Essentials of Excel, Excel VBA, SAS and Minitab for Statistical and Financial Analyses

Cheng-Few Lee • John Lee • Jow-Ran Chang • Tzu Tai

Essentials of Excel, Excel VBA, SAS and Minitab for Statistical and Financial Analyses

Springer

Cheng-Few Lee
Department of Finance, Rutgers University
Piscataway Township
NJ, USA

John Lee
Center for PBBEF Research
Morris Plains
NJ, USA

Jow-Ran Chang
Department of Quantitative Finance
National Tsing Hua University
Hsinchu, Taiwan

Tzu Tai
Mezocliq, LLC
New York
NY, USA

ISBN 978-3-319-38865-6 ISBN 978-3-319-38867-0 (eBook)
DOI 10.1007/978-3-319-38867-0

Library of Congress Control Number: 2016954950

Printed on acid-free paper

This Springer imprint is published by Springer Nature
The registered company is Springer International Publishing AG
The registered company address is: Gewerbestrasse 11, 6330 Cham, Switzerland

Preface

There are 32 chapters in this book, and it is divided into three parts. Part I includes the first 21 chapters which were proposed to supplement the textbook entitled *Statistics for Business and Financial Economics, 3rd edition* (by Cheng-Few Lee, John C. Lee, and Alice C. Lee). There are nine chapters (Chaps. 22–30) in Part II addressing the advanced applications of Microsoft Excel Programs. There are two chapters in Part III addressing the applications of SAS programs in financial analysis and research. These two chapters are applications of simultaneous equations in finance research.

Part I not only can be used to supplement the abovementioned statistics book, it also can be used to teach basic statistics courses by itself. In addition, Part I can be combined with Part II to learn VBA programming in Excel. Parts II and III are the extensions of Part I: Chaps. 22–24 introduce how to write Macro Recorder and VBA programs in Excel, Chaps. 25 and 26 illustrate how to use Excel to evaluate options, Chaps. 27, 28, 29, and 30 show how to use Excel to evaluate Greek letters, implied variance, and portfolio and use simulation to do the analysis, and Chaps. 31 and 32 show how to use SAS to deal with simultaneous equations and hedge ratios for financial data, respectively. Every chapter in this book uses the financial data of individual company, stock index, options, and futures. Therefore, this book will be the most comprehensive type of computer application book which shows how to use financial data to do empirical analysis.

There are three possible applications of this book as follows:

A. To supplement business statistics books, especially for a book such as *Statistics for Business and Financial Economics, 3rd edition* (by Cheng-Few Lee, John C. Lee, and Alice C. Lee).
B. Chapters 1–21 can be used independently to teach statistics courses. Under this approach, the instructor can use data and computer programs to teach a course. In other words, this approach can be regarded as a data analysis approach to teach statistics.
C. This book can be used to learn how to apply Microsoft Excel, SAS, and Minitab to financial data analysis.
D. This book can be used to supplement courses such as portfolio management and futures and options.

In sum, this book can be used by academic courses and for practitioners in the financial industry. Finally, we appreciate the extensive help of our assistants Yaqing Xiao and Alex McCullough.

Piscataway Township, NJ	Cheng-Few Lee
Morris Plains, NJ	John Lee
Hsinchu, Taiwan	Jow-Ran Chang
New York, NY	Tzu Tai

Contents

Part A Statistical Analysis

Part A
Statistical Analysis

Chapter 1
Introduction

1.1 Introduction

A person studies and uses *statistics* to learn how to study, analyze, and understand a data set of particular interest. But one problem in studying statistics is that it is a very intense computational discipline. We can get a sense of the computational intensity of statistics by looking at the average statistical formula and the correlation coefficient statistical formula shown below.

$$\bar{x} = \frac{\sum_{i=1}^{n} x_i}{n} \qquad r = \frac{\frac{1}{n-1}\sum_{i=1}^{n}(x_i-\bar{x})(y_i-\bar{y})}{\left[\frac{1}{n-1}\sum_{i=1}^{n}(x_i-\bar{x})^2\right]^{1/2}\left[\frac{1}{n-1}\sum_{i=1}^{n}(y_i-\bar{y})^2\right]^{1/2}}$$

Average Correlation coefficient

The average formula looks fairly simple, but it becomes more and more difficult as n increases. The correlation coefficient formula is obviously more complex and tedious to perform—and other computational formulas are even more complicated. Therefore one of the biggest costs in both time and money in studying and using statistics is the implementation costs of implementing statistical concepts on data sets that we are interested in studying, analyzing, and understanding.

To combat these implementation costs, statistical computer programs have been developed. Some of the more popular statistical programs are SAS, SPSS, and Minitab. In this book we will look at Minitab and SAS. One of the main reasons to use Minitab is that it is the easiest to use among the popular statistical programs. We will look at SAS because it's the leading statistical package used in industry.

Traditionally, schools use a statistical program to accompany the study of statistics. One problem with this is that most students will not become professional statisticians. This means that all the time and effort they spent learning on how to use the statistical program has been basically wasted. And, if they do ever need to do some statistical analysis, the statistical program that they did use in school is probably not available.

One solution to the above problem is to use Microsoft Excel as the computer program to do the statistical analysis. Excel can do statistical analysis. While Excel's statistical features are not as strong or sophisticated as the traditional statistical programs, it does take care 90 % of the statistical analysis done in the business world. One of the most important characteristics about Excel is that it is

C.-F. Lee et al., *Essentials of Excel, Excel VBA, SAS and Minitab for Statistical and Financial Analyses*, DOI 10.1007/978-3-319-38867-0_1

virtually everywhere. This is important because it allows a person to perform the statistical analysis right away instead of spending time, effort, and money to acquire a traditional statistical program. Many times it could be weeks before they get the statistical program.

The benefits of using Excel have been recognized in the academic world. This is why you will find more and more statistical books using Excel to do the statistical analysis.

In this book we will use Microsoft Excel, Minitab, and SAS to do our statistical analysis. We use Microsoft Excel because of the benefits stated above. We will use Minitab and SAS to double-check our work in Excel and to do some statistical analysis that is not possible or practical to do in Excel.

In this book we will use Microsoft Excel 2013 and Minitab 13.

1.2 Programming!

Among the business students and business people the word *programming* is a dirty word. This is why many business applications do not mention its programming capabilities. This is also why many statistical books do not emphasize or even mention the programming aspects of the application.

Contrary to this attitude, business people and business students should study the programming capabilities of the statistical application and the statistical application should mention its programming capabilities and the statistical books should show and discuss the computer programs of the application.

One of the most compelling reasons is it should allow the business person to make more *money*! The reason being is that knowing how to program allows them to accomplish more things which makes them more valuable. The following *New York Times* article talks how programming has help people make more money:

As Tech Booms, Workers Turn to Coding for Career Change: http://www.nytimes.com/2015/07/29/technology/code-academy-as-career-game-changer.html

One might ask, "Why not ask for a programmer to do it?" There lies the problem. It is not easy to ask for programming resource in the business world and it is also expensive to ask for a programming resource.

In Minitab we will find out that beyond the basic statistical analysis, we will require some programming in Minitab. It is interesting to note that many statistical features in Minitab are accomplished by using Minitab macros! The user can study these Minitab macros to learn how to program Minitab macros.

Excel gives the user a professional quality programming language and a professional programming environment. Because of this there are many commercial applications developed in Excel. The professional quality programming language is called Visual Basic for Application or VBA.

This book will publish all Excel macros for users to study.

1.3 Statistical Environment of Microsoft Excel 2013

Excel has many built-in statistical features. One statistical feature that Excel has is it provides a lot of statistical functions. To view the statistical functions, go to the *Formulas* tab and choose the *Statistical* item as shown below.

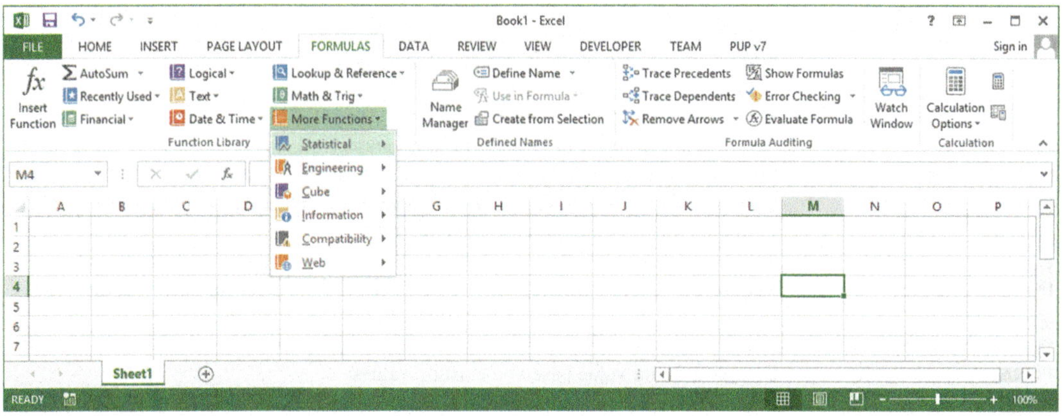

Below is a list of all the available statistical functions in Excel:

Statistical functions	
AVEDEV	Returns the average of the absolute deviations of data points from their mean
AVERAGE	Returns the average of its arguments
AVERAGEA	Returns the average of its arguments, including numbers, text, and logical values
AVERAGEIF	Returns the average (arithmetic mean) of all the cells in a range that meet a given criteria
AVERAGEIFS	Returns the average (arithmetic mean) of all cells that meet multiple criteria
BETA.DIST	Returns the beta cumulative distribution function
BETA.INV	Returns the inverse of the cumulative distribution function for a specified beta distribution
BINOM.DIST	Returns the individual term binomial distribution probability
BINOM.DIST.RANGE	Returns the probability of a trial result using a binomial distribution
BINOM.INV	Returns the smallest value for which the cumulative binomial distribution is less than or equal to a criterion value
CHISQ.DIST	Returns the cumulative beta probability density function
CHISQ.DIST.RT	Returns the one-tailed probability of the chi-squared distribution
CHISQ.INV	Returns the cumulative beta probability density function
CHISQ.INV.RT	Returns the inverse of the one-tailed probability of the chi-squared distribution
CHISQ.TEST	Returns the test for independence
CONFIDENCE.NORM	Returns the confidence interval for a population mean
CONFIDENCE.T	Returns the confidence interval for a population mean, using a Student's t-distribution
CORREL	Returns the correlation coefficient between two data sets
COUNT	Counts how many numbers are in the list of arguments
COUNTA	Counts how many values are in the list of arguments
COUNTBLANK	Counts the number of blank cells within a range
COUNTIF	Counts the number of cells within a range that meet the given criteria
COUNTIFS	Counts the number of cells within a range that meet multiple criteria
COVARIANCE.P	Returns covariance, the average of the products of paired deviations
COVARIANCE.S	Returns the sample covariance, the average of the products' deviations for each data point pair in two data sets
DEVSQ	Returns the sum of squares of deviations
EXPON.DIST	Returns the exponential distribution
F.DIST	Returns the F probability distribution
F.DIST.RT	Returns the F probability distribution
F.INV	Returns the inverse of the F probability distribution

(continued)

Statistical functions	
F.INV.RT	Returns the inverse of the F probability distribution
F.TEST	Returns the result of an F-test
FISHER	Returns the Fisher transformation
FISHERINV	Returns the inverse of the Fisher transformation
FORECAST	Returns a value along a linear trend
FORECAST.ETS	Returns a future value based on existing (historical) values by using the AAA version of the Exponential Smoothing (ETS) algorithm
FORECAST.ETS.CONFINT	Returns a confidence interval for the forecast value at the specified target date
FORECAST.ETS. SEASONALITY	Returns the length of the repetitive pattern Excel detects for the specified time series
FORECAST.ETS.STAT	Returns a statistical value as a result of time series forecasting
FORECAST.LINEAR	Returns a future value based on existing values
FREQUENCY	Returns a frequency distribution as a vertical array
GAMMA	Returns the gamma function value
GAMMA.DIST	Returns the gamma distribution
GAMMA.INV	Returns the inverse of the gamma cumulative distribution
GAMMALN	Returns the natural logarithm of the gamma function, Γ(x)
GAMMALN.PRECISE	Returns the natural logarithm of the gamma function, Γ(x)
GAUSS	Returns 0.5 less than the standard normal cumulative distribution
GEOMEAN	Returns the geometric mean
GROWTH	Returns values along an exponential trend
HARMEAN	Returns the harmonic mean
HYPGEOM.DIST	Returns the hypergeometric distribution
INTERCEPT	Returns the intercept of the linear regression line
KURT	Returns the kurtosis of a data set
LARGE	Returns the kth largest value in a data set
LINEST	Returns the parameters of a linear trend
LOGEST	Returns the parameters of an exponential trend
LOGNORM.DIST	Returns the cumulative lognormal distribution
LOGNORM.INV	Returns the inverse of the lognormal cumulative distribution
MAX	Returns the maximum value in a list of arguments
MAXA	Returns the maximum value in a list of arguments, including numbers, text, and logical values
MEDIAN	Returns the median of the given numbers
MIN	Returns the minimum value in a list of arguments
MINA	Returns the smallest value in a list of arguments, including numbers, text, and logical values
MODE.MULT	Returns a vertical array of the most frequently occurring or repetitive values in an array or range of data
MODE.SNGL	Returns the most common value in a data set
NEGBINOM.DIST	Returns the negative binomial distribution
NORM.DIST	Returns the normal cumulative distribution
NORM.INV	Returns the inverse of the normal cumulative distribution
NORM.S.DIST	Returns the standard normal cumulative distribution
NORM.S.INV	Returns the inverse of the standard normal cumulative distribution
PEARSON	Returns the Pearson product moment correlation coefficient
PERCENTILE.EXC	Returns the kth percentile of values in a range, where k is in the range 0.0.1, exclusive
PERCENTILE.INC	Returns the kth percentile of values in a range
PERCENTRANK.EXC	Returns the rank of a value in a data set as a percentage (0.0.1, exclusive) of the data set
PERCENTRANK.INC	Returns the percentage rank of a value in a data set
PERMUT	Returns the number of permutations for a given number of objects
PERMUTATIONA	Returns the number of permutations for a given number of objects (with repetitions) that can be selected from the total objects

(continued)

Statistical functions	
PHI	Returns the value of the density function for a standard normal distribution
POISSON.DIST	Returns the Poisson distribution
PROB	Returns the probability that values in a range are between two limits
QUARTILE.EXC	Returns the quartile of the data set, based on percentile values from 0.0.1, exclusive
QUARTILE.INC	Returns the quartile of a data set
RANK.AVG	Returns the rank of a number in a list of numbers
RANK.EQ	Returns the rank of a number in a list of numbers
RSQ	Returns the square of the Pearson product moment correlation coefficient
SKEW	Returns the skewness of a distribution
SKEW.P	Returns the skewness of a distribution based on a population: a characterization of the degree of asymmetry of a distribution around its mean
SLOPE	Returns the slope of the linear regression line
SMALL	Returns the kth smallest value in a data set
STANDARDIZE	Returns a normalized value
STDEV.P	Calculates standard deviation based on the entire population
STDEV.S	Estimates standard deviation based on a sample
STDEVA	Estimates standard deviation based on a sample, including numbers, text, and logical values
STDEVPA	Calculates standard deviation based on the entire population, including numbers, text, and logical values
STEYX	Returns the standard error of the predicted y-value for each x in the regression
T.DIST	Returns the percentage points (probability) for the Student's t-distribution
T.DIST.2T	Returns the percentage points (probability) for the Student's t-distribution
T.DIST.RT	Returns the Student's t-distribution
T.INV	Returns the t-value of the Student's t-distribution as a function of the probability and the degrees of freedom
T.INV.2T	Returns the inverse of the Student's t-distribution
T.TEST	Returns the probability associated with a Student's t-test
TREND	Returns values along a linear trend
TRIMMEAN	Returns the mean of the interior of a data set
VAR.P	Calculates variance based on the entire population
VAR.S	Estimates variance based on a sample
VARA	Estimates variance based on a sample, including numbers, text, and logical values
VARPA	Calculates variance based on the entire population, including numbers, text, and logical values
WEIBULL.DIST	Returns the Weibull distribution
Z.TEST	Returns the one-tailed probability value of a z-test

1.4 Environment of Minitab

Minitab consists of two main parts, the *data window* and the *session window*.

The data window is where the statistical data are entered. The session window is where all statistical analysis is shown and where statistical commands called session commands can be issued. When you start Minitab, the data window and session window are displayed by default. This is shown below.

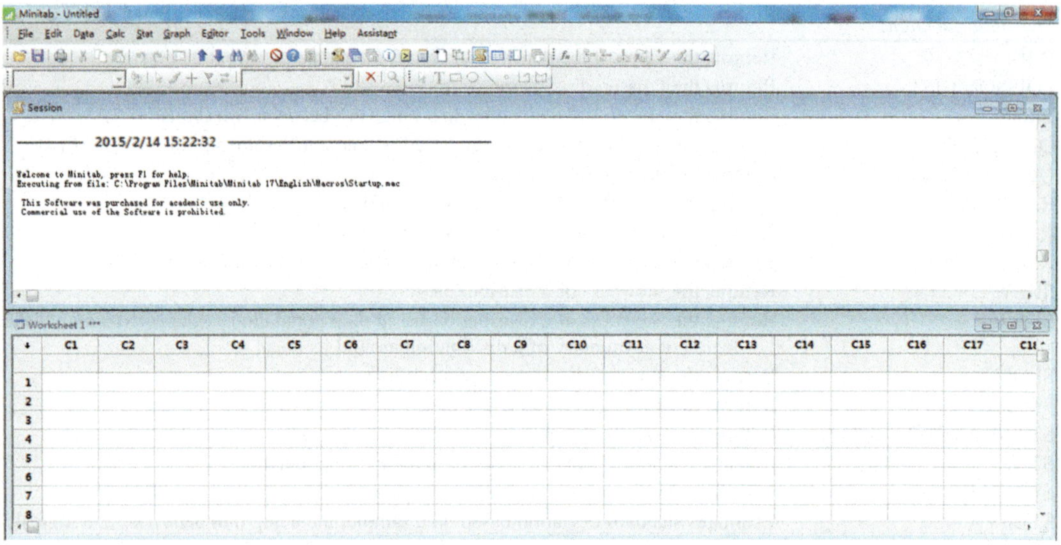

The session window is where statistical commands and reports are displayed. Many statistical calculations require several commands. With Minitab you can group commands together and give a name to the group of commands. Such a group is called a *macro*. The user can execute the group of commands just by typing the group name. Grouping commands together is computer programming.

By default only the statistical reports are shown in the session window. The default way to do statistical calculations is through Minitab's menus. Learning and becoming familiar with the Minitab commands are very important in getting things done in Minitab. A lot of the sophisticated analysis done by Minitab is done through Minitab macros.

To enable the statistical commands in the session window, first select the session window as shown below.

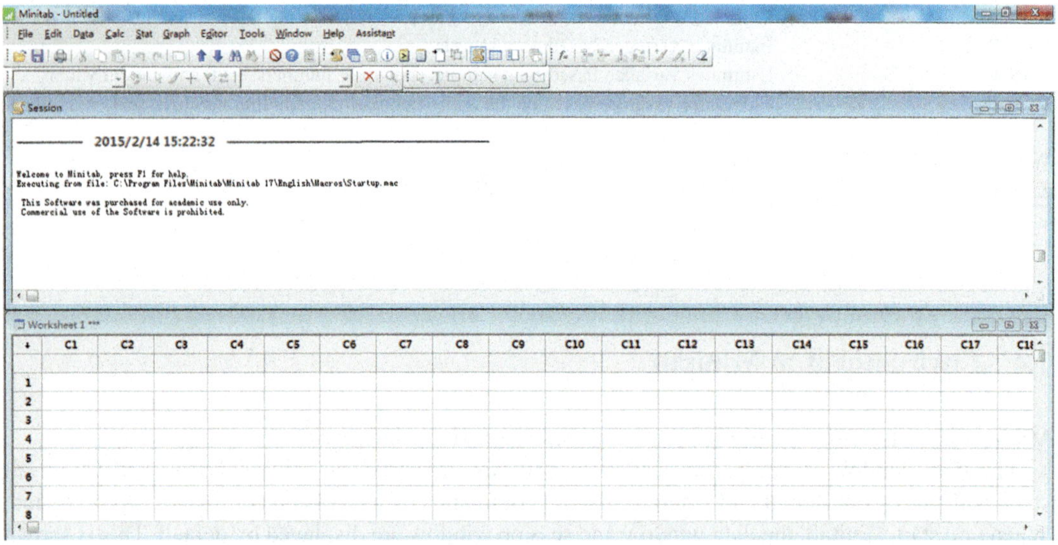

You will know that the session window is activated because the session window will be bright while the data session window will be dimmed.

Next, choose **Editor** → **Enable Commands** to enable the commands in the session window. This is shown below.

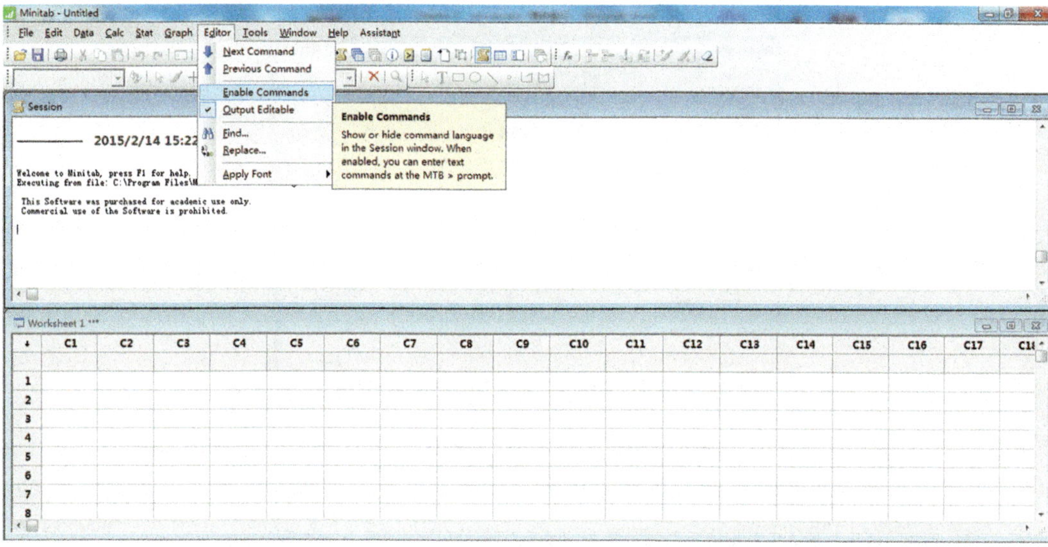

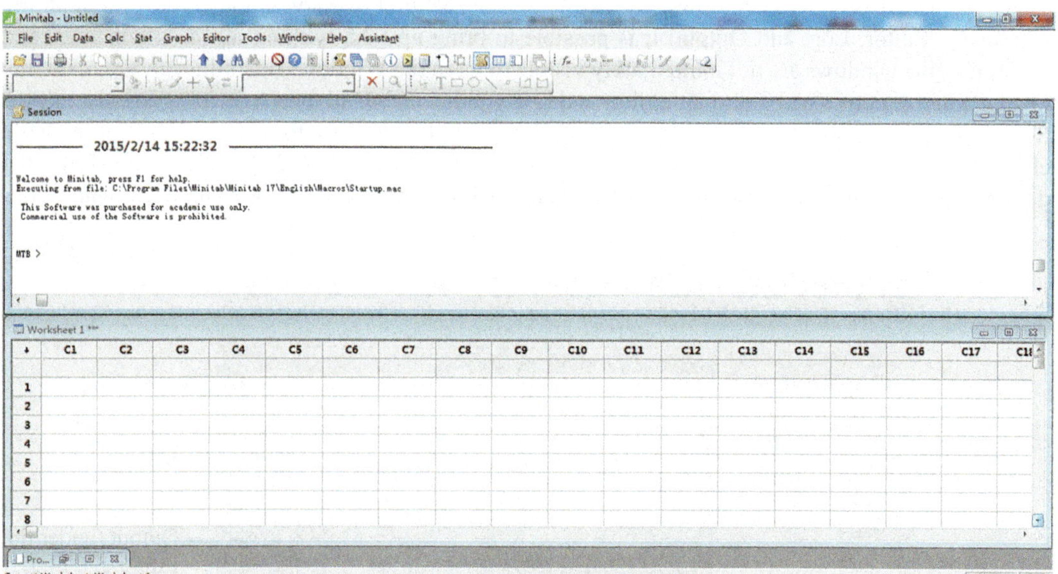

Minitab will show the prompt, `MTB>` to indicate that session commands are enabled. This is shown above.

You will have to do this every time you start Minitab. To set it so Minitab command prompt is shown every time when you start Minitab, do the following:

Choose Tools > **Options** > **Session Window** > **Submitting Commands.**
Under **Command Language**, choose **Enable**.
Click **OK**. as shown below.

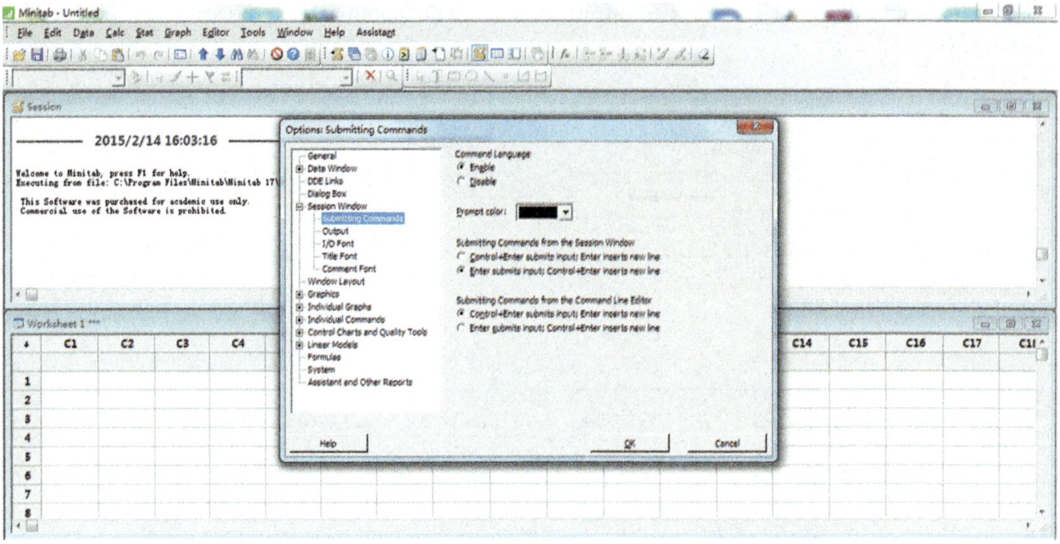

1.5 Environment of SAS

There are five basic SAS windows: the Results and Explorer windows and three programming windows – Editor, Log, and Output. It is possible to bring up SAS without all these windows, and sometimes the windows are not immediately visible (e.g., in the Windows operating environment, the Output window comes up behind the Editor and Log windows), but all these windows do exist in your SAS session. There are also many other SAS windows that you may use for tasks such as getting help, changing SAS system options, and customizing your SAS session. The following figure shows the default view for a Microsoft Windows SAS session, with pointers to the five main SAS windows.

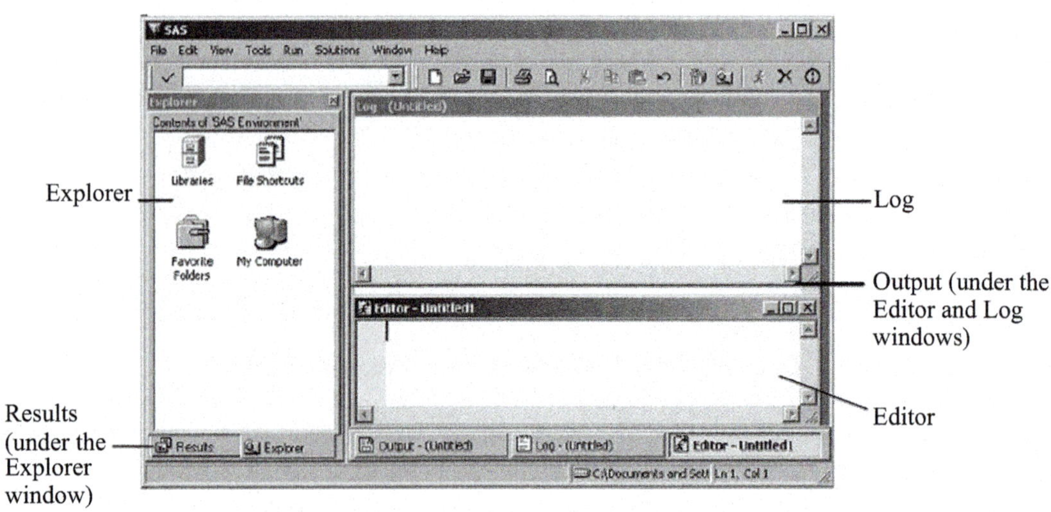

Editor This window is a text editor. You can use it to type in, edit, and submit SAS programs as well as edit other text files such as raw data files. In Windows operating environments, the default editor is the Enhanced Editor. The Enhanced Editor is syntax sensitive and color codes your programs making it easier to read them and find mistakes. The Enhanced Editor also allows you to collapse and expand the various steps in your program. For other operating environments, the default editor is the Program Editor whose features vary with the version of SAS and operating environment.

Log The Log window contains notes about your SAS session, and after you submit a SAS program, any notes, errors, or warnings associated with your program as well as the program statements themselves will appear in the Log window.

Output If your program generates any printable results, then they will appear in the Output window.

Results The Results window is like a table of contents for your Output window; the results tree lists each part of your results in an outline form.

Explorer The Explorer window gives you easy access to your SAS files and libraries.

1.6 SAS Commands

There are SAS commands for performing a variety of tasks. You may have up to three ways to issue commands: menus, the toolbar, or the SAS command bar (or command line). The following figure shows the location of these three methods of issuing SAS commands in the Windows operating environment default view.

Menus Most operating environments will have pull-down menus located either at the top of each window or at the top of your screen. If your menus are at the top of your screen, then the menus will change when you activate the different windows.

Toolbar The toolbar, if you have one, gives you quick access to commands that are already accessible through the pull-down menus. Not all operating environments have a toolbar.

SAS Command Bar The command bar is a place that you can type in SAS commands. Most of the commands that you can type in the command bar are also accessible through the pull-down menus or the toolbar.

Controlling Your Windows The Window pull-down menu gives you choices on how the windows are placed on your screen. You can also activate any of the programming windows by selecting it from the Window pull-down menu, typing the name of the window in the command line area of your SAS session or simply clicking on the window.

1.7 Implementation and Case Study Approach

Microsoft Excel is the overwhelming favorite program of business user. Business users try to use Excel to solve all their business needs. But most of the business users only use a fraction of the features available in Excel and many business users use Excel in an inefficient way. The main problem is that Excel is a large sophisticated program. This is shown by the many large "encyclopedia"-like Excel books available. Most people do not read an encyclopedia from cover to cover. For the same reason, most people do not read these "encyclopedia" Excel books from cover to cover.

The best way to learn Excel is by applying Excel to problems. "Learning by doing" is the best way to learn Excel. Therefore this book will "kill two birds with one stone." This book will discuss the concepts of statistics and simultaneously discuss how to use Excel to implement the concepts of statistics.

The approach, perspective, and emphasis on how to use Excel will be different from other statistics and Excel books. Most of the other authors of the statistics that use Excel come from an academic perspective. This book will come from a business perspective. By a quirk of fate, I make my living as an Excel professional. Because I work in the business environment, I am very concerned about the *implementation* of the statistical concepts. Those authors who come from an academic background are concerned mainly with the concepts of statistics. These authors mainly present reports generated from the statistical program or statistical reports generated from Excel. There are a lot of issues that go into implementing these reports. Because of my background, I will discuss the implementation issues in both Excel and Minitab. Therefore, I will be talking about issues and concepts in Excel and Minitab that have nothing to do with statistics.

This book will present every Excel program in its entirety in this book. It is very annoying to study a program code that is partially presented in a book and then for the book to say to go to the CD to study the rest of the program.

Bibliography

Lee CF, Lee JC, Lee AC (2013) Statistics for business and financial economics. Springer, New York
Microsoft Inc., Excel 2013. Microsoft Inc., Redmond
Minitab Inc. Minitab 17. Minitab Inc., State College
SAS Institute Inc. SAS 2014. SAS Institute Inc., Cary

Chapter 2
Data Collection, Presentation, and Yahoo Finance

2.1 Data Presentation

Statistics is mainly associated with numbers. A complete statistical analysis of a data set consists of both numbers and graphs. Graphs many times allow users to understand a data set that is very hard to capture with number.

Prior to Microsoft Excel, it has been very hard to create professional-looking graphs. There is a feature in Excel called a *chart wizard* that has made creating graphs easy. We will look at the chart wizard in detail in this chapter. In addition, we also present four ways to read a data into SAS. Then we use SAS code called PROC SGPLOT to create color graphics.

Until very recently it's been real hard to get financial data to use for data analysis. Now with the advent of the Internet, it has been much easier to get financial data for data analysis. One of the most popular places to get financial data is Yahoo Finance. Yahoo Finance's web link is https://www.yahoo.com/finance. We will also look at how to retrieve financial data from Google Finance. Google Finance's web link is https://www.google.com/finance. Finally we would look at a Quandl.com.

2.2 Yahoo Finance

In this chapter we will look at financial data that spans through time. The most popular technique to analyze financial data that spans through time is to use line graphs.

In this chapter we will analyze the stock price of JPM from 1984 to 2013.

One of the most popular places to get financial data is Yahoo Finance. To get JPM's historical stock price, you would go to http://finance.yahoo.com/q/hp?s=JPM+Historical+Prices. It is interesting to note that to get historical stock price of IBM, you would go to http://finance.yahoo.com/q/hp?s=IBM+Historical+Prices. To get the historical price of MSFT, you would go to http://finance.yahoo.com/q/hp?s=MFST+Historical+Prices. Note the web link pattern.

The following are the steps to get JPM's historical stock price from 1984 to 2013:

1. Using a web browser like Internet Explorer, Firefox, or Chrome, go to the location http://finance.yahoo.com/q/hp?s=JPM+Historical+Prices. Yahoo Finance will show you the available date ranges. The earliest available historical data available for JPM is December 30, 1983. This example will use Firefox.

C.-F. Lee et al., *Essentials of Excel, Excel VBA, SAS and Minitab for Statistical and Financial Analyses*, DOI 10.1007/978-3-319-38867-0_2

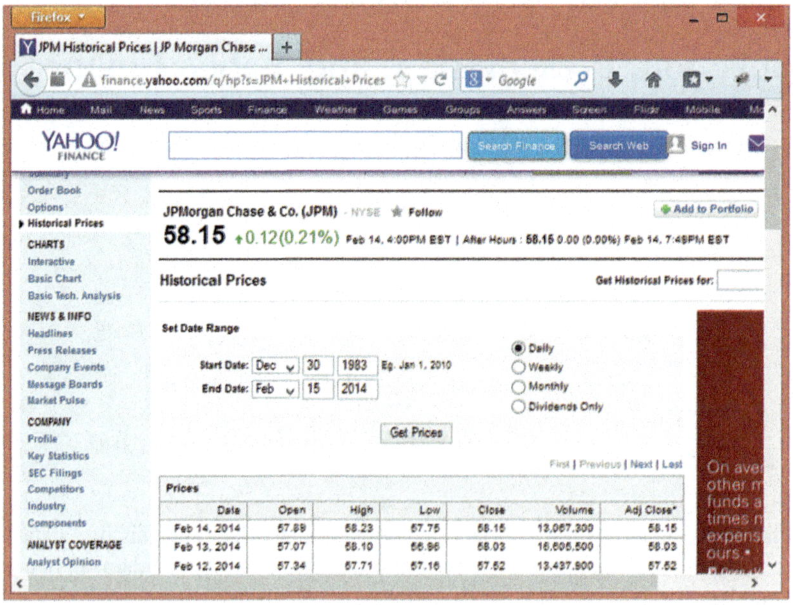

2. At the top of the web page, set the date range of interest. Let's set it from January 1, 1984, to December 31, 2013.

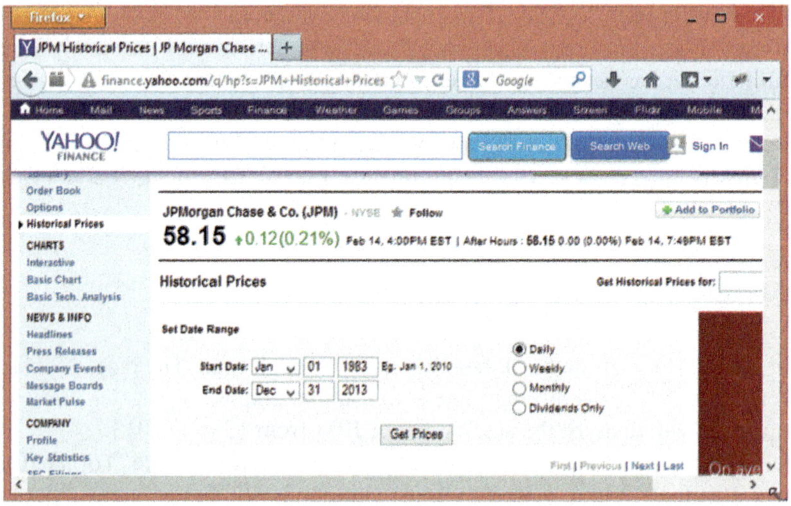

3. Click on the *Get Prices* button.
4. At the bottom of the web page, click on the *Download to Spreadsheet* link.

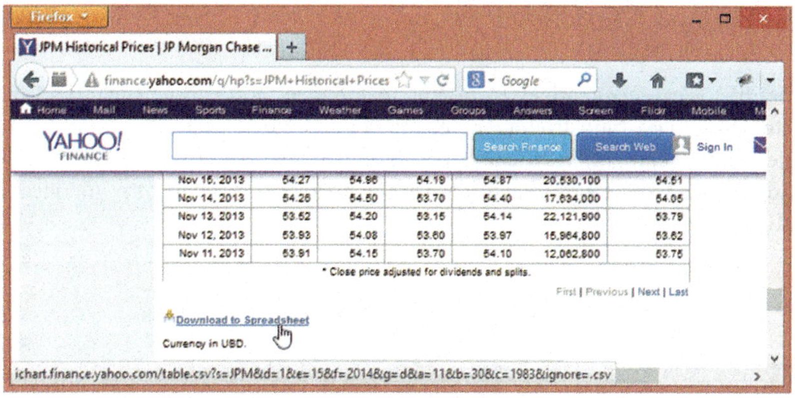

5. Firefox will prompt you and suggest you open JPM's historical data in Microsoft Excel.

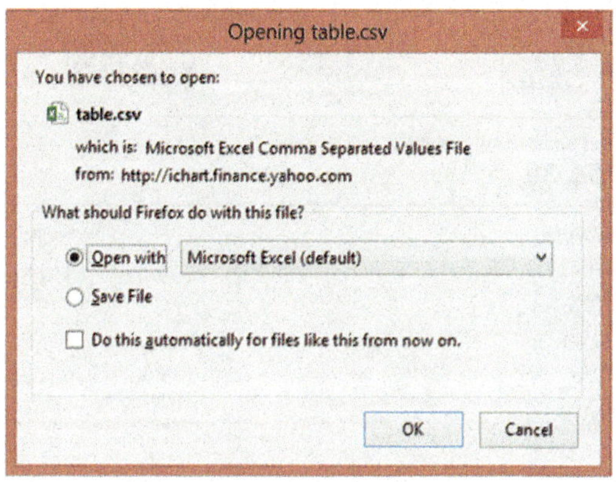

If you go to the bottom of the worksheet you will notice that there are 7566 historical prices for JPM from 1984 to 2013.

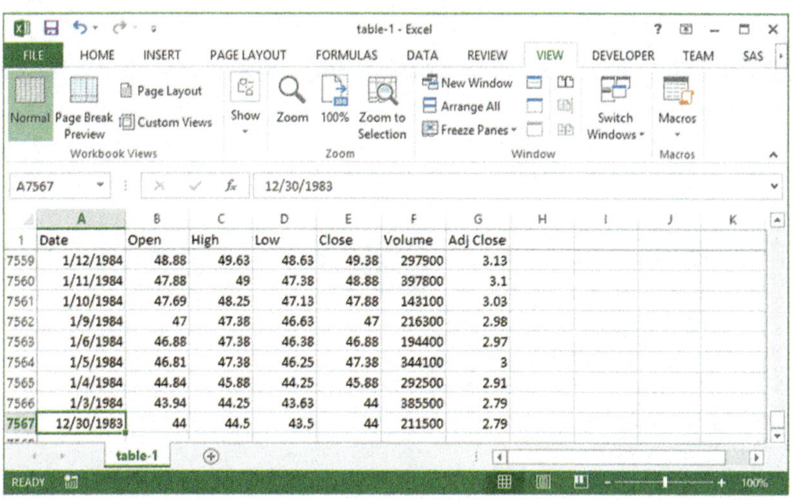

2.3 Yahoo Finance Market Data

Besides getting historical stock prices from Yahoo Finance, you can get historical prices of some indexes like the Dow Jones Industrial average, SP500, and the NASDAQ. You can get prices of the 10-year T-note and you can get historical prices. You can also get historical prices of mutual funds.

The web link for the Dow Jones industrial average historical prices is http://finance.yahoo.com/q/hp?s=%5EDJI+Historical+Prices

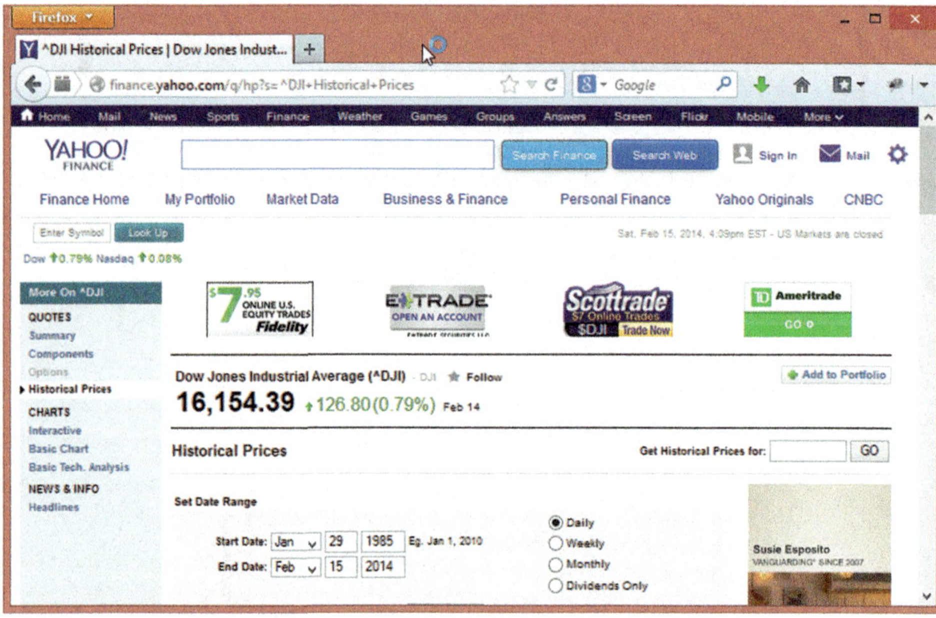

The web link for the S&P 500 historical prices is http://finance.yahoo.com/q/hp?s=^GSPC+Historical+Prices

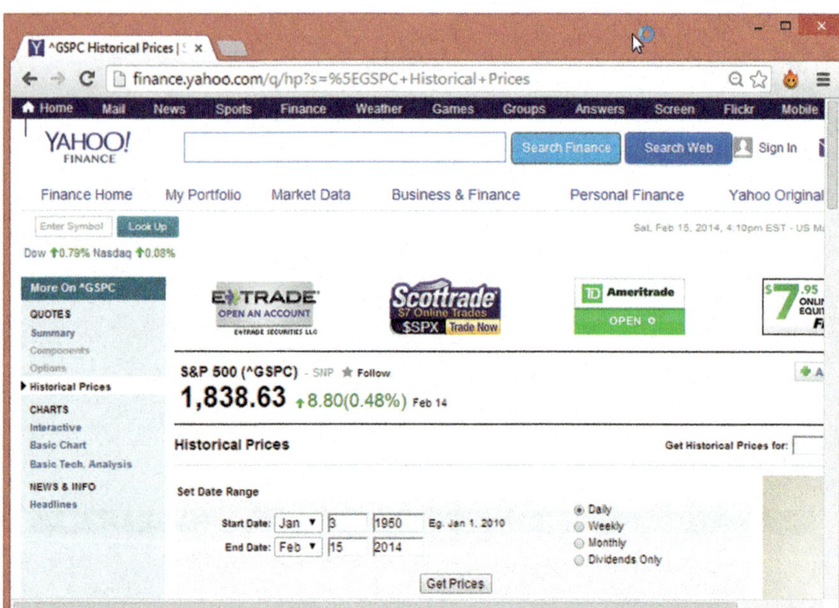

The web link for NASDAQ's historical prices is http://finance.yahoo.com/q/hp?s=^IXIC+Historical+Prices

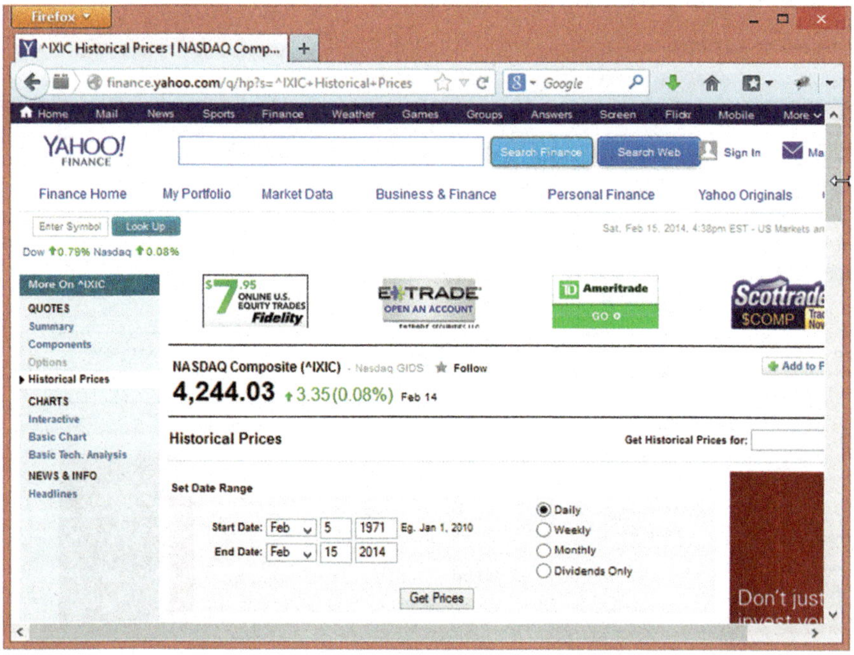

The web link for the 10-year note historical prices is http://finance.yahoo.com/q/hp?s=^TNX+Historical+Prices

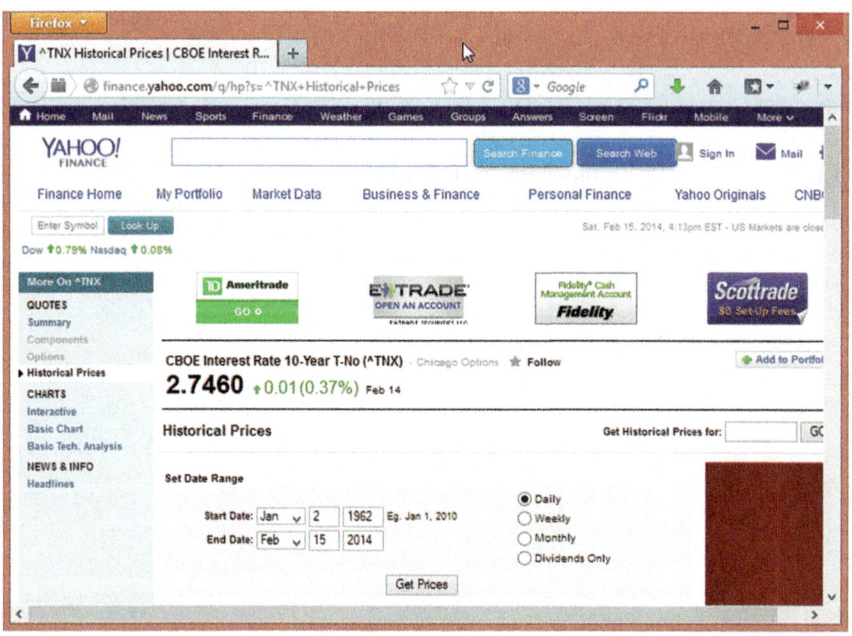

The web link for the Franklin Biotechnology historical prices is http://finance.yahoo.com/q/hp?s=FTDZX+Historical+Prices

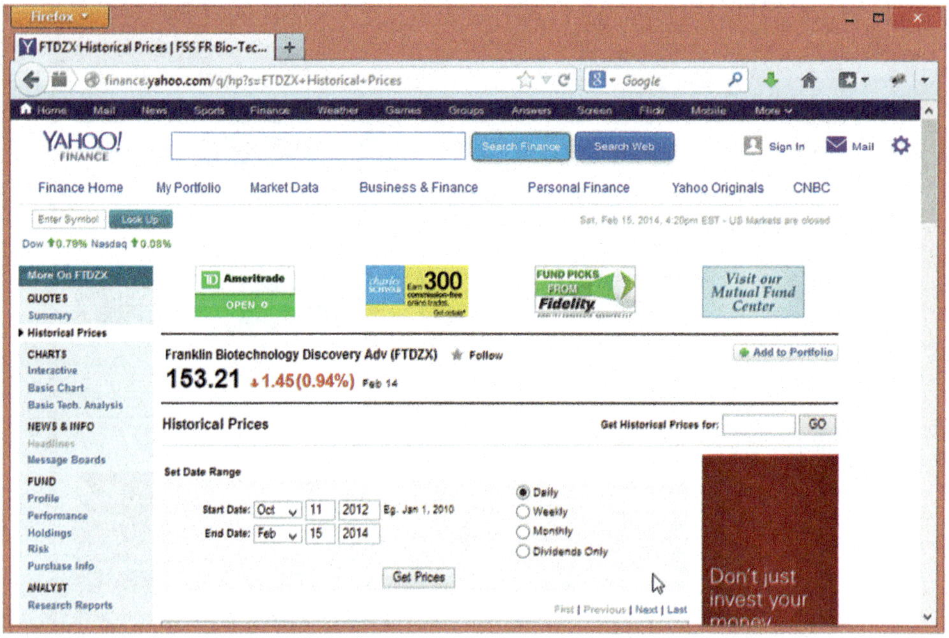

2.4 Components of Dow Jones Industrial Average

The web link for the components of the Dow Jones Industrial Average is http://money.cnn.com/data/dow30/

As of October 12, 2015 the components of DJIA is

Companies in the Dow Jones Industrial Average

	Company	Price	Change	% Change	Volume	YTD change
1	MMM 3M	149.11	-0.95	-0.63%	2,027,618	-9.26%
2	AXP American Express	76.6	-0.71	-0.92%	3,247,383	-17.67%
3	AAPL Apple	111.79	0.19	0.17%	33,049,256	1.28%
4	BA Boeing	140.29	-0.39	-0.28%	2,339,688	7.93%
5	CAT Caterpillar	70.17	-0.33	-0.47%	4,724,800	-23.34%
6	CVX Chevron	88.38	-0.36	-0.41%	8,012,541	-21.22%
7	CSCO Cisco	27.85	-0.11	-0.39%	15,107,154	0.13%
8	KO Coca-Cola	41.65	-0.35	-0.83%	12,903,557	-1.35%
9	DIS Disney	106.59	0.24	0.23%	8,342,685	13.16%
10	DD E I du Pont de Nemours and Co	55.76	0.1	0.18%	4,506,809	-20.62%
11	XOM Exxon Mobil	79.16	-0.14	-0.18%	12,326,598	-14.38%
12	GE General Electric	27.87	-0.22	-0.78%	43,203,188	10.29%
13	GS Goldman Sachs	180.97	0.74	0.41%	2,918,822	-6.63%
14	HD Home Depot	121.61	-0.29	-0.24%	2,971,353	15.85%
15	IBM IBM	149.62	-1.52	-1.01%	3,915,707	-6.74%
16	INTC Intel	32.04	-0.17	-0.53%	35,588,607	-11.71%
17	JNJ Johnson & Johnson	95.45	-0.54	-0.56%	12,761,886	-8.72%
18	JPM JPMorgan Chase	61.55	-0.17	-0.28%	15,621,860	-1.65%
19	MCD McDonald's	103.38	0.14	0.14%	4,697,279	10.33%
20	MRK Merck	49.47	-1.24	-2.45%	14,905,931	-12.89%
21	MSFT Microsoft	46.89	-0.11	-0.23%	19,987,816	0.95%
22	NKE Nike	125.81	-0.62	-0.49%	3,177,677	30.85%
23	PFE Pfizer	32.98	-0.24	-0.72%	19,446,103	5.87%
24	PG Procter & Gamble	74.11	-0.22	-0.30%	8,390,248	-18.64%
25	TRV Travelers Companies Inc	103.31	-0.41	-0.40%	1,287,612	-2.40%
26	UTX United Technologies	94.28	-1.15	-1.21%	4,079,277	-18.02%
27	UNH UnitedHealth	123.99	1.48	1.21%	6,655,729	22.65%
28	VZ Verizon	44.36	0.06	0.14%	11,566,364	-5.17%
29	V Visa	75	0.01	0.01%	7,439,403	14.42%
30	WMT Wal-Mart	66.73	-0.2	-0.30%	8,858,374	-22.30%

2.5 Retrieving Market Data From Microsoft Excel

It's possible to retrieve market data directly from Microsoft Excel. To do this we need the web link to the historical stock price. To get the web link to the data file, go to *Download to Spreadsheet* link. At the link right-mouse click and choose the *Copy Link Location* item.

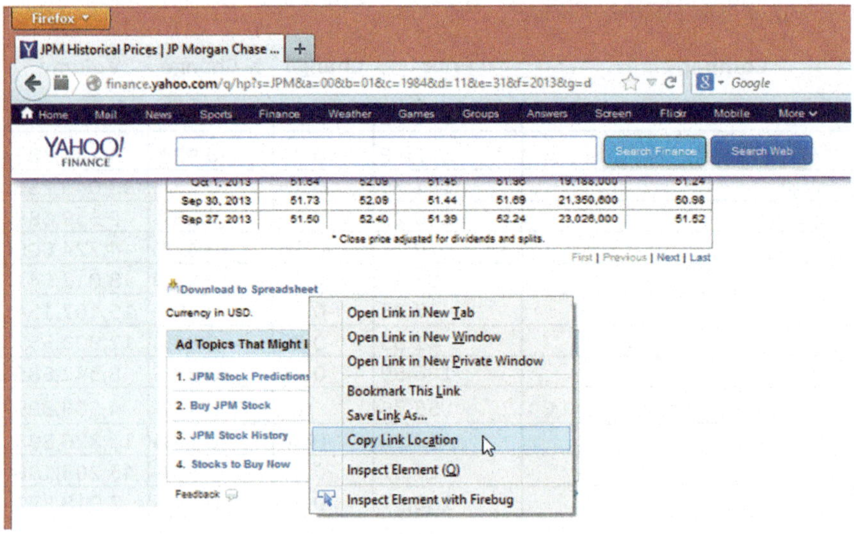

To see the value of the web link, paste the link to an empty cell in Microsoft Excel.

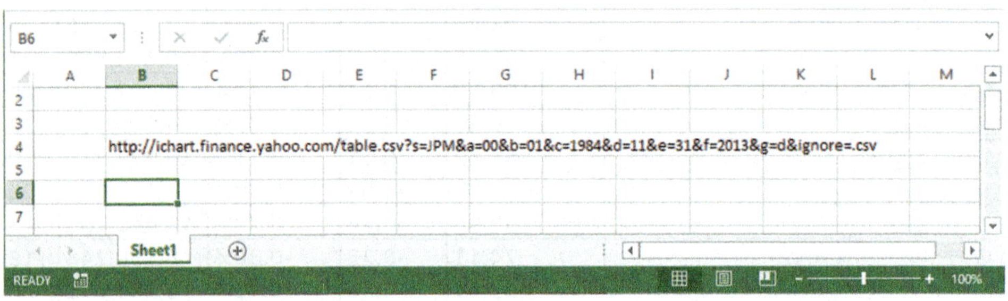

Do the following to retrieve JPM's historical data directly from Microsoft Excel:

1. Click on the *File* menu.

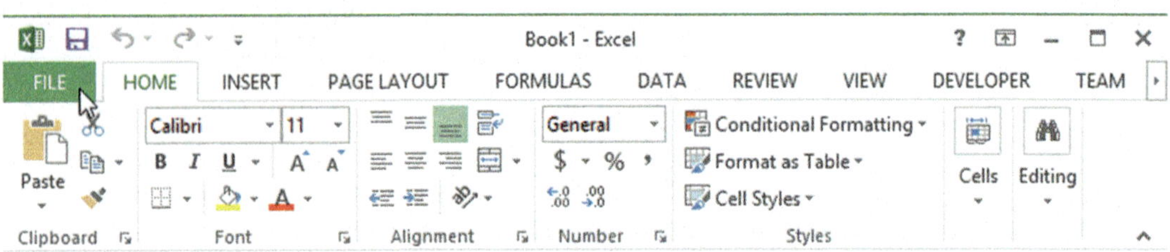

2. (a) Click on *Open*.
   (b) Click on *Other Web Locations*.
   (c) Click on the *Browser*.

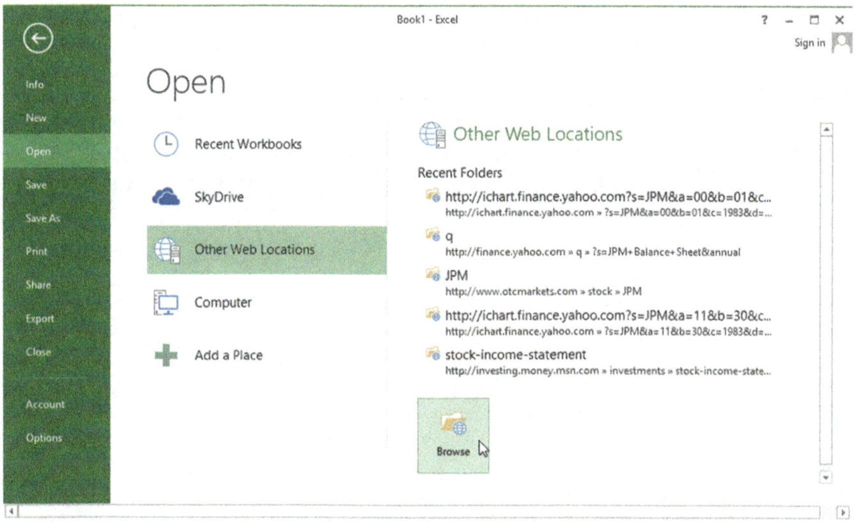

3. Paste the JPM data web link in the *Open* dialog box and then press on *Open* button to retrieve the JPM data to Microsoft Excel.

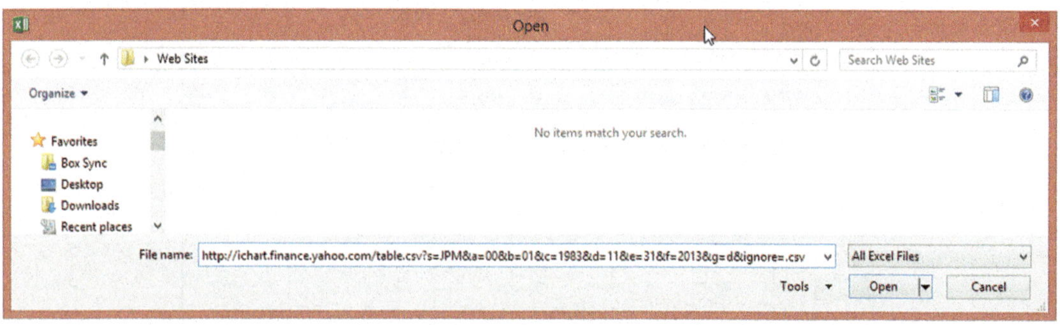

The reason this process is of interest is that you can manually modify the JPM web link to customize the data you want to retrieve. For example, if you want to get Microsoft's 2012 historical stock price, you modify the data link to the following: http://ichart.finance.yahoo.com/table.csv?s=MSFT&a=00&b=01&c=2012&d=11&e=31&f=2012&g=d&ignore=.csv

Notice that there is a logic to modifying the data link. The *s* parameter is for the ticker. The *a* parameter is for the month of the starting period. In this parameter you actually need to adjust by subtracting one from the month. For example, to indicate the month of January, you would indicate 00 instead of 01. The *b* parameter is for the day of the start period. The *c* is for the year of the starting period. The *d* parameter is for the month of the ending period. Notice that 11 is indicated for December. The *e* parameter is for the day of the ending period. The *f* parameter is for the year of the ending period. The *g* parameter indicates the frequency of the data will by daily.

2.6 Charting JPM's Data Using Excel's Chart Wizard

We will graph JPM's stock price by using Excel's chart wizard.

The first thing that we will need to do is to retrieve JPM's historical stock price from Excel. Use the method shown above to retrieve JPM's historical stock price. The link to get JPM's historical price from 1983 to 2013 is http://ichart.finance.yahoo.com/table.csv?s=JPM&a=00&b=01&c=1983&d=11&e=31&f=2013&g=d&ignore=.csv

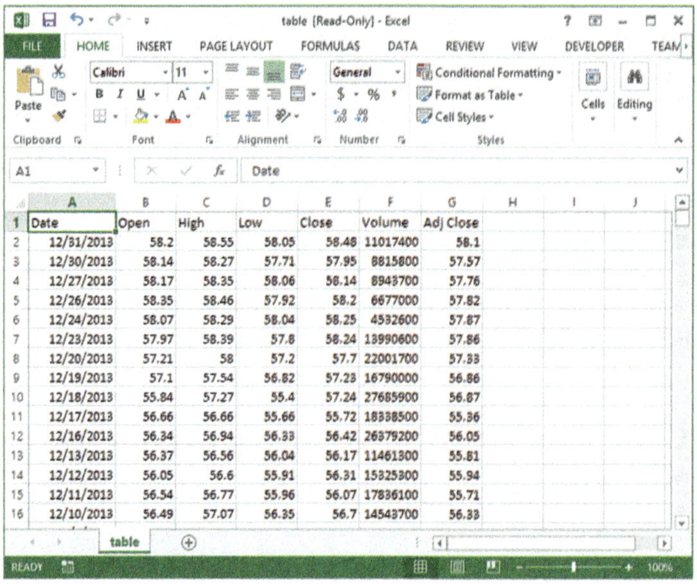

	A	B	C	D	E	F	G
1	Date	Open	High	Low	Close	Volume	Adj Close
2	12/31/2013	58.2	58.55	58.05	58.48	11017400	58.1
3	12/30/2013	58.14	58.27	57.71	57.95	8815800	57.57
4	12/27/2013	58.17	58.35	58.06	58.14	8943700	57.76
5	12/26/2013	58.35	58.46	57.92	58.2	6677000	57.82
6	12/24/2013	58.07	58.29	58.04	58.25	4532600	57.87
7	12/23/2013	57.97	58.39	57.8	58.24	13990600	57.86
8	12/20/2013	57.21	58	57.2	57.7	22001700	57.33
9	12/19/2013	57.1	57.54	56.82	57.23	16790000	56.86
10	12/18/2013	55.84	57.27	55.4	57.24	27685900	56.87
11	12/17/2013	56.66	56.66	55.66	55.72	18338500	55.36
12	12/16/2013	56.34	56.94	56.33	56.42	26379200	56.05
13	12/13/2013	56.37	56.56	56.04	56.17	11461300	55.81
14	12/12/2013	56.05	56.6	55.91	56.31	15325300	55.94
15	12/11/2013	56.54	56.77	55.96	56.07	17836100	55.71
16	12/10/2013	56.49	57.07	56.35	56.7	14543700	56.33

Next, save the Excel file as JPM.

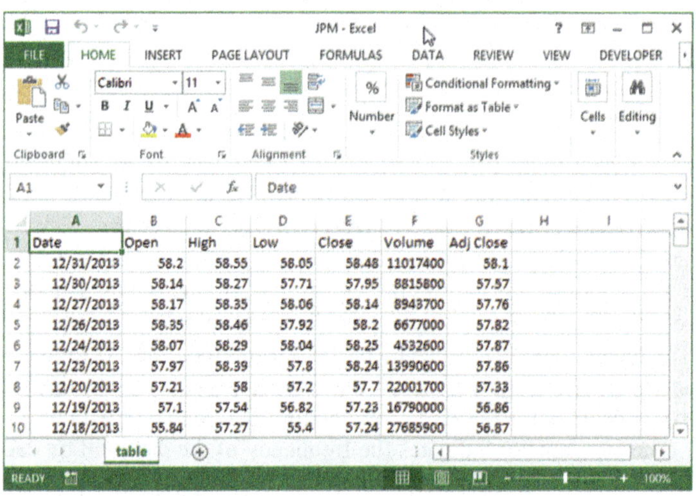

	A	B	C	D	E	F	G
1	Date	Open	High	Low	Close	Volume	Adj Close
2	12/31/2013	58.2	58.55	58.05	58.48	11017400	58.1
3	12/30/2013	58.14	58.27	57.71	57.95	8815800	57.57
4	12/27/2013	58.17	58.35	58.06	58.14	8943700	57.76
5	12/26/2013	58.35	58.46	57.92	58.2	6677000	57.82
6	12/24/2013	58.07	58.29	58.04	58.25	4532600	57.87
7	12/23/2013	57.97	58.39	57.8	58.24	13990600	57.86
8	12/20/2013	57.21	58	57.2	57.7	22001700	57.33
9	12/19/2013	57.1	57.54	56.82	57.23	16790000	56.86
10	12/18/2013	55.84	57.27	55.4	57.24	27685900	56.87

Put the cursor inside the table; then in the *Insert* tab, click on the *Recommend Charts* item.

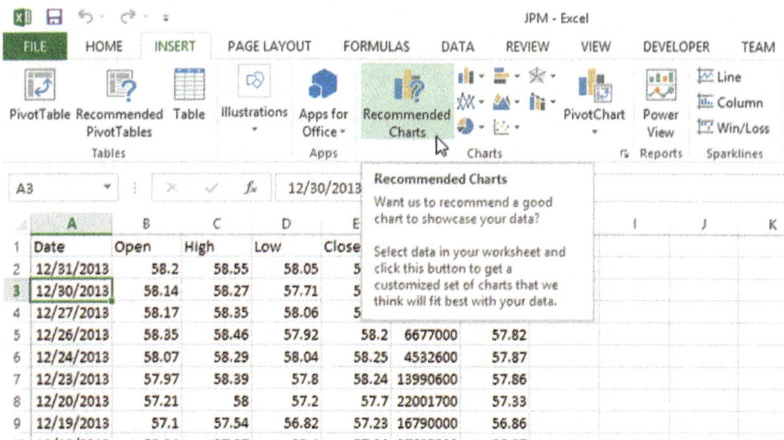

Microsoft Excel will show you a suggested chart.

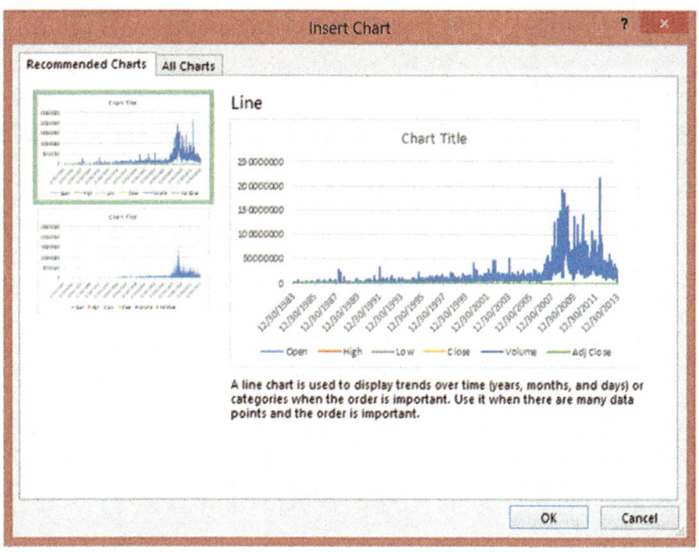

After clicking on the *OK* button, Microsoft Excel will display a chart next to the table.

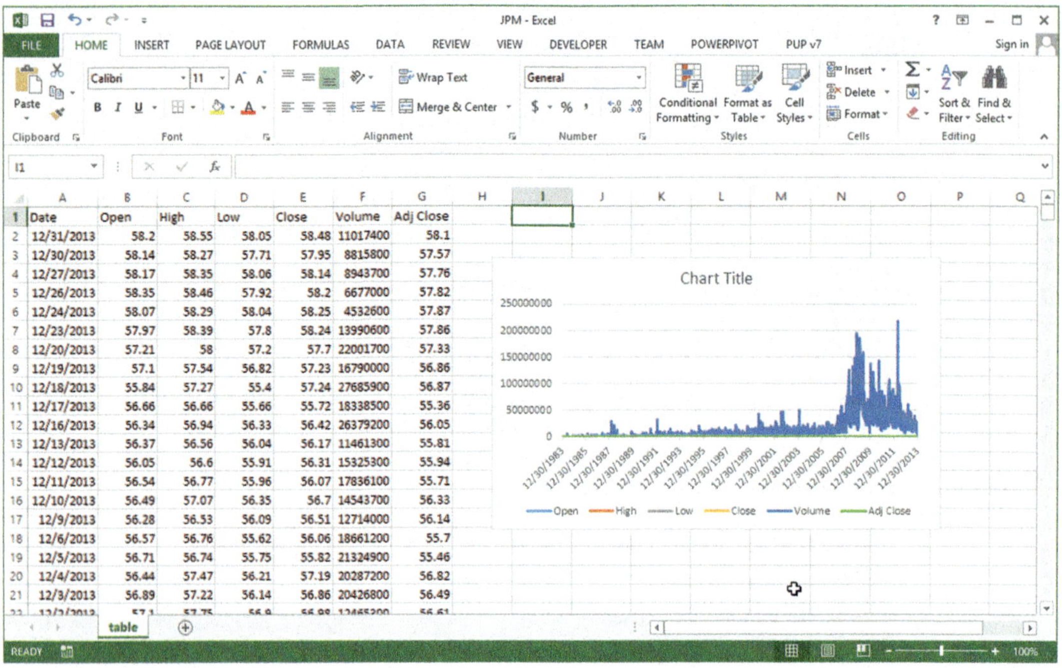

This chart needs to be modified. We are interested in the y-axis to be the *Adj Close* column and the x-axis to be the *Date* column. To do this right-mouse click over the chart and choose the *Select Data* menu item.

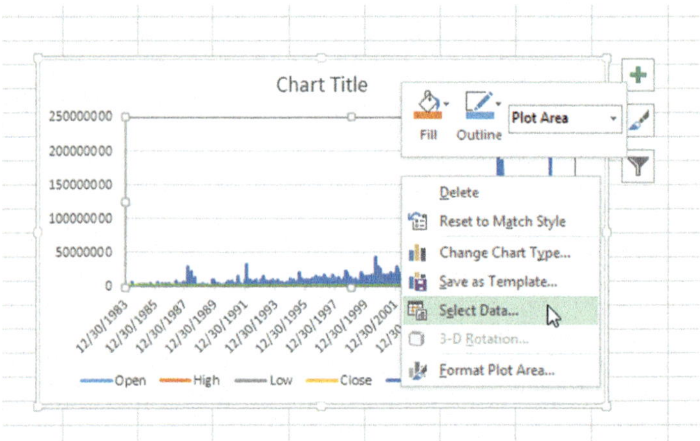

In the *Select Data Source* dialog box, unselect everything except for *Adj Close* and then press the *OK* button.

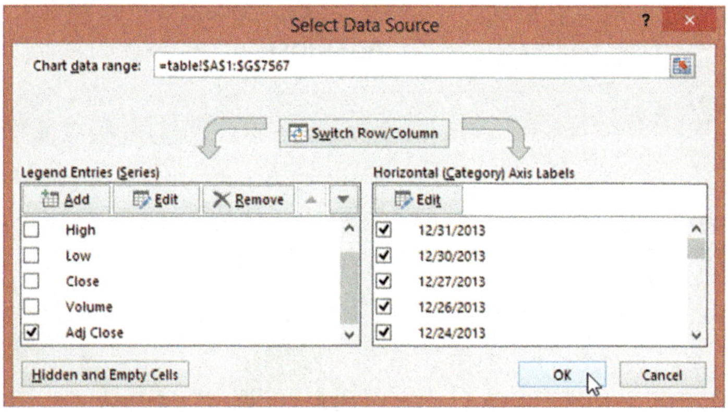

Right-mouse click on the resulting chart and choose *Move Chart*.

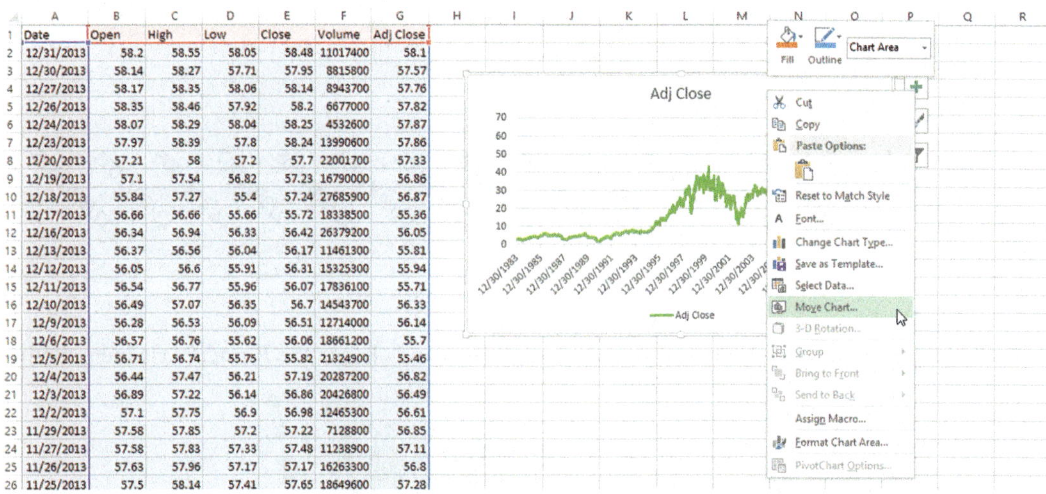

In the *Move Chart* dialog box, indicate the name of the new sheet as *JPM*.

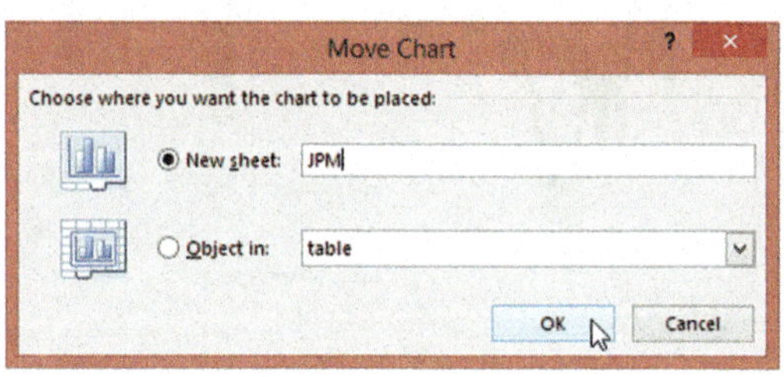

The resulting chart looks like the following. One problem with the chart is that the date axis looks too crowded.

To format the date axis, right-mouse click on the dates and choose the *Format Axis* menu item.

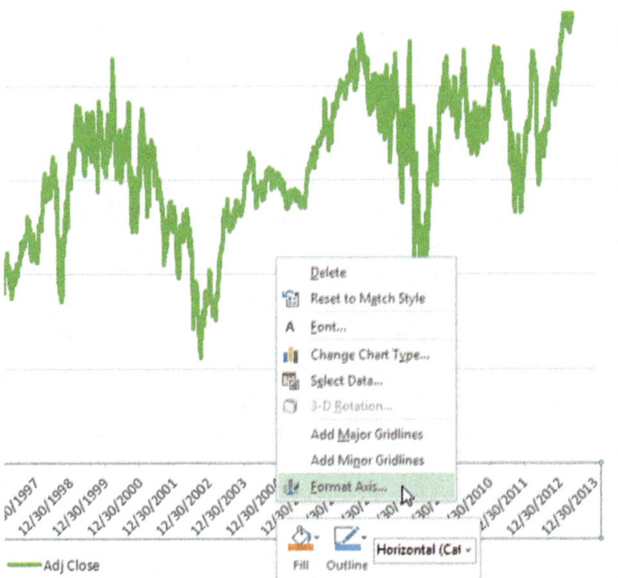

To the right of the chart, a *Format Axis* will appear.

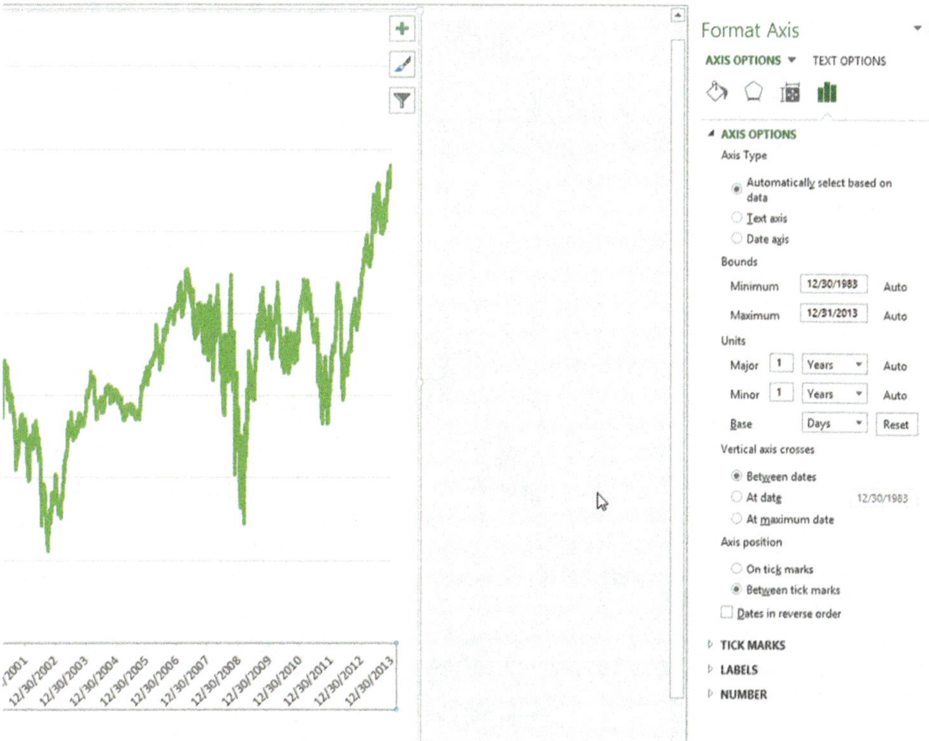

To make the date axis less cluttered, change the *Major* units to 5.

The resulting chart would look like the following:

Many large companies have offices around the world. The above date format could be confusing to other countries because they list date first, then month, then year. To make it less confusing, let's reformat the date that shows mmm-dd-yyy. To do this go the number section *number* section of the *Format Axis* panel. In the *Category* option, choose "custom." Next in the *Format Code* option, type in "mmm-dd-yyy." Next press the *Add* button. Then choose "mmm-dd-yyy."

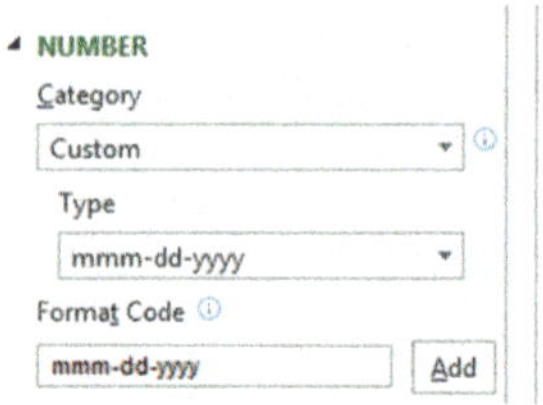

2.7 Read Data into SAS and Create a Chart Graph

In this chapter, we'll focus on creating SAS data set from raw data files. There are mainly two methods for reading raw data files in SAS:

- Use the Import Wizard.
- Use the DATA step. This method works for almost any type of raw data file.

We will talk about these methods for most of this chapter.

2.7.1 Reading Files With the Import Wizard

The Import Wizard can read delimited raw data files, including comma-separated values (CSV) files, which are produced by Excel and many other applications.

We'll use this method to read the following file:

d:\Minitab and Microsoft Excel Book\Ch2\Data used in Excel, Minitab, and SAS book.csv

In SAS, select File/ Import data . . . /

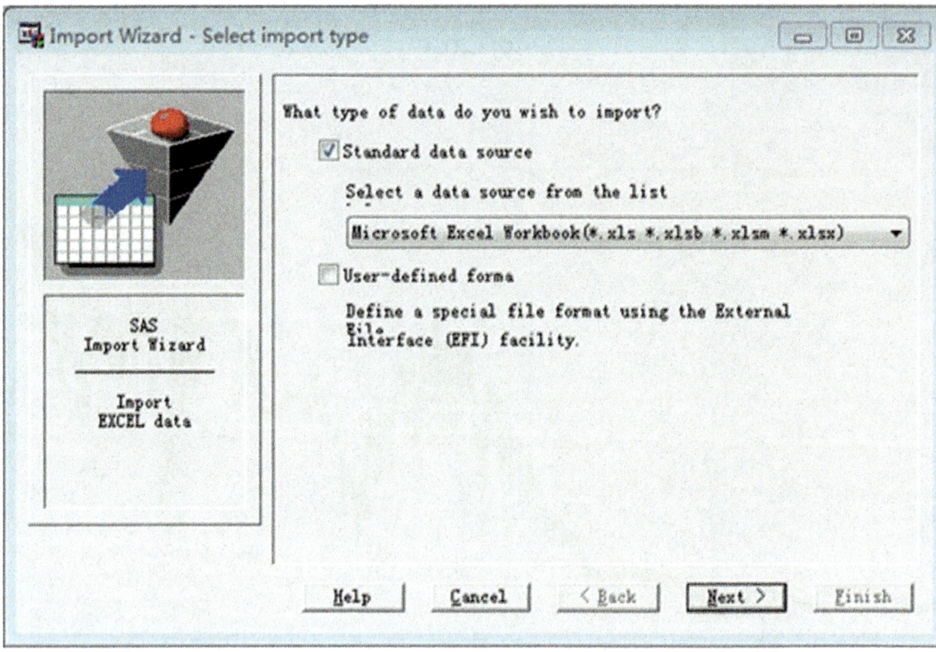

For the menu "Select a data source from the list," select "Comma Separated Values (*.csv)."

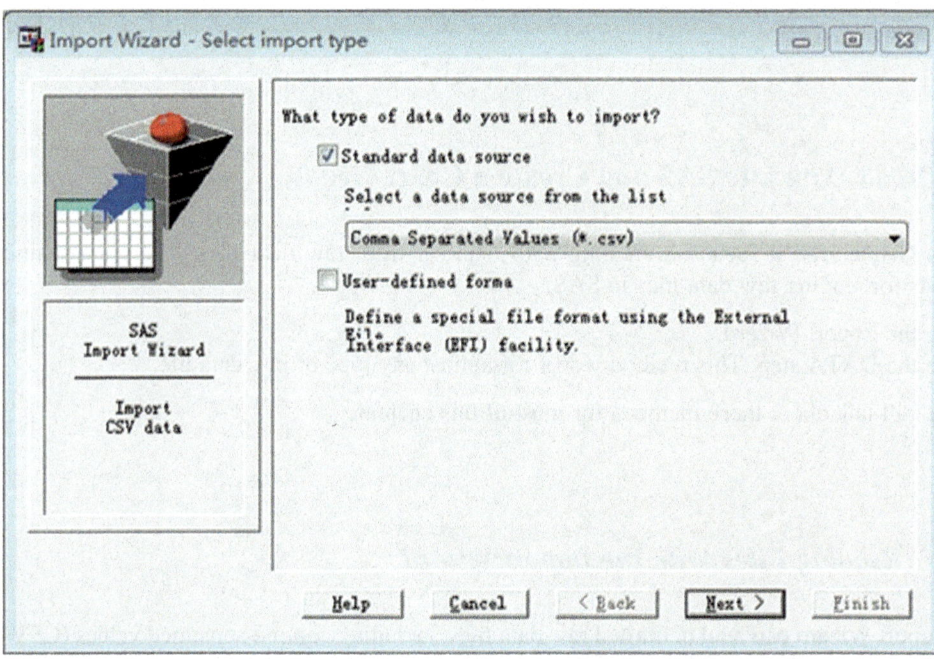

Select "Next."
Browse to the directory that has the CSV file (d:\Minitab and Microsoft Excel Book\Ch2).

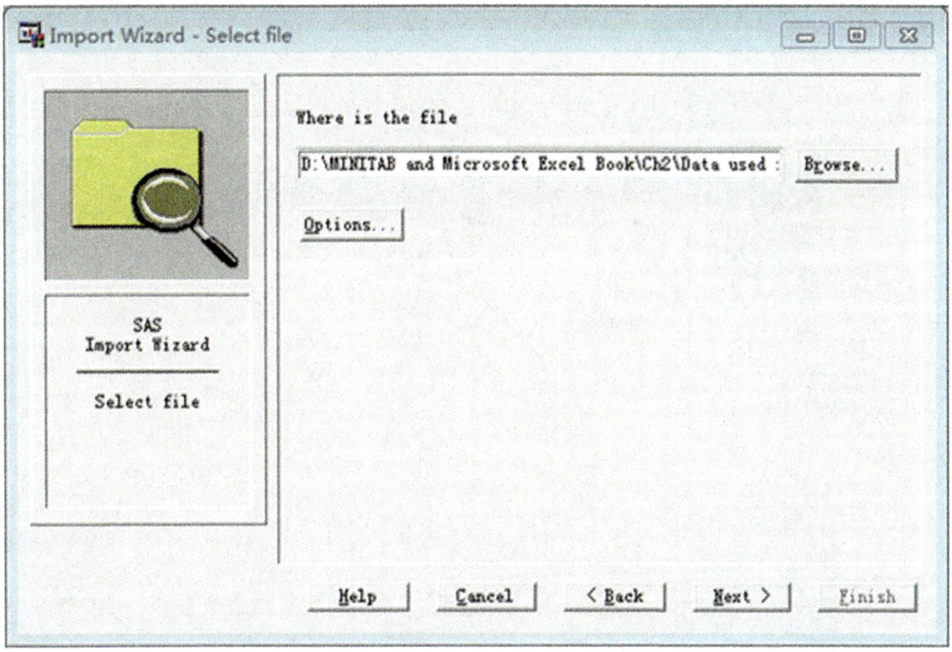

Select “Next.”

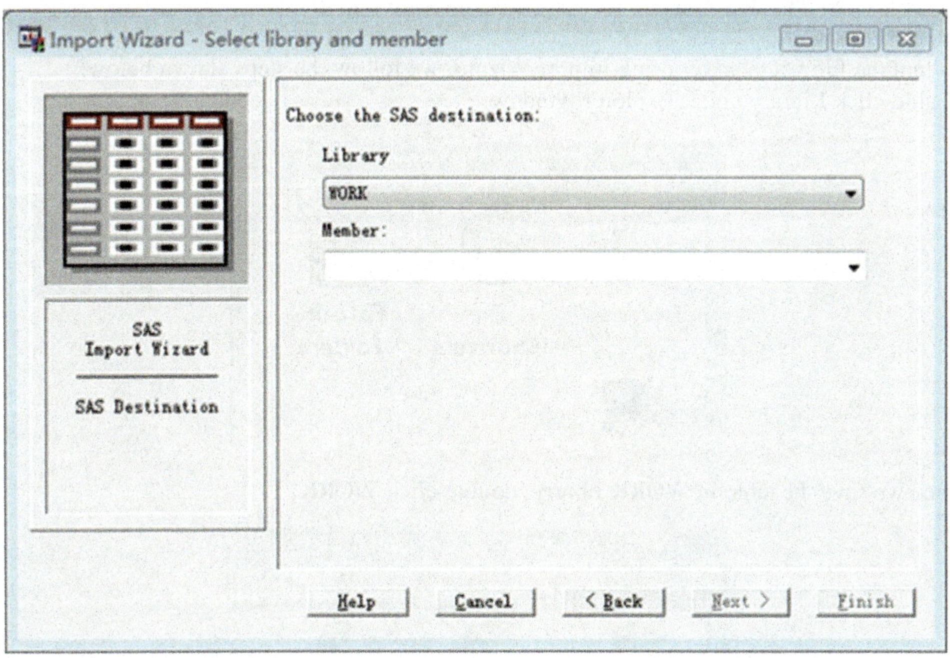

Accept the WORK library as the destination.
In the “member” field, give the name you want SAS to use for the data set, such as “JNJ.”
Select “Next.”

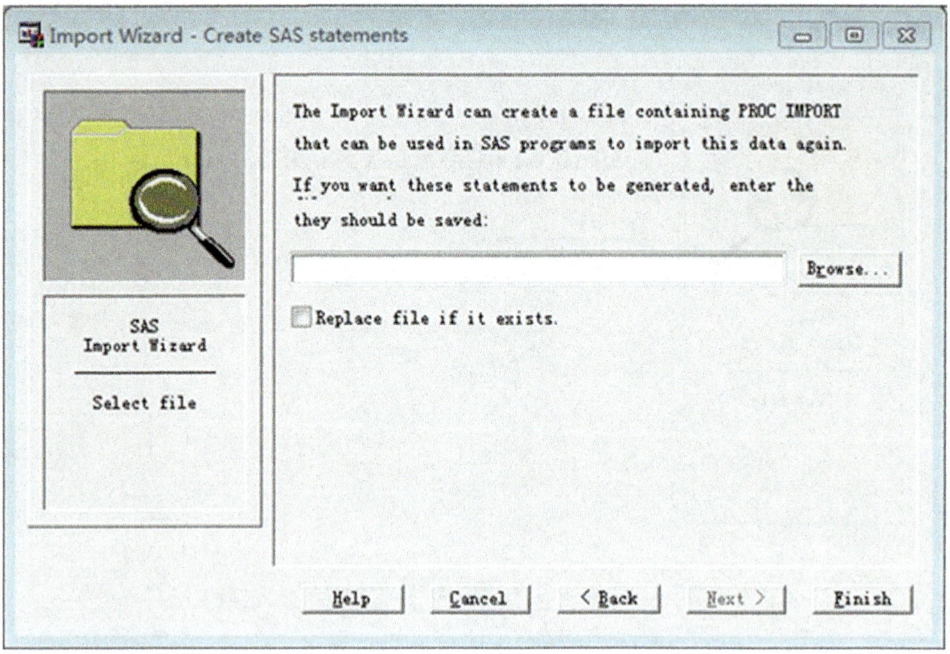

The Import Wizard will offer to create a file containing the import SAS commands. I suggest you create the file, so that later you can see how the data were imported.

Select “Finish.”

To find the file you read by using Import Wizard, we follow the steps shown below:

Double-click Library under Explorer window.

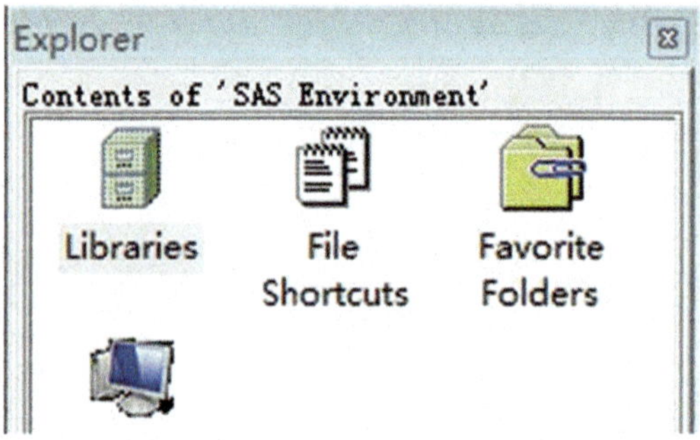

Since we save the table in WORK library, double-click WORK.

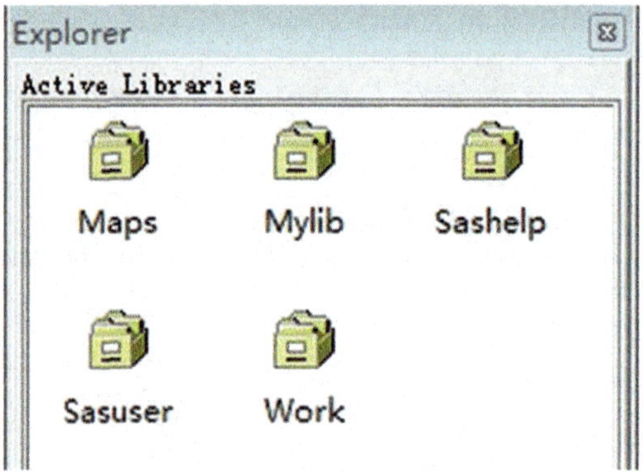

Double-click the file name you defined earlier. Here, we click Jnj. Then we get the View table on the right.

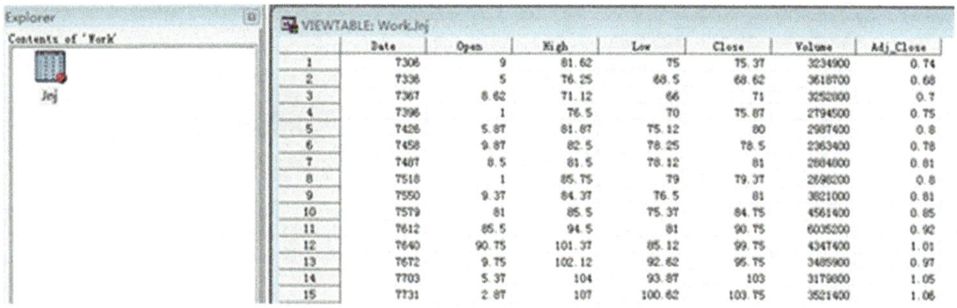

By default, View table uses variable labels for column headings, or if a variable does not have a label, the variable name is displayed. Sometimes you may want to see the actual variable names instead of the labels. To do this, select Column Names from the View menu. You can also control colors, fonts, and view the column attributes (Delwiche 2008).

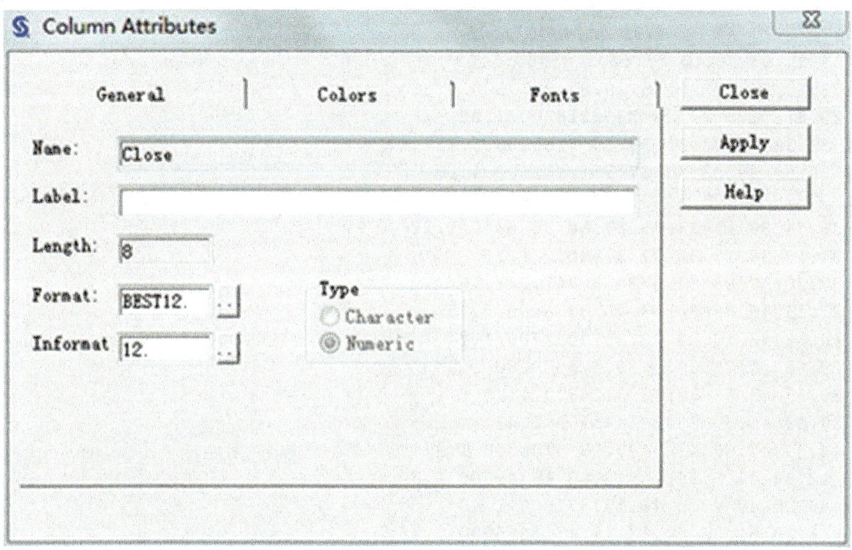

2.7.2 Read Internal Raw Data Into SAS

If you include your data in the body of your SAS program, then the data are internal raw data. This method is convenient if you have a small amount of data. Use the DATALINES statement to indicate raw internal data. In the following example, the data set is to be named JNJ. The INPUT statement tells SAS how to read the data, specifically the variable names. There are seven variables in the data. We input date in the form of yyyy/mm/dd.*/

```
Data JNJ;
   INPUT Date yymmdd10. Open High Low Close Volume AdjClose;
/*Define the name of each column and keep Date in form yyyy/mm/dd*/
   DATALINES;
1980/1/2 79 81.62 75 75.37 3234900 0.74
1980/2/1 75 76.25 68.5 68.62 3618700 0.68
1980/3/3 68.62 71.12 66 71 3252800 0.7
1980/4/1 71 76.5 70 75.87 2794500 0.75
1980/5/1 75.87 81.87 75.12 80 2987400 0.8
1980/6/2 79.87 82.5 78.25 78.5 2363400 0.78
1980/7/1 78.5 81.5 78.12 81 2884800 0.81
1980/8/1 81 85.75 79 79.37 2698200 0.8
1980/9/2 79.37 84.37 76.5 81 3821000 0.81
1980/10/1 81 85.5 75.37 84.75 4561400 0.85
1980/11/3 85.5 94.5 81 90.75 6035200 0.92
1980/12/1 90.75 101.37 85.12 99.75 4347400 1.01
1981/1/2 99.75 102.12 92.62 95.75 3485900 0.97
1981/2/2 95.37 104 93.87 103 3179800 1.05
1981/3/2 102.87 107 100.62 103.75 3521400 1.06
1981/4/1 103.75 112.37 101 106.62 3808200 1.09
1981/5/1 106.62 109 35.25 38.25 2763200 1.18
1981/6/1 38.25 39.38 34.5 35 2694800 1.08
1981/7/1 35 35.13 30.12 33.63 2201400 1.03
1981/8/3 33.63 34.13 29.75 29.87 2657800 0.92
1981/9/1 29.87 33.5 28.25 33 3218200 1.02
1981/10/1 33 36.75 32.38 36.38 3166200 1.13
1981/11/2 36.63 38 33.75 37.75 2981500 1.18
1981/12/1 37.75 37.75 34 37.13 1914400 1.16
1982/1/4 36.75 38.13 33.75 37.88 3420600 1.18
1982/2/1 37.13 39.13 36 37 3769000 1.16
1982/3/1 37.38 37.88 32.5 37 3674200 1.16
1982/4/1 37.13 40.38 36.63 40 3576400 1.25
1982/5/3 40.38 42.13 38 39.25 3093700 1.24
1982/6/1 39.63 40.75 37.5 39.5 3339200 1.25
1982/7/1 39.25 43 38 39.63 3124300 1.25
1982/8/2 40.88 45.5 38 45 5445100 1.43
1982/9/1 44.38 47.88 42.5 43.13 5700000 1.37
1982/10/1 42.88 44.5 38.75 43.88 18349200 1.39
1982/11/1 44.25 48 43.5 46.5 7377900 1.48
1982/12/1 46.25 51.25 44.63 49.63 5563500 1.58
1983/1/3 49.75 51.38 46 48.38 4785200 1.54
1983/2/1 48.38 48.88 45 47.38 6238700 1.52
1983/3/1 48.5 50.63 44.63 47.38 8630800 1.52
1983/4/4 46.5 50.63 44.75 50.13 7923600 1.61
1983/5/2 49.5 51.5 45.75 46 6312300 1.48
1983/6/1 46 49.5 45.38 49.25 8221300 1.59
1983/7/1 49.5 49.75 42.75 44 6167600 1.42
1983/8/1 43.88 44.63 40.13 41.13 6729500 1.34
1983/9/1 41.38 45.88 40.38 45.13 5903700 1.47
1983/10/3 44.75 48 43.25 43.88 5479600 1.42
1983/11/1 43.75 44.75 41.5 42.38 4831400 1.38
1983/12/1 42.38 42.38 39 40.88 7203800 1.34
1984/1/3 40.5 42.88 37.25 38.38 8445400 1.25
1984/2/1 38.38 40 34.13 35.88 7448200 1.18
1984/3/1 35.88 36.38 33.38 35 5349700 1.15
1984/4/2 35.13 38.63 34.88 37.38 5069200 1.23
```

```
1984/5/1 37.38 38 30.12 30.62 7257600 1.02
1984/6/1 30.62 33.13 29.37 30 8657200 1
1984/7/2 30 30.87 28 29.25 6648200 0.97
1984/8/1 29.25 34.25 29.25 33.75 8312600 1.13
1984/9/4 33.63 34.88 31.75 32.5 5879800 1.09
1984/10/1 32.13 33.88 30.25 32.5 5378300 1.09
1984/11/1 32.5 36.63 32.25 35.5 7493000 1.2
1984/12/3 35.38 37.25 33.75 36.13 6300800 1.22
1985/1/2 35.88 39.25 35.13 39 6820000 1.32
1985/2/1 38.25 40.13 37 39.38 7219500 1.34
1985/3/1 39 42.13 38.75 42 7756800 1.43
1985/4/1 41.75 44.75 40.38 43.88 9078000 1.49
1985/5/1 43.63 47.38 42.88 47 6525600 1.61
1985/6/3 47.38 47.88 44.88 46.38 5812800 1.59
1985/7/1 46.5 49.13 45 46.13 7084200 1.58
1985/8/1 46.75 47.5 44.25 46.75 4921300 1.61
1985/9/3 46.75 48 42.75 44.38 6103400 1.53
1985/10/1 44.25 47.75 43.13 47 5933800 1.62
1985/11/1 46.88 50.63 46.63 49.88 11448100 1.73
1985/12/2 49.63 55.25 48.25 52.63 8357500 1.83
1986/1/2 52.5 54 49.25 53.75 6325600 1.87
1986/2/3 53.75 55 45.75 49.75 31687300 1.74
1986/3/3 49.88 58.5 48.13 57.38 14274400 2.01
1986/4/1 57.75 69.37 52.88 65.87 14886700 2.31
1986/5/1 64.87 70.5 64 68.37 7941500 2.41
1986/6/2 68.12 73.75 65.62 72.62 7095500 2.55
1986/7/1 72.62 72.87 64.37 68.12 8141600 2.4
1986/8/1 68.25 74.5 67 73.87 7158000 2.61
1986/9/2 73.5 73.87 60.25 64 9497900 2.26
1986/10/1 64.5 70.12 61.88 69.75 5719400 2.47
1986/11/3 70 70.87 64.75 68.5 6965800 2.43
1986/12/1 67.5 72 65.25 65.62 8123400 2.33
1987/1/2 65.62 78.87 65.62 75.62 12411500 2.69
1987/2/2 75.37 88.5 74.75 86.25 10537000 3.08
1987/3/2 87 94.5 85.37 88.5 9923800 3.16
1987/4/1 87.75 92.12 83.12 87.12 11077100 3.11
1987/5/1 86.75 90.37 80.25 86 7792400 3.08
1987/6/1 86.37 95.25 85 91.87 6439400 3.29
1987/7/1 92 96.75 89.12 95.25 7525800 3.42
1987/8/3 95 105.37 92.5 99.75 8788400 3.59
1987/9/1 100 100.5 89.5 95.5 10951500 3.44
1987/10/1 96 99.5 55.25 80 16290900 2.88
1987/11/2 79.25 80.62 70.75 73 7791200 2.64
1987/12/1 73.75 80.37 70 74.87 7306400 2.71
1988/1/4 75.87 79.87 69.62 78.37 12605600 2.84
1988/2/1 79 83.5 76 83.25 7739200 3.03
1988/3/1 83.37 86.62 77.37 79.37 7203600 2.89
1988/4/4 78.87 84.12 75.37 75.75 7738700 2.76
1988/5/2 76 78 69.25 77.75 7497100 2.85
1988/6/1 78.25 81.75 76.87 79.37 8363700 2.91
1988/7/1 79.5 80.37 76 80 5805400 2.93
1988/8/1 80 82.37 78.75 81 5395800 2.99
1988/9/1 80.62 86.75 79.75 85.87 6199900 3.17
1988/10/3 85.75 88.12 84.25 87 5440600 3.21
1988/11/1 87 88 80.62 86 5783500 3.19
1988/12/1 86 87.5 84.25 85.12 4323200 3.16
```

```
1989/1/3 85 92 83 91.62 6595200 3.4
1989/2/1 91.62 92.25 85.5 86.12 6282500 3.21
1989/3/1 86.75 93.5 85.12 90.75 6739500 3.39
1989/4/3 90.87 96.87 89.87 96.12 5653600 3.59
1989/5/1 95.5 97.25 47.13 49.38 6451800 3.71
1989/6/1 49.75 53.75 47.13 47.88 6226600 3.59
1989/7/3 48.75 54.25 47.88 53.38 5582100 4.01
1989/8/1 53.25 53.38 49.38 52.38 6416200 3.95
1989/9/1 52.63 54.88 50 54.75 5751400 4.13
1989/10/2 54.25 56.63 50.5 55.38 6351500 4.18
1989/11/1 55 57.25 52.38 56.5 4700300 4.29
1989/12/1 57 59.5 55.5 59.38 4862400 4.51
1990/1/2 58.75 60.75 51.13 52.38 6447200 3.97
1990/2/1 54.25 56.63 52 53.63 5187400 4.09
1990/3/1 53.88 57.5 53.75 56.13 4443300 4.28
1990/4/2 55.75 59.38 55.5 56.63 5000700 4.32
1990/5/1 57 64.37 56.75 63.88 5984100 4.9
1990/6/1 64.5 68.25 62.88 67.75 5422400 5.2
1990/7/2 67.75 74.12 67.5 70.62 5833500 5.42
1990/8/1 70.37 70.75 60 65.62 7473600 5.06
1990/9/4 64.62 66.75 60.5 63.25 5779200 4.88
1990/10/1 64 69.5 63.38 64.37 6154100 4.96
1990/11/1 64.25 70.37 63.75 69.75 6097000 5.41
1990/12/3 70.75 72.12 69.5 71.75 4826800 5.56
1991/1/2 71.37 76.62 65.37 76.5 7080600 5.93
1991/2/1 76.25 82.87 75 82.12 6873900 6.39
1991/3/1 81.12 99.5 81 95.87 9212300 7.46
1991/4/1 95.62 100.37 93.12 94.5 7249400 7.35
1991/5/1 94.25 96.25 88.87 90.62 6019000 7.08
1991/6/3 89.25 90 82 83.5 7708900 6.53
1991/7/1 84.5 91.62 81.75 91.62 5131900 7.16
1991/8/1 91.62 93.37 85 92.87 4132800 7.29
1991/9/3 93.12 93.62 88 88.37 4013300 6.94
1991/10/1 88.5 95.5 84.75 95.12 5340500 7.47
1991/11/1 94.87 100.12 92.5 96.37 4997000 7.6
1991/12/2 95.25 116.25 95.25 114.5 5877000 9.02
1992/1/2 112.5 117.37 102.25 107.5 6822000 8.47
1992/2/3 106.75 109.75 96.5 101.25 6128900 8.01
1992/3/2 100.75 102.62 94 97.5 5168400 7.71
1992/4/1 97 102.75 88.87 94.87 8204400 7.51
1992/5/1 94.37 97.75 92 96.87 5255800 7.7
1992/6/1 96.5 98 43 44.75 6779600 7.12
1992/7/1 44.63 49.63 44.63 49.25 3981100 7.83
1992/8/3 48.88 51.25 48.13 49.38 3353600 7.89
1992/9/1 49.25 52.88 46.5 47.13 4022900 7.53
1992/10/1 47.38 51 43.25 50 4469400 7.99
1992/11/2 49.75 51.88 47.63 50.38 3571100 8.09
1992/12/1 50.25 54.75 49.75 50.5 3778000 8.11
1993/1/4 50.25 50.38 43.5 44 5943400 7.06
1993/2/1 44.38 46.13 37.63 42.5 8792100 6.86
1993/3/1 42.88 44.75 38.63 42.63 7387700 6.88
1993/4/1 42.63 43.88 38.38 43.63 5828400 7.04
1993/5/3 44 45.75 42.25 44.63 4794600 7.24
1993/6/1 44.5 45.75 40.5 41.5 5156000 6.73
1993/7/1 41.25 41.38 36.13 36.5 6572500 5.92
1993/8/2 36.63 41 35.63 40.75 6467000 6.66
```

```
1993/9/1 40.63 41.75 37.63 39.38 5682700 6.44
1993/10/1 39.25 42.63 38.5 42.13 4521600 6.89
1993/11/1 42.13 45.63 41.88 43.63 4521300 7.17
1993/12/1 43.75 45.5 42.38 44.88 4086300 7.38
1994/1/3 44.75 45.75 40.13 42.38 4914800 6.97
1994/2/1 42.25 43.13 40 40.13 4753500 6.64
1994/3/1 40.13 40.88 36 37.75 4976400 6.24
1994/4/4 36.5 41.88 36.25 41.38 5611300 6.85
1994/5/2 41.38 44.88 40.5 44.25 4487100 7.37
1994/6/1 44.13 44.25 42.25 42.88 3707700 7.14
1994/7/1 43.13 47.63 42.25 47 5284200 7.83
1994/8/1 47 50.63 46.63 50.13 6292700 8.4
1994/9/1 50.13 52.38 48.38 51.75 4807300 8.67
1994/10/3 51.88 56.13 49.5 54.63 5166900 9.16
1994/11/1 54.25 56.5 51.25 53.38 5489800 9
1994/12/1 53.63 55.75 52.13 54.75 4274200 9.23
1995/1/3 54.63 59.5 53.63 58.13 4788300 9.8
1995/2/1 58.38 58.38 55.5 56.75 4734100 9.61
1995/3/1 56.63 63 56.38 59.5 4601900 10.08
1995/4/3 59.38 65.12 58.38 65 5082500 11.01
1995/5/1 64.75 67.62 61.63 66.12 5342000 11.25
1995/6/1 66 71.25 65.37 67.5 3554600 11.49
1995/7/3 67.25 71.75 64.37 71.75 4475000 12.21
1995/8/1 71.5 71.5 66.87 69 2942700 11.8
1995/9/1 69 74.87 68.37 74.12 3586200 12.68
1995/10/2 73.62 84.12 73.12 81.5 4341000 13.94
1995/11/1 81.25 87.25 79.62 86.62 3630800 14.87
1995/12/1 86.87 92.37 85.12 85.5 4993700 14.68
1996/1/2 85.5 96.5 83.12 96 5841500 16.48
1996/2/1 96.25 100.37 92.37 93.5 8224400 16.11
1996/3/1 94.25 98.25 91 92.25 5989000 15.89
1996/4/1 93.12 95.25 85.62 92.5 5838900 15.94
1996/5/1 92 98.37 88.37 97.37 4684800 16.84
1996/6/3 97 100.37 47.5 49.5 4315300 17.13
1996/7/1 49.75 50 44.13 47.75 3736300 16.52
1996/8/1 47.75 53.38 47.75 49.25 3133800 17.1
1996/9/3 48.63 51.75 48.25 51.25 3278800 17.8
1996/10/1 51.25 52.88 47.13 49.25 4234300 17.1
1996/11/1 49.75 54 48.88 53.25 4170900 18.56
1996/12/2 53.13 53.75 48 49.75 4130100 17.34
1997/1/2 49.88 58.5 48.63 57.75 5994500 20.13
1997/2/3 58 62.75 56.63 57.5 5223600 20.1
1997/3/3 57 61.38 52.75 52.88 5688800 18.49
1997/4/1 52.88 61.5 51.13 61.13 5022700 21.37
1997/5/1 60.63 62.63 58.25 60 4496500 21.05
1997/6/2 60.25 66.5 58.75 64.37 5156000 22.59
1997/7/1 64.37 65.87 60 62.13 5637000 21.8
1997/8/1 62 62.5 55.94 56.69 6240000 19.97
1997/9/2 56.75 60.56 55.13 57.69 5979500 20.32
1997/10/1 58.13 62.13 53 57.38 5845500 20.21
1997/11/3 58.44 65 58.19 62.94 5023100 22.25
1997/12/1 63.38 67.31 63.13 65.87 4470100 23.28
1998/1/2 65.06 69.44 63.38 66.94 5345800 23.66
1998/2/2 68.5 75.5 68.06 75.37 4606200 26.73
1998/3/2 75.19 76.44 71.06 73.44 3975200 26.04
1998/4/1 73.31 77.5 68.06 71.5 5082900 25.36
```

```
1998/5/1 72.5 72.69 68.12 69.06 3633500 24.58
1998/6/1 69.87 77.87 67.06 74 4660500 26.33
1998/7/1 73.75 79.37 70.81 77.25 5129600 27.49
1998/8/3 77.19 77.81 68.87 69 4583000 24.63
1998/9/1 69.12 80.81 68.12 78.25 5777300 27.94
1998/10/1 77.12 84.69 72.62 81.5 5827800 29.1
1998/11/2 81.44 89.75 80.62 81.25 4600300 29.09
1998/12/1 81.06 84.37 76.19 83.87 6107700 30.03
1999/1/4 82.94 86.31 77 85.12 6118700 30.48
1999/2/1 84.87 88.12 81.62 85.37 4678900 30.66
1999/3/1 84.75 94 83.69 93.5 4473600 33.58
1999/4/1 93.25 103 88.87 97.5 5234400 35.02
1999/5/3 97.44 98 87.81 92.62 5109600 33.36
1999/6/1 92.44 98 89.25 98 4073300 35.3
1999/7/1 97.37 99.5 91.06 91.06 4256900 32.8
1999/8/2 91.94 105.87 90.31 102.25 4936300 36.94
1999/9/1 102.25 102.62 90 91.87 5154800 33.19
1999/10/1 92.12 106.44 92.12 104.75 6326000 37.84
1999/11/1 105.25 106.87 101.81 103.75 4536500 37.58
1999/12/1 103.25 103.44 90.12 93.25 6547700 33.78
2000/1/3 93.12 96.94 80.5 86.06 8133900 31.18
2000/2/1 85.44 86.31 70.06 72 9840400 26.18
2000/3/1 72.25 82.25 66.12 70.25 12519600 25.54
2000/4/3 70.62 84.44 70 82.5 9266200 29.99
2000/5/1 82.25 90 81.87 89.5 5671200 32.66
2000/6/1 89.25 101.98 82.5 101.87 7550100 37.17
2000/7/3 101.25 101.44 90.25 93.06 7025700 33.96
2000/8/1 92.5 98.87 91.25 91.95 4573000 33.66
2000/9/1 92.5 97.81 90.31 93.94 5096500 34.39
2000/10/2 94.19 97.5 90.31 92.12 6778300 33.72
2000/11/1 92 103 89.19 100 7308700 36.73
2000/12/1 99.25 105.94 95.25 105.06 7247600 38.59
2001/1/2 104 104.69 90.37 93.13 8434400 34.21
2001/2/1 93.13 98.25 93.13 97.33 6520200 35.87
2001/3/1 97.8 99.85 80.5 87.47 14132500 32.24
2001/4/2 87.5 96.7 85.19 96.48 10913100 35.56
2001/5/1 96.6 102 95.82 96.95 8773800 35.86
2001/6/1 96.96 103.5 49.13 49.96 9688900 36.96
2001/7/2 50.1 54.98 50 54.1 8229200 40.02
2001/8/1 54.1 57.6 52.3 52.71 7394900 39.12
2001/9/4 55.35 57.16 50.2 55.4 12474300 41.12
2001/10/1 55.4 59.66 53.05 57.91 8206900 42.98
2001/11/1 57.58 60.97 57.51 58.25 7557100 43.37
2001/12/3 57.82 60.18 55.42 59.1 8706900 44
2002/1/2 58.9 60.1 56.6 57.51 6601800 42.81
2002/2/1 57.26 60.97 54.7 60.9 7271200 45.48
2002/3/1 60.9 65.89 60.6 64.95 8019000 48.5
2002/4/1 64.2 65.29 61.78 63.86 7653200 47.69
2002/5/1 63.97 64.63 59.16 61.35 7646900 45.97
2002/6/3 61.1 61.74 52 52.26 10014800 39.16
2002/7/1 51.7 54.85 41.4 52.6 16378700 39.41
2002/8/1 53 56.5 49.75 54.31 7507900 40.85
2002/9/3 53.88 56.25 51.45 54.08 7827500 40.67
2002/10/1 54.95 61.3 54.3 58.75 10150900 44.18
2002/11/1 58.33 61.24 56.3 57.02 6937200 43.03
2002/12/2 56.59 57.07 53 53.71 7122900 40.53
```

```
2003/1/2 54.25 57.55 51.54 53.61 7956100 40.46
2003/2/3 53.5 53.95 49.1 52.45 7655000 39.74
2003/3/3 53.16 58.68 51.5 57.87 8083400 43.85
2003/4/1 57.87 59.08 53.95 56.36 7919500 42.71
2003/5/1 56.36 57.52 52.55 54.35 8083600 41.36
2003/6/2 54.35 55.57 50.75 51.7 10821600 39.35
2003/7/1 51.7 54.24 50.55 51.79 8639900 39.41
2003/8/1 50.95 52.25 49 49.58 7795900 37.91
2003/9/2 49.78 52.5 49.09 49.52 9837000 37.86
2003/10/1 49.71 51.15 48.05 50.33 9170000 38.48
2003/11/3 50.08 52.89 48.26 49.3 8957100 37.88
2003/12/1 50.3 51.74 48.79 51.66 8703400 39.69
2004/1/2 51.66 54.04 50.43 53.42 8932100 41.04
2004/2/2 53.41 54.9 52.79 53.91 6945300 41.6
2004/3/1 54.49 54.49 49.25 50.72 7831500 39.14
2004/4/1 51.05 54.77 49.9 54.03 8295300 41.69
2004/5/3 54.22 56.39 54.02 55.71 7309700 43.21
2004/6/1 55.23 57.28 54.5 55.7 6259700 43.21
2004/7/1 55.35 58.14 54.43 55.27 6996800 42.87
2004/8/2 55 58.1 54.37 58.1 5129900 45.3
2004/9/1 58 58.8 56.09 56.33 5635800 43.92
2004/10/1 56.39 58.48 54.81 58.38 6985400 45.52
2004/11/1 58.35 61.49 57.81 60.32 6310600 47.25
2004/12/1 60.33 64.25 59.6 63.42 8040900 49.68
2005/1/3 63.35 64.98 61.2 64.7 7118000 50.68
2005/2/1 64.61 66.89 64.52 65.6 6020500 51.61
2005/3/1 66.47 68.68 66.12 67.16 7756800 52.84
2005/4/1 67.32 69.99 66.65 68.63 7865400 53.99
2005/5/2 68.68 68.87 66.76 67.1 6294600 53.05
2005/6/1 67.1 67.61 64.43 65 7791300 51.39
2005/7/1 64.82 65.43 62.67 63.96 7802500 50.57
2005/8/1 64.25 65.35 61.65 63.39 6349400 50.37
2005/9/1 63.18 65.33 62.85 63.28 6909100 50.29
2005/10/3 63.18 64.6 60.53 62.62 11139300 49.76
2005/11/1 62.48 63.78 60 61.75 11850700 49.33
2005/12/1 61.85 62.06 59.76 60.1 9291900 48.01
2006/1/3 61.07 63.1 57.45 57.54 12062900 45.96
2006/2/1 57.67 59.71 56.7 57.65 10120200 46.31
2006/3/1 57.62 61.1 57.45 59.22 8300200 47.57
2006/4/3 59.49 59.76 57.32 58.61 9458100 47.08
2006/5/1 58.71 60.93 58.19 60.22 9312000 48.68
2006/6/1 60.25 62 58.97 59.92 9531300 48.44
2006/7/3 60.13 63 59.68 62.55 7101400 50.56
2006/8/1 62.38 65.13 62.29 64.66 6743000 52.57
2006/9/1 64.87 65.04 63.25 64.94 6809600 52.8
2006/10/2 65 69.41 64.5 67.4 9157300 54.8
2006/11/1 67.38 69.03 65.22 65.91 9708100 53.89
2006/12/1 65.91 67.25 65.29 66.02 8170400 53.98
2007/1/3 66.13 68.22 65.9 66.8 9188300 54.62
2007/2/1 66.8 67.15 62.72 62.93 11003400 51.75
2007/3/1 62.35 62.93 59.87 60.26 15064600 49.56
2007/4/2 60.26 65.45 60.03 64.22 13144300 52.81
2007/5/1 64.53 64.94 61.65 63.27 13292500 52.37
2007/6/1 63.35 63.77 60.73 61.62 12951700 51.01
2007/7/2 61.94 63.88 59.72 60.5 14868200 50.08
2007/8/1 60.51 62.98 60.31 61.79 14975300 51.49
```

```
2007/9/4 61.63 65.75 61.26 65.7 12074400 54.75
2007/10/1 65.66 66.8 63.55 65.17 10584000 54.31
2007/11/1 64.8 68.75 63.73 67.74 14358000 56.8
2007/12/3 67.62 68.5 66.55 66.7 10284200 55.93
2008/1/2 66.56 68.85 61.72 63.14 17202500 52.94
2008/2/1 63.35 64.07 61.36 61.96 11547600 52.3
2008/3/3 62 65.81 61.17 64.87 14414500 54.76
2008/4/1 65.13 67.74 64.51 67.09 10684100 56.63
2008/5/1 67.13 68.32 64.54 66.74 10997900 56.73
2008/6/2 66.55 67.1 63.1 64.34 11466800 54.69
2008/7/1 63.75 69.25 63.74 68.47 14655400 58.2
2008/8/1 68.5 71.9 67.93 70.43 11163400 60.26
2008/9/2 71.09 72.76 66.9 69.28 18434600 59.27
2008/10/1 69.07 69.07 52.06 61.34 28704800 52.48
2008/11/3 61.49 63.01 54.49 58.58 20175800 50.53
2008/12/1 57.66 60.25 54.95 59.83 14868800 51.61
2009/1/2 60.13 61 55.09 57.69 18162400 49.77
2009/2/2 57.25 59.13 49.94 50 17801100 43.49
2009/3/2 49.03 53.7 46.25 52.6 20083500 45.75
2009/4/1 52.3 54.13 50.12 52.36 15542900 45.54
2009/5/1 52.53 56.48 51.67 55.16 13488300 48.4
2009/6/1 55.75 57.23 54.54 56.8 14441400 49.84
2009/7/1 56.82 62.47 55.71 60.89 12890300 53.43
2009/8/3 61.16 61.49 59.5 60.44 10573800 53.47
2009/9/1 60.35 61.54 59.28 60.89 9928500 53.86
2009/10/1 60.93 62.74 58.78 59.05 13693800 52.24
2009/11/2 59.15 63.44 58.8 62.84 12211600 56.03
2009/12/1 62.95 65.41 62.95 64.41 9965800 57.43
2010/1/4 64.71 65.95 62.22 62.86 13840500 56.05
2010/2/1 63.26 64.7 61.89 63 11721000 56.6
2010/3/1 63.27 65.49 63.02 65.2 11386100 58.58
2010/4/1 65.37 66.2 64.21 64.3 11910300 57.77
2010/5/3 65.16 65.62 58.19 58.3 19229500 52.86
2010/6/1 58.21 60.15 57.55 59.06 16961100 53.55
2010/7/1 59.07 61.73 56.86 58.09 15413600 52.67
2010/8/2 58.5 60.25 56.99 57.02 11543000 52.18
2010/9/1 57.67 62.7 57.44 61.96 10575400 56.71
2010/10/1 62.09 64.23 61.3 63.74 10382400 58.33
2010/11/1 63.88 64.92 61.25 61.55 11144100 56.82
2010/12/1 62.18 63.16 61.56 61.85 9869700 57.09
2011/1/3 62.63 63.54 59.69 59.77 13815400 55.17
2011/2/1 59.7 61.45 59.36 61.44 11577000 57.23
2011/3/1 61.25 61.5 57.5 59.25 11487600 55.19
2011/4/1 59.63 65.97 59.06 65.72 14148000 61.21
2011/5/2 65.92 67.37 64.8 67.29 13684000 63.22
2011/6/1 67.02 67.36 64.69 66.52 10978000 62.5
2011/7/1 66.48 68.05 64.13 64.79 10972100 60.87
2011/8/1 65.28 66.23 59.08 65.8 22001900 62.37
2011/9/1 65.66 66.14 61 63.69 13305600 60.37
2011/10/3 63.65 65.86 60.83 64.39 12633800 61.03
2011/11/1 63.68 65.49 61.05 64.72 11997400 61.92
2011/12/1 64.68 66.32 62.85 65.58 10249700 62.74
2012/1/3 65.59 66 64.34 65.91 11284900 63.05
```

```
2012/2/1 65.97 66.32 64.02 65.08 10192900 62.81
2012/3/1 65.11 66.18 64.12 65.96 9975900 63.66
2012/4/2 66.04 66.3 62.76 65.1 12051400 62.83
2012/5/1 64.91 65.49 62 62.43 13241200 60.84
2012/6/1 62.04 67.7 61.71 67.56 24223300 65.84
2012/7/2 67.42 69.7 67 69.22 13507600 67.46
2012/8/1 69.48 69.75 66.85 67.43 10184900 66.31
2012/9/4 67.49 69.64 66.97 68.91 11079800 67.76
2012/10/1 68.92 72.74 67.8 70.82 13195100 69.64
2012/11/1 71.1 71.9 68.51 69.73 10721100 69.18
2012/12/3 69.8 71.62 69.36 69.48 12080200 68.93
;
run;
```

2.7.3 *Reading Text Files Into SAS*

If you put your data in a file to be read into SAS, it is considered external raw data. Use the INFILE statement to tell SAS the filename and path of the external file containing the data.

Suppose that you have a file named "Data used in Excel, Minitab, and SAS book.dat" in the Windows directory d:\Minitab and Microsoft Excel Book\Ch2. After the INFILE keyword, the file path and name must be in quotation marks. Under Windows, the path is specified in this way:

```
    INFILE 'd:\MINITAB and Microsoft Excel Book\Ch2\Data used in Excel, MINITAB, and SAS
book.dat';
```

Submit the program to read the file. Remember to check the Log window. Things often go wrong at this step.

```
    libname Mydata 'd:\SASdata\MySASdata';
    /*Setting new library*/
    data Mydata.JNJ;
       INFILE 'd:\MINITAB and Microsoft Excel Book\Ch2\Data used in Excel, MINITAB, and
SAS book.dat';
    /*Change the INFILE statement for your directory*/
       INPUT Date yymmdd10. Open High Low Close Volume Adj_Close;
    /*Define the name of each column*/
    RUN;

    /* To print the file, We use the following code:*/

    proc print data=Mydata.JNJ (obs=20);
    /*Print the first 20 records from data Mydata.JNJ*/
    format Date date9.;
    /*Print Column Date in form DDMMMYYYY*/
    run;
```

2.7.4 Reading Excel Files Into SAS

Using the Import Wizard is an easy way to import data into SAS. Although the Import Wizard is easy, it can be time consuming if used repeatedly. Here, we introduce an alternative approach to import data in excel files.

The **out=option** in the **PROC IMPORT** tells SAS what the name should be for the newly created SAS data file and where to store the data set once it is imported.

Next the **DATAFILE=option** tells SAS where to find the file we want to import.

The **DBMS=option** is used to identify the type of file being imported.

The **replace** option will overwrite an existing file.

To specify which sheet SAS should import, use the **sheet="sheetname"** statement. The default is for SAS to read the first sheet.

The **getnames=yes** is the default setting and SAS will automatically use the first row of data as variable names. If the first row of your sheet does not contain variable names, use the **getnames=no**.

```
    libname Mydata 'd:\SASdata\MySASdata';
    proc import out=Mydata.JNJ
          datafile="d:\MINITAB and Microsoft Excel Book\Ch2\Data used in Excel, MINITAB,
and SAS book.xls"
          DBMS=excel2000 replace;
          sheet="JNJ Monthly stock price";
          getnames=yes;
            run;
```

2.7.5 Charting JNJ's Data

In this section, we will talk about the ways how to print your SAS data. We mainly focus on **PROC GPLOT** which could create "publication quality" color graphics.

The **PROC SGPLOT** statement plots a line to show the close price of each month. Reference lines will be drawn automatically. **AXIS** specifies the axis that contains the reference line values. **LABEL=option** specifies one or more text strings to be used as labels for the reference lines.

```
Goptions ftext='Times New Roman' htext=4 gunit=pct;
/*Select font and size*/
title  height=8 "JNJ's Monthly Stock Price-Jan. 1980 to Dec. 2000";
/*Set the size of title*/
axis1  label=('Date');
/*Label the horizontal axis*/
axis2  label=(angle=90 '$')
       order=(20 to 120 by 10);
/*Label the vertical axis, rotate 90 degrees, and scale from 20 to 120 by 10*/
Proc gplot data=Mydata.JNJ;
plot  close*date/haxis=axis1 vaxis=axis2 autovref;
/*Plot close on date, and show horizontal reference lines automatically */
symbol v=none i=join c=black;
/*Decide the type of points(v=none or square or heart..) and
the color of the line(c=black or red or..)*/
Run;
```

The following graph is achieved by using the code above:

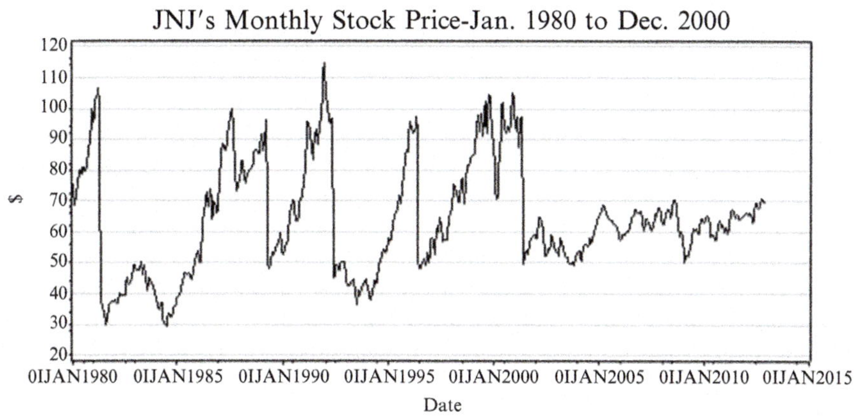

2.8 JPM's Financial Statements

Financial accounting statements are used to understand a firm's financial condition. The three required financial accounting statements for publicly traded companies are the *balance sheet*, *income statement*, and *statement of cash flows*. We can use Yahoo Finance to see a company's financial statement for the past 3 years.

The web link to JPM's balance sheet for the last 3 years is http://finance.yahoo.com/q/bs?s=JPM+Balance+Sheet&annual

The web link to JPM's income statement for the last 3 years is http://finance.yahoo.com/q/is?s=JPM+Income+Statement&annual

The web link to JPM's statement of cash flow for the last 3 years is http://finance.yahoo.com/q/cf?s=JPM+Cash+Flow&annual

Once you know the link to these financial statements for one company it is easy to get the financial statements for another company. For example, the following are the links to Merck's financial statements:

The web link to Merck's balance sheet for the last 3 years is http://finance.yahoo.com/q/bs?s=MRK+Balance+Sheet&annual

The web link to Merck's income statement for the last 3 years is http://finance.yahoo.com/q/is?s=MRK+Income+Statement&annual

The web link to Merck's statement of cash flow for the last 3 years is http://finance.yahoo.com/q/cf?s=MRK+Cash+Flow&annual

We can use Yahoo Finance to see a company's financial statement for the past four quarters.

The web link to JPM's balance sheet for the last four quarters is http://finance.yahoo.com/q/bs?s=JPM+Balance+Sheet

The web link to JPM's income statement for the last 3 years is http://finance.yahoo.com/q/is?s=JPM+Income+Statement

The web link to JPM's statement of cash flow for the last 4 years is http://finance.yahoo.com/q/cf?s=JPM+Cash+Flow

The web link to Merck's balance sheet for the last four quarters is http://finance.yahoo.com/q/bs?s=MRK+Balance+Sheet&annual

The web link to Merck's income statement for the last four quarters is http://finance.yahoo.com/q/is?s=MRK+Income+Statement&annual

The web link to Merck's statement of cash flow for the last four quarters is http://finance.yahoo.com/q/cf?s=MRK+Cash+Flow&annual

2.9 Retrieving Stock Prices From Google Finance

Another source to get historical stock price is from Google Finance. Below are examples of URLs to retrieve stock prices. Note that the logical pattern in the URL construction.

The following is the URL to retrieve historical stock prices for Proctor from Google Finance, http://finance.google.com/finance/historical?q=GE&startdate=Jan+1+2014&enddate=Dec+31+2014&output=csv

The following is the URL to retrieve historical stock prices for Walmart from Google Finance, http://finance.google.com/finance/historical?q=WMT&startdate=Jan+1+2014&enddate=Dec+31+2014&output=csv

The following is the URL to retrieve stock prices for historical Walmart from Google Finance, http://finance.google.com/finance/historical?q=MMM&startdate=Jan+1+2014&enddate=Dec+31+2014&output=csv

2.10 Retrieving Stock Prices from QUANDL

QUANDL offers free and premium data sets. The following two articles discuss in more detail what QUANDL is: http://www.computerworld.com/article/2474710/business-intelligence/quandl--wikipedia-for-data-.html

http://blog.revolutionanalytics.com/2013/02/quandl-a-wikipedia-for-time-series-data.html

The following is the URL to retrieve stock prices for Apple from QUANDL: https://www.quandl.com/api/v3/datasets/WIKI/AAPL.csv?start_date=2014-01-01&end_date=2014-12-31

The following is the URL to retrieve stock prices for Goldman Sachs from QUANDL: https://www.quandl.com/api/v3/datasets/WIKI/GS.csv?start_date=2014-01-01&end_date=2014-12-31

The following is the URL to retrieve stock prices for McDonald's from QUANDL: https://www.quandl.com/api/v3/datasets/WIKI/MCD.csv?start_date=2014-01-01&end_date=2014-12-31

One thing you can do at QUANDL that you can't do with the other sites is to request data in XML or JASON format.

The following is the URL to retrieve stock prices for Apple from QUANDL in XML format: https://www.quandl.com/api/v3/datasets/wiki/AAPL.xml?start_date=2014-01-01&end_date=2014-12-31

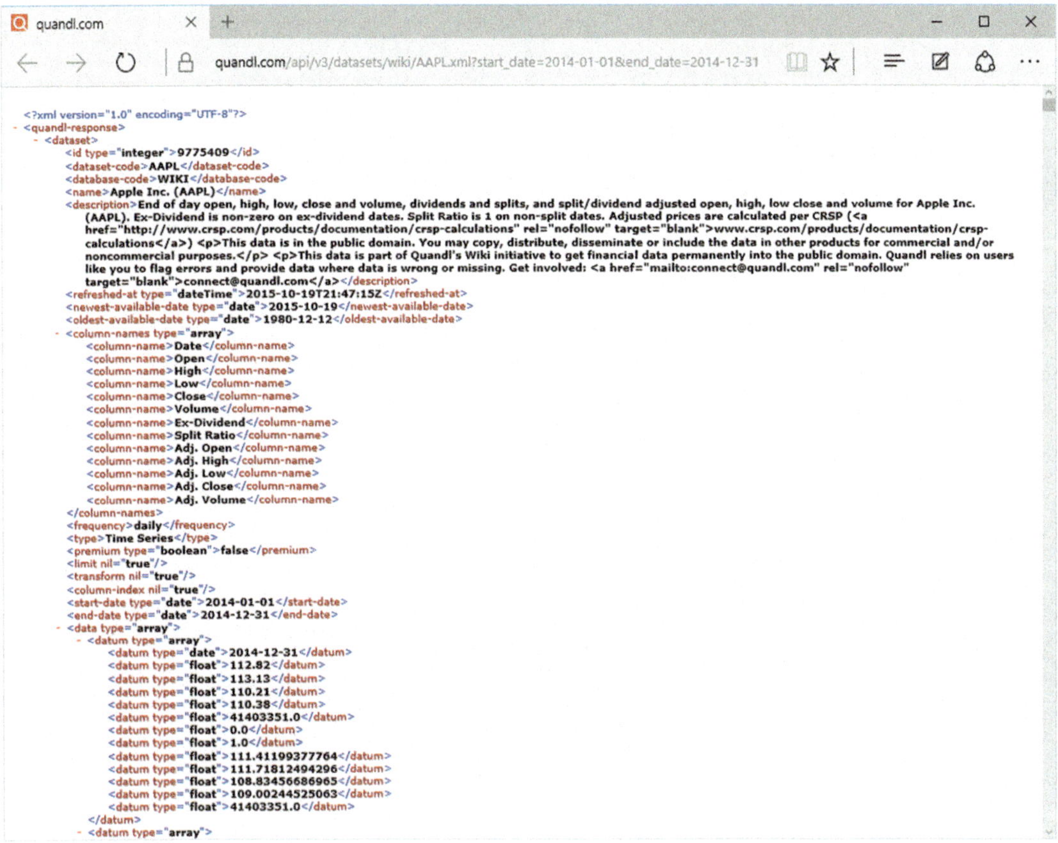


```
quandl.com/api/v3/datasets/wiki/AAPL.xml?start_date=2014-01-01&end_date=2014-12-31

<?xml version="1.0" encoding="UTF-8"?>
- <quandl-response>
  - <dataset>
      <id type="integer">9775409</id>
      <dataset-code>AAPL</dataset-code>
      <database-code>WIKI</database-code>
      <name>Apple Inc. (AAPL)</name>
      <description>End of day open, high, low, close and volume, dividends and splits, and split/dividend adjusted open, high, low close and volume for Apple Inc.
        (AAPL). Ex-Dividend is non-zero on ex-dividend dates. Split Ratio is 1 on non-split dates. Adjusted prices are calculated per CRSP (<a
        href="http://www.crsp.com/products/documentation/crsp-calculations" rel="nofollow" target="blank">www.crsp.com/products/documentation/crsp-
        calculations</a>) <p>This data is in the public domain. You may copy, distribute, disseminate or include the data in other products for commercial and/or
        noncommercial purposes.</p> <p>This data is part of Quandl's Wiki initiative to get financial data permanently into the public domain. Quandl relies on users
        like you to flag errors and provide data where data is wrong or missing. Get involved: <a href="mailto:connect@quandl.com" rel="nofollow"
        target="blank">connect@quandl.com</a></description>
      <refreshed-at type="dateTime">2015-10-19T21:47:15Z</refreshed-at>
      <newest-available-date type="date">2015-10-19</newest-available-date>
      <oldest-available-date type="date">1980-12-12</oldest-available-date>
    - <column-names type="array">
        <column-name>Date</column-name>
        <column-name>Open</column-name>
        <column-name>High</column-name>
        <column-name>Low</column-name>
        <column-name>Close</column-name>
        <column-name>Volume</column-name>
        <column-name>Ex-Dividend</column-name>
        <column-name>Split Ratio</column-name>
        <column-name>Adj. Open</column-name>
        <column-name>Adj. High</column-name>
        <column-name>Adj. Low</column-name>
        <column-name>Adj. Close</column-name>
        <column-name>Adj. Volume</column-name>
      </column-names>
      <frequency>daily</frequency>
      <type>Time Series</type>
      <premium type="boolean">false</premium>
      <limit nil="true"/>
      <transform nil="true"/>
      <column-index nil="true"/>
      <start-date type="date">2014-01-01</start-date>
      <end-date type="date">2014-12-31</end-date>
    - <data type="array">
      - <datum type="array">
          <datum type="date">2014-12-31</datum>
          <datum type="float">112.82</datum>
          <datum type="float">113.13</datum>
          <datum type="float">110.21</datum>
          <datum type="float">110.38</datum>
          <datum type="float">41403351.0</datum>
          <datum type="float">0.0</datum>
          <datum type="float">1.0</datum>
          <datum type="float">111.41199377764</datum>
          <datum type="float">111.71812494296</datum>
          <datum type="float">108.83456686965</datum>
          <datum type="float">109.00244525063</datum>
          <datum type="float">41403351.0</datum>
        </datum>
      - <datum type="array">
```


2.11 Statistical Summary

Statistics is mainly associated with numbers. A complete statistical analysis of a data set consists of both numbers and graphs. In this chapter we showed how to graph a data set in Excel, Minitab, and SAS. In Excel we showed in detail on how to use the chart wizard to graph a data set. In Minitab, there is no chart wizard. Because of this, it is very hard to create good-looking charts in Minitab. In SAS, we show four ways to read data into SAS and present how to use PROC GPLOT commend to create a chart. We also showed how to retrieve data from Yahoo.com, Google.com, and Quandl.com.

Bibliography

Delwiche LD (2008) The Little SAS®Book: A Primer, 4th edn

Lee CF, Lee JC, Lee AC (2013) Statistics for business and financial economics. Springer, New York

Microsoft Inc., Excel 2013. Microsoft Inc., Redmond

Minitab Inc. Minitab 17. Minitab Inc., State College, Pennsylvania

SAS Institute Inc. SAS 2014. SAS Institute Inc., Cary

Chapter 3
Histograms and the Rate of Returns of JPM and JNJ

3.1 Introduction

In this chapter we are interested in the dispersion and the general location of JPM's annual returns. In this chapter we will use *histograms* to analyze the dispersion and location of JPM's annual stock price returns.

3.2 Calculating the Rate of Return of JPM

Suppose P1 is the price of a stock on the last business day of the previous year and that P2 is the price of the stock on the last business day of the current year. Then, the annual rate of return will be calculated as follows:

$$\text{Annual rate of return} = (\text{P2} - \text{P1})/\text{P1}.$$

The above annual rate of returns calculation ignores the effects of dividends.

To calculate the annual rate of return from JPM, we will need the December ending price for each year. How do we do this? Do we scroll through the daily prices and then write down the December price? In Excel there is an easier way. Below shows how to get only the December prices in Excel.

We will use the *month* function in Excel to indicate the month of a date. The *month* function is illustrated in cell H2. Notice that in cell H2, the *month* function indicates that the date in cell A2 is in the first month of the year.

The web link to JPM's month data on Yahoo Finance is

http://ichart.finance.yahoo.com/table.csv?s=JPM&a=00&b=01&c=1984&d=00&e=2&f=2014&g=m&ignore=.csv

C.-F. Lee et al., *Essentials of Excel, Excel VBA, SAS and Minitab for Statistical and Financial Analyses*, DOI 10.1007/978-3-319-38867-0_3

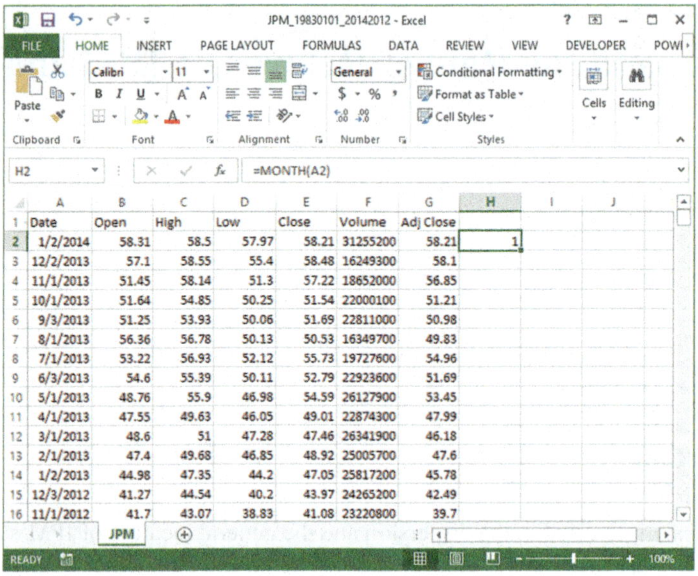

Note that for JPM monthly data is given at the beginning of the month. For this reason we will be looking for the month of January instead of the month of December.

Next we will need to apply the *month* function to the rest of the data set. Instead of typing in the *month* function for every data item, we will use a special Excel feature. To perform this special Excel feature, we will need to put the cursor on the right bottom corner of cell H2 as shown below.

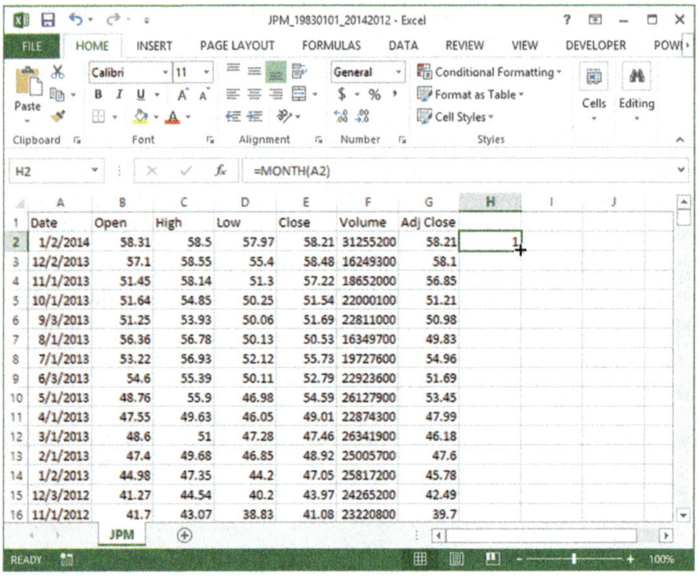

Notice that the cursor turns to a black cross when it's at the right bottom corner of cell H2. Double-clicking the right bottom corner of cell H2 will automatically copy the *month* function to the rest of the other data items and then adjust the formula accordingly. This process is shown below.

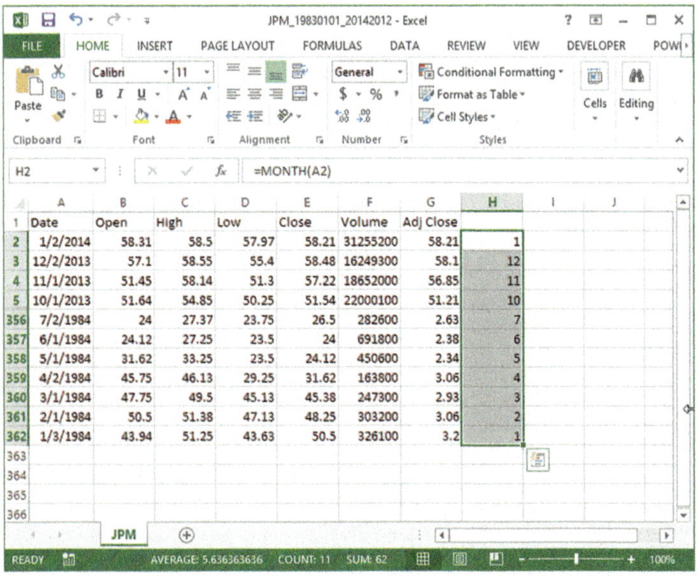

Let's now filter the data. To do this, first select cell A1 and then choose the *Data* tab and then choose the *Filter* item as shown below.

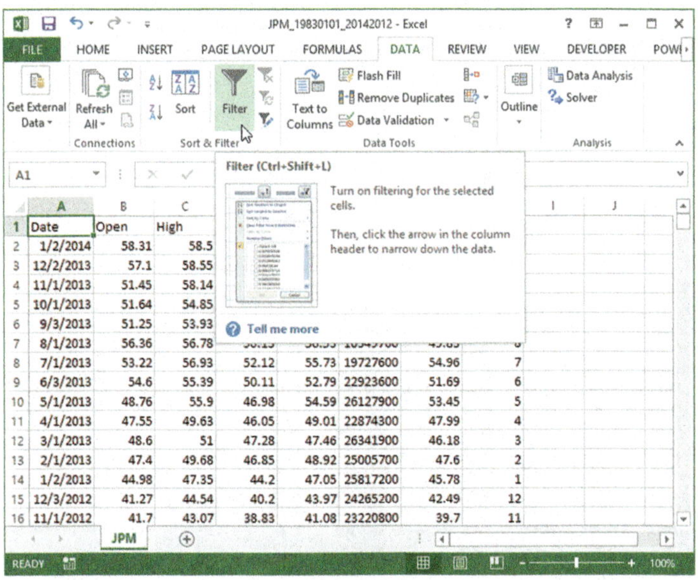

Choosing the *Filter* will result in drop-down boxes in cells A1 to H1. Set it that only “12” is selected.

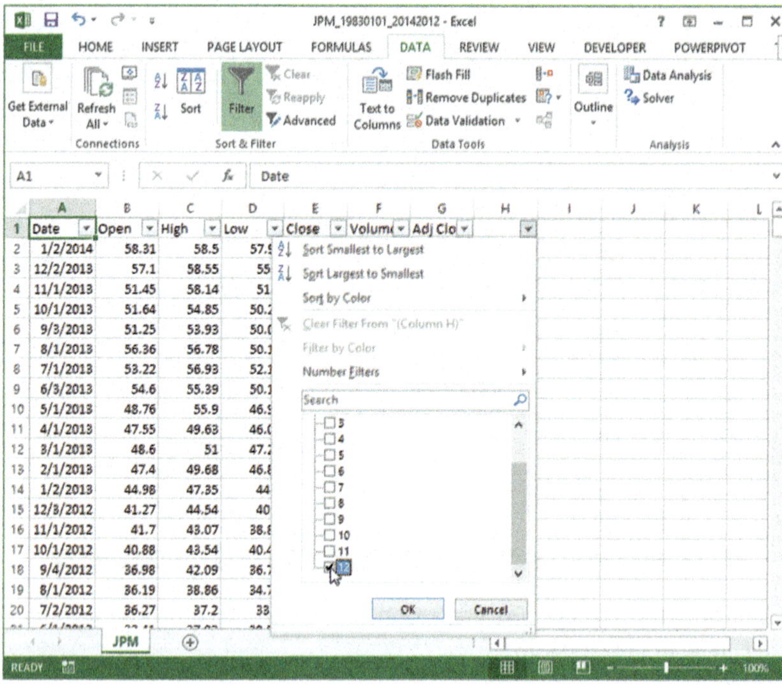

Choosing "12" will cause Excel to show only January data. This is shown below.

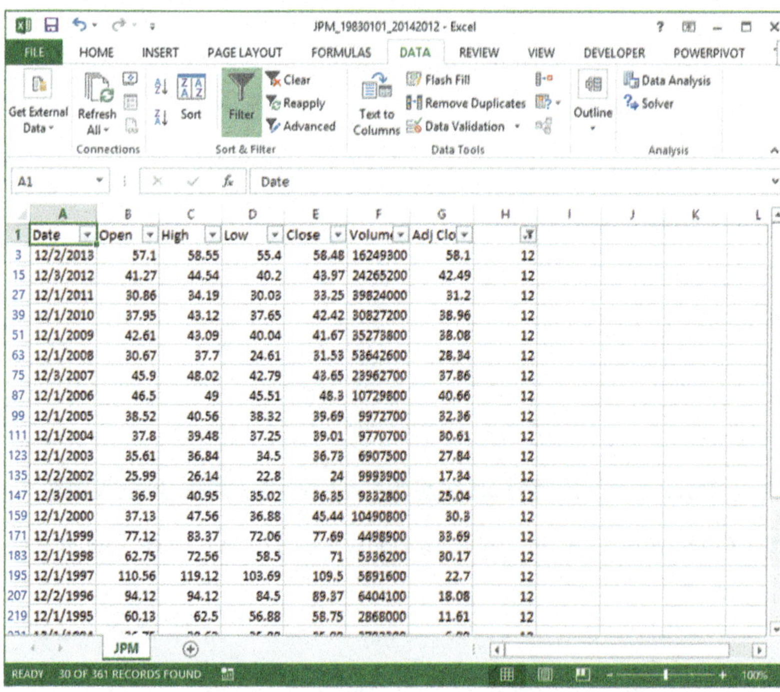

Notice that Excel hid all non-January data. Now let's get the January data to a new worksheet in a new workbook. We will first need to select the whole data range. A fancy way to do this is to first select cell A1 and then press the *F5* key.

Pressing the *F5* key will result in the following *Go To* dialog box.

In the *Go To* dialog box press the *Special* dialog box to get the following dialog box as shown below.

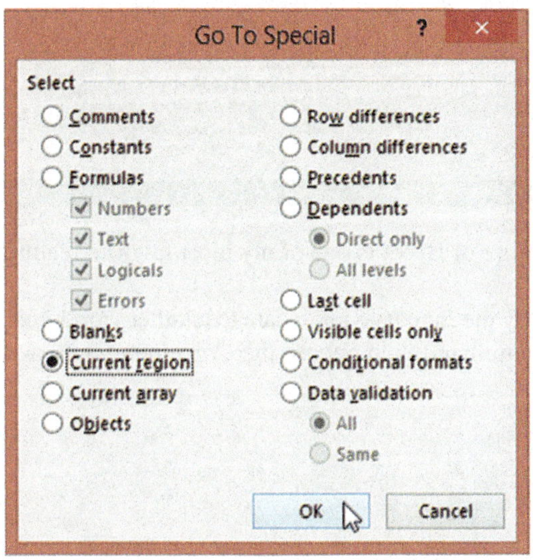

In the *Go To Special* dialog box, choose the *Current region* option and then press the *OK* button. This will result in all the data to be automatically selected.

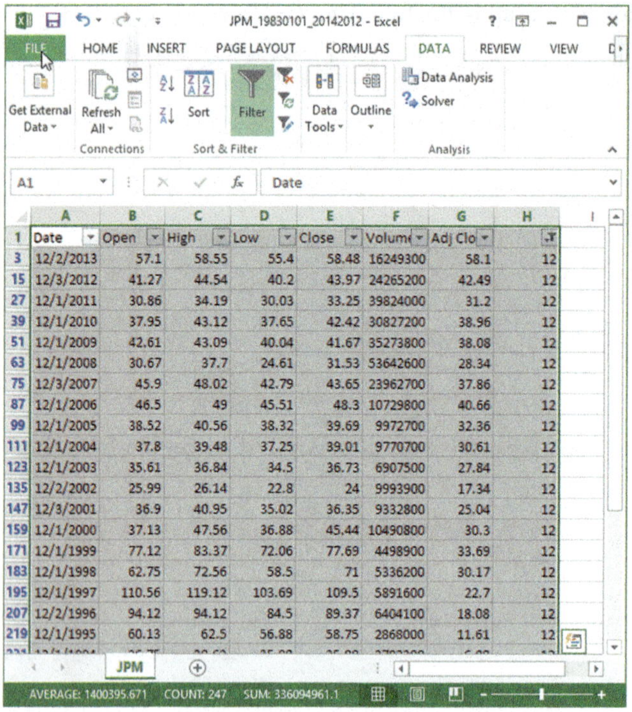

The *Current Region* feature of Excel is one of my most favorite features and one of my most used feature in Excel.

Now we will want to copy the January month data to another workbook. To do this, we need to first click on the right-mouse button and then choose the *Copy* menu item as shown below.

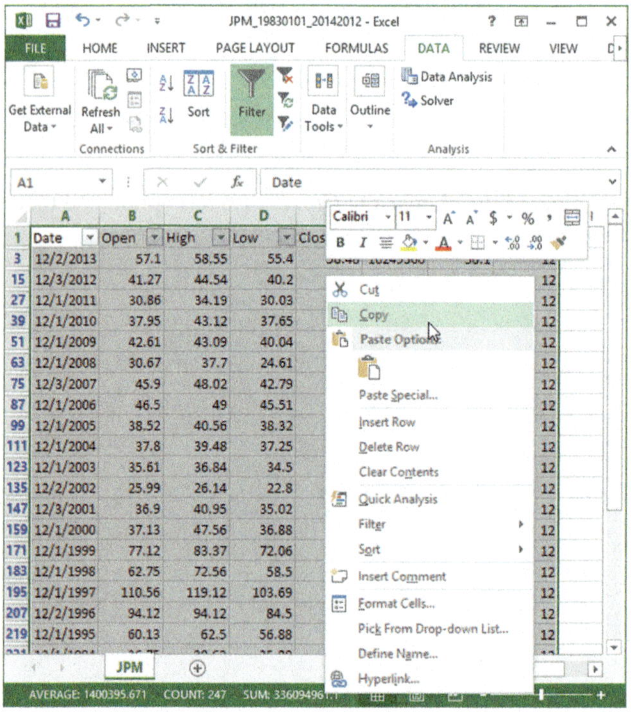

Choosing the *Copy* feature will result in a marquee around the data. This is shown below. The marquee indicates that the data is ready to be copied.

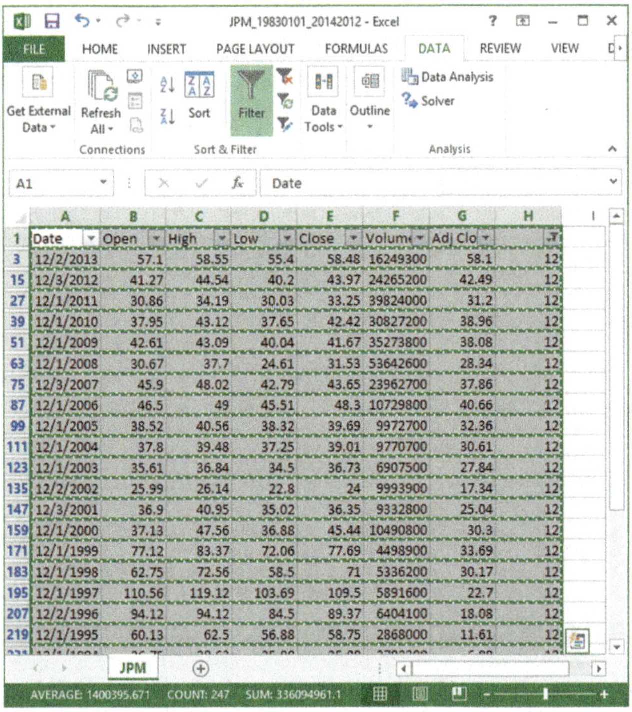

To create a new workbook, press the *Ctrl-N* key combination.
In cell A1 of sheet1 in the new workbook, right-mouse click and choose the *Paste* as shown below.

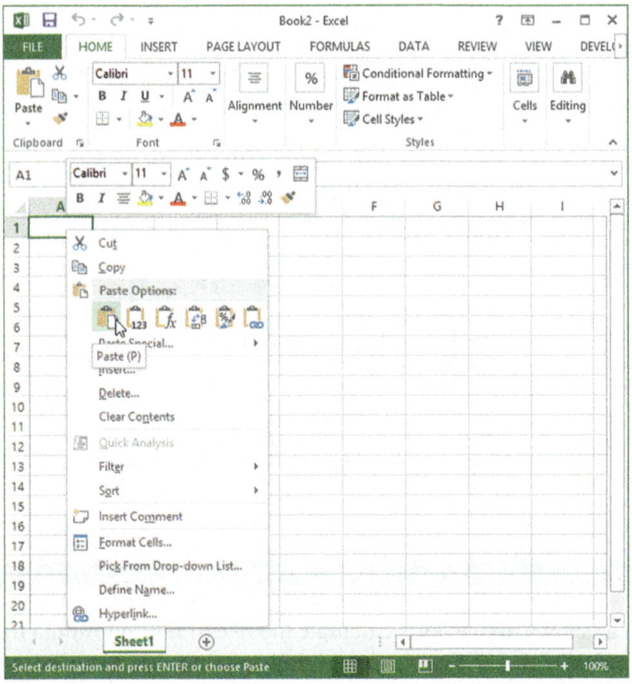

Choosing *Paste* will result in the January data being pasted to the new workbook as shown below.

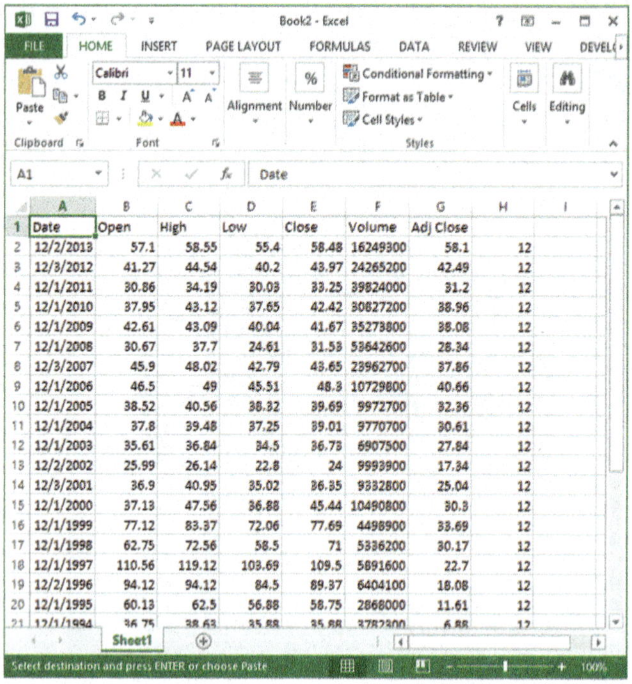

We can now calculate the rate of returns. Below shows the formula in cell I2 to calculate the rate of return for JPM for the year 2013.

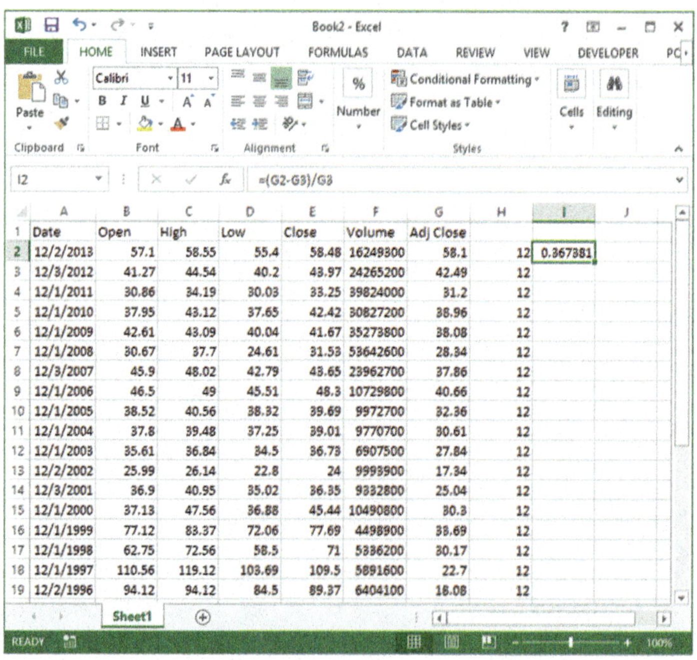

Using the concept as shown above, let's calculate the rate of return formula from cell I2 to the cells from I2 to I31. The rate of return for each year is shown below.

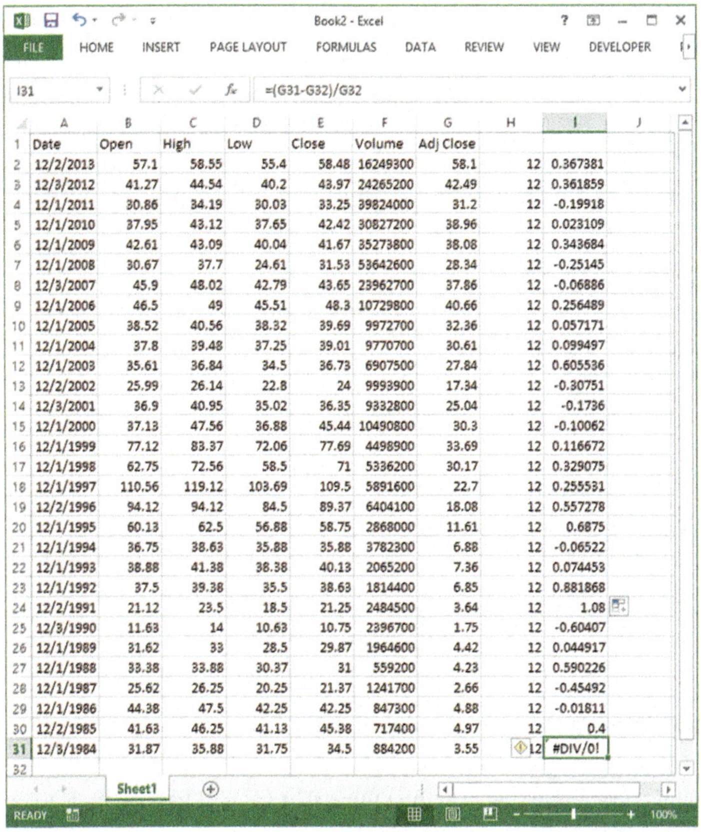

	A	B	C	D	E	F	G	H	I
1	Date	Open	High	Low	Close	Volume	Adj Close		
2	12/2/2013	57.1	58.55	55.4	58.48	16249300	58.1	12	0.367381
3	12/3/2012	41.27	44.54	40.2	43.97	24265200	42.49	12	0.361859
4	12/1/2011	30.86	34.19	30.03	33.25	39824000	31.2	12	-0.19918
5	12/1/2010	37.95	43.12	37.65	42.42	30827200	38.96	12	0.023109
6	12/1/2009	42.61	43.09	40.04	41.67	35273800	38.08	12	0.343684
7	12/1/2008	30.67	37.7	24.61	31.53	53642600	28.34	12	-0.25145
8	12/3/2007	45.9	48.02	42.79	43.65	23962700	37.86	12	-0.06886
9	12/1/2006	46.5	49	45.51	48.3	10729800	40.66	12	0.256489
10	12/1/2005	38.52	40.56	38.32	39.69	9972700	32.36	12	0.057171
11	12/1/2004	37.8	39.48	37.25	39.01	9770700	30.61	12	0.099497
12	12/1/2003	35.61	36.84	34.5	36.73	6907500	27.84	12	0.605536
13	12/2/2002	25.99	26.14	22.8	24	9993900	17.34	12	-0.30751
14	12/3/2001	36.9	40.95	35.02	36.35	9332800	25.04	12	-0.1736
15	12/1/2000	37.13	47.56	36.88	45.44	10490800	30.3	12	-0.10062
16	12/1/1999	77.12	83.37	72.06	77.69	4498900	33.69	12	0.116672
17	12/1/1998	62.75	72.56	58.5	71	5336200	30.17	12	0.329075
18	12/1/1997	110.56	119.12	103.69	109.5	5891600	22.7	12	0.255531
19	12/2/1996	94.12	94.12	84.5	89.37	6404100	18.08	12	0.557278
20	12/1/1995	60.13	62.5	56.88	58.75	2868000	11.61	12	0.6875
21	12/1/1994	36.75	38.63	35.88	35.88	3782300	6.88	12	-0.06522
22	12/1/1993	38.88	41.38	38.38	40.13	2065200	7.36	12	0.074453
23	12/1/1992	37.5	39.38	35.5	38.63	1814400	6.85	12	0.881868
24	12/2/1991	21.12	23.5	18.5	21.25	2484500	3.64	12	1.08
25	12/3/1990	11.63	14	10.63	10.75	2396700	1.75	12	-0.60407
26	12/1/1989	31.62	33	28.5	29.87	1964600	4.42	12	0.044917
27	12/1/1988	33.38	33.88	30.37	31	559200	4.23	12	0.590226
28	12/1/1987	25.62	26.25	20.25	21.37	1241700	2.66	12	-0.45492
29	12/1/1986	44.38	47.5	42.25	42.25	847300	4.88	12	-0.01811
30	12/2/1985	41.63	46.25	41.13	45.38	717400	4.97	12	0.4
31	12/3/1984	31.87	35.88	31.75	34.5	884200	3.55	12	#DIV/0!
32									

Notice that "#DIV/0!" is displayed in cell I31. This makes sense because the formula is trying to reference cell E32 and there is no data for cell G32. From our exercise above, we have calculated 29 rates of returns.

One thing we are interested in any data set is how the data set is distributed. Let's first use Minitab to analyze the distribution of the returns of JPM's stock price. Let's save the above Excel workbook so we can use it in Minitab. Before we save the above data file, let's delete cell I31 and type "return" in cell I1. After doing this, click on the *Save* icon as shown below.

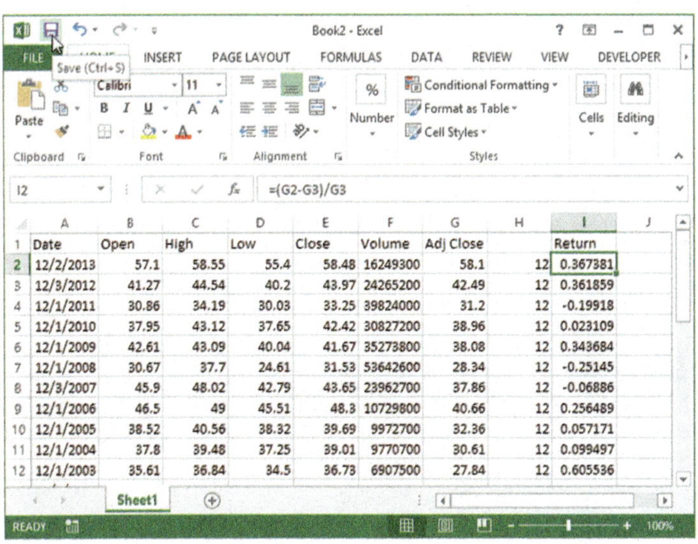

	A	B	C	D	E	F	G	H	I
1	Date	Open	High	Low	Close	Volume	Adj Close		Return
2	12/2/2013	57.1	58.55	55.4	58.48	16249300	58.1	12	0.367381
3	12/3/2012	41.27	44.54	40.2	43.97	24265200	42.49	12	0.361859
4	12/1/2011	30.86	34.19	30.03	33.25	39824000	31.2	12	-0.19918
5	12/1/2010	37.95	43.12	37.65	42.42	30827200	38.96	12	0.023109
6	12/1/2009	42.61	43.09	40.04	41.67	35273800	38.08	12	0.343684
7	12/1/2008	30.67	37.7	24.61	31.53	53642600	28.34	12	-0.25145
8	12/3/2007	45.9	48.02	42.79	43.65	23962700	37.86	12	-0.06886
9	12/1/2006	46.5	49	45.51	48.3	10729800	40.66	12	0.256489
10	12/1/2005	38.52	40.56	38.32	39.69	9972700	32.36	12	0.057171
11	12/1/2004	37.8	39.48	37.25	39.01	9770700	30.61	12	0.099497
12	12/1/2003	35.61	36.84	34.5	36.73	6907500	27.84	12	0.605536

In the *Save As* dialog box, navigate to the folder that you want to save in the workbook and then save the file as "JPM return" as shown below.

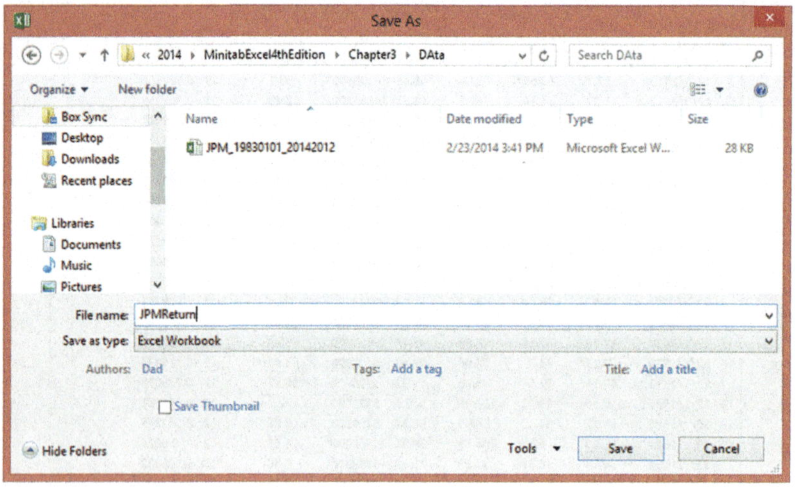

3.3 Frequency Histograms in Excel

Let's now create a histogram in Excel. The first thing that we will have to do is to open JPM "returns" file that we created above. Let's then copy the return data to sheet2 of the workbook. We do the following steps to copy the return data to sheet2.

1. Select cells I1 to I30 in sheet1.
2. Right-mouse click and choose *Copy*.

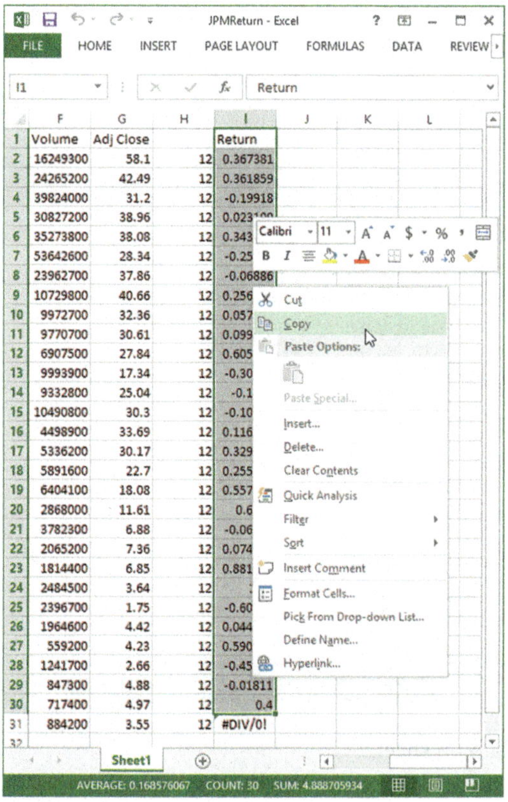

3. Click the "plus" sign next to sheet1 to add another sheet.

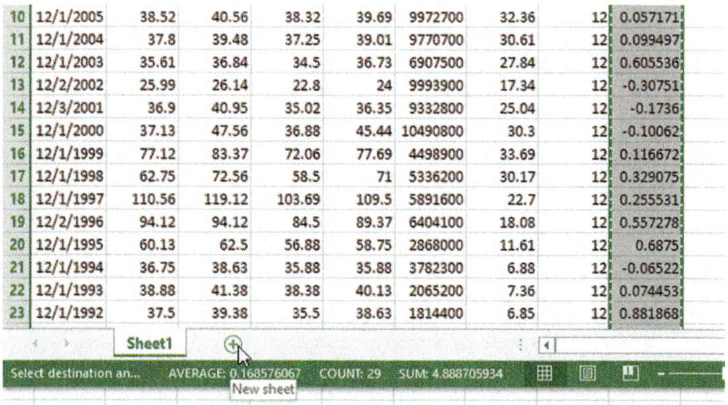

4. At cell A1 on sheet2, click the right-mouse button and choose the *Paste Value* icon as shown below.

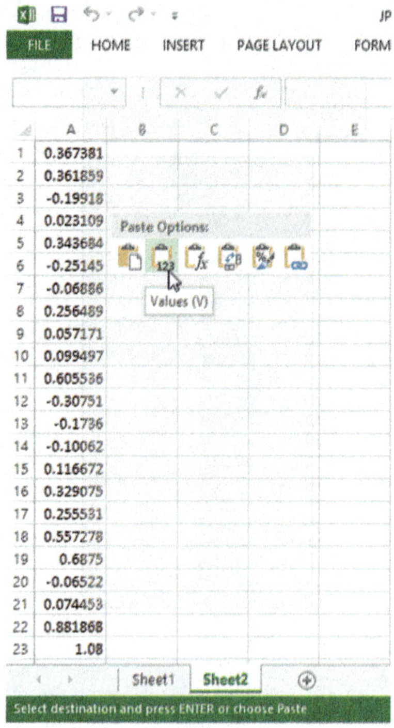

Below is the "return" data on sheet2.

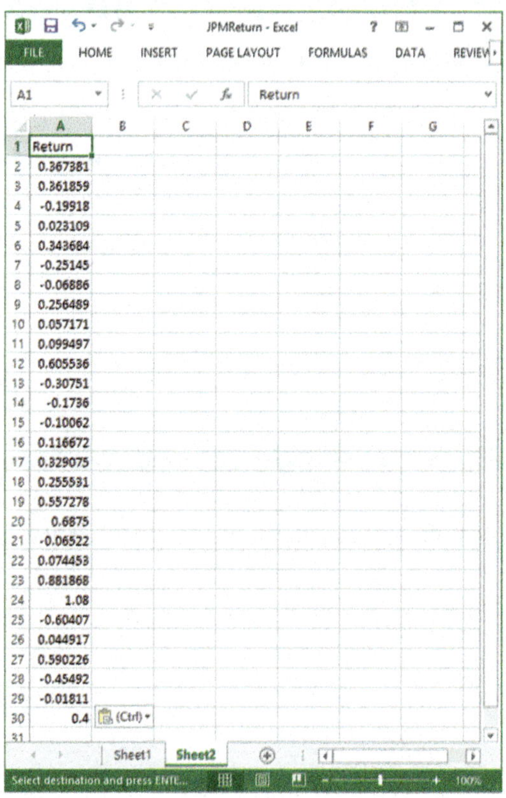

It would be helpful if the "return" data was in some logical order. Let's sort the return data in descending order. To do this first, choose the cell A1 and then click on the right mouse and choose *Sort A to Z* as shown below.

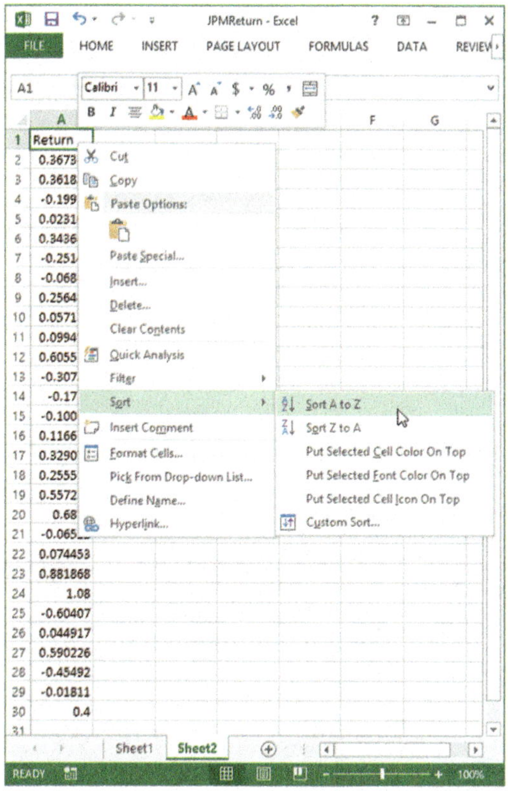

The next thing we will need to do is to create the intervals for the histograms. Excel calls the intervals *bins*. Below shows the *bins* created in column B.

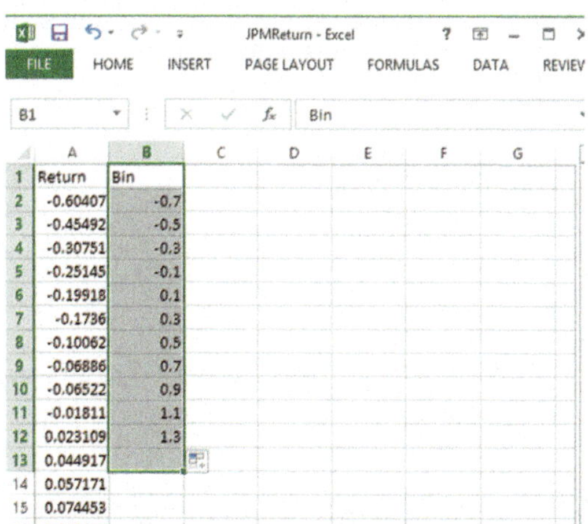

The interpretations of histograms in Excel are different than histograms in Minitab. Notice that the bins of the histograms are in cells B2 to B13. Notice that the last bin is empty. Before we can interpret what the bins means, we need to create the *frequencies* for each *bin*. We will use the Excel *frequency* function to calculate the frequency of each bin.

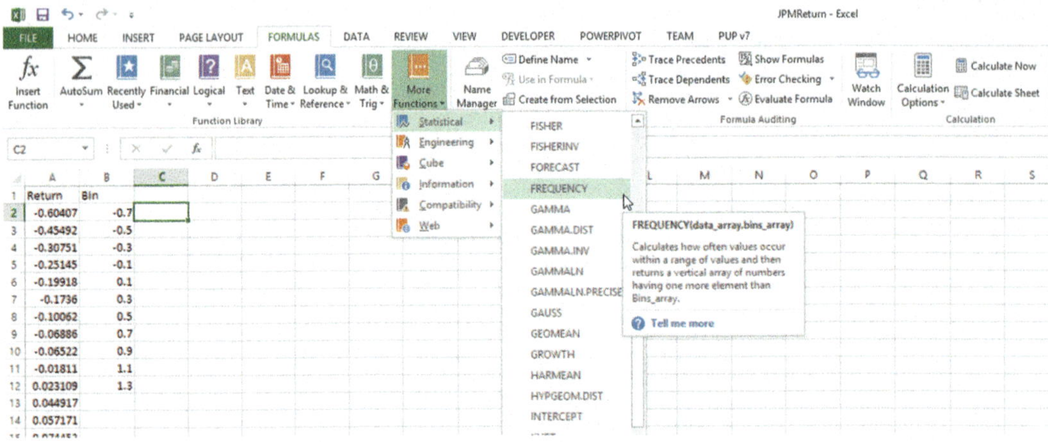

Let's first select cell C2 and then choose the *Formulas* tab and then choose the *Frequency* function as shown above.

It will result in the following.

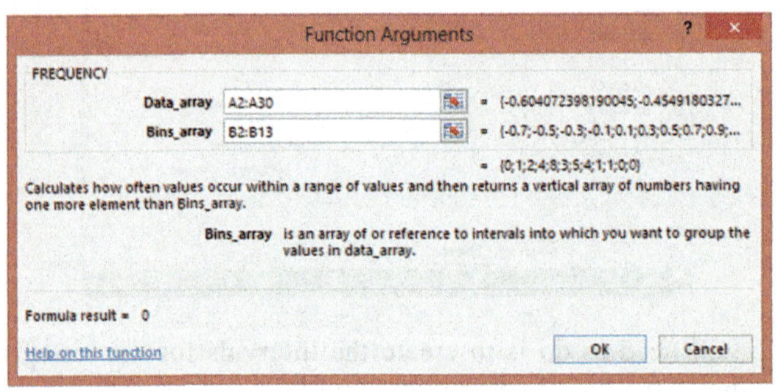

Fill out the *Function Arguments* dialog box as shown above. Notice that the bin's array argument in the dialog box includes cell B13, an empty cell. Pressing the *OK* button after entering the arguments will get you the following.

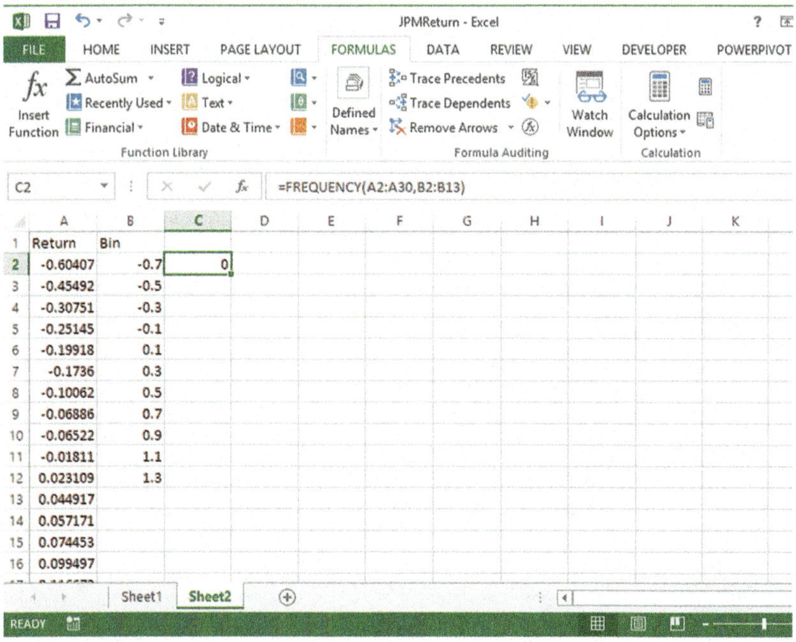

The frequency function is an *array formula*. Array formulas require special actions. To finish the frequency function, we need to select the range that the frequency applies to. This would be cells C2 to C13. This is shown below.

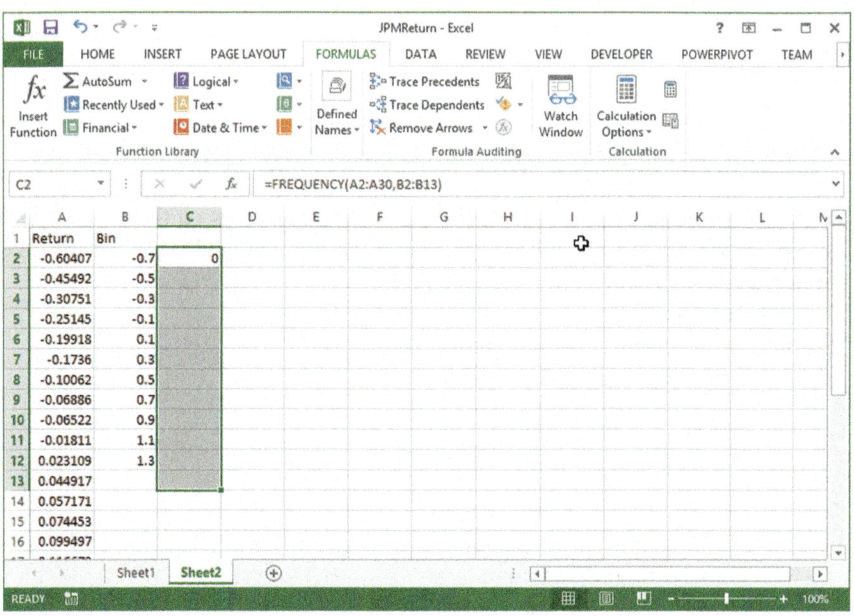

Notice that in the selection range, that cell C2 is the active cell. The next thing that we need to do is to press the *F2* key on the keyboard. The result is shown below.

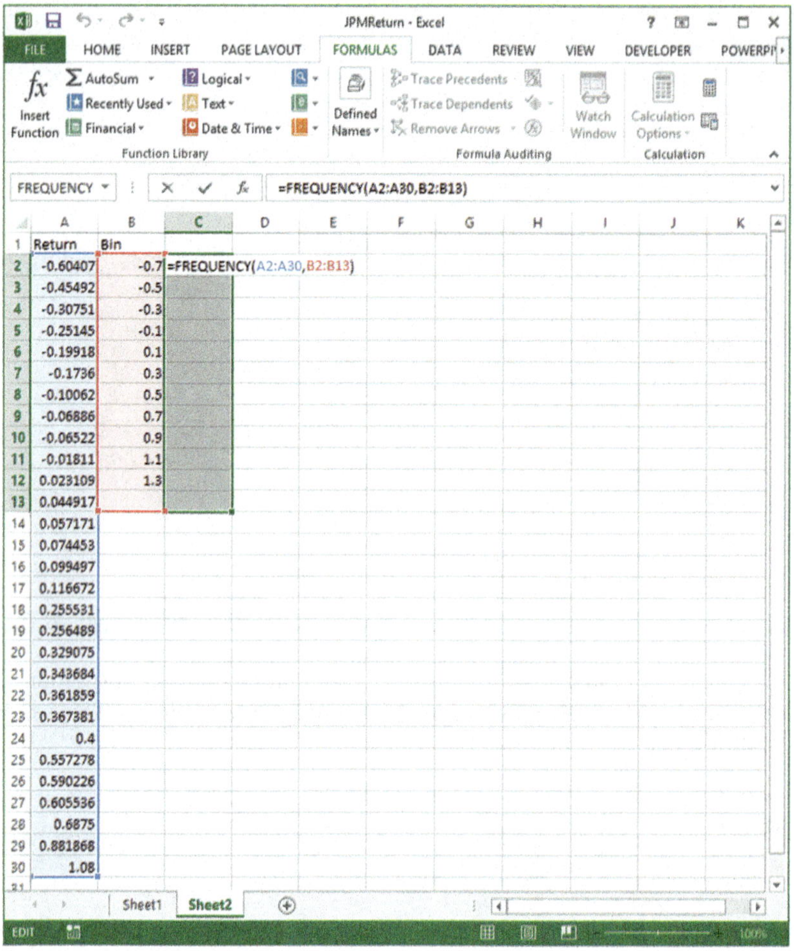

The next thing we need to do is first press the *Shift-Ctrl-Enter* keys on the keyboard. This means that we would press the second key without releasing the first key and press the third key without releasing the first and second key.

Below shows the end result.

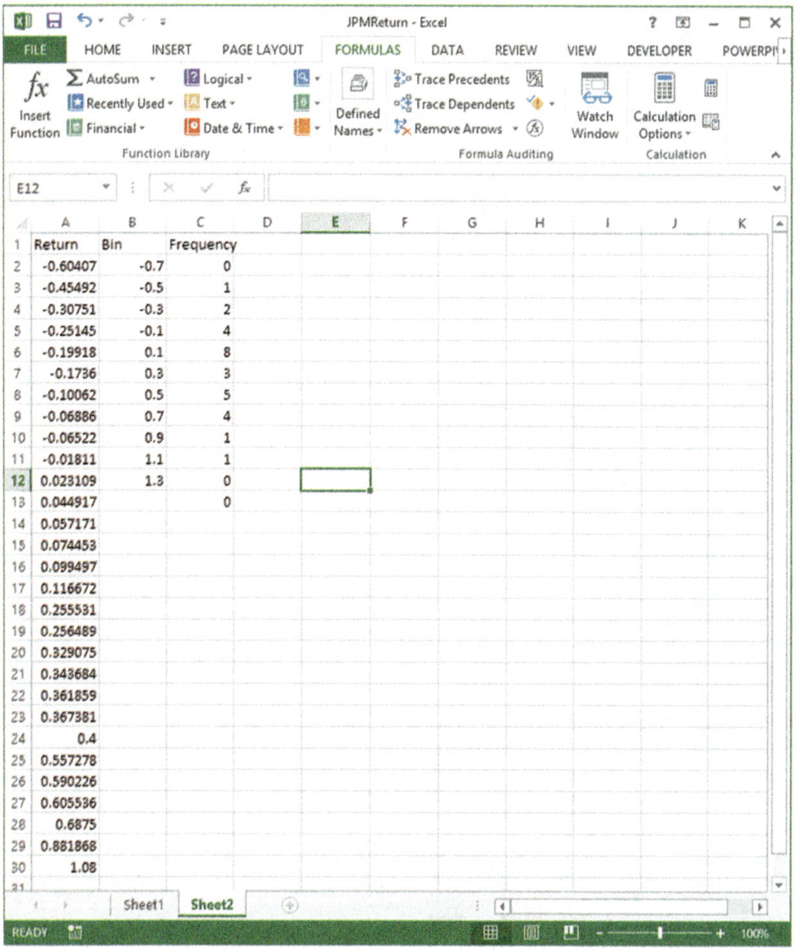

Now let's chart the "bin" and "frequency" column. To do this select the "bin" and "frequency" column as shown below. Then click on the *Quick Analysis* icon as shown blow

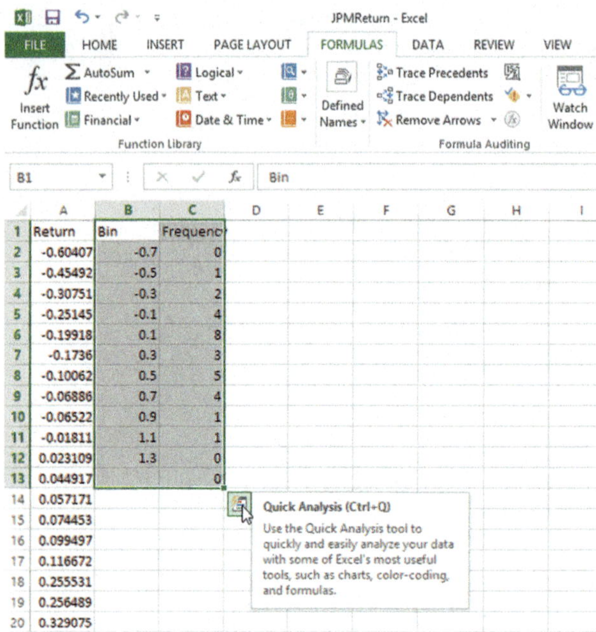

Then click on the *Charts* as shown below.

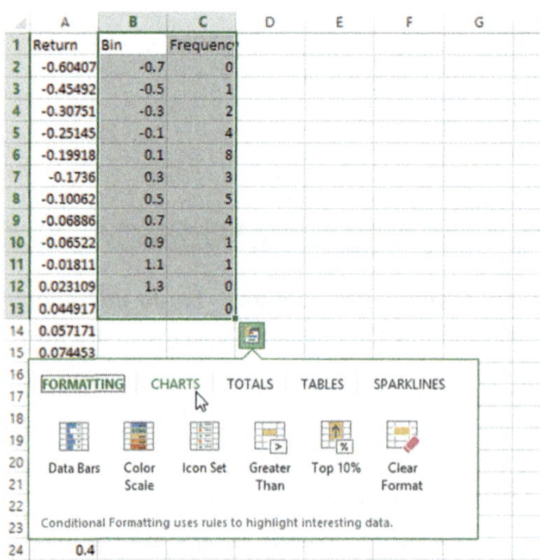

Then click on the *Clustered Column*

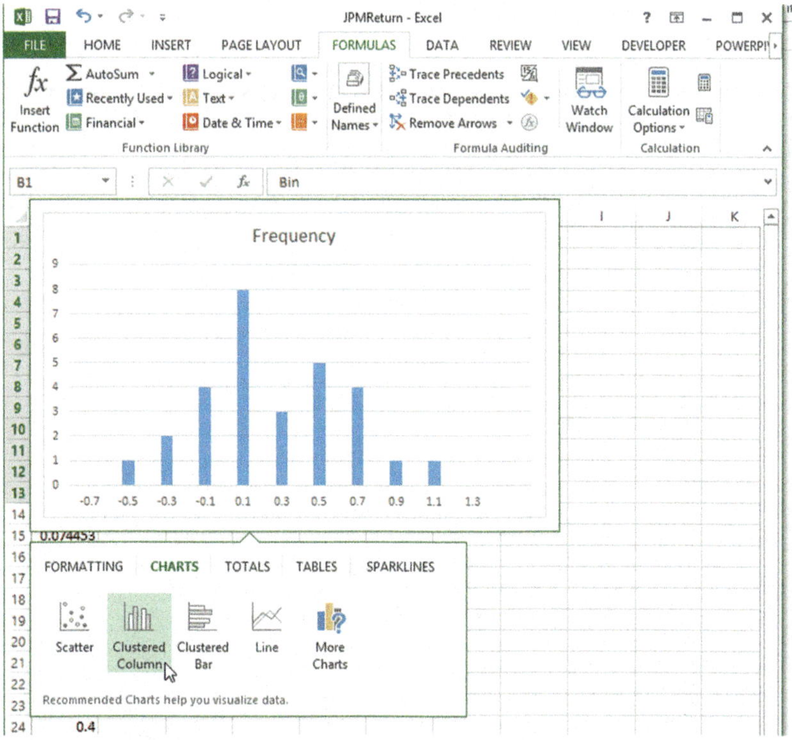

The resulting histogram is shown below.

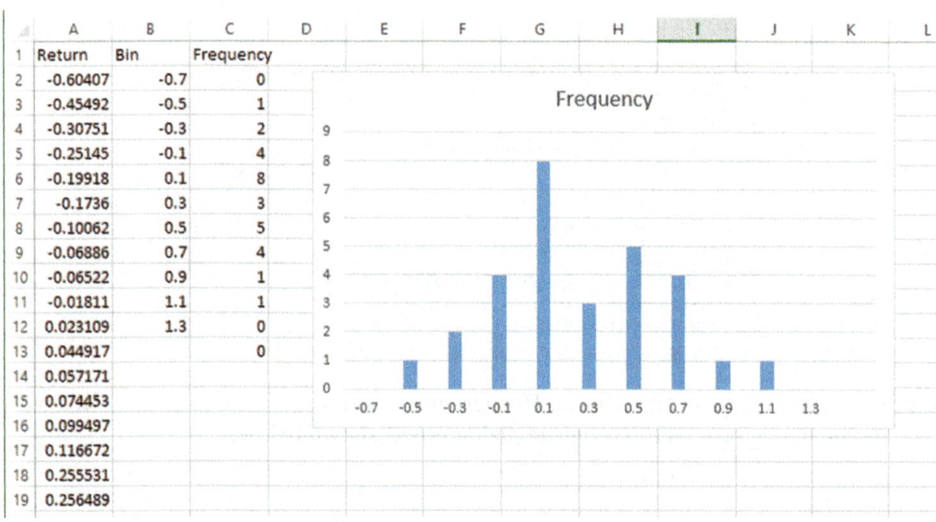

3.4 Excel's Data Analysis Tool: Histograms

In Sect. 3.4 we manually created a histogram in Excel. In this section we will use Excel's *Data Analysis* tool to create a histogram. Go to the *Data* tab and then choose the *Data Analysis* tool in Excel as shown below.

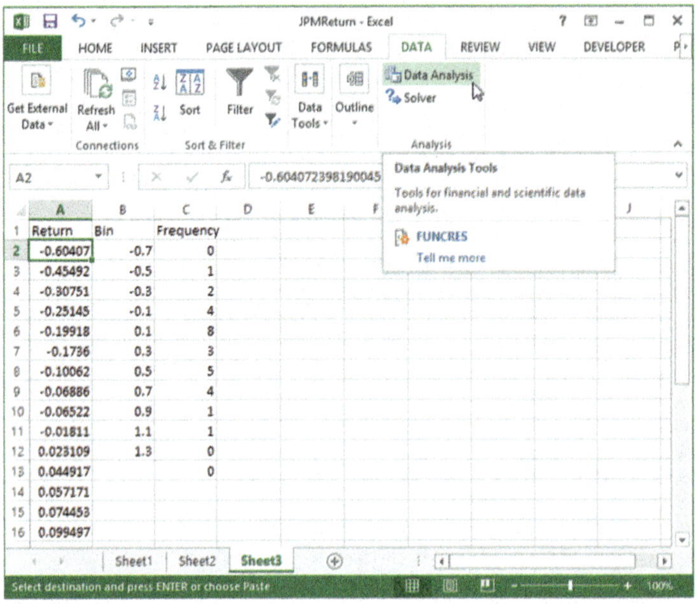

Choosing the *Data Analysis* menu item will result in the *Data Analysis* dialog box shown below.

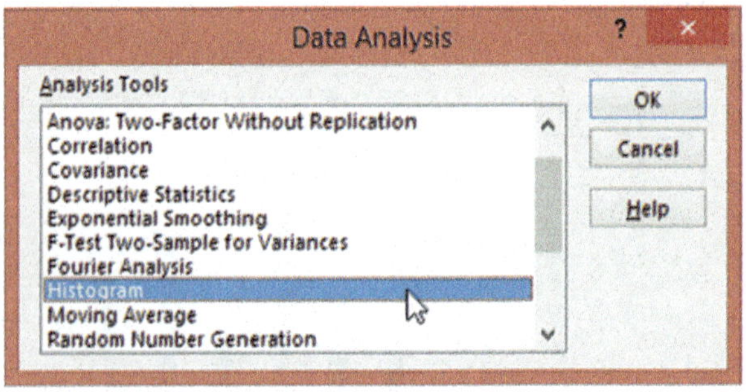

In the *Data Analysis* dialog box, choose *Histogram* and then press the *OK* button. This will result in the *Histogram* dialog box. The *Histogram* dialog box is shown below.

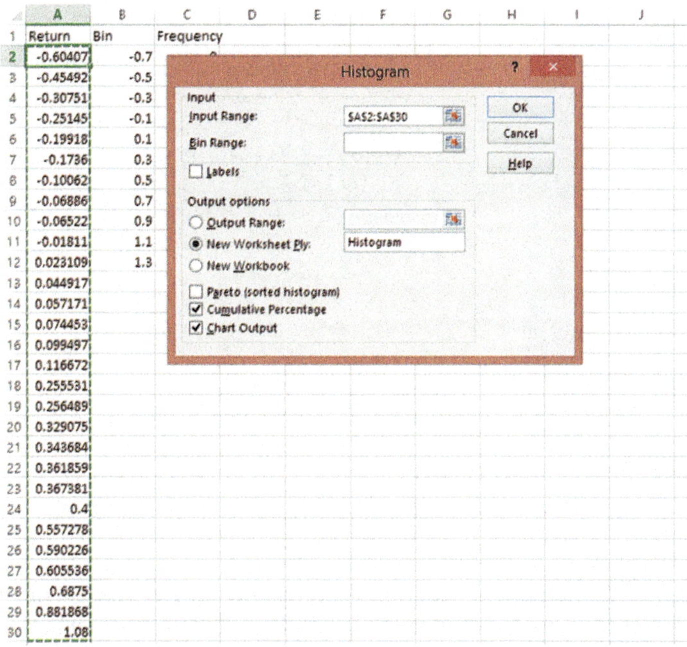

Fill out the *Histogram* dialog box as shown above. Notice that the *Bin Range* is left empty. The reason that we do this is that most of the time, we do not know what the correct bins are for a histogram. Leaving the *Bin Range* empty will let Excel set the bins. Pressing the *OK* button will result in the histogram shown below.

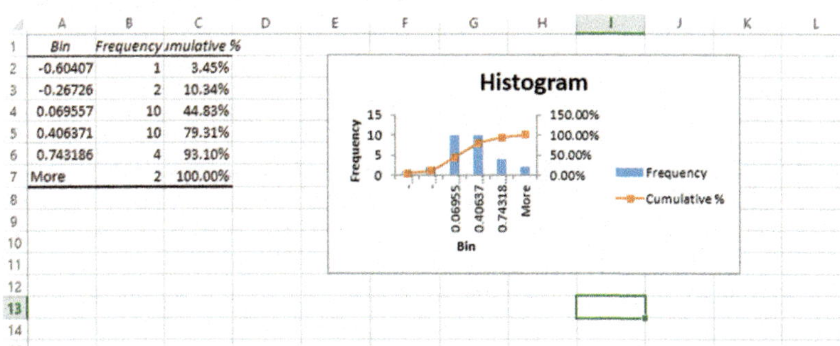

Notice that the small histogram is included in the worksheet "Histogram". This histogram is not legible. To solve this problem, we will put the histogram in its own sheet. To do this, move the cursor over the histogram and then click on the right-mouse button. A shortcut menu will show up.

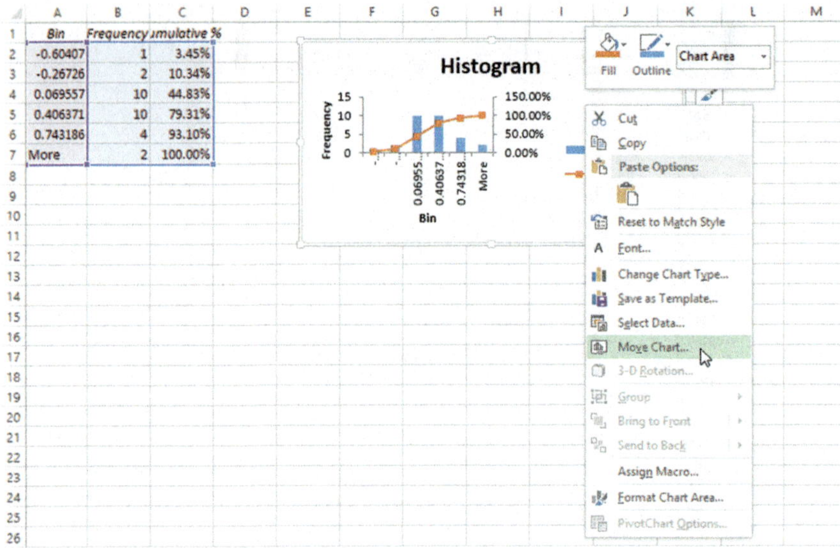

Next, choose the *Move Chart* menu item in the shortcut menu. The shortcut menu is shown above. Choosing the *Move Chart* menu item will result in the *Move Chart* dialog box shown below.

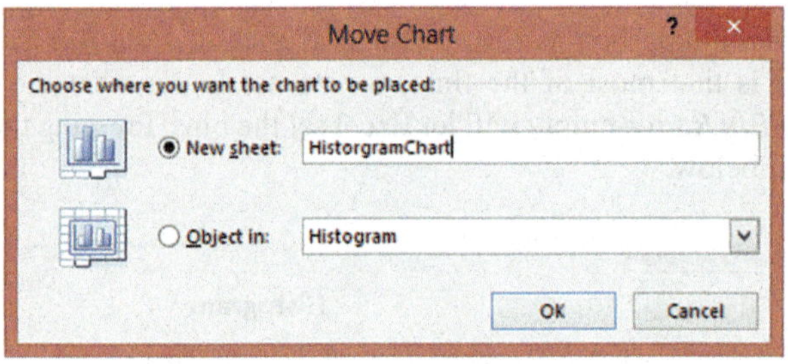

Next, fill out the *Move Chart* dialog box as shown above. Pressing the *OK* button will result in the histogram shown below.

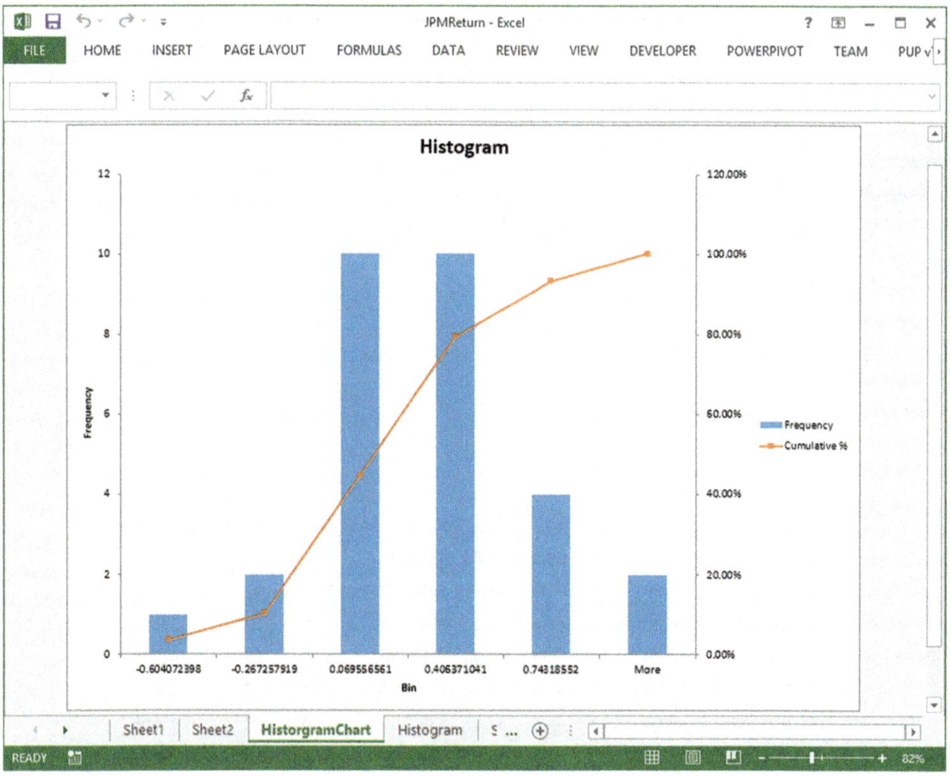

3.5 Histogram of JNJ's Return Using SAS

The annual stock rate of return, P_t, at year t, is defined as

$$R_t = \frac{P_t - P_{t-1}}{P_{t-1}},$$

where P_t and P_{t-1} are the stock price at year t and t − 1, respectively. Here, we use adjusted close price of the beginning of December from 1980 to 2012 to calculate the annual stock rate of return for JNJ.

To calculate the annual rate of return form JNJ, we will need the December ending price for each year. To achieve it, we read the data from external raw file and add a new column called *Month*. The function *Month()* returns the month of the year from a SAS date value. The IF statement deletes all the records with the value of Month not equal to 12.

```
libname Mydata 'd:\SASdata\MySASdata';
data Mydata.JNJ;
   INFILE 'd:\MINITAB and Microsoft Excel Book\Ch2\Data used in Excel, MINITAB, and
SAS book.dat';
   INPUT Date yymmdd10. Open High Low Close Volume Adj_Close;
   Month=Month(Date);
   if Month=12;
run;
```

Before calculating the value of annual rate of return, we introduce function *LAGn()*. A *LAGn* *(n = 1–100)* function returns the value of the *n*th previous execution of the function. It is easy to assume that the **LAGn** functions return values of the *n*th previous observation. Based on the formula given above, annual rate of return equals to *(Close-lag(Close))/lag(Close)*.

```
data Mydata.Rate_of_return;
   set Mydata.JNJ;
   Close2=lag(Close);
   Rate_of_return=(Close-Close2)/Close2;
run;

Print the data using following commands:
proc print data=Mydata.Rate_of_return (drop=Close2);
format Date date9.;
run;
```

Obs	Date	Open	High	Low	Close	Volume	Adj_Close	Month	Rate_of_return
1	1-Dec-80	90.75	101	85.1	99.8	4347400	1.01	12	.
2	1-Dec-81	37.75	37.8	34	37.1	1914400	1.16	12	-0.62777
3	1-Dec-82	46.25	51.3	44.6	49.6	5563500	1.58	12	0.33665
4	1-Dec-83	42.38	42.4	39	40.9	7203800	1.34	12	-0.1763
5	3-Dec-84	35.38	37.3	33.8	36.1	6300800	1.22	12	-0.11619
6	2-Dec-85	49.63	55.3	48.3	52.6	8357500	1.83	12	0.45668
7	1-Dec-86	67.5	72	65.3	65.6	8123400	2.33	12	0.24682
8	1-Dec-87	73.75	80.4	70	74.9	7306400	2.71	12	0.14096
9	1-Dec-88	86	87.5	84.3	85.1	4323200	3.16	12	0.1369
10	1-Dec-89	57	59.5	55.5	59.4	4862400	4.51	12	-0.3024
11	3-Dec-90	70.75	72.1	69.5	71.8	4826800	5.56	12	0.20832
12	2-Dec-91	95.25	116	95.3	115	5877000	9.02	12	0.59582
13	1-Dec-92	50.25	54.8	49.8	50.5	3778000	8.11	12	-0.55895
14	1-Dec-93	43.75	45.5	42.4	44.9	4086300	7.38	12	-0.11129
15	1-Dec-94	53.63	55.8	52.1	54.8	4274200	9.23	12	0.21992
16	1-Dec-95	86.87	92.4	85.1	85.5	4993700	14.68	12	0.56164
17	2-Dec-96	53.13	53.8	48	49.8	4130100	17.34	12	-0.41813
18	1-Dec-97	63.38	67.3	63.1	65.9	4470100	23.28	12	0.32402
19	1-Dec-98	81.06	84.4	76.2	83.9	6107700	30.03	12	0.27327
20	1-Dec-99	103.25	103	90.1	93.3	6547700	33.78	12	0.11184
21	1-Dec-00	99.25	106	95.3	105	7247600	38.59	12	0.12665
22	3-Dec-01	57.82	60.2	55.4	59.1	8706900	44	12	-0.43746
23	2-Dec-02	56.59	57.1	53	53.7	7122900	40.53	12	-0.0912
24	1-Dec-03	50.3	51.7	48.8	51.7	8703400	39.69	12	-0.03817
25	1-Dec-04	60.33	64.3	59.6	63.4	8040900	49.68	12	0.22764
26	1-Dec-05	61.85	62.1	59.8	60.1	9291900	48.01	12	-0.05235
27	1-Dec-06	65.91	67.3	65.3	66	8170400	53.98	12	0.0985
28	3-Dec-07	67.62	68.5	66.6	66.7	1E+07	55.93	12	0.0103
29	1-Dec-08	57.66	60.3	55	59.8	1.5E+07	51.61	12	-0.103
30	1-Dec-09	62.95	65.4	63	64.4	9965800	57.43	12	0.07655
31	1-Dec-10	62.18	63.2	61.6	61.9	9869700	57.09	12	-0.03975
32	1-Dec-11	64.68	66.3	62.9	65.6	1E+07	62.74	12	0.06031
33	3-Dec-12	69.8	71.6	69.4	69.5	1.2E+07	68.93	12	0.05947

3.5.1 Frequency Histograms in SAS

Let's now use SAS to see how the rate of return data of JNJ is distributed. A good procedure to use at this point is PROC UNIVARIATE.

PROC UNIVARIATE, which is part of Base SAS software, produces statistics and graphs describing the distribution of a single variable. Using PROC UNIVARIATE is fairly simple. After the PROC statement, you specify one or more numeric variables in a HISTOGRAM statement.

```
Goptions ftext='Times New Roman' htext=6 gunit=pct;
title height=7 'Histogram of the Rate of Return';
proc capability data=Mydata.Rate_of_return noprint;
histogram Rate_of_return;
run;
```

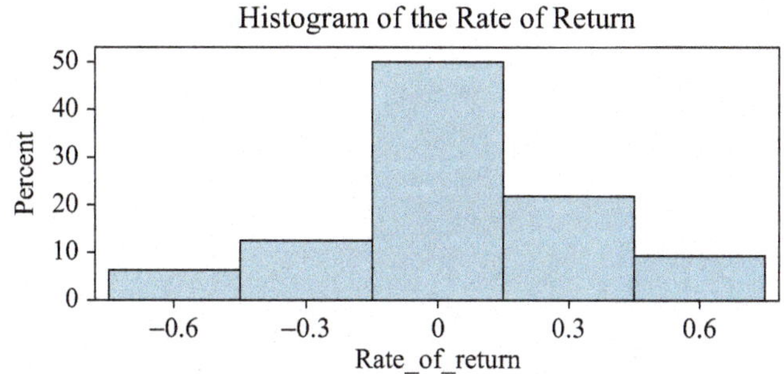

3.5.2 Cumulative Frequency Histograms in SAS

To plot the cumulative frequency histograms in SAS, we introduce another procedure: PROC GCHART. The GCHART procedure produces six types of charts: block charts, horizontal and vertical bar charts, pie and donut charts, and star charts. The charted variables can be either numeric or character.

Using PROC GCHART is simple as well. After the PROC statement, you specify one or more numeric variables in a VBAR statement. The statement TYPE=option specify the statistic represented in the chart

```
proc gchart data=Mydata.Rate_of_return;
vbar Rate_of_return /TYPE=CFREQ;
format Rate_of_return;
run;
```

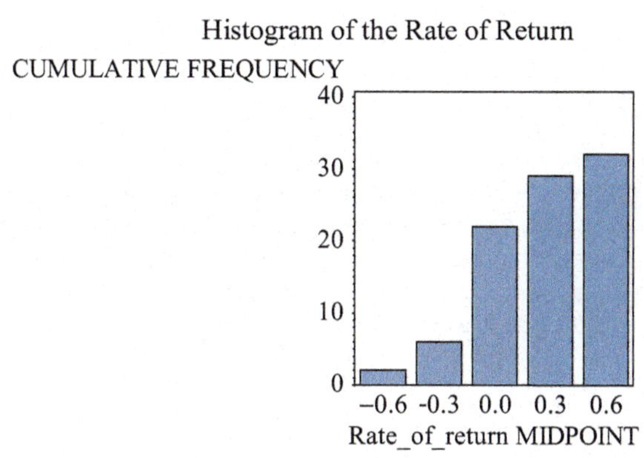

3.5.3 Frequency Histograms vs. Cumulative Percentage Line Plot

What if people want to plot the frequency histograms with line plot overlays? PROC GBARLINE provides a convenient and elegant method of doing just that, i.e., producing bar-line charts. Bar-line charts are vertical bar charts with one or more plot overlays.

Since Rate_of_Return is a continuous variable. We set up a format that assigns values to ranges in a subsequent section and also delete the missing values:

```
proc format;
value range
   -0.75--0.45= '-0.60'
   -0.45--0.15= '-0.30'
   -0.15-0.15= '0'
   0.15-0.45= '0.30'
   0.45-high= '0.60';
 run;

data Mydata.Rate_of_return;
set Mydata.Rate_of_return;
if Rate_of_return=. then delete;
run;

Then we use Proc Freq to count the frequency of each section:

proc freq data=Mydata.Rate_of_return;
tables Rate_of_return/out=Return_Freq;
format Rate_of_return range.;
run;

data Return_Freq;
set Return_Freq;
Cum_Percent+PERCENT;
run;
```

The SAS GRAPH procedure GBARLINE may provide an effective solution to the bar charts and line plots on one graph for data with limited groupings on the X-axis. The following procedure shows how a bar chart and a line plot can be displayed on the same output.

```
Goptions ftext='Times New Roman' htext=5 gunit=pct;
title h=6 'Frequency Histograms vs Cumulative Percentage Line Graph';
axis1 label=('Frequency Count') minor=none; /* left */
axis2 label=('Cumulative Percent') minor=none; /* right */
axis3 label=('Rate of Return'); /* bottom */
/* Bar legend */
legend1 position=(middle right) across=1
        label=(position=(top) j=l);
/* Line plot legend */
legend2 position=(bottom right) across=1 repeat=1
        label=(position=(top) j=l "Lines");
symbol1 c=black value=none;
proc gbarline data=Return_Freq;
        bar Rate_of_return/discrete sumvar=COUNT
        raxis=axis1 maxis=axis3 legend=legend;
        plot /sumvar=Cum_Percent legend=legend2 axis=axis2;
run;
quit;
```

The BAR statement generates a bar chart of the Frequency Count output on the left axis using COUNT as the summary variable (SUMVAR=COUNT) and the PLOT statement displays a line plot of the cumulative percentage values on the right axis using Cum_Percent as the summary variable (SUMVAR=Cum_Percent).

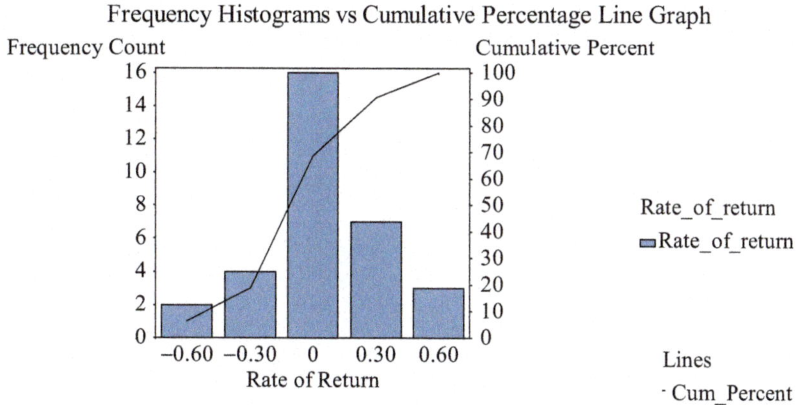

3.6 Statistical Summary

In this chapter we look at the histogram of the rate of return of JP Morgan and Johnson and Johnson. The histogram provided information about the shape and form of the distribution of the rate of return of JP Morgan and Johnson and Johnson.

Bibliography

Lee CF, Lee JC, Lee AC (2013) Statistics for business and financial economics. Springer, New York
Microsoft Inc., Excel 2013. Microsoft Inc., Redmond
Minitab Inc. Minitab 17. Minitab Inc., State College
SAS Institute Inc. SAS 2014. SAS Institute Inc., Cary

Chapter 4
Numerical Summary Measures on Rate of Returns of Amazon, Walmart, and the S&P 500

4.1 Introduction

In this chapter we will calculate numeric summary measures on the rate of return of the S&P 500, Amazon, and Walmart. Statistical numerical numbers look at data in five different ways. The five different ways are *measure of central location, measure of variability, measure of relative standing, measure of share, and measure of linear relationship*. We will look at the weekly, monthly, and annual rate of return from 2008 to 2015. 2008 was the year of the great recession. Since 2008 we have seen the market improve tremendously.

The formula to calculate the rate of return for a period is easy, but what is usually difficult to do is to get data to calculate the rate of return. We will look www.quandl.com. This website has made it easy to retrieve stock price data and other financial data.

4.2 Retrieving Stock Prices from Quandl.com

The following is the URL to retrieve weekly pricing data of Amazon from January 2008 to December 2014:

https://www.quandl.com/api/v3/datasets/yahoo/AMZN.csv?start_date=Jan-01-2008&end_date=Dec-31-2014&collapse=monthly&api_key=uBSmYjQ3xghF8sqmTjf-

The following is the URL to retrieve weekly pricing data of Walmart from January 2008 to December 2014:

https://www.quandl.com/api/v3/datasets/yahoo/WMT.csv?start_date=Jan-01-2008&end_date=Dec-31-2014&collapse=monthly&api_key=uBSmYjGExghF8sqmTjf-

The following is the URL to retrieve weekly pricing data of the S&P 500 from January 2008 to December 2014. Note that the Quandl.com ticker for the S&P 500 is *INDEX_GSPC*:

https://www.quandl.com/api/v3/datasets/yahoo/INDEX_GSPC.csv?start_date=Jan-01-2008&end_date=Dec-31-2014&collapse=monthly&api_key=uBSmYjGExghF8sqmTjf-

The URL has four parameters. The *collapse* parameter indicates the frequency of the data. The valid values are daily, weekly, monthly, quarterly, and annual. Qunadl.com allows the first 50 requests per day to be anonymous; after that Quandl.com requires users to register to use www.quandl.com.

C.-F. Lee et al., *Essentials of Excel, Excel VBA, SAS and Minitab for Statistical and Financial Analyses*, DOI 10.1007/978-3-319-38867-0_4

Registration is free. Quandl.com gives users a unique ID. Users will put the unique ID in the *api_key* argument.

Putting the above URL in a web browser would result the web browser returning Amazon's stock price data from Quandl.com.

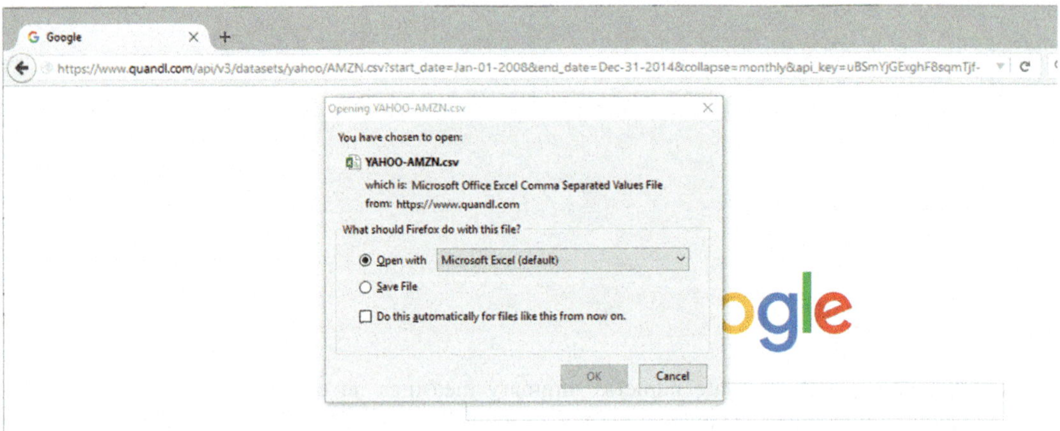

You can also the URL in Excel.

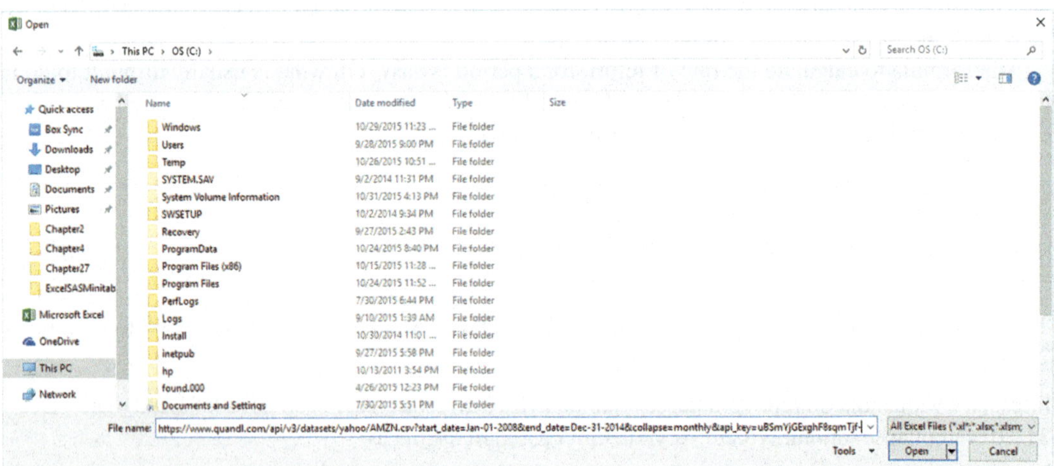

4.3 Numeric Measures of the S&P 500, Amazon, and Walmart

Below shows the numerical weekly, monthly, and annual statistics of the rate of returns of the S&P 500, Amazon, and Walmart from January 2008 to December 2014. The spreadsheets were created using Quandl.com, Excel VBA, and Excel functions. Notice the weekly rate of return statistics and 366 data points per stock ticker.

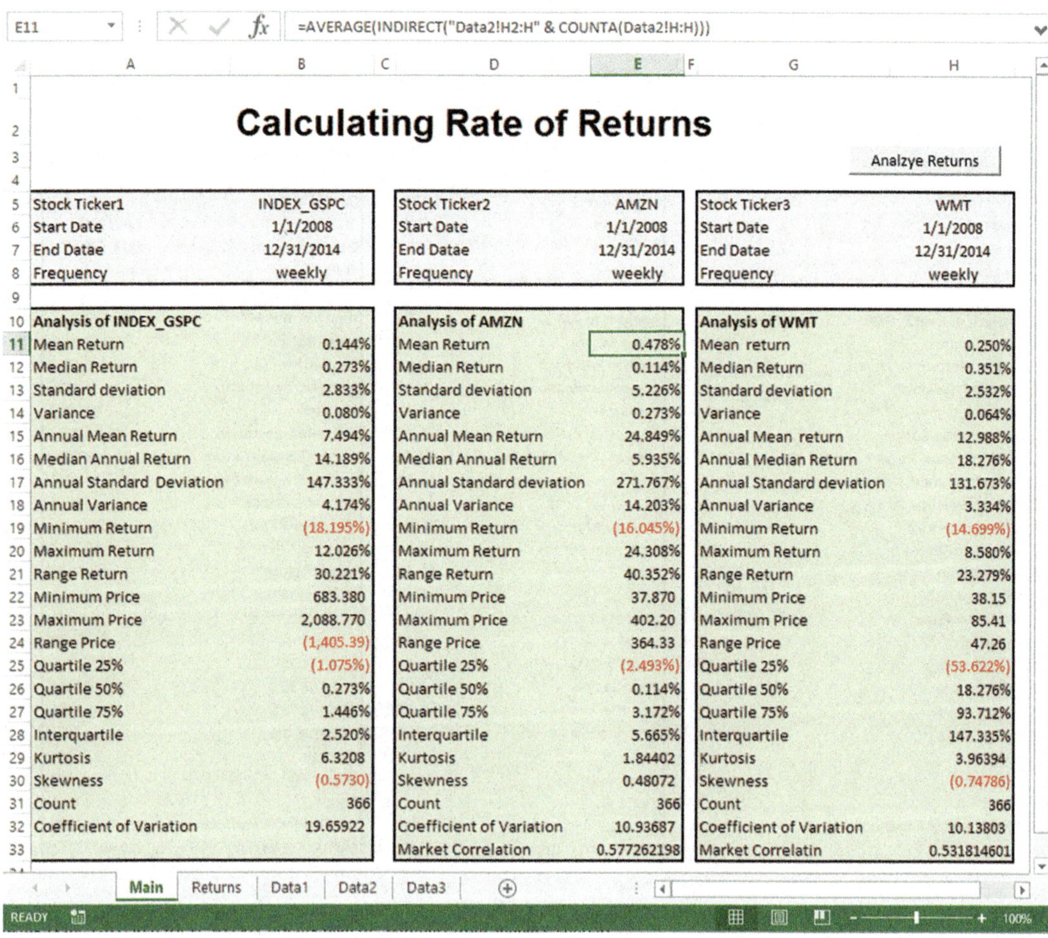

E11 =AVERAGE(INDIRECT("Data2!H2:H" & COUNTA(Data2!H:H)))

Calculating Rate of Returns

Analzye Returns

Stock Ticker1	INDEX_GSPC	Stock Ticker2	AMZN	Stock Ticker3	WMT
Start Date	1/1/2008	Start Date	1/1/2008	Start Date	1/1/2008
End Datae	12/31/2014	End Datae	12/31/2014	End Datae	12/31/2014
Frequency	weekly	Frequency	weekly	Frequency	weekly

Analysis of INDEX_GSPC		Analysis of AMZN		Analysis of WMT	
Mean Return	0.144%	Mean Return	0.478%	Mean return	0.250%
Median Return	0.273%	Median Return	0.114%	Median Return	0.351%
Standard deviation	2.833%	Standard deviation	5.226%	Standard deviation	2.532%
Variance	0.080%	Variance	0.273%	Variance	0.064%
Annual Mean Return	7.494%	Annual Mean Return	24.849%	Annual Mean return	12.988%
Median Annual Return	14.189%	Median Annual Return	5.935%	Annual Median Return	18.276%
Annual Standard Deviation	147.333%	Annual Standard deviation	271.767%	Annual Standard deviation	131.673%
Annual Variance	4.174%	Annual Variance	14.203%	Annual Variance	3.334%
Minimum Return	(18.195%)	Minimum Return	(16.045%)	Minimum Return	(14.699%)
Maximum Return	12.026%	Maximum Return	24.308%	Maximum Return	8.580%
Range Return	30.221%	Range Return	40.352%	Range Return	23.279%
Minimum Price	683.380	Minimum Price	37.870	Minimum Price	38.15
Maximum Price	2,088.770	Maximum Price	402.20	Maximum Price	85.41
Range Price	(1,405.39)	Range Price	364.33	Range Price	47.26
Quartile 25%	(1.075%)	Quartile 25%	(2.493%)	Quartile 25%	(53.622%)
Quartile 50%	0.273%	Quartile 50%	0.114%	Quartile 50%	18.276%
Quartile 75%	1.446%	Quartile 75%	3.172%	Quartile 75%	93.712%
Interquartile	2.520%	Interquartile	5.665%	Interquartile	147.335%
Kurtosis	6.3208	Kurtosis	1.84402	Kurtosis	3.96394
Skewness	(0.5730)	Skewness	0.48072	Skewness	(0.74786)
Count	366	Count	366	Count	366
Coefficient of Variation	19.65922	Coefficient of Variation	10.93687	Coefficient of Variation	10.13803
		Market Correlation	0.577262198	Market Correlatin	0.531814601

Main Returns Data1 Data2 Data3

READY

B13 =STDEV.P(INDIRECT("Data1!H2:H" & COUNTA(Data1!H:H)))

Calculating Rate of Returns

Analzye Returns

Stock Ticker1	INDEX_GSPC	Stock Ticker2	AMZN	Stock Ticker3	WMT
Start Date	1/1/2008	Start Date	1/1/2008	Start Date	1/1/2008
End Datae	12/31/2014	End Datae	12/31/2014	End Datae	12/31/2014
Frequency	monthly	Frequency	monthly	Frequency	monthly

Analysis of INDEX_GSPC		Analysis of AMZN		Analysis of WMT	
Mean Return	0.603%	Mean Return	2.148%	Mean return	0.943%
Median Return	1.260%	Median Return	2.623%	Median Return	1.174%
Standard deviation	4.804%	Standard deviation	9.650%	Standard deviation	4.705%
Variance	0.231%	Variance	0.931%	Variance	0.221%
Annual Mean Return	7.234%	Annual Mean Return	25.772%	Annual Mean return	11.313%
Median Annual Return	15.117%	Median Annual Return	31.471%	Annual Median Return	14.089%
Annual Standard Deviation	57.650%	Annual Standard deviation	115.801%	Annual Standard deviation	56.458%
Annual Variance	2.770%	Annual Variance	11.175%	Annual Variance	2.656%
Minimum Return	(16.942%)	Minimum Return	(25.402%)	Minimum Return	(15.947%)
Maximum Return	10.772%	Maximum Return	27.260%	Maximum Return	14.776%
Range Return	27.715%	Range Return	52.662%	Range Return	30.724%
Minimum Price	735.090	Minimum Price	42.700	Minimum Price	40.00
Maximum Price	2,067.560	Maximum Price	398.79	Maximum Price	85.41
Range Price	(1,332.47)	Range Price	356.09	Range Price	45.40
Quartile 25%	(1.530%)	Quartile 25%	(5.297%)	Quartile 25%	(27.389%)
Quartile 50%	1.260%	Quartile 50%	2.623%	Quartile 50%	14.089%
Quartile 75%	3.586%	Quartile 75%	8.398%	Quartile 75%	46.233%
Interquartile	5.115%	Interquartile	13.696%	Interquartile	73.622%
Kurtosis	1.4937	Kurtosis	0.36573	Kurtosis	1.65691
Skewness	(0.8584)	Skewness	(0.11874)	Skewness	(0.01673)
Count	84	Count	84	Count	84
Coefficient of Variation	7.96963	Coefficient of Variation	4.49335	Coefficient of Variation	4.99043
		Market Correlation	0.466328654	Market Correlatin	0.402730597

Main | Returns | Data1 | Data2 | Data3

READY 100%

E15 fx =AVERAGE(INDIRECT("Data2!H2:H" & COUNTA(Data2!H:H)))* IF(E8="weekly",52,1) * IF(E8="monthly",12,1)

Calculating Rate of Returns

Analzye Returns

	A	B	D	E	G	H
5	Stock Ticker1	INDEX_GSPC	Stock Ticker2	AMZN	Stock Ticker3	WMT
6	Start Date	1/1/2008	Start Date	1/1/2008	Start Date	1/1/2008
7	End Datae	12/31/2014	End Datae	12/31/2014	End Datae	12/31/2014
8	Frequency	annual	Frequency	annual	Frequency	annual
9						
10	Analysis of INDEX_GSPC		Analysis of AMZN		Analysis of WMT	
11	Mean Return	15.105%	Mean Return	45.669%	Mean return	10.251%
12	Median Return	13.094%	Median Return	39.368%	Median Return	12.858%
13	Standard deviation	9.400%	Standard deviation	59.133%	Standard deviation	7.502%
14	Variance	0.884%	Variance	34.967%	Variance	0.563%
15	Annual Mean Return	15.105%	Annual Mean Return	45.669%	Annual Mean return	10.251%
16	Median Annual Return	13.094%	Median Annual Return	39.368%	Annual Median Return	12.858%
17	Annual Standard Deviation	9.400%	Annual Standard deviation	59.133%	Annual Standard deviation	7.502%
18	Annual Variance	0.884%	Annual Variance	34.967%	Annual Variance	0.563%
19	Minimum Return	(0.003%)	Minimum Return	(22.177%)	Minimum Return	(2.583%)
20	Maximum Return	29.601%	Maximum Return	162.325%	Maximum Return	18.164%
21	Range Return	29.604%	Range Return	184.502%	Range Return	20.747%
22	Minimum Price	903.250	Minimum Price	51.280	Minimum Price	46.36
23	Maximum Price	2,058.900	Maximum Price	398.79	Maximum Price	84.26
24	Range Price	(1,155.65)	Range Price	347.51	Range Price	37.89
25	Quartile 25%	11.739%	Quartile 25%	5.577%	Quartile 25%	5.382%
26	Quartile 50%	13.094%	Quartile 50%	39.368%	Quartile 50%	12.858%
27	Quartile 75%	20.942%	Quartile 75%	55.454%	Quartile 75%	16.203%
28	Interquartile	9.203%	Interquartile	49.877%	Interquartile	10.821%
29	Kurtosis	0.0595	Kurtosis	2.19607	Kurtosis	(0.76743)
30	Skewness	0.0304	Skewness	1.27436	Skewness	(0.86984)
31	Count	7	Count	7	Count	7
32	Coefficient of Variation	0.62232	Coefficient of Variation	1.29481	Coefficient of Variation	0.73188
33			Market Correlation	0.658255655	Market Correlatin	-0.113165047

Main | Returns | Data1 | Data2 | Data3

Below shows the annual return of the S&P 500, Amazon, and Walmart.

	A	B	C	D	E
1	Date	INDEX_GSPC	AMZN	WMT	Frequency
2	12/31/2014	11.391%	(22.177%)	11.871%	annual
3	12/31/2013	29.601%	58.963%	18.164%	annual
4	12/31/2012	13.406%	44.928%	16.989%	annual
5	12/31/2011	(0.003%)	(3.833%)	13.845%	annual
6	12/31/2010	12.783%	33.809%	3.220%	annual
7	12/31/2009	23.454%	162.325%	(2.583%)	annual
8	12/31/2008				annual
9					
10					

Main | Returns | Data1 | Data2 ...

READY

Below shows the Excel formulas to calculate each numeric measures. Note that the Excel function *FormulaText* is used to show the formula of a specific cell.

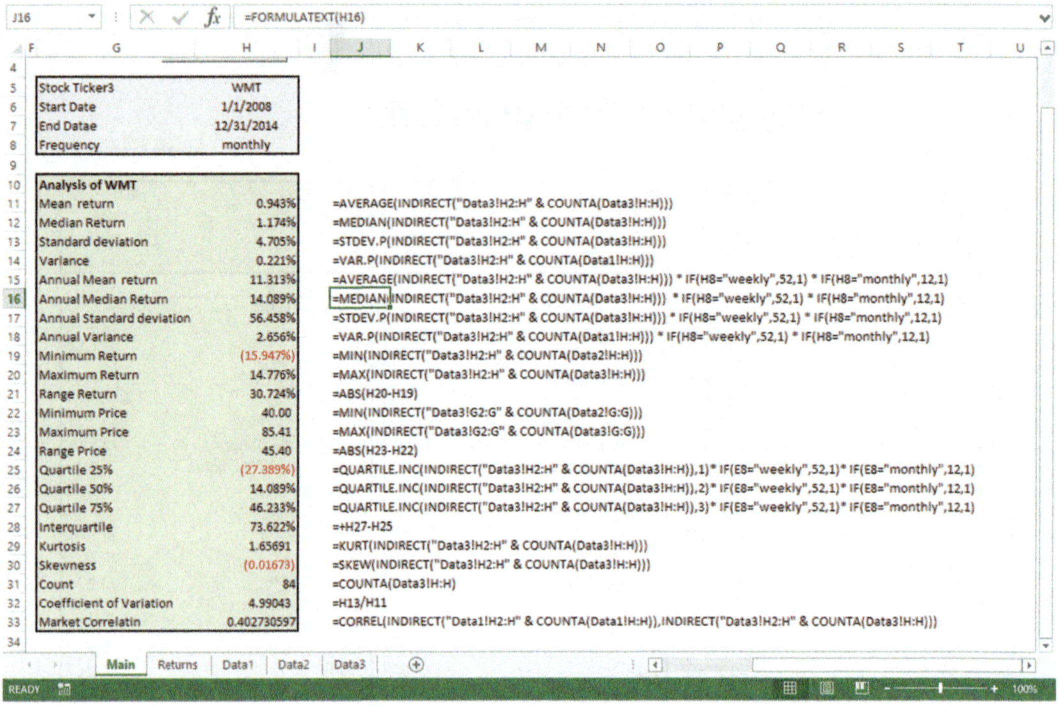

The VBA to retrieve the stock prices from www.Quandl.com is shown in Appendix 4.2

4.4 Arithmetic Mean (Average)

We will look at two measures of the central of tendency of a data set. The first one is the arithmetic mean or average. Mathematically it is defined as

$$\overline{x} = \frac{\sum_{i=1}^{n} x_i}{n}$$

The *Average* Excel formula is used to calculate the arithmetic mean.

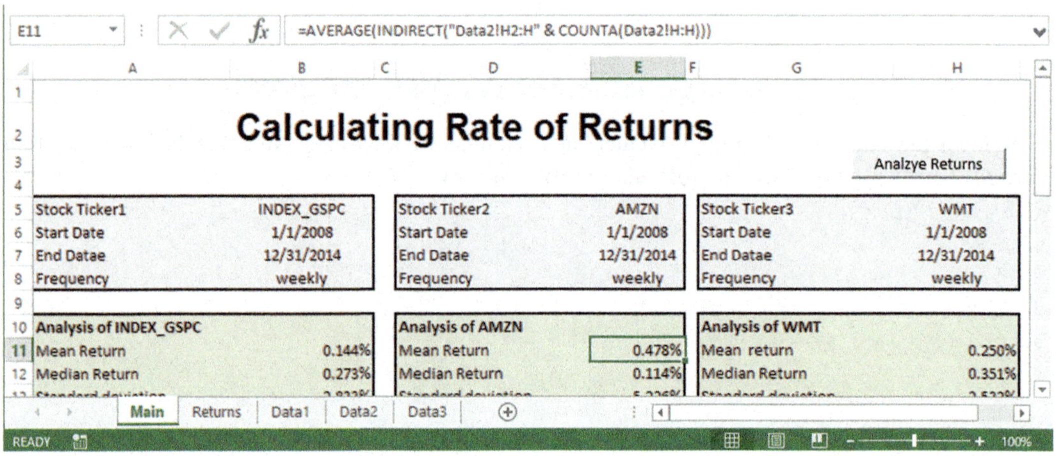

We would multiply weekly returns by 52 to annualize weekly returns. We would multiply returns by 12 to annualize monthly returns. This is a quick way to approximate the annualize returns. Below shows that the approximation worked for both the S&P 500 and Walmart but not Amazon. This approximation only works when the variance of the returns are not very large.

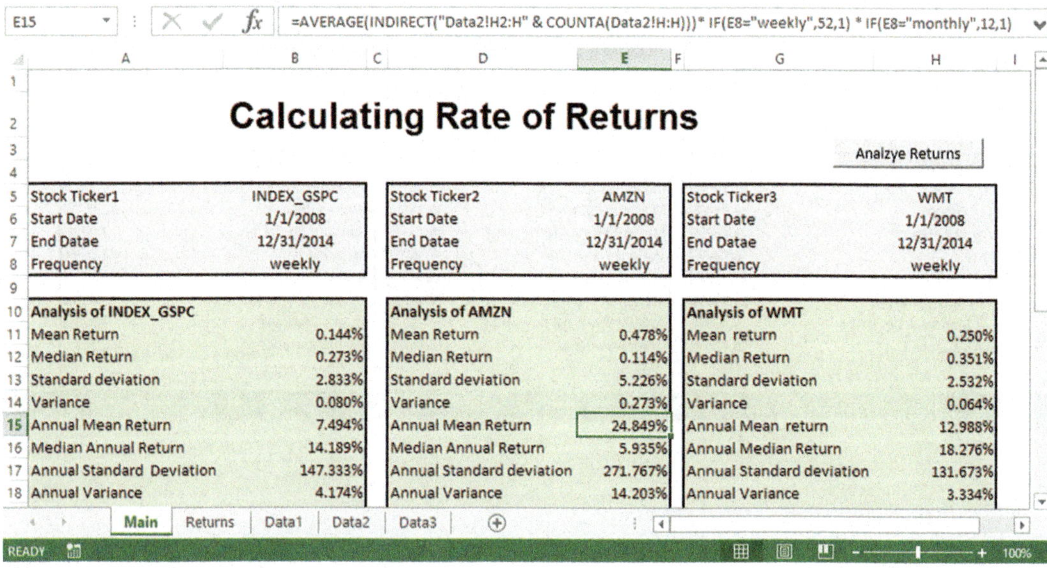

E15 =AVERAGE(INDIRECT("Data2!H2:H" & COUNTA(Data2!H:H)))* IF(E8="weekly",52,1) * IF(E8="monthly",12,1)

Calculating Rate of Returns

Analzye Returns

Stock Ticker1	INDEX_GSPC	Stock Ticker2	AMZN	Stock Ticker3	WMT
Start Date	1/1/2008	Start Date	1/1/2008	Start Date	1/1/2008
End Datae	12/31/2014	End Datae	12/31/2014	End Datae	12/31/2014
Frequency	weekly	Frequency	weekly	Frequency	weekly
Analysis of INDEX_GSPC		**Analysis of AMZN**		**Analysis of WMT**	
Mean Return	0.144%	Mean Return	0.478%	Mean return	0.250%
Median Return	0.273%	Median Return	0.114%	Median Return	0.351%
Standard deviation	2.833%	Standard deviation	5.226%	Standard deviation	2.532%
Variance	0.080%	Variance	0.273%	Variance	0.064%
Annual Mean Return	7.494%	Annual Mean Return	24.849%	Annual Mean return	12.988%
Median Annual Return	14.189%	Median Annual Return	5.935%	Annual Median Return	18.276%
Annual Standard Deviation	147.333%	Annual Standard deviation	271.767%	Annual Standard deviation	131.673%
Annual Variance	4.174%	Annual Variance	14.203%	Annual Variance	3.334%

Main | Returns | Data1 | Data2 | Data3

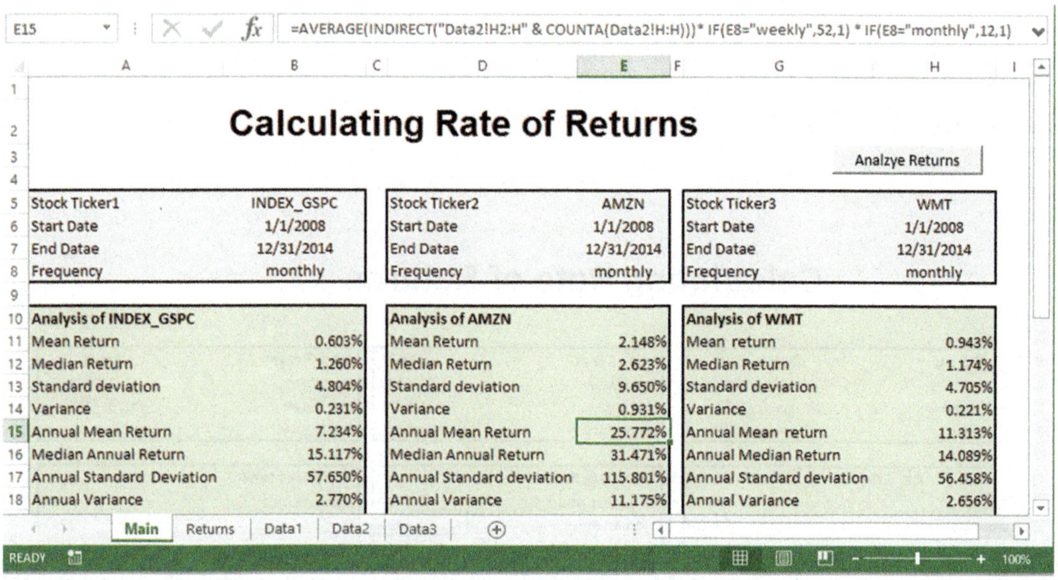

E15 =AVERAGE(INDIRECT("Data2!H2:H" & COUNTA(Data2!H:H)))* IF(E8="weekly",52,1) * IF(E8="monthly",12,1)

Calculating Rate of Returns

Analzye Returns

Stock Ticker1	INDEX_GSPC	Stock Ticker2	AMZN	Stock Ticker3	WMT
Start Date	1/1/2008	Start Date	1/1/2008	Start Date	1/1/2008
End Datae	12/31/2014	End Datae	12/31/2014	End Datae	12/31/2014
Frequency	monthly	Frequency	monthly	Frequency	monthly
Analysis of INDEX_GSPC		**Analysis of AMZN**		**Analysis of WMT**	
Mean Return	0.603%	Mean Return	2.148%	Mean return	0.943%
Median Return	1.260%	Median Return	2.623%	Median Return	1.174%
Standard deviation	4.804%	Standard deviation	9.650%	Standard deviation	4.705%
Variance	0.231%	Variance	0.931%	Variance	0.221%
Annual Mean Return	7.234%	Annual Mean Return	25.772%	Annual Mean return	11.313%
Median Annual Return	15.117%	Median Annual Return	31.471%	Annual Median Return	14.089%
Annual Standard Deviation	57.650%	Annual Standard deviation	115.801%	Annual Standard deviation	56.458%
Annual Variance	2.770%	Annual Variance	11.175%	Annual Variance	2.656%

Main | Returns | Data1 | Data2 | Data3

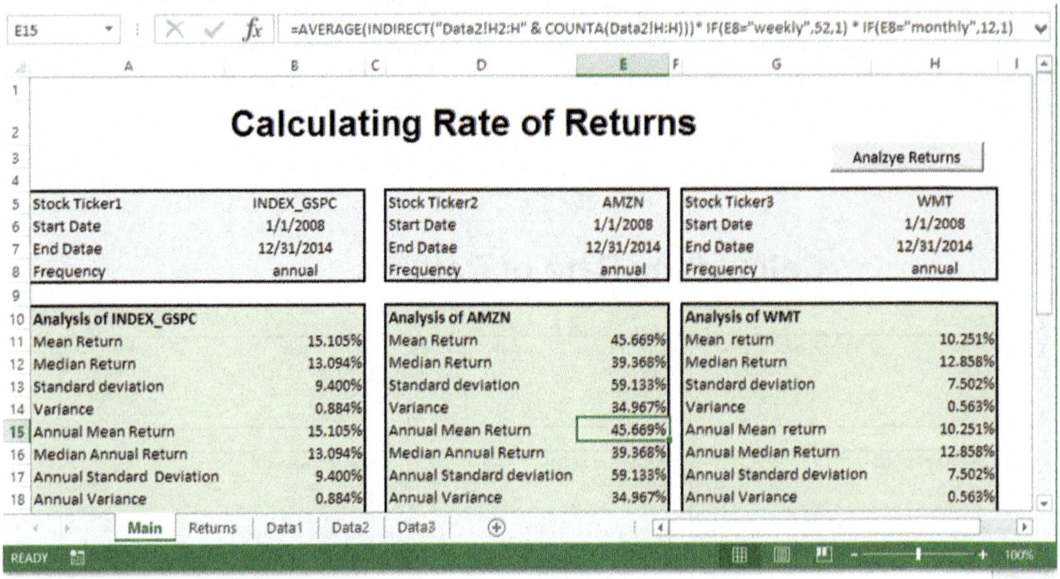

4.5 Median

Another measure of the central tendency of a data set is the median. The median is the middle observation of a data set. It is important to note that unlike the mean, the median is not influenced by *extreme* values.

The *Median* Excel function is used to calculate the median.

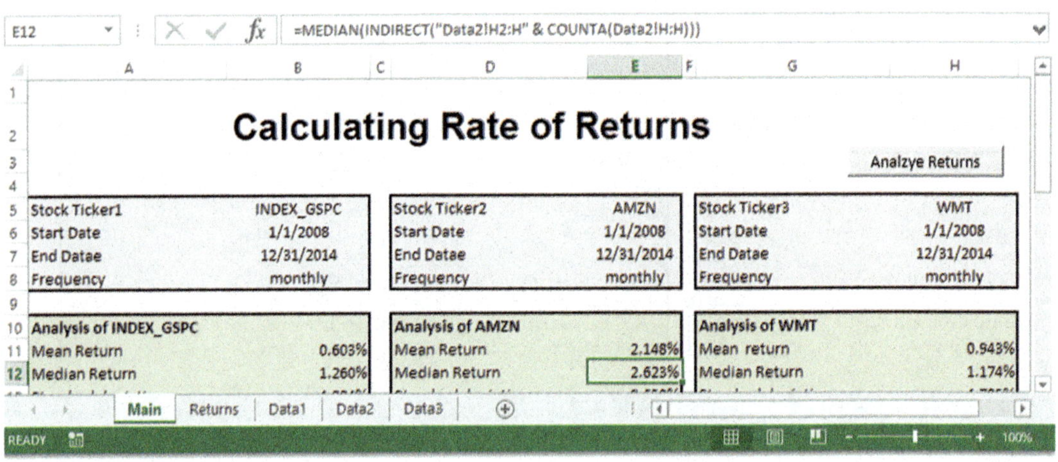

4.6 Standard Deviation

We will now look at the concept of the how tightly clustered a data set is. This concept is called the *measure of dispersion*. The standard deviation is a measure of dispersion. The mathematical formula for the standard deviation is

$$s = \sqrt{\frac{\sum_{i=1}^{n}(x_i - \overline{x})^2}{n-1}}$$

The *STDEV.P* Excel function is used to calculate the standard deviation.

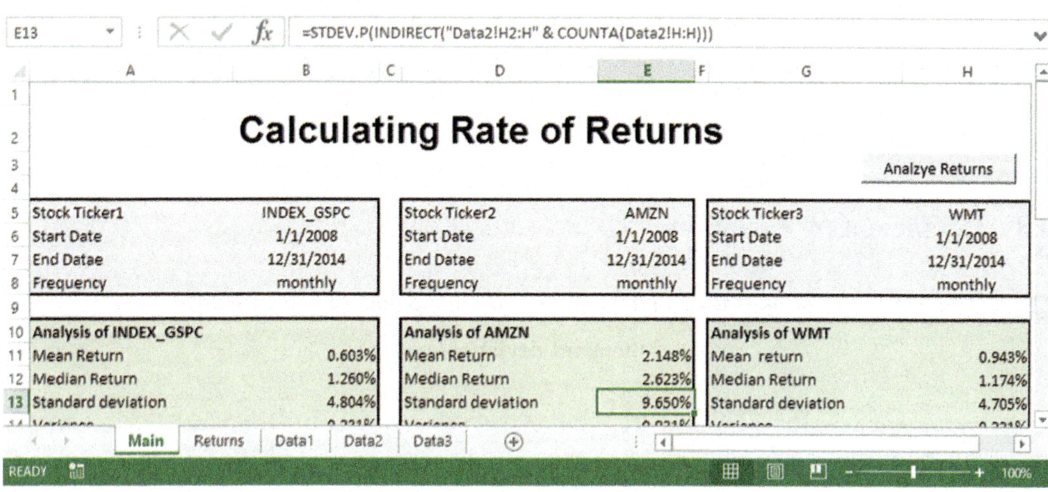

4.7 Variance

The variance is another measure of the spread, or dispersion, of a data set. Mathematically it looks like the following:

$$\sigma^2 = \sqrt{\frac{\sum_{i=1}^{n}(x_i - \overline{x})^2}{n-1}} * \sqrt{\frac{\sum_{i=1}^{n}(x_i - \overline{x})^2}{n-1}} = \frac{\sum_{i=1}^{n}(x_i - \overline{x})^2}{n-1}$$

The *VAR.P* Excel function is used to calculate the variance.

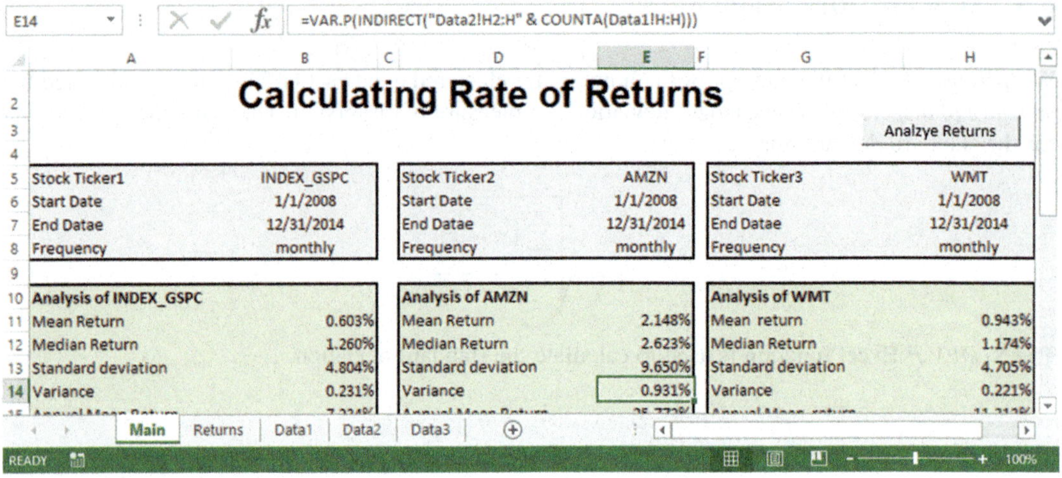

4.8 Coefficient of Variation

The *Coefficient of Variation* is used to help variations among data sets. It is defined as

Standard deviation/mean

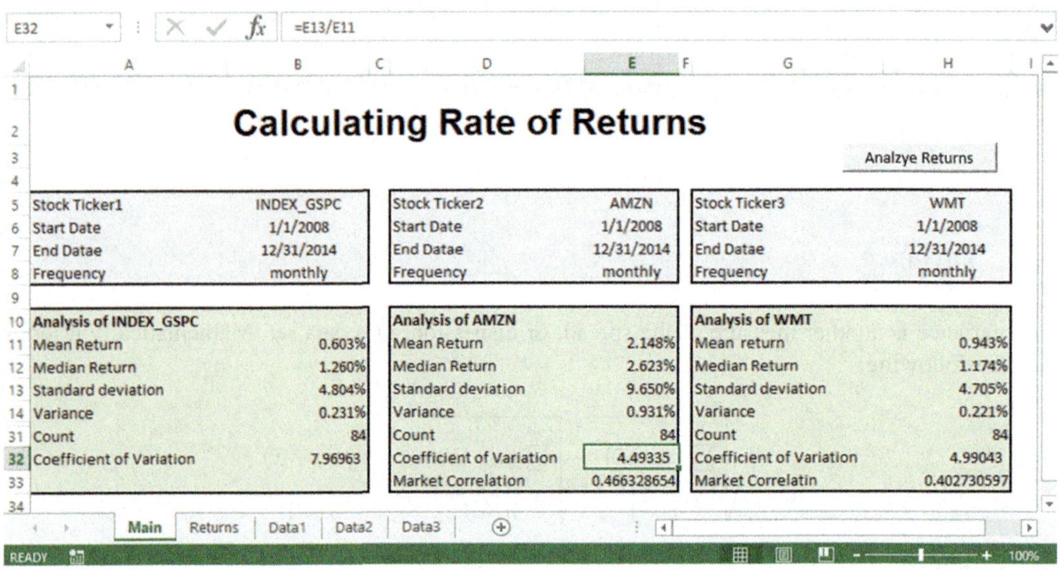

4.9 Minimum, Maximum, and Range

Another measure of spread of a data set is the *range*. The range is the difference between the highest and lowest value.

The *Min* Excel function is used to calculate the minimum. The *Max* Excel function is used to calculate the minimum.

J19 =FORMULATEXT(H19)

	G	H	I	J
5	Stock Ticker3	WMT		
6	Start Date	1/1/2008		
7	End Datae	12/31/2014		
8	Frequency	monthly		
9				
10	Analysis of WMT			
11	Mean return	0.943%		
12	Median Return	1.174%		
13	Standard deviation	4.705%		
14	Variance	0.221%		
15	Annual Mean return	11.313%		
16	Annual Median Return	14.089%		
17	Annual Standard deviation	56.458%		
18	Annual Variance	2.656%		
19	Minimum Return	(15.947%)		=MIN(INDIRECT("Data3!H2:H" & COUNTA(Data2!H:H)))
20	Maximum Return	14.776%		=MAX(INDIRECT("Data3!H2:H" & COUNTA(Data3!H:H)))
21	Range Return	30.724%		=ABS(H20-H19)
22	Minimum Price	40.00		=MIN(INDIRECT("Data3!G2:G" & COUNTA(Data2!G:G)))
23	Maximum Price	85.41		=MAX(INDIRECT("Data3!G2:G" & COUNTA(Data3!G:G)))
24	Range Price	45.40		=ABS(H23-H22)
25	Quartile 25%	(27.389%)		
26	Quartile 50%	14.089%		
27	Quartile 75%	46.233%		
28	Interquartile	73.622%		
29	Kurtosis	1.65691		
30	Skewness	(0.01673)		
31	Count	84		
32	Coefficient of Variation	4.99043		
33	Market Correlatin	0.402730597		
34				

Main | Returns | Data1 | Data2 | D ...

READY 100%

4.10 Quartiles

We will now look at the concept of the *measure of relative position.* This concept indicates where a data value is in respect to all of the other data values in the data set. When we are analyzing relative positions we first assume that the data set is sorted in ascending order.

Quartile is a measure of relative position. When we are talking about quartiles, we are breaking up the data set into four equal parts. The first quartile is the first 25 % of the data set. The second quartile is the first 50 % of the data set after the first quartile. The third quartile is the first 75 % of the data set after the first quartile.

The *Quartile.Inc* Excel function is used to calculate the quartiles. The function uses a one to indicate the first quartile, a two to indicate the second quartile, and a three to indicate the third quartile.

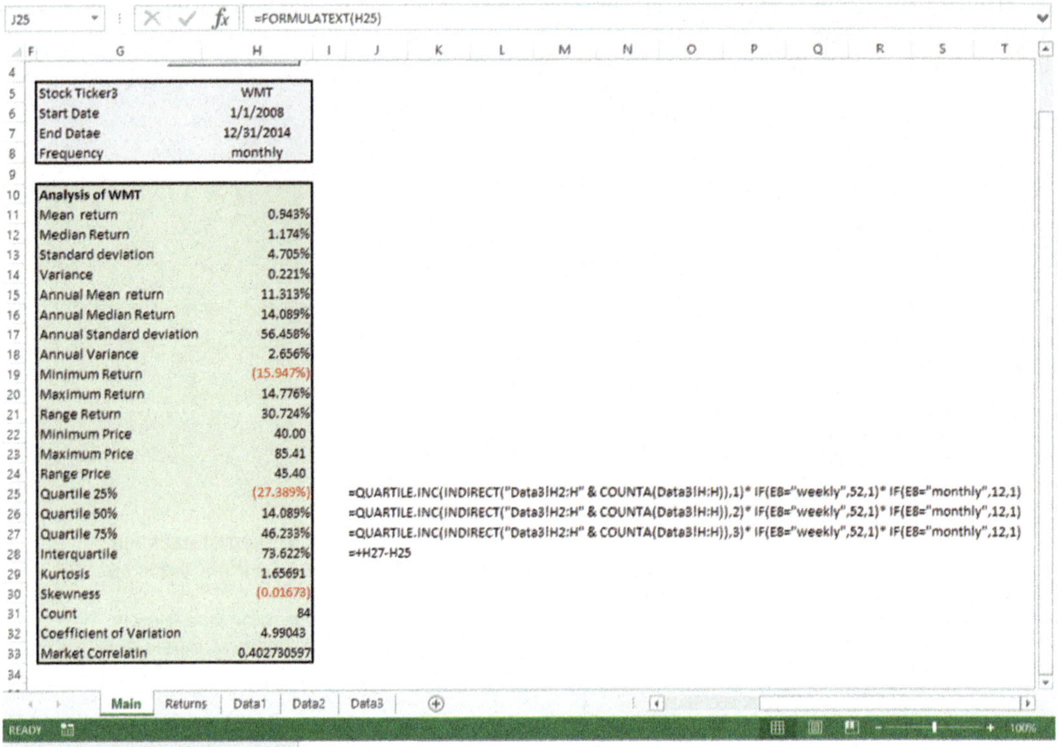

4.11 Excel Macro: Box Plot

We will now run the program in the workbook *BoxPlot.xls*m to create box plots in Excel. The Excel program for the box plot is shown in Appendix 4.1.

The rate of return data should be put in the worksheet "data" as shown below.

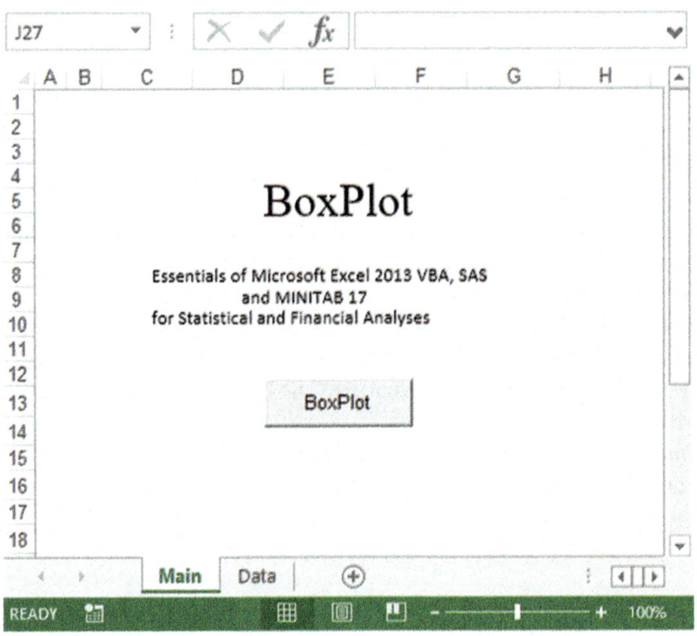

Now go to worksheet "main" and push the button *boxplot*. Below shows the resulting box plot chart.

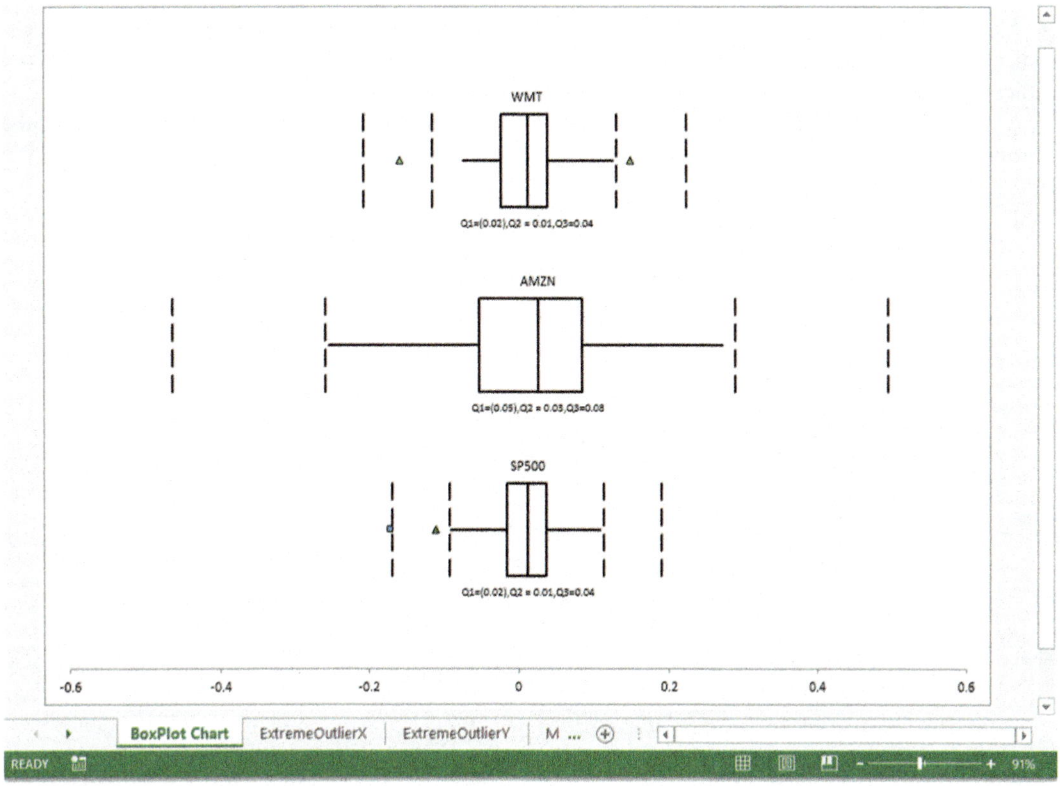

Above shows the box plots of the rate of returns of the S&P 500, Amazon, and Walmart. Each box plot shows the inner and outer fences of a box plot. The four inner fences are defined as follows:

Upper Inner Fence: Q_3 + 1.5 * Interquartile
Lower Inner Fence: Q_1 Quartile − 1.5 * Interquartile
Upper Outer Fence: Q_3 + 3 * Interquartile
Lower Outer Fence: Q_1 − 3 * Interquartile

Showing the fences of a box plot allows us to show that a whisker is the lines leaving the interquartile to the last data point before the inner fences. Showing the fences allows us to show mild and extreme values of a data set. Data between the inner and upper fences are defined as *mild outliers*, and data beyond the outer fences are defined as extreme outliers.

4.12 *Z* Score

Another measure of relative position is the *Z* score. The *Z* score expresses the relative position of any particular data item in terms of the number of standard deviations above or below the mean. The score is a very important concept that will be used extensively in future chapters. It is defined as

$$Z = \frac{x - \overline{x}}{s}$$

A zero Z score indicates that a particular data value is equal to the mean. A positive 1 Z score indicates that a data value is one standard deviation to the right of the mean. A negative 1 Z score indicates that a data value is one standard deviation to the left of the mean.

A very important feature of mound-shaped data sets is that approximately 68 % of the observations have a Z score between -1 and 1, and approximately 95 % of the observations have a Z score between -2 and 2.

This is graphically shown below.

Z SCORE, 2 STANDARD DEVIATIONS

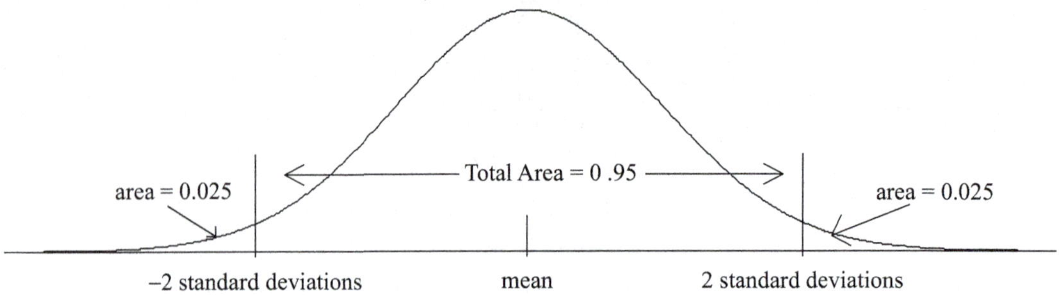

Z SCORE, 1 STANDARD DEVIATION

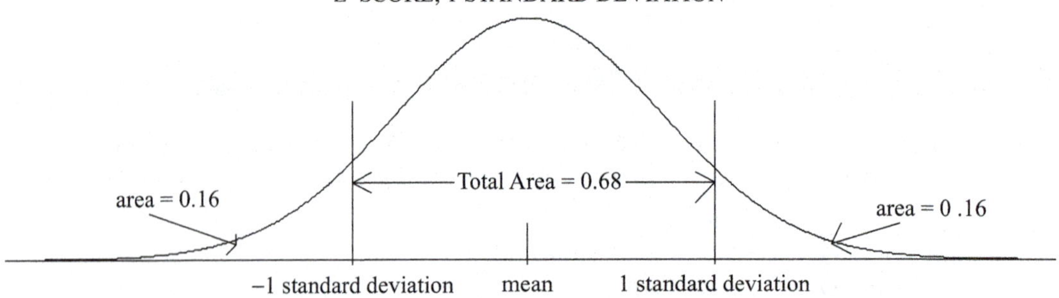

In the above section we used box plots to identify outliers. Another way to identify outliers is to use Z scores. In most data sets 5 % of the data items are more than two standard deviations away from the mean, and 3 % of the data items are more than three standard deviations away. Lets classify those data items that are more than two standard deviations but less than three standard deviations as *mild outliers* and those data items that are more than three standard deviations as *extreme outliers*.

4.13 Z Score in Excel

We will use the Excel *standardize* function to calculate the Z scores. The *standardize* function requires two arguments. The first argument is the value of the data item of interest. The second argument is the mean of the data set. The third argument is the standard deviation of the argument. Below, in cell J13, shows the Z score for Amazon for January 2014 is -1.26455. This indicates that the -10.055 % return for this month is 1.26455 standardize deviation to the left of average monthly return of 2.148 %.

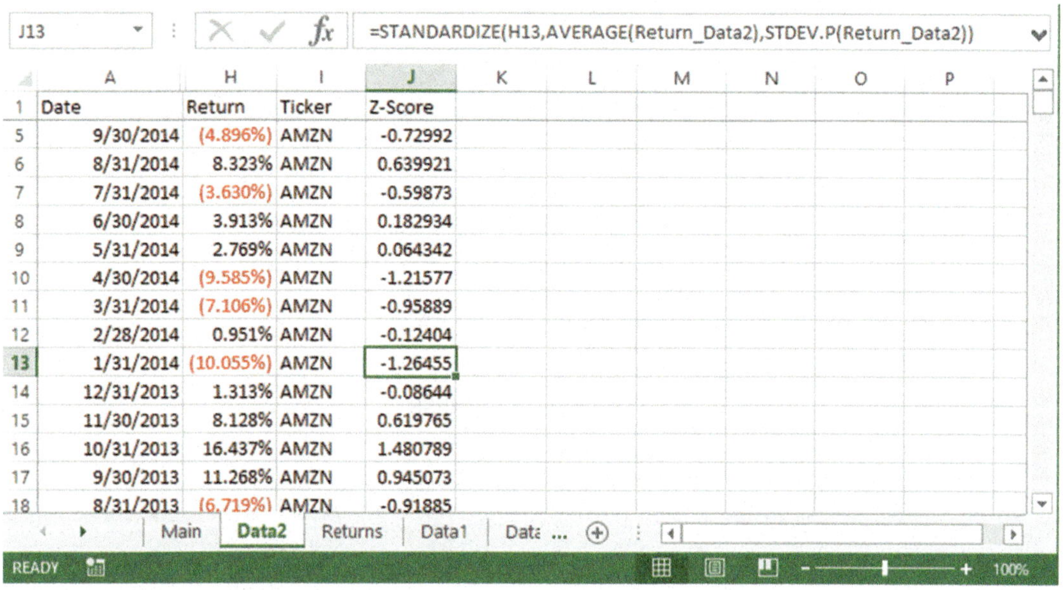

J13 =STANDARDIZE(H13,AVERAGE(Return_Data2),STDEV.P(Return_Data2))

	A	H	I	J
1	Date	Return	Ticker	Z-Score
5	9/30/2014	(4.896%)	AMZN	-0.72992
6	8/31/2014	8.323%	AMZN	0.639921
7	7/31/2014	(3.630%)	AMZN	-0.59873
8	6/30/2014	3.913%	AMZN	0.182934
9	5/31/2014	2.769%	AMZN	0.064342
10	4/30/2014	(9.585%)	AMZN	-1.21577
11	3/31/2014	(7.106%)	AMZN	-0.95889
12	2/28/2014	0.951%	AMZN	-0.12404
13	1/31/2014	(10.055%)	AMZN	-1.26455
14	12/31/2013	1.313%	AMZN	-0.08644
15	11/30/2013	8.128%	AMZN	0.619765
16	10/31/2013	16.437%	AMZN	1.480789
17	9/30/2013	11.268%	AMZN	0.945073
18	8/31/2013	(6.719%)	AMZN	-0.91885

4.14 Skewness

We will now look at the concept of *measure of shape*. Sometimes we want to know whether a data set exhibits a symmetric pattern. The *coefficient of skewness* answers this question. The formula for the coefficient of skewness is

$$CS = \frac{\frac{\sum_{i=1}^{N}(x_i - \mu)^3}{N}}{\sigma^3}$$

We interpret skew numbers as follows:

Zero skewness coefficient means that the distribution is symmetric, with the mean equal to the median.

Positive skewness coefficient means that the distribution is skewed to the right and that the median lies below or to the left of the mean.

Negative skewness coefficient means that the distribution is skewed to the left and the median is above or to the right of the mean.

These concepts are graphically illustrated below.

ZERO SKEWNESS

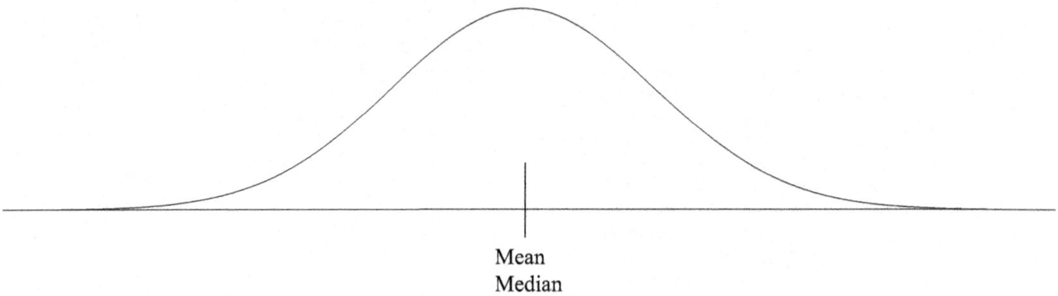

POSITIVE SKEWNESS

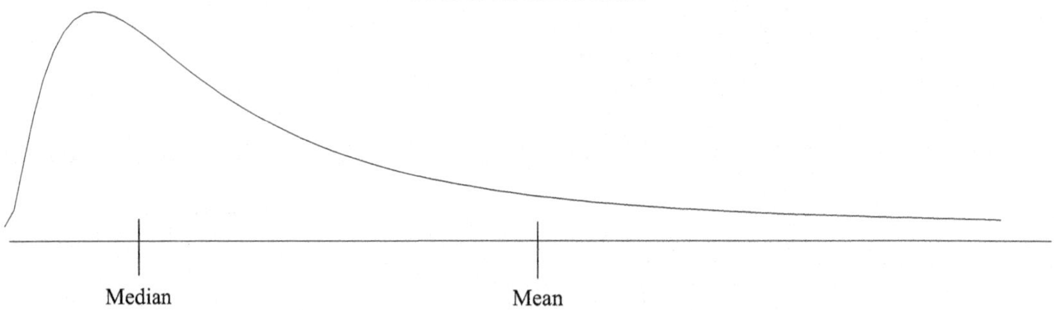

NEGATIVE SKEWNESS

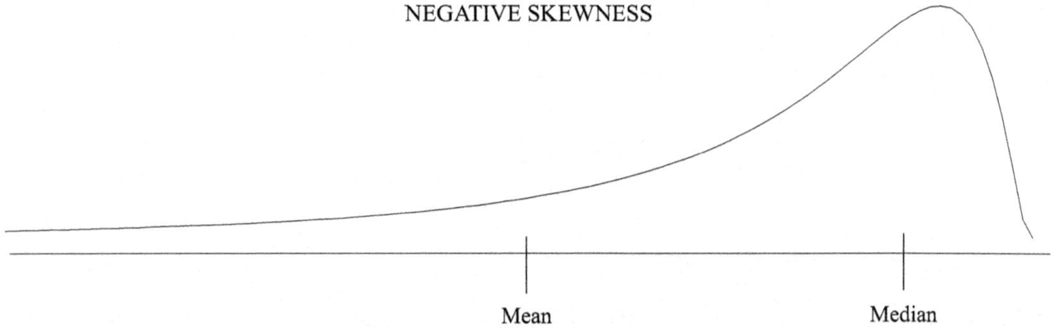

Notice that for the zero skewness, the hump of the distribution is in the middle. The hump of the positive skewness is toward the left. The hump of the negative skewness is toward the right.

4.15 Skewness in Excel

The *Skew* Excel function is used to calculate the skewness of a data set.

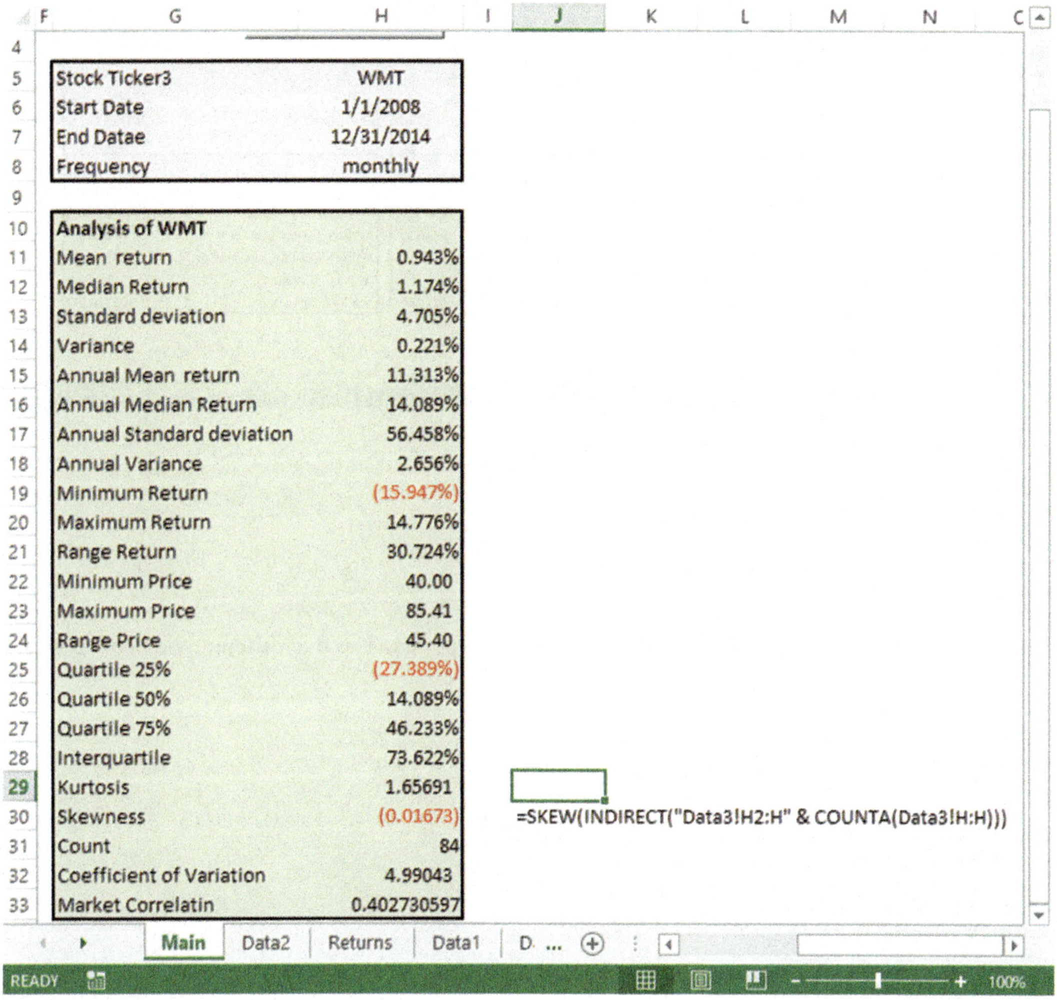

Stock Ticker3	WMT
Start Date	1/1/2008
End Datae	12/31/2014
Frequency	monthly

Analysis of WMT	
Mean return	0.943%
Median Return	1.174%
Standard deviation	4.705%
Variance	0.221%
Annual Mean return	11.313%
Annual Median Return	14.089%
Annual Standard deviation	56.458%
Annual Variance	2.656%
Minimum Return	(15.947%)
Maximum Return	14.776%
Range Return	30.724%
Minimum Price	40.00
Maximum Price	85.41
Range Price	45.40
Quartile 25%	(27.389%)
Quartile 50%	14.089%
Quartile 75%	46.233%
Interquartile	73.622%
Kurtosis	1.65691
Skewness	(0.01673)
Count	84
Coefficient of Variation	4.99043
Market Correlatin	0.402730597

4.16 Coefficient of Correlation

The *coefficient of correlation* is a measurement of linear relationship between two variables. The coefficient of correlation is a number between −1 and 1. A 1 means a perfect positive correlation and a −1 means a perfect negative correlation.

The *Correl* Excel function is used to calculate the skewness of a dataset.

The shows the correlation between Amazon and the S&P 500 and Walmart and the S&P 500.

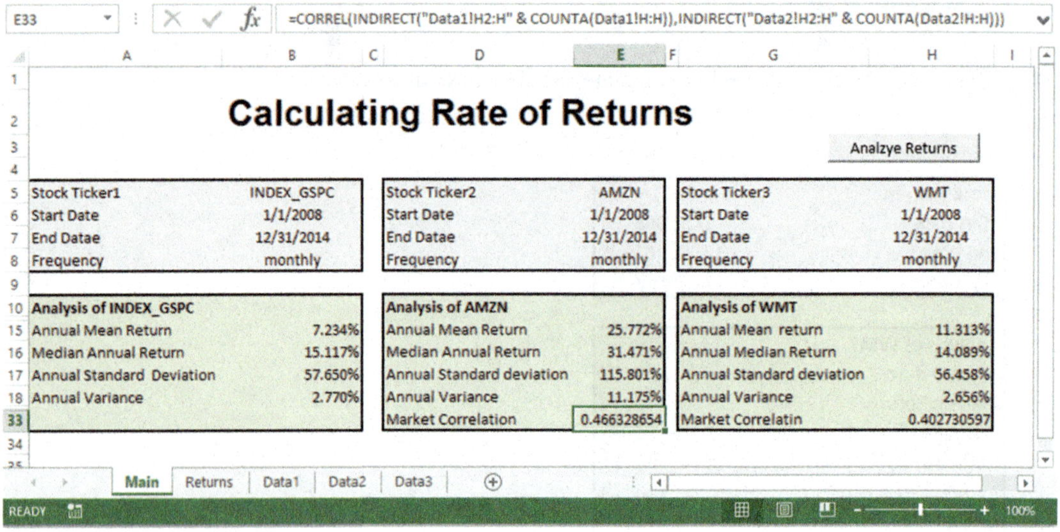

4.17 Statistical Summary

Numeric measures look at a data set from different viewpoints. The five different viewpoints are:

1. Measure of central tendency or measure of location:

   (a) Arithmetic mean
   (b) Median

   This group tells us the central location of a data set.
2. Measure of spread:

   (a) Standard deviation
   (b) Variance

   This group tells us how spread out the data set is.
3. Measure of relative position:

   (a) Quartiles
   (b) Z score

   This group tells us the relative position of a data value in a data set.
4. Measure of shape:

   (a) Skewness

   This measure tells us if a data set is symmetrically relative to the mean or median.
5. Measure of linear relationship:

   (a) Coefficient of correlation

   This measure tells us the linear relationship between two variables.

4.18 SAS Programming Code Instructions and Examples

4.18.1 Descriptive Statistics Report in SAS

In this section, we will analyze the rate of return for JNJ, MRK, and the S&P 500. The S&P 500 is an index of stocks that many people consider as a proxy for the economy. The rate of returns of JNJ, MRK, and the S&P 500 is read from three different documents and is shown below.

```
    /*Input the data from JNJ*/
    libname Mydata 'd:\SASdata\MySASdata';
    data Mydata.JNJ;
       INFILE 'd:\MINITAB and Microsoft Excel Book\Ch4\Ch4JNJ.txt' dlm='09'x firstobs=2;
       INPUT CUSIP_hand        CUSIP $ Year EPS_Basic DPS_Fiscal Close_Fiscal
Rate_of_Return;
    run;

    /*Input the data from MRK*/
    data Mydata.MRK;
       INFILE 'd:\MINITAB and Microsoft Excel Book\Ch4\Ch4MRK.txt' dlm='09'x firstobs=2;
       INPUT CUSIP_hand        CUSIP $ Year EPS_Basic DPS_Fiscal Close_Fiscal
Rate_of_Return;
    run;


    /*Input the data from SP500*/
    data Mydata.SP500;
       INFILE 'd:\MINITAB and Microsoft Excel Book\Ch4\Ch4SP500.txt' dlm='09'x firstobs=2;
       INPUT Year Close Rate_of_Return;
    run;

    /*Merge the previous table into one*/
    data Mydata.Return (keep=Year JNJ MRK SP500);
       merge Mydata.JNJ (rename=(Rate_of_return=JNJ))
             Mydata.MRK (rename=(Rate_of_Return=MRK))
                Mydata.SP500 (rename=(Rate_of_Return=SP500));
       by Year;
    run;

    proc print data=Mydata.Return;
    run;
```

The rate of returns of JNJ, MRK, and the S&P 500 is shown below.

Obs	Year	JNJ	MRK	S&P 500
1	1971	0.72807	0.25253	0.18216
2	1972	0.32487	−0.28125	0.10995
3	1973	−0.13602	−0.09397	−0.01549
4	1974	−0.28270	−0.17802	−0.22952
5	1975	0.10974	0.04331	0.04107
6	1976	−0.13092	−0.01625	0.18392
7	1977	−0.01603	−0.18532	−0.03773
8	1978	−0.03909	0.21847	−0.02108
9	1979	0.07458	0.06839	0.07169
10	1980	0.25867	0.17301	0.15252

(continued)

Obs	Year	JNJ	MRK	S&P 500
11	1981	−0.62782	0.00000	0.07859
12	1982	0.33670	−0.00147	−0.06506
13	1983	−0.17632	0.06795	0.34049
14	1984	−0.11621	0.04011	−0.00006
15	1985	0.45675	0.45745	0.16422
16	1986	0.24703	−0.09580	0.26540
17	1987	0.14095	0.27951	0.21410
18	1988	0.13689	−0.63565	−0.07359
19	1989	−0.30250	0.34199	0.21502
20	1990	0.20842	0.15968	0.03585
21	1991	0.59582	0.85257	0.12420
22	1992	−0.55895	−0.73949	0.10516
23	1993	−0.11139	−0.20749	0.08625
24	1994	0.22006	0.10909	0.01951
25	1995	0.56164	0.72131	0.17658
26	1996	−0.41813	0.21333	0.23772
27	1997	0.32412	0.33124	0.30265
28	1998	0.27324	0.39151	0.24280
29	1999	0.11177	−0.54449	0.22278
30	2000	0.12668	0.39349	0.07526
31	2001	−0.43748	−0.37196	−0.16328
32	2002	−0.09120	−0.03724	−0.16768
33	2003	−0.03817	−0.18389	−0.02889
34	2004	0.22764	−0.30433	0.17138
35	2005	−0.05235	−0.01027	0.06773
36	2006	0.09850	0.37064	0.08551
37	2007	0.01030	0.33280	0.12723
38	2008	−0.10300	−0.47685	−0.17408
39	2009	0.07655	0.20197	−0.22294
40	2010	−0.03975	−0.01368	0.20244
41	2011	0.06031	0.04606	0.11199

We used **PROC UNIVARIATE** to produce the descriptive statistics. Below is the SAS command to produce a descriptive statistics report of JNJ, MRK, and the S&P 500.

```
PROC UNIVARIATE DATA=Mydata.Return;
    VAR JNJ MRK SP500;
RUN;
```

Here, we only show the output of JNJ below. The statistics of MRK and S&P 500 is reported in the same form.

Moments			
N	41	Sum weights	41
Mean	0.04954403	Sum observations	2.03130537
Std deviation	0.29352041	Variance	0.08615423
Skewness	−0.0923951	Kurtosis	0.32605592
Uncorrected SS	3.54680823	Corrected SS	3.44616917
Coeff variation	592.443504	Std error mean	0.04584019

Basic statistical measures			
Location		Variability	
Mean	0.049544	Std deviation	0.29352
Median	0.074576	Variance	0.08615
Mode		Range	1.35589
		Interquartile range	0.33903

Tests for location: Mu0 = 0				
Test	Statistics		p value	
Student's t	t	1.080799	Pr > |t|	0.2863
Sign	M	2.5	Pr ≥ |M|	0.5327
Signed rank	S	94.5	Pr ≥ |S|	0.2250

Quantiles (definition 5)	
Quantile	Estimate
100 % max	0.7280684
99 %	0.7280684
95 %	0.5616438
90 %	0.3367003
75 % Q3	0.2276423
50 % median	0.0745763
25 % Q1	−0.1113861
10 %	−0.3024963
5 %	−0.4374777
1 %	−0.6278192
0 % min	−0.6278192

Extreme observations			
Lowest		Highest	
Value	Obs	Value	Obs
−0.627819	11	0.336700	12
−0.558952	22	0.456747	15
−0.437478	31	0.561644	25
−0.418129	26	0.595819	21
−0.302496	19	0.728068	1

We can also use **ods tagsets.excelxp** to export the output from SAS into Excel (note: This command is compatible with SAS 9.1.3 and above). The example here will create a new worksheet for each table we get previously by using **PROC UNIVARIATE**. The file is named as Statistics under "d:\".

```
PROC UNIVARIATE DATA=Mydata.Return;
    VAR JNJ;
RUN;
```

Below are forms in file statistics from different worksheets:

Moments			
N	41	Sum weights	41
Mean	0.04954403	Sum observations	2.03130537
Std deviation	0.29352041	Variance	0.08615423
Skewness	−0.0923951	Kurtosis	0.32605592
Uncorrected SS	3.54680823	Corrected SS	3.44616917
Coeff variation	592.443504	Std error mean	0.04584019

Basic statistical measures			
Location		Variability	
Mean	0.049544	Std deviation	0.29352
Median	0.074576	Variance	0.08615
Mode		Range	1.35589
		Interquartile range	0.33903

Tests for location: Mu0 = 0				
Test	Statistics		p value	
Student's t	t	1.080799	Pr > |t|	0.2863
Sign	M	2.5	Pr ≥ |M|	0.5327
Signed rank	S	94.5	Pr ≥ |S|	0.2250

Quantiles (definition 5)	
Quantile	Estimate
100 % max	0.7280684
99 %	0.7280684
95 %	0.5616438
90 %	0.3367003
75 % Q3	0.2276423
50 % median	0.0745763
25 % Q1	−0.1113861
10 %	−0.3024963
5 %	−0.4374777
1 %	−0.6278192
0 % min	−0.6278192

Extreme observations			
Lowest		Highest	
Value	Obs	Value	Obs
−0.627819	11	0.336700	12
−0.558952	22	0.456747	15
−0.437478	31	0.561644	25
−0.418129	26	0.595819	21
−0.302496	19	0.728068	1

4.18.2 Arithmetic Mean (Average)

```
PROC MEANS MEAN DATA=Mydata.Return;
    VAR JNJ MRK SP500;
RUN;
```

Variable	Mean
JNJ	0.0495440
MRK	0.0411947
S&P 500	0.0784090

4.18.3 Median

```
PROC MEANS MEDIAN DATA=Mydata.Return;
    VAR JNJ MRK SP500;
RUN;
```

Variable	Median
JNJ	0.0745763
MRK	0.0433145
S&P 500	0.0862538

4.18.4 Standard Deviation

```
PROC MEANS STD DATA=Mydata.Return;
    VAR JNJ MRK SP500;
RUN;
```

Variable	Std dev
JNJ	0.2935204
MRK	0.3345002
S&P 500	0.1422031

4.18.5 Variance

```
PROC MEANS VAR DATA=Mydata.Return;
    VAR JNJ MRK SP500;
RUN;
```

Variable	Variance
JNJ	0.0861542
MRK	0.1118904
S&P 500	0.0202217

4.18.6 Range

```
PROC MEANS RANGE DATA=Mydata.Return;
    VAR JNJ MRK SP500;
RUN;
```

Variable	Range
JNJ	1.3558876
MRK	1.5920625
S&P 500	0.5700130

4.18.7 SAS Box Plot

In this section, we will mainly use the **BOXPLOT** procedure to create box plots in SAS. The **BOXPLOT** procedure creates side-by-side box plots of measurements organized in groups. A box plot displays the mean, quartiles, and minimum and maximum observations for a group.

```
data Mydata.Return;
      set Mydata.Return;
      Company='JNJ';
run;

Goptions ftext='Times New Roman' htext=6 gunit=pct;
title  height=7 'Box Plot for JNJ';
proc boxplot data=Mydata.Return;
 plot JNJ*Company/BOXWIDTH=6 boxstyle=schematic IDSYMBOL=square IDCOLOR=blue;
run;
```

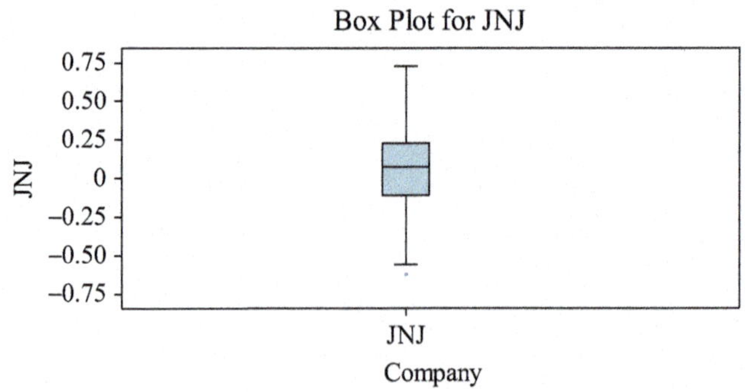

```
data Mydata.Return;
      set Mydata.Return;
      Company='MRK';
run;

Goptions ftext='Times New Roman' htext=7 gunit=pct;
title  height=6 'Box Plot for MRK';
axis1  label=(h=6pct;
axis2  label=(h=6pct);
symbol w=2;
proc boxplot data=Mydata.Return;
     plot MRK*Company/haxis=axis1 vaxis=axis2 BOXWIDTH=6
                         boxstyle=schematic IDSYMBOL=square
                         IDCOLOR=blue;
run;
```

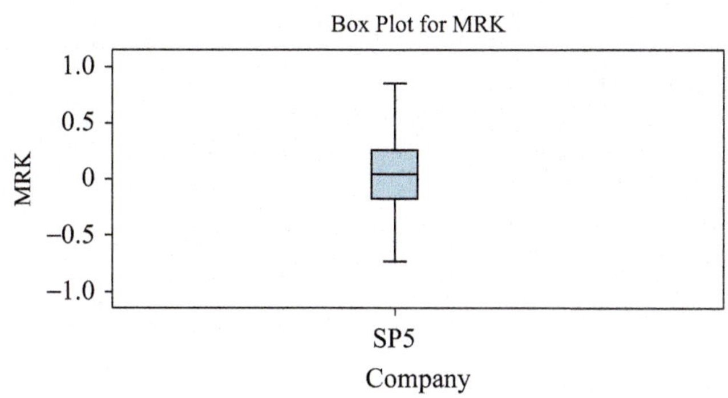


```
data Mydata.Return;
      set Mydata.Return;
      Company='SP500';
run;

Goptions ftext='Times New Roman' htext=7 gunit=pct;
title  height=6 'Box Plot for SP500';
axis1  label=(h=6pct);
axis2  label=(h=6pct);
symbol w=2;
proc boxplot data=Mydata.Return;
     plot SP500*Company/haxis=axis1 vaxis=axis BOXWIDTH=6
                            boxstyle=schematic IDSYMBOL=square
                            IDCOLOR=blue;
run;
```

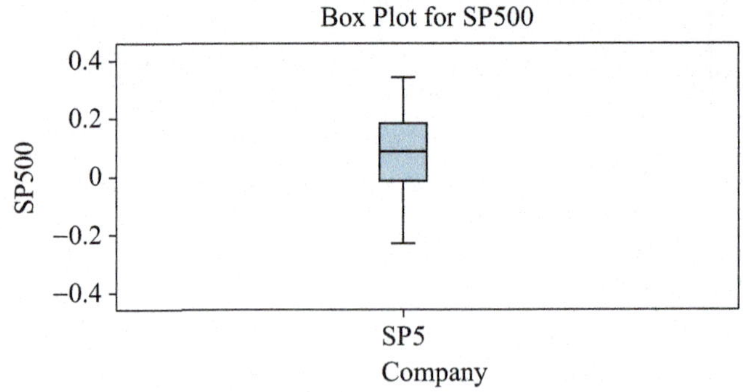

When **BOXSTYLE = SCHEMATIC** is specified, the whiskers are drawn to the most extreme points in the group that lie within the fences. The upper fence is defined as the third quartile (represented by the upper edge of the box) +1.5 times the interquartile range (IQR). The lower fence is defined as the first quartile (represented by the lower edge of the box) −1.5 times the interquartile range. Observations outside the fences are identified with a special symbol. The default symbol is a square, and you can specify the shape and color for this symbol with the **IDSYMBOL=**and **IDCOLOR=**options.[1]

IDCOLOR=color specifies the color of the symbol marker used to identify outliers in schematic box plots. **IDSYMBOL=symbol** specifies the symbol marker used to identify outliers in schematic box plots.

4.18.8 Extreme Outliers

The second procedure for generating box plots is **PROC UNIVARIATE**. To create a box plot, the **PLOT** option is first included. This causes **PROC UNIVARIATE** to create a stem-and-leaf plot, a box plot, following the default statistics output by the procedure for the variable listed in the **VAR** statement.

```
proc univariate data=Mydata.Return plot;
      var JNJ MRK SP500;
run;
```

[1] http://support.sas.com/documentation/cdl/en/qcug/63922/HTML/default/viewer.htm#qcug_shewhart_a0000003887.htm

```
                      The UNIVARIATE Procedure
                         Variable:  JNJ

                       Extreme Observations

         ------Lowest------          ------Highest-----

             Value       Obs              Value       Obs

         -0.627819        11           0.336700        12
         -0.558952        22           0.456747        15
         -0.437478        31           0.561644        25
         -0.418129        26           0.595819        21
         -0.302496        19           0.728068         1

     Stem Leaf                     #             Boxplot
        7 3                        1                |
        6 0                        1                |
        5 6                        1                |
        4 6                        1                |
        3 224                      3                |
        2 123567                   6             +-----+
        1 011344                   6             |     |
        0 1678                     4             *--+--*
       -0 954442                   6             |     |
       -1 843210                   6             +-----+
       -2 8                        1                |
       -3 0                        1                |
       -4 42                       2                |
       -5 6                        1                |
       -6 3                        1                0
          ----+----+----+----+
      Multiply Stem.Leaf by 10**-1
```

```
                      The UNIVARIATE Procedure
                         Variable:  MRK

                       Extreme Observations

         ------Lowest------          ------Highest-----

             Value       Obs              Value       Obs

         -0.739489        22           0.391509        28
         -0.635646        18           0.393488        30
         -0.544492        29           0.457447        15
         -0.476854        38           0.721311        25
         -0.371963        31           0.852573        21

     Stem Leaf                     #             Boxplot
        8 5                        1                |
        7 2                        1                |
        6                                           |
        5                                           |
        4 6                        1                |
        3 334799                   6                |
        2 01258                    5             +-----+
        1 167                      3             |     |
        0 044577                   6             *--+--*
       -0 942110                   6             |     |
       -1 9880                     4             +-----+
       -2 81                       2                |
       -3 70                       2                |
       -4 8                        1                |
       -5 4                        1                |
       -6 4                        1                |
       -7 4                        1                |
          ----+----+----+----+
      Multiply Stem.Leaf by 10**-1
```

```
                 The UNIVARIATE Procedure
                   Variable:  SP500

                  Extreme Observations

     ------Lowest------         ------Highest-----

         Value      Obs             Value      Obs

     -0.229523        4          0.237724       26
     -0.222935       39          0.242801       28
     -0.174081       38          0.265403       16
     -0.167680       32          0.302655       27
     -0.163282       31          0.340490       13

   Stem Leaf                       #             Boxplot
      3 04                         2                |
      2 7                          1                |
      2 012244                     6                |
      1 567888                     6             +-----+
      1 11123                      5             |     |
      0 778899                     6             *--+--*
      0 244                        3             |     |
     -0 43220                      5             +-----+
     -0 77                         2                |
     -1                                             |
     -1 776                        3                |
     -2 32                         2                |
        ----+----+----+----+
    Multiply Stem.Leaf by 10**-1
```

Notice that the above character box plot indicates the extreme box plot with a “0.”

4.18.9 Z Score in SAS

We will use the **PROC STANDARD** to calculate the Z score for our rate of return data sets. This is shown below.

```
PROC STANDARD DATA=Mydata.Return MEAN=0 STD=1 OUT=ZSCORES;
      VAR JNJ MRK SP500;
RUN;
PROC PRINT DATA=ZSCORES;
RUN;
```

Obs	Year	JNJ	MRK	S&P 500
1	1971	2.31168	0.63178	0.72958
2	1972	0.93802	−0.96396	0.22178
3	1973	−0.63219	−0.40408	−0.66029
4	1974	−1.13194	−0.65535	−2.16544
5	1975	0.20507	0.00634	−0.26256
6	1976	−0.61482	−0.17172	0.74196
7	1977	−0.22339	−0.67718	−0.81674
8	1978	−0.30196	0.52997	−0.69965
9	1979	0.08528	0.08131	−0.04726
10	1980	0.71249	0.39407	0.52119
11	1981	−2.30772	−0.12315	0.00131
12	1982	0.97832	−0.12756	−1.00889
13	1983	−0.76951	0.07998	1.84300

(continued)

Obs	Year	JNJ	MRK	S&P 500
14	1984	−0.56470	−0.00324	−0.55183
15	1985	1.38731	1.24440	0.60341
16	1986	0.67282	−0.40956	1.31498
17	1987	0.31142	0.71247	0.95417
18	1988	0.29760	−2.02344	−1.06888
19	1989	−1.19937	0.89924	0.96069
20	1990	0.54128	0.35421	−0.29931
21	1991	1.86111	2.42564	0.32199
22	1992	−2.07310	−2.33388	0.18812
23	1993	−0.54828	−0.74346	0.05517
24	1994	0.58092	0.20298	−0.41420
25	1995	1.74468	2.03323	0.69034
26	1996	−1.59332	0.51461	1.12034
27	1997	0.93546	0.86710	1.57694
28	1998	0.76213	1.04728	1.15604
29	1999	0.21201	−1.75093	1.01526
30	2000	0.26278	1.05319	−0.02217
31	2001	−1.65924	−1.23515	−1.69962
32	2002	−0.47951	−0.23450	−1.73055
33	2003	−0.29883	−0.67290	−0.75451
34	2004	0.60677	−1.03296	0.65378
35	2005	−0.34714	−0.15385	−0.07509
36	2006	0.16680	0.98488	0.04993
37	2007	−0.13370	0.87176	0.34332
38	2008	−0.51970	−1.54873	−1.77556
39	2009	0.09201	0.48065	−2.11911
40	2010	−0.30420	−0.16406	0.87219
41	2011	0.03667	0.01454	0.23618

4.18.10 Skewness

```
PROC MEANS Skewness DATA=Mydata.Return;
    VAR JNJ MRK SP500;
RUN;
```

Variable	Skewness
JNJ	−0.0923951
MRK	−0.1154083
S&P 500	−0.4929031

4.18.11 Pivot Tables

In some situations, the rate of return data would look like the data shown below.

Obs	Return	Company
1	-0.73949	MRK
2	-0.63565	MRK
3	-0.62782	JNJ
4	-0.55895	JNJ
5	-0.54449	MRK
6	-0.47685	MRK
7	-0.43748	JNJ
8	-0.41813	JNJ
9	-0.37196	MRK
10	-0.30433	MRK
11	-0.30250	JNJ
12	-0.28270	JNJ
13	-0.28125	MRK
14	-0.22952	SP500
15	-0.22294	SP500
16	-0.20749	MRK
17	-0.18532	MRK
18	-0.18389	MRK
19	-0.17802	MRK
20	-0.17632	JNJ
21	-0.17408	SP500
22	-0.16768	SP500
23	-0.16328	SP500
24	-0.13602	JNJ
25	-0.13092	JNJ
26	-0.11621	JNJ
27	-0.11139	JNJ
28	-0.10300	JNJ
29	-0.09580	MRK
30	-0.09397	MRK
31	-0.09120	JNJ
32	-0.07359	SP500
33	-0.06506	SP500
34	-0.05235	JNJ
35	-0.03975	JNJ
36	-0.03909	JNJ
37	-0.03817	JNJ
38	-0.03773	SP500
39	-0.03724	MRK
40	-0.02889	SP500
41	-0.02108	SP500
42	-0.01625	MRK
43	-0.01603	JNJ
44	-0.01549	SP500

```
45    -0.01368    MRK
46    -0.01027    MRK
47    -0.00147    MRK
48    -0.00006    SP500
49     0.00000    MRK
50     0.01030    JNJ
51     0.01951    SP500
52    0.03585     SP500
53    0.04011     MRK
54    0.04107     SP500
55    0.04331     MRK
56    0.04606     MRK
57    0.06031     JNJ
58    0.06773     SP500
59    0.06795     MRK
60    0.06839     MRK
61    0.07169     SP500
62    0.07458     JNJ
63    0.07526     SP500
64    0.07655     JNJ
65    0.07859     SP500
66    0.08551     SP500
67    0.08625     SP500
68    0.09850     JNJ
69    0.10516     SP500
70    0.10909     MRK
71    0.10974     JNJ
72    0.10995     SP500
73    0.11177     JNJ
74    0.11199     SP500
75    0.12420     SP500
76    0.12668     JNJ
77    0.12723     SP500
78    0.13689     JNJ
79    0.14095     JNJ
80    0.15252     SP500
81    0.15968     MRK
82    0.16422     SP500
83    0.17138     SP500
84    0.17301     MRK
85    0.17658     SP500
86    0.18216     SP500
87    0.18392     SP500
88    0.20197     MRK
89    0.20244     SP500
90    0.20842     JNJ
```

91	0.21333	MRK
92	0.21410	SP500
93	0.21502	SP500
94	0.21847	MRK
95	0.22006	JNJ
96	0.22278	SP500
97	0.22764	JNJ
98	0.23772	SP500
99	0.24280	SP500
100	0.24703	JNJ
101	0.25253	MRK
102	0.25867	JNJ
103	0.26540	SP500
104	0.27324	JNJ
105	0.27951	MRK
106	0.30265	SP500
107	0.32412	JNJ
108	0.32487	JNJ
109	0.33124	MRK
110	0.33280	MRK
111	0.33670	JNJ
112	0.34049	SP500
113	0.34199	MRK
114	0.37064	MRK
115	0.39151	MRK
116	0.39349	MRK
117	0.45675	JNJ
118	0.45745	MRK
119	0.56164	JNJ
120	0.59582	JNJ
121	0.72131	MRK
122	0.72807	JNJ
123	0.85257	MRK

To calculate statistic of each company, we sort the data by company first and then use **PROC MEANS**. Remember to specify **by Company** when using **PROC MEANS**, which informs SAS to do the calculation by different companies.

```
data Mydata.Pivot;
   INFILE 'd:\MINITAB and Microsoft Excel Book\Ch4\Pivot.txt' dlm='09'x;
   INPUT Return Company $;
run;

proc sort data=Mydata.Pivot;
by Company;
run;

PROC MEANS MEAN STD VAR MIN MAX DATA=Mydata.Pivot;
    var Return;
    by Company;
run;
```

Analysis variable: return				
Mean	Std dev	Variance	Minimum	Maximum
0.0495432	0.2935205	0.0861543	−0.6278200	0.7280700

Analysis variable: return				
Mean	Std dev	Variance	Minimum	Maximum
0.0411949	0.3344999	0.1118902	−0.7394900	0.8525700

Analysis variable: return				
Mean	Std dev	Variance	Minimum	Maximum
0.0784090	0.1422029	0.0202217	−0.2295200	0.3404900

```
goption ftext='Times New Roman' htext=6 gunit=pct;
title height=7 'Box Plot for JNJ, MRK, and SP500';
proc boxplot data=Mydata.Pivot;
    plot Return*Company/boxwidth=5;
run;
```

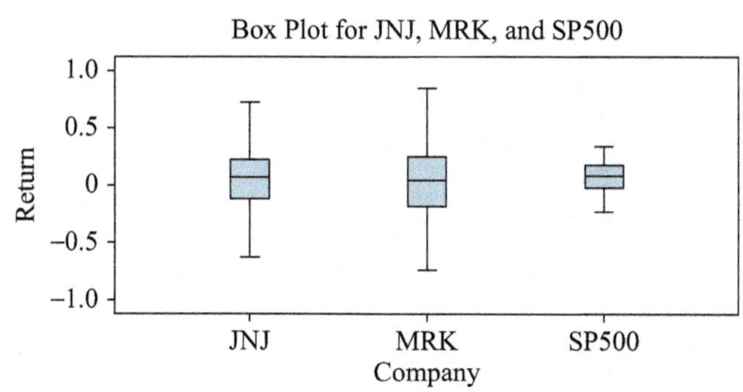

Appendix 4.1: Excel Code—Box Plot

```
'/**************************************************************************
'/ Essentials of Microsoft Excel 2013, SAS and MINITAB 17
'/ for Statistical and Financial Analyses
'/
'/**************************************************************************

'/**************************************************************************
'/Purpose:  The purpose of this program is to create as many box plots as data
'/          series in the worksheet 'Data'. In worksheet 'Data' the first row
'/          is assumed to be the label of the data series.
'/          Each box plot will have
'/               1) Inner Fences     Q1 - 1.5(IQR) and Q3 + 1.5(IQR)
'/               2) Outer Fences     Q1 - 3(IQR) and Q3 + 3(IQR)
'/               3) Whiskers
'/
'/          The whiskers of a boxplot are the lines that are leaving the boxplot
'/          and end at the last data point before the inner fence.
'/
'/          Values lying between the inner and outer fence are consider mild outliers
'/          Values lying outside the outer fence is consider extreme outliers
Option Explicit

Sub BoxPlot()
    Dim ChrtBox As Chart
    Dim SeriesCurrent As Series
    Dim rngData As Range
    Dim rDataset As Range
    Dim rCell As Range
    Dim wsMildOutlierY As Worksheet
    Dim wsMildOutlierX As Worksheet
    Dim wsExtremeOutlierY As Worksheet
    Dim wsExtremeOutlierX As Worksheet

    Dim wbReport As Workbook

    'Check to see if there is data to plot

    If wsData.Cells(1).CurrentRegion.Rows.Count = 1 Then

        MsgBox "There is no data to box plot!", vbCritical, "BoxPlot"
        Exit Sub
    End If

    Set wbReport = Workbooks.Add

    'Store outlier calculations
    Set wsMildOutlierY = wbReport.Worksheets.Add
    Set wsMildOutlierX = wbReport.Worksheets.Add
    Set wsExtremeOutlierY = wbReport.Worksheets.Add
    Set wsExtremeOutlierX = wbReport.Worksheets.Add
    wsMildOutlierY.Name = "MildOutlierY"
    wsMildOutlierX.Name = "MildOutlierX"
    wsExtremeOutlierY.Name = "ExtremeOutlierY"
    wsExtremeOutlierX.Name = "ExtremeOutlierX"

    'Create Chart
    Set ChrtBox = wbReport.Charts.Add

    'Format Chart
    If ChrtBox.SeriesCollection.Count = 0 Then
        Set SeriesCurrent = ChrtBox.SeriesCollection.NewSeries
    Else
        Set SeriesCurrent = ChrtBox.SeriesCollection(1)
    End If
```

```
    With ChrtBox
            .ChartType = xlXYScatterLinesNoMarkers
            With .Axes(xlCategory)
                  .HasMajorGridlines = False
                  .HasMinorGridlines = False
            End With
            With .Axes(xlValue)
                  .HasMajorGridlines = False
                  .HasMinorGridlines = False
                  .Delete
            End With

            .PlotArea.Interior.ColorIndex = xlNone
            .HasLegend = False
            .Name = "BoxPlot Chart"
    End With

    'Dynamically calculate the data range in 'data worksheet
    Set rDataset = ThisWorkbook.Worksheets("Data").Range("a1").CurrentRegion

    'Loop through the data labels
    For Each rCell In rDataset.Rows(1).Cells
        'Get data series with no label
        Set rngData = rDataset.Columns(rCell.Column).Resize(rDataset.Rows.Count - 1,
1).Offset(1, 0)
        'Name the chart series as the data label
        SeriesCurrent.Name = rngData.Cells(1).Offset(-1, 0)
        'Draw boxplot for particular data series
        DrawBoxPlot nMultiplier:=rCell.Column - 1, ChrtBox:=ChrtBox, _
        SeriesCurrent:=SeriesCurrent, rngData:=rngData, _
        wsMildOutlierY:=wsMildOutlierY, wsMildOutlierX:=wsMildOutlierX, _
        wsExtremeOutlierY:=wsExtremeOutlierY, wsExtremeOutlierX:=wsExtremeOutlierX
        'Add new chart series for next boxplot
        Set SeriesCurrent = ChrtBox.SeriesCollection.NewSeries
    Next

    'Delete unused chart series
    SeriesCurrent.Delete

    On Error Resume Next
    Application.DisplayAlerts = False
    With wbReport
        .Worksheets("Sheet1").Delete
        .Worksheets("Sheet2").Delete
        .Worksheets("Sheet3").Delete
    End With
    Application.DisplayAlerts = True
End Sub
Sub DrawBoxPlot(nMultiplier As Integer, _
                ChrtBox As Chart, _
                SeriesCurrent As Series, _
                rngData As Range, _
                wsMildOutlierY As Worksheet, _
                wsMildOutlierX As Worksheet, _
                wsExtremeOutlierY As Worksheet, _
                wsExtremeOutlierX As Worksheet)
    Dim sngX(0 To 13) As Single
    Dim sngY(0 To 13) As Single
    Dim sngLower_Inner_fence As Single
    Dim sngUpper_Inner_fence As Single
    Dim sngLower_Outer_fence As Single
    Dim sngUpper_Outer_fence As Single
    Dim sngLowerWisker As Single
    Dim sngUpperWisker As Single
    Dim sngInterQuartile
    Dim sngQuartile(4) As Variant
    Dim lngCounter As Long
    Dim rngNumber As Range
```

```
    Dim nYoffset As Integer
    Dim SeriesOutliers As Series
    Dim rMildOutlierX As Range
    Dim rMildOutlierY As Range
    Dim rExtremeOutlierX As Range
    Dim rExtremeOutlierY As Range

    nYoffset = 2


    If rngData.Rows.Count < 3 Then
        Call MsgBox("You must have a minimum of three data items in column A!", ,
"Boxplot")
        Exit Sub
    End If

    'Calcualte Quartiles
    sngQuartile(1) = Application.Quartile(rngData, 1)
    sngQuartile(2) = Application.Quartile(rngData, 2)
    sngQuartile(3) = Application.Quartile(rngData, 3)


    sngInterQuartile = sngQuartile(3) - sngQuartile(1)

    'calculate the inner and outer fences of a box plot
    With ChrtBox
            'Lower Inner Fence
            sngLower_Inner_fence = (sngQuartile(1) - (1.5) * sngInterQuartile)
            Set SeriesOutliers = .SeriesCollection.NewSeries
            With SeriesOutDim nYoffset As Integer
                    .XValuDim SeriesOutliers As Series ce, sngLower_Inner_fence)
                    .ValueDim rMildOutlierX As Range   tiplier, 2 + nYoffset *
nMultiplier)
                           Dim rMildOutlierY As Range
                    With .Dim rExtremeOutlierX As Range
                        .LDim rExtremeOutlierY As Range
                        .ColorIndex = 1
                    End With
                    .Name = "Lower Inner Fence" & rngData.Cells(1).Offset(-1, 0)
            End With

            'Upper Inner Fence
            sngUpper_Inner_fence = (sngQuartile(3) + (1.5) * sngInterQuartile)

            Set SeriesOutliers = .SeriesCollection.NewSeries
            With SeriesOutliers
                    .XValues = Array(sngUpper_Inner_fence, sngUpper_Inner_fence)
                    .Values = Array(1 + nYoffset * nMultiplier, 2 + nYoffset *
nMultiplier)
                    With .Border
                        .LineStyle = xlDash
                        .ColorIndex = 1
                    End With
                    .Name = "Upper Inner Fence" & rngData.Cells(1).Offset(-1, 0)
            End With

            'Lower outer Fence
            sngLower_Outer_fence = (sngQuartile(1) - (3) * sngInterQuartile)
            Set SeriesOutliers = .SeriesCollection.NewSeries
            With SeriesOutliers
                    .XValues = Array(sngLower_Outer_fence, sngLower_Outer_fence)
                    .Values = Array(1 + nYoffset * nMultiplier, 2 + nYoffset *
nMultiplier)
                    With .Border
                        .LineStyle = xlDash
                        .ColorIndex = 1
                    End With
                    .Name = "Lower Outer Fence" & rngData.Cells(1).Offset(-1, 0)
```

```
            End With

            'Upper outer Fence
            sngUpper_Outer_fence = (sngQuartile(3) + (3) * sngInterQuartile)

            Set SeriesOutliers = .SeriesCollection.NewSeries
            With SeriesOutliers
                    .XValues = Array(sngUpper_Outer_fence, sngUpper_Outer_fence)
                    .Values = Array(1 + nYoffset * nMultiplier, 2 + nYoffset *
nMultiplier)
                    With .Border
                        .LineStyle = xlDash
                        .ColorIndex = 1
                    End With
                    .Name = "Upper Outer Fence" & rngData.Cells(1).Offset(-1, 0)
            End With
    End With

    'Caculate the upper whisker
    sngUpperWisker = sngQuartile(3)
    For Each rngNumber In rngData.Cells
        If (rngNumber.Value > sngUpperWisker) And (rngNumber.Value <
sngUpper_Inner_fence) Then
            sngUpperWisker = rngNumber.Value
        End If

    Next

    'Caculate the lower whisker
    sngLowerWisker = sngQuartile(1)
    For Each rngNumber In rngData.Cells
        If (rngNumber.Value < sngLowerWisker) And (rngNumber.Value >
sngLower_Inner_fence) Then
            sngLowerWisker = rngNumber.Value
        End If

    Next

    'Draw the box and whiskers
    sngX(0) = sngQuartile(1)
    sngY(0) = 1 + nYoffset * nMultiplier
    sngX(1) = sngQuartile(2)
    sngY(1) = 1 + nYoffset * nMultiplier
    sngX(2) = sngQuartile(3)
    sngY(2) = 1 + nYoffset * nMultiplier
    sngX(3) = sngQuartile(3)
    sngY(3) = 1.5 + nYoffset * nMultiplier
    sngX(4) = sngUpperWisker
    sngY(4) = 1.5 + nYoffset * nMultiplier
    sngX(5) = sngQuartile(3)
    sngY(5) = 1.5 + nYoffset * nMultiplier
    sngX(6) = sngQuartile(3)
    sngY(6) = 2 + nYoffset * nMultiplier
    sngX(7) = sngQuartile(2)
    sngY(7) = 2 + nYoffset * nMultiplier
    sngX(8) = sngQuartile(2)
    sngY(8) = 1 + nYoffset * nMultiplier
    sngX(9) = sngQuartile(2)
    sngY(9) = 2 + nYoffset * nMultiplier
    sngX(10) = sngQuartile(1)
    sngY(10) = 2 + nYoffset * nMultiplier
    sngX(11) = sngQuartile(1)
```

```
sngY(11) = 1 + nYoffset * nMultiplier
sngX(12) = sngQuartile(1)
sngY(12) = 1.5 + nYoffset * nMultiplier
sngX(13) = sngLowerWisker
sngY(13) = 1.5 + nYoffset * nMultiplier

With SeriesCurrent
    .XValues = sngX
    .Values = sngY
End With


'Calculate and plot the mild outliers
Set rMildOutlierY = wsMildOutlierY.Columns(rngData.Column).Cells(1)
Set rMildOutlierX = wsMildOutlierX.Columns(rngData.Column).Cells(1)

lngCounter = 0
For Each rngNumber In rngData.Cells
    If ((rngNumber.Value >= sngUpper_Inner_fence) And _
        (rngNumber.Value <= sngUpper_Outer_fence)) Or _
        ((rngNumber.Value <= sngLower_Inner_fence) And _
         (rngNumber.Value >= sngLower_Outer_fence)) Then

        rMildOutlierX = rngNumber.Value
        rMildOutlierY = 1.5 + nYoffset * nMultiplier

        Set rMildOutlierX = rMildOutlierX.Offset(1, 0)
        Set rMildOutlierY = rMildOutlierY.Offset(1, 0)
        lngCounter = lngCounter + 1
    End If
Next

On Error Resume Next
If Not lngCounter = 0 Then

    With ChrtBox
        Set SeriesOutliers = .SeriesCollection.NewSeries
        With SeriesOutliers
                .XValues = wsMildOutlierX.Columns(rMildOutlierX.Column). _
                           Cells(1).Resize(WorksheetFunction. _
                           CountA(rMildOutlierX.EntireColumn, 1))
                .Values = wsMildOutlierY.Columns(rMildOutlierY.Column). _
                          Cells(1).Resize(WorksheetFunction. _
                          CountA(rMildOutlierY.EntireColumn, 1))
                With .Border
                    .LineStyle = xlNone
                End With

                .Name = "MildOutliers" & rngData.Cells(1).Offset(-1, 0)
                .MarkerStyle = xlTriangle
                .MarkerSize = 5
                .MarkerForegroundColorIndex = 1
        End With


    End With
End If


'Calculate and plot the Extreme outliers
Set rExtremeOutlierY = wsExtremeOutlierY.Columns(rngData.Column).Cells(1)
Set rExtremeOutlierX = wsExtremeOutlierX.Columns(rngData.Column).Cells(1)
```

```
lngCounter = 0
For Each rngNumber In rngData.Cells
    If (rngNumber.Value > sngUpper_Outer_fence) Or _
         (rngNumber.Value < sngLower_Outer_fence) Then

        rExtremeOutlierX = rngNumber.Value
        rExtremeOutlierY = 1.5 + nYoffset * nMultiplier

        Set rExtremeOutlierX = rExtremeOutlierX.Offset(1, 0)
        Set rExtremeOutlierY = rExtremeOutlierY.Offset(1, 0)
        lngCounter = lngCounter + 1
    End If
Next

On Error Resume Next
If Not lngCounter = 0 Then

    With ChrtBox
        Set SeriesOutliers = .SeriesCollection.NewSeries
        With SeriesOutliers
                .XValues = wsExtremeOutlierX.Columns(rMildOutlierX.Column). _
                          Cells(1).Resize(WorksheetFunction. _
                          CountA(rMildOutlierX.EntireColumn, 1))
                .Values = wsExtremeOutlierY.Columns(rMildOutlierY.Column). _
                         Cells(1).Resize(WorksheetFunction. _
                         CountA(rMildOutlierY.EntireColumn, 1))
                With .Border
                    .LineStyle = xlNone
                End With

                .Name = "ExtremeOutliers" & rngData.Cells(1).Offset(-1, 0)
                .MarkerStyle = xlSquare
                .MarkerSize = 5
                .MarkerForegroundColorIndex = 1
        End With

    End With
End If


'Put the data series label on the box
With SeriesCurrent
    .Name = rngData.Cells(1).Offset(-1, 0)
    .Border.ColorIndex = 1
    With .Points(10)
    .ApplyDataLabels
        With .DataLabel
            .Characters.Text = rngData.Cells(1).Offset(-1, 0).Value
            .Position = xlLabelPositionAbove
        End With
    End With
    With .Points(2)
        .ApplyDataLabels
            With .DataLabel
                'Show Q1, Q2, Q3
                With .Characters
                    .Text = "Q1=" & Format(sngQuartile(1), "#,##0.00;(#,##0.00)") _
                    & ",Q2 = " & Format(sngQuartile(2), "#,##0.00;(#,##0.00)") _
                    & ",Q3=" & Format(sngQuartile(3), "#,##0.00;(#,##0.00)")
                    .Font.Size = 8

                    End With
                    .Position = xlLabelPositionBelow
                End With
        End With
    End With

End Sub
```

Appendix 4.2: Excel Code—Rate of Return

```
'/*****************************************************************************************
'/ Essentials of Microsoft Excel 2013, SAS and MINITAB 17
'/ for Statistical and Financial Analyses
'/
'/*****************************************************************************************


Sub Main()

    Application.StatusBar = "Calculating Return for Ticker 1"
    'Ticker1 Returns
    GetReturns ticker:=Worksheets("Main").Range("b5"), _
               Start_Date:=Worksheets("Main").Range("b6"), _
               End_Date:=Worksheets("Main").Range("b7"), _
               frequency:=Worksheets("Main").Range("b8"), _
               shtData:=Worksheets("Data1")

     Application.StatusBar = "Calculating Return for Ticker 2"

    'Ticker2 Returns
    GetReturns ticker:=Worksheets("Main").Range("e5"), _
               Start_Date:=Worksheets("Main").Range("e6"), _
               End_Date:=Worksheets("Main").Range("e7"), _
               frequency:=Worksheets("Main").Range("e8"), _
               shtData:=Worksheets("Data2")

     Application.StatusBar = "Calculating Return for Ticker 3"
    'Ticker3 Returns
    GetReturns ticker:=Worksheets("Main").Range("h5"), _
               Start_Date:=Worksheets("Main").Range("h6"), _
               End_Date:=Worksheets("Main").Range("h7"), _
               frequency:=Worksheets("Main").Range("h8"), _
               shtData:=Worksheets("Data3")


   CompileReturn

    Application.StatusBar = False

End Sub

Sub GetReturns(Start_Date As Date, _
              End_Date As Date, _
              ticker As String, _
              frequency As String, _
              shtData As Worksheet)



    Dim URL As String
    Dim rReturn As Excel.Range

    shtData.UsedRange.ClearContents

    'Build URL for ticker
    URL = "https://www.quandl.com/api/v3/datasets/yahoo/" & ticker & ".csv?"
    URL = URL & "start_date=" & Format(Start_Date, "mmm-dd-yyyy") & "&end_date=" &
Format(End_Date, "mmm-dd-yyyy")
    URL = URL & "&collapse=" & frequency
    URL = URL & "&api_key=uBSmYjGExghF8sqmTjf-"


    'Use qurytables to return URL info
    With shtData.QueryTables.Add(Connection:="URL;" & URL,
Destination:=shtData.Range("a1"))
        .BackgroundQuery = True
        .TablesOnlyFromHTML = False
        .Refresh BackgroundQuery:=False
        .SaveData = True
    End With
```

```
    'data is return as a comma list
    'parse list into column
    shtData.Range("a1").CurrentRegion.TextToColumns _
                        Destination:=shtData.Range("a1"), DataType:=xlDelimited, _
                        TextQualifier:=xlDoubleQuote, ConsecutiveDelimiter:=False,
Tab:=True, _
                        Semicolon:=False, Comma:=True, Space:=False, other:=False

    shtData.Range("a1").CurrentRegion.ColumnWidth = 14


    Set rReturn = shtData.Range("a1").CurrentRegion
    'Delete empty row
    rReturn.Cells(1).Offset(rReturn.Rows.Count - 1, 0).EntireRow.Delete Shift:=xlUp

    Set rReturn = shtData.Range("a1").CurrentRegion
    Set rReturn = rReturn.Columns(1).Offset(0, rReturn.Columns.Count)
    rReturn.Name = "Return"

    shtData.Names.Add Name:="Return_" & shtData.Name, _
                RefersToR1C1:=rReturn.Offset(1, 0).Resize(rReturn.Rows.Count - 2, 1)

    'Calculate rate of return
    'Use R1C1 address
    rReturn.FormulaR1C1 = "=(RC[-1]-R[1]C[-1])/R[1]C[-1]"
    'format percentage
    rReturn.NumberFormat = "0.000%;[Red](0.000%)"
    rReturn.Cells(1) = "Return"
    rReturn.Cells(rReturn.Cells.Count).ClearContents
    rReturn.Offset(0, 1).Value = ticker
    rReturn.Offset(0, 1).Cells(1).Value = "Ticker"

    'Calculate Z-Score
    rReturn.Offset(0, 2).Formula = "=STANDARDIZE(RC[-2],AVERAGE(Return_" & _
                                   rReturn.Parent.Name & "),STDEV.P(Return_" & _
                                   rReturn.Parent.Name & "))"
    rReturn.Offset(0, 2).Cells(1).Value = "Z-Score"

End Sub

Sub CompileReturn()

    Dim sht As Excel.Worksheet
    Dim rData As Excel.Range
    Dim rTmp As Excel.Range

    Set sht = ThisWorkbook.Worksheets("Returns")

    sht.UsedRange.ClearContents

    'Datasheet1
    Set rData = ThisWorkbook.Worksheets("Data1").Cells(1).CurrentRegion
    Set rTmp = rData.Columns(1)
    rTmp.Copy Destination:=sht.Range("a1")

    'Copy over return
    'Column 8 is column H
     Set rTmp = ThisWorkbook.Worksheets("Data1").Columns(8)
     sht.Range("b1").Resize(rTmp.Rows.Count, 1).Value = rTmp.Value
     sht.Range("b1").Resize(rTmp.Rows.Count, 1).NumberFormat = "0.000%;[Red](0.000%)"
     sht.Range("b1").Value = ThisWorkbook.Worksheets("Main").Range("B5")
```

```
    'Datasheet2
    Set rData = ThisWorkbook.Worksheets("Data2").Cells(1).CurrentRegion
    Set rTmp = rData.Columns(8)
    sht.Range("c1").Resize(rTmp.Rows.Count, 1).Value = rTmp.Value
    sht.Range("c1").Resize(rTmp.Rows.Count, 1).NumberFormat = "0.000%;[Red](0.000%)"
    sht.Range("c1").Value = ThisWorkbook.Worksheets("Main").Range("e5")


    'Datasheet3
    Set rData = ThisWorkbook.Worksheets("Data3").Cells(1).CurrentRegion
    Set rTmp = ThisWorkbook.Worksheets("Data3").Columns(8)
    sht.Range("d1").Resize(rTmp.Rows.Count, 1).Value = rTmp.Value
    sht.Range("d1").Resize(rTmp.Rows.Count, 1).NumberFormat = "0.000%;[Red](0.000%)"
    sht.Range("d1").Value = ThisWorkbook.Worksheets("Main").Range("h5")

    'populate ticker
    Set rTmp = sht.Range("a1").CurrentRegion
    Set rTmp = rTmp.Columns(1).Offset(0, rTmp.Columns.Count)
    rTmp.Value = ThisWorkbook.Worksheets("Main").Range("b8")
    rTmp.Cells(1).Value = "Frequency"


End Sub
```

Appendix 4.3: SAS Macro Code—Rate of Return, Q–Q Graph, and Descriptive Statistics

```
/*This macro is designed to read excel or csv files, do some basic statistical analysis,
and output the
results in RTF or HTML Format. There three macro variables: FileName is to defined to
file name you want
 to read. Format is used to decide what kind of output you prefer. There are two chioce:
RTF or HTML.
FileType has two chioce as well: xls or csv.*/


%macro SASprogram(FileName, Format, FileType);

/*Two IF-THEN statement are used to decide the output file.*/
%If &Format = RTF %then %do;
ODS RTF path="d:\" body="&FileName.RTF.RTF";/*File Path*/
%end;

%If &Format = HTML %then %do;
 ODS LISTING CLOSE;/*Eliminates the output to appear in the output window*/
ODS HTML path="d:\" body="&FileName.html.html";/*File Path*/
%end;

/*Two IF-THEN statement are used to define what type of file will be read.*/
 %If &FileType = xls %then %do;
/*Read Data in Excel*/
libname Finance 'd:\SASFinance'; /*Define library name*/
proc import out=Finance.&FileName /*Define data name in SAS*/
      datafile="d:\&FileName..xls" /*Indicate file path in local*/
       DBMS=excel2000 replace;
      sheet="&FileName. Monthly stock price"; /*Define which sheet to read*/
      getnames=yes; /*If the first row is the names of variables*/
   run;
%end;
```

```
%If &FileType = csv %then %do;
 /*Read Data in csv*/
libname Finance 'd:\SASFinance';
proc import out=Finance.JNJ
      datafile="d:\JNJ.csv"
      DBMS=excel2000 replace;
      sheet="&FileName. Monthly stock price";
       getnames=yes;
   run;
%end;

/*Monthly Rate of return*/
data Finance.&FileName;
   set Finance.&FileName;
   Close2=lag(Close); /*LAG function is used to return values from a queue*/
   Rate_of_return=(Close-Close2)/Close2;
 run;

/*Histogram*/
proc capability data=Finance.&FileName noprint;
histogram Rate_of_return;
title 'Histogram of the Rate of Return';
run;

/*Scatter Plot*/
Goptions reset=global;
 Goptions htext=5 gunit=pct;
title  height=8 "&FileName.'s Monthly Stock Returns";
title2 height=6 "Jan. 1980 to Dec. 2000";
axis1  label=('Date');
axis2  label=(angle=90 'Monthly Stock Returns');

 Proc gplot data=Finance.&FileName;
plot  close*date/haxis=axis1 vaxis=axis2 autovref;
symbol v=none i=join c=black;
Run;

/*QQ Plot*/
/*Specifying MU=EST and SIGMA=EST with the NORMAL primary option requests the reference
line for
 which mu and sigma are estimated by the sample mean and standard deviation. The COLOR=
and L= options
specify the color and type of the line, and the SQUARE option displays the plot in a
square format.*/
Goptions reset=global;
 proc univariate data=Finance.&FileName noprint;
      qqplot Rate_of_return/ normal(mu=est sigma=est color=red l=2)
square;
   run;
/*http://support.sas.com/documentation/cdl/en/procstat/63104/HTML/default/viewer.htm#proc
stat_univariate_sect084.htm*/
```

Bibliography

Lee CF, Lee JC, Lee AC (2013) Statistics for business and financial economics. Springer, New York
Microsoft Inc., Excel 2013. Microsoft Inc., Redmond
Minitab Inc. Minitab 17. Minitab Inc., State College
SAS Institute Inc. SAS 2014. SAS Institute Inc., Cary

Chapter 5
Probability Concepts and Their Analysis

5.1 Introduction

In the previous chapters we studied on how to analyze and describe a data set. It was implied that we had all the data items that we need. We will now deal with the situations where this is not the case. We will be unable to get all the data items that we are interested in analyzing because of prohibitive costs. Cost could be time, money, or effort.

Because we had all the data items that we need we could make definitive statements like the mean of the data is 50. But now since we will have change, our definitive statement to some like there is a 90 % chance that the mean of the data set is 50. Interesting enough it is not necessary to have a definitive answer before making a decision.

What we will be doing is in a systematic way to obtain a partial set of the data set of interest and then from the partial data set make statements about the complete data set. Loosely we can call the partial data set of interest as the *sample data*. Also we can loosely call the complete data set as the *population data*. The study in how to do this is called *inferential statistics*.

The first thing that we will need to study is to study *probability*. Interesting probability is actually the opposite of inferential statistics. In probability we have the population data. We are interested in knowing what is the probability in getting a specific sample data.

So inferential statistics is basically

$$\text{Sample Data} \rightarrow \text{Population data.}$$

And probability is basically

$$\text{Population Data} \rightarrow \text{Sample Data.}$$

5.2 Probability

One of the most fundamental concepts in probability is that the probability of the total number of possible outcomes equals 100 % (many times referred to as 1) and the probability of an item is between 0 and 100 % (i.e., between 0 and 1).

As a corollary, the sum of the probability of all the items of the total number of possible outcomes equals one.

C.-F. Lee et al., *Essentials of Excel, Excel VBA, SAS and Minitab for Statistical and Financial Analyses*, DOI 10.1007/978-3-319-38867-0_5

Suppose there are only two items, A and B. Then

$$\text{probability (A)} + \text{probability (B)} = 1$$

Therefore, if we know the probability of B, we should know the probability of A by rearranging the above formula to

$$\text{probability (A)} = 1 - \text{probability (B)}$$

An underlying issue is to find out the total number of possible outcomes and knowing what items we are interested in and finding the percentage of these items compared to the total number of possible outcomes.

$$\text{Probability of an event} = \frac{\text{items of interest}}{\text{total number of possible outcomes}}$$

For example, the total number of possible outcomes of tossing a die is six because there are only six possible numbers. The probability of tossing the number three therefore would be 1/6 = 0.17. If we change to two dice, then the total number of possible outcomes would be 12. Then, if we ask what is the chance of getting a three when we throw the two dice, the probability would be 2/12 = 0.17.

What does a probability of an event mean? There is a good chance if we roll a dice ten times the number three never shows up or the number three shows up five times. This is not in line with our expectation that the probability of number three from a dice is 0.17.

What is important to note is that the number three will approach 0.17 as we increase the number of times we roll the dice. This is best shown by a computer simulation. We will do a computer simulation in Excel.

5.3 Excel Macro: Probability Simulation

We will do some computer simulation to illustrate the probability concept. For example, if we were to flip a coin 5, 10, 50, 100, 500, and 5000 times, will we see fifty percent heads for each try? We will use the workbook *probabilitysimulator.xls* that is included in the CD that accompanies this book to simulate the flipping of the coins and chart the results. The Excel code in shown in Appendix 5.1.

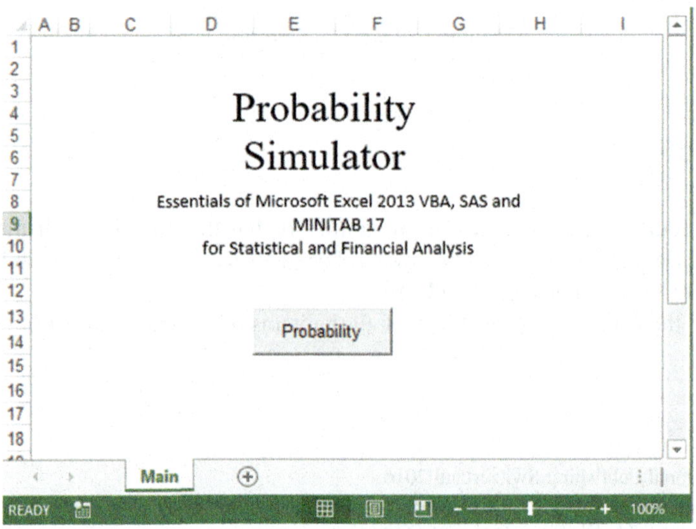

Pressing the *Probability* button will show the *Probability Simulator* dialog box.

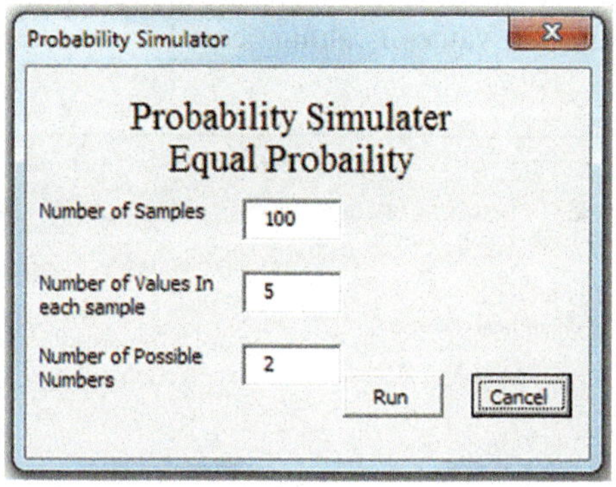

We will first generate 100 samples and in each sample have five values. To do this, fill out the *Probability Simulator* as shown above and then press *Run* button to get the following chart.

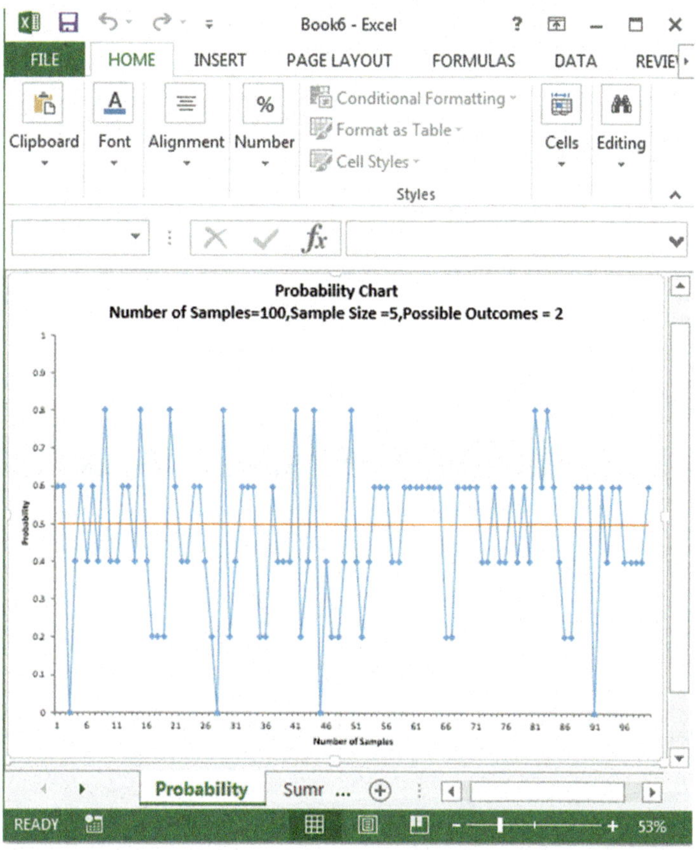

In the above chart there is a line across the value 0.5 because that is the theoretical probability when there are two possible numbers with equal values.

Below shows the probability chart for 100 samples with each sample having 10, 50, 100, 500, and 5000 values. Notice that as the number of values increase, the variability of the graphs decreases. The last graph showing a value of 5000 values is almost equal to the theoretical value of 0.5 or 50 % heads.

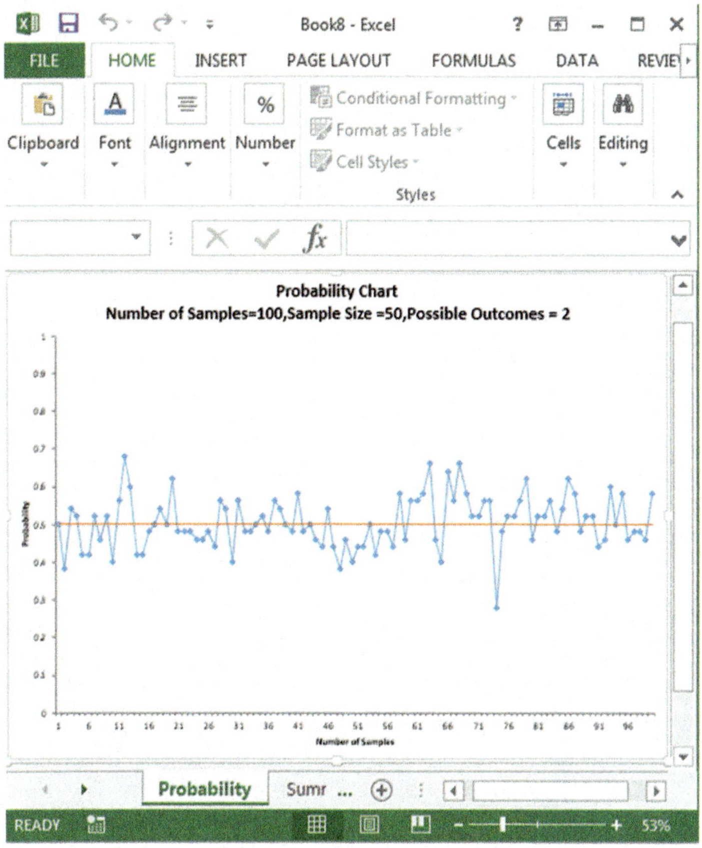
Book8 - Excel
FILE
HOME
INSERT
PAGE LAYOUT
FORMULAS
DATA
Clipboard
Font
Alignment
Number
Conditional Formatting
Format as Table
Cell Styles
Styles
Cells
Editing
Probability Chart
Number of Samples=100,Sample Size =50,Possible Outcomes = 2
Probability
Sumr ...
READY
53%

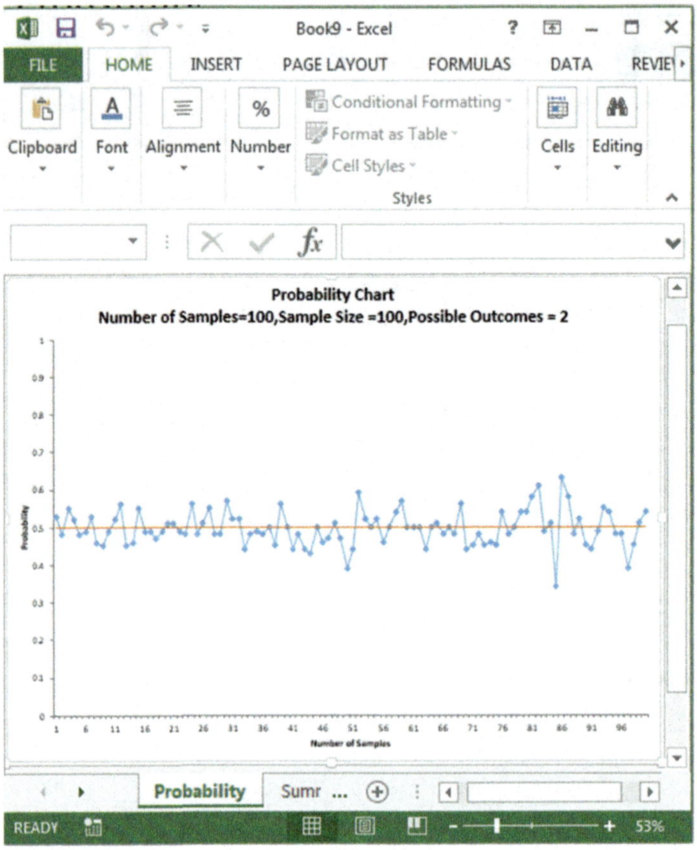

Below is another chart with 100 samples with a sample size of 1000 values. This chart will have three possible values. Notice that the chart data hovers around 0.333 (1/3).

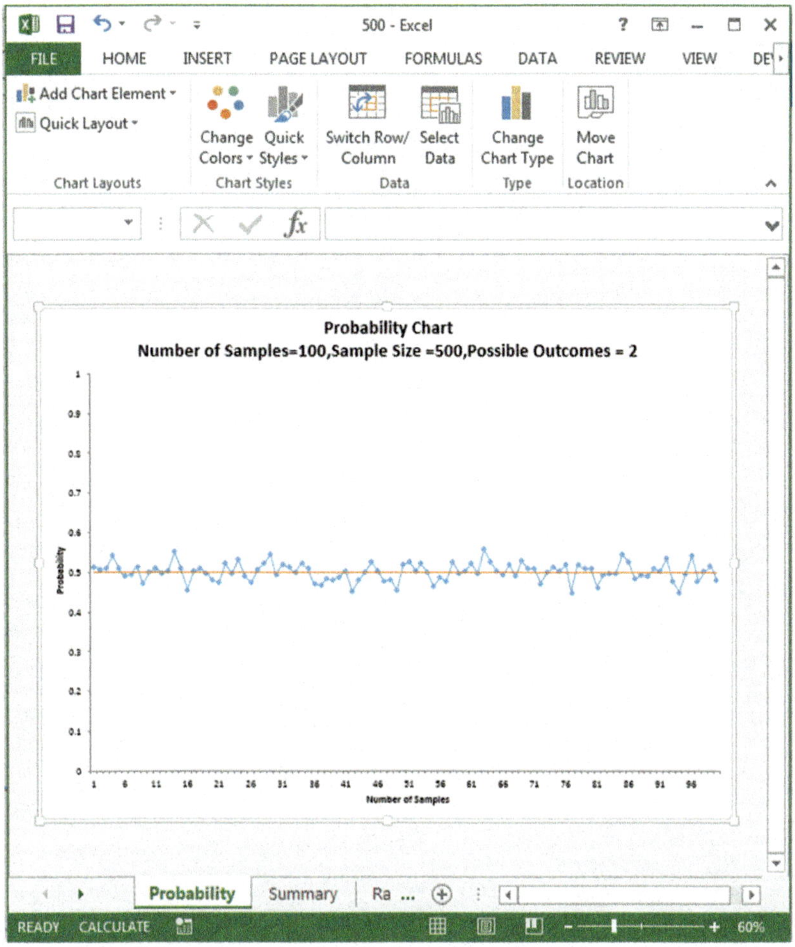
500 - Excel
FILE
HOME
INSERT
PAGE LAYOUT
FORMULAS
DATA
REVIEW
VIEW
Add Chart Element
Quick Layout
Change Colors
Quick Styles
Switch Row/ Column
Select Data
Change Chart Type
Move Chart
Chart Layouts
Chart Styles
Data
Type
Location
Probability Chart
Number of Samples=100,Sample Size =500,Possible Outcomes = 2
Probability
Summary
Ra ...
READY
CALCULATE
60%

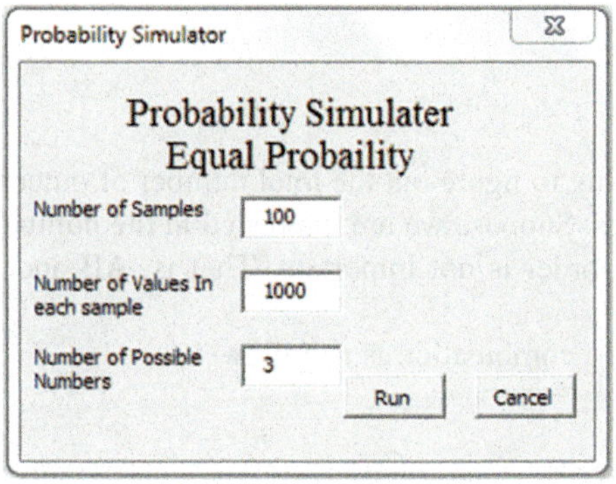
Probability Simulator
Probability Simulater
Equal Probaility
Number of Samples
100
Number of Values In each sample
1000
Number of Possible Numbers
3
Run
Cancel

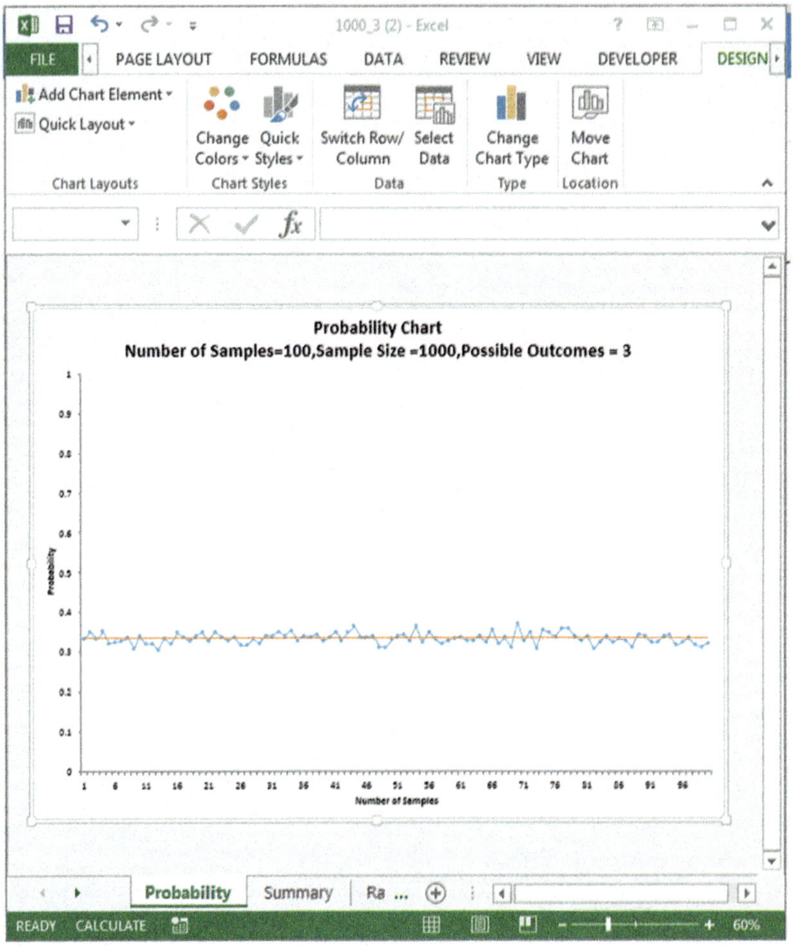

5.4 Combinations

Many times when we are trying to figure out the total number of outcomes we need to figure out the *combination* of a set of values. Suppose we are interested in the number of possible arrangements of the letters A, B, and C, and order is not important. That is, AB and BA are considered the same arrangements.

Mathematical formula for a combination is as follows:

$$nC_r = \frac{n!}{r!(n-r)!}$$

where:

$n =$ total number of elements in the list
$r =$ the number of items chosen for the combination

Suppose we are interested in finding out how many combinations of three names are from a list of four names.

Excel has a built-in function called *combin* to calculate the number of combinations. This is shown below.

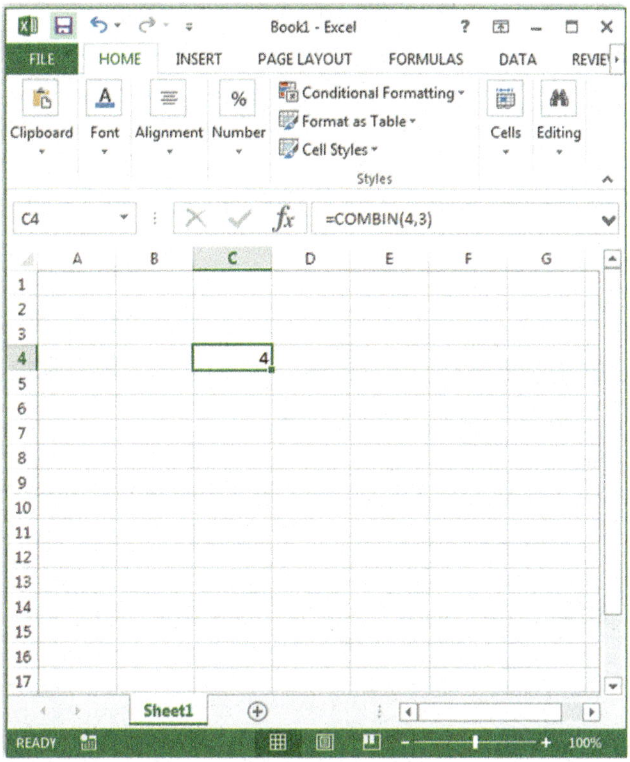

Excel calculates that there are four combinations of three names from a list of four names. Many times we are interested in the list of combinations. In the next section we use an Excel macro to list the combination of interest.

5.5 Excel Macro: Combination

We will now look at the combination program in the workbook *combination.xls* that is in the CD that is included in this workbook. This program will list all combinations of interest. We will use the program to find all combinations of three names from a list of four names. The Excel VBA combination program is shown in Appendix 5.2.

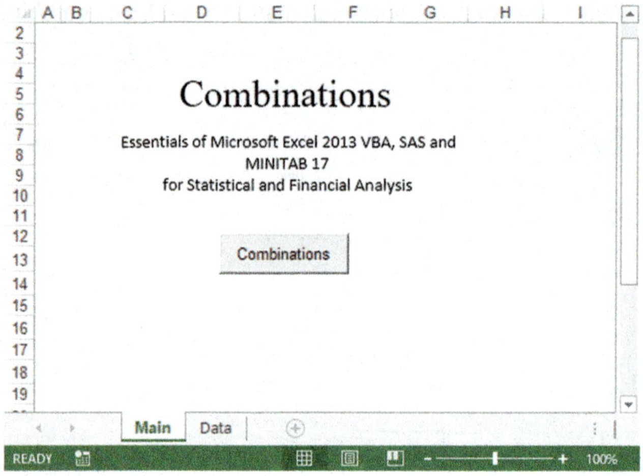

The combination list of four names will be as follows:

Bob, Tom, Wendy, Mary

We will need to put the list of names in the worksheet *Data* of the workbook *combination.xls* as shown below.

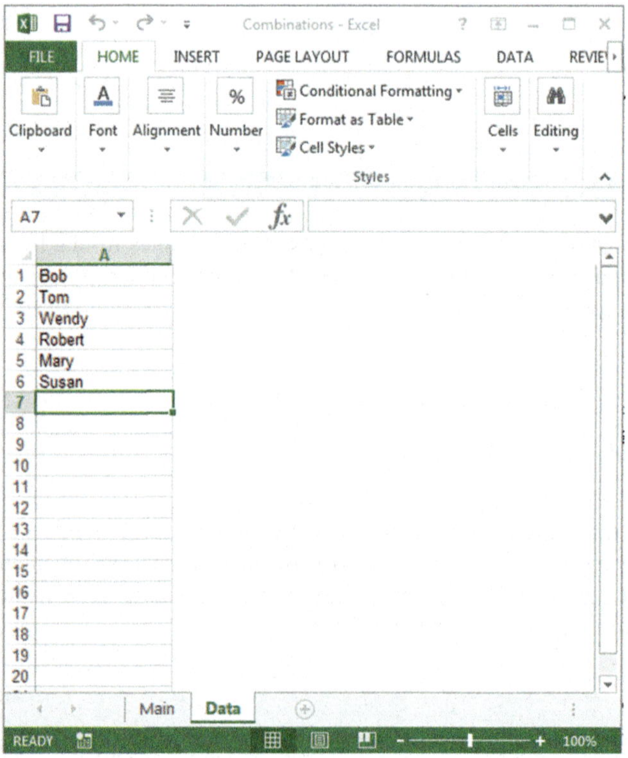

To run the program we need to push the *Combinations* button on the worksheet *main*. Pushing the button will get the dialog box shown below.

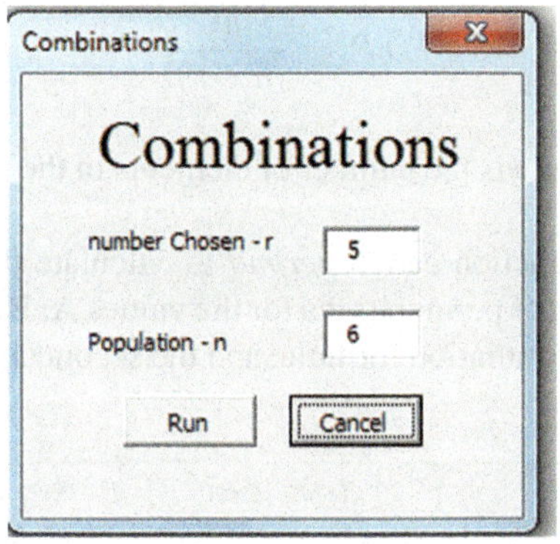

After filling out the dialog box as shown above, we should then press the *Run* button. The combination list is shown below.

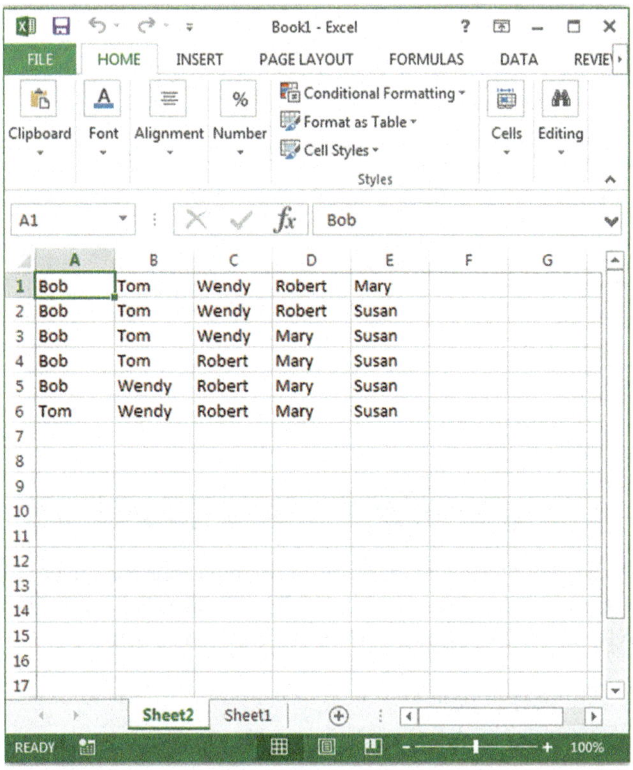

5.6 Permutations

Many times when we are trying to figure out the total number of outcomes we need to figure out the *permutation* of a set of values. Suppose we are interested in the number of possible arrangements of the letters A, B, and C, and order is important. That is, AB and BA are two different arrangements. The concept of permutations can help us answer our question. The mathematical notation for permutation is

$$nP_r = \frac{n!}{(n-r)!}$$

where n in the permutation formula is the number of elements in the population and r is the number of elements in the sample.

In Excel there is a built-in function called *permut* to calculate the permutation of a sequence of values. Below shows the number of permutations for the values A, B, and C. The first argument of the function indicates the n of the permutation formula, and the second argument of the function indicates the r of the permutation formula.

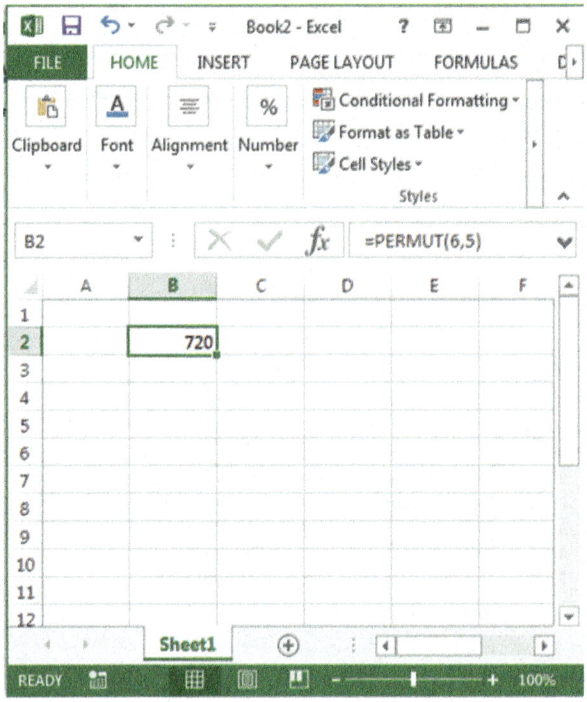

As shown above, Excel calculates that there are 24 permutations for a list of three names from a list of four names. Many times we are interested in the list of permutations. In the next section we will used an Excel macro to list all 24 permutations for a list of three names.

5.7 Excel Macro: Permutation

We will now use the *permutation.xls* workbook that is included in the CD that accompanies this book to generate the list of all possible permutations. The program in this workbook will show all the possible permutations where n is equal to r in the permutation formula. The Excel permutation program is shown in Appendix 5.3.

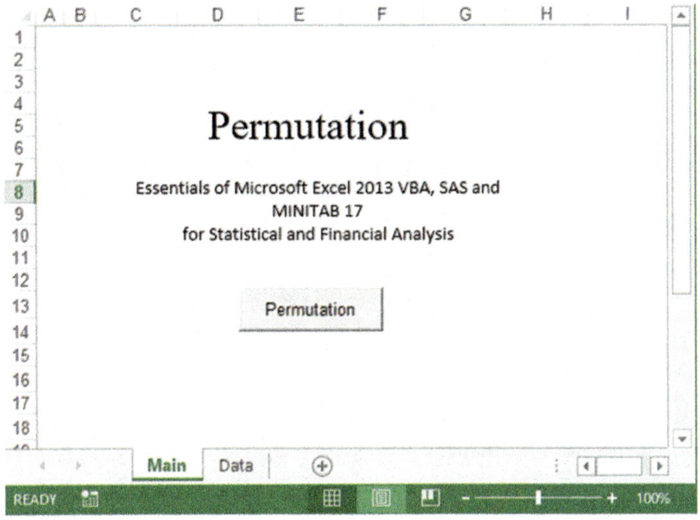

In the worksheet *Data*, put in the population that you want to populate. We will generate all the possible permutations of the four names listed in the previous section. Enter the names as shown below.

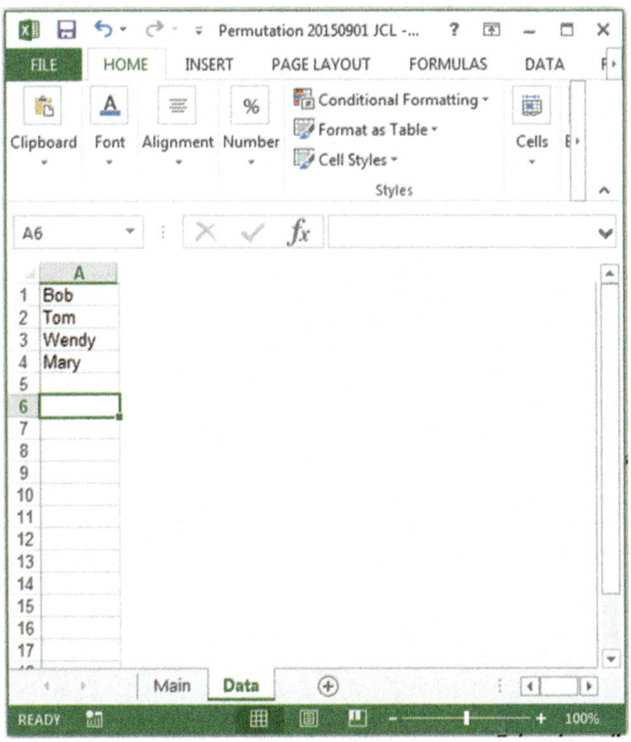

Pushing the *Permutation* button will get you the dialog box shown below.

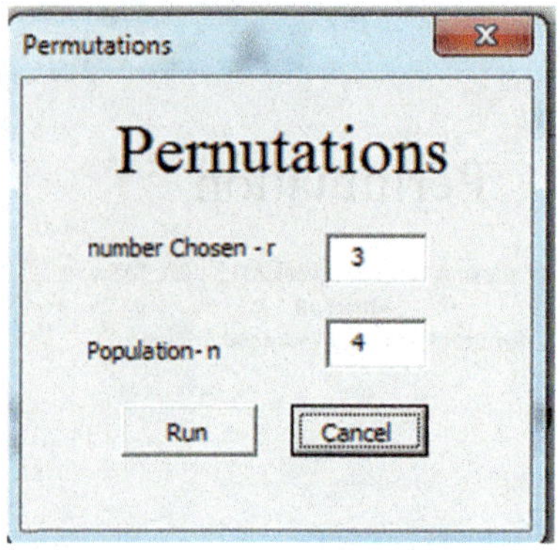

Below shows all the possible permutations of three names from a four names list after pressing the *Permutation* button on the *main* worksheet.

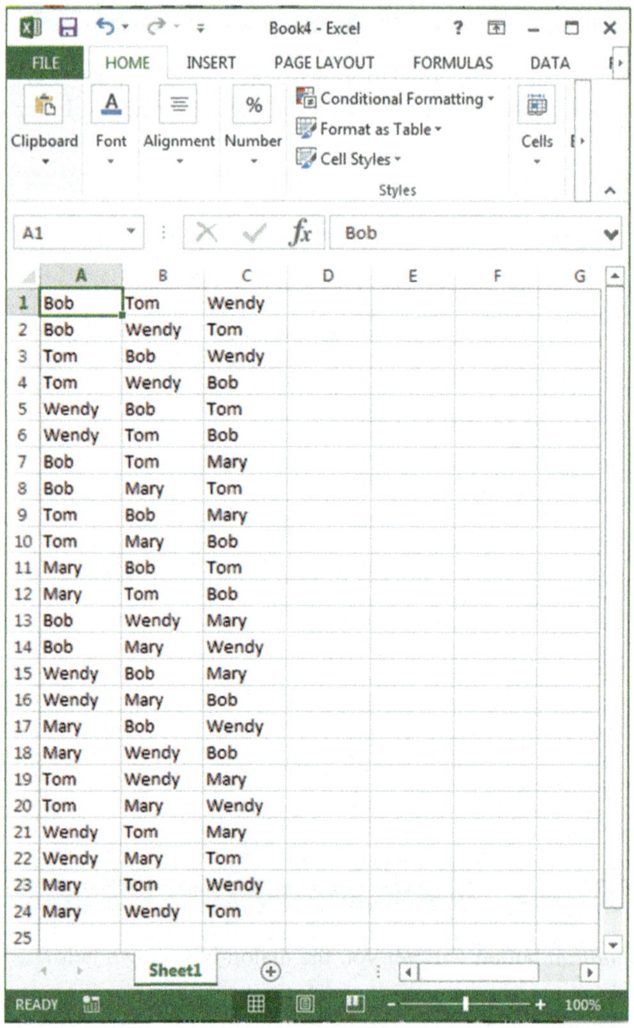

5.8 SAS Programming Code Instructions and Examples

5.8.1 SAS Probability Simulation

In this section, we use Do loop to simulate the process of tossing a coin. The function **UNIFORM()** will generate a random number between 0 and 1 each time when the loop runs. We consider the value larger than 0.5 as head and otherwise as tail. In this example, we toss 1000 times for each sample and then count the number of heads.

```
DATA Random;
do sample=1 to 100;
do toss=1 to 1000;
retain head;
coin=UNIFORM(123456);
if coin>0.5 then head=sum(head, 1);
end;
percent=head/1000;
output;
end;
run;
```

The following commands calculate the percentage of head for each sample.

```
data random;
set random;
percent1=lag(percent);
if percent1=. then percent1=0;
percent2=percent-percent1;
run;
```

Below shows the probability chart for 100 samples with each sample having 1000 values. Notice that it is almost equal to the theoretical value of 0.5 or 50 % heads.

```
Goptions ftext='Times New Roman' htext=4 gunit=pct;
title  height=8 "Probability Chart";
axis1  label=('Sample')
       order=(0 to 100 by 10);
axis2  label=('Probability')
       order=(0.4 to 0.6 by 0.02);
symbol v=dot i=join c=black;
Proc gplot data=random;
plot  percent2*sample/haxis=axis1 vaxis=axis2 vref=0.5;
Run;
```

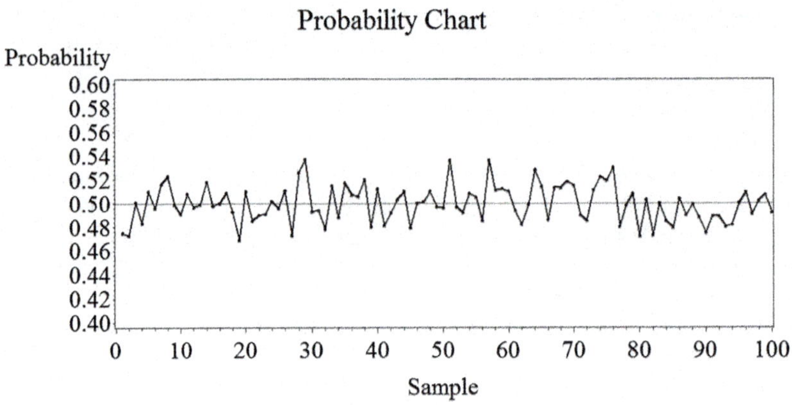

5.8.2 Combinations

Function **COMB**(n, r) is used to compute the number of combinations of n elements taken r at a time. n is a nonnegative integer that represents the total number of elements from which the sample is chosen. r is a nonnegative integer that represents the number of chosen elements.[1]

```
data Comb;
ncomb=comb(4,2);
run;
```

Obs	Ncomb
1	6

We will use function **lexcomb** to generate the combination list of four names shown below: Bob, Tom, Wendy, Mary

```
data Name;
   array x[4] $ ('Bob' 'Tom' 'Wendy' 'Mary');
   n=dim(x);
   k=3;
   ncomb=comb(n,k);
   do j=1 to ncomb;
      rc=lexcomb(j, k, of x[*]);
      put j 5. +3 x1-x3 +3 rc=;
      if rc<0 then leave;
   end;
run;
```

SAS writes the following output to the log:

```
1    Bob Mary Tom     rc=1
2    Bob Mary Wendy     rc=3
3    Bob Tom Wendy    rc=2
4    Mary Tom Wendy     rc=1
```

[1] http://support.sas.com/documentation/cdl/en/lrdict/64316/HTML/default/viewer.htm#a003112051.htm

To put this function in a more general way, we will use SAS macro to rewrite it. In order to create a macro program, you must first define it. You begin a macro definition with a %MACRO statement, and you end the definition with a %MEND statement. You can use positional parameters to create the macro variables k in the com macro definition as follows:

```
%macro com(k);
      data Name;
      array x[4] $ ('Bob' 'Tom' 'Wendy' 'Mary');
      n=4;
      ncomb=comb(n,&k);
      do j=1 to ncomb;
            rc=lexcomb(j, &k, of x[*]);
            output Name;
            if rc<0 then leave;
end;
data Name (keep=x1-x&k);
set Name;
run;
proc print data=name;
run;
%mend;
```

In order to substitute the value of a macro variable in the program, you must reference the macro variable. A macro variable reference is created by preceding the macro variable name with an ampersand (&). The reference causes the macro processor to search for the named variable in the symbol table and to return the value of the variable if the variable exists.

In this case, when you call the com macro you assign values to the macro variables that are created in the parameters. In the following example, the value two is assigned to the macro variable k.[2]

```
%com(k=2)
```

Obs	x1	x2
1	Bob	Mary
2	Bob	Tom
3	Bob	Wendy
4	Mary	Tom
5	Mary	Wendy
6	Tom	Wendy

5.8.3 Permutation

Function **PERM**(n, r) is used to compute the number of permutations of n items that are taken r at a time. n is an integer that represents the total number of elements from which the sample is chosen. r is an integer value that represents the number of chosen elements.

[2] http://www.ifeique.com/R3oa9Kd1pI/60477/index.htm

```
data Perm;
nperm=perm(4,2);
run;
```

Obs	Nperm
1	12

We will use function **lexperm** to generate all the possible permutations of the four names listed in the previous section.[3]

```
data Name;
   array x[4] $ ('Bob' 'Tom' 'Wendy' 'Mary');
   array y[3] $ y1-y3;
   k=3;
   /*Function fact() is used to calculate the factorial of k.*/
   nfact=fact(k);
   /*Function dim() is used to get the dimention of array x.*/
   n=dim(x);
   ncomb=comb(n,k);
   do j=1 to ncomb;
      rc=lexcomb(j, k, of x[*]);
      if rc<0 then leave;
   /*Let array y equal to the first three elements of array x.*/
      do m=1 to 3;
      y[m]=x[m];
      end;
   do i=1 to nfact;
      mc=lexperm(i, of y[*]);
      output Name;
      if mc<0 then leave;
   end;
   end;
run;
proc print data=name (keep=y1 y2 y3);
run;
```

Obs	y1	y2	y3
1	Bob	Mary	Tom
2	Bob	Tom	Mary
3	Mary	Bob	Tom
4	Mary	Tom	Bob
5	Tom	Bob	Mary
6	Tom	Mary	Bob
7	Bob	Mary	Wendy
8	Bob	Wendy	Mary
9	Mary	Bob	Wendy
10	Mary	Wendy	Bob
11	Wendy	Bob	Mary
12	Wendy	Mary	Bob

(continued)

[3] http://support.sas.com/documentation/cdl/en/lrdict/64316/HTML/default/viewer.htm#a003112233.htm

Obs	y1	y2	y3
13	Bob	Tom	Wendy
14	Bob	Wendy	Tom
15	Tom	Bob	Wendy
16	Tom	Wendy	Bob
17	Wendy	Bob	Tom
18	Wendy	Tom	Bob
19	Mary	Tom	Wendy
20	Mary	Wendy	Tom
21	Tom	Mary	Wendy
22	Tom	Wendy	Mary
23	Wendy	Mary	Tom
24	Wendy	Tom	Mary

A general function is generated by using SAS macro as follows:

```
%macro per(k);
data Name;
        array x[4] $ ('Bob' 'Tom' 'Wendy' 'Mary');
        array y[3] $ y1-y3;
        nfact=fact(&k);
   n=dim(x);
        ncomb=comb(n,&k);
          do j=1 to ncomb;
             rc=lexcomb(j, &k, of x[*]);
             if rc<0 then leave;
                do m=1 to &k;
                    y[m]=x[m];
                end;
                do i=1 to nfact;
                    mc=lexperm(i, of y[*]);
                    output Name;
                    if mc<0 then leave;
                end;
         end;
run;
proc print data=name (keep=y1-y&k);
run;
%mend;
%per(k=2)
```

Obs	y1	y2
1	Bob	Mary
2	Mary	Bob
3	Bob	Tom
4	Tom	Bob
5	Bob	Wendy
6	Wendy	Bob
7	Mary	Tom
8	Tom	Mary
9	Mary	Wendy
10	Wendy	Mary
11	Tom	Wendy
12	Wendy	Tom

5.9 Statistical Summary

This chapter begins our study of inferential statistics. We first look at the very basic concepts of probability, because inferential statistics needs to use probability to make inferences about a population from sample data.

We are studying inferential statistics to do the following:

$$\text{Sample Data} \rightarrow \text{Population Data}$$

We study probability to do the following:

$$\text{Population Data} \rightarrow \text{Sample Data}$$

In this chapter we did a probability simulation to illustrate the concept of probability. We simulated the chance of getting flipping a coin and getting a head. Logically we concluded that the probability should be 50 %. We saw in the simulation that the more times we flipped in a trial the closer the result was to 50 %

We also looked at Excel and SAS programs to list the combination and permutations of a list.

Appendix 5.1: Excel Code—Probability Simulator

```
Option Explicit
'/**************************************************************************
'/ Essentials of Microsoft Excel 2013 VBA, SAS and
'/              MINITAB 17
'/ for Statistical and Financial Analyses
'/
'/**************************************************************************

Private Sub cmdCancel_Click()
    Unload Me
End Sub

Private Sub cmdRun_Click()


    If Me.txtSamples < 1 Or Me.txtSamples > 256 Then
        MsgBox "Please enter a number for the Samples between 1 and 256", _
         vbExclamation, "Probability Simulator"
         Exit Sub
    ElseIf Me.txtValues < 1 Or Me.txtValues > 65536 Then
        MsgBox "Please enter a number for the Values between 1 and 65536", _
         vbExclamation, "Probability Simulator"
         Exit Sub
    ElseIf Me.txtPossibleNum < 0 Then
        MsgBox "Please enter a number greater than 0 for the PossibleNum.", _
         vbExclamation, "Probability Simulator"
        Exit Sub
    End If
    Me.Hide
    ProbabilitySimulation
    Application.StatusBar = False
    Unload Me
End Sub
```

```
'/*******************************************************************
'/Purpose:This procedure randomlly chooses between 0 and 1 and then
'/        graphs the percentage of 1s in each sample
'/*******************************************************************
Sub ProbabilitySimulation()

    Dim lSamples As Long
    Dim lValues As Long
    Dim lBegin As Long
    Dim lEnd As Long
    Dim wbProbability As Workbook
    Dim wsRandom As Worksheet
    Dim wsSummary As Worksheet
    Dim rCell As Range
    Dim rRandom As Range
    Dim lRow As Long
    Dim lColumn As Long
    Dim vProb As Variant
    Dim dtStart As Date


    lSamples = Me.txtSamples
    lValues = Me.txtValues
    lBegin = 1
    lEnd = Me.txtPossibleNum


    Set wbProbability = Workbooks.Add
    Set wsRandom = wbProbability.Worksheets(1)
    Set wsSummary = wbProbability.Worksheets(2)

    wsRandom.Name = "Random"
    wsSummary.Name = "Summary"
    Application.Calculation = xlCalculationManual
    With wsRandom
        .Name = "Random"
        '.Range("a1").Value = "=RANDBETWEEN(" & lBegin & "," & lEnd & ")"
        Set rRandom = .Range("a1").Resize(lValues, lSamples)
    End With

    dtStart = VBA.Now()
    ReDim vProb(1 To lValues, 1 To lSamples)

    For lColumn = 1 To lSamples
        For lRow = 1 To lValues
            Application.StatusBar = "Generating Sample Number: " & lColumn & _
                                    ", Sample Item Number: " & lRow & _
                                    " Number of Samples:" & lSamples & _
                                    " Sample Size:" & lValues & _
                                    " Start Time: " & VBA.Format(dtStart, "hh:mm:ss am/pm")
            vProb(lRow, lColumn) = Evaluate("=RANDBETWEEN(" & lBegin & "," & lEnd & ")")

        Next
    Next

    rRandom = vProb

    With wsSummary
            .Range("a1").Value = "Summary"
            .Range("a2").Value = 1
            .Range("B2").Value = 2
            .Range("a2:b2").AutoFill Destination:=.Range("a2").Resize(1, lSamples),
Type:=xlFillDefault
            .Range("A4").Value = "PerCent Head"
    End With

    Set rRandom = wbProbability.Worksheets("Random").Range("a1").CurrentRegion


    For Each rCell In wsSummary.Range("a5").Resize(1, lSamples).Cells

        rCell.Value = "=SUMif(random!" & rRandom.Columns(rCell.Column).Address & ",1)/" & lValues
        rCell.Offset(1, 0) = (1 / Me.txtPossibleNum)
    Next
```

```
    Dim chtProb As Chart
    Set chtProb = Charts.Add
    With chtProb
        .ChartType = xlLineMarkers
        .SetSourceData Source:=Sheets("Summary").Range("A5").Resize(2, lSamples), PlotBy _
        :=xlRows
        .Location Where:=xlLocationAsNewSheet
        .HasTitle = True
        .ChartTitle.Characters.Text = "Probability Chart" & Chr(10) _
                                      & "Number of Samples=" & lSamples & "," _
                                      & "Sample Size =" & lValues & "," _
                                      & "Possible Outcomes = " & Me.txtPossibleNum
        .ChartTitle.Characters(Start:=1, Length:=17).Font.Size = 18
        .Axes(xlCategory, xlPrimary).HasTitle = True
        .Axes(xlCategory, xlPrimary).AxisTitle.Characters.Text = "Number of Samples"
        .Axes(xlValue, xlPrimary).HasTitle = True
        .Axes(xlValue, xlPrimary).AxisTitle.Characters.Text = "Probability"
        .Axes(xlCategory).TickLabelSpacing = 5
        .Axes(xlValue).MaximumScale = 1
        .Axes(xlValue).MajorGridlines.Delete
        .SeriesCollection(2).MarkerStyle = xlNone
        .SeriesCollection(1).XValues = Sheets("Summary").Range("A2").Resize(2, lSamples)
        .SeriesCollection(2).XValues = Sheets("Summary").Range("A2").Resize(2, lSamples)
        .HasLegend = False
        .PlotArea.ClearFormats
        .Name = "Probability"
    .PlotArea.Interior.ColorIndex = xlNone
    End With

    Dim chtSeries As Series

    On Error Resume Next
    Application.DisplayAlerts = False
    wbProbability.Worksheets("Sheet3").Delete
    Application.DisplayAlerts = True
    Application.Calculation = xlCalculationAutomatic

End Sub

    Private Sub UserForm_Initialize()

        With Me
            .txtSamples = 100
            .txtValues = 50
            .txtPossibleNum = 2
        End With
    End Sub
```

Appendix 5.2: Excel Code—Combination

```
'/*****************************************************************************
'/ Essentials of Microsoft Excel 2013, SAS and
'/              MINITAB 17
'/ for Statistical and Financial Analyses
'/
'/*****************************************************************************
Option Explicit
Option Base 1
Dim mvComboList As Variant
Dim mwsList As Worksheet
Property Let ComboList(c As Variant)
    mvComboList = c
End Property
Property Get ComboList() As Variant
    ComboList = mvComboList
End Property
Property Set Listcombo(w As Worksheet)
    Set mwsList = w
End Property
Property Get Listcombo() As Worksheet
    Set Listcombo = mwsList
End Property

Private Sub cmdCancel_Click()
    Unload Me
End Sub
```

```
Private Sub cmdRun_Click()
    Me.Hide
    ComboHeader
    Application.StatusBar = False
    Unload Me
End Sub


Private Sub UserForm_Initialize()
    With Me
        .ComboList = WorksheetFunction.Transpose(wsData.Cells(1, 1).CurrentRegion)
        If WorksheetFunction.CountA(wsData.Columns(1)) > 0 Then
            .TxtComboN = UBound(.ComboList)
            .txtComboR = .TxtComboN - 1
        Else
            .TxtComboN = 0
            .txtComboR.Enabled = False
        End If
        .Hide
    End With
End Sub

Sub Combinations(vCombo As Variant, lStart As Long, lN As Long, lR As Long, lMaxR As Long, rPtr As
Range)
    Dim i As Long
    Dim ws As Worksheet

    If lR > lMaxR Then
        i = 1
        While i <= lMaxR
            Application.StatusBar = "Combination on Row " & rPtr.Row & " of " _
                & WorksheetFunction.Combin(Me.TxtComboN, Me.txtComboR)
            rPtr.Offset(0, i - 1) = Me.ComboList(vCombo(i))
            i = i + 1
        Wend
        If Not (rPtr.Row = 655356) Then
            Set rPtr = rPtr.Offset(1, 0)
        Else
            'Put comb on new worksheet because its more than
            '655356 rows
            Set ws = rPtr.Parent.Parent.Worksheets.Add
            Set rPtr = ws.Cells(1)
        End If
    Else
        i = lStart
        While i <= lN
            vCombo(lR) = i
            Combinations vCombo:=vCombo, lStart:=i + 1, lN:=lN, lR:=lR + 1, _
                lMaxR:=lMaxR, rPtr:=rPtr
            i = i + 1
        Wend
    End If
End Sub


Sub ComboHeader()
    Dim vCombo As Variant
    Dim rPtr As Range
    Dim wb As Workbook

    ReDim vCombo(256)  ' Max number of columns

    Set wb = Workbooks.Add
    Set Me.Listcombo = wb.Worksheets.Add
    Set rPtr = Me.Listcombo.Cells(1)

    Combinations vCombo:=vCombo, lStart:=1, lN:=Me.TxtComboN, lR:=1, _
        lMaxR:=Me.txtComboR, rPtr:=rPtr
End Sub
```

Appendix 5.3: Excel Code—Permutation

```
'/**************************************************************************
'/ Essentials of Microsoft Excel 2013, SAS and
'/              MINITAB 17
'/ for Statistical and Financial Analyses
'/
'/**************************************************************************
Option Explicit
Option Base 1
'/****************Permuatation Variables
Dim mwbPerm As Workbook
Dim mwsPerm As Worksheet
Dim mrData As Range
Dim mrPerm As Range
'/****************Combination Variables
Dim mvComboList As Variant
Dim mwsList As Worksheet
Property Let ComboList(c As Variant)
    mvComboList = c
End Property
Property Get ComboList() As Variant
    ComboList = mvComboList
End Property
Property Set Listcombo(w As Worksheet)
    Set mwsList = w
End Property
Property Get Listcombo() As Worksheet
    Set Listcombo = mwsList
End Property

Private Sub cmdCancel_Click()
    Unload Me
End Sub

Private Sub cmdRun_Click()
    Me.Hide
    ComboHeader
    Me.Listcombo.Name = "Combination"
    MainPermCaller
    Me.Listcombo.Parent.Close False
    Application.StatusBar = False
    Unload Me
End Sub

Private Sub UserForm_Initialize()
    With Me
        .ComboList = WorksheetFunction.Transpose(wsData.Cells(1, 1).CurrentRegion)
        If WorksheetFunction.CountA(wsData.Columns(1)) > 0 Then
            .TxtComboN = UBound(.ComboList)
            .txtComboR = .TxtComboN - 1
        Else
            .TxtComboN = 0
            .txtComboR.Enabled = False
        End If
        .Hide
    End With
End Sub

Sub Combinations(vCombo As Variant, lStart As Long, lN As Long, lR As Long, lMaxR As Long, rPtr As
Range)
    Dim i As Long
    Dim ws As Worksheet

    If lR > lMaxR Then
        i = 1
        While i <= lMaxR
            Application.StatusBar = "Combination on Row " & rPtr.Row & " of " _
                & WorksheetFunction.Combin(Me.TxtComboN, Me.txtComboR)
            rPtr.Offset(0, i - 1) = Me.ComboList(vCombo(i))
            i = i + 1
        Wend
        If Not (rPtr.Row = 655356) Then
            Set rPtr = rPtr.Offset(1, 0)
        Else
            'Put comb on new worksheet because its more than
            '655356 rows
            Set ws = rPtr.Parent.Parent.Worksheets.Add
            Set rPtr = ws.Cells(1)
        End If
```

```
    Else

        i = lStart
        While i <= lN
            vCombo(lR) = i
            Combinations vCombo:=vCombo, lStart:=i + 1, lN:=lN, lR:=lR + 1, _
                lMaxR:=lMaxR, rPtr:=rPtr
            i = i + 1
        Wend
    End If
End Sub


Sub ComboHeader()
    Dim vCombo As Variant
    Dim rPtr As Range
    Dim wb As Workbook

    ReDim vCombo(256)   ' Max number of columns

    Set wb = Workbooks.Add
    Set Me.Listcombo = wb.Worksheets.Add
    Set rPtr = Me.Listcombo.Cells(1)

    Combinations vCombo:=vCombo, lStart:=1, lN:=Me.TxtComboN, lR:=1, _
        lMaxR:=Me.txtComboR, rPtr:=rPtr
End Sub


'/************************Permutation Section
Property Set PermuteSheet(w As Worksheet)
    Set mwsPerm = w
End Property
Property Get PermuteSheet() As Worksheet
    Set PermuteSheet = mwsPerm
End Property

Sub MainPermCaller()
    Dim vList As Variant
    Dim iCount As Long
    Dim sList As String
    Dim rCell As Range

    Set mwbPerm = Workbooks.Add
    Set Me.PermuteSheet = mwbPerm.Worksheets(1)
    Set mrPerm = Me.PermuteSheet.Cells(1)


    Set mrData = Me.Listcombo.Cells(1).CurrentRegion
    For Each rCell In mrData.Columns(1).Cells
        Set vList = Nothing
        ReDim vList(mrData.Columns.Count)
        For iCount = 1 To mrData.Columns.Count
            vList(iCount) = rCell.Resize(1, mrData.Columns.Count).Cells(iCount)
        Next
        Permutation vList:=vList, sList:=sList
    Next


    Application.StatusBar = False
End Sub

'/**************************************************************
'/Purpose:  To create a permuation list recursively
'/          Each permutation is stored in a string list
'/**************************************************************
Sub Permutation(ByVal vList As Variant, ByVal sList As String)
    Dim lCount As Long
    Dim nCount As Long
    Dim iArray As Long
    Dim vNewList As Variant
```

```
    Dim sNewList As String

    sNewList = sList
    If UBound(vList) <> 1 Then
        For lCount = 1 To UBound(vList)
            iArray = 1
            ReDim vNewList(UBound(vList) - 1)
            For nCount = 1 To UBound(vList)
                If nCount <> lCount Then
                    vNewList(iArray) = vList(nCount)
                    iArray = iArray + 1
                Else
                    If Not (Len(Trim(sList)) = 0) Then
                        sNewList = sNewList & ";" & vList(lCount)
                    Else
                        sNewList = vList(lCount)
                    End If
                End If
            Next
            Permutation vList:=vNewList, sList:=sNewList
            sNewList = sList
        Next

    Else

        sList = sList & ";" & vList(1) & ";"
        mrPerm.Resize(1, Me.txtComboR).Value = ArrayList(sList:=sList, sSeperator:=";")
        Set mrPerm = mrPerm.Offset(1, 0)
    End If

End Sub


'/*******************************************************************
'/Purpose:  To convert a string list to an array
'/*******************************************************************
Function ArrayList(sList As String, sSeperator As String) As Variant
    Dim nFound As Integer
    Dim Arrayconvert As Variant

    ReDim Arrayconvert(1)
    Do
        nFound = InStr(1, sList, sSeperator)
        If nFound > 0 And Len(sList) > 1 Then
            If Not IsEmpty(Arrayconvert(1)) Then
                ReDim Preserve Arrayconvert(UBound(Arrayconvert) + 1)
            Else
                ReDim Arrayconvert(1)

            End If
            Arrayconvert(UBound(Arrayconvert)) = Mid(sList, 1, nFound - 1)
            sList = Mid(sList, nFound + 1, Len(sList) - (nFound))
        End If

    Loop While nFound > 0

    ArrayList = Arrayconvert
End Function
```

Bibliography

Lee CF, Lee JC, Lee AC (2013) Statistics for business and financial economics. Springer, New York
Microsoft Inc., Excel 2013. Microsoft Inc., Redmond
Minitab Inc. Minitab 17. Minitab Inc., State College
SAS Institute Inc. SAS 2014. SAS Institute Inc., Cary

Chapter 6
Discrete Random Variables and Probability Distributions

6.1 Introduction and Probability Distribution

In this chapter and the next three chapters, we will begin to study specific data sets of interest in statistics. With all of these data sets, we are interested in how the data items are distributed and what is the likelihood of getting a specific data item from a data set.

For example, let's show a distribution of a die. In a die there are six possible numbers, namely, 1, 2, 3, 4, 5, and 6. We can think of these possible numbers as the population. If we roll a die, we have a 16.66 % chance of getting the number 3. We also have a 16.66 % chance of getting the numbers 1, 4, 5, or 6. Now let's list out the probability of getting each specific data item as shown below.

Probability Distribution of a Die

#	probability	
1	16.667%	
2	16.667%	
3	16.667%	
4	16.667%	
5	16.667%	
6	16.667%	
Total Prob	100%	*

* number rounded

The list of all the probability distribution of each data item is called a *probability distribution.* Frequently we graph the probability distribution of a data set. Below is the graph of the probability distribution of a data set.

C.-F. Lee et al., *Essentials of Excel, Excel VBA, SAS and Minitab for Statistical and Financial Analyses*, DOI 10.1007/978-3-319-38867-0_6

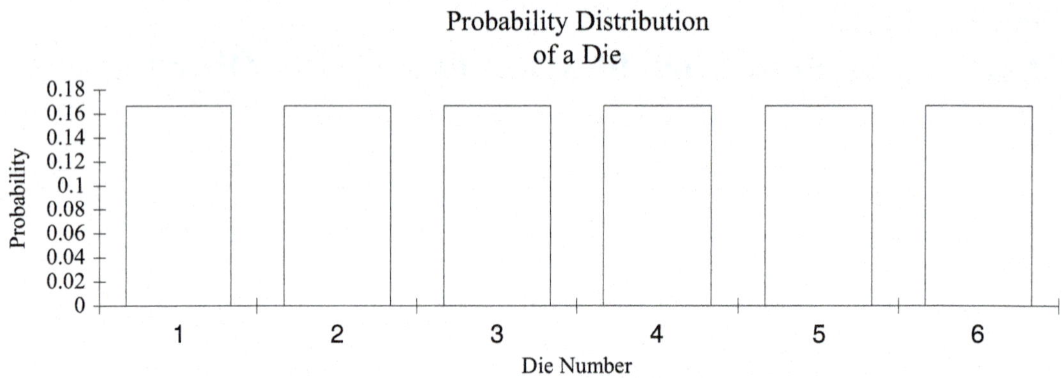

So far we have been looking at data sets that have countable data items. In this chapter we will look at countable data sets of interest in statistics. Statistics model each data set of interest with a mathematical formula. The mathematical formula for our die would be

$$F(x) = .1666, \text{ where } 1 \leq x \leq 6$$

In this function the variable x can be any values between and including 1–6. Statistics call this type of x variable a *discrete random* variable. Minitab likes to call the above function a *probability density function*.

6.2 Cumulative Probability Distribution

In the previous section, we were interested in getting the probability of getting a specific data item. In statistics we are also interested in getting the probability of getting more than one specific data item. For our die we would be interested in the following questions. What is the probability of getting either a 1 or 2? What is the probability of getting a 1 or 2 or 3? What is the probability of getting either a 1 or 2 or 3 or 4? What is the probability of getting a 1 or 2 or 3 or 4 or 5? What is the probability of getting a 1 or 2 or 3 or 4 or 5 or 6? The function models these questions as

$$F(x) = .1666x, \text{ where } 1 \leq x \leq 6$$

Minitab likes to call the above function a *cumulative density function*. The Minitab below tries to create the cumulative density function for our die probability density function. It flags an error that does not make sense.

```
MTB > name c1 "number"
MTB > name c2 "probability"
MTB > name c3 "CDF"
MTB > set "number"

* ERROR * Argument is a constant or matrix, but a column was expected.

MTB > help set
MTB > set 'number'
DATA> .
DATA> .
DATA> end
No new data recorded.
* NOTE * Text found in data line.
MTB > print 'number'
```

Data Display

```
number

* NOTE * One or more variables are undefined.

MTB > name c1 "number"
MTB > name c2 "probability"
MTB > name c3 "CDF"
MTB > set 'number'
DATA> 1 2 3 4 5 6
DATA> end
MTB > set 'probability'
DATA> .1666 .1666 .1666 .1666 .1666 .1666
DATA> end
MTB > print 'number' 'probability' 'CDF'
```

Data Display

```
Row  number  probability  CDF
  1       1       0.1666
  2       2       0.1666
  3       3       0.1666
  4       4       0.1666
  5       5       0.1666
  6       6       0.1666

* NOTE * One or more variables are undefined.

MTB > cdf 'number' 'cdf';
SUBC> discrete 'number' 'probability'.
* ERROR * Probabilities must sum to one.
```

The following Minitab commands reset the probabilities and verify that the sum of the probabilities is equal to 1.

```
MTB > set 'probability'
DATA> .1666 .1666 .1666 .1666 .1666 .167
DATA> end
MTB > sum 'probability'
```

Sum of Probability

```
Sum of probability = 1.0000
```

The following Minitab commands create and print the cumulative density function.

```
MTB > cdf 'number' 'cdf';
SUBC> discrete 'number' 'probability'.
MTB > print 'number' 'probability' 'cdf'
```

Data Display

Row	number	probability	CDF
1	1	0.1666	0.1666
2	2	0.1666	0.3332
3	3	0.1666	0.4998
4	4	0.1666	0.6664
5	5	0.1666	0.8330
6	6	0.1670	1.0000

```
MTB > plot 'cdf'*'number';
SUBC> connect;
SUBC> title 'Die Cummulative Distribution'.
```

Scatterplot of CDF vs. Number

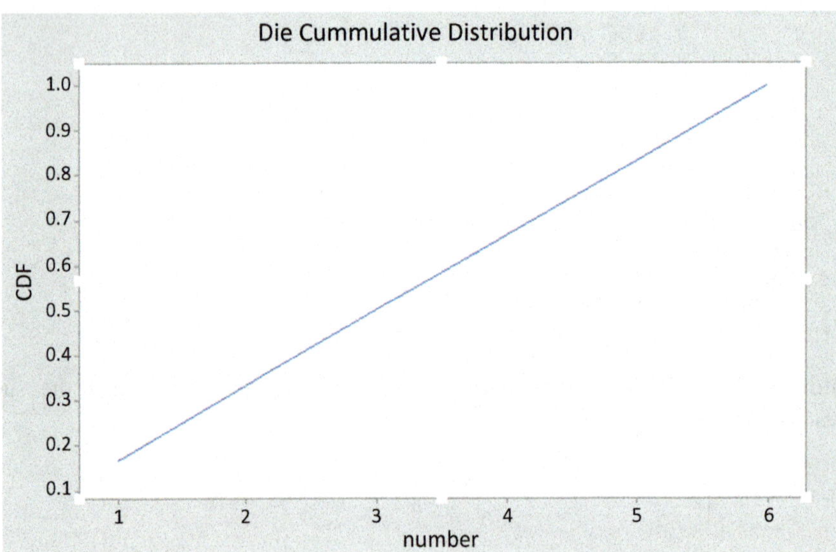

```
MTB > plot 'probability'*'number';
SUBC> connect;
SUBC> title 'Die Probability Distribution'.
```

Scatterplot of Probability vs. Number

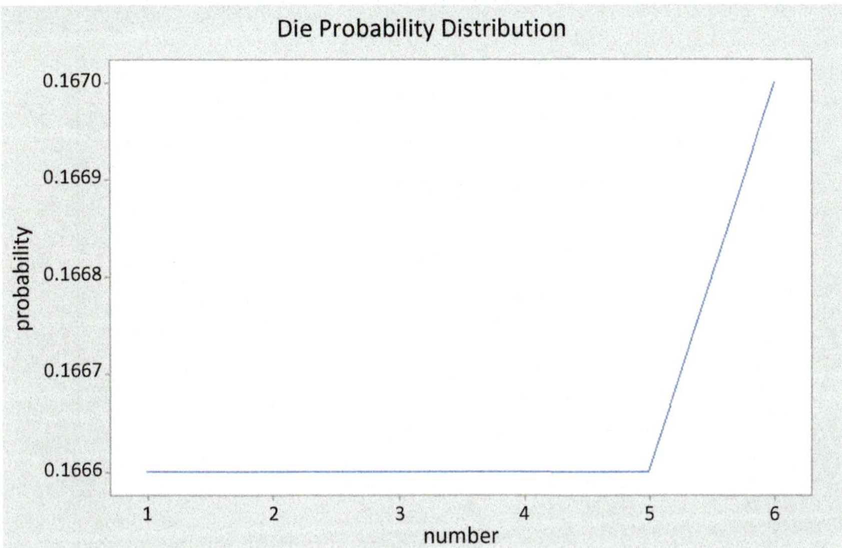

6.3 Binomial Distribution

The first important distribution that we will look at is the *binomial distribution*. There are two important characteristics of the binomial distribution. The first important characteristic is that the probability of something happening is independent of what happen previously. For example, if I flip a coin, there is a 50 % chance that I will get a head. If I then flip it one more time, there is still a 50 % chance that I will get a head. Now suppose I randomly put in ten numbers into a hat. The first time I pull out a number, each number has a 10 % chance of being chosen. The second time I pull out a number, the probability for each of the remaining numbers of being chosen is not 10 % anymore. It is now 1/9 or 11 %.

The second important characteristic about the binomial distribution is there are only two possible values possible. These values could represent True, False, or Success, Failure or 0, 1.

The binomial distribution is mathematically defined as the following:

$$\frac{n!}{r!(n-r)!}p^r q^{n-r}$$

In the formula n is how many times we flip the coin, r is how many times we want the outcome to be tails, p is the probability of success, and q is the probability of failure.

As an example, suppose that we are playing in a friendly betting game. The game consists of betting how many times we will get heads when we flip a coin ten times. If we think about this game carefully, it follows a binomial distribution. In this game we will have to pay $1 to play it, and we will receive $2 if we guess how many times the coin will turn up heads. What number should we choose? Should we choose 0, 3, 5, 7, or another number? We can just randomly choose one of the six numbers (0–5) or we can make an educated bet.

6.4 Excel Macro: Binomial Distribution

To make an educated bet, let's create a binomial distribution table for the ten coin flips and also graph the binomial distribution for ten coin flips. At the same time, we will also create a table and graph the cumulative binomial distribution. To create the tables and graphs, we will use the workbook *binomial.xlsm*. The code to produce the binomial distribution is shown in Appendix 6.1.

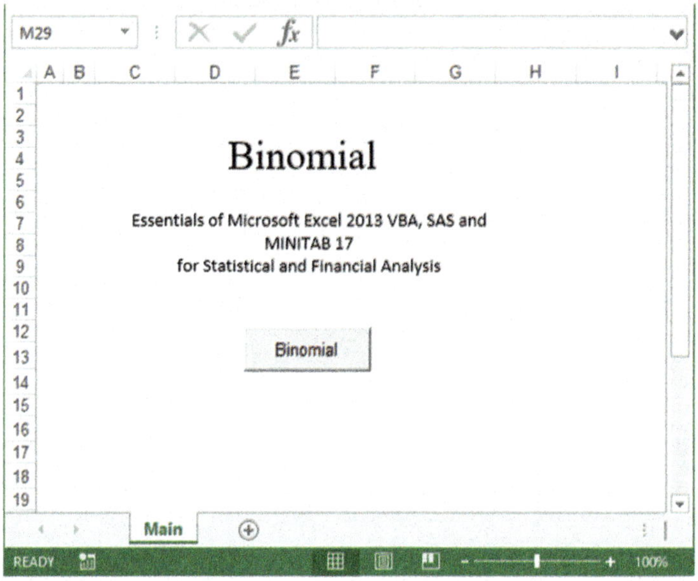

Pushing the *binomial* button will produce the following dialog box.

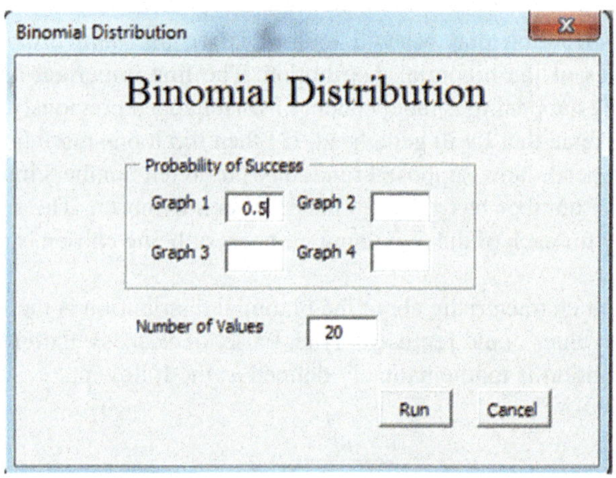

Next fill out the dialog box as shown above. After doing this, press the *Run* button. Pushing the *Run* button produces the following four things:

1. Binomial CDF graph
2. Binomial PDF graph
3. Binomial CDF table
4. Binomial PDF table

The table and graphs are shown below.

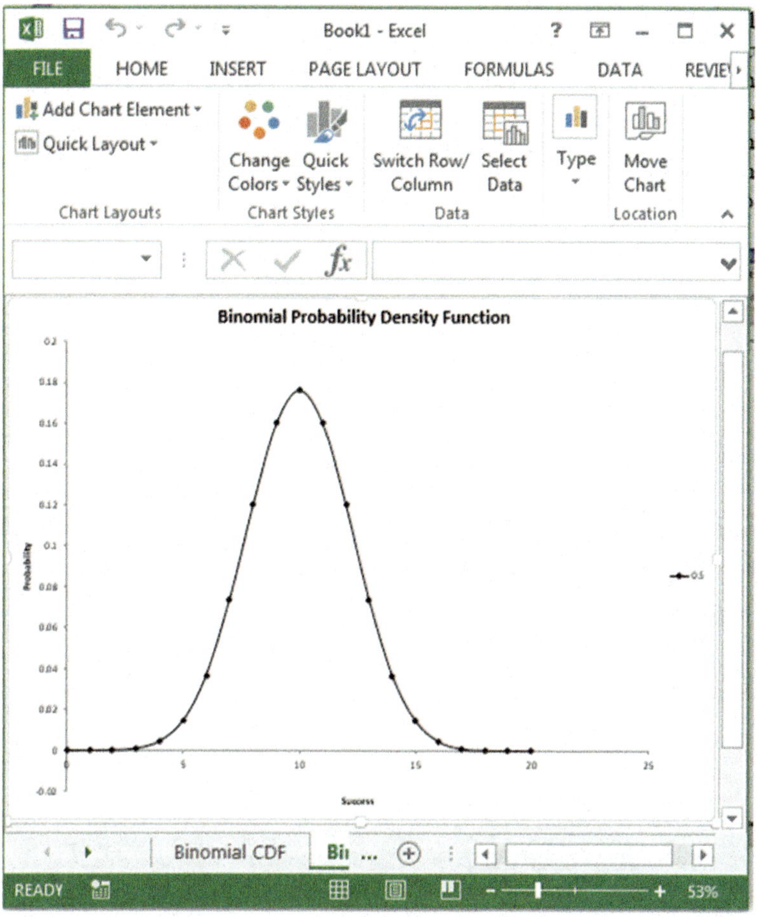

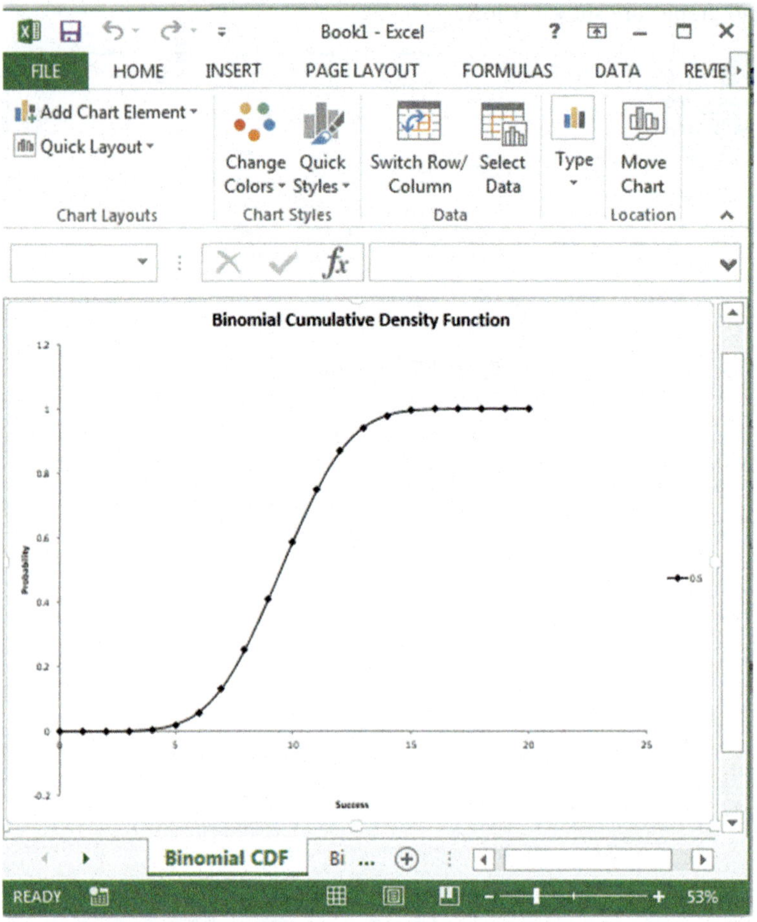
Book1 - Excel
FILE
HOME
INSERT
PAGE LAYOUT
FORMULAS
DATA
REVIE
Add Chart Element
Quick Layout
Chart Layouts
Change Colors
Quick Styles
Chart Styles
Switch Row/ Column
Select Data
Data
Type
Move Chart
Location
Binomial Cumulative Density Function
Probability
Success
0.5
Binomial CDF
Bi ...
READY
53%

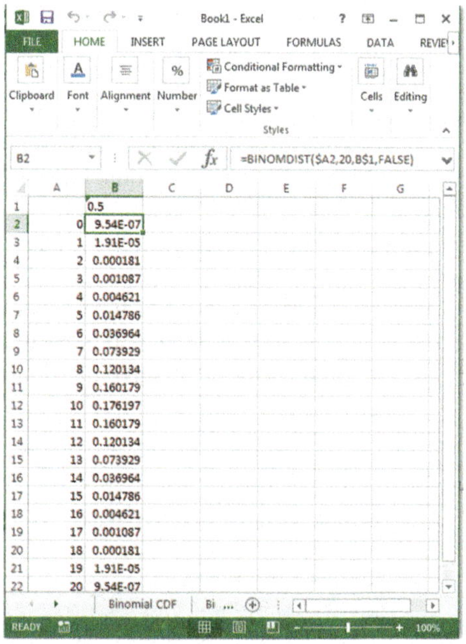

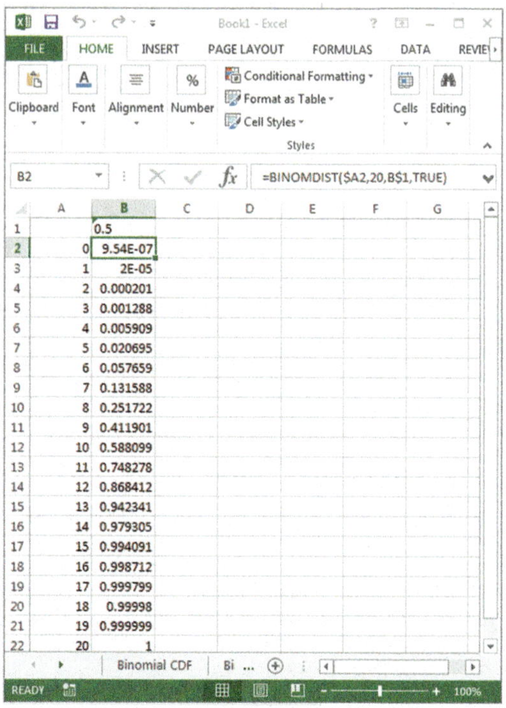

We used the *binomdist* Excel function to create the binomial PDF and CDF distribution. The first parameter is the number of success. The second parameter is the number of independent trials. The third parameter is the probability of success on each trial. The fourth parameter indicates if it's a cumulative binomial distribution.

From the binomial distribution, we can see that the best choice would be betting that we would see five heads when flipping a coin ten times. There will be a 24.61 % chance that we will see five heads after flipping a coin ten times.

Now suppose in our friendly betting game that there is another game where if we bet \$5, we can choose three numbers, and if we are correct, we will win \$10. Which three numbers should we choose? The most logical choice would be 3, 4, or 5. We would have a 20 + 24 + 20 = 64 % chance of winning.

Let's now create three binomial distributions of ten coin flips. The first distribution will have it so in each flip there is only a 25 % chance of getting a head. The second binomial distribution will have it so that there would be a 50 % chance of getting head. The third binomial distribution will have it so that there would be a 75 % chance of getting a head. How will these three distributions compare with each other? Below shows the table and graphs for these three distributions.

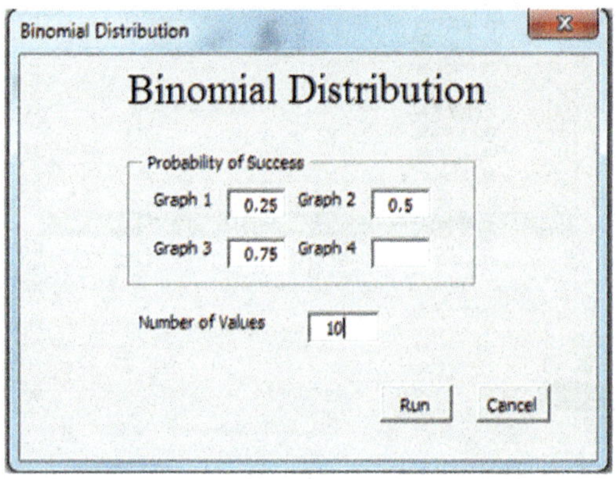

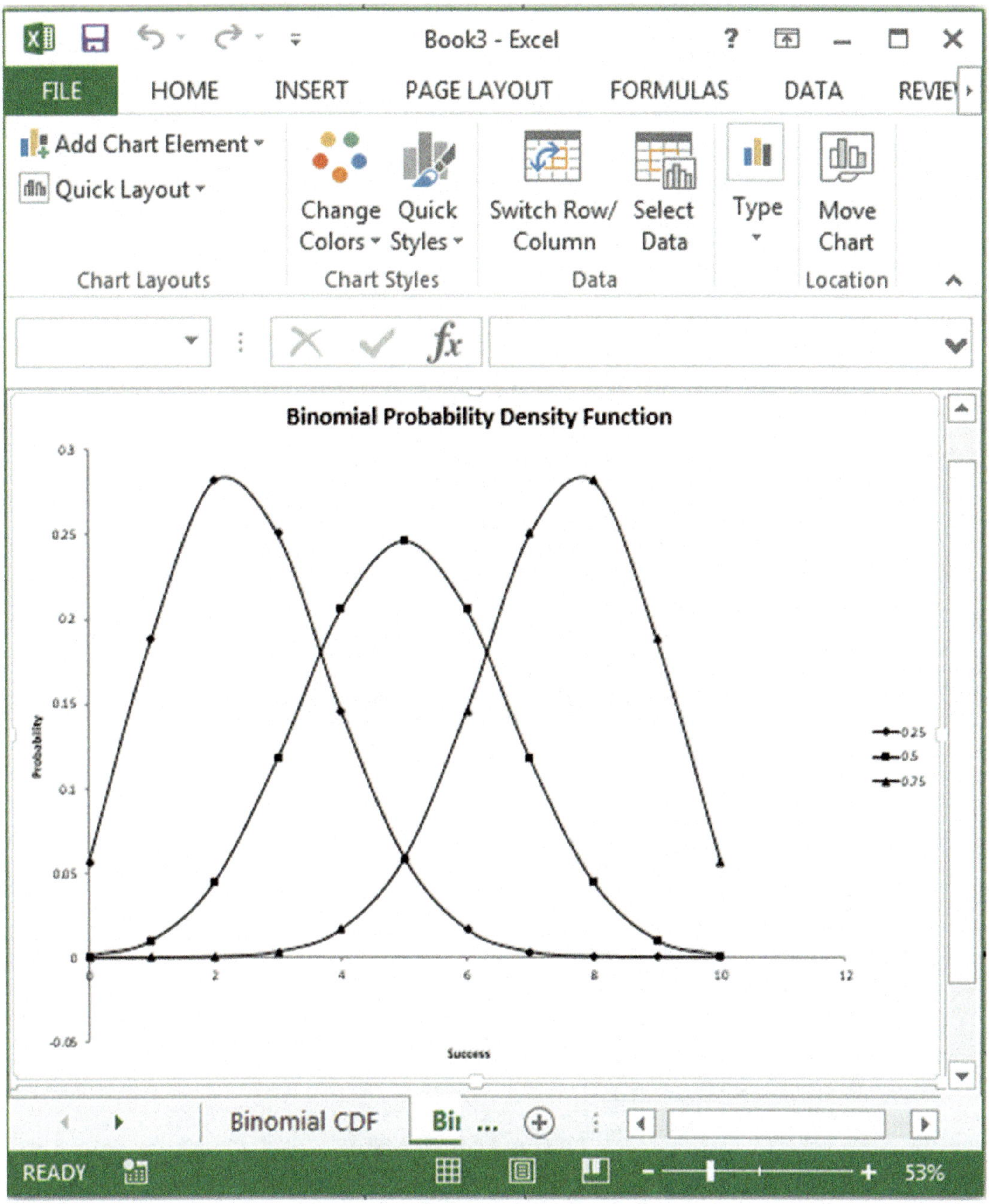
Book3 - Excel
FILE
HOME
INSERT
PAGE LAYOUT
FORMULAS
DATA
Add Chart Element
Quick Layout
Change Colors
Quick Styles
Switch Row/ Column
Select Data
Type
Move Chart
Chart Layouts
Chart Styles
Data
Location
Binomial Probability Density Function
Probability
Success
0.25
0.5
0.75
Binomial CDF
READY
53%

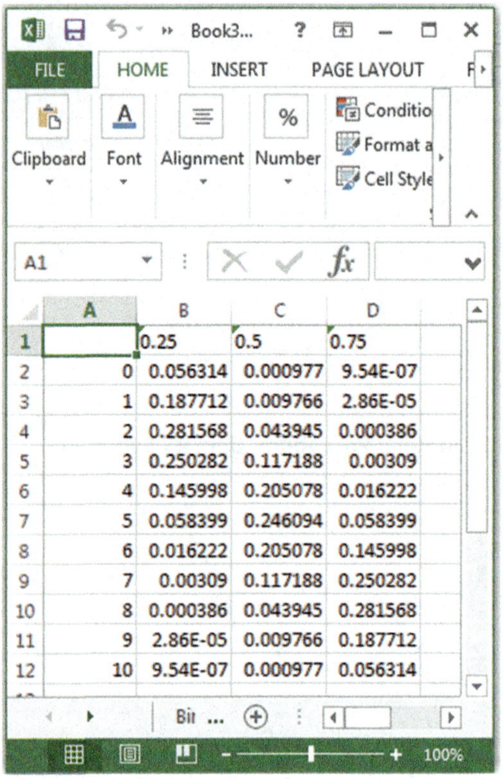

	A	B	C	D
1		0.25	0.5	0.75
2	0	0.056314	0.000977	9.54E-07
3	1	0.187712	0.009766	2.86E-05
4	2	0.281568	0.043945	0.000386
5	3	0.250282	0.117188	0.00309
6	4	0.145998	0.205078	0.016222
7	5	0.058399	0.246094	0.058399
8	6	0.016222	0.205078	0.145998
9	7	0.00309	0.117188	0.250282
10	8	0.000386	0.043945	0.281568
11	9	2.86E-05	0.009766	0.187712
12	10	9.54E-07	0.000977	0.056314

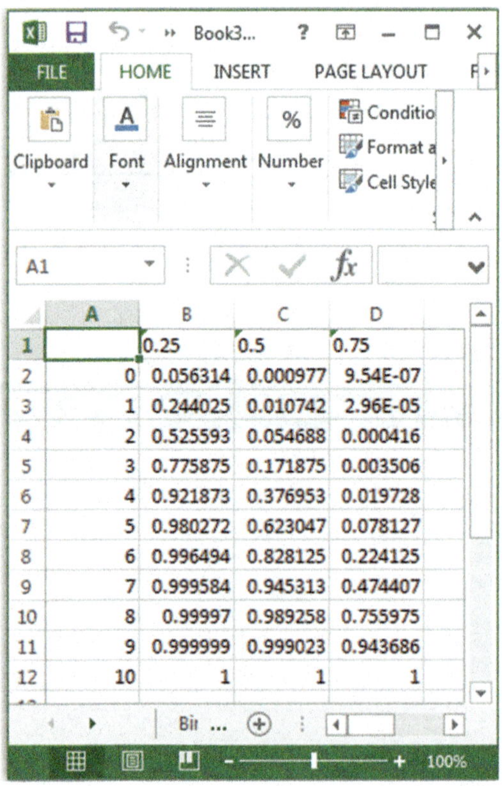

	A	B	C	D
1		0.25	0.5	0.75
2	0	0.056314	0.000977	9.54E-07
3	1	0.244025	0.010742	2.96E-05
4	2	0.525593	0.054688	0.000416
5	3	0.775875	0.171875	0.003506
6	4	0.921873	0.376953	0.019728
7	5	0.980272	0.623047	0.078127
8	6	0.996494	0.828125	0.224125
9	7	0.999584	0.945313	0.474407
10	8	0.99997	0.989258	0.755975
11	9	0.999999	0.999023	0.943686
12	10	1	1	1

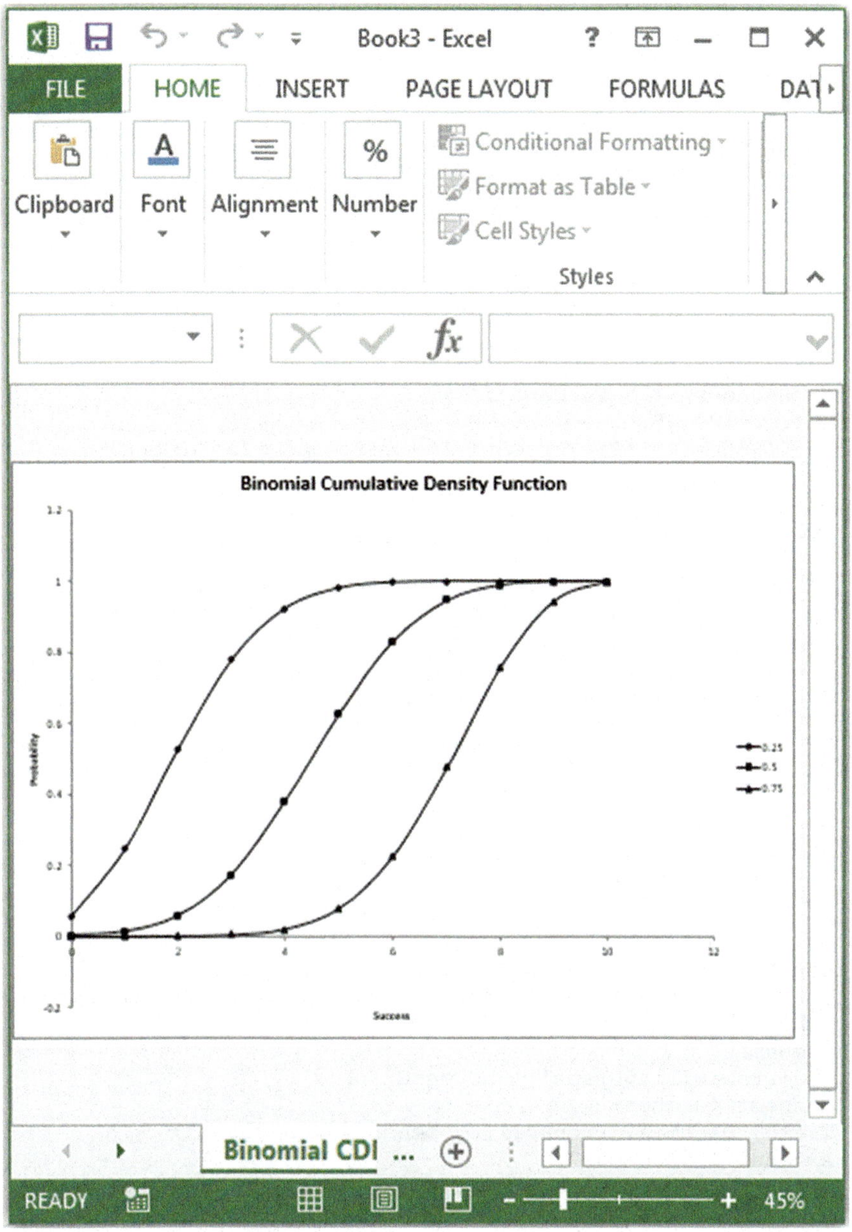

6.5 Binomial Distribution in Minitab

Below shows the Minitab commands to create the binomial PDFs and binomial CDFs shown in the previous section.

```
MTB > name c1 'Success'
MTB > name c2 '25 Percent'
MTB > name c3 '50 Percent'
MTB > name c4 '75 Percent'
MTB > name c5 'ProbabilityGraph'
MTB > name c6 'Group'
MTB > name c7 'SuccessGraph'
MTB > set 'Success'
DATA> 0:10
DATA> end
MTB > pdf 'Success' '25 Percent';
SUBC> binomial 10 .25.
MTB > pdf 'Success' '50 Percent';
SUBC> binomial 10 .50.
MTB > pdf 'Success' '75 Percent';
SUBC> binomial 10 .75.
MTB > print 'Success' '25 Percent' '50 Percent' '75 Percent'.
```

Data Display

Row	Success	25 Percent	50 Percent	75 Percent
1	0	0.056314	0.000977	0.000001
2	1	0.187712	0.009766	0.000029
3	2	0.281568	0.043945	0.000386
4	3	0.250282	0.117188	0.003090
5	4	0.145998	0.205078	0.016222
6	5	0.058399	0.246094	0.058399
7	6	0.016222	0.205078	0.145998
8	7	0.003090	0.117188	0.250282
9	8	0.000386	0.043945	0.281568
10	9	0.000029	0.009766	0.187712
11	10	0.000001	0.000977	0.056314

```
MTB > Stack  'Success' 'Success' 'Success' 'SuccessGraph'.
MTB > Stack  '25 Percent' '50 Percent' '75 Percent' 'ProbabilityGraph';
SUBC>   Subscripts 'group';
SUBC>   UseNames.
MTB > Plot 'ProbabilityGraph'*'SuccessGraph';
SUBC>   Connect 'group';
SUBC>   Title "Binomial Probability Density Functions";
SUBC>   Axlabel 1 "Success";
SUBC>   AXLabel 2 "Probability".
```

Plot ProbabilityGraph * SuccessGraph

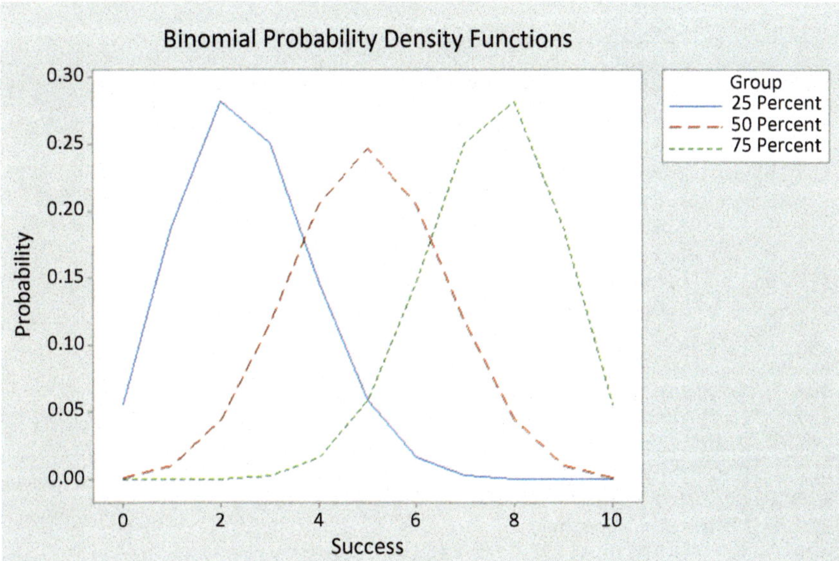


```
MTB > name c1 'Success'
MTB > name c2 '25 Percent'
MTB > name c3 '50 Percent'
MTB > name c4 '75 Percent'
MTB > name c5 'ProbabilityGraph'
MTB > name c6 'Group'
MTB > name c7 'SuccessGraph'
MTB > set 'Success'
DATA> 0:10
DATA> end
MTB > cdf 'Success' '25 Percent';
SUBC> binomial 10 .25.
MTB > cdf 'Success' '50 Percent';
SUBC> binomial 10 .50.
MTB > cdf 'Success' '75 Percent';
SUBC> binomial 10 .75.
MTB > print 'Success' '25 Percent' '50 Percent' '75 Percent'.
```

Data Display

```
 Row  Success  25 Percent  50 Percent  75 Percent

   1        0     0.05631     0.00098     0.00000
   2        1     0.24403     0.01074     0.00003
   3        2     0.52559     0.05469     0.00042
   4        3     0.77588     0.17188     0.00351
   5        4     0.92187     0.37695     0.01973
   6        5     0.98027     0.62305     0.07813
   7        6     0.99649     0.82813     0.22412
   8        7     0.99958     0.94531     0.47441
   9        8     0.99997     0.98926     0.75597
  10        9     1.00000     0.99902     0.94369
  11       10     1.00000     1.00000     1.00000
```

```
MTB > Stack  'Success' 'Success' 'Success' 'SuccessGraph'.
MTB > Stack  '25 Percent' '50 Percent' '75 Percent' 'ProbabilityGraph';
      SUBC>   Subscripts 'group';
      SUBC>   UseNames.
      MTB > Plot 'ProbabilityGraph'*'SuccessGraph';
      SUBC>   Connect 'group';
      SUBC>   Title "Binomial Cumulative Density Functions";
      SUBC>   Axlabel 1 "Success";
      SUBC>   AXLabel 2 "Probability".
```

Plot ProbabilityGraph * SuccessGraph

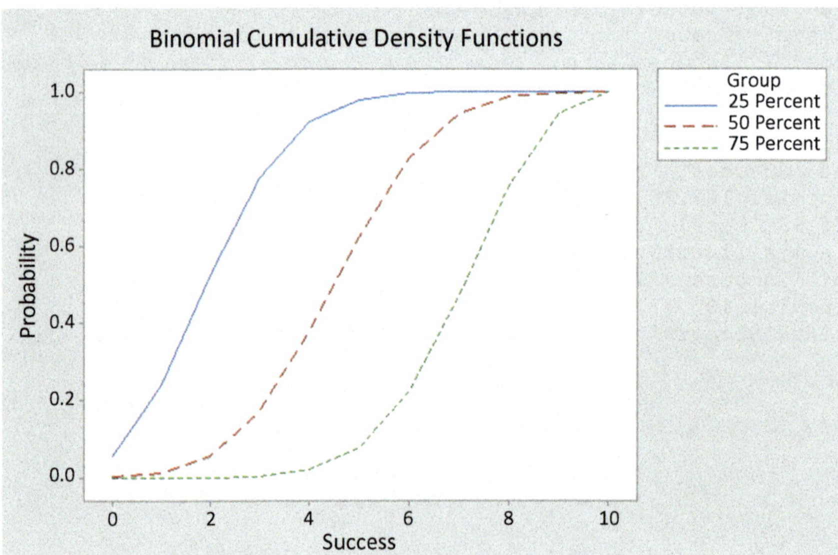

6.6 Hypergeometric Distribution

In the previous section, we talked about the binomial distribution. One of the biggest assumptions of the binomial distribution is in the computation of probability where the trials are independent. If the experiment consists of randomly drawing n elements (samples), with replacement, from a set of N elements, then the trials are independent.

In most practical situations, however, sampling is carried out *without replacement*, and the number sampled is not extremely small relative to the total number in the population. For example, when a sample of accounts receivable is drawn from a firm's accounting records for a sample audit, sampling units are ordinarily not replaced before the selection of subsequent units.

If the number sampled is extremely small relative to the total number of items, then the trial is almost independent even if the sampling is without replacement. Under such circumstances, and in sampling with replacement, the binomial distribution can be used in the analysis.

We will look at the *hypergeometric* distribution where there are samplings without replacement. The hypergeometric probability density function is defined as follows:

$$P[(x \text{ successes and } (n-x) \text{ failures})] = \frac{\binom{h}{n}\binom{N-h}{n-x}}{\binom{N}{n}}$$

The *hypergeometric* formula gives the probability of x successes when a random sample of n is drawn without replacement from a population of N within which h units have the characteristic denoting success.

Consider a group of ten students in which four are Democrats and six are Republicans. A sample of size six has been selected. What is the probability that there will be only one Democrat in this sample? Below shows how to solve this problem in Excel.

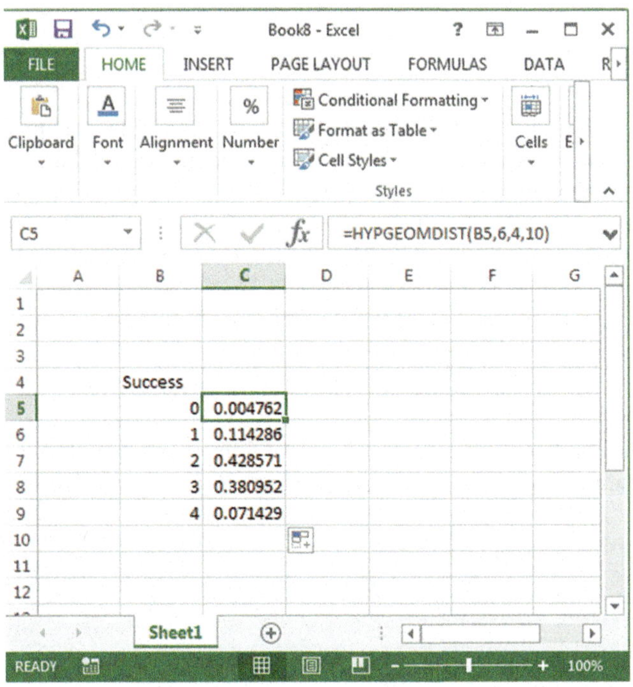

As shown above Microsoft Excel has a hypergeometric probability function called *hypgeomdist*. This function has four parameters. The first parameter is the number of success in the sample. The second parameter is the size of the sample. The third parameter is the number of success in the population. The fourth parameter is the population size.

As shown above in the hypergeometric table, the probability that there will be only one Democrat is .1143.

Below shows the Minitab commands in creating a hypergeometric distribution.

```
MTB > name c1 'Success'
MTB > name c2 'Probability'
MTB > pdf 'Success' 'Probability';
SUBC> hypergeometric 10,4,6.
MTB > print 'Success' 'Probability'
```

Data Display

```
Row   Success   Probability

  1         0      0.004762
  2         1      0.114286
  3         2      0.428571
  4         3      0.380952
  5         4      0.071429
  6         5      0.000000
  7         6      0.000000
```

6.7 Poisson Random Variable

In the previous two sections, we have discussed two major types of discrete probability distributions, one for binomial random variables and the other for hypergeometric random variables. Both of these random variables were defined in terms of the number of success, and these successes were *obtained within a fixed number of trials* of some random experiment. In this section, we will discuss a distribution called the *Poisson distribution*. This distribution can be used to deal with a single type of outcome or "event."

The formula for the density function of the Poisson distribution is

$$e^{-\lambda}\lambda^{x}/x! \text{ for } x = 0, 1, 2, 3, \ \ldots \text{ and } \lambda > 0$$

where x represents the number of rare events in a unit of time, space, or volume; λ is the mean value of x; e is the base of natural logarithms and is approximately equal to 2.71828; and ! is the factorial symbol.

6.8 Excel Macro: Poisson Distribution

To analyze the Poisson distribution, we will look at the workbook *Poisson.xls*.

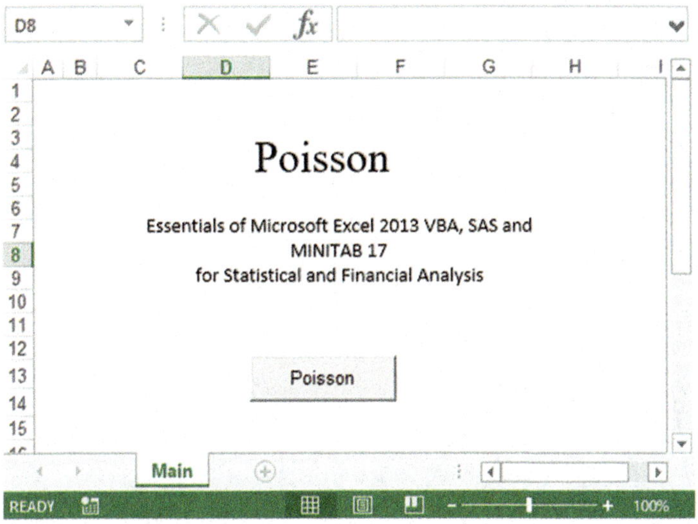

We will now analyze four different Poisson distributions with occurrences of 4, 6, 8, and 10. Press the *Poisson* button above to create these four Poisson distributions. Pressing the *Poisson* button will bring the *Poisson Distribution* dialog box shown below.

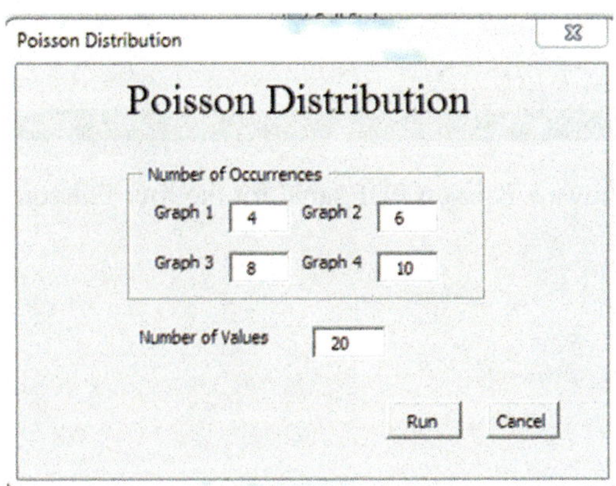

Fill out the dialog box above and then press the *Run* button to the four Poisson distributions. The dialog box below shows the four PDF Poisson distributions.

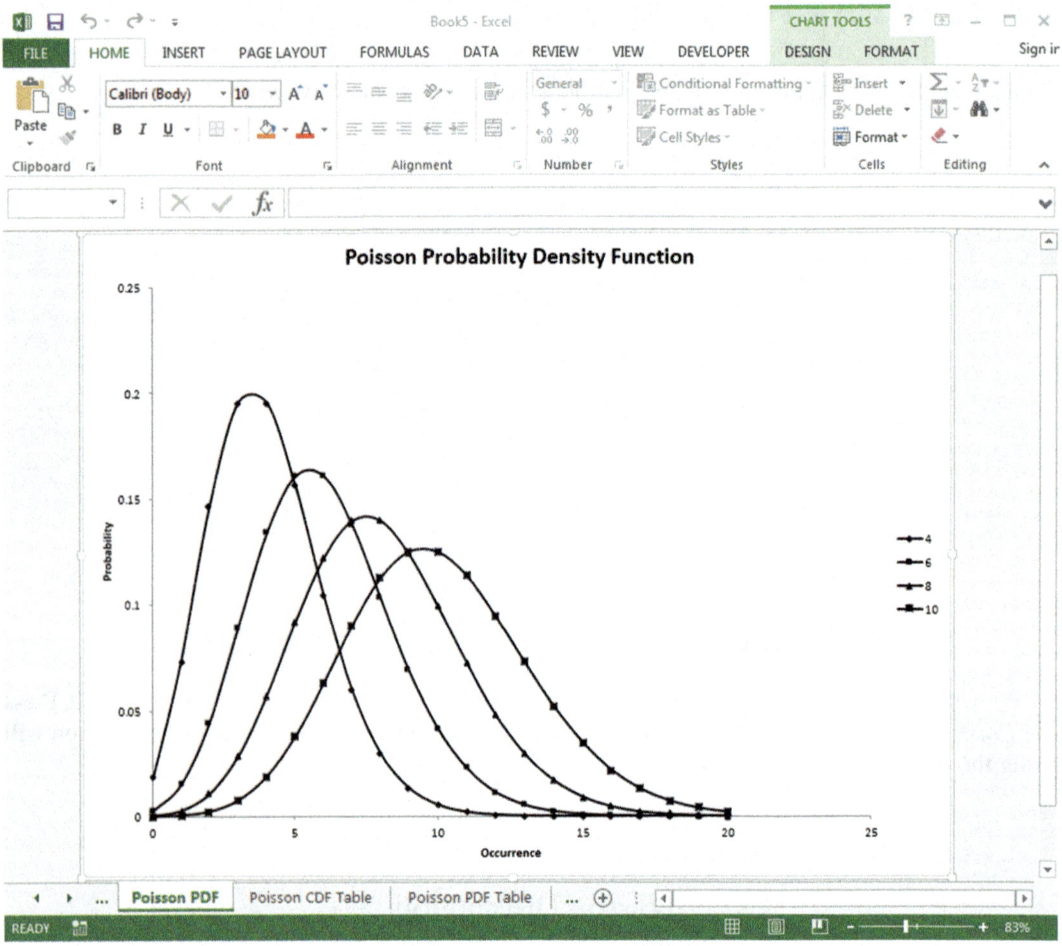

The workbook below shows a Poisson PDF table for the four Poisson distributions.

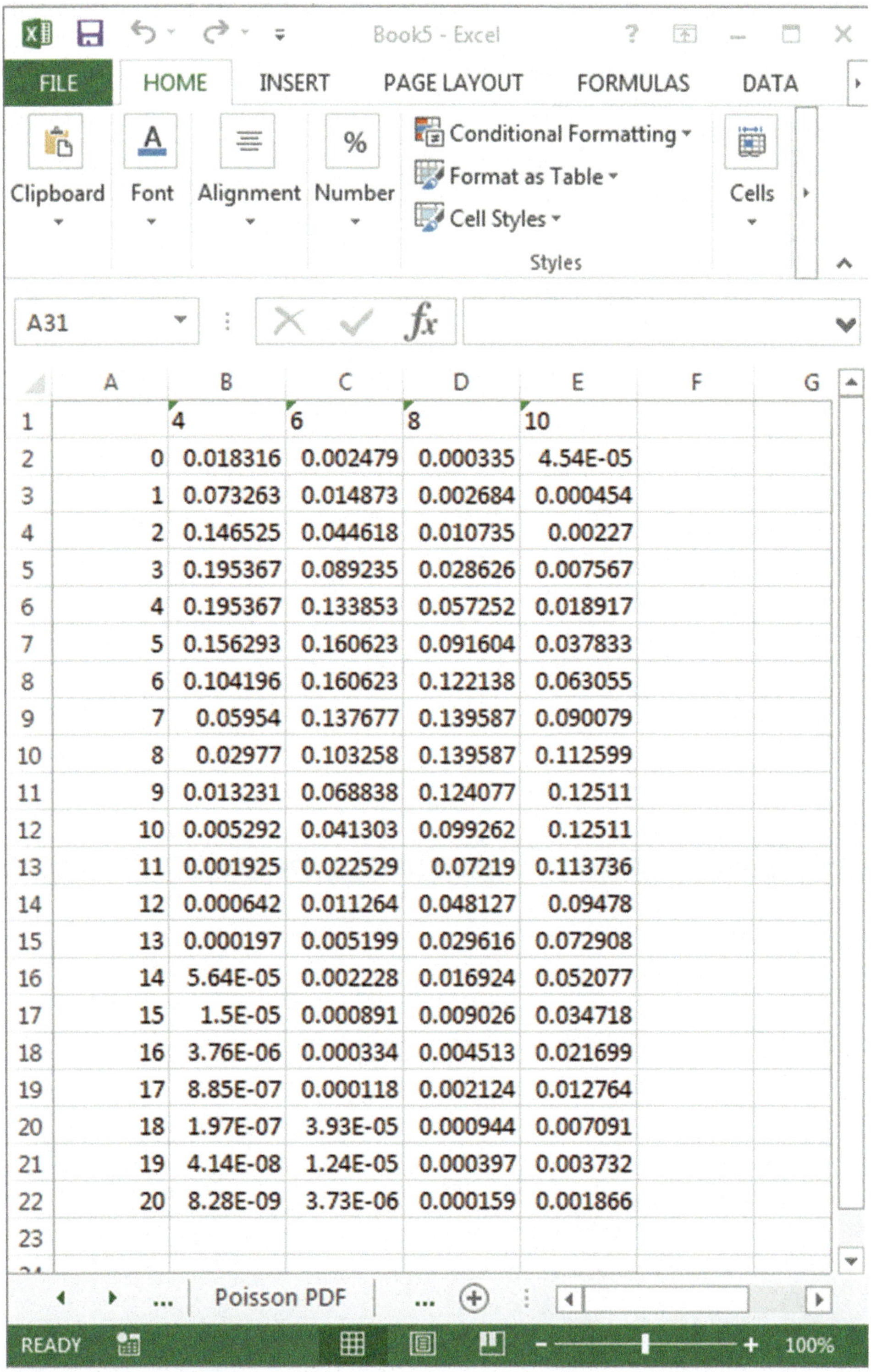
Book5 - Excel
FILE HOME INSERT PAGE LAYOUT FORMULAS DATA
Clipboard Font Alignment Number
Conditional Formatting
Format as Table
Cell Styles
Styles
Cells
A31
A B C D E F G
1 4 6 8 10
2 0 0.018316 0.002479 0.000335 4.54E-05
3 1 0.073263 0.014873 0.002684 0.000454
4 2 0.146525 0.044618 0.010735 0.00227
5 3 0.195367 0.089235 0.028626 0.007567
6 4 0.195367 0.133853 0.057252 0.018917
7 5 0.156293 0.160623 0.091604 0.037833
8 6 0.104196 0.160623 0.122138 0.063055
9 7 0.05954 0.137677 0.139587 0.090079
10 8 0.02977 0.103258 0.139587 0.112599
11 9 0.013231 0.068838 0.124077 0.12511
12 10 0.005292 0.041303 0.099262 0.12511
13 11 0.001925 0.022529 0.07219 0.113736
14 12 0.000642 0.011264 0.048127 0.09478
15 13 0.000197 0.005199 0.029616 0.072908
16 14 5.64E-05 0.002228 0.016924 0.052077
17 15 1.5E-05 0.000891 0.009026 0.034718
18 16 3.76E-06 0.000334 0.004513 0.021699
19 17 8.85E-07 0.000118 0.002124 0.012764
20 18 1.97E-07 3.93E-05 0.000944 0.007091
21 19 4.14E-08 1.24E-05 0.000397 0.003732
22 20 8.28E-09 3.73E-06 0.000159 0.001866
23
Poisson PDF
READY
100%

The above table uses the Excel function *Poisson* function to create the Poisson distribution. The *Poisson* function has three arguments. The first argument is the number of events. The second argument is the expected numeric value. The third argument indicates if the cumulative Poisson distribution should be calculated or not.

The chart below shows the four CDF Poisson distributions.

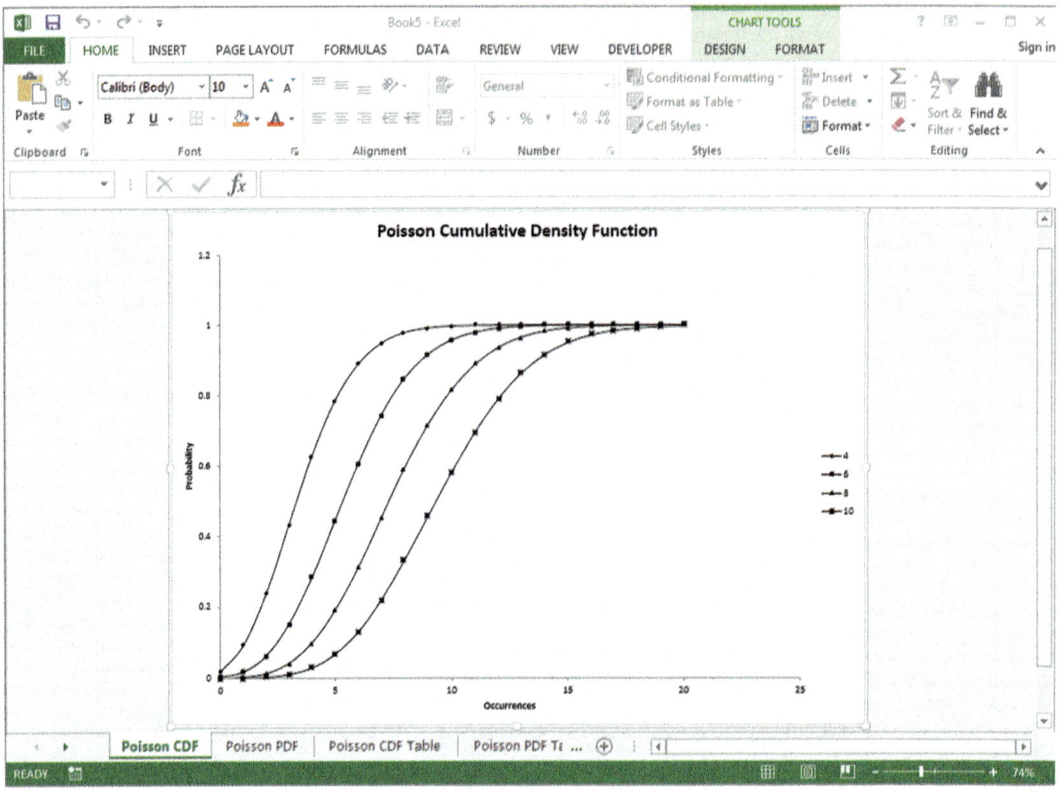

The chart below shows a Poisson PDF table for the four Poisson distributions.

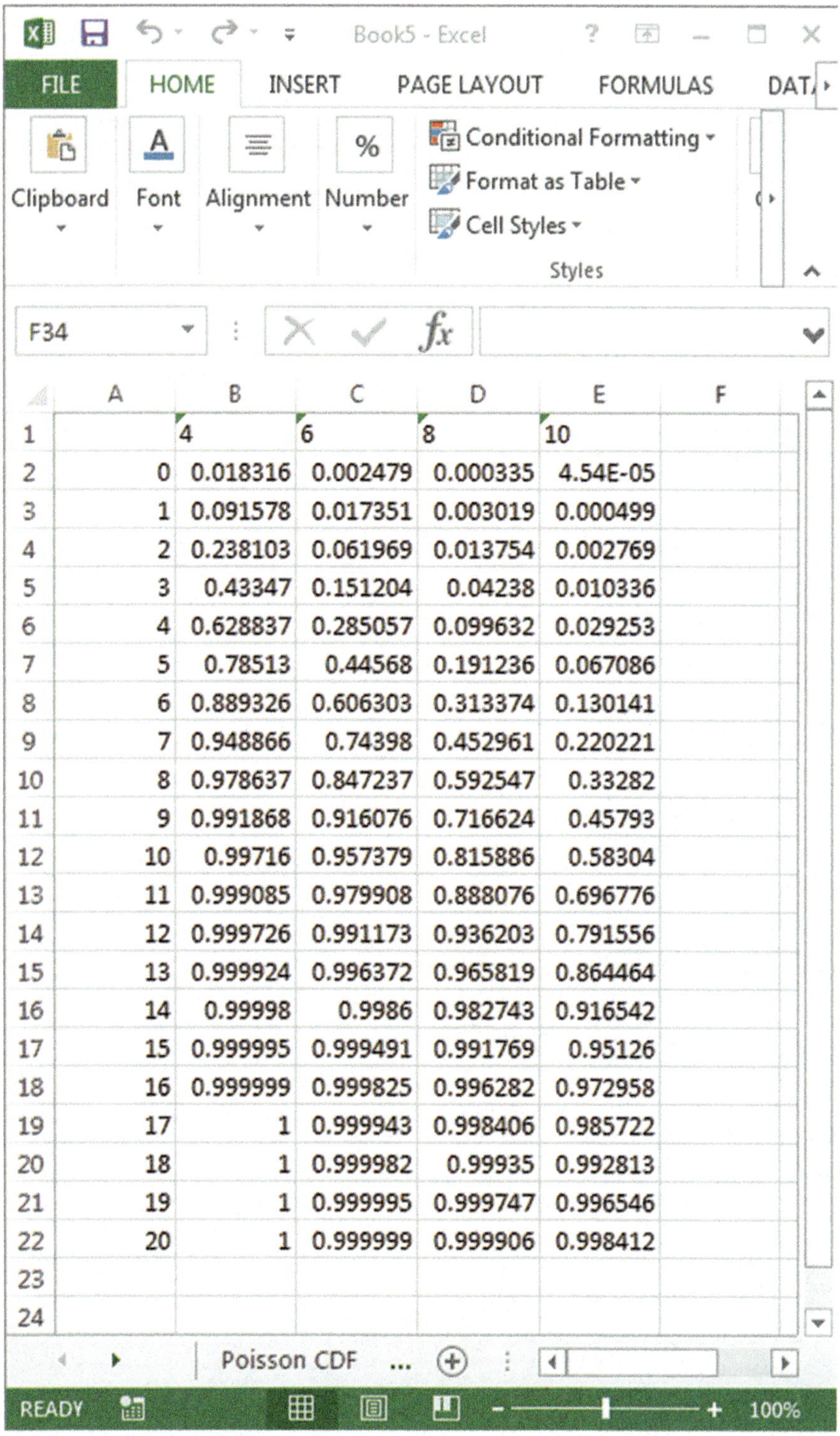

	4	6	8	10
0	0.018316	0.002479	0.000335	4.54E-05
1	0.091578	0.017351	0.003019	0.000499
2	0.238103	0.061969	0.013754	0.002769
3	0.43347	0.151204	0.04238	0.010336
4	0.628837	0.285057	0.099632	0.029253
5	0.78513	0.44568	0.191236	0.067086
6	0.889326	0.606303	0.313374	0.130141
7	0.948866	0.74398	0.452961	0.220221
8	0.978637	0.847237	0.592547	0.33282
9	0.991868	0.916076	0.716624	0.45793
10	0.99716	0.957379	0.815886	0.58304
11	0.999085	0.979908	0.888076	0.696776
12	0.999726	0.991173	0.936203	0.791556
13	0.999924	0.996372	0.965819	0.864464
14	0.99998	0.9986	0.982743	0.916542
15	0.999995	0.999491	0.991769	0.95126
16	0.999999	0.999825	0.996282	0.972958
17	1	0.999943	0.998406	0.985722
18	1	0.999982	0.99935	0.992813
19	1	0.999995	0.999747	0.996546
20	1	0.999999	0.999906	0.998412

6.9 Poisson Distribution in Minitab

Below shows the Minitab commands to create the Poisson PDFs and Poisson CDFs shown in the previous section.

```
MTB > name c1 'x-axis'
MTB > name c2 'm=4'
MTB > name c3 'm=6'
MTB > name c4 'm=8'
MTB > name c5 'm=10'
MTB > set 'x-axis'
DATA>    0:20
DATA> end
MTB > pdf 'x-axis' 'm=4';
SUBC> poisson 4.
MTB > pdf 'x-axis' 'm=6';
SUBC> poisson 6.
MTB > pdf 'x-axis' 'm=8';
SUBC> poisson 8.
MTB > pdf 'x-axis' 'm=10';
SUBC> poisson 10.
MTB > name c6 'graphx'
MTB > name c7 'graphy'
MTB > name c8 'group'
MTB > Stack 'x-axis' 'x-axis' 'x-axis' 'x-axis' 'graphx'.
MTB > Stack  'm=4' 'm=6' 'm=8'  'm=10' 'graphy';
SUBC>   Subscripts 'group';
SUBC>   UseNames.
MTB > Plot 'graphy'*'graphx';
SUBC>   Connect 'group';
SUBC>   Title "Poisson Distributions";
SUBC>   Axlabel 1 "Success";
SUBC>   AXLabel 2 "Probability".
```

Scatterplot of graphy vs. graphx

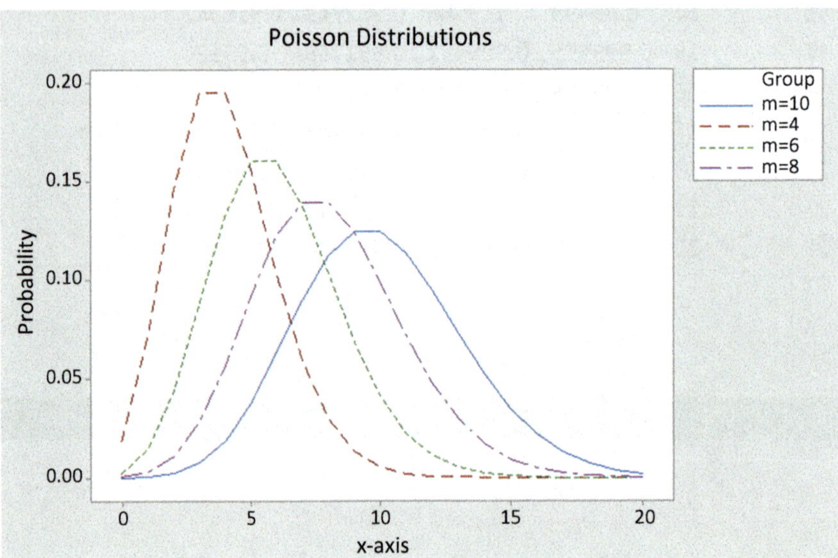

```
MTB > name c1 'x-axis'
MTB > name c2 'm=4'
MTB > name c3 'm=6'
MTB > name c4 'm=8'
MTB > name c5 'm=10'
MTB > set 'x-axis'
DATA>    0:20
DATA> end
MTB > cdf 'x-axis' 'm=4';
SUBC> poisson 4.
MTB > cdf 'x-axis' 'm=6';
SUBC> poisson 6.
MTB > cdf 'x-axis' 'm=8';
SUBC> poisson 8.
MTB > cdf 'x-axis' 'm=10';
SUBC> poisson 10.
MTB > name c6 'graphx'
MTB > name c7 'graphy'
MTB > name c8 'group'
MTB > Stack 'x-axis' 'x-axis' 'x-axis' 'x-axis' 'graphx'.
MTB > Stack  'm=4' 'm=6' 'm=8'  'm=10' 'graphy';
SUBC>   Subscripts 'group';
SUBC>   UseNames.
MTB > Plot 'graphy'*'graphx';
SUBC>   Connect 'group';
SUBC>   Title "Cumulative Poisson Distributions";
SUBC>   Axlabel 1 "Success";
SUBC>   AXLabel 2 "Probability".
```

Scatterplot of graphy vs. graphx

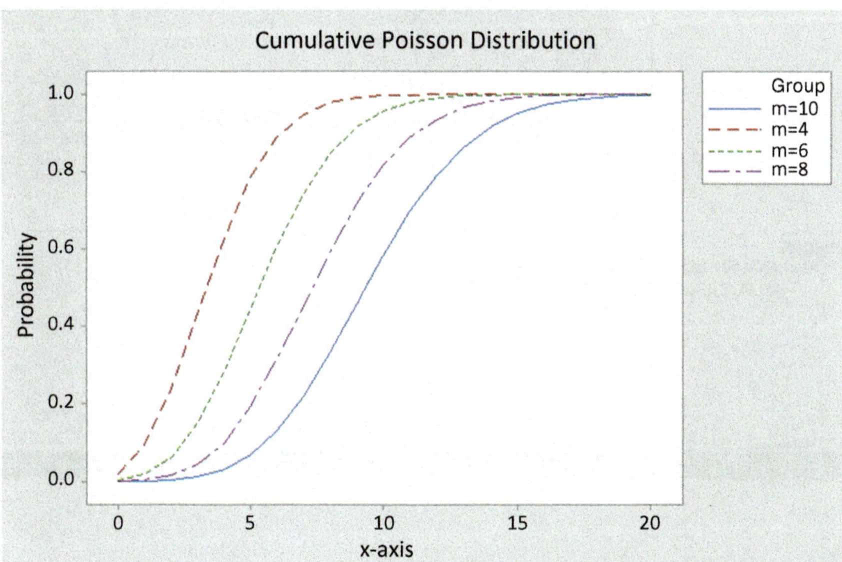

6.10 Stephen Bullen's Charting Method

We used Excel's VBA programming language to chart the binomial and Poisson distributions. In this section we will show Stephen Bullen's charting method. His charting method requires no VBA programming. Stephen Bullen is one of the world's leading Microsoft Excel experts. Go to his web site *http://BMSLtd.co.uk* the file *chtfrmla.xls*. We will use this file to chart the binomial and Poisson distribution. The chtfrmla.xls file is shown below.

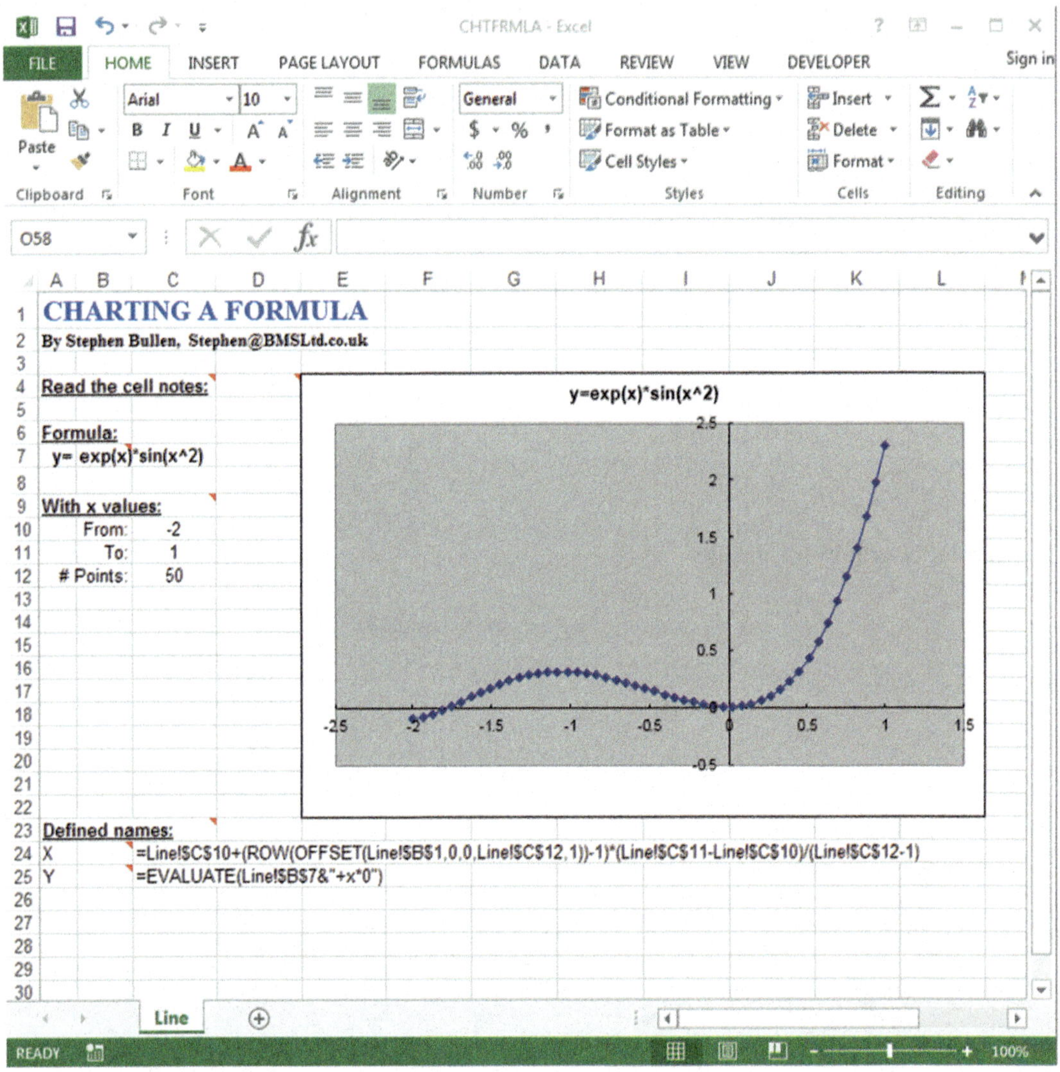

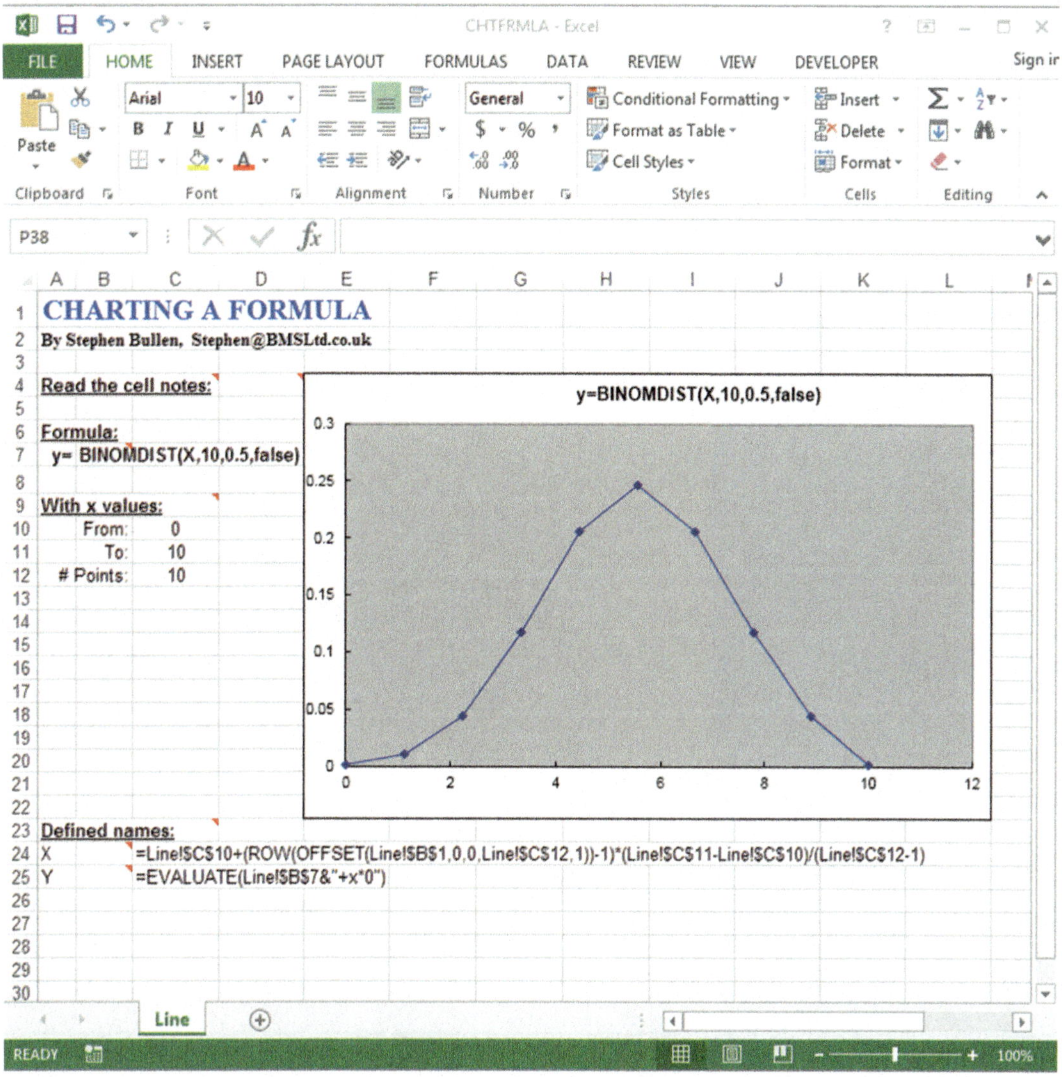

Above shows the binomial distribution using Stephen Bullen's charting method. This was accomplished by entering the binomial function in cell B7. There are two rules in entering functions in this cell. The first rule is that the first character should be a single apostrophe. The second rule is to enter an '*x*' in the argument where it represents the independent variable of a chart. The independent variable is the first argument of the binomial distribution. It is also important to note that cell C10 indicates where the chart will start and that cell C11 is where the chart will end and cell C12 indicates how many points there are in the chart.

Stephen Bullen uses two Excel features to create his charts. These two features are *defined names* and Excel 4.0 macros. In Excel, there are many uses for defined names. One of the uses of defined names is to give names to formulas and then to use the formulas by referencing the name. Below shows how to see the define names for the above workbook.

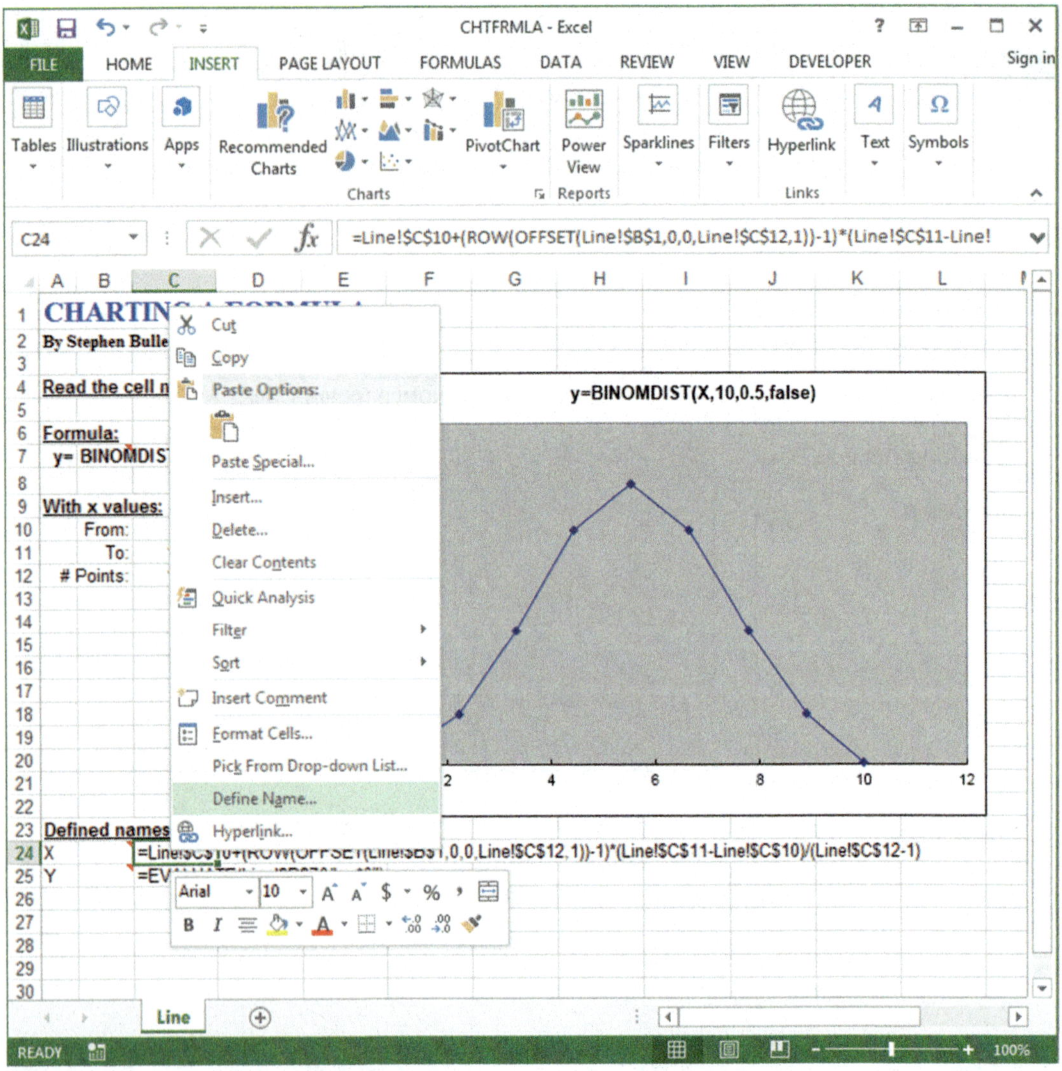
CHTFRMLA - Excel
FILE HOME INSERT PAGE LAYOUT FORMULAS DATA REVIEW VIEW DEVELOPER
Sign in
Tables Illustrations Apps Recommended Charts PivotChart Power View Sparklines Filters Hyperlink Text Symbols
Charts Reports Links
C24
=Line!$C$10+(ROW(OFFSET(Line!$B$1,0,0,Line!$C$12,1))-1)*(Line!$C$11-Line!
CHARTIN
By Stephen Bulle
Read the cell n
Formula:
y= BINOMDIS
With x values:
From:
To:
Points:
Defined names
X
Y
Cut
Copy
Paste Options:
Paste Special...
Insert...
Delete...
Clear Contents
Quick Analysis
Filter
Sort
Insert Comment
Format Cells...
Pick From Drop-down List...
Define Name...
Hyperlink...
y=BINOMDIST(X,10,0.5,false)
2
4
6
8
10
12
,0,0,Line!$C$12,1))-1)*(Line!$C$11-Line!$C$10)/(Line!$C$12-1)
Line
READY
100%

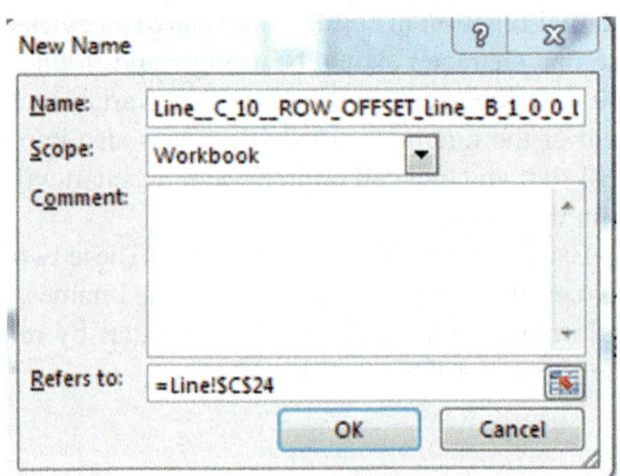
New Name
Name:
Line__C_10__ROW_OFFSET_Line__B_1_0_0_l
Scope:
Workbook
Comment:
Refers to:
=Line!$C$24
OK
Cancel

As shown above, the Stephen Bullen's workbook has two defined names. The above dialog box shows the formula for the defined name *y*. Notice that in the defined name *y*, there is the word "evaluate." "Evaluate" is part of the Excel 4.0 macro language.

To see how Excel uses these defined names, click on the binomial distribution as shown below.

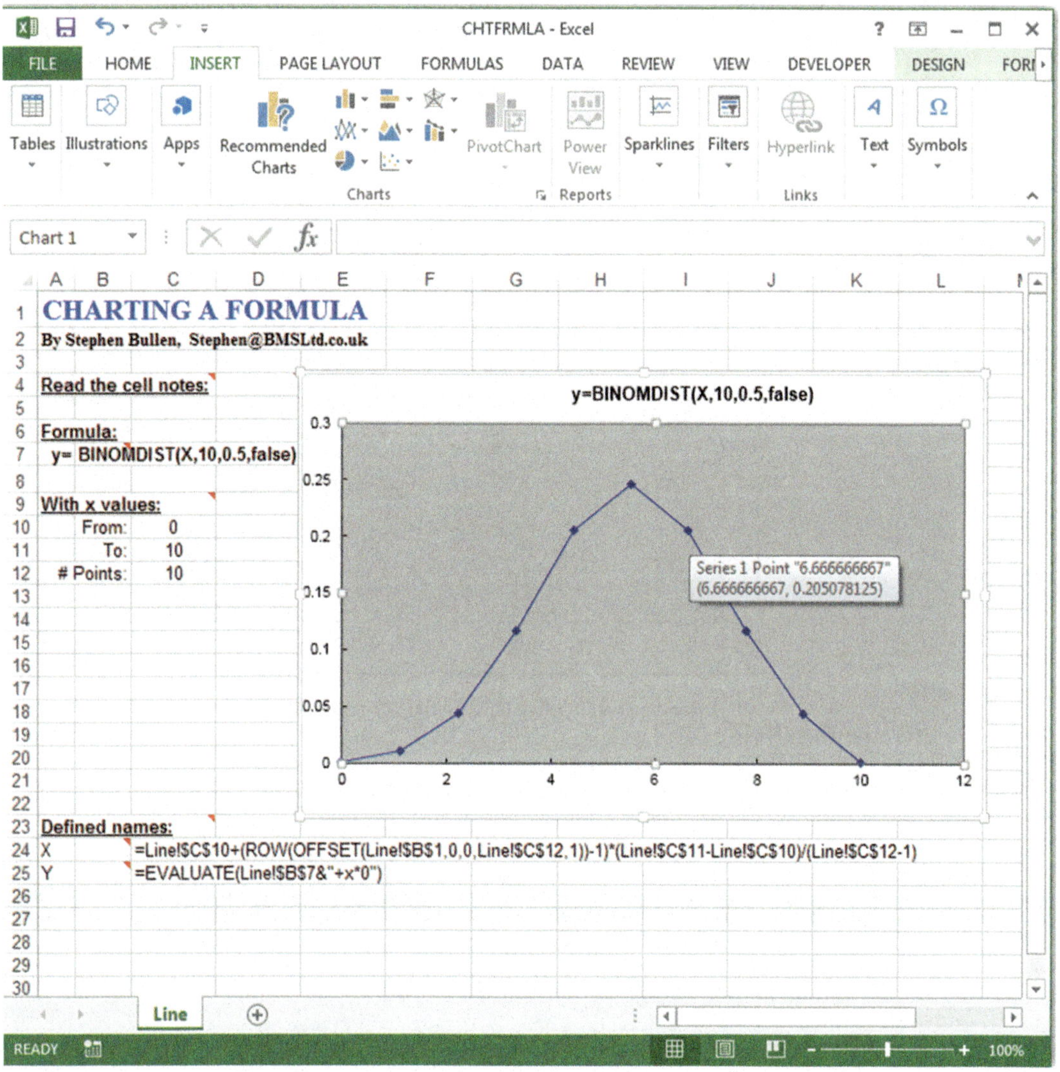

Clicking the binomial distribution will show in the formula bar the following formula:

$$= \text{SERIES}(\text{Line!x}, \text{Line!y}, 1)$$

This formula uses the defined names *x* and *y*. If it wasn't for the defined names, it would be necessary to put the formula indicated in cells C24 and C25.

Below shows the cumulative binomial distribution, the Poisson distribution, and the cumulative Poisson distribution.

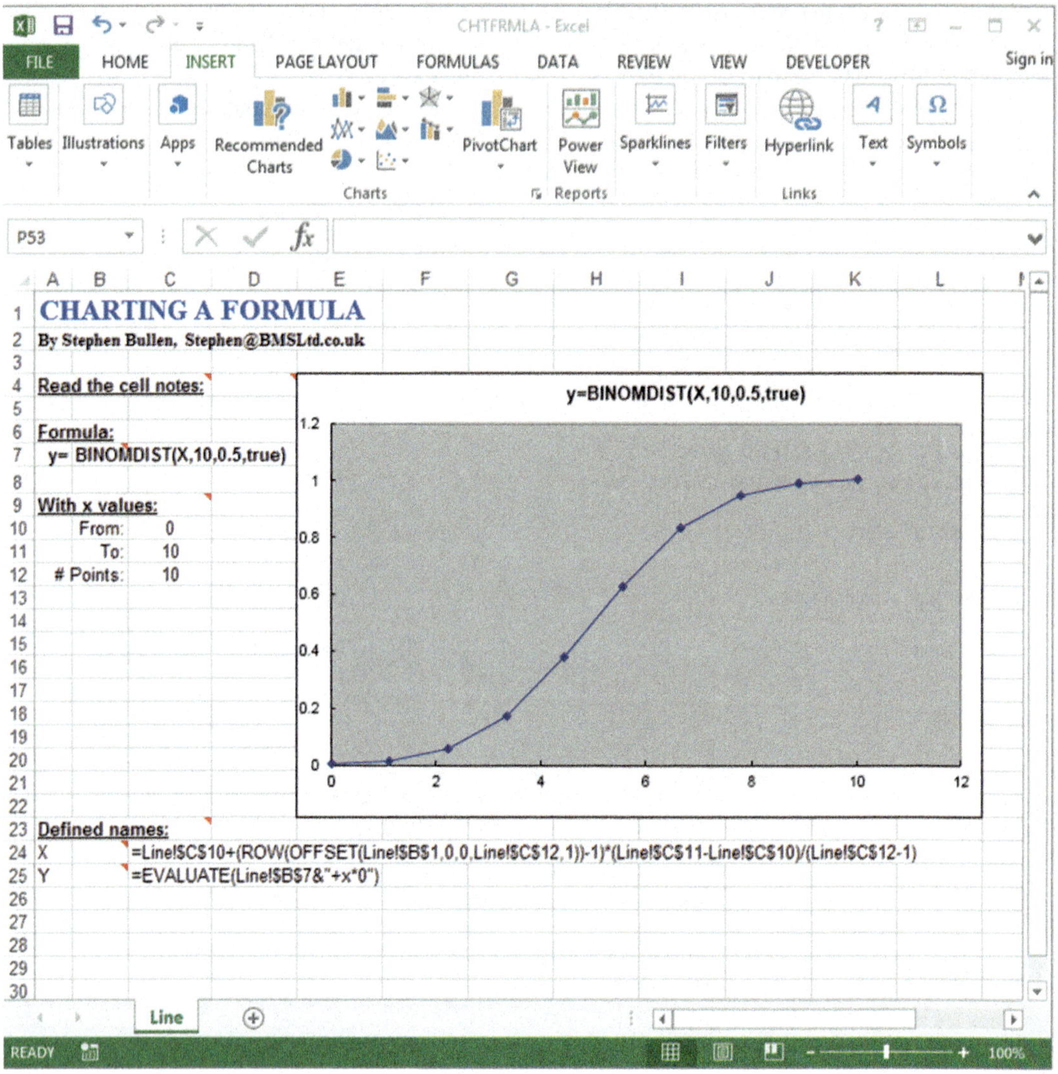
CHTFRMLA - Excel
FILE HOME INSERT PAGE LAYOUT FORMULAS DATA REVIEW VIEW DEVELOPER
Sign in
Tables Illustrations Apps Recommended Charts PivotChart Power View Sparklines Filters Hyperlink Text Symbols
Charts Reports Links
P53
CHARTING A FORMULA
By Stephen Bullen, Stephen@BMSLtd.co.uk
Read the cell notes:
Formula:
y= BINOMDIST(X,10,0.5,true)
With x values:
From: 0
To: 10
# Points: 10
y=BINOMDIST(X,10,0.5,true)
Defined names:
X =Line!$C$10+(ROW(OFFSET(Line!$B$1,0,0,Line!$C$12,1))-1)*(Line!$C$11-Line!$C$10)/(Line!$C$12-1)
Y =EVALUATE(Line!$B$7&"+x*0")
Line
READY
100%

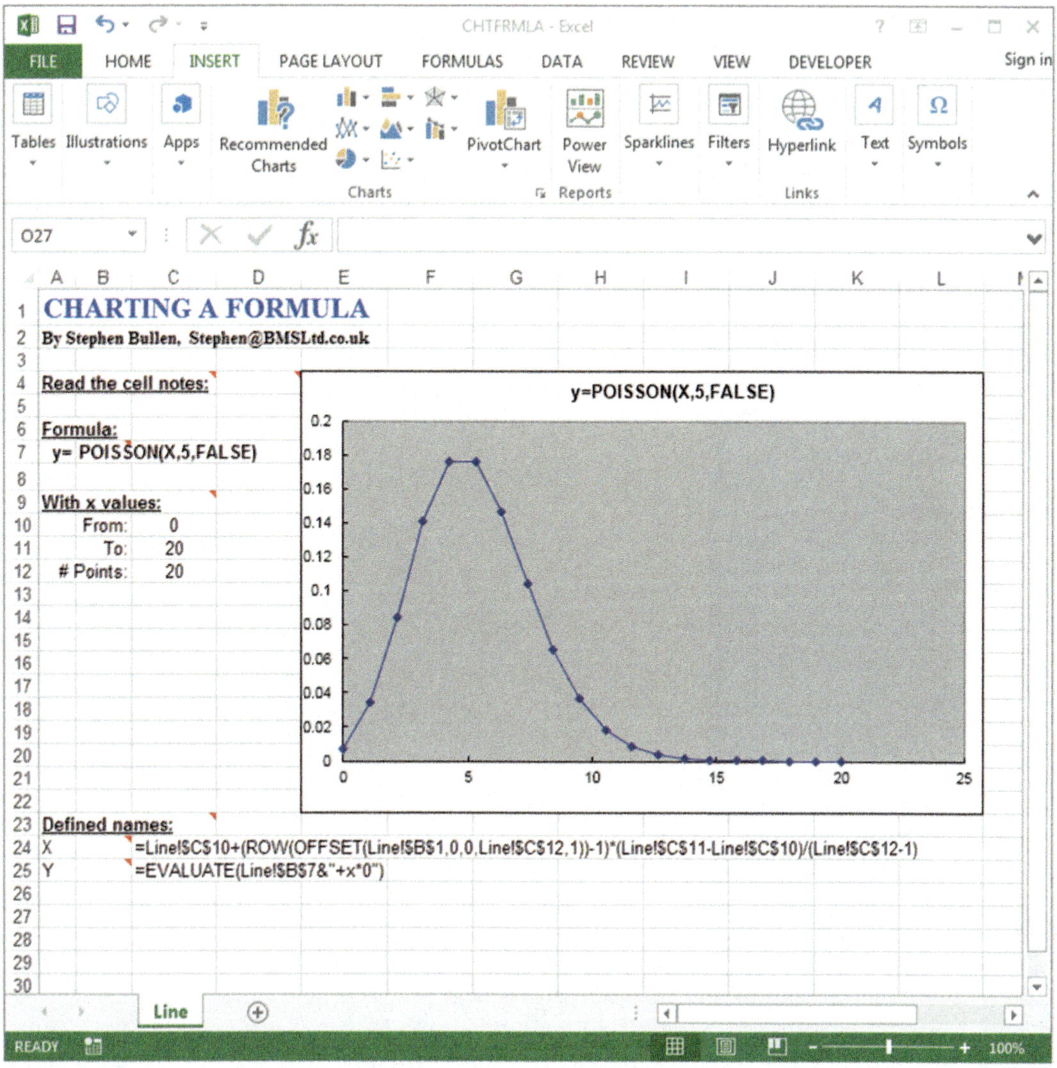
CHTFRMLA - Excel
FILE HOME INSERT PAGE LAYOUT FORMULAS DATA REVIEW VIEW DEVELOPER
Sign in
Tables Illustrations Apps Recommended Charts PivotChart Power View Sparklines Filters Hyperlink Text Symbols
Charts Reports Links
O27
CHARTING A FORMULA
By Stephen Bullen, Stephen@BMSLtd.co.uk
Read the cell notes:
Formula:
y= POISSON(X,5,FALSE)
With x values:
From: 0
To: 20
# Points: 20
y=POISSON(X,5,FALSE)
Defined names:
X =Line!$C$10+(ROW(OFFSET(Line!$B$1,0,0,Line!$C$12,1))-1)*(Line!$C$11-Line!$C$10)/(Line!$C$12-1)
Y =EVALUATE(Line!$B$7&"+x*0")
Line
READY
100%

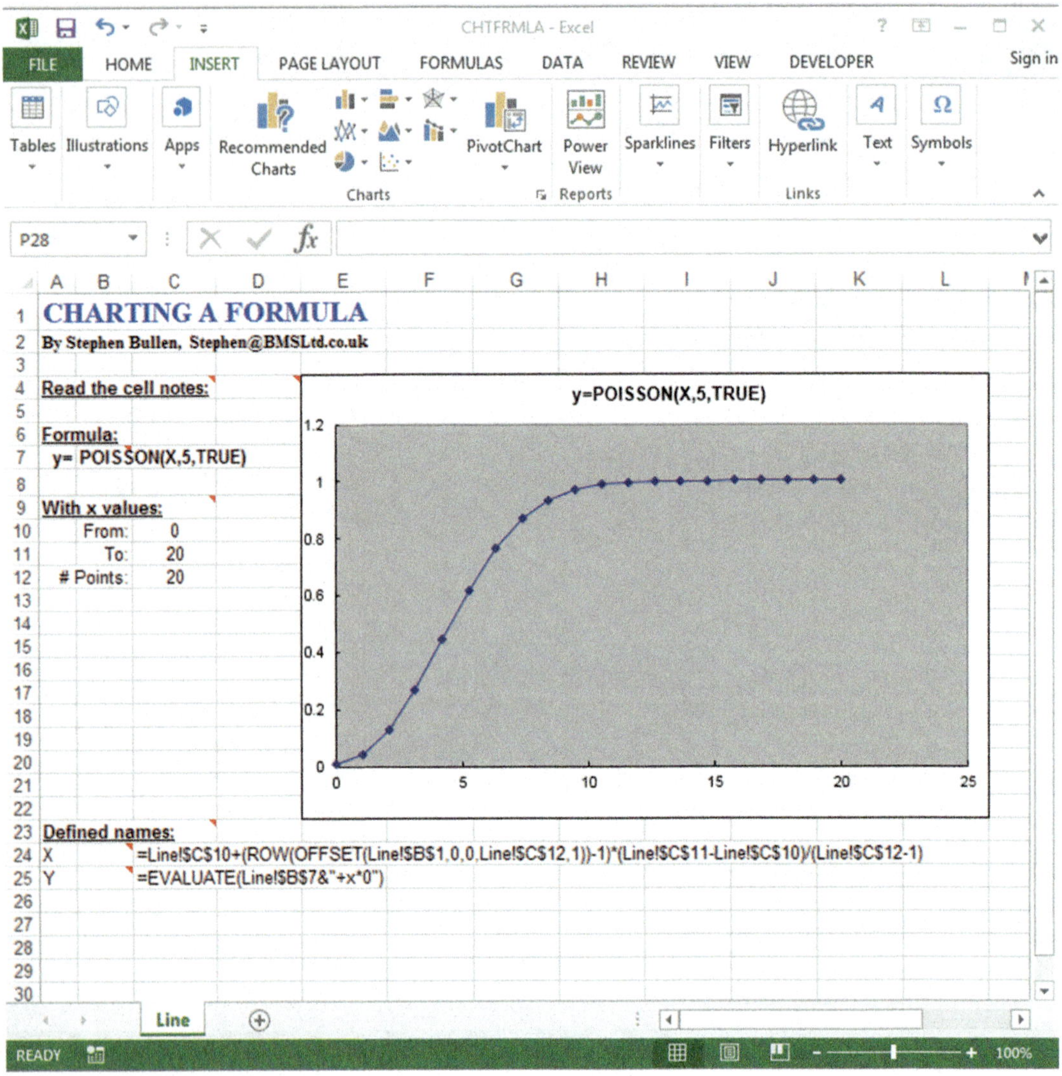

6.11 SAS Programming Code Instructions and Examples

6.11.1 Introduction and Probability Distribution

Since we have equal chance to get each number in a single roll, we use the following loop to generate the data.

```
data Die;
do num=1 to 6 by 1;
probability=1/6;
output;
end;
run;
```

Let's list out the probability of getting each specific data item as shown below.

Obs	Num	Probability
1	1	0.16667
2	2	0.16667
3	3	0.16667
4	4	0.16667
5	5	0.16667
6	6	0.16667

The graph of the probability distribution is shown below.

```
Goptions ftext='Times New Roman' htext=6 gunit=pct;
title height=7 'Binomial Probability Density Function';
axis1  label=('Probability') order=(0 to 0.18 by 0.03);
axis2  label=('Number');
proc gchart data=die;
vbar  num/ RAXIS=axis1 MAXIS=axis2 discrete sumvar=probability WIDTH=8;
run;
```

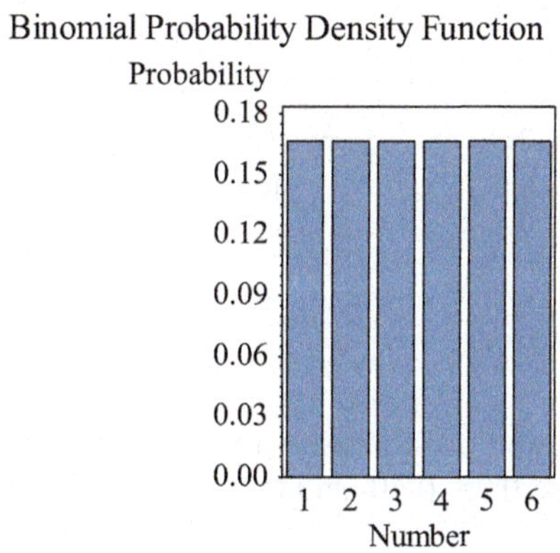

6.11.2 Cumulative Probability Distribution

The following SAS commands create and print the cumulative density function.

```
data Die;
set Die;
retain CDF;
CDF=sum(CDF, probability);
run;
```

Obs	Num	Probability	CDF
1	1	0.16667	0.16667
2	2	0.16667	0.33333
3	3	0.16667	0.50000
4	4	0.16667	0.66667
5	5	0.16667	0.83333
6	6	0.16667	1.00000

```
Goptions ftext='Times New Roman' htext=4 gunit=pct;
title  height=7 'Die Cumulative Distribution';
axis1  label=('Number');
axis2  label=(angle=90 'CDF')
       order=(0 to 1 by 0.1);
symbol v=none i=join c=black;
Proc gplot data=die;
plot  CDF*num/haxis=axis1 vaxis=axis2;
Run;
```

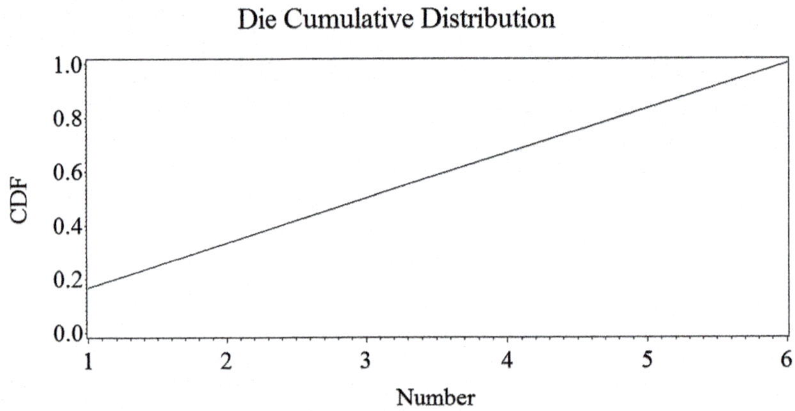

6.11.3 Binomial Distribution

In this section, we will use function **pdf** to create a binomial distribution table for the ten coin flips with three different probabilities: 0.25, 0.50, and 0.75.

```
data binom;
do success = 0 to 10 by 1;
percent25=pdf('binomial', success, 0.25, 10);
percent50=pdf('binomial', success, 0.50, 10);
percent75=pdf('binomial', success, 0.75, 10);
output binom;
end;
run;
```

Obs	Success	Percent25	Percent50	Percent75
1	0	0.05631	0.00098	0.00000
2	1	0.18771	0.00977	0.00003
3	2	0.28157	0.04395	0.00039
4	3	0.25028	0.11719	0.00309
5	4	0.14600	0.20508	0.01622
6	5	0.05840	0.24609	0.05840
7	6	0.01622	0.20508	0.14600

(continued)

Obs	Success	Percent25	Percent50	Percent75
8	7	0.00309	0.11719	0.25028
9	8	0.00039	0.04395	0.28157
10	9	0.00003	0.00977	0.18771
11	10	0.00000	0.00098	0.05631

Let's graph the binomial distribution for ten coin flips by using **PROC GPLOT**. We specify option **OVERLAY** to superimpose the second and third plots onto the first.

```
Goptions ftext='Times New Roman' htext=6 gunit=pct;
title  height=7 'Binomial Probability Density Function';
legend1 label=none value=('0.25' '0.50' '0.75') shape=symbol(5,1);
axis1  label=('Success');
axis2  label=(angle=90 'Probability')
       order=(-0.05 to 0.30 by 0.05);
symbol1 i=join  v=dot L=1 c=black w=2;
symbol2 i=join  v=dot L=5 c=black w=2;
symbol3 i=join  v=dot L=3 c=black w=2;
proc gplot data=binom;
plot (percent25 percent50 percent75)*success/haxis=axis1 vaxis=axis2 overlay
legend=legend1;
run;
```

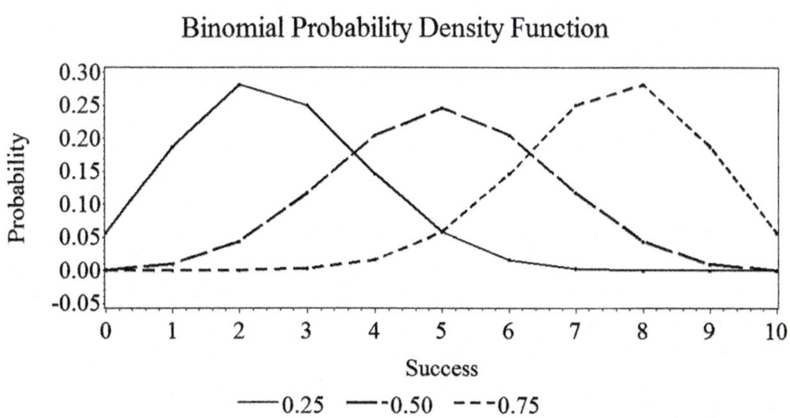

Let's now create cumulative binomial distributions of ten coin flips.

```
data binom;
set binom;
retain CDF25 CDF50 CDF75;
CDF25=sum(CDF25, percent25);
CDF50=sum(CDF50, percent50);
CDF75=sum(CDF75, percent75);
run;
```

Obs	Success	Percent25	Percent50	Percent75	CDF25	CDF50	CDF75
1	0	0.05631	0.00098	0.00000	0.05631	0.00098	0.00000
2	1	0.18771	0.00977	0.00003	0.24403	0.01074	0.00003
3	2	0.28157	0.04395	0.00039	0.52559	0.05469	0.00042
4	3	0.25028	0.11719	0.00309	0.77588	0.17188	0.00351
5	4	0.14600	0.20508	0.01622	0.92187	0.37695	0.01973
6	5	0.05840	0.24609	0.05840	0.98027	0.62305	0.07813
7	6	0.01622	0.20508	0.14600	0.99649	0.82813	0.22412
8	7	0.00309	0.11719	0.25028	0.99958	0.94531	0.47441
9	8	0.00039	0.04395	0.28157	0.99997	0.98926	0.75597
10	9	0.00003	0.00977	0.18771	1.00000	0.99902	0.94369
11	10	0.00000	0.00098	0.05631	1.00000	1.00000	1.00000

```
Goptions ftext='Times New Roman' htext=6 gunit=pct;
title  height=7 'Binomial Cumulative Density Function';
legend1 label=none value=('0.25' '0.50' '0.75') shape=symbol(6,1);
axis1  label=('Success');
axis2  label=(angle=90 'Probability')
       order=(0 to 1 by 0.2);
symbol1 i=join  v=dot L=1 c=black w=2;
symbol2 i=join  v=dot L=5 c=black w=2;
symbol3 i=join  v=dot L=3 c=black w=2;
proc gplot data=binom;
plot (CDF25 CDF50 CDF75)*success/haxis=axis1 vaxis=axis2 overlay
legend=legend1;
run;
```

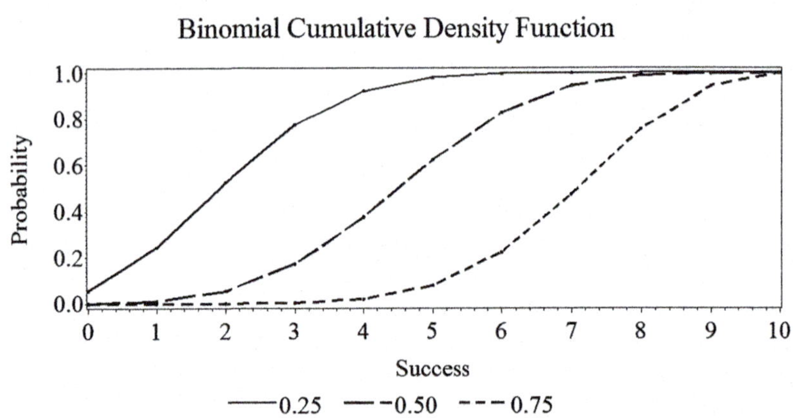

6.11.4 Hypergeometric Distribution

The hypergeometric probability density function is defined as follows:

$$P[(x \text{ successes and } (n-x) \text{ failures})] = \frac{\binom{h}{n}\binom{N-h}{n-x}}{\binom{N}{n}}$$

Consider a group of ten students in which four are Democrats and six are Republicans. A sample of size six has been selected. What is the probability that there will be only one Democrat in this sample? We use function **PDF** to generate the probability distribution as mentioned in the previous section.

```
data hyper;
do success = 0 to 4 by 1;
percent=pdf('hyper', success, 10, 4, 6);
output hyper;
end;
run;
```

Obs	Success	Percent
1	0	0.00476
2	1	0.11429
3	2	0.42857
4	3	0.38095
5	4	0.07143

6.11.5 Poisson Random Variable

```
data poisson;
do success = 0 to 20 by 1;
percent4=pdf('POISSON', success, 4);
percent6=pdf('POISSON', success, 6);
percent8=pdf('POISSON', success, 8);
percent10=pdf('POISSON', success, 10);
output poisson;
end;
run;
```

The table below shows a Poisson PDF table for the four Poisson distributions.

Obs	Success	Percent4	Percent6	Percent8	Percent10
1	0	0.01832	0.00248	0.00034	0.00005
2	1	0.07326	0.01487	0.00268	0.00045
3	2	0.14653	0.04462	0.01073	0.00227
4	3	0.19537	0.08924	0.02863	0.00757
5	4	0.19537	0.13385	0.05725	0.01892

(continued)

Obs	Success	Percent4	Percent6	Percent8	Percent10
6	5	0.15629	0.16062	0.09160	0.03783
7	6	0.10420	0.16062	0.12214	0.06306
8	7	0.05954	0.13768	0.13959	0.09008
9	8	0.02977	0.10326	0.13959	0.11260
10	9	0.01323	0.06884	0.12408	0.12511
11	10	0.00529	0.04130	0.09926	0.12511
12	11	0.00192	0.02253	0.07219	0.11374
13	12	0.00064	0.01126	0.04813	0.09478
14	13	0.00020	0.00520	0.02962	0.07291
15	14	0.00006	0.00223	0.01692	0.05208
16	15	0.00002	0.00089	0.00903	0.03472
17	16	0.00000	0.00033	0.00451	0.02170
18	17	0.00000	0.00012	0.00212	0.01276
19	18	0.00000	0.00004	0.00094	0.00709
20	19	0.00000	0.00001	0.00040	0.00373
21	20	0.00000	0.00000	0.00016	0.00187

```
Goptions ftext='Times New Roman' htext=4 gunit=pct;
title  height=8 'Poisson Probability Distribution Function';
legend1 label=none value=('4' '6' '8' '10') shape=symbol(5,1);
axis1  label=('success');
axis2  label=(angle=90 'Probability')
       order=(0 to 0.25 by 0.05);
symbol1 i=join  v=dot L=1 c=black;
symbol2 i=join  v=dot L=2 c=black;
symbol3 i=join  v=dot L=3 c=black;
symbol4 i=join  v=dot L=4 c=black;
proc gplot data=poisson;
plot (percent4 percent6 percent8 percent10)*success/haxis=axis1
vaxis=axis2 overlay legend=legend1;
run;
```

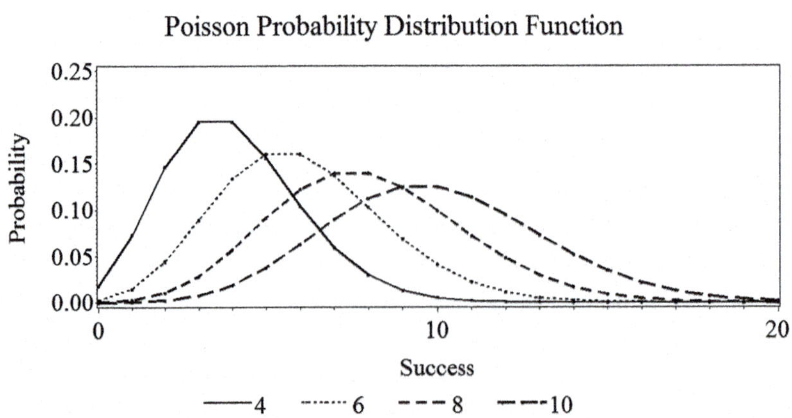

The code below generates the four CDF Poisson distributions.

```
data poisson;
set poisson;
retain CDF4 CDF6 CDF8 CDF10;
CDF4=sum(CDF4, percent4);
CDF6=sum(CDF6, percent6);
CDF8=sum(CDF8, percent8);
CDF10=sum(CDF10, percent10);
run;
```

Obs	Success	Percent4	Percent6	Percent8	Percent10	CDF4	CDF6	CDF8	CDF10
1	0	0.01832	0.00248	0.00034	0.00005	0.01832	0.00248	0.00034	0.00005
2	1	0.07326	0.01487	0.00268	0.00045	0.09158	0.01735	0.00302	0.00050
3	2	0.14653	0.04462	0.01073	0.00227	0.23810	0.06197	0.01375	0.00277
4	3	0.19537	0.08924	0.02863	0.00757	0.43347	0.15120	0.04238	0.01034
5	4	0.19537	0.13385	0.05725	0.01892	0.62884	0.28506	0.09963	0.02925
6	5	0.15629	0.16062	0.09160	0.03783	0.78513	0.44568	0.19124	0.06709
7	6	0.10420	0.16062	0.12214	0.06306	0.88933	0.60630	0.31337	0.13014
8	7	0.05954	0.13768	0.13959	0.09008	0.94887	0.74398	0.45296	0.22022
9	8	0.02977	0.10326	0.13959	0.11260	0.97864	0.84724	0.59255	0.33282
10	9	0.01323	0.06884	0.12408	0.12511	0.99187	0.91608	0.71662	0.45793
11	10	0.00529	0.04130	0.09926	0.12511	0.99716	0.95738	0.81589	0.58304
12	11	0.00192	0.02253	0.07219	0.11374	0.99908	0.97991	0.88808	0.69678
13	12	0.00064	0.01126	0.04813	0.09478	0.99973	0.99117	0.93620	0.79156
14	13	0.00020	0.00520	0.02962	0.07291	0.99992	0.99637	0.96582	0.86446
15	14	0.00006	0.00223	0.01692	0.05208	0.99998	0.99860	0.98274	0.91654
16	15	0.00002	0.00089	0.00903	0.03472	1.00000	0.99949	0.99177	0.95126
17	16	0.00000	0.00033	0.00451	0.02170	1.00000	0.99983	0.99628	0.97296
18	17	0.00000	0.00012	0.00212	0.01276	1.00000	0.99994	0.99841	0.98572
19	18	0.00000	0.00004	0.00094	0.00709	1.00000	0.99998	0.99935	0.99281
20	19	0.00000	0.00001	0.00040	0.00373	1.00000	0.99999	0.99975	0.99655
21	20	0.00000	0.00000	0.00016	0.00187	1.00000	1.00000	0.99991	0.99841

```
Goptions ftext='Times New Roman' htext=6 gunit=pct;
title  height=7 'Poisson Cumulative Density Function';
legend1 label=none value=('4' '6' '8' '10') shape=symbol(6,1);
axis1  label=('Success');
axis2  label=(angle=90 'Probability')
       order=(0 to 1 by 0.2);
symbol1 i=join  v=dot L=1 c=black w=2;
symbol2 i=join  v=dot L=2 c=black w=2;
symbol3 i=join  v=dot L=3 c=black w=2;
symbol4 i=join  v=dot L=4 c=black w=2;
proc gplot data=poisson;
plot (CDF4 CDF6 CDF8 CDF10)*success/haxis=axis1 vaxis=axis2 overlay
legend=legend1;
run;
```

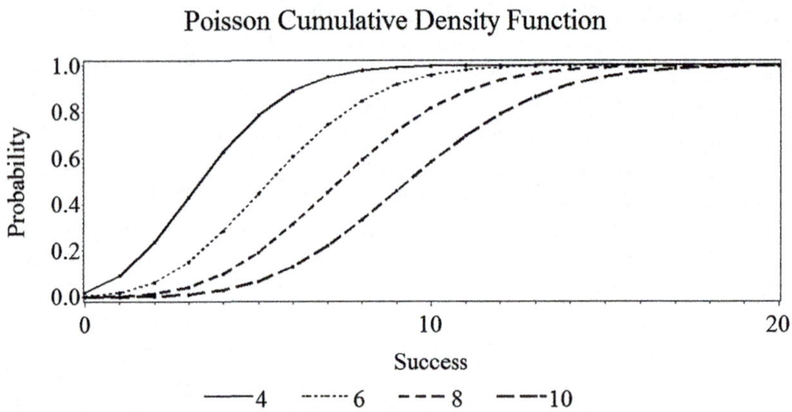

6.11.6 Charting Method in SAS

```
data func;
do x = -2 to 1 by 0.06;
y=exp(x)*sin(x*x);
output func;
end;
run;

Goptions ftext='Times New Roman' htext=6 gunit=pct;
title  height=7 'y=exp(x)*sin(x^2)';
axis1  label=('x')
   order=(-2 to 1 by 0.5);
axis2  label=('y');
symbol1 i=join  v=dot L=1 c=black w=2 ;
proc gplot data=func;
plot y*x/haxis=axis1 vaxis=axis2 vref=0 wvref=3 href=0 whref=3;
run;
```

We use option **vref=** and **href=** to specify position of reference lines perpendicular to vertical and horizontal axis respectively.

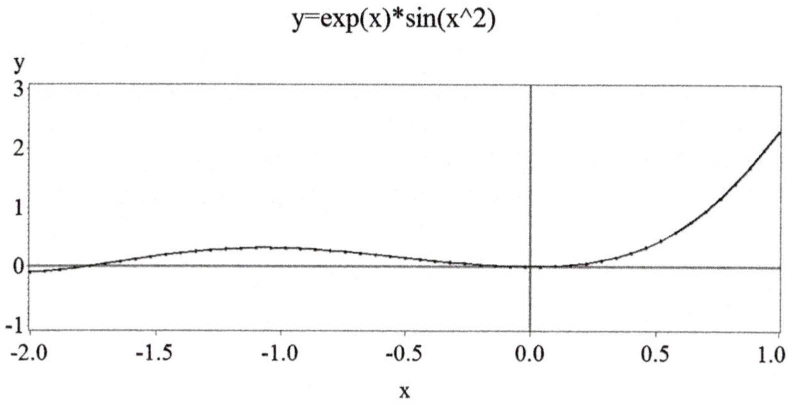


```
data die;
set die;
prob1=lag(CDF);
if prob1=. then prob1=0;
prob2=CDF-prob1;
run;

Goptions ftext='Times New Roman' htext=4 gunit=pct;
title  height=8 'Die Probability Distribution';
axis1  label=('Number');
axis2  label=(angle=90 'Probability')
       order=(0.1660 to 0.1670 by 0.0002);
symbol v=none i=join c=black;
Proc gplot data=die;
plot  prob2*num/haxis=axis1 vaxis=axis2;
Run;
```

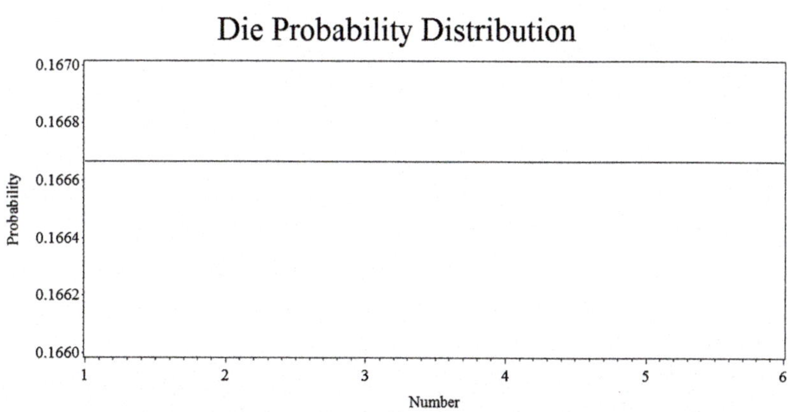

6.12 Statistical Summary

In this chapter we saw that a *discrete random variable* is a variable that has a countable number of outcomes. We also saw that a *probability distribution* is a list of all the probabilities for the outcomes of a variable. We study the binomial, hypergeometric, and Poisson distributions. We saw that as the probability of success increases for a binomial distribution, the more the binomials' distribution will shift to the right. For the Poisson distribution, we saw that as the number of occurrences increases, the more the Poisson shifts to the right. The Poisson also resembles more and more the normal distribution as the number of occurrences increases. The normal distribution is a very important distribution, and we will talk about it a lot more in the next couple of chapters.

Appendix 6.1: Excel Code—Binomial Distribution

```
'/*************************************************************************
'/
'/ Essentials of Microsoft Excel 2013, SAS and
'/              MINITAB 17
'/ for Statistical and Financial Analysis
'/
'/*************************************************************************
Option Explicit
Dim mwbReport As Workbook
Dim mrPDFTable As Range
Dim mrCDFTable As Range
Property Set ReportWb(w As Workbook)
    Set mwbReport = w
End Property
Property Get ReportWb() As Workbook
    Set ReportWb = mwbReport
End Property
Property Set BinomialPDFTable(r As Range)
    Set mrPDFTable = r
End Property
Property Get BinomialPDFTable() As Range
    Set BinomialPDFTable = mrPDFTable
End Property
Property Set BinomialCDFTable(r As Range)
    Set mrCDFTable = r
End Property
Property Get BinomialCDFTable() As Range
    Set BinomialCDFTable = mrCDFTable
End Property
Private Sub cmdCancel_Click()
    Unload Me
End Sub
Sub Binomial()
    Me.Hide
    'Me.txtNumPoints = Me.txtNumPoints + 1
    GeneratePDFTable
    GenerateCDFTable
    CreatePDFBinomialChart
    CreateCDFBinomialChart
End Sub
```

```
'/**************************************************************
'/Purpose: To create the Binomial CDF Distribution from the
'/         the binomial CDF table
'/*************************************************************
Sub CreateCDFBinomialChart()
    Dim chrt As Chart
    Dim obj As Object

    Set chrt = Me.ReportWb.Charts.Add

    With chrt
        .ChartType = xlXYScatterSmooth
        .SetSourceData Source:=Me.BinomialCDFTable, PlotBy:=xlColumns
        .Location Where:=xlLocationAsNewSheet
        .HasTitle = True
        .ChartTitle.Characters.Text = " Binomial Cumulative Density Function"
        .ChartTitle.Characters.Font.Size = 18
        .Axes(xlCategory, xlPrimary).HasTitle = True
        .Axes(xlCategory, xlPrimary).AxisTitle.Characters.Text = "Success"
        .Axes(xlValue, xlPrimary).HasTitle = True
        .Axes(xlValue, xlPrimary).AxisTitle.Characters.Text = "Probability"
        .Axes(xlValue).MajorGridlines.Delete
        .HasLegend = True
        .PlotArea.ClearFormats
        .Name = "Binomial CDF"
    .PlotArea.Interior.ColorIndex = xlNone
    End With

    'make every series color black.  This is done
    'because some color do not print well on a
    'laser printer
    For Each obj In chrt.SeriesCollection
        With obj
            .Border.ColorIndex = 1
            .MarkerBackgroundColorIndex = 1
            .MarkerForegroundColorIndex = 1
        End With
    Next
End Sub
'/**************************************************************
'/Purpose: To create the Binomial Distribution from the
'/         the binomial table
'/**************************************************************
Sub CreatePDFBinomialChart()
    Dim chrt As Chart
    Dim obj As Object
    Set chrt = Me.ReportWb.Charts.Add
    With chrt
        .ChartType = xlXYScatterSmooth
        .SetSourceData Source:=Me.BinomialPDFTable, PlotBy:=xlColumns
        .Location Where:=xlLocationAsNewSheet
        .HasTitle = True
        .ChartTitle.Characters.Text = " Binomial Probability Density Function"
        .ChartTitle.Characters.Font.Size = 18
        .Axes(xlCategory, xlPrimary).HasTitle = True
        .Axes(xlCategory, xlPrimary).AxisTitle.Characters.Text = "Success"
        .Axes(xlValue, xlPrimary).HasTitle = True
        .Axes(xlValue, xlPrimary).AxisTitle.Characters.Text = "Probability"
        .Axes(xlValue).MajorGridlines.Delete
        .HasLegend = True
        .PlotArea.ClearFormats
        .Name = "Binomial PDF"
    .PlotArea.Interior.ColorIndex = xlNone
    End With

    'make every series color black.  This is done
    'because some color do not print well on a
    'laser printer
    For Each obj In chrt.SeriesCollection
        With obj
            .Border.ColorIndex = 1
            .MarkerBackgroundColorIndex = 1
            .MarkerForegroundColorIndex = 1
        End With
    Next
End Sub
```

```
'/*****************************************************************
'/Purpose:  Generate Binomial table based on user input on the form
'/          This form will be used to generate the binomial
'/          distribution chart.
'/CallBy:   Binomial
'/******************************************************************
Sub GeneratePDFTable()
    Dim wsTable As Worksheet
    Dim rTable As Range
    Dim rCell As Range

    Set Me.ReportWb = Workbooks.Add
    Set wsTable = Me.ReportWb.Worksheets.Add
    Set rTable = wsTable.Cells(2, 1)
    For Each rCell In rTable.Resize(Me.txtNumPoints + 1, 1).Cells
        rCell.Value = rCell.Row - 2
    Next
    If Len(Trim(Me.txtGraph1)) <> 0 Then
        wsTable.Cells(1, 2) = Me.txtGraph1
        rTable.Resize(Me.txtNumPoints + 1, 1).Offset(0, 1) = _
        "=BINOMDIST(RC1," & Me.txtNumPoints & ",R1C,FALSE)"

    End If
    If Len(Trim(Me.txtGraph2)) <> 0 Then
        wsTable.Cells(1, [counta(wstable.Rows(1))] + 2) = Me.txtGraph2
        With rTable.Resize(Me.txtNumPoints + 1, 1)
            .Offset(0, 1).Copy Destination:=.Offset(0, [counta(wstable.Rows(1))+1])
        End With
    End If

    If Len(Trim(Me.txtGraph3)) <> 0 Then
        wsTable.Cells(1, WorksheetFunction.CountA(wsTable.Rows(1)) + 2) = Me.txtGraph3
        With rTable.Resize(Me.txtNumPoints + 1, 1)
            .Offset(0, 1).Copy Destination:=.Offset(0, WorksheetFunction.CountA(wsTable.Rows(1)))
        End With

    End If
    If Len(Trim(Me.txtGraph4)) <> 0 Then
        wsTable.Cells(1, WorksheetFunction.CountA(wsTable.Rows(1)) + 2) = Me.txtGraph4
        With rTable.Resize(Me.txtNumPoints + 1, 1)
            .Offset(0, 1).Copy Destination:=.Offset(0, WorksheetFunction.CountA(wsTable.Rows(1)))
        End With
    End If
    Set Me.BinomialPDFTable = wsTable.Cells(1).CurrentRegion
    wsTable.Name = "Binomial PDF Table"
End Sub
'/******************************************************************
'/Purpose: To generate a Binomial CDF table.  This is done by
'/         copying the Binomial PDF table to a new sheet and then
'/         replacing the false value in the binomial function to true
'/****************************************************************
Sub GenerateCDFTable()
    Dim wsCDF As Worksheet

    Set wsCDF = Me.ReportWb.Worksheets.Add
    Me.BinomialPDFTable.Copy Destination:=wsCDF.Cells(1)
    wsCDF.Cells.Replace What:="false", Replacement:="true", LookAt:=xlPart, _
        SearchOrder:=xlByRows, MatchCase:=False
    Set Me.BinomialCDFTable = wsCDF.Cells(1).CurrentRegion
    wsCDF.Name = "Binomial CDF Table"
End Sub

Private Sub cmdRun_Click()

    Binomial
    Unload Me
End Sub

'/*******************************************************************
'/Purpose:  This procedure runs everytime the form is loaded.  This
'/          procedure sets some initial values.
'/*******************************************************************
Private Sub UserForm_Initialize()
    Me.txtNumPoints = 20
    Me.txtGraph1 = 0.5
End Sub
```

Appendix 6.2: Excel Code—Poisson Distribution

```
'/***********************************************************************
'/ Essentials of Microsoft Excel 2013, SAS and
'/                MINITAB 17
'/ for Statistical and Financial Analysis
'/
'/***********************************************************************
Option Explicit
Dim mwbReport As Workbook
Dim mrPDFTable As Range
Dim mrCDFTable As Range

Property Set ReportWb(w As Workbook)
    Set mwbReport = w
End Property
Property Get ReportWb() As Workbook
    Set ReportWb = mwbReport
End Property
Property Set PDFTable(r As Range)
    Set mrPDFTable = r
End Property
Property Get PDFTable() As Range
    Set PDFTable = mrPDFTable
End Property
Property Set CDFTable(r As Range)
    Set mrCDFTable = r
End Property
Property Get CDFTable() As Range
    Set CDFTable = mrCDFTable
End Property
Private Sub cmdCancel_Click()
    Unload Me
End Sub
Sub Poisson()
    Me.Hide
    GeneratePDFTable
    GenerateCDFTable
    CreatePDFChart
    CreateCDFChart
End Sub


'/************************************************************
'/Purpose: To create the Poisson CDF Distribution from the
'/         the Poisson CDF table
'/************************************************************
Sub CreateCDFChart()
    Dim chrt As Chart
    Dim obj As Object

    Set chrt = Me.ReportWb.Charts.Add

    With chrt
        .ChartType = xlXYScatterSmooth
        .SetSourceData Source:=Me.CDFTable, PlotBy:=xlColumns
        .Location Where:=xlLocationAsNewSheet
        .HasTitle = True
        .ChartTitle.Characters.Text = " Poisson Cumulative Density Function"
        .ChartTitle.Characters.Font.Size = 18
        .Axes(xlCategory, xlPrimary).HasTitle = True
        .Axes(xlCategory, xlPrimary).AxisTitle.Characters.Text = "Occurrences"
        .Axes(xlValue, xlPrimary).HasTitle = True
        .Axes(xlValue, xlPrimary).AxisTitle.Characters.Text = "Probability"
        '.Axes(xlValue).MaximumScale = 1
        '.Axes(xlValue).MinimumScale = 0
        .Axes(xlValue).MajorGridlines.Delete
        .HasLegend = True
        .PlotArea.ClearFormats
        .Name = "Poisson CDF"
    .PlotArea.Interior.ColorIndex = xlNone
    End With
```

```
    'make every series color black.  This is done
    'because some color do not print well on a
    'laser printer
    For Each obj In chrt.SeriesCollection
        With obj
            .Border.ColorIndex = 1
            .MarkerBackgroundColorIndex = 1
            .MarkerForegroundColorIndex = 1
        End With
    Next
End Sub

'/***************************************************************
'/Purpose: To create the Poisson Distribution from the
'/         the binomial table
'/**************************************************************
Sub CreatePDFChart()
    Dim chrt As Chart
    Dim obj As Object
    Set chrt = Me.ReportWb.Charts.Add

    With chrt
        .ChartType = xlXYScatterSmooth
        .SetSourceData Source:=Me.PDFTable, PlotBy:=xlColumns
        .Location Where:=xlLocationAsNewSheet
        .HasTitle = True
        .ChartTitle.Characters.Text = " Poisson Probability Density Function"
        .ChartTitle.Characters.Font.Size = 18
        .Axes(xlCategory, xlPrimary).HasTitle = True
        .Axes(xlCategory, xlPrimary).AxisTitle.Characters.Text = "Occurrence"
        .Axes(xlValue, xlPrimary).HasTitle = True
        .Axes(xlValue, xlPrimary).AxisTitle.Characters.Text = "Probability"
        '.Axes(xlValue).MaximumScale = 1
        '.Axes(xlValue).MinimumScale = 0
        .Axes(xlValue).MajorGridlines.Delete
        .HasLegend = True
        .PlotArea.ClearFormats
        .Name = "Poisson PDF"
    .PlotArea.Interior.ColorIndex = xlNone
    End With

    'make every series color black.  This is done
    'because some color do not print well on a
    'laser printer
    For Each obj In chrt.SeriesCollection
        With obj
            .Border.ColorIndex = 1
            .MarkerBackgroundColorIndex = 1
            .MarkerForegroundColorIndex = 1
        End With
    Next
End Sub


'/******************************************************************
'/Purpose:  Generate Poisson table based on user input on the form
'/          This form will be used to generate the Poisson
'/          distribution chart.
'/CallBy:   Poisson
'/*******************************************************************
Sub GeneratePDFTable()
    Dim wsTable As Worksheet
    Dim rTable As Range
    Dim rCell As Range

    Set Me.ReportWb = Workbooks.Add
    Set wsTable = Me.ReportWb.Worksheets.Add

    Set rTable = wsTable.Cells(2, 1)

    For Each rCell In rTable.Resize(Me.txtNumPoints + 1, 1).Cells
        rCell.Value = rCell.Row - 2
    Next
```

```
If Len(Trim(Me.txtGraph1)) <> 0 Then
    wsTable.Cells(1, 2) = Me.txtGraph1
    rTable.Resize(Me.txtNumPoints + 1, 1).Offset(0, 1).Formula = _
    "=POISSON(RC1,R1C,FALSE)"

End If

If Len(Trim(Me.txtGraph2)) <> 0 Then
    wsTable.Cells(1, [counta(wstable.Rows(1))] + 2) = Me.txtGraph2
    With rTable.Resize(Me.txtNumPoints + 1, 1)
        .Offset(0, 1).Copy Destination:=.Offset(0, [counta(wstable.Rows(1))+1])
    End With
End If
If Len(Trim(Me.txtGraph3)) <> 0 Then
    wsTable.Cells(1, WorksheetFunction.CountA(wsTable.Rows(1)) + 2) = Me.txtGraph3
    With rTable.Resize(Me.txtNumPoints + 1, 1)
        .Offset(0, 1).Copy Destination:=.Offset(0, WorksheetFunction.CountA(wsTable.Rows(1)))
    End With
End If
If Len(Trim(Me.txtGraph4)) <> 0 Then
    wsTable.Cells(1, WorksheetFunction.CountA(wsTable.Rows(1)) + 2) = Me.txtGraph4
    With rTable.Resize(Me.txtNumPoints + 1, 1)
            .Offset(0, 1).Copy Destination:=.Offset(0, WorksheetFunction.CountA(wsTable.Rows(1)))
        End With
    End If
    Set Me.PDFTable = wsTable.Cells(1).CurrentRegion
    wsTable.Name = "Poisson PDF Table"
End Sub


'/**********************************************************************
'/Purpose: To generate a Binomial CDF table.  This is done by
'/         copying the Binomial PDF table to a new sheet and then
'/         replacing the false value in the binomial function to true
'/*******************************************************************
Sub GenerateCDFTable()
    Dim wsCDF As Worksheet
    Set wsCDF = Me.ReportWb.Worksheets.Add
    Me.PDFTable.Copy Destination:=wsCDF.Cells(1)
    wsCDF.Cells.Replace What:="false", Replacement:="true", LookAt:=xlPart, _
        SearchOrder:=xlByRows, MatchCase:=False
    Set Me.CDFTable = wsCDF.Cells(1).CurrentRegion
    wsCDF.Name = "Poisson CDF Table"
End Sub
Private Sub cmdRun_Click()
    Poisson
    Unload Me
End Sub
'/**********************************************************************
'/Purpose:  This procedure runs everytime the form is loaded.  This
'/          procedure sets some initial values.
'/**********************************************************************
Private Sub UserForm_Initialize()
    Me.txtNumPoints = 20
    Me.txtGraph1 = 5
End Sub
```

Bibliography

Lee CF, Lee JC, Lee AC (2013) Statistics for business and financial economics. Springer, New York

Microsoft Inc., Excel 2013. Microsoft Inc., Redmond

Minitab Inc. Minitab 17. Minitab Inc., State College

SAS Institute Inc. SAS 2014. SAS Institute Inc., Cary

Chapter 7
The Normal and Lognormal Distributions

7.1 Introduction

In the last chapter, we examined the probability distribution of discrete random variables. In this chapter, we will look at the probability distribution of *continuous random variables*.

Unlike the values of discrete random variables, which are limited to a finite or countable number of distinct (integer) values, values of continuous variables are not limited to being integers; theoretically, they are infinitely divisible. A continuous random variable may take on any value within an interval.

7.2 Uniform Distribution

One interesting characteristic of continuous random variables is that the entire range of probabilities under a continuous random variable has a total area equal to 1. This is most clearly shown in the uniform continuous distribution. The formula for the uniform distribution is

$$f(x) = \frac{1}{d - c}, c < x < d$$

The c is the beginning point and the d is the ending point of the uniform distribution. This is a uniform distribution because any two points between the points c and d will have the same probability.

7.3 Uniform Distribution in Minitab

Below creates the uniform distribution graph using Minitab.

C.-F. Lee et al., *Essentials of Excel, Excel VBA, SAS and Minitab for Statistical and Financial Analyses*, DOI 10.1007/978-3-319-38867-0_7

```
MTB >  Set c1
DATA>   1( 0 : 10 / .01 )1
DATA>   End.
MTB > pdf c1 c2;
SUBC> uniform 2 7.
MTB > Plot c2*c1;
SUBC>   Connect;
SUBC>   Title "Uniform Probability Density Function";
SUBC>   Axlabel 1 "X-Axis";
SUBC>   AXLabel 2 "f(x)".
```

Scatterplot of C2 vs C1

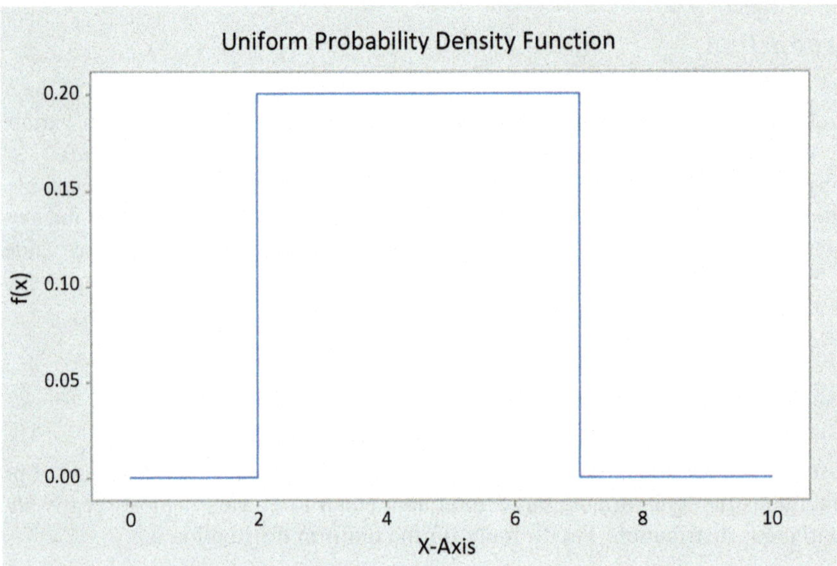

We can confirm that this plot has an area that is equal to 1 by seeing that the height is 0.2 and the width is 5. If we multiply these two numbers together, we get 1.

From the above graph, we can see the probability of any point of the uniform distribution. For example, the probability is .2 when the uniform distribution is 4. The probability is also .2 when the uniform distribution is 6. This graph and the function that defines this graph are commonly known as the probability density function (PDF) of the uniform distribution or the uniform probability density function.

Another important concept is to know the total probability of a distribution from left to right to a certain point in the distribution. The graph or function that answers this question is called the cumulative density function (CDF). For example, the CDF for when the uniform distribution is equal to 4 is $(4 - 2) * .2 = .4$.

Below creates the cumulative uniform distribution graph using Minitab.

```
MTB > Set c1
DATA>   1( 1 : 10 / .01 )1
DATA>   End.
MTB > cdf c1 c3;
SUBC> uniform 2 7.
MTB > Plot c3*c1;
SUBC>   Connect;
SUBC>   Title "Uniform Cumulative Density Function";
SUBC>   Axlabel 1 "X-Axis";
SUBC>   AXLabel 2 "f(x)".
```

Scatterplot of C3 vs C1

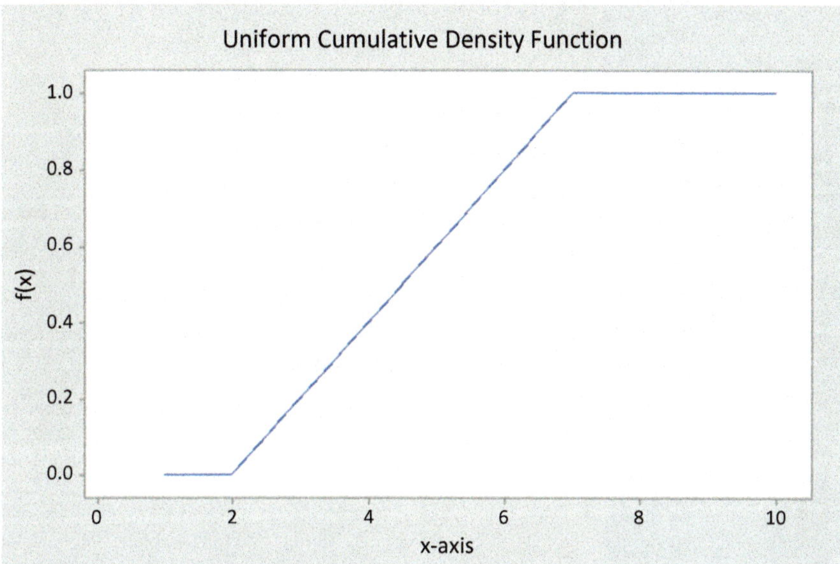

Below shows the *c* and *d* points of the uniform distribution that influence the uniform distribution. We will do this by graphing three distributions together. One distribution will be longer than the above distribution, and one distribution will be shorter than the above distribution.

```
MTB > name c1 'x-axis'
MTB > name c2 '2to7'
MTB > name c3 '1to9'
MTB > name c4 '3to5'
MTB > Set 'x-axis'
DATA>   1( -1 : 12 / .01 )1
DATA>   End.
MTB > cdf 'x-axis' '2to7';
SUBC> uniform 2 7.
MTB > cdf 'x-axis' '1to9';
SUBC> uniform 1 9.
MTB > cdf 'x-axis' '3to5';
SUBC> uniform 3 5.
MTB > name c5 'graphX'
MTB > name c6 'graphY'
MTB > name c7 'group'
MTB > Stack 'x-axis' 'x-axis' 'x-axis' 'graphX'.
MTB > Stack  '2to7' '1to9' '3to5' 'graphY';
SUBC>   Subscripts 'group';
SUBC>   UseNames.
MTB > Plot 'graphY'*'graphX';
SUBC>   Connect 'group';
SUBC>   Title "Cumulative Uniform Density Functions";
SUBC>   Axlabel 1 "X-Axis";
SUBC>   AXLabel 2 "f(x)".
```

Plot graphY * graphX

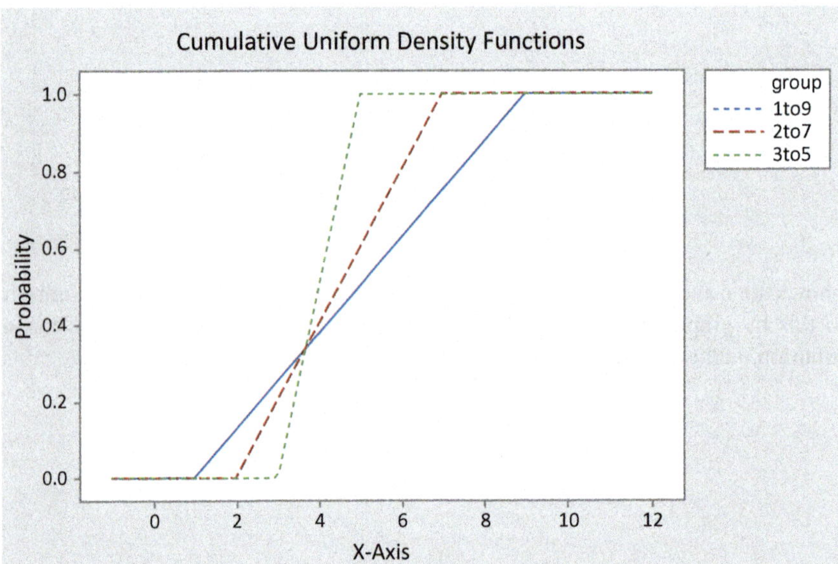

If we do the geometry calculation on all the graphs, we will find that all the graphs have an area equal to 1.

Below shows the cumulative uniform density functions for the above probability density functions.

```
MTB > name c1 'x-axis'
MTB > name c2 '2to7'
MTB > name c3 '1to9'
MTB > name c4 '3to5'
MTB > Set 'x-axis'
DATA>   1( -1 : 12 / .01 )1
DATA>   End.
MTB > cdf 'x-axis' '2to7';
SUBC> uniform 2 7.
MTB > cdf 'x-axis' '1to9';
SUBC> uniform 1 9.
MTB > cdf 'x-axis' '3to5';
SUBC> uniform 3 5.
MTB > name c5 'graphX'
MTB > name c6 'graphY'
MTB > name c7 'group'
MTB > Stack 'x-axis' 'x-axis' 'x-axis' 'graphX'.
MTB > Stack  '2to7' '1to9' '3to5' 'graphY';
SUBC>   Subscripts 'group';
SUBC>   UseNames.
MTB > Plot 'graphY'*'graphX';
SUBC>   Connect 'group';
SUBC>   Title "Uniform Probability Density Functions";
SUBC>   Axlabel 1 "X-Axis";
SUBC>   AXLabel 2 " Probability ".
```

Plot graphY * graphX

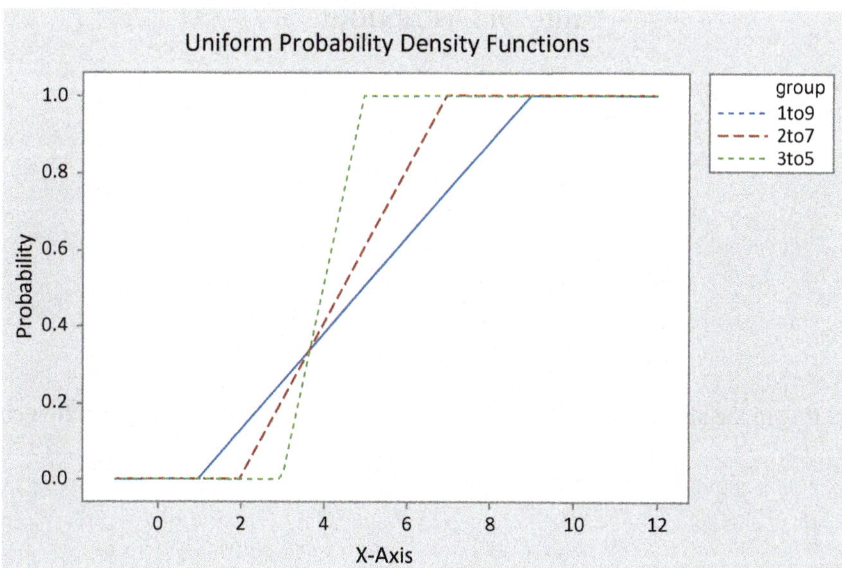

7.4 Excel Macro: Uniform Distribution

We will now use Excel to illustrate the uniform distribution. To illustrate this, we will look at the workbook *uniform.xls* that is included in the CD that accompanies this book. Uniform.xls will only create the uniform probability density function. The Excel code to produce the uniform probability density functions is shown in Appendix 7.2.

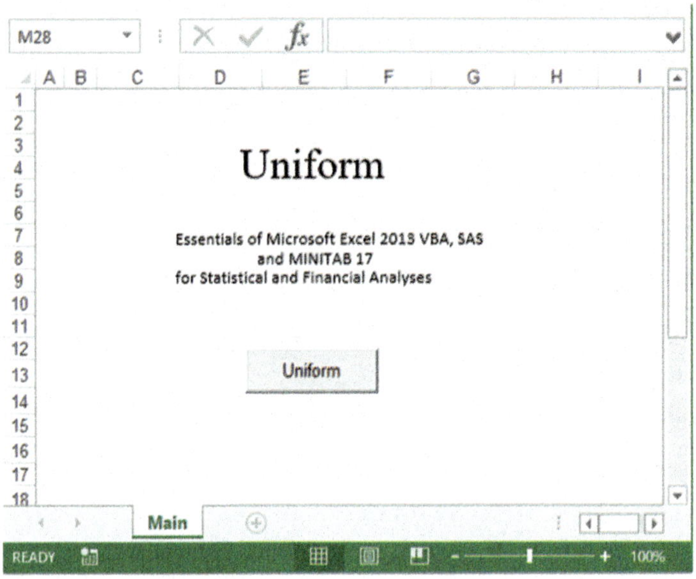

To create a uniform distribution, push the *Uniform* button shown above to get the dialog box below.

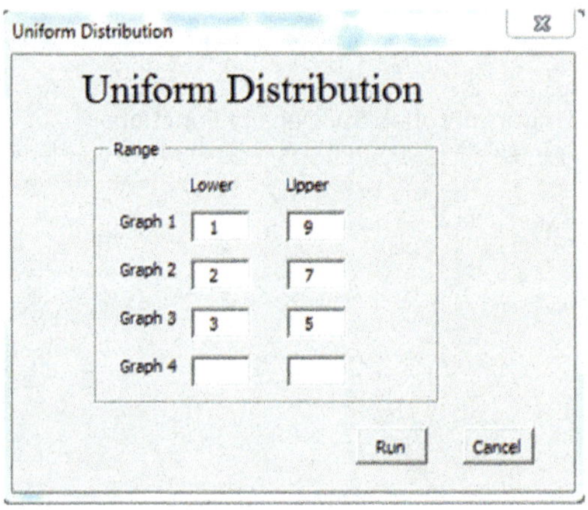

Push the *Run* in the above dialog box to get the uniform probability density function chart shown below.

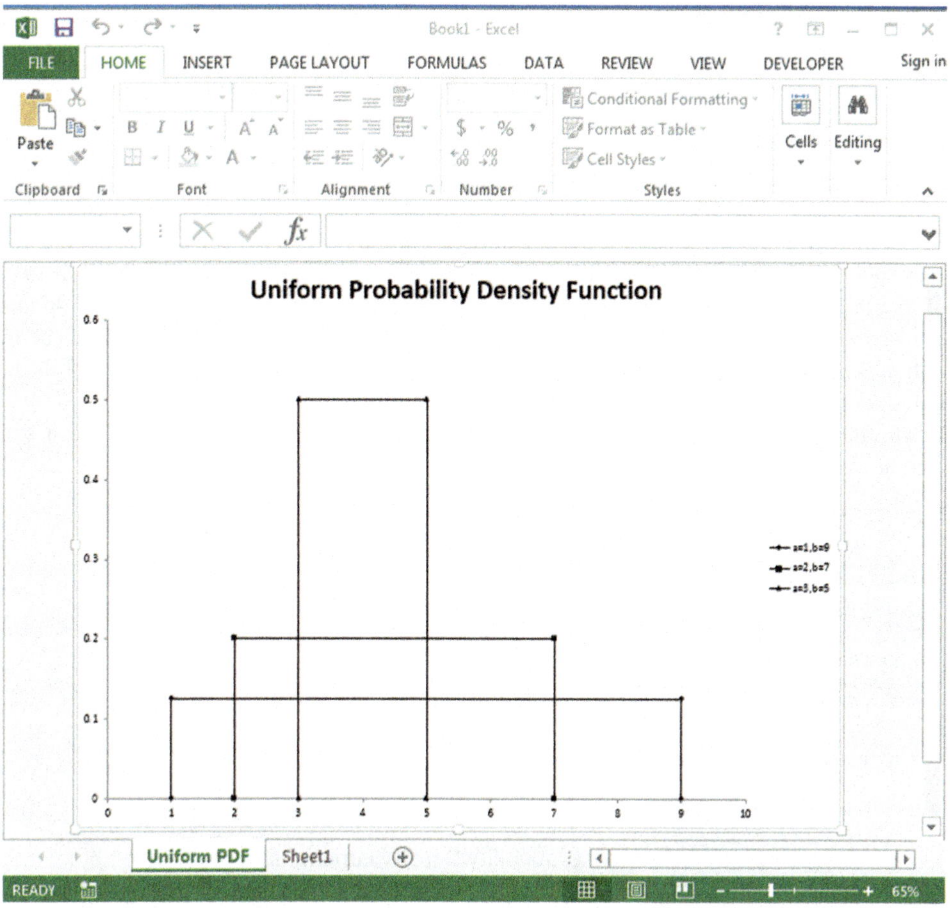

7.5 Normal Distribution

Continuous variables have many shapes, but one of the most important is the normal random variable. The probability density function of the normal distribution looks like the following.

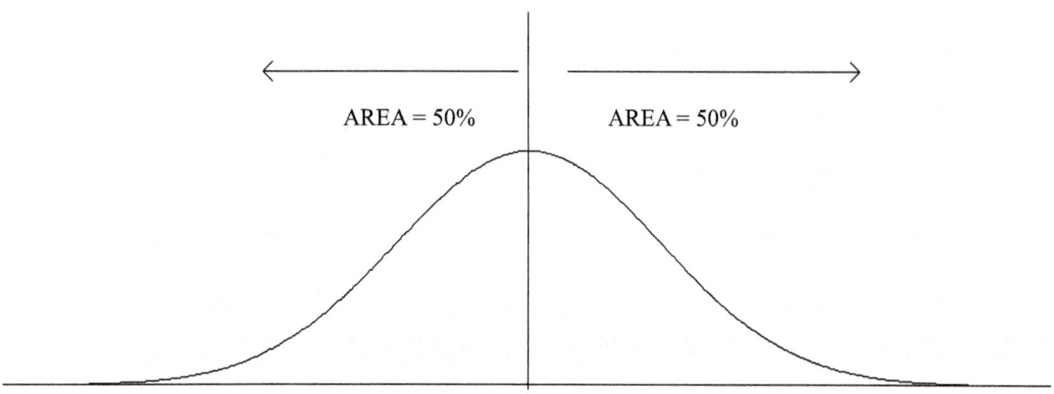

The formula for this function is

$$f(x) = \frac{1}{\sqrt{2\pi}\sigma} e^{-(x-\mu)^2/2\sigma^2}, \quad -\infty < x < \infty$$

We can see that this could be very complicated and tedious to calculate, but with Minitab and Excel, it is no longer a problem.

We should keep in mind some of the characteristics of the normal distribution. First, the distribution goes indefinitely in both directions. The area under the normal distribution is equal to 1. The normal distribution has a bell shape where the greatest probability density is at the midpoint. The normal distribution is symmetrical at its midpoint. That is, the probability to the left of the normal distribution is equal to 0.5 and the probability to the right is also equal to 0.5. The shape of the normal distribution is influenced by the standard deviation, σ, and the mean, μ.

The Minitab commands for creating a normal distribution with a mean of 15 and a standard deviation are as follows:

```
MTB > Set c1
DATA>   1( -10 : 40 / 1 )1
DATA>   End.
MTB > PDF c1 c2;
SUBC>   Normal 15 5.
MTB > Plot C2*c1;
SUBC>   Connect;
SUBC>   Title "Uniform Probability Density Functions";
SUBC>   Axlabel 1 "X-Axis";
SUBC>   AXLabel 2 " Y-Axis ".
```

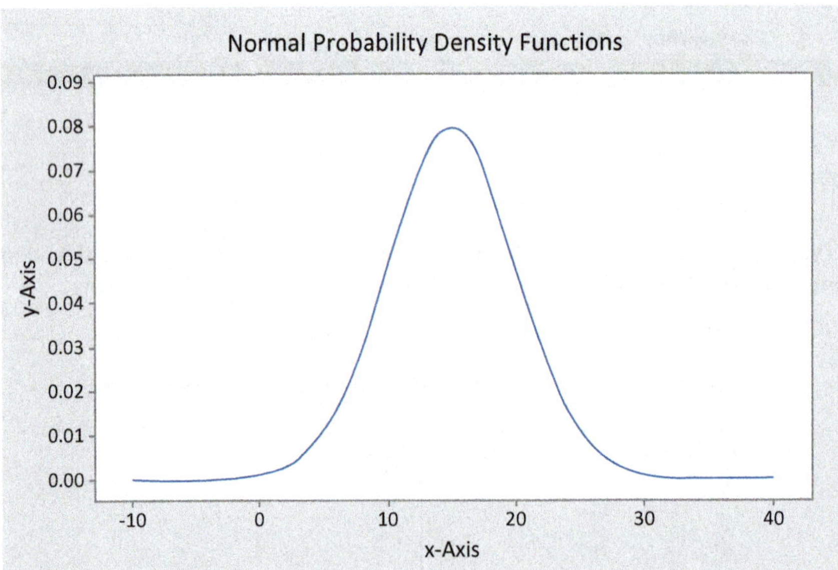

We will now calculate and plot the cumulative density function of the normal distribution to demonstrate that the area or total probability of a normal distribution is equal to 1.

```
MTB > cdf c1 c3;
SUBC> normal 15 5.
MTB > Plot c3*c1;
SUBC>   Connect;
SUBC>   Title "Cumulative Normal Distribution Function";
SUBC>   Axlabel 1 "X-Axis";
SUBC>   AXLabel 2 "f(x)".
```

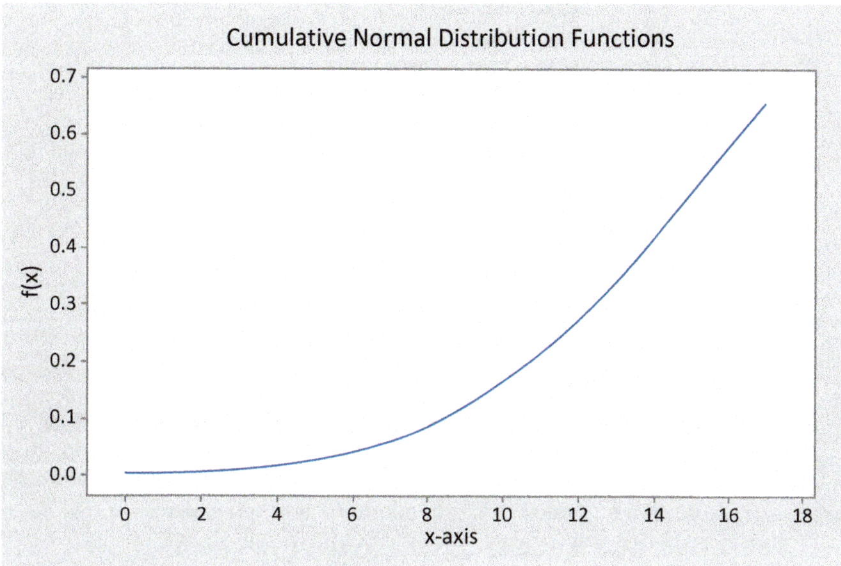

7.6 Excel Macro: Normal Distribution

We will now use Excel to illustrate that the normal distribution is influenced by σ and μ. To illustrate this, we will look at the workbook *normal*.xls that is distributed with the CD that accompanies this book.

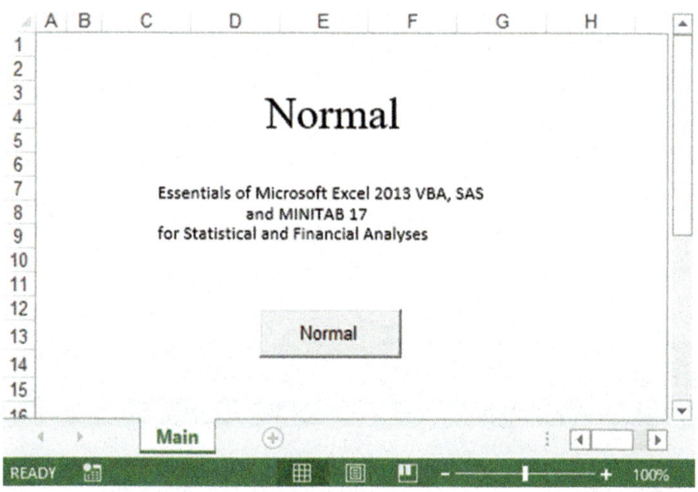

We will first show three normal distributions in the same graph with the same standard deviation and different means. We will then create three normal distributions in the same graph with the same mean but different standard deviations.

To create normal distributions, push the *Normal* button shown above to get the dialog box below.

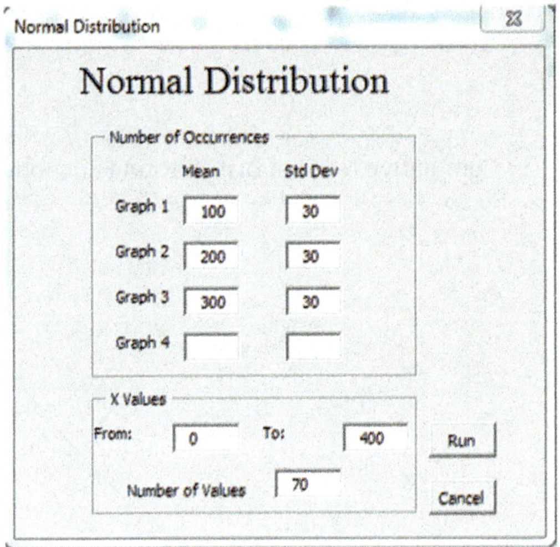

Push the *Run* in the above dialog box to get the following two charts shown below.

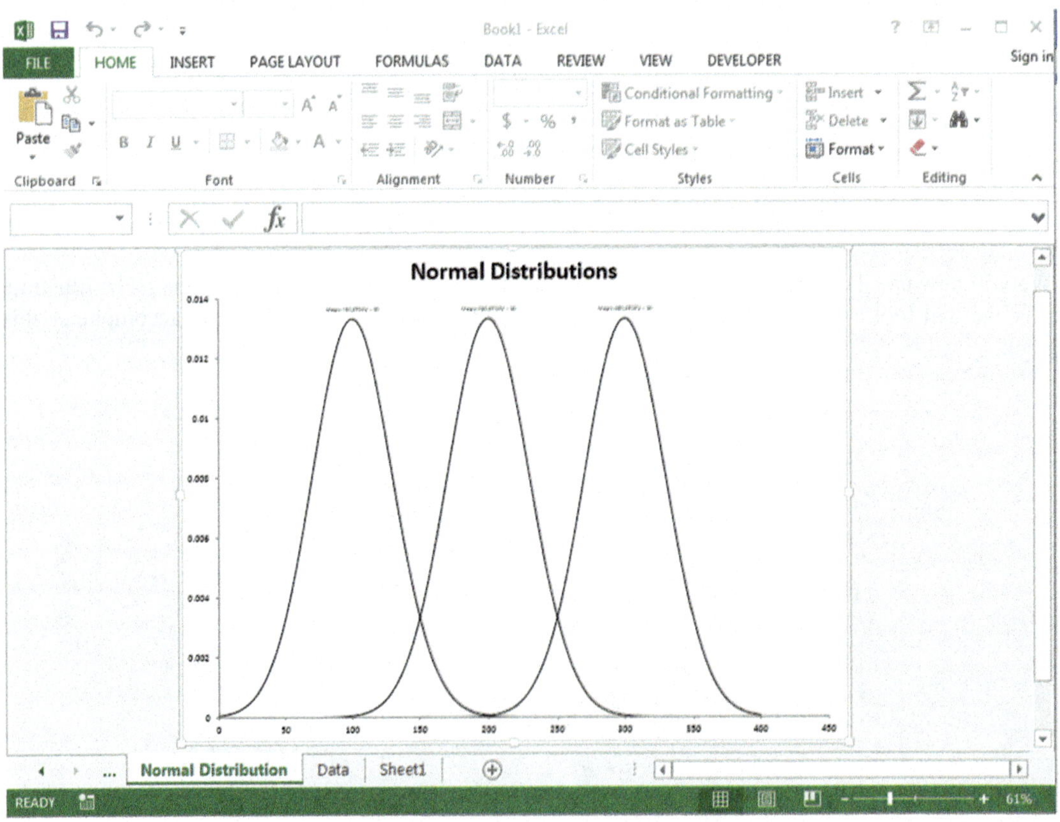

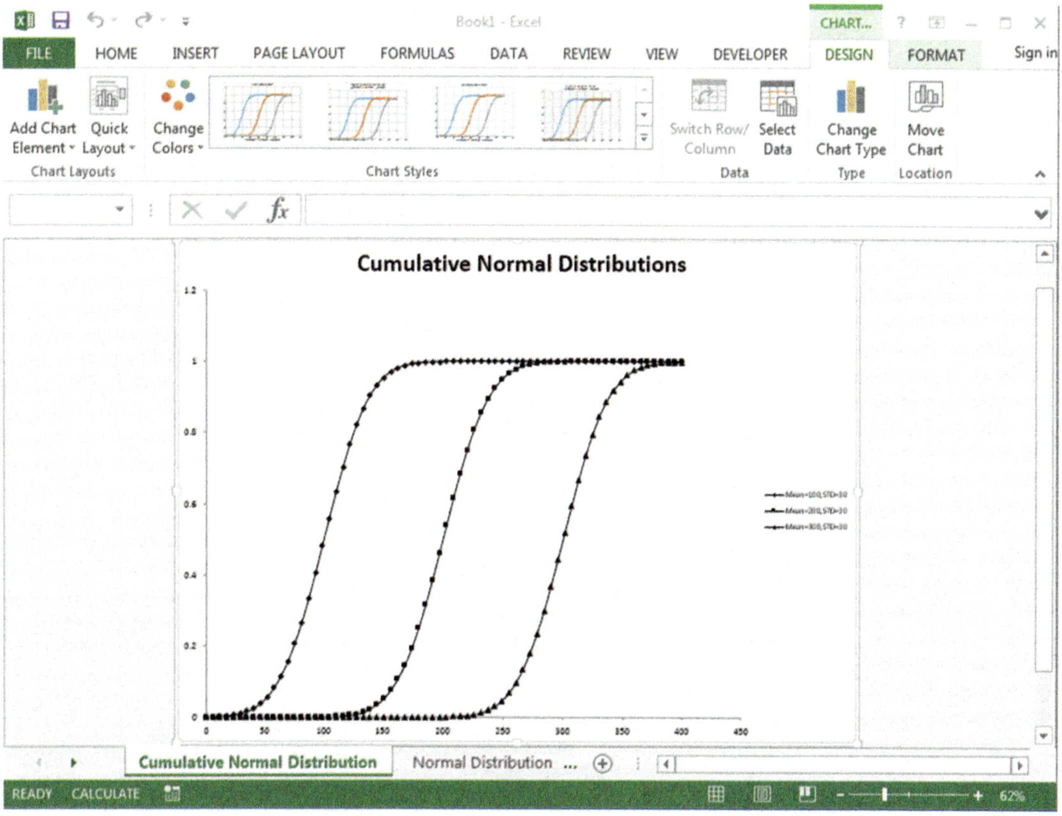

We will now create three normal distributions in the same graph with the same mean but different standard deviations.

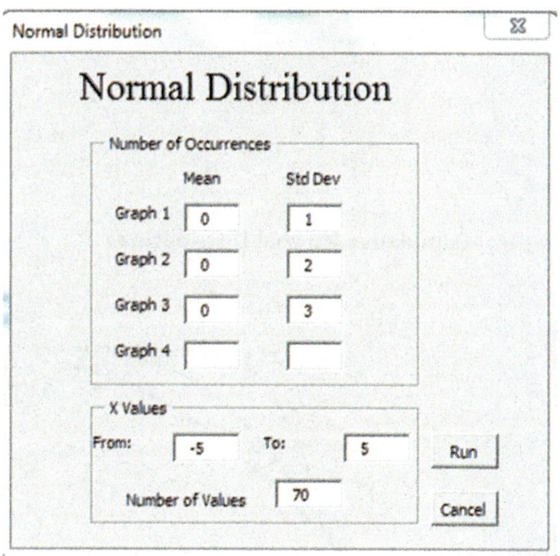

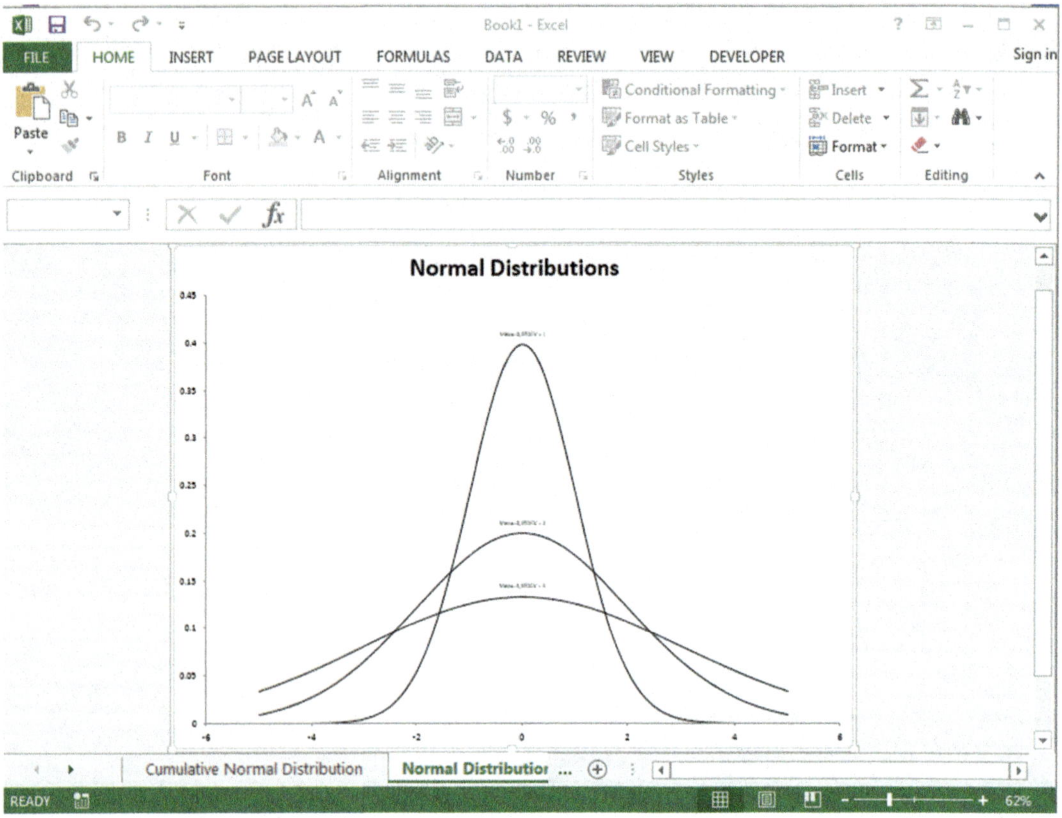
Book1 - Excel
FILE
HOME
INSERT
PAGE LAYOUT
FORMULAS
DATA
REVIEW
VIEW
DEVELOPER
Sign in
Paste
Clipboard
Font
Alignment
Number
Conditional Formatting
Format as Table
Cell Styles
Styles
Insert
Delete
Format
Cells
Editing
Normal Distributions
Cumulative Normal Distribution
Normal Distributior ...
READY
62%

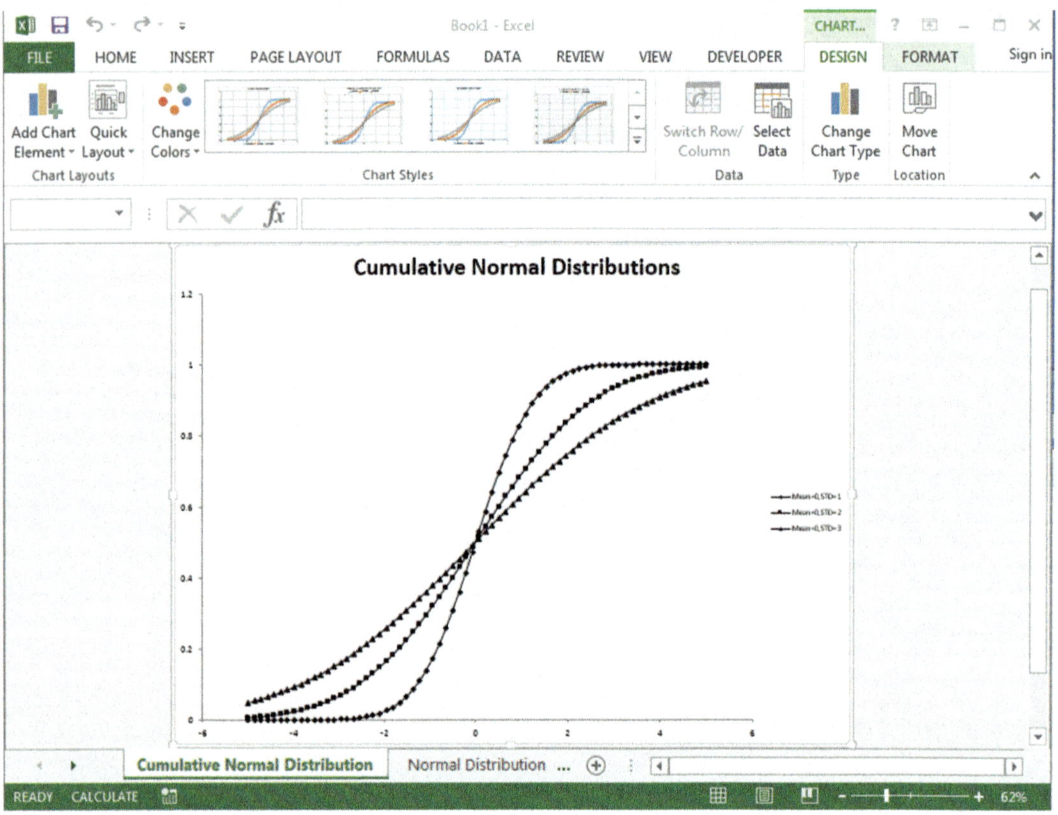
Book1 - Excel
CHART...
FILE
HOME
INSERT
PAGE LAYOUT
FORMULAS
DATA
REVIEW
VIEW
DEVELOPER
DESIGN
FORMAT
Sign in
Add Chart Element
Quick Layout
Change Colors
Chart Layouts
Chart Styles
Switch Row/ Column
Select Data
Data
Change Chart Type
Type
Move Chart
Location
Cumulative Normal Distributions
Cumulative Normal Distribution
Normal Distribution ...
READY
CALCULATE
62%

The program in the workbook *normal.xls* uses Excel's *normdist* function. The Excel code for the workbook *normal.xls* is shown in Appendix 7.1. The *normdist* function is Excel's normal probability density function. This function takes four parameters. The first parameter is the x value. The second parameter is the mean of the normal distribution. The third parameter is the standard deviation of the normal distribution. The fourth parameter indicates if the function is a cumulative normal distribution or not. Enter either a *true* or *false* value for the fourth parameter.

Below shows the Minitab commands to create the PDF and CDF normal distributions shown in the previous section.

```
MTB > name c1 'x-axis'
MTB > name c2 'm=100'
MTB > name c3 'm=200'
MTB > name c4 'm=300'
MTB > set 'x-axis'
DATA>     1( -10 : 500 / 1 )1
DATA> end
MTB > pdf 'x-axis' 'm=100';
SUBC> normal 100 30.
MTB > pdf 'x-axis' 'm=200';
SUBC> normal 200 30.
MTB > pdf 'x-axis' 'm=300';
SUBC> normal 300 30.
MTB > name c5 'graphx'
MTB > name c6 'graphy'
MTB > name c7 'group'
MTB > Stack 'x-axis' 'x-axis' 'x-axis' 'graphx'.
MTB > Stack  'm=100' 'm=200' 'm=300' 'graphy';
SUBC>   Subscripts 'group';
SUBC>   UseNames.
MTB > Plot 'graphy'*'graphx';
SUBC>   Connect 'group';
SUBC>   Title "Normal Distributions";
SUBC>   Axlabel 1 "X-Axis";
SUBC>   AXLabel 2 " Probability ".
```

Scatterplot of graphY vs graphX

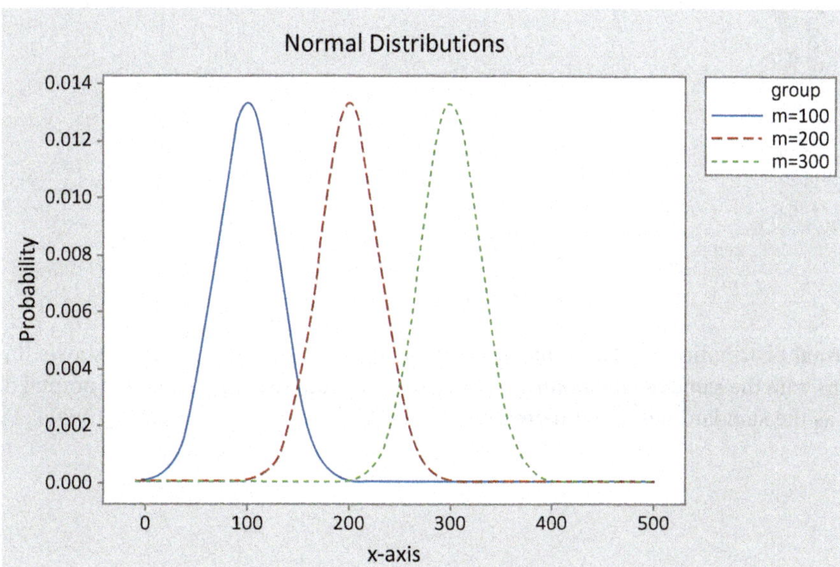

```
MTB > name c1 'x-axis'
MTB > name c2 'm=100'
MTB > name c3 'm=200'
MTB > name c4 'm=300'
MTB > set 'x-axis'
DATA>    1( -10 : 500 / 1 )1
DATA> end
MTB > cdf 'x-axis' 'm=100';
SUBC> normal 100 30.
MTB > cdf 'x-axis' 'm=200';
SUBC> normal 200 30.
MTB > cdf 'x-axis' 'm=300';
SUBC> normal 300 30.
MTB > name c5 'graphx'
MTB > name c6 'graphy'
MTB > name c7 'group'
MTB > Stack 'x-axis' 'x-axis' 'x-axis' 'graphx'.
MTB > Stack  'm=100' 'm=200' 'm=300' 'graphy';
SUBC>   Subscripts 'group';
SUBC>   UseNames.
MTB > Plot 'graphy'*'graphx';
SUBC>   Connect 'group';
SUBC>   Title "Cumulative Normal Distributions";
SUBC>   Axlabel 1 "X-Axis";
SUBC>   AXLabel 2 " Probability ".
```

Scatterplot of graphY vs graphX

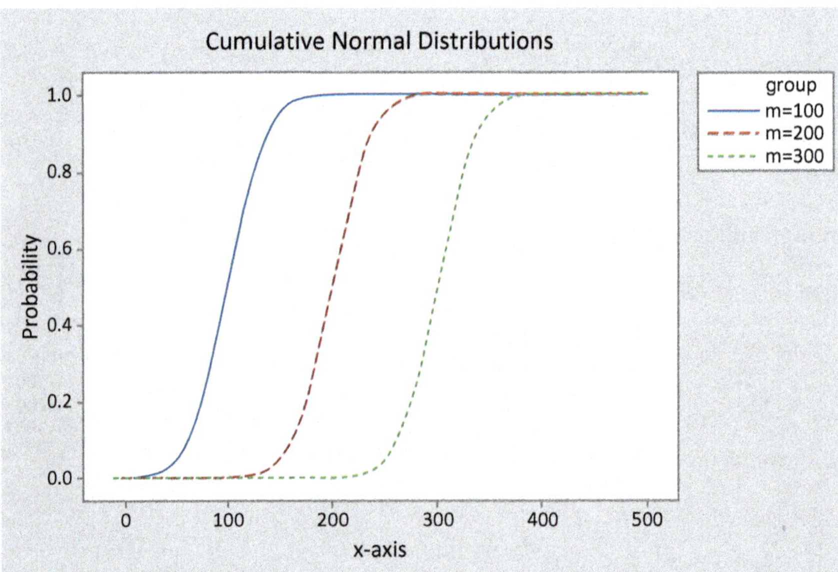

The normal distribution is also influenced by the standard deviation. Below shows three normal distributions with the same mean but different standard deviations. Notice that the normal distribution gets fatter as the standard deviation increases.

```
MTB > name c1 'x-axis'
MTB > name c2 's=100'
MTB > name c3 's=200'
MTB > name c4 's=300'
MTB > set 'x-axis'
DATA>    1( -10 : 500 / 1 )1
DATA> end
MTB > pdf 'x-axis' 's=100';
SUBC> normal 200 20.
MTB > pdf 'x-axis' 's=200';
SUBC> normal 200 30.
MTB > pdf 'x-axis' 's=300';
SUBC> normal 200 40.
MTB > name c5 'graphx'
MTB > name c6 'graphy'
MTB > name c7 'group'
MTB > Stack 'x-axis' 'x-axis' 'x-axis' 'graphx'.
MTB > Stack  's=100' 's=200' 's=300' 'graphy';
SUBC>   Subscripts 'group';
SUBC>   UseNames.
MTB > Plot 'graphy'*'graphx';
SUBC>   Connect 'group';
SUBC>   Title "Normal Distributions";
SUBC>   Axlabel 1 "X-Axis";
SUBC>   AXLabel 2 "Probability ".
```

Scatterplot of graphY vs graphX

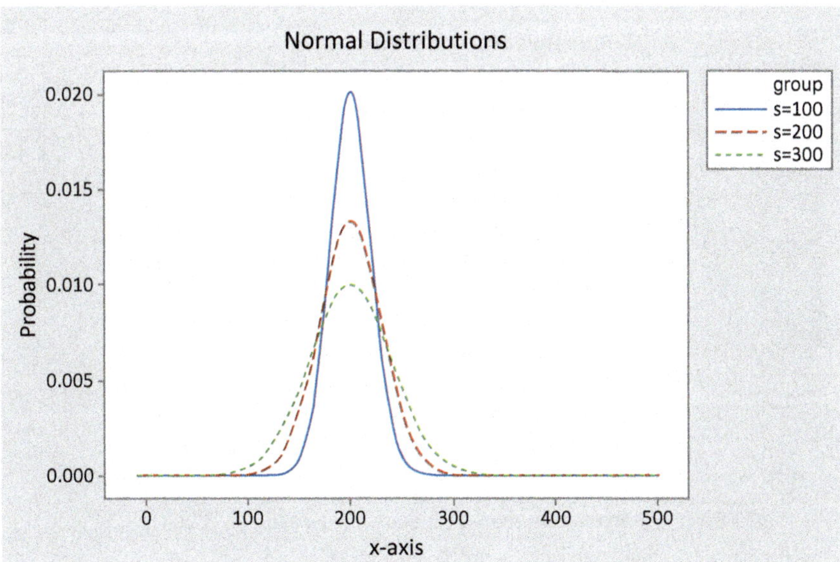

```
MTB > name c1 'x-axis'
MTB > name c2 's=100'
MTB > name c3 's=200'
MTB > name c4 's=300'
MTB > set 'x-axis'
DATA>    1( -10 : 500 / 1 )1
DATA> end
MTB > cdf 'x-axis' 's=100';
SUBC> normal 200 20.
MTB > cdf 'x-axis' 's=200';
SUBC> normal 200 30.
MTB > cdf 'x-axis' 's=300';
SUBC> normal 200 40.
MTB > name c5 'graphx'
MTB > name c6 'graphy'
MTB > name c7 'group'
MTB > Stack 'x-axis' 'x-axis' 'x-axis' 'graphx'.
MTB > Stack  's=100' 's=200' 's=300' 'graphy';
SUBC>   Subscripts 'group';
SUBC>   UseNames.
MTB > Plot 'graphy'*'graphx';
SUBC>   Connect 'group';
SUBC>   Title "Cumulative Normal Distributions";
SUBC>   Axlabel 1 "X-Axis";
SUBC>   AXLabel 2 "Probability ".
```

Scatterplot of graphY vs graphX

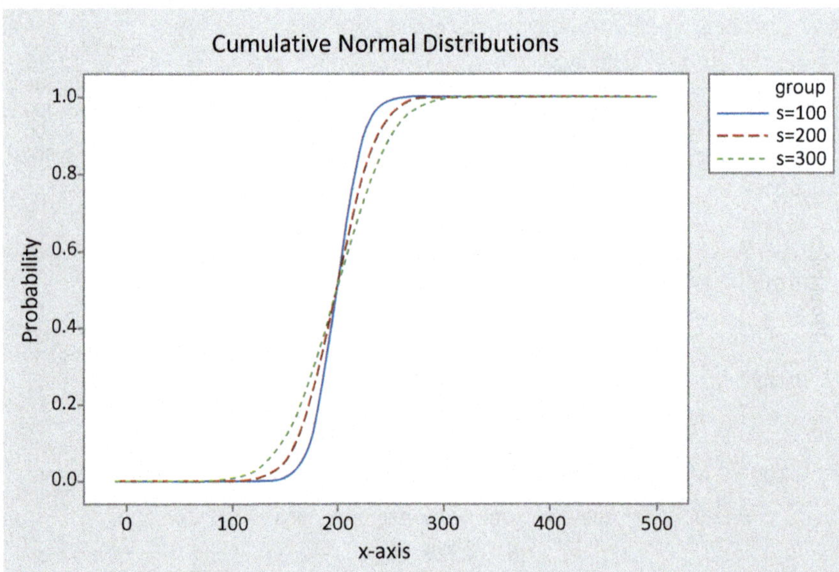

Once we know that a variable follows a normal distribution, we can make a decision based on this knowledge. Many times for a problem we know the distribution and the cumulative probability of the problem, and we are interested in finding the x value of the distribution. In statistics, the inverse cumulative distribution function answers the above question.

A Dunkin' Donuts shop located in New Brunswick, New Jersey, sells dozens of fresh donuts. Any donuts remaining unsold at the end of the day are either discarded or sold elsewhere at a loss.

The demand for the Dunkin' Donuts at this shop has followed a normal distribution with $\mu = 50$ and $\sigma = 5$ dozen. Use Minitab to figure out how many dozen donuts this Dunkin' Donuts shop should make each day so that it can meet the demand 95 % of the time.

```
MTB > invcdf .95;
SUBC> normal 50 5.
```

Inverse Cumulative Distribution Function

```
Normal with mean = 50.0000 and standard deviation = 5.00000

  P( X <= x)            x
     0.9500         58.2243
```

From the above output, we can see that the Dunkin' Donuts shop should make 58.22 dozen to meet the demand 95 % of the time.

This is illustrated graphically below.

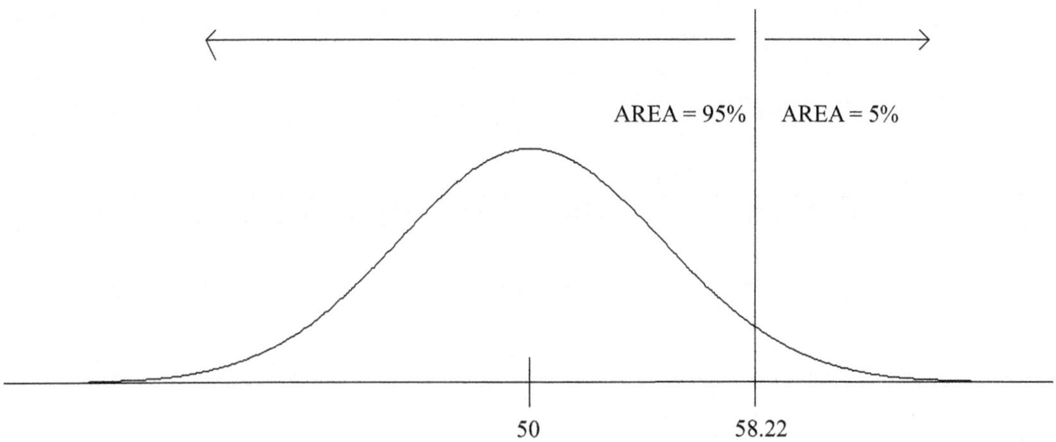

Below shows the calculation done in Excel using the *normalinv* function. This function takes three parameters. The first parameter is the probability. The second parameter is the mean. The third parameter is the standard deviation.

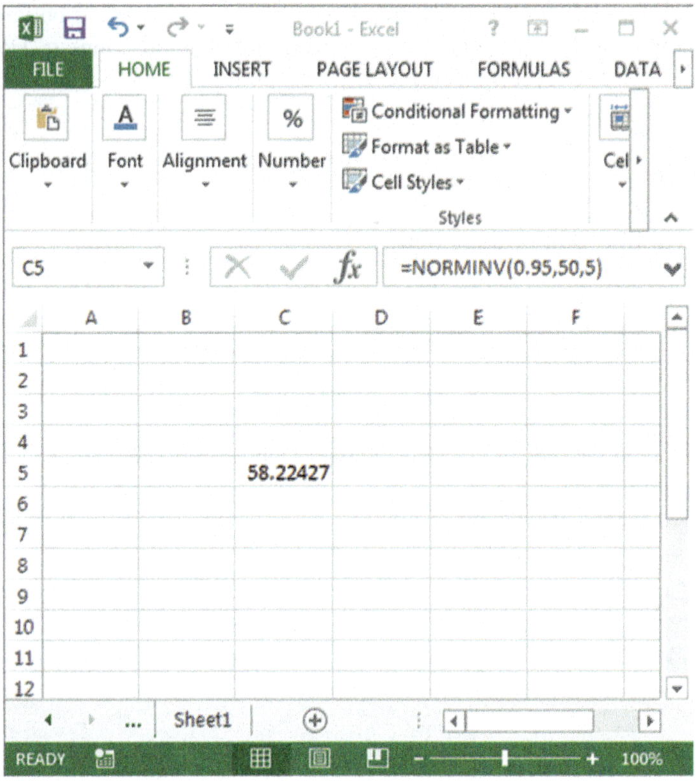

7.7 Standard Normal Distribution

We will find that the normal distribution is one of the most widely used continuous distributions in statistics. In fact, there is an infinite number of normal curves in statistics. To analyze normal curves more easily, most statistical analysis transforms normal curves to a *standard normal distribution*. This is done by using the Z score as discussed in Chap. 4. The standard normal distribution is a normal curve that has a mean of 0 and a standard deviation of 1.

7.8 Lognormal Distribution

Below are the Minitab commands to produce a lognormal distribution graph. In the *lognormal* subcommand below, the first argument is the mean of the lognormal distribution, and the second argument is the standard deviation of the lognormal distribution.

```
MTB > Set c1
DATA>   1( .006 : 17 / .01 )1
DATA>   End.
MTB > pdf c1 c2;
SUBC> lognormal 1 1.
MTB > Plot c2*c1;
SUBC>   Connect;
SUBC>   Title "Lognormal Density Function";
SUBC>   Axlabel 1 "x-Axis";
SUBC>   AXLabel 2 " f(x)".
```

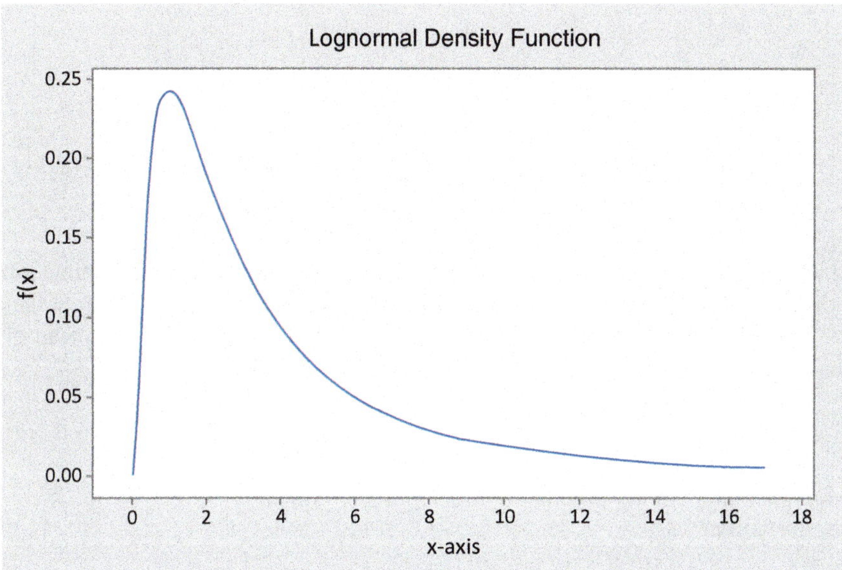

Now let us calculate and plot the lognormal cumulative distribution function to illustrate that the total area under the density function of the lognormal distribution is equal to 1.

```
MTB > cdf c1 c3;
SUBC> lognormal 1 1.
MTB > Plot c3*c1;
SUBC>   Connect;
SUBC>   Title "Cumulative Lognormal Distribution";
SUBC>   Axlabel 1 "x-Axis";
SUBC>   AXLabel 2 " f(x)".
```

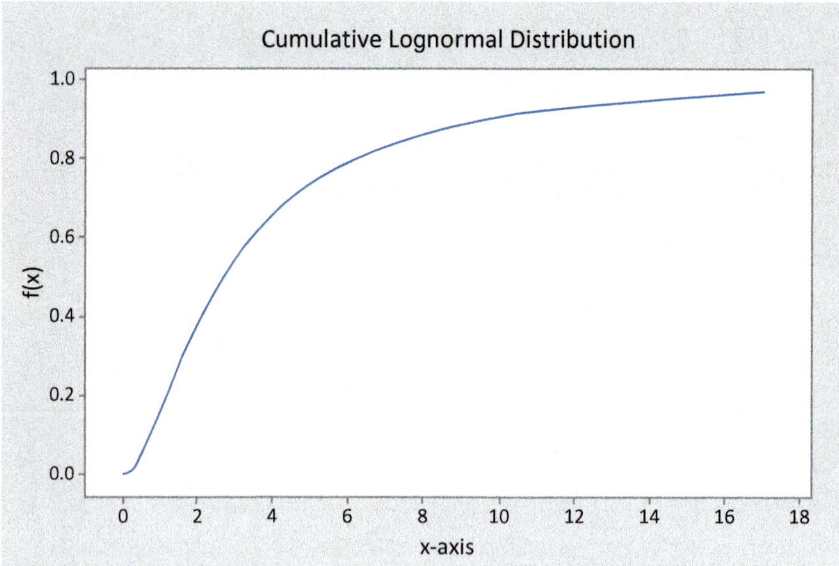

We will use Minitab to demonstrate how the mean and the standard deviation influence the lognormal distribution. The first thing that we will show is how the mean influences the lognormal distribution. We will use three lognormal distributions. One of them will have a mean of 1. One of them will have a mean of 2. One of them will have a mean of 3.

```
MTB > name c1 'x-axis'
MTB > name c2 'm=1'
MTB > name c3 'm=2'
MTB > name c4 'm=3'
MTB > set 'x-axis'
DATA>    1( 0 : 20 / .01 )1
DATA> end
MTB > pdf 'x-axis' 'm=1';
SUBC> lognormal 1 1.
MTB > pdf 'x-axis' 'm=2';
SUBC> lognormal 2 1.
MTB > pdf 'x-axis' 'm=3';
SUBC> lognormal 3 1.
MTB > name c5 'graphx'
MTB > name c6 'graphy'
MTB > name c7 'group'
MTB > Stack 'x-axis' 'x-axis' 'x-axis' 'graphx'.
MTB > Stack  'm=1' 'm=2' 'm=3' 'graphy';
SUBC>   Subscripts 'group';
SUBC>   UseNames.
MTB > Plot 'graphy'*'graphx';
SUBC>   Connect 'group';
SUBC>   Title "LogNormal Distributions";
SUBC>   Axlabel 1 "x-Axis";
SUBC>   AXLabel 2 " Probability ".
```

Scatterplot of graphY vs graphX

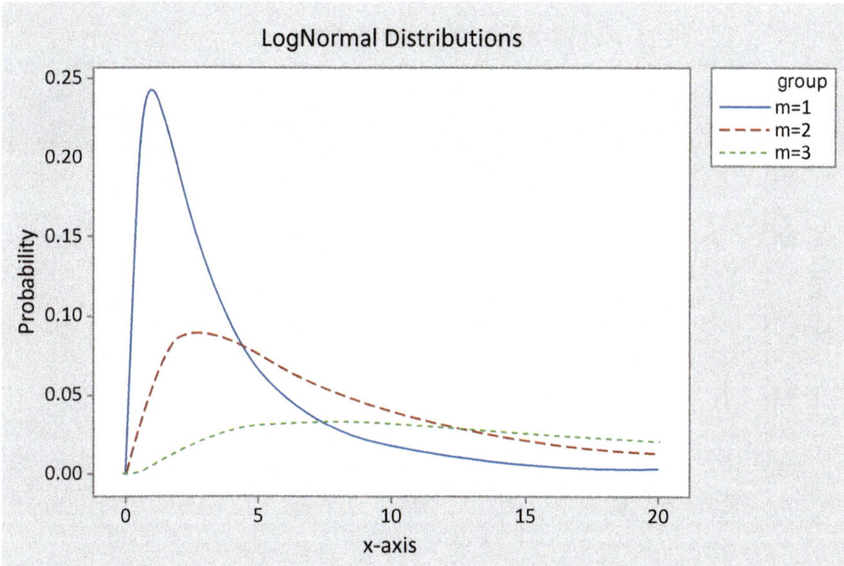

From the above exercise, we can see that increasing the mean flattens the lognormal distribution.

```
MTB > name c1 'x-axis'
MTB > name c2 'm=1'
MTB > name c3 'm=2'
MTB > name c4 'm=3'
MTB > set 'x-axis'
DATA>    1( 0 : 100 / 1 )1
DATA> end
MTB > cdf 'x-axis' 'm=1';
SUBC> lognormal 1 1.
MTB > cdf 'x-axis' 'm=2';
SUBC> lognormal 2 1.
MTB > cdf 'x-axis' 'm=3';
SUBC> lognormal 3 1.
MTB > name c5 'graphx'
MTB > name c6 'graphy'
MTB > name c7 'group'
MTB > Stack 'x-axis' 'x-axis' 'x-axis' 'graphx'.
MTB > Stack  'm=1' 'm=2' 'm=3' 'graphy';
SUBC>   Subscripts 'group';
SUBC>   UseNames.
MTB > Plot 'graphy'*'graphx';
SUBC>   Connect 'group';
SUBC>   Title "LogNormal Distributions";
SUBC>   Axlabel 1 "x-Axis";
SUBC>   AXLabel 2 " Probability ".
```

Scatterplot of graphY vs graphX

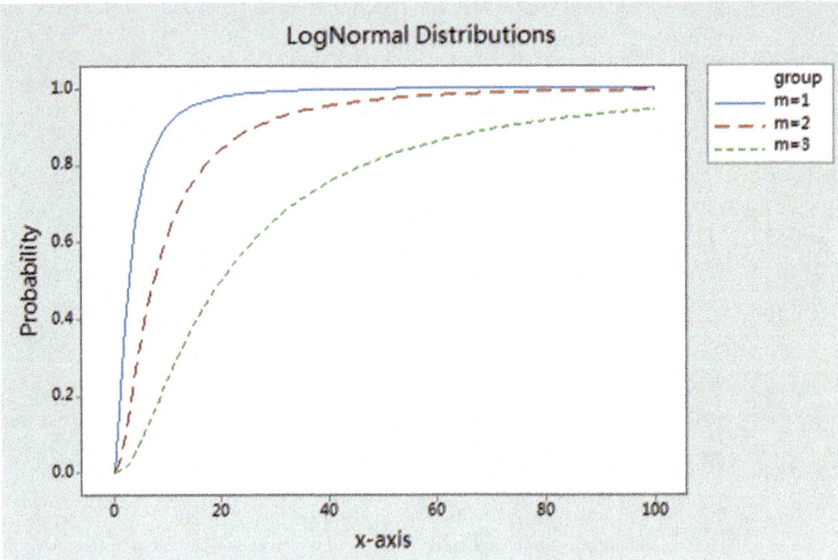

In Excel the function *lognorm.dist* returns the lognormal distribution. Below shows three lognormal distributions with a mean of 1, 2, and 3. The three lognormal distributions have a standard deviation of 1.

E10 =LOGNORM.DIST($B10,3,1,FALSE)

Lognormal Distribution

	Mean = 1	Mean = 2	Mean = 3	
1	0.2420	0.0540	0.004432	=LOGNORM.DIST($B10,3,1,FALSE)
2	0.1903	0.0849	0.013942	=LOGNORM.DIST($B11,3,1,FALSE)
3	0.1323	0.0886	0.021814	=LOGNORM.DIST($B12,3,1,FALSE)
4	0.0926	0.0826	0.027126	=LOGNORM.DIST($B13,3,1,FALSE)
5	0.0663	0.0739	0.030342	=LOGNORM.DIST($B14,3,1,FALSE)
6	0.0486	0.0651	0.032045	=LOGNORM.DIST($B15,3,1,FALSE)
7	0.0364	0.0569	0.032699	=LOGNORM.DIST($B16,3,1,FALSE)
8	0.0278	0.0497	0.032644	=LOGNORM.DIST($B17,3,1,FALSE)
9	0.0216	0.0435	0.032116	=LOGNORM.DIST($B18,3,1,FALSE)
10	0.0171	0.0381	0.031282	=LOGNORM.DIST($B19,3,1,FALSE)
11	0.0137	0.0335	0.030255	=LOGNORM.DIST($B20,3,1,FALSE)
12	0.0110	0.0296	0.029115	=LOGNORM.DIST($B21,3,1,FALSE)
13	0.0090	0.0262	0.027917	=LOGNORM.DIST($B22,3,1,FALSE)
14	0.0074	0.0232	0.026699	=LOGNORM.DIST($B23,3,1,FALSE)
15	0.0062	0.0207	0.025487	=LOGNORM.DIST($B24,3,1,FALSE)
16	0.0052	0.0185	0.024297	=LOGNORM.DIST($B25,3,1,FALSE)
17	0.0044	0.0166	0.023143	=LOGNORM.DIST($B26,3,1,FALSE)
18	0.0037	0.0149	0.022031	=LOGNORM.DIST($B27,3,1,FALSE)
19	0.0032	0.0134	0.020965	=LOGNORM.DIST($B28,3,1,FALSE)
20	0.0027	0.0122	0.019947	=LOGNORM.DIST($B29,3,1,FALSE)

LogNormal

Now we will see how the lognormal distribution behaves as the standard deviation decreases.

```
MTB > name c1 'x-axis'
MTB > name c2 's=1'
MTB > name c3 's=.6'
MTB > name c4 's=.4'
MTB > set 'x-axis'
DATA>    1( 0 : 20 /.01 )1
DATA> end
MTB > pdf 'x-axis' 's=1';
SUBC> lognormal 1 1.
MTB > pdf 'x-axis' 's=.6';
SUBC> lognormal 1 .6.
MTB > pdf 'x-axis' 's=.4';
SUBC> lognormal 1 .4.
MTB > name c5 'graphx'
MTB > name c6 'graphy'
MTB > name c7 'group'
MTB > Stack 'x-axis' 'x-axis' 'x-axis' 'graphx'.
MTB > Stack  's=1' 's=.6' 's=.4' 'graphy';
SUBC>   Subscripts 'group';
SUBC>   UseNames.
MTB > Plot 'graphy'*'graphx';
SUBC>   Connect 'group';
SUBC>   Title "LogNormal Distributions";
SUBC>   Axlabel 1 "x-Axis";
SUBC>   AXLabel 2 " Probability ".
```

Plot graphy * graphx

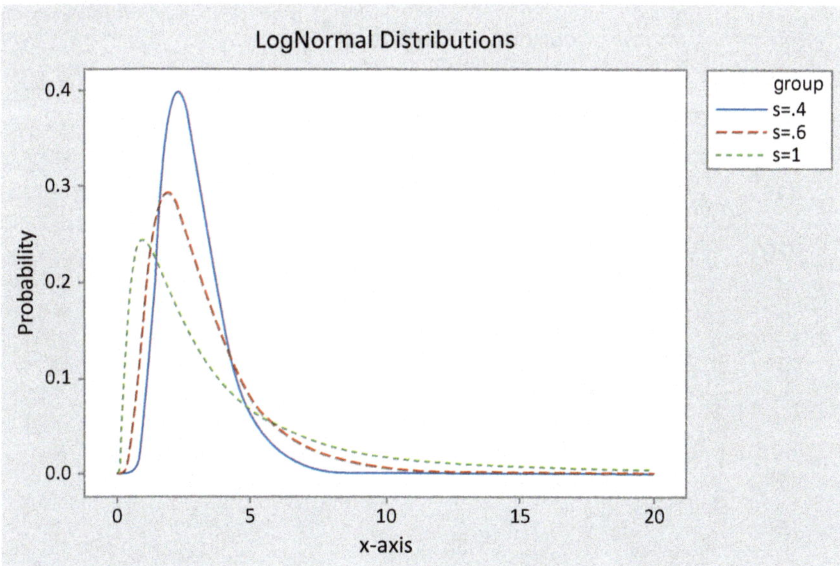

We can see as the standard deviation decreases while keeping the mean constant, it causes the lognormal distribution to peak.

Now we will see how the lognormal distribution behaves as the standard deviation increases.

```
MTB > name c1 'x-axis'
MTB > name c2 's=1'
MTB > name c3 's=1.6'
MTB > name c4 's=1.2'
MTB > set 'x-axis'
DATA>    1( 0 : 20 / .01 )1
DATA> end
MTB > pdf 'x-axis' 's=1';
SUBC> lognormal 1 1.
MTB > pdf 'x-axis' 's=1.6';
SUBC> lognormal 1 1.6.
MTB > pdf 'x-axis' 's=1.2';
SUBC> lognormal 1 1.2.
MTB > name c5 'graphx'
MTB > name c6 'graphy'
MTB > name c7 'group'
MTB > Stack 'x-axis' 'x-axis' 'x-axis' 'graphx'.
MTB > Stack  's=1' 's=1.6' 's=1.2' 'graphy';
SUBC>   Subscripts 'group';
SUBC>   UseNames.
MTB > Plot 'graphy'*'graphx';
SUBC>   Connect 'group';
SUBC>   Title "LogNormal Distributions";
SUBC>   Axlabel 1 "x-Axis";
SUBC>   AXLabel 2 " Probability ".
```

Scatterplot of graphY vs graphX

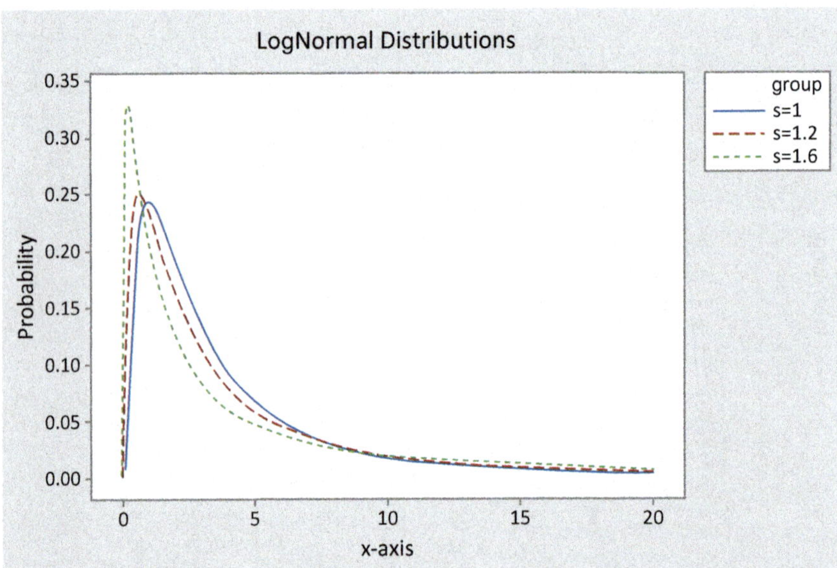

From the above graph, we can see that as the standard deviation increases, the lognormal peaks and shifts to the left.

Microsoft Excel does not have a lognormal PDF function.

7.9 Normal Approximating the Binomial

We can approximate a binomial distribution using a normal distribution if n is sufficiently large. A rule of thumb would be n equal to or greater than 30.

Why would we want to use the normal distribution to approximate the binomial distribution? The answer is with version 13 of Minitab, there is really no reason to. But with previous versions of Minitab and older software, it was necessary because they were unable to calculate the binomial distribution if n was too large. Below shows the result of Minitab 12.

```
MTB > let k1=sqrt(134.4)
MTB > cdf 185;
SUBC> normal 160 k1.
```

Cumulative Distribution Function

```
Normal with mean = 160.000 and standard deviation = 11.5931

         x        P( X <= x)
  185.0000           0.9845

MTB > cdf 185;
SUBC> binomial 1000 .16.

Cumulative Distribution Function

Binomial with n = 1000 and p = 0.160000

         x        P( X <= x)

* ERROR * Completion of computation impossible.
```

The reason this was a problem for older versions of Minitab is the $n!$ in the binomial formula. One of the calculations that Minitab would have to make would be the factorial of 185, which has a number of more than 18 integers! Unless specifically programmed to handle this large a number, a computer program cannot calculate the binomial distribution. In Minitab 17, this is no longer a problem.

The reason that we can use a normal distribution to approximate the binomial distribution is because our sample size is very large and because of the central limit theorem, which will be discussed in Chap. 8.

Example Suppose a very bumpy conveyor belt in a brewery transports beer bottles from the point where they are capped to the point where they are boxed for shipping. There is a 16 % chance that each beer bottle will fall off the conveyor belt. In 1 h, exactly 1000 beer bottles travel from one end of the belt to the other. Use Minitab to find the probability that 185 beer bottles or fewer will fall off the conveyor belt.

The mean of this problem is

$$1,000*0.16 = 160$$

The variance of this problem is

$$1,000*0.16*(1 - 0.16) = 134.4$$

Minitab's efforts to perform the normal and binomial calculations are illustrated below.

```
MTB > let k1=sqrt(134.4)
MTB > cdf 185;
SUBC> normal 160 k1.
```

Cumulative Distribution Function

```
Normal with mean = 160.000 and standard deviation = 11.5931

          x     P( X <= x )
   185.0000          0.9845
```

```
MTB > cdf 185;
SUBC> binomial 1000 .16.
```

Cumulative Distribution Function

```
Binomial with n = 1000 and p = 0.160000
          x     P( X <= x )
     185.00          0.9848
```

Below is a graphical illustration of the above problem.

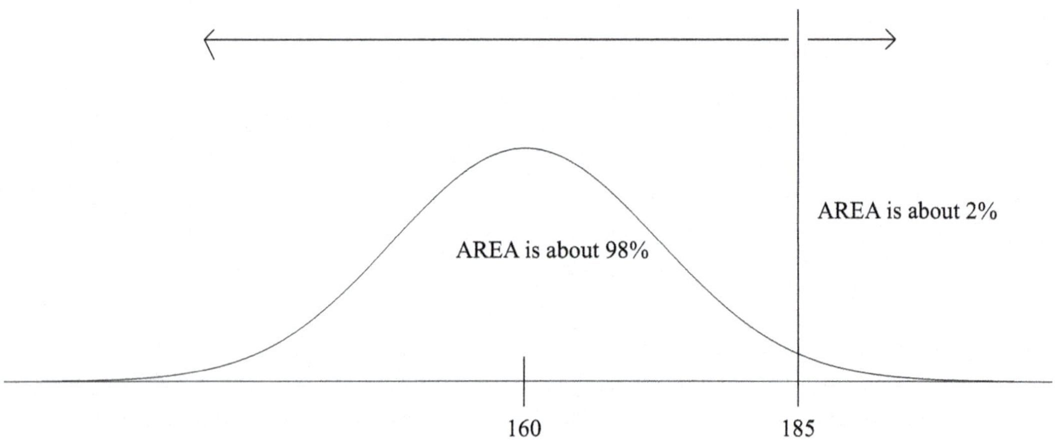

7.10 Normal Approximating the Poisson

The mean and variance of a Poisson distribution are both λ.

Example Assume that during a 20-min period, a bank has an average of 50 customers entering it. Suppose that we would like to find the probability of 57 customers or fewer coming into the bank within a 20-min period. Assume that this is a Poisson random variable situation.

Our mean and variance would be 50.

```
MTB > let k1=sqrt(50)
MTB > cdf 57;
SUBC> normal 50 k1.
```

Cumulative Distribution Function

```
Normal with mean = 50.0000 and standard deviation = 7.07107

          x        P( X <= x)
   57.0000            0.8389
   MTB > cdf 57;
   SUBC> poisson 50.
```

Cumulative Distribution Function

```
Poisson with mu = 50.0000

          x        P( X <= x)
     57.00            0.8551
```

From the normal distribution calculations, we can see that the probability of getting 57 customers or fewer in a 20-min period is 0.8389. From the Poisson distribution calculations, we can see that the probability of getting 57 or fewer in a 20-min period is 0.8551. Even though these two results are not the same, the difference is so small that making a decision on either number will in general result in the same conclusion.

The graph of the probability of 57 customers or fewer in a 20-min period is as follows.

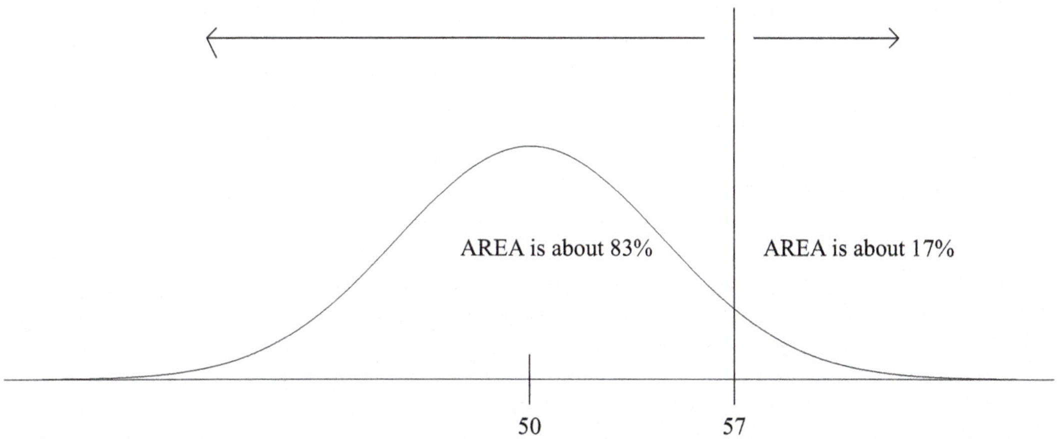

7.11 SAS Programming Code Instructions and Examples

As we mention in previous chapters, we mainly use function **PDF** to compute probability density (mass) functions. In this chapter, we will also introduce function **CDF** which returns a value from a cumulative probability distribution.

1. **Uniform Distribution**
   The **PDF** function for the **uniform** distribution returns the probability density function of a uniform distribution, with left location parameter **l** and right location parameter **r**, which is evaluated at the value **x**. It follows the form[1] **PDF('UNIFORM',x,l,r)**.

[1] The detailed information about SAS PDF function can be found in the link http://v8doc.sas.com/sashtml/lgref/z0270634.htm#z0226403

```
data unif;
do X=2 to 7 by 0.1;
pdf=pdf('uniform', X, 2, 7);
output unif;
end;
run;

Goptions ftext='Times New Roman' htext=6 gunit=pct;
title height=7 'Uniform Probability Density Function';
axis1  label=('f(x)') order=(-0.1 to 0.3 by 0.1);
axis2  label=('X-Axis') order=(1 to 8 by 1);
symbol1 i=join  v=none L=1 c=black w=2;
proc gplot data=unif;
plot  pdf*X/vaxis=axis1 haxis=axis2;
run;
```

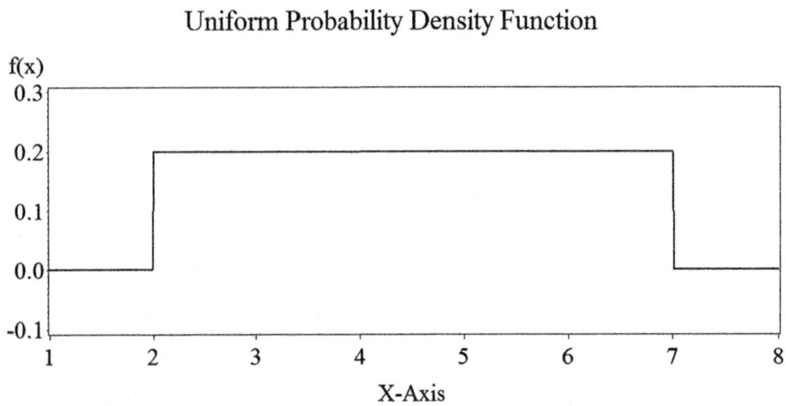

The **CDF** function for the uniform distribution returns the probability that an observation from a uniform distribution, with the left location parameter **l** and the right location parameter **r**, is less than or equal to **x**. The form follows **CDF('UNIFORM',x,l,r)**.

```
data unif;
do X=1 to 8 by 0.01;
CDF=cdf('uniform', X, 2, 7);
output unif;
end;
run;

Goptions ftext='Times New Roman' htext=6 gunit=pct;
title height=7 'Uniform Cumulative Density Function';
axis1  label=('f(x)') order=(0 to 1.1 by 0.1);
axis2  label=('X-Axis') order=(1 to 8 by 1);
symbol1 i=join  v=none L=1 c=black w=2;
proc gplot data=unif;
plot  CDF*X/vaxis=axis1 haxis=axis2;
run;
```

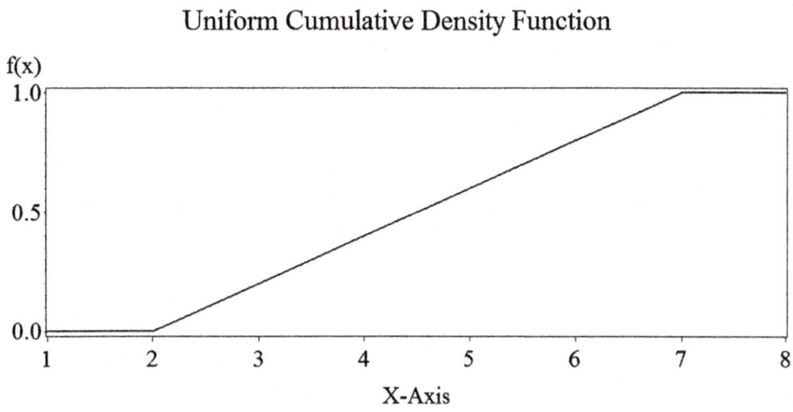

Below shows how the value of l and r influences the uniform distribution. We will graph three distributions with different **l** and **r** values.

```
data unif;
do X=0 to 10 by 0.01;
probability1=pdf('uniform', X, 1, 9);
probability2=pdf('uniform', X, 2, 7);
probability3=pdf('uniform', X, 3, 5);
output unif;
end;
run;

Goptions ftext='Times New Roman' htext=6 gunit=pct;
title  height=7 'Uniform Probability Density Function';
legend1 label=none value=('1to9' '2to7' '3to5') shape=symbol(5,1);
axis1  label=('f(x)') order=(0 to 0.6 by 0.1);
axis2  label=('X-Axis') order=(0 to 10 by 1);
symbol1 i=join  v=none L=1 c=black w=2;
symbol2 i=join  v=none L=5 c=black w=2;
symbol3 i=join  v=none L=3 c=black w=2;
proc gplot data=unif;
plot (probability1 probability2 probability3)*X/vaxis=axis1 haxis=axis2
overlay legend=legend1;
run;
```

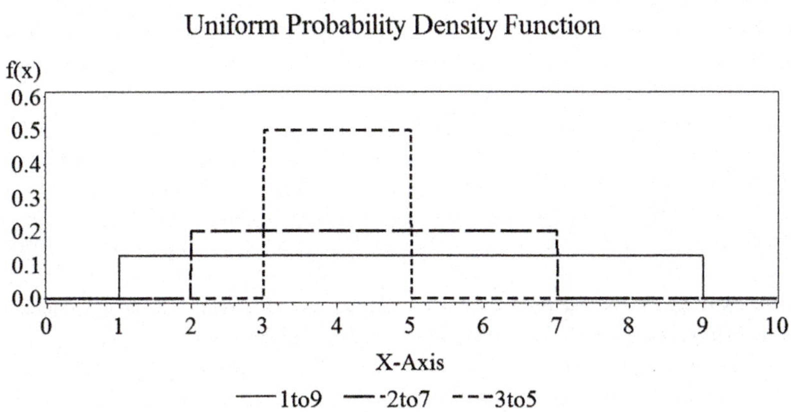

```
data unif;
do X=0 to 10 by 0.01;
CDF1=cdf('uniform', X, 1, 9);
CDF2=cdf('uniform', X, 2, 7);
CDF3=cdf('uniform', X, 3, 5);
output unif;
end;
run;

Goptions ftext='Times New Roman' htext=6 gunit=pct;
title  height=7 'Uniform Cumulative Distribution Function';
legend1 label=none value=('1to9' '2to7' '3to5') shape=symbol(5,1);
axis1  label=('Probability') order=(-0.1 to 1.1 by 0.2);
axis2  label=('X-Axis') order=(0 to 10 by 1);
symbol1 i=join  v=none L=1 c=black w=2;
symbol2 i=join  v=none L=5 c=black w=2;
symbol3 i=join  v=none L=3 c=black w=2;
proc gplot data=unif;
plot (CDF1 CDF2 CDF3)*X/vaxis=axis1 haxis=axis2 overlay legend=legend1;
run;
```

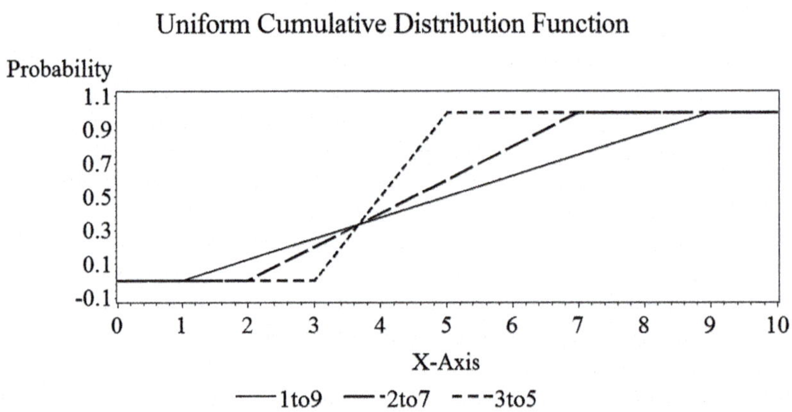

2. **Normal Distribution**
   The SAS commands for creating a normal distribution with a mean of $\boldsymbol{\theta}$ and a standard deviation of $\boldsymbol{\sigma}$ are as follows[2]: **PDF('NORMAL',x,$\boldsymbol{\theta}$,$\boldsymbol{\sigma}^2$)**.

   We will first show three normal distributions in the same graph with the same standard deviation and different means.

[2] The detailed information about SAS PDF function can be found in the link http://support.sas.com/documentation/cdl/en/lefunctionsref/63354/HTML/default/viewer.htm#n0n7cce4a3gfqkn1vr0p1x0of99s.htm

```
data norm;
do X=0 to 400 by 1;
probability1=pdf('normal', X, 100, 30);
probability2=pdf('normal', X, 200, 30);
probability3=pdf('normal', X, 300, 30);
output norm;
end;
run;

Goptions ftext='Times New Roman' htext=6 gunit=pct;
title  height=7 'Normal Distribution';
legend1 label=none value=('Mean 100' 'Mean 200' 'Mean 300')
shape=symbol(5,1);
axis1  label=('Probability') order=(0 to 0.014 by 0.002);
axis2  label=('X-Axis') order=(0 to 400 by 100);
symbol1 i=join  v=none L=1 c=black w=2;
symbol2 i=join  v=none L=5 c=black w=2;
symbol3 i=join  v=none L=3 c=black w=2;
proc gplot data=norm;
plot (probability1 probability2 probability3)*X/vaxis=axis1 haxis=axis2
overlay legend=legend1;
run;
```

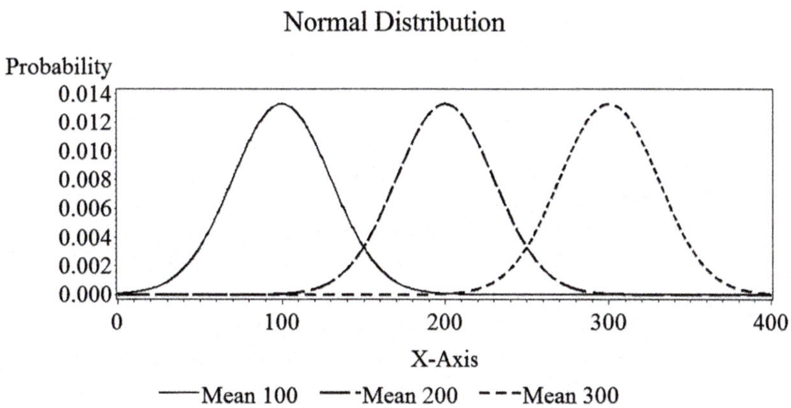

The CDF function for the normal distribution can be generated by replacing **PDF** with **CDF** in previous commands.

```
data norm;
do X=0 to 400 by 1;
CDF1=cdf('NORMAL', X, 100, 30);
CDF2=cdf('NORMAL', X, 200, 30);
CDF3=cdf('NORMAL', X, 300, 30);
output norm;
end;
run;

Goptions ftext='Times New Roman' htext=6 gunit=pct;
title  height=7 'Cumulative Normal Distribution';
legend1 label=none value=('STD 10' 'STD 20' 'STD 30') shape=symbol(5,1);
axis1  label=('Probability') order=(0 to 1 by 0.2);
axis2  label=('X-Axis') order=(0 to 400 by 100);
symbol1 i=join  v=none L=1 c=black w=2;
symbol2 i=join  v=none L=5 c=black w=2;
symbol3 i=join  v=none L=3 c=black w=2;
proc gplot data=norm;
plot (CDF1 CDF2 CDF3)*X/vaxis=axis1 haxis=axis2 overlay legend=legend1;
run;
```

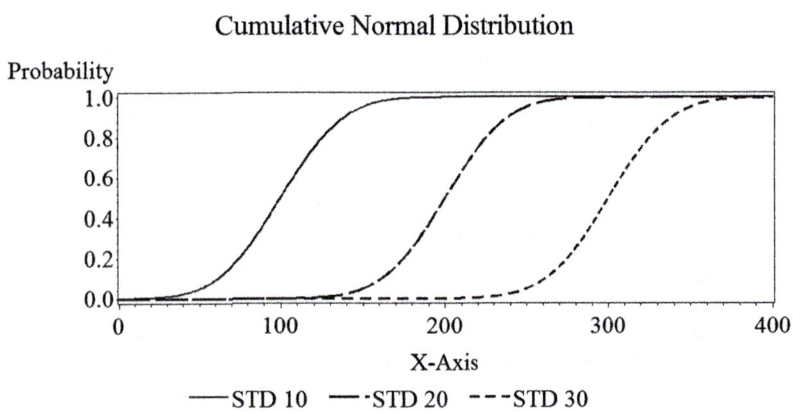

We will now create three normal distributions in the same graph with the same mean but different standard deviations.

```
data norm;
do X=0 to 400 by 1;
probability1=pdf('normal', X, 200, 10);
probability2=pdf('normal', X, 200, 20);
probability3=pdf('normal', X, 200, 30);
output norm;
end;
run;

Goptions ftext='Times New Roman' htext=6 gunit=pct;
title  height=7 'Normal Distribution';
legend1 label=none value=('STD 10' 'STD 20' 'STD 30') shape=symbol(5,1);
axis1  label=('Probability') order=(0 to 0.05 by 0.01);
axis2  label=('X-Axis') order=(0 to 400 by 100);
symbol1 i=join  v=none L=1 c=black w=2;
symbol2 i=join  v=none L=5 c=black w=2;
symbol3 i=join  v=none L=3 c=black w=2;
proc gplot data=norm;
plot (probability1 probability2 probability3)*X/vaxis=axis1 haxis=axis2
overlay legend=legend1;
run;
```

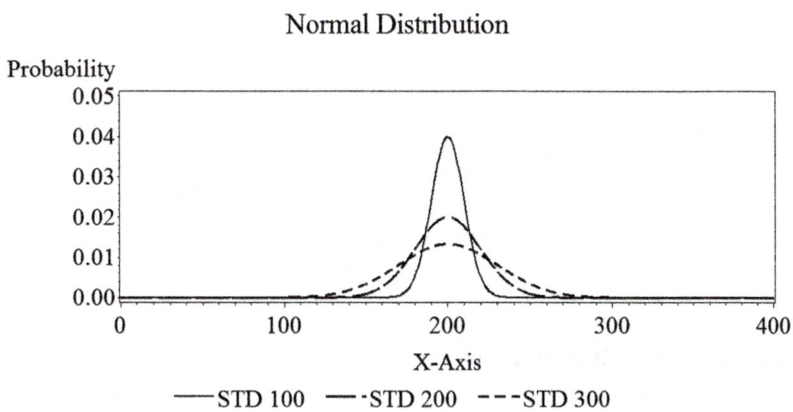

```
data norm;
do X=0 to 400 by 1;
CDF1=cdf('NORMAL', X, 200, 10);
CDF2=cdf('NORMAL', X, 200, 20);
CDF3=cdf('NORMAL', X, 200, 30);
output norm;
end;
run;

Goptions ftext='Times New Roman' htext=6 gunit=pct;
title  height=7 'Cumulative Normal Distribution';
legend1 label=none value=('STD 10' 'STD 20' 'STD 30') shape=symbol(5,1);
axis1  label=('Probability') order=(0 to 1 by 0.2);
axis2  label=('X-Axis') order=(0 to 400 by 100);
symbol1 i=join  v=none L=1 c=black w=2;
symbol2 i=join  v=none L=5 c=black w=2;
symbol3 i=join  v=none L=3 c=black w=2;
proc gplot data=norm;
plot (CDF1 CDF2 CDF3)*X/vaxis=axis1 haxis=axis2 overlay legend=legend1;
run;
```

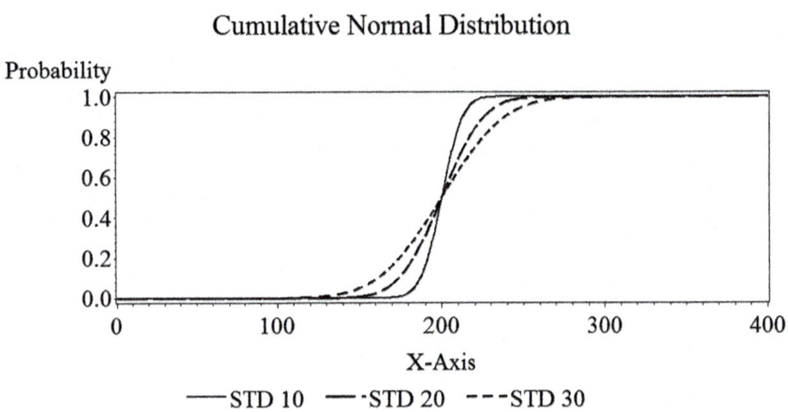

3. **Inverse Cumulative Distribution Function**
   In SAS, function probit(p) computes the inverse of the standard normal cdf, and $\boldsymbol{\sigma}$ ***probit(p)+**$\boldsymbol{\theta}$ will give you the inverse of the normal cdf with parameters $\boldsymbol{\theta}$ and $\boldsymbol{\sigma}$.

```
data x;
p=0.95;
mean=50;
STD=5;
x=STD*probit(p)+mean;
run;
```

Obs	p	Mean	STD	x
1	0.95	50	5	58.2243

To illustrate it graphically, we use the following command:

```
data norm1;
do X=0 to 100 by 1;
probability=pdf('normal', X, 50, 5);
output norm1;
end;
run;

Goptions ftext='Times New Roman' htext=6 gunit=pct;
title  height=7 'Normal Distribution';
axis1  label=('Probability') order=(0 to 0.1 by 0.02);
axis2  label=('X-Axis') order=(0 to 100 by 20);
symbol1 i=join  v=none L=1 c=black w=2;
proc gplot data=norm1;
plot probability*X/vaxis=axis1 haxis=axis2 overlay href=58.2243 whref=
run;
```

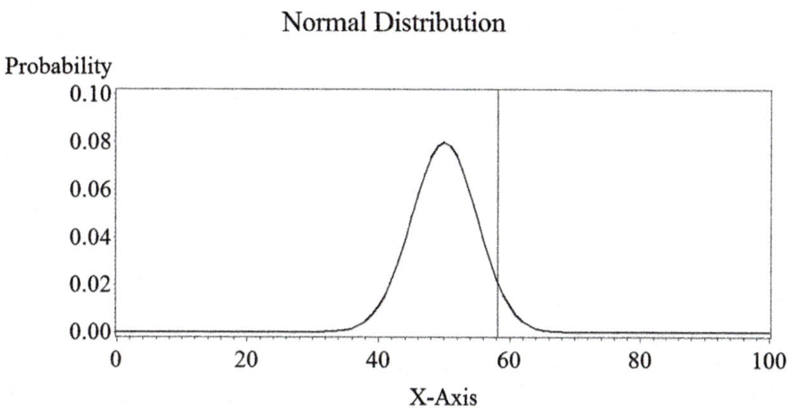

4. **Lognormal Distribution**
The PDF function for the lognormal distribution returns the probability density function of a lognormal distribution, with location parameter $\boldsymbol{\theta}$ and scale parameter $\boldsymbol{\lambda}$, which is evaluated at the value **x**. The equation follows **PDF('LOGNORMAL',x,$\boldsymbol{\theta}$,$\boldsymbol{\lambda}$)**.

```
data lognorm;
do X=0 to 20 by 0.1;
probability1=pdf('LOGNORMAL', X, 1, 1);
probability2=pdf('LOGNORMAL', X, 2, 1);
probability3=pdf('LOGNORMAL', X, 3, 1);
output lognorm;
end;
run;

Goptions ftext='Times New Roman' htext=6 gunit=pct;
title  height=7 'Lognormal Distribution';
legend1 label=none value=('Mean=1' 'Mean=2' 'Mean=3') shape=symbol(5,1);
axis1  label=('Probability') order=(0 to 0.25 by 0.05);
axis2  label=('X-Axis') order=(0 to 20 by 10);
symbol1 i=join  v=none L=1 c=black w=2;
symbol2 i=join  v=none L=5 c=black w=2;
symbol3 i=join  v=none L=3 c=black w=2;
proc gplot data=lognorm;
plot (probability1 probability2 probability3)*X/vaxis=axis1 haxis=axis2
overlay legend=legend1;
run;
```

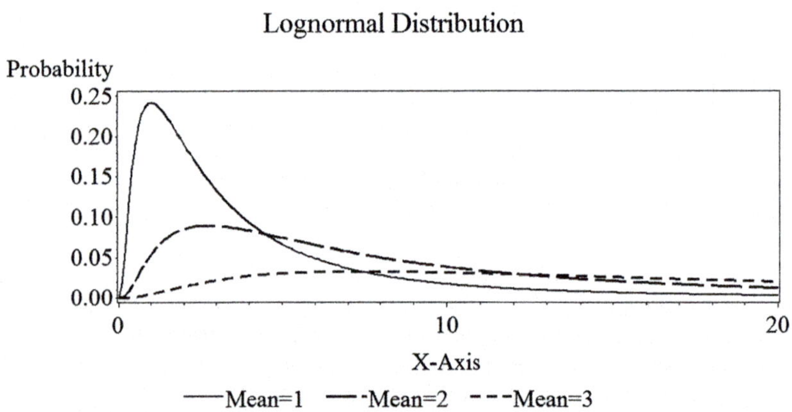

The CDF function for the lognormal distribution can be generated by replacing **PDF** with **CDF** in previous commands.

```
data lognorm;
do X=0 to 100 by 0.1;
CDF1=cdf('LOGNORMAL', X, 1, 1);
CDF2=cdf('LOGNORMAL', X, 2, 1);
CDF3=cdf('LOGNORMAL', X, 3, 1);
output lognorm;
end;
run;

Goptions ftext='Times New Roman' htext=6 gunit=pct;
title  height=7 'Cumulative Lognormal Distribution';
legend1 label=none value=('Mean=1' 'Mean=2' 'Mean=3') shape=symbol(5,1);
axis1  label=('Probability') order=(0 to 1 by 0.5);
axis2  label=('X-Axis') order=(0 to 100 by 50);
symbol1 i=join  v=none L=1 c=black w=2;
symbol2 i=join  v=none L=5 c=black w=2;
symbol3 i=join  v=none L=3 c=black w=2;
proc gplot data=lognorm;
plot (CDF1 CDF2 CDF3)*X/vaxis=axis1 haxis=axis2 overlay legend=legend1;
run;
```

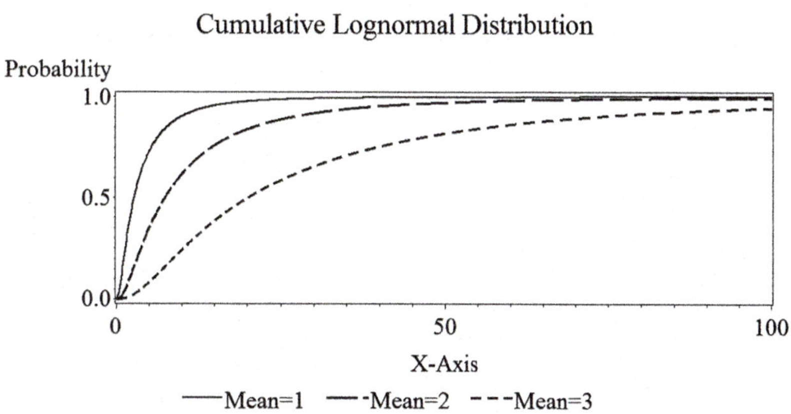

(5) **Normal Approximating the Binomial**

In this section, we repeat the example we concerned in the previous chapter:

Suppose a very bumpy conveyor belt in a brewery transports beer bottles from the point where they are capped to the point where they are boxed for shipping. There is a 16 % chance that each beer bottle will fall off the conveyor belt. In 1 h, exactly 1000 beer bottles travel from one end of the belt to the other. Use Minitab to find the probability that 185 beer bottles or fewer will fall off the conveyor belt.

The mean of this problem is $1000 * 0.16 = 160$.

The variance of this problem is $1000 * 0.16 * (1 - 0.16) = 134.4$.

We can use the following commands to approximate a binomial distribution using a normal distribution if the sample size is sufficiently large.

```
data norml;
ref=1-cdf('normal', 185, 160, sqrt(134.4));
do X=100 to 220 by 1;
probability=pdf('normal', X, 160, sqrt(134.4));
output norml;
end;
run;

Goptions ftext='Times New Roman' htext=6 gunit=pct;
title  height=7 'Normal Distribution';
axis1  label=('Probability') order=(0 to 0.1 by 0.02);
axis2  label=('X-Axis') order=(100 to 220 by 20);
symbol1 i=join  v=none L=1 c=black w=2;
proc gplot data=norml;
plot probability*X/vaxis=axis1 haxis=axis2 overlay href=180 whref=2;
run;
```

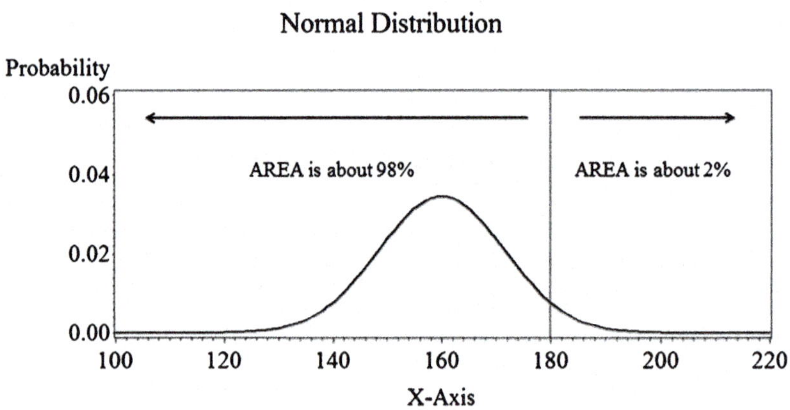

7.12 Statistical Summary

In this chapter, we looked at the uniform, normal, and lognormal continuous distributions. We saw that the normal distribution is influenced by the mean and standard deviation. For the normal distribution, we saw that changing the mean of the normal distribution causes the normal distribution to shift. We saw that changing the standard deviation of the normal distribution causes the normal distribution to become fatter or skinnier.

We saw that the lognormal distribution is influenced by the mean and the standard deviation. We saw that increasing the mean causes the lognormal distribution to become flatter. We saw that increasing the standard deviation causes the lognormal to peak and approach 0. We saw that decreasing the standard deviation causes the lognormal distribution to peak.

We also saw that the normal distribution can approximate a binomial and Poisson distribution.

Appendix 7.1: Excel Code—Normal Distribution

```
'/**************************************************************************
'/ Essentials of Microsoft Excel 2013, SAS and
'/                  MINITAB 17
'/ for Statistical and Financial Analysis
'/
'/**************************************************************************
Option Explicit
Dim mwbReport As Workbook
Dim mwsReportData As Worksheet

Property Set ReportWb(w As Workbook)
    Set mwbReport = w
End Property
Property Get ReportWb() As Workbook
    Set ReportWb = mwbReport
End Property
Property Set ReportData(w As Worksheet)
    Set mwsReportData = w
End Property
Property Get ReportData() As Worksheet
    Set ReportData = mwsReportData
End Property
Private Sub cmdCancel_Click()
    Unload Me
End Sub

'/********************************************************************
'/Purpose:  This is the procedure that is run after pressing the
'/          the run button on the form.  This is the main procedure
'/********************************************************************
Private Sub cmdRun_Click()
    Me.Hide
    Normal
    Unload frmNormal
End Sub

'/**********************************************************************
'/Purpose:  This procedure runs everytime the form is loaded.  This
'/          procedure sets some initial values.
'/***********************************************************************
Private Sub UserForm_Initialize()
    With Me
        .txtNumPoints = 70
        .txtMeanGraph1 = 0
        .TxtSTDDEVGraph1 = 1
        .txtXFrom = -5
        .txtXto = 5
    End With
End Sub

Sub Normal()
    Dim Chrt As Chart
    Dim ChrtCDF As Chart
    Dim Ser As Series

    Set Me.ReportWb = Workbooks.Add
    Set Me.ReportData = Me.ReportWb.Worksheets.Add
    Me.ReportData.Name = "Data"
    With Me
        .ReportWb.Names.Add Name:="Data!xFrom", RefersToR1C1:=Me.txtXFrom
        .ReportWb.Names.Add Name:="Data!xTo", RefersToR1C1:=Me.txtXto
        .ReportWb.Names.Add Name:="Data!NumPoints", RefersToR1C1:=Me.txtNumPoints
    End With

Set Chrt = Me.ReportWb.Charts.Add
Set ChrtCDF = Me.ReportWb.Charts.Add
Chrt.ChartType = xlXYScatterSmoothNoMarkers
ChrtCDF.ChartType = xlXYScatterSmoothNoMarkers
If Chrt.SeriesCollection.Count = 1 Then
    Chrt.SeriesCollection(1).Delete
End If
If ChrtCDF.SeriesCollection.Count = 1 Then
    Chrt.SeriesCollection(1).Delete
End If
```

```
If Len(Trim(Me.txtMeanGraph1)) <> 0 And _
   Len(Trim(Me.TxtSTDDEVGraph1)) <> 0 Then
        PDFChartGraph1 Chrt:=Chrt
        CDFChartGraph1 Chrt:=ChrtCDF
End If
If Len(Trim(Me.txtMeanGraph2)) <> 0 And _
   Len(Trim(Me.txtSTDDEVGraph2)) <> 0 Then
        PDFChartGraph2 Chrt:=Chrt
        CDFChartGraph2 Chrt:=ChrtCDF
End If
If Len(Trim(Me.txtMeanGraph3)) <> 0 And _
   Len(Trim(Me.txtSTDDEVGraph3)) <> 0 Then
        PDFChartGraph3 Chrt:=Chrt
        CDFChartGraph3 Chrt:=ChrtCDF
End If
If Len(Trim(Me.txtMeanGraph4)) <> 0 And _
   Len(Trim(Me.txtSTDDEVGraph4)) <> 0 Then
        PDFChartGraph4 Chrt:=Chrt
        CDFChartGraph4 Chrt:=ChrtCDF
End If
With Chrt

    .PlotArea.ClearFormats
    .Axes(xlValue).MajorGridlines.Delete
    .Axes(xlCategory).CrossesAt = .Axes(xlCategory).MinimumScale
    .Name = "Normal Distribution"
    .HasTitle = True
    If .SeriesCollection.Count > 1 Then
        .ChartTitle.Characters.Text = "Normal Distributions"
    Else
        .ChartTitle.Characters.Text = "Normal Distribution"
    End If
    .ChartTitle.Font.Size = 24
    .Legend.Font.Size = 8
    .HasLegend = False
End With

With ChrtCDF

    .PlotArea.ClearFormats
    .Axes(xlValue).MajorGridlines.Delete
    .Axes(xlCategory).CrossesAt = .Axes(xlCategory).MinimumScale
    .Name = "Cumulative Normal Distribution"
    .HasTitle = True
    If .SeriesCollection.Count > 1 Then
        .ChartTitle.Characters.Text = " Cumulative Normal Distributions"
    Else
        .ChartTitle.Characters.Text = " Cumulative Normal Distribution"
    End If
    .ChartTitle.Font.Size = 24
    .Legend.Font.Size = 8
    For Each Ser In .SeriesCollection
        With Ser
            .MarkerStyle = xlMarkerStyleAutomatic
            .Border.ColorIndex = 1
            .MarkerForegroundColorIndex = 1
            .MarkerBackgroundColorIndex = 1
        End With
    Next

    End With

End Sub

Sub PDFChartGraph1(Chrt As Chart)
    Dim Ser As Series
    Dim SerLabel As Series

    With Me
        .ReportWb.Names.Add Name:="Data!Mean1", _
                RefersToR1C1:=Me.txtMeanGraph1
        .ReportData.Names.Add Name:="Data!StdDev1", _
                RefersToR1C1:=Me.TxtSTDDEVGraph1

        .ReportWb.Names.Add Name:="Data!yOne", RefersToR1C1:= _
        "=EVALUATE(""Normdist(x,Data!Mean1,Data!StdDev1,false)+x*0"")"
        .ReportWb.Names.Add Name:="Data!x", RefersToR1C1:= _
        "=Data!xFrom+(ROW(OFFSET(Data!R1C2,0,0,NumPoints,1))-1)*(Data!xTo-(Data!xFrom))/(NumPoints-1)"
    End With
```

```
    Set Ser = Chrt.SeriesCollection.NewSeries
    With Ser
        .XValues = "=Data!x"
        .Values = "=Data!yOne"
        .Name = "Mean=" & Me.txtMeanGraph1 & ",STD=" & Me.TxtSTDDEVGraph1
        .Border.ColorIndex = 1
    End With

    ChartLegend Chrt:=Chrt, txtMean:=Me.txtMeanGraph1, txtStv:=Me.TxtSTDDEVGraph1


End Sub
Sub CDFChartGraph1(Chrt As Chart)
    Dim Ser As Series

    With Me
        .ReportWb.Names.Add Name:="Data!yCDFOne", RefersToR1C1:= _
        "=EVALUATE(""Normdist(x,Data!Mean1,Data!StdDev1,True)+x*0"")"
    End With

    Set Ser = Chrt.SeriesCollection.NewSeries
    With Ser
        .XValues = "=Data!x"
        .Values = "=Data!yCDFOne"
        .Name = "Mean=" & Me.txtMeanGraph1 & ",STD=" & Me.TxtSTDDEVGraph1
    End With

End Sub
Sub CDFChartGraph2(Chrt As Chart)
    Dim Ser As Series

    With Me
        .ReportWb.Names.Add Name:="Data!yCDFTwo", RefersToR1C1:= _
        "=EVALUATE(""Normdist(x,Data!Mean2,Data!StdDev2,True)+x*0"")"
    End With

    Set Ser = Chrt.SeriesCollection.NewSeries
    With Ser
        .XValues = "=Data!x"
        .Values = "=Data!yCDFTwo"
        .Name = "Mean=" & Me.txtMeanGraph2 & ",STD=" & Me.txtSTDDEVGraph2
    End With

End Sub
Sub CDFChartGraph3(Chrt As Chart)
    Dim Ser As Series
    With Me
        .ReportWb.Names.Add Name:="Data!yCDFThree", RefersToR1C1:= _
        "=EVALUATE(""Normdist(x,Data!Mean3,Data!StdDev3,True)+x*0"")"
    End With

    Set Ser = Chrt.SeriesCollection.NewSeries
    With Ser
        .XValues = "=Data!x"
        .Values = "=Data!yCDFThree"
        .Name = "Mean=" & Me.txtMeanGraph3 & ",STD=" & Me.txtSTDDEVGraph3
    End With

End Sub
Sub CDFChartGraph4(Chrt As Chart)
    Dim Ser As Series

    With Me
        .ReportWb.Names.Add Name:="Data!yCDFFour", RefersToR1C1:= _
        "=EVALUATE(""Normdist(x,Data!Mean4,Data!StdDev4,True)+x*0"")"
    End With


    Set Ser = Chrt.SeriesCollection.NewSeries
    With Ser
        .XValues = "=Data!x"
        .Values = "=Data!yCDFFour"
        .Name = "Mean=" & Me.txtMeanGraph2 & ",STD=" & Me.txtSTDDEVGraph2
    End With
```

```
End Sub
Sub PDFChartGraph2(Chrt As Chart)
    Dim Ser As Series

    With Me
        .ReportWb.Names.Add Name:="Data!Mean2", _
                RefersToR1C1:=Me.txtMeanGraph2
        .ReportData.Names.Add Name:="Data!StdDev2", _
                RefersToR1C1:=Me.txtSTDDEVGraph2

        .ReportWb.Names.Add Name:="Data!yTwo", RefersToR1C1:= _
        "=EVALUATE(""Normdist(x,Data!Mean2,Data!StdDev2,false)+x*0"")"
        .ReportWb.Names.Add Name:="Data!x", RefersToR1C1:= _
        "=Data!xFrom+(ROW(OFFSET(Data!R1C2,0,0,NumPoints,1))-1)*(Data!xTo-(Data!xFrom))/(NumPoints-1)"
    End With


    Set Ser = Chrt.SeriesCollection.NewSeries
    With Ser
        .XValues = "=Data!x"
        .Values = "=Data!yTwo"
        .Name = "Mean=" & Me.txtMeanGraph2 & ",STD=" & Me.txtSTDDEVGraph2
        .Border.ColorIndex = 1
    End With

    ChartLegend Chrt:=Chrt, txtMean:=Me.txtMeanGraph2, txtStv:=Me.txtSTDDEVGraph2

End Sub
Sub PDFChartGraph3(Chrt As Chart)
    Dim Ser As Series

    With Me
        .ReportWb.Names.Add Name:="Data!Mean3", _
                RefersToR1C1:=Me.txtMeanGraph3
        .ReportData.Names.Add Name:="Data!StdDev3", _
                RefersToR1C1:=Me.txtSTDDEVGraph3

        .ReportWb.Names.Add Name:="Data!yThree", RefersToR1C1:= _
        "=EVALUATE(""Normdist(x,Data!Mean3,Data!StdDev3,false)+x*0"")"
        .ReportWb.Names.Add Name:="Data!x", RefersToR1C1:= _
        "=Data!xFrom+(ROW(OFFSET(Data!R1C2,0,0,NumPoints,1))-1)*(Data!xTo-(Data!xFrom))/(NumPoints-1)"
    End With

    Set Ser = Chrt.SeriesCollection.NewSeries
    With Ser
        .XValues = "=Data!x"
        .Values = "=Data!yThree"
        .Name = "Mean=" & Me.txtMeanGraph3 & ",STD=" & Me.txtSTDDEVGraph3
        .Border.ColorIndex = 1
    End With

    ChartLegend Chrt:=Chrt, txtMean:=Me.txtMeanGraph3, txtStv:=Me.txtSTDDEVGraph3


End Sub
Sub PDFChartGraph4(Chrt As Chart)
    Dim Ser As Series

    With Me
        .ReportWb.Names.Add Name:="Data!Mean4", _
                RefersToR1C1:=Me.txtMeanGraph4
        .ReportData.Names.Add Name:="Data!StdDev4", _
                RefersToR1C1:=Me.txtSTDDEVGraph4

        .ReportWb.Names.Add Name:="Data!yFour", RefersToR1C1:= _
        "=EVALUATE(""Normdist(x,Data!Mean3,Data!StdDev4,false)+x*0"")"
        .ReportWb.Names.Add Name:="Data!x", RefersToR1C1:= _
        "=Data!xFrom+(ROW(OFFSET(Data!R1C2,0,0,NumPoints,1))-1)*(Data!xTo-(Data!xFrom))/(NumPoints-1)"
    End With

    Set Ser = Chrt.SeriesCollection.NewSeries
    With Ser
        .XValues = "=Data!x"
        .Values = "=Data!yFour"
        .Name = "Mean=" & Me.txtMeanGraph4 & ",STD=" & Me.txtSTDDEVGraph4
        .Border.ColorIndex = 1
    End With
```

```
    ChartLegend Chrt:=Chrt, txtMean:=Me.txtMeanGraph4, txtStv:=Me.txtSTDDEVGraph4

End Sub

Sub ChartLegend(Chrt As Chart, txtMean As Object, txtStv As Object)
    Dim SerLabel As Series
    Set SerLabel = Chrt.SeriesCollection.NewSeries
    Chrt.Activate
    With SerLabel

        .Values = Array(0, WorksheetFunction.NormDist(CDbl(txtMean), CDbl(txtMean), CDbl(txtStv),
False))
        .XValues = Array(CDbl(txtMean), CDbl(txtMean))

        With .Points(2)
            .ApplyDataLabels
            With .DataLabel
                .Characters.Text = "Mean=" & txtMean & ",STDEV = " & txtStv
                .Position = xlLabelPositionAbove
                .Font.Size = 6
            End With
        End With


        .Border.LineStyle = xlNone

    End With

End Sub
```

Appendix 7.2: Excel Code—Uniform Distribution

```
'/**********************************************************************
'/ Essentials of Microsoft Excel 2013, SAS and
'/               MINITAB 17
'/ for Statistical and Financial Analysis
'/
'/**********************************************************************
Option Explicit
Dim mwbReport As Workbook
Dim mwsReportData As Worksheet

Property Set ReportWb(w As Workbook)
    Set mwbReport = w
End Property
Property Get ReportWb() As Workbook
    Set ReportWb = mwbReport
End Property
Private Sub cmdCancel_Click()
    Unload Me
End Sub

'/*******************************************************************
'/Purpose:  This is the procedure that is run after pressing the
'/          the run button on the form.  This is the main procedure
'/*******************************************************************
Private Sub cmdRun_Click()
    Me.Hide
    Uniform
    Unload Me
End Sub
```

```
Sub Uniform()
    Dim ChrtPDF As Chart
    Dim ChrtCDF As Chart
    Dim nCount As Integer
    Dim Ser As Series

    Set Me.ReportWb = Workbooks.Add
    Set ChrtPDF = Me.ReportWb.Charts.Add

    For nCount = 1 To 4

        If Len(Trim(Me.Controls("txtUpper" & nCount).Text)) <> 0 And _
           Len(Trim(Me.Controls("txtLower" & nCount).Text)) <> 0 Then
                UniformPDF chrtBox:=ChrtPDF, _
                lLowerNumber:=Me.Controls("txtLower" & nCount).Text, _
                lUpperNumber:=Me.Controls("txtUpper" & nCount).Text
        End If
    Next

    FormatChart chrtBox:=ChrtPDF

    With ChrtPDF
            .Activate
            .HasTitle = True
            With .ChartTitle.Characters
                .Text = "Uniform Probability Density Function"
                .Font.Size = 24
            End With
            .Name = "Uniform PDF"
             For Each Ser In .SeriesCollection
                With Ser
                    .MarkerStyle = xlMarkerStyleAutomatic
                    .Border.ColorIndex = 1
                    .MarkerForegroundColorIndex = 1
                    .MarkerBackgroundColorIndex = 1
                End With
            Next
            .Axes(xlCategory).CrossesAt = .Axes(xlCategory).MinimumScale

    End With
End Sub

'/*********************************************************************
'/Purpose: Creates both the PDF and CDF of the Uniform distribution.
'          The upper and lower numbers are prompted.  Used scatter plot
'/*********************************************************************
Sub UniformPDF(chrtBox As Chart, lLowerNumber As Long, lUpperNumber As Long)
    Dim seriesCurrent As Series
    Dim sngX(1 To 4) As Single
    Dim sngY(1 To 4) As Single

    chrtBox.Activate

    'Creat the Uniform Probability Density Function
    sngX(1) = lLowerNumber
    sngX(2) = lLowerNumber
    sngX(3) = lUpperNumber
    sngX(4) = lUpperNumber
    sngY(1) = 0
    sngY(2) = 1 / (lUpperNumber - lLowerNumber)
    sngY(3) = 1 / (lUpperNumber - lLowerNumber)
    sngY(4) = 0

    Set seriesCurrent = chrtBox.SeriesCollection.NewSeries

    With seriesCurrent
        .XValues = sngX
        .Values = sngY
        .Name = "a=" & lLowerNumber & ",b=" & lUpperNumber
    End With

End Sub
```

```
'/*****************************************************************
'/Purpose:  Format a basic Chart
'/*****************************************************************
Sub FormatChart(chrtBox As Chart)

    With chrtBox
            .Activate
            .ChartType = xlXYScatterLinesNoMarkers
            '.HasTitle = True
            With .Axes(xlCategory)
                  .HasMajorGridlines = False
                  .HasMinorGridlines = False
            End With
            With .Axes(xlValue)
                  .HasMajorGridlines = False
                  .HasMinorGridlines = False
            End With
            With .PlotArea
                .Interior.ColorIndex = xlNone
                .ClearFormats
            End With
            .HasLegend = True
    End With

End Sub

Private Sub UserForm_Initialize()
    Me.txtLower1 = 2
    Me.txtUpper1 = 7
End Sub
```

Appendix 7.3: QQ Plot

A QQ plot is a probability plot of quantiles of two distributions against each other. It is a graphical method for comparing two probability distributions. If the two distributions are similar, the points on the QQ plot will approximately lie on the line Y = X.

Below is an example to show you how to use Excel to graph a QQ plot. Assume we have gotten the JNJ stock return data. Data period is 1967–2011. We would like to compare this data and normal distribution.

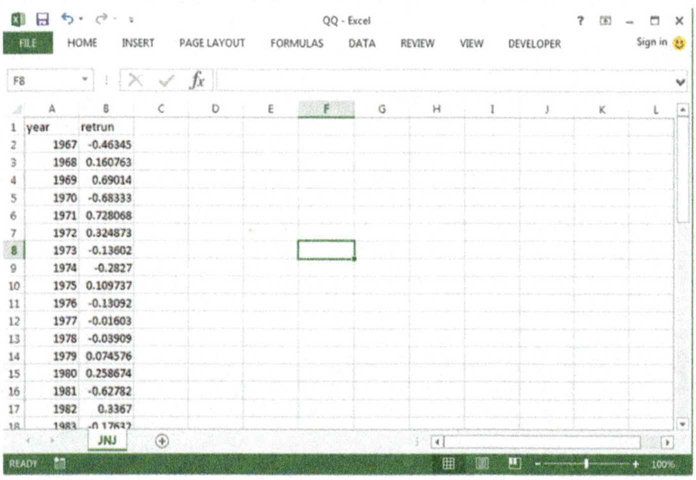

Here are the main steps to calculate the QQ plot:
Step 1: Select column B and choose Data → Sort A to Z.

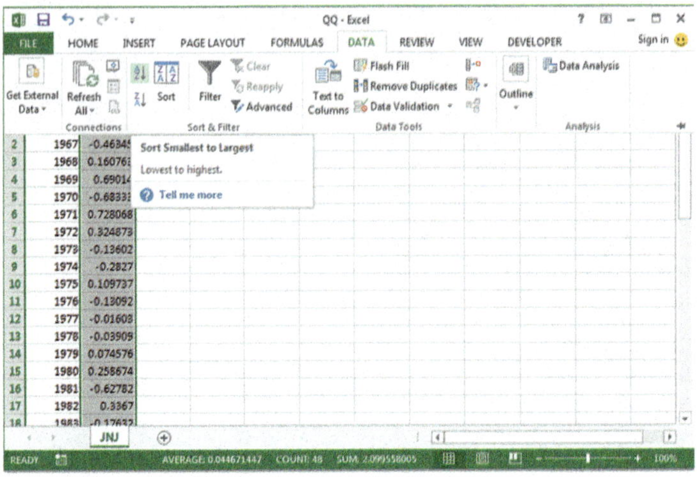

After pressing the A to Z button and expanding the selection, we get the result like below:

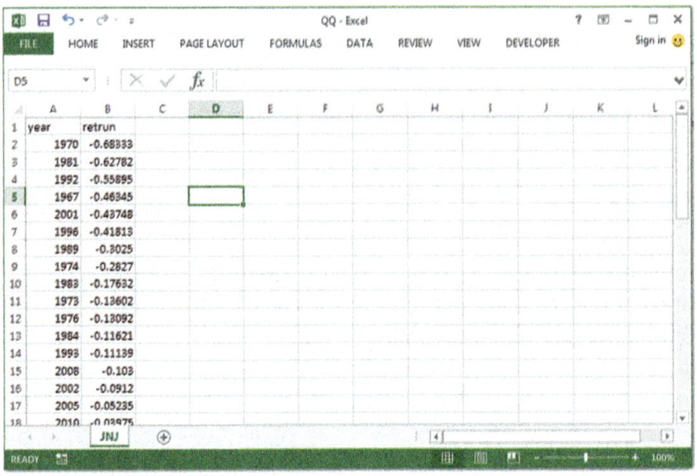

Step 2: Standardize the return data.

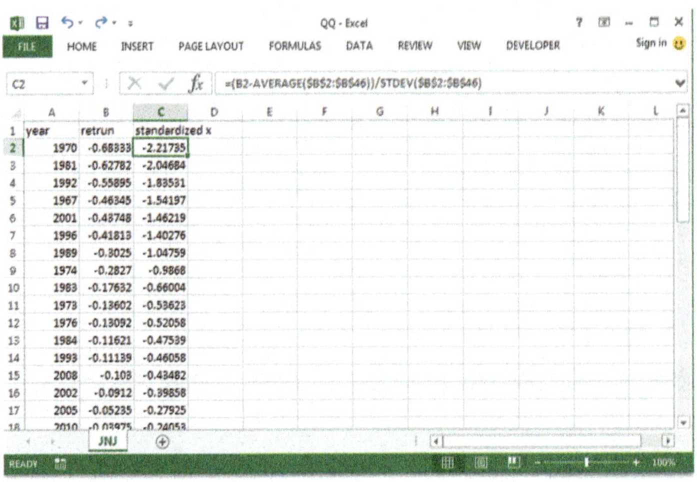

The formula for standardized return in cell C4 is

$$= \frac{(\text{B2} - \text{AVERAGE}(\$\text{B}\$2 : \$\text{B}\$46))}{\text{STDEV}(\$\text{B}\$2 : \$\text{B}\$46)}$$

Step 3: Construct the sample number column E.

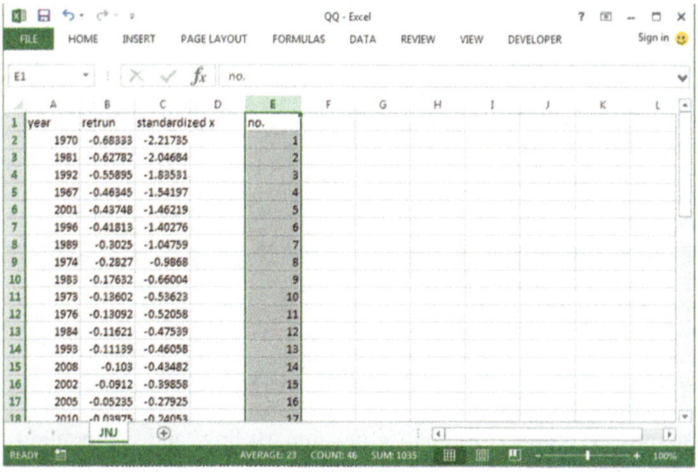

Step 4: Use the number of Step 3 to calculate the cumulative distribution function (CDF).

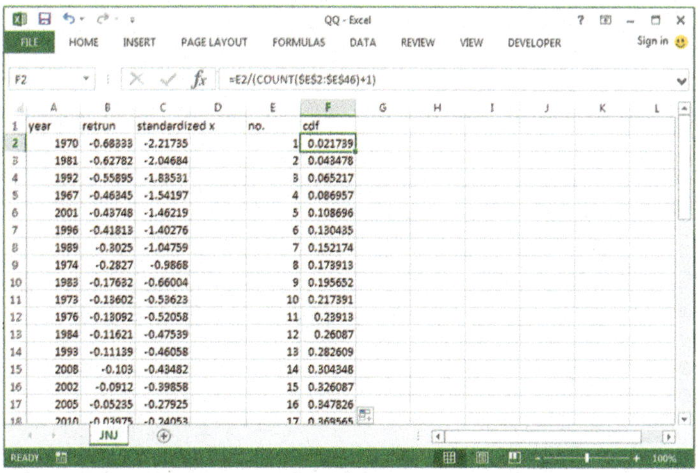

The CDF formula in the cell F2 is

$$= \frac{\text{E2}}{(\text{COUNT}(\$\text{E}\$2 : \$\text{E}\$46) + 1)}$$

Step 5: Use NORMSINV function to z of column F.

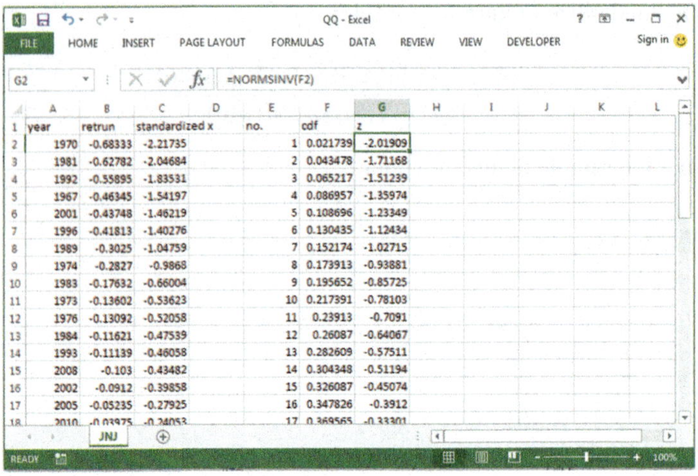

QQ - Excel

G2 =NORMSINV(F2)

	A	B	C	D	E	F	G
1	year	retrun	standardized x		no.	cdf	z
2	1970	-0.68333	-2.21735		1	0.021739	-2.01909
3	1981	-0.62782	-2.04684		2	0.043478	-1.71168
4	1992	-0.55895	-1.83531		3	0.065217	-1.51239
5	1967	-0.46345	-1.54197		4	0.086957	-1.35974
6	2001	-0.43748	-1.46219		5	0.108696	-1.23349
7	1996	-0.41813	-1.40276		6	0.130435	-1.12434
8	1989	-0.3025	-1.04759		7	0.152174	-1.02715
9	1974	-0.2827	-0.9868		8	0.173913	-0.93881
10	1983	-0.17632	-0.66004		9	0.195652	-0.85725
11	1973	-0.13602	-0.53623		10	0.217391	-0.78103
12	1976	-0.13092	-0.52058		11	0.23913	-0.7091
13	1984	-0.11621	-0.47539		12	0.26087	-0.64067
14	1993	-0.11139	-0.46058		13	0.282609	-0.57511
15	2008	-0.103	-0.43482		14	0.304348	-0.51194
16	2002	-0.0912	-0.39858		15	0.326087	-0.45074
17	2005	-0.05235	-0.27925		16	0.347826	-0.3912
18	2010	-0.03975	-0.24053		17	0.369565	-0.33301

JNJ

The formula of z in cell G2 is

$$= \text{NORMSINV(F2)}$$

Step 6: Use columns G, z and z, to chart a benchmark line, and use column C and column G, standard return and z, to graph a comparative line.

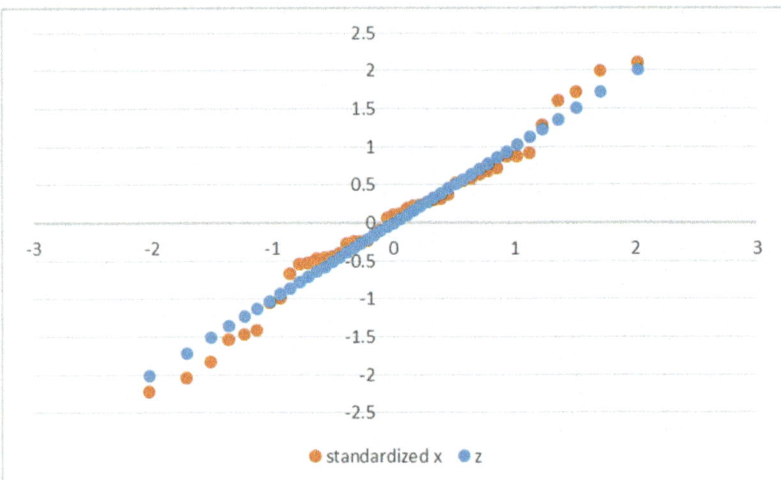

Bibliography

Microsoft Inc. Excel 2013. Microsoft Inc., Redmond
Lee CF, Lee JC, Lee AC (2013) Statistics for business and financial economics. Springer, New York
Minitab Inc. Minitab 17. Minitab Inc., State College
SAS Institute Inc. SAS 2014. SAS Institute Inc., Cary

Chapter 8
Sampling Distributions and Central Limit Theorem

8.1 Introduction

Many times it is impossible or too costly to analyze the population data. Because of this we are only able to analyze a sample from the population. After analyzing the sample data, are we able to understand the population data from the sample data? The answer is yes. In this chapter we will study why the answer is yes.

8.2 Sample Distribution

To study this issue, we will sample from a very small population. The concepts that we learn here will apply to large populations. To make things simple, let's suppose that the population that we are interested in analyzing has the following six elements:

$$1, 2, 3, 4, 5, 6$$

The mean of the above population would be 3.5.

Suppose we are interested in learning about the mean of the population. Suppose we are restricted in only randomly sample two items from the population at a time. One possible sample that we might get is a sample that consists of the elements 1 and 2. The mean of this sample would be 1.5. What can we say about the population mean from this sample? Not much. Let's say we randomly get another sample from the population. What is the chance of us getting another sample that has the elements 1 and 2 again? The chance would be small but greater than zero. Our second sample might contain the elements 1 and 3. This leads us to the question of what are all the possible combinations for a sample size of 2 from the population of interests. Below shows all the possible combinations and the mean of each possible combination.

C.-F. Lee et al., *Essentials of Excel, Excel VBA, SAS and Minitab for Statistical and Financial Analyses*, DOI 10.1007/978-3-319-38867-0_8

Possible sample combinations		
		Mean
1	2	1.5
1	3	2
1	4	2.5
1	5	3
1	6	3.5
2	3	2.5
2	4	3
2	5	3.5
2	6	4
3	4	3.5
3	5	4
3	6	4.5
4	5	4.5
4	6	5
5	6	5.5

Looking at the above table, we will notice that there are more occurrences of samples with the mean of 3.5 than 2. This mean there would be a greater chance for use to get a sample that has a mean of 3.5. In the above table, there are 15 elements. Let's suppose that there is an equal chance to randomly select any of the samples. If this is the case, we can then create the table below that shows the probability of getting each specific mean.

Sampling distribution	
1.5	6.67 %
2	6.67 %
2.5	13.33 %
3	13.33 %
3.5	20.00 %
4	13.33 %
4.5	13.33 %
5	6.67 %
5.5	6.67 %

The table above is often called a sampling distribution because it's the possible distributions of all the means from the samples.

Now when we get a sample that has the elements 1 and 2, we can say that there was a 6.67 % chance of getting a sample that has a mean of 1.5. And if we get a sample that contains the elements 1 and 3, we can say that there was 20 % chance of getting a sample that has a mean of 3.5.

It is important to note that the mean of this sample distribution is

$$
\begin{aligned}
&(1.5^*.067) + (2^*.067) + (2.5^*.1333) \\
&+(3^*.1333) + (3.5^*.2) + (4^*.1333) \\
&+(4.5^*.1333) + (5^*.067) + (5.5^*.067) = 3.5
\end{aligned}
$$

which is the same as the population mean. Below is a histogram of the sample distribution.

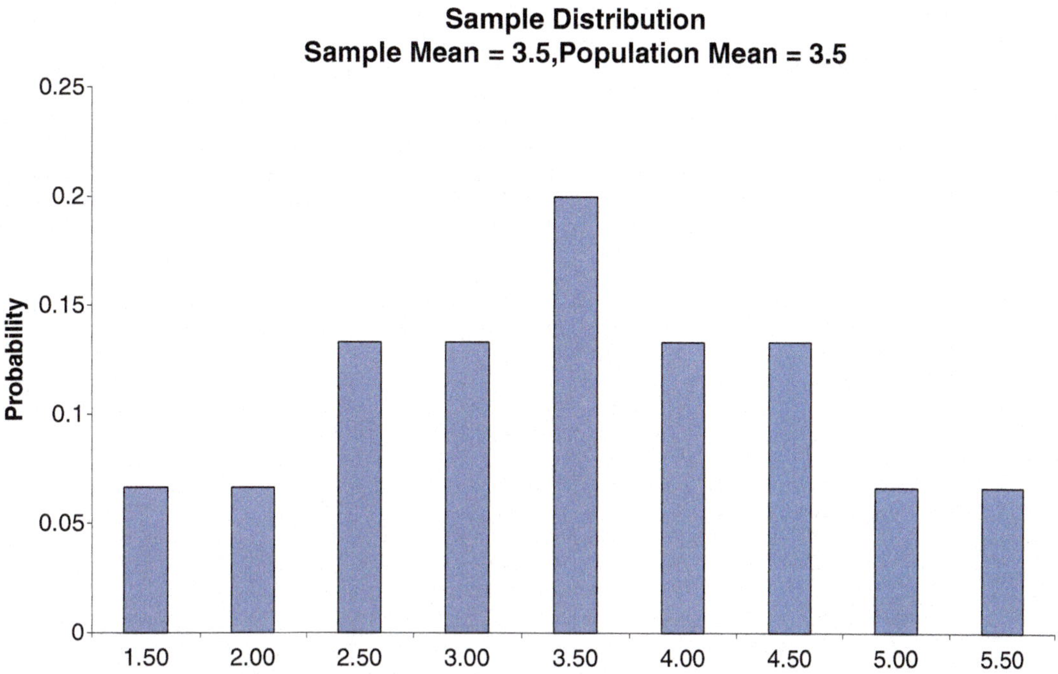

Let's now create a sample distribution with a sample size of 3 and 4.

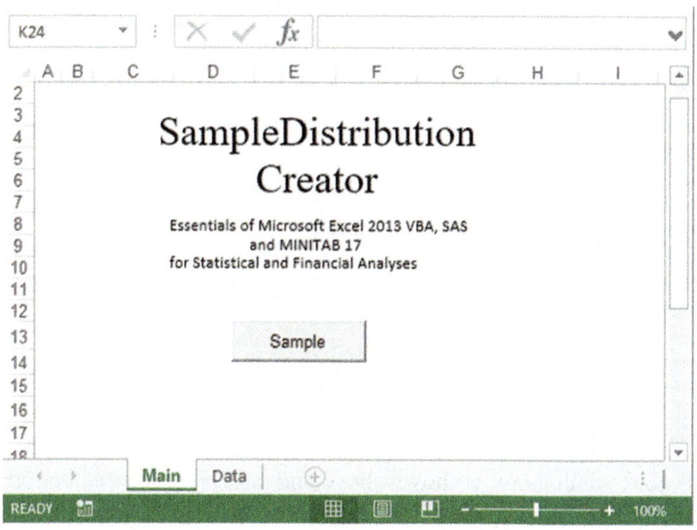

The workbook *sampledistribution.xlsm* will consist of two worksheets. The *Main* worksheet is where we push the button to create a sample distribution from the population distribution. The *Data* worksheet is where we indicate the population data set. Let's put the population data set in the worksheet *Data* as shown below.

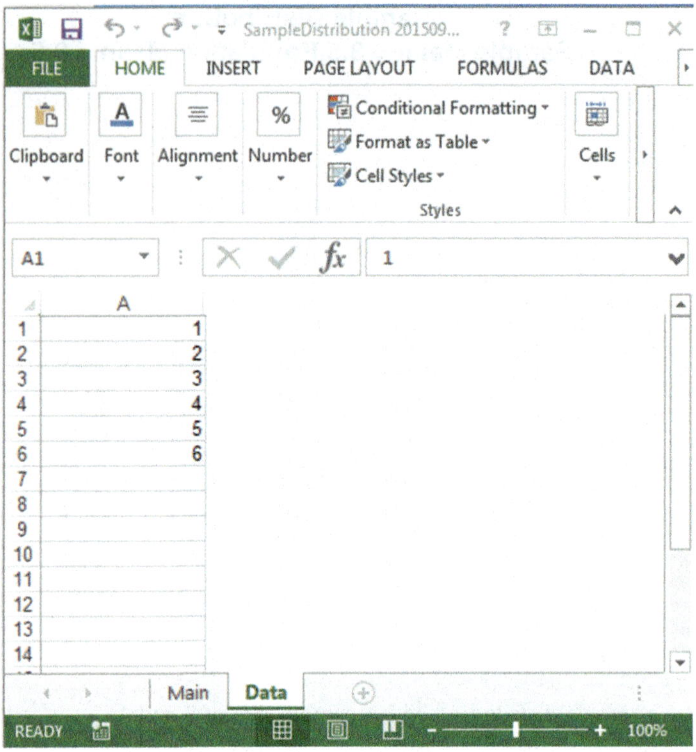

Now we will create the sample distribution when the same size is 3 by pressing the button on the *Main* worksheet. Pressing the button will result in the following dialog box.

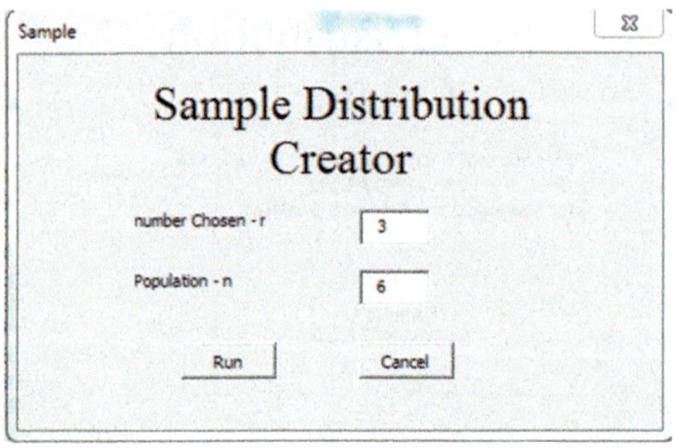

If we fill out the above dialog box as shown above and then press the *Run* button, a workbook with the following sheets will be created.

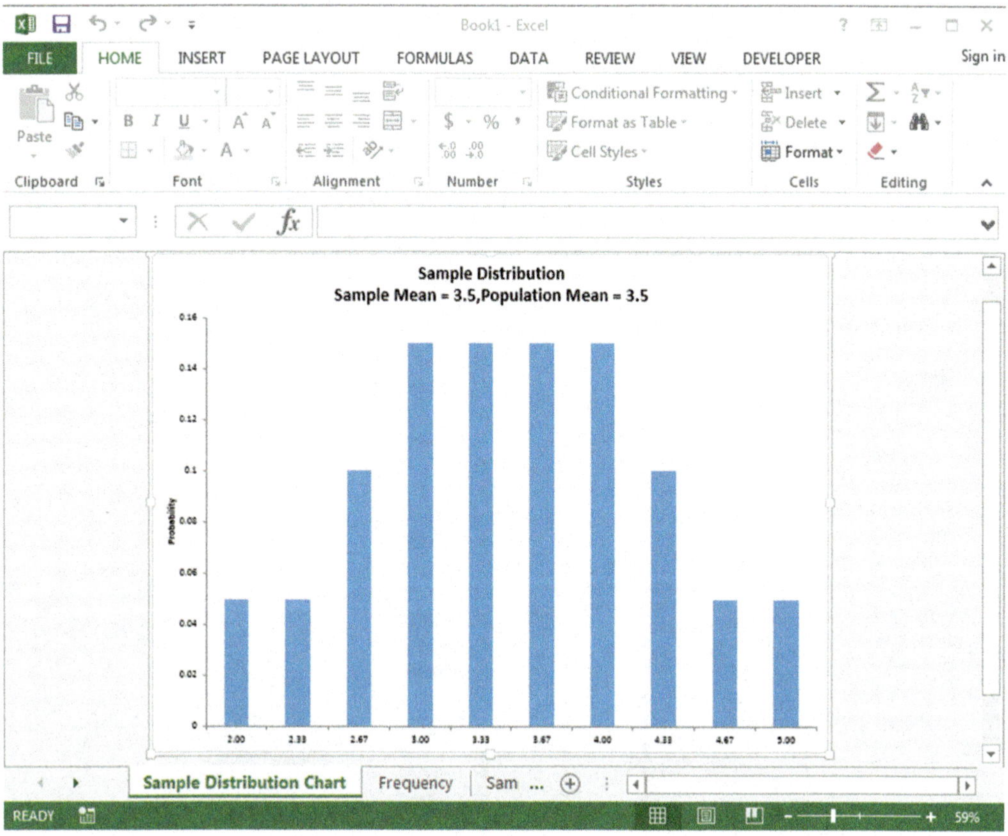

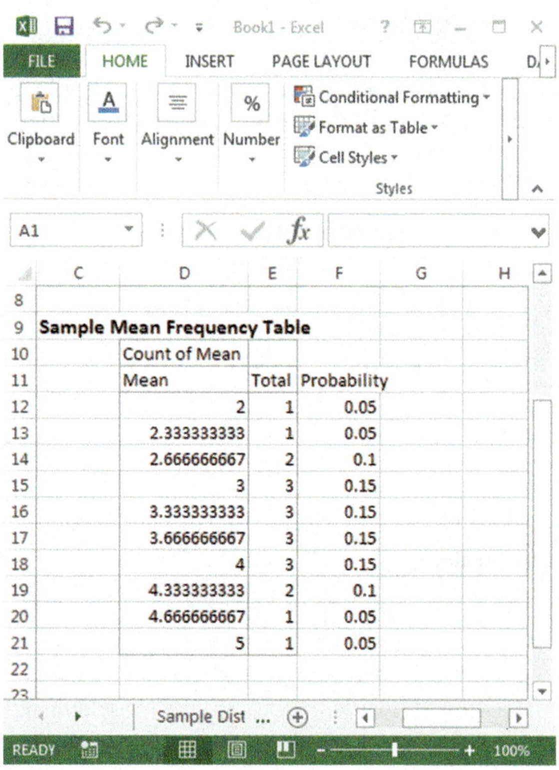

Sample Mean Frequency Table

Count of Mean		
Mean	Total	Probability
2	1	0.05
2.333333333	1	0.05
2.666666667	2	0.1
3	3	0.15
3.333333333	3	0.15
3.666666667	3	0.15
4	3	0.15
4.333333333	2	0.1
4.666666667	1	0.05
5	1	0.05

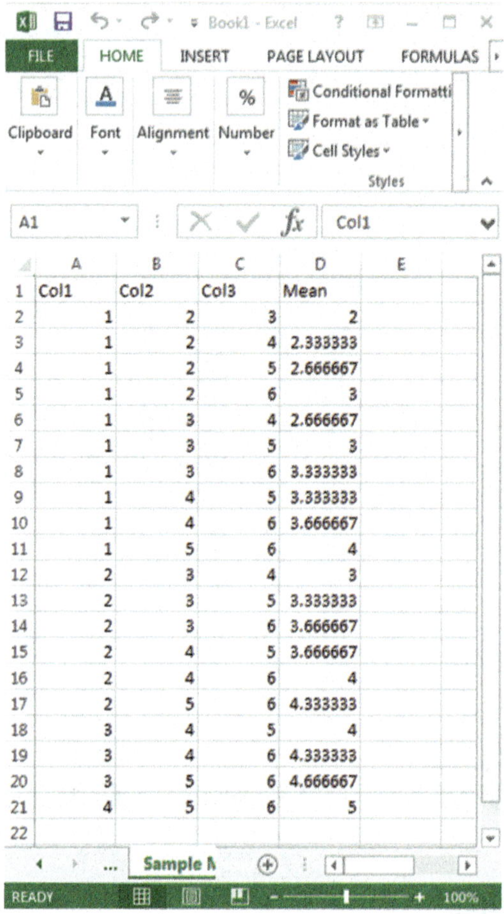

Now we will create the sample distribution when the same size is 4 by pressing the button on the *Main* worksheet. Pressing the button will result in the following dialog box.

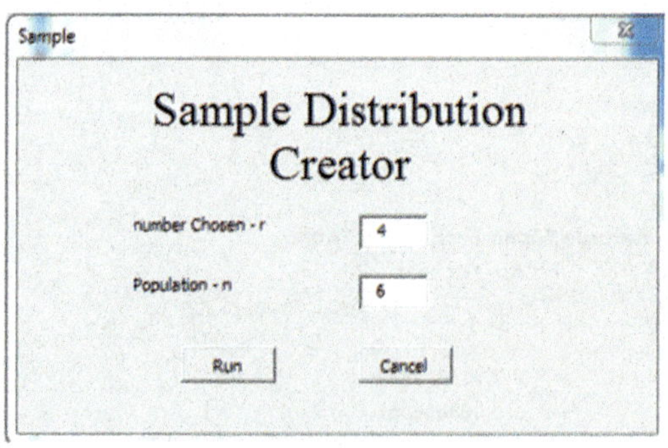

If we fill out the above dialog box as shown above and then press the *Run* button, a workbook with the following sheets will be created.

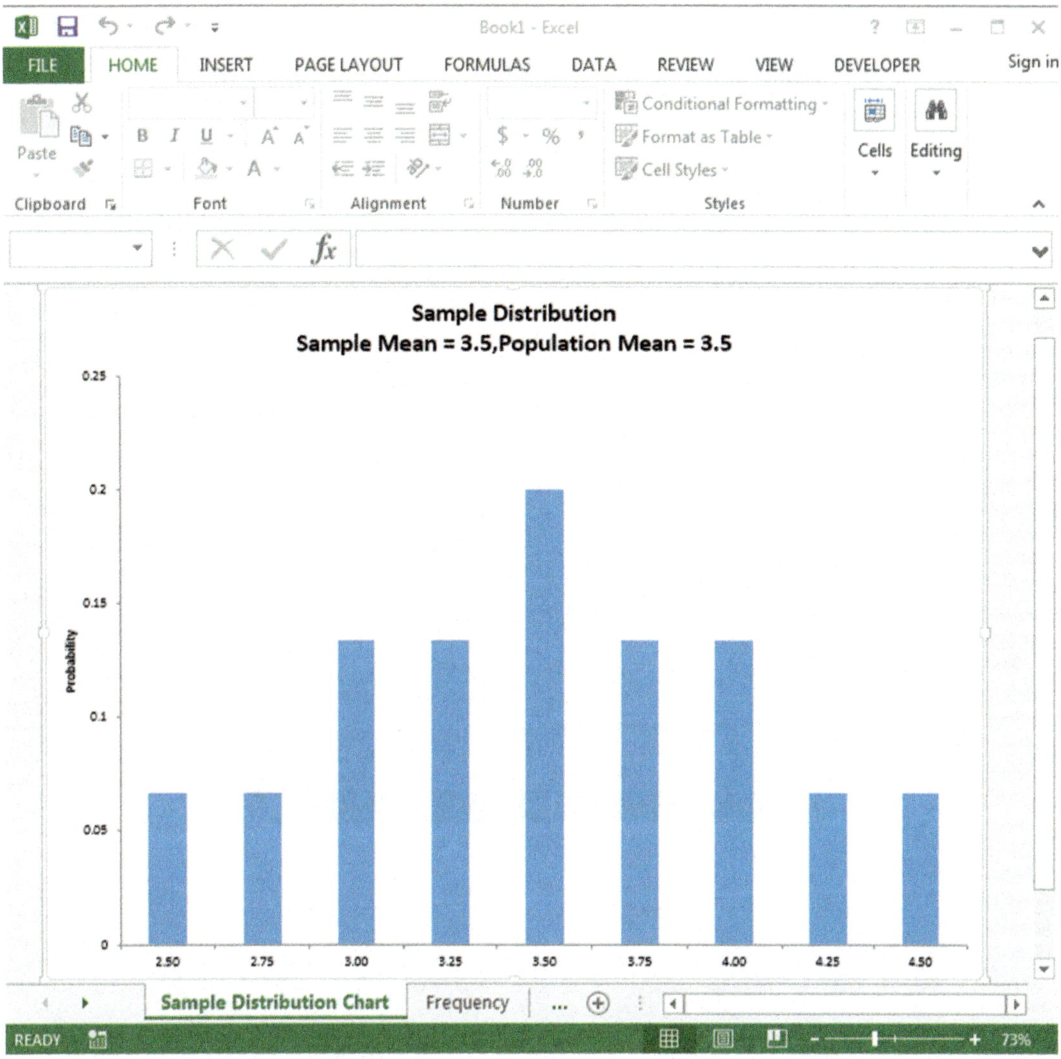
Sample Distribution
Sample Mean = 3.5,Population Mean = 3.5
Probability
0.25
0.2
0.15
0.1
0.05
0
2.50
2.75
3.00
3.25
3.50
3.75
4.00
4.25
4.50
Sample Distribution Chart
Frequency

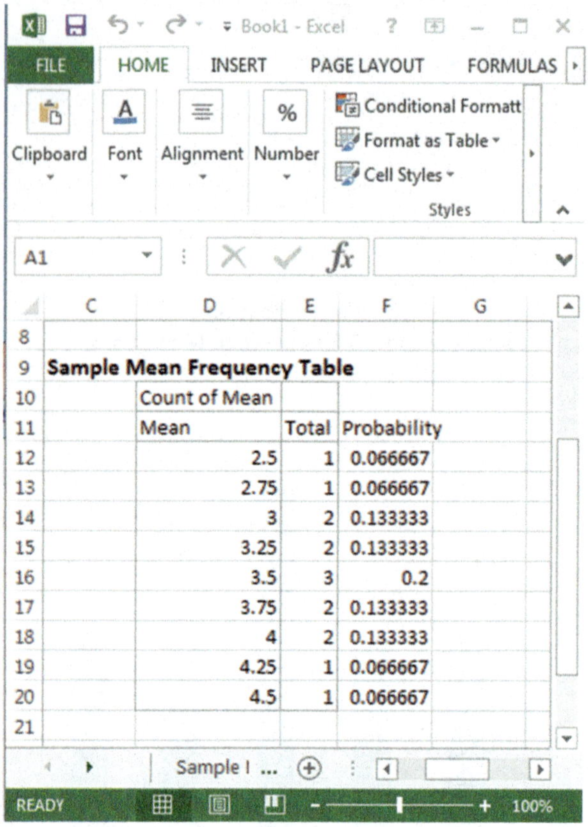

Sample Mean Frequency Table

Count of Mean		
Mean	Total	Probability
2.5	1	0.066667
2.75	1	0.066667
3	2	0.133333
3.25	2	0.133333
3.5	3	0.2
3.75	2	0.133333
4	2	0.133333
4.25	1	0.066667
4.5	1	0.066667

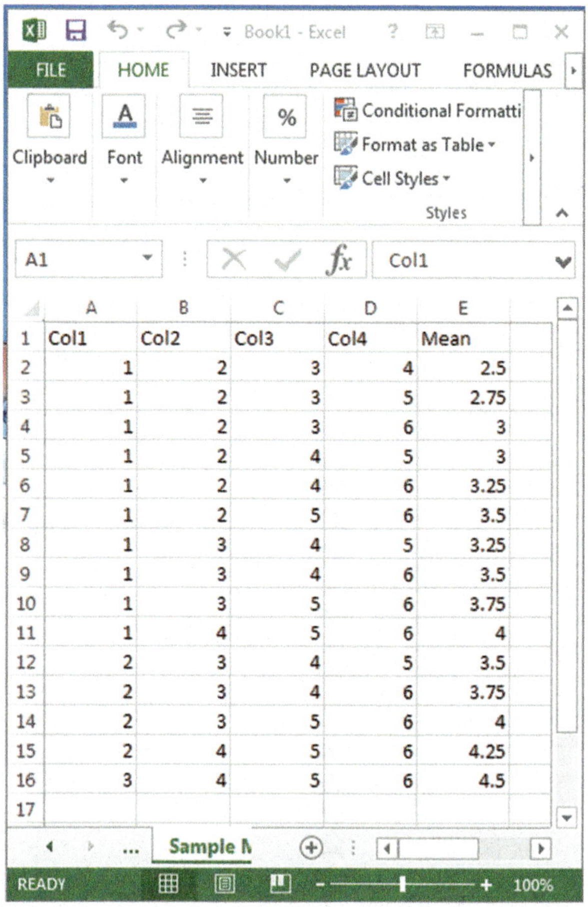

	A	B	C	D	E
1	Col1	Col2	Col3	Col4	Mean
2	1	2	3	4	2.5
3	1	2	3	5	2.75
4	1	2	3	6	3
5	1	2	4	5	3
6	1	2	4	6	3.25
7	1	2	5	6	3.5
8	1	3	4	5	3.25
9	1	3	4	6	3.5
10	1	3	5	6	3.75
11	1	4	5	6	4
12	2	3	4	5	3.5
13	2	3	4	6	3.75
14	2	3	5	6	4
15	2	4	5	6	4.25
16	3	4	5	6	4.5
17					

Notice that as the sample size increases, the resulting sample distribution approaches a normal distribution.

8.3 Mean of Sample Distribution Equals Mean of Population

It is important to note that the mean of a sample distribution is equal to the mean of the population. We will demonstrate this with the following population.

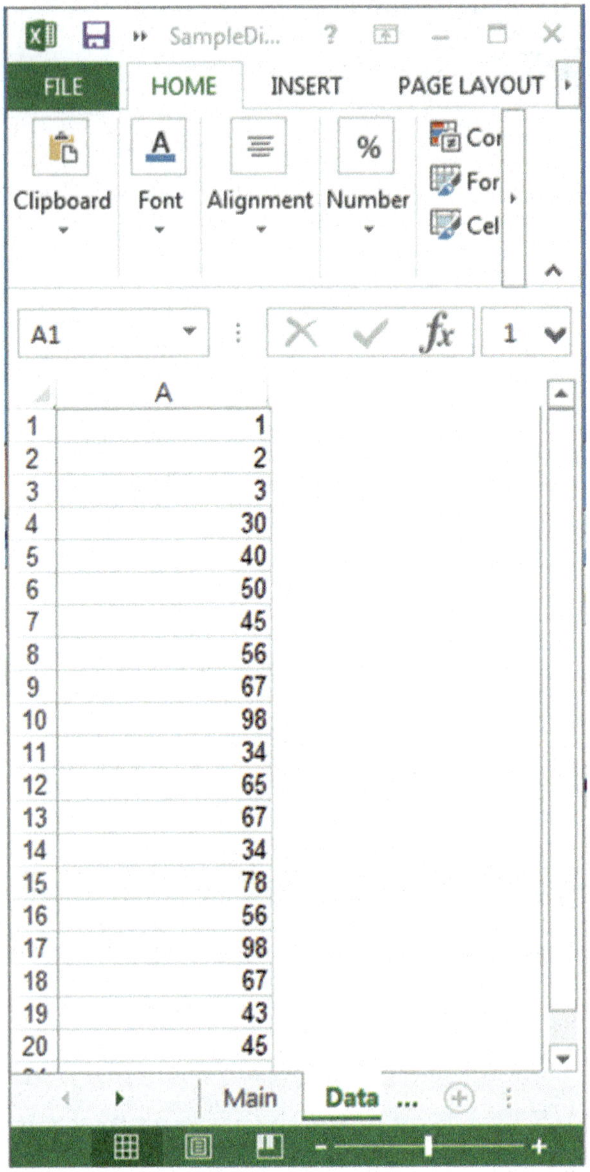

With this population we will create the sample distributions with the sizes 5 and 15. The following is the sample distribution when the sample size is 5.

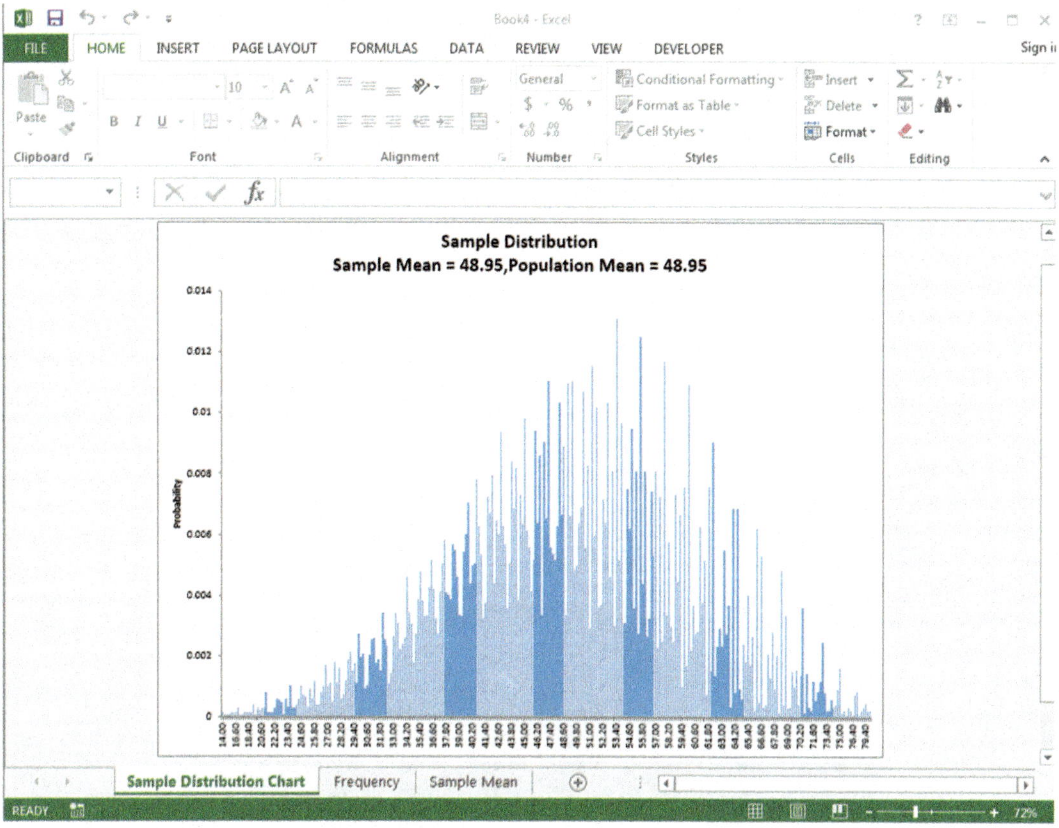

There are 15,504 possible sample combinations for this distribution.

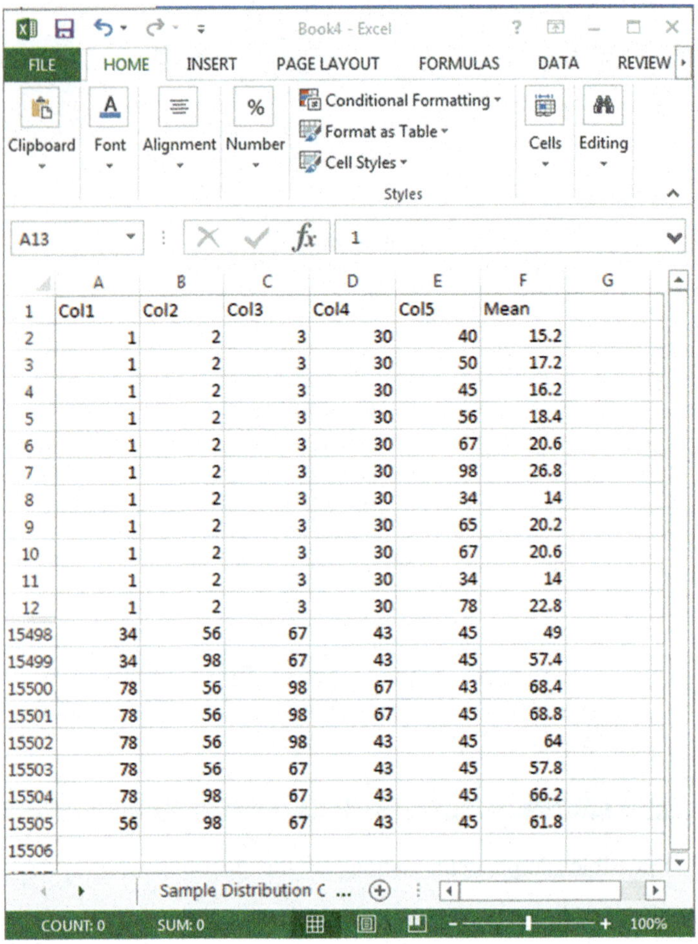

	A	B	C	D	E	F	G
1	Col1	Col2	Col3	Col4	Col5	Mean	
2	1	2	3	30	40	15.2	
3	1	2	3	30	50	17.2	
4	1	2	3	30	45	16.2	
5	1	2	3	30	56	18.4	
6	1	2	3	30	67	20.6	
7	1	2	3	30	98	26.8	
8	1	2	3	30	34	14	
9	1	2	3	30	65	20.2	
10	1	2	3	30	67	20.6	
11	1	2	3	30	34	14	
12	1	2	3	30	78	22.8	
15498	34	56	67	43	45	49	
15499	34	98	67	43	45	57.4	
15500	78	56	98	67	43	68.4	
15501	78	56	98	67	45	68.8	
15502	78	56	98	43	45	64	
15503	78	56	67	43	45	57.8	
15504	78	98	67	43	45	66.2	
15505	56	98	67	43	45	61.8	
15506							

There are 297 possible means for this distribution.

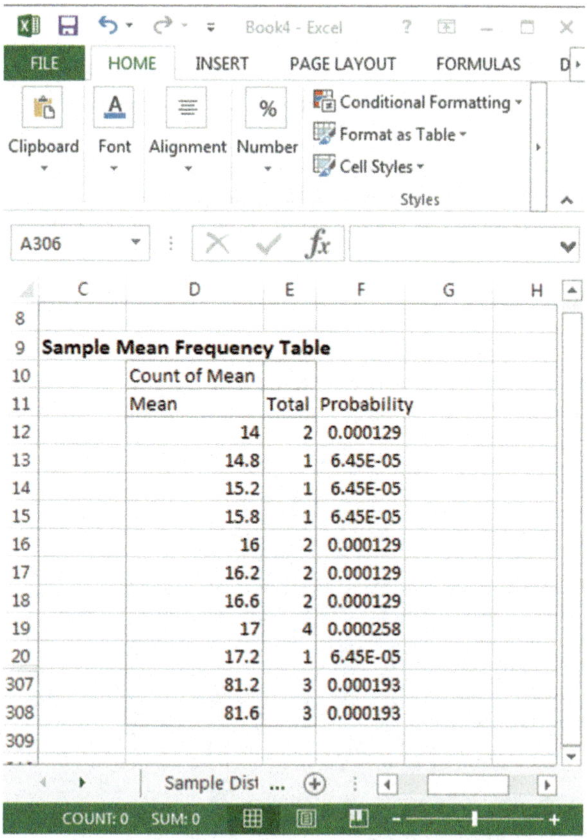

Sample Mean Frequency Table

Count of Mean		
Mean	Total	Probability
14	2	0.000129
14.8	1	6.45E-05
15.2	1	6.45E-05
15.8	1	6.45E-05
16	2	0.000129
16.2	2	0.000129
16.6	2	0.000129
17	4	0.000258
17.2	1	6.45E-05
81.2	3	0.000193
81.6	3	0.000193

The following is the sample distribution when the sample size is 15.

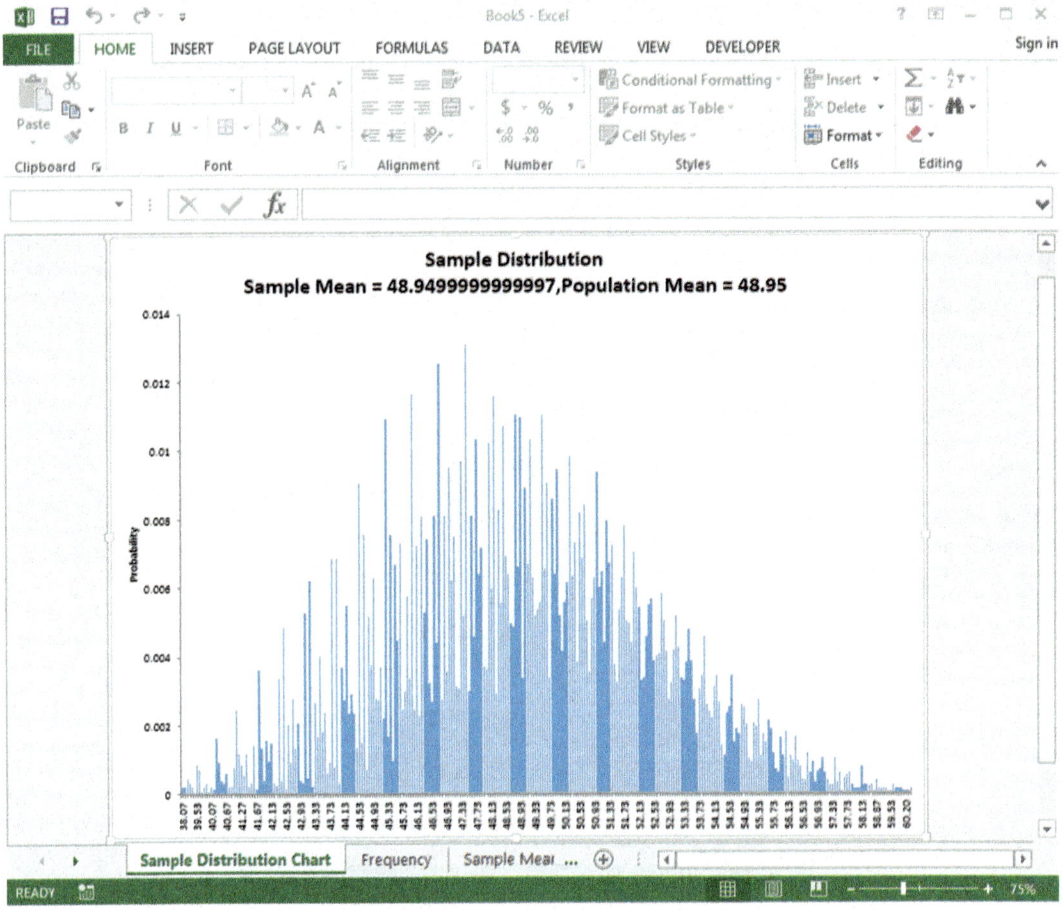

There are 15,504 possible sample combinations for this distribution.

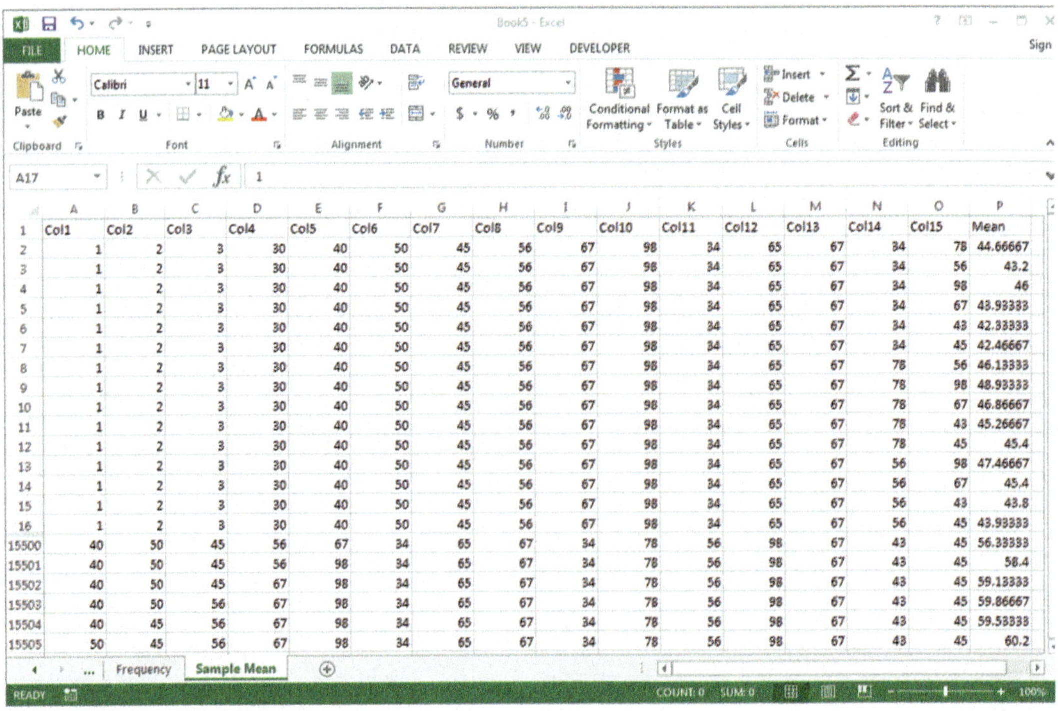

	A	B	C	D	E	F	G	H	I	J	K	L	M	N	O	P
1	Col1	Col2	Col3	Col4	Col5	Col6	Col7	Col8	Col9	Col10	Col11	Col12	Col13	Col14	Col15	Mean
2	1	2	3	30	40	50	45	56	67	98	34	65	67	34	78	44.66667
3	1	2	3	30	40	50	45	56	67	98	34	65	67	34	56	43.2
4	1	2	3	30	40	50	45	56	67	98	34	65	67	34	98	46
5	1	2	3	30	40	50	45	56	67	98	34	65	67	34	67	43.93333
6	1	2	3	30	40	50	45	56	67	98	34	65	67	34	43	42.33333
7	1	2	3	30	40	50	45	56	67	98	34	65	67	34	45	42.46667
8	1	2	3	30	40	50	45	56	67	98	34	65	67	78	56	46.13333
9	1	2	3	30	40	50	45	56	67	98	34	65	67	78	98	48.93333
10	1	2	3	30	40	50	45	56	67	98	34	65	67	78	67	46.86667
11	1	2	3	30	40	50	45	56	67	98	34	65	67	78	43	45.26667
12	1	2	3	30	40	50	45	56	67	98	34	65	67	78	45	45.4
13	1	2	3	30	40	50	45	56	67	98	34	65	67	56	98	47.46667
14	1	2	3	30	40	50	45	56	67	98	34	65	67	56	67	45.4
15	1	2	3	30	40	50	45	56	67	98	34	65	67	56	43	43.8
16	1	2	3	30	40	50	45	56	67	98	34	65	67	56	45	43.93333
15500	40	50	45	56	67	34	65	67	34	78	56	98	67	43	45	56.33333
15501	40	50	45	56	98	34	65	67	34	78	56	98	67	43	45	58.4
15502	40	50	45	67	98	34	65	67	34	78	56	98	67	43	45	59.13333
15503	40	50	56	67	98	34	65	67	34	78	56	98	67	43	45	59.86667
15504	40	45	56	67	98	34	65	67	34	78	56	98	67	43	45	59.53333
15505	50	45	56	67	98	34	65	67	34	78	56	98	67	43	45	60.2

There are 297 possible means for this distribution.

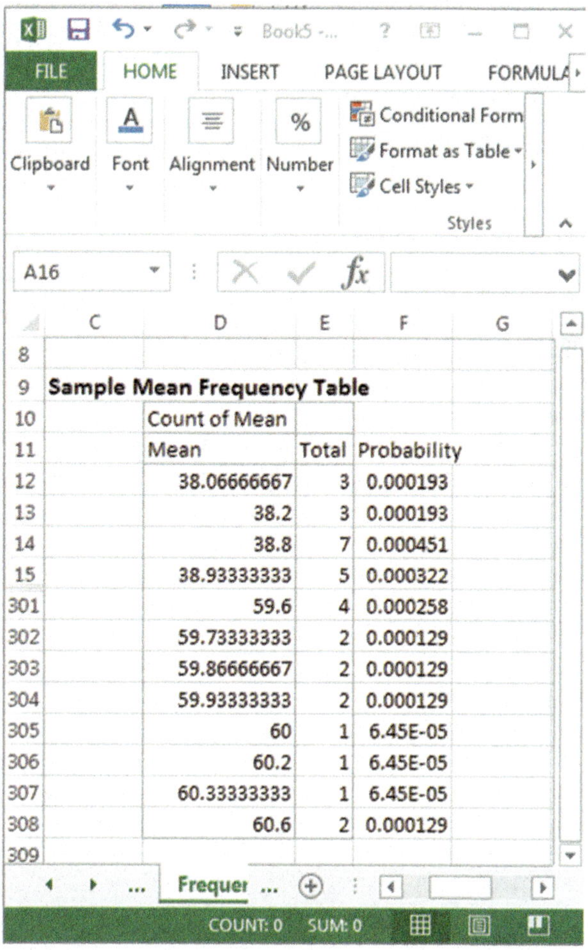

	C	D	E	F	G
8					
9	**Sample Mean Frequency Table**				
10		Count of Mean			
11		Mean	Total	Probability	
12		38.06666667	3	0.000193	
13		38.2	3	0.000193	
14		38.8	7	0.000451	
15		38.93333333	5	0.000322	
301		59.6	4	0.000258	
302		59.73333333	2	0.000129	
303		59.86666667	2	0.000129	
304		59.93333333	2	0.000129	
305		60	1	6.45E-05	
306		60.2	1	6.45E-05	
307		60.33333333	1	6.45E-05	
308		60.6	2	0.000129	
309					

8.4 Central Limit Theorem

In the above sections, we calculated the sample distribution. The sample distributions that we created in Sect. 8.2 were relatively simple to calculate. We saw that the sample distributions that we created in Sect. 8.3 were more complicated. In general most sample distributions that we are interested in are very hard to calculate. Fortunately because of the *central limit theorem*, we do not have to calculate sample distributions. The *central limit theorem* is stated below.

> As the sample size (n) from a given population gets 'large enough,' the sampling distribution of the mean, $\bar{x}$, can be approximated by a normal distribution with mean, μ, and standard deviation, $\sigma/\sqrt{n}$, regardless of the distribution of the individual values in the population.

Notice that the *central limit theorem* does not indicate what the sample size should be before it approximates a normal distribution. Some population distributions have sample distributions that approximate a normal distribution when the sample size is small, while other population distributions require a larger sample size before the sample distribution approximates a normal distribution. By convention most people use a rule of thumb of having the sample size greater than 30 to guarantee that a sample distribution approximates a normal distribution.

To illustrate that *central limit theorem*, we will randomly select 300 samples for the distributions that we have studied.

For each distribution we will randomly select:

1. 300 samples of size 5
2. 300 samples of size 10
3. 300 samples of size 30
4. 300 samples of size 50

From this exercise, we will see that for every distribution shown in this chapter, the sample mean distribution approximates a normal distribution. Some sample mean distribution will have an approximate normal distribution with a sample size as small as 5, whereas other sample mean distributions will approach a normal distribution as the sample size increases.

We will use the *random* command to generate random numbers from any distribution given by its subcommand. It should be noted that since these numbers are randomly generated, an exact reproduction of the histogram is not possible, but the general behavior of each histogram shown below should result.

8.5 Uniform Distribution

Sample Size = 5, 300 Samples

```
MTB > random 300 c1-c5;
SUBC> uniform 5 6.
MTB > rmean c1-c5 c10
MTB > Histogram c10;
SUBC>   MidPoint;
SUBC>   NInterval 100;
SUBC>   Bar;
SUBC>   Axlabel 1 "X-Axis".
```

Histogram of C10

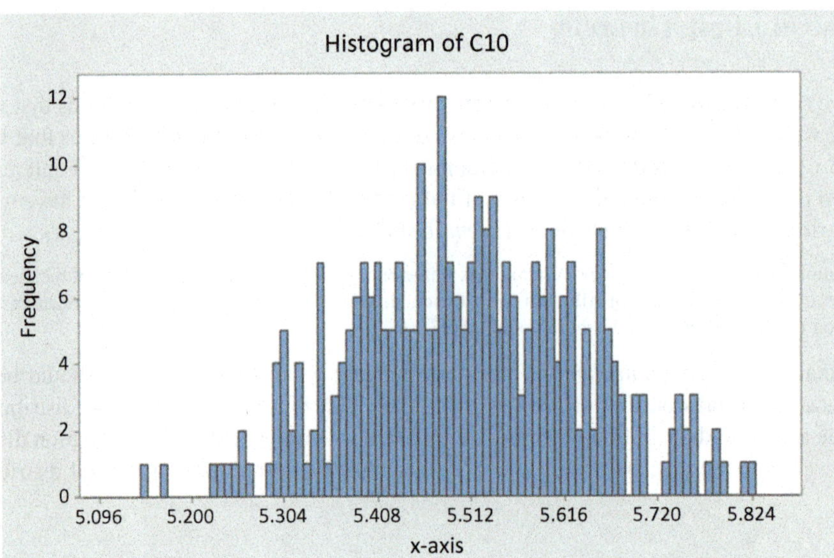

Sample Size = 10, 300 Samples

```
MTB > random 300 c1-c10;
SUBC> uniform 5 6.
MTB > rmean c1-c10 c15
MTB > Histogram c15;
SUBC>   MidPoint;
SUBC>   NInterval 100;
SUBC>   Bar;
SUBC>   Axlabel 1 "X-Axis".
```

Histogram of C15

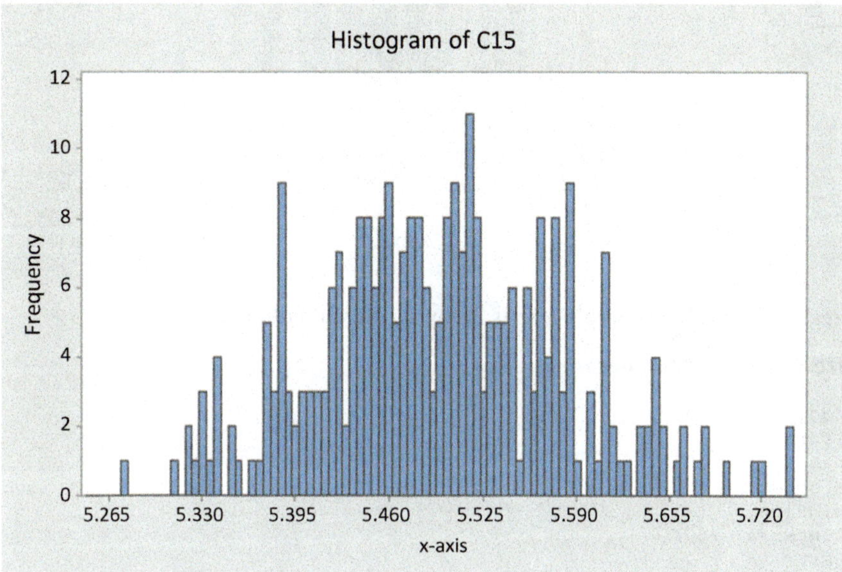

We represent the fact that sample size is 10 by having ten columns.

Sample Size = 30, 300 Samples

```
MTB > random 300 c1-c30;
SUBC> uniform 5 6.
MTB > rmean c1-c30 c35
MTB > Histogram c35;
SUBC>   MidPoint;
SUBC>   NInterval 100;
SUBC>   Bar;
SUBC>   Axlabel 1 "X-Axis".
```

Histogram of C35

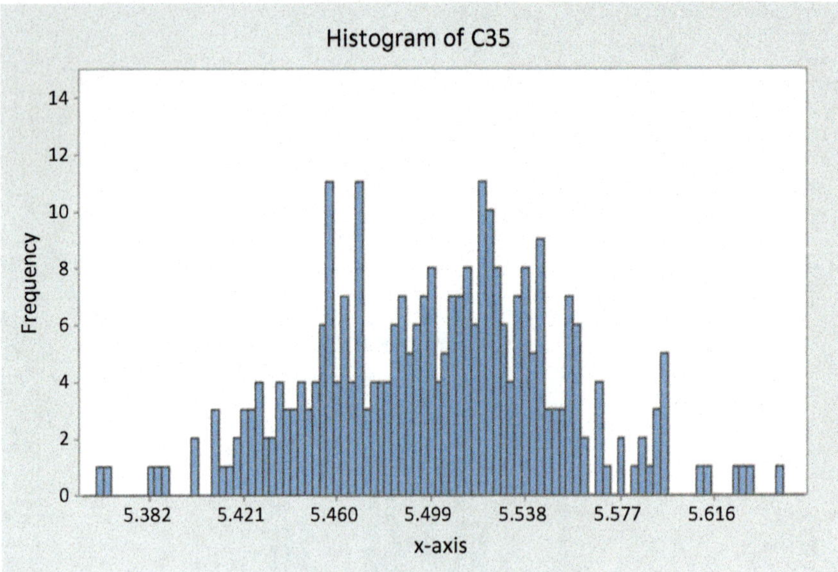

We represent the fact that sample size is 30 by having 30 columns.

Sample Size = 50, 300 Samples

```
MTB > random 300 c1-c50;
SUBC> uniform 5 6.
MTB > rmean c1-c50 c55
MTB > Histogram c55;
SUBC>   MidPoint;
SUBC>   NInterval 100;
SUBC>   Bar;
SUBC>   Axlabel 1 "X-Axis".
```

Histogram of C55

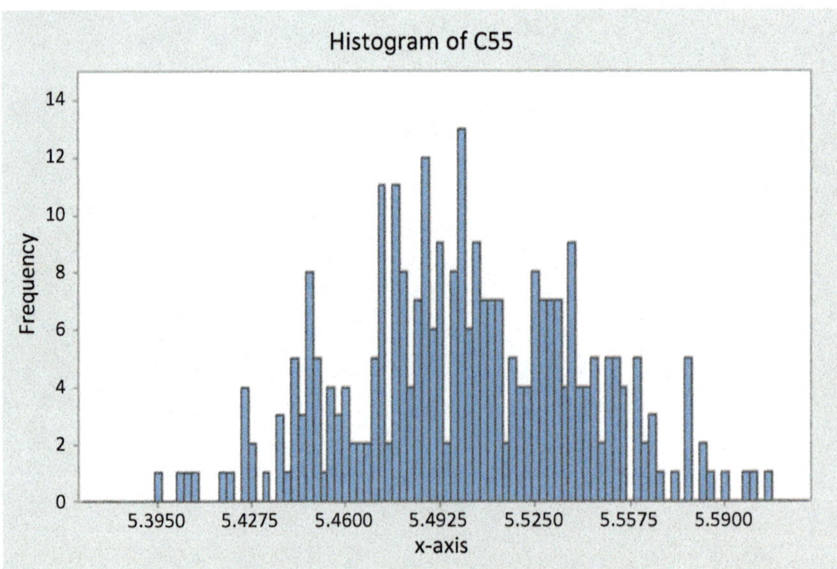

We represent the fact that sample size is 50 by having 50 columns.

8.6 Normal Distribution

In the commands issued below, the *normal* subcommand tells the *random* command to generate random numbers from the normal distribution. The first argument of the normal distribution is the mean of the distribution. The second argument is the standard deviation of the distribution.

Sample Size = 5, 300 Samples

```
MTB > random 300 c1-c5;
SUBC> normal 5 2.
MTB > rmean c1-c5 c10
MTB > Histogram C10;
SUBC>   MidPoint;
SUBC>   NInterval 100;
SUBC>   Bar;
SUBC>   Axlabel 1 "X-Axis".
```

Histogram of C10

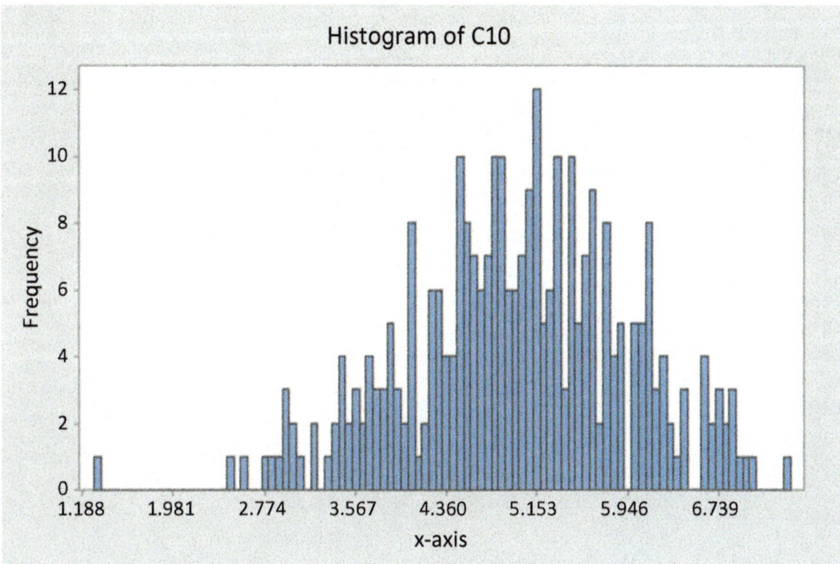

Sample Size = 10, 300 Samples

```
MTB > random 300 c1-c10;
SUBC> normal 5 2.
MTB > rmean c1-c10 c15
MTB > Histogram c15;
SUBC>   MidPoint;
SUBC>   NInterval 100;
SUBC>   Bar;
SUBC>   Axlabel 1 "X-Axis".
```

Histogram of C15

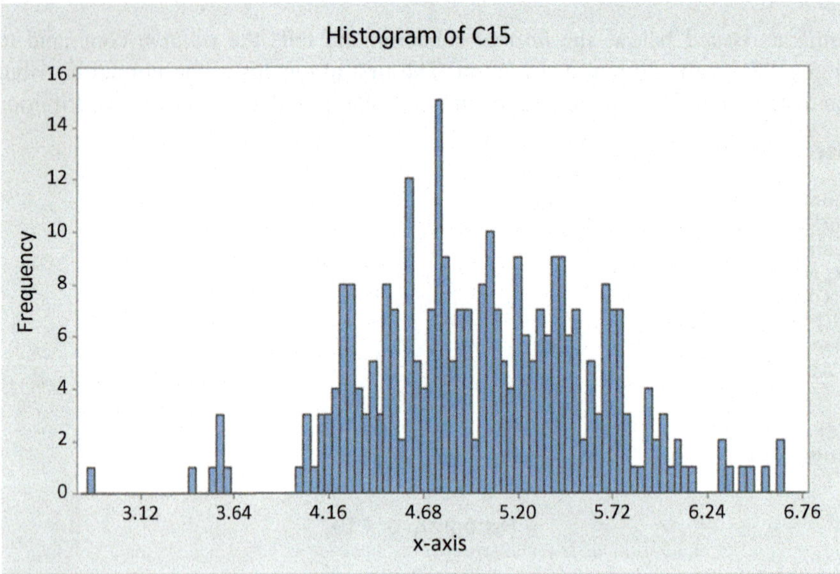

Sample Size = 30, 300 Samples

```
MTB > random 300 c1-c30;
SUBC> normal 5 2.
MTB > rmean c1-c30 c35
MTB > Histogram c35;
SUBC>   MidPoint;
SUBC>   NInterval 100;
SUBC>   Bar;
SUBC>    Axlabel 1 "X-Axis".
```

Histogram of C35

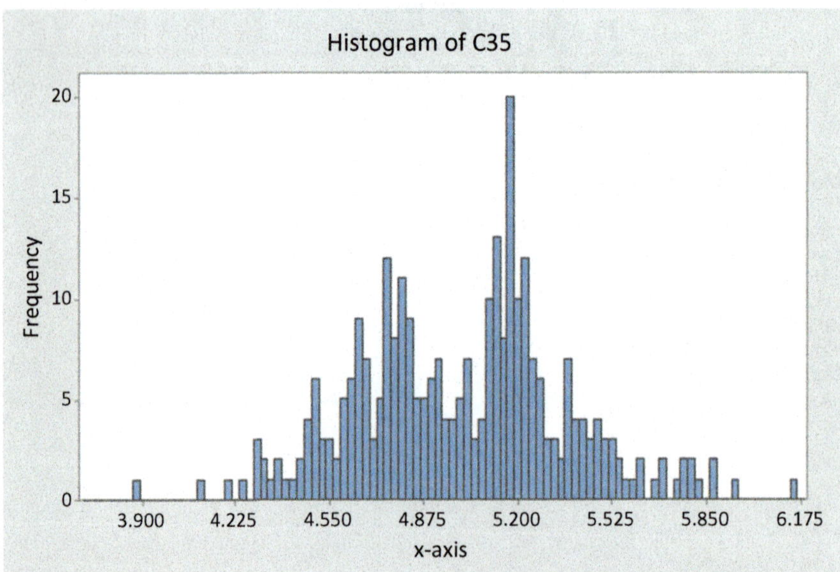

Sample Size = 50, 300 Samples

```
MTB > random 300 c1-c50;
SUBC> normal 5 2.
MTB > rmean c1-c50 c55
MTB > Histogram c55;
SUBC>   MidPoint;
SUBC>   NInterval 100;
SUBC>   Bar;
SUBC>    Axlabel 1 "X-Axis".
```

Histogram of C55

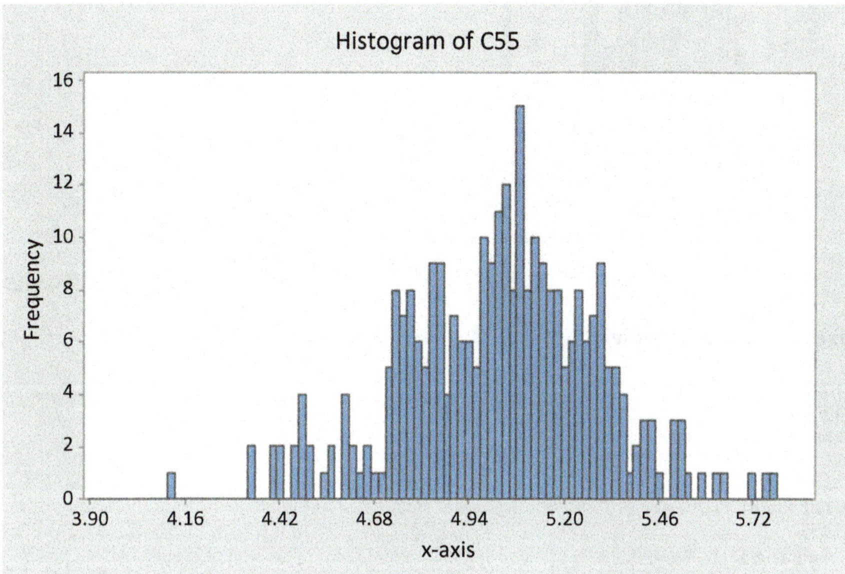

8.7 Lognormal Distribution

In the commands issued below, the *lognormal* subcommand tells the *random* command to generate numbers randomly from the lognormal distribution. The first argument of the *lognormal* subcommand is the mean. The second argument is the standard deviation.

Sample Size = 5, 300 Samples

```
MTB > random 300 c1-c5;
SUBC> lognormal 1 1.
MTB > rmean c1-c5 c10
MTB > Histogram C10;
SUBC>   MidPoint;
SUBC>   NInterval 100;
SUBC>   Bar;
SUBC>    Axlabel 1 "X-Axis".
```

Histogram of C10

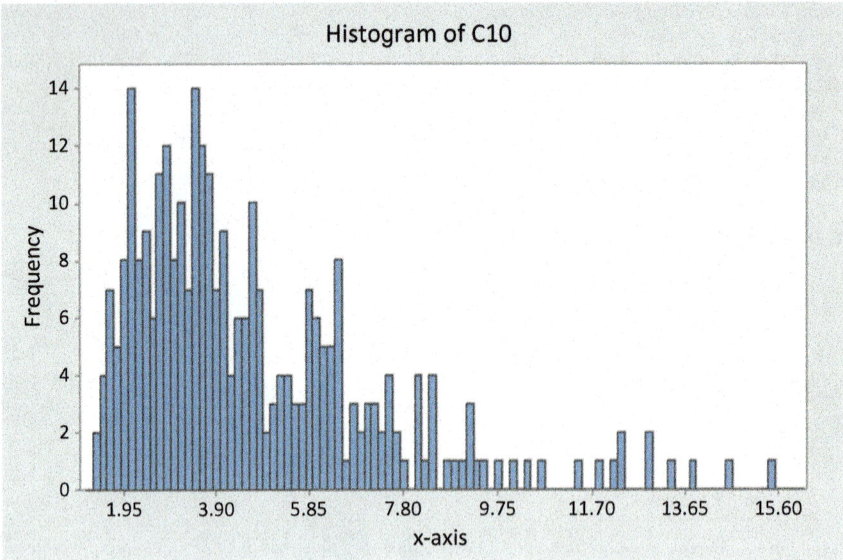

Sample Size = 10, 300 Samples

```
MTB > random 300 c1-c10;
SUBC> lognormal 1 1.
MTB > rmean c1-c10 c15
MTB > Histogram C15;
SUBC>   MidPoint;
SUBC>   NInterval 100;
SUBC>   Bar;
SUBC>    Axlabel 1 "X-Axis".
```

Histogram of C15

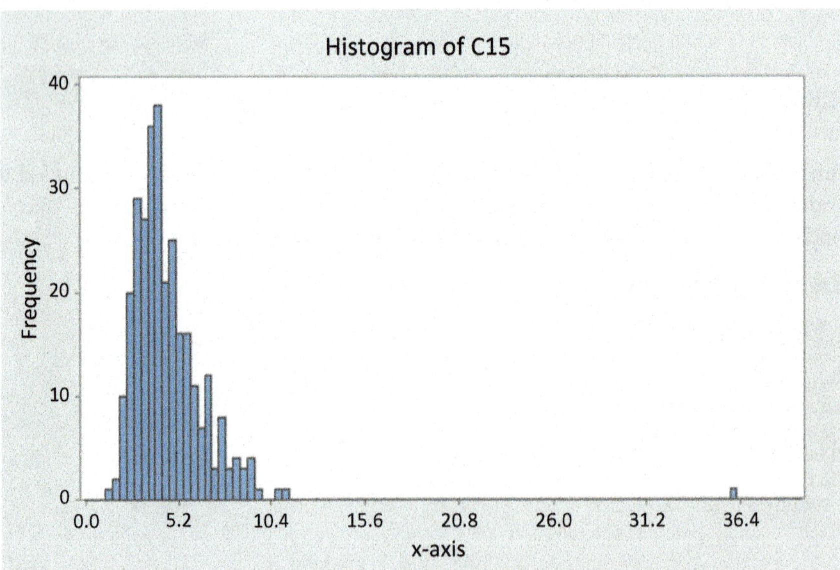

Sample Size = 30, 300 Samples

```
MTB > random 300 c1-c30;
SUBC> lognormal 1 1.
MTB > rmean c1-c30 c35
MTB > Histogram C35;
SUBC>   MidPoint;
SUBC>   NInterval 100;
SUBC>   Bar;
SUBC>   Axlabel 1 "X-Axis".
```

Histogram of C35

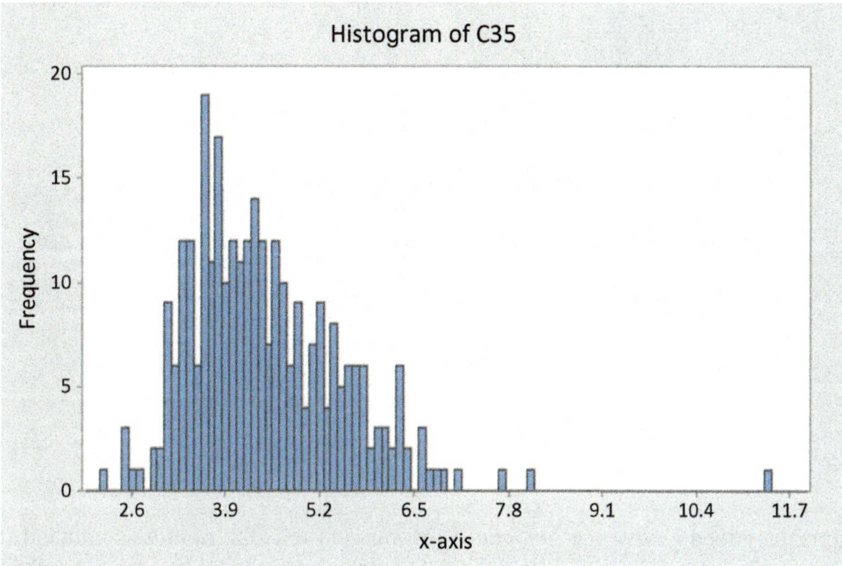

Sample Size = 50, 300 Samples

```
MTB >  random 300 c1-c50;
SUBC> lognormal 1 1.
MTB > rmean c1-c50 c55
MTB > Histogram c55;
SUBC>   MidPoint;
SUBC>   NInterval 100;
SUBC>   Bar;
SUBC>   Axlabel 1 "X-Axis".
```

Histogram of C55

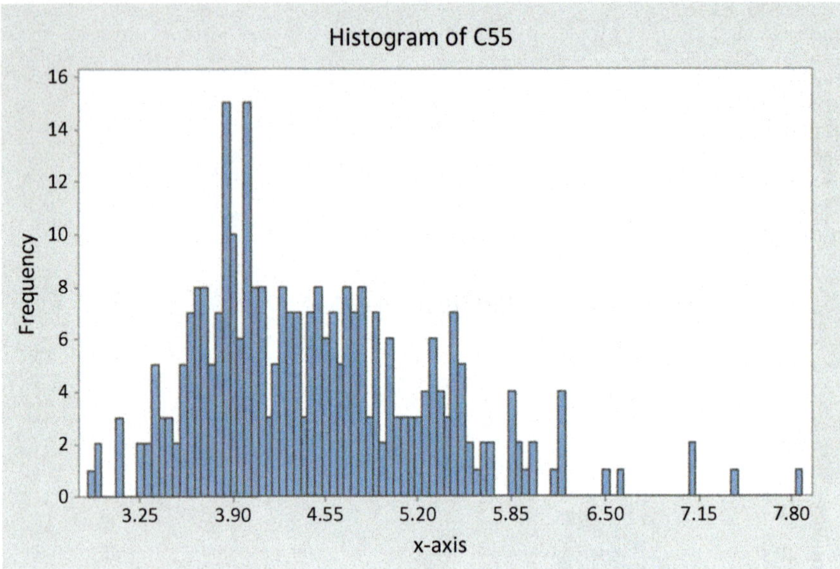

8.8 Binomial Distribution

In the commands issued below, the *binomial* subcommand tells the *random* command to generate random numbers from the binomial distribution. The first argument of the *binomial* subcommand is the number of trials. The second argument is the probability.

Sample Size = 10, 300 Samples

```
MTB > random 300 c1-c5;
SUBC> binomial 100 .3.
MTB > rmean c1-c5 c10
MTB > Histogram c10;
SUBC>   MidPoint;
SUBC>   NInterval 100;
SUBC>   Bar;
SUBC>   Axlabel 1 "X-Axis".
```

Histogram of C10

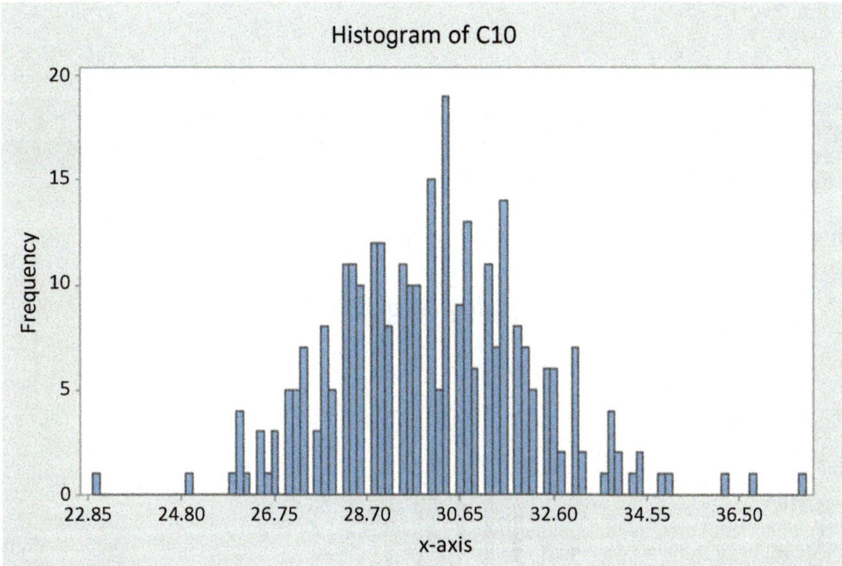

Sample Size = 10, 300 Samples

```
MTB > random 300 c1-c10;
SUBC> binomial 100 .3.
MTB > rmean c1-c10 c15
MTB > Histogram c15;
SUBC>   MidPoint;
SUBC>   NInterval 100;
SUBC>   Bar;
SUBC>   Axlabel 1 "X-Axis".
```

Histogram of C15

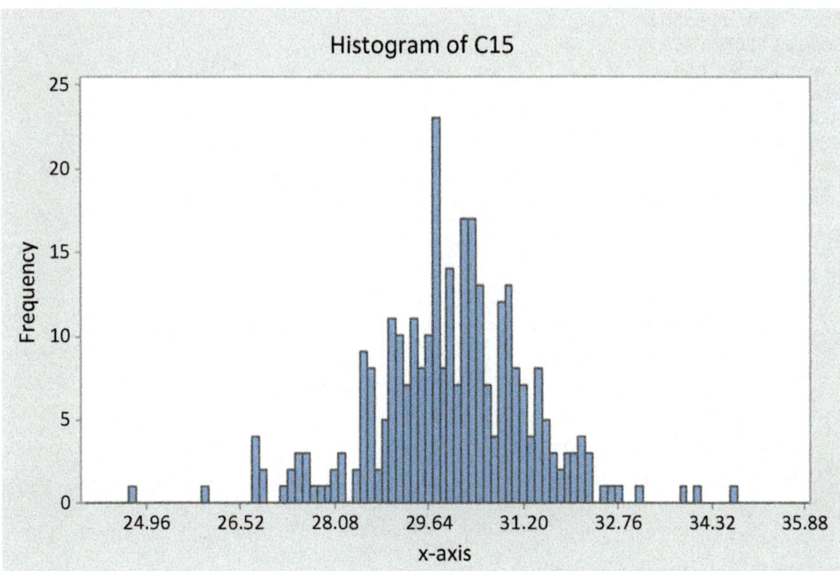

Sample Size = 30, 300 Samples

```
MTB > random 300 c1-c30;
SUBC> binomial 100 .3.
MTB > rmean c1-c30 c35
MTB > Histogram c35;
SUBC>   MidPoint;
SUBC>   NInterval 100;
SUBC>   Bar;
SUBC>    Axlabel 1 "X-Axis".
```

Histogram of C35

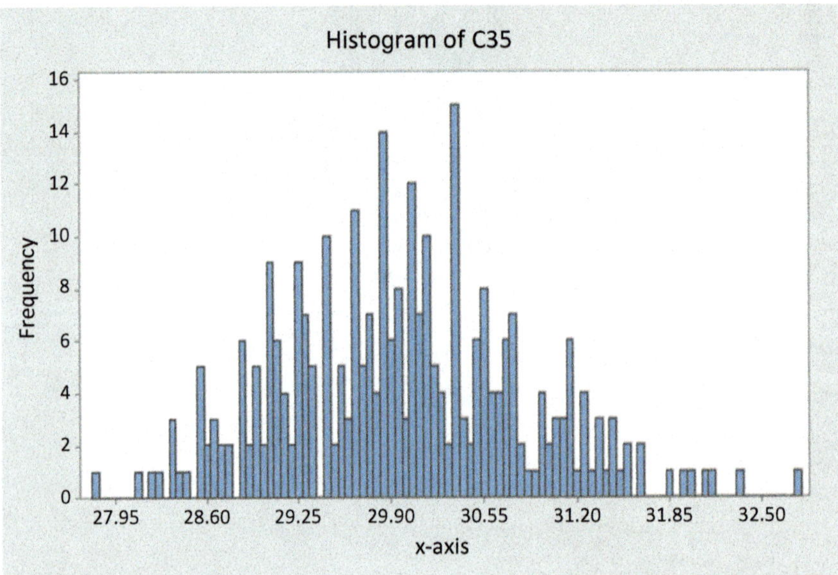

Sample Size = 50, 300 Samples

```
MTB > random 300 c1-c50;
SUBC> binomial 100 .3.
MTB > rmean c1-c50 c55
MTB > Histogram c55;
SUBC>   MidPoint;
SUBC>   NInterval 100;
SUBC>   Bar;
SUBC>    Axlabel 1 "X-Axis".
```

Histogram of C55

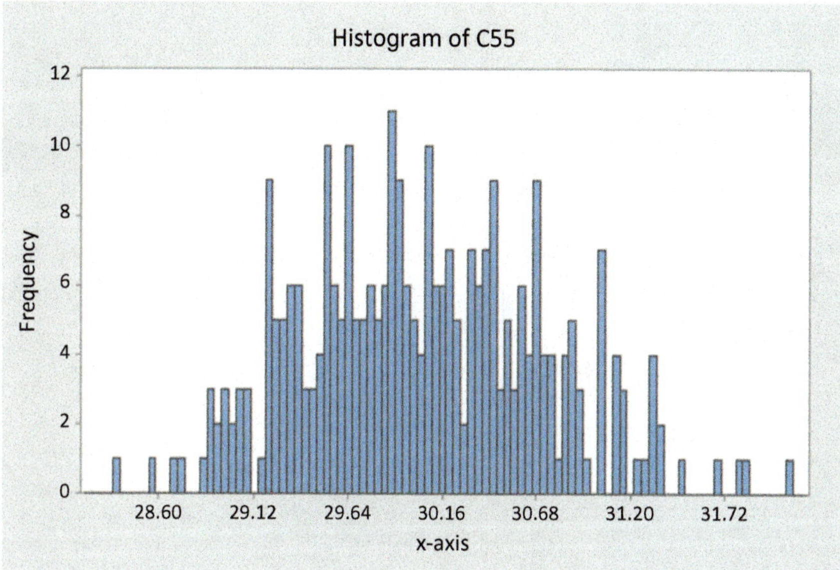

8.9 Poisson Distribution

The *poisson* subcommand tells the *random* command to randomly generate numbers from the Poisson distribution. The argument of the *poisson* subcommand is the mean.

Sample Size = 5, 300 Samples

```
MTB > random 300 c1-c5;
SUBC> poisson 5.
MTB > rmean c1-c5 c10
MTB > Histogram c10;
SUBC>   MidPoint;
SUBC>   NInterval 100;
SUBC>   Bar;
SUBC>   Axlabel 1 "X-Axis".
```

Histogram of C10

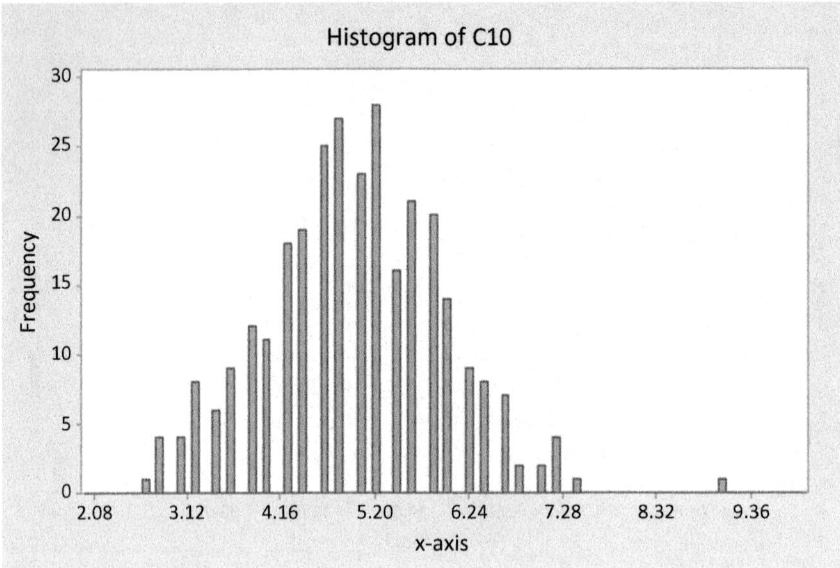

Sample Size = 10, 300 Samples

```
MTB > random 300 c1-c10;
SUBC> poisson 5.
MTB > rmean c1-c10 c15
MTB > Histogram C15;
SUBC>   MidPoint;
SUBC>   NInterval 100;
SUBC>   Bar;
SUBC>   Axlabel 1 "X-Axis".
```

Histogram of C15

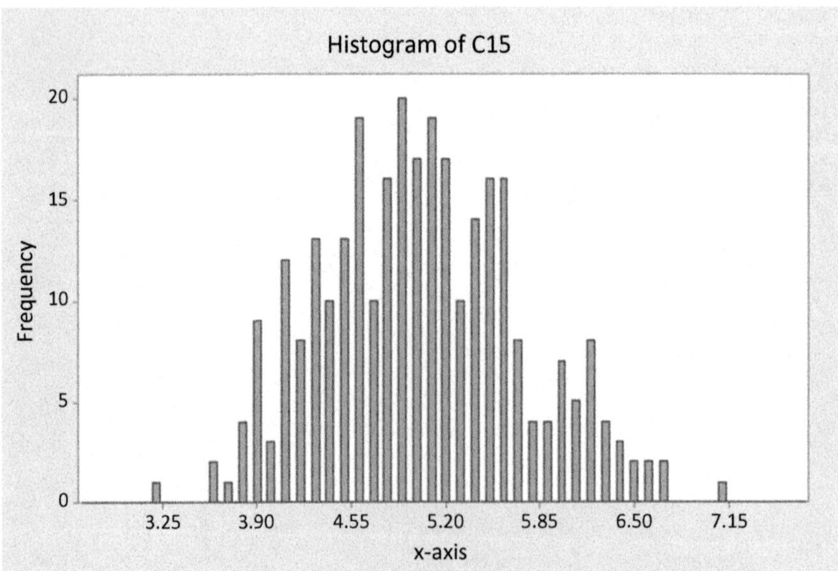

Sample Size = 30, 300 Samples

```
MTB > random 300 c1-c30;
SUBC> poisson 5.
MTB > rmean c1-c30 c35
MTB >  Histogram C35;
SUBC>   MidPoint;
SUBC>   NInterval 100;
SUBC>   Bar;
SUBC>    Axlabel 1 "X-Axis".
```

Histogram of C35

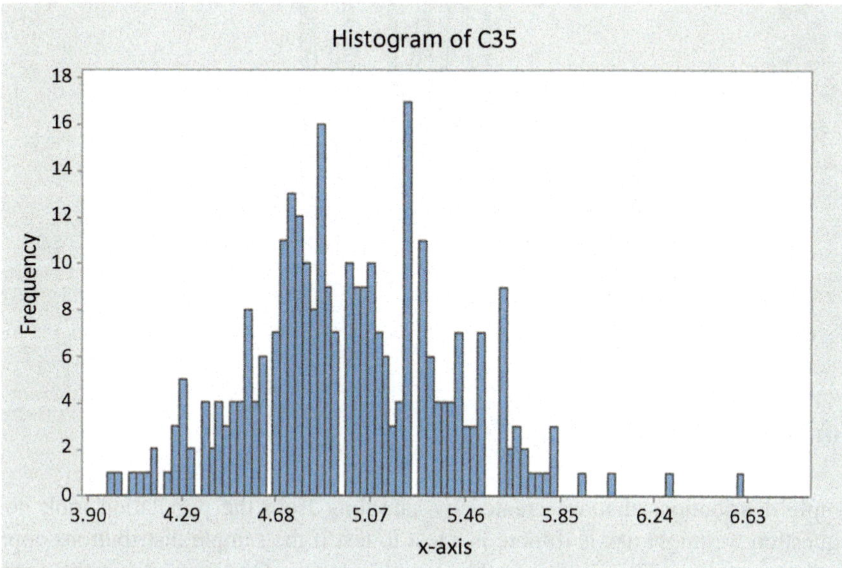

Sample Size = 50, 300 Samples

```
MTB > random 300 c1-c50;
SUBC> poisson 5.
MTB > rmean c1-c50 c55
MTB > Histogram c55;
SUBC>   MidPoint;
SUBC>   NInterval 100;
SUBC>   Bar;
SUBC>    Axlabel 1 "X-Axis".
```

Histogram of C55

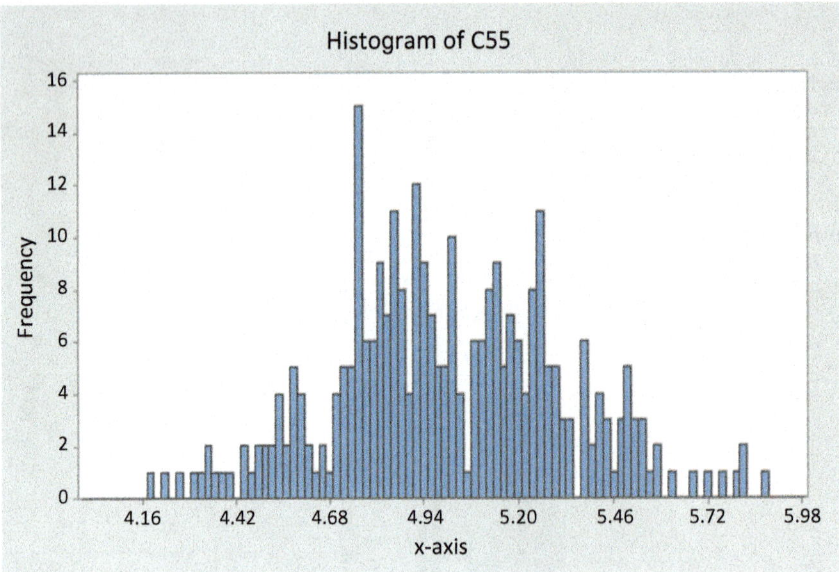

8.10 Normal Probability Plot

All the sample distributions that we created by sampling from the population look normal. One important question we might ask is if there is a test to test if the sample distributions approximate a normal distribution or not. The answer to this question is yes. One way to test the normality of a sample distribution is to create a normal probability plot. This is a chart with the probability on the y-axis and the x-values of the sample distribution on the x-axis. This chart indicates that a distribution is approximately normal if the chart displays an approximate straight line.

We will test the normality of the sample distributions from a lognormal distribution and a uniform distribution. We will first test the normality of a sample distribution from a lognormal distribution when the sample size is 5.

8.11 Lognormal Sample Distribution: Normality Test

We will first test the normality of a sample distribution from a lognormal distribution when the sample size is 5. The following works only in Minitab 13. It does not work in Minitab 17.

```
MTB > random 300 c1-c5;
SUBC> lognormal 1 1.
MTB > rmean c1-c5 c10
MTB > Histogram C10;
SUBC>   MidPoint;
SUBC>   NInterval 100;
SUBC>   Bar;
SUBC>   Axlabel 1 "X-Axis".
```

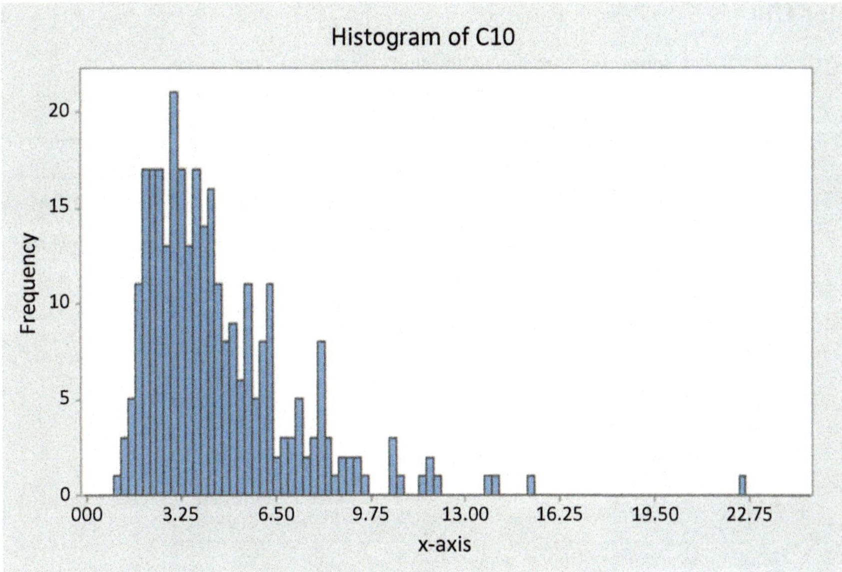


```
MTB > %Qqplot C10;
SUBC>   Normal;
SUBC>   MLE;
SUBC>   Allpts;
SUBC>   Title "Lognormal Sample Distribution - Normality Test";
SUBC>   Conf 95.
Executing from file: C:\Program Files\Mtbwin\MACROS\Qqplot.MAC
```

Distribution function analysis

```
Normal Dist. Parameter Estimates (ML)


Variable: C10


Mean     4.26138
StDev    2.12938

Goodness of Fit

Anderson-Darling (adjusted) = 5.951
```

Prob plot for C10

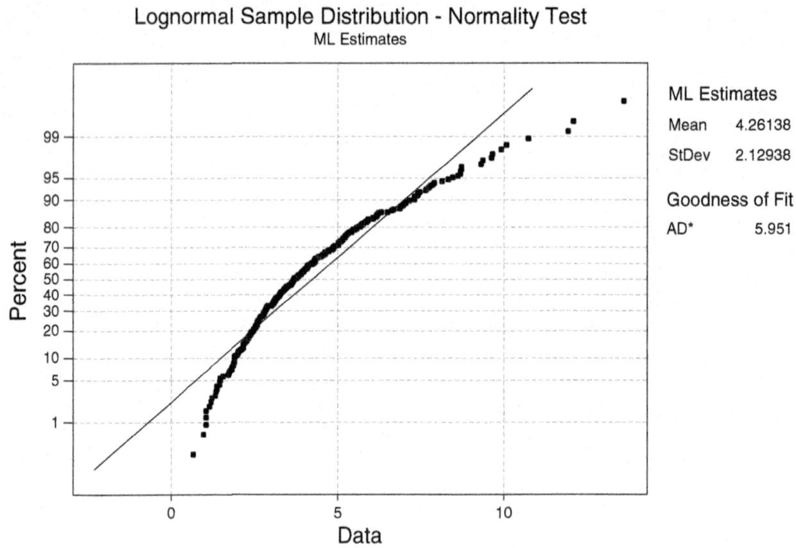

Above is the normal probability plot for the lognormal sample distribution with a sample size of 5. The plot will indicate that the sample distribution will be normal or approximately normal if the points lie or nearly line on a straight line. From the above plot, we have doubts that the lognormal sample distribution of sample size 5 is or approximates a normal distribution.

We will now test the normality of a sample distribution from a lognormal distribution when the sample size is 10.

```
MTB > random 300 c1-c10;
SUBC> binomial 100 .3.
MTB > rmean c1-c10 c15
MTB > Histogram c15;
SUBC>   MidPoint;
SUBC>   NInterval 100;
SUBC>   Bar;
SUBC>   Axis 1;
SUBC>     Label "x-axis";
SUBC>   Axis 2.
```

Histogram C15

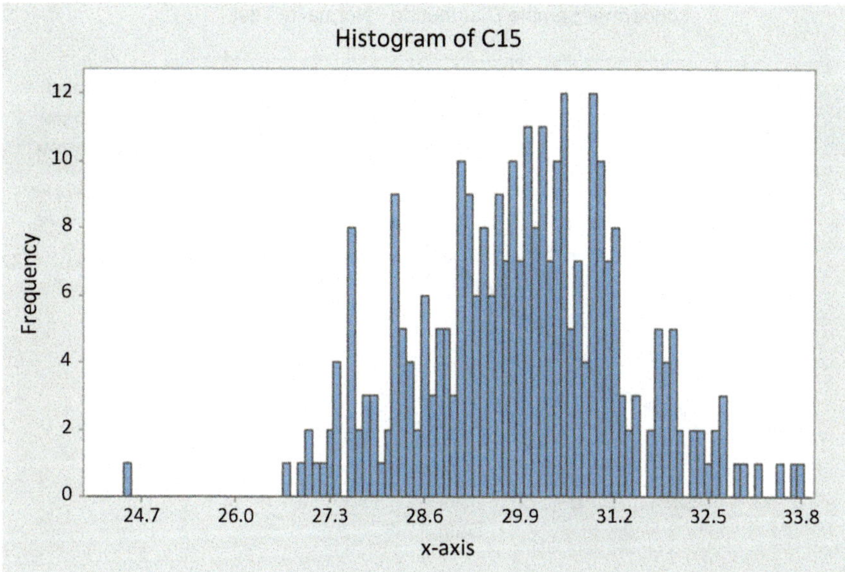


```
MTB > %Qqplot c15;
SUBC>   Normal;
SUBC>   MLE;
SUBC>   Allpts;
SUBC>   Title "Lognormal Sample Distribution - Normality Test";
SUBC>   Conf 95.
Executing from file: C:\Program Files\Mtbwin\MACROS\Qqplot.MAC
```

Distribution function analysis

```
Normal Dist. Parameter Estimates (ML)


Variable: C15


Mean     30.008
StDev    1.47544

Goodness of Fit

Anderson-Darling (adjusted) = 0.243
```

Prob plot for C15

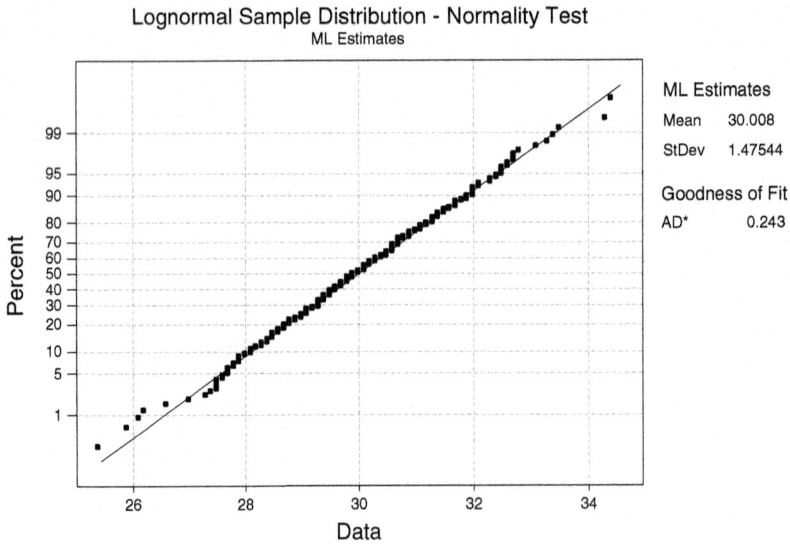

Above is the normal probability plot for the lognormal sample distribution with a sample size of 10. The above normal probability plot does not cast doubt that the lognormal sample distribution approximates a normal distribution.

We will first test the normality of a sample distribution from a uniform distribution when the sample size is 5.

```
MTB > random 300 c1-c5;
SUBC> uniform 5 6.
MTB > rmean c1-c5 c10
MTB > Histogram c10;
SUBC>   MidPoint;
SUBC>   NInterval 100;
SUBC>   Bar;
SUBC>   Axis 1;
SUBC>     Label "x-axis";
SUBC>   Axis 2;
SUBC>   Tick 1;
SUBC>   Tick 2.
```

Histogram C10

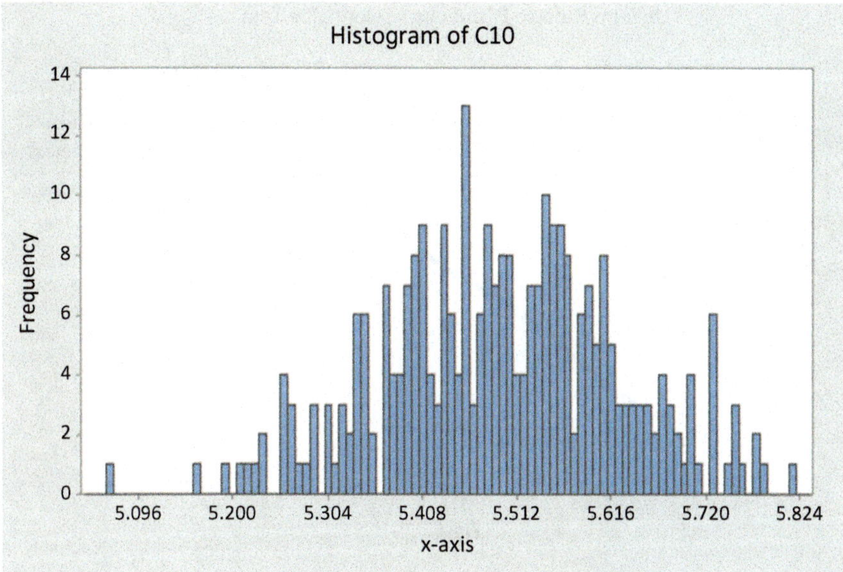


```
MTB > %Qqplot c10;
SUBC>   Normal;
SUBC>   MLE;
SUBC>   Allpts;
SUBC>   Title "Uniform Sample Distribution - Normality Test";
SUBC>   Conf 95.
Executing from file: C:\Program Files\Mtbwin\MACROS\Qqplot.MAC
```

Distribution function analysis

```
Normal Dist. Parameter Estimates (ML)


Variable: C10


Mean     5.49947
StDev    0.134898

Goodness of Fit

Anderson-Darling (adjusted) = 0.561
```

Prob plot for C10

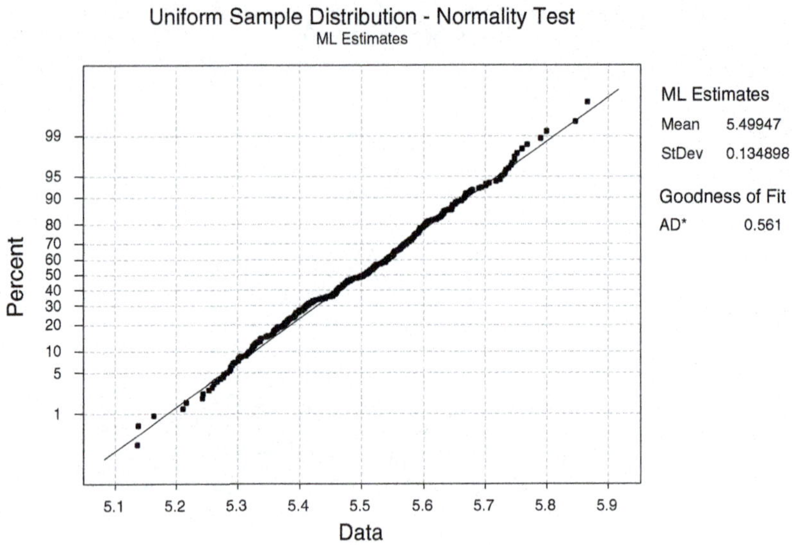

From the above normal probability plot, we can see that a sample size as small as 5 is all that is needed to create a sample distribution that is approximately normal.

8.12 SAS Programming Code Instructions and Examples

In this section, we use SAS function to generate random numbers from a specified distrubtion.[1] We also write SAS macro function to generate sample distribution.

8.12.1 Sample Distribution

Let's suppose that the population that we are interested in analyzing has the following six elements:

$$1, 2, 3, 4, 5, 6$$

We use SAS macro to generate all the possible combinations and the mean of this population. Please check the related section in Chap. 5 on how to construct it.

[1] For detailed information of SAS, a random function for different distributions can be found in http://support.sas.com/documentation/cdl/en/lrdict/64316/HTML/default/viewer.htm#a001466748.htm

```
%macro com(k);
    data Name;
    array x[6] (1,2,3,4,5,6);
    n=6;
    ncomb=comb(n,&k);
do j=1 to ncomb;
        rc=lexcomb(j, &k, of x[*]);
        output Name;
        if rc<0 then leave;
end;
run;

data Name (keep=x1-x&k meanvar);
set Name;
meanvar=sum(of x1-x&k)/&k;
run;
%mend;

%com(k=2)
```

Below shows all the possible combinations and the mean of each possible combination.

Obs	x1	x2	Meanvar
1	1	2	1.5
2	1	3	2.0
3	1	4	2.5
4	1	5	3.0
5	1	6	3.5
6	2	3	2.5
7	2	4	3.0
8	2	5	3.5
9	2	6	4.0
10	3	4	3.5
11	3	5	4.0
12	3	6	4.5
13	4	5	4.5
14	4	6	5.0
15	5	6	5.5

```
Goptions ftext='Times New Roman' htext=3 gunit=pct;
title height=7 'Sample Distribution';
axis1  label=('Probability');
axis2  label=('Mean') order=(1.5 to 5.5 by 0.5);
proc gchart data=Name;
vbar  meanvar/ RAXIS=axis1 MAXIS=axis2 discrete WIDTH=6 type=pct;
run;
```

Below is a histogram of the sample distribution.

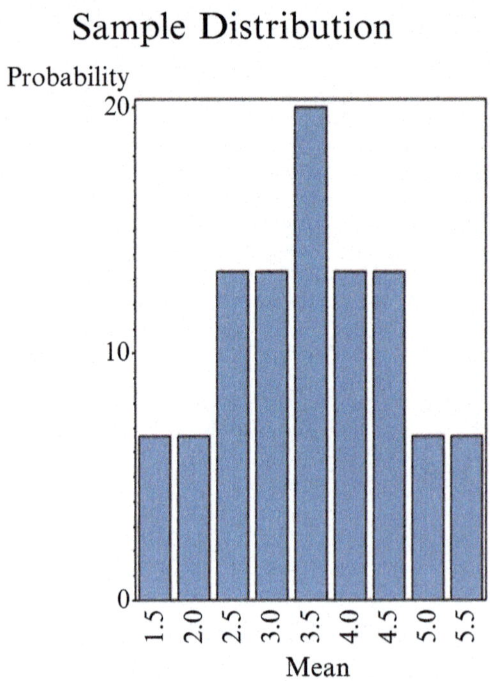

Now we will create the sample distribution when the sample size is 3 by changing the value of k to 3.

```
%com(k=3)

Goptions ftext='Times New Roman' htext=4 gunit=pct;
title height=7 'Sample Distribution';
axis1  label=('Probability');
axis2  label=('Mean') value=(angle=90 rotate=0);
proc gchart data=Name;
vbar  meanvar/ RAXIS=axis1 MAXIS=axis2 discrete width=6  space=1.5 type=pct;
run;
```

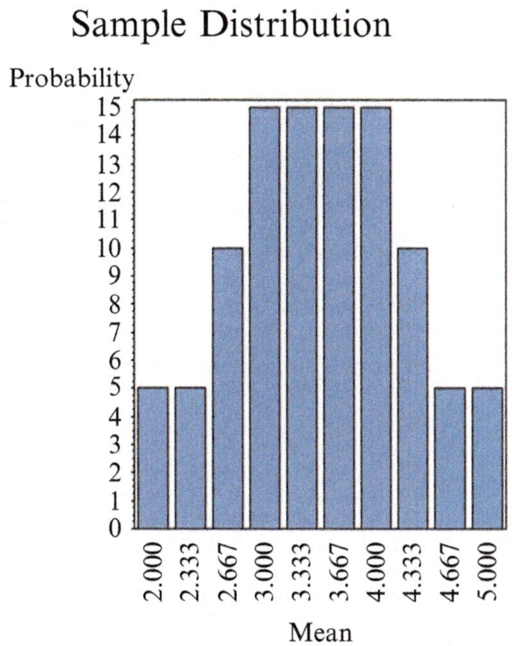

Now we will create the sample distribution when the sample size is 4 by changing the value of k to 4.

```
%com(k=4)

Goptions ftext='Times New Roman' htext=4 gunit=pct;
title height=7 'Sample Distribution';
axis1  label=('Probability');
axis2  label=('Mean') value=(angle=90 rotate=0);
proc gchart data=Name;
vbar  meanvar/ RAXIS=axis1 MAXIS=axis2 discrete width=6  space=1.5 type=pct;
run;
```

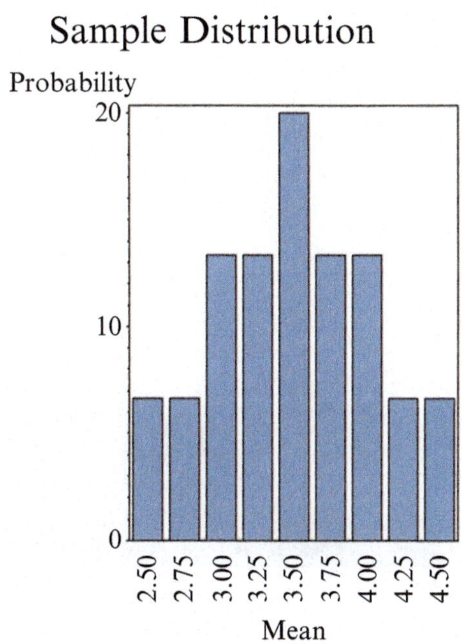

8.12.2 Mean of Sample Distribution Equals Mean of Population

Let's suppose that the population that we are interested in analyzing has the following 20 elements:

$$1, 2, 3, 30, 40, 50, 45, 56, 67, 98, 34, 65, 67, 34, 78, 56, 98, 67, 43, 45$$

Notice that element in this population is not unique. Thus, **ALLCOMB** function is used to generate all combinations of the values of n variables taken k at a time.

```
%macro sample(k);
 data Name;
 array x[20] (1,2,3,30,40,50,45,56,67,98,34,65,67,34,78,56,98,67,43,45);
 n=20;
 ncomb=comb(n,&k);
 do j=1 to ncomb;
       rc=allcomb(j, &k, of x[*]);
       output Name;
       if rc<0 then leave;
end;

data Name (keep=meanvar);
set Name;
meanvar=sum(of x1-x&k)/&k;
%mend;

%sample(k=5)

Goptions ftext='Times New Roman' htext=6 gunit=pct;
title height=7 'Sample Distribution';
axis1  label=('Probability');
axis2  label=('Mean') value=none;
proc gchart data=Name;
vbar  meanvar/ RAXIS=axis1 MAXIS=axis2 discrete type=pct levels=all;
run;
```

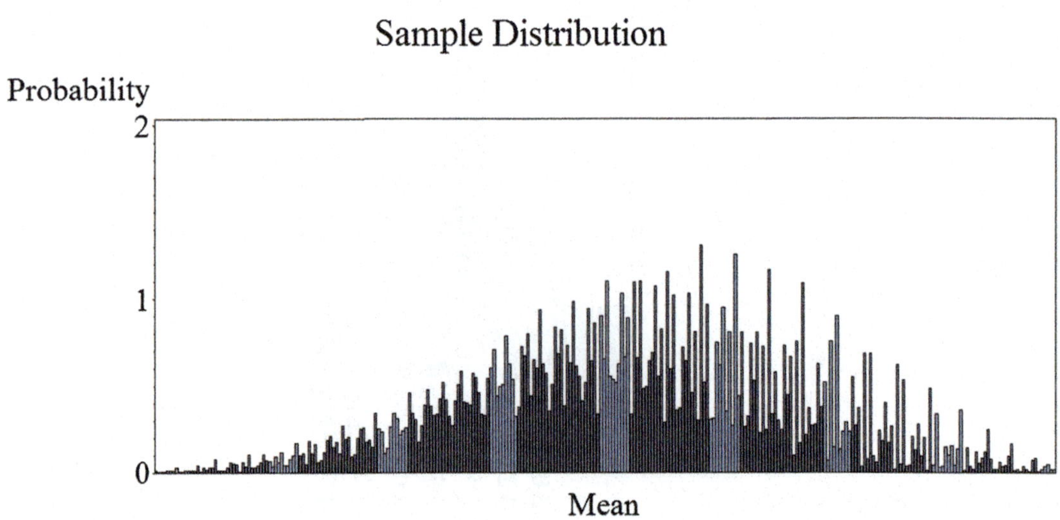

The following is the sample distribution when the sample size is 15.

```
%sample(k=15)

Goptions ftext='Times New Roman' htext=6 gunit=pct;
title height=7 'Sample Distribution';
axis1  label=('Probability');
axis2  label=('Mean') value=none;
proc gchart data=Name;
vbar  meanvar/ RAXIS=axis1 MAXIS=axis2 discrete type=pct levels=all;
run;
```

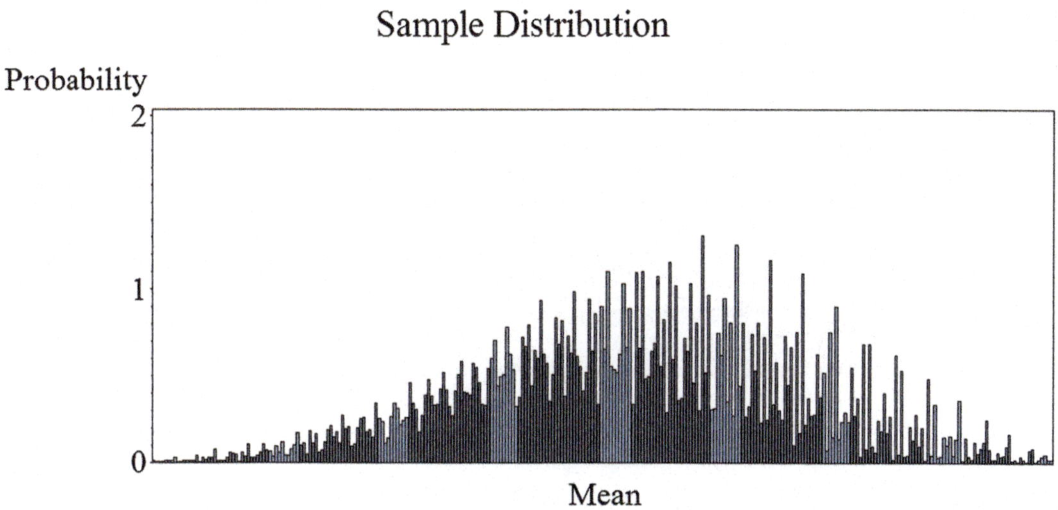

8.12.3 Uniform Distribution

```
%macro Unif(k);
data Unif;
array u[&k];
call streaminit(12345); /* set random number seed */
do j = 1 to 300;
do i = 1 to &k;
   u[i] = rand("Uniform")+5; /* u ~ U[0,1] */
end;
output Unif;
end;
run;

data Unif (keep=meanvar);
set Unif;
meanvar=sum(of u1-u&k)/&k;
format meanvar 4.2;
run;

Goptions ftext='Times New Roman' htext=4 gunit=pct;
title height=7 'Sample Distribution';
axis1  label=('Frequency');
axis2  label=('Mean') value=(angle=90 rotate=0);
proc gchart data=Unif;
vbar  meanvar/ RAXIS=axis1 MAXIS=axis2 discrete type=freq levels=all;
run;
%mend;
```

Sample Size = 5, 300 Samples

```
%Unif(k=5);
```

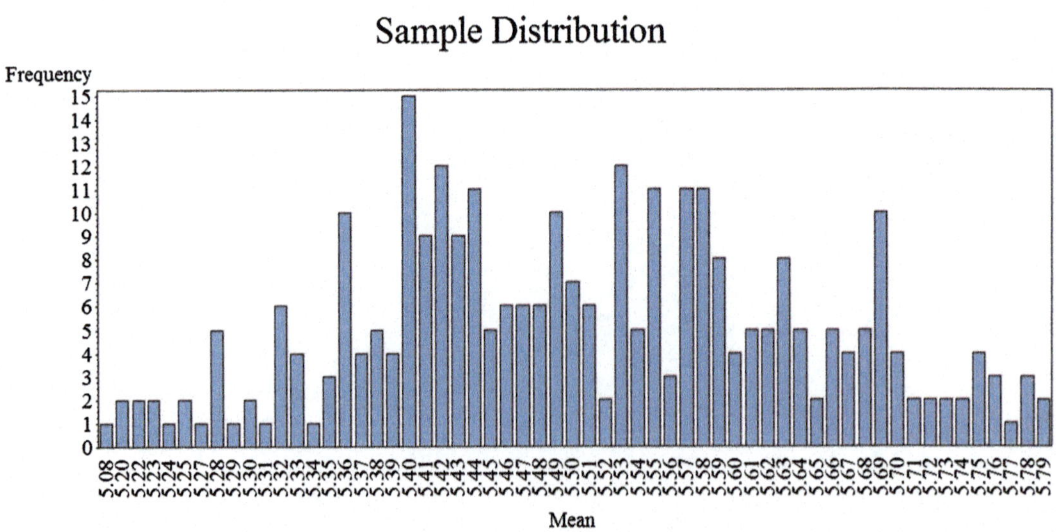

Sample Size = 10, 300 Samples

```
%Unif(k=10);
```

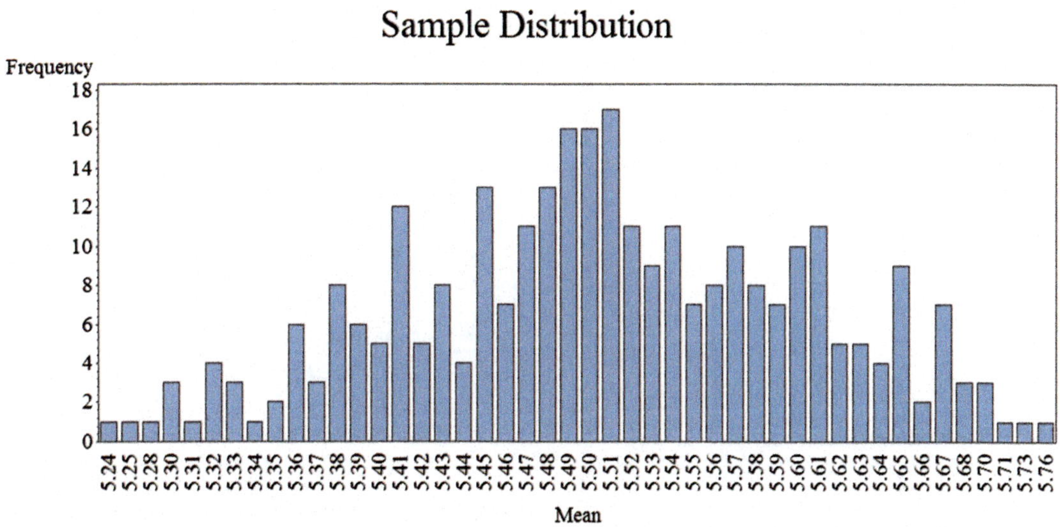

Sample Size = 30, 300 Samples

```
%Unif(k=30);
```

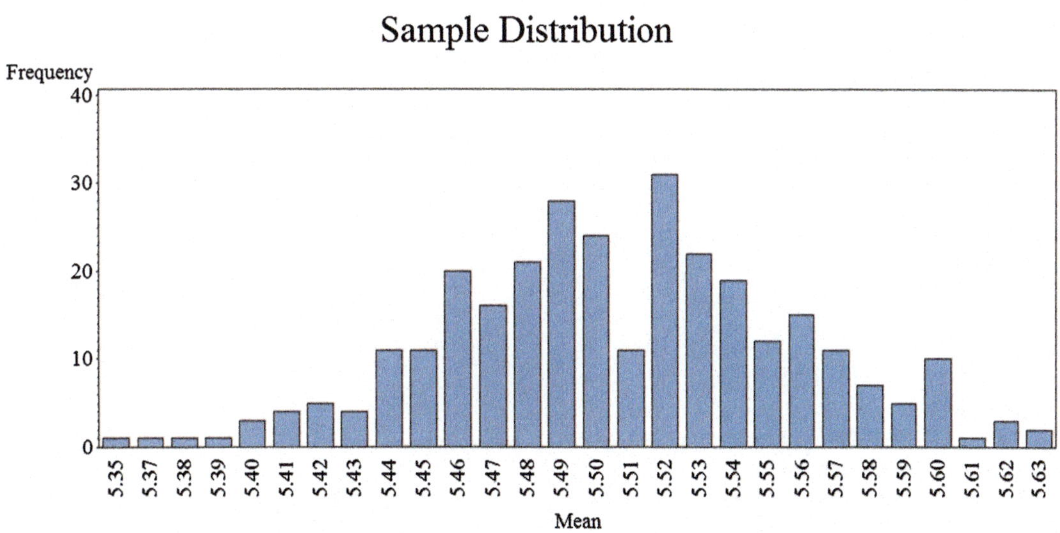

Sample Size = 50, 300 Samples

```
%Unif(k=50);
```

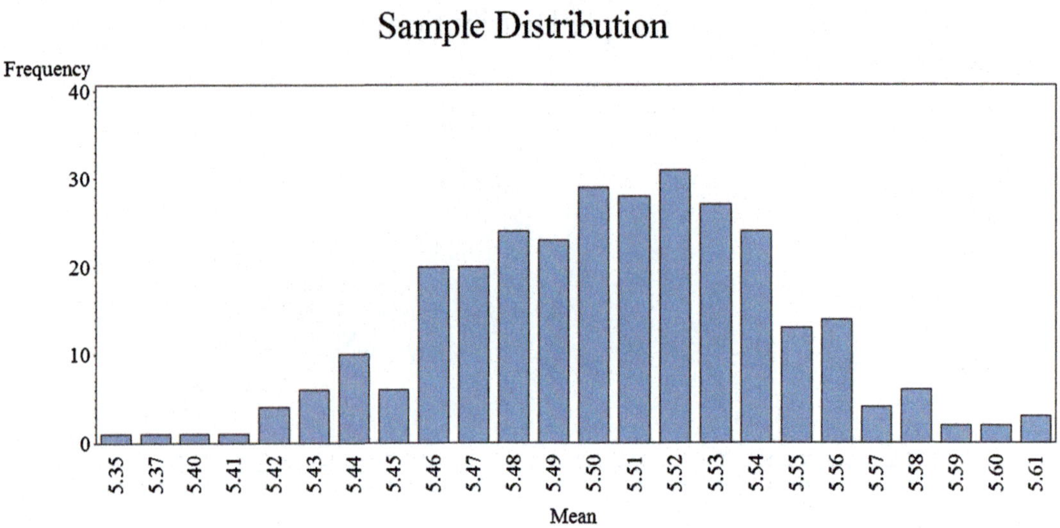

8.12.4 Normal Distribution

```
%macro Norm(k);
data Norm;
array u[&k];
call streaminit(12345); /* set random number seed */
do j = 1 to 300;
do i = 1 to &k;
   u[i] = rand("Normal",5,2); /* u ~ N[5,2] */
end;
output Norm;
end;
run;

data Norm (keep=meanvar);
set Norm;
meanvar=sum(of u1-u&k)/&k;
format meanvar 3.1;
run;
Goptions ftext='Times New Roman' htext=4 gunit=pct;
title height=7 'Sample Distribution';
axis1  label=('Frequency');
axis2  label=('Mean') value=(angle=90 rotate=0);
proc gchart data=Norm;
vbar  meanvar/ RAXIS=axis1 MAXIS=axis2 discrete type=freq levels=all;
run;
%mend;
```

Sample Size = 5, 300 Samples

```
%Norm(k=5);
```

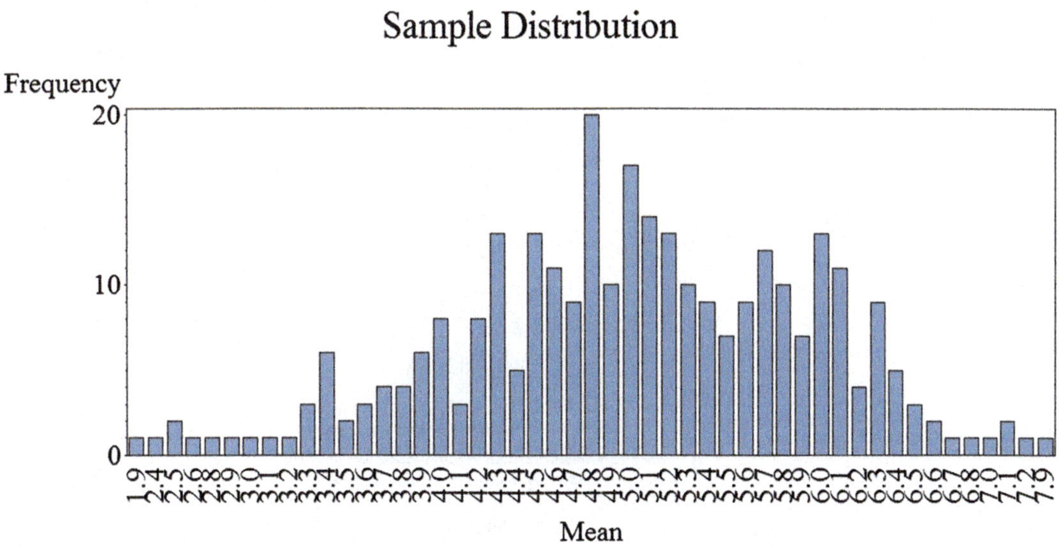

Sample size = 10, 300 samples

```
%Norm(k=10);
```

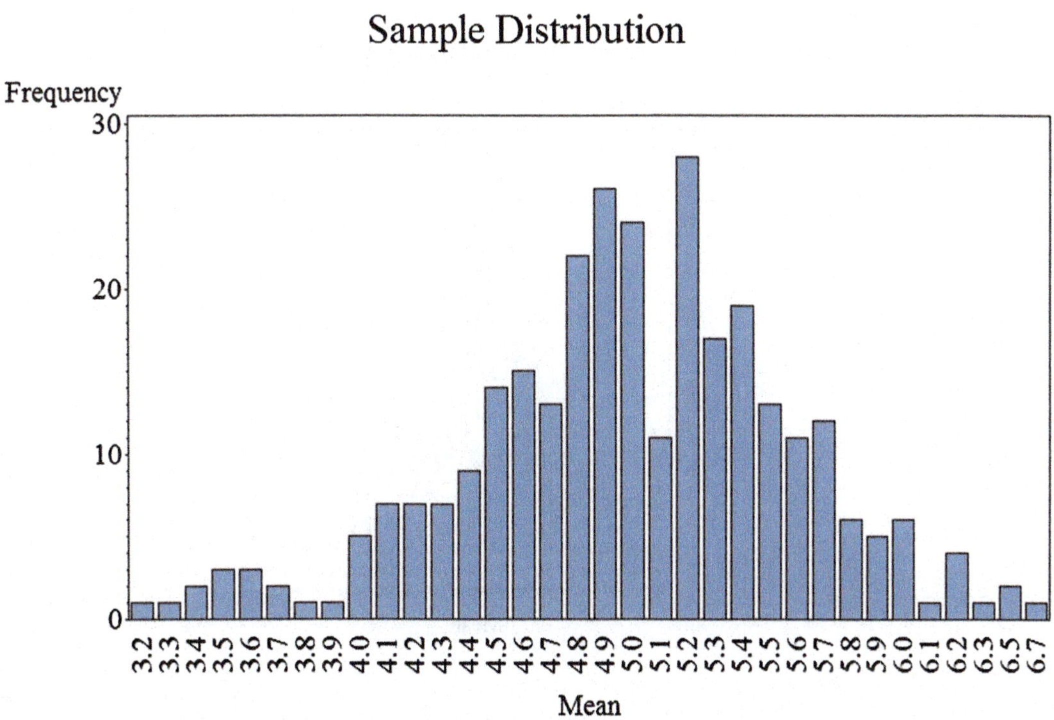

Sample Size = 30, 300 Samples

```
%Norm(k=30);
```

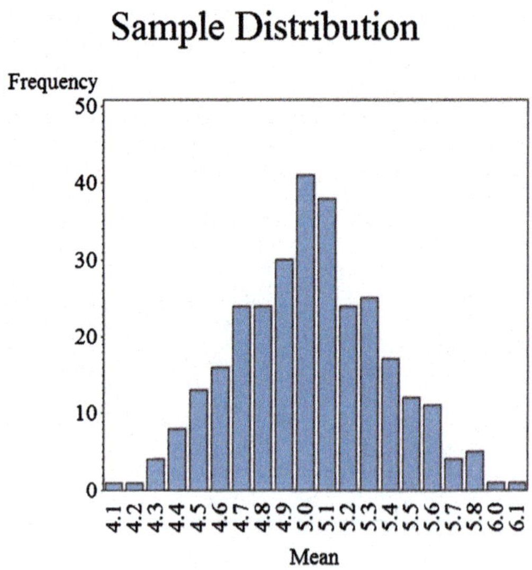

Sample Size = 50, 300 Samples

```
%Norm(k=50);
```

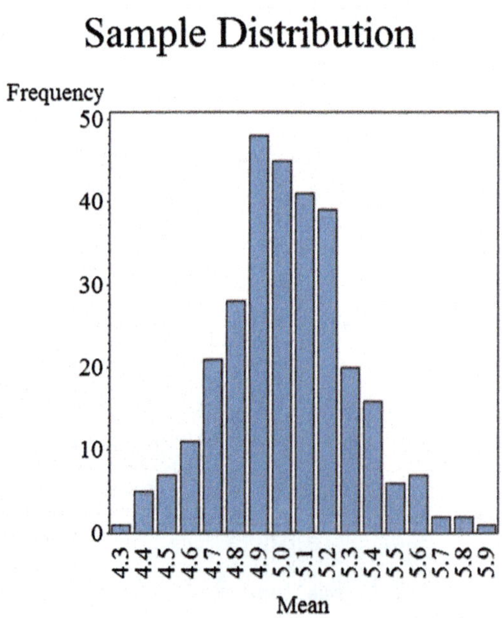

8.12.5 Lognormal Distribution

```
%macro LogNorm(k);
data LogNorm;
array u[&k];
call streaminit(12345); /* set random number seed */
do j = 1 to 300;
do i = 1 to &k;
   u[i] = rand("LOGNORMAL"); /* u ~ Lognormal(0,1) */
end;
output LogNorm;
end;
run;

data LogNorm (keep=meanvar);
set LogNorm;
meanvar=sum(of u1-u&k)/&k;
format meanvar 3.1;
run;
Goptions ftext='Times New Roman' htext=4 gunit=pct;
title height=7 'Sample Distribution';
axis1  label=('Frequency');
axis2  label=('Mean') value=(angle=90 rotate=0);
proc gchart data=LogNorm;
vbar  meanvar/ RAXIS=axis1 MAXIS=axis2 discrete type=freq levels=
run;
%mend;
```

Sample Size = 5, 300 Samples

```
%LogNorm(k=5);
```

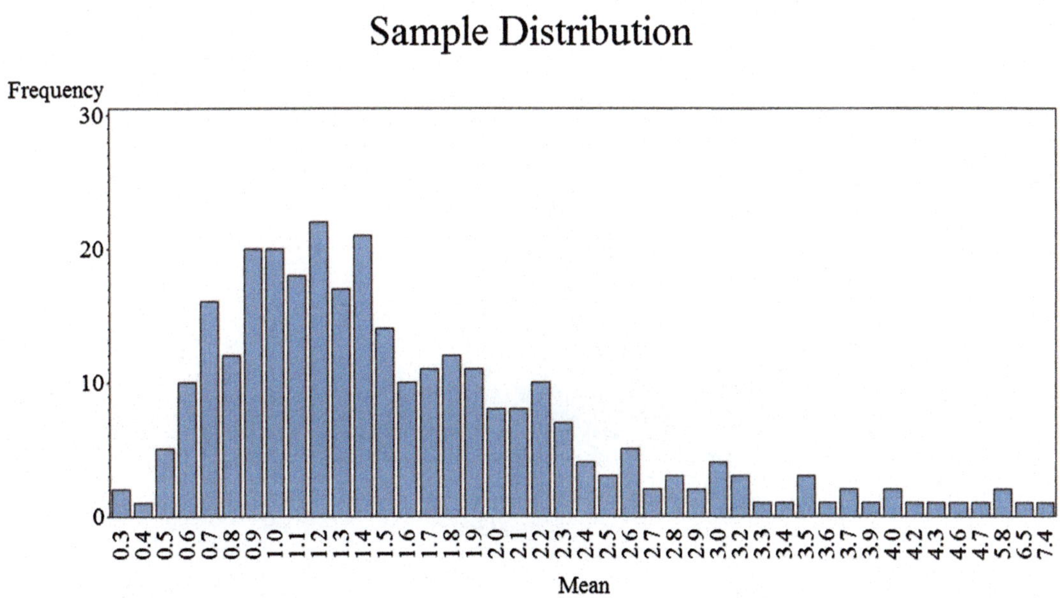

Sample size = 10, 300 samples

```
%LogNorm(k=10);
```

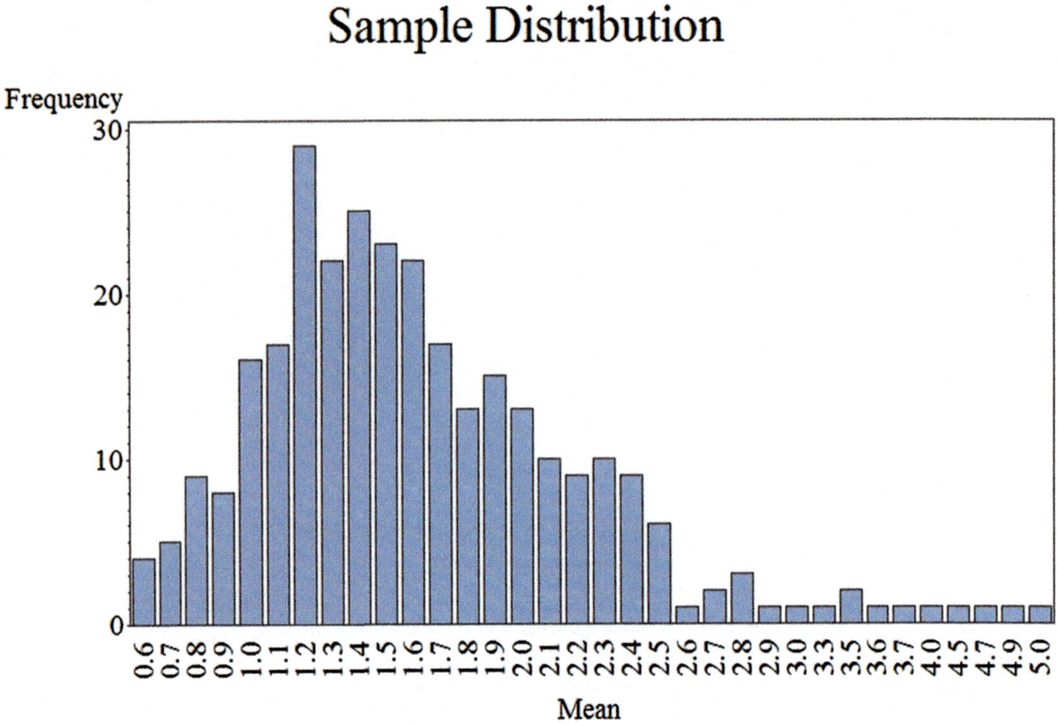

Sample Size = 30, 300 Samples

```
%LogNorm(k=30);
```

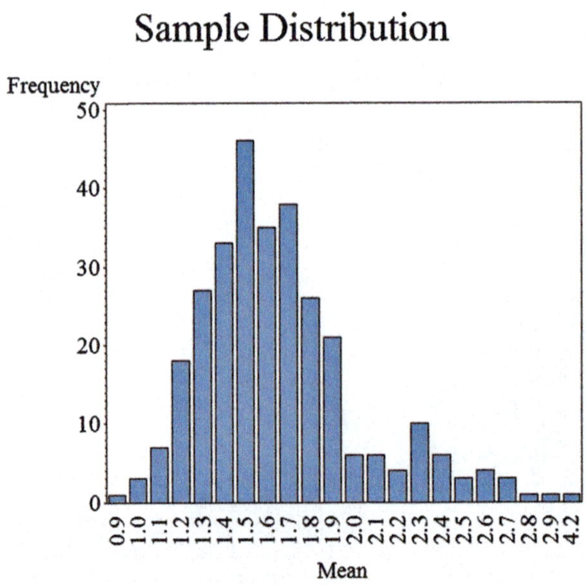

Sample Size = 50, 300 Samples

```
%LogNorm(k=50);
```

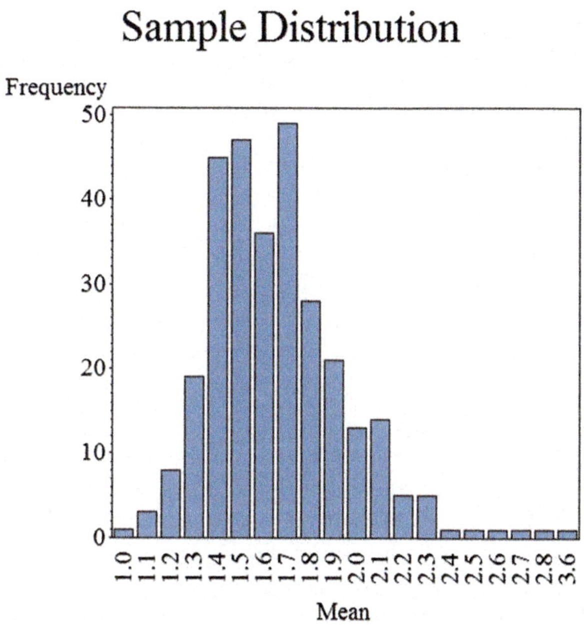

8.12.6 Binomial Distribution

```
%macro Bino(k);
data Bino;
array u[&k];
call streaminit(12345); /* set random number seed */
do j = 1 to 300;
do i = 1 to &k;
   u[i] = rand("Binomial",0.3,100); /* u ~ Binomial[0.3,100] */
end;
output Bino;
end;
run;

data Bino (keep=meanvar);
set Bino;
meanvar=sum(of u1-u&k)/&k;
format meanvar 4.1;
run;
Goptions ftext='Times New Roman' htext=4 gunit=pct;
title height=7 'Sample Distribution';
axis1  label=('Frequency');
axis2  label=('Mean') value=(angle=90 rotate=0);
proc gchart data=Bino;
vbar  meanvar/ RAXIS=axis1 MAXIS=axis2 discrete type=freq levels=all;
run;
%mend;
```

Sample Size = 5, 300 Samples

```
%Bino(k=5);
```

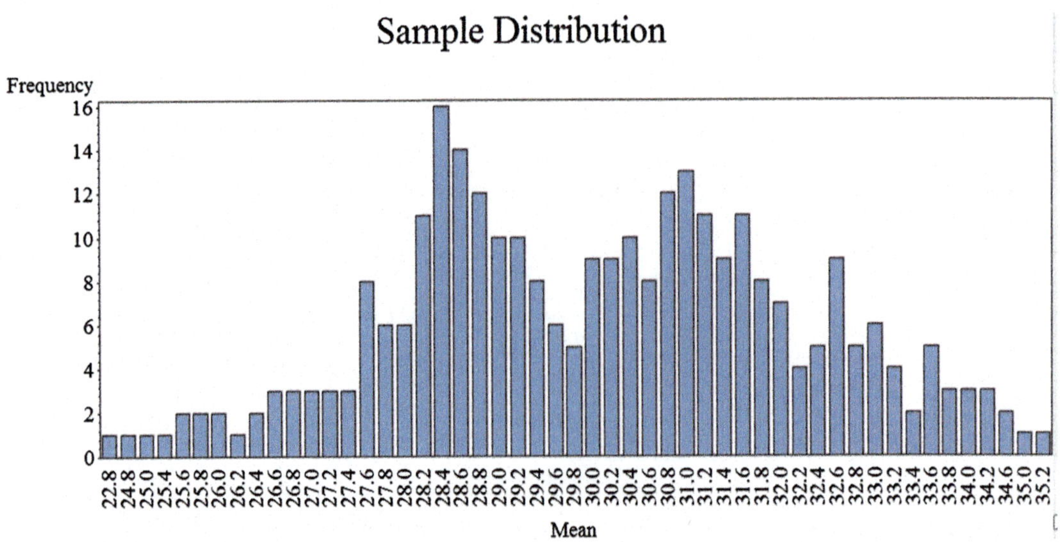

Sample Size = 10, 300 Samples

```
%Bino(k=10);
```

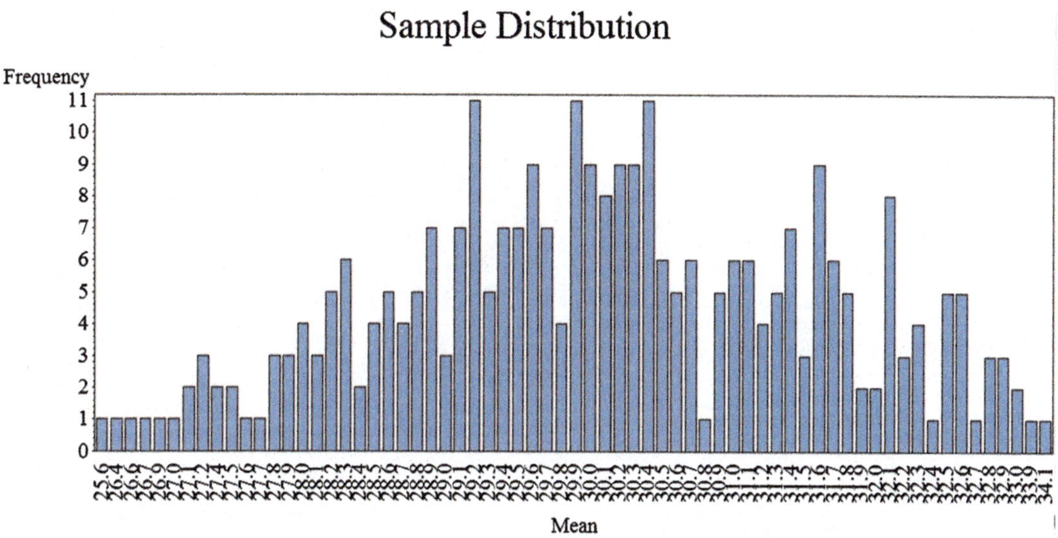

Sample size = 30, 300 samples

```
%Bino(k=30);
```

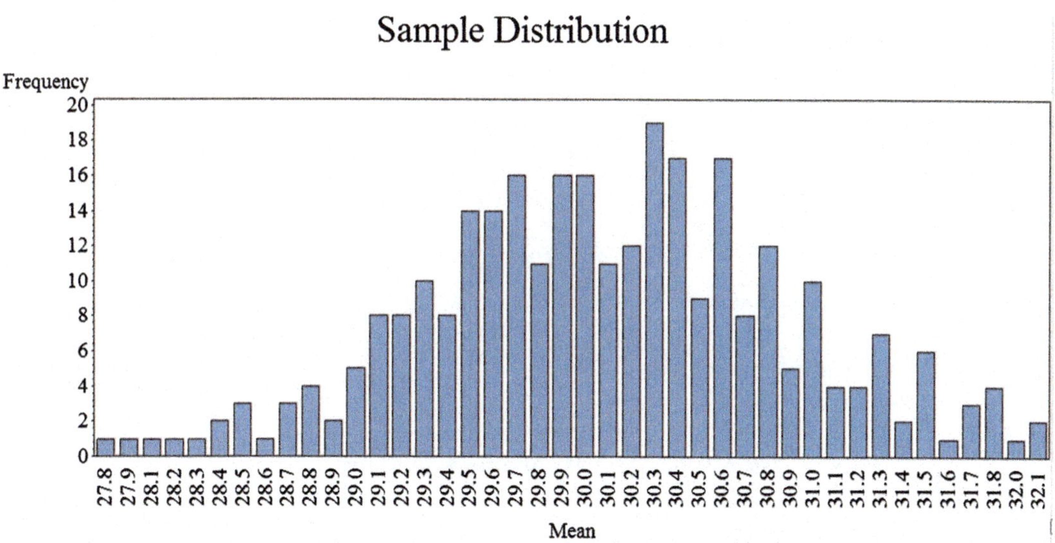

Sample size = 50, 300 samples

```
%Bino(k=50);
```

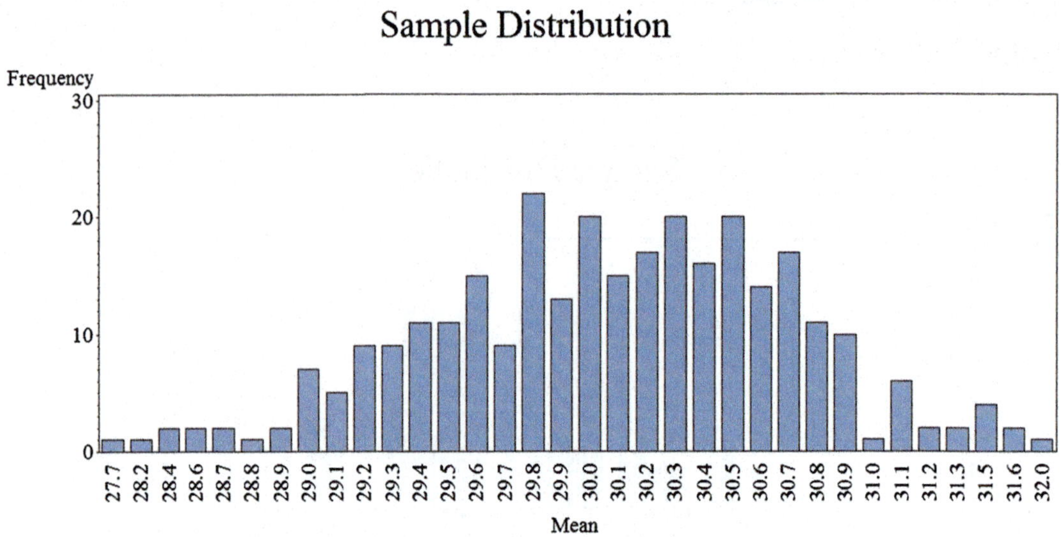

8.12.7 Poisson Distribution

```
%macro Poiss(k);
data Poiss;
array u[&k];
call streaminit(12345); /* set random number seed */
do j = 1 to 300;
do i = 1 to &k;
   u[i] = rand("Poisson",5);
end;
output Poiss;
end;
run;

data Poiss (keep=meanvar);
set Poiss;
meanvar=sum(of u1-u&k)/&k;
format meanvar 3.1;
run;
Goptions ftext='Times New Roman' htext=4 gunit=pct;
title height=7 'Sample Distribution';
axis1  label=('Frequency');
axis2  label=('Mean') value=(angle=90 rotate=0);
proc gchart data=Poiss;
vbar  meanvar/ RAXIS=axis1 MAXIS=axis2 discrete type=freq levels=all;
run;
%mend;
```

Sample Size = 5, 300 Samples

```
%Poiss(k=5);
```

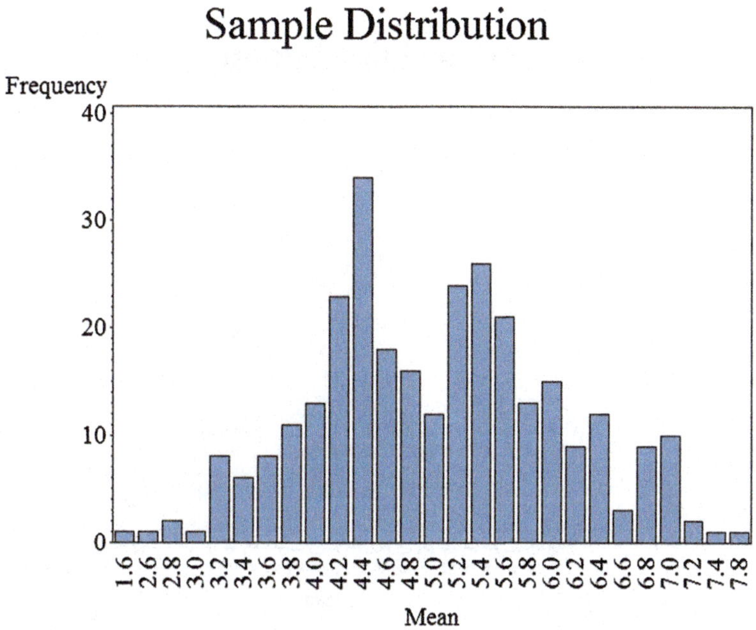

Sample Size = 10, 300 Samples

```
%Poiss(k=10);
```

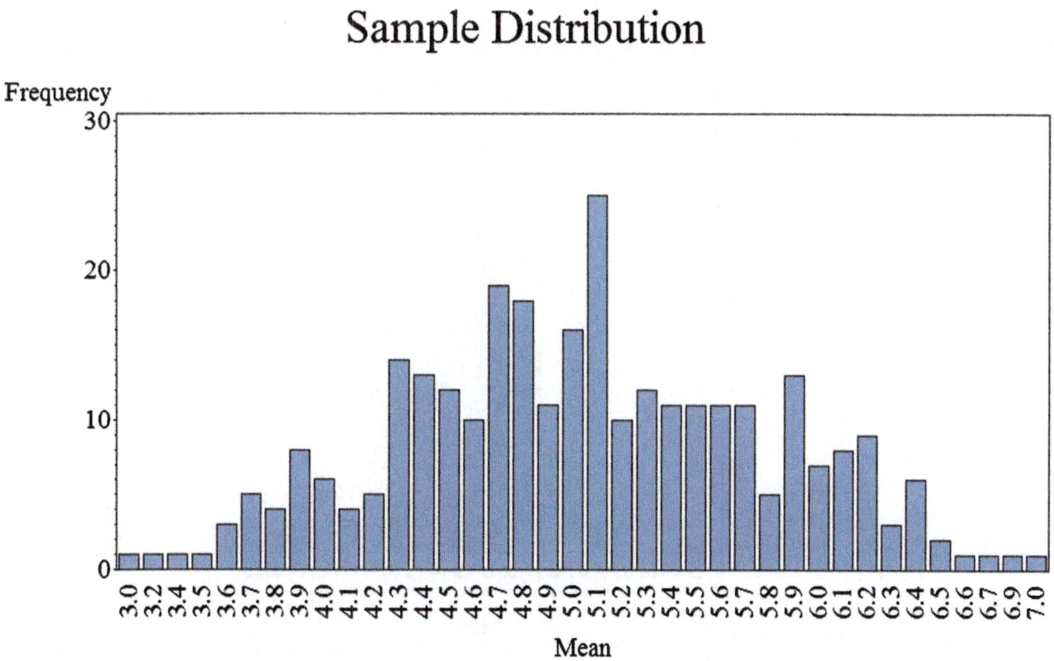

Sample size = 30, 300 samples

```
%Poiss(k=30);
```

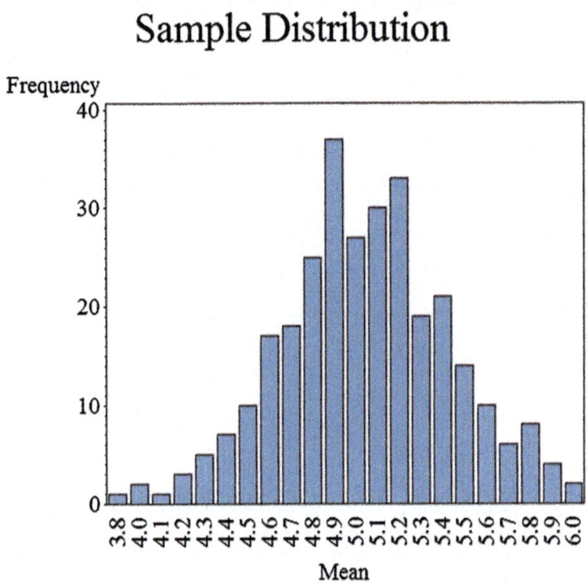

Sample Size = 50, 300 Samples

```
%Poiss(k=50);
```

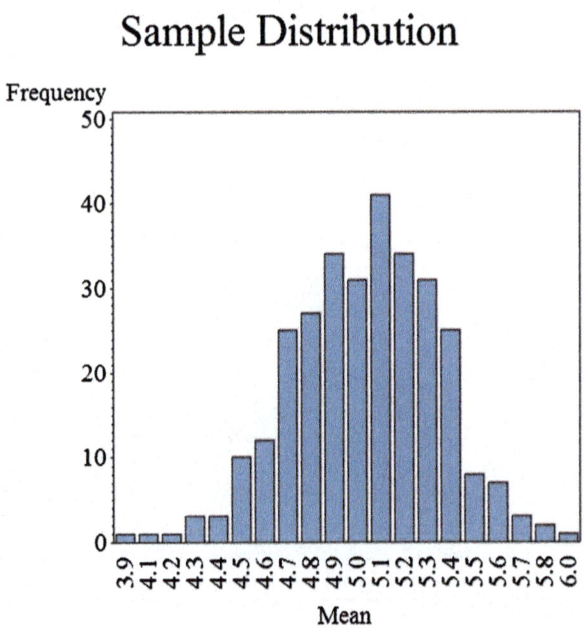

8.12.8 Lognormal Sample Distribution: Normality Test

Let's test the normality of a sample distribution from a lognormal distribution when the sample size is 5.

```
%LogNorm(k=5);
```

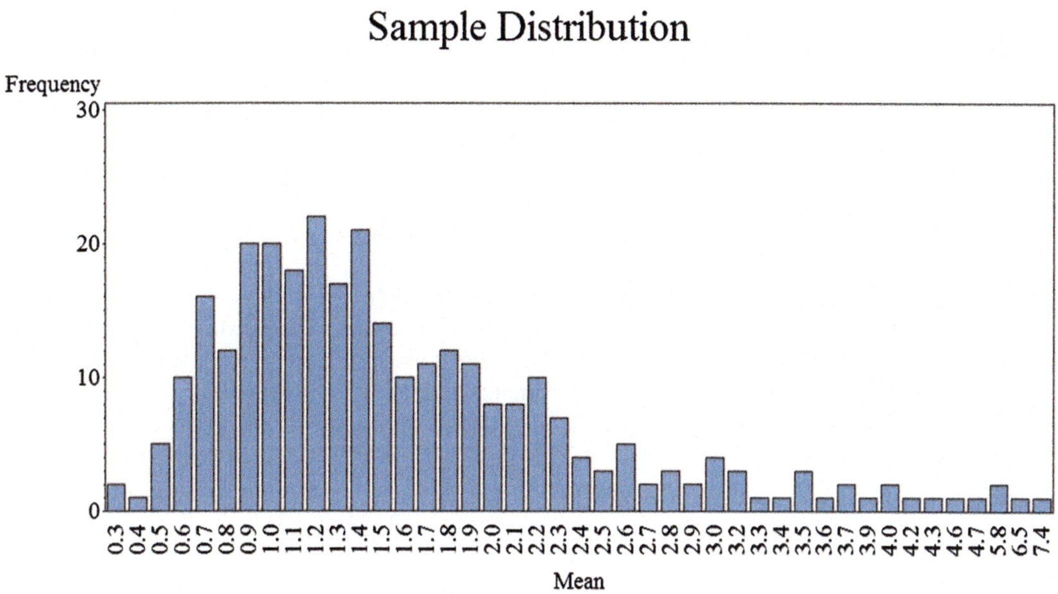


```
Goptions ftext='Times New Roman' htext=4 gunit=pct;
title height=7 'LogNormal Sample Distribution';
proc capability data=LogNorm;
   qqplot meanvar/normal(mu=est sigma=est color=red l=1);
run;
```

Specifying MU=EST and SIGMA=EST with the NORMAL option requests the reference line (alternatively, you can specify numeric values for μ_0 and σ_0 with the MU = and SIGMA = options). The COLOR = and L = options specify the color of the line and the line type. The SQUARE option displays the plot in a square format, and the NOSPECLEGEND option suppresses the legend for the specification lines.

Basic statistical measures			
Location		Variability	
Mean	1.634086	Std deviation	0.98384
Median	1.398539	Variance	0.96794
Mode		Range	7.15096
		Interquartile range	1.01261

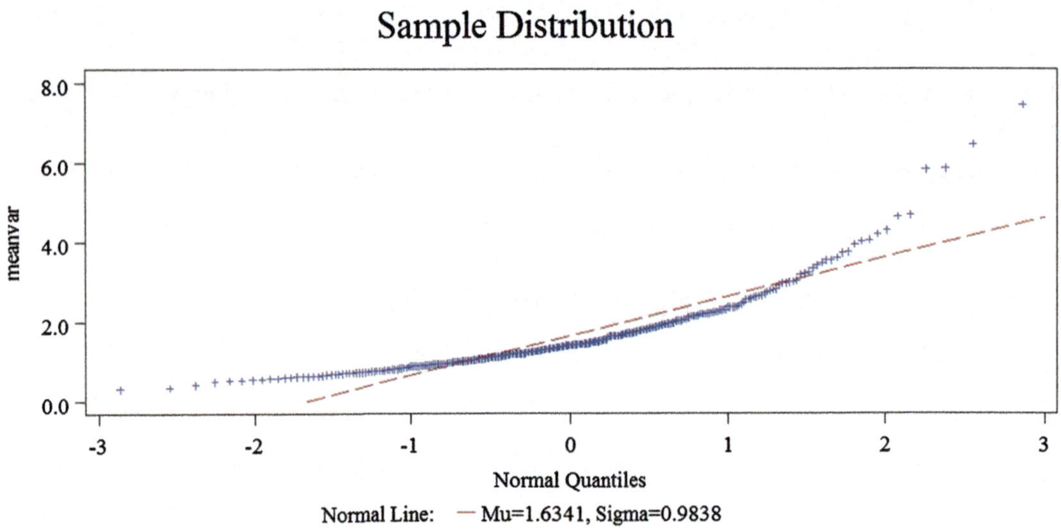

Above is the normal probability plot for the lognormal sample distribution with a sample size of 5. The plot will indicate that the sample distribution will be normal or approximately normal if the points lie or nearly line on a straight line. From the above plot, we have doubts that the lognormal sample distribution of sample size 5 is or approximates a normal distribution.

We will now test the normality of a sample distribution from a lognormal distribution when the sample size is 10.

```
%LogNorm(k=10)
```

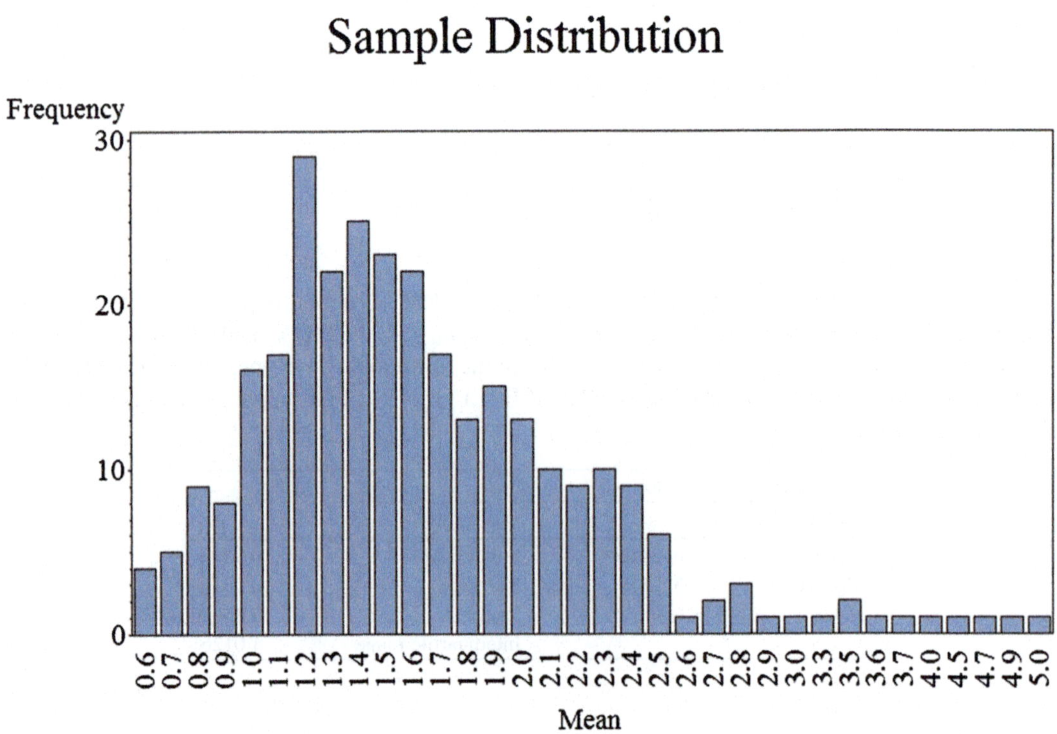

```
Goptions ftext='Times New Roman' htext=4 gunit=pct;
title height=7 'LogNormal Sample Distribution';
proc capability data=LogNorm;
   qqplot meanvar/normal(mu=est sigma=est color=red l=1);
run;
```

Basic statistical measures			
Location		Variability	
Mean	1.641284	Std deviation	0.67217
Median	1.514105	Variance	0.45182
Mode		Range	4.43876
		Interquartile range	0.72819

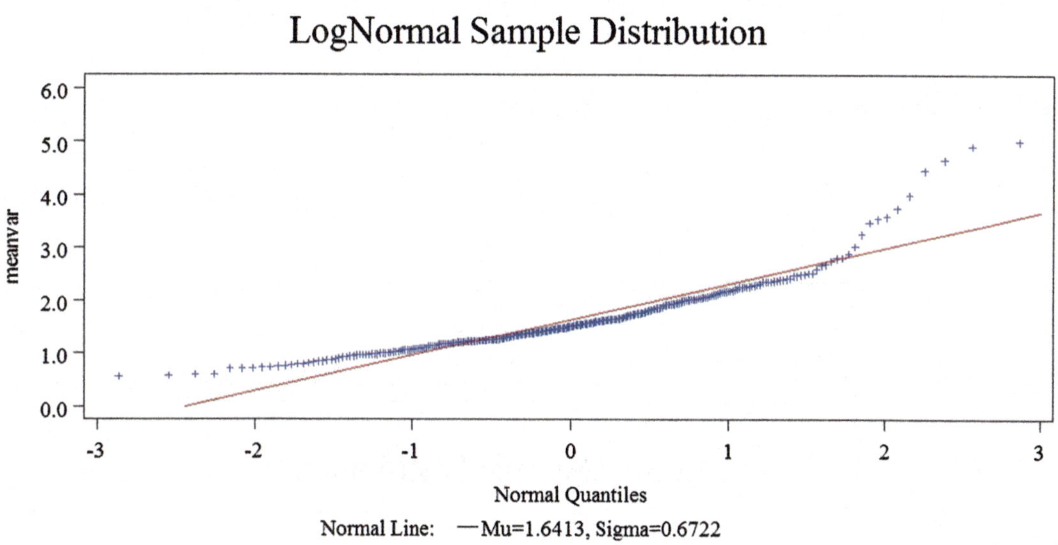

8.12.9 Uniform Sample Distribution: Normality Test

```
%Unif(k=5)
```

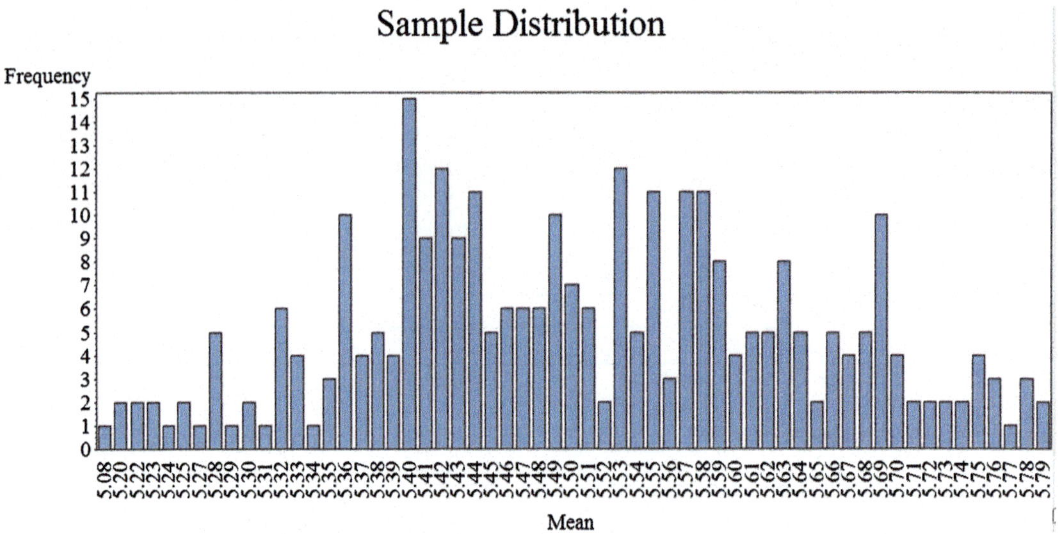


```
Goptions ftext='Times New Roman' htext=4 gunit=pct;
title height=7 'LogNormal Sample Distribution';
proc capability data=Unif;
   qqplot meanvar/normal(mu=est sigma=est color=red l=1);
run;
```

Basic statistical measures			
Location		Variability	
Mean	5.506585	Std deviation	0.13492
Median	5.500363	Variance	0.01820
Mode		Range	0.70500
		Interquartile range	0.19212

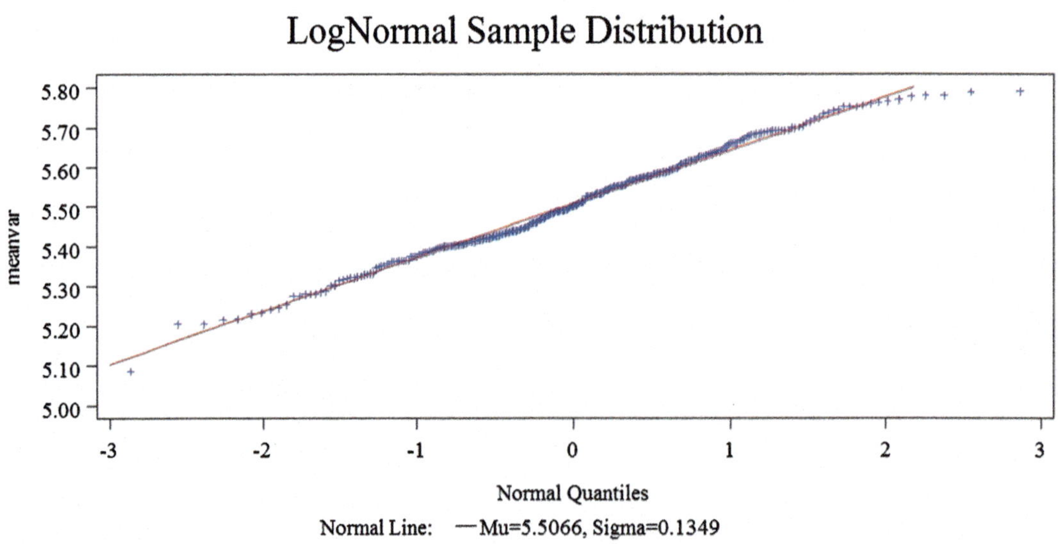

From the above normal probability plot, we can see that a sample size as small as 5 is all that is needed to create a sample distribution that is approximately normal.

8.13 Statistical Summary

In this chapter we first calculated the sample distribution for a small population. We saw that once we calculated the sample distribution, we can make educated statements about the population. We noticed that as the sample size increases, the sample distribution approaches a normal distribution.

Most of the time, it is very hard to calculate the sample distribution for a specific population distribution. So instead of calculating the sample distribution, we can derive the sample distribution by randomly sampling a population distribution many times. We saw that the resulting sample distribution is approximately a normal distribution if the sample size is greater than 30.

We created normal probability plots to test to see if a sample distribution is approximately normal. We saw that we had doubts with the normality of a sample distribution from a sample size 5 from a lognormal distribution. We saw that the sample distribution of a lognormal distribution was approximately normal when the sample size became 10. We saw that the sample distribution of a uniform distribution was normal when the size of the sample distribution was 5.

Appendix 8.1: Excel Code—Sample Distribution Creator

```
'/*************************************************************************
'           Essentials of Microsoft Excel 2013 VBA, SAS
'                     and MINITAB 17
'           for Statistical and Financial Analyses
'/*************************************************************************

Option Explicit
Option Base 1
Dim mvComboList As Variant
Dim mwsList As Worksheet
Property Let ComboList(c As Variant)
    mvComboList = c
End Property
Property Get ComboList() As Variant
    ComboList = mvComboList
End Property
Property Set Listcombo(w As Worksheet)
    Set mwsList = w
End Property
Property Get Listcombo() As Worksheet
    Set Listcombo = mwsList
End Property

Private Sub cmdCancel_Click()
    Unload Me
End Sub
Private Sub cmdRun_Click()
    Dim ws As Worksheet
    Me.Hide
    ComboHeader
    ChartMean
    Application.DisplayAlerts = False
    For Each ws In Me.Listcombo.Parent.Worksheets
        If Left(ws.Name, 5) = "Sheet" Then
            ws.Delete
        End If
    Next
    Application.DisplayAlerts = True
    Application.StatusBar = False
    Unload Me
End Sub
```

```
Private Sub UserForm_Initialize()
    With Me
        .ComboList = WorksheetFunction.Transpose(wsData.Cells(1, 1).CurrentRegion)
        If WorksheetFunction.CountA(wsData.Columns(1)) > 0 Then
            .TxtComboN = UBound(.ComboList)
            .txtComboR = .TxtComboN - 1
        Else
            .TxtComboN = 0
            .txtComboR.Enabled = False
        End If
        .Hide
    End With
End Sub


Sub Combinations(vCombo As Variant, lStart As Long, lN As Long, lR As Long, lMaxR As
Long, rPtr As Range)
    Dim i As Long
    Dim ws As Worksheet

    If lR > lMaxR Then
        i = 1
        While i <= lMaxR
            Application.StatusBar = "Combination on Row " & rPtr.Row & " of " _
                & WorksheetFunction.Combin(Me.TxtComboN, Me.txtComboR)
            rPtr.Offset(0, i - 1) = Me.ComboList(vCombo(i))
            i = i + 1
        Wend
        If Not (rPtr.Row = 655356) Then
            Set rPtr = rPtr.Offset(1, 0)
        Else
            'Put comb on new worksheet because its more than
            '655356 rows
            Set ws = rPtr.Parent.Parent.Worksheets.Add
            Set rPtr = ws.Cells(1)
        End If
    Else


        i = lStart
        While i <= lN
            vCombo(lR) = i
            Combinations vCombo:=vCombo, lStart:=i + 1, lN:=lN, lR:=lR + 1, _
                lMaxR:=lMaxR, rPtr:=rPtr
            i = i + 1
        Wend
    End If

End Sub


Sub ComboHeader()
    Dim vCombo As Variant
    Dim rPtr As Range
    Dim wb As Workbook

    ReDim vCombo(256)  ' Max number of columns

    Set wb = Workbooks.Add
    Set Me.Listcombo = wb.Worksheets.Add
    Set rPtr = Me.Listcombo.Cells(2, 1)
    Me.Listcombo.Name = "Sample Mean"
    Combinations vCombo:=vCombo, lStart:=1, lN:=Me.TxtComboN, lR:=1, _
        lMaxR:=Me.txtComboR, rPtr:=rPtr
End Sub
```

```
Sub ChartMean()
    Dim rSample As Range
    Dim rHeader As Range
    Dim rMean As Range
    Dim rCell As Range
    Dim pivReport As PivotTable
    Dim wsPiv As Worksheet
    Dim rPivData As Range
    Dim Chrt As Chart

    Set rSample = Me.Listcombo.Cells(2, 1).CurrentRegion
    Set rHeader = rSample.Rows(1).Offset(-1, 0)

    For Each rCell In rHeader.Cells
        rCell.Value = "Col" & rCell.Column
    Next

    Set rMean = rSample.Columns(1).Offset(, rSample.Columns.Count)
    rMean.Cells(1).Offset(-1, 0).Value = "Mean"

    rMean.Value = "=Sum(" & rSample.Rows(1).Address(False, False) _
                          & ") /" & rSample.Columns.Count
    rMean.Formula = rMean.Cells(1).Formula

    Set wsPiv = Me.Listcombo.Parent.Worksheets.Add
    Set pivReport = wsPiv.PivotTableWizard(SourceType:=xlDatabase, _
        SourceData:=Union(rSample, rHeader, rMean, rMean.Cells(1). _
        Offset(-1, 0)), tabledestination:=wsPiv.Cells(10, 4), _
        tablename:="Central")

        With pivReport
            .SmallGrid = False
            .AddFields RowFields:=rMean.Cells(1).Offset(-1, 0).Value
            With .PivotFields(rMean.Cells(1).Offset(-1, 0).Value)
                .Orientation = xlDataField
                .Caption = "Count"
                .Position = 1
                .Function = xlCount
            End With
            .ColumnGrand = False
            .Parent.Name = "Frequency"
            Set rPivData = .DataBodyRange
            With rPivData.Cells(1).Offset(-3, -2)
                    .Value = "Sample Mean Frequency Table"
                    .Font.Size = 12
                    .Font.Bold = True
            End With

            .DataBodyRange.CurrentRegion.Copy
            .DataBodyRange.CurrentRegion.PasteSpecial xlPasteValues
        End With

    'Calculate probability for chart
    With rPivData
        .Offset(0, 1) = "=RC[-1]/" & WorksheetFunction.Sum(rPivData)
        .Cells(1).Offset(-1, 1) = "Probability"
        .Parent.Cells(1).Select
    End With
```

```
    Set Chrt = Me.Listcombo.Parent.Charts.Add

    With Chrt
        .SetSourceData Source:=rPivData.Offset(0, 1), PlotBy:=xlColumns
        .SeriesCollection(1).XValues = rPivData.Offset(0, -1)
        .HasLegend = False
        .Axes(xlValue, xlPrimary).HasTitle = True
        .Axes(xlValue, xlPrimary).AxisTitle.Characters.Text = "Probability"
        .PlotArea.ClearFormats
        .Axes(xlValue).MajorGridlines.Delete
        .HasTitle = True
        With .ChartTitle
            .Characters.Text = "Sample Distribution" & Chr(10) & "Sample Mean = " & _
                                 WorksheetFunction.Sum(rMean) /
WorksheetFunction.Sum(rPivData) & _
                                 ",Population Mean = " & _
                                 WorksheetFunction.Sum(Me.ComboList) /
UBound(Me.ComboList)
            .Font.Size = 18

        End With
        .Axes(xlCategory).TickLabels.NumberFormat = "0.00"
        .Name = "Sample Distribution Chart"
      End With

End Sub
```

Bibliography

Microsoft Inc., Excel 2013. Microsoft Inc., Redmond

Lee CF, Lee JC, Lee AC (2013) Statistics for business and financial economics. Springer, New York

Minitab Inc. Minitab 17. Minitab Inc., State College

SAS Institute Inc. SAS 2014. SAS Institute Inc., Cary

Chapter 9
Other Continuous Distributions

9.1 Introduction

We will now look at other continuous distributions that are commonly used in statistics. It is important to study all these distributions because later we will make statistical inferences based on the particular distribution we are working with. We should remember that the area under the density curve of each of these distributions is equal to 1.

9.2 *t* Distribution

The *t* distribution is a symmetric distribution with mean 0.

The mathematical formula for the *t* distribution is

$$t = \frac{\overline{x} - \mu}{\frac{s_x}{\sqrt{n}}}$$

The *t* distribution is influenced by the degrees of freedom. As the degrees of freedom increase, the *t* distribution more closely resembles a standard normal distribution. In fact, the standard normal distribution is a *t* distribution with an infinite number of degrees of freedom. The characteristics of the *t* distribution are illustrated below. Below are the Minitab commands to produce *t* distribution and a standard normal distribution.

C.-F. Lee et al., *Essentials of Excel, Excel VBA, SAS and Minitab for Statistical and Financial Analyses*, DOI 10.1007/978-3-319-38867-0_9

```
MTB > name c1 'x-axis'
MTB > name c2 '1deg'
MTB > name c3 '4deg'
MTB > name c4 'stdnormal'
MTB > set 'x-axis'
DATA>    1(-5:5/.01)1
DATA> end
MTB > pdf 'x-axis' '1deg';
SUBC> t 1.
MTB > pdf 'x-axis' '4deg';
SUBC> t 4.
MTB > pdf 'x-axis' 'stdnormal';
SUBC> normal 0 1.
MTB > name c5 'graphx'
MTB > name c6 'graphy'
MTB > name c7 'group'
MTB > Stack 'x-axis' 'x-axis' 'x-axis' 'graphx'.
MTB > Stack  '1deg' '4deg' 'stdnormal' 'graphy';
SUBC>   Subscripts 'group';
SUBC>   UseNames.
MTB > Plot 'graphy'*'graphx';
SUBC>   Connect 'group';
SUBC>   Title "t Distributions and";
SUBC>   Title "Standard Normal Distribution";
SUBC>   Axlabel 1 "x-Axis";
SUBC>   AXLabel 2 " Probability ".
```

9.2.1 *Scatterplot of graphy vs. graphx*

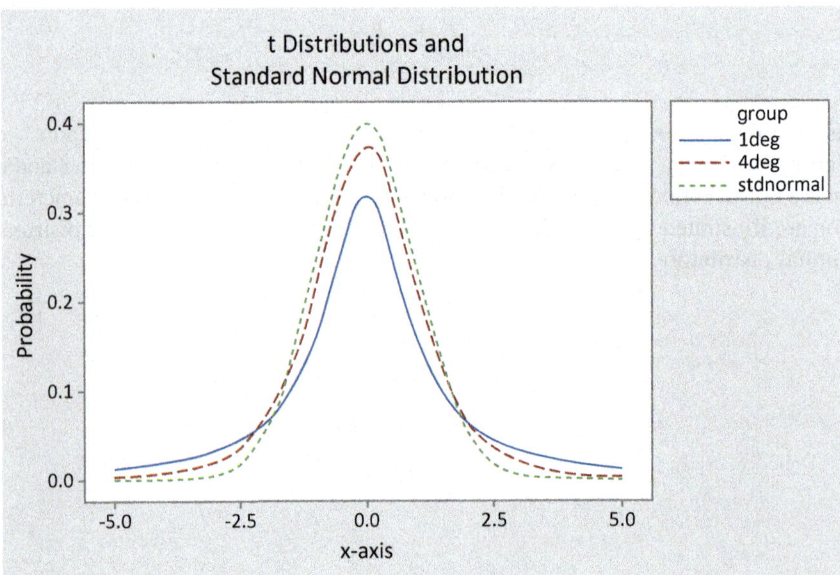

We will now calculate and plot the *t* cumulative density function for both the one degree of freedom t distribution and the four degrees of freedom t distribution to illustrate that the total probability of a *t* distribution equals 1.

```
MTB > name c1 'x-axis'
MTB > name c2 '1deg'
MTB > name c3 '4deg'
MTB > name c4 'stdnormal'
MTB > set 'x-axis'
DATA>    1(-5:5/.01)1
DATA> end
MTB > cdf 'x-axis' '1deg';
SUBC> t 1.
MTB > cdf 'x-axis' '4deg';
SUBC> t 4.
MTB > cdf 'x-axis' 'stdnormal';
SUBC> normal 0 1.
MTB > name c5 'graphx'
MTB > name c6 'graphy'
MTB > name c7 'group'
MTB > Stack 'x-axis' 'x-axis' 'x-axis' 'graphx'.
MTB > Stack  '1deg' '4deg' 'stdnormal' 'graphy';
SUBC>   Subscripts 'group';
SUBC>   UseNames.
MTB > Plot 'graphy'*'graphx';
SUBC>   Connect 'group';
SUBC>   Title "Cumulative t Distributions";
SUBC>   Axlabel 1 "x-Axis";
SUBC>   AXLabel 2 " Probability ".
```

9.2.2 Scatterplot of graphy vs. graphx

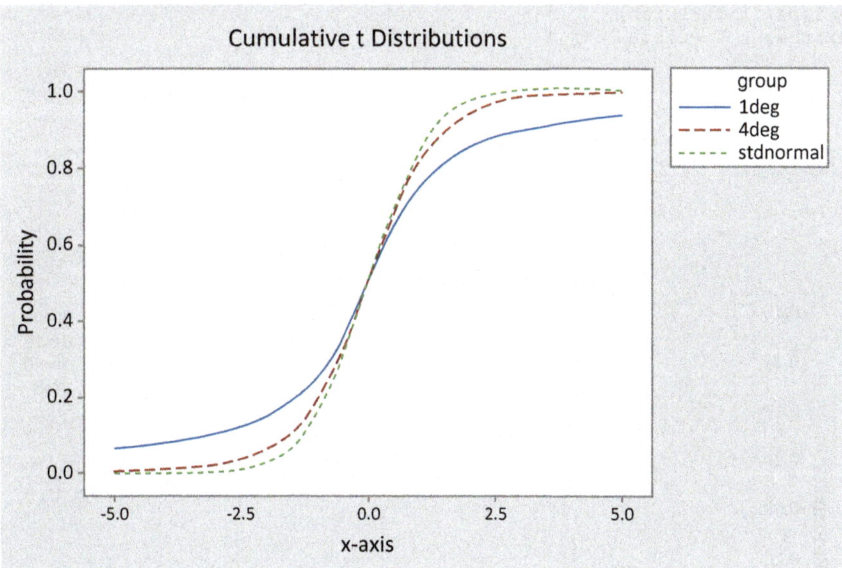

9.3 Chi-Square (χ^2) Distribution

Another distribution often used in statistics is the chi-square (χ^2) distribution. The mathematical formula for the chi-square is

$$\chi^2 = \sum_{i=1}^{n} \left(\frac{X_i - \mu}{\sigma_X} \right)^2$$

This probability density function is greatly influenced by the degrees of freedom.

The Minitab session commands below will show that as the degree of freedom increases, the distribution shifts to the right. We will create three chi-square distributions with 5, 10, and 30 degrees of freedom to show that the chi-square distribution does in fact shift to the right as the degree of freedom increases.

```
MTB > name c1 'x-axis'
MTB > name c2 'df=5'
MTB > name c3 'df=10'
MTB > name c4 'df=30'
MTB > set 'x-axis'
DATA>    1(0:60/.01)1
DATA> end
MTB > pdf 'x-axis' 'df=5';
SUBC> chisquare 5.
MTB > pdf 'x-axis' 'df=10';
SUBC> chisquare 10.
MTB > pdf 'x-axis' 'df=30';
SUBC> chisquare 30.
MTB > name c5 'graphx'
MTB > name c6 'graphy'
MTB > name c7 'group'
MTB > Stack 'x-axis' 'x-axis' 'x-axis' 'graphx'.
MTB > Stack  'df=5' 'df=10' 'df=30' 'graphy';
SUBC>   Subscripts 'group';
SUBC>   UseNames.
MTB > Plot 'graphy'*'graphx';
SUBC>   Connect 'group';
SUBC>   Title "Chi-Square Distributions";
SUBC>   Axlabel 1 "x-Axis";
SUBC>   AXLabel 2 " Probability ".
```

9.3.1 *Scatterplot of graphy vs. graphx*

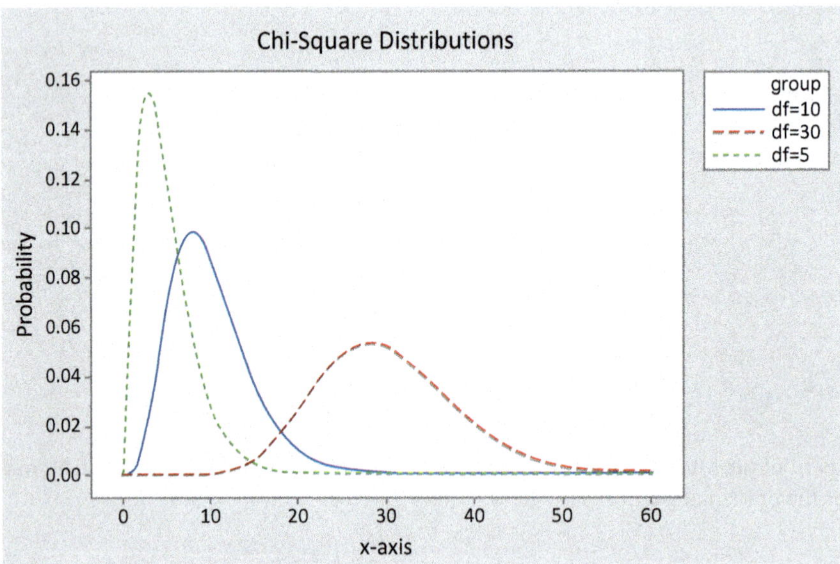

Below are the Minitab commands to graph the cumulative chi-square distribution with degrees of freedom of 5, 10, and 30.

```
MTB > name c1 'x-axis'
MTB > name c2 'df=5'
MTB > name c3 'df=10'
MTB > name c4 'df=30'
MTB > set 'x-axis'
DATA>    1(0:60/.01)1
DATA> end
MTB > cdf 'x-axis' 'df=5';
SUBC> chisquare 5.
MTB > cdf 'x-axis' 'df=10';
SUBC> chisquare 10.
MTB > cdf 'x-axis' 'df=30';
SUBC> chisquare 30.
MTB > name c5 'graphx'
MTB > name c6 'graphy'
MTB > name c7 'group'
MTB > Stack 'x-axis' 'x-axis' 'x-axis' 'graphx'.
MTB > Stack  'df=5' 'df=10' 'df=30' 'graphy';
SUBC>   Subscripts 'group';
SUBC>   UseNames.
MTB > Plot 'graphy'*'graphx';
SUBC>   Connect 'group';
SUBC>   Title "Chi-Square Distributions";
SUBC>   Axlabel 1 "x-Axis";
SUBC>   AXLabel 2 " Probability ".
```

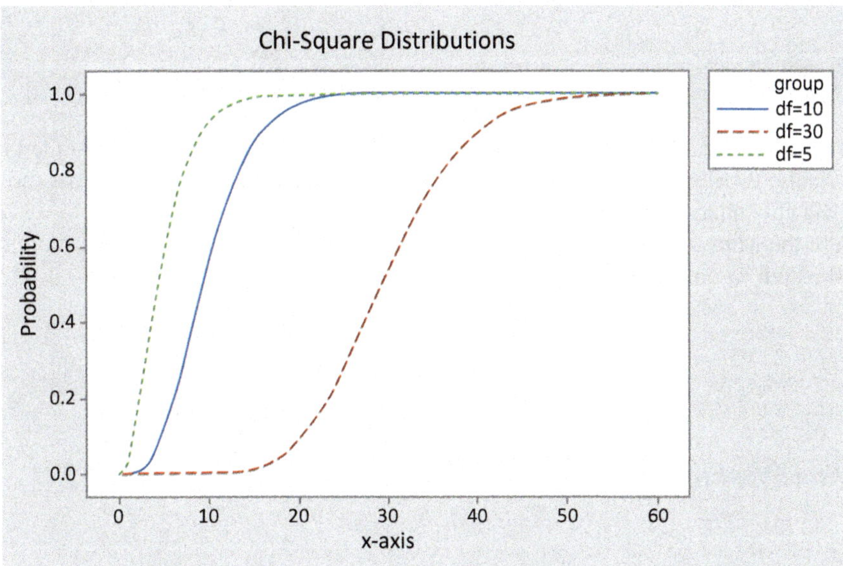

Excel does not have a PDF or a CDF for the chi-square distribution. Excel does have a function called *chisq.dist.rt*. This function is shown below.

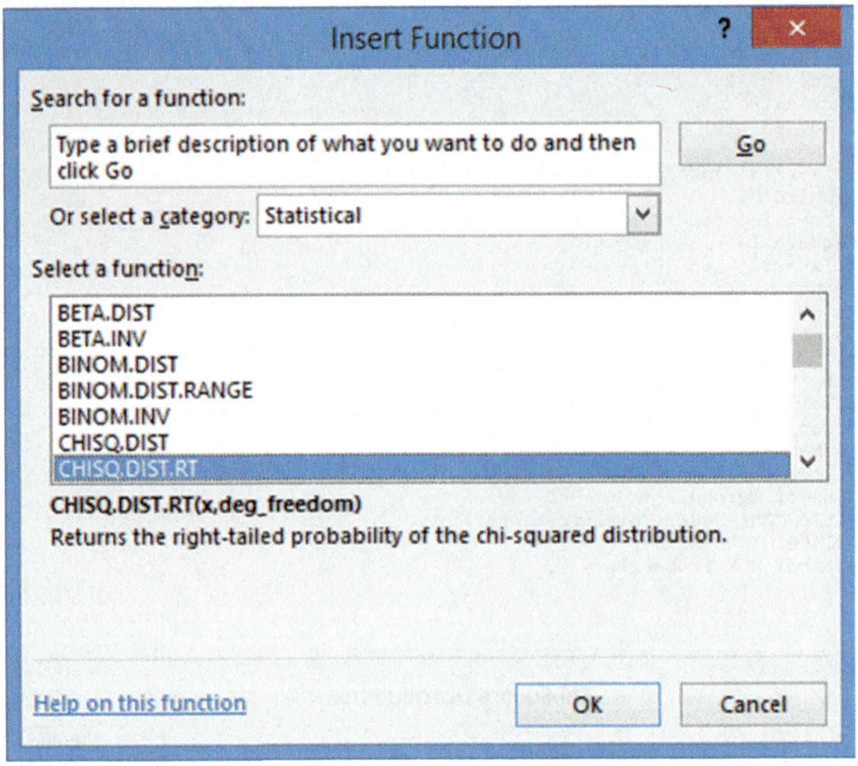

Excel's function wizard indicates that the *chisq.dist.rt* function returns the right-tailed probability of the chi-square distribution. What this actually means is that this function returns the one-tailed *p-value* of the chi-square distribution.

Below are the Minitab commands to calculate the *p-value* of a chi-square distribution with one degree of freedom when x is equal to 2.

```
MTB > cdf 2 k1;
SUBC> chisquare 1.
MTB > print k1
```

9.3.2 Data Display

```
K1     0.842701
MTB > let k2 = 1 - k1
MTB > print k2
```

9.3.3 Data Display

```
K2     0.157299
```

Below shows the above *p-value* calculation in Excel using the *chi.dist.rt* distribution.

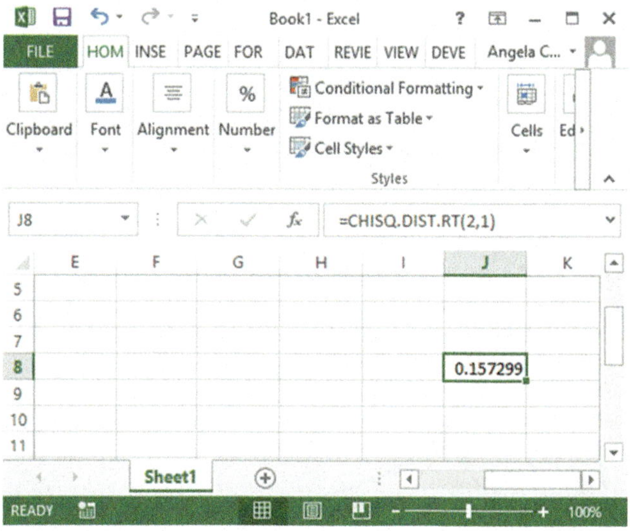

9.4 *F* Distribution

Another distribution often used in statistics is the *F* distribution. Its mathematical formula is

$$F = \frac{s_X^2/\sigma_X^2}{s_Y^2/\sigma_Y^2}$$

Each *F* distribution is influenced by two different degrees of freedom. There is a degree of freedom for the numerator and a degree of freedom for the denominator of the formula.

The session commands below will show that as the two different degrees of freedom increase, the more the *F* distribution shifts to the right and the more peaked the *F* distribution becomes. We will generate and plot three *F* distributions to illustrate this characteristic of the *F* distribution.

```
MTB > name c1 'x-axis'
MTB > name c2 'df=5,5'
MTB > name c3 'df=10,10'
MTB > name c4 'df=30,30'
MTB > set 'x-axis'
DATA> 1(0:5/.01)1
DATA> end
MTB > pdf 'x-axis' 'df=5,5';
SUBC> f 5 5.
MTB > pdf 'x-axis' 'df=10,10';
SUBC> f 10 10.
MTB > pdf 'x-axis' 'df=30,30';
SUBC> f 30 30.
MTB > name c5 'graphx'
MTB > name c6 'graphy'
MTB > name c7 'group'
MTB > Stack 'x-axis' 'x-axis' 'x-axis' 'graphx'.
MTB > Stack  'df=5,5' 'df=10,10' 'df=30,30' 'graphy';
SUBC>   Subscripts 'group';
SUBC>   UseNames.
MTB > Plot 'graphy'*'graphx';
SUBC>   Connect 'group';
SUBC>   Title "F Distributions";
SUBC>   Axlabel 1 "x-Axis";
SUBC>   AXLabel 2 " Probability ".
```

9.4.1 Scatterplot of graphy vs. graphx

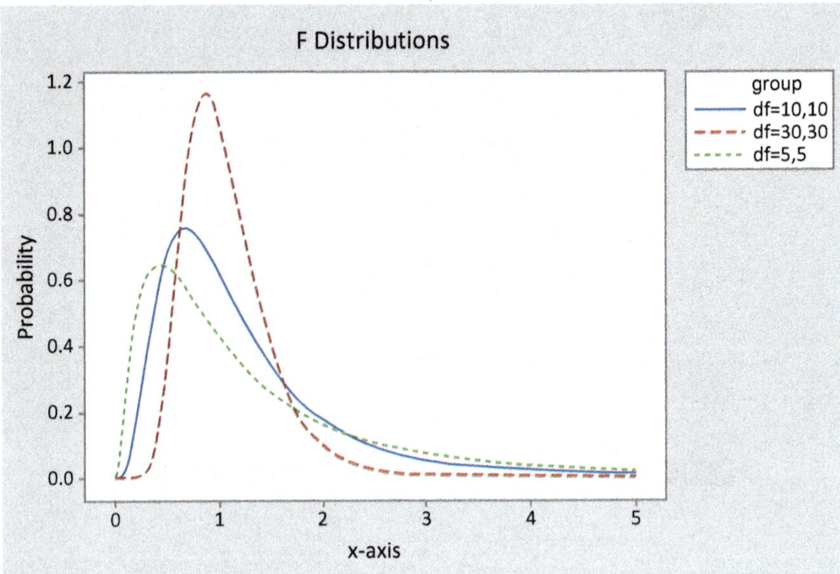

Below are the commands to calculate the cumulative *F* distribution for graphs with 5,5 and 10,10 and 30,30 degrees of freedom.

```
MTB > name c1 'x-axis'
MTB > name c2 'df=5,5'
MTB > name c3 'df=10,10'
MTB > name c4 'df=30,30'
MTB > set 'x-axis'
DATA> 1(0:5/.01)1
DATA> end
MTB > cdf 'x-axis' 'df=5,5';
SUBC> f 5 5.
MTB > cdf 'x-axis' 'df=10,10';
SUBC> f 10 10.
MTB > cdf 'x-axis' 'df=30,30';
SUBC> f 30 30.
MTB > name c5 'graphx'
MTB > name c6 'graphy'
MTB > name c7 'group'
MTB > Stack 'x-axis' 'x-axis' 'x-axis' 'graphx'.
MTB > Stack  'df=5,5' 'df=10,10' 'df=30,30' 'graphy';
SUBC>   Subscripts 'group';
SUBC>   UseNames.
MTB > Plot 'graphy'*'graphx';
SUBC>   Connect 'group';
SUBC>   Title "Cumulative F Distributions";
SUBC>   Axlabel 1 "x-Axis";
SUBC>   AXLabel 2 " Probability ".
```

9.4.2 *Scatterplot of graphy vs. graphx*

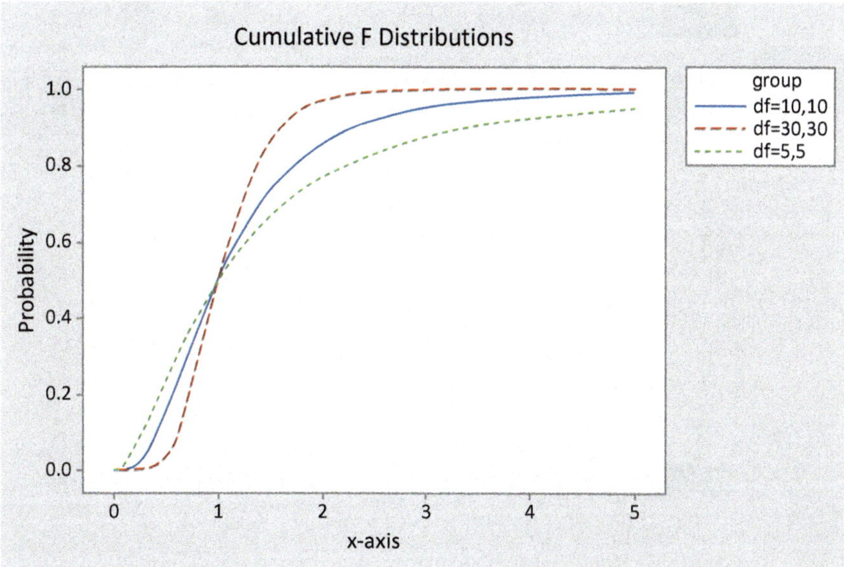

The only thing that Excel calculates for the *F* distribution is the one-tail *p-value*. The function that calculates the *p-value* is the *fdist*.

Below are the Minitab commands to calculate the *p-value* of an *F* distribution that has a degrees of freedom of 5,5 and when x is 2.

```
MTB > cdf 2 k1;
SUBC> f 5 5.
MTB > print k1
```

9.4.3 *Data Display*

```
K1     0.767489
MTB > let k2 = 1-k1
MTB > print k2
```

9.4.4 *Data Display*

```
K2     0.232511
```

Below is the above one-tail *p-value* calculation done in Excel using the *f.dist.rt* function.

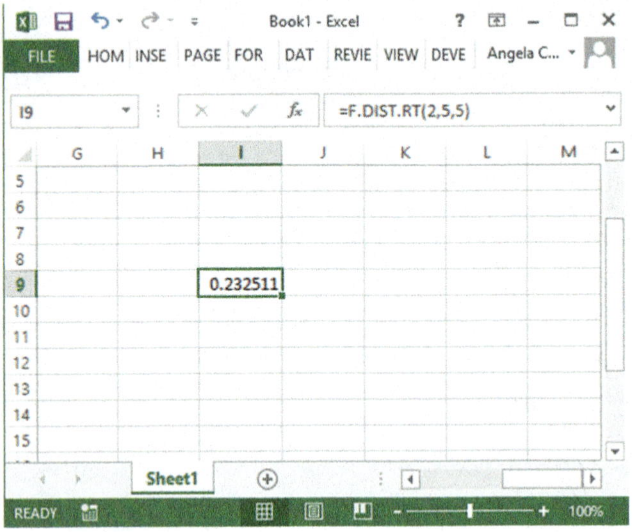

9.5 Exponential Distribution

The last continuous distribution that we will look at is the exponential distribution. The formula for it in Minitab is

$$f(t) = \frac{1}{\lambda} e^{-t/\lambda}, \quad t > 0$$

Some books use an alternative formula:

$$f(t) = \lambda e^{-\lambda t}, \quad t \geq 0$$

This second way of expressing the exponential formula has the interesting property that when $t = 0$ the exponential function will be equal to λ.

9.6 Exponential Distribution in Minitab

The following Minitab session commands create exponential distributions with lambda equal to 1, 2, and 3.

```
MTB > name c1 'x-axis'
MTB > name c2 'lamda=1'
MTB > name c3 'lamda=2'
MTB > name c4 'lamda=3'
MTB > set 'x-axis'
DATA> 1(0:4/.01)1
DATA> end
MTB > pdf 'x-axis' 'lamda=1';
SUBC> exponential 1.
MTB > pdf 'x-axis' 'lamda=2';
SUBC> exponential 2.
MTB > pdf 'x-axis' 'lamda=3';
SUBC> exponential 3.
MTB > name c5 'graphx'
MTB > name c6 'graphy'
MTB > name c7 'group'
MTB > Stack 'x-axis' 'x-axis' 'x-axis' 'graphx'.
MTB > Stack  'lamda=1' 'lamda=2' 'lamda=3' 'graphy';
SUBC>   Subscripts 'group';
SUBC>   UseNames.
MTB > Plot 'graphy'*'graphx';
SUBC>   Connect 'group';
SUBC>   Title "Exponential Distributions";
SUBC>   Axlabel 1 "x-Axis";
SUBC>   AXLabel 2 " Probability ".
```

9.6.1 *Scatterplot of graphy vs. graphx*

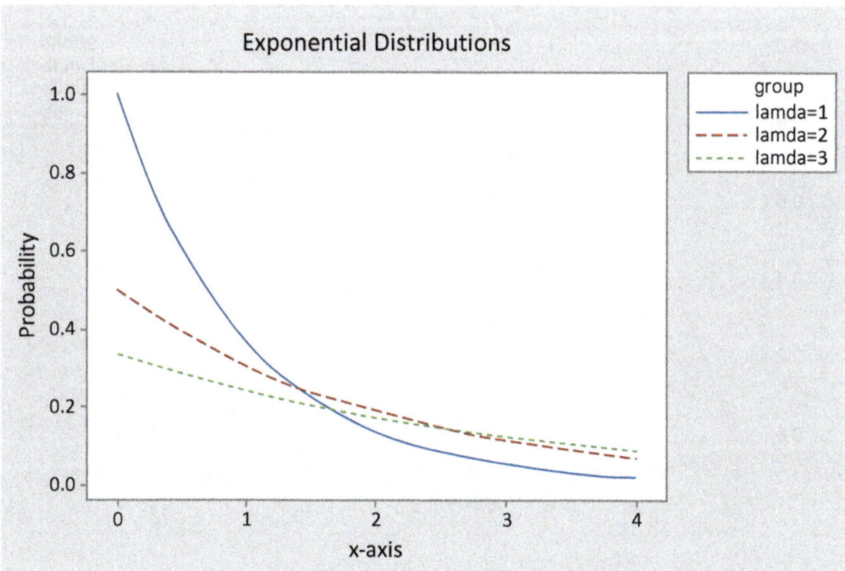

Note in the above graph that the left side of the exponential distribution approaches 1 divided by lambda. For example, in the above graph, the exponential distribution with a lambda of 2, the left side of the distribution approaches .5.

```
MTB > name c1 'x-axis'
MTB > name c2 'lamda=1'
MTB > name c3 'lamda=2'
MTB > name c4 'lamda=3'
MTB > set 'x-axis'
DATA> 1(0:4/.01)1
DATA> end
MTB > cdf 'x-axis' 'lamda=1';
SUBC> exponential 1.
MTB > cdf 'x-axis' 'lamda=2';
SUBC> exponential 2.
MTB > cdf 'x-axis' 'lamda=3';
SUBC> exponential 3.
MTB > name c5 'graphx'
MTB > name c6 'graphy'
MTB > name c7 'group'
MTB > Stack 'x-axis' 'x-axis' 'x-axis' 'graphx'.
MTB > Stack  'lamda=1' 'lamda=2' 'lamda=3' 'graphy';
SUBC>   Subscripts 'group';
SUBC>   UseNames.
MTB > Plot 'graphy'*'graphx';
SUBC>   Connect 'group';
SUBC>   Title "Cumulative Exponential Distributions";
SUBC>   Axlabel 1 "x-Axis";
SUBC>   AXLabel 2 " Probability ".
```

9.6.2 Scatterplot of graphy vs. graphx

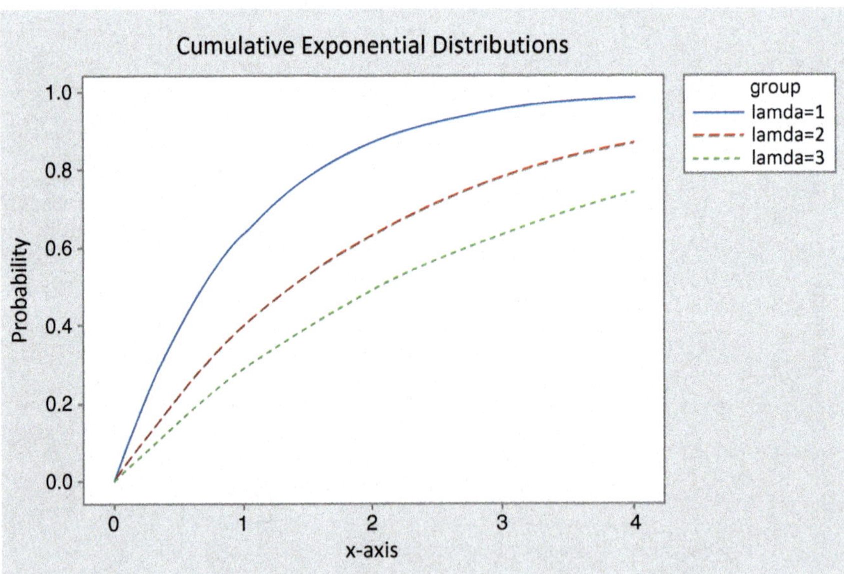

9.7 Excel Macro—Exponential Distribution

The Excel code that produces the above exponential distribution is shown in Appendix 9.1.

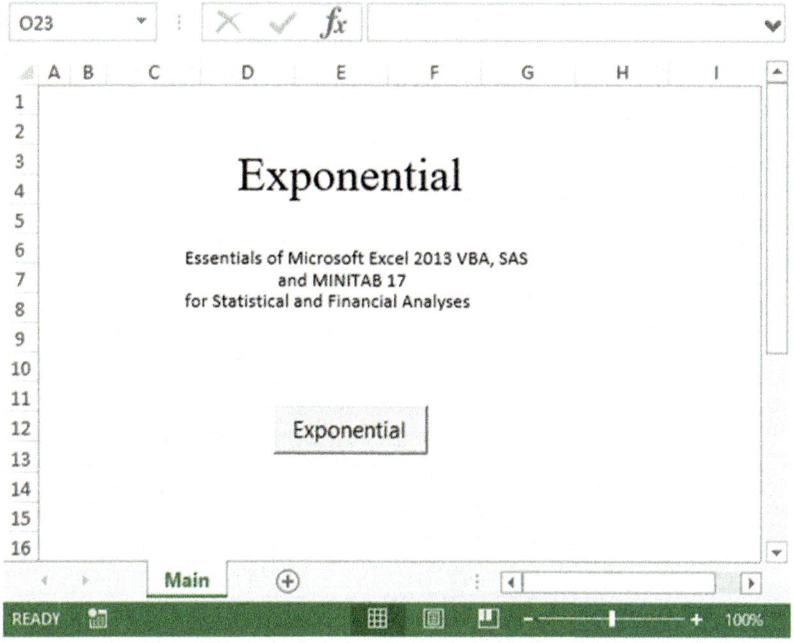

Push the *Exponential* button shown above to get the dialog box below.

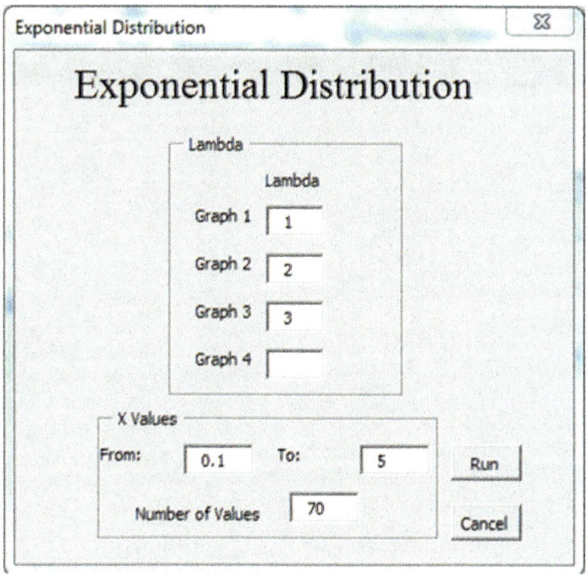

The next thing to do is to fill out the *Exponential Distribution* dialog box as shown above. Then press the *Run* button to get the two charts shown below.

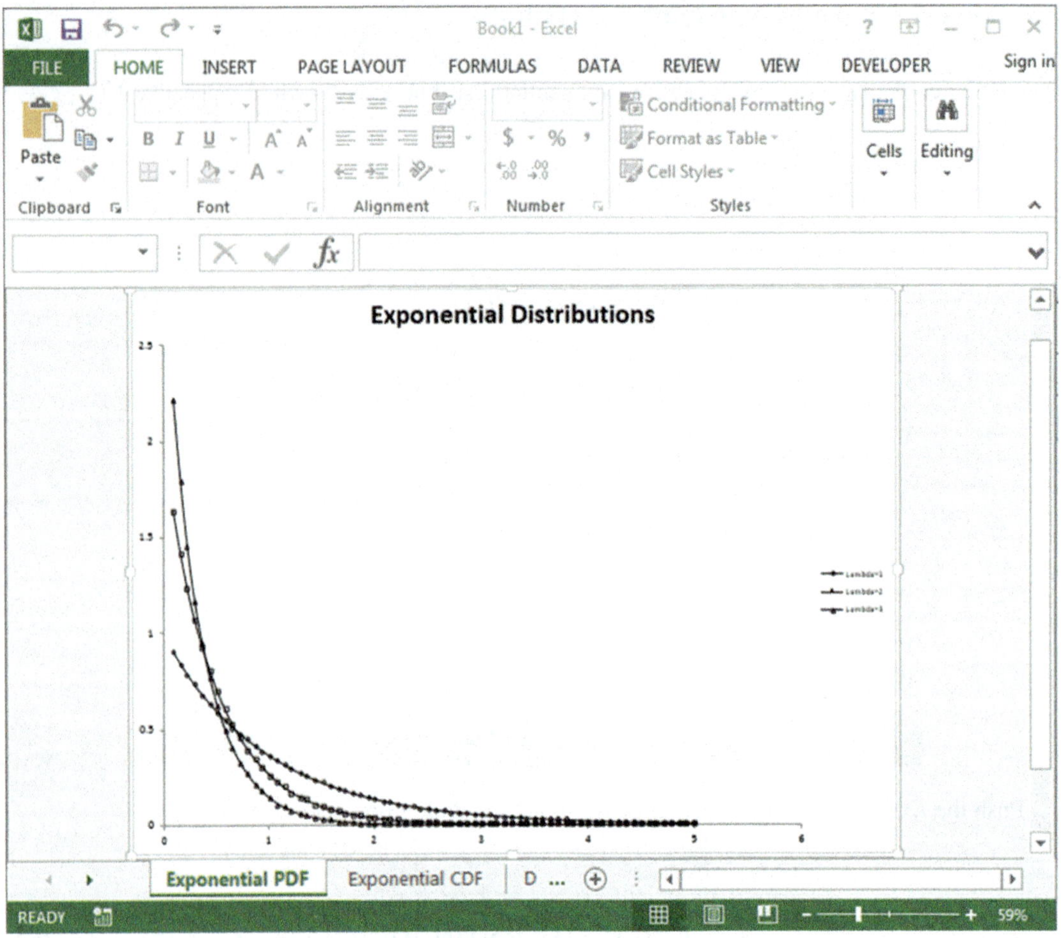
Book1 - Excel
FILE
HOME
INSERT
PAGE LAYOUT
FORMULAS
DATA
REVIEW
VIEW
DEVELOPER
Sign in
Paste
Clipboard
Font
Alignment
Number
Conditional Formatting
Format as Table
Cell Styles
Styles
Cells
Editing
Exponential Distributions
Exponential PDF
Exponential CDF
READY
59%

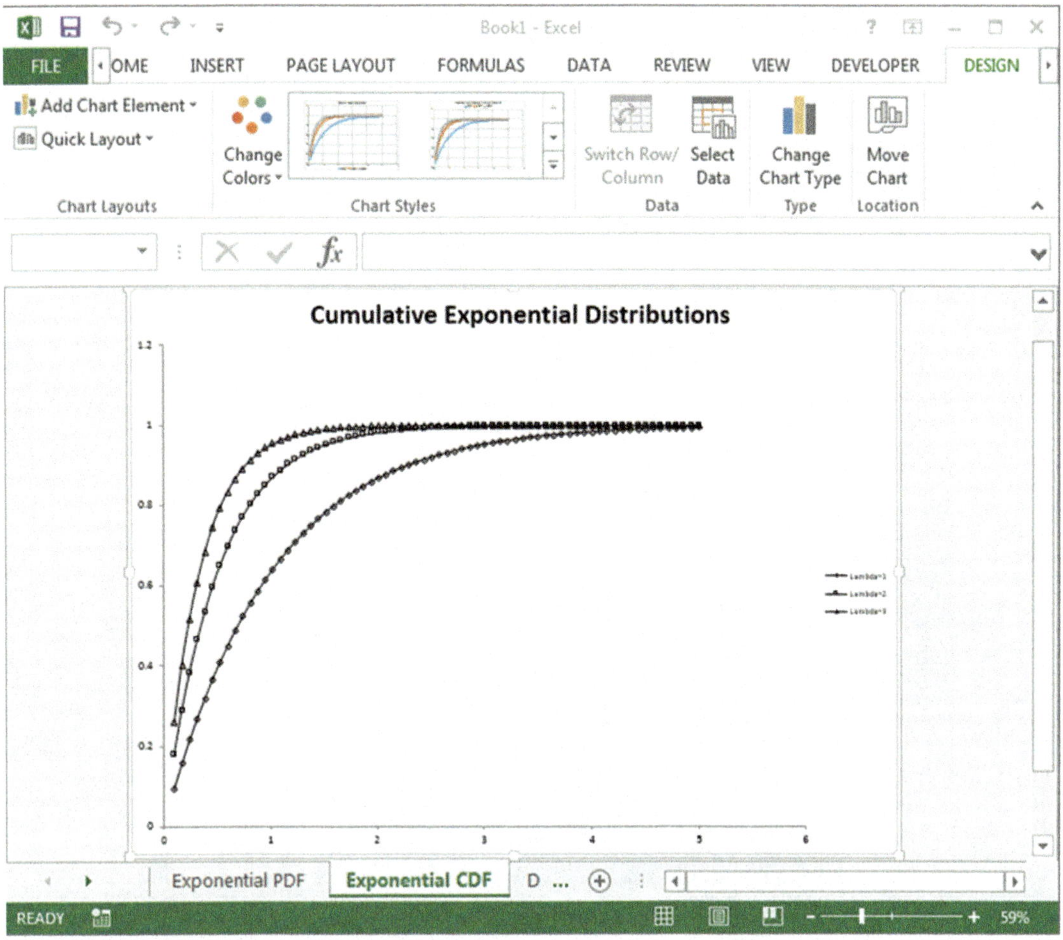

9.8 Central Limit Theorem: Other Distributions

We will now illustrate the central limit theorem with the distributions we have examined in this chapter. As in Chap. 8, the distribution of the sample mean generated from the four distributions, which have been discussed in this chapter, will approach a normal distribution if the sample size is large enough.

9.8.1 Exponential Distribution

The *exponential* subcommand tells the *random* command to generate numbers at random from the exponential distribution. The argument of the exponential distribution indicates the λ of the exponential formula.

Sample Size = 5, 300 Samples

```
MTB > random 300 c1-c5;
SUBC> exponential 1.
MTB > rmean c1-c5 c10
MTB > Histogram C10;
SUBC>   MidPoint;
SUBC>   NInterval 100;
SUBC>   Bar;
SUBC>   Axis 1;
SUBC>   Axlabel 1 "x-Axis";
SUBC>   AXLabel 2 " Probability ".
```

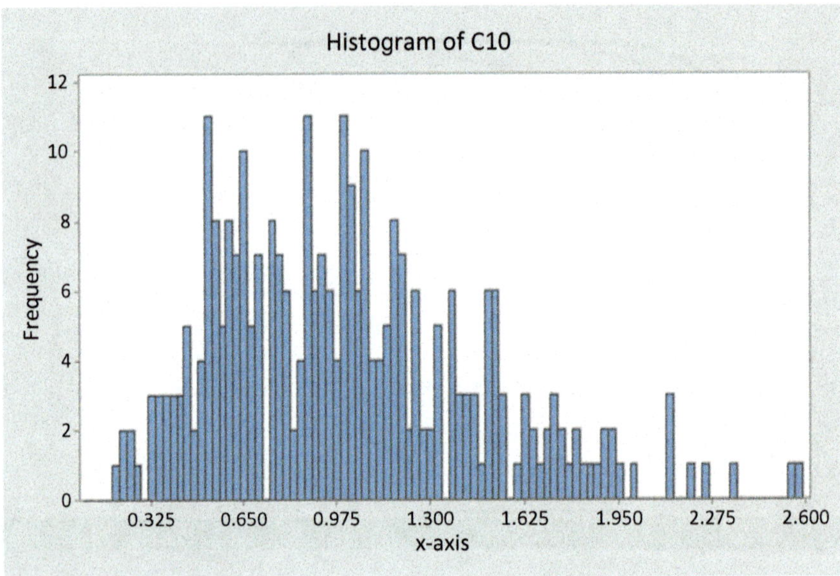

Sample Size = 10, 300 Samples

```
MTB >  random 300 c1-c10;
SUBC> exponential 1.
MTB > rmean c1-c10 c15
MTB > Histogram C15;
SUBC>   MidPoint;
SUBC>   NInterval 100;
SUBC>   Bar;
SUBC>   Axlabel 1 "x-Axis";
SUBC>   AXLabel 2 " Probability ".
```

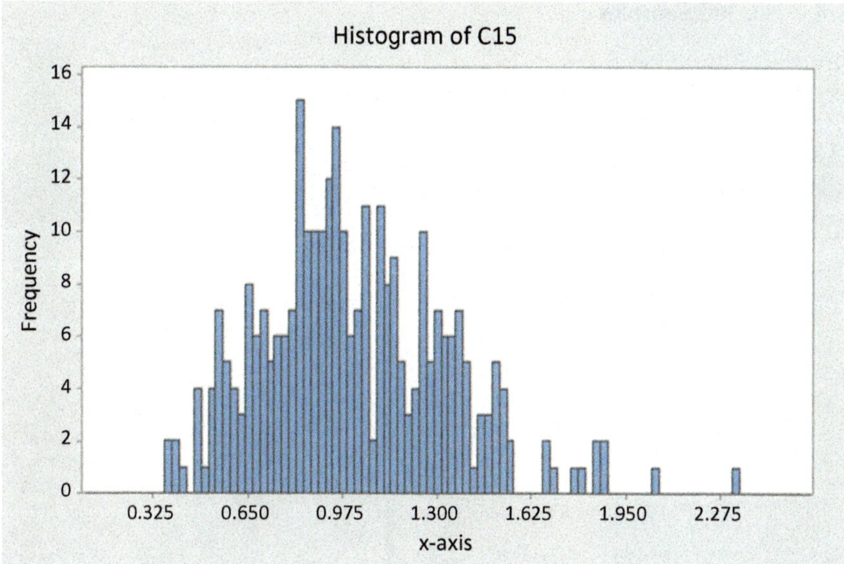

Sample Size = 30, 300 Samples

```
MTB > random 300 c1-c30;
SUBC> exponential 1.
MTB > rmean c1-c30 c35
MTB > Histogram C35;
SUBC>   MidPoint;
SUBC>   NInterval 100;
SUBC>   Bar;
SUBC>   Axlabel 1 "x-Axis";
SUBC>   AXLabel 2 " Probability ".
```

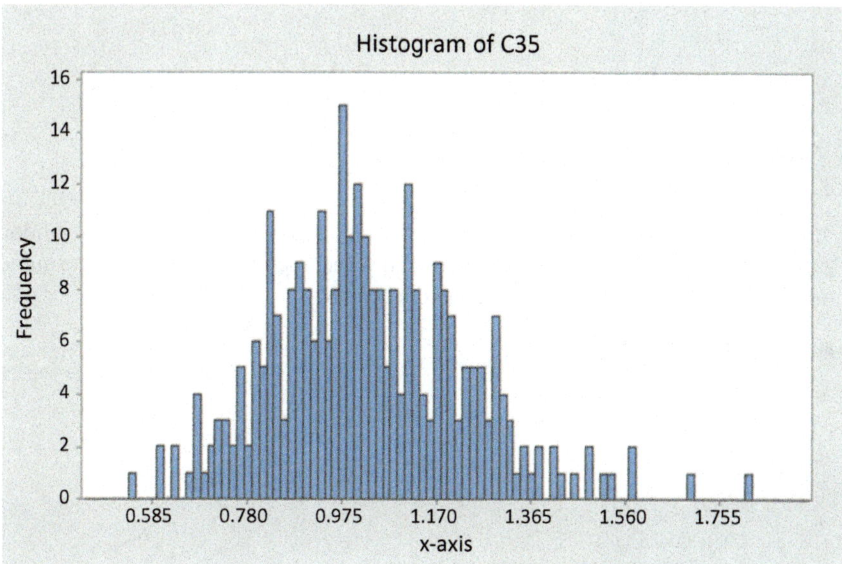

Sample Size = 50, 300 Samples

```
MTB > random 300 c1-c50;
SUBC> exponential 1.
MTB > rmean c1-c50 c55
MTB > Histogram C55;
SUBC>   MidPoint;
SUBC>   NInterval 100;
SUBC>   Bar;
SUBC>   Axlabel 1 "x-Axis";
SUBC>   AXLabel 2 " Probability ".
```

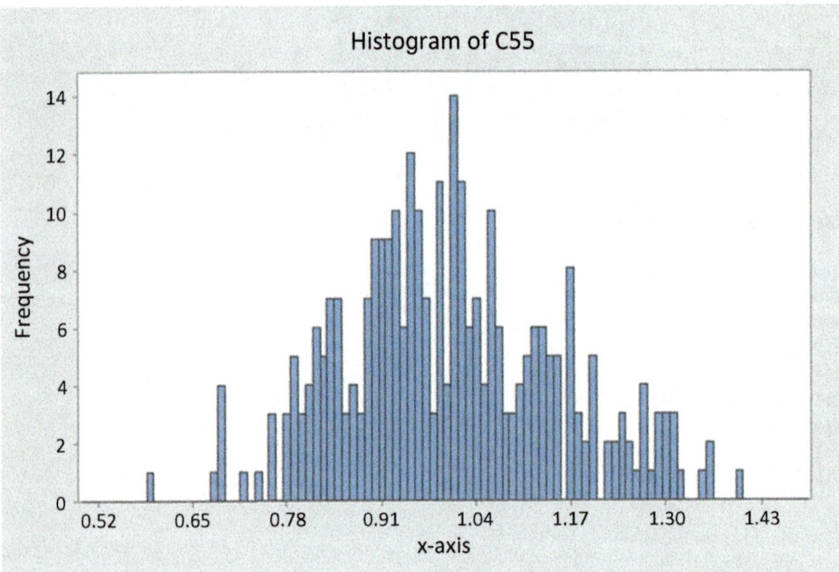

9.8.2 F Distribution

The *f* subcommand in Minitab tells the *random* subcommand to generate numbers at random from the *F* distribution. The first argument of the *f* subcommand is the degree of freedom of the numerator. The second argument is the degree of freedom of the denominator.

Sample Size = 5, 300 Samples

```
MTB > random 300 c1-c5;
SUBC> f 5 5.
MTB > rmean c1-c5 c10
MTB > Histogram C10;
SUBC>   MidPoint;
SUBC>   NInterval 100;
SUBC>   Bar;
SUBC>   Axlabel 1 "x-Axis";
SUBC>   AXLabel 2 " Probability ".
```

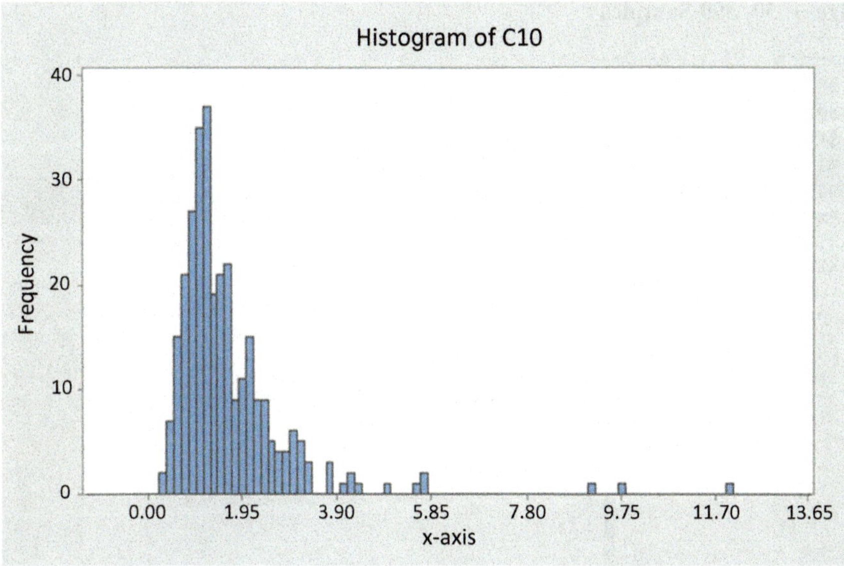

Sample Size = 10, 300 Samples

```
MTB > random 300 c1-c10;
SUBC> f 5 5.
MTB > rmean c1-c10 c15
MTB > Histogram C15;
SUBC>   MidPoint;
SUBC>   NInterval 100;
SUBC>   Bar;
SUBC>   Axlabel 1 "x-Axis";
SUBC>   AXLabel 2 " Probability ".
```

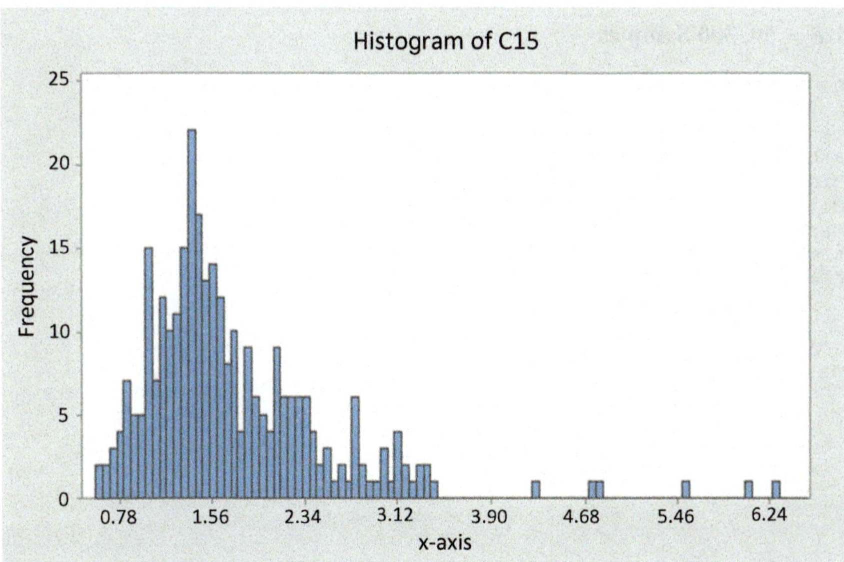

Sample Size = 30, 300 Samples

```
MTB >  random 300 c1-c30;
SUBC> f 5 5.
MTB > rmean c1-c30 c35
MTB > Histogram c35;
SUBC>   MidPoint;
SUBC>   NInterval 100;
SUBC>   Bar;
SUBC>   Axlabel 1 "x-Axis";
SUBC>   AXLabel 2 " Probability ".
```

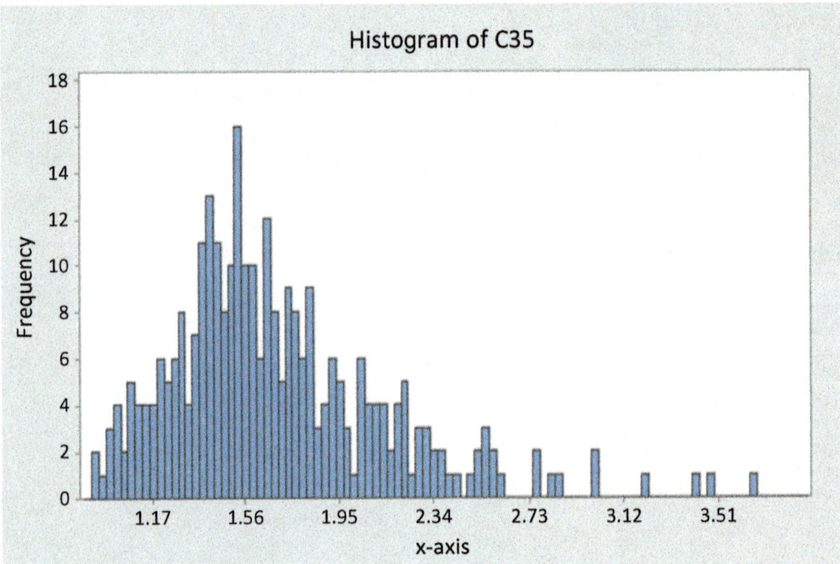

Sample Size = 50, 300 Samples

```
MTB > random 300 c1-c50;
SUBC> f 5 5.
MTB > rmean c1-c50 c55
MTB > Histogram C55;
SUBC>   MidPoint;
SUBC>   NInterval 100;
SUBC>   Bar;
SUBC>   Axlabel 1 "x-Axis";
SUBC>   AXLabel 2 " Probability ".
```

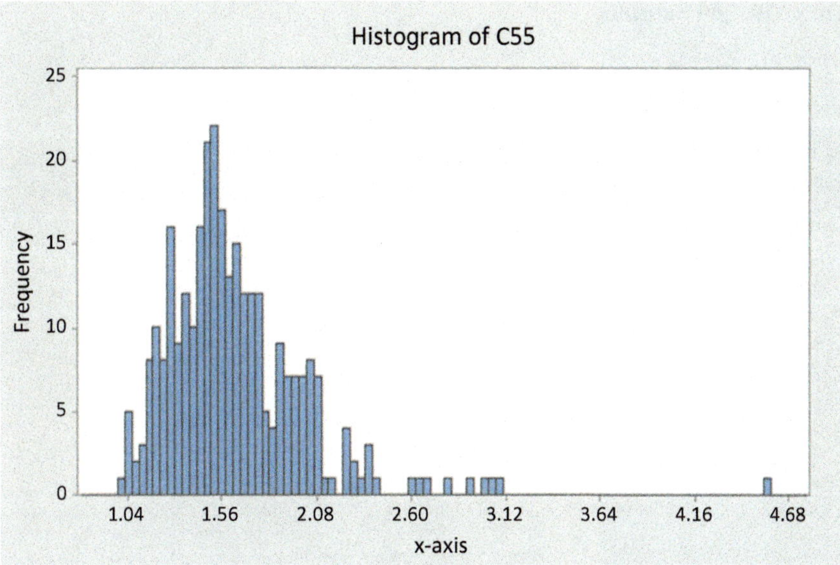

9.8.3 Chi-Square Distribution

The *chisquare* subcommand tells the *random* subcommand to generate numbers at random from the chi-square distribution. The first argument of the *chisquare* subcommand is degrees of freedom.

Sample Size = 5, 300 Samples

```
MTB >  random 300 c1-c5;
SUBC> chisquare 5.
MTB > rmean c1-c5 c10
MTB > Histogram C10;
SUBC>   MidPoint;
SUBC>   NInterval 100;
SUBC>   Bar;
SUBC>  Axlabel 1 "x-Axis";
SUBC>  AXLabel 2 " Probability ".
```

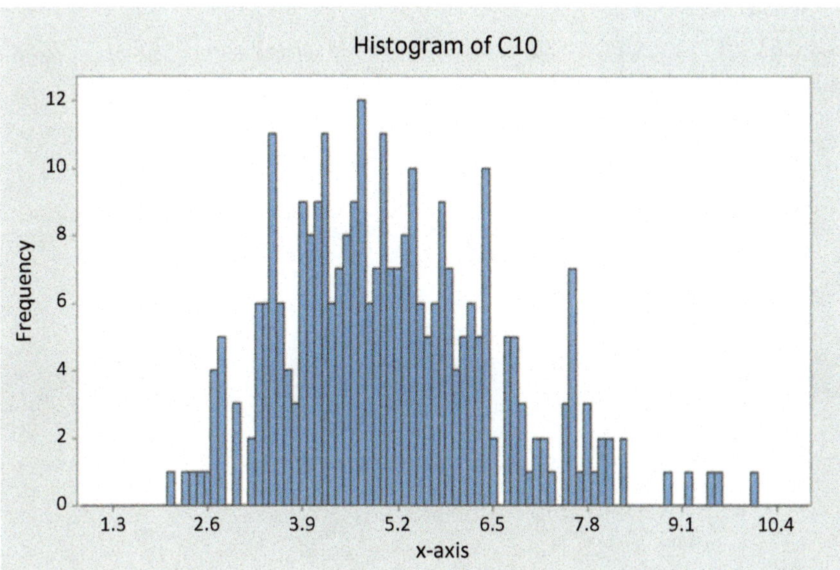

Sample Size = 10, 300 Samples

```
MTB > random 300 c1-c10;
SUBC> chisquare 5.
MTB > rmean c1-c10 c15
MTB > Histogram C15;
SUBC>   MidPoint;
SUBC>   NInterval 100;
SUBC>   Bar;
SUBC>   Axlabel 1 "x-Axis";
SUBC>   AXLabel 2 " Probability ".
```

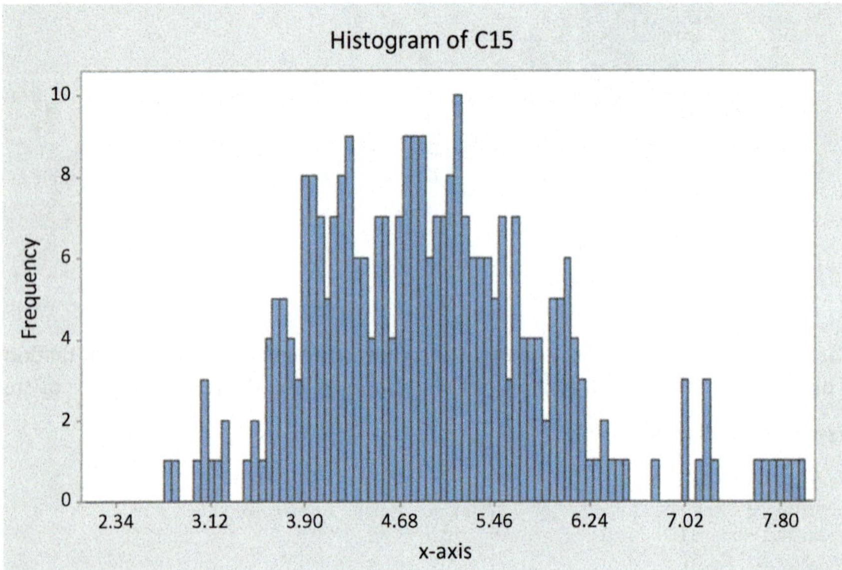

Sample Size = 30, 300 Samples

```
MTB > random 300 c1-c30;
SUBC> chisquare 5.
MTB > rmean c1-c30 c35
MTB > Histogram C35;
SUBC>   MidPoint;
SUBC>   NInterval 100;
SUBC>   Bar;
SUBC>   Axlabel 1 "x-Axis";
SUBC>   AXLabel 2 " Probability ".
```

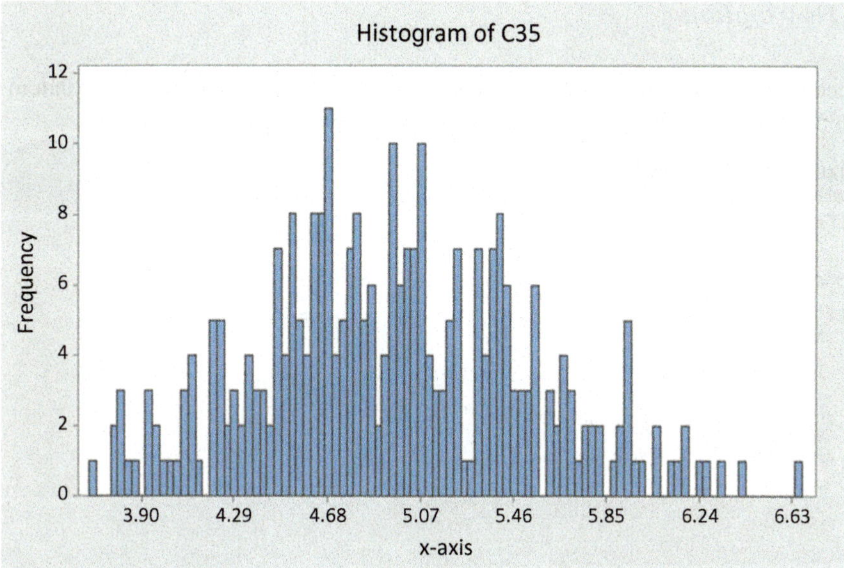

Sample Size = 50, 300 Samples

```
MTB > random 300 c1-c50;
SUBC> chisquare 5.
MTB > rmean c1-c50 c55
MTB > Histogram C55;
SUBC>   MidPoint;
SUBC>   NInterval 100;
SUBC>   Bar;
SUBC>   Axlabel 1 "x-Axis";
SUBC>   AXLabel 2 " Probability ".
```

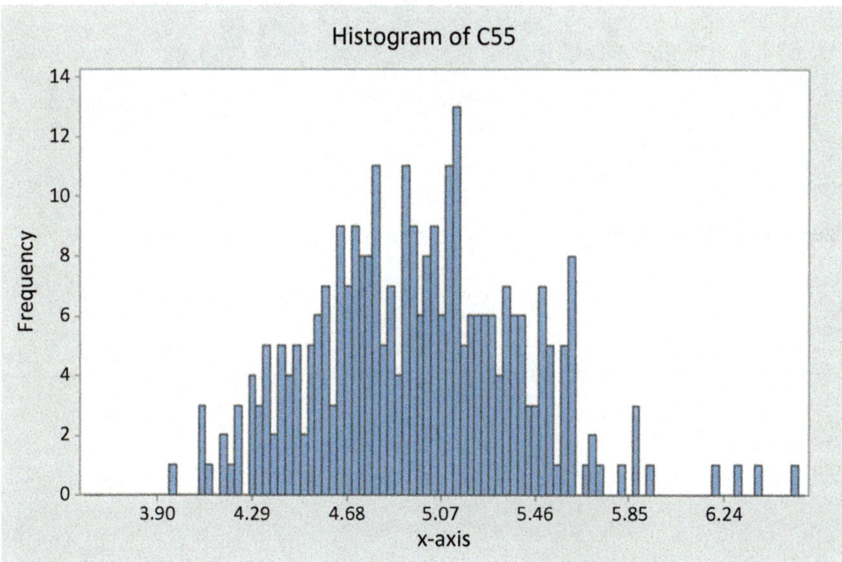

9.8.4 t Distribution

The *t* subcommand tells the *random* subcommand to generate numbers at random from the *t* distribution. The first argument of the *t* subcommand is the degree of freedom.

Sample Size = 5, 300 Samples

```
MTB >  random 300 c1-c5;
SUBC> t 10.
MTB > rmean c1-c5 c10
MTB > Histogram c10;
SUBC>   MidPoint;
SUBC>   NInterval 100;
SUBC>   Bar;
SUBC>   Axlabel 1 "x-Axis";
SUBC>   AXLabel 2 " Probability ".
```

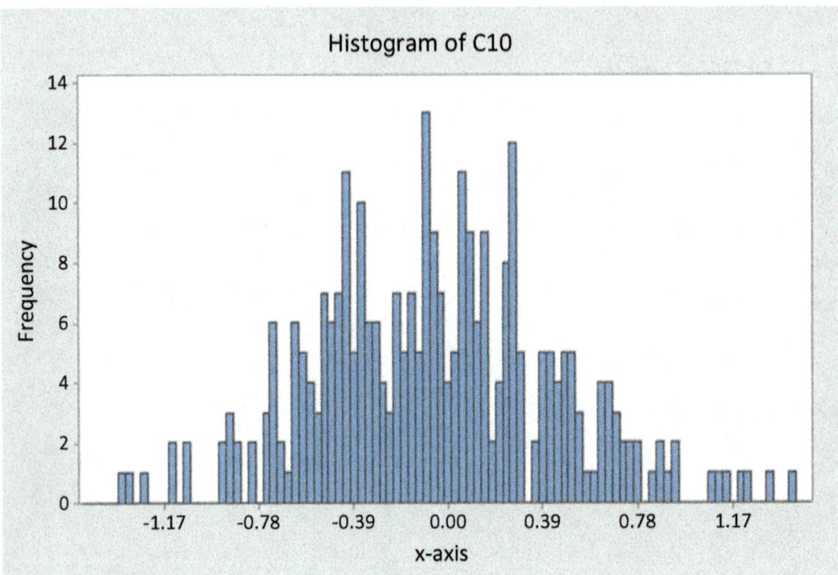

Sample Size = 10, 300 Samples

```
MTB > random 300 c1-c10;
SUBC> t 10.
MTB > rmean c1-c10 c15
MTB > Histogram C15;
SUBC>   MidPoint;
SUBC>   NInterval 100;
SUBC>   Bar;
SUBC>   Axlabel 1 "x-Axis";
SUBC>   AXLabel 2 " Probability ".
```

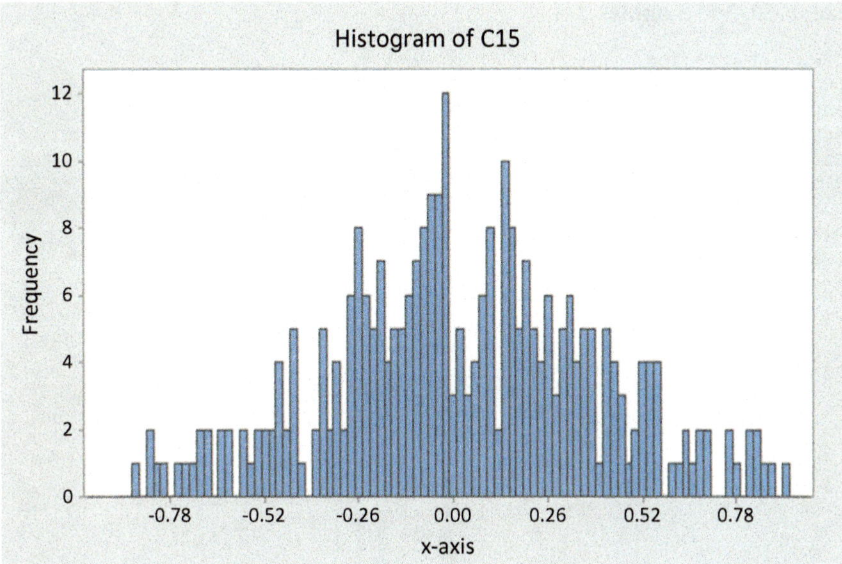

Sample Size = 30, 300 Samples

```
MTB > random 300 c1-c30;
SUBC> t 10.
MTB > rmean c1-c30 c35
MTB > Histogram C35;
SUBC>   MidPoint;
SUBC>   NInterval 100;
SUBC>   Bar;
SUBC>   Axlabel 1 "x-Axis";
SUBC>   AXLabel 2 " Probability ".
```

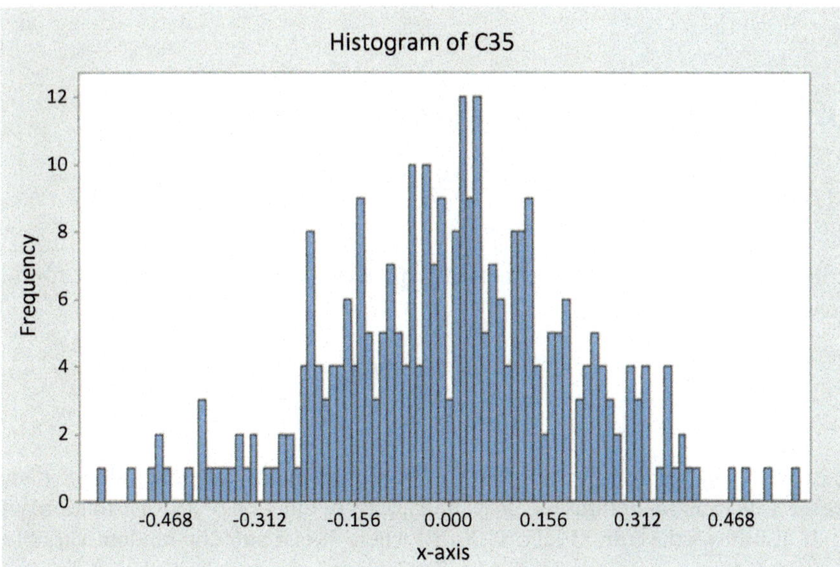

Sample Size = 50, 300 Samples

```
MTB > random 300 c1-c50;
SUBC> t 10.
MTB > rmean c1-c50 c55
MTB > Histogram C55;
SUBC>   MidPoint;
SUBC>   NInterval 100;
SUBC>   Bar;
SUBC>   Axlabel 1 "x-Axis";
SUBC>   AXLabel 2 " Probability ".
```

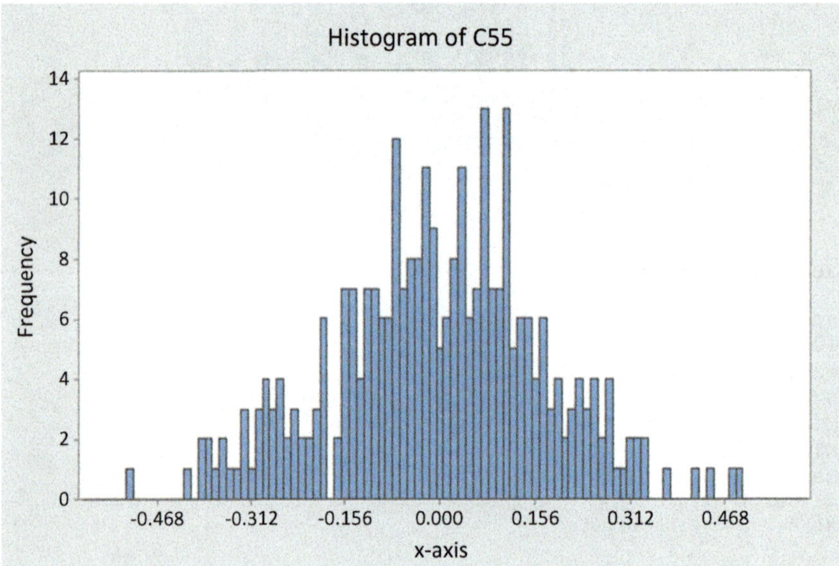

9.9 SAS Programming Code Instructions and Examples

9.9.1 t Distribution

The t distribution is a symmetric distribution with mean 0. The mathematical formula for the t distribution is

$$t = \frac{\overline{x} - \mu}{\frac{s_x}{\sqrt{n}}}$$

The same as the way we describe in Chap. 7, function pdf is used to generate the data. The PDF function for the T distribution returns the probability density function of a T distribution, with degrees of freedom df. It follows the form: PDF('T',X,df), where X is a numeric random variable.

Below are the SAS commands to produce t distribution and a standard normal distribution.

```
data Tdist;
do X=-5 to 5 by 0.01;
PDF1=pdf('T', X, 1);
PDF2=pdf('T', X, 4);
PDF3=pdf('normal', X, 0, 1);
output Tdist;
end;
run;

Goptions ftext='Times New Roman' htext=6 gunit=pct;
title height=7 'T Distributions and Standard Normal Distribution';
legend1 label=none value=('1 deg' '4 deg' 'STDNormal') shape=symbol(5,1);
axis1  label=('Probability') order=(0 to 0.4 by 0.1);
axis2  label=('X-Axis') order=(-5 to 5 by 5);
symbol1 i=join  v=none L=1 c=black w=2;
symbol2 i=join  v=none L=5 c=black w=2;
symbol3 i=join  v=none L=3 c=black w=2;
proc gplot data=Tdist;
plot (PDF1 PDF2 PDF3)*X/vaxis=axis1 haxis=axis2 overlay legend=legend1;
run;
```

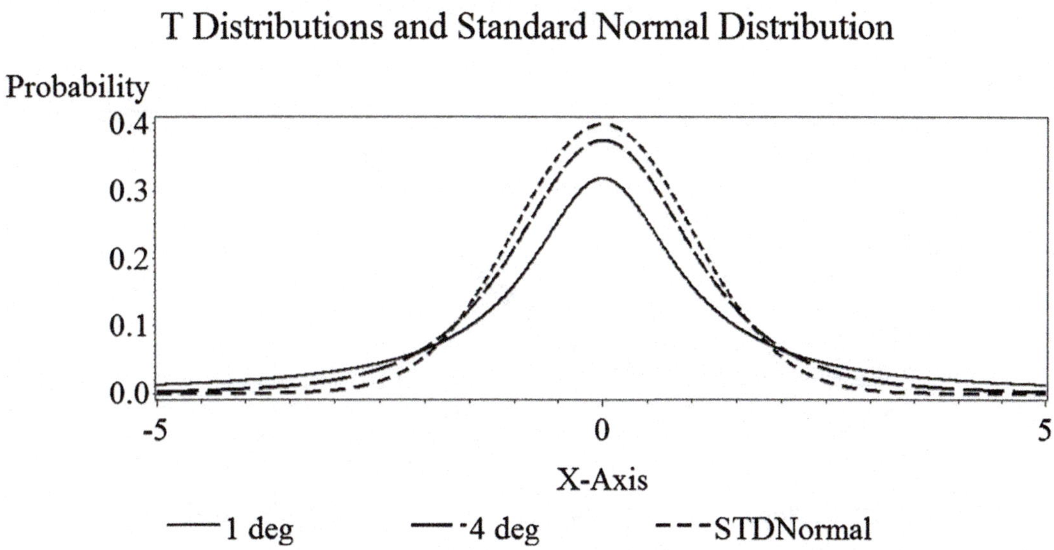

We will now calculate and plot the t cumulative density function for both the one degree of freedom t distribution and the four degrees of freedom t distribution to illustrate that the total probability of a t distribution equals 1.

```
data Tdist;
do X=-5 to 5 by 0.01;
CDF1=cdf('T', X, 1);
CDF2=cdf('T', X, 4);
CDF3=cdf('normal', X, 0, 1);
output Tdist;
end;
run;
Goptions ftext='Times New Roman' htext=6 gunit=pct;
title  height=7 'Cumulative T Distributions';
legend1 label=none value=('1 deg' '4 deg' 'STDNormal') shape=symbol(5,1);
axis1  label=('Probability') order=(0 to 1 by 0.5);
axis2  label=('X-Axis') order=(-5 to 5 by 5);
symbol1 i=join  v=none L=1 c=black w=2;
symbol2 i=join  v=none L=5 c=black w=2;
symbol3 i=join  v=none L=3 c=black w=2;
proc gplot data=Tdist;
plot (CDF1 CDF2 CDF3)*X/vaxis=axis1 haxis=axis2 overlay legend=legend1;
run;
```

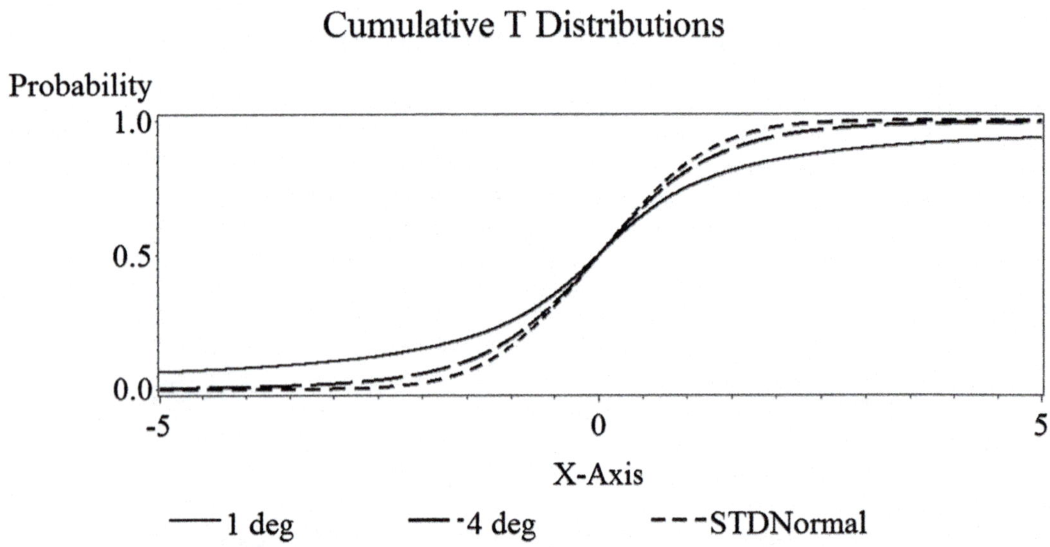

Below shows the SAS commands to calculate the cumulative distribution of a t distribution with one degree of freedom when x is equal to 2.

```
data Tdist;
PDF=pdf('T', 2, 1, 1);
run;

proc print data=Tdist;
run;
```

Cumulative T Distributions

Obs	CDF1
1	0.14380

9.9.2 Chi-Square (χ^2) Distribution

Another distribution often used in statistics is the chi-square (χ^2) distribution. The mathematical formula for the chi-square is

$$\chi^2 = \sum_{i=1}^{n} \left(\frac{X_i - \mu}{\sigma_X} \right)^2$$

The PDF function for the chi-square distribution returns the probability density function of a chi-square distribution, with df degrees of freedom. It follows the form: PDF('CHISQUARE',X,df), where X is a numeric random variable. We will create three chi-square distributions with 5, 10, and 30 degrees of freedom. The SAS commands below will show that as the degree of freedom increases, the distribution shifts to the right.

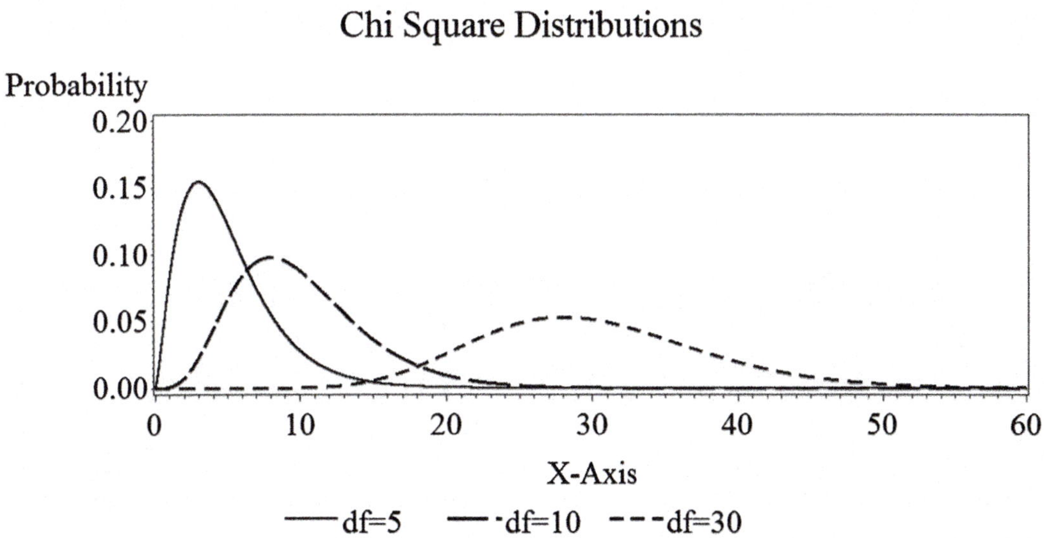

Below are the SAS commands to graph the cumulative chi-square distribution with degrees of freedom of 5, 10, and 30.

```
data Chidist;
do X=0 to 60 by 0.01;
CDF1=cdf('CHISQUARE', X, 5);
CDF2=cdf('CHISQUARE', X, 10);
CDF3=cdf('CHISQUARE', X, 30);
output Chidist;
end;
run;

Goptions ftext='Times New Roman' htext=6 gunit=pct;
title  height=7 'Cumulative Chi Square Distributions';
legend1 label=none value=('df=5' 'df=10' 'df=30') shape=symbol(5,1);
axis1  label=('Probability') order=(0 to 1 by 0.5);
axis2  label=('X-Axis') order=(0 to 60 by 10);
symbol1 i=join  v=none L=1 c=black w=2;
symbol2 i=join  v=none L=5 c=black w=2;
symbol3 i=join  v=none L=3 c=black w=2;
proc gplot data=Chidist;
plot (CDF1 CDF2 CDF3)*X/vaxis=axis1 haxis=axis2 overlay legend=legend1;
run;
```

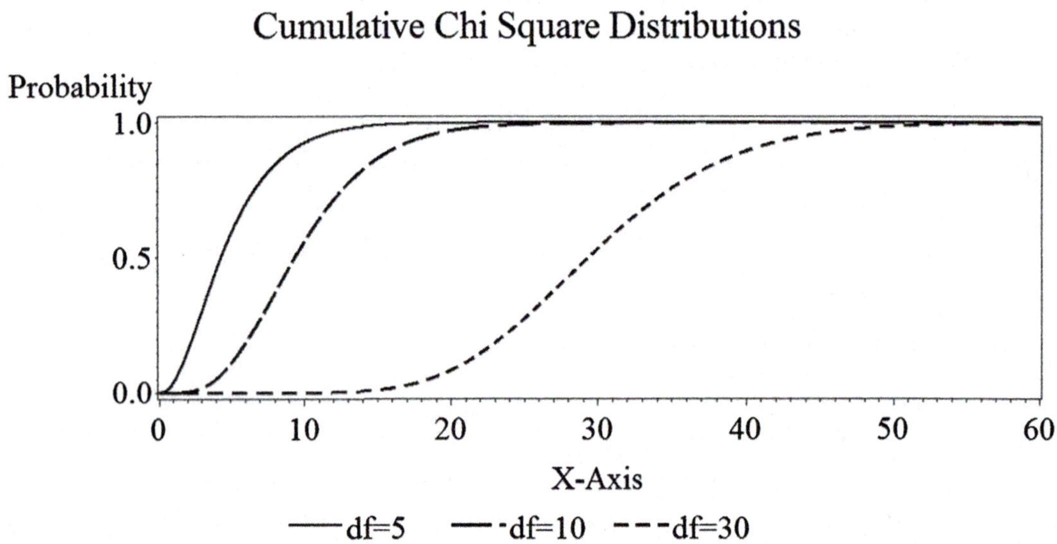

Below are the SAS commands to calculate the p-value of a chi-square distribution with one degree of freedom when x is equal to 2.

```
data Chidist;
K1=1-cdf('CHISQUARE', 2, 1);
run;

proc print data=Chidist;
run;
```

Chi Square Distributions

Obs	K1
1	0.15730

9.9.3 F Distribution

Another distribution often used in statistics is the F distribution. Its mathematical formula is

$$F = \frac{s_X^2/\sigma_X^2}{s_Y^2/\sigma_Y^2}$$

Each F distribution is influenced by two different degrees of freedom. The PDF function for the F distribution follows the form: PDF('F', X, ndf, ddf), where X is a numeric random variable. It returns the probability density function of an F distribution, with ndf numerator degrees of freedom and ddf denominator degrees of freedom.

The commands below will show that as the two different degrees of freedom increase, the more the F distribution shifts to the right and the more peaked the F distribution becomes.

```
data Fdist;
do X=0 to 5 by 0.01;
PDF1=pdf('F', X, 5,5);
PDF2=pdf('F', X, 10,10);
PDF3=pdf('F', X, 30,30);
output Fdist;
end;
run;

Goptions ftext='Times New Roman' htext=6 gunit=pct;
title  height=7 'F Distributions';
legend1 label=none value=('df=5,5' 'df=10,10' 'df=30,30') shape=symbol(5,1);
axis1  label=('Probability') order=(0 to 1.5 by 0.5);
axis2  label=('X-Axis') order=(0 to 5 by 1);
symbol1 i=join  v=none L=1 c=black w=2;
symbol2 i=join  v=none L=5 c=black w=2;
symbol3 i=join  v=none L=3 c=black w=2;
proc gplot data=Fdist;
plot (PDF1 PDF2 PDF3)*X/vaxis=axis1 haxis=axis2 overlay legend=legend1;
run;
```

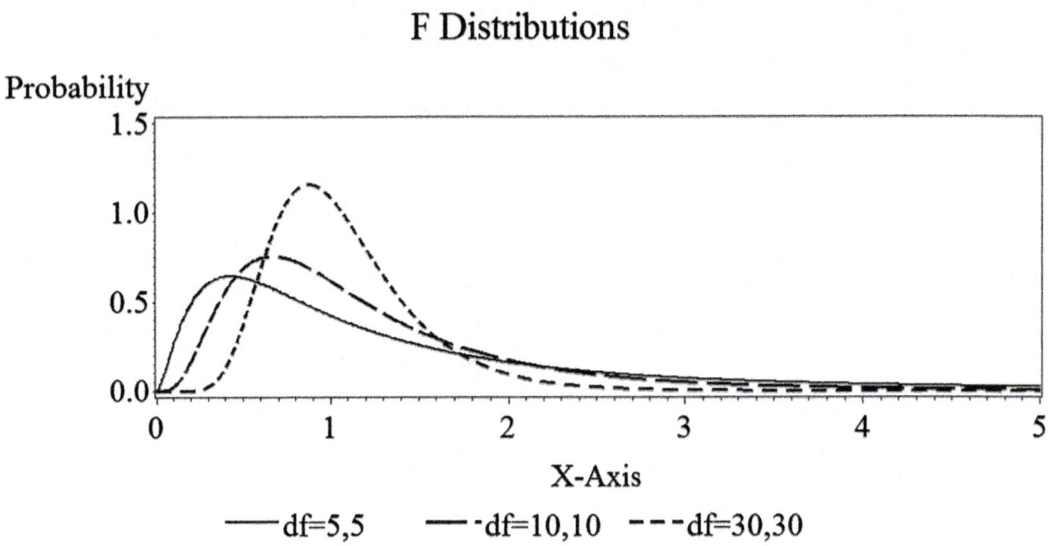

Below are the commands to calculate the cumulative *F* distribution for graphs with 5,5 and 10,10 and 30,30 degrees of freedom.

```
data Fdist;
do X=0 to 5 by 0.01;
CDF1=cdf('F', X, 5,5);
CDF2=cdf('F', X, 10,10);
CDF3=cdf('F', X, 30,30);
output Fdist;
end;
run;

Goptions ftext='Times New Roman' htext=6 gunit=pct;
title  height=7 'Cumulative F Distributions';
legend1 label=none value=('df=5,5' 'df=10,10' 'df=30,30') shape=symbol(5,1);
axis1  label=('Probability') order=(0 to 1 by 0.5);
axis2  label=('X-Axis') order=(0 to 5 by 1);
symbol1 i=join  v=none L=1 c=black w=2;
symbol2 i=join  v=none L=5 c=black w=2;
symbol3 i=join  v=none L=3 c=black w=2;
proc gplot data=Fdist;
plot (CDF1 CDF2 CDF3)*X/vaxis=axis1 haxis=axis2 overlay legend=legend1;
run;
```

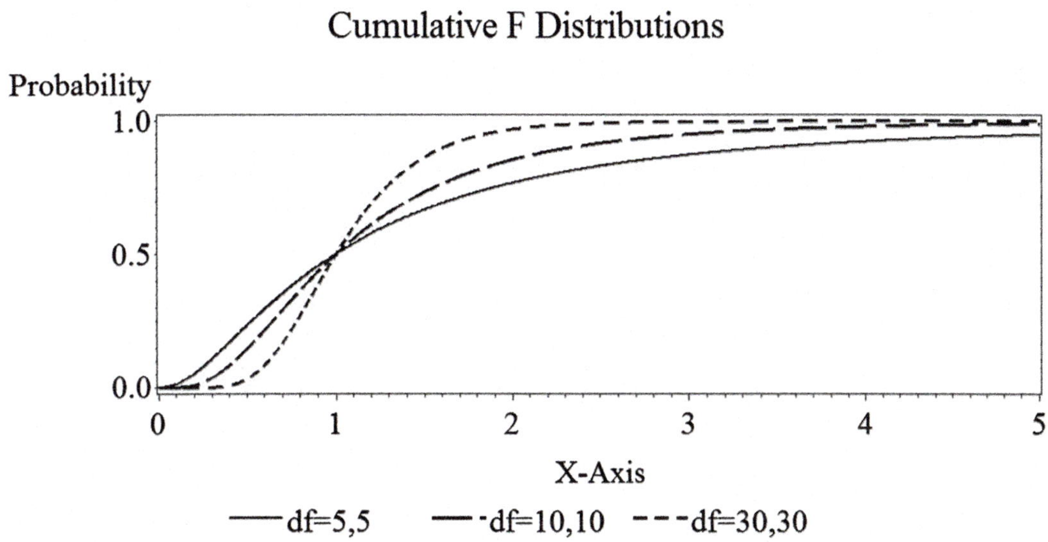

Below is the above one-tail *p-value* calculation done in SAS program:

```
data Fdist;
K1=1-cdf('F', 2, 5, 5);
run;

proc print data=Fdist;
run;
```

Cumulative F Distributions

Obs	K1
1	0.23251

9.9.4 Exponential Distribution

The last continuous distribution that we will look at is the exponential distribution

$$f(t) = \frac{1}{\lambda} e^{-t/\lambda}, \quad t > 0.$$

The following SAS commands create exponential distributions with lambda equal to 1, 2, and 3.

```
data Expdist;
do X=0 to 4 by 0.01;
PDF1=pdf('EXPONENTIAL', X, 1);
PDF2=pdf('EXPONENTIAL', X, 2);
PDF3=pdf('EXPONENTIAL', X, 3);
output Expdist;
end;
run;

Goptions ftext='Times New Roman' htext=6 gunit=pct;
title  height=7 'Exponential Distributions';
legend1 label=none value=('lamda=1' 'lamda=2' 'lamda=3') shape=symbol(5,1);
axis1  label=('Probability') order=(0 to 1 by 0.2);
axis2  label=('X-Axis') order=(0 to 4 by 1);
symbol1 i=join  v=none L=1 c=black w=2;
symbol2 i=join  v=none L=5 c=black w=2;
symbol3 i=join  v=none L=3 c=black w=2;
proc gplot data=Expdist;
plot (PDF1 PDF2 PDF3)*X/vaxis=axis1 haxis=axis2 overlay legend=legend1;
run;
```

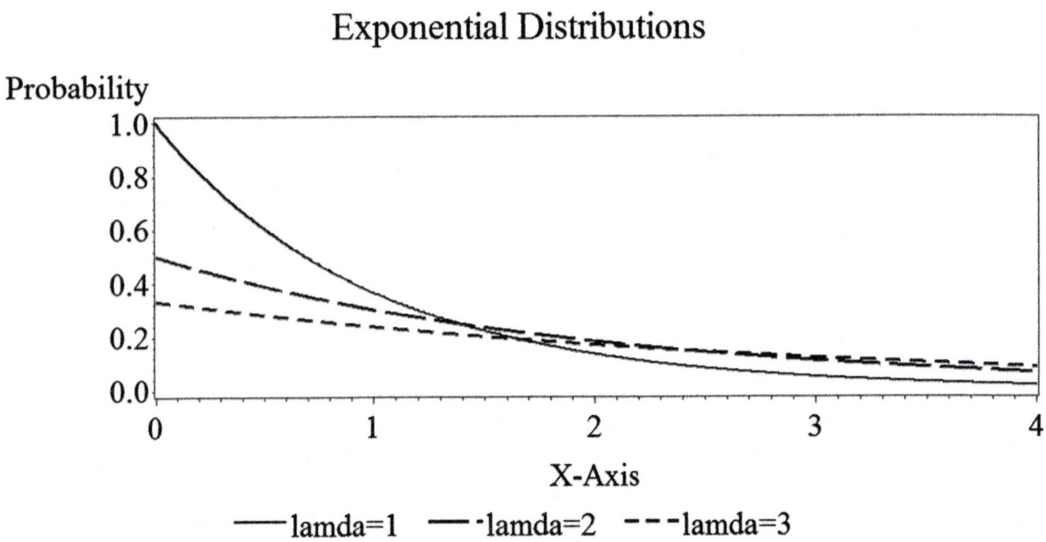

```
data Expdist;
do X=0 to 4 by 0.01;
CDF1=cdf('EXPONENTIAL', X, 1);
CDF2=cdf('EXPONENTIAL', X, 2);
CDF3=cdf('EXPONENTIAL', X, 3);
output Expdist;
end;
run;

Goptions ftext='Times New Roman' htext=6 gunit=pct;
title  height=7 'Cumulative Exponential Distributions';
legend1 label=none value=('lamda=1' 'lamda=2' 'lamda=3') shape=symbol(5,1);
axis1  label=('Probability') order=(0 to 1 by 0.2);
axis2  label=('X-Axis') order=(0 to 4 by 1);
symbol1 i=join  v=none L=1 c=black w=2;
symbol2 i=join  v=none L=5 c=black w=2;
symbol3 i=join  v=none L=3 c=black w=2;
proc gplot data=Expdist;
plot (CDF1 CDF2 CDF3)*X/vaxis=axis1 haxis=axis2 overlay legend=legend1;
run;
```

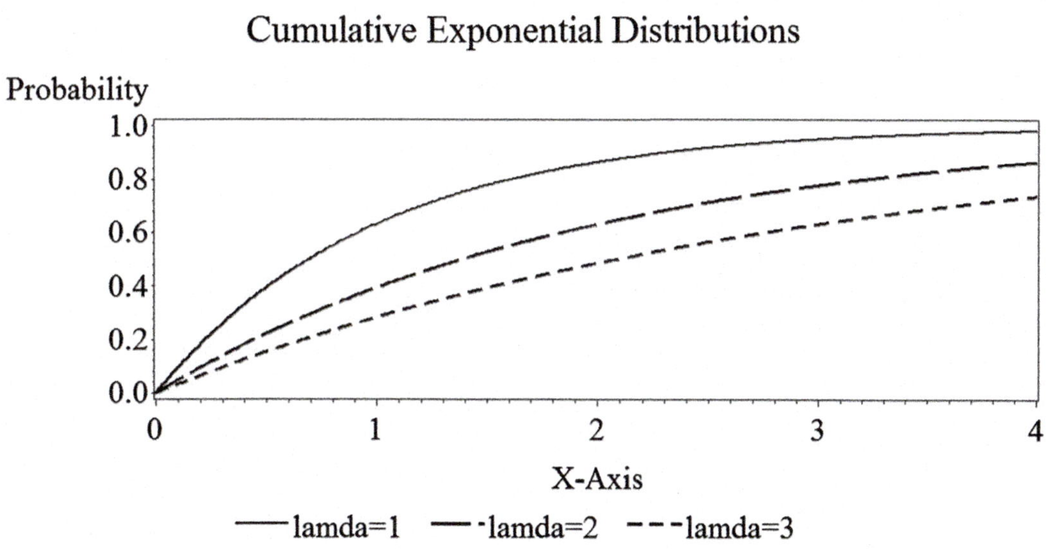

9.9.5 Central Limit Theorem: Other Distributions

We will now illustrate the central limit theorem with the distributions we have examined in this chapter.

9.9.5.1 Exponential Distribution

The *exponential* subcommand tells the *random* command to generate numbers at random from the exponential distribution.

```
%macro Exp(k);
data Exp;
array u[&k];
call streaminit(12345);
do j = 1 to 300;
do i = 1 to &k;
   u[i] = rand("EXPONENTIAL");
end;
output Exp;
end;
run;

data Exp (keep=meanvar);
set Exp;
meanvar=sum(of u1-u&k)/&k;
format meanvar 3.1;
run;
%mend;
```

Sample Size = 5, 300 Samples

```
%Exp(k=5);
Goptions ftext='Times New Roman' htext=4 gunit=pct;
title height=7 'Exponential Distribution';
axis1  label=('Frequency');
axis2  label=('Mean') value=(angle=90 rotate=0);
proc gchart data=Exp;
vbar  meanvar/RAXIS=axis1 MAXIS=axis2 type=freq levels=all midpoints=(0.3 to
3.1 by 0.1);
run;
```

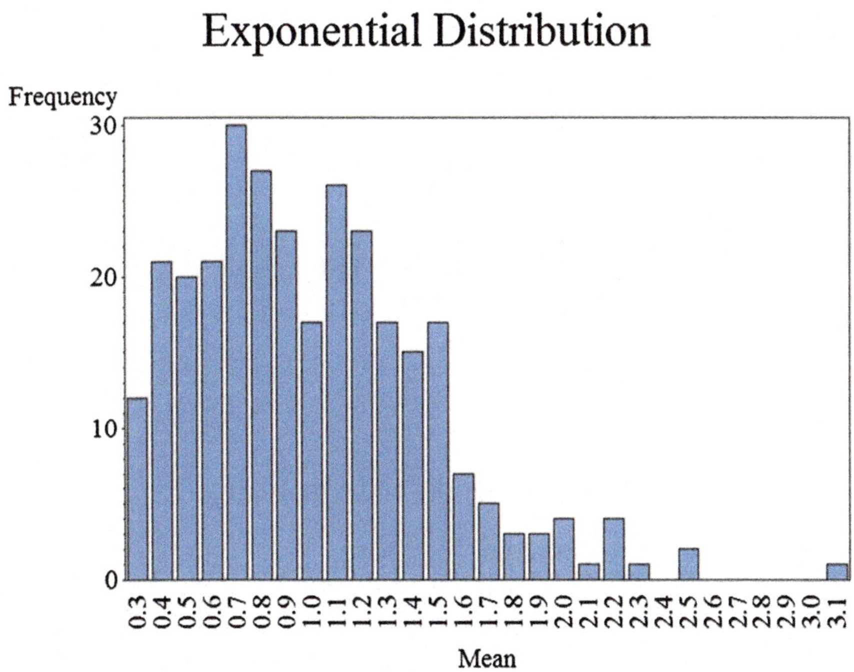

Sample Size = 10, 300 Samples

```
%Exp(k=10);
Goptions ftext='Times New Roman' htext=4 gunit=pct;
title height=7 'Exponential Distribution';
axis1  label=('Frequency');
axis2  label=('Mean') value=(angle=90 rotate=0);
proc gchart data=Exp;
vbar  meanvar/RAXIS=axis1 MAXIS=axis2 type=freq levels=all midpoints=(0.3 to
2.1 by 0.1);
run;
```

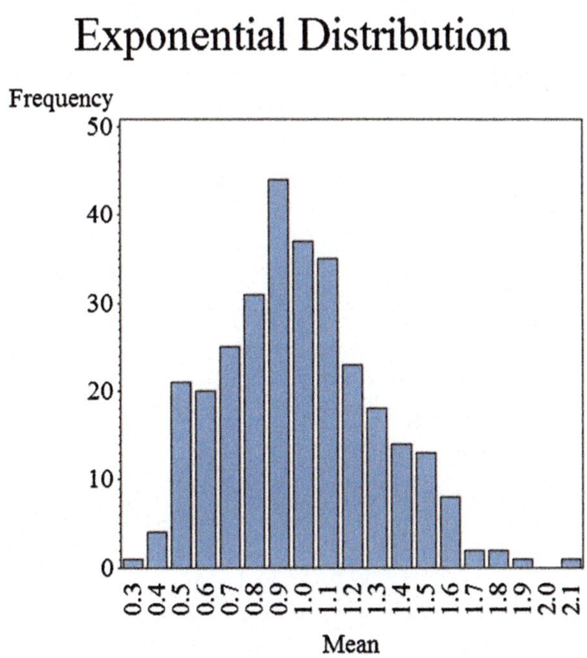

Sample Size = 30, 300 Samples

```
Sample Size = 30, 300 Samples
%Exp(k=30);
Goptions ftext='Times New Roman' htext=4 gunit=pct;
title height=7 'Exponential Distribution';
axis1  label=('Frequency');
axis2  label=('Mean') value=(angle=90 rotate=0);
proc gchart data=Exp;
vbar  meanvar/RAXIS=axis1 MAXIS=axis2 type=freq levels=all midpoints=(0.6 to
1.5 by 0.1);
run;
```

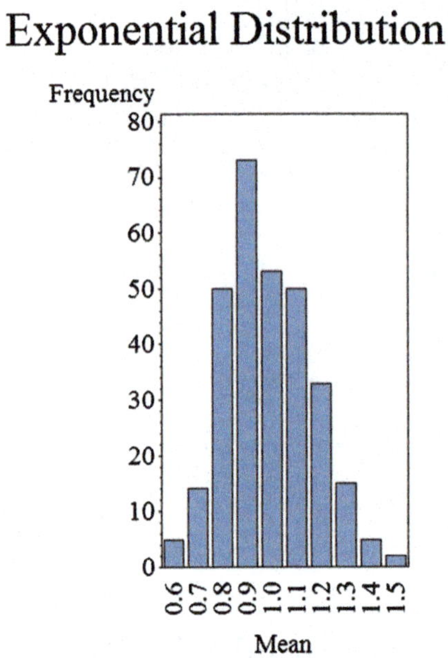

Sample Size = 50, 300 Samples

```
%Exp(k=50);
Goptions ftext='Times New Roman' htext=4 gunit=pct;
title height=7 'Exponential Distribution';
axis1  label=('Frequency');
axis2  label=('Mean') value=(angle=90 rotate=0);
proc gchart data=Exp;
vbar  meanvar/RAXIS=axis1 MAXIS=axis2 type=freq levels=all midpoints=(0.7 to
1.5 by 0.1);
run;
```

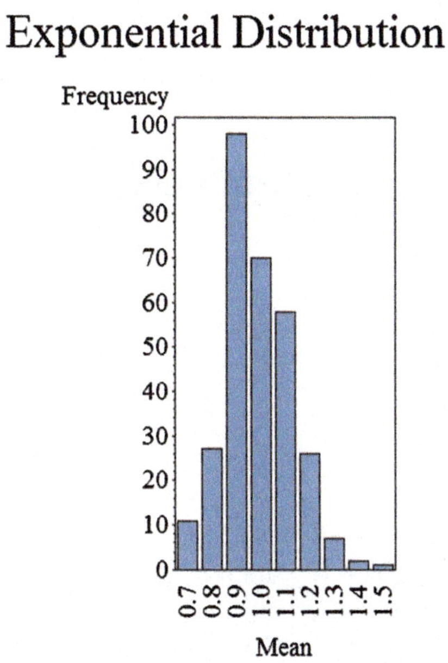

9.9.5.2 *F* Distribution

The *f* subcommand in SAS tells the random subcommand to generate numbers at random from the F distribution.

```
%macro F(k);
data F;
array u[&k];
call streaminit(12345);
do j = 1 to 300;
do i = 1 to &k;
   u[i] = rand("F", 5, 5);
end;
output F;
end;
run;

data F (keep=meanvar);
set F;
meanvar=sum(of u1-u&k)/&k;
format meanvar 3.1;
run;
Goptions ftext='Times New Roman' htext=4 gunit=pct;
title height=7 'F Distribution';
axis1  label=('Frequency');
axis2  label=('Mean') value=(angle=90 rotate=0);
proc gchart data=F;
vbar  meanvar/ RAXIS=axis1 MAXIS=axis2 discrete type=freq levels=all;
run;
%mend;
```

Sample Size = 5, 300 Samples

```
%F(k=5);
```

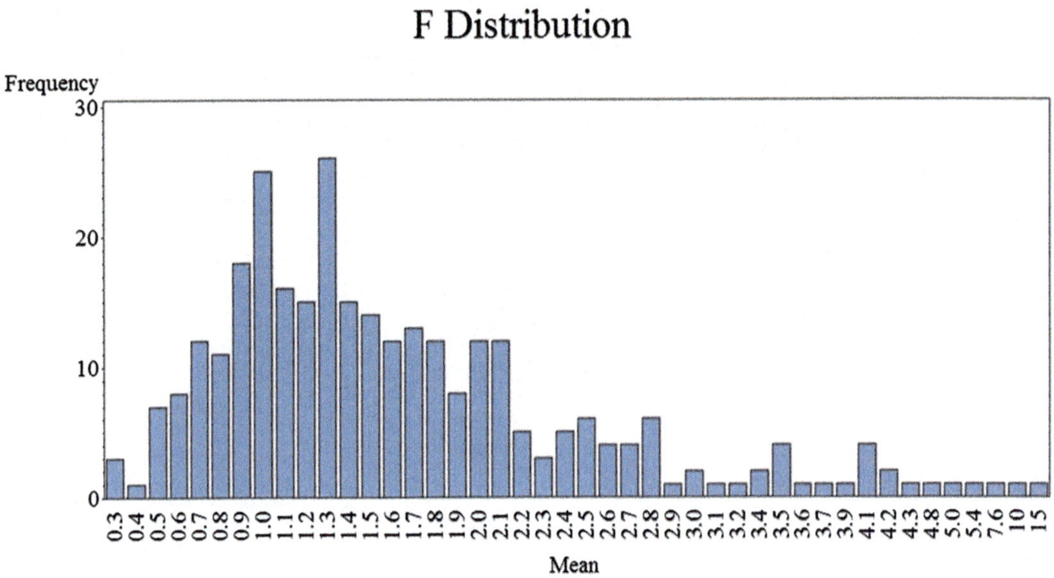

Sample Size = 10, 300 Samples

```
%F(k=10);
```

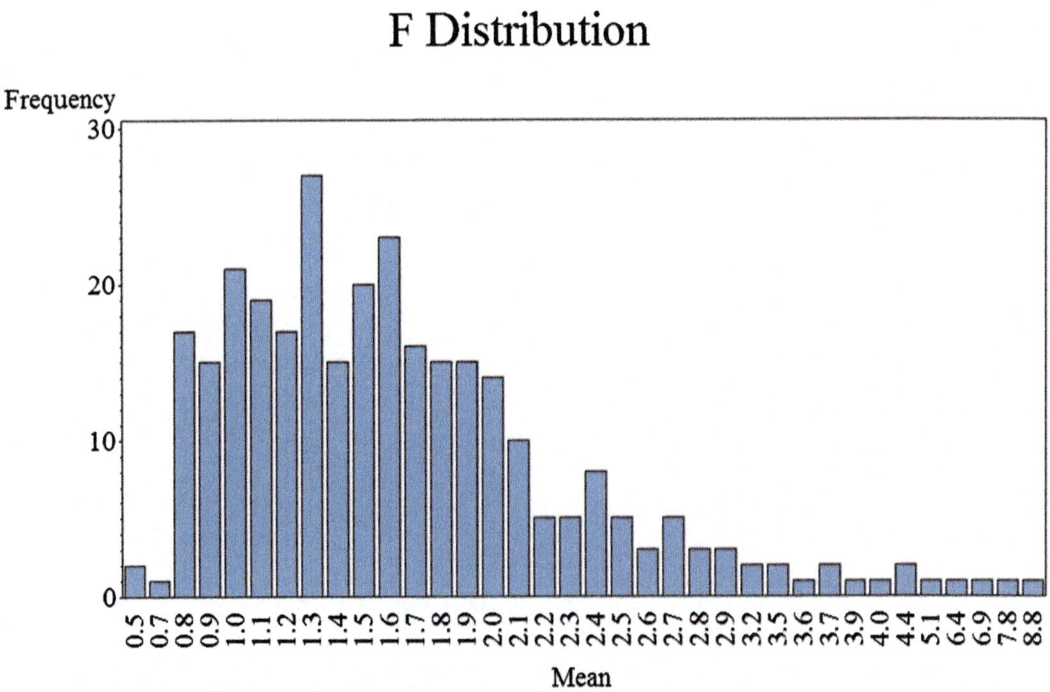

Sample Size = 30, 300 Samples

```
%F(k=30);
```

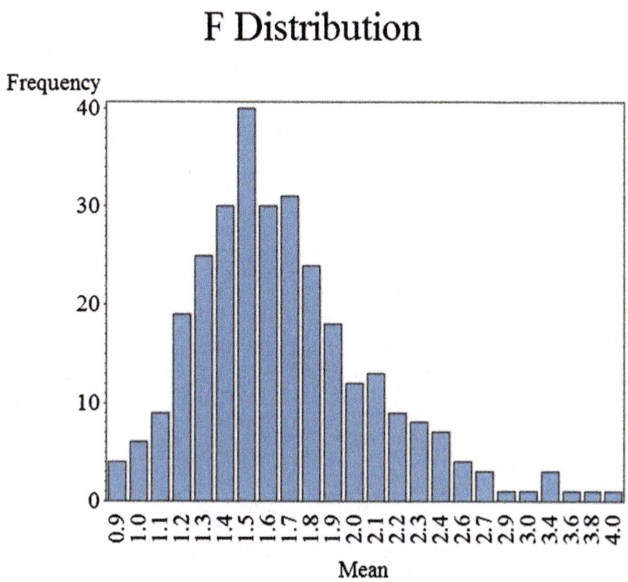

Sample Size = 50, 300 Samples

```
%F(k=50);
```

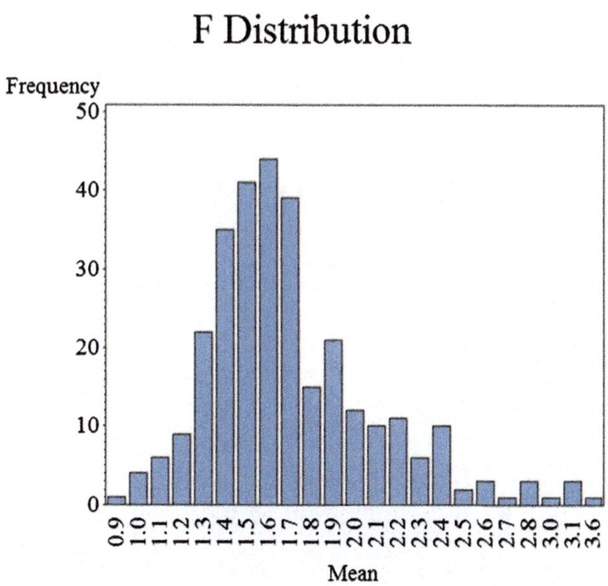

9.9.5.3 Chi-Square Distribution

The *chisquare* subcommand tells the *random* subcommand to generate numbers at random from the chi-square distribution.

```
%macro Chisquare(k);
data Chisquare;
array u[&k];
call streaminit(12345);
do j = 1 to 300;
do i = 1 to &k;
   u[i] = rand("CHISQUARE", 5);
end;
output Chisquare;
end;
run;

data Chisquare (keep=meanvar);
set Chisquare;
meanvar=sum(of u1-u&k)/&k;
format meanvar 3.1;
run;
Goptions ftext='Times New Roman' htext=4 gunit=pct;
title height=7 'Chi Square Distribution';
axis1  label=('Frequency');
axis2  label=('Mean') value=(angle=90 rotate=0);
proc gchart data=Chisquare;
vbar  meanvar/ RAXIS=axis1 MAXIS=axis2 discrete type=freq levels=all;
run;
%mend;
```

Sample Size = 5, 300 Samples

```
%Chisquare(k=5);
```

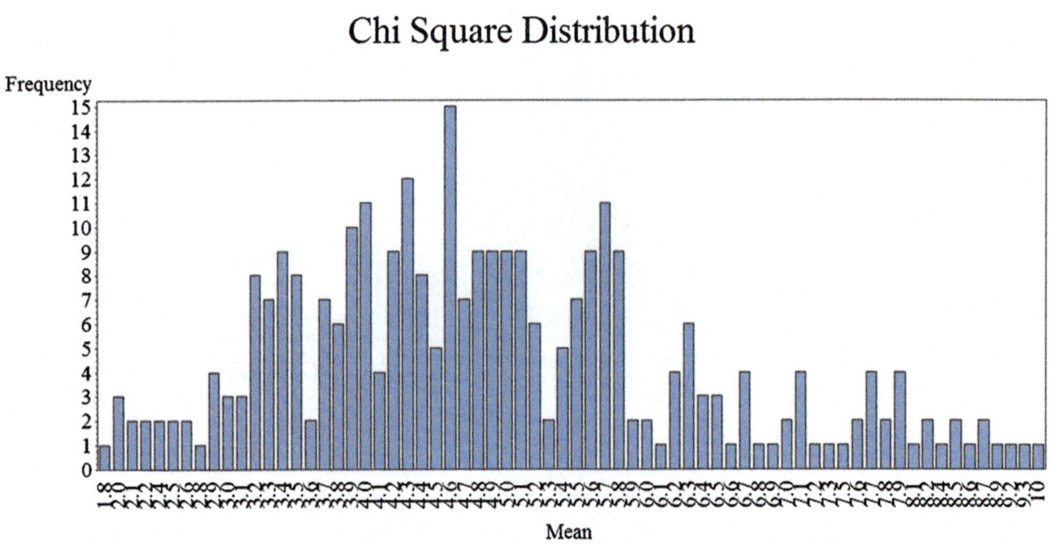

Sample Size = 10, 300 Samples

```
%Chisquare(k=10);
```

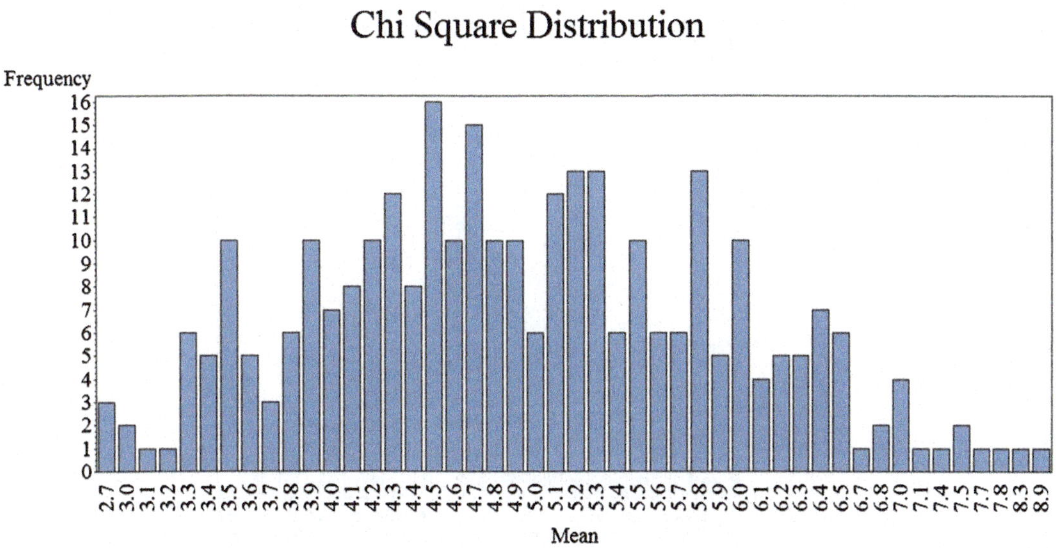

Sample Size = 30, 300 Samples

```
%Chisquare(k=30);
```

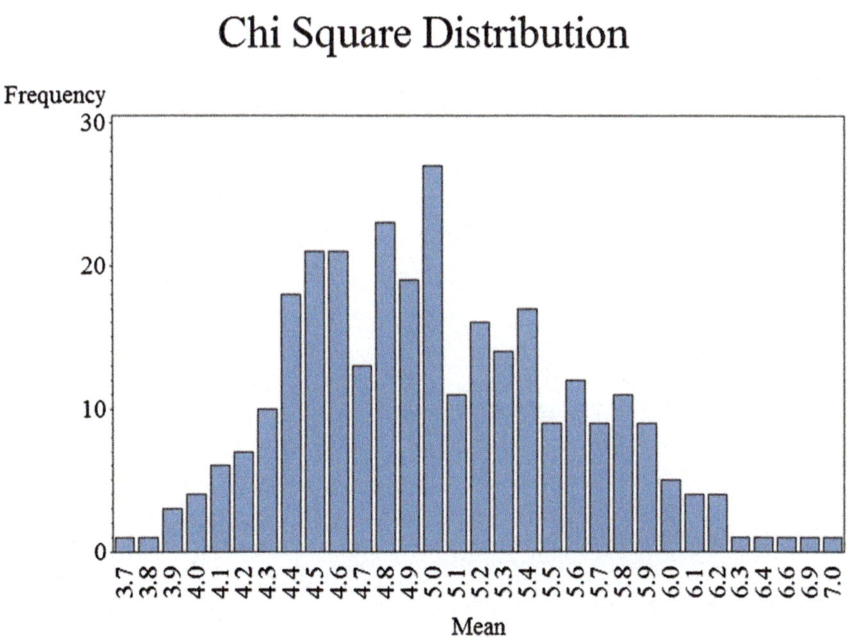

Sample Size = 50, 300 Samples

```
%Chisquare(k=50);
```

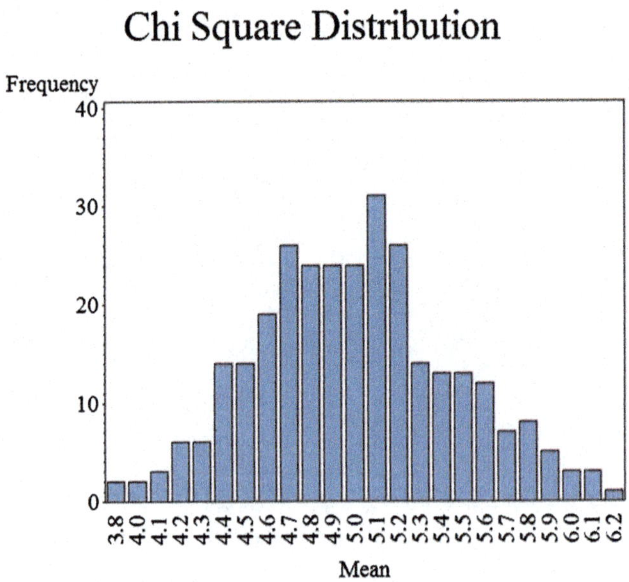

9.9.5.4 *t* Distribution

The *t* subcommand tells the *random* subcommand to generate numbers at random from the *t* distribution.

```
%macro T(k);
data T;
array u[&k];
call streaminit(12345);
do j = 1 to 300;
do i = 1 to &k;
   u[i] = rand("T", 5);
end;
output T;
end;
run;

data T (keep=meanvar);
set T;
meanvar=sum(of u1-u&k)/&k;
format meanvar 3.1;
run;
Goptions ftext='Times New Roman' htext=4 gunit=pct;
title height=7 'T Distribution';
axis1  label=('Frequency');
axis2  label=('Mean') value=(angle=90 rotate=0);
proc gchart data=T;
vbar  meanvar/ RAXIS=axis1 MAXIS=axis2 discrete type=freq levels=all;
run;
%mend;
```

Sample Size = 5, 300 Samples

```
%T(k=5);
```

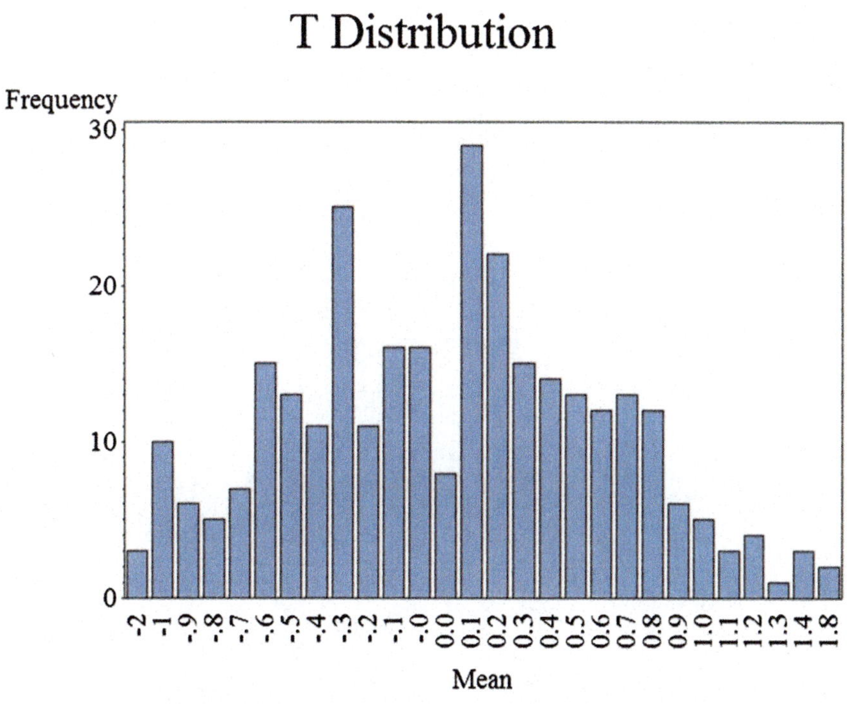

Sample Size = 10, 300 Samples

```
%T(k=10);
```

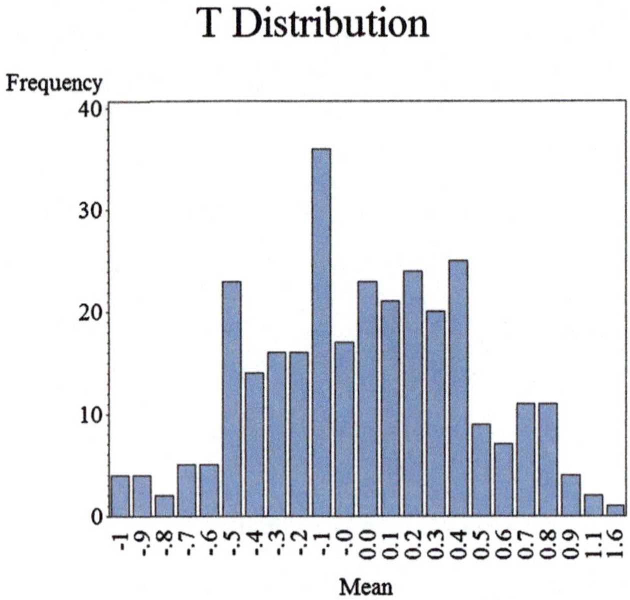

Sample Size = 30, 300 Samples

```
%T(k=30);
```

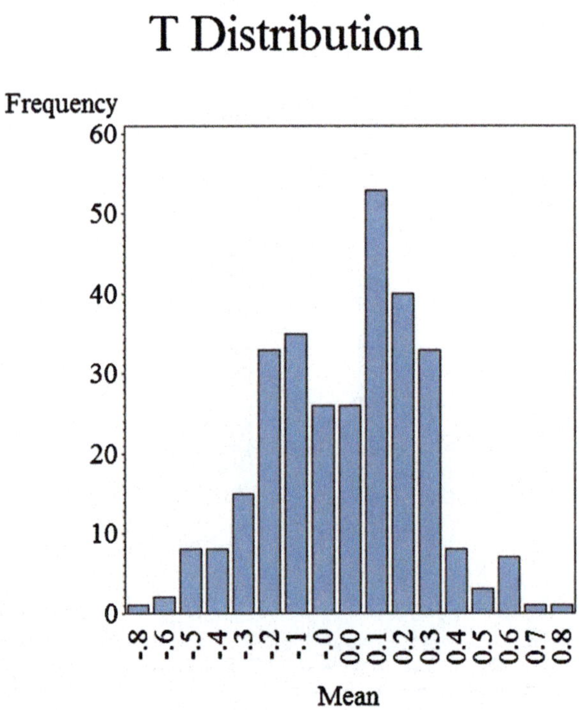

Sample Size = 50, 300 Samples

```
%T(k=50);
```

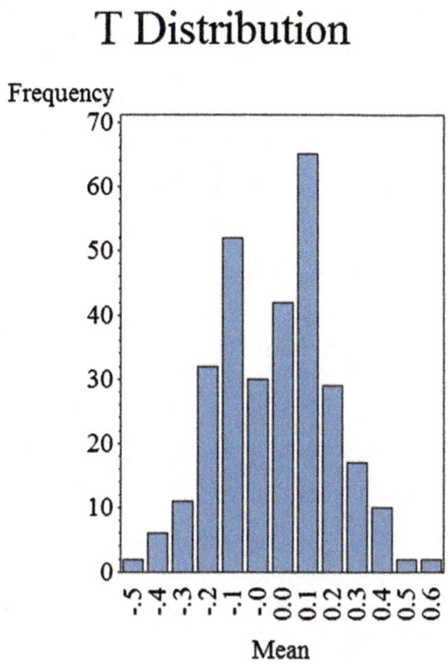

9.10 Statistical Summary

In this chapter we completed our survey of important continuous variables. We looked at the t, chi-square, F, and exponential distributions. We have demonstrated that the t distribution approaches a standard normal distribution as the degrees of freedom increase. For the chi-square distribution, we have demonstrated that as the degrees of freedom increase, the farther right the density function shifts. For the F distribution, we have demonstrated that as the degrees of freedom increase, the more peaked its density function becomes. For the exponential distribution, we have shown how the shape can be affected by the value of the parameter λ.

We also illustrated the central limit theorem with the distributions discussed in this chapter. We have demonstrated that the sample mean of all these distributions approaches a normal distribution as the sample size increases.

Appendix 9.1: Excel Code—Exponential Distribution

```
'/*************************************************************************
'/ Essentials of Microsoft Excel 2013, SAS and
'/                     MINITAB 17
'/ for Statistical and Financial Analysis
'/
'/*************************************************************************
Option Explicit
Dim mwbReport As Workbook
Dim mwsReportData As Worksheet

Property Set ReportWb(w As Workbook)
    Set mwbReport = w
End Property
Property Get ReportWb() As Workbook
    Set ReportWb = mwbReport
End Property
Property Set ReportData(w As Worksheet)
    Set mwsReportData = w
End Property
Property Get ReportData() As Worksheet
    Set ReportData = mwsReportData
End Property
Private Sub cmdCancel_Click()
    Unload Me
End Sub

'/*******************************************************************
'/Purpose:  This is the procedure that is run after pressing the
'/          the run button on the form.  This is the main procedure
'/*******************************************************************
Private Sub cmdRun_Click()
    Me.Hide
    Exponential
    Unload Me
End Sub


'/*********************************************************************
'/Purpose:  This procedure runs everytime the form is loaded.  This
'/          procedure sets some initial values.
'/*********************************************************************
Private Sub UserForm_Initialize()
    With Me
        .txtNumPoints = 70
        .txtLambda1 = 1
        .txtXFrom = 0.1
        .txtXto = 5
    End With
End Sub

Sub Exponential()
    Dim Chrt As Chart
    Dim ChrtCDF As Chart
    Dim Ser As Series
    Dim nCount As Integer

    Set Me.ReportWb = Workbooks.Add
    Set Me.ReportData = Me.ReportWb.Worksheets.Add
    Me.ReportData.Name = "Data"
    With Me
```

```
        .ReportWb.Names.Add Name:="Data!xFrom", RefersToR1C1:=Me.txtXFrom
        .ReportWb.Names.Add Name:="Data!xTo", RefersToR1C1:=Me.txtXto
        .ReportWb.Names.Add Name:="Data!NumPoints", RefersToR1C1:=Me.txtNumPoints
    End With

    Set ChrtCDF = Me.ReportWb.Charts.Add
    Set Chrt = Me.ReportWb.Charts.Add

    Chrt.ChartType = xlXYScatterSmoothNoMarkers
    ChrtCDF.ChartType = xlXYScatterSmoothNoMarkers
    If Chrt.SeriesCollection.Count = 1 Then
        Chrt.SeriesCollection(1).Delete
    End If
    If ChrtCDF.SeriesCollection.Count = 1 Then
        Chrt.SeriesCollection(1).Delete
    End If
    For nCount = 1 To 4
        If Len(Trim(Me.Controls("txtLambda" & nCount))) <> 0 Then
                PDFChartGraph Chrt:=Chrt, objLambda:=Me.Controls("txtLambda" & nCount)
                CDFChartGraph Chrt:=ChrtCDF, objLambda:=Me.Controls("txtLambda" & nCount)
        End If
    Next
    With Chrt
        .PlotArea.ClearFormats
        .Axes(xlValue).MajorGridlines.Delete
        .Axes(xlCategory).CrossesAt = .Axes(xlCategory).MinimumScale
        .Name = "Exponential PDF"
        .HasTitle = True
        If .SeriesCollection.Count > 1 Then
            .ChartTitle.Characters.Text = "Exponential Distributions"
        Else
            .ChartTitle.Characters.Text = "Exponential Distribution"
        End If
        .ChartTitle.Font.Size = 24
        .Legend.Font.Size = 8
        For Each Ser In .SeriesCollection
            With Ser
                .MarkerStyle = xlMarkerStyleAutomatic
                .Border.ColorIndex = 1
                .MarkerForegroundColorIndex = 1
                .MarkerBackgroundColorIndex = xlNone
            End With
        Next
    End With

    With ChrtCDF
        .PlotArea.ClearFormats
        .Axes(xlValue).MajorGridlines.Delete
        .Axes(xlCategory).CrossesAt = .Axes(xlCategory).MinimumScale
        .Name = "Exponential CDF"
        .HasTitle = True
        If .SeriesCollection.Count > 1 Then
            .ChartTitle.Characters.Text = " Cumulative Exponential Distributions"
        Else
            .ChartTitle.Characters.Text = " Cumulative Exponential Distribution"
        End If
        .ChartTitle.Font.Size = 24
        .Legend.Font.Size = 8
        For Each Ser In .SeriesCollection
            With Ser
                .MarkerStyle = xlMarkerStyleAutomatic
                .Border.ColorIndex = 1
                .MarkerForegroundColorIndex = 1
                .MarkerBackgroundColorIndex = xlNone
            End With
        Next

    End With

End Sub
```

```
Sub PDFChartGraph(Chrt As Chart, objLambda As Object)
    Dim Ser As Series
    Dim SerLabel As Series
    Dim sName As String

    sName = "Data!Lambda" & Right(objLambda.Name, 1)
    With Me
        .ReportWb.Names.Add Name:=sName, RefersToR1C1:=objLambda
        .ReportWb.Names.Add Name:="Data!y" & objLambda.Name, RefersToR1C1:= _
        "=EVALUATE(""EXPONDIST(x," & sName & ",false)+x*0"")"
        .ReportWb.Names.Add Name:="Data!x", RefersToR1C1:= _
        "=Data!xFrom+(ROW(OFFSET(Data!R1C2,0,0,NumPoints,1))-1)*(Data!xTo-
(Data!xFrom))/(NumPoints-1)"
    End With


    Set Ser = Chrt.SeriesCollection.NewSeries
    With Ser
        .XValues = "=Data!x"
        .Values = "=Data!y" & objLambda.Name
        .Name = "Lambda=" & objLambda.Text
        .Border.ColorIndex = 1

    End With

     With Selection.Border
        .ColorIndex = 1
        .Weight = xlThin
        .LineStyle = xlDash
    End With


End Sub

Sub CDFChartGraph(Chrt As Chart, objLambda As Object)
    Dim Ser As Series
    Dim sName As String

    sName = "Data!Lambda" & Right(objLambda.Name, 1)
    With Me
        .ReportWb.Names.Add Name:="Data!yCDF" & objLambda.Name, RefersToR1C1:= _
        "=EVALUATE(""EXPONDIST(x," & sName & ",TRUE)+x*0"")"
    End With


    Set Ser = Chrt.SeriesCollection.NewSeries
    With Ser
        .XValues = "=Data!x"
        .Values = "=Data!yCDF" & objLambda.Name
        .Name = "Lambda=" & objLambda.Text
    End With
 End Sub
```

Bibliography

Microsoft Inc., Excel 2013. Microsoft Inc., Redmond
Lee CF, Lee JC, Lee AC (2013) Statistics for business and financial economics. Springer, New York
Minitab Inc. Minitab 17. Minitab Inc., State College
SAS Institute Inc. SAS 2014. SAS Institute Inc., Cary

Chapter 10
Estimation

10.1 Introduction

In this chapter we use the continuous variables that we studied in Chaps 7 and 9 to estimate the mean, variance, and proportion of populations.

Often we are interested in making inferences about a population parameter based on a sample parameter. One method for doing this is to make a confidence interval that might contain the population parameter.

Confidence intervals are intervals that have a chance of containing the population parameter. For example, a 90 % confidence interval means that 90 % of the confidence intervals that we create will contain the true population parameter. The concept of confidence intervals is illustrated below.

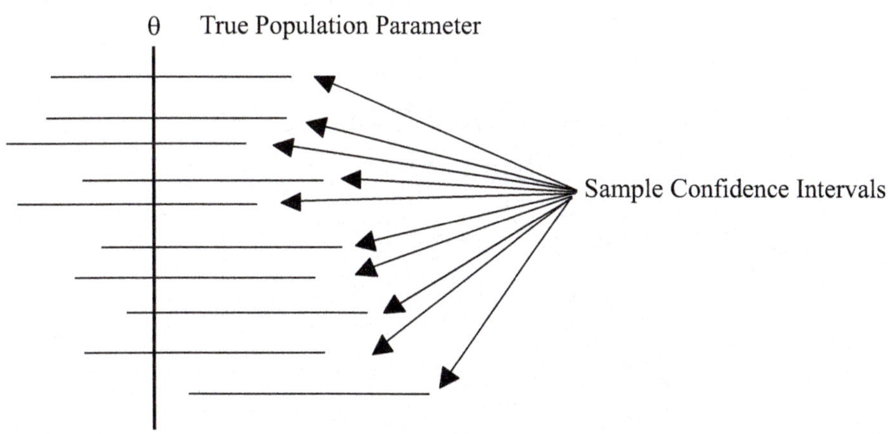

From this diagram of ten 90 % confidence intervals, we can see that nine of the intervals contain the true population parameter while one of them does not contain the true mean.

C.-F. Lee et al., *Essentials of Excel, Excel VBA, SAS and Minitab for Statistical and Financial Analyses*, DOI 10.1007/978-3-319-38867-0_10

10.2 Excel Macro: Confidence Interval Simulation

The Excel code for the program in the workbook is shown in Appendix 10.1. The Excel program in this workbook will randomly generate 100 confidence intervals from a given normal distribution.

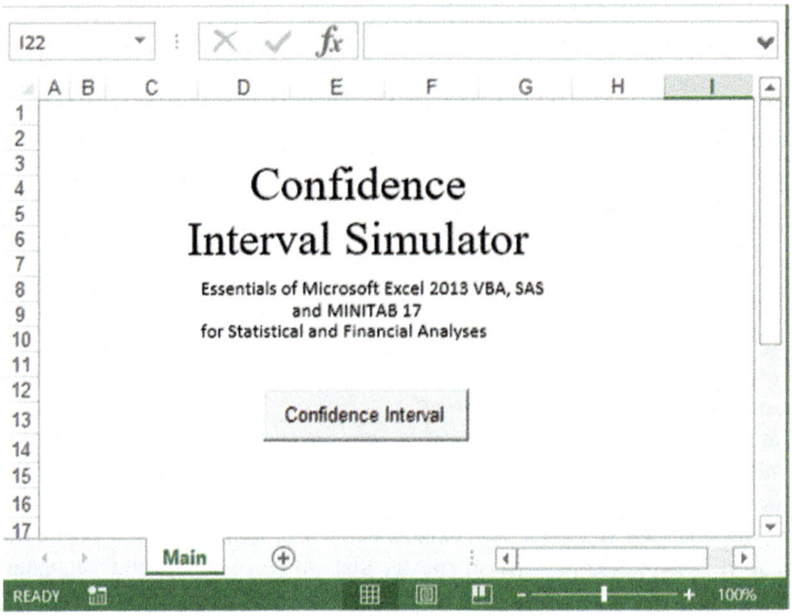

Push the *Confidence Interval* button shown above to get the following dialog box shown below.

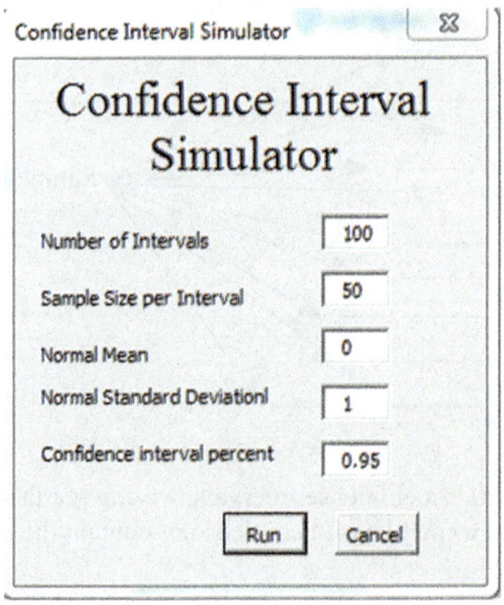

Pushing the *Run* button above will show 95 % confidence intervals as shown below. Below also shows 85 and 75 % confidence intervals.

Below is the result of simulating a 95 % confidence interval. We would expect in a 95 % confidence interval simulation that out of 100 confidence intervals five of the confidence intervals will not contain the population mean. In the simulation, four of the confidence intervals did not contain thepopulation mean.

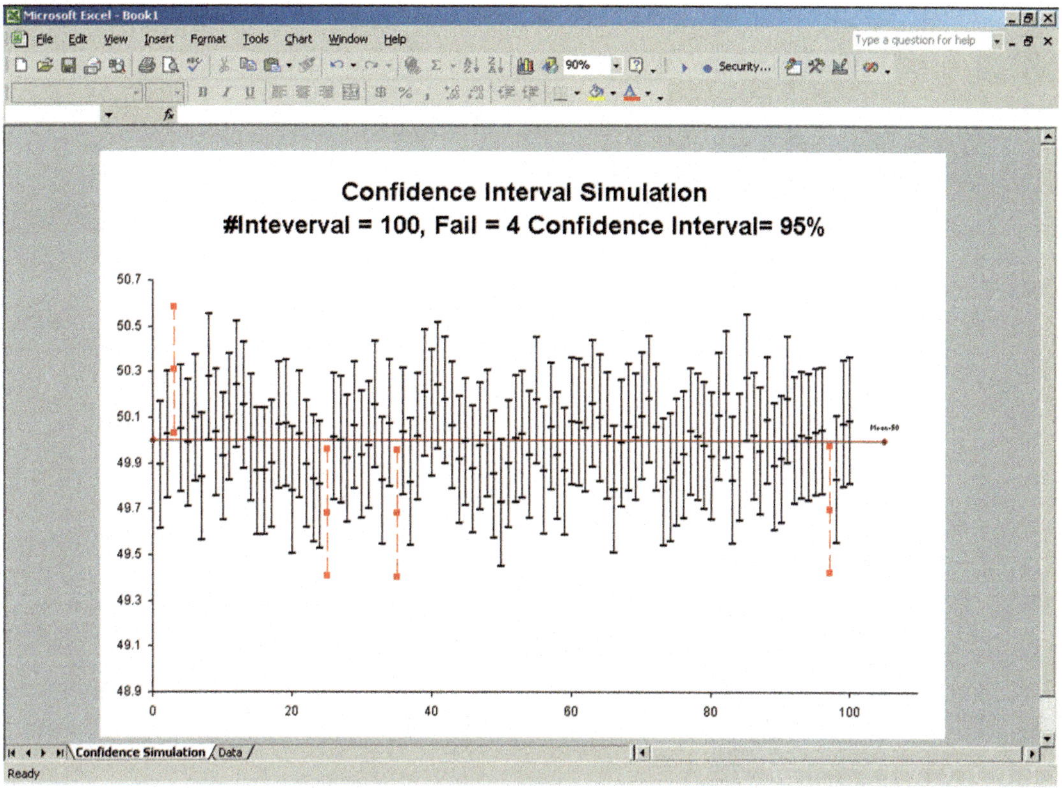

Below is the result of simulating an 85 % confidence interval. We would expect in an 85 % confidence interval simulation that out of 100 confidence intervals 15 of the confidence intervals will not contain the population mean. In the simulation, 18 of the confidence intervals did not contain the population mean.

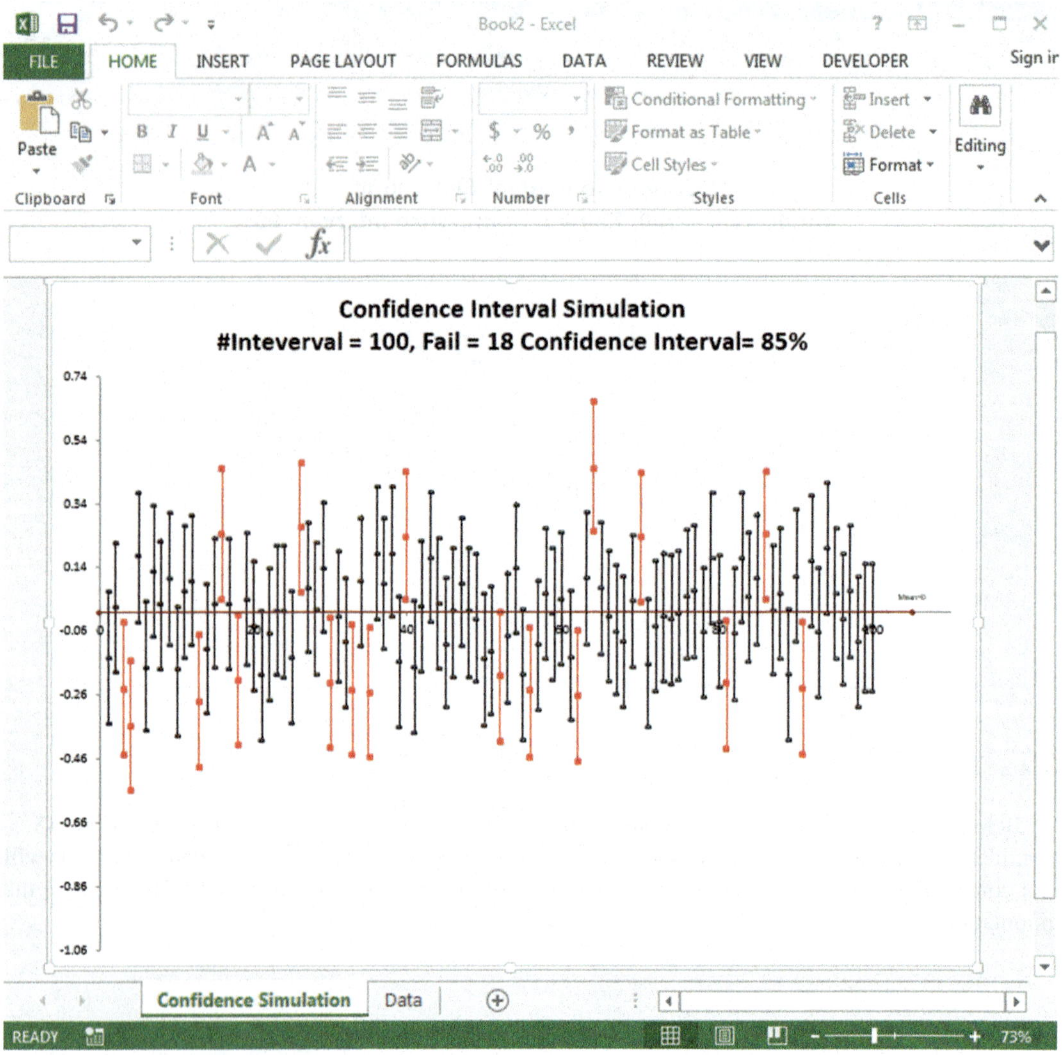
Book2 - Excel
FILE
HOME
INSERT
PAGE LAYOUT
FORMULAS
DATA
REVIEW
VIEW
DEVELOPER
Sign in
Paste
Clipboard
Font
Alignment
Number
Conditional Formatting
Format as Table
Cell Styles
Styles
Insert
Delete
Format
Cells
Editing
Confidence Interval Simulation
#Inteverval = 100, Fail = 18 Confidence Interval= 85%
0.74
0.54
0.34
0.14
-0.06
-0.26
-0.46
-0.66
-0.86
-1.06
Confidence Simulation
Data
READY
73%

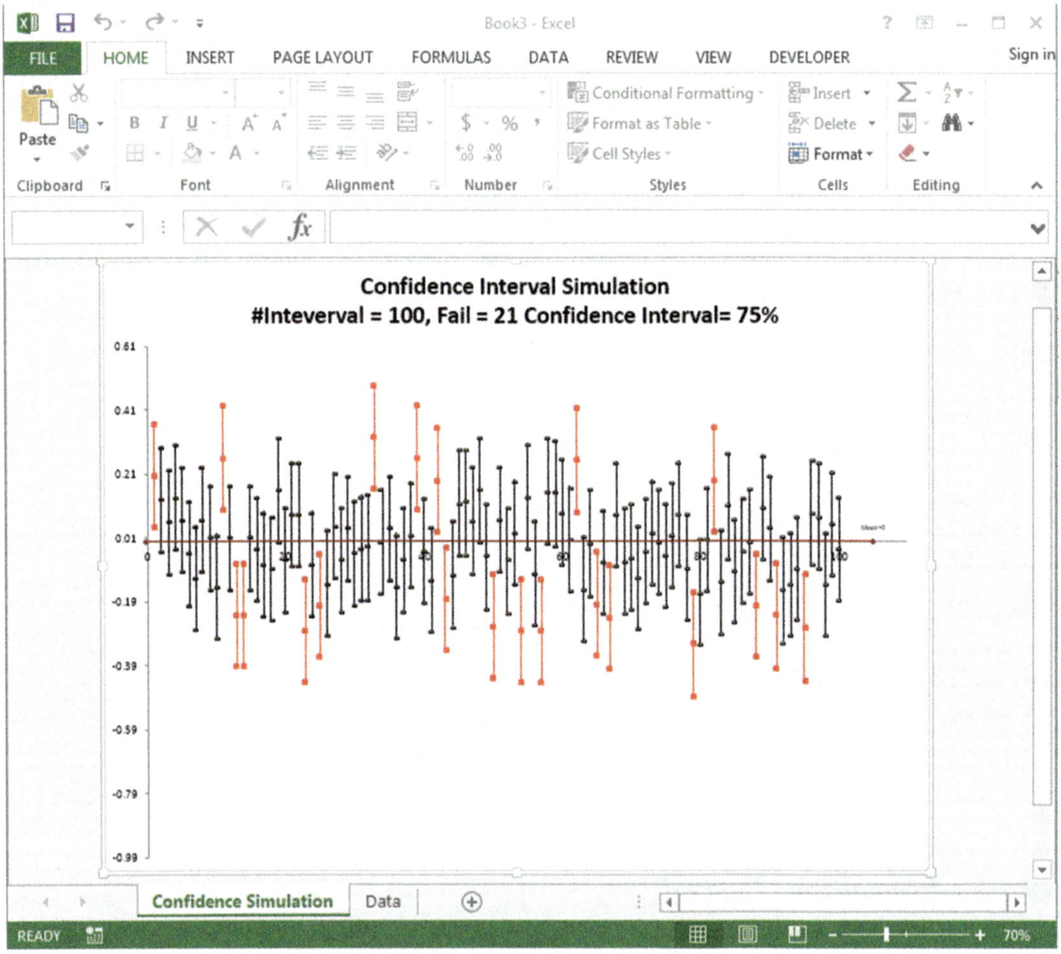

Above is the result of simulating a 75 % confidence interval. We would expect in a 75 % confidence interval simulation that out of 100 confidence intervals 25 of the confidence intervals will not contain the population mean. In the simulation, 21 of the confidence intervals did not contain the population mean.

10.3 Interval Estimates for μ When σ^2 Is Known

In general, if the sample size is greater than 30, we know that the sample distribution will approximate a normal distribution. We can substitute the sample variance for the population variance when the sample size is greater than 30. We can do this because of the central limit theorem. Below is the formula to create the confidence interval for the population mean.

$$\overline{x} - z_{\alpha/2}\left(\sigma/\sqrt{n}\right) < \mu < \overline{x} + z_{\alpha/2}\left(\sigma/\sqrt{n}\right)$$

Example A Suppose we are interested in creating a 95 % confidence interval for the following starting salaries for accountants.

$24,000	$20,000	$18,000	$26,000	$23,750	$32,000	$36,789	$25,300
$27,500	$35,000	$31,000	$25,500	$25,000	$50,000	$23,750	$30,500
$28,000	$29,000	$27,000	$27,750	$25,000	$41,000	$28,900	$32,000
$22,000	$26,000	$34,000	$26,550	$33,000	$27,900	$23,750	$31,000
$32,000	$21,430	$42,014	$36,450	$24,750	$34,000	$35,000	$27,000

We first need to put the data in column c1 of the worksheet.

To find the 95 % confidence interval in Minitab, we will use the *zinterval* command to calculate the above formula. First, however, we need to know the standard deviation of our accounting salary data set. For this we will use the *describe* command.

```
MTB > describe c1
```

10.3.1 Descriptive Statistics

```
Variable              N        Mean      Median      TrMean      StDev
SE Mean
C1                   40       29240       27825       28877       6352
1004

Variable       Minimum     Maximum          Q1          Q3
C1               18000       50000       25000       32750
```

```
MTB > zinterval 95 6352 c1
```

10.3.2 Z Confidence Intervals

```
The assumed sigma = 6352

Variable      N       Mean     StDev   SE Mean        95.0 % CI
C1           40      29240      6352      1004   (  27271,    31208)
```

The *describe* command describes the data in column c1 using several descriptive statistical measures. We need to use the *describe* command because we need to know the standard deviation of the data set for the *zinterval* command. Here the standard deviation is shown as 6352.

The *zinterval* command is used to calculate a confidence interval when σ is known. The first argument of this command, "95," indicates that we are interested in a 95 % confidence interval. The second argument, "6352," is the standard deviation of the data set. We got this information by issuing the *describe* command. The last argument, "c1," is the location of the data set.

Minitab calculated that the 95 % confidence interval for the accounting salaries is between $27,271 and $31,208 and that the mean is $29,240.

Often we are given the sample size, mean, and standard deviation but not the data set. Without the data set we cannot use the *zinterval* command to create the confidence interval. To solve this problem, we will create a Minitab macro to calculate the confidence interval. The macro will be called *zint*. The *zint* macro is shown in Appendix 10.2.

Example B Suppose a real estate agent in Connecticut is interested in the mean home price in the state. A random sample of 50 homes shows a mean price of $175,622, assuming a population sample standard deviation of \$37,221. What is the 95 % confidence interval for the mean home price for Connecticut?

In Excel, we will use the *confidence.norm* function to calculate the confidence interval for the mean home price. The *confidence.norm* function has three parameters. They are

Alpha: The significance level used to compute the confidence level.
Standard_dev: The standard deviation for the data range.
Size: The sample size.

How to set up an Excel spreadsheet to calculate the confidence interval is shown below.

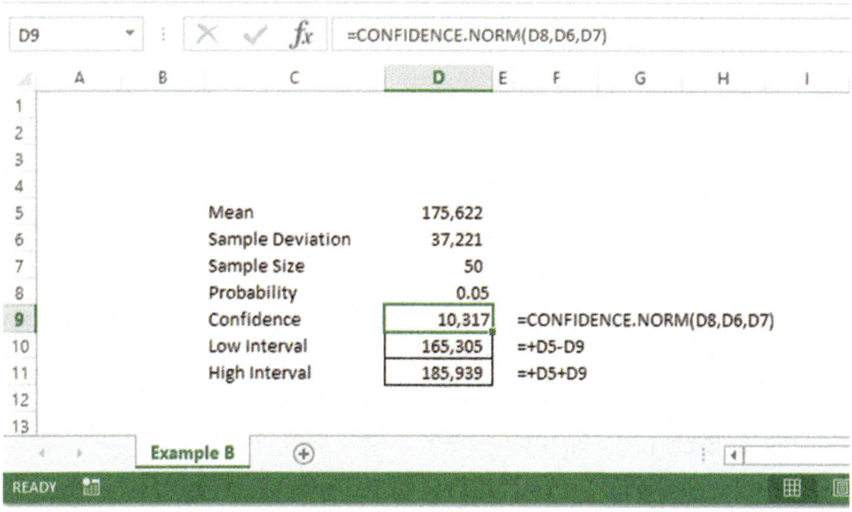

Our output tells us that the 95 % confidence interval for the mean home price in Connecticut is from \$165,305 to \$185,939.

This is graphically illustrated below.

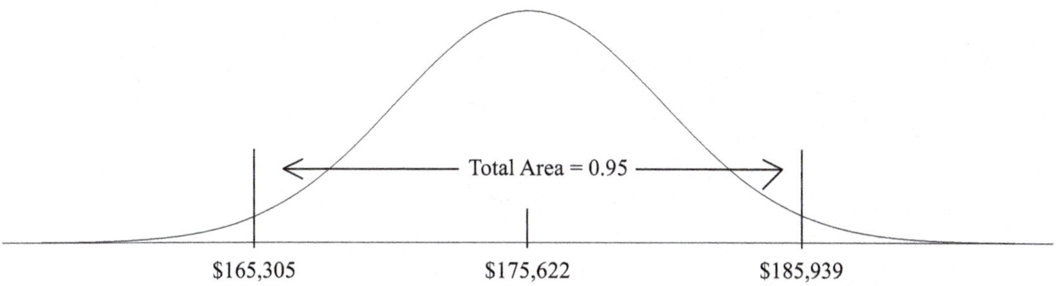

Below uses the Minitab *zint* macro to demonstrate the above problem.

```
MTB > %zint
Executing from file: C:\Program Files\Mtbwin\MACROS\zint.MAC
calculates a mean confidence interval using a normal distribution
What is the mean?
DATA> 175622
What is the standard deviation?
DATA> 37221
What is the sample size of the data set?
DATA> 50
What is the desired percentage confidence interval?
DATA> .95
The confidence interval is:
```

10.3.3 Data Display

```
Row   Clowint  Chighint

  1    165305    185939
```

Example C Suppose a machine dispenses sand into bags. The population standard deviation is 9.0 pounds, and the weights are normally distributed. A random sample of 100 bags is taken, and the sample mean is 105 pounds. Calculate a 95 % confidence interval for the mean weight of the bags using the *zint* macro.

The calculation in Excel is shown below.

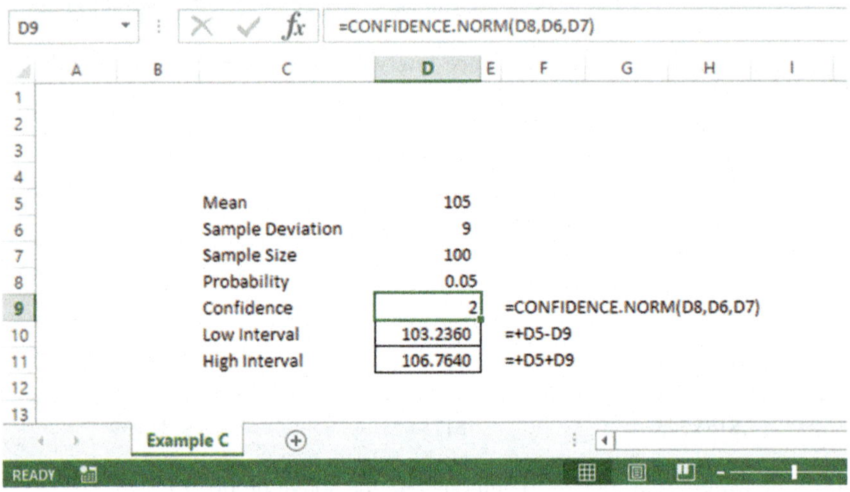

From the output we can see that the 95 % confidence interval for the mean weight of the bags is 103.236 to 106.764 pounds.

This is graphically illustrated below.

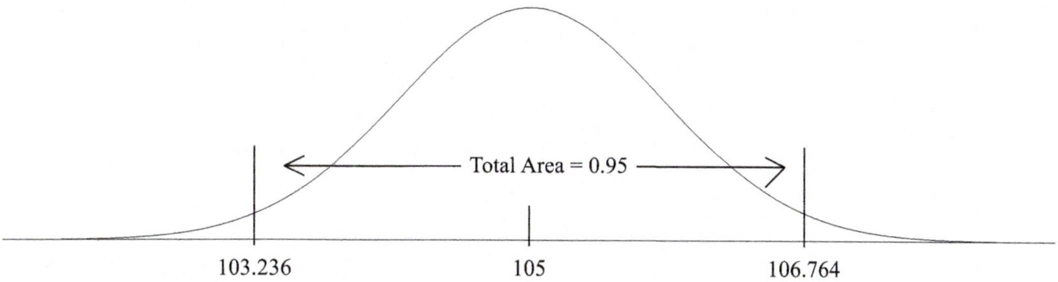

Below uses the Minitab *zint* macro to demonstrate the above problem.

```
MTB > %zint
Executing from file: C:\Program Files\Mtbwin\MACROS\zint.MAC
calculates a mean confidence interval using a normal distribution
What is the mean?
DATA> 105
What is the standard deviation?
DATA> 9
What is the sample size of the data set?
DATA> 100
What is the desired percentage confidence interval?
DATA> .95
The confidence interval is:
```

10.3.4 Data Display

```
Row   Clowint  Chighint

  1   103.236   106.764
```

Example D To study the effect of internal audit departments on external audit fees, W. A. Wallace conducted a survey of the audit department of 32 diverse companies (*Harvard Business Review*, March–April 1984). She found that the mean annual external audit paid by the 32 companies was $779,030 and the standard deviation was $1,083,162. Use Minitab to find the 95 % confidence interval for the mean external audit fees.

The calculation in Excel is shown below.

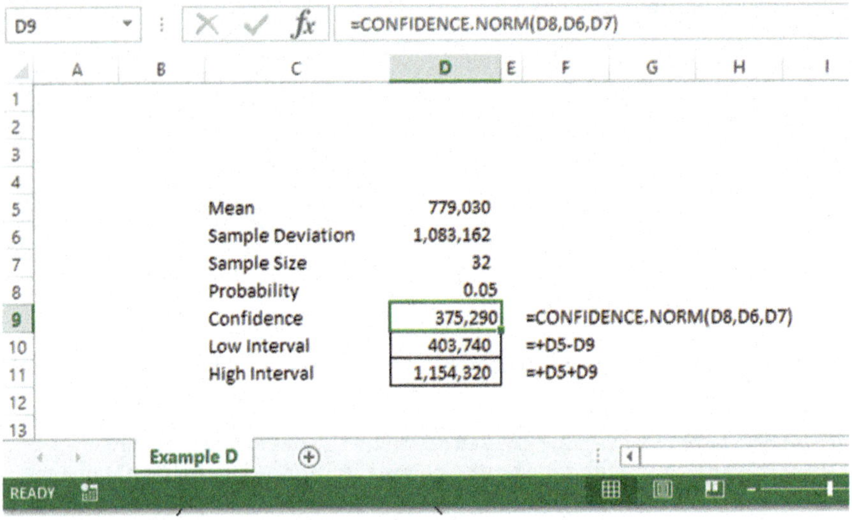

From the output we can see that the 95 % confidence interval for the mean external audit fees is from $403,740 to $1,154,320.

This is graphically illustrated below.

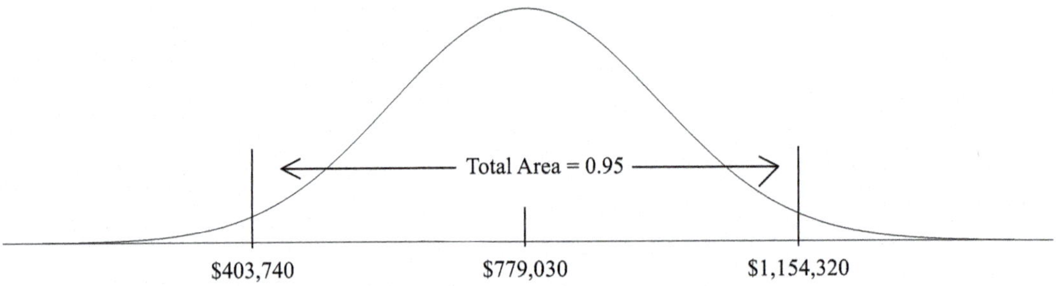

Below uses the Minitab *zint* macro to demonstrate the above problem.

```
MTB > %zint
Executing from file: C:\Program Files\Mtbwin\MACROS\zint.MAC
calculates a mean confidence interval using a normal distribution
What is the mean?
DATA> 779030
What is the standard deviation?
DATA> 1083162
What is the sample size of the data set?
DATA> 32
What is the desired percentage confidence interval?
DATA> .95
The confidence interval is:
```

10.3.5 Data Display

```
Row   Clowint   Chighint

  1    403740    1154320
```

10.4 Confidence Intervals for μ When σ^2 is Unknown

When the sample size is small, less than 30, and the data have a normal distribution, we will have to use the *t* test. The *t* test uses a *t* distribution. The formula to create the confidence interval would be:

$$\overline{x} - t_{n-1,\alpha/2}\left(s/\sqrt{n}\right) < \mu < \overline{x} + t_{n-1,\alpha/2}\left(s/\sqrt{n}\right)$$

Example E Suppose managers at the Smooth Ride Car Rental Company are interested in the mean number of miles that customers drive per day. A random sample of six car rentals shows that the customers drove the following numbers of miles:

152, 222, 300, 84, 90, 122

Construct a 99 % confidence interval for the mean number of miles driven.

```
MTB > tinterval 99 c1
```

10.4.1 T Confidence Intervals

```
Variable       N       Mean     StDev   SE Mean          99.0 % CI
C1             6      161.7      84.4      34.5   (    22.7,     300.6)
```

We can see from the Minitab output that the 99 % confidence interval is between 22.7 and 300.6 miles.

We will now calculate the interval in Excel. Below shows how to set up the calculation in Excel.

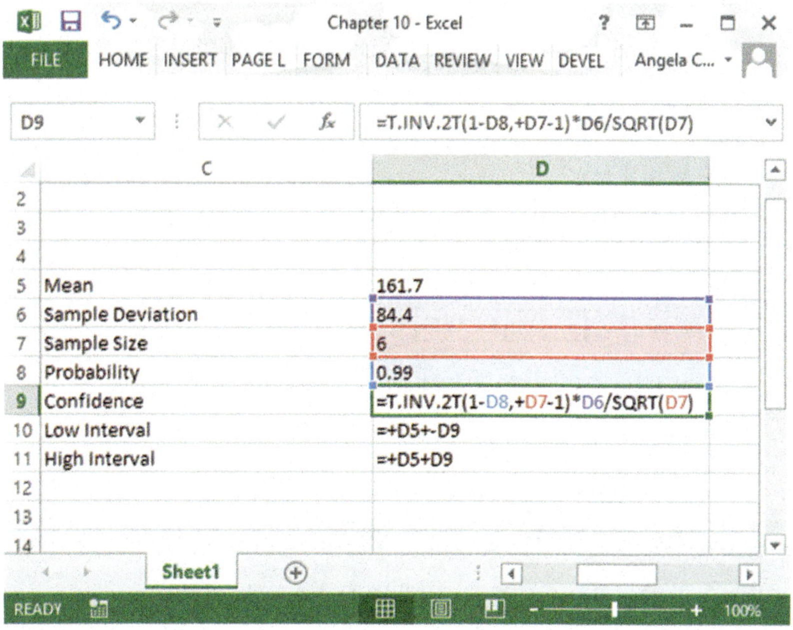

Below shows the calculations in Excel.

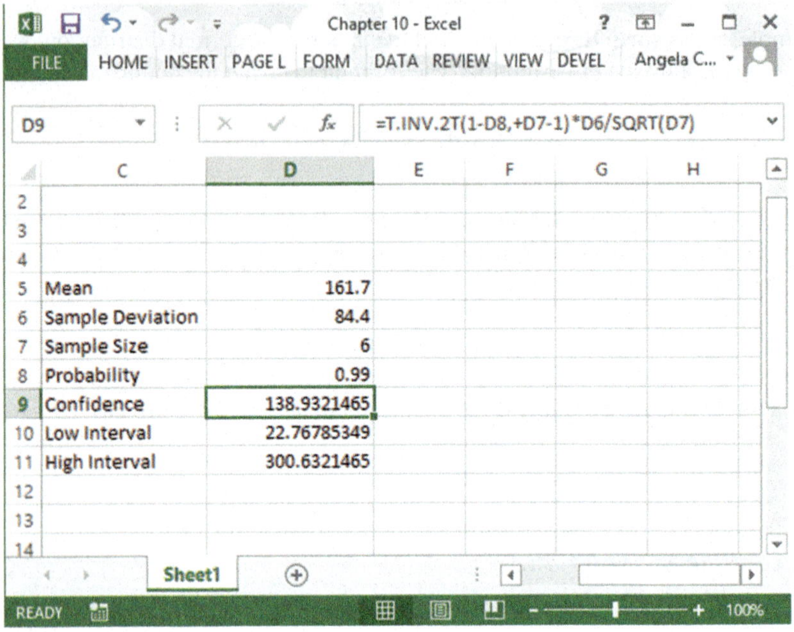

Example F Suppose after randomly selecting eight football players, we found that the mean weight was 212.75 and the sample standard deviation was 23.34. Calculate the 95 % confidence interval.

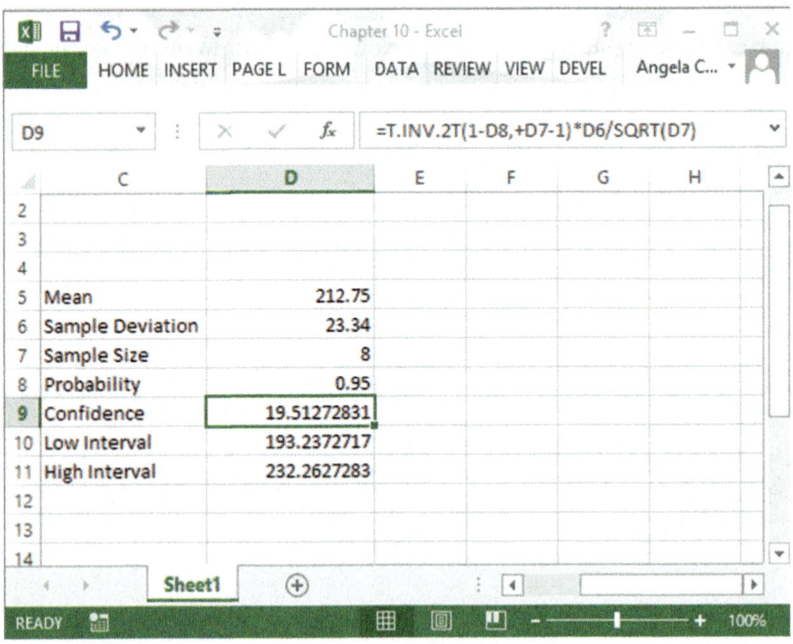

The output then indicates that the 95 % confidence interval for the average weight of football players is from 193.237 to 232.263 pounds.

This is illustrated below.

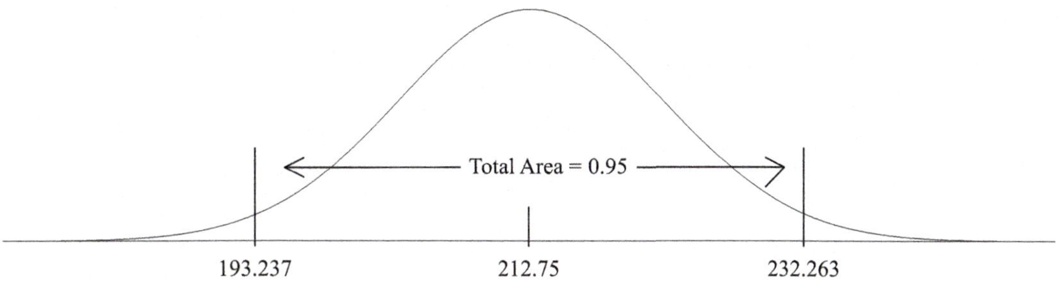

Below uses the Minitab *tint* macro to demonstrate the above problem. The Minitab *tint* macro is shown in Appendix 10.3.

```
MTB > %tint
Executing from file: C:\Program Files\Mtbwin\MACROS\tint.MAC
calculates a mean confidence interval using a normal distribution
What is the mean?
DATA> 212.75
What is the standard deviation?
DATA> 23.34
What is the sample size of the data set?
DATA> 8
What is the desired percentage confidence interval?
DATA> .95
The confidence interval is:
```

10.4.2 Data Display

```
Row   Clowint  Chighint

  1   193.237   232.263
```

10.5 Confidence Intervals for the Population Proportion

Because of the central limit theorem, whenever we calculate the confidence interval for the population proportion, we will use the normal distribution if the sample is large enough. The formula to calculate the confidence interval for the population proportion is

$$\hat{P} - z_{\alpha/2}\sqrt{\frac{\hat{P}\left(1-\hat{P}\right)}{n}} < P < \hat{P} + z_{\alpha/2}\sqrt{\frac{\hat{P}\left(1-\hat{P}\right)}{n}}$$

Example G A marketing firm discovers that 65 % of the 30 customers who participated in a blind taste test prefer brand A to brand B. Develop a 95 % confidence interval for the number of people who prefer brand A.

Below shows how to set up Excel to calculate a confidence interval for a population proportion.

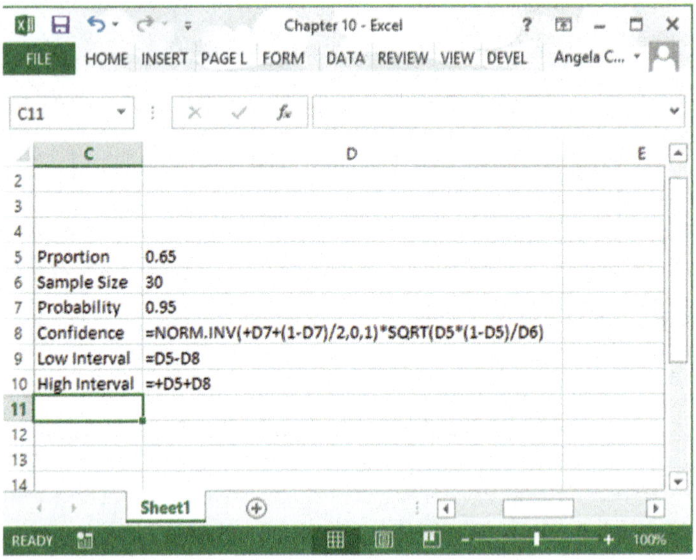

Below shows the calculation done in Excel.

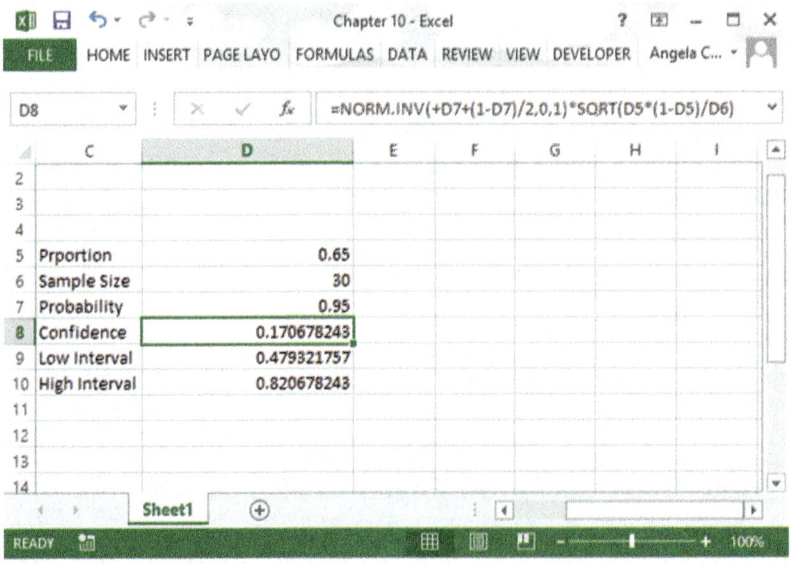

From the above calculation, the confidence interval for those who preferred brand A lies between 47.9322 and 82.0678 %.

Graphically the confidence interval would look like the following.

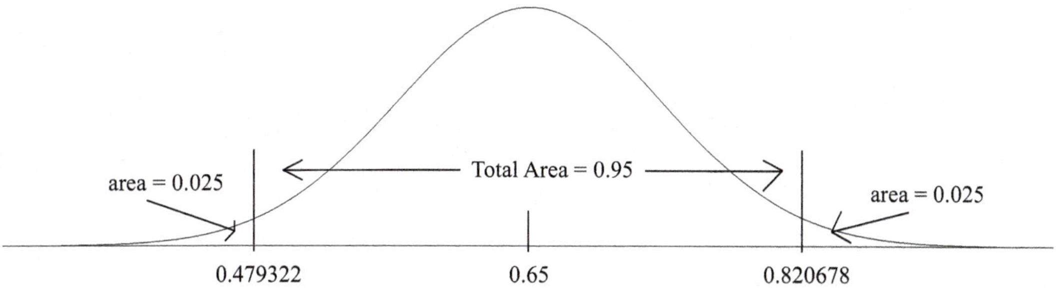

Below uses the Minitab *pint* macro to demonstrate the above problem. The Minitab *pint* macro is shown in Appendix 10.4.

```
MTB > %pint
Executing from file: C:\Program Files\Mtbwin\MACROS\pint.MAC
calculates a proportion confidence interval using a normal distribution
What is the proportion?
DATA> .65
What is the sample size of the data set?
DATA> 30
What is the desired percentage confidence interval?
DATA> .95
The confidence interval is:
```

10.5.1 Data Display

```
Row    Clowint   Chighint

  1   0.479322   0.820678
```

Example H Suppose that a random sample of 100 voters is taken, and 55 % of the sample supports the incumbent candidate. Construct a 90 % confidence interval for the proportion that supports the incumbent candidate.

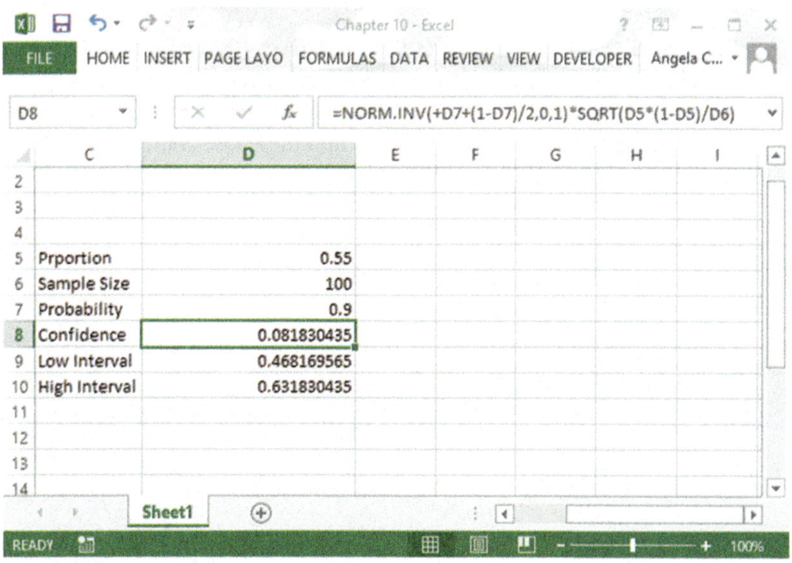

From the above calculation, we learn that the confidence interval for those who support the incumbent candidate is from 46.817 to 63.183 %.

Graphically the confidence interval looks like the following.

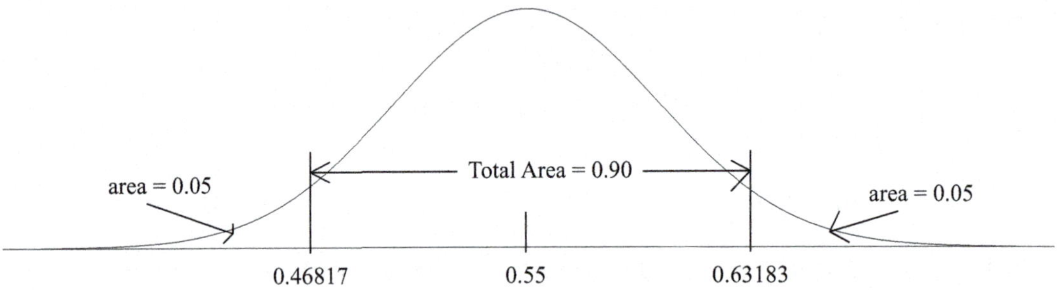

Below uses the Minitab *Pint* macro to demonstrate the above problem.

```
MTB > %pint
Executing from file: C:\Program Files\Mtbwin\MACROS\pint.MAC
calculates a proportion confidence interval using a normal distribution
What is the proportion?
DATA> .55
What is the sample size of the data set?
DATA> 100
What is the desired percentage confidence interval?
DATA> .90
The confidence interval is:
```

10.5.2 Data Display

```
Row    Clowint   Chighint
```

1 0.468170 0.631830

Example I A study in the *Journal of Advertising Research* (April/May 1984) to find the proportion of working adults using computer equipment on the job employed the random sample approach to survey 616 working adults. The survey revealed that 184 of the adults, or 29.9 %, regularly used computer equipment on the job. Use the *pint* macro to find the 95 % confidence interval for working adults' computer usage.

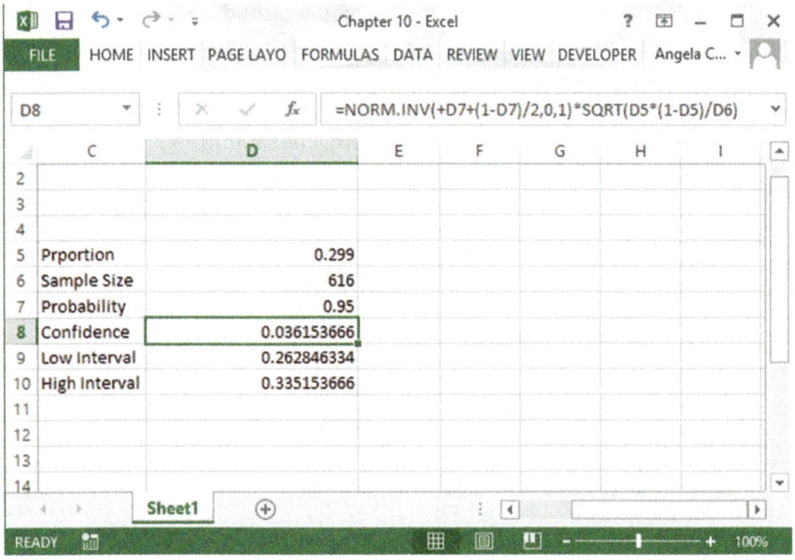

From the above calculation, we can see that the 95 % confidence interval for working adults' computer usage was from 26.2846 to 33.5154 %.

This is graphically shown below.

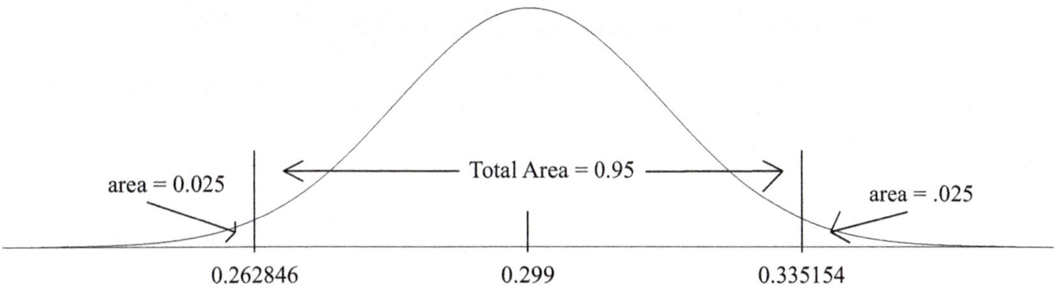

Below uses the Minitab *Pint* macro to demonstrate the above problem.

```
MTB > %pint
Executing from file: C:\Program Files\Mtbwin\MACROS\pint.MAC
calculates a proportion confidence interval using a normal distribution
What is the proportion?
DATA> .299
What is the sample size of the data set?
DATA> 616
What is the desired percentage confidence interval?
DATA> .95
The confidence interval is:
```

10.5.3 Data Display

```
Row    Clowint    Chighint

  1   0.262846   0.335154
```

Example J Guffey, Harris, and Laumer (1979) studied the attitudes of shoppers toward shoplifting and devices for its prevention. They sampled 403 shopping center patrons. Twenty-four percent of this sample expressed awareness of and discomfort with the use of TV cameras as a device to prevent shoplifting. Use the *pint* macro to find the 95 % confidence interval of the population in order to find the proportion that dislikes the use of TV devices to prevent shoplifting.

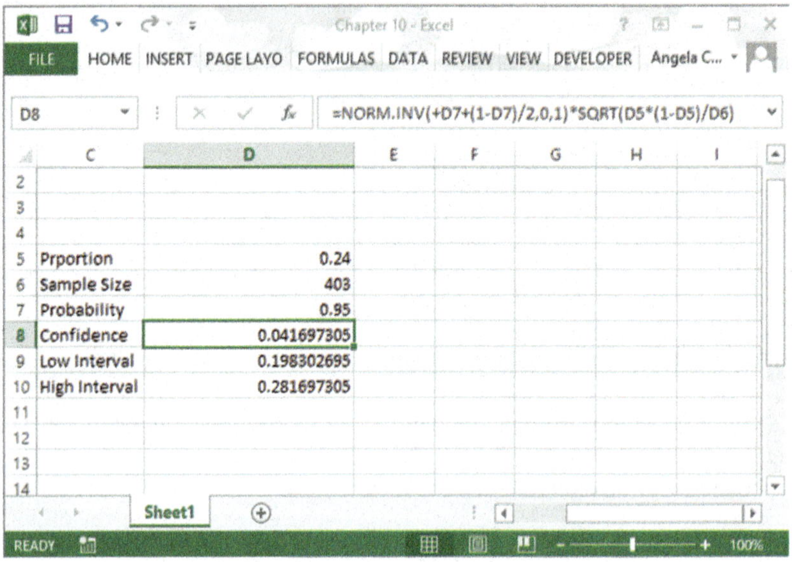

From the above calculations, we can see that the 95 % confidence interval proportion that dislikes using TV cameras to prevent shoplifting is from 19.8303 to 28.1697 %.

This is graphically shown below.

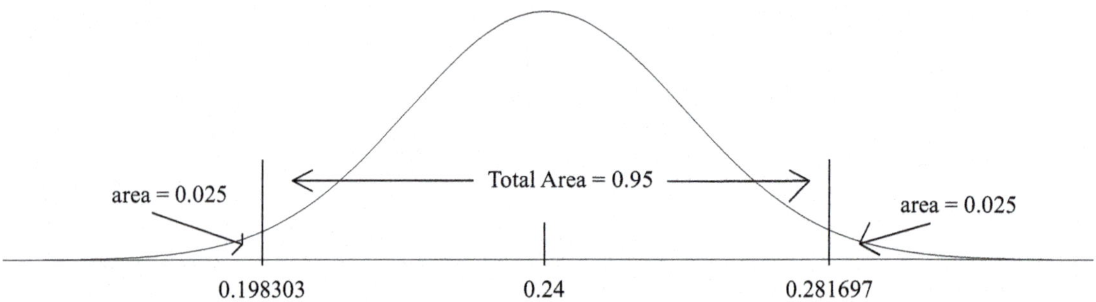

Below uses the Minitab *Pint* macro to demonstrate the above problem.

```
MTB > %pint
Executing from file: C:\Program Files\Mtbwin\MACROS\pint.MAC
calculates a proportion confidence interval using a normal distribution
What is the proportion?
DATA> .24
What is the sample size of the data set?
DATA> 403
What is the desired percentage confidence interval?
DATA> .95
The confidence interval is:
```

10.5.4 Data Display

```
Row    Clowint    Chighint

  1   0.198303   0.281697
```

10.6 Confidence Intervals for the Variance

The confidence interval for variances uses a chi-square distribution. The formula for the confidence interval for a variance is:

$$\frac{(n-1)s_x^2}{\chi^2_{\nu,\,\alpha/2}} < \sigma^2 < \frac{(n-1)s_x^2}{\chi^2_{\nu,\,1-\alpha/2}}$$

Example K Suppose a random sample of 30 bags of sand is taken and the sample variance of weights is 5.5. Find a 95 % confidence interval for the population variance.

Below shows how to set up Excel to calculate a confidence interval for a population variance.

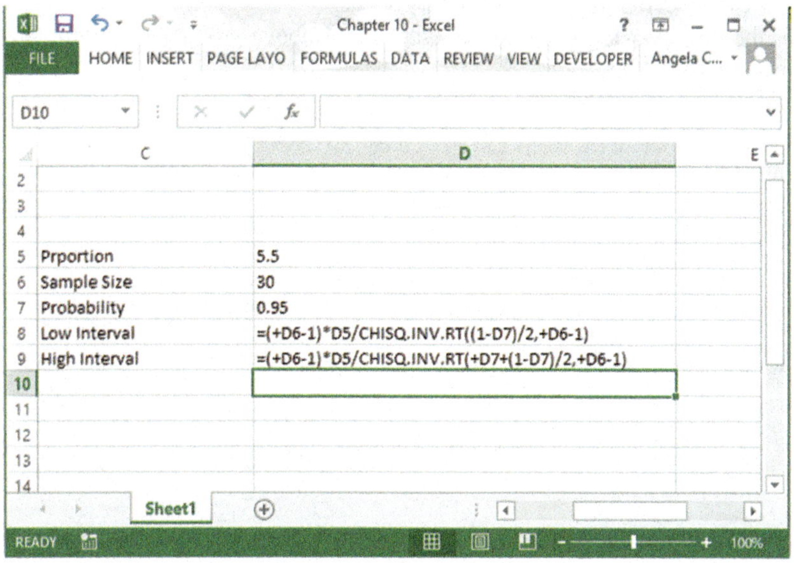

Below shows the calculation done in Excel.

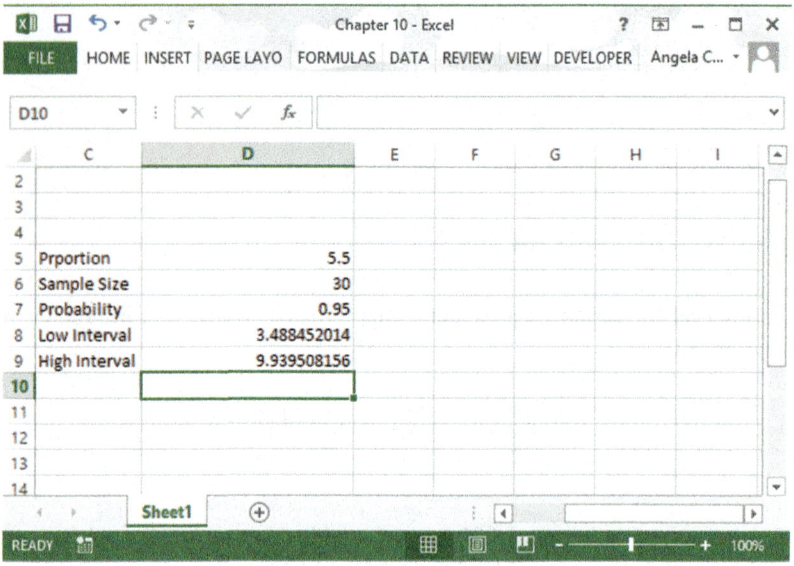

From the above calculation, we learn that the confidence interval for the sample variance of sandbag weight is between 3.4885 and 9.93951 pounds.

Below uses the Minitab *Xint* macro to demonstrate the above problem. The Minitab *pint* macro is shown in Appendix 10.5.

```
MTB > %xint
Executing from file: C:\Program Files\Mtbwin\MACROS\xint.MAC
calculates a variance confidence interval using chisquare distribution
What is the sample variance?
DATA> 5.5
What is the sample size of the data set?
DATA> 30
What is the desired percentage confidence interval?
DATA> .95
The confidence interval is:
```

10.6.1 Data Display

```
Row   Clowint   Chighint

  1   3.48845   9.93951
```

10.7 SAS Programming Code Instructions and Examples

10.7.1 Interval Estimates for μ When σ^2 is Known

Example A First, we'll read in the data and compute 95 % confidence intervals for the mean.

```
Data ACCOUNTANTS;
   INPUT Salary @@;
   /*uses an @@ so SAS does not automatically go to a new line of raw data
for each observation*/
   DATALINES;
24000 20000 18000 26000 23750 32000 36789 25300
27500 35000 31000 25500 25000 50000 23750 30500
28000 29000 27000 27750 25000 41000 28900 32000
22000 26000 34000 26550 33000 27900 23750 31000
32000 21430 42014 36450 24750 34000 35000 27000
;
run;
```

We need to know the mean, standard deviation, and number of records from our accounting salary data set. We use **output out** = option to read such information into file Accmeans.

```
proc univariate data=ACCOUNTANTS;
output out = Accmeans mean=Mean STD=Std N=n;
   var Salary;
run;
```

For a 95 % confidence interval z = 1.96, since 95 % of the area under a standard normal density is between −1.96 and 1.96. Let's calculate the confidence interval base on the formula we mentioned.

```
data Acclimits;
set Accmeans;
lowlim95 = mean - 1.96 * (Std/sqrt(N));
uplim95 = mean + 1.96 * (Std/sqrt(N));
run;

proc print data= Acclimits;
run;
```

Obs	n	Mean	Std	lowlim95	uplim95
1	40	29239.58	6351.90	27271.10	31208.05

Example B Below use SAS to demonstrate the problem.

```
data homeprice;
mean=175622;
Std=37221;
N=50;
lowlim95 = mean - 1.96 * (Std/sqrt(N));
uplim95 = mean + 1.96 * (Std/sqrt(N));
run;
```

Obs	Mean	Std	*N*	lowlim95	uplim95
1	175622	37221	50	165304.87	185939.13

Example C Below use SAS to demonstrate the problem.

```
data bagsand;
mean=105;
Std=9;
N=100;
lowlim95 = mean - 1.96 * (Std/sqrt(N));
uplim95 = mean + 1.96 * (Std/sqrt(N));
run;
```

Obs	Mean	Std	N	lowlim95	uplim95
1	105	9	100	103.236	106.764

Example D Below use SAS to demonstrate the problem.

```
data department;
mean=779030;
Std=1083162;
N=32;
lowlim95 = mean - 1.96 * (Std/sqrt(N));
uplim95 = mean + 1.96 * (Std/sqrt(N));
run;
```

Obs	Mean	Std	N	lowlim95	uplim95
1	779030	1083162	32	403733.51	1154326.49

10.7.2 Interval Estimates for μ When σ^2 is Unknown

Example E We will now calculate the interval in SAS. Below show how to set up the calculation by using **PROC MEAN**.

```
Data CarRental;
   INPUT Miles @@;
   DATALINES;
152 222 300 84 90 122
;
run;

proc means data=CarRental alpha=0.01 clm mean std;
var Miles;
run;
```

Analysis variable: miles			
Lower 99 % CL for mean	Upper 99 % CL for mean	Mean	Std dev
22.7451826	300.5881507	161.6666667	84.3935227

Example F For a 95 % confidence interval, t value is 2.365 with degree of freedom equal 7. Let's calculate the confidence interval base on the formula we mentioned.

```
data Players;
mean=212.75;
Std=23.34;
N=8;
lowlim95 = mean - 2.365 * (Std/sqrt(N));
uplim95 = mean + 2.365 * (Std/sqrt(N));
run;
```

Obs	Mean	Std	N	lowlim95	uplim95
1	212.75	23.34	8	193.234	232.266

10.7.3 Confidence Intervals for the Population Proportion

Example G Below shows how to set up SAS to calculate a confidence interval for a population proportion.

```
data customers;
p=0.65;
N=30;
lowInterval = p - 1.96 * sqrt(p*(1-p)/N);
upInterval = p + 1.96 * sqrt(p*(1-p)/N);
run;
```

Obs	p	N	lowInterval	upInterval
1	0.65	30	0.47932	0.82068

Example H Below shows how to set up SAS to calculate a confidence interval for a population proportion.

```
data voters;
p=0.55;
N=100;
lowInterval = p - 1.645 * sqrt(p*(1-p)/N);
upInterval = p + 1.645 * sqrt(p*(1-p)/N);
run;
```

Obs	p	N	lowInterval	upInterval
1	0.55	100	0.46816	0.63184

Example I Below shows how to set up SAS to calculate a confidence interval for a population proportion.

```
data Survey;
p=0.299;
N=616;
lowInterval = p - 1.96 * sqrt(p*(1-p)/N);
upInterval = p + 1.96 * sqrt(p*(1-p)/N);
run;
```

Obs	p	N	lowInterval	upInterval
1	0.299	616	0.26285	0.33515

Example J Below shows how to set up SAS to calculate a confidence interval for a population proportion.

```
data Shoppers;
p=0.24;
N=403;
lowInterval = p - 1.96 * sqrt(p*(1-p)/N);
upInterval = p + 1.96 * sqrt(p*(1-p)/N);
run;
```

Obs	p	N	lowInterval	upInterval
1	0.24	403	0.19830	0.28170

10.7.4 Confidence Intervals for the Variance

Example K For chi-square distribution, $\chi^2_{N-1,(1-\alpha)/2}$ and $\chi^2_{N-1,\ \alpha/2}$ are equal to 16.074 and 45.722, respectively. Below show how to set up SAS to calculate a confidence interval for variance.

```
data Bags;
s=5.5;
chi1=16.047;
chi2=45.722;
N=30;
lowInterval = (N-1)*s/chi2;
upInterval = (N-1)*s/chi1;
run;
```

Obs	s	chi1	chi2	N	lowInterval	upInterval
1	5.5	16.047	45.722	30	3.48847	9.93955

10.8 Statistical Summary

In this chapter we created the confidence interval for the population mean when the population variance was known and unknown. We also found the confidence interval for the population proportion and the confidence interval for the population variance.

In this chapter we calculated the confidence interval for the population mean. Most of the time the population variance is unknown but because of the central limit theorem, we can use the sample variance to approximate the population variance and use the normal distribution to calculate the confidence interval. We can substitute the sample variance for the population variance when the sample size is 30 or greater.

If the sample size is smaller than 30, then we will need to use the t distribution to calculate the confidence interval for the population mean.

Also because of the central limit theorem, we can calculate the confidence interval for the population proportion using the normal distribution.

We used the chi-square distribution, in calculating the confidence interval for the variance of a normal distribution.

Appendix 10.1: Excel Code—Confidence Interval Simulator

```
'/***************************************************************************
'/ Essentials of Microsoft Excel 2013 VBA, SAS
'/              and MINITAB 17
'/ for Statistical and Financial Analyses
'/
'/***************************************************************************
Option Explicit

Private Sub cmdCancel_Click()
    Unload Me
End Sub
```

```
'/*****************************************************************
'/Purpose:  This is the procedure that is run after pressing the
'/          the run button on the form.  This is the main procedure
'/*****************************************************************
Private Sub cmdRun_Click()
    Me.Hide
    ConfIntSimNorm
    Unload Me
End Sub


'/*******************************************************************
'/Purpose:  This procedure runs everytime the form is loaded.  This
'/          procedure sets some initial values.
'/*******************************************************************
Private Sub UserForm_Initialize()
        With Me
            .txtInterval = 100
            .txtPercent = 0.95
            .txtSample = 50
            .txtMean = 0
            .txtStandDev = 1
        End With
End Sub

Sub ConfIntSimNorm()
    Dim chrtBox As Chart
    Dim lCount As Long
    Dim seriesCurrent As Series
    Dim rX As Range
    Dim rY As Range
    Dim wbData As Workbook
    Dim wsData As Worksheet
    Dim lTimes As Long
    Dim dblMaxY As Double
    Dim rData As Range
    Dim rAvg As Range

    Application.DisplayAlerts = False
    On Error Resume Next
    Set chrtBox = ThisWorkbook.Sheets("Confidence Simulation").Delete

    Set wbData = Workbooks.Add
    Set wsData = wbData.Worksheets.Add
    wsData.Name = "Data"


    Err = 0
    On Error GoTo 0

    Set chrtBox = wbData.Charts.Add
    chrtBox.Name = "Confidence Simulation"
    CreatChart chrtBox:=chrtBox, seriesCurrent:=seriesCurrent
    ConfIntGraph chrtBox:=chrtBox, rData:=rData, wsData:=wsData,
seriesCurrent:=seriesCurrent
    Application.DisplayAlerts = False
    For Each wsData In wbData.Worksheets
        If Left(wsData.Name, 5) = "Sheet" Then
            wsData.Delete
        End If
    Next
    Application.DisplayAlerts = True
End Sub
```

```
Sub ConfIntGraph(chrtBox As Chart, wsData As Worksheet, _
                rData As Range, seriesCurrent As Series)
    Dim rCell As Range
    Dim lCount As Long
    Dim ncount As Integer
    Dim nInterval As Integer
    Dim nSampSize As Integer
    Dim dblMean As Double
    Dim dblSTDV As Double
    Dim sngPercent As Single
    Dim sngZpercent As Single
    Dim rAvg As Range
    Dim lFail As Long


    nInterval = Me.txtInterval
    nSampSize = Me.txtSample
    dblMean = Me.txtSample

dblSTDV = Me.txtStandDev
sngPercent = Me.txtPercent

sngZpercent = WorksheetFunction.NormInv(1 - ((1 - sngPercent)) / 2, dblMean, dblSTDV)

sngZpercent = (sngZpercent - dblMean) / dblSTDV

Application.Run "ATPVBAEN.XLA!Random", wsData.Range("$A$1"), nSampSize, nInterval _
    , 2, , dblMean, dblSTDV

Set rData = wsData.Cells(1).CurrentRegion

Set rAvg = rData.Columns(1).Offset(0, rData.Columns.Count + 3)
With rAvg
    .Formula = "=AVERAGE(RC[-53]:RC[-4])"
    .Offset(0, -1).Formula = "=+RC[1]-(" & sngZpercent * _
        dblSTDV & "* (1/sqrt(" & nSampSize & ")))"
    .Offset(0, 1).Formula = "=+rc[-1]+(" & sngZpercent * _
        dblSTDV & "* (1/sqrt(" & nSampSize & ")))"
End With

lCount = 1
lFail = 0
For Each rCell In rAvg.Offset(0, -1).Cells

    With seriesCurrent
        .XValues = "={" & lCount & "," & lCount & "," & lCount & "}"
        .Values = rCell.Resize(1, 3)

        If rCell.Value < dblMean And rCell.Offset(0, 2) > dblMean Then

                With .Border
                        .ColorIndex = 1
                        .LineStyle = xlContinuous
                End With
                .MarkerStyle = xlMarkerStyleDash
                .MarkerSize = 5
                .MarkerForegroundColorIndex = 1

        Else
                lFail = lFail + 1
                With .Border
                        .ColorIndex = 3
                        .LineStyle = 2
                End With
```

```
                    .MarkerBackgroundColorIndex = 3
                    .MarkerForegroundColorIndex = 3
                    .MarkerStyle = xlSquare
                    .MarkerSize = 5

                End If
            End With
            lCount = lCount + 1
            Set seriesCurrent = chrtBox.SeriesCollection.NewSeries
        Next

        'Draw the mean
        With seriesCurrent
                .XValues = "={0," & nInterval + 5 & "}"
                .Values = "={" & dblMean & "," & dblMean & "}"
                        .Border.ColorIndex = 30
                        .MarkerBackgroundColorIndex = 30
                        .MarkerForegroundColorIndex = 30
                        .MarkerStyle = xlDiamond
                        .MarkerSize = 5
                With .Points(2)
                    .ApplyDataLabels
                    With .DataLabel
                        .Characters.Text = "Mean=" & dblMean
                        .Position = xlLabelPositionAbove
                        .Font.Size = 6
                        End With
                    End With
    End With

    With chrtBox
            With .ChartTitle.Characters
                .Text = "Confidence Interval Simulation" & Chr(10) _
                            & "#Inteverval = " & nInterval & ", Fail = " _
                            & lFail & " Confidence Interval= " & Format(sngPercent, "0%")
                .Font.Size = 20
            End With
            .Axes(xlValue).MinimumScale = _
                Round(WorksheetFunction.Min(rAvg.Offset(0, -1)) - dblSTDV / 2, 2)
                With .Axes(xlCategory)
                    .MinimumScale = 0
                    .MaximumScale = nInterval + 10

                 End With
            .Activate
    End With
End Sub
Sub CreatChart(chrtBox As Chart, seriesCurrent As Series)

    If chrtBox.SeriesCollection.Count = 0 Then
        Set seriesCurrent = chrtBox.SeriesCollection.NewSeries
    Else
        Set seriesCurrent = chrtBox.SeriesCollection(1)
    End If
    With chrtBox
            .ChartType = xlXYScatterSmooth
            .HasTitle = True
            With .Axes(xlCategory)
                  .HasMajorGridlines = False
                  .HasMinorGridlines = False
            End With
            With .Axes(xlValue)
                  .HasMajorGridlines = False
                  .HasMinorGridlines = False
            End With
            With .PlotArea
                .Interior.ColorIndex = xlNone
                .ClearFormats
            End With
            .HasLegend = False
    End With
End Sub
```

Appendix 10.2: Minitab Code—*Zint*

```
macro
Zint
mconstant Kmean Kstdv Ksize Kpercent Kinvcdf KinvcdfP KinvcdfS Kupper Klower
mcolumn   Cmean Cstdv Csize CPercent Chighint Clowint
noecho
#
note calculates a mean confidence interval using a normal distribution
note What is the mean?
set 'terminal' Cmean;
nobs 1.
end
note What is the standard deviation?
set 'terminal' Cstdv;
nobs 1.
End

note What is the sample size of the data set?
set 'terminal' Csize;
nobs 1.
End

note What is the desired percentage confidence interval?
set 'terminal' Cpercent;
nobs 1.
End

let Kmean = Cmean
let Kstdv = Cstdv
let Ksize = Csize
let Kpercent = Cpercent
let KinvcdfP = (1 + Kpercent)/2
invcdf kinvcdfP KinvcdfS;
normal 0 1.





#calculating the upper and lower end of interval
let Chighint= kmean + Kinvcdfs *(Kstdv/sqrt(Ksize))
let Clowint= kmean - Kinvcdfs *(Kstdv/sqrt(Ksize))


note The confidence interval is:
print clowint chighint
endmacro
```

Appendix 10.3: Minitab Code—*Tint*

```
macro
Tint
mconstant Kmean Kstdv Ksize Kpercent Kinvcdf KinvcdfP KinvcdfS Kupper Klower KFree
mcolumn   Cmean Cstdv Csize CPercent Chighint Clowint
noecho
#
note calculates a mean confidence interval using a normal distribution
note What is the mean?
set 'terminal' Cmean;
nobs 1.
end

note What is the standard deviation?
set 'terminal' Cstdv;
nobs 1.
End

note What is the sample size of the data set?
set 'terminal' Csize;
nobs 1.
End

note What is the desired percentage confidence interval?
set 'terminal' Cpercent;
nobs 1.
End

let Kmean = Cmean
let Kstdv = Cstdv
let Ksize = Csize
let Kpercent = Cpercent
let KinvcdfP = (1 + Kpercent)/2
let Kfree = KSize -1
invcdf kinvcdfP KinvcdfS;
t Kfree.

#calculating the upper and lower end of interval
let Chighint= kmean + Kinvcdfs *(Kstdv/sqrt(Ksize))
let Clowint= kmean - Kinvcdfs *(Kstdv/sqrt(Ksize))


note The confidence interval is:
print clowint chighint

endmacro
```

Appendix 10.4: Minitab Code—*Pint*

```
macro
Pint
mconstant Kporport Ksize Kpercent Kinvcdf KinvcdfP KinvcdfS Kupper Klower
mcolumn   Cporport Csize CPercent Chighint Clowint
noecho
#
note calculates a proportion confidence interval using a normal distribution
note What is the proportion?
set 'terminal' Cporport;
nobs 1.
end

note What is the sample size of the data set?
set 'terminal' Csize;
nobs 1.
End

note What is the desired percentage confidence interval?
set 'terminal' Cpercent;
nobs 1.
End

let Kporport = Cporport
let Ksize = Csize
let Kpercent = Cpercent
let KinvcdfP = (1 + Kpercent)/2
invcdf kinvcdfP KinvcdfS;
normal 0 1.


#calculating the upper and lower end of interval
let Chighint= Kporport+ Kinvcdfs * (sqrt((Kporport*(1-Kporport))/Ksize))
let Clowint= Kporport - Kinvcdfs * (sqrt((Kporport*(1-Kporport))/Ksize))


note The confidence interval is:
print clowint chighint

endmacro
```

Appendix 10.5: Minitab Code—*Xint*

```
macro
Xint
mconstant Kvar Ksize Kpercent Kinvcdf KinvcdfP KinvcdfS Kupper Klower Kfree
mcolumn   Cvar Csize CPercent Chighint Clowint
noecho
#
note calculates a variance confidence interval using chisquare distribution
note What is the sample variance?
set 'terminal' Cvar;
nobs 1.
end


note What is the sample size of the data set?
set 'terminal' Csize;
nobs 1.
End

note What is the desired percentage confidence interval?
set 'terminal' Cpercent;
nobs 1.
End

let Kvar = Cvar
let Ksize = Csize
let Kpercent = Cpercent
let Kfree = Ksize -1
let KinvcdfP = (1 + Kpercent)/2
invcdf kinvcdfP KinvcdfS;
chisquare Kfree.

let Clowint = (Kfree * Kvar) /KinvcdfS

let KinvcdfP = 1 - KinvcdfP
invcdf kinvcdfP KinvcdfS;
chisquare Kfree.

let Chighint= (Kfree * Kvar) /KinvcdfS

note The confidence interval is:
print clowint chighint

endmacro
```

Bibliography

Microsoft Inc., Excel 2013. Microsoft Inc., Redmond
Lee CF, Lee JC, Lee AC (2013) Statistics for business and financial economics. Springer, New York
Minitab Inc. Minitab 17. Minitab Inc., State College
SAS Institute Inc. SAS 2014. SAS Institute Inc., Cary

Chapter 11
Hypothesis Testing

11.1 Introduction

In the last chapter, we made inferences about a population parameter by creating a confidence interval from a sample. We will now look at another method, called *hypothesis testing*, for making inferences about a population parameter. In hypothesis testing we infer that the stated null hypothesis (H_0) is true until there is convincing but *not perfect* evidence that the null hypothesis is not true. Our evidence is from the sample that we obtain. We conclude that there is convincing evidence when the *p*-value is less than the alpha value. The *p*-value will be discussed later in this chapter.

Because the sample evidence is not perfect, we can make four kinds of decisions:

1. We can make a correct decision by rejecting H_0 when in fact the true H_0 is false.
2. We can make a correct decision by failing to reject H_0 when in fact the true H_0 is true.
3. We can make an incorrect decision by rejecting H_0 when in fact the true H_0 is true.
4. We can make an incorrect decision by accepting H_0 when in fact the true H_0 is false.

In statistics we call situation 3 a type I error.
We will make a type I error alpha percent of the time.
In statistics we call situation 4 a type II error.
Below is a diagram of the four possible resulting decisions of hypothesis testing.

	Actual H_0 True	Actual H_0 False
Fail to Reject H_0	Correct Decision	Type II Error
Reject H_0	Type I Error	Correct Decision

11.2 One-Tailed Tests of Mean for Large Samples

We begin by examining the case where only one sample is drawn and that sample is large. Using a large sample offers two important advantages. First, we can apply the central limit theorem. Second, the large sample enables us, through our choice of significance level, alpha, to reduce our chance of making a type II error.

C.-F. Lee et al., *Essentials of Excel, Excel VBA, SAS and Minitab for Statistical and Financial Analyses*, DOI 10.1007/978-3-319-38867-0_11

Example 11.1 Suppose an accountant is interested in finding out if the average starting salary for accountants is greater than $25,000 or not. Test at the 5 % alpha level. The accountant collected the following sample of 40 salaries.

$24,000	$20,000	$18,000	$26,000	$23,750	$32,000	$36,789	$25,300
$27,500	$35,000	$31,000	$25,500	$25,000	$50,000	$23,750	$30,500
$28,000	$29,000	$27,000	$27,750	$25,000	$41,000	$28,900	$32,000
$22,000	$26,000	$34,000	$26,550	$33,000	$27,900	$23,750	$31,000
$32,000	$21,430	$42,014	$36,450	$24,750	$34,000	$35,000	$27,000

The null and alternative hypotheses would then be

$$\mathrm{H}_0 = \$25,000$$

$$\mathrm{H}_1 > \$25,000$$

The test statistic for this would be

$$\frac{\overline{x} - \mu_0}{\sigma/\sqrt{n}} > Z_\alpha$$

To do this test, we should first put the salary data in column c1 of the Minitab worksheet, and then we will use the *ztest* command to perform the calculation.

```
MTB > set c1
DATA> 24000 20000 18000 26000 23750 32000 36789 25300
DATA> 27500 35000 31000 25500 25000 50000 23750 30500
DATA> 28000 29000 27000 27750 25000 41000 28900 32000
DATA> 22000 26000 34000 26550 33000 27900 23750 31000
DATA> 32000 21430 42014 36450 24750 34000 35000 27000
DATA> end
MTB > describe c1
```

11.2.1 Descriptive Statistics

```
Variable           N       Mean     Median     TrMean      StDev    SE Mean
C1                40      29240      27825      28877       6352       1004

Variable     Minimum    Maximum         Q1         Q3
C1             18000      50000      25000      32750

MTB > ztest 25000 6352 c1;
SUBC> alternate 1.
```

11.2.2 Z Test

```
Test of mu = 25000 vs mu > 25000
The assumed sigma = 6352

Variable      N      Mean     StDev   SE Mean        Z          P
C1           40     29240      6352      1004     4.22     0.0000

MTB > invcdf .95;
SUBC> normal 0 1.
```

11.2.3 Inverse Cumulative Distribution Function

```
Normal with mean = 0 and standard deviation = 1.00000

 P( X <= x)          x
    0.9500      1.6449
```

From the *ztest* command output, the calculated sample *Z* score for the data set is 4.22. We are testing at a 5 % alpha level. For a one-sided right-tailed test, the 5 % alpha level begins at the value 1.6449. We will reject the null hypothesis when the calculated *Z* value is greater than 1.6449 or when the calculated *Z* value lies in the 5 % alpha area. The test statistic which we calculated, 4.22, lies in the reject region.

Graphically we are seeing the following:

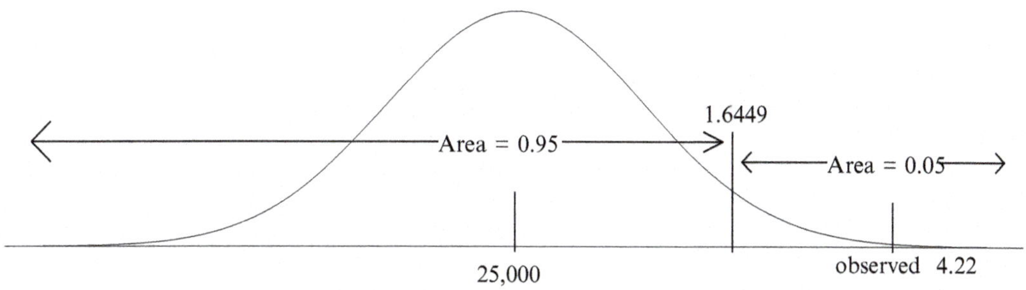

11.2.3.1 Hypothesis Testing and the *p*-Value

Another way to test the null hypothesis is to use the *p*-value. A *p*-value is the probability of getting a value more extreme than the calculated statistic. If we see that the *p*-value is less than the alpha value, we know that there is significant evidence to reject the null hypothesis because the *p*-value would lie in the region of the alpha value as shown in the above graph. In our example, the *p*-value is the area to the right of the calculated value of 4.22. This area to the right of 4.22 is the probability of getting a value more extreme than the calculated statistic.

With the advent of computer statistical software, the use of the *p*-value method for accepting or rejecting the null hypothesis is easier than using the test statistic method. In the latter method, we would also have to look up the value that begins the alpha value. In our example, the value 1.6449 is the value where an alpha value of 5 % begins. We then have to see if the calculated value is greater than the value of 1.6449. If it is, we conclude that there is enough evidence to reject the null hypothesis. If not, we conclude there is not enough evidence to reject the null hypothesis.

In the p-value method, we only need to see if the p-value is greater or less than the alpha. If it is greater or equal to the alpha value, we fail to reject the null hypothesis. If it is less than the alpha value, we reject the null hypothesis.

The calculated p-value in our example is 0.00. Since this p-value is less than the alpha value of 5 %, there is enough evidence to reject the null hypothesis that the mean accounting salary is $25,000 and accept the alternative hypothesis that the mean accounting salary is greater than $25,000.

We should remember that we have strong but not perfect evidence to reject the null hypothesis. There is a 5 % chance that we should not be rejecting the null hypothesis. As stated above, this kind of error in judgment is called a type I error.

Example 11.2 Say we want to test whether the average weight of 60-ounce bags of cat food is equal to, or smaller than, 60 ounces at significance level $\alpha = .05$. The null and alternative hypotheses can be stated as

$$H_{0\mu} = 60$$

$$H_{1\mu} < 60$$

In addition, suppose we know that the sample size $n = 100$, sample mean $\overline{X} = 59$, and standard deviation $s_x = 5$.

The test statistic for this would be

$$\frac{\overline{x} - \mu_0}{\sigma/\sqrt{n}} < Z_\alpha$$

The calculation is shown below. Based on the calculation shown below, we would reject the null hypothesis because for a left-tailed test, we reject the null hypothesis when the p-value is less than the alpha value.

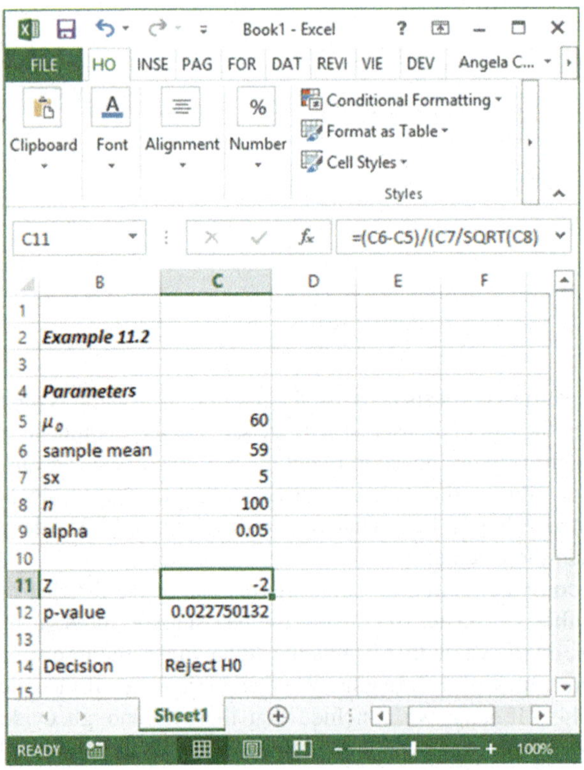

In Excel, if we press *ctrl-‘*, Excel will flip from showing the values as above or formulas as below. Showing the formulas in Excel in a worksheet illustrates how the worksheet is set up.

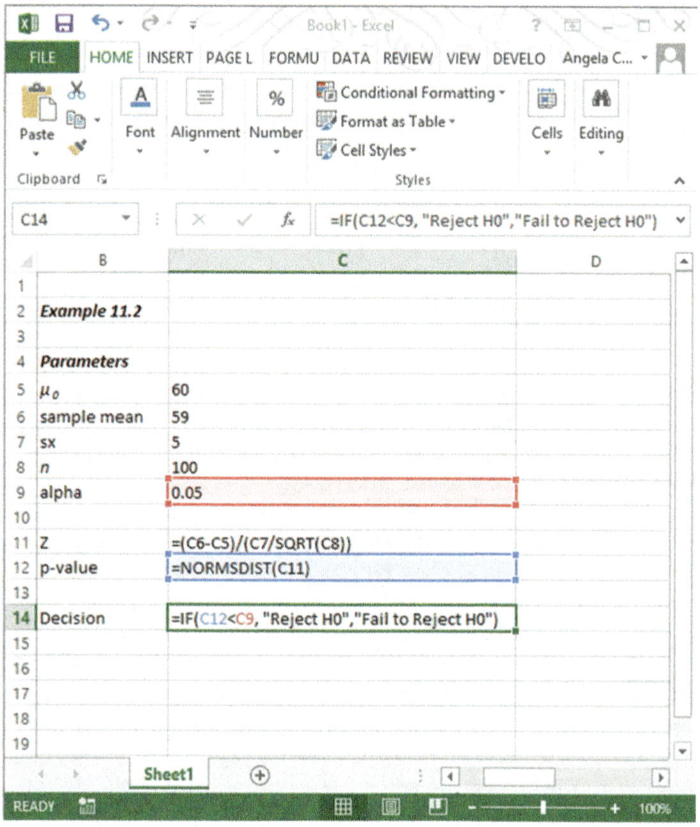

11.3 One-Tailed Tests of Mean for Large Samples: Two-Sample Test of Means

Another important issue is how to test the difference between two population means, μ_1 and μ_1, of two normally distributed populations with variances $\sigma_1{}^2$ and $\sigma_2{}^2$. Because we will use large sample, the assumption of normality is not necessary.

The test statistic for a right-tailed test testing two-sample test of means would be

$$Z_\alpha > \frac{(\bar{x}_1 - \bar{x}_2) - (\mu_1 - \mu_2)}{\sqrt{\frac{s_1^2}{n_1} + \frac{s_2^2}{n_2}}}.$$

Example 11.3 David Smith conducts a market survey to compare the prices of unleaded gasoline at Texaco stations and Shell stations. A random sample of 32 Texaco stations and 38 Shell stations in central New Jersey is used. The cost of 1 gallon of unleaded gasoline is recorded and the resulting data are summarized here.

Sample A (Texaco)					
1.06	0.97	0.97	0.96	1.02	1.09
1.08	1.04	1.11	1.12	1.19	1.07
1.14	1.17	1.22	0.97	1.08	
1.05	1.21	0.95	0.99	1.18	
1.05	1.21	1.03	1.14	1.14	
1.13	1.00	1.16	0.96	0.98	

Sample B (Shell)				
1.08	0.96	1.06	1.11	1.07
1.17	1.01	1.05	1.04	1.09
1.05	1.06	1.14	1.04	0.94
1.01	0.99	1.07	1.18	0.94
1.08	1.13	1.16	1.00	0.94
1.13	0.91	1.13	0.96	0.95
1.00	1.09	1.15	1.13	
0.98	1.04	1.03	1.17	

Test the following hypotheses:

$$H_0: \mu_1 = \mu_2$$
$$H_1: \mu_1 > \mu_2$$

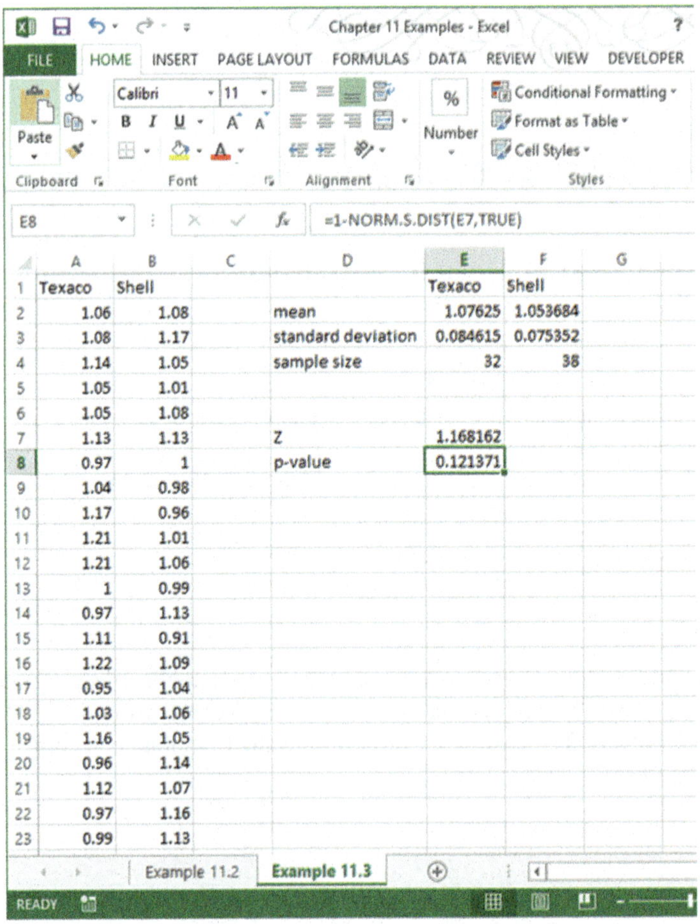

From the above calculations done in Excel, we can see that we cannot reject the null hypothesis that the average price per gallon of unleaded gasoline from Texaco stations is the same as that from Shell stations.

Below is the formula view of the above calculations.

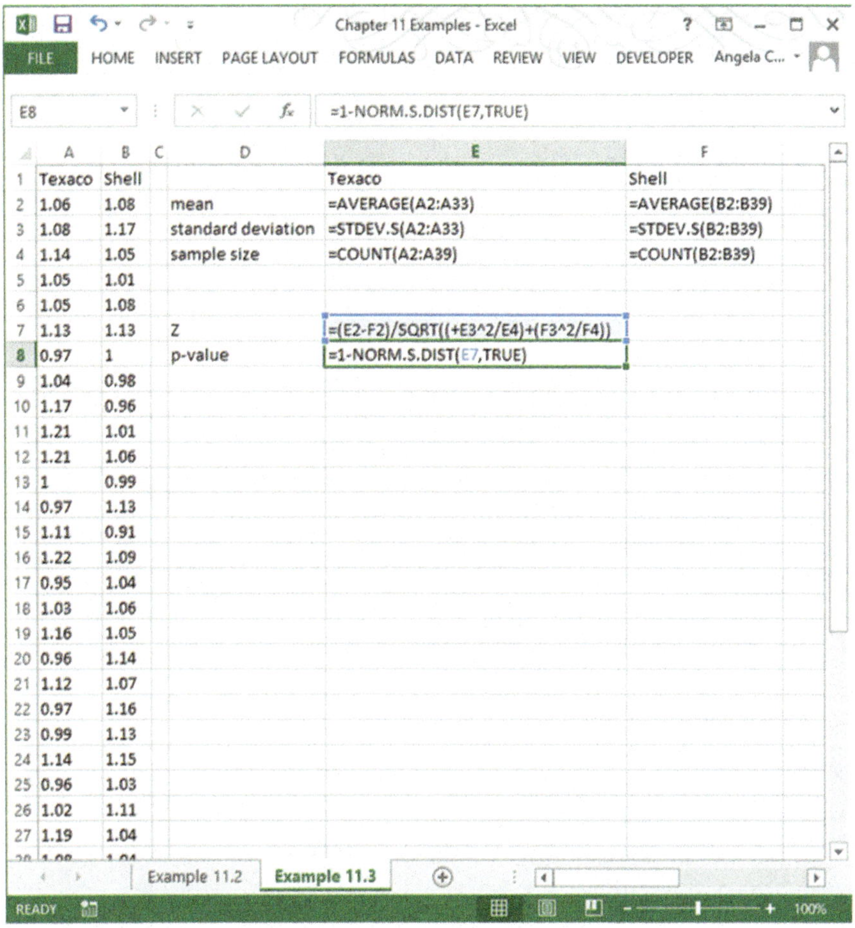

	A	B	C	D	E	F
1	Texaco	Shell			Texaco	Shell
2	1.06	1.08		mean	=AVERAGE(A2:A33)	=AVERAGE(B2:B39)
3	1.08	1.17		standard deviation	=STDEV.S(A2:A33)	=STDEV.S(B2:B39)
4	1.14	1.05		sample size	=COUNT(A2:A39)	=COUNT(B2:B39)
5	1.05	1.01				
6	1.05	1.08				
7	1.13	1.13		z	=(E2-F2)/SQRT((+E3^2/E4)+(F3^2/F4))	
8	0.97	1		p-value	=1-NORM.S.DIST(E7,TRUE)	
9	1.04	0.98				
10	1.17	0.96				
11	1.21	1.01				
12	1.21	1.06				
13	1	0.99				
14	0.97	1.13				
15	1.11	0.91				
16	1.22	1.09				
17	0.95	1.04				
18	1.03	1.06				
19	1.16	1.05				
20	0.96	1.14				
21	1.12	1.07				
22	0.97	1.16				
23	0.99	1.13				
24	1.14	1.15				
25	0.96	1.03				
26	1.02	1.11				
27	1.19	1.04				

11.4 Two-Tailed Tests of Mean for Large Samples

In a two-tailed test, we are testing the following hypotheses:

$$H_{0\,\mu} = D$$
$$H_{1\,\mu} \neq D$$

Then we do the same calculations for the two-tailed test as for the one-tailed test. The main difference is that we make decision based on half of the stated alpha. There are two reasons for this:

1. We are testing based on a large sample, and because of the central limit theorem, we can use the normal distribution.
2. Because we can use the normal distribution, we need to split the alpha into half because the normal distribution is symmetrical.

Example 11.4 Instead of checking if the salary was larger than $25,000, we can test to see if the mean is equal to $25,000 or not. The null and the alternative hypotheses would then be

$$\mathrm{H}_0 = \$25,000$$
$$\mathrm{H}_1 \neq \$25,000$$

This would represent a two-tailed test. The mathematical formula to calculate this would be

$$\frac{\bar{x} - \mu_0}{\sigma/\sqrt{n}} > Z_{\alpha/2} \quad \text{or} \quad \frac{\bar{x} - \mu_0}{\sigma/\sqrt{n}} < -Z_{\alpha/2}$$

The following shows how Minitab would test the hypothesis:

```
MTB > describe c1
```

11.4.1 Descriptive Statistics

```
Variable        N      Mean    Median    TrMean     StDev   SE Mean
C1             40     29240     27825     28877      6352      1004

Variable     Minimum   Maximum        Q1        Q3
C1             18000     50000     25000     32750

MTB > ztest 25000 6352 c1
```

11.4.2 Z Test

```
Test of mu = 25000 vs mu not = 25000
The assumed sigma = 6352

Variable     N      Mean    StDev   SE Mean       Z         P
C1          40     29240     6352      1004    4.22    0.0000


MTB > invcdf .975;
SUBC> normal 0 1.
```

11.4.3 Inverse Cumulative Distribution Function

```
Normal with mean = 0 and standard deviation = 1.00000

  P( X <= x)          x
    0.9750       1.9600
```

From the *ztest* command output, the calculated Z score for the data set is 4.22. We will reject the null hypothesis if the sample Z score is greater than the standard Z score of 1.96. Since this is the case, we can reject the null hypothesis that the mean salary is $25,000 and conclude that the mean salary is different from $25,000.

Using the *p*-value method, we can see that the calculated *p*-value is 0.0. Since this is less than the alpha value of 0.05, indicating that the *p*-value is in the alpha area, we would reject the null hypothesis and accept the alternative hypothesis.

Graphically we are seeing the following:

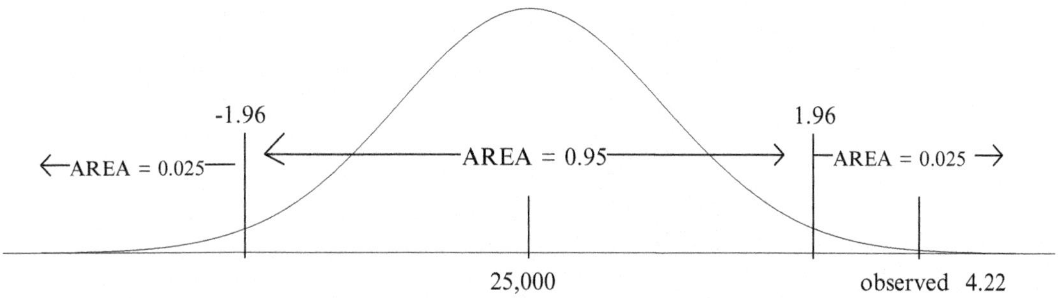

Example 11.5 C. S. Patterson's study of a sample of 47 large public electric utilities with revenues of $300 million or more according to Moody's Manual (*Financial Management*, Summer 1984). Patterson's study focused on the financing practices and policies of these regulated utilities. Compilation of the actual debt ratios, or long-term debt divided by total capital, of the companies yielded the following results:

$$\overline{X} = 0.485, \quad s_x = 0.29$$

Test the following two-tailed hypotheses:

$$H_{0:\mu} = .459$$
$$H_{1:\mu} \neq .459$$

Below shows the calculations done in Excel. Note that the p-value 3.96184E-10 is in scientific notation. This value is the same as $3.96184 * 10^{-10}$. From the calculation below, we would reject the null hypothesis H_0.

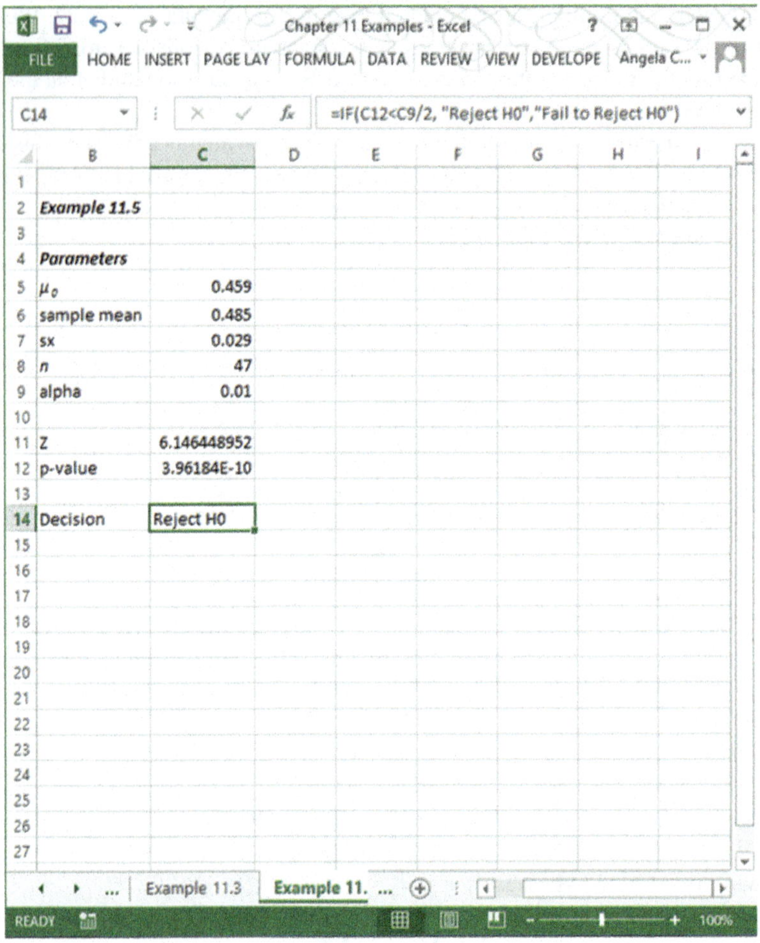

Note that in the above calculation for the two-tailed test, the format is the same as the one-tailed test. The only difference in the worksheet format is in the decision-making process made in cell C14. Note that in cell C14, the function divides cell C9, which contains the alpha value, by 2.

Below is the formula view of the above calculations.

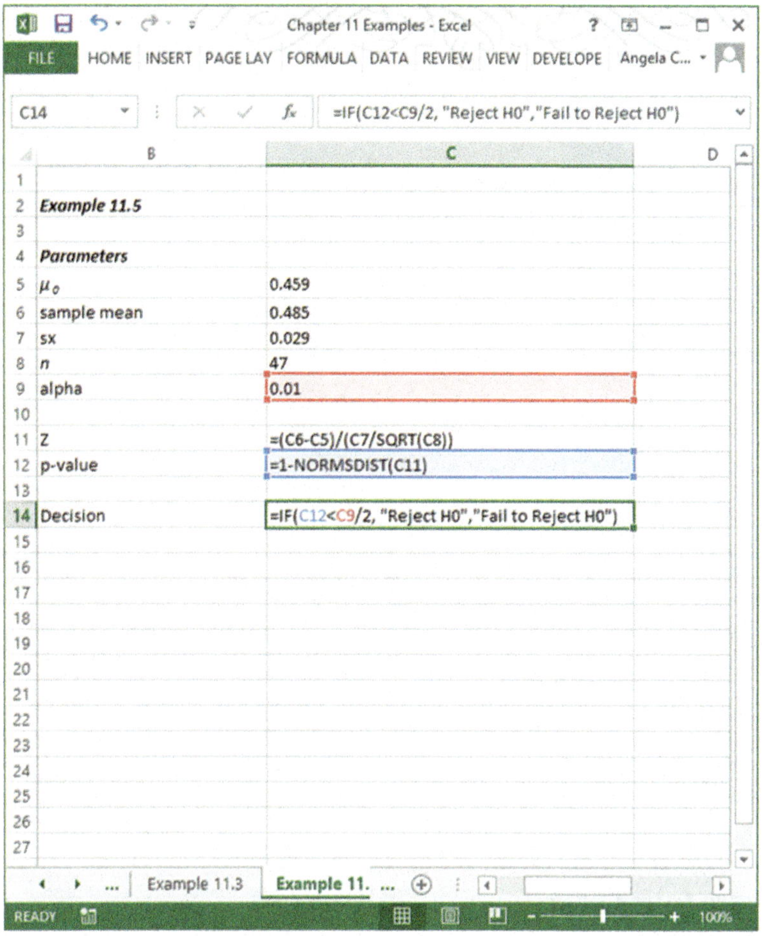

11.5 One-Tailed Tests of Mean for Small Samples

For one-tailed tests of mean for samples of less than 30, the *t*-test should be used. The *t*-test, which uses a *t* distribution, looks like the following:

$$\frac{\overline{x} - \mu_0}{s/\sqrt{n}} = t$$

We will use the *t*-test in the following example:

Example 11.6 United Van Lines Company is considering purchasing a large, new moving van. The sales agency has agreed to lease the truck to United Van Lines for 4 weeks (24 working days) on a trial basis. The main concern of United Van Lines is the miles per gallon (mpg) of gasoline that the van obtains on a typical moving day. The mpg values for the 24 trial days are

8.5	9.5	8.7	8.9	9.1	10.1	12	11.5	10.5	9.6
8.7	11.6	10.9	9.8	8.8	8.6	9.4	10.8	12.3	11.1
10.2	9.7	9.8	8.1						

They want to know if the average was greater than 9.5. Our hypothesis would then be

$$H_0 = 9.5$$
$$H_1 > 9.5$$

```
MTB > set c1
DATA> 8.5 9.5 8.7 8.9 9.1 10.1 12 11.5 10.5 9.6
DATA> 8.7 11.6 10.9 9.8 8.8 8.6 9.4 10.8 12.3 11.1
DATA> 10.2 9.7 9.8 8.1
DATA> end
MTB > ttest 9.5 c1;
SUBC> alternative 1.
```

11.5.1 One-Sample T: C1

```
Test of mu = 9.5 vs mu > 9.5

Variable            N       Mean     StDev   SE Mean
C1                 24      9.925     1.189     0.243

Variable        95.0% Lower Bound         T        P
C1                          9.509      1.75    0.047


MTB > invcdf .95;
SUBC> t 23.
```

11.5.2 Inverse Cumulative Distribution Function

```
Student's t distribution with 23 DF

P( X <= x )          x
    0.9500      1.7139
```

The critical value for 95 % was 1.7109. The calculated value for the sample was 1.75. Since the calculated value was larger than the critical value, we reject the null hypothesis and accept the alternative hypothesis.

Using the p-value method, we can see that the calculated p-value of 0.047 is less than the alpha value of 0.05. Therefore, we can conclude that there is enough evidence to reject the null hypothesis and accept the alternative hypothesis.

Graphically this situation would look like the following:

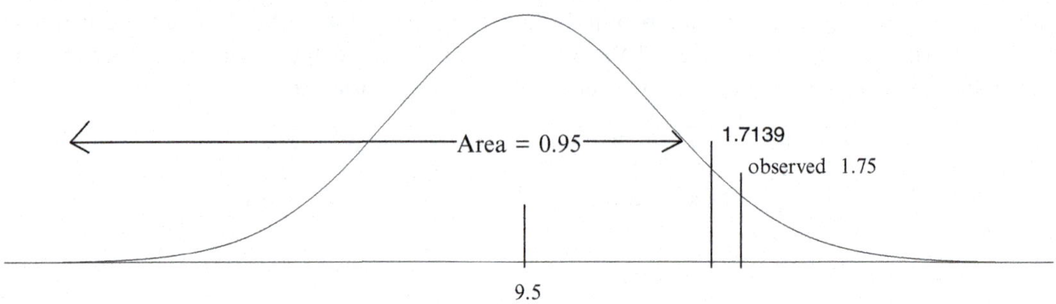

Below shows Example 11.6's calculations in Excel.

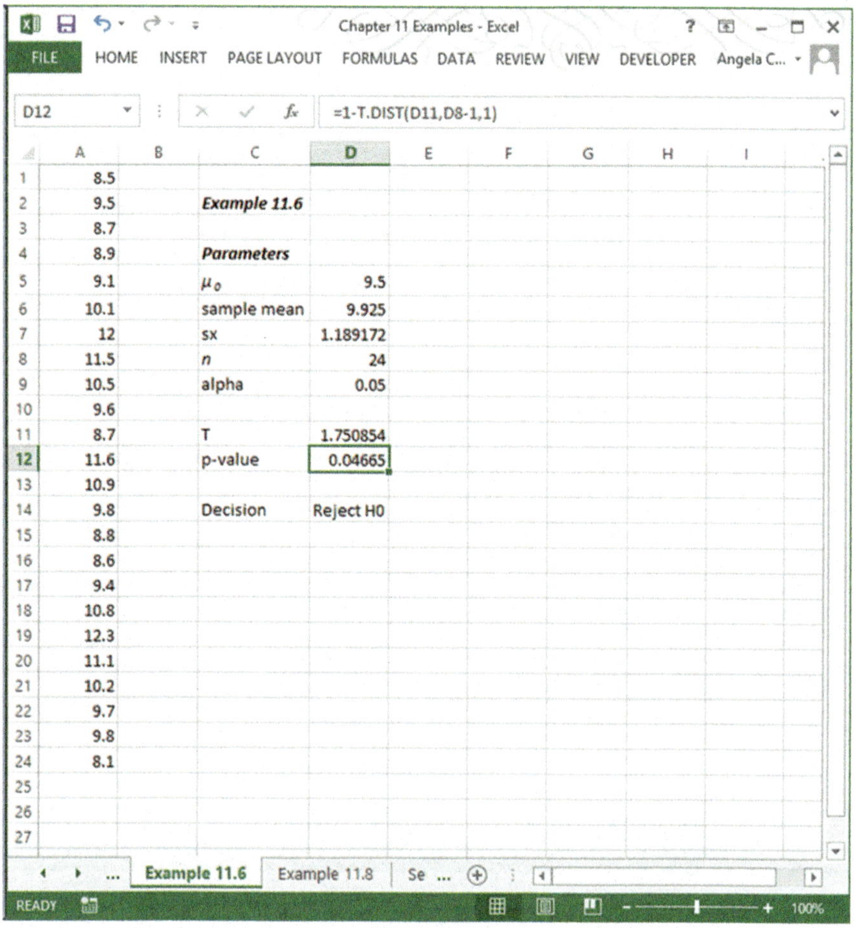

Below shows the formula view on how Example 11.6 is set up in Excel.

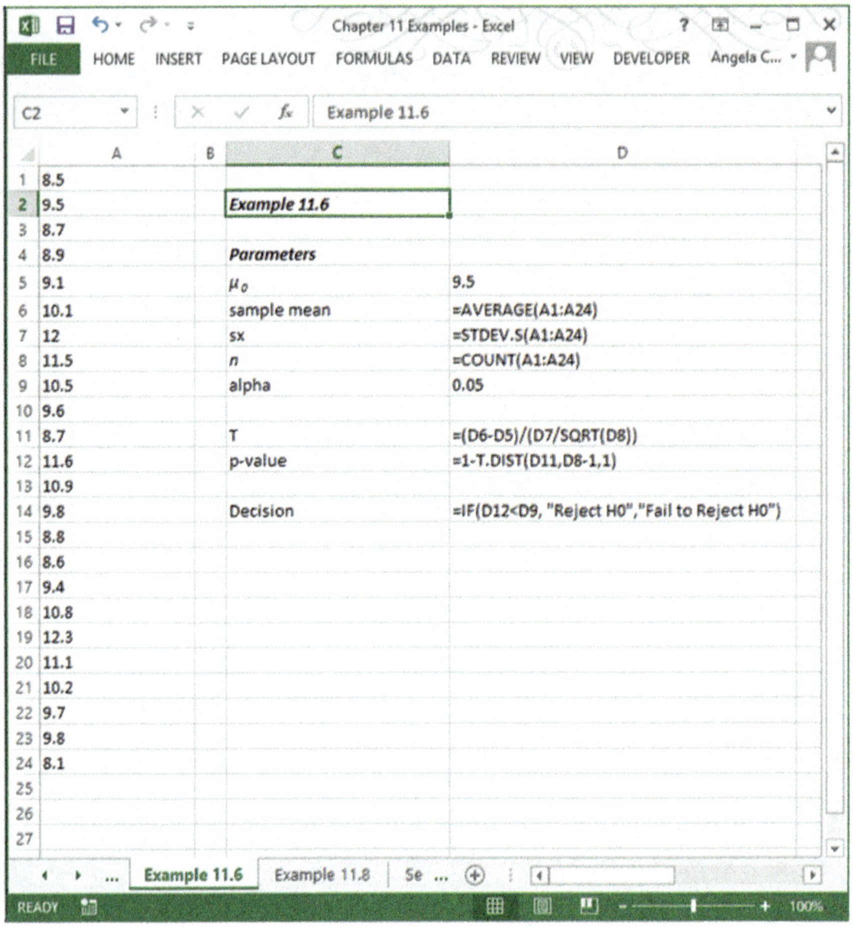

	A	B	C	D
1	8.5			
2	9.5		*Example 11.6*	
3	8.7			
4	8.9		***Parameters***	
5	9.1		μ_0	9.5
6	10.1		sample mean	=AVERAGE(A1:A24)
7	12		sx	=STDEV.S(A1:A24)
8	11.5		*n*	=COUNT(A1:A24)
9	10.5		alpha	0.05
10	9.6			
11	8.7		T	=(D6-D5)/(D7/SQRT(D8))
12	11.6		p-value	=1-T.DIST(D11,D8-1,1)
13	10.9			
14	9.8		Decision	=IF(D12<D9, "Reject H0","Fail to Reject H0")
15	8.8			
16	8.6			
17	9.4			
18	10.8			
19	12.3			
20	11.1			
21	10.2			
22	9.7			
23	9.8			
24	8.1			
25				
26				
27				

11.6 Difference of Two Means: Small Samples

We will test the difference between two means when the population variances are unknown and the samples are small.

To illustrate this concept, we will test to see if the average EPS for GM and Ford is equal or not.

Example 11.7 The hypotheses to be tested are

$$H_0 : \mu_{GM} - \mu_{FORD} = 0$$
$$H_1 : \mu_{GM} - \mu_{FORD} \neq 0$$

The data for the EPS for GM and Ford are shown below.

EPS FOR GM AND FORD

YEAR	GM	FORD
1970	2.09	4.77
1971	6.72	6.18
1972	7.51	8.52
1973	8.34	9.13
1974	3.27	3.86
1975	4.32	3.46
1976	10.08	10.45
1977	11.62	14.61
1978	12.24	13.35
1979	10.04	9.75
1980	-2.65	-12.83
1981	1.07	-8.81
1982	3.09	-5.46
1983	11.84	10.29
1984	14.22	15.79
1985	12.28	13.63
1986	8.22	12.32
1987	10.06	18.10
1988	13.64	10.96
1989	6.33	8.22
1990	-4.09	1.86

To do this analysis, we will put the GM data in column c1 in the Minitab worksheet and the Ford data in column c2. We will also name column c1 *gm* and column c2 *ford*.

When testing to see if the samples of two means are equal or not, it is helpful to graph the data. In this case we will dotplot the GM and Ford data, as shown below.

```
MTB > name c1 'GM'
MTB > name c2 'FORD'
MTB > set 'GM'
DATA> 2.09 6.72 7.51 8.34 3.27 4.32 10.08 11.62 12.24 10.04
DATA> -2.65 1.07 3.09 11.84 14.22 12.28 8.22 10.06 13.64 6.33 -4.09
DATA> end
MTB > set 'FORD'
DATA> 4.77 6.18 8.52 9.13 3.86 3.46 10.45 14.61 13.35 9.75
DATA> -12.83 -8.81 -5.46 10.29 15.79 13.63 12.32 18.10 10.96 8.22 1.86
DATA> END
MTB > dotplot c1 c2;
SUBC> same.
```

11.6.1 Dotplot

```
                                         .  .
                 .  .     .. : .   : .:  : .:  ..
 ---+---------+---------+---------+---------+---------+---GM

   .     .     .           .  : . .   :..:.  .... .   .
 ---+---------+---------+---------+---------+---------+---FORD
-12.0      -6.0       0.0       6.0      12.0      18.0
```

The *same* subcommand tells the *dotplot* command to use the same scale for both plots.

The resulting dotplots show that there is visually not much difference between the mean EPS of GM and that of Ford. Now let us use Minitab to mathematically calculate whether there is a significant difference.

```
MTB > twosample .95 c1 c2;
SUBC> pooled.
```

11.6.2 Two-Sample T-Test and Confidence Interval

```
Two sample T for GM vs FORD

          N       Mean      StDev   SE Mean
GM       21       7.15       5.19       1.1
FORD     21       7.05       7.99       1.7

95% CI for mu GM - mu FORD: ( -4.1,  4.3)
T-Test mu GM = mu FORD (vs not =): T = 0.05  P = 0.96  DF = 40
Both use Pooled StDev = 6.74
```

We use the *two-sample* command to see if the two means are equal or not. The first argument, ".95," of the *two-sample* command is the confidence level we want. The second argument, "c1," is the location of the data in the first sample. The third argument, "c2," is the location of the second data set. The *pooled* subcommand lets the *two-sample* command assume that the two populations have equal variances.

The calculated p-value is 0.96. So at the 0.05 alpha level, we cannot reject the hypothesis that the average EPS for GM and Ford are equal.

11.7 Hypothesis Testing for a Population Proportion

If the sample size is large, the Z test can be used to test the hypothesis for a population proportion. If the sample size is small, then the t-test should be used.

The test statistic for a two-tailed population proportion is

$$-z_{\alpha/2} > \frac{\hat{P} - P_0}{\sqrt{P_0(1 - P_0)/n}} \quad \text{or} \quad z_{\alpha/2} > \frac{\hat{P} - P_0}{\sqrt{P_0(1 - P_0)/n}}$$

Example 11.8 Francis Company is evaluating the promotability of its employees. That is, it is determining the proportion of employees whose ability, training, and supervisory experience qualify them for promotion to the next level of management. The human resources director of Francis Company tells the president that 80 % of the employees in the company are "promotable." However, a special committee appointed by the president finds that only 75 % of the 200 employees who have been interviewed are qualified for promotion. Use this information to do a two-tailed hypothesis test at $\alpha = 5\ \%$.

The null and alternative hypotheses would be

$$
\begin{aligned}
H_0 &: p = 0.80 \\
H_1 &: p \neq 0.80
\end{aligned}
$$

The calculations are shown below on the left. The formula view is shown on the right.

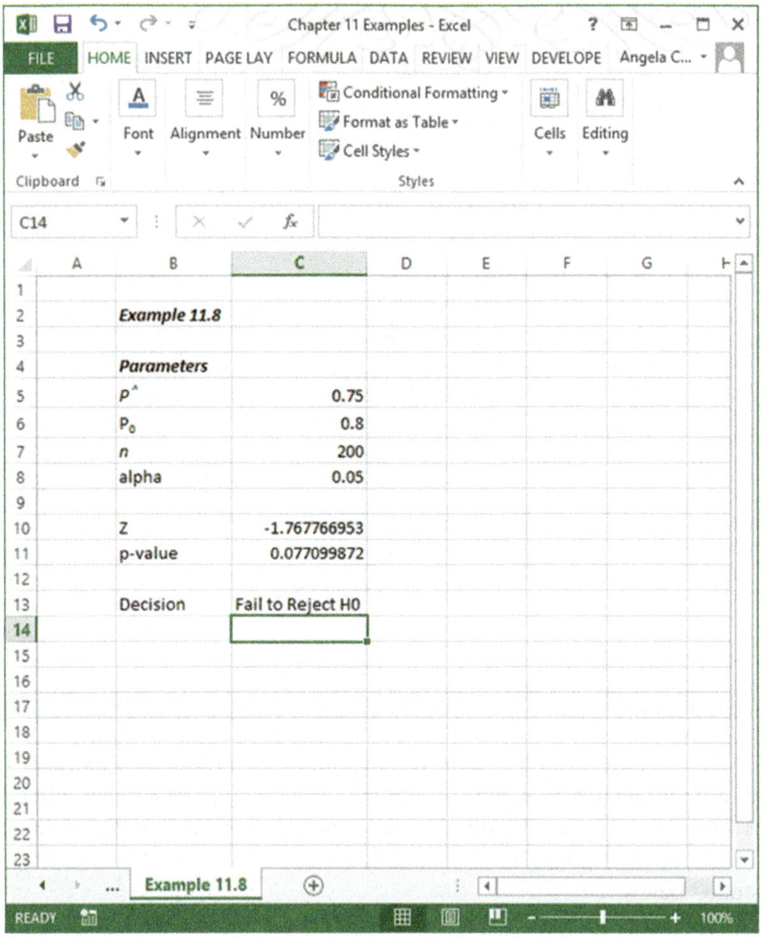

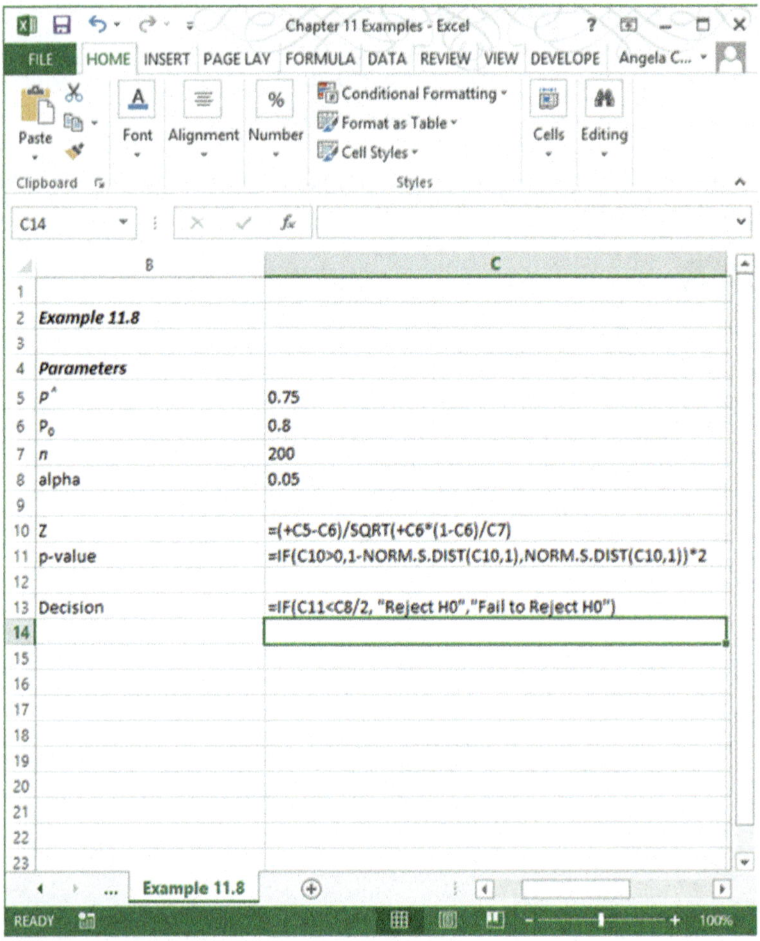

Since the sample proportion is less than the hypothesized proportion, the test statistic will be negative. Because it will be negative, we will use the lower calculated Z value. We cannot reject the null hypothesis because the lower Z value for the sample is greater than the Z value for the given alpha.

This is illustrated below.

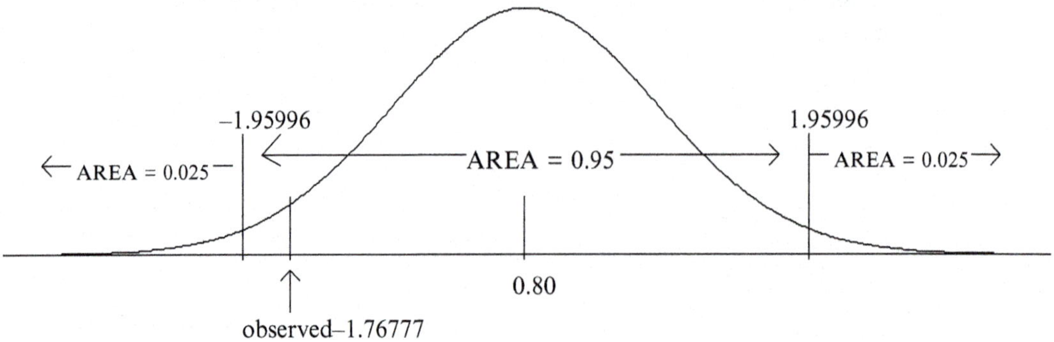

11.8 The Power of a Test and Power Function

The *power of a test* is the probability of rejecting H_0 when it is false. The probability is equal to $(1-\beta)$, where β denotes the probability of a type II error. Other things being equal, the greater the power of the test, the better the test.

Below are the power functions for a right-tailed, left-tailed, and two-tailed test.

Right-Tailed Power Function

$$Power = P\left(Z > \frac{\mu_o - \mu_1}{\sigma_x/\sqrt{n}} + z_a\right)$$

Left-Tailed Power Function

$$Power = P\left(Z > \frac{\mu_o - \mu_1}{\sigma_x/\sqrt{n}} - z_a\right)$$

Two-Tailed Power Function

$$Power = P\left(Z > \frac{\mu_o - \mu_1}{\sigma_x/\sqrt{n}} + z_{a/2}\right) + P\left(Z > \frac{\mu_o - \mu_1}{\sigma_x/\sqrt{n}} - z_{a/2}\right)$$

Testing the Average Weight of Cat Food per Bag

Say we want to test whether the average weight of 60-ounce bags of cat food is equal to, or smaller than, 60 ounces at significance level $\alpha = .05$. The null and alternative hypotheses can be stated as

$$H_{0\mu} = 60$$
$$H_{1\mu} > 60$$

In addition, suppose we know that the sample size $n = 100$, sample mean $\overline{X} = 59$, and standard deviation $s_x = 5$. Below shows the calculation for the power of μ_1 when $\mu_1 = 60, 60.5, 61, 61.5, 62$.

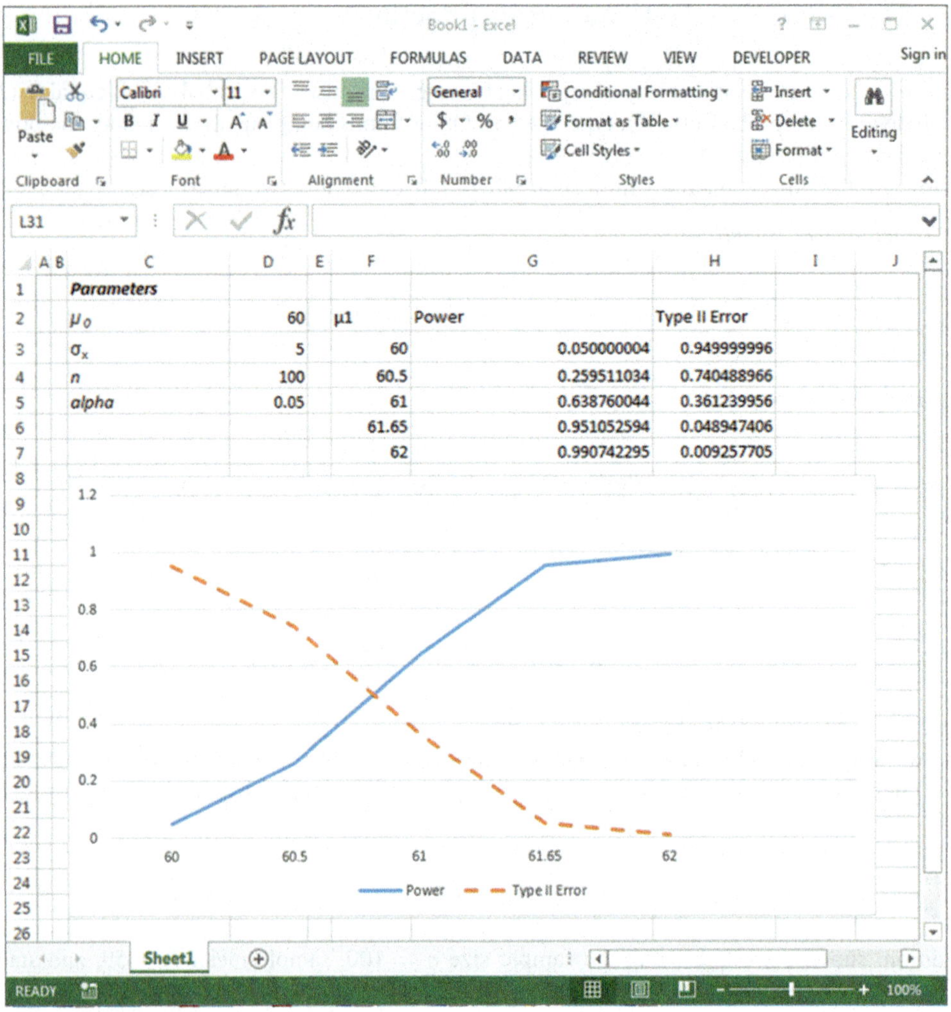

Pressing the *ctrl-'* key will show how the above worksheet is formatted. The format of the spreadsheet is shown below.

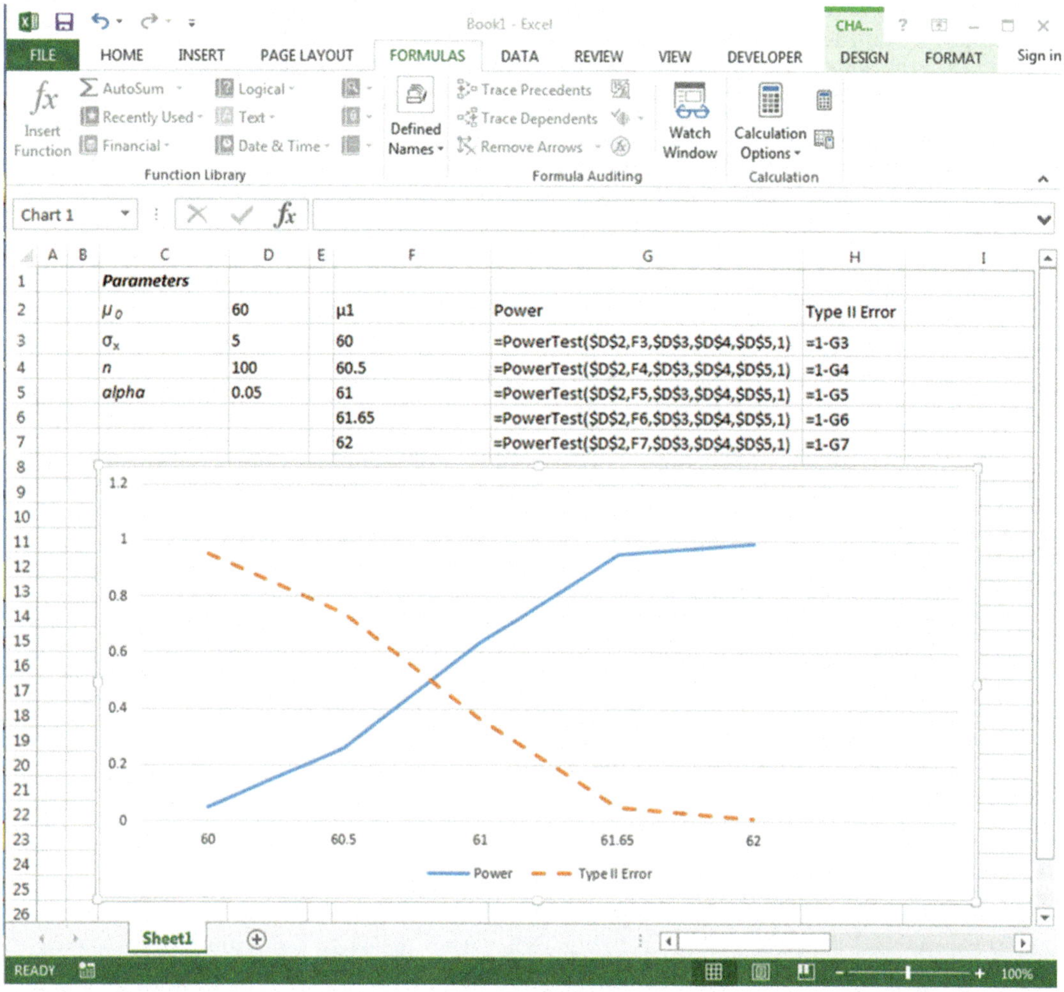

Notice above that there is a function called *PowerTest*. This function is a custom function. Custom functions can be programmed and used in a worksheet. The *PowerTest* function is shown below.

```
'/*****************************************************************
'/Purpose: Calculates the Power of as test. The PowerType argument
'/         indicates if the power test is for a right tail, left tail
'/         or a two tail test.
'/*****************************************************************
Function PowerTest(Mean0 As Double, Mean1 As Double, Stdev As Double, _
                    Size As Double, Alpha As Single, PowerType As Integer) As Double

    Dim dblRatio As Double
    Dim Z As Single

    If PowerType = 3 Then
        Z = WorksheetFunction.NormInv(1 - Alpha / 2, 0, 1)
    Else

        Z = WorksheetFunction.NormInv(1 - Alpha, 0, 1)
    End If

    dblRatio = (Mean0 - Mean1) / (Stdev / Sqr(Size))

    If PowerType = 1 Then
        Z = dblRatio + Z
    ElseIf PowerType = 2 Then
        Z = dblRatio - Z
    ElseIf PowerType = 3 Then
        Z = dblRatio + Z + dblRatio - Z
    End If

    If PowerType = 1 Then
        PowerTest = 1 - WorksheetFunction.NormDist(Z, 0, 1, True)
    ElseIf PowerType = 2 Then
        PowerTest = WorksheetFunction.NormDist(Z, 0, 1, True)
    ElseIf PowerType = 3 Then
        PowerTest = WorksheetFunction.NormDist(Z, 0, 1, True) * 2
    End If

End Function
```

11.9 Power and Sample Size

In this section we will demonstrate how the sample size affects the power of a test. Below shows how the sample size affects the power of the test. Below shows that as the sample size increases, the better the power test gets.

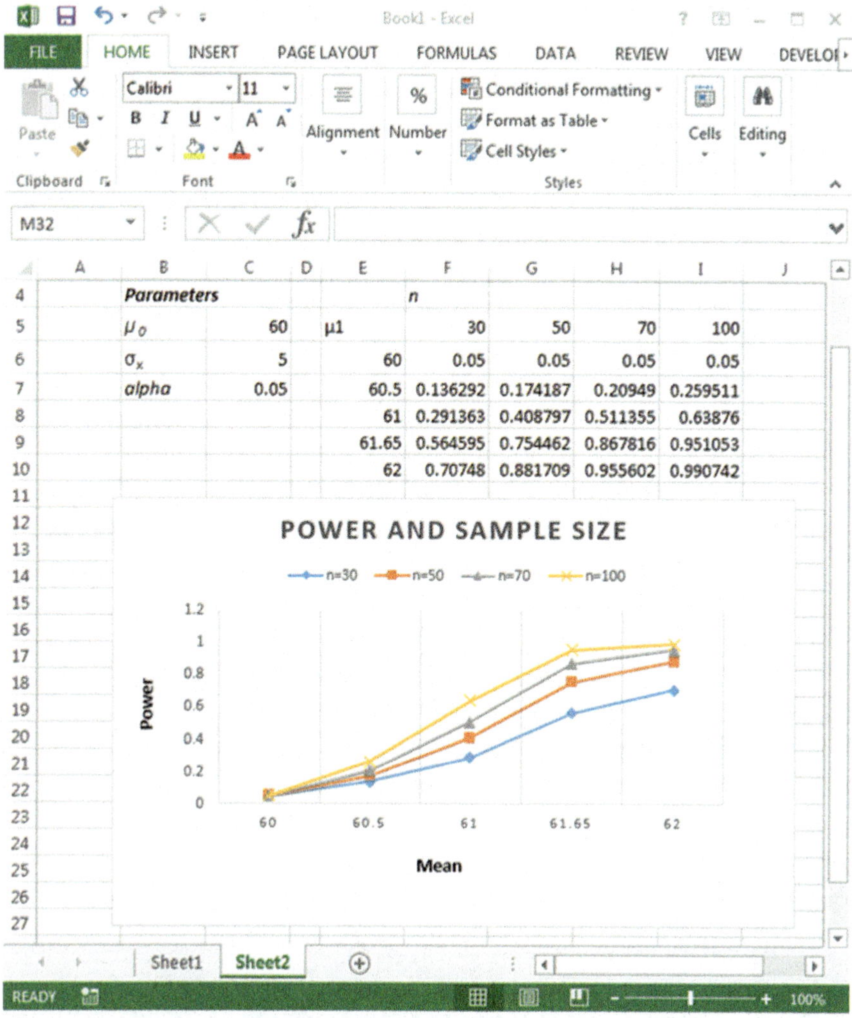

11.10 Power and Alpha Size

In this section, we will demonstrate how the sample size affects the power of a test. Below shows how the alpha size affects the power of the test. Below shows that as the alpha size increases, the better the power of the test.

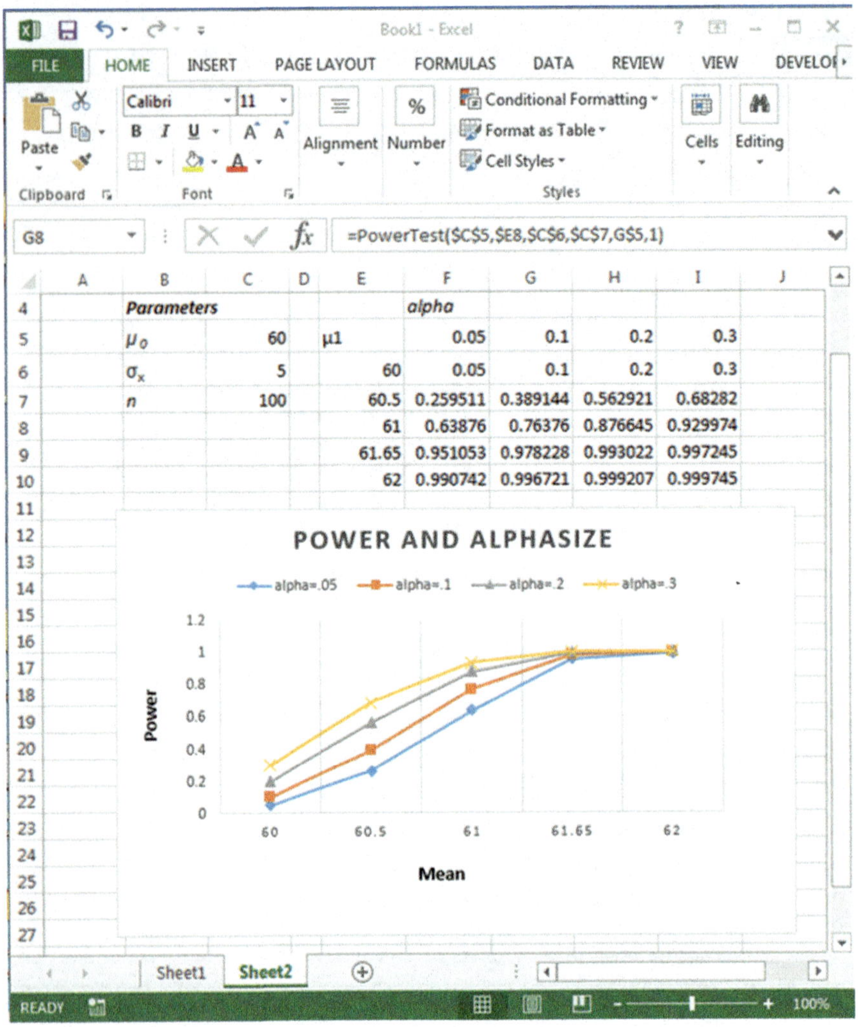

11.11 SAS Programming Code Instructions and Examples

11.11.1 One-Tailed Tests of Mean for Large Samples

Example 11.1 To do the same test, we should first put the salary data into SAS and then we will use the **PROC TTEST** command to perform the calculation. **SIDES=L** specifies lower one-sided tests (SIDES=U specifies upper one-sided tests).

```
Data ACCOUNTANTS;
    INPUT Salary @@;
    /*uses an @@ so SAS does not automatically go to a new line of raw data
for each observation*/
    DATALINES;
24000 20000 18000 26000 23750 32000 36789 25300
27500 35000 31000 25500 25000 50000 23750 30500
28000 29000 27000 27750 25000 41000 28900 32000
22000 26000 34000 26550 33000 27900 23750 31000
32000 21430 42014 36450 24750 34000 35000 27000
;
run;

proc ttest data=ACCOUNTANTS Sides=U alpha=0.05 H0=25000;
/*Tests if the mean score is larger than 15650*/;
var Salary;
run;
```

N	Mean	Std dev	Std err	Minimum	Maximum
40	29239.6	6351.9	1004.3	18000.0	50000.0

Mean	95 % CL	Mean	Std dev	95 % CL	Std dev
29239.6	27547.4	Infty	6351.9	5203.2	8156.1

DF	t value	Pr > t
39	4.22	<0.0001

Example 11.2 Based on the calculation shown below, we would reject the null hypothesis because for a left-tailed test, we reject the null hypothesis when the p-value is less than the alpha value.

```
data Bags;
H0=60;
Mean=59;
STD=5;
N=100;
alpha=0.05;
Z=(Mean-H0)/(STD/sqrt(100));
t=probit(1-alpha);
/*Function probit(p) computes the inverse of the standard normal cdf */
Pvalue=cdf('Normal',Z);
if Pvalue<alpha then
        Decision='Reject H0';
    else if Decision='Accept H0';
run;
```

Obs	H0	Mean	STD	N	Alpha	Z	t	P-value	Decision
1	60	59	5	100	0.05	−2	1.64485	0.022750	Reject H0

Example 11.3 Before calculation, we should first put the gasoline data into SAS. **PROC APPEND** is used to **add the observations from one SAS data set to the end of another SAS data set.**

```
data Gasoline;
   INPUT Cost @@;
   Brand="Texaco";
   DATALINES;
1.06      0.97  0.97  0.96  1.02  1.09
1.08      1.04  1.11  1.12  1.19  1.07
1.14      1.17  1.22  0.97  1.08
1.05      1.21  0.95  0.99  1.18
1.05      1.21  1.03  1.14  1.14
1.13      1.00  1.16  0.96  0.98
;
run;

data GasolineAdd;
INPUT Cost @@;
Brand="Shell";
DATALINES;
1.08      0.96  1.06  1.11  1.07
1.17      1.01  1.05  1.04  1.09
1.05      1.06  1.14  1.04  0.94
1.01      0.99  1.07  1.18  0.94
1.08      1.13  1.16  1.00  0.94
1.13      0.91  1.13  0.96  0.95
1.00      1.09  1.15  1.13
0.98      1.04  1.03  1.17
;
run;

proc append base=Gasoline  data=GasolineAdd;
run;
```

We will use the **PROC TTEST** command to perform the same calculation in SAS.

```
proc ttest data=Gasoline sides=L;
class brand;
var Cost;
run;
```

Brand	N	Mean	Std dev	Std err	Minimum	Maximum
Shell	38	1.0537	0.0754	0.0122	0.9100	1.1800
Texaco	32	1.0763	0.0846	0.0150	0.9500	1.2200
Diff (1–2)		−0.0226	0.0797	0.0191		

Brand	Method	Mean	95 % CL mean		Std dev	95 % CL Std dev	
Shell		1.0537	1.0289	1.0785	0.0754	0.0614	0.0975
Texaco		1.0763	1.0457	1.1068	0.0846	0.0678	0.1125
Diff (1–2)	Pooled	−0.0226	−Infty	0.00933	0.0797	0.0683	0.0958
Diff (1–2)	Satterthwaite	−0.0226	−Infty	0.00968			

Method	Variances	DF	t value	Pr < t
Pooled	Equal	68	−1.18	0.1211
Satterthwaite	Unequal	62.774	−1.17	0.1236

Equality of variances				
Method	Num DF	Den DF	F value	Pr > F
Folded F	31	37	1.26	0.4964

Example 11.4

```
proc ttest data=ACCOUNTANTS Sides=U alpha=0.05 H0=25000;
/*Tests if the mean score is 15650*/;
var Salary;
run;
```

N	Mean	Std dev	Std err	Minimum	Maximum
40	29239.6	6351.9	1004.3	18000.0	50000.0

Mean	95 % CL Mean		Std dev	95 % CL Std dev	
29239.6	27547.4	Infty	6351.9	5203.2	8156.1

DF	t value	Pr > t
39	4.22	< 0.0001

Example 11.5

```
data Utilites;
H0=0.459;
Mean=0.485;
STD=0.029;
N=47;
alpha=0.01;
Z=(Mean-H0)/(STD/sqrt(N));
t=probit(1-alpha/2);
Pvalue=1-cdf('Normal',Z);
if Pvalue<(alpha/2) or Pvalue>(1-alpha/2)  then
      Decision='Reject H0';
   else Decision='Accept H0';
run;
```

Obs	H0	Mean	STD	N	Alpha	Z	t	P-value	Decision
1	0.459	0.485	0.029	47	0.01	6.14645	2.57583	3.9618E-10	Reject H0

Example 11.6 Below shows Example 11.6's calculations in SAS.

```
Data Trials;
   INPUT MPG @@;
   DATALINES;
8.5 9.5 8.7 8.9 9.1 10.1 12 11.5 10.5 9.6
8.7 11.6 10.9 9.8 8.8 8.6 9.4 10.8 12.3 11.1
10.2 9.7 9.8 8.1
;
run;

proc ttest data=Trials Sides=U alpha=0.05 H0=9.5;
var MPG;
run;
```

N	Mean	Std dev	Std err	Minimum	Maximum
24	9.9250	1.1892	0.2427	8.1000	12.3000

Mean	95 % CL Mean		Std dev	95 % CL Std dev	
9.9250	9.5090	Infty	1.1892	0.9242	1.6681

DF	t value	Pr > t
23	1.75	0.0466

Example 11.7 The hypotheses to be tested are

$$H_0 : \mu_{JNJ} - \mu_{MRK} = 0$$
$$H_1 : \mu_{JNJ} - \mu_{MRK} \neq 0$$

The data for the EPS for JNJ and MRK are shown below.

EPS for JNJ and MRK		
Year	JNJ	MRK
1980	6.500	5.540
1981	2.510	5.360
1982	2.790	5.610
1983	2.570	6.100
1984	2.750	6.710
1985	3.360	7.580
1986	1.850	4.850
1987	4.830	6.680
1988	5.720	3.050
1989	3.250	3.780
1990	3.430	4.560
1991	4.390	5.490
1992	2.460	2.120
1993	2.740	1.870

(continued)

EPS for JNJ and MRK		
Year	JNJ	MRK
1994	3.120	2.380
1995	3.720	2.700
1996	2.170	3.200
1997	2.470	3.830
1998	2.270	4.410
1999	3.000	2.510
2000	3.450	2.960
2001	1.870	3.180
2002	2.200	3.170
2003	2.420	2.950
2004	2.870	2.620
2005	3.500	2.110
2006	3.760	2.040
2007	3.670	1.510
2008	4.620	3.660
2009	4.450	5.670
2010	4.850	0.280
2011	3.540	2.040

Before calculation, we should first put the EPS data into SAS and change it in the following form.

Obs	Year	EPS	Company
1	1980	6.50	JNJ
2	1981	2.51	JNJ
3	1982	2.79	JNJ
4	1983	2.57	JNJ
5	1984	2.75	JNJ
6	1985	3.36	JNJ
7	1986	1.85	JNJ
8	1987	4.83	JNJ
9	1988	5.72	JNJ
10	1989	3.25	JNJ
11	1990	3.43	JNJ
12	1991	4.39	JNJ
13	1992	2.46	JNJ
14	1993	2.74	JNJ
15	1994	3.12	JNJ
16	1995	3.72	JNJ
17	1996	2.17	JNJ
18	1997	2.47	JNJ
19	1998	2.27	JNJ
20	1999	3.00	JNJ
21	2000	3.45	JNJ
22	2001	1.87	JNJ
23	2002	2.20	JNJ
24	2003	2.42	JNJ
25	2004	2.87	JNJ
26	2005	3.50	JNJ
27	2006	3.76	JNJ
28	2007	3.67	JNJ
29	2008	4.62	JNJ

(continued)

Obs	Year	EPS	Company
30	2009	4.45	JNJ
31	2010	4.85	JNJ
32	2011	3.54	JNJ
33	1980	5.54	MRK
34	1981	5.36	MRK
35	1982	5.61	MRK
36	1983	6.10	MRK
37	1984	6.71	MRK
38	1985	7.58	MRK
39	1986	4.85	MRK
40	1987	6.68	MRK
41	1988	3.05	MRK
42	1989	3.78	MRK
43	1990	4.56	MRK
44	1991	5.49	MRK
45	1992	2.12	MRK
46	1993	1.87	MRK
47	1994	2.38	MRK
48	1995	2.70	MRK
49	1996	3.20	MRK
50	1997	3.83	MRK
51	1998	4.41	MRK
52	1999	2.51	MRK
53	2000	2.96	MRK
54	2001	3.18	MRK
55	2002	3.17	MRK
56	2003	2.95	MRK
57	2004	2.62	MRK
58	2005	2.11	MRK
59	2006	2.04	MRK
60	2007	1.51	MRK
61	2008	3.66	MRK
62	2009	5.67	MRK
63	2010	0.28	MRK
64	2011	2.04	MRK

```
Data EPS;
   INPUT Year JNJ MRK;
   DATALINES;
1980        6.500  5.540
1981        2.510  5.360
1982        2.790  5.610
1983        2.570  6.100
1984        2.750  6.710
1985        3.360  7.580
1986        1.850  4.850
1987        4.830  6.680
1988        5.720  3.050
1989        3.250  3.780
1990        3.430  4.560
1991        4.390  5.490
1992        2.460  2.120
1993        2.740  1.870
1994        3.120  2.380
1995        3.720  2.700
1996        2.170  3.200
1997        2.470  3.830
1998        2.270  4.410
1999        3.000  2.510
2000        3.450  2.960
2001        1.870  3.180
2002        2.200  3.170
2003        2.420  2.950
2004        2.870  2.620
2005        3.500  2.110
2006        3.760  2.040
2007        3.670  1.510
2008        4.620  3.660
2009        4.450  5.670
2010        4.850  0.280
2011        3.540  2.040
;
run;

data JNJ(rename=(JNJ=EPS));
set EPS;
Company='JNJ';
drop MRK;
run;

data MRK(rename=(MRK=EPS));
set EPS;
Company='MRK';
drop JNJ;
run;
```

```
proc append base=JNJ data=MRK;
run;

proc ttest data=JNJ sides=2;
class Company;
var EPS;
run;
```

Company	N	Mean	Std dev	Std err	Minimum	Maximum
JNJ	32	3.3469	1.1056	0.1954	1.8500	6.5000
MRK	32	3.7663	1.7608	0.3113	0.2800	7.5800
Diff (1–2)		−0.4194	1.4702	0.3675		

Company	Method	Mean	95 % CL Mean		Std dev	95 % CL Std dev	
JNJ		3.3469	2.9483	3.7455	1.1056	0.8864	1.4699
MRK		3.7663	3.1314	4.4011	1.7608	1.4117	2.3410
Diff (1–2)	Pooled	−0.4194	−1.1541	0.3153	1.4702	1.2508	1.7836
Diff (1–2)	Satterthwaite	−0.4194	−1.1569	0.3181			

Method	Variances	DF	t value	Pr > |t|
Pooled	Equal	62	−1.14	0.2583
Satterthwaite	Unequal	52.156	−1.14	0.2591

Equality of variances				
Method	Num DF	Den DF	F value	Pr > F
Folded F	31	31	2.54	0.0115

Example 11.8 The calculation is shown below by using SAS.

```
data Employees;
P=0.75;
P0=0.8;
N=200;
alpha=0.05;
Z=(P-P0)/sqrt(P0*(1-P0)/N);
if Z>0 then
      Pvalue=1-cdf('Normal',Z);
     else Pvalue=cdf('Normal',Z)*2;
if Pvalue<(alpha/2) then
      Decision='Reject H0';
      else Decision='Accept H0';
run;
```

Obs	P	P0	N	Alpha	Z	P-value	Decision
1	0.75	0.8	200	0.05	−1.76777	0.077100	Accept H0

Testing the Average Weight of Cat Food per Bag

Use the **ONESAMPLEMEANS** statement in the **PROC POWER** to compute the power. Indicate power as the result parameter by specifying the **POWER=** option with a missing value (.). Specify your conjectures for the mean and standard deviation by using the **MEAN=** and **STDDEV=** options and for the sample size by using the **NTOTAL=** option. The statements required to perform this analysis are as follows:

```
proc power;
   onesamplemeans SIDES=L
      mean   = 60
      ntotal = 100
      stddev = 5
     NULLMEAN=60 60.5 61 61.65 62
      power  = .;
run;
```

Fixed scenario elements	
Distribution	Normal
Method	Exact
Number of sides	L
Mean	60
Standard deviation	5
Total sample size	100
Alpha	0.05

Computed power

Index	Null mean	Power
1	60.0	0.050
2	60.5	0.257
3	61.0	0.634
4	61.7	0.949
5	62.0	0.990

11.12 Statistical Summary

In this chapter we made inferences about a population parameter by checking if there is enough evidence to reject the null hypothesis. A type I mistake is rejecting the null hypothesis when in fact it is true. A type II mistake is accepting the null hypothesis when in fact it is false.

Bibliography

Microsoft Inc., Excel 2013. Microsoft Inc., Redmond

Lee CF, Lee JC, Lee AC (2013) Statistics for business and financial economics. Springer, New York

Minitab Inc. Minitab 17. Minitab Inc., State College

SAS Institute Inc. SAS 2014. SAS Institute, Cary

Chapter 12
Analysis of Variance and Chi-Square Tests

12.1 Introduction

Often in statistics we are interested in whether two or more samples have the same mean from the same population. Statistics uses the concept of analysis of variance to answer this question.

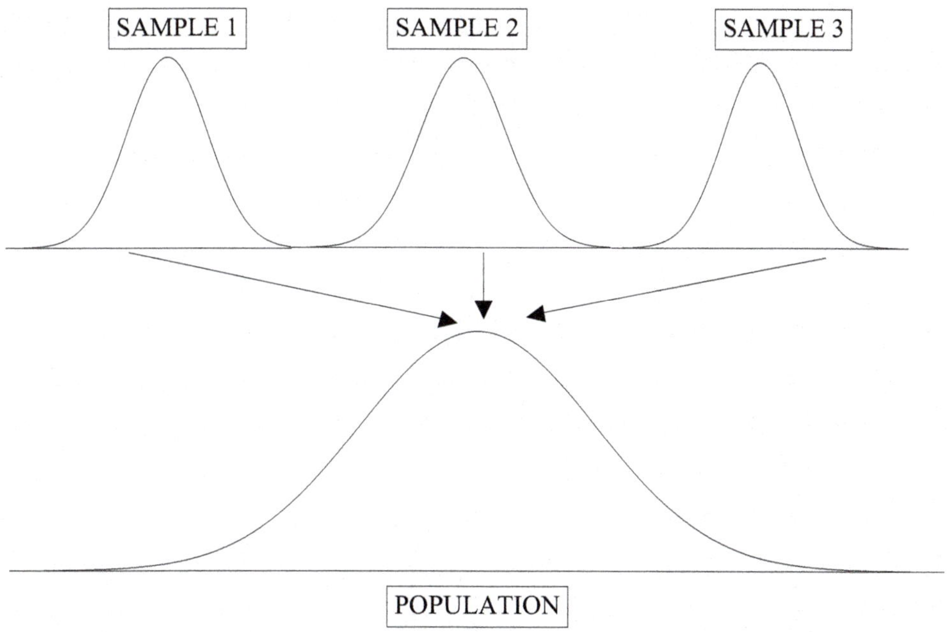

The first thing we must find out is whether the group of samples comes from the same population as shown in the above diagram or whether the group of samples does not come from the same population, as shown below.

C.-F. Lee et al., *Essentials of Excel, Excel VBA, SAS and Minitab for Statistical and Financial Analyses*, DOI 10.1007/978-3-319-38867-0_12

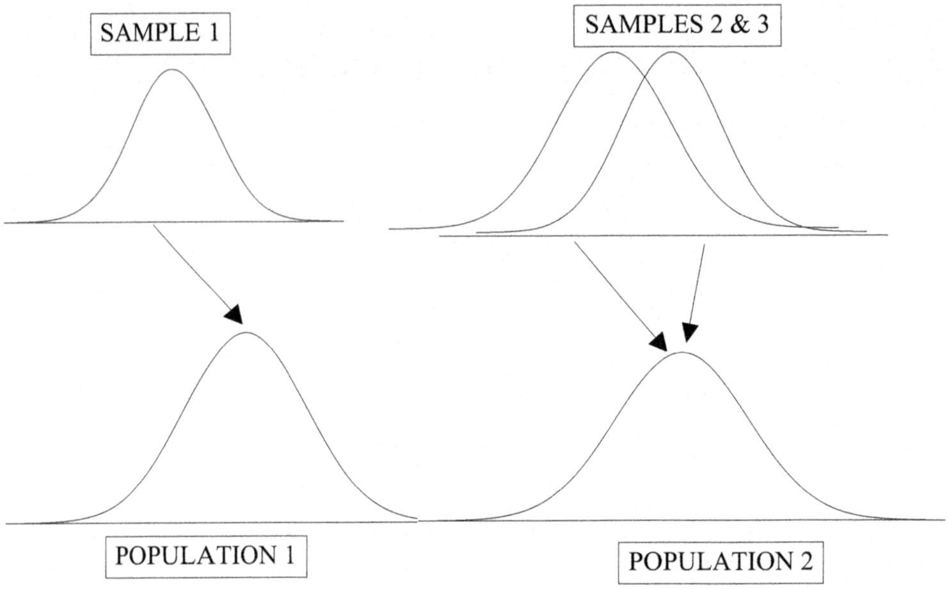

Analysis of variance, which is based on the concept of variation among samples, can be separated into two components, between-group variation and within-group variation. This is shown below.

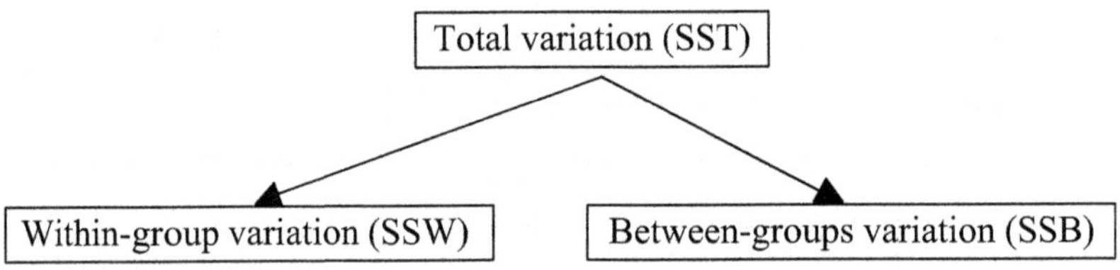

Total variation is the variation of all the samples about the overall mean of the samples.

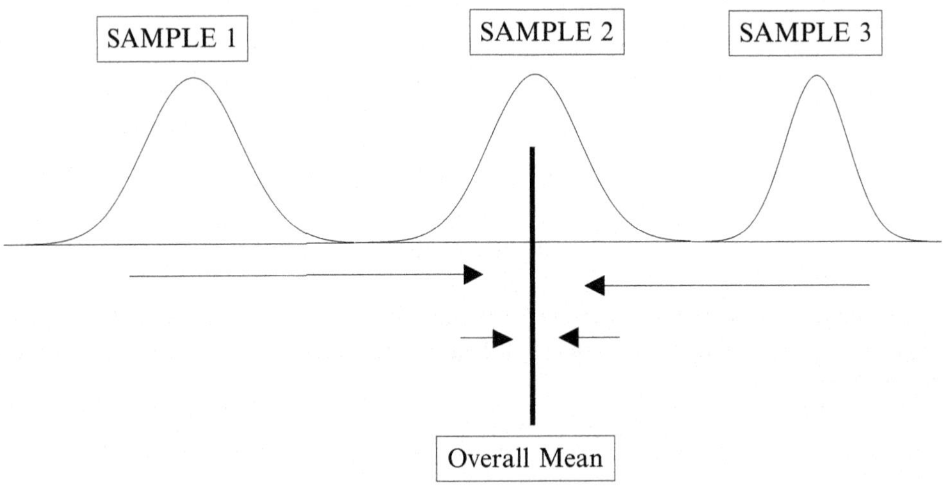

Within-group variation is variation about the mean of each individual sample.

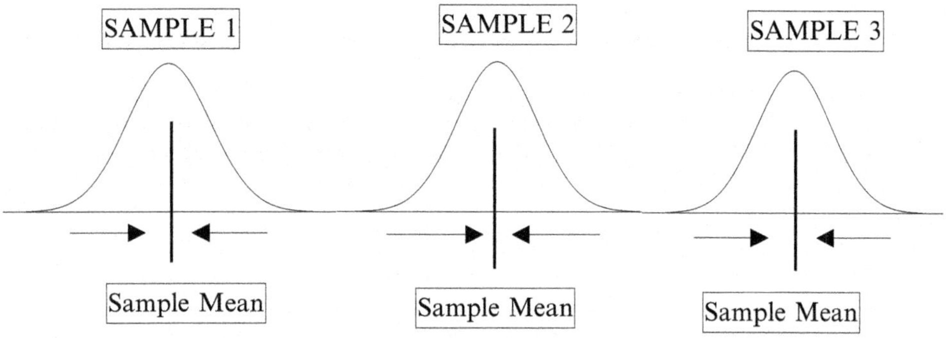

Between-group variation is variation of group means about the overall mean.

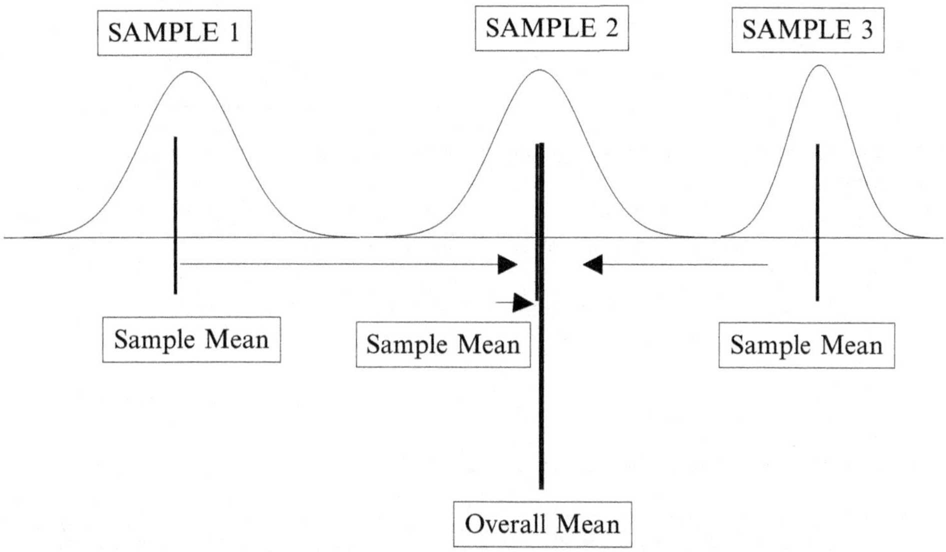

If the between-group variation is significantly larger than the within-group variation, we can conclude that all the samples did not come from the same population. If the between-group variation is not significantly larger, we cannot conclude that all the samples did not come from the same population.

Therefore, based on the above concepts, the test statistic for the analysis of variance is

$$\frac{\text{between-group variation}}{\text{within-group variation}} = \frac{\text{between-group mean square}}{\text{within-group mean square}}$$

The above ratio follows an F distribution.

In the analysis of variance, the general null and alternative hypotheses are

H_0: all the samples have the same mean

H_1: not all the samples have equal means.

12.2 One-Way Analysis of Variance

One-way analysis of variance studies the mean of two or more samples based on one classification.

Example 12.1 A research institution says that gasoline is gasoline. There is no difference among different brands of gasoline in terms of mileage per gallon. An independent consumer rights organization did an experiment on three different brands of gasoline. Twenty cars of the same condition were divided into three groups, and each group was filled with a different brand of gasoline. The test results show:

Gasoline		
A	B	C
34	29	32
28	32	34
29	31	30
37	43	42
42	31	32
27	29	33
29	28	

Is there any difference among the three brands of gasoline? Is there any significant difference at the 5 % alpha level?

We will use the analysis of variance to see if there are any significant differences among the means. The null hypothesis and the alternative hypothesis would be

H_0: gasoline mean A = gasoline mean B = gasoline mean C

H_1: not all three gasoline means are equal

To do this analysis, let us put the data for gasoline A in column c1, gasoline B in column c2, and gasoline C in column c3.

When finished the worksheet should look like the following.

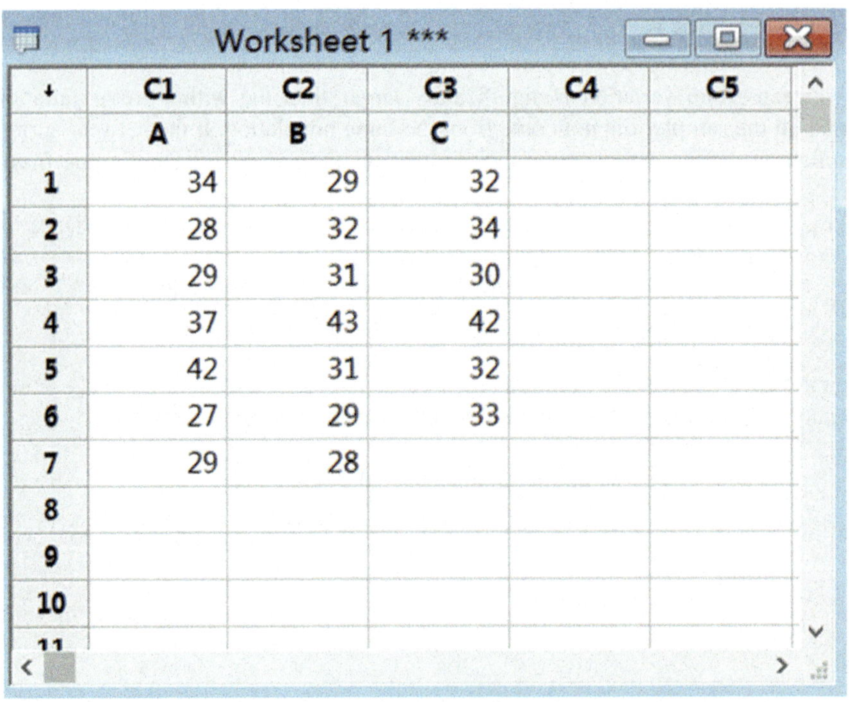

Let us first have a look at the dotplot of the three gasolines to get an overall picture of the data.

```
MTB > dotplot c1-c3;
SUBC> start 27 43.
```

12.2.1 Dotplot

```
       .  .   :                .         .              .
       +---------+---------+---------+---------+---------+-----A

          .   :     :    .                                 .
       +---------+---------+---------+---------+---------+-----B

                    .      :   .  .                     .
       +---------+---------+---------+---------+---------+-----C
     27.0      30.0      33.0      36.0      39.0      42.0
```

From the dotplots we can see that the three gasolines get similar mileage. This suggests to us that the three gasoline brands might even have the same mean. We will test this by doing a one-way analysis of variance test. We will use the *aovoneway* Minitab command to calculate the one-way analysis of variance.

```
MTB > aovoneway c1-c3
```

12.2.2 One-Way Analysis of Variance

```
Analysis of Variance
Source      DF        SS        MS        F        P
Factor       2      13.7       6.8     0.27    0.768
Error       17     433.1      25.5
Total       19     446.8
                                   Individual 95% CIs For Mean
                                   Based on Pooled StDev
Level        N      Mean     StDev --------+---------+---------+--------
A            7    32.286     5.589    (-------------*------------)
B            7    31.857     5.113   (------------*-------------)
C            6    33.833     4.215        (--------------*-------------)
                                   --------+---------+---------+--------
Pooled StDev =     5.048                 30.0      33.0      36.0
```

From the Minitab output, we can see that the p-value for the one-way analysis is 0.768. Since this p-value is greater than an alpha value of 0.05, we cannot reject the null hypothesis that the three brands of gasoline have the same mean.

We will now do the one-way analysis of variance in Excel.

Enter the data in Excel as shown below.

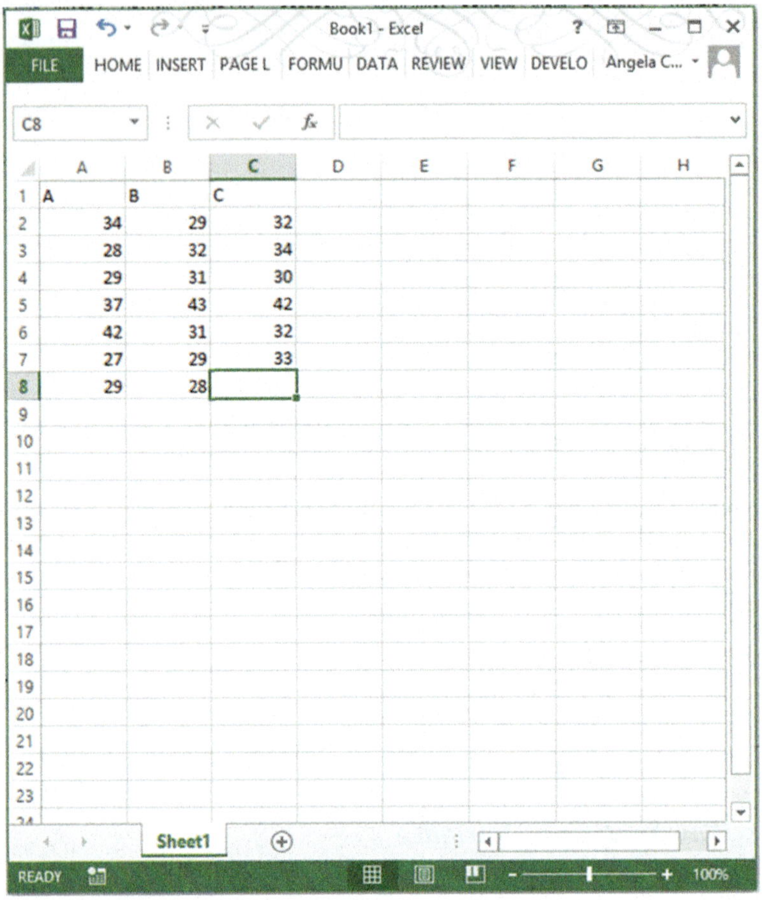

To do the one-way analysis of variance, we will use the *Analysis Tools Pack* add-in.

The Analysis Tools Pack will add *Data Analysis* menu item under the *DATA* menu. The *Data Analysis* menu item brings up the following dialog box.

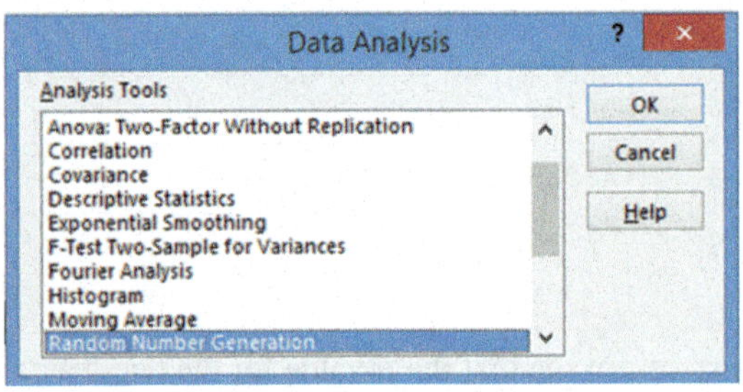

We will choose the *ANOVA*: *Single Factor* item to do the one-way analysis of variance. Fill out the ANOVA: Single Factor dialog box as shown below.

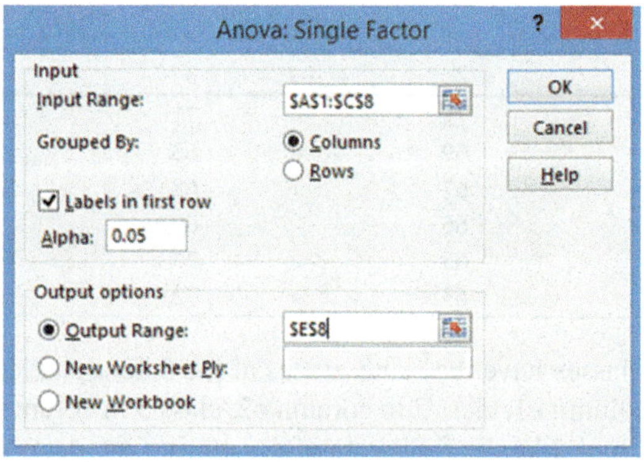

The Excel worksheet should look like the following after the analysis.

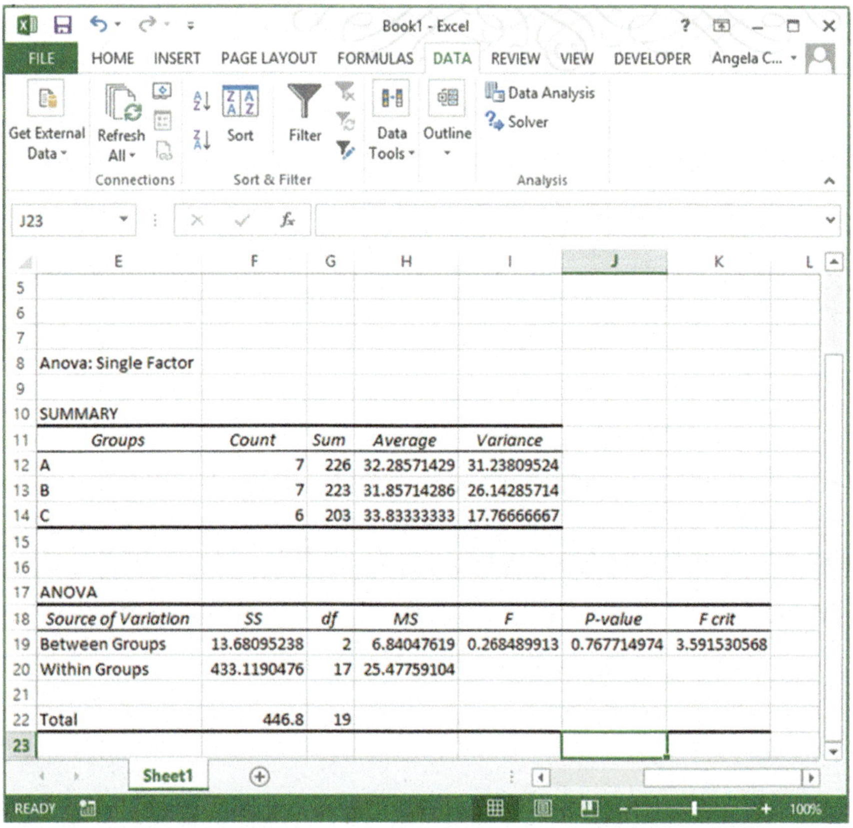

Example 12.2 There are many books to help people learn computer software packages. An instructor checked these books and found that they are all of similar quality. He picked four books and used a different book in each of his four classes. If the students all have the same average grades, he will use the cheapest book. The test results for students using the four books are:

Classes			
1	2	3	4
43	77	72	72
45	72	73	74

(continued)

Classes			
1	2	3	4
67	75	71	75
68	69	65	65
73	67	68	66
72	66	69	68
55	65	73	74
62	63	72	81

Test to see if the four classes have the same grades at the 5 % alpha level.

We will put class 1 in column c1, class 2 in column c2, class 3 in column c3, and class 4 in column c4. The worksheet should look like the following after entering the data.

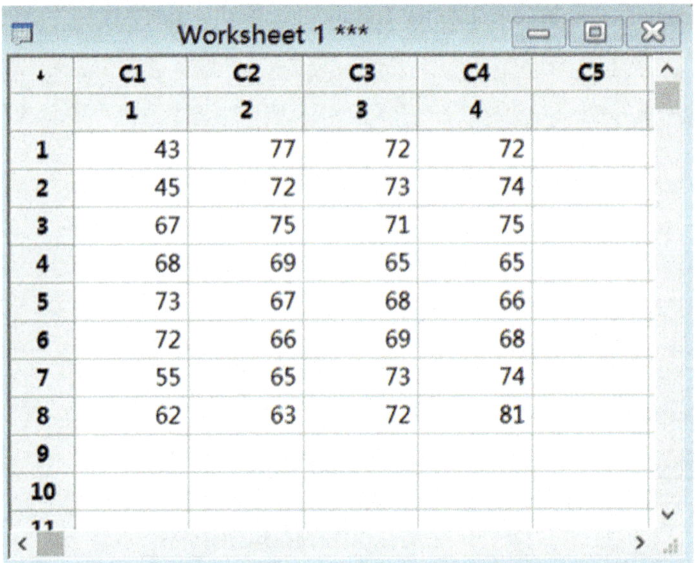

Worksheet 1 ***

↓	C1	C2	C3	C4	C5
	1	2	3	4	
1	43	77	72	72	
2	45	72	73	74	
3	67	75	71	75	
4	68	69	65	65	
5	73	67	68	66	
6	72	66	69	68	
7	55	65	73	74	
8	62	63	72	81	
9					
10					

Let us look at a dotplot of the four classes.

```
MTB > dotplot c1-c4;
SUBC> start 41 81.
```

12.2.3 Dotplot

```
       . .            .        .     ..    ..
   +---------+---------+---------+---------+---------+---1

                                . . .. .   .   . .
   +---------+---------+---------+---------+---------+---2

                                  .   ..  .::
   +---------+---------+---------+---------+---------+---3

                                  . . .    .  :.      .
   +---------+---------+---------+---------+---------+---4
40.0      48.0      56.0      64.0      72.0      80.0
```

From the dotplot of the four classes, we can see that the plots are somewhat similar. This suggests to us that the four classes might have the same average grade. We will test this by doing a one-way analysis of variance calculation.

Our null and alternative hypotheses would be

H_0: class 1 grade average = class 2 grade average = class 3 grade average = class 4 grade average

H_1: not all class averages are equal

```
MTB > aovoneway c1-c4
```

12.2.4 One-Way Analysis of Variance

```
Analysis of Variance
Source      DF        SS        MS        F        P
Factor       3     612.8     204.3     4.10    0.016
Error       28    1394.1      49.8
Total       31    2007.0
                                     Individual 95% CIs For Mean
                                     Based on Pooled StDev
Level        N      Mean     StDev   --------+---------+---------+-------
-
1            8    60.625    11.747    (-------*--------)
2            8    69.250     4.979                 (-------*--------)
3            8    70.375     2.825                   (-------*--------)
4            8    71.875     5.330                (--------*-------
)
                                     --------+---------+---------+-------
-
Pooled StDev =    7.056                  60.0      66.0      72.0
```

From the Minitab output, we can see that the calculated p-value is 0.016. Therefore, since the p-value is less than the alpha value of 0.05, we reject the null hypothesis that every class has the same average grade.

The Minitab output also shows the calculated mean grade for each class and the 95 % confidence interval of mean grade for each class. From the diagram we conclude that the class with a statistically different mean grade is class 1. This conclusion stems from a graphical approach.

There are also numerical approaches. The advantage of numerical approaches is that they are more precise. We will now look at numerical approaches to test which class has statistically different grades.

To do this we will have to use the *oneway* Minitab command. For the *oneway* command, we are going to put all the data in column c1 and indicate the different classes in column c2.

After entering the class data, the worksheet should look like the following.

Worksheet 1 ***

↓	C1	C2	C3	C4
1	43	1		
2	45	1		
3	67	1		
4	68	1		
5	73	1		
6	72	1		
7	55	1		
8	62	1		
9	77	2		
10	72	2		
11	75	2		
12	69	2		
13	67	2		
14	66	2		
15	65	2		
16	63	2		

Worksheet 1 ***

↓	C1	C2	C3	C4
1	72	1		
2	73	1		
3	71	1		
4	65	1		
5	68	1		
6	69	1		
7	73	1		
8	72	1		
9	72	2		
10	74	2		
11	75	2		
12	65	2		
13	66	2		
14	68	2		
15	74	2		
16	81	2		

Notice that every data item in column c1 is assigned to a specific class in column c2.

We use the Tukey method to determine which samples are statistically different from the other samples. For this method, Minitab uses the *tukey* subcommand.

```
MTB > oneway c1 c2;
SUBC> tukey.
```

12.2.5 One-Way Analysis of Variance

```
Analysis of Variance for C1
Source     DF        SS          MS          F        P
C2          3     612.8       204.3       4.10    0.016
Error      28    1394.1        49.8
Total      31    2007.0
                                    Individual 95% CIs For Mean
                                    Based on Pooled StDev
Level       N      Mean       StDev  --------+---------+---------+--------
1           8    60.625      11.747   (-------*--------)
2           8    69.250       4.979                (-------*--------)
3           8    70.375       2.825                  (-------*--------)
4           8    71.875       5.330                    (--------*-------)
                                     --------+---------+---------+--------
Pooled StDev =    7.056                    60.0      66.0      72.0

Tukey's pairwise comparisons

    Family error rate = 0.0500
Individual error rate = 0.0108

Critical value = 3.86

Intervals for (column level mean) - (row level mean)

                1            2            3

    2     -18.255
            1.005

    3     -19.380      -10.755
           -0.120        8.505

    4     -20.880      -12.255      -11.130
           -1.620        7.005        8.130
```

The Tukey method calculates the confidence interval between two groups. To determine which two groups are statistically different, we see if the interval contains zero or not. If it does, then it is likely that the two groups are not significantly different. For example, the 95 % confidence interval for groups 2 and 3 is

$$-10.755 < \mu_2 - \mu_3 < 8.505$$

Since this interval contains zero, we conclude that we cannot reject that class 2 and class 3 are statistically different.

From the Tukey matrix, we can also see that class 1 and class 3 are statistically different because their interval does not contain 0. Also class 1 and class 4 are statistically different because their interval does not contain 0.

This illustrates that we might not get the same answer using both a numerical method and a graphical method. Graphically we saw that class 1 was statistically different from every other class, but numerically we saw that it was not statistically different from class 2.

So far we have looked at examples based on a single classification. The first example analyzed the mean based on gasoline. The second example analyzed the mean based on classes. These classifications are called *factors*. In the gasoline example, we had three different types of gasoline. In the classroom example, we had four classes. The individual types of gasoline and types of class are called *levels*.

12.3 Two-Way Analysis of Variance

Doing an analysis of variance with one factor is called a one-way analysis of variance. We will now expand the analysis of variance and test to see if the means from different samples are the same or not based on two factors. Analysis of variance of two factors is called a two-way analysis of variance.

We can think of factors as influencing the value of the mean of each sample. For example, in the gasoline example, we were interested in miles per gallon. The factor that we thought might be influencing the mileage per gallon was the different brands. Now we are interested in two-factor analysis, in two things affecting the mean that we are interested in.

Example 12.3 Suppose we are interested to know if years of work experience and geographical location influence annual salary. The data, given below, show salaries in thousands of dollars.

	Years of work experience		
Region	1	2	3
1. West	16, 16.5	19, 17	24, 25
2. Midwest	21, 20.5	20, 19	21, 22.5
3. Northeast	18, 19	21, 20.9	22, 21
4. South	13, 13.5	20, 20.8	25, 23

Here we are interested in the mean of the annual salary. We believe that two factors influence the mean: years of work experience and region of the country. It turns out that we should also look at the influence of the interaction of these two factors.

The variability of annual salary can thus be broken down in the following way:

$$\begin{aligned}\text{Total Variation} = &\ \text{Variation of Factor A} + \text{Variation of Factor B} \\ &+ \text{Variation of Interaction of Factors AB} + \text{Random Error}\end{aligned}$$

In a two-way analysis of variance, there are three tests in which we are interested. In our salary example, they would be the following:

1. To test the hypothesis of no difference due to years of working experience:

$$\begin{aligned} &\text{H}_0: \text{mean year 1} = \text{mean year 2} = \text{mean year 3} \\ &\text{H}_1: \text{not all mean years are equal} \end{aligned}$$

The test would follow an F distribution and would have the following ratio:

$$F = \frac{\text{variation of factor A (years of working experience)}}{\text{random error}}.$$

2. To test the hypothesis of no difference due to geographic region:

$$\begin{aligned} &\text{H}_0: \text{mean region 1} = \text{mean region 2} = \text{mean region 3} \\ &\text{H}_1: \text{not all mean regions are equal} \end{aligned}$$

The test would follow an F distribution and would have the following ratio:

$$F = \frac{\text{variation of factor B (geographic region)}}{\text{random error}}.$$

3. To test the hypothesis of no difference due to interaction of work experience and region:

$$\begin{aligned} &\text{H}_0: \text{interaction of work experience and geographic region has no effect} \\ &\text{H}_1: \text{interaction of work experience and geographic region has an effect} \end{aligned}$$

The test would follow an F distribution and would have the following ratio:

$$F = \frac{\text{variation of interaction of factor A and factor B (experience and region)}}{\text{random error}}$$

We will now test our annual salary example using the *twoway* command. The Minitab commands to do an ANOVA analysis as shown below.

```
MTB > name c1 'salary'
MTB > name c2 'region'
MTB > name c3 'work'
MTB > set 'salary'
DATA> 16 16.5 19 17 24 25 21 20.5 20 19 21 22.5
DATA> 18 19 21 20.9 22 21 13 13.5 20 20.8 25 23
DATA> end
MTB > set 'region'
DATA> 1 1 1 1 1 1 2 2 2 2 2 2 3 3 3 3 3 3 4 4 4 4 4 4
DATA> end
set 'work'
DATA> 1 1 2 2 3 3 1 1 2 2 3 3 1 1 2 2 3 3 1 1 2 2 3 3
DATA> end
MTB > twoway c1 c2 c3;
SUBC> mean c2 c3.
```

12.3.1 Two-Way Analysis of Variance

```
Analysis of Variance for salary
Source          DF         SS          MS         F        P
region           3      7.921       2.640      4.05    0.033
work             2    132.903      66.452    101.91    0.000
Interaction      6     77.730      12.955     19.87    0.000
Error           12      7.825       0.652
Total           23    226.380

                             Individual 95% CI
region          Mean     ------+---------+---------+---------+-----
1              19.58           (---------*---------)
2              20.67                          (---------*---------)
3              20.32                     (---------*---------)
4              19.22     (----------*---------)
                         ------+---------+---------+---------+-----
                             18.90     19.60     20.30     21.00



                             Individual 95% CI
work            Mean     --------+---------+---------+---------+---
1              17.19      (--*--)
2              19.71                  (---*--)
3              22.94                                  (--*--)
                         --------+---------+---------+---------+---
                             18.00     20.00     22.00     24.00
```

Looking at the statistics report, we have the following conclusions:

1. We conclude that the geographic region does affect annual salary because the calculated p-value 0.033 is less than the alpha value of interest, which is .05.
2. We conclude that work experience does affect annual salary because the calculated p-value 0.00 is less than the alpha value of interest, which is .05.
3. We conclude that the interaction of work experience and geographic regions does affect annual salary because the calculated p-value of 0.00 is less than the alpha value of interest which is .05.

12.4 ANOVA Interaction Plot

We will now create an interaction plot to see how the interaction of work experience and region affect salary. Below uses the Minitab macro "Interact.Mac" to create the interaction plot. The following works in Minitab 13 and not in Minitab 17.

```
MTB > %Interact 'region' 'work';
SUBC>   Response 'salary'.
Executing from file: C:\Program Files\Mtbwin\MACROS\Interact.MAC
```

Macro is running ... please wait

12.4.1 Interaction Plot for Salary

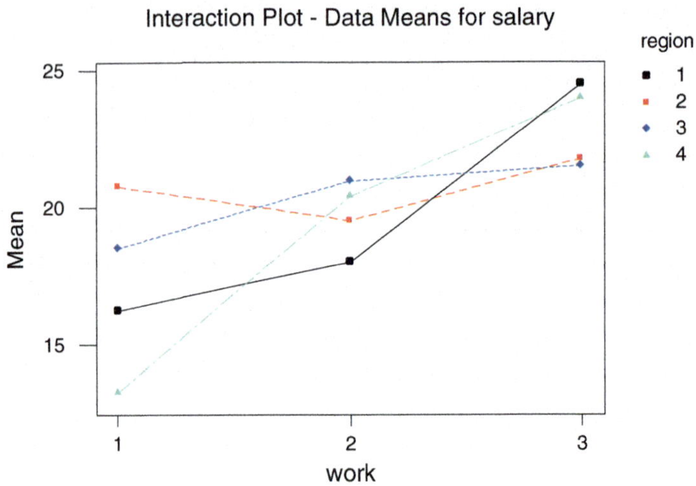

12.5 Chi-Square Test

In this section, we will use chi-square test to test goodness of fit and independence of data set.

12.5.1 Goodness of Fit

In this section, we will show how the chi-square statistic can be used to the appropriateness of a distribution—its goodness of fit for a set of data. Goodness-of-fit tests are designed to study the frequency distribution to determine whether a set of data are generated from a certain probability distribution, such as the uniform, binomial, Poisson, or normal distribution.

The chi-square statistic for determining whether the data follow a specific probability distribution is

$$\chi^2_{k-1} = \sum_{i=1}^{k} \frac{\left(f_i^0 - f_i^e\right)^2}{f_i^e}$$

where:

f_i^0 = observed frequency

f_i^e = expected frequency

k = number of groups

χ^2_{k-1} = chi-square statistic with $(k - 1)$ degrees of freedom

Example 12.4 A marketing manager wants to test his belief that four different categories of cars share the auto market equally. These four categories of cars are brand A, brand B, brand C, and imported cars. He sends out 2000 questionnaires to car owners throughout the nations and receives the following responses:

Brand	Number of owners
A	475
B	505
C	495
Import	525
Total	2000

Test the following hypotheses:

$$H0: \text{Same market shares (uniform distribution)}$$

$$H1: \text{Different market Shares}$$

Below are the calculations that are done in Excel. We fail to reject the null hypothesis because the calculated p-value, .457489552, is greater than the alpha value, .05.

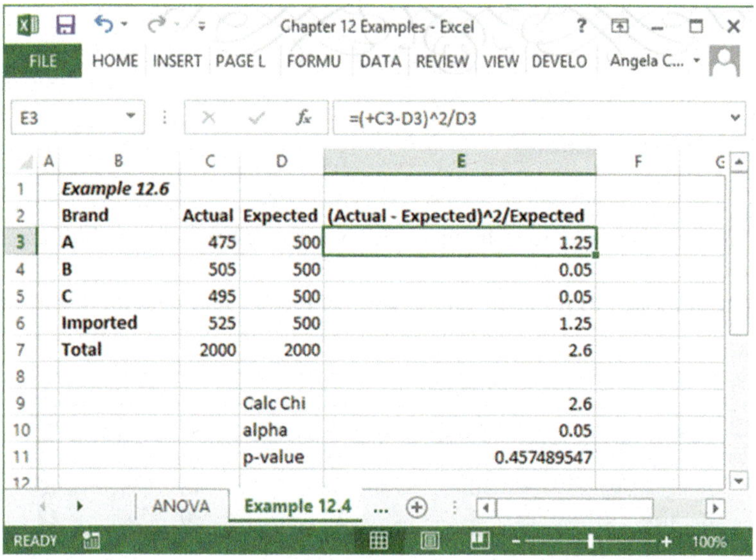

Below shows how the above worksheet is set up.

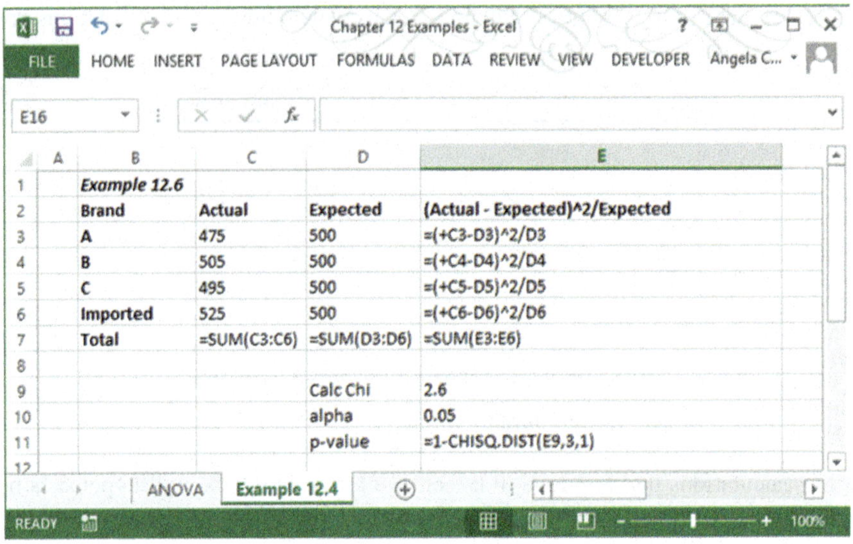

	A	B	C	D	E
1		*Example 12.6*			
2		**Brand**	**Actual**	**Expected**	**(Actual - Expected)^2/Expected**
3		**A**	475	500	=(+C3-D3)^2/D3
4		**B**	505	500	=(+C4-D4)^2/D4
5		**C**	495	500	=(+C5-D5)^2/D5
6		**Imported**	525	500	=(+C6-D6)^2/D6
7		**Total**	=SUM(C3:C6)	=SUM(D3:D6)	=SUM(E3:E6)
8					
9				Calc Chi	2.6
10				alpha	0.05
11				p-value	=1-CHISQ.DIST(E9,3,1)

12.5.2 Test of Independence

Example 12.5 We are now interested in checking to see whether two variables are independent of each other or not.

Suppose someone has asked us if majors and grades are related. The data for this question are shown below:

	Grades			
Major	A	B	C	F
Science	12	36	34	8
Humanities	10	24	46	10
Business	8	30	70	12

The hypotheses would be

$$H_0 : \text{major and grades are independent}$$
$$H_1 : \text{major and grades are not independent}$$

For this problem, we would follow a chi-square test.

```
MTB > chisquare c1-c4
```

12.5.3 Chi-Square Test

```
Expected counts are printed below observed counts

              A        B        C        F    Total
    1        12       36       34        8       90
           9.00    27.00    45.00     9.00

    2        10       24       46       10       90
           9.00    27.00    45.00     9.00

    3         8       30       70       12      120
          12.00    36.00    60.00    12.00

Total        30       90      150       30      300

Chi-Sq =  1.000 +  3.000 +  2.689 +  0.111 +
          0.111 +  0.333 +  0.022 +  0.111 +
          1.333 +  1.000 +  1.667 +  0.000 = 11.378
DF = 6, P-Value = 0.077

MTB > invcdf .95;
SUBC> chisquare 6.
```

12.5.4 Inverse Cumulative Distribution Function

```
Chi-Square with 6 DF

 P( X <= x)           x
    0.9500      12.5916
```

We conclude that grades and major are independent, because the calculated chi-square statistic of 11.378 is less than the critical value of 0.05 alpha, which is 12.5916.

Below are the calculations done in Excel.

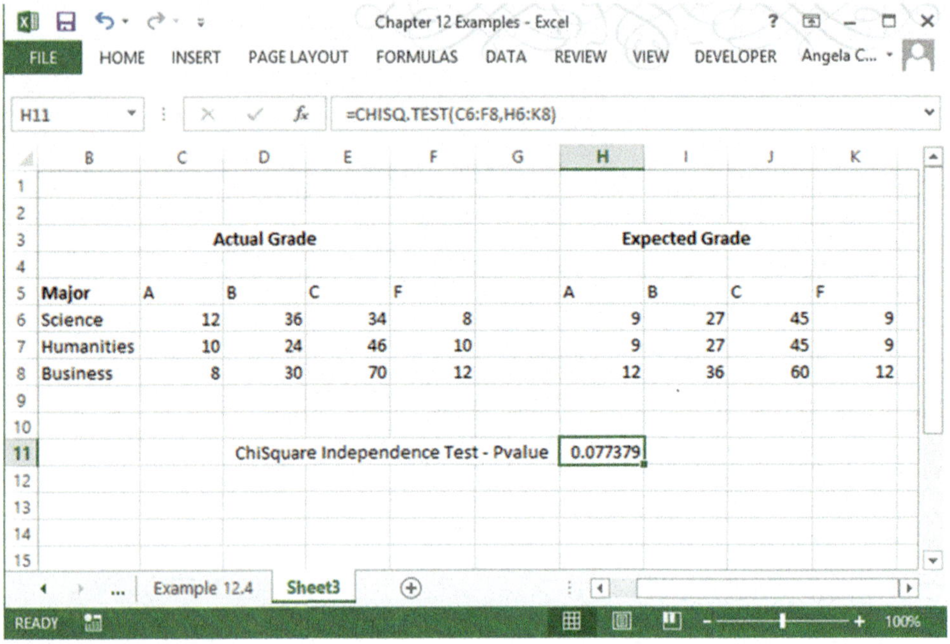

Below shows how the above worksheet is set up in Excel.

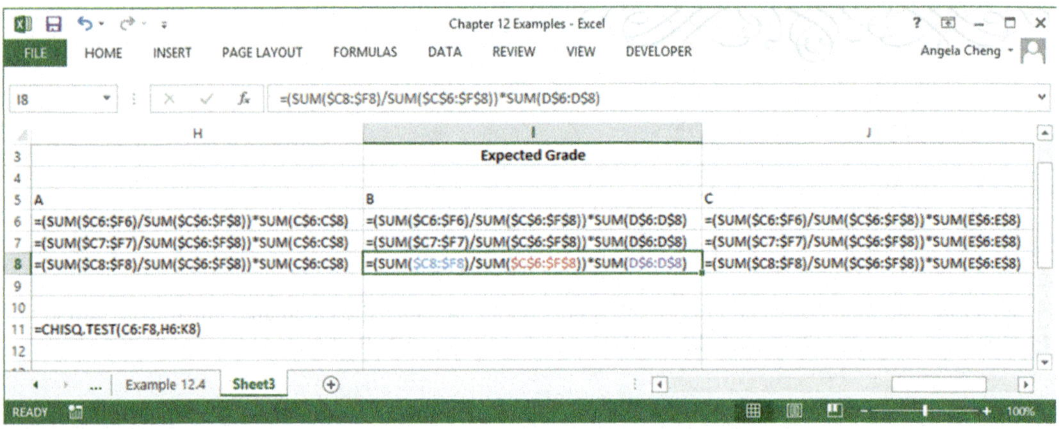

In doing the calculations in Excel, we use the *chisq.test* function. This function returns a p-value. The calculated p-value in Excel is 0.07739. Since the p-value is greater than the alpha value of .05, we cannot reject the null hypothesis that major and grades are independent.

12.6 SAS Programming Code Instructions and Examples

12.6.1 One-Way Analysis of Variance

Example 12.2A To do this analysis, let us put the data for gasoline A, B, and C in column Gasolines and the types of gasolines in column Group.

```
Data Gasoline;
   INPUT Gasolines Group $ @@;
   DATALINES;
34 A  29 B  32 C
28 A  32 B  34 C
29 A  31 B  30 C
37 A  43 B  42 C
42 A  31 B  32 C
27 A  29 B  33 C
29 A  28 B   . C
;
run;
```

Let us first have a look at the dotplot of the three gasolines to get an overall picture of the data.

```
  Goptions ftext='Times New Roman' htext=3;
  title  height=4 'Dotplot of Three Gasolines';
  symbol1 v=dot  c=black L=1;
  proc gplot data=Gasoline;
     plot Group*Gasolines;
run;
```

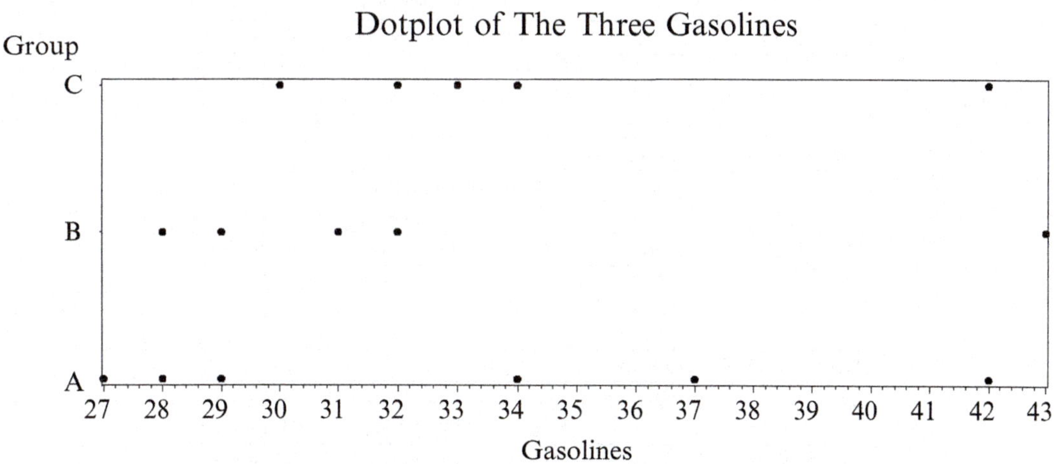

We will use the **PROC ANOVA** command to calculate the one-way analysis of variance.

```
proc anova data=Gasoline;
   class Group;
   model Gasolines=Group;
   means Group /TUKEY CLDIFF;
run;
```

Source	DF	Sum of squares	Mean square	F value	Pr > F
Model	2	13.6809524	6.8404762	0.27	0.7677
Error	17	433.1190476	25.4775910		
Corrected total	19	446.8000000			

R-square	Coeff var	Root MSE	Gasoline mean
0.030620	15.48323	5.047533	32.60000

```
PROC SORT DATA =Gasoline;
BY Group;
RUN;

PROC MEANS MEAN STD N CLM DATA =Gasoline;
    BY Group;
    VAR Gasolines;
RUN;
```

Group = A

Analysis variable: gasolines				
Mean	Std dev	N	Lower 95 % CL for mean	Upper 95 % CL for mean
32.2857143	5.5891050	7	27.1166542	37.4547743

Group = B

Analysis variable : gasolines				
Mean	Std dev	N	Lower 95 % CL for mean	Upper 95 % CL for mean
31.8571429	5.1130086	7	27.1283986	36.5858871

Group = C

Analysis variable: gasolines				
Mean	Std dev	N	Lower 95 % CL for mean	Upper 95 % CL for mean
33.8333333	4.2150524	6	29.4099071	38.2567595

Example 12.6 To do this analysis, let us put the grades for classes 1, 2, 3, and 4 in column Grade and the types of class in column Class.

```
Data Books;
   INPUT Grade Class $ @@;
   DATALINES;
43 1 77 2 72 3 72 4
45 1 72 2 73 3 74 4
67 1 75 2 71 3 75 4
68 1 69 2 65 3 65 4
73 1 67 2 68 3 66 4
72 1 66 2 69 3 68 4
55 1 65 2 73 3 74 4
62 1 63 2 72 3 81 4
;
run;
```

Let us look at a dotplot of the four classes.

```
Goptions ftext='Times New Roman' htext=3;
title  height=4 'Dotplot of Different Classes';
symbol1 v=dot  c=black L=1;
proc gplot data=Books;
  plot Class*Grade;
run;
```

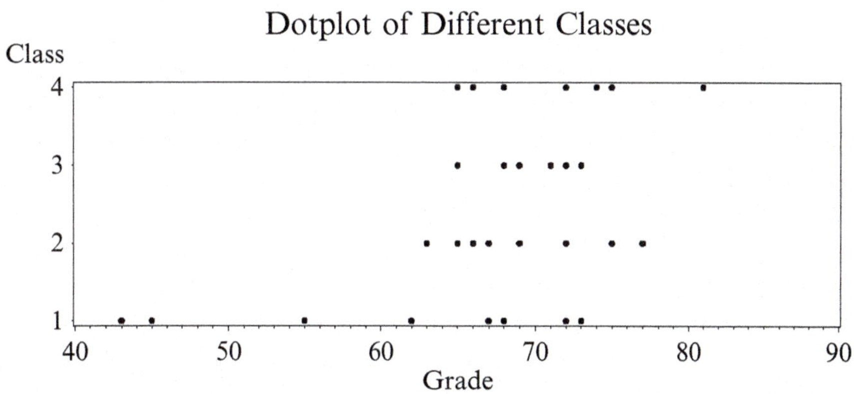

From the dotplot of the four classes, we can see that the plots are somewhat similar. This suggests to us that the four classes might have the same average grade. We will test this by doing a one-way analysis of variance calculation.

```
proc anova data=Books;
   class Class;
   model Grade=Class;
   means Class /TUKEY CLDIFF;
run;
```

Source	DF	Sum of squares	Mean square	F value	Pr > F
Model	3	612.843750	204.281250	4.10	0.0156
Error	28	1394.125000	49.790179		
Corrected total	31	2006.968750			

R-square	Coeff var	Root MSE	Grade mean
0.305358	10.37202	7.056216	68.03125

Alpha	0.05
Error degrees of freedom	28
Error mean square	49.79018
Critical value of studentized range	3.86124
Minimum significant difference	9.6328

Comparisons significant at the 0.05 level are indicated by ***

Class comparison	Difference between means	Simultaneous 95 % confidence limits		
4-3	1.5	−8.133	11.133	
4-2	2.625	−7.008	12.258	
4-1	11.25	1.617	20.883	***
3-4	−1.5	−11.133	8.133	
3-2	1.125	−8.508	10.758	
3-1	9.75	0.117	19.383	***
2-4	−2.625	−12.258	7.008	
2-3	−1.125	−10.758	8.508	
2-1	8.625	−1.008	18.258	
1-4	−11.25	−20.883	−1.617	***
1-3	−9.75	−19.383	−0.117	***
1-2	−8.625	−18.258	1.008	

```
PROC SORT DATA=Books;
BY Class;
RUN;

PROC MEANS MEAN STD N CLM DATA=books;
    BY Class;
    VAR Grade;
RUN;
```

Class = 1

Analysis variable: grade

Mean	Std dev	N	Lower 95 % CL for mean	Upper 95 % CL for mean
60.6250000	11.7465800	8	50.8046133	70.4453867

Class = 2

Analysis variable: grade				
Mean	Std dev	N	Lower 95 % CL for mean	Upper 95 % CL for mean
69.2500000	4.9785253	8	65.0878487	73.4121513

Class = 3

Analysis variable: grade				
Mean	Std dev	N	Lower 95 % CL for mean	Upper 95 % CL for mean
70.3750000	2.8252686	8	68.0130163	72.7369837

Class = 4

Analysis variable: grade				
Mean	Std dev	N	Lower 95 % CL for mean	Upper 95 % CL for mean
71.8750000	5.3301702	8	67.4188662	76.3311338

12.6.2 Two-Way Analysis of Variance

Example 12.3A

```
Data Salary;
     INPUT Salary Year Region $ @@;
     DATALINES;
  16 1 West  19 2 West  24 3  West
  16.5 1 West  17 2 West  25 3 West
  21 1 Midwest  20 2 Midwest  21 3 Midwest
  20.5 1 Midwest  19 2 Midwest  22.5 3 Midwest
  18 1 Northeast  21 2 Northeast  22 3 Northeast
  19 1 Northeast  20.9 2 Northeast  21 3 Northeast
  13 1 South  20 2 South  25 3 South
  13.5 1 South  20.8 2 South  23 3 South
  ;
run;

  PROC SORT DATA=Salary;
  BY Region Year;
RUN;

  proc anova data=Salary;
     class Region Year;
     model Salary=Region | Year;
     means Region Year/TUKEY CLDIFF;
  run;
```

Source	DF	ANOVA SS	Mean square	F value	Pr > F
Region	3	7.9212500	2.6404167	4.05	0.0334
Year	2	132.9033333	66.4516667	101.91	< 0.0001
Region*year	6	77.7300000	12.9550000	19.87	< 0.0001

R-square	Coeff var	Root MSE	Salary mean
0.965434	4.048549	0.807517	19.94583

```
  PROC SORT DATA=Salary;
  BY Region;
RUN;
PROC MEANS MEAN STD N CLM DATA=Salary;
    BY Region;
    VAR Salary;
RUN;
```

Region = Midwest

Analysis variable: salary				
Mean	Std dev	N	Lower 95 % CL for mean	Upper 95 % CL for mean
20.6666667	1.1690452	6	19.4398290	21.8935044

Region = Northeast

Analysis variable: salary				
Mean	Std dev	N	Lower 95 % CL for mean	Upper 95 % CL for mean
20.3166667	1.4972196	6	18.7454310	21.8879023

Region = South

Analysis variable: salary				
Mean	Std dev	N	Lower 95 % CL for mean	Upper 95 % CL for mean
19.2166667	4.9438514	6	14.0284128	24.4049205

Region = West

Analysis variable: salary				
Mean	Std dev	N	Lower 95 % CL for mean	Upper 95 % CL for mean
19.5833333	3.9549547	6	15.4328629	23.7338037

```
  PROC SORT DATA=Salary;
  BY Year;
RUN;
PROC MEANS MEAN STD N CLM DATA=Salary;
    BY Year;
    VAR Salary;
RUN;
```

Year = 1

Analysis variable: salary				
Mean	Std dev	N	Lower 95 % CL for mean	Upper 95 % CL for mean
17.1875000	2.9873244	8	14.6900343	19.6849657

Year = 2

Analysis variable: salary				
Mean	Std dev	N	Lower 95 % CL for mean	Upper 95 % CL for mean
19.7125000	1.3516524	8	18.5824903	20.8425097

Year = 3

Analysis variable: salary				
Mean	Std dev	N	Lower 95 % CL for mean	Upper 95 % CL for mean
22.9375000	1.6132819	8	21.5887626	24.2862374

12.7 ANOVA Interaction Plot

To create the interaction plot, we first have to calculate the mean salary for different regions in each year. PROC SQL is used to solve this problem.

The SQL procedure is a SAS procedure. This procedure begins with the declaration **PROC SQL;** and ends with a **QUIT;** statement (not **RUN;**). In between these two declarations, SQL code may be used. SQL code obeys the rules of the SAS system regarding maximum lengths of variable names, reserved words, and the use of the semi colon as the line delimiter. The simplest and most commonly used SQL statement is SELECT, which is used to extract data from a table.

```
proc sql;
    create table SalaryMean as
     select distinct Year,Region,
           mean(Salary) as Salary_mean
           from Salary
           group by Region, Year;
quit;

Data SalaryMean;
   INPUT Midwest Northeast South West Year @@;
   DATALINES;
20.75 18.50 13.25 16.25 1
19.50 20.95 20.40 18.00 2
21.75 21.50 24.00 24.50 3
;
run;
```

Below uses the SAS command **PROC GPLOT** to create the interaction plot.

```
   Goptions ftext='Times New Roman' htext=6 gunit=pct;
   title  height=7 'Interaction Plot-Data Means for Salary';
   legend1 label=none value=('Midwest' 'Northeast' 'South' 'West')
shape=symbol(5,1);
   axis1  label=('Mean') order=(10 to 26 by 5);
   axis2  label=('Work') ;
   symbol1 i=join  v=none L=1 c=black w=2;
   symbol2 i=join  v=none L=1 c=red w=2;
   symbol3 i=join  v=none L=1 c=blue w=2;
   symbol4 i=join  v=none L=1 c=green w=2;
   proc gplot data=SalaryMean;
   plot (Midwest Northeast South West)*Year/vaxis=axis1 haxis=axis2 overlay
legend=legend1;
   run;
```

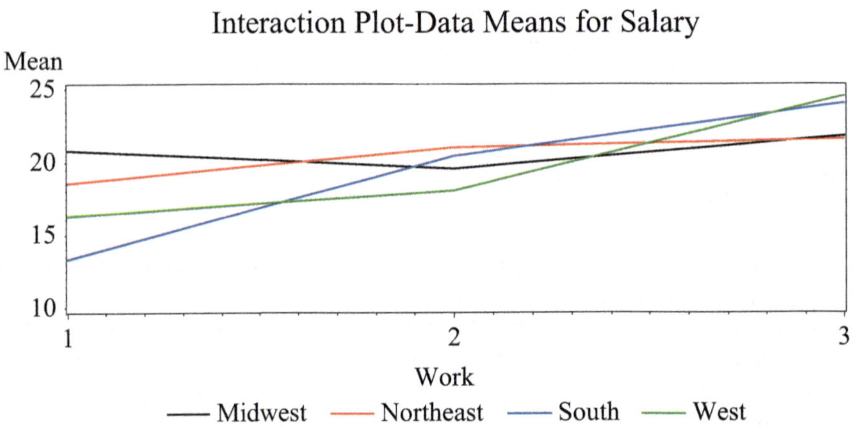

Example 12.4 Below are the calculations that are done in SAS. We fail to reject the null hypothesis because the calculated p-value, .54251, is greater than the alpha value .05.

```
  Data ChiSquare;
  set Cars;
  Expected=500;
  CalcChi=(Actual-Expected)**2/Expected;
  run;

  proc sql;
  select sum(CalcChi) as ChiSquare, CDF('CHISQUARE',2.6,3) as Pvalue
  from ChiSquare
quit;
```

Chi-square	P-value
2.6	0.54251

Example 12.7 To solve this problem in SAS, we would follow a chi-square test.

```
  DATA Grades;
       INPUT Major $ Grade $ Count @@;
       DATALINES;
   Science A 12 Science B 36 Science C 34 Science F 8
   Humanities A 10 Humanities B 24 Humanities C 46 Humanities F 10
   Business A 8 Business B 30 Business C 70 Business F 12
  ;
run;
  PROC FREQ ORDER=DATA DATA=Grades;
       TABLES Major*Grade / CMH NOPERCENT NOCOL NOROW;
       WEIGHT Count;
       TITLE1 'Chi Square Test';
RUN;
```

Table of major by grade					
Major	Grade				
Frequency	A	B	C	F	Total
Science	12	36	34	8	90
Humanities	10	24	46	10	90
Business	8	30	70	12	120
Total	30	90	150	30	300

Cochran-Mantel-Haenszel statistics (based on table scores)				
Statistic	Alternative hypothesis	DF	Value	Prob
1	Nonzero correlation	1	6.7708	0.0093
2	Row mean scores differ	2	7.0424	0.0296
3	General association	6	11.3782	0.0774

We conclude that grades and major are independent, because the calculated p-value in SAS is 0.07739. Since the p-value is greater than the alpha value of .05, we cannot reject the null hypothesis that major and grades are independent.

12.8 Statistical Summary

In this chapter we use either one-way or two-way ANOVA to see if several sample means are statistically different from each other. In addition, we discussed how the chi-square statistic can be used to test whether two variables are independent of each other and where a set of data comes from a specific distribution (goodness of fit).

Bibliography

Microsoft Inc., Excel 2013. Microsoft Inc., Redmond
Lee CF, Lee JC, Lee AC (2013) Statistics for business and financial economics. Springer, New York
Minitab Inc. Minitab 17. Minitab Inc., State College
SAS Institute Inc. SAS 2014. SAS Institute Inc., Cary

Chapter 13
Simple Linear Regression and the Correlation Coefficient

13.1 Introduction

In the previous chapters, we have been primarily interested in analyzing single variables. In this chapter, we are interested in analyzing two variables and the relationship between them. The two techniques that we will look at are regression analysis and correlation analysis.

13.2 Regression Analysis

There are many types of relationships between two or more variables. Some of the types of relationships between two variables are linear, logarithmic, and polynomial. In the next four chapters, we will be studying the linear relationships between two or more variables. We will use regression analysis in studying linear relationships.

Mathematically, a linear equation looks like the following:

$$y = a + bx$$

The two variables of interest in this equation are x and y. In this equation x is called the independent variable and y is called the dependent variable.

In regression analysis, we are modeling the relationship between two variables and then using the model to predict the value of the dependent variable from the value of the independent variable.

C.-F. Lee et al., *Essentials of Excel, Excel VBA, SAS and Minitab for Statistical and Financial Analyses*, DOI 10.1007/978-3-319-38867-0_13

Graphically, the relationship between the two variables is believed to look like the following:

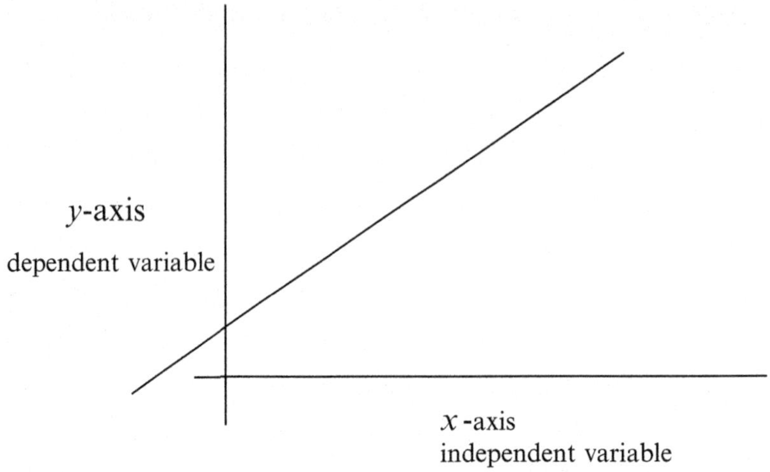

Example 13.1 The Organization of Petroleum Exporting Countries (OPEC) has tried to control the price of crude oil since 1973. The price of crude oil rose dramatically from the mid-1970s to the mid-1980s. As a result, motorists were confronted with a similar upward spiral of gasoline prices. The following table presents a gallon of return leaded gasoline and a barrel of crude oil in terms of the average value at the point of production during 1975–1988 (data from the *Statistical Abstract of the United States*, 1990, pp. 483, 485).

Price of gasoline and crude oil

Year	Gasoline y (cents/gallon)	Crude oil x ($/barrel)
1975	57	7.67
1976	59	8.19
1977	62	8.57
1978	63	9.00
1979	86	12.64
1980	119	21.59
1981	133	31.77
1982	122	28.52
1983	116	26.19
1984	113	25.88
1985	112	24.09
1986	86	12.51
1987	90	15.40
1988	90	12.57

Let us first investigate the relationship between the price of a gallon of gasoline and the price of a barrel of crude oil by graphing it. We will enter the data into an Excel worksheet as shown below:

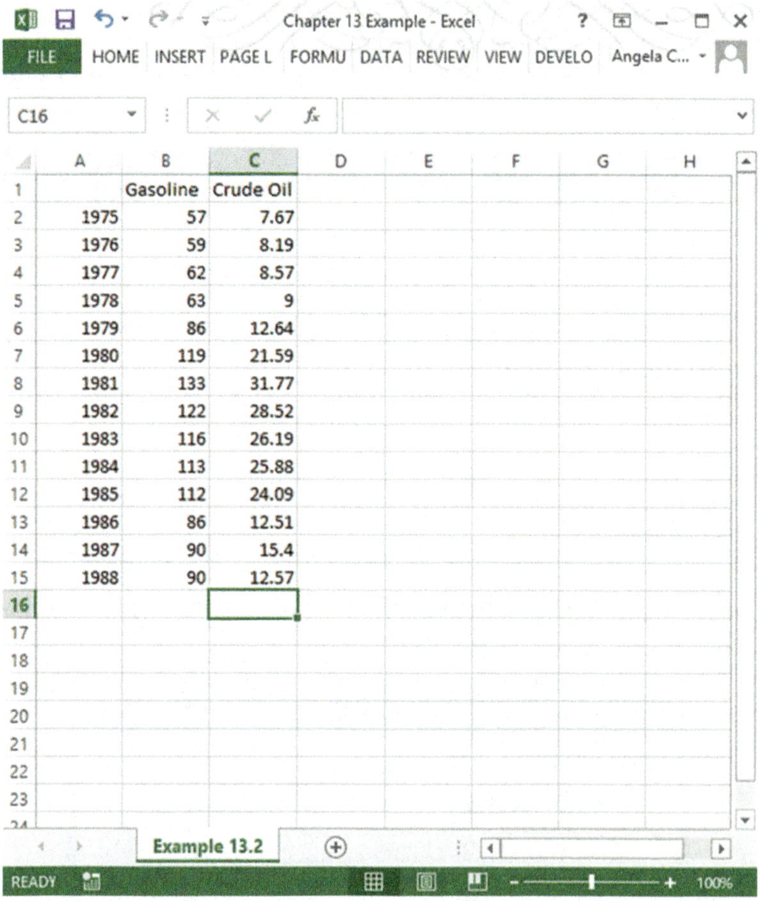

	A	B	C
1		Gasoline	Crude Oil
2	1975	57	7.67
3	1976	59	8.19
4	1977	62	8.57
5	1978	63	9
6	1979	86	12.64
7	1980	119	21.59
8	1981	133	31.77
9	1982	122	28.52
10	1983	116	26.19
11	1984	113	25.88
12	1985	112	24.09
13	1986	86	12.51
14	1987	90	15.4
15	1988	90	12.57

To graph the above data, first we highlight everything then choose, **Insert→Scatter Chart**.

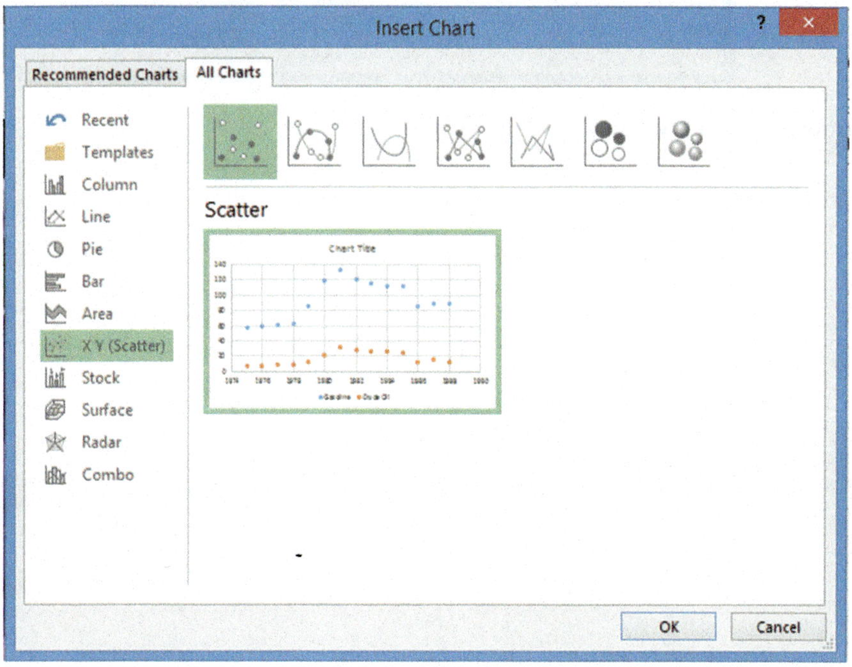

As shown above, Excel guessed the years as the x-axis data and the gasoline and crude oil prices as the y-axis data, which is incorrect. The gasoline data should be the y-axis and the crude oil data should be the x-axis. To change the axis values go under **Chart Tools→Design→Select Data**. Below is the dialog box that is shown.

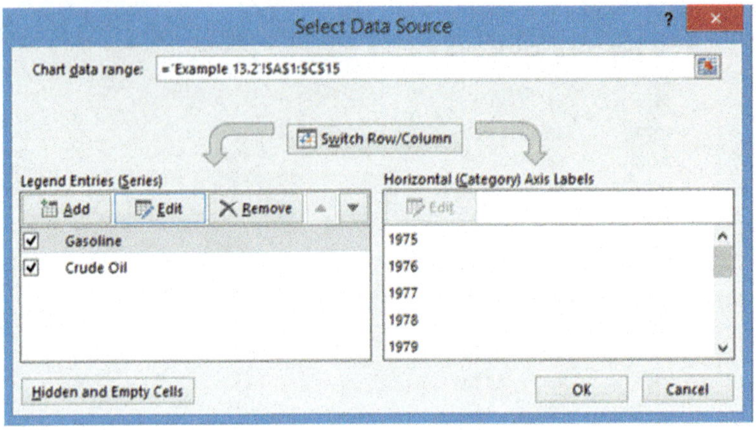

Select *Edit* under Legend Entries. The following dialog box shows the correct settings for the series tab:

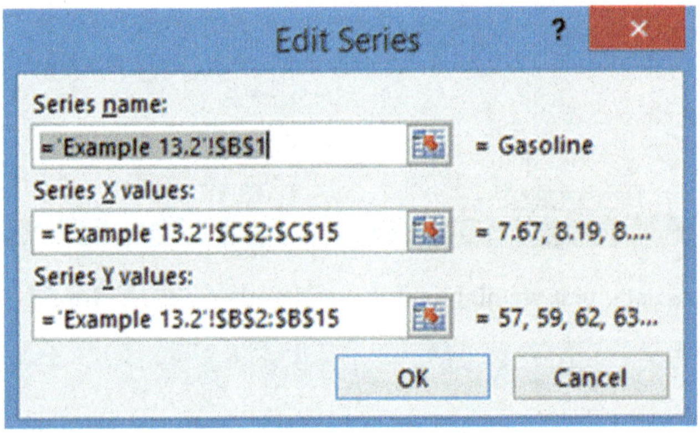

The gasoline and crude oil data look like the following:

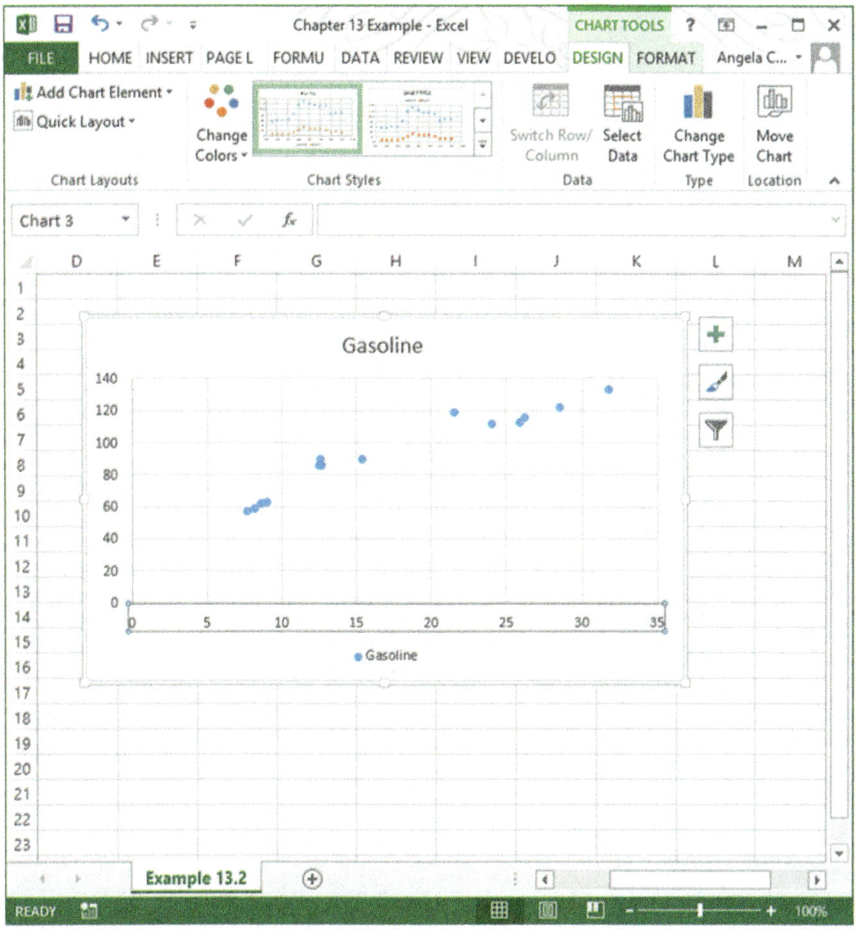

From the above graph, the gasoline and crude oil data look linear. One question often asked is, "What is the best linear line that fits the data?" To answer this question, we can user a feature in Excel to draw the best linear line that fits the above data. To do this, choose **Chart Tools→Design→Add Chart Elements→Trendline**, which is shown below:

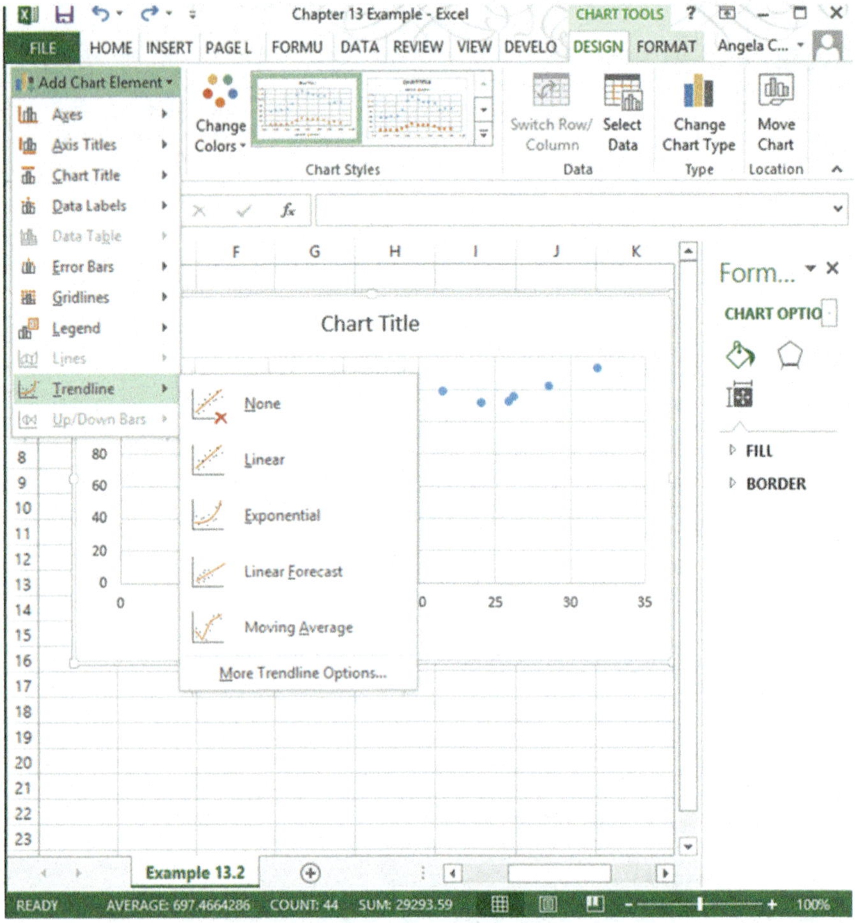

Choosing the *Linear* menu item as shown above will result in the *Add Trendline* as shown below:

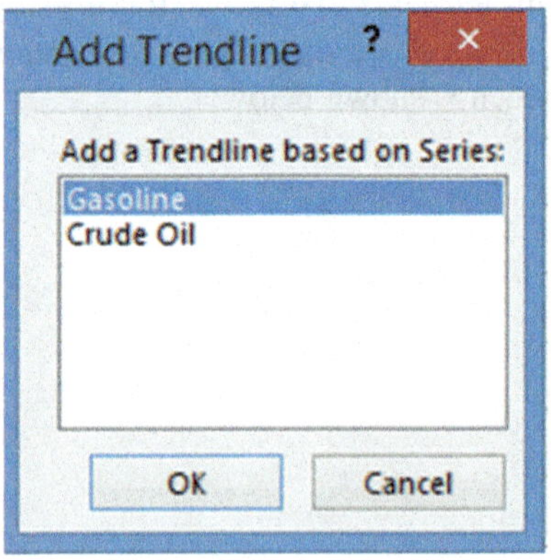

The *Add Trendline* dialog box illustrates that there are many points to fit the gasoline and crude oil data. In this chapter and the next couple of chapters we will concentrate on the linear regression method.

Next, double click on the chart, and format options will appear on the right-hand side. Choose *TRENDLINE OPTION* and fill out the options as below:

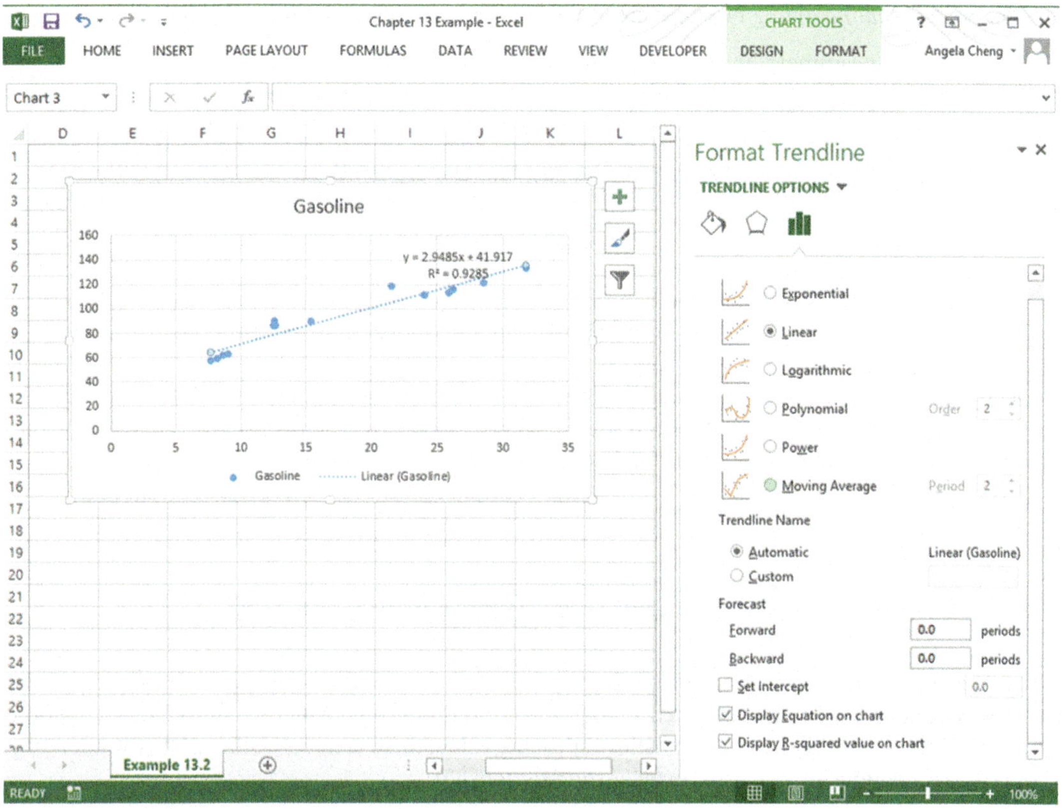

The main things that we did was to check,

(1) *Display equation on chart*
(2) *Display R-squared value on chart*

The gasoline/crude oil data look like the following after we press the *OK* button of the *Add Trendline* dialog box:

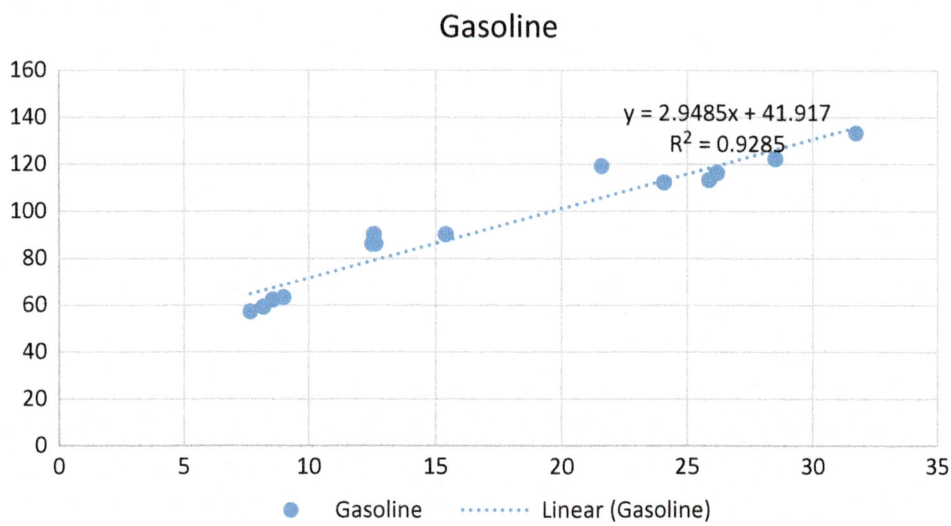

The *Add-Trendline* dialog box did three things to the gasoline and crude oil graph:

1. It drew in the best linear line that fitted the gasoline and crude oil graph.
2. It gave the equation of the best linear line that fitted the gasoline and crude oil graph.
3. It calculated the R^2 for the best linear line that fitted the gasoline and crude oil graph.

Excel gives the best linear line that fits the gasoline and crude oil data as,

$$y = 2.9485x + 41.917$$

It is important to note this linear equation produces an expected value of y. For example, the way to interpret the above equation is when crude oil increases by one dollar, the expected increase of gasoline is 44.8655 (2.9485 * 1 + 41.917).

R^2 is a number from 0 to 1. R^2 represents how well the linear line explains the relationship between gasoline and crude oil. Excel calculated the R^2 number as 0.9285. This means that the linear line, given by Excel, explains the relationship between gasoline and crude oil pretty well.

13.3 Deterministic Relationship and Stochastic Relationship

When we study linear equations in algebra, we expect that for each x value there is only one value of y. This kind of relationship is called a *deterministic relationship*. When we study linear equations in statistics, this is not the case. In statistics for each x value, there is a probability of distribution of y. This kind of relationship is called a *stochastic relationship*.

13.4 Standard Assumptions for Linear Regression

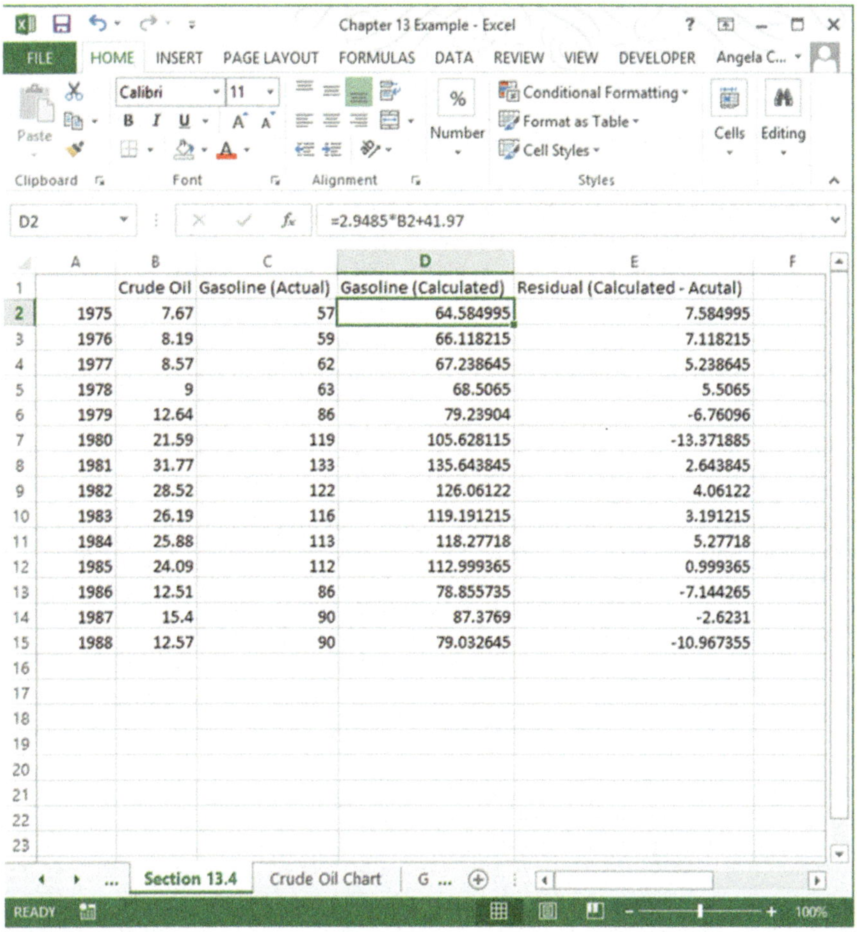

	Crude Oil	Gasoline (Actual)	Gasoline (Calculated)	Residual (Calculated - Acutal)
1975	7.67	57	64.584995	7.584995
1976	8.19	59	66.118215	7.118215
1977	8.57	62	67.238645	5.238645
1978	9	63	68.5065	5.5065
1979	12.64	86	79.23904	-6.76096
1980	21.59	119	105.628115	-13.371885
1981	31.77	133	135.643845	2.643845
1982	28.52	122	126.06122	4.06122
1983	26.19	116	119.191215	3.191215
1984	25.88	113	118.27718	5.27718
1985	24.09	112	112.999365	0.999365
1986	12.51	86	78.855735	-7.144265
1987	15.4	90	87.3769	-2.6231
1988	12.57	90	79.032645	-10.967355

The best line (regression line) that fitted the gasoline and crude oil data did not fit the data perfectly; there were differences or residuals. The residuals are calculated above for the regression line of the gasoline and crude oil data. Also it was noted that the regression line was to be interpreted as an expected value for the dependent variable.

It is also important to note that there are five assumptions for the regression line. There assumptions must be true for us to properly interpret the regression line. The five assumptions are:

1. The independent variables (x) are independent of the residuals.
2. The residuals are normally distributed.
3. The residuals have a mean of zero.
4. The residuals are independent of each other.
5. The residuals have constant variances.

The above five assumptions led us to interpret the resulting regressing line as the expected or average regression line. It is also important to note from the above five assumptions that for any give x value of a regression line, the corresponding y value has a normal distribution. In our regression analysis, we are assuming that the data population comes from a normal distribution, or because of a large sample size we can use the central limit theorem to indicate an approximation of a normal distribution.

For example in our gasoline/crude oil example, the regression equation $y = 2.9485x + 41.917$ indicates that a crude oil price of \$24.09 should result in an expected gasoline price of 112 cents. The actual observed gasoline price, when crude oil is \$24.09, is 113 cents.

13.5 The Coefficient of Determination

One important question to ask about the regression line is how well the regression line explains the data set. This question is measured by the *coefficient of determination*, which is denoted as R^2. In our gasoline and crude oil example, the R^2 was calculated as 0.9285.

It is beneficial to discuss some terminology about R^2.

The following will be a simplistic view of how to derive R^2. Statisticians like to talk about the following equation when they talked about the explanatory power of a regression equation,

$$\begin{aligned} &\text{SST} = \text{SSE} + \text{SSR} \\ &\text{Total} \rightarrow \text{Unexplained} \rightarrow \text{Explained} \end{aligned}$$

The above equation indicates that the explanatory power of a regression equals the explained plus the unexplained. Let us now manipulate the above equation to get R^2. From the above equation, we can get

$$\text{SST}/\text{SST} = 1 = \text{SSE}/\text{SST} + \text{SSR}/\text{SST}$$

Finally, we can get

$$R^2 = \text{SSR}/\text{SST} = 1 - \text{SSE}/\text{SST}$$

13.6 Regression Analysis in Minitab

```
MTB > name  c1 'year'
MTB > name c2 'gasoline'
MTB > name c3 'crude oil'
MTB > set 'year'
DATA> 1975 1976 1977 1978 1979 1980 1981 1982 1983 1984 1985 1986 1987 1988
DATA> end
MTB > set 'gasoline'
DATA> 57 59 62 63 86 119 133 122 116 113 112 86 90 90
DATA> end
MTB > set 'crude oil'
DATA> 7.67 8.19 8.57 9.00 12.64 21.59 31.77 28.52 26.19 25.88 24.09 12.51 15.40 12.57
DATA> end.
MTB > Regress;
SUBC>   Response 'gasoline';
SUBC>   Nodefault;
SUBC>   Continuous 'crude oil';
SUBC>   Terms C3;
SUBC>   Constant;
SUBC>   Unstandardized;
SUBC>   TExpand;
SUBC>   Tmethod;
SUBC>   Tanova;
SUBC>   Tsummary;
SUBC>   Tcoefficients;
SUBC>   Tequation;
SUBC>   TDiagnostics 1.
```

13.6.1 Regression Analysis: Gasoline Versus Crude Oil

Analysis of Variance

Source	DF	Seq SS	Contribution	Adj SS	Adj MS	F-Value	P-Value
Regression	1	8202.3	92.85%	8202.3	8202.27	155.95	0.000
crude oil	1	8202.3	92.85%	8202.3	8202.27	155.95	0.000
Error	12	631.2	7.15%	631.2	52.60		
Total	13	8833.4	100.00%				

Model Summary

S	R-sq	R-sq(adj)	PRESS	R-sq(pred)
7.25237	92.85%	92.26%	818.588	90.73%

Coefficients

Term	Coef	SE Coef	95% CI	T-Value	P-Value	VIF
Constant	41.92	4.56	(31.99, 51.85)	9.20	0.000	
crude oil	2.948	0.236	(2.434, 3.463)	12.49	0.000	1.00

Regression Equation

gasoline = 41.92 + 2.948 crude oil

Fits and Diagnostics for All Observations

Obs	gasoline	Fit	SE Fit	95% CI	Resid	Std Resid	Del Resid	HI
1	57.00	64.53	3.02	(57.95, 71.11)	-7.53	-1.14	-1.16	0.173235
2	59.00	66.06	2.93	(59.69, 72.44)	-7.06	-1.06	-1.07	0.162719
3	62.00	67.19	2.86	(60.96, 73.41)	-5.19	-0.78	-0.76	0.155396
4	63.00	68.45	2.79	(62.38, 74.52)	-5.45	-0.81	-0.80	0.147479
5	86.00	79.19	2.25	(74.29, 84.09)	6.81	0.99	0.99	0.096162
6	119.00	105.57	2.17	(100.85, 110.30)	13.43	1.94	2.24	0.089413
7	133.00	135.59	3.89	(127.11, 144.07)	-2.59	-0.42	-0.41	0.288144
8	122.00	126.01	3.25	(118.93, 133.09)	-4.01	-0.62	-0.60	0.200827
9	116.00	119.14	2.83	(112.98, 125.30)	-3.14	-0.47	-0.45	0.152008
10	113.00	118.22	2.77	(112.18, 124.27)	-5.22	-0.78	-0.77	0.146380
11	112.00	112.95	2.49	(107.52, 118.37)	-0.95	-0.14	-0.13	0.117868
12	86.00	78.80	2.26	(73.87, 83.74)	7.20	1.04	1.05	0.097511
13	90.00	87.32	2.00	(82.97, 91.68)	2.68	0.38	0.37	0.075973
14	90.00	78.98	2.26	(74.06, 83.90)	11.02	1.60	1.73	0.096884

Obs	Cook's D	DFITS
1	0.14	-0.530183
2	0.11	-0.472175
3	0.06	-0.327854
4	0.06	-0.333622
5	0.05	0.322048
6	0.18	0.702534
7	0.04	-0.259747
8	0.05	-0.301478
9	0.02	-0.192193
10	0.05	-0.317189
11	0.00	-0.048612
12	0.06	0.344850
13	0.01	0.106068
14	0.14	0.565288

From the above, we can see that Minitab calculated R^2 as 0.929. As discussed above on how to get R^2, we can also derive R^2 from the Minitab report. The R^2 calculation is shown below:

$$\begin{aligned} R^2 &= 1 - \text{Error/Total} \\ &= 1 - 631.12/8833.4 \\ &= 1 - 0.071447 \\ &= 0.929 \end{aligned}$$

The Minitab regression output indicates that the line that best fit the gasoline data set is

```
gasoline = 41.9 + 2.95 crude oil
```

The bottom half of the Minitab output analyzes each data item in the data set. The *Fit* column indicates the calculated gasoline value from the linear equation that best fits the gasoline data. For example, the following calculates the gasoline value for the first observed data item.

$$\text{Gasoline} = 41.9 + 2.95*(57.00) = 64.53$$

The *Residual* column indicates the difference between the actual gasoline values versus the calculated gasoline value. For example, for the first observed data item, the residual is

$$\text{Residual} = 57.00 - 64.53 = -7.53$$

As indicated above, in regression analysis it is assumed that the residuals are normally distributed. The *St Resid* transforms the residual differences to a standard normal distribution. This column indicates that the first observation is −1.14 standard deviations away from the expected residual mean.

13.7 Regression Analysis in Excel

When we first looked at the gasoline and crude oil example, we graph the data and let Excel chart calculated the regression line and R^2. We will now use Excel's Data analysis add-in to a regression analysis; to use the *Data Analysis* feature, choose **Data→Data Analysis** and then choose the regression option. The data should be entered in Excel the same way as in Sect. 13.2.

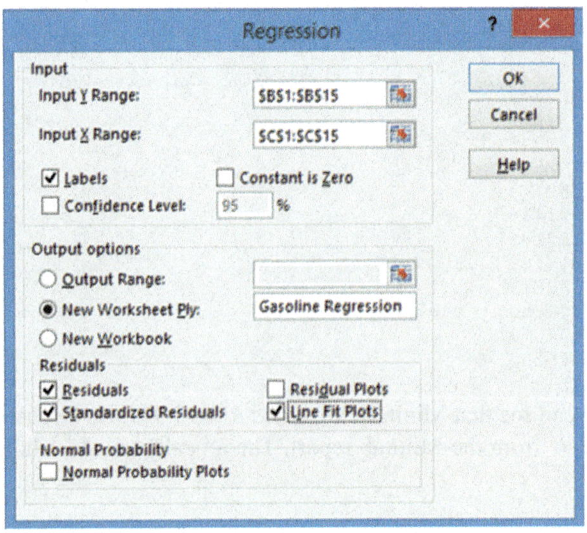

After filling out the *Regression* dialog box as shown above, press the *OK* button. After pressing the *OK* button, you will get the following regression analysis in Excel:

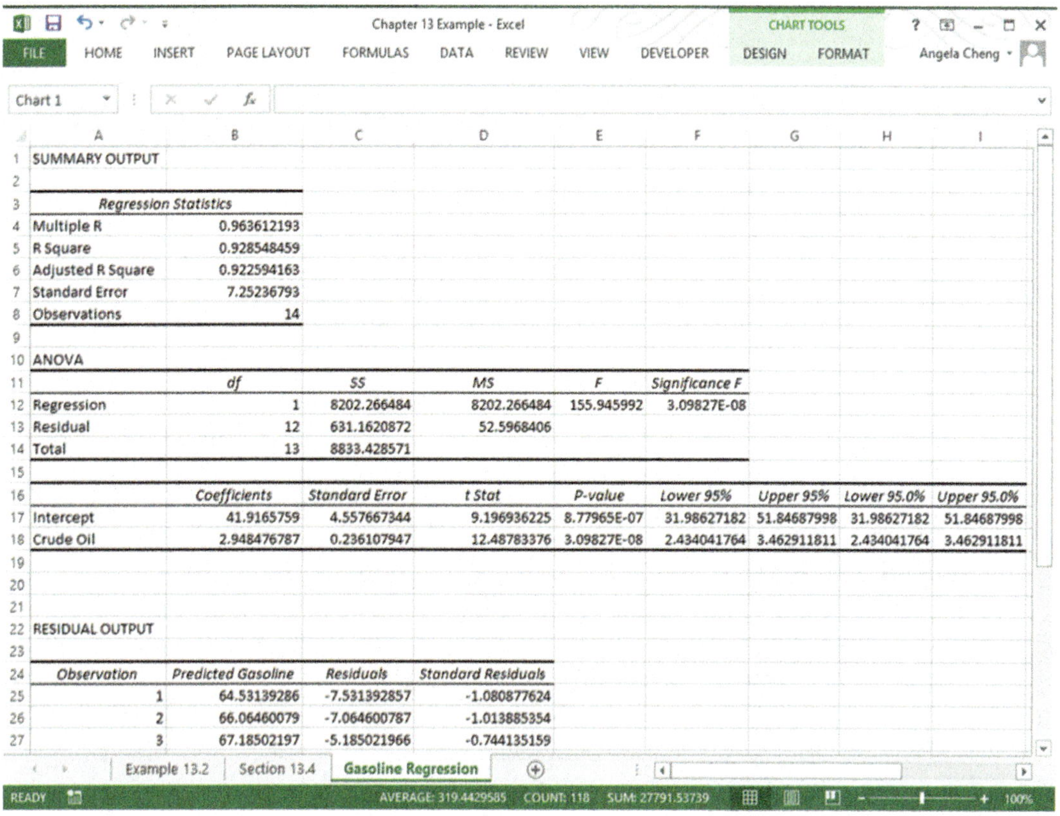

SUMMARY OUTPUT

Regression Statistics	
Multiple R	0.963612193
R Square	0.928548459
Adjusted R Square	0.922594163
Standard Error	7.25236793
Observations	14

ANOVA

	df	SS	MS	F	Significance F
Regression	1	8202.266484	8202.266484	155.945992	3.09827E-08
Residual	12	631.1620872	52.5968406		
Total	13	8833.428571			

	Coefficients	Standard Error	t Stat	P-value	Lower 95%	Upper 95%	Lower 95.0%	Upper 95.0%
Intercept	41.9165759	4.557667344	9.196936225	8.77965E-07	31.98627182	51.84687998	31.98627182	51.84687998
Crude Oil	2.948476787	0.236107947	12.48783376	3.09827E-08	2.434041764	3.462911811	2.434041764	3.462911811

RESIDUAL OUTPUT

Observation	Predicted Gasoline	Residuals	Standard Residuals
1	64.53139286	-7.531392857	-1.080877624
2	66.06460079	-7.064600787	-1.013885354
3	67.18502197	-5.185021966	-0.744135159

A chart was also produced because we chose the *Line Fit Plots* option in the *Regression* dialog box. The chart is shown below:

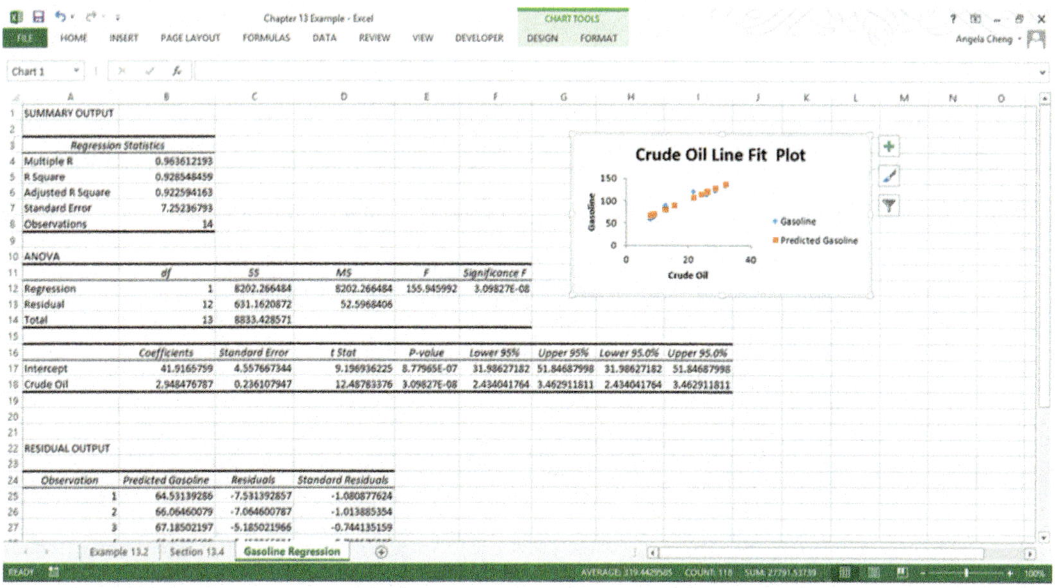

The chart does not look very good because it is too small. To solve this problem, we will put the chart on its own sheet. To do this, move the cursor over the chart and click on the right mouse button. This will result in a short-cut menu shown below. In the short-cut menu, choose the *Move Chart* menu-item.

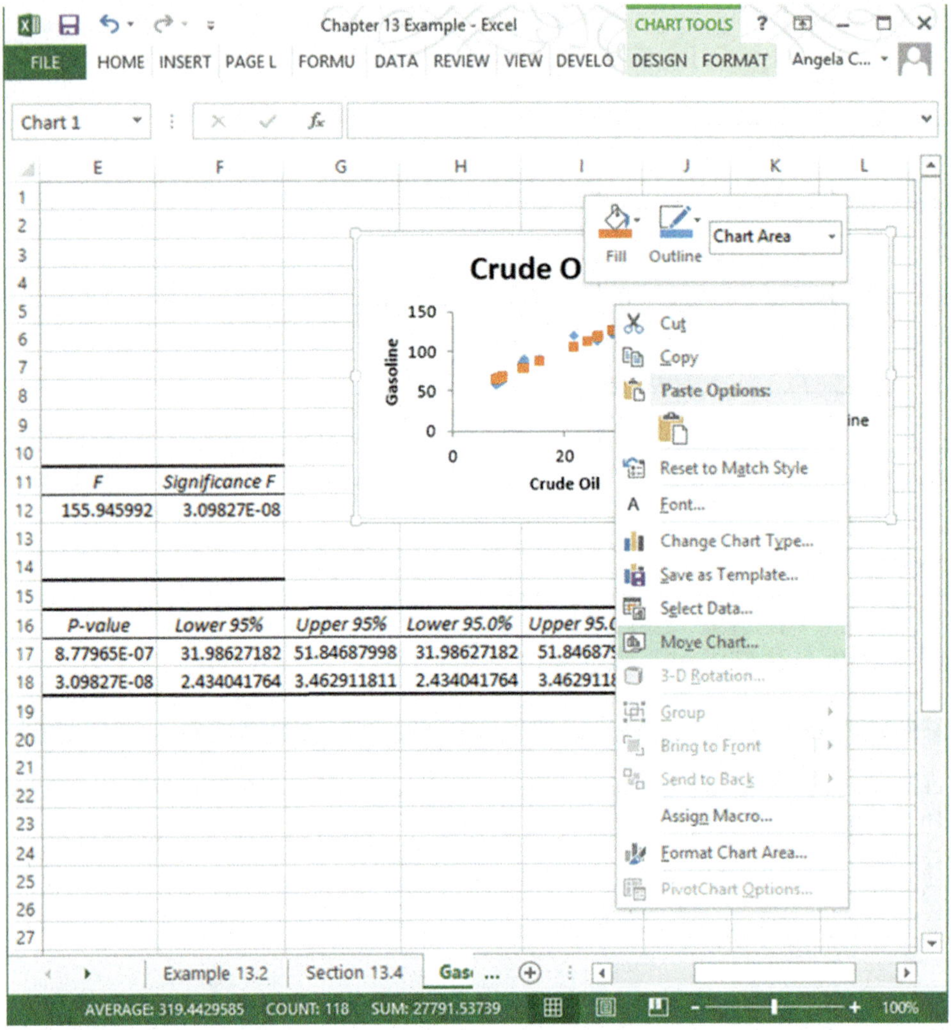

Choosing the *Move Chart* menu-item will result in the *Move Chart* dialog box.

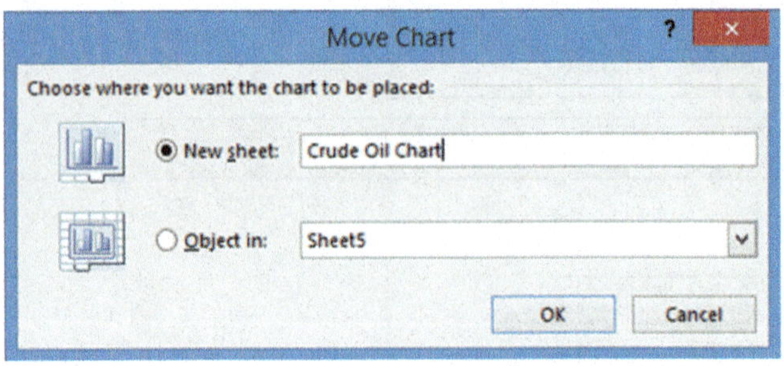

Fill out the *Move Chart* dialog box as shown above. Pressing the *OK* butt[...]n the following chart:

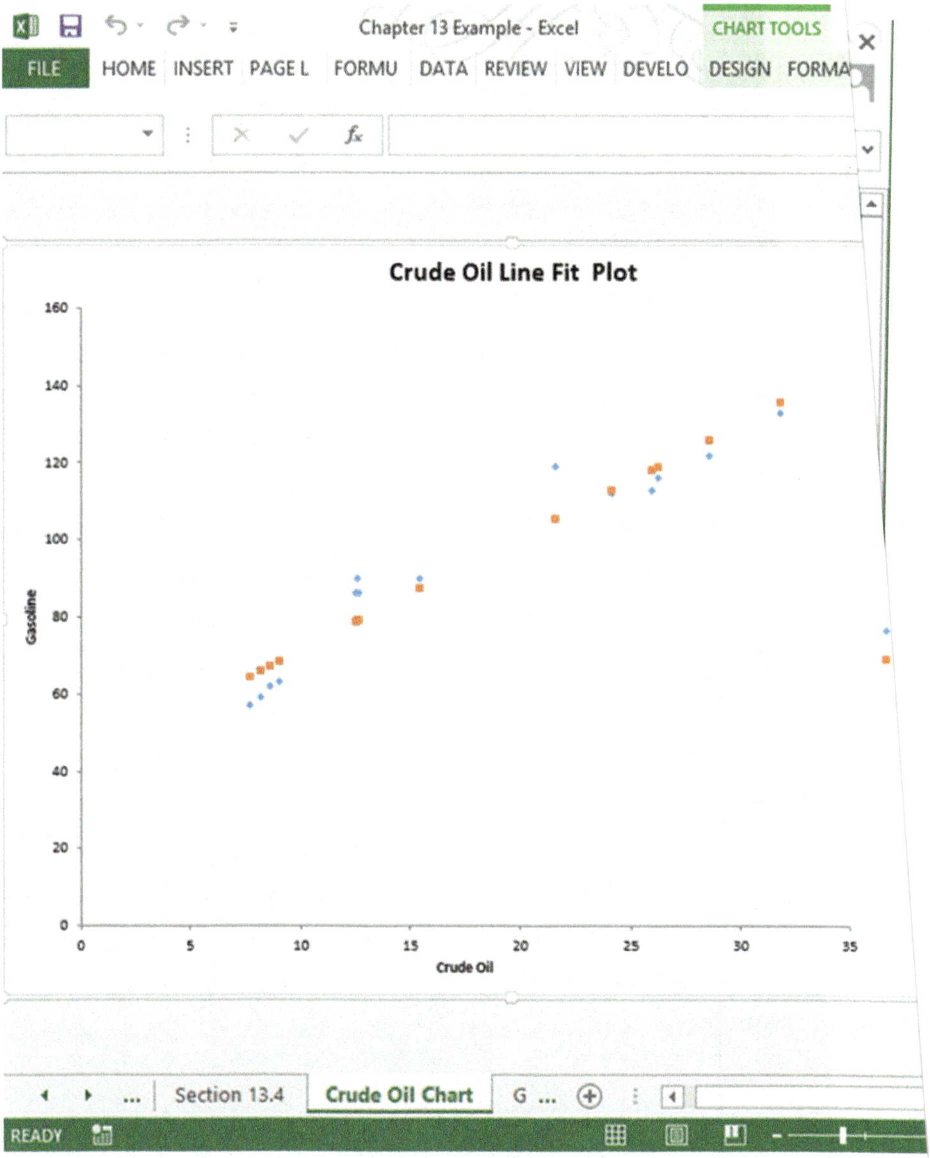

Excel produces the chart with a white background.

13.8 Correlation Coefficient

Correlation analysis is another way to study the relationship between two variables. In pa[...] measures the strength of association between two variables. Correlation is expressed [...] between −1 and +1. When $r = +1$, then x and y exist in a perfectly positive linear rel[...]

$r = -1$ mean riables x and y exist in a perfectly negative linear relationship; and $r = 0$ means that x nearly related. These concepts are graphically illustrated below:

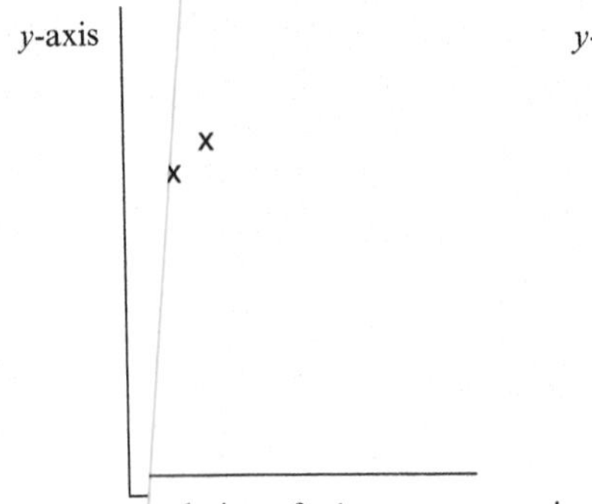

rrelation of +1

$r = 1$

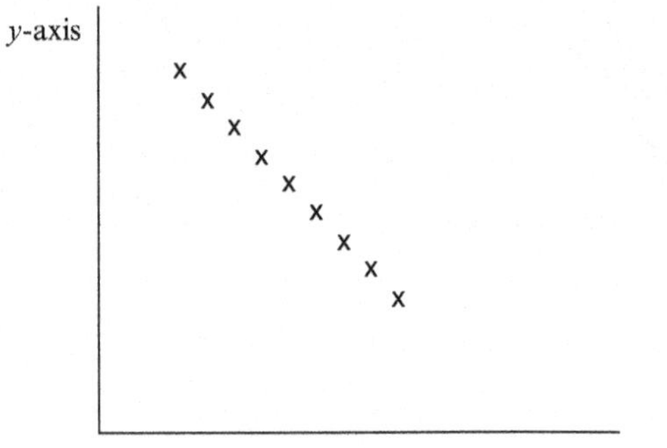

Perfect Correlation of –1

$r = -1$

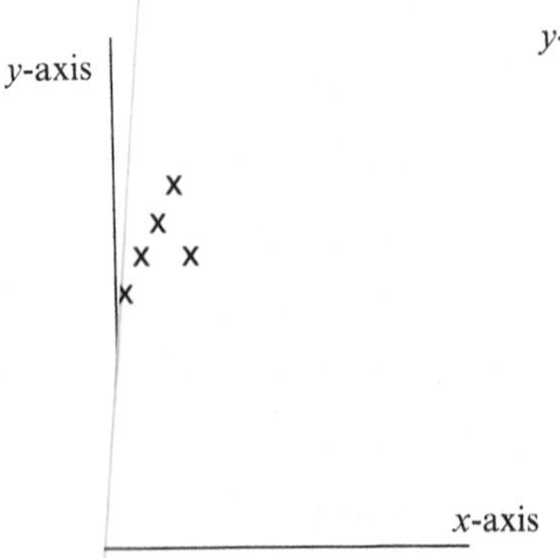

, correlation less than +1

$< r < 1$

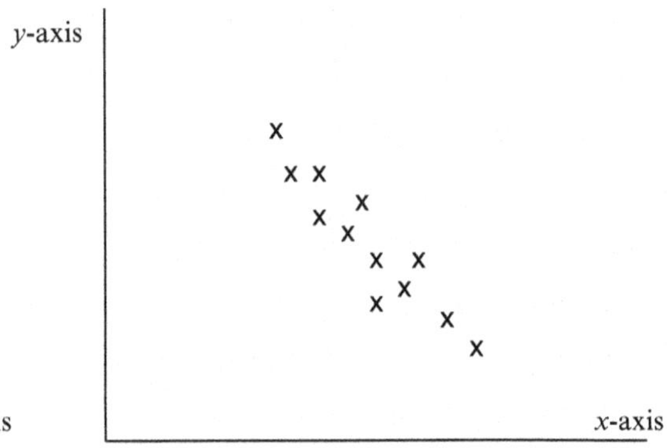

Line Deteriorates, correlation greater than –1

$-1 < r < 0$

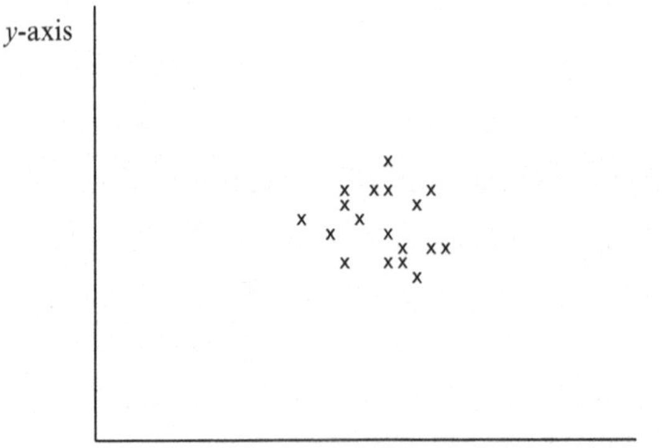

No linear relationship

$r = 0$

The formula for the correlation coefficient is

$$r = \frac{\text{sample covariance}}{(\text{sample standard deviation of } x)\ (\text{sample standard deviation of } y)}$$

$$= \frac{\frac{1}{n-1}\sum_{i=1}^{n}(x_i - \bar{x})(y_i - \bar{y})}{\left[\frac{1}{n-1}\sum_{i=1}^{n}(x_i - \bar{x})^2\right]^{1/2}\left[\frac{1}{n-1}\sum_{i=1}^{n}(y_i - \bar{y})^2\right]^{1/2}} = \frac{s_{xy}}{s_x s_y}$$

Let us look again at the gasoline/crude oil problem, using the Minitab *correlation* command to calculate the correlation coefficient.

```
MTB > correlation 'gasoline' 'crude oil'
```

13.8.1 Correlations: Gasoline, Crude Oil

```
Pearson correlation of gasoline and crude oil = 0.964
P-Value = 0.000
```

In Excel there is the *correl* worksheet function to calculate the correlation coefficient. The use of this function is shown below:

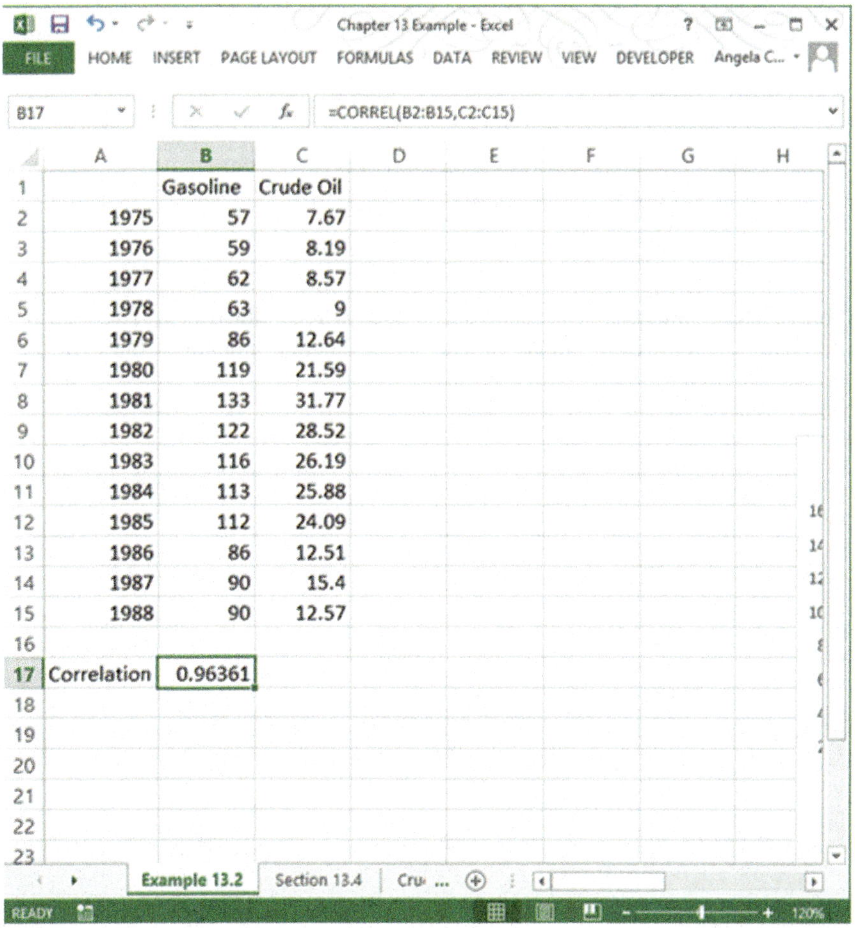

B17 =CORREL(B2:B15,C2:C15)

	A	B	C
1		Gasoline	Crude Oil
2	1975	57	7.67
3	1976	59	8.19
4	1977	62	8.57
5	1978	63	9
6	1979	86	12.64
7	1980	119	21.59
8	1981	133	31.77
9	1982	122	28.52
10	1983	116	26.19
11	1984	113	25.88
12	1985	112	24.09
13	1986	86	12.51
14	1987	90	15.4
15	1988	90	12.57
16			
17	Correlation	0.96361	

From the correlation coefficient value, 0.964, we can see that there is a strong positive relationship between pounds and inches.

If we look carefully at the graphical representation of regression analysis and correlation analysis, we might find that they are very similar. Because of these similarities, we might ask if there is any relationship between regression analysis and correlation analysis. In fact, there is a relationship, as expressed by the following equation:

$$\text{R-square} = r^2$$

We can check this relationship by squaring our calculated correlation coefficient of 0.964. This would be 0.964 * 0.964 = 0.928, which equals our R-square.

13.9 Regression Examples

Example 13.2 Suppose we have random samples of ten households showing the numbers of cars per household, as shown below:

Numbers of cars per household		
Household	Cars, y	People, x
1	4	6
2	1	2
3	3	4
4	2	3
5	2	4
6	3	4
7	4	6
8	1	3
9	2	2
10	2	2

In this example, Excel functions *intercept* and *slope* will be illustrated. The *intercept* function calculates the a in the regression equation. The *slope* function calculates the b in the regression equation.

The calculations in Excel are shown below:

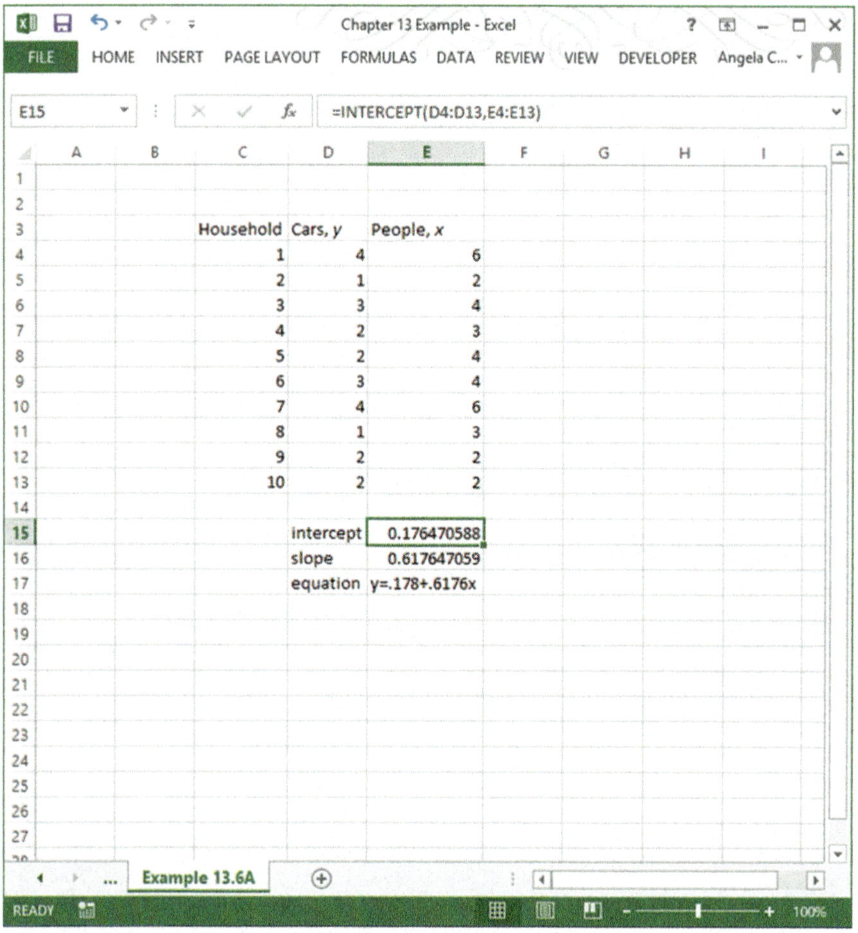

E15 =INTERCEPT(D4:D13,E4:E13)

	C	D	E
3	Household	Cars, *y*	People, *x*
4	1	4	6
5	2	1	2
6	3	3	4
7	4	2	3
8	5	2	4
9	6	3	4
10	7	4	6
11	8	1	3
12	9	2	2
13	10	2	2
15		intercept	0.176470588
16		slope	0.617647059
17		equation	y=.178+.6176x

How the above worksheet is setup is shown below:

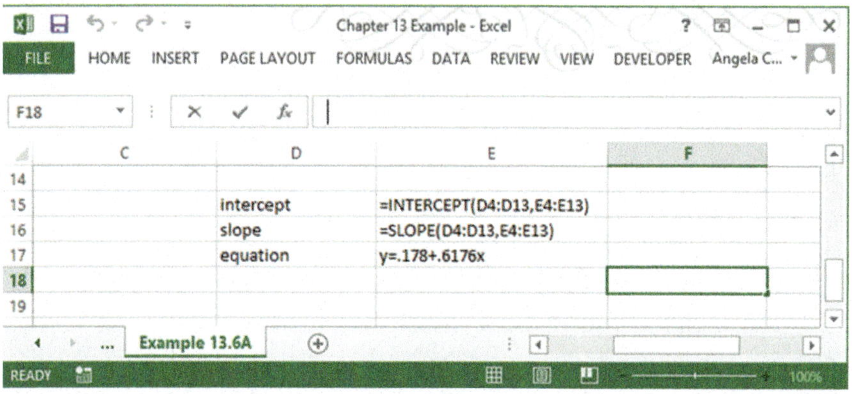

	D	E
15	intercept	=INTERCEPT(D4:D13,E4:E13)
16	slope	=SLOPE(D4:D13,E4:E13)
17	equation	y=.178+.6176x

Below shows the use of Minitab to do a regression analysis to investigate the relationship between the number of people in a household and the number of cars in a household:

```
MTB > name c1 'household'
MTB > name c2 'cars'
MTB > name c3 'people'
MTB > set 'household'
DATA> 1 2 3 4 5 6 7 8 9 10
DATA> end
MTB > set 'cars'
DATA> 4 1 3 2 2 3 4 1 2 2
DATA> end
MTB > set 'people'
DATA> 6 2 4 3 4 4 6 3 2 2
DATA> end
MTB > Regress;
SUBC>   Response 'cars';
SUBC>   Nodefault;
SUBC>   Continuous 'people';
SUBC>   Terms C3;
SUBC>   Constant;
SUBC>   Unstandardized;
SUBC>   TExpand;
SUBC>   Tmethod;
SUBC>   Tanova;
SUBC>   Tsummary;
SUBC>   Tcoefficients;
SUBC>   Tequation;
SUBC>   TDiagnostics 1.
```

13.9.1 Regression Analysis: Cars Versus People

Analysis of Variance

Source	DF	Seq SS	Contribution	Adj SS	Adj MS	F-Value	P-Value
Regression	1	7.7824	74.83%	7.7824	7.7824	23.78	0.001
people	1	7.7824	74.83%	7.7824	7.7824	23.78	0.001
Error	8	2.6176	25.17%	2.6176	0.3272		
Lack-of-Fit	2	0.7843	7.54%	0.7843	0.3922	1.28	0.344
Pure Error	6	1.8333	17.63%	1.8333	0.3056		
Total	9	10.4000	100.00%				

Model Summary

S	R-sq	R-sq(adj)	PRESS	R-sq(pred)
0.572019	74.83%	71.68%	3.71012	64.33%

Coefficients

Term	Coef	SE Coef	95% CI	T-Value	P-Value	VIF
Constant	0.176	0.491	(-0.955, 1.308)	0.36	0.728	
people	0.618	0.127	(0.326, 0.910)	4.88	0.001	1.00

Regression Equation

cars = 0.176 + 0.618 people

Fits and Diagnostics for All Observations

Obs	cars	Fit	SE Fit	95% CI	Resid	Std Resid	Del Resid	HI	Cook's D
1	4.000	3.882	0.354	(3.067, 4.698)	0.118	0.26	0.25	0.382353	0.02
2	1.000	1.412	0.272	(0.785, 2.038)	-0.412	-0.82	-0.80	0.225490	0.10
3	3.000	2.647	0.188	(2.214, 3.080)	0.353	0.65	0.63	0.107843	0.03
4	2.000	2.029	0.196	(1.577, 2.482)	-0.029	-0.05	-0.05	0.117647	0.00
5	2.000	2.647	0.188	(2.214, 3.080)	-0.647	-1.20	-1.24	0.107843	0.09
6	3.000	2.647	0.188	(2.214, 3.080)	0.353	0.65	0.63	0.107843	0.03
7	4.000	3.882	0.354	(3.067, 4.698)	0.118	0.26	0.25	0.382353	0.02
8	1.000	2.029	0.196	(1.577, 2.482)	-1.029	-1.92	-2.44	0.117647	0.24
9	2.000	1.412	0.272	(0.785, 2.038)	0.588	1.17	1.20	0.225490	0.20
10	2.000	1.412	0.272	(0.785, 2.038)	0.588	1.17	1.20	0.225490	0.20

Obs	DFITS
1	0.193434
2	-0.431265
3	0.218350
4	-0.018700
5	-0.429927
6	0.218350
7	0.193434
8	-0.889513
9	0.647617
10	0.647617

Example 13.3 Wallin and Gilman (1986) used a simple linear regression analysis to investigate the effect of research and development (R&D) spending on a company's value. Data for the 20 largest R&D spenders in terms of the 1981–1982 averages are presented below. In the table below, y and x represent the price/earnings (P/E) ratio and R&D expenditures (R/S) ratio, respectively.

P/E ratio and R/S ratio for top 20 R&D spenders (1981–1982)

Company	P/E ratio	R/S ratio
1	5.6	0.003
2	7.2	0.004
3	8.1	0.009
4	9.9	0.021
5	6	0.023
6	8.2	0.03
7	6.3	0.035
8	10	0.037
9	8.5	0.044
10	13.2	0.051
11	8.4	0.058
12	11.1	0.058
13	11.1	0.067
14	13.2	0.08
15	13.4	0.08
16	11.5	0.083
17	9.8	0.091
18	16.1	0.092
19	7	0.064
20	5.9	0.028

Source: C. C. Wallin and J.J. Gilman (1986) "Determining the Optimum Level for R&D Spending," Research Management, Vol. 14, No. 5, Sept./Oct. 1986, 19–24 (adapted from Fig. 1, p. 20)

Below is the regression analysis done in Minitab:

```
MTB > name c1 'P/E'
MTB > name c2 'R/S'
MTB > read 'P/E' 'R/S'
DATA> 5.6 .003
DATA> 7.2 .004
DATA> 8.1 .009
DATA> 9.9 .021
DATA> 6.0 .023
DATA> 8.2 .030
DATA> 6.3 .035
DATA> 10.0 .037
DATA> 8.5 .044
DATA> 13.2 .051
DATA> 8.4 .058
DATA> 11.1 .058
DATA> 11.1 .067
DATA> 13.2 .080
DATA> 13.4 .080
DATA> 11.5 .083
DATA> 9.8 .091
DATA> 16.1 .092
DATA> 7.0 .064
DATA> 5.9 .028
DATA> end
MTB > correlation 'P/E' 'R/S'
```

13.9.2 Correlations: P/E, R/S

```
Pearson correlation of P/E and R/S = 0.726
P-Value = 0.000

MTB > plot 'P/E' * 'R/S'
```

13.9.3 Plot P/E * R/S

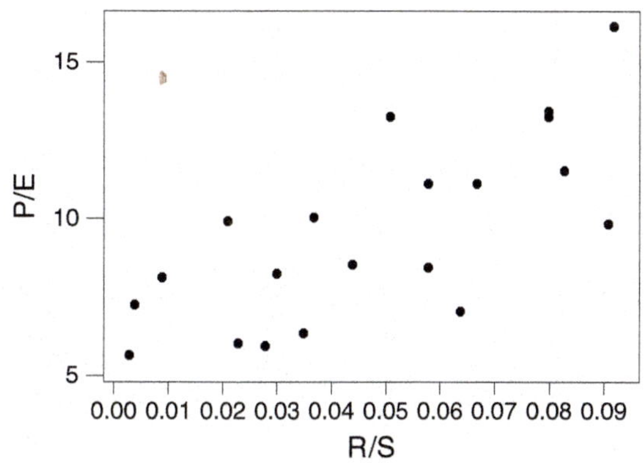


```
MTB > Regress;
SUBC>   Response 'P/E';
SUBC>   Nodefault;
SUBC>   Continuous 'R/S';
SUBC>   Terms C2;
SUBC>   Constant;
SUBC>   Unstandardized;
SUBC>   TExpand;
SUBC>   Tmethod;
SUBC>   Tanova;
SUBC>   Tsummary;
SUBC>   Tcoefficients;
SUBC>   Tequation;
SUBC>   Tdw;
SUBC>   TDiagnostics 1.
```

13.9.4 Regression Analysis: P/E versus R/S

Analysis of Variance

Source	DF	Seq SS	Contribution	Adj SS	Adj MS	F-Value	P-Value
Regression	1	86.404	52.74%	86.404	86.404	20.09	0.000
R/S	1	86.404	52.74%	86.404	86.404	20.09	0.000
Error	18	77.414	47.26%	77.414	4.301		
Lack-of-Fit	16	73.749	45.02%	73.749	4.609	2.52	0.322
Pure Error	2	3.665	2.24%	3.665	1.832		
Total	19	163.817	100.00%				

Model Summary

S	R-sq	R-sq(adj)	PRESS	R-sq(pred)
2.07383	52.74%	50.12%	96.2706	41.23%

Coefficients

Term	Coef	SE Coef	95% CI	T-Value	P-Value	VIF
Constant	5.977	0.917	(4.050, 7.904)	6.52	0.000	
R/S	74.1	16.5	(39.4, 108.8)	4.48	0.000	1.00

Regression Equation

P/E = 5.977 + 74.1 R/S

Fits and Diagnostics for All Observations

Obs	P/E	Fit	SE Fit	95% CI	Resid	Std Resid	Del Resid	HI
1	5.600	6.199	0.875	(4.361, 8.038)	-0.599	-0.32	-0.31	0.178002
2	7.200	6.273	0.861	(4.465, 8.082)	0.927	0.49	0.48	0.172364
3	8.100	6.644	0.793	(4.979, 8.309)	1.456	0.76	0.75	0.146078
4	9.900	7.533	0.642	(6.183, 8.882)	2.367	1.20	1.22	0.095944
5	6.000	7.681	0.620	(6.378, 8.983)	-1.681	-0.85	-0.84	0.089366
6	8.200	8.199	0.550	(7.044, 9.355)	0.001	0.00	0.00	0.070344
7	6.300	8.570	0.510	(7.497, 9.642)	-2.270	-1.13	-1.14	0.060566
8	10.000	8.718	0.497	(7.673, 9.763)	1.282	0.64	0.63	0.057544
9	8.500	9.236	0.468	(8.253, 10.220)	-0.736	-0.36	-0.36	0.050966
10	13.200	9.755	0.467	(8.774, 10.735)	3.445	1.71	1.81	0.050610
11	8.400	10.273	0.493	(9.238, 11.309)	-1.873	-0.93	-0.93	0.056477
12	11.100	10.273	0.493	(9.238, 11.309)	0.827	0.41	0.40	0.056477
13	11.100	10.940	0.561	(9.761, 12.118)	0.160	0.08	0.08	0.073163
14	13.200	11.903	0.705	(10.422, 13.383)	1.297	0.67	0.65	0.115424
15	13.400	11.903	0.705	(10.422, 13.383)	1.497	0.77	0.76	0.115424
16	11.500	12.125	0.743	(10.565, 13.685)	-0.625	-0.32	-0.31	0.128224
17	9.800	12.717	0.850	(10.932, 14.503)	-2.917	-1.54	-1.61	0.167945
18	16.100	12.791	0.864	(10.977, 14.606)	3.309	1.75	1.87	0.173482
19	7.000	10.717	0.535	(9.594, 11.841)	-3.717	-1.86	-2.00	0.066458
20	5.900	8.051	0.568	(6.857, 9.245)	-2.151	-1.08	-1.08	0.075144

Obs	Cook's D	DFITS
1	0.01	-0.144570
2	0.03	0.219283
3	0.05	0.310455
4	0.08	0.396305
5	0.04	-0.263896
6	0.00	0.000108
7	0.04	-0.289034
8	0.01	0.154705
9	0.00	-0.082364
10	0.08	0.417811
11	0.03	-0.226592
12	0.01	0.098062
13	0.00	0.021928
14	0.03	0.236437
15	0.04	0.274033
16	0.01	-0.120608
17	0.24	-0.722773
18	0.32	0.858187
19	0.12	-0.534922
20	0.05	-0.308920

Durbin-Watson Statistic

Durbin-Watson Statistic = 2.57937

Below is the regression report from Excel.
Summary Output

Regression statistics	
Multiple R	0.726249237
R square	0.527437955
Adjusted R square	0.501184508
Standard error	2.073830766
Observations	20

ANOVA

	df	SS	MS	*F*	Significance *F*
Regression	1	86.40356719	86.40356719	20.09023638	0.000288085
Residual	18	77.41393281	4.300774045		
Total	19	163.8175			

	Coefficients	Standard error	*t* Stat	*P*-value	Lower 95 %	Upper 95 %	Lower 95.0 %	Upper 95.0 %
Intercept	5.977161615	0.917371118	6.515532807	3.98998E-06	4.049834923	7.904488307	4.049834943	7.904488307
R/S ratio	74.06760721	16.524784	4.482213335	0.000288085	39.35029742	108.784917	39.35029742	108.784917

13.10 SAS Programming Code Instructions and Examples

13.10.1 Regression Analysis

Example 13.2 Let us read the data into SAS first.

```
Data Gasoline;
   INPUT Year Y X @@;
   DATALINES;
1975 57 7.67
1976 59 8.19
1977 62 8.75
1978 63 9.00
1979 86 12.64
1980 119 21.59
1981 133 31.77
1982 122 28.52
1983 116 26.19
1984 113 25.88
1985 112 24.09
1986 86 12.51
1987 90 15.40
1988 90 12.57
 ;
run;
```

To graph the data above, **PROC GPLOT** function is used as we mentioned in the previous chapter.

```
title  height=7 'Scatter Plot of Gasoline  and Crude Oil';
legend1 label=none value=('Gasoline' 'Crude Oil') shape=symbol(5,1);
axis1  label=('Dollar') ;
axis2  label=('Year') ;
symbol1 i=none  v=PLUS L=1 c=blue h=3;
symbol2 i=none  v=STAR L=1 c=red h=3;
proc gplot data=Gasoline;
plot (Y X)*Year/vaxis=axis1 haxis=axis2 overlay legend=legend1 autovref;
run;
```

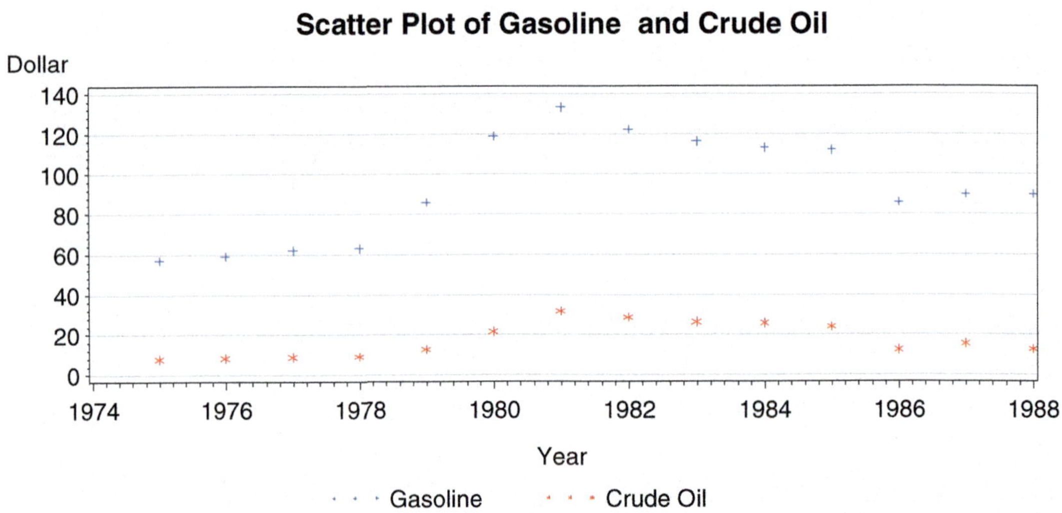

To detect the relationship between the gasoline and crude oil, we use the following command:

```
title  height=7 'Gasoline & Crude Oil';
axis1  label=('Gasoline') ;
axis2  label=('Crude Oil') ;
symbol1 i=none  v=PLUS L=1 c=blue h=3;
symbol2 i=none  v=STAR L=1 c=red h=3;
proc gplot data=Gasoline;
plot Y*X/vaxis=axis1 haxis=axis2 overlay autovref;
run;
```

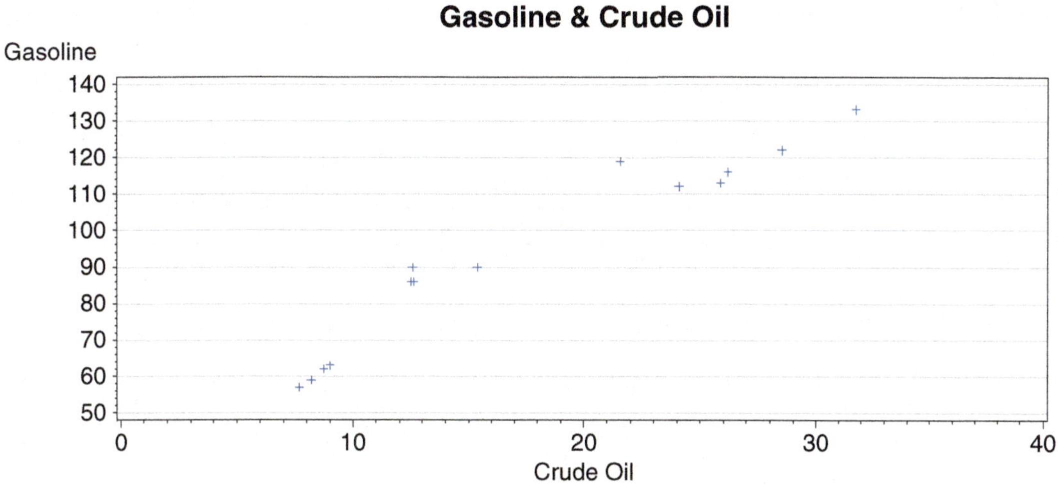

REGEGN option is specified to add the best linear line that fitted the gasoline and crude oil graph.

```
Goptions ftext='Times New Roman' htext=6 gunit=pct;
title  height=7 'Gasoline & Crude Oil';
symbol1 v=dot  c=black L=1 h=2 i=rl;
axis1  label=('Cents/gallon');
axis2  label=('dollar/barrel');
Proc gplot data=Gasoline;
plot  Y*X /vaxis=axis1 haxis=axis2 regeqn;
Run;
```

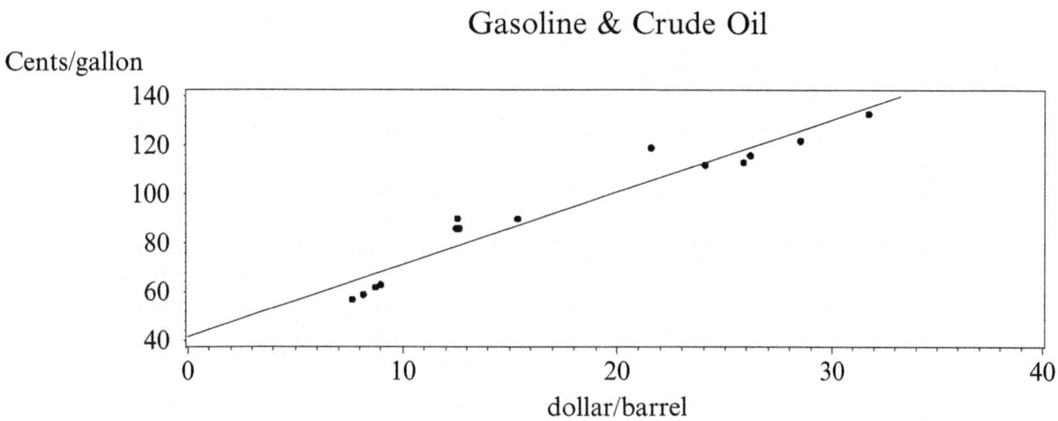

Regression Equation:
Y = 41.80984 + 2.952414*X

13.10.2 Regression Analysis in SAS

In SAS, PROC REG is used to do regression analysis. CLM option produces confidence limits for a mean predicted value for each observation. P option is specified to display observed, predicted, and residual values for each observation that does not contain missing values for independent variables.

```
PROC REG DATA = Gasoline;
 MODEL Y = X /P CLM;
 output out=Gasoline_out p=predict;
RUN;
```

Analysis of variance

Source	DF	Sum of squares	Mean square	F value	Pr > F
Model	1	8196.51586	8196.51586	154.43	< 0.0001
Error	12	636.91271	53.07606		
Corrected total	13	8833.42857			

Root MSE	7.28533	R-square	0.9279
Dependent mean	93.42857	Adj R-Sq	0.9219
Coeff var	7.79776		

Parameter estimates

Variable	DF	Parameter estimate	Standard error	t value	Pr > $\|t\|$
Intercept	1	41.80984	4.58747	9.11	< 0.0001
X	1	2.95241	0.23758	12.43	< 0.0001

Output statistics

Obs	Dependent variable	Predicted value	Std error Mean predict	95 % CL mean		Residual
1	57.0000	64.4548	3.0376	57.8364	71.0733	−7.4548
2	59.0000	65.9901	2.9439	59.5760	72.4042	−6.9901
3	62.0000	67.6435	2.8454	61.4438	73.8431	−5.6435
4	63.0000	68.3816	2.8024	62.2756	74.4875	−5.3816
5	86.0000	79.1283	2.2617	74.2005	84.0562	6.8717
6	119.0000	105.5524	2.1778	100.8074	110.2975	13.4476
7	133.0000	135.6080	3.9130	127.0823	144.1337	−2.6080
8	122.0000	126.0127	3.2659	118.8968	133.1285	−4.0127
9	116.0000	119.1335	2.8407	112.9441	125.3230	−3.1335
10	113.0000	118.2183	2.7876	112.1447	124.2919	−5.2183
11	112.0000	112.9335	2.5009	107.4844	118.3825	−0.9335
12	86.0000	78.7445	2.2776	73.7821	83.7070	7.2555
13	90.0000	87.2770	2.0090	82.8997	91.6543	2.7230
14	90.0000	78.9217	2.2702	73.9753	83.8681	11.0783

Sum of residuals	0
Sum of squared residuals	636.91271
Predicted residual SS (PRESS)	826.05572

```
   title  height=7 'Crude Oil Line Fit Plot';
   legend1 label=none value=('Gasoline' 'Predicted Gasoline')
shape=symbol(5,1);
   axis1  label=('Gasoline') ;
   axis2  label=('Crude Oil') ;
   symbol1 i=none  v=PLUS L=1 c=blue h=3;
   symbol2 i=none  v=STAR L=1 c=red h=3;
   proc gplot data=Gasoline_out;
   plot (Y predict)*X/vaxis=axis1 haxis=axis2 overlay legend=legend1 autovref;
   run;
```

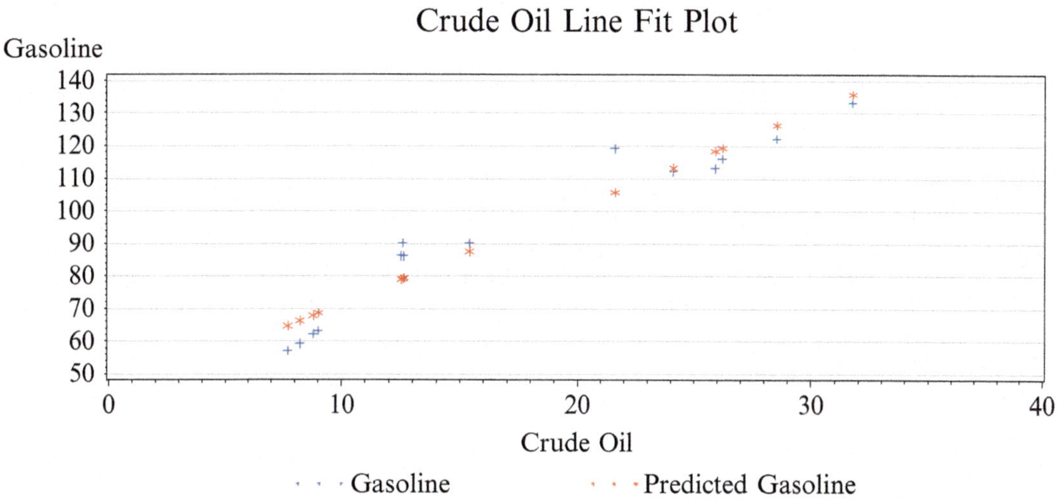

13.10.3 Correlation Coefficient

Let us look again at the gasoline/crude oil problem, using the SAS command **PROC CORR** to calculate the correlation coefficient.

```
proc corr data=Gasoline cov;
     var Y X;
   run ;
```

Pearson correlation coefficients, $N = 14$		
Prob > $\|r\|$ under H0: Rho = 0		
	Y	*X*
Y	1.00000	0.96327 < 0.0001
X	0.96327 < 0.0001	1.00000

In this form,

r = correlation between two variables in the sample

Rho = correlation between the same two variables in the population

A common assumption is that there is NO relationship between X and Y in the population: Rho = 0. The simplest formula for computing the appropriate t value to test significance of a correlation coefficient employs the t distribution:

$$t = r\sqrt{\frac{n-2}{1-r^2}}$$

The degree of freedom for entering the t-distribution is $n-2$. Based on the value of t, we could calculate the significance of the correlation coefficient.

13.10.4 Regression Examples

Example 13.4 Below shows the use of SAS to do a regression analysis to investigate the relationship between the number of people in a household and the number of cars in a household.

```
Data CarsPerHousehold;
    INPUT Household Cars People @@;
    DATALINES;
1 4 6
2 1 2
3 3 4
4 2 3
5 2 4
6 3 4
7 4 6
8 1 3
9 2 2
10 2 2
;
run;

PROC REG DATA=CarsPerHousehold;
     MODEL Cars = People / P CLM;
RUN;
```

Analysis of variance					
Source	DF	Sum of squares	Mean square	F value	Pr > F
Model	1	7.78235	7.78235	23.78	0.0012
Error	8	2.61765	0.32721		
Corrected total	9	10.40000			

Root MSE	0.57202	R-square	0.7483
Dependent mean	2.40000	Adj R-Sq	0.7168
Coeff var	23.83413		

Parameter estimates					
Variable	DF	Parameter estimate	Standard error	t value	Pr > $\|t\|$
Intercept	1	0.17647	0.49050	0.36	0.7283
People	1	0.61765	0.12665	4.88	0.0012

Output statistics						
Obs	Dependent variable	Predicted value	Std error Mean predict	95 % CL mean		Residual
1	4.0000	3.8824	0.3537	3.0667	4.6980	0.1176
2	1.0000	1.4118	0.2716	0.7854	2.0381	−0.4118
3	3.0000	2.6471	0.1878	2.2139	3.0802	0.3529
4	2.0000	2.0294	0.1962	1.5770	2.4819	−0.0294
5	2.0000	2.6471	0.1878	2.2139	3.0802	−0.6471
6	3.0000	2.6471	0.1878	2.2139	3.0802	0.3529
7	4.0000	3.8824	0.3537	3.0667	4.6980	0.1176
8	1.0000	2.0294	0.1962	1.5770	2.4819	−1.0294
9	2.0000	1.4118	0.2716	0.7854	2.0381	0.5882
10	2.0000	1.4118	0.2716	0.7854	2.0381	0.5882

Sum of residuals	0
Sum of squared residuals	2.61765
Predicted residual SS (PRESS)	3.71012

Example 13.5 Below is the regression report from SAS.

```
Data RandD;
   INPUT Company PE RS @@;
   DATALINES;
1  5.6 .003
2  7.2 .004
3  8.1 .009
4  9.9 .021
5  6.0 .023
6  8.2 .030
7  6.3 .035
8  10.0 .037
9  8.5 .044
10  13.2 .051
11  8.4 .058
12  11.1 .058
13  11.1 .067
14  13.2 .080
15  13.4 .080
16  11.5 .083
17  9.8 .091
18  16.1 .092
19  7.0 .064
20  5.9 .028
;
run;

PROC REG DATA=RandD;
    MODEL PE = RS / P CLM;
RUN;
```

Analysis of variance

Source	DF	Sum of squares	Mean square	*F* value	Pr > *F*
Model	1	86.40357	86.40357	20.09	0.0003
Error	18	77.41393	4.30077		
Corrected total	19	163.81750			

Root MSE	2.07383	R-square	0.5274
Dependent mean	9.52500	Adj R-Sq	0.5012
Coeff var	21.77250		

Parameter estimates

Variable	DF	Parameter estimate	Standard error	*t* value	Pr > |*t*|
Intercept	1	5.97716	0.91737	6.52	< 0.0001
RS	1	74.06761	16.52478	4.48	0.0003

Output statistics						
Obs	Dependent variable	Predicted value	Std error Mean predict	95 % CL mean		Residual
1	5.6000	6.1994	0.8750	4.3612	8.0376	−0.5994
2	7.2000	6.2734	0.8610	4.4646	8.0823	0.9266
3	8.1000	6.6438	0.7926	4.9785	8.3090	1.4562
4	9.9000	7.5326	0.6424	6.1830	8.8821	2.3674
5	6.0000	7.6807	0.6200	6.3782	8.9832	−1.6807
6	8.2000	8.1992	0.5500	7.0436	9.3548	0.000810
7	6.3000	8.5695	0.5104	7.4973	9.6418	−2.2695
8	10.0000	8.7177	0.4975	7.6725	9.7628	1.2823
9	8.5000	9.2361	0.4682	8.2525	10.2197	−0.7361
10	13.2000	9.7546	0.4665	8.7744	10.7348	3.4454
11	8.4000	10.2731	0.4928	9.2377	11.3085	−1.8731
12	11.1000	10.2731	0.4928	9.2377	11.3085	0.8269
13	11.1000	10.9397	0.5609	9.7612	12.1182	0.1603
14	13.2000	11.9026	0.7046	10.4223	13.3828	1.2974
15	13.4000	11.9026	0.7046	10.4223	13.3828	1.4974
16	11.5000	12.1248	0.7426	10.5646	13.6849	−0.6248
17	9.8000	12.7173	0.8499	10.9318	14.5028	−2.9173
18	16.1000	12.7914	0.8638	10.9767	14.6061	3.3086
19	7.0000	10.7175	0.5346	9.5943	11.8407	−3.7175
20	5.9000	8.0511	0.5685	6.8567	9.2454	−2.1511

Sum of residuals	0
Sum of squared residuals	77.41393
Predicted residual SS (PRESS)	96.27062

13.11 Statistical Summary

In this chapter, we look at two methods of analyzing the relationship between two variables. The two methods are the simple linear regression method and the correlation method. The simple linear regression analysis can be used to measure how much one variable is explained by another variable.

It is important to note that in algebra linear equations have a *deterministic relationship*, while in statistics linear equations have a *stochastic relationship*.

The correlation analysis can be used to measure the strength of the linear relationship between two variables.

Bibliography

Microsoft Inc., Excel 2013. Microsoft Inc., Redmond
Lee CF, Lee JC, Lee AC (2013) Statistics for business and financial economics. Springer, New York
Minitab Inc. Minitab 17. Minitab Inc., State College
SAS Institute Inc. SAS 2014. SAS Institute Inc., Cary

Chapter 14
Simple Linear Regression and Correlation: Analyses and Applications

14.1 Introduction

When we do regression analysis, there are five underlying assumptions:

1. The dependent and independent variables have a linear relationship.
2. The expected value of the error term is zero.
3. The variance of the error term is constant.
4. The error terms are independent.
5. The errors are normally distributed.

In general, we use sample instead of population data to estimate regression coefficients. If the regression model follows the above five assumptions, then we can make inferences about the population regression coefficients from the sample regression coefficients. In the last chapter, we used regression and correlation analysis to relate two variables on the basis of sample information. But data from a sample represent only part of the total population. Because of this, we may think of our estimated sample regression line as an estimate of a true, but unknown population regression line of the form:

$$y = \alpha + \beta x$$

Therefore, what we are doing is using the sample regression line

$$y = a + bx$$

to estimate the population regression line.

14.2 Two-Tail t Test for β

One statistical inference that we are interested in is to test the hypothesis about the value of β. Specifically, we want to know whether β is equal to zero. If β does equal to zero, then this would mean that the independent variable actually does not explain any part of the variation of the dependent variable. The null hypothesis (H_0) and the alternative hypothesis (H_1) would be the following.

C.-F. Lee et al., *Essentials of Excel, Excel VBA, SAS and Minitab for Statistical and Financial Analyses*, DOI 10.1007/978-3-319-38867-0_14

$$H_0: \beta = 0$$
$$H_1: \beta \neq 0$$

We will use the t distribution to test the above hypothesis. The test statistic is shown below.

$$t = \frac{\dfrac{\sum_{i=1}^{n}(x_i-\bar{x})(y_i-\bar{y})}{\sum_{i=1}^{n}(x_i-\bar{x})^2} - 0}{\dfrac{s_e}{\sqrt{\sum_{i=1}^{n}(x_i-\bar{x})^2}}} = \frac{b-0}{s_b}$$

where s_e and s_b are the sample standard deviation of the error terms and standard error of the slope, respectively.

Example 14.1

Sample Height and Weight data for Children

x, inches	y, pounds
55	92
56	95
57	99
58	97
59	102
60	104

Below is the regression analysis in Minitab.

```
MTB > name c1 'height'
MTB > name c2 'weight'
MTB > read 'height' 'weight'
DATA> 55 92
DATA> 56 95
DATA> 57 99
DATA> 58 97
DATA> 59 102
DATA> 60 104
DATA> end
6 rows read.

MTB > Regress;
SUBC>   Response 'weight';
SUBC>   Nodefault;
SUBC>   Continuous 'height';
SUBC>   Terms height;
SUBC>   Constant;
SUBC>   Unstandardized;
SUBC>   TExpand;
SUBC>   Tmethod;
SUBC>   Tanova;
SUBC>   Tsummary;
SUBC>   Tcoefficients;
SUBC>   Tequation;
SUBC>   TDiagnostics 1.
```

Regression Analysis: Weight Versus Height

Analysis of Variance

Source	DF	Seq SS	Contribution	Adj SS	Adj MS	F-Value	P-Value
Regression	1	89.157	90.21%	89.157	89.157	36.86	0.004
height	1	89.157	90.21%	89.157	89.157	36.86	0.004
Error	4	9.676	9.79%	9.676	2.419		
Total	5	98.833	100.00%				

Model Summary

S	R-sq	R-sq(adj)	PRESS	R-sq(pred)
1.55533	90.21%	87.76%	15.4607	84.36%

Coefficients

Term	Coef	SE Coef	95% CI	T-Value	P-Value	VIF
Constant	-31.6	21.4	(-91.0, 27.8)	-1.48	0.213	
height	2.257	0.372	(1.225, 3.289)	6.07	0.004	1.00

Regression Equation

weight = -31.6 + 2.257 height

Fits and Diagnostics for All Observations

Obs	weight	Fit	SE Fit	95% CI	Resid	Std Resid	Del Resid	HI
1	92.000	92.524	1.126	(89.398, 95.649)	-0.524	-0.49	-0.44	0.523810
2	95.000	94.781	0.845	(92.435, 97.127)	0.219	0.17	0.15	0.295238
3	99.000	97.038	0.662	(95.201, 98.875)	1.962	1.39	1.68	0.180952
4	97.000	99.295	0.662	(97.458, 101.132)	-2.295	-1.63	-2.44	0.180952
5	102.000	101.552	0.845	(99.206, 103.899)	0.448	0.34	0.30	0.295238
6	104.000	103.810	1.126	(100.684, 106.935)	0.190	0.18	0.15	0.523810

Obs	Cook's D	DFITS
1	0.13	-0.45711
2	0.01	0.09437
3	0.21	0.79111
4	0.29	-1.14633
5	0.02	0.19505
6	0.02	0.16183

The *regress* command calculated the t test statistic as 6.07. Usually we would have to check a t table to find the associated t value for an alpha value of 0.05. Then we would test to see if the calculated sample t value was greater or less than the value in the t table.

With the advent of computer software, however, the decision process is much simpler. Most statistical software also calculates the p-value associated with the calculated t value. To reject or accept the null hypothesis, we only need to compare the calculated p-value with the alpha value that we are interested in.

The *p-value* of a one-tail test is area to the area to the right or left calculated t test statistics, depending if it a right or left tail test. Because the t distribution is symmetrical, the *p-value* for a two-tail test is twice the area of a one-tail test.

If we want to test at the 5 % alpha level, we can reject the null hypothesis that $\beta = 0$ because the p-value associated with the test statistic is 0.004, which is smaller than the 0.05 alpha value.

We will reject the null hypothesis any time the *p*-value is less than the alpha value.

In the above regression report, the heading `SE Coef` is called the standard error of the coefficient. The numbers represent the standard deviation of α and β of the regression line. For example, 0.3718 is the standard deviation of the coefficient for independent variable height.

14.3 Standard Error of the Regression Line

Regression output also includes a number that represents the variability of actual observation of the y values versus the expected y values of the regression line. This number is called the *standard error of the regression line*. In the previous example the standard error of the regression line was 1.555. Below shows the best regression line that fits the above data set.

```
MTB > Fitline 'weight' 'height';
SUBC>   Confidence 95.0.
```

Regression Analysis: Weight Versus Height

```
The regression equation is
weight = - 31.62 + 2.257 height


S = 1.55533    R-Sq = 90.2%    R-Sq(adj) = 87.8%


Analysis of Variance

Source       DF       SS       MS      F       P
Regression    1  89.1571  89.1571  36.86   0.004
Error         4   9.6762   2.4190
Total         5  98.8333
```

Fitted Line: Weight Versus Height

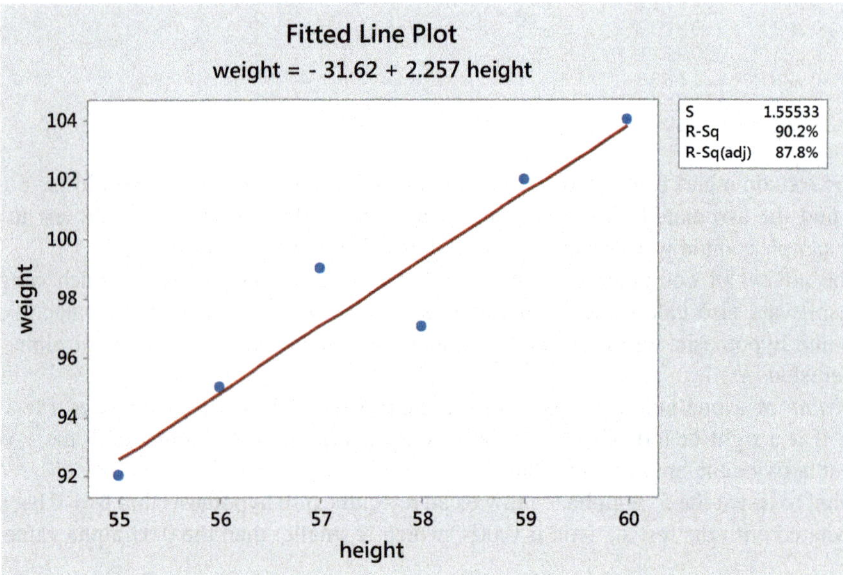

14.4 Two-Tail t Test for α

We can also test whether the population intercept α is significantly different from zero or not. As with the test for the population slope, β, we can use the regression output to test the significance of α of a linear equation. We will use the t statistic defined on the previous page to test whether α is significantly different from zero or not. The null and alternative hypotheses for α would be:

$$H_0: \alpha = 0$$
$$H_1: \alpha \neq 0$$

The part of the regression output that is necessary for our analysis is

```
Predictor          Coef          Stdev      t-ratio          p
Constant         -31.62          21.39        -1.48      0.213
```

We can see that the calculated t value for α is -1.48 and the associated p-value is 0.213. If the alpha value that we are interested in is 0.05, we cannot reject the null hypothesis because the p-value is larger than the alpha value of 0.05.

14.5 Confidence Interval of β

Another way to make inferences about the population parameter is to create a confidence interval. If we are interested in a 95 % confidence interval, this would mean that 95 % of the confidence intervals created will contain the true population parameter. The concept of confidence intervals was discussed in Chap. 10.

The formula for the β confidence interval is

$$b - t_{\alpha/2}s_b \leq \beta \leq b + t_{\alpha/2}s_b$$

where β is the population slope and b is the sample slope. Below uses Excel to calculate the β confidence interval.

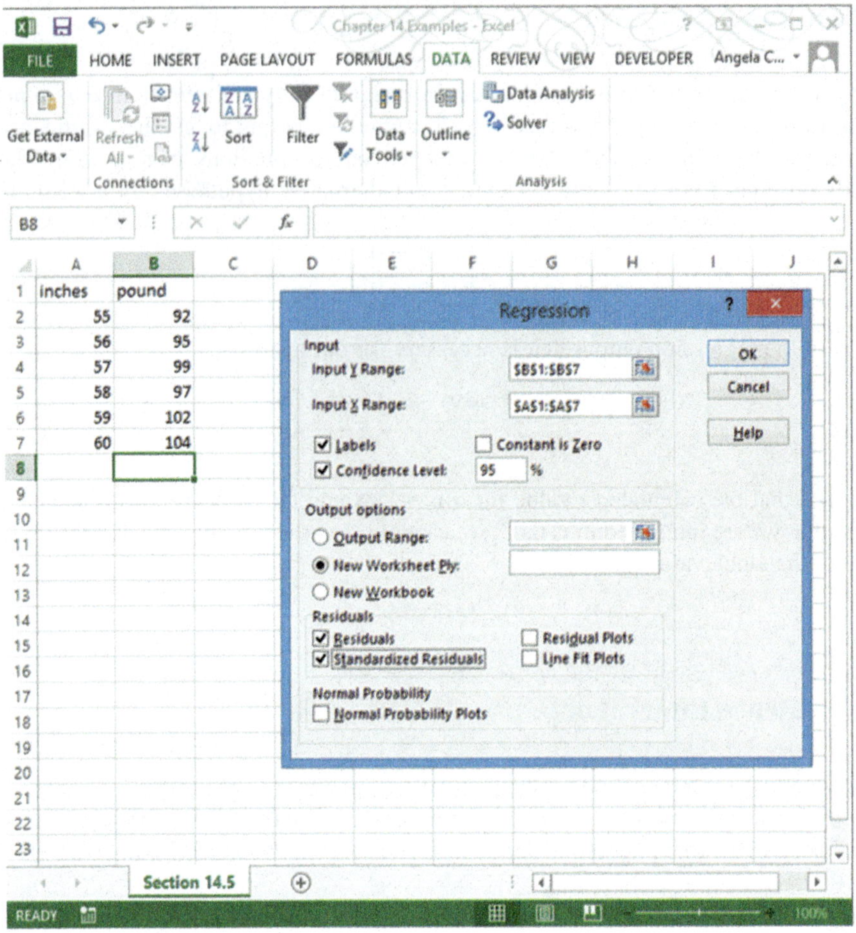
Chapter 14 Examples - Excel
FILE
HOME
INSERT
PAGE LAYOUT
FORMULAS
DATA
REVIEW
VIEW
DEVELOPER
Angela C...
Get External Data
Refresh All
Connections
Sort
Filter
Sort & Filter
Data Tools
Outline
Data Analysis
Solver
Analysis
B8
inches
pound
55 92
56 95
57 99
58 97
59 102
60 104
Regression
Input
Input Y Range: $B$1:$B$7
Input X Range: $A$1:$A$7
Labels
Constant is Zero
Confidence Level: 95 %
OK
Cancel
Help
Output options
Output Range:
New Worksheet Ply:
New Workbook
Residuals
Residuals
Standardized Residuals
Residual Plots
Line Fit Plots
Normal Probability
Normal Probability Plots
Section 14.5
READY

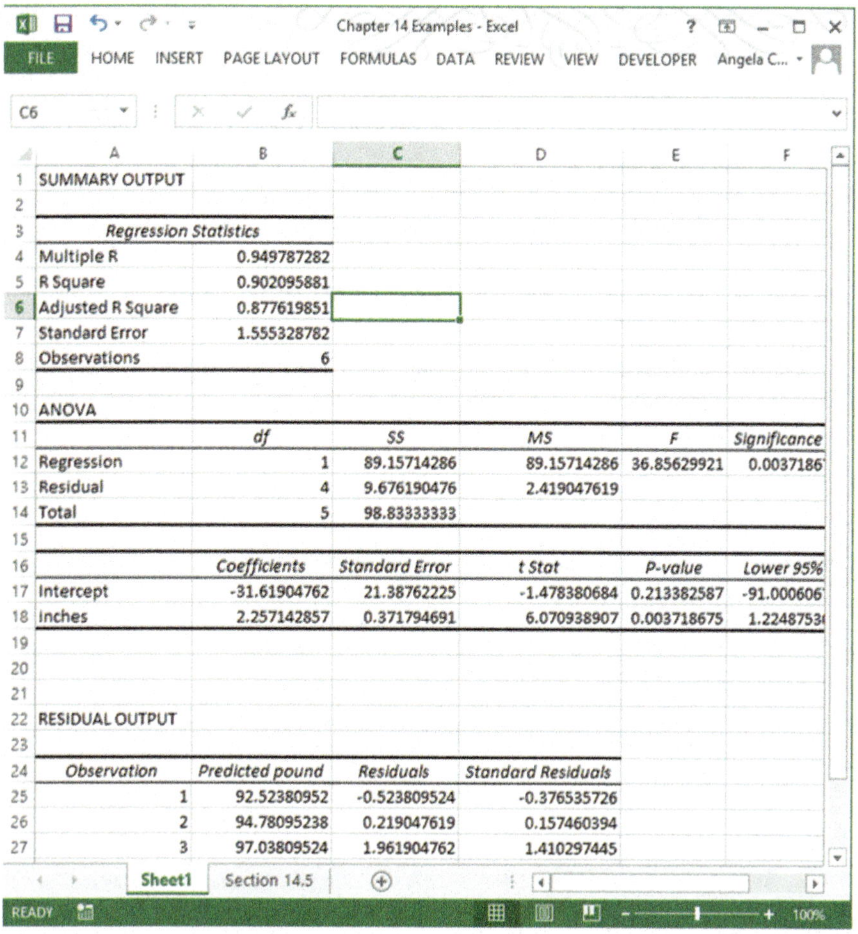

	A	B	C	D	E	F
1	SUMMARY OUTPUT					
2						
3	*Regression Statistics*					
4	Multiple R	0.949787282				
5	R Square	0.902095881				
6	Adjusted R Square	0.877619851				
7	Standard Error	1.555328782				
8	Observations	6				
9						
10	ANOVA					
11		*df*	*SS*	*MS*	*F*	*Significance*
12	Regression	1	89.15714286	89.15714286	36.85629921	0.0037186
13	Residual	4	9.676190476	2.419047619		
14	Total	5	98.83333333			
15						
16		*Coefficients*	*Standard Error*	*t Stat*	*P-value*	*Lower 95%*
17	Intercept	-31.61904762	21.38762225	-1.478380684	0.213382587	-91.000606
18	Inches	2.257142857	0.371794691	6.070938907	0.003718675	1.2248753
19						
20						
21						
22	RESIDUAL OUTPUT					
23						
24	*Observation*	*Predicted pound*	*Residuals*	*Standard Residuals*		
25	1	92.52380952	-0.523809524	-0.376535726		
26	2	94.78095238	0.219047619	0.157460394		
27	3	97.03809524	1.961904762	1.410297445		

From the above regression analysis, we can see that the 95 % confidence interval for the inches independent variable constant is $1.225 \leq \beta \leq 3.2892$

14.6 *F* Test

We used the *t* test to see if β is significantly different from zero. We can also use the *F* test to see if β is significantly different from zero. Below is a reproduction of the part of the regression output that we will need to use the *F* test.

```
Analysis of Variance

Source            DF       SS        MS        F        P
Regression         1   89.157    89.157    36.86    0.004
Residual Error     4    9.676     2.419
Total              5   98.833
```

From the regression output we can see that the calculated *F* value was 36.86 for the height/weight problem and the associated *p*-value was 0.004. If we use an alpha value of 0.05, we can conclude that β is significantly different from zero and that the height does indeed help explain the variation in weight.

The difference between the *F* test and the *t* test is that the *F* test asks if all the coefficients of the independent variables are equal to zero or not. Unlike the *F* test, the *t* test is calculated for each individual coefficient of each independent variable. We will see the difference between the *F* test and the *t* test in the next chapter when we analyze regression lines with more than one independent variable.

14.7 The Relationship Between the *F* Test and the *t* Test

One question that we might ask is, "If we are using both the *F* test and the *t* test to test the same hypothesis, is there any relationship between the *F* test and the *t* test?" The answer is yes. For a simple regression, the mathematical relationship is

$$F = t^2$$

Our *t* value is 6.07 and our *F* value is 36.86. If we square our *t* value, we get

$$t^2 = 6.07^2 = 36.8449$$

The slight discrepancy in this *F* and *t* relationship is due to rounding errors.

14.8 Predicting

One of the important uses of a simple regression line, as we have seen, is to obtain predictions about the dependent variable based on the value of the independent variable.

We will use the *predict* subcommand of the *regress* command to predict the value of the dependent variable in the following problem.

Ralph Farmer of the Department of Agriculture is interested in the relationship between the amount of fertilizer used and the number of bushels of wheat harvested. He collects the following information on six farmers.

x pounds of fertilizer	y bushels of wheat
100	1000
150	1250
180	1710
200	2100
222	2500

He wants to predict the number of bushels of wheat that would be harvested if 160 lb of fertilizer were used. Put pounds of fertilizer in column c1 and bushels of wheat in column c2. The worksheet should look like the following.

Next we will analyze the data using regression analysis and a *predict* subcommand.

```
MTB > name c1 'fertilizer'
MTB > name c2 'wheat'
MTB > read 'fertilizer' 'wheat'
DATA> 100 1000
DATA> 150 1250
DATA> 180 1710
DATA> 200 2100
DATA> 222 2500
DATA> end

MTB > Predict 'wheat';
SUBC>   Nodefault;
SUBC>   KPredictors 160;
SUBC>   TEquation;
SUBC>   TPrediction.
```

Prediction for Wheat

```
Regression Equation

wheat = -402 + 12.40 fertilizer


Variable    Setting
fertilizer      160


   Fit    SE Fit          95% CI                  95% PI
1583.00  86.8152  (1306.71, 1859.28)  (922.413, 2243.58)
```

The output indicates that the predicted number of bushels is 1583 for 160 lb of fertilizer. The output gives both the confidence and prediction intervals. The 95 % confidence interval for 160 lb of fertilizer is from 1306.7 bushels to 1859.3 bushels. The 95 % prediction interval for 160 lb of fertilizer is from 922.4 bushels to 2243.6 bushels.

14.9 Regression Examples

Example 14.2 Healthy Hamburgers has a chain of 12 stores in northern Illinois. Sales figures and profits are shown below. Our task is to obtain a regression line for the data and predict what the profit will be if a store does $10 million in sales.

sales, x (millions)	profits, y (millions)
7.00	0.15
2.00	0.10
6.00	0.13
4.00	0.15
14.00	0.25
15.00	0.27
16.00	0.24
12.00	0.20
14.00	0.27
20.00	0.44
15.00	0.34
7.00	0.17

To do this analysis, we will put sales in column c1 and name the column *sales*. We will put profits in column c2 and name the column *profits*.

We will begin our analysis by plotting the data.

```
MTB > name c1 'sales'
MTB > name c2 'profits'
MTB > read 'sales' 'profits'
DATA> 7 .15
DATA> 2 .1
DATA> 6 .13
DATA> 4 .15
DATA> 14 .25
DATA> 15 .27
DATA> 16 .24
DATA> 12 .2
DATA> 14 .27
DATA> 20 .44
DATA> 15 .34
DATA> 7 .17
DATA> end

MTB >  Plot 'profits' * 'sales'
```

Scatterplot of Profits Versus Sales

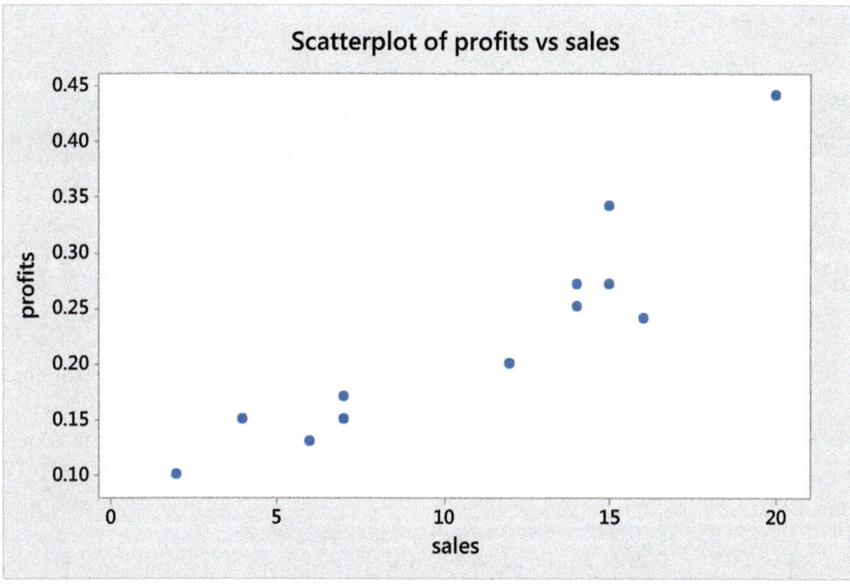

Next we will analyze the data by finding a regression line for the data.

```
MTB > Regress;
SUBC>   Response 'profits';
SUBC>   Nodefault;
SUBC>   Continuous 'sales';
SUBC>   Terms sales;
SUBC>   Constant;
SUBC>   Unstandardized;
SUBC>   Tmethod;
SUBC>   Tanova;
SUBC>   Tsummary;
SUBC>   Tcoefficients;
SUBC>   Tequation;
SUBC>   TDiagnostics 0.
```

Regression Analysis: Profits Versus Sales

```
Analysis of Variance

Source          DF    Adj SS    Adj MS  F-Value  P-Value
Regression       1  0.087298  0.087298    52.61    0.000
  sales          1  0.087298  0.087298    52.61    0.000
Error           10  0.016594  0.001659
  Lack-of-Fit    7  0.013744  0.001963     2.07    0.296
  Pure Error     3  0.002850  0.000950
Total           11  0.103892


Model Summary

        S    R-sq  R-sq(adj)  R-sq(pred)
0.0407357  84.03%     82.43%      74.08%


Coefficients

Term         Coef  SE Coef  T-Value  P-Value   VIF
Constant   0.0506   0.0269     1.88    0.089
sales     0.01593  0.00220     7.25    0.000  1.00


Regression Equation

profits = 0.0506 + 0.01593 sales


Fits and Diagnostics for Unusual Observations

                                   Std
Obs  profits     Fit   Resid  Resid
 10   0.4400  0.3692  0.0708   2.11  R

R  Large residual

MTB > Predict 'profits';
SUBC>   Nodefault;
SUBC>   KPredictors 10;
SUBC>   TEquation;
SUBC>   TPrediction.
```

Prediction for Profits

```
Regression Equation

profits = 0.0506 + 0.01593 sales


Variable  Setting
sales          10


     Fit     SE Fit              95% CI                    95% PI
0.209903  0.0119627  (0.183248, 0.236558)  (0.115305, 0.304501)
```

From the regression output we can see that the regression line is

```
profits = 0.0506 + 0.0159 sales
```

This line indicates that for every \$1 million increase in sales, there will be an expected \$0.0159 million dollar increase in profits.

The regression model predicts that \$10 million in sales will result in an expected \$0.2099 million in profits. The 95 % prediction interval for this profit is from \$0.1153 million to \$0.3045 million.

The *regress* command calculated the *p*-value for the sales coefficient as zero. Therefore, we can conclude that the sales coefficient does not equal zero, which means that sales do explain profits.

Below is the regression done in Excel. On the Excel ribbon choose Data → Data Analysis → Regression.

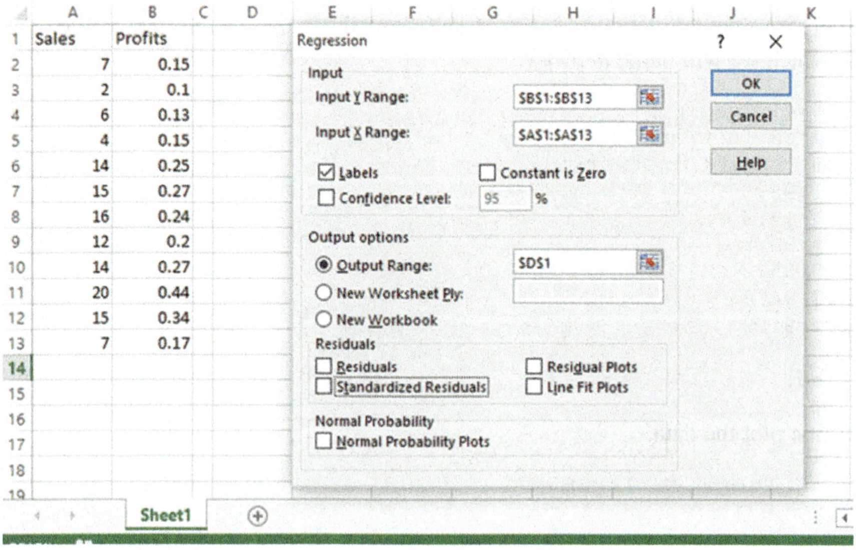

	A	B	C	D	E	F	G	H	I	J	K	L
1	Sales	Profits		SUMMARY OUTPUT								
2	7	0.15										
3	2	0.1		*Regression Statistics*								
4	6	0.13		Multiple R	0.9167							
5	4	0.15		R Square	0.8403							
6	14	0.25		Adjusted R Square	0.8243							
7	15	0.27		Standard Error	0.0407							
8	16	0.24		Observations	12							
9	12	0.2										
10	14	0.27		ANOVA								
11	20	0.44			*df*	*SS*	*MS*	*F*	*Significance F*			
12	15	0.34		Regression	1	0.0873	0.0873	52.6080	0.000027			
13	7	0.17		Residual	10	0.0166	0.0017					
14				Total	11	0.1039						
15												
16					*Coefficients*	*Standard Error*	*t Stat*	*P-value*	*Lower 95%*	*Upper 95%*	*Lower 95.0%*	*Upper 95.0%*
17				Intercept	0.0506	0.0269	1.8832	0.0891	(0.0093)	0.1105	(0.0093)	0.1105
18				Sales	0.0159	0.0022	7.2531	0.0000	0.0110	0.0208	0.0110	0.0208

Example 14.3 Krugman and Rust (1987) investigated the effects of cable TV penetration on network share of advertising revenue by using simple regression analysis. Use Minitab to find a regression line for the following data, and find out the predicted network share of TV advertising for the 54 % of US households that subscribe to cable.

Year	U.S. Cable TV Households (%), x	Network Share of Television Revenues, y
1980	21.1	98.9
1981	23.7	97.9
1982	25.8	96.5
1983	30.0	95.2
1984	35.7	94.0
1985	41.1	91.9

We will put the network data in column c1 and name the column *cable*. The cable data will go in column c2, which we will name *network*.

```
MTB > name c1 'cable'
MTB > name c2 'network'
MTB > read 'cable' 'network'
DATA> 21.1 98.9
DATA> 23.7 97.9
DATA> 25.8 96.5
DATA> 30.0 95.2
DATA> 35.7 94.0
DATA> 41.1 91.9
DATA> end
6 rows read.
```

We will first plot the data.

```
MTB > plot 'network'*'cable'
```

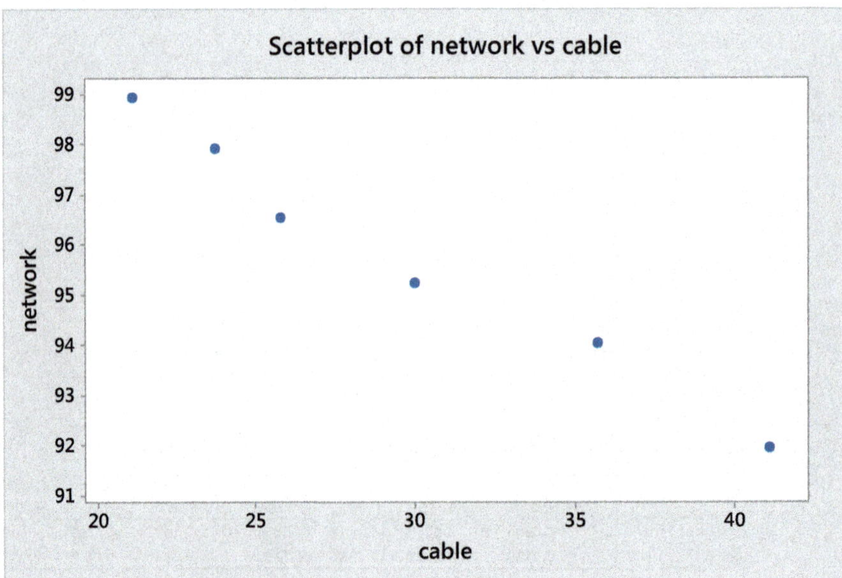

We will now regress the data set.

```
MTB > Regress;
SUBC>   Response 'network';
SUBC>   Nodefault;
SUBC>   Continuous 'cable';
SUBC>   Terms cable;
SUBC>   Constant;
SUBC>   Unstandardized;
SUBC>   Tmethod;
SUBC>   Tanova;
SUBC>   Tsummary;
SUBC>   Tcoefficients;
SUBC>   Tequation;
SUBC>   TDiagnostics 0.
```

Regression Analysis: Network Versus Cable

```
Analysis of Variance

Source       DF   Adj SS   Adj MS  F-Value  P-Value
Regression    1  32.6436  32.6436   200.98    0.000
  cable       1  32.6436  32.6436   200.98    0.000
Error         4   0.6497   0.1624
Total         5  33.2933


Model Summary

       S    R-sq  R-sq(adj)  R-sq(pred)
0.403016  98.05%     97.56%      96.32%


Coefficients

Term         Coef  SE Coef  T-Value  P-Value   VIF
Constant  105.634    0.718   147.22    0.000
cable     -0.3349   0.0236   -14.18    0.000  1.00


Regression Equation

network = 105.634 - 0.3349 cable

MTB > Predict 'network';
SUBC>   Nodefault;
SUBC>   KPredictors 54;
SUBC>   TEquation;
SUBC>   TPrediction.
```

Prediction for Network

```
Regression Equation

network = 105.634 - 0.3349 cable


Variable  Setting
cable          54


    Fit    SE Fit         95% CI                95% PI
87.5515  0.600125  (85.8853, 89.2177)  (85.5444, 89.5586)  XX

XX denotes an extremely unusual point relative to predictor levels used to fit
the model.
```

From the regression output we can see that the regression line for the data is

```
network = 106 - 0.335 cable
```

This regression line indicates that for every 1 % increase in households with cable, the networks lose 0.335 % of their share of television advertising revenues.

The regression commands predict that if the cable industry gained 54 % of the households, then the networks would take in 87.551 % of the total revenues. The prediction interval would be from 85.544 to 89.559 %.

14.10 Market Model

In finance there is a model called the *market model*. This model indicates that the return of a stock is linearly related to the return of the market. The equation of the market model is

$$\boldsymbol{R} = \boldsymbol{a} + \boldsymbol{\beta} \boldsymbol{R}_m + \boldsymbol{\varepsilon}$$

Where R is the return of a particular stock and R_m is the return of the market. The coefficient $\boldsymbol{\beta}$ is called the stock's *beta* coefficient. The coefficient indicates how sensitive a stock's return is to the movement of the market. It is common to use the S&P500 index as a proxy for the market. A beta of one indicates the stock's return moves in tangent with the market. A beta of zero indicates that the stock's return is not affected by the market. A beta of two indicates that a stock's return is twice as sensitive as the market. A negative beta indicates that a stock's return moves in the opposite direction of the market.

Yahoo Finance has list of betas for over 8000 stocks. The URL to this list of beta is http://screener.finance.yahoo.com/b?beta=0/&s=beta&db=stocks&vw=1&b=41.

Yahoo Finance calculates beta over a 36-month period and uses the S&P 500 index to approximate the market. The following URL examples how Yahoo Finance calculates beta: https://help.yahoo.com/kb/finance/SLN2347.html?impressions=true

Below shows some companies with betas <-10.

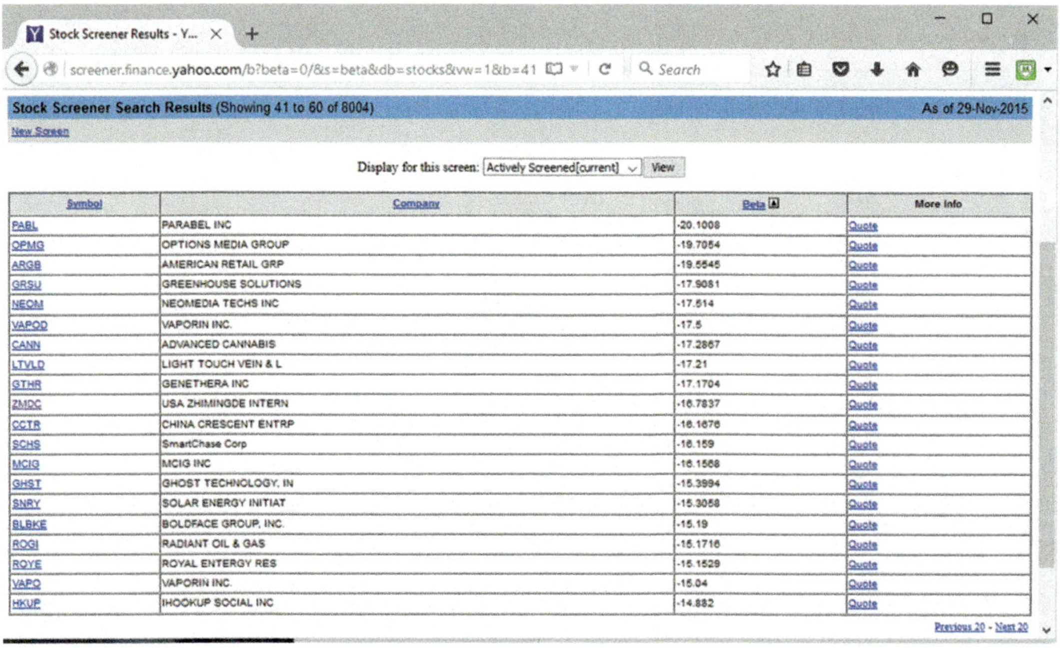
Stock Screener Results - Y...

screener.finance.yahoo.com/b?beta=0/&s=beta&db=stocks&vw=1&b=41

Stock Screener Search Results (Showing 41 to 60 of 8004) As of 29-Nov-2015

New Screen

Display for this screen: Actively Screened[current] View

Symbol	Company	Beta	More Info
PABL	PARABEL INC	-20.1008	Quote
OPMG	OPTIONS MEDIA GROUP	-19.7054	Quote
ARGB	AMERICAN RETAIL GRP	-19.5545	Quote
GRSU	GREENHOUSE SOLUTIONS	-17.9081	Quote
NEOM	NEOMEDIA TECHS INC	-17.514	Quote
VAPOD	VAPORIN INC.	-17.5	Quote
CANN	ADVANCED CANNABIS	-17.2867	Quote
LTVLD	LIGHT TOUCH VEIN & L	-17.21	Quote
GTHR	GENETHERA INC	-17.1704	Quote
ZMDC	USA ZHIMINGDE INTERN	-16.7837	Quote
CCTR	CHINA CRESCENT ENTRP	-16.1676	Quote
SCHS	SmartChase Corp	-16.159	Quote
MCIG	MCIG INC	-16.1568	Quote
GHST	GHOST TECHNOLOGY, IN	-15.3994	Quote
SNRY	SOLAR ENERGY INITIAT	-15.3058	Quote
BLBKE	BOLDFACE GROUP, INC.	-15.19	Quote
ROGI	RADIANT OIL & GAS	-15.1716	Quote
ROYE	ROYAL ENTERGY RES	-15.1529	Quote
VAPO	VAPORIN INC.	-15.04	Quote
HKUP	IHOOKUP SOCIAL INC	-14.882	Quote

Previous 20 - Next 20

Below shows companies with betas >10.

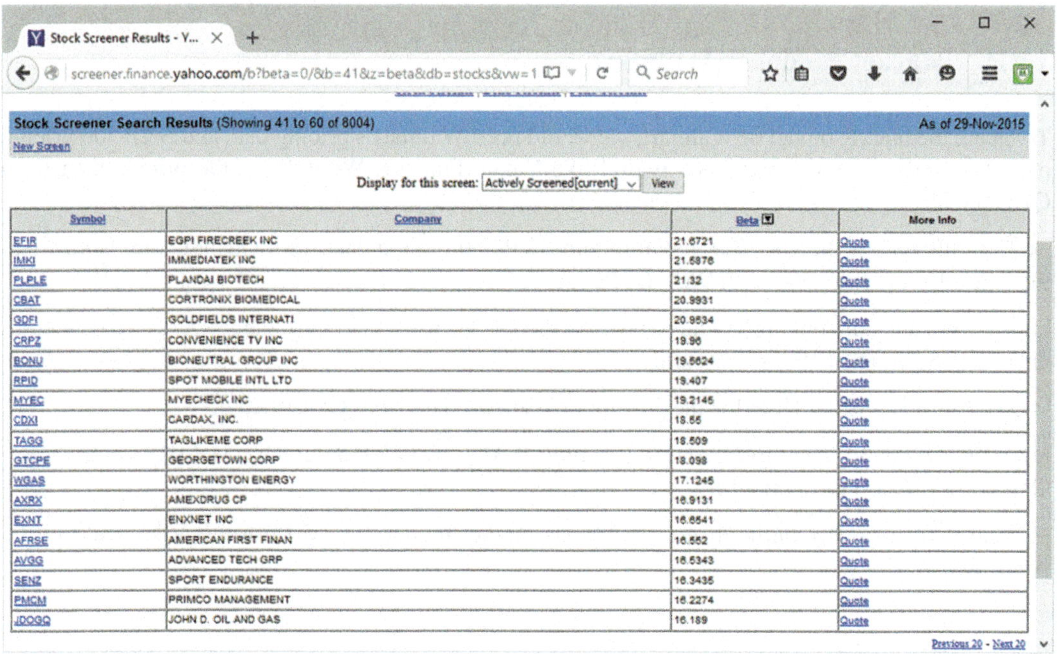
Stock Screener Results - Y...

screener.finance.yahoo.com/b?beta=0/&b=41&z=beta&db=stocks&vw=1

Stock Screener Search Results (Showing 41 to 60 of 8004) As of 29-Nov-2015

New Screen

Display for this screen: Actively Screened[current] View

Symbol	Company	Beta	More Info
EFIR	EGPI FIRECREEK INC	21.6721	Quote
IMKI	IMMEDIATEK INC	21.5876	Quote
PLPLE	PLANDAI BIOTECH	21.32	Quote
CBAT	CORTRONIX BIOMEDICAL	20.9931	Quote
GDFI	GOLDFIELDS INTERNATI	20.9534	Quote
CRPZ	CONVENIENCE TV INC	19.96	Quote
BONU	BIONEUTRAL GROUP INC	19.5624	Quote
RPID	SPOT MOBILE INTL LTD	19.407	Quote
MYEC	MYECHECK INC	19.2145	Quote
CDXI	CARDAX, INC.	18.55	Quote
TAGG	TAGLIKEME CORP	18.509	Quote
GTCPE	GEORGETOWN CORP	18.098	Quote
WGAS	WORTHINGTON ENERGY	17.1245	Quote
AXRX	AMEXDRUG CP	16.9131	Quote
EXNT	ENXNET INC	16.6541	Quote
AFRSE	AMERICAN FIRST FINAN	16.552	Quote
AVGG	ADVANCED TECH GRP	16.5343	Quote
SENZ	SPORT ENDURANCE	16.3435	Quote
PMCM	PRIMCO MANAGEMENT	16.2274	Quote
JDOGQ	JOHN D. OIL AND GAS	16.189	Quote

Previous 20 - Next 20

Below shows companies with betas of zero.

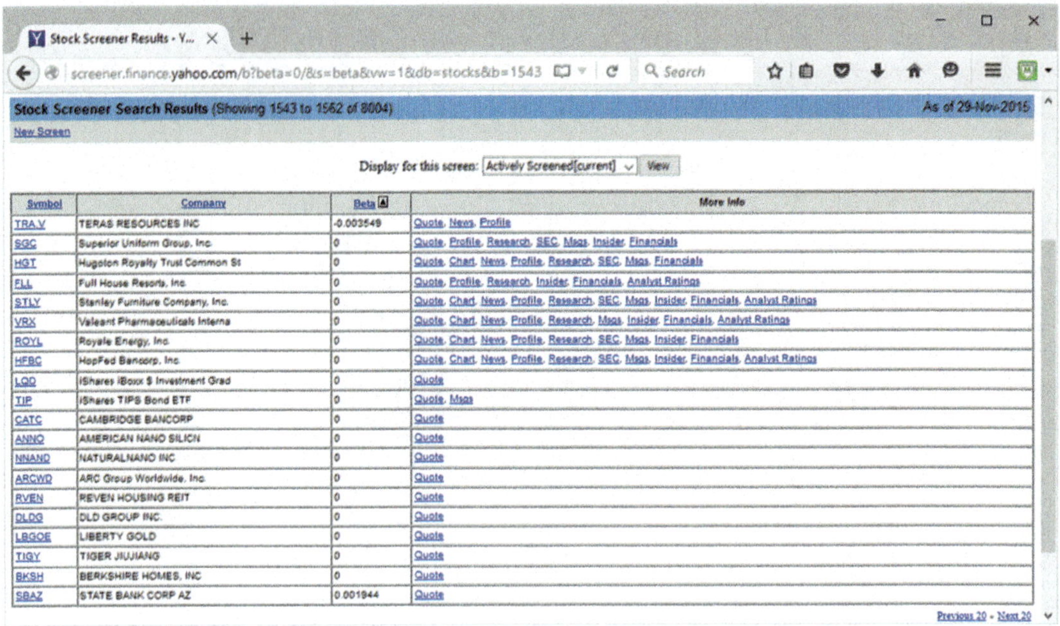
Stock Screener Results - Y...

screener.finance.yahoo.com/b?beta=0/&s=beta&vw=1&db=stocks&b=1543

Stock Screener Search Results (Showing 1543 to 1562 of 8004) As of 29-Nov-2015

New Screen

Display for this screen: Actively Screened[current] View

Symbol	Company	Beta	More Info
TRA.V	TERAS RESOURCES INC	-0.003549	Quote, News, Profile
SGC	Superior Uniform Group, Inc	0	Quote, Profile, Research, SEC, Msgs, Insider, Financials
HGT	Hugoton Royalty Trust Common St	0	Quote, Chart, News, Profile, Research, SEC, Msgs, Financials
FLL	Full House Resorts, Inc	0	Quote, Profile, Research, Insider, Financials, Analyst Ratings
STLY	Stanley Furniture Company, Inc.	0	Quote, Chart, News, Profile, Research, SEC, Msgs, Insider, Financials, Analyst Ratings
VRX	Valeant Pharmaceuticals Interna	0	Quote, Chart, News, Profile, Research, Msgs, Insider, Financials, Analyst Ratings
ROYL	Royale Energy, Inc	0	Quote, Chart, News, Profile, Research, SEC, Msgs, Insider, Financials
HFBC	HopFed Bancorp, Inc	0	Quote, Chart, News, Profile, Research, SEC, Msgs, Insider, Financials, Analyst Ratings
LQD	iShares iBoxx $ Investment Grad	0	Quote
TIP	iShares TIPS Bond ETF	0	Quote, Msgs
CATC	CAMBRIDGE BANCORP	0	Quote
ANNO	AMERICAN NANO SILICN	0	Quote
NNAND	NATURALNANO INC	0	Quote
ARCWD	ARC Group Worldwide, Inc	0	Quote
RVEN	REVEN HOUSING REIT	0	Quote
DLDG	DLD GROUP INC.	0	Quote
LBGOE	LIBERTY GOLD	0	Quote
TIGY	TIGER JIUJIANG	0	Quote
BKSH	BERKSHIRE HOMES, INC	0	Quote
SBAZ	STATE BANK CORP AZ	0.001944	Quote

Previous 20 - Next 20

14.11 Market Model of Walmart

To create the market model of Walmart, we would need Walmart's pricing data and S&P 500 pricing data. We will use a 36-month time period like Yahoo Finance. We will get the pricing data from Quandl.com

The following is the URL for the 36-month S&P 500 pricing data from Quandl.com. Notice that the ticker for the S&P 500 is *INDEX_GSPC*.

https://www.quandl.com/api/v3/datasets/yahoo/INDEX_GSPC.csv?start_date=Oct-31-2012&end_date=Oct-31-2015&collapse=monthly

The following is the URL for the 36-month Walmart data from Quandl.com:

https://www.quandl.com/api/v3/datasets/yahoo/WMT.csv?start_date=Oct-31-2012&end_date=Oct-31-2015&collapse=monthly

Instead of manually building the Walmart's market model, we will use the VBA code in the Appendix 14.1.

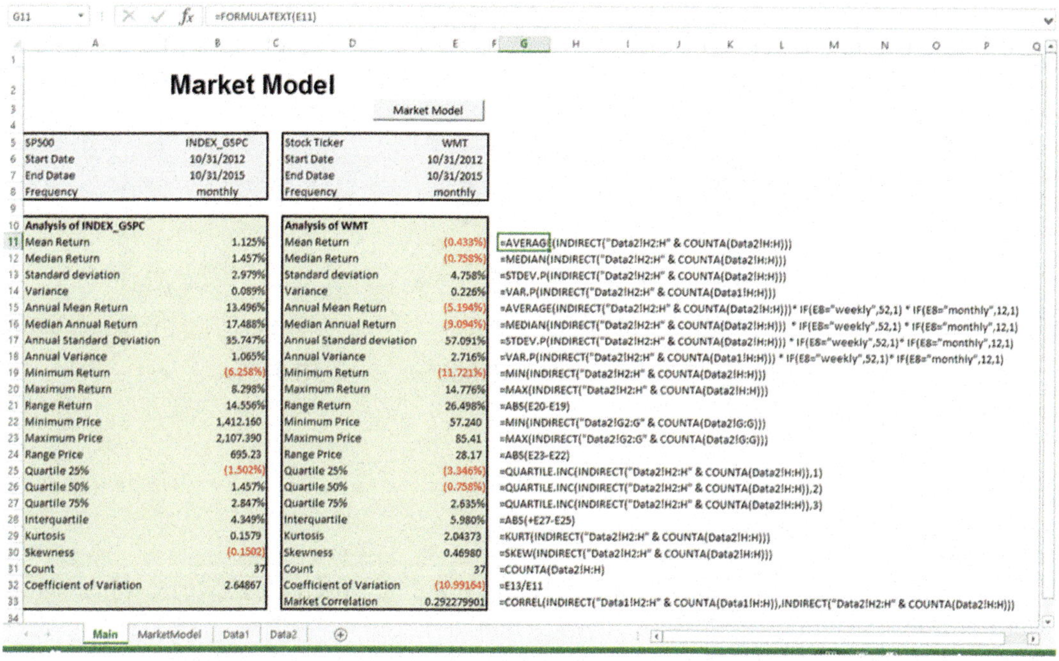

G11 =FORMULATEXT(E11)

Market Model

Market Model

SP500	INDEX_GSPC	Stock Ticker	WMT	
Start Date	10/31/2012	Start Date	10/31/2012	
End Datae	10/31/2015	End Datae	10/31/2015	
Frequency	monthly	Frequency	monthly	
Analysis of INDEX_GSPC		**Analysis of WMT**		
Mean Return	1.125%	Mean Return	(0.433%)	=AVERAGE(INDIRECT("Data2!H2:H" & COUNTA(Data2!H:H)))
Median Return	1.457%	Median Return	(0.758%)	=MEDIAN(INDIRECT("Data2!H2:H" & COUNTA(Data2!H:H)))
Standard deviation	2.979%	Standard deviation	4.758%	=STDEV.P(INDIRECT("Data2!H2:H" & COUNTA(Data2!H:H)))
Variance	0.089%	Variance	0.226%	=VAR.P(INDIRECT("Data2!H2:H" & COUNTA(Data1!H:H)))
Annual Mean Return	13.496%	Annual Mean Return	(5.194%)	=AVERAGE(INDIRECT("Data2!H2:H" & COUNTA(Data2!H:H)))* IF(E8="weekly",52,1) * IF(E8="monthly",12,1)
Median Annual Return	17.488%	Median Annual Return	(9.094%)	=MEDIAN(INDIRECT("Data2!H2:H" & COUNTA(Data2!H:H))) * IF(E8="weekly",52,1) * IF(E8="monthly",12,1)
Annual Standard Deviation	35.747%	Annual Standard deviation	57.091%	=STDEV.P(INDIRECT("Data2!H2:H" & COUNTA(Data2!H:H))) * IF(E8="weekly",52,1)* IF(E8="monthly",12,1)
Annual Variance	1.065%	Annual Variance	2.716%	=VAR.P(INDIRECT("Data2!H2:H" & COUNTA(Data1!H:H))) * IF(E8="weekly",52,1)* IF(E8="monthly",12,1)
Minimum Return	(6.258%)	Minimum Return	(11.721%)	=MIN(INDIRECT("Data2!H2:H" & COUNTA(Data2!H:H)))
Maximum Return	8.298%	Maximum Return	14.776%	=MAX(INDIRECT("Data2!H2:H" & COUNTA(Data2!H:H)))
Range Return	14.556%	Range Return	26.498%	=ABS(E20-E19)
Minimum Price	1,412.160	Minimum Price	57.240	=MIN(INDIRECT("Data2!G2:G" & COUNTA(Data2!G:G)))
Maximum Price	2,107.390	Maximum Price	85.41	=MAX(INDIRECT("Data2!G2:G" & COUNTA(Data2!G:G)))
Range Price	695.23	Range Price	28.17	=ABS(E23-E22)
Quartile 25%	(1.502%)	Quartile 25%	(3.346%)	=QUARTILE.INC(INDIRECT("Data2!H2:H" & COUNTA(Data2!H:H)),1)
Quartile 50%	1.457%	Quartile 50%	(0.758%)	=QUARTILE.INC(INDIRECT("Data2!H2:H" & COUNTA(Data2!H:H)),2)
Quartile 75%	2.847%	Quartile 75%	2.635%	=QUARTILE.INC(INDIRECT("Data2!H2:H" & COUNTA(Data2!H:H)),3)
Interquartile	4.349%	Interquartile	5.980%	=ABS(+E27-E25)
Kurtosis	0.1579	Kurtosis	2.04373	=KURT(INDIRECT("Data2!H2:H" & COUNTA(Data2!H:H)))
Skewness	(0.1502)	Skewness	0.46980	=SKEW(INDIRECT("Data2!H2:H" & COUNTA(Data2!H:H)))
Count	37	Count	37	=COUNTA(Data2!H:H)
Coefficient of Variation	2.64867	Coefficient of Variation	(10.99164)	=E13/E11
		Market Correlation	0.292279901	=CORREL(INDIRECT("Data1!H2:H" & COUNTA(Data1!H:H)),INDIRECT("Data2!H2:H" & COUNTA(Data2!H:H)))

Main MarketModel Data1 Data2

The VBA code in the Appendix assumes that there are four worksheets. The four worksheets are *Main*, *MarketModel*, *Data1*, and *Data2*. The *Data1* worksheet will contain the data for the S&P 500 data. The *Data2* worksheet will contain the data for the stock. The *MarketModel* worksheet will contain the return data for both the S&P 500 and the stock of interest.

On the *Main* worksheet cell E5 should be the stock ticker of interest. Pushing the *Market Model* button runs the procedure *Main* in the Appendix.

Pushing the *Market Model* button results in the following regression for Walmart:

H	I	J	K	L	M	N
SUMMARY OUTPUT						
Regression Statistics						
Multiple R	0.292279901					
R Square	0.085427541					
Adjusted R Square	0.058528351					
Standard Error	0.046817267					
Observations	36					
ANOVA						
	df	*SS*	*MS*	*F*	*Significance F*	
Regression	1	0.006960987	0.006960987	3.175840643	0.083668107	
Residual	34	0.07452312	0.002191856			
Total	35	0.081484106				
	Coefficients	*Standard Error*	*t Stat*	*P-value*	*Lower 95%*	*Upper 95%*
Intercept	-0.009578308	0.008340477	-1.148412482	0.258814123	-0.026528197	0.007371581
INDEX_GSPC	0.466796717	0.26193796	1.782088843	0.083668107	-0.065525265	0.999118699

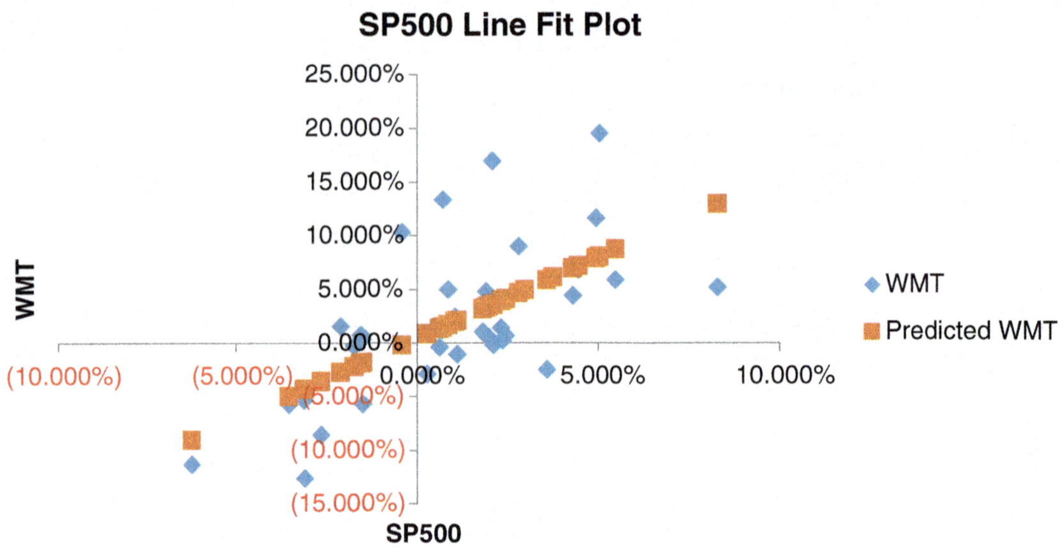

The market model indicates that Walmart's beta is 0.466 on November 27, 2015, which agrees with Yahoo Finance.

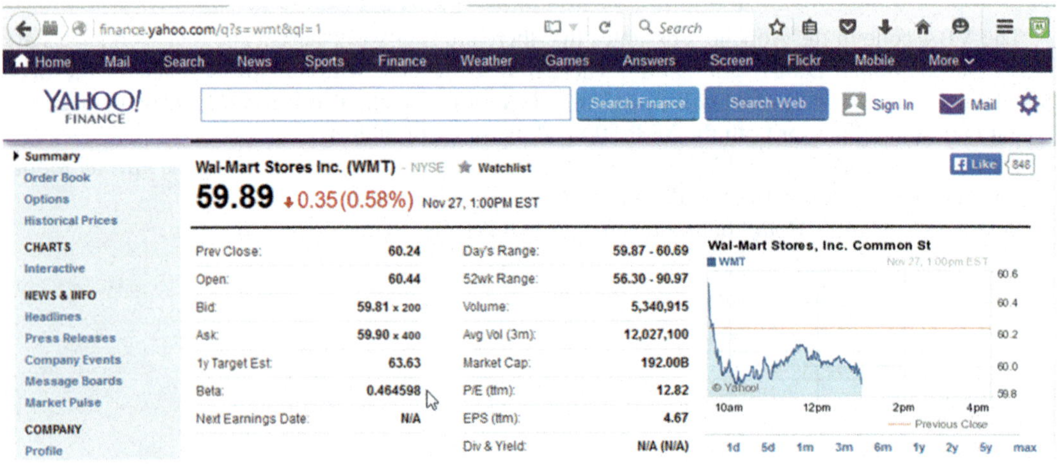

14.12 Market Model of Morgan Stanley

We will now create a market model of Morgan Stanley. The URL for Morgan Stanley's pricing data is

https://www.quandl.com/api/v3/datasets/yahoo/MS.csv?start_date=Oct-31-2012&end_date=Oct-31-2015&collapse=monthly

Pushing the *Market Model* button results in the following regression for Morgan Stanley:

H	I	J	K	L	M	N
SUMMARY OUTPUT						
Regression Statistics						
Multiple R	0.650972215					
R Square	0.423764825					
Adjusted R Square	0.406816732					
Standard Error	0.053970098					
Observations	36					
ANOVA						
	df	*SS*	*MS*	*F*	*Significance F*	
Regression	1	0.072830026	0.072830026	25.00368717	1.7159E-05	
Residual	34	0.09903423	0.002912771			
Total	35	0.171864256				
	Coefficients	*Standard Error*	*t Stat*	*P-value*	*Lower 95%*	*Upper 95%*
Intercept	0.004257987	0.009614751	0.442859809	0.660672065	-0.015281538	0.023797512
INDEX_GSPC	1.509898037	0.301957341	5.000368703	1.7159E-05	0.896246889	2.123549185

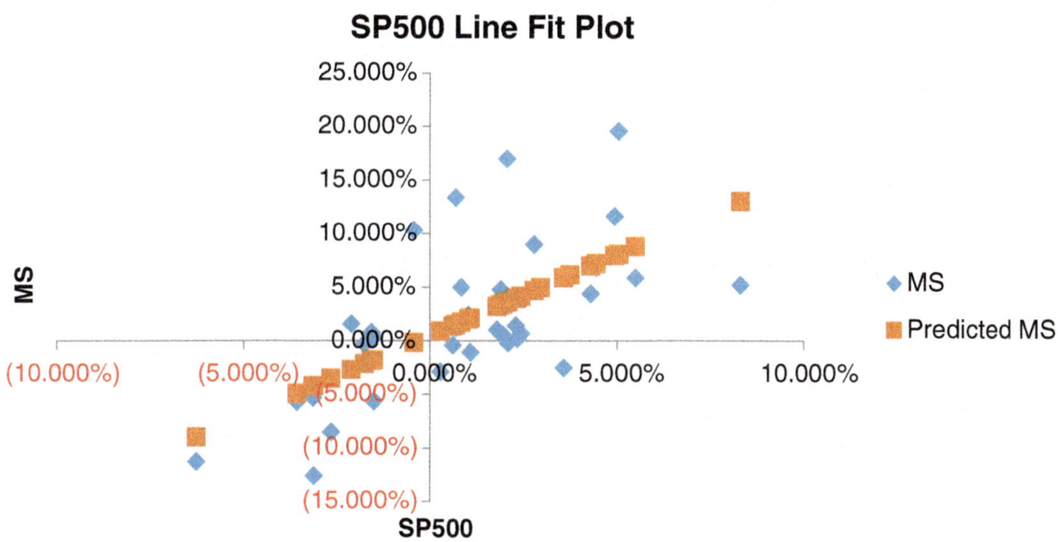

The market model indicates that Morgan Stanley's beta is 1.509 on November 27, 2015, which agrees with Yahoo Finance.

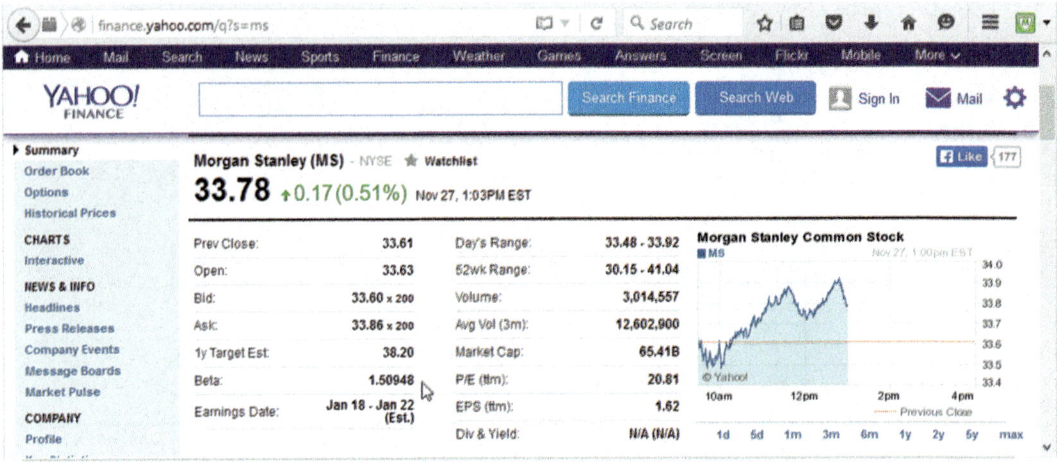

14.13 SAS Programming Code Instructions and Examples

14.13.1 Two-Tail t Test for β

Example 14.4 Below is the regression analysis in SAS.

```
Data Children;
   INPUT Height Weight @@;
   DATALINES;
55 92
56 95
57 99
58 97
59 102
60 104
;
run;

PROC REG DATA=Children;
    MODEL  Weight=Height/ P CLM;
RUN;
```

Analysis of variance					
Source	DF	Sum of squares	Mean square	F Value	Pr > F
Model	1	89.15714	89.15714	36.86	0.0037
Error	4	9.67619	2.41905		
Corrected total	5	98.83333			

Root MSE	1.55533	R-Square	0.9021
Dependent mean	98.16667	Adj R-sq	0.8776
Coeff var	1.58438		

Parameter estimates

Variable	DF	Parameter estimate	Standard error	t Value	Pr > |t|
Intercept	1	−31.61905	21.38762	−1.48	0.2134
Height	1	2.25714	0.37179	6.07	0.0037

Output statistics

Obs	Dependent variable	Predicted value	Std error mean predict	95 % CL mean		Residual
1	92.0000	92.5238	1.1257	89.3985	95.6492	−0.5238
2	95.0000	94.7810	0.8451	92.4346	97.1273	0.2190
3	99.0000	97.0381	0.6616	95.2012	98.8750	1.9619
4	97.0000	99.2952	0.6616	97.4583	101.1322	−2.2952
5	102.0000	101.5524	0.8451	99.2060	103.8988	0.4476
6	104.0000	103.8095	1.1257	100.6842	106.9349	0.1905

Sum of residuals	0
Sum of squared residuals	9.67619
Predicted residual SS (PRESS)	15.46072

14.13.2 Standard Error of the Regression Line

Below shows the command to graph the best regression line that fits the above data set in SAS.

```
Goptions ftext='Times New Roman' htext=6 gunit=pct;
title  height=7 'Regression Plot';
symbol1 v=dot  c=black L=1 h=2 i=rl;
axis1  label=('Height');
axis2  label=('Weight');
Proc gplot data=Children;
plot  Weight*Height /vaxis=axis1 haxis=axis2 regeqn;
Run;
```

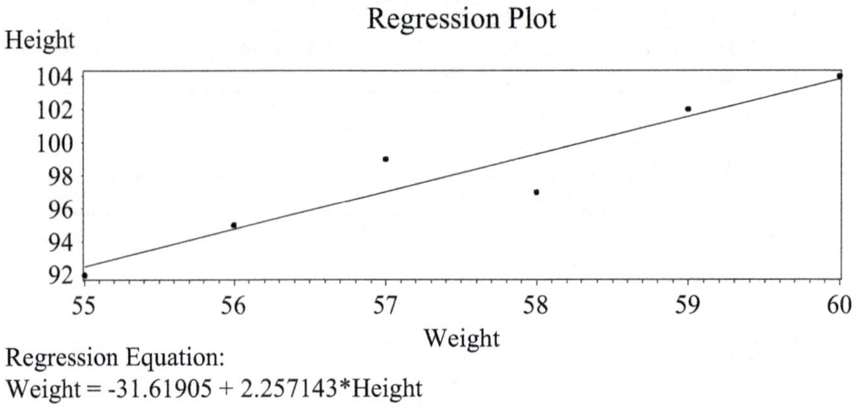

14.13.3 Confidence Interval of β

CLPARM option produces confidence limits for the parameter estimates (if the SOLUTION option is also specified) and for the results of all ESTIMATE statements.

```
PROC GLM DATA = Children;
    MODEL Weight=Height/CLPARM;
RUN;
```

Parameter	Estimate	Standard error	t Value	Pr > |t|	95 % Confidence limits	
Intercept	−31.61904762	21.38762225	−1.48	0.2134	−91.00060673	27.76251149
Height	2.25714286	0.37179469	6.07	0.0037	1.22487531	3.28941041

14.13.4 F Test

```
PROC REG DATA=Children;
  MODEL Weight=Height;
RUN;
```

Analysis of variance					
Source	DF	Sum of squares	Mean square	F Value	Pr > F
Model	1	89.15714	89.15714	36.86	0.0037
Error	4	9.67619	2.41905		
Corrected total	5	98.83333			

14.13.5 Predicting

We will use the **p** option of the **PROC REG** command to predict the value of the dependent variable in the following problem.

```
DATA Farmers;
    INPUT fertilizer wheat @@;
    DATALINES;
100 1000
150 1250
180 1710
200 2100
222 2500
;
run;
```

For example, we want to predict the number of bushels of wheat that would be harvested if 160 lb of fertilizer were used. The basic idea is to insert a new record (**160 .**) to the original data and then do the regression as we described previously.

```
data Farmers_predict;
input fertilizer wheat;
cards;
160 .
;
run;

data Farmers_predict;
set Farmers_predict Farmers;
run;
```

The **P option** causes **PROC REG** to display the observation number, the actual value, the predicted value, and the residual. The **OUTPUT** statement creates a new SAS data set that saves all the results. In this case, the new data set is named Farmers_out. R = res and P = predict option are specified to rename residuals and predicted values.

```
proc reg data=Farmers_predict;
model wheat=fertilizer/P clm cli;
output out=Farmers_out r=res p=predict;
run;
```

Output statistics

Obs	Dependent variable	Predicted value	Std error mean predict	95 % CL mean		95 % CL predict		Residual
1		1583	86.8152	1307	1859	922.4131	2244	
2	1000	838.7560	163.3375	318.9431	1359	44.8765	1633	161.2440
3	1250	1459	93.5575	1161	1757	789.1148	2129	−208.9577
4	1710	1831	86.4506	1556	2106	1171	2491	−121.0787
5	2100	2079	102.8072	1752	2406	1396	2763	20.8406
6	2500	2352	132.7512	1930	2775	1618	3086	147.9518

14.13.6 Regression Examples

Example 14.5 Below is the regression done in SAS.

```
DATA Sale;
    INPUT Sales Profits @@;
    DATALINES;
7 0.15
2 0.10
6 0.13
4 0.15
14 0.25
15 0.27
16 0.24
12 0.20
14 0.27
20 0.44
15 0.34
7 0.17
;
run;

Goptions ftext='Times New Roman' htext=6 gunit=pct;
title  height=7 'Profits and Sales';
symbol1 v=dot  c=black L=1 h=2 i=rl;
axis1  label=('Sales') ;
axis2  label=('Profits') ;
proc gplot data=Sale;
plot Profits*Sales/vaxis=axis1 haxis=axis2 overlay autovref regeqn;
run;
```

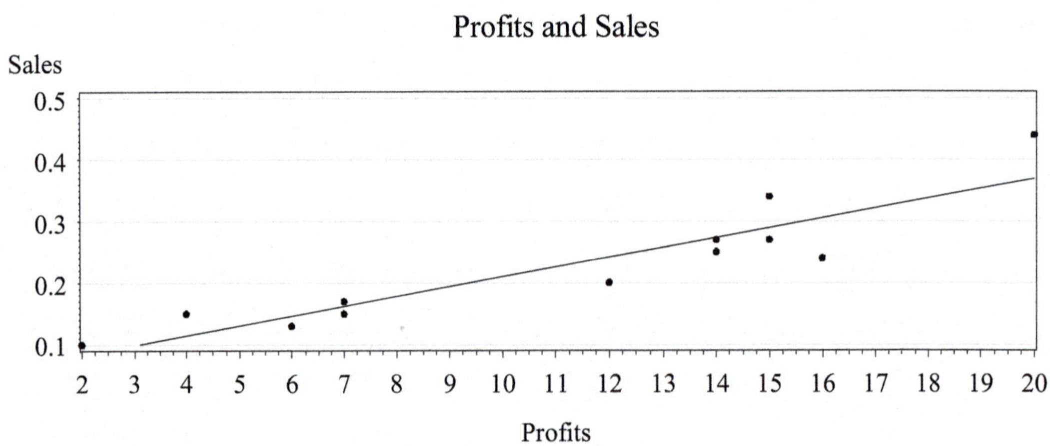

Regression Equation:
Profits = 0.050601 + 0.01593*Sales

```
PROC GLM DATA = Sale;
    MODEL Profits=Sales/CLPARM;
RUN;
```

Source	DF	Sum of squares	Mean square	F Value	Pr > F
Model	1	0.08729767	0.08729767	52.61	<0.0001
Error	10	0.01659399	0.00165940		
Corrected total	11	0.10389167			

R-Square	Coeff var	Root MSE	Profits mean
0.840276	18.03796	0.040736	0.225833

Parameter	Estimate	Standard error	t Value	Pr > |t|	95 % Confidence limits	
Intercept	0.0506007752	0.02686945	1.88	0.0891	−0.0092680884	0.1104696388
Sales	0.0159302326	0.00219632	7.25	<0.0001	0.0110365198	0.0208239453

Example 14.6 Below is the regression done in SAS.

```
DATA TV;
    INPUT Year Cable Network @@;
    DATALINES;
1980 21.1 98.9
1981 23.7 97.9
1982 25.8 96.5
1983 30.0 95.2
1984 35.7 94.0
1985 41.1 91.9
;
run;

PROC GLM DATA = TV;
    MODEL Network=Cable;
RUN;
```

Source	DF	Sum of squares	Mean square	F Value	Pr > F
Model	1	32.64364490	32.64364490	200.98	0.0001
Error	4	0.64968844	0.16242211		
Corrected total	5	33.29333333			

R-Square	Coeff var	Root MSE	Network mean
0.980486	0.420978	0.403016	95.73333

Parameter	Estimate	Standard error	t Value	Pr > |t|
Intercept	105.6341413	0.71750239	147.22	<0.0001
Cable	−0.3348639	0.02362063	−14.18	0.0001

```
title  height=7 'Relationship Between Cable and Network';
axis1  label=('Cable') ;
axis2  label=('Network') ;
proc gplot data=TV;
plot Network*Cable/vaxis=axis1 haxis=axis2 overlay autovref;
run;
```

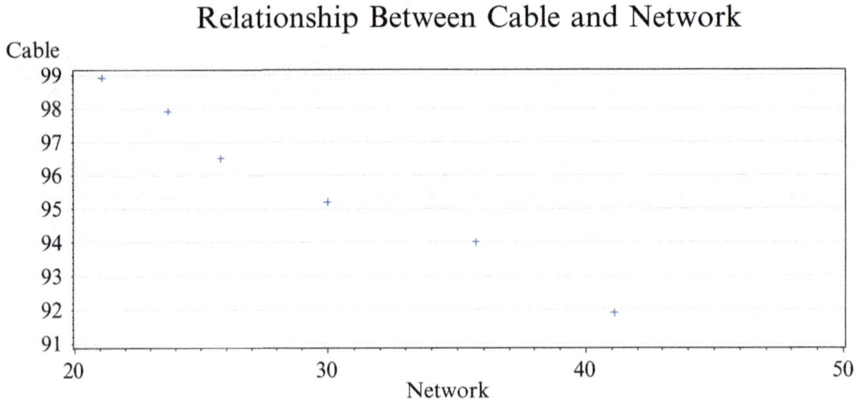

14.14 Statistical Summary

In this chapter we extended our use of regression analysis. We determined whether the β coefficient would be significantly different from zero or not. We tested this by seeing if the p-value calculated for the b coefficient was less than the alpha value that we were interested in. We calculated the confidence interval for β. We found the regression line for a given data set. We made predictions using the regression line and found the confidence interval for those predictions.

Appendix 14.1: Market Model VBA Code

```
Option Explicit

'Build Market model for any stock.  The data source is from Quandl.com
'The URL for Walmart is
'https://www.quandl.com/api/v3/datasets/yahoo/WMT.csv?start_date=Oct-31-
2012&end_date=Oct-31-2015&collapse=monthly
'The URL for the S&P 500is
'https://www.quandl.com/api/v3/datasets/yahoo/INDEX_GSPC.csv?start_date=Oct-31-
2012&end_date=Oct-31-2015&collapse=monthly
'Yahoo uses 36 months to calculate Beta
```

```
Sub Main()

    Application.StatusBar = "Calculating Return for SP500"
    'SP500 Returns
    GetReturns ticker:=Worksheets("Main").Range("b5"), _
               Start_Date:=Worksheets("Main").Range("b6"), _
               End_Date:=Worksheets("Main").Range("b7"), _
               frequency:=Worksheets("Main").Range("b8"), _
               shtData:=Worksheets("Data1")

     Application.StatusBar = "Calculating Return for Ticker 2"

    'Ticker2 Returns
    GetReturns ticker:=Worksheets("Main").Range("e5"), _
               Start_Date:=Worksheets("Main").Range("e6"), _
               End_Date:=Worksheets("Main").Range("e7"), _
               frequency:=Worksheets("Main").Range("e8"), _
               shtData:=Worksheets("Data2")

   CompileReturn
   MarketModel

    Application.StatusBar = False

End Sub

Sub GetReturns(Start_Date As Date, _
              End_Date As Date, _
              ticker As String, _
              frequency As String, _
              shtData As Worksheet)



    Dim URL As String
    Dim rReturn As Excel.Range

    shtData.UsedRange.ClearContents

    'Build URL for ticker
    URL = "https://www.quandl.com/api/v3/datasets/yahoo/" & ticker & ".csv?"
    URL = URL & "start_date=" & Format(Start_Date, "mmm-dd-yyyy") & "&end_date=" &
Format(End_Date, "mmm-dd-yyyy")
    URL = URL & "&collapse=" & frequency
    URL = URL & "&api_key=uBSmYjGExghF8sqmTjf-"


    'Use qurytables to return URL info
    With shtData.QueryTables.Add(Connection:="URL;" & URL,
Destination:=shtData.Range("a1"))
         .BackgroundQuery = True
         .TablesOnlyFromHTML = False
         .Refresh BackgroundQuery:=False
         .SaveData = True
    End With
```

```
    'data is return as a comma list
    'parse list into column
    shtData.Range("a1").CurrentRegion.TextToColumns _
                        Destination:=shtData.Range("a1"), DataType:=xlDelimited, _
                        TextQualifier:=xlDoubleQuote, ConsecutiveDelimiter:=False,
Tab:=True, _
                        Semicolon:=False, Comma:=True, Space:=False, other:=False

    shtData.Range("a1").CurrentRegion.ColumnWidth = 14

    Set rReturn = shtData.Range("a1").CurrentRegion
    'Delete empty row
    rReturn.Cells(1).Offset(rReturn.Rows.Count - 1, 0).EntireRow.Delete Shift:=xlUp

    Set rReturn = shtData.Range("a1").CurrentRegion
    Set rReturn = rReturn.Columns(1).Offset(0, rReturn.Columns.Count)
    rReturn.Name = "Return"

    shtData.Names.Add Name:="Return_" & shtData.Name, _
                    RefersToR1C1:=rReturn.Offset(1, 0).Resize(rReturn.Rows.Count - 2, 1)

    'Calculate rate of return
    'Use R1C1 address
    rReturn.FormulaR1C1 = "=(RC[-1]-R[1]C[-1])/R[1]C[-1]"
    'format percentage
    rReturn.NumberFormat = "0.000%;[Red](0.000%)"
    rReturn.Cells(1) = "Return"
    rReturn.Cells(rReturn.Cells.Count).ClearContents
    rReturn.Offset(0, 1).Value = ticker
    rReturn.Offset(0, 1).Cells(1).Value = "Ticker"

    'Calculate Z-Score
    rReturn.Offset(0, 2).Formula = "=STANDARDIZE(RC[-2],AVERAGE(Return_" & _
                                    rReturn.Parent.Name & "),STDEV.P(Return_" & _
                                    rReturn.Parent.Name & "))"
    rReturn.Offset(0, 2).Cells(1).Value = "Z-Score"

End Sub

Sub CompileReturn()

    Dim sht As Excel.Worksheet
    Dim rData As Excel.Range
    Dim rTmp As Excel.Range

    Set sht = ThisWorkbook.Worksheets("MarketModel")

    sht.UsedRange.ClearContents

    'Datasheet1
    Set rData = ThisWorkbook.Worksheets("Data1").Cells(1).CurrentRegion
    Set rTmp = rData.Columns(1)
    rTmp.Copy Destination:=sht.Range("a1")

    'Copy over return
    'Column 8 is column H
     Set rTmp = ThisWorkbook.Worksheets("Data1").Columns(8)
     sht.Range("b1").Resize(rTmp.Rows.Count, 1).Value = rTmp.Value
     sht.Range("b1").Resize(rTmp.Rows.Count, 1).NumberFormat = "0.000%;[Red](0.000%)"
     sht.Range("b1").Value = ThisWorkbook.Worksheets("Main").Range("B5")
```

```
    'Datasheet2
    Set rData = ThisWorkbook.Worksheets("Data2").Cells(1).CurrentRegion
    Set rTmp = rData.Columns(8)
    sht.Range("c1").Resize(rTmp.Rows.Count, 1).Value = rTmp.Value
    sht.Range("c1").Resize(rTmp.Rows.Count, 1).NumberFormat = "0.000%;[Red](0.000%)"
    sht.Range("c1").Value = ThisWorkbook.Worksheets("Main").Range("e5")


    'populate ticker
    Set rTmp = sht.Range("a1").CurrentRegion
    Set rTmp = rTmp.Columns(1).Offset(0, rTmp.Columns.Count)
    rTmp.Value = ThisWorkbook.Worksheets("Main").Range("b8")
    rTmp.Cells(1).Value = "Frequency"


End Sub

Sub MarketModel()

        Dim vnt As Variant
        Dim rStock As Excel.Range
        Dim rSP As Excel.Range
        Dim rReturn As Excel.Range

        'Delete all active charts
        For Each vnt In ThisWorkbook.Worksheets("MarketModel").Shapes

                vnt.Delete
        Next

        Set rReturn = ThisWorkbook.Worksheets("MarketModel").Cells(1).CurrentRegion
        Set rReturn = rReturn.Resize(rReturn.Rows.Count - 1, rReturn.Columns.Count)
        Set rStock = rReturn.Columns(3)
        Set rSP = rReturn.Columns(2)

         Application.Run "ATPVBAEN.XLAM!Regress", rStock, _
        rSP, False, True, , ThisWorkbook.Worksheets("MarketModel").Range("$h$1") _
        , True, True, False, True, , False

        ThisWorkbook.Worksheets("main").Range("h1").Select
        ThisWorkbook.Worksheets("MarketModel").Shapes(1).ScaleWidth 1.9348958333,
msoFalse, _
       msoScaleFromTopLeft
       ThisWorkbook.Worksheets("MarketModel").Shapes(1).ScaleHeight 1.9901477833,
msoFalse, _
       msoScaleFromTopLeft

      Dim chartMarket As ChartObject
       Set chartMarket = ThisWorkbook.Worksheets("MarketModel").ChartObjects(1)
```

```
        With chartMarket.Chart
            .HasTitle = True
            .ChartTitle.Text = "SP500 Line Fit Plot"
            .Axes(xlCategory).AxisTitle.Text = "SP500"

        End With

        ThisWorkbook.Worksheets("MarketModel").Shapes(1).Top = _
                ThisWorkbook.Worksheets("marketmodel").Range("O2").Top
        ThisWorkbook.Worksheets("MarketModel").Shapes(1).Left = _
                ThisWorkbook.Worksheets("marketmodel").Range("O2").Left

        ThisWorkbook.Worksheets("marketmodel").Activate
        ThisWorkbook.Worksheets("marketmodel").Range("h1").Activate


End Sub
```

Bibliography

Excel 2013 is published by Microsoft Inc., Redmond
Krugman DM, Rust RT (1987) The impact of cable penetration on network viewing. J Mark Res 27(9):9–12
Lee CF, Lee JC, Lee AC (2013) Statistics for business and financial economics. Springer, New York
Minitab 17, Minitab Inc., State College
SAS 2014 is published by SAS Institute Inc., Cary

Chapter 15
Multiple Linear Regression

15.1 Introduction

In the previous two chapters, we examined the relationship between two variables. In this chapter we will look at the relationship among three or more variables, using regression analysis.

When we looked at the relationship between two variables using regression analysis, one variable explained the other one, the variable of interest. Now we will have two or more variables explaining the variable of interest. The variables that are doing the explaining are called independent variables. The variable that is being explained is called the dependent variable. The mathematical formula for the regressed equation looks like the following:

$$y = a + b_1x_1 + b_2x_2 + b_3x_3 + \ldots + b_nx_n$$

The x's are the independent variables. The y is the dependent variable. The b's are the coefficients of the independent variables.

A regression that contains more than one independent variable is called a multiple linear regression. We will see that many of the analyses and tests done in simple linear regression extend over to multiple linear regression.

Suppose we believe that years of education and years of work experience can explain the amount of annual salary. To test this, we have the following data.

y	$x1$	$x2$
Annual salary (thousands)	Years of education	Years of work experience
15	5	7
17	10	5
26	9	14
24	13	8
27	15	6

To test our belief, we will use Minitab to do a regression analysis on the above data.

C.-F. Lee et al., *Essentials of Excel, Excel VBA, SAS and Minitab for Statistical and Financial Analyses*, DOI 10.1007/978-3-319-38867-0_15

Then we use the *regress* command and get the following results.

```
MTB > name c1 'salary'
MTB > name c2 'YearsEducation'
MTB > name c3 'YearsWork'
MTB > read 'salary' 'YearsEducation' 'YearsWork'
DATA> 15 5 7
DATA> 17 10 5
DATA> 26 9 14
DATA> 24 13 8
DATA> 27 15 6
DATA> end

MTB > Regress;
SUBC>   Response 'salary';
SUBC>   Nodefault;
SUBC>   Continuous 'YearsEducation' 'YearsWork';
SUBC>   Terms YearsEducation YearsWork;
SUBC>   Constant;
SUBC>   Unstandardized;
SUBC>   TExpand;
SUBC>   Tmethod;
SUBC>   Tanova;
SUBC>   Tsummary;
SUBC>   Tcoefficients;
SUBC>   Tequation;
SUBC>   Tdw;
SUBC>   TDiagnostics 1.
```

Regression Analysis: Salary Versus YearsEducation, YearsWork

Analysis of Variance

Source	DF	Seq SS	Contribution	Adj SS	Adj MS	F-Value	P-Value
Regression	2	113.009	95.13%	113.009	56.504	19.51	0.049
YearsEducation	1	65.773	55.36%	87.089	87.089	30.08	0.032
YearsWork	1	47.236	39.76%	47.236	47.236	16.31	0.056
Error	2	5.791	4.87%	5.791	2.896		
Total	4	118.800	100.00%				

Model Summary

S	R-sq	R-sq(adj)	PRESS	R-sq(pred)
1.70165	95.13%	90.25%	35.5200	70.10%

Coefficients

Term	Coef	SE Coef	95% CI	T-Value	P-Value	VIF
Constant	0.98	3.44	(-13.82, 15.78)	0.29	0.802	
YearsEducation	1.238	0.226	(0.267, 2.210)	5.48	0.032	1.04
YearsWork	0.992	0.246	(-0.065, 2.050)	4.04	0.056	1.04

Regression Equation

salary = 0.98 + 1.238 YearsEducation + 0.992 YearsWork

Fits and Diagnostics for All Observations

Obs	salary	Fit	SE Fit	95% CI	Resid	Std Resid	Del Resid	HI	Cook's D
1	15.00	14.12	1.50	(7.67, 20.57)	0.88	1.09	1.22	0.776259	1.38
2	17.00	18.33	1.08	(13.70, 22.96)	-1.33	-1.01	-1.01	0.399789	0.23
3	26.00	26.02	1.63	(19.00, 33.04)	-0.02	-0.04	-0.03	0.920113	0.01
4	24.00	25.02	0.96	(20.88, 29.16)	-1.02	-0.73	-0.60	0.319056	0.08
5	27.00	25.51	1.30	(19.91, 31.11)	1.49	1.36	3.41	0.584783	0.86

Obs	DFITS
1	2.27126
2	-0.82728
3	-0.10444
4	-0.40978
5	4.04598

Durbin-Watson Statistic

Durbin-Watson Statistic = 2.39454

This Minitab output shows that the regression equation is

salary = 0.98 + 1.238 YearsEducation + 0.992 YearsWork

This equation, which gives the relationship between annual salary and years of education and years of work experience, suggests that for every additional year of education, there will be a \$1238 increase in annual salary. It also suggests that for every year of additional work experience, there will be a \$992 increase in annual salary.

Below shows the regression analysis performed in Excel.

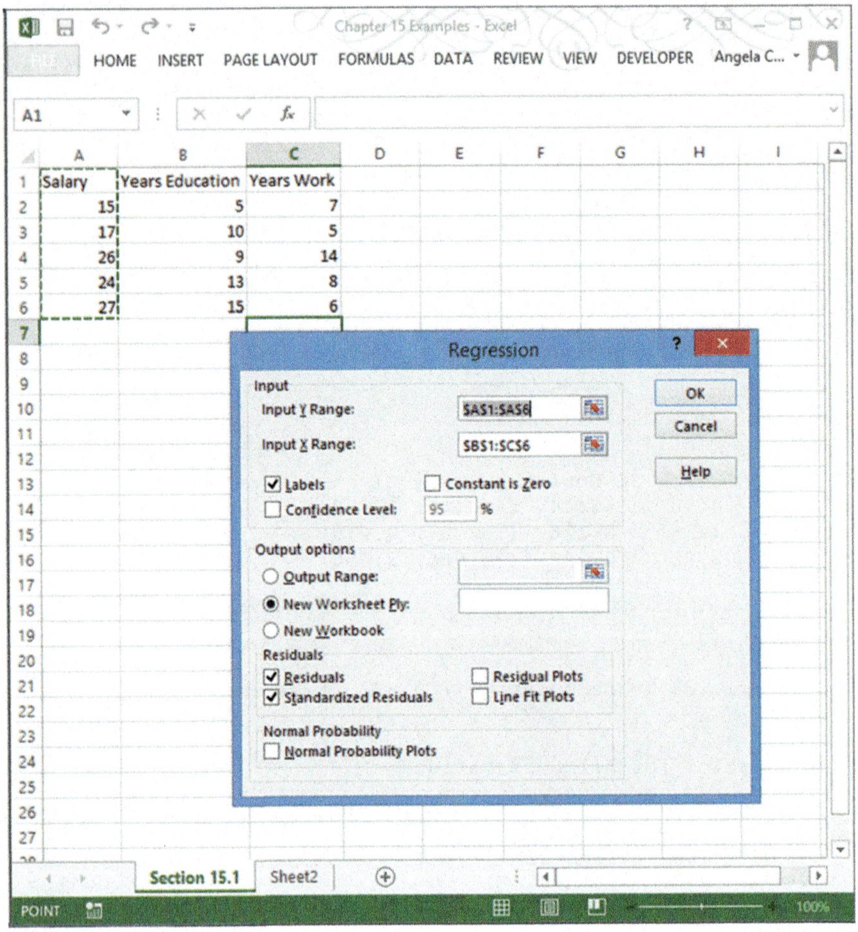

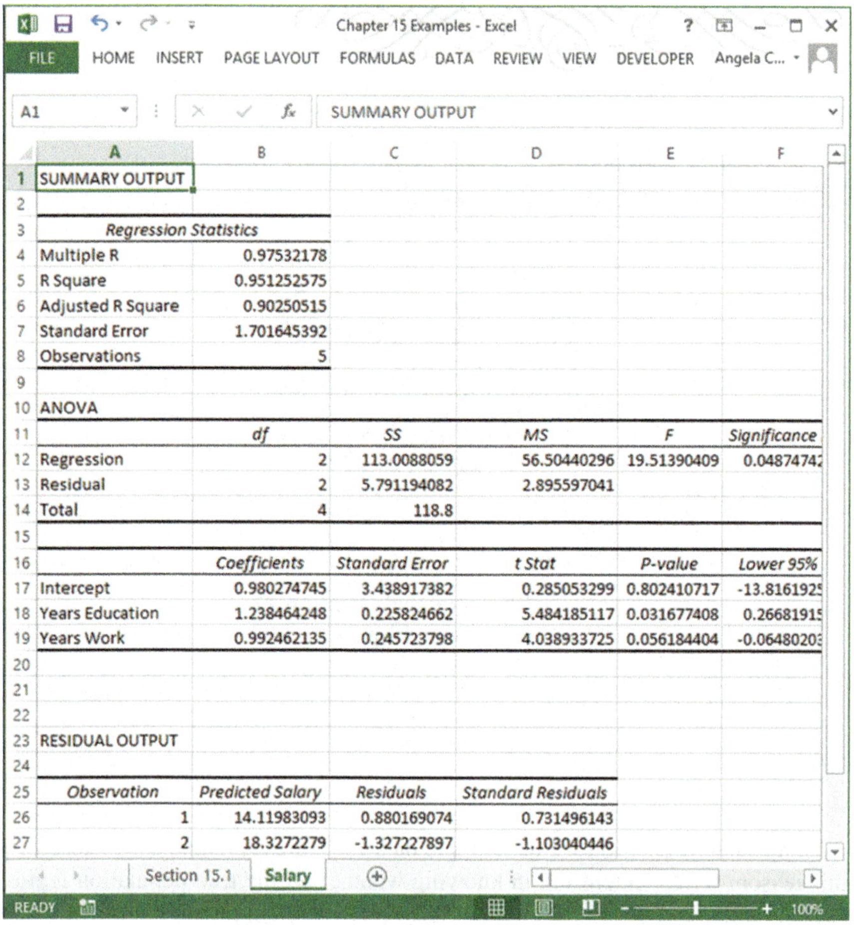

	A	B	C	D	E	F
1	SUMMARY OUTPUT					
2						
3	*Regression Statistics*					
4	Multiple R	0.97532178				
5	R Square	0.951252575				
6	Adjusted R Square	0.90250515				
7	Standard Error	1.701645392				
8	Observations	5				
9						
10	ANOVA					
11		*df*	*SS*	*MS*	*F*	*Significance*
12	Regression	2	113.0088059	56.50440296	19.51390409	0.0487474
13	Residual	2	5.791194082	2.895597041		
14	Total	4	118.8			
15						
16		*Coefficients*	*Standard Error*	*t Stat*	*P-value*	*Lower 95%*
17	Intercept	0.980274745	3.438917382	0.285053299	0.802410717	-13.816192
18	Years Education	1.238464248	0.225824662	5.484185117	0.031677408	0.2668191
19	Years Work	0.992462135	0.245723798	4.038933725	0.056184404	-0.0648020
20						
21						
22						
23	RESIDUAL OUTPUT					
24						
25	*Observation*	*Predicted Salary*	*Residuals*	*Standard Residuals*		
26	1	14.11983093	0.880169074	0.731496143		
27	2	18.3272279	-1.327227897	-1.103040446		

15.2 *R*-Square

To see how well the independent variables (education and experience) explain the variance of the dependent variable (salary) in the preceding section, we should look at the *R*-square. The following part of the Minitab output shows that the independent variables explain 95.13 % of the variation of the dependent variable.

```
s = 1.70165          R-sq = 95.13%       R-sq(adj) = 90.25%
```

The *R*-square can also be calculated using the following part of the Minitab output.

```
Analysis of Variance

Source               DF   Seq SS  Contribution   Adj SS   Adj MS  F-Value  P-Value
Regression            2  113.009        95.13%  113.009   56.504    19.51    0.049
  YearsEducation      1   65.773        55.36%   87.089   87.089    30.08    0.032
  YearsWork           1   47.236        39.76%   47.236   47.236    16.31    0.056
Error                 2    5.791         4.87%    5.791    2.896
Total                 4  118.800       100.00%
```

$$R\text{-square} = \frac{113.009}{118.8} = 0.951$$

A Venn diagram of the R-square would look like the following.

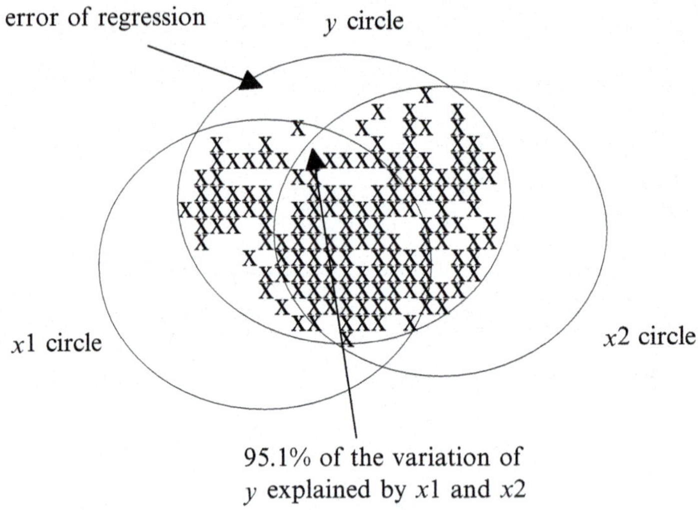

15.3 *F* Test

Many times in multiple regression we use sample data for regression analysis instead of population data. For this reason, we are interested in knowing whether all the true population regression slope coefficients equal zero. To test this, we will use the F test. The F statistic looks like the following:

$$F_{k,n-k-1} = \frac{\dfrac{\sum_{i=1}^{n}(\hat{y}_i - \bar{y})^2}{k}}{\dfrac{\sum_{i=1}^{n}(\hat{y}_i - \hat{y}_i)^2}{n-k-1}}$$

The hypotheses would be

$$H_0: \beta_1 = \beta_2 = 0$$
$$H_1: \text{at least one of } \beta_1 \text{ or } \beta_2 \text{ is not equal to zero}$$

where β_1 and β_2 are population regression parameters.

Below is a reproduction of the part of the regression analysis (from Sect. 15.1) that we will need for the *F* test.

```
Analysis of Variance

Source               DF   Seq SS  Contribution   Adj SS   Adj MS  F-Value  P-Value
Regression            2  113.009        95.13%  113.009   56.504    19.51    0.049
  YearsEducation      1   65.773        55.36%   87.089   87.089    30.08    0.032
  YearsWork           1   47.236        39.76%   47.236   47.236    16.31    0.056
Error                 2    5.791         4.87%    5.791    2.896
Total                 4  118.800       100.00%
```

If we test the hypothesis at the 0.05 alpha level, there are two ways to see if we can reject or accept the null hypothesis.

The first way to test the null hypothesis is to find the critical value for a significant level of alpha = 0.05. In Minitab we can use the *invcdf* command to find this value, as illustrated below:

```
MTB > invcdf .95;
SUBC> f 2 2.
```

Inverse Cumulative Distribution Function

```
F distribution with 2 DF in numerator and 2 DF in denominator

 P( X <= x)            x
    0.9500       19.0000
```

If we have an alpha of 0.05, then the area of interest is 1 − 0.05 = 0.95. The *invcdf* command works with the *f* subcommand to find the *F* value that corresponds with an area of 0.95. From the Minitab output after the *invcdf* command, we can see that the area of 0.95 corresponds with an *F* value of 19. The first argument, "0.95," of the *invcdf* indicates the area of interest. The first argument, "2," of the *f* subcommand indicates the first degree of freedom for the *F* test. This degree of freedom is equal to the number of independent variables. The second argument, "2," of the *f* subcommand indicates the second degree of freedom for the *F* test. This degree of freedom is obtained by taking the number of data items minus the number of independent variables minus one or, mathematically, $n - k - 1$. In this case we have 5 data items minus 2 independent variables minus 1 equals 2.

We will reject the null hypothesis if the calculated *F* value is greater than the *F* value that corresponds with an alpha value of 0.05.

Since the F value that corresponds with an alpha value of 0.05 is 19 and Minitab calculated the F value as 19.51, we can conclude that at least one of the regression coefficients is significantly different from zero. This is shown below.

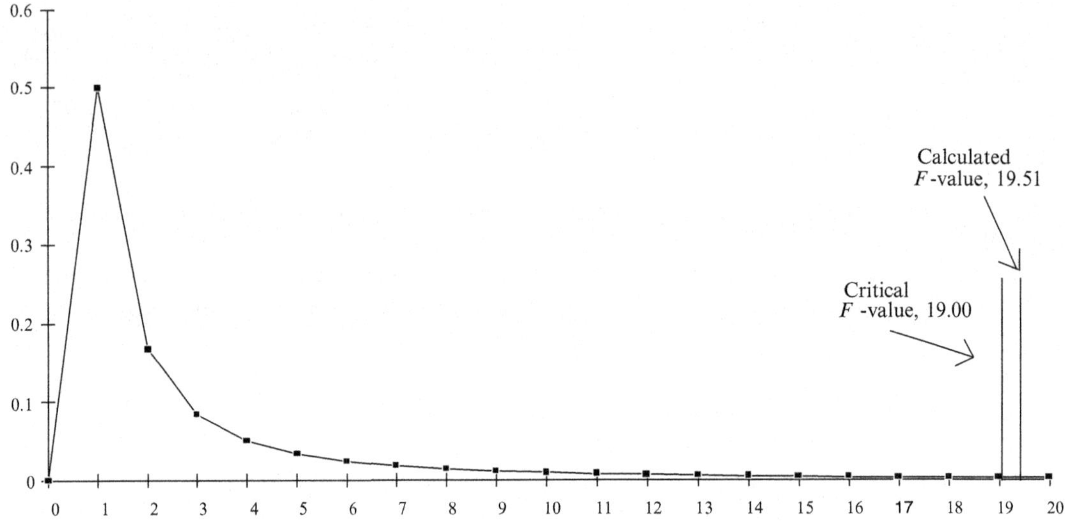

The second way to test the null hypothesis is to find the p-value for the calculated F value. In a one-tail test like the F test, the p-value is the area to the right of the calculated F value. Minitab calculated the p-value as 0.049. This means that the area to the right of the calculated F value is 0.049. An alpha value of 0.05 means that there is an F value that has an area to the right of 0.05. We reject the null hypothesis and accept the alternative hypothesis when the p-value is less than the alpha value. Since our p-value is less than the alpha value of 0.05, we reject the null hypothesis that all the population coefficients are equal to zero.

15.4 Confidence Interval of β

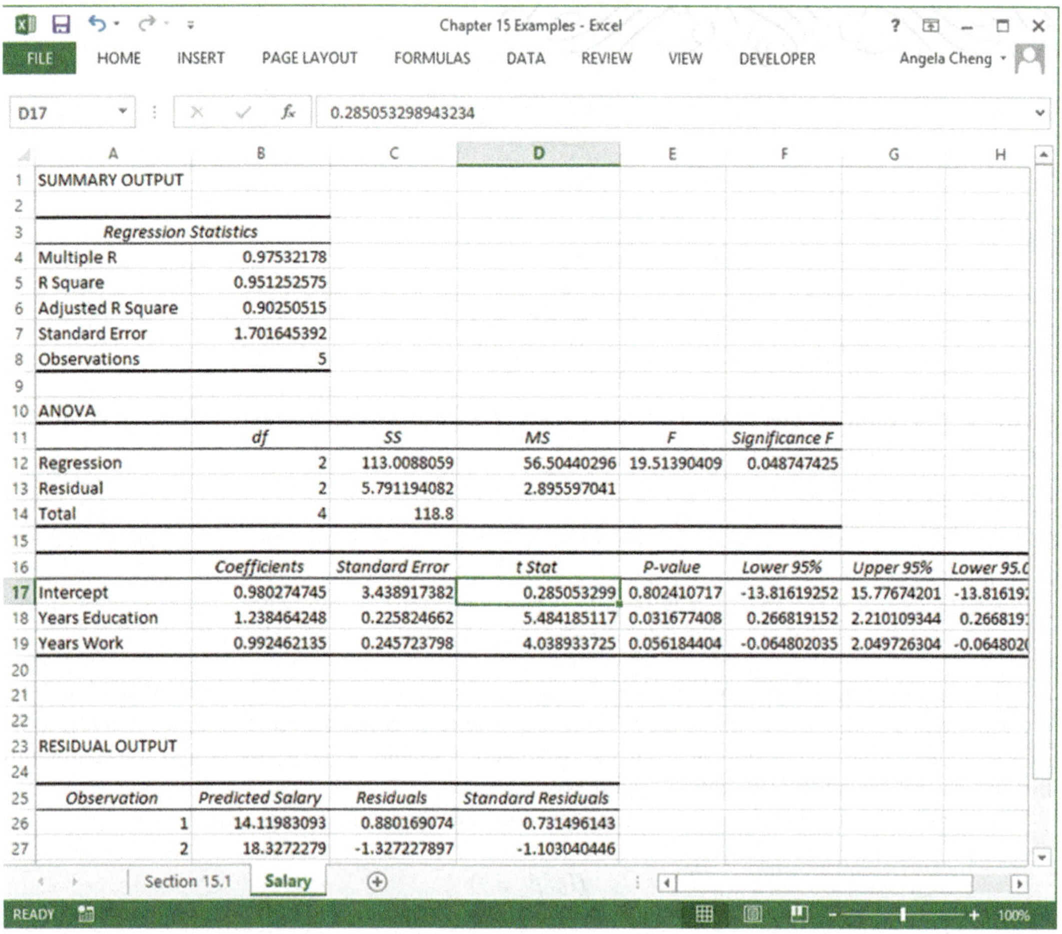

SUMMARY OUTPUT

Regression Statistics	
Multiple R	0.97532178
R Square	0.951252575
Adjusted R Square	0.90250515
Standard Error	1.701645392
Observations	5

ANOVA

	df	SS	MS	F	Significance F
Regression	2	113.0088059	56.50440296	19.51390409	0.048747425
Residual	2	5.791194082	2.895597041		
Total	4	118.8			

	Coefficients	Standard Error	t Stat	P-value	Lower 95%	Upper 95%	Lower 95.0
Intercept	0.980274745	3.438917382	0.285053299	0.802410717	-13.81619252	15.77674201	-13.81619
Years Education	1.238464248	0.225824662	5.484185117	0.031677408	0.266819152	2.210109344	0.266819
Years Work	0.992462135	0.245723798	4.038933725	0.056184404	-0.064802035	2.049726304	-0.064802

RESIDUAL OUTPUT

Observation	Predicted Salary	Residuals	Standard Residuals
1	14.11983093	0.880169074	0.731496143
2	18.3272279	-1.327227897	-1.103040446

From the above regression analysis, we can see that the 95 % confidence interval for the *YearsEducation* independent variable constant is $0.2668 \leq \beta \leq 2.2101$.

From the above regression analysis, we can see that the 95 % confidence interval for the *YearsWork* independent variable constant is $-0.0648 \leq \beta \leq 2.0497$.

15.5 *t* Test

From the *F* test we know that at least one of the two independent variables has a significant effect on the dependent variable. We will now use the *t* test to see which of the independent variables has a significant effect.

Let us begin by testing if x_1, years of education, has a significant effect on annual salary. The t test statistic is the following:

$$t = \frac{\frac{\sum_{i=1}^{n}(x_i-\overline{x})(y_i-\overline{y})}{\sum_{i=1}^{n}(x_i-\overline{x})^2} - 0}{\frac{s_e}{\sqrt{\sum_{i=1}^{n}(x_i-\overline{x})^2}}} = \frac{b-0}{s_b}$$

The null and alternative hypotheses would be

$$H_0: \beta_1 = 0$$
$$H_1: \beta_1 \neq 0$$

Below is a reproduction of the calculated t value from the regression output in Sect. 15.1. Minitab calculates the t statistic as a two-tail test.

```
Coefficients

Term              Coef  SE Coef          95% CI         T-Value  P-Value   VIF
Constant          0.98     3.44   (-13.82, 15.78)          0.29    0.802
YearsEducation   1.238    0.226   ( 0.267, 2.210)          5.48    0.032  1.04
YearsWork        0.992    0.246   (-0.065, 2.050)          4.04    0.056  1.04
```

From this we can see that the calculated t statistic for x_1, or years of education, is 5.48 and the corresponding p-value 0.032. Since the p-value is less than the alpha value of 0.05, we will reject the null hypothesis that x_1 has no effect on the dependent variable. That is, we would reject the null hypothesis and accept the alternative hypothesis.

The hypotheses for analyzing the independent variable x_2, years of work experience, are

$$H_0: \beta_2 = 0$$
$$H_1: \beta_2 \neq 0$$

From the Minitab output we can see that the calculated p-value equals 0.056. Since the p-value is greater than the alpha value, we cannot reject the null hypothesis that this independent variable has no significant effect on the dependent variable.

Calculating t statistics is one area where we can see that using Minitab is very easy compared to doing the statistical analysis manually, which would take a very long time.

15.6 Predicting

If we look more carefully at the regression equation for the annual salary problem,

`salary = 0.98 + 1.24 YearsEducation + 0.992 YearsWork`, we realize that this equation is also a model of the relationship between all annual salaries and the years of experience and years of education. We can use this model to *predict* the value of annual salaries, y, which we have not observed.

Suppose we are interested in predicting what the annual salary would be for someone who has 12 years of education and 12 years of work experience. As in Chap. 14, we will use the *predict* subcommand of the *regress* command.

```
MTB > Predict 'salary';
SUBC>   Nodefault;
SUBC>   KPredictors 12 12;
SUBC>   TEquation;
SUBC>   TPrediction.
```

Prediction for Salary

```
Regression Equation

salary = 0.98 + 1.238 YearsEducation + 0.992 YearsWork


Variable          Setting
YearsEducation         12
YearsWork              12


    Fit   SE Fit          95% CI                 95% PI
27.7514  1.34884  (21.9478, 33.5550)  (18.4086, 37.0942)
```

From the above Minitab output, we can see that the predicted annual salary is $27,751 and the 95 % prediction interval is from $18,409 to $37,094.

15.7 Regression Assumptions

It was indicated in Chaps. 13 and 14 that there were five underlying regression assumptions:

1. The dependent and independent variables have a linear relationship.
2. The expected value of the error term is zero.
3. The variance of the error term is constant.
4. The error terms are independent.
5. The errors are normally distributed.

When doing multiple linear regression analysis, we have to add the following assumption:

The independent variables are uncorrelated.

The normal probability plot is one way to determine if the errors are normally distributed. If the residuals basically follow a straight line, it indicates the errors are normally distributed. The Minitab *Gnormal* produces the normal probability plot.

```
MTB >  Regress;
SUBC>   Response 'salary';
SUBC>   Nodefault;
SUBC>   Continuous 'YearsEducation' 'YearsWork';
SUBC>   Terms YearsEducation YearsWork;
SUBC>   Constant;
SUBC>   Gnormal;
SUBC>   Unstandardized;
SUBC>   TExpand;
SUBC>   Tmethod;
SUBC>   Tanova;
SUBC>   Tsummary;
SUBC>   Tcoefficients;
SUBC>   Tequation;
SUBC>   Tdw;
SUBC>   TDiagnostics 1.
```

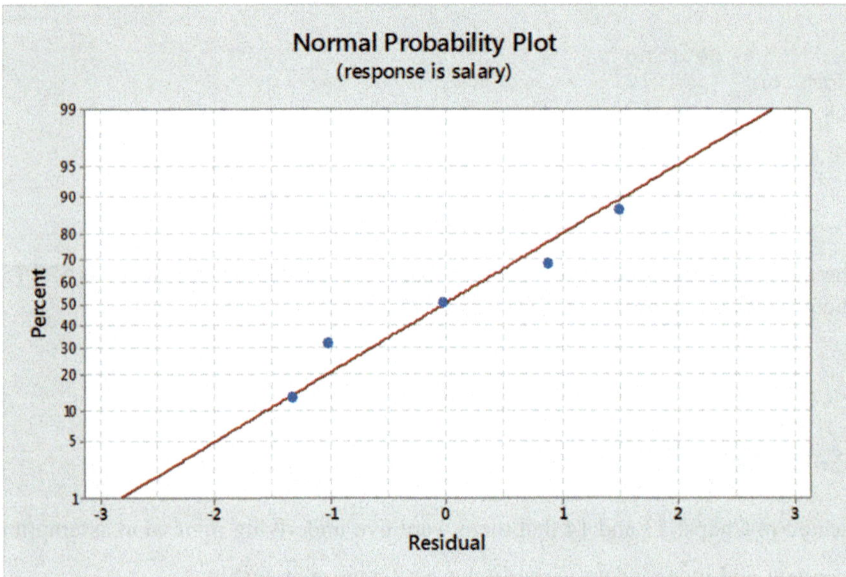

15.8 Another Regression Example

To show how the multiple regression technique can be used by real estate appraisers, Andrews and Ferguson (1986) use the data below to do a multiple regression analysis. They used this formula:

$$y_i = a + b_1 x_{1i} + b_2 x_{2i} + e_i$$

where
y_i = sale price for the ith house
x_{1i} = home size for the ith house
x_{2i} = condition rating for the ith house

Sale Price, y (thousands of dollars)	Home Size, $x1$ (hundreds of sq. ft.)	Condition Rating, $x2$ (1 to 10)
60.0	23	5
32.7	11	2
57.7	20	9
45.5	17	3
47.0	15	8
55.3	21	4
64.5	24	7
42.6	13	6
54.5	19	7
57.5	25	2

Use Minitab to do a regression on the data and predict what the sale price of a house would be if the home size was 1800 square feet and the condition rating was five.

```
MTB > name c1 'price'
MTB > name c2 'size'
MTB > name c3 'rating'
MTB > read 'price' 'size' 'rating'
DATA> 60 23 5
DATA> 32.7 11 2
DATA> 57.7 20 9
DATA> 45.5 17 3
DATA> 47.0 15 8
DATA> 55.3 21 4
DATA> 64.5 24 7
DATA> 42.6 13 6
DATA> 54.5 19 7
DATA> 57.5 25 2
DATA> end
10 rows read.
MTB > Regress;
SUBC>   Response 'price';
SUBC>   Nodefault;
SUBC>   Continuous 'size' 'rating';
SUBC>   Terms size rating;
SUBC>   Constant;
SUBC>   Unstandardized;
SUBC>   Tmethod;
SUBC>   Tanova;
SUBC>   Tsummary;
SUBC>   Tcoefficients;
SUBC>   Tequation;
SUBC>   TDiagnostics 0.
```

Regression Analysis: Price Versus Size, Rating

```
Analysis of Variance

Source       DF   Adj SS   Adj MS  F-Value  P-Value

Regression    2  819.328  409.664   350.87    0.000
  size        1  704.360  704.360   603.27    0.000
  rating      1   91.476   91.476    78.35    0.000
Error         7    8.173    1.168
Total         9  827.501


Model Summary

      S    R-sq  R-sq(adj)  R-sq(pred)
1.08055  99.01%     98.73%      97.45%


Coefficients

Term          Coef  SE Coef  T-Value  P-Value   VIF
Constant      9.78     1.63     6.00    0.001
size        1.8709   0.0762    24.56    0.000  1.00
rating       1.278    0.144     8.85    0.000  1.00


Regression Equation

price = 9.78 + 1.8709 size + 1.278 rating


Fits and Diagnostics for Unusual Observations

Obs   price     Fit   Resid  Std Resid
 10  57.500  59.112  -1.612      -2.11  R

R  Large residual

MTB > Predict 'price';
SUBC>   Nodefault;
SUBC>   KPredictors 18 5;
SUBC>   TEquation;
SUBC>   TPrediction.
```

Prediction for Price

```
Regression Equation

price = 9.78 + 1.8709 size + 1.278 rating


Variable  Setting
size           18
rating          5


    Fit    SE Fit          95% CI                   95% PI
49.8498  0.349456  (49.0235, 50.6761)  (47.1644, 52.5352)
```

From the Minitab output we can see that the regression equation is

```
price = 9.78 + 1.87 size + 1.28 rating
```

This regression equation says that for every 100 square foot increase in size, a house's price will increase by \$1870. For every increase in rating, there will be a \$1290 increase in the price of a house.

From the regression output we can see that the coefficients for size and rating do not equal zero at the 5 % alpha level. We conclude this because the *p*-value calculated for both coefficients is zero and this would be less than the 5 % alpha level.

The regression predicts that the sale price would be \$49,850 for a house with 1800 square feet and a rating of 5. The prediction interval for this prediction is from \$47,164 to \$52,536.

15.9 SAS Programming Code Instructions and Examples

15.9.1 Introduction

Below shows the regression analysis performed in SAS.

```
DATA Education;
    INPUT salary YearsEducation YearsWork @@;
    DATALINES;
15 5 7
17 10 5
26 9 14
24 13 8
27 15 6
;
run;
```

In **PROC GLM**, ALPHA = p specifies the level of significance p for 100(1 − p) % confidence intervals. The value must be between 0 and 1; the default value of p results in 95 % intervals.

```
PROC GLM DATA =Education;
    MODEL salary = YearsEducation YearsWork /P CLM SS1 CLPARM ALPHA=0.05;
RUN;
```

Source	DF	Sum of squares	Mean square	F Value	Pr > F
Model	2	113.0088059	56.5044030	19.51	0.0487
Error	2	5.7911941	2.8955970		
Corrected total	4	118.8000000			

R-Square	Coeff var	Root MSE	Salary mean
0.951253	7.805713	1.701645	21.80000

Source	DF	Type I SS	Mean square	F Value	Pr > F
YearsEducation	1	65.77297297	65.77297297	22.71	0.0413
YearsWork	1	47.23583294	47.23583294	16.31	0.0562

Parameter	Estimate	Standard error	t Value	Pr > $\lvert t \rvert$	95 % Confidence limits	
Intercept	0.980274745	3.43891738	0.29	0.8024	−13.81619252	15.776742007
YearsEducation	1.238464248	0.22582466	5.48	0.0317	0.266819152	2.210109344
YearsWork	0.992462135	0.24572380	4.04	0.0562	−0.064802035	2.049726304

Sum of residuals	−0.00000000
Sum of squared residuals	5.79119408
Sum of squared residuals—error SS	0.00000000
PRESS statistic	35.52000728
First order autocorrelation	−0.45531768
Durbin–Watson D	2.39454006

15.9.2 R-Square

R-Square	Coeff var	Root MSE	Salary mean
0.951253	7.805713	1.701645	21.80000

15.9.3 F Test

Source	DF	Sum of squares	Mean square	F Value	Pr > F
Model	2	113.0088059	56.5044030	19.51	0.0487
Error	2	5.7911941	2.8955970		
Corrected total	4	118.8000000			

15.9.4 Confidence Interval of β

Parameter	Estimate	Standard error	t Value	Pr > $\lvert t \rvert$	95 % Confidence limits	
Intercept	0.980274745	3.43891738	0.29	0.8024	−13.81619252	15.776742007
YearsEducation	1.238464248	0.22582466	5.48	0.0317	0.266819152	2.210109344
YearsWork	0.992462135	0.24572380	4.04	0.0562	−0.064802035	2.049726304

15.9.5 t Test

Parameter	Estimate	Standard error	t Value	Pr > |t|	95 % Confidence limits	
Intercept	0.980274745	3.43891738	0.29	0.8024	−13.81619252	15.776742007
YearsEducation	1.238464248	0.22582466	5.48	0.0317	0.266819152	2.210109344
YearsWork	0.992462135	0.24572380	4.04	0.0562	−0.064802035	2.049726304

15.9.6 Predicting

We will use the **p** option of the **PROC REG** command to predict the value of the dependent variable in the following problem.

```
data Education_Predict;
input salary YearsEducation YearsWork;
cards;
. 12 12
;
run;
data Education_Predict;
set Education_Predict Education;
run;
proc reg data=Education_Predict;
model salary=YearsEducation YearsWork/noint P clm cli;
output out=Education_Out r=res p=predict;
run;
```

Output statistics								
Obs	Dependent variable	Predicted value	Std error mean predict	95 % CL mean		95 % CL predict		Residual
1		27.7514	0.8935	25.1410	30.8280	22.6524	33.3165	
2	15.0000	13.7445	0.5974	11.8433	15.6458	8.8497	18.6394	1.2555
3	17.0000	18.1095	0.6312	16.1006	20.1184	13.1718	23.0472	−1.1095
4	26.0000	26.1992	1.2558	22.2029	30.1956	20.1729	32.2255	−0.1992
5	24.0000	25.1056	0.7605	22.6855	27.5258	19.9868	30.2245	−1.1056
6	27.0000	25.6010	1.0522	22.2526	28.9495	19.9834	31.2187	1.3990

15.9.7 Another Regression Example

Use SAS to do a regression on the data and predict what the sale price of a house would be if the home size was 1800 square feet and the condition rating was five.

```
DATA RealEstate;
    INPUT price size rating @@;
    DATALINES;
60 23 5
32.7 11 2
57.7 20 9
45.5 17 3
47.0 15 8
55.3 21 4
64.5 24 7
42.6 13 6
54.5 19 7
57.5 25 2
;
run;

PROC GLM DATA =RealEstate;
    MODEL price = size rating / P CLM SS1 CLPARM;
RUN;

data RealEstate_Predict;
input price size rating;
cards;
. 18 5
;
run;
data RealEstate_Predict;
set RealEstate_Predict RealEstate;
run;
```

Just two additional lines of code are needed to generate a simple in RTF file, a form of output that is read Microsoft Word. **ods rtf file="output.rtf"** statement opens the RTF destination; all output generated will be sent to the file output.rtf. If this file already exists, it will be overwritten. It is only when **ODS RTF CLOSE** statements are passed that the RTF file is saved to disk.

```
ods rtf file='c:\name.rtf';
proc reg data=RealEstate_Predict;
model price = size rating/P clm cli;
output out=RealEstate_Out r=res p=predict;
run;
ods rtf close;
```

Analysis of variance					
Source	DF	Sum of squares	Mean square	F Value	Pr > F
Model	2	819.32795	409.66398	350.87	<0.0001
Error	7	8.17305	1.16758		
Corrected total	9	827.50100			

Root MSE	1.08055	R-Square	0.9901
Dependent mean	51.73000	Adj R-Sq	0.9873
Coeff var	2.08882		

Parameter estimates

Variable	DF	Parameter estimate	Standard error	t Value	Pr > |t|
Intercept	1	9.78227	1.63048	6.00	0.0005
Size	1	1.87094	0.07617	24.56	<0.0001
Rating	1	1.27814	0.14440	8.85	<0.0001

Output statistics

Obs	Dependent variable	Predicted value	Std error mean predict	95 % CL mean		95 % CL predict		Residual
1		49.8498	0.3495	49.0235	50.6761	47.1644	52.5352	
2	60.0000	59.2045	0.4714	58.0899	60.3191	56.4169	61.9921	0.7955
3	32.7000	32.9188	0.8200	30.9799	34.8578	29.7114	36.1263	−0.2188
4	57.7000	58.7042	0.6375	57.1969	60.2116	55.7377	61.6708	−1.0042
5	45.5000	45.4226	0.4919	44.2595	46.5856	42.6153	48.2299	0.0774
6	47.0000	48.0714	0.6019	46.6481	49.4948	45.1466	50.9962	−1.0714
7	55.3000	54.1845	0.4276	53.1735	55.1955	51.4366	56.9323	1.1155
8	64.5000	63.6317	0.5705	62.2826	64.9808	60.7423	66.5211	0.8683
9	42.6000	41.7733	0.5710	40.4231	43.1235	38.8834	44.6632	0.8267
10	54.5000	54.2770	0.4206	53.2824	55.2717	51.5352	57.0189	0.2230
11	57.5000	59.1119	0.7657	57.3013	60.9226	55.9803	62.2435	−1.6119

Sum of residuals	0
Sum of squared residuals	8.17305
Predicted residual SS (PRESS)	21.09523

15.10 Statistical Summary

In this chapter we looked at the relationship among three or more variables. We saw that many of the concepts of the relationship between two variables extend to the situations of three or more variables.

Bibliography

Andrews RL, Ferguson JT (1986) Integrating judgment with a regression appraisal. Real Estate Appraiser Anal 52(2):71–74

Lee CF, Lee JC, Lee AC (2013) Statistics for business and financial economics. Springer, New York

Microsoft Inc., Excel 2013. Microsoft Inc., Redmond

Minitab Inc. Minitab 17. Minitab Inc., State College

SAS Institute Inc. SAS 2014. SAS Institute Inc., Cary

Chapter 16
Residual and Regression Assumption Analysis

16.1 Introduction

When we do regression analysis, there are six underlying assumptions:

1. The dependent and independent variables have a linear relationship.
2. The expected value of the residual term is zero.
3. The variance of the residual term is constant.
4. The residual terms are independent.
5. The independent variables are uncorrelated.
6. The residuals are normally distributed.

To make the proper statistical inferences, these six assumptions must not be materially violated. In this chapter we will look at the six major assumptions by studying an article that was published in *Accounting Review* (Benston 1966). G. J. Benston did a multiple regression that had the following dependent variable and independent variables:

y = hours of labor in ith week
$x1$ = thousands of pounds shipped
$x2$ = percentage of units shipped by truck
$x3$ = average number of pounds per shipment

From the above, we can conclude that Benston felt that thousands of pounds shipped, percentage of units shipped by truck, and average number of pounds per shipment could explain the hours of labor in a particular week. The data for his model are displayed below.

C.-F. Lee et al., *Essentials of Excel, Excel VBA, SAS and Minitab for Statistical and Financial Analyses*, DOI 10.1007/978-3-319-38867-0_16

week	y	x1	x2	x3
1	100	5.1	90	20
2	85	3.8	99	22
3	108	5.3	58	19
4	116	7.5	16	15
5	92	4.5	54	20
6	63	3.3	42	26
7	79	5.3	12	25
8	101	5.9	32	21
9	88	4.0	56	24
10	71	4.2	64	29
11	122	6.8	78	10
12	85	3.9	90	30
13	50	3.8	74	28
14	114	7.5	89	14
15	104	4.5	90	21
16	111	6.0	40	20
17	110	8.1	55	16
18	100	2.9	64	19
19	82	4.0	35	23
20	85	4.8	58	25

y = hours of labor in *i* th week
x1 = thousands of pounds shipped
x2 = percentage of units shipped by truck
x3 = average number of pounds per shipmen

We will now regress the above data using the *regress* command.

```
MTB > name c1 'labor'
MTB > name c2 'thousand'
MTB > name c3 'percent'
MTB > name c4 'average'
MTB > read 'labor' 'thousand' 'percent' 'average'
DATA> 100 5.1 90 20
DATA> 85   3.8 99  22
DATA> 108  5.3 58  19
DATA> 116  7.5 16  15
DATA> 92   4.5 54  20
DATA> 63   3.3 42  26
DATA> 79   5.3 12  25
DATA> 101  5.9 32  21
DATA> 88   4.0 56  24
DATA> 71   4.2 64  29
DATA> 122  6.8 78  10
DATA> 85   3.9 90  30
DATA> 50   3.8 74  28
DATA> 114  7.5 89  14
DATA> 104  4.5 90  21
DATA> 111  6.0 40  20
DATA> 110  8.1 55  16
```

```
DATA> 100   2.9 64   19
DATA> 82    4.0 35   23
DATA> 85    4.8 58   25
DATA> end
20 rows read.
MTB > Name C5 "FITS" C6 "RESI" C7 "SRES".
MTB > Regress;
SUBC>   Response 'labor';
SUBC>   Nodefault;
SUBC>   Continuous 'thousand' 'percent' 'average'
SUBC>   Terms thousand percent average;
SUBC>   Constant;
SUBC>   Unstandardized;
SUBC>   TExpand;
SUBC>   Tmethod;
SUBC>   Tanova;
SUBC>   Tsummary;
SUBC>   Tcoefficients;
SUBC>   Tequation;
SUBC>   Tdw;
SUBC>   TDiagnostics 1;
SUBC>   Fits 'FITS';
SUBC>   Residuals 'RESI';
SUBC>   Sresiduals 'SRES'.
```

Regression Analysis: Labor Vs. Thousand, Percent, and Average

```
Analysis of Variance

Source      DF  Seq SS  Contribution   Adj SS   Adj MS  F-Value  P-Value
Regression   3  5158.3        77.01%  5158.31  1719.44    17.87    0.000
  thousand   1  3400.6        50.77%   138.19   138.19     1.44    0.248
  percent    1   198.4         2.96%    24.62    24.62     0.26    0.620
  average    1  1559.3        23.28%  1559.32  1559.32    16.20    0.001
Error       16  1539.9        22.99%  1539.89    96.24
Total       19  6698.2       100.00%


Model Summary

      S    R-sq  R-sq(adj)    PRESS  R-sq(pred)
9.81035  77.01%     72.70%  2475.94      63.04%


Coefficients

Term        Coef  SE Coef          95% CI         T-Value  P-Value   VIF
Constant   131.9     25.7  (   77.5,   186.4)      5.13    0.000
thousand    2.73     2.28  (  -2.10,    7.55)      1.20    0.248  2.25
percent   0.0472   0.0933  (-0.1507,  0.2451)      0.51    0.620  1.09
average   -2.587    0.643  ( -3.950,  -1.225)     -4.03    0.001  2.17


Regression Equation

labor = 131.9 + 2.73 thousand + 0.0472 percent - 2.587 average
```

Fits and Diagnostics for All Observations

Obs	labor	Fit	SE Fit	95% CI	Resid	Std Resid	Del Resid	HI
1	100.00	98.33	3.53	(90.84, 105.81)	1.67	0.18	0.18	0.129547
2	85.00	90.03	4.40	(80.70, 99.37)	-5.03	-0.57	-0.56	0.201606
3	108.00	99.95	2.50	(94.64, 105.26)	8.05	0.85	0.84	0.065142
4	116.00	114.31	5.47	(102.71, 125.91)	1.69	0.21	0.20	0.311128
5	92.00	94.99	3.12	(88.39, 101.60)	-2.99	-0.32	-0.31	0.100936
6	63.00	75.63	4.08	(66.98, 84.28)	-12.63	-1.42	-1.47	0.172984
7	79.00	82.25	5.13	(71.37, 93.13)	-3.25	-0.39	-0.38	0.273634
8	101.00	95.18	3.46	(87.84, 102.52)	5.82	0.63	0.62	0.124558
9	88.00	83.37	2.80	(77.43, 89.32)	4.63	0.49	0.48	0.081641
10	71.00	71.36	4.39	(62.05, 80.67)	-0.36	-0.04	-0.04	0.200197
11	122.00	128.27	5.71	(116.16, 140.38)	-6.27	-0.79	-0.78	0.339071
12	85.00	69.18	5.63	(57.24, 81.12)	15.82	1.97	2.19	0.329760
13	50.00	73.33	3.92	(65.01, 81.64)	-23.33	-2.59	-3.30	0.159821
14	114.00	120.35	5.48	(108.74, 131.96)	-6.35	-0.78	-0.77	0.311611
15	104.00	94.10	3.53	(86.62, 101.59)	9.90	1.08	1.09	0.129665
16	111.00	98.42	3.04	(91.98, 104.86)	12.58	1.35	1.39	0.096014
17	110.00	115.20	5.35	(103.86, 126.55)	-5.20	-0.63	-0.62	0.297673
18	100.00	93.69	6.40	(80.13, 107.25)	6.31	0.85	0.84	0.424990
19	82.00	84.97	3.94	(76.62, 93.32)	-2.97	-0.33	-0.32	0.161271
20	85.00	83.06	2.92	(76.87, 89.26)	1.94	0.21	0.20	0.088751

Obs	Cook's D	DFITS	
1	0.00	0.06830	
2	0.02	-0.28235	
3	0.01	0.22197	
4	0.00	0.13494	
5	0.00	-0.10471	
6	0.10	-0.67027	
7	0.01	-0.23232	
8	0.01	0.23442	
9	0.01	0.14313	
10	0.00	-0.01987	
11	0.08	-0.55607	
12	0.48	1.53673	
13	0.32	-1.43939	R
14	0.07	-0.51801	
15	0.04	0.41968	
16	0.05	0.45203	
17	0.04	-0.40405	
18	0.13	0.72244	
19	0.01	-0.14083	
20	0.00	0.06261	

R Large residual

Durbin-Watson Statistic

Durbin-Watson Statistic = 2.42977

From the Minitab output, we can see that the regressed equation is

```
labor = 131.9 + 2.73 thousand + 0.0472 percent - 2.587 average
```

This equation tells us that:

1. For every thousand pound increase, there will be 2.73 h increase in labor.
2. For every 1 % increase in units shipped, there will be 0.0472 h increase in labor.
3. For every average pound increase per shipment, there will be 2.59 h decrease in labor.

From the *R*-square calculation of 72.7 %, we can see that the independent variables explain 72.7 % of the variability of the dependent variable.

16.2 Linearity

One way to check linearity is to plot the *y* variable with each *x* variable to check if the plot looks linear. If one of these plots does not look linear, then the linearity assumption might be violated. Below are three plots of *y* with each of the three *x* variables. For each plot we will also look at the strength of the relationship between *y* and *x*.

```
MTB > plot 'labor'*'thousand'
```

Scatterplot of Labor vs. Thousand

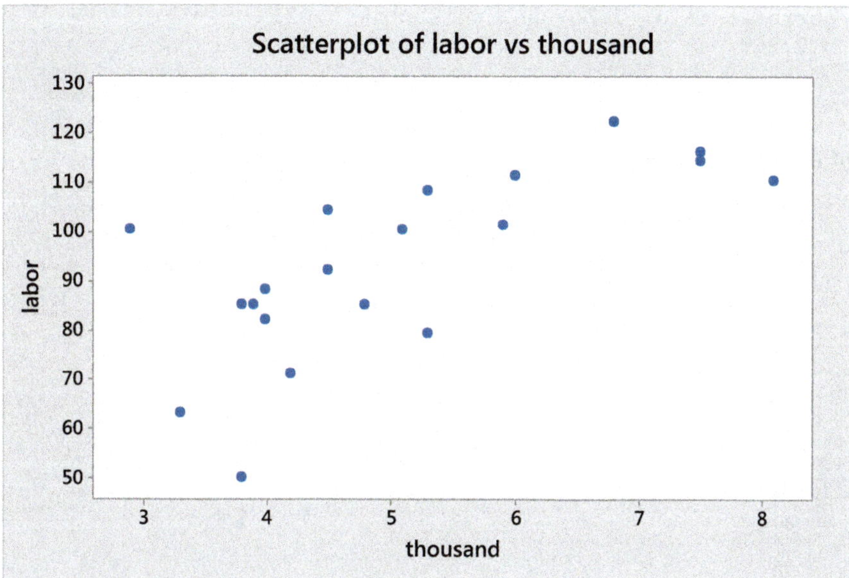


```
MTB > correlation 'labor' 'thousand'
```

Correlation: Labor, Thousand

```
Pearson correlation of labor and thousand = 0.713
P-Value = 0.000
```

```
MTB > plot 'labor' * 'percent'
```

Scatterplot of Labor vs. Percent

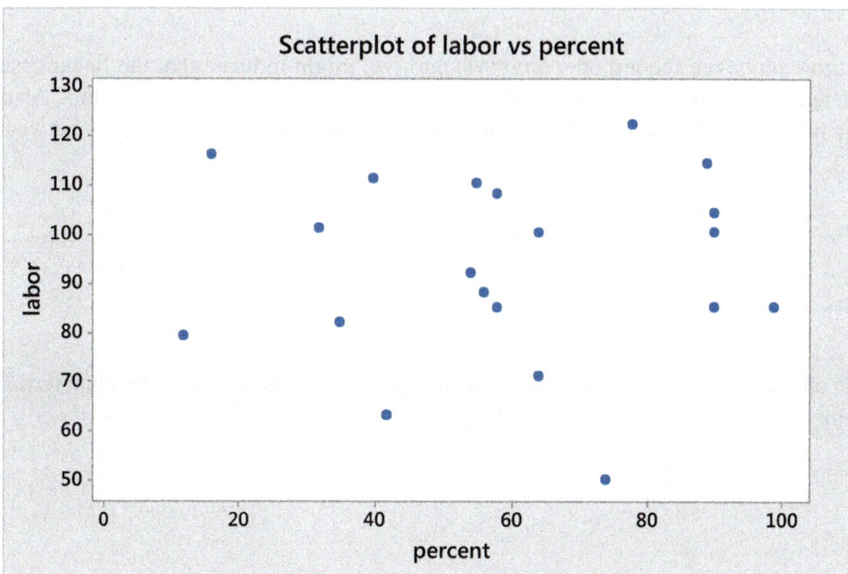

```
MTB > correlation 'labor' 'percent'
```

Correlations: Labor, Percent

```
Pearson correlation of labor and percent = 0.031
P-Value = 0.898
```

```
MTB > plot 'labor' * 'average'
```

Scatterplot of Labor vs. Average

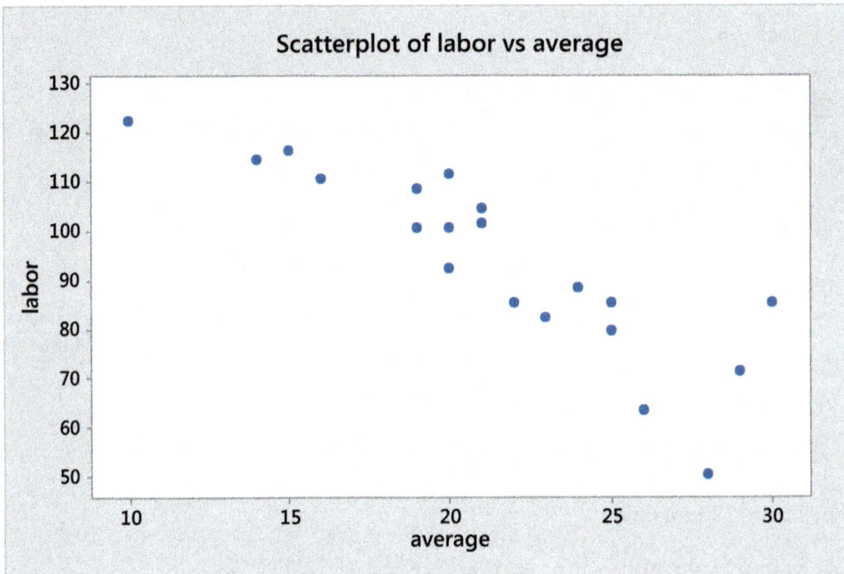


```
MTB > correlation 'labor' 'average'
```

Correlations: Labor, Average

```
Pearson correlation of labor and average = -0.865
P-Value = 0.000
```

Of the three plots, the second one, *labor* vs. *percent*, might indicate that the linear assumption is being violated. In this plot, we can see that there is no general linear relationship. Also from the correlation number of 0.031, we can see that there is only a weak association between the two variables.

16.3 The Expected Value of the Residual Term Is Zero

One way to check this assumption is to produce a dotplot and a boxplot of the residual terms and see if the midpoint is zero.

```
MTB > dotplot resi
```

Dotplot of RESI

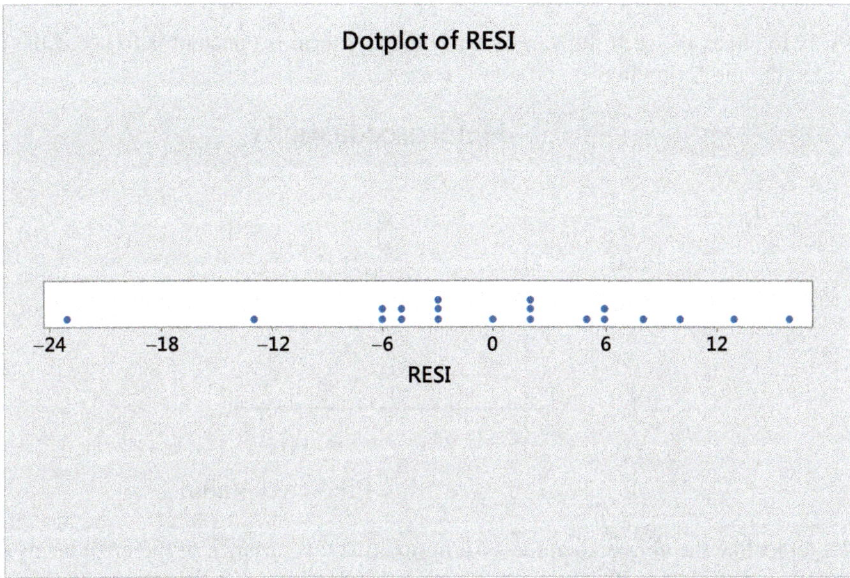


```
MTB > boxplot 'resi';
SUBC> transpose.
```

Boxplot of RESI

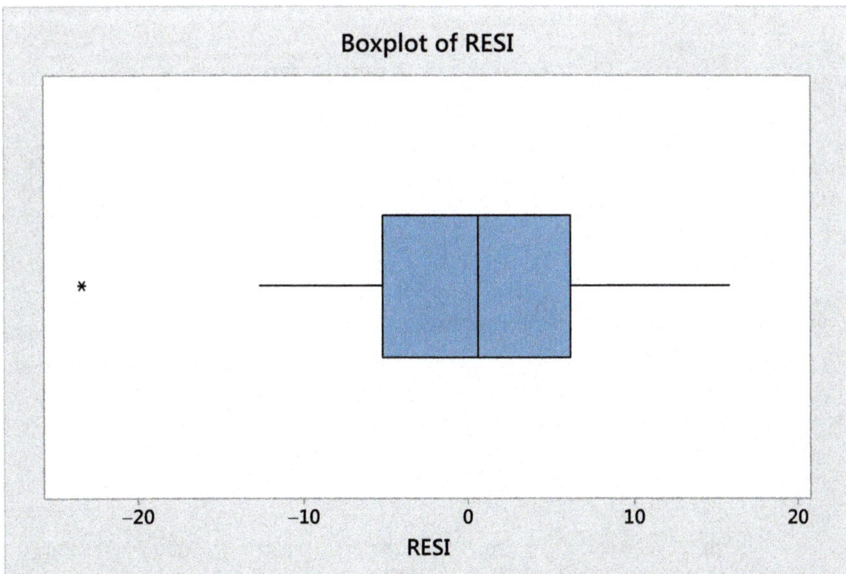

From the above dotplot and boxplot, we can see that the midpoint is very close to zero.

16.4 The Variance of the Error Term Is Constant

One quick way to check to see if the variance of the error term is constant is to see if the plot of the residuals looks like the following.

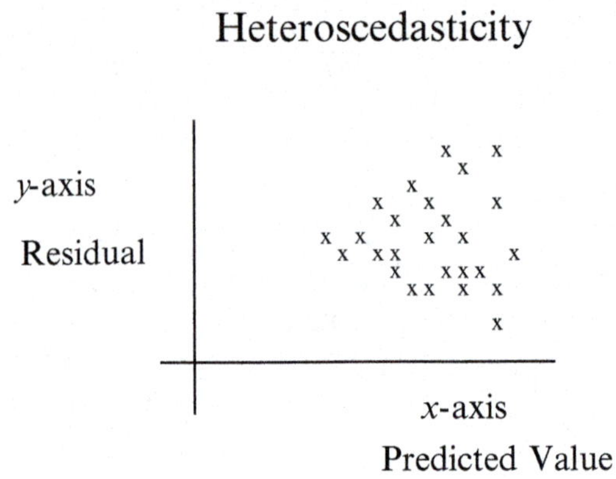

If the plot looks like the above, then the assumption that the variance of the error terms is constant is violated. This problem is called *heteroscedasticity*. Below we will plot the residuals against the independent variables and the plot the residuals against the expected values. Below shows the plot in Minitab.

```
MTB > plot resi * fits
```

Scatterplot of RESI vs. FITS

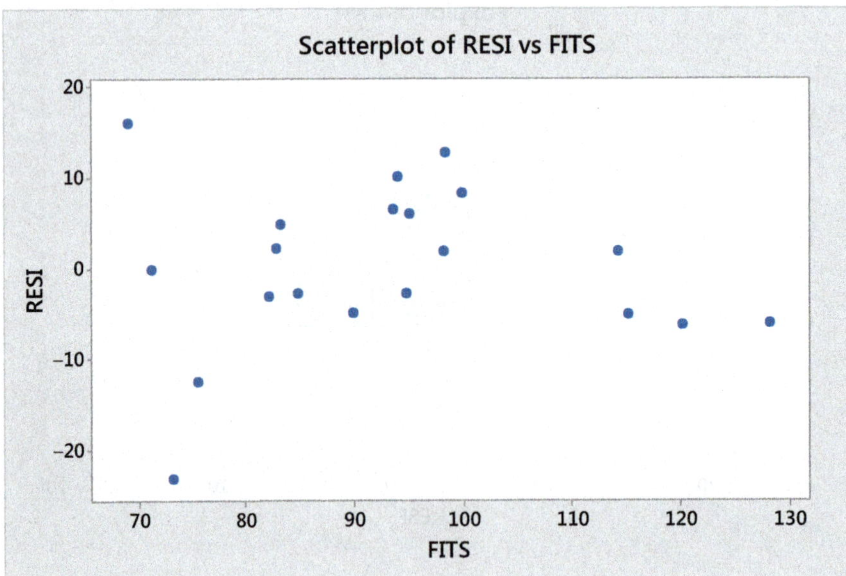


```
MTB > plot resi * labor
```

Scatterplot of RESI vs. Labor

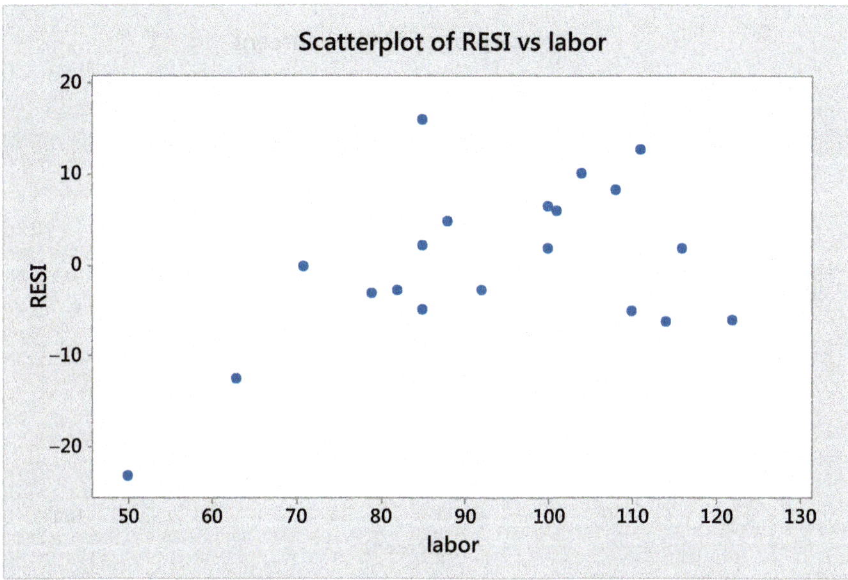


```
MTB > plot resi * thousand
```

Scatterplot of RESI vs. Thousand

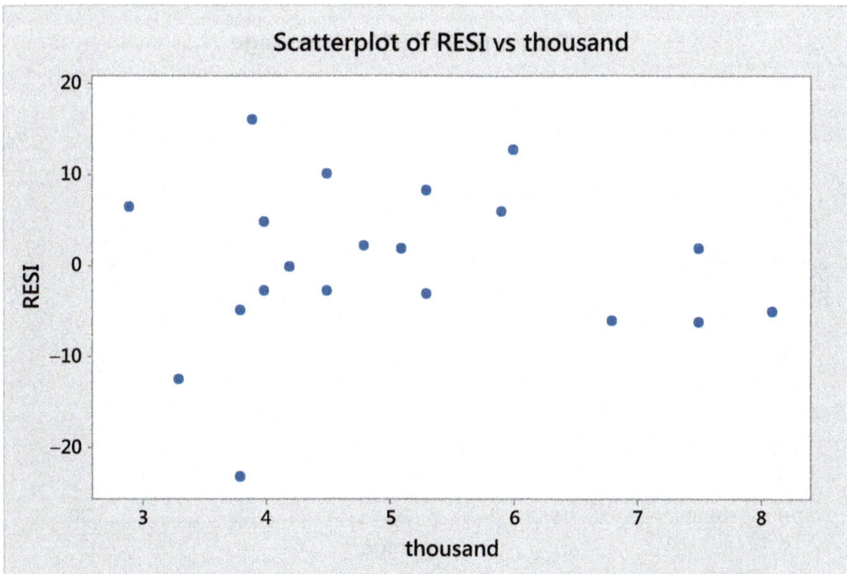


```
MTB > plot resi * percent
```

Scatterplot of RESI vs. Percent

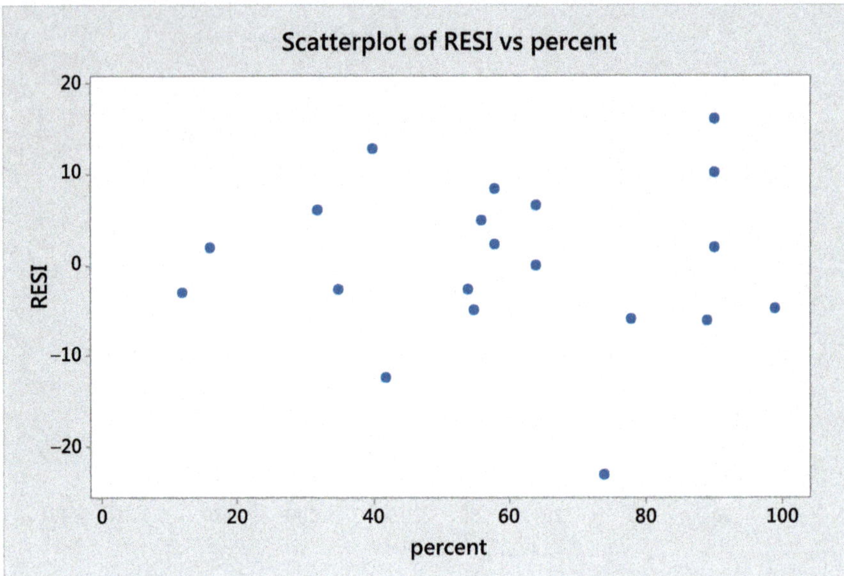


```
MTB > plot resi * average
```

Scatterplot of RESI vs. Average

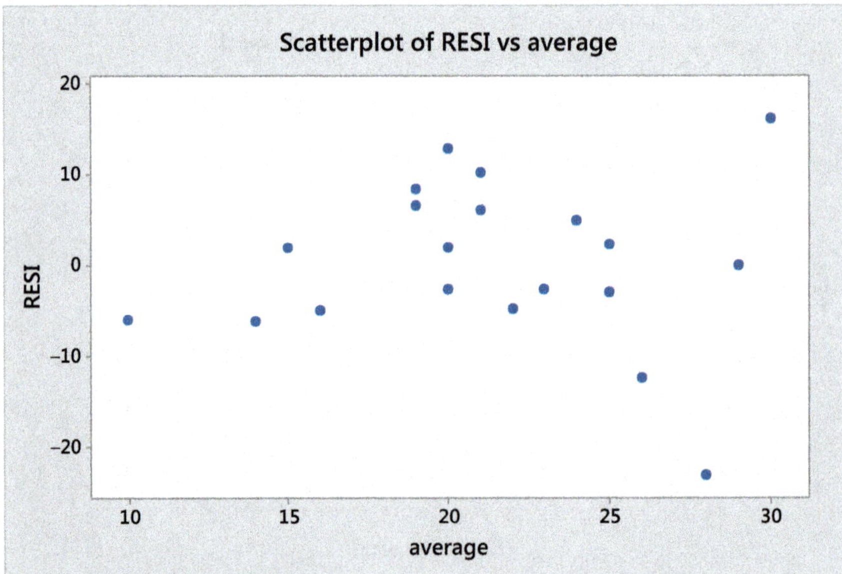

We can also test for heteroscedasticity by calculating the nR^2 statistics known as the variance inflation factor (VIF).

The following are the steps to calculate for the existence of heteroscedasticity:

(1) Run a standard regression.
(2) Calculate the residuals, residual = expected—actual.

(3) Run a regression using the square of the residuals as the dependant variable and the estimated dependent variable as the independent variable.
(4) Estimate nR^2, where n is the sample size and R^2 is the coefficient of determination.
(5) Use χ^2 statistics with one degree of freedom to test whether nR^2 is significantly different from zero.

16.5 Excel Macro: Regression Assumption Tests

The code to produce these regression assumptions is shown in Appendix 16.1.

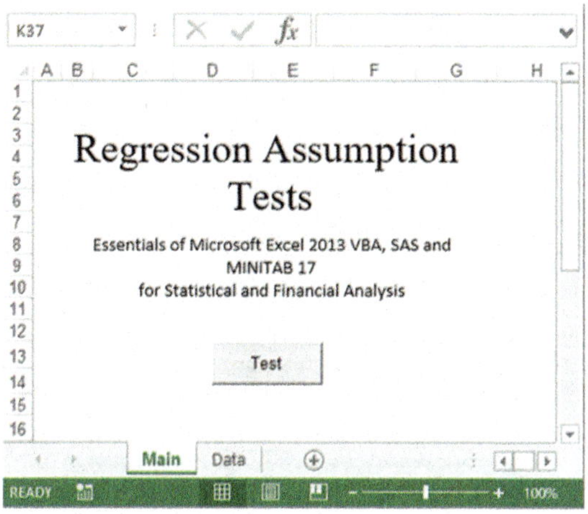

Before pressing the *Test* button, the regression data should be entered in the *data* worksheet as shown below. In the *data* sheet, the first row should be the name of the variable and the first column should be the dependant variable.

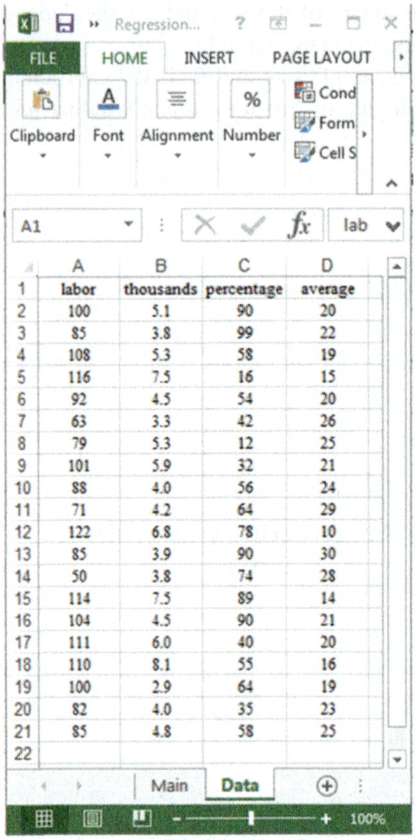

	A	B	C	D
1	labor	thousands	percentage	average
2	100	5.1	90	20
3	85	3.8	99	22
4	108	5.3	58	19
5	116	7.5	16	15
6	92	4.5	54	20
7	63	3.3	42	26
8	79	5.3	12	25
9	101	5.9	32	21
10	88	4.0	56	24
11	71	4.2	64	29
12	122	6.8	78	10
13	85	3.9	90	30
14	50	3.8	74	28
15	114	7.5	89	14
16	104	4.5	90	21
17	111	6.0	40	20
18	110	8.1	55	16
19	100	2.9	64	19
20	82	4.0	35	23
21	85	4.8	58	25

Below shows the regression assumption test report.

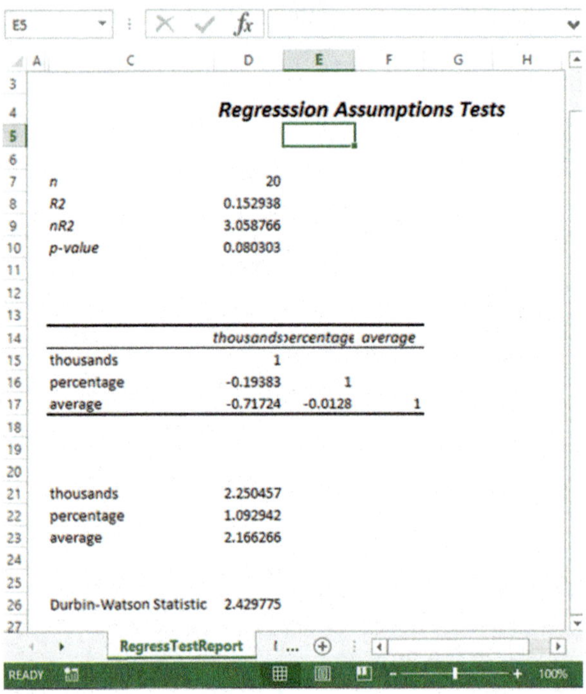

Regresssion Assumptions Tests

n	20
R2	0.152938
nR2	3.058766
p-value	0.080303

	thousands	*percentage*	*average*
thousands	1		
percentage	-0.19383	1	
average	-0.71724	-0.0128	1

thousands	2.250457
percentage	1.092942
average	2.166266

Durbin-Watson Statistic 2.429775

16.6 The Residual Terms Are Independent

If the residual terms are not independent, we have a situation called *autocorrelation*. The Durbin–Watson test is a test statistic for detecting autocorrelation. The Durbin–Watson calculation from the Minitab regression output is reproduced below.

```
Durbin-Watson statistic = 2.43
```

The DW statistic has a value between 0 and 4. A 0 means very strong positive autocorrelation. A 4 means very strong negative autocorrelation. A value of 2 means no or very little autocorrelation. Our test statistic of 2.43, therefore, suggests that our data set has very little autocorrelation.

According to James Stevens:

> This is a very important assumption. . . . . if independence is violated only mildly then the probability of a type one error will be *several* times greater than the level the experimenter thinks he or she is working at. Thus, instead of rejecting falsely 5 % of the time, the experimenter may be rejecting falsely 25 % or 30 % of the time. (Stevens 1986)

The regression analysis in Excel did not calculate the Durbin–Watson statistics. We can calculate it easily by using Excel regression report done in Sect. 16.1. We will create the DW statistics based on its definition. The definition of the DW statistics is

$$DW = \frac{\sum_{t=2}^{n} (e_t - e_{t-1})^2}{\sum_{t=1}^{n} e_t^2}$$

Below is the Durban–Watson report that is produce in the workbook *regressionassumptiontests.xls*.

AutoCorrelation Test

Durbin-Watson Statistic	2.429775

16.7 The Independent Variables Are Uncorrelated: Multicollinearity

The next thing we would like to see is how correlated the independent variables are. The more correlated the independent variables, the less sure we are about how much each independent variable explains the dependent variable. *Multicollinearity* can cause problems like, inflating regression coefficients, producing a t statistic that is too small (Kvanli et al. 1992). The ideal situation would be no correlation at all between the independent variables. This would look like the following, using a Venn diagram.

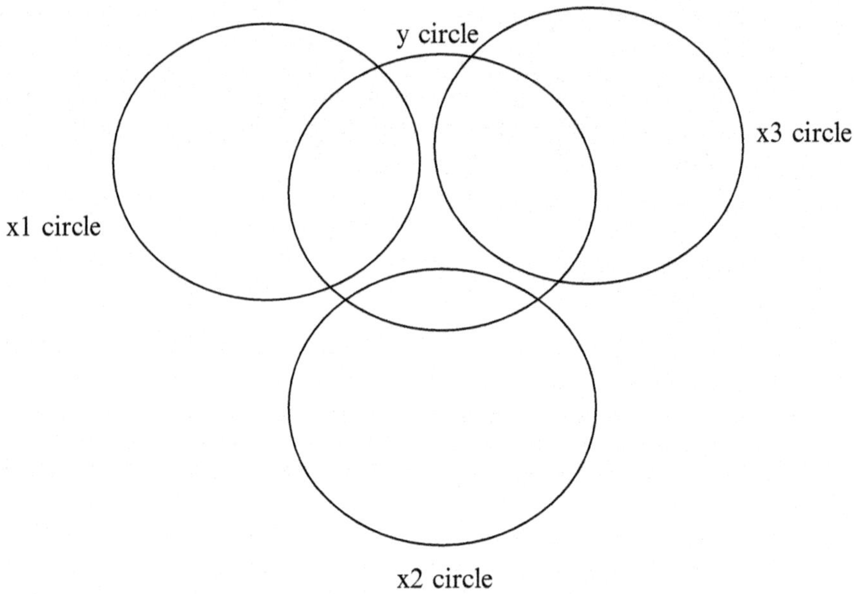

In this diagram none of the independent variables intersect any other independent variable. But most of the time, this is not the case. Usually there are intersections among the independent variables, indicating some degree of correlation. This would look like the following in a Venn diagram.

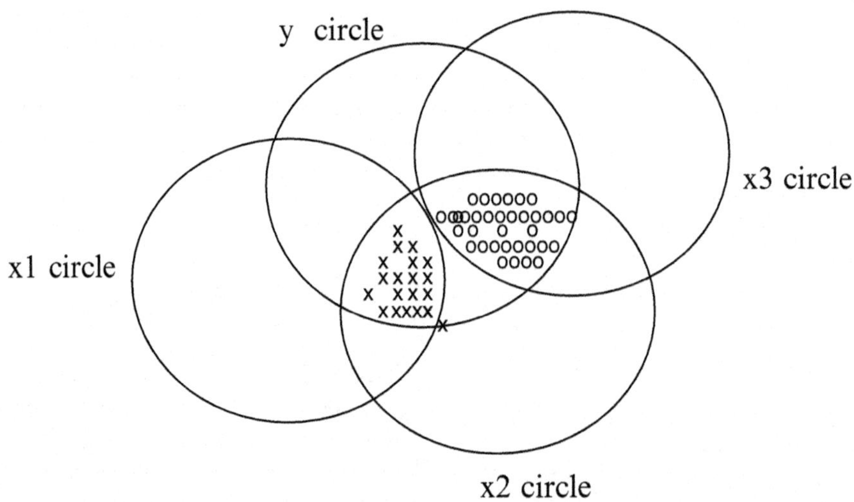

The area with small x's indicates multicollinearity between circles $x1$ and $x2$. This means that $x1$ and $x2$ are both explaining the same part of y. Similarly, we can see that there is multicollinearity between $x2$ and $x3$ by the small o's. This means that $x2$ and $x3$ are explaining the same part of y. We can see therefore that circle $x2$ is not very useful in the regression because the part of y that it is explaining is also being explained by $x1$ and $x3$.

```
MTB > correlation 'thousand'-'average'
```

Correlations: Thousand, Percent, Average

```
        thousand  percent
percent   -0.194
           0.413

average   -0.717   -0.013
           0.000    0.957

Cell Contents: Pearson correlation
               P-Value
```

From this Minitab multicollinearity table, we can see that there is strong multicollinearity between thousands of pounds shipped and the average number of pounds per shipment.

16.8 Variance Infactionary Factor (VIF)

The correlation matrix only measures the correlation between one independent variable with another independent variable and not the possible correlation between any one independent variable and all others taken as a group. A statistics called the *variance inflationary factor* (*VIF*) measures the correlation between any one independent variable and all others taken as a group. Researchers have used VIF = 10 as a critical value rule of thumb to determine whether too much correlation exists between an independent variable and the other independent variables taken as a group. The VIF is defined for each independent variable as

$$VIF = \frac{1}{1 - R^2}$$

The steps to calculate the VIF for each independent variable would be to regress each of the independent variables against the other independent variables. In our example we would do three independent variable regressions. The next step then would be to find the R^2 for each regression and then calculate the VIF.

Calculating the VIF in Minitab is easy. There is a VIF subcommand for the regress command. The VIF value for each independent variable is shown below using Minitab.

Regression Analysis: Labor Vs. Thousand, Percent, Average

```
The regression equation is
labor = 132 + 2.73 thousand + 0.0472 percent - 2.59 average

Predictor         Coef     SE Coef        T       P      VIF
Constant        131.92       25.69     5.13   0.000
thousand         2.726       2.275     1.20   0.248      2.3
percent        0.04722     0.09335     0.51   0.620      1.1
average        -2.5874      0.6428    -4.03   0.001      2.2

S = 9.810       R-Sq = 77.0%      R-Sq(adj) = 72.7%

Analysis of Variance

Source            DF        SS        MS        F        P
Regression         3    5158.3    1719.4    17.87    0.000
Residual Error    16    1539.9      96.2
Total             19    6698.2
```

Based on the VIF calculation for our example, we can conclude that there is no problem with each of our independent variables correlating with the other independent variables as a group.

Below is the multicollinearity report that is produced in the workbook *regressionassumptiontests.xlsm*.

Multicoliniearity Test

Correlation Maxtrix

	thousands	*percentage*	*average*
thousands	1		
percentage	-0.193825	1	
average	-0.717235	-0.012804	1

Variance Influence Factor - VIF

thousands	2.250457
percentage	1.092942
average	2.166266

16.9 Testing the Normality of the Residuals

The following works in Minitab 13 but not in Minitab 17. Minitab gives three kinds of normality test. The three types of normality tests are:

(1) Anderson–Darling test
(2) Ryan–Joiner test
(3) Kolmogorov–Smirnov test

All three normality tests have the following hypotheses:

Ho: Residuals follow a normal distribution

H1: Residuals do not follow a normal distribution

The following is the Anderson–Darling normality test.

```
MTB > %NormPlot 'residual'.
Executing from file: C:\Program Files\Mtbwin\MACROS\NormPlot.MAC
Macro is running ... please wait
```

Normal Prob Plot: Residual

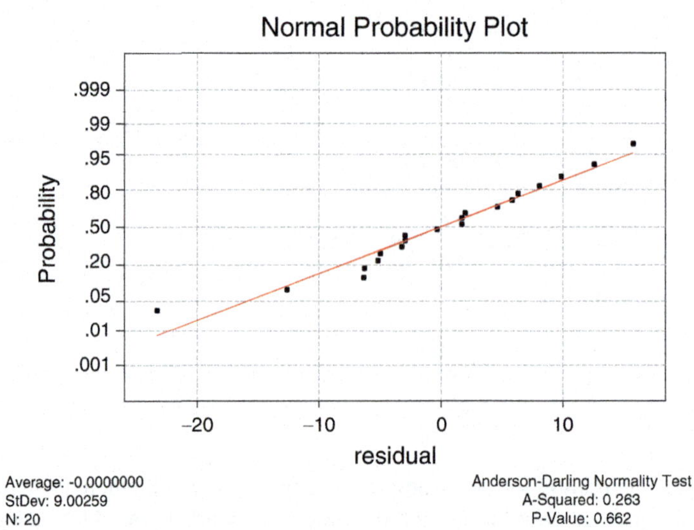

The following is the Ryan–Joiner normality test.

```
MTB > %NormPlot 'residual';
SUBC>   SWTest.
Executing from file: C:\Program Files\Mtbwin\MACROS\NormPlot.MAC
Macro is running ... please wait
```

Normal Prob Plot: Residual

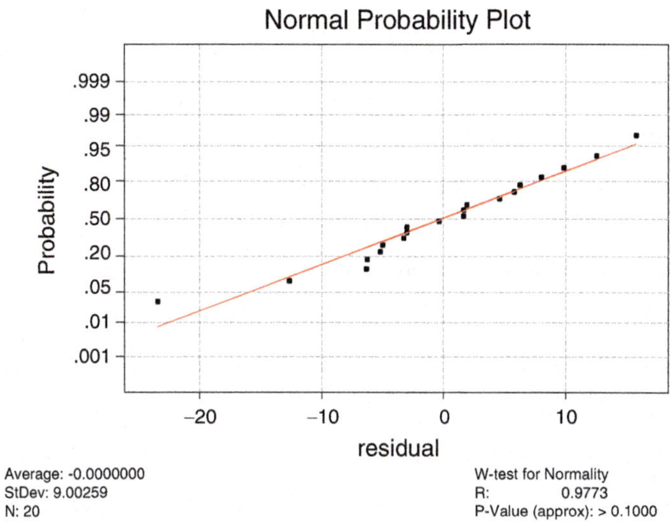

The following is the Kolmogorov–Smirnov normality test.

```
MTB > %NormPlot 'residual';
SUBC>   Kstest.
Executing from file: C:\Program Files\Mtbwin\MACROS\NormPlot.MAC
Macro is running ... please wait
```

Normal Prob Plot: Residual

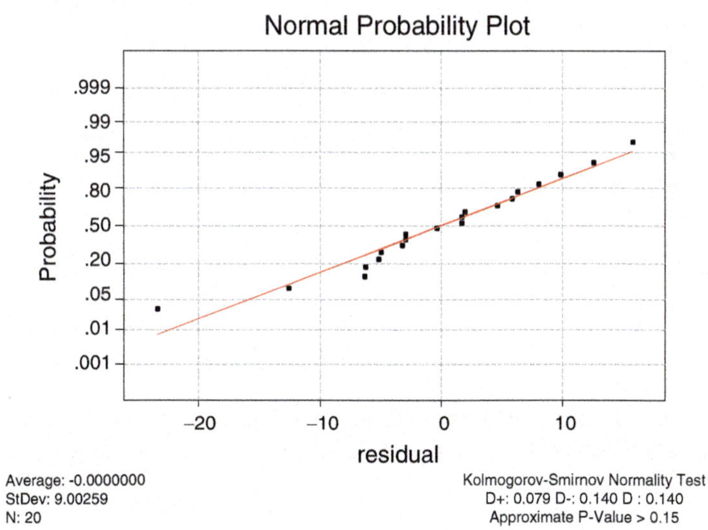

From the three different normality tests, we can see that we cannot reject the null hypothesis that the residuals follow a normal distribution.

It is interesting to note that the normality test done using a Minitab macro was written by Minitab. The Minitab macro that was used to calculate the normality test was called *normplot.mac*. *Normplot.mac* is stored under the folder called *macros*. The macro is available for view for people who want to study how Minitab macros are written. To view this Minitab macro and all other Minitab, use Microsoft Word or NOTEPAD.

16.10 SAS Programming Code Instructions and Examples

16.10.1 Introduction

Below shows the regression analysis performed in SAS.

```
Data Benston;
   INPUT labor thousand percent average @@;
   DATALINES;
100 5.1 90 20
85   3.8 99  22
108  5.3 58  19
116  7.5 16  15
92   4.5 54  20
63   3.3 42  26
79   5.3 12  25
101  5.9 32  21
88   4.0 56  24
71   4.2 64  29
122  6.8 78  10
85   3.9 90  30
50   3.8 74  28
114  7.5 89  14
104  4.5 90  21
111  6.0 40  20
110  8.1 55  16
100  2.9 64  19
82   4.0 35  23
85   4.8 58  25
;
run;

PROC REG DATA = Benston;
 MODEL labor = thousand percent average /P CLM;
    output out=Benston_Out r=res p=predict;
RUN;
```

Analysis of variance					
Source	DF	Sum of squares	Mean square	F Value	Pr > F
Model	3	5158.31383	1719.43794	17.87	<0.0001
Error	16	1539.88617	96.24289		
Corrected total	19	6698.20000			

Root MSE	9.81035	R-square	0.7701
Dependent mean	93.30000	Adj R-Sq	0.7270
Coeff var	10.51484		

Parameter estimates					
Variable	DF	Parameter estimate	Standard error	t Value	Pr > |t|
Intercept	1	131.92425	25.69321	5.13	<0.0001
Thousand	1	2.72609	2.27500	1.20	0.2483
Percent	1	0.04722	0.09335	0.51	0.6199
Average	1	−2.58744	0.64282	−4.03	0.0010

Output statistics						
Obs	Dependent variable	Predicted value	Std error mean predict	95 % CL mean		Residual
1	100.0000	98.3281	3.5310	90.8427	105.8135	1.6719
2	85.0000	90.0343	4.4049	80.6963	99.3722	−5.0343
3	108.0000	99.9498	2.5039	94.6418	105.2578	8.0502
4	116.0000	114.3138	5.4721	102.7134	125.9141	1.6862
5	92.0000	94.9926	3.1168	88.3853	101.5999	−2.9926
6	63.0000	75.6300	4.0803	66.9802	84.2797	−12.6300
7	79.0000	82.2531	5.1318	71.3741	93.1320	−3.2531
8	101.0000	95.1828	3.4623	87.8430	102.5227	5.8172
9	88.0000	83.3742	2.8031	77.4319	89.3165	4.6258
10	71.0000	71.3599	4.3895	62.0546	80.6652	−0.3599
11	122.0000	128.2703	5.7125	116.1602	140.3803	−6.2703
12	85.0000	69.1823	5.6336	57.2397	81.1250	15.8177
13	50.0000	73.3291	3.9219	65.0150	81.6433	−23.3291
14	114.0000	120.3481	5.4763	108.7388	131.9575	−6.3481
15	104.0000	94.1050	3.5326	86.6162	101.5938	9.8950
16	111.0000	98.4206	3.0399	91.9765	104.8648	12.5794
17	110.0000	115.2035	5.3525	103.8568	126.5502	−5.2035
18	100.0000	93.6905	6.3955	80.1326	107.2483	6.3095
19	82.0000	84.9700	3.9397	76.6183	93.3218	−2.9700
20	85.0000	83.0621	2.9226	76.8664	89.2577	1.9379

Sum of residuals	0
Sum of squared residuals	1539.88617
Predicted residual SS (PRESS)	2475.94271

16.10.2 Linearity

Below are three plots of y with each of the three x variables. For each plot we will also look at the strength of the relationship between y and x.

```
title  height=7 'Scatter Plot: labor & thousand';
proc gplot data=Benston;
plot labor*thousand;
run;
```

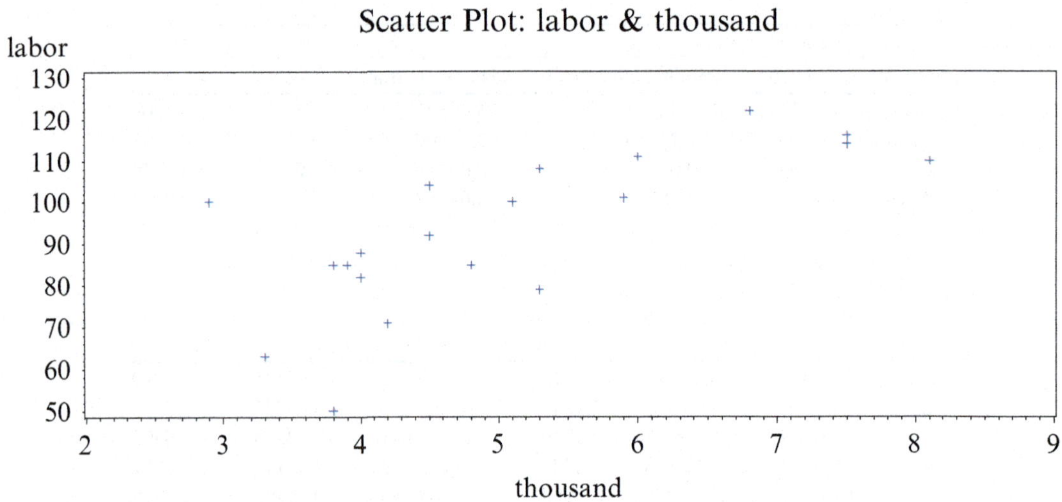


```
title  height=7 'Scatter Plot: labor & percent';
proc gplot data=Benston;
plot labor*percent;
run;
```

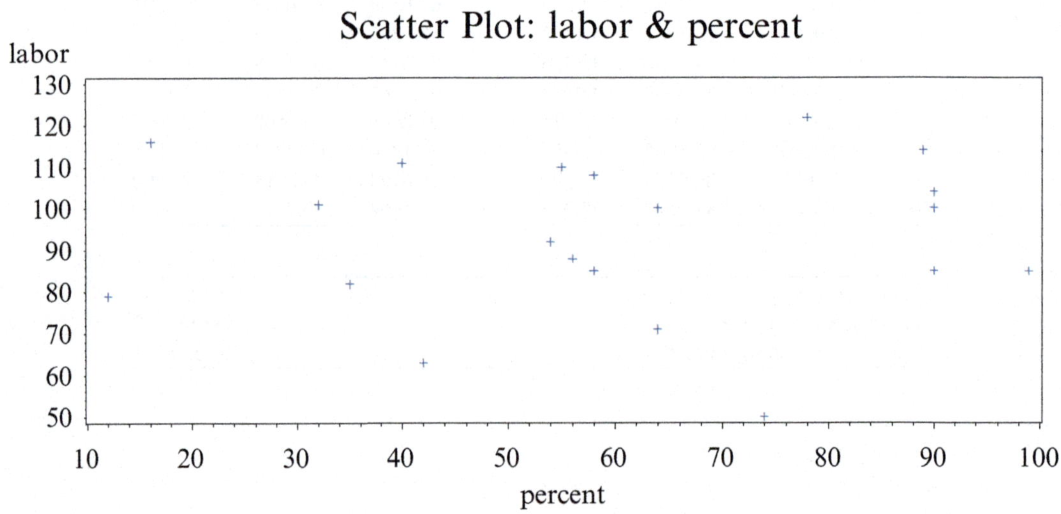

```
title  height=7 'Scatter Plot: labor & average';
proc gplot data=Benston;
plot labor*average;
run;
```

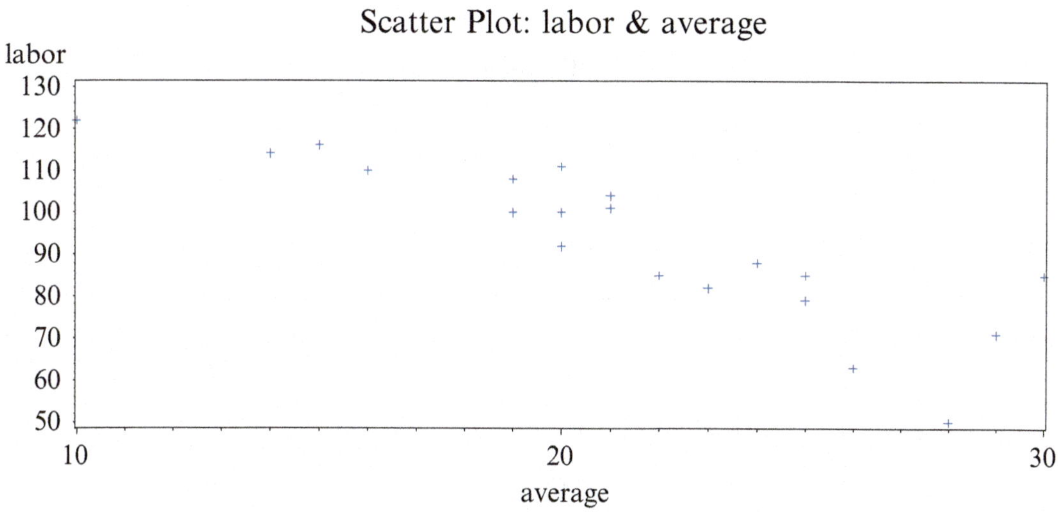


```
proc corr data=Benston cov;
run;
```

Pearson correlation coefficients, N = 20				
Prob > |r| under H0: Rho = 0				
	Labor	Thousand	Percent	Average
Labor	1.00000	0.71253	0.03072	−0.86550
		0.0004	0.8977	<0.0001
Thousand	0.71253	1.00000	−0.19383	−0.71724
	0.0004		0.4129	0.0004
Percent	0.03072	−0.19383	1.00000	−0.01280
	0.8977	0.4129		0.9573
Average	−0.86550	−0.71724	−0.01280	1.00000
	<0.0001	0.0004	0.9573	

16.10.3 The Expected Value of the Residual Term Is Zero

One way to check this assumption is to produce a dotplot and a boxplot of the residual terms and see if the midpoint is zero.

```
Goptions ftext='Times New Roman' htext=6;
title  height=7 'Dotplot residual';
symbol1 v=dot  c=black L=1;
proc gplot data=Benston_Out;
   plot group*res;
run;
```

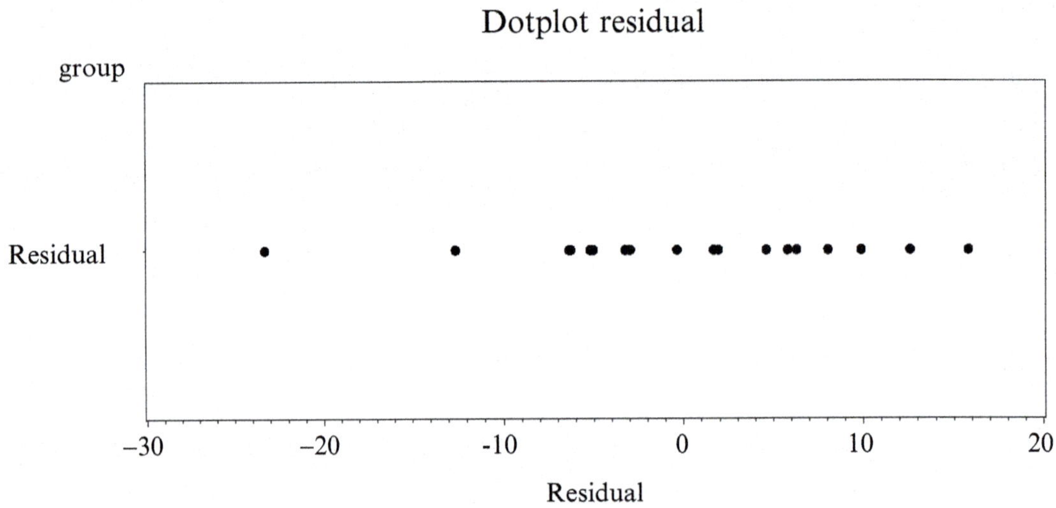


```
Goptions ftext='Times New Roman' htext=6;
title  height=7 'Boxplot residual';
proc boxplot data=Benston_Out;
   plot res*group/BOXWIDTH=6 boxstyle=schematic IDSYMBOL=square;
run;
```

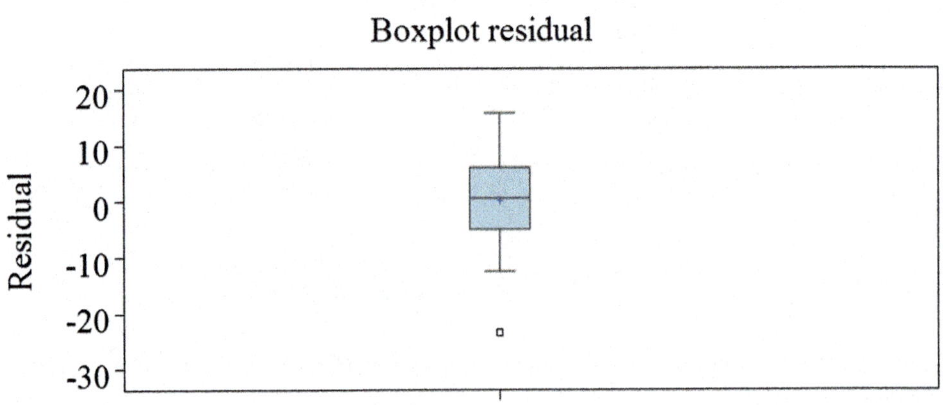

16.10.4 The Variance of the Error Term Is Constant

```
Goptions ftext='Times New Roman' htext=6 gunit=pct;
title  height=7 'Residual Versus the Fitted Values';
symbol1 v=dot  c=black L=1 h=2 i=rl;
axis1  label=('Residual');
axis2  label=('Fitted Value');
Proc gplot data=Benston_Out;
plot  res*predict /vaxis=axis1 haxis=axis2;
Run;
```

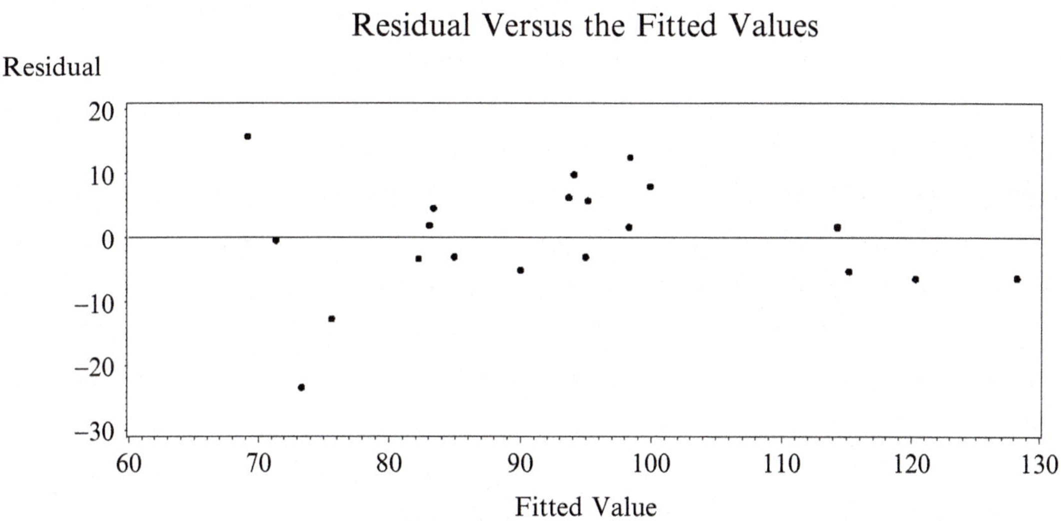


```
Goptions ftext='Times New Roman' htext=6 gunit=pct;
title  height=7 'Residual Versus Thousand';
symbol1 v=dot  c=black L=1 h=2 i=rl;
axis1  label=('Residual');
axis2  label=('Thousand');
Proc gplot data=Benston_Out;
plot  res*thousand /vaxis=axis1 haxis=axis2;
Run;
```

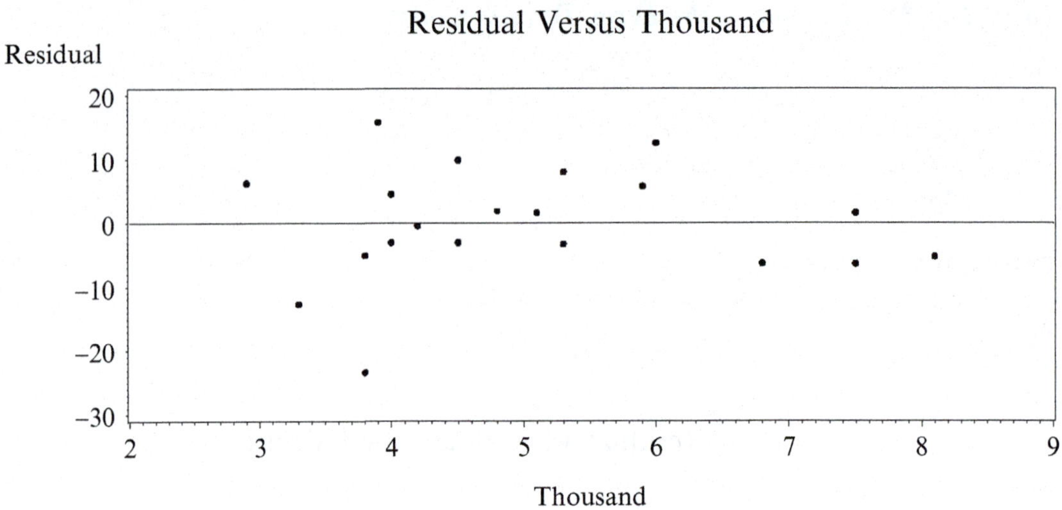


```
Goptions ftext='Times New Roman' htext=6 gunit=pct;
title  height=7 'Residual Versus Percent';
symbol1 v=dot  c=black L=1 h=2 i=rl;
axis1  label=('Residual');
axis2  label=('Percent');
Proc gplot data=Benston_Out;
plot  res*percent /vaxis=axis1 haxis=axis2;
Run;
```

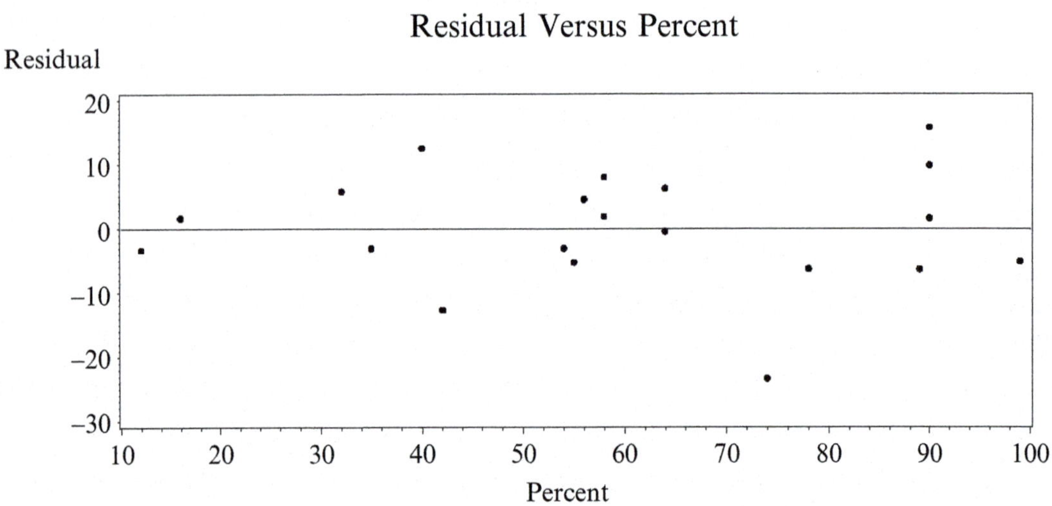

```
Goptions ftext='Times New Roman' htext=6 gunit=pct;
title  height=7 'Residual Versus Average';
symbol1 v=dot  c=black L=1 h=2 i=rl;
axis1  label=('Residual');
axis2  label=('Average');
Proc gplot data=Benston_Out;
plot  res*average /vaxis=axis1 haxis=axis2;
Run;
```

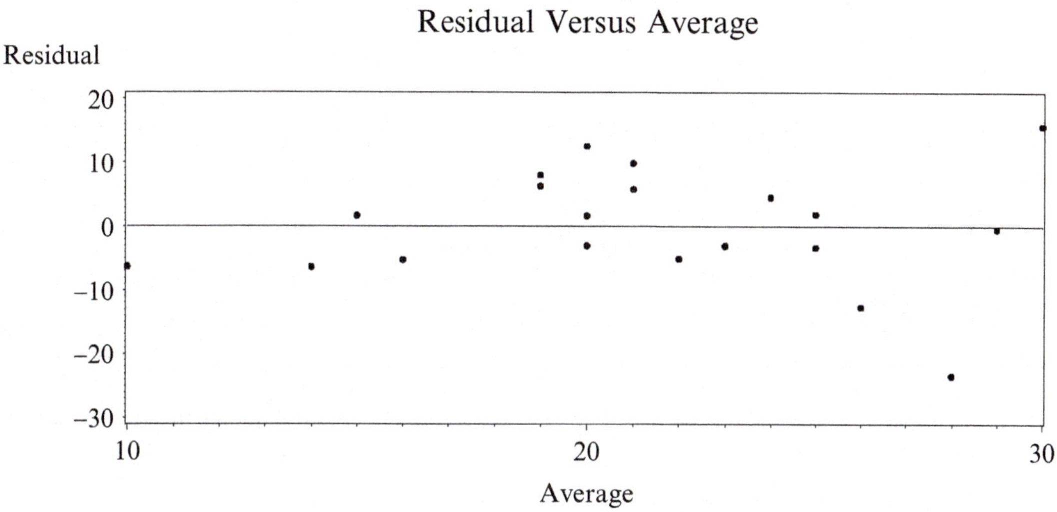

16.10.5 The Residual Terms Are Independent Linearity

The Durbin–Watson test statistic can be computed in **PROC REG** by using option **DW** after the model statement.

```
PROC REG DATA = Benston;
    MODEL labor = thousand percent average /DW;
RUN;
```

Durbin–Watson D	2.430
Number of observations	20
First-order autocorrelation	−0.217

16.10.6 SAS Macro: Regression Assumption Tests

The best way to check the regression assumption is to create a macro, so that you can get everything in the same time.

```
%macro reg (name, dv, x1, xlast);
proc corr data=&name cov;
     var &dv &x1 -- &xlast;
     run;
proc reg data=&name;
    model &dv = &x1 -- &xlast /vif dw;
run;
%mend;
```

The following error message might be generated when attempting to define a list using dissimilar variable names in a VAR statement.

ERROR: <variable> does not have a numeric suffix.

For example, a list is defined in the VAR statement below.

```
proc transpose data=sashelp.class;
   var age-height;
run;
```

To circumvent the problem, use two hyphens instead of one when defining the list, as shown below:

```
proc transpose data=sashelp.class;
   var age--height;
run;
```

That's the reason why we code two hyphens in the macro above. The macro is used in the following form:

```
%reg(Benston, labor, thousand, average);
```

Pearson correlation coefficients, N = 20				
Prob > |r| under H0: Rho = 0				
	Labor	Thousand	Percent	Average
Labor	1.00000	0.71253	0.03072	−0.86550
		0.0004	0.8977	<0.0001
Thousand	0.71253	1.00000	−0.19383	−0.71724
	0.0004		0.4129	0.0004
Percent	0.03072	−0.19383	1.00000	−0.01280
	0.8977	0.4129		0.9573
Average	−0.86550	−0.71724	−0.01280	1.00000
	<0.0001	0.0004	0.9573	

Analysis of variance					
Source	DF	Sum of squares	Mean square	F Value	Pr > F
Model	3	5158.31383	1719.43794	17.87	<0.0001
Error	16	1539.88617	96.24289		
Corrected total	19	6698.20000			

Root MSE	9.81035	R-square	0.7701
Dependent mean	93.30000	Adj R-Sq	0.7270
Coeff var	10.51484		

Parameter estimates						
Variable	DF	Parameter estimate	Standard error	t Value	Pr > |t|	Variance inflation
Intercept	1	131.92425	25.69321	5.13	<0.0001	0
Thousand	1	2.72609	2.27500	1.20	0.2483	2.25046
Percent	1	0.04722	0.09335	0.51	0.6199	1.09294
Average	1	−2.58744	0.64282	−4.03	0.0010	2.16627

Durbin–Watson D	2.430
Number of observations	20
First-order autocorrelation	−0.217

16.10.7 The Independent Variables Are Uncorrelated: Multicollinearity

The eigenvalues can also be used to measure the presence of multicollinearity. We will only be using the eigenvalue for diagnosing collinearity in multiple regression. If one or more of the eigenvalues are small (close to zero) and the corresponding condition number is large, then it indicates multicollinearity.

Let λ1. . . . . . . .λp be the eigenvalues of correlation matrix. The condition number of correlation matrix is defined as

$$\mathrm{Kj} = \sqrt{\lambda_{\max}/\lambda \mathrm{j}},\ \mathrm{j} = 1, 2, \ldots\ldots, \mathrm{p}.$$

```
PROC REG DATA = Benston;
    MODEL labor = thousand percent average /collin;
RUN;
```

Collinearity diagnostics						
			Proportion of variation			
Number	Eigenvalue	Condition index	Intercept	Thousand	Percent	Average
1	3.76061	1.00000	0.00051083	0.00219	0.00847	0.00161
2	0.13562	5.26582	0.00132	0.10796	0.51565	0.00179
3	0.09872	6.17201	0.00161	0.06635	0.29841	0.13737
4	0.00505	27.29270	0.99656	0.82351	0.17747	0.85922

16.10.8 Variance Infactionary Factor (VIF)

The VIF option in the MODEL statement provides the variance inflation factors (VIF). These factors measure the inflation in the variances of the parameter estimates due to collinearities that exist among the regressor variables.

```
PROC REG DATA = Benston;
    MODEL labor = thousand percent average /vif;
RUN;
```

Parameter estimates						
Variable	DF	Parameter estimate	Standard error	t Value	Pr > |t|	Variance inflation
Intercept	1	131.92425	25.69321	5.13	<0.0001	0
Thousand	1	2.72609	2.27500	1.20	0.2483	2.25046
Percent	1	0.04722	0.09335	0.51	0.6199	1.09294
Average	1	−2.58744	0.64282	−4.03	0.0010	2.16627

16.10.9 Testing the Normality of the Residuals

Option **NORMAL** provides three EDF goodness-of-fit tests for the lognormal distribution: the Anderson–Darling, the Cramér–von Mises, and the Kolmogorov–Smirnov tests.

```
PROC UNIVARIATE DATA = Benston_out;
    VAR res;
   HISTOGRAM res / NORMAL;
RUN;
```

Goodness-of-fit tests for normal distribution				
Test	Statistic		p Value	
Kolmogorov–Smirnov	D	0.14035940	Pr > D	>0.150
Cramer–von Mises	W-Sq	0.03553969	Pr > W-Sq	>0.250
Anderson–Darling	A-Sq	0.26332128	Pr > A-Sq	>0.250

The PROBPLOT statement[1] creates a probability plot, which compares ordered variable values with the percentiles of a specified theoretical distribution. If the data distribution matches the theoretical distribution, the points on the plot form a linear pattern. Consequently, you can use a probability plot to determine how well a theoretical distribution models a set of measurements.

The **MU**= and **SIGMA**= normal-options display a distribution reference line that corresponds to the normal distribution with mean $\mu_0 = 0$ and standard deviation $\sigma_0 = 9.0026$, and the **COLOR**= normal-option specifies the color for the line.

```
title  height=7 'Normal Probability Plot';
proc capability data=Benston_out noprint;
   probplot res;
run;
```

[1] http://support.sas.com/documentation/cdl/en/procstat/63104/HTML/default/viewer.htm#procstat_univariate_sect018.htm

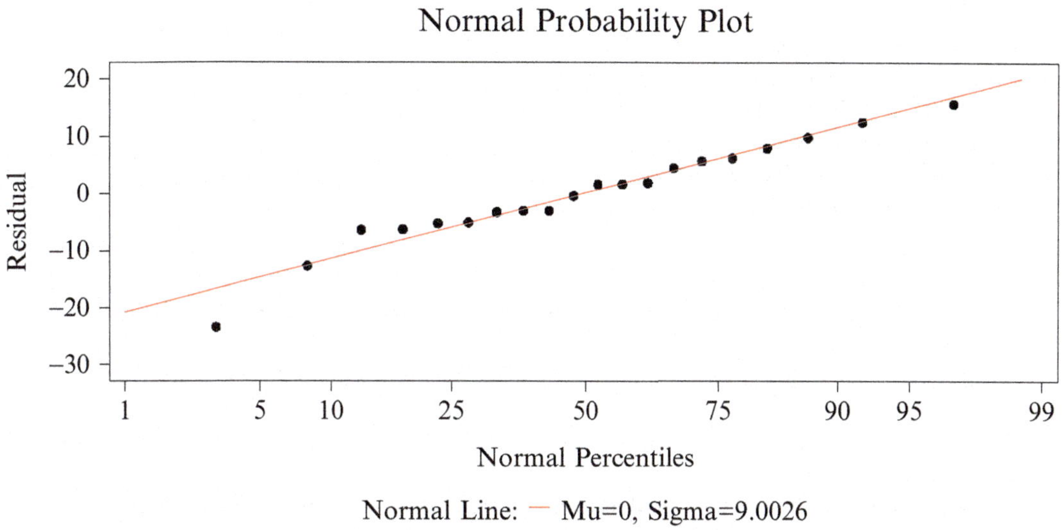

Below is the normal probability chart that is produced using **PROC UNIVARIATE** in SAS.

```
Goptions ftext='Times New Roman' htext=5;
title  height=7 'Normal Probability Chart for Regression Residuals';
proc univariate data=Benston_out noprint;
      qqplot res/normal(mu=est sigma=est color=red l=2) square;
run;
```

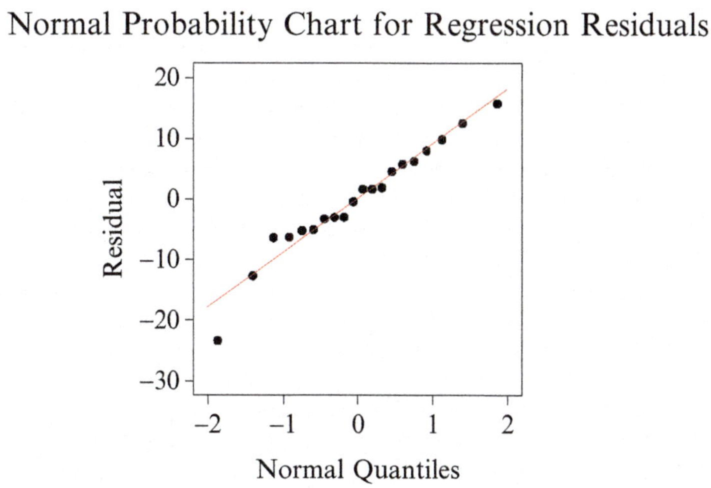

16.11 Statistical Summary

In this chapter we analyzed the six basic assumptions of linear regression. Analyzing them is important in every regression situation because when we make inferences about the regressed equation, we are assuming that the six basic assumptions are not materially violated. If they are violated and we are not aware of that, then our inferences could be seriously flawed.

The six assumptions were:

1. The dependent and independent variables have a linear relationship.
2. The expected value of the residual term is zero.
3. The variance of the residual term is constant.
4. The residual terms are independent.
5. The independent variables are uncorrelated.
6. The residuals are normally distributed.

We checked assumption 1 by plotting each individual independent variable with the dependent variable to see if the plot looked linear.

We checked assumption 2 by seeing if the dotplot and boxplot of the residuals have midpoints around zero.

We checked assumption 3 by plotting the variance of the residuals to see if the variance was constant or not.

We checked assumption 4 by calculating the Durbin–Watson statistic to see if there was any autocorrelation.

We checked assumption 5 by calculating the correlation of the independent variables and by calculating the *variance inflationary factor* (*VIF*) for each independent variable.

We checked assumption 6 by using Minitab to calculate the following three normality tests:

(1) Anderson–Darling test
(2) Ryan–Joiner test
(3) Kolmogorov–Smirnov test

Appendix 16.1: Excel Code—Regression Test

```
'/*************************************************************************
'/      Essentials of Microsoft Excel 2013 VBA, SAS
'/                    and MINITAB 17
'/        for Statistical and Financial Analysis
'/
'/*************************************************************************
Option Explicit
Dim mrData As Range
Dim mrReport_Ptr As Range
Dim mwsReport As Worksheet
Dim mrResidual As Range
Dim mwbTest As Workbook
Dim mwsData As Worksheet
Dim mwsNormPlot As Worksheet
Dim mChrt As Chart
Dim mSeries As Series

Private Sub CommandButton1_Click()
Dim ws As Worksheet
    Set mwbTest = Workbooks.Add
    Set mwsData = mwbTest.Worksheets(1)
    ThisWorkbook.Worksheets("Data").Cells(1).CurrentRegion.Copy
Destination:=mwsData.Cells(1)

    mwsData.Name = "Data"

    Set mrData = mwsData.Cells(1).CurrentRegion

    Set mwsReport = mwbTest.Worksheets.Add
```

```
    mwsReport.Name = "RegressTestReport"
    Set mrReport_Ptr = mwsReport.Range("c7")

    With mrReport_Ptr
        With .Offset(-3, 1)
            .Value = "Regresssion Assumptions Tests"
            With .Font
                    .Size = 16
                    .Italic = True
                    .Bold = True
            End With
        End With
    End With

    Heteroscedasticity

    Set mrReport_Ptr = mrReport_Ptr.Offset(5, 0)

    Multicolinearity

    Set mrReport_Ptr = mrReport_Ptr.Offset(1, 0)

    DurbinWatson

    With mrReport_Ptr
        .Offset(0, -1).EntireColumn.ColumnWidth = 0.1
        .EntireColumn.AutoFit
    End With

    NormProbPlot
    mrReport_Ptr.Parent.Move before:=Sheets(1)

    mwsReport.Activate
    ActiveWindow.DisplayGridlines = False

End Sub


'/**********************************************************************
'/Purpose:  To test for Heteroscedasticity
'/          To run this macro the first row should be the header
'/          The first column should be the independent variable data
'/          All other adjacent columns should be the independent variable
'/**********************************************************************
Sub Heteroscedasticity()
    Dim rRegress As Range
    Dim wsRegress As Worksheet
    Dim wsResidRegress As Worksheet
    Dim rResidRegress As Range
    Dim lDataSize As Long
    Dim wsReport As Worksheet
    Dim sngRSquare As Single
    Dim rRegressHetro As Range
    Dim rRSquare As Range

    'Initialize Variables
    Application.DisplayAlerts = False
    On Error Resume Next
    Set wsRegress = Worksheets("Regression")
    wsRegress.Delete
    Set wsResidRegress = Worksheets("ResidualRegression")
    wsResidRegress.Delete

    Err = 0
    On Error GoTo 0
    Application.DisplayAlerts = True
```

```
    lDataSize = mrData.Rows.Count - 1

    'Run regression on active sheet.
    'Assumes dependent variable in columns 1
    'Assumes other columns are independent variables
    'assumes that first row is a header
    Application.Run "ATPVBAEN.XLAM!Regress", mrData.Columns(1), _
        mrData.Resize(mrData.Rows.Count, mrData.Columns.Count - 1).Offset(0, 1), False,
True, , "regression", True, False _
        , False, False, , False

    'EXCEL will automatically selection regression report
    Set rRegress = Selection
    With rRegress
            .CurrentRegion.EntireColumn.AutoFit
            .Parent.Move after:=mrData.Parent
    End With

    'Find the calcualted residual column
    Set mrResidual = rRegress.Find("Residuals")
    Set mrResidual = mrResidual.CurrentRegion.Columns(3)


    'Calculate Residual square next to residual
    With mrResidual
        .Offset(0, 1).Formula = "=RC[-1]*RC[-1]"
        .Cells(1).Offset(0, 1).Value = "eSquare"
    End With

    'Run a regression on the residual square
    Application.Run "ATPVBAEN.XLAM!Regress", mrResidual.Offset(0, 1), _
        mrResidual.Offset(0, -1), False, True, , "ResidualRegression", True, False _
        , False, False, , False

    'Residual regression
    Set rRegressHetro = Selection

    With rRegressHetro
            .EntireColumn.AutoFit
            .Parent.Move after:=mrData.Parent

    End With
    Set rRSquare = rRegressHetro.Find("R Square")

    'Labels for report

    With mrReport_Ptr
        .Value = "n"
        .Offset(1, 0).Value = "R2"
        .Offset(2, 0).Value = "nR2"
        .Offset(3, 0).Value = "p-value"
        .Resize(4, 1).Font.Italic = True
        With .Offset(-1, -1)
            .Value = "Heteroscedasticity Test"
            With .Font
                    .Size = 12
                    .Italic = True
                    .Bold = True
            End With
        End With
    End With
    'values
    With mrReport_Ptr.Offset(0, 1)
            .Value = lDataSize
            .Offset(1, 0) = rRSquare.Offset(0, 1)
            .Offset(2, 0) = lDataSize * rRSquare.Offset(0, 1)
            .Offset(3, 0) = WorksheetFunction.ChiDist(.Offset(2, 0), 1)
    End With
End Sub
```

```
Sub NormProbPlot()
    Dim lResidualSize As Long

    Set mwsNormPlot = mwbTest.Worksheets.Add
    mwsNormPlot.Name = "NormPlotData"
    lResidualSize = mrResidual.Rows.Count - 1

    mrResidual.Resize(lResidualSize, 1).Offset(1, 0).Copy
Destination:=mwsNormPlot.Cells(4)
    mwsNormPlot.Cells(4).Sort key1:=mwsNormPlot.Cells(4), order1:=xlAscending
    With mwsNormPlot
        .Cells(1, 1).Value = 1
        .Cells(2, 1).Value = 2
        .Cells(1, 1).Resize(2, 1).AutoFill Destination:=.Cells(1).Resize(lResidualSize,
1)
        .Cells(1, 2).Resize(lResidualSize, 1).Formula = "=RC[-1]/(COUNTA(C[-1])+1)"
        .Cells(1, 3).Resize(lResidualSize, 1).Formula = "=NORMSINV(RC[-1])"
    End With

    Set mChrt = Nothing
    CreatChart chrtBox:=mChrt, seriesCurrent:=mSeries, wbWorkbook:=mwbTest

    With mSeries
        .Values = mwsNormPlot.Cells(1).CurrentRegion.Columns(4)
        .XValues = mwsNormPlot.Cells(1).CurrentRegion.Columns(3)
    End With

    With mChrt
        .Name = "Normal Probability Chart"
        .ChartTitle.Text = "Normal Probability Chart for Regression Residuals"
        With .Axes(xlValue)
            .CrossesAt = .MinimumScale
        End With
        With .Axes(xlCategory)
            .CrossesAt = .MinimumScale
        End With
        .Axes(xlCategory, xlPrimary).HasTitle = True
        .Axes(xlCategory, xlPrimary).AxisTitle.Characters.Text = _
        "Standard Normal Distribution"
        .Axes(xlValue, xlPrimary).HasTitle = True
        .Axes(xlValue, xlPrimary).AxisTitle.Characters.Text = "Residuals"

    End With


End Sub

Sub DurbinWatson()
    Dim sDWNumerator As String
    Dim sDWDenominator As String

    sDWNumerator = "SumXMY2(" & mrResidual.Resize(mrResidual.Rows.Count - 1, 1) _
                   .Offset(1, 0).Address(external:=True) & "," _
                   & mrResidual.Resize(mrResidual.Rows.Count - 1,
1).Address(external:=True) & ")"

    sDWDenominator = "SumSq(" & mrResidual.Address(external:=True) & ")"

    With mrReport_Ptr
        With .Offset(0, -1)
            .Value = "AutoCorrelation Test"
            With .Font
                    .Size = 12
                    .Italic = True
                    .Bold = True
            End With
        End With
    End With
```

```
    Set mrReport_Ptr = mrReport_Ptr.Offset(1, 0)

    With mrReport_Ptr
        .Value = "Durbin-Watson Statistic"
        .Offset(0, 1).Formula = "=" & sDWNumerator & "/" & sDWDenominator
    End With

End Sub


'/*******************************************************************
'/Purpose: Procdues the correlation matrix and VIF statistics for
'/         each independant variables.
'/*******************************************************************
Sub Multicolinearity()
    Dim rY As Range
    Dim rX As Range
    Dim rCell As Range

    With mrReport_Ptr
        With .Offset(0, -1)
            .Value = "Multicoliniearity Test"
            With .Font
                    .Size = 12
                    .Italic = True
                    .Bold = True
            End With
        End With
    End With

    Set mrReport_Ptr = mrReport_Ptr.Offset(1, 0)
    With mrReport_Ptr
        With .Offset(0, -1)
            .Value = "Correlation Maxtrix"
            With .Font
                    .Size = 11
                    .Italic = True

            End With
        End With
    End With
    Set mrReport_Ptr = mrReport_Ptr.Offset(1, 0)

    'produces the correlation matric
    'using the Excel's statistical add-in
    Application.Run "ATPVBAEN.XLAM!Mcorrel", mrData.Columns(2).Resize(,
mrData.Columns.Count - 1), _
        mrReport_Ptr, "C", True

    Set mrReport_Ptr = mrReport_Ptr.Offset(Selection.Rows.Count + 2, 0)

    With mrReport_Ptr
        With .Offset(0, -1)
            .Value = "Variance Influence Factor - VIF"
            With .Font
                    .Size = 11
                    .Italic = True
                    .Bold = True
            End With
        End With
    End With

    Set mrReport_Ptr = mrReport_Ptr.Offset(1, 0)

    'check to see if there is more than one independant variable
    If mrData.Columns.Count > 2 Then

        Set rX = mrData.Columns(2).Resize(, mrData.Columns.Count - 1)

        Dim rCell2 As Range
        Dim rXregress As Range
        Dim rXReport As Range
        Dim wsTemp As Worksheet
```

```
For Each rCell In rX.Rows(1).Cells

    For Each rCell2 In rX.Rows(1).Cells

      If Not (rCell2 = rCell) Then

        If Not (rXregress Is Nothing) Then
            Set rXregress = Union(rXregress, rCell2.Resize(mrData.Rows.Count, 1))

        Else

            Set rXregress = rCell2.Resize(mrData.Rows.Count, 1)
        End If
      End If
        Next
        Set wsTemp = mwbTest.Worksheets.Add
        rXregress.Copy
        wsTemp.Cells(1, 1).PasteSpecial xlAll
        Set rXregress = Selection
    Application.Run "ATPVBAEN.XLAM!Regress", rCell.Resize(mrData.Rows.Count, 1),
    rXregress, False, True, , "MultiCol-" & rCell, True, False _
      , False, False, , False
    Application.DisplayAlerts = False
            wsTemp.Delete
    Application.DisplayAlerts = True
    Set rXReport = Selection
    rXReport.EntireColumn.AutoFit
    Set rXregress = Nothing

    With mrReport_Ptr
        .Value = rCell
        .Offset(0, 1) = 1 / (1 - rXReport.Cells(5, 2))
    End With
    Set mrReport_Ptr = mrReport_Ptr.Offset(1, 0)

    Next
    End If

End Sub



Sub CreatChart(chrtBox As Chart, seriesCurrent As Series, wbWorkbook As Workbook)
    Set chrtBox = wbWorkbook.Charts.Add

    If chrtBox.SeriesCollection.Count = 0 Then
        Set seriesCurrent = chrtBox.SeriesCollection.NewSeries
    Else
        Set seriesCurrent = chrtBox.SeriesCollection(1)
    End If
    With chrtBox
            .ChartType = xlXYScatter
            .HasTitle = True
            With .Axes(xlCategory)
                   .HasMajorGridlines = False
                   .HasMinorGridlines = False
            End With
            With .Axes(xlValue)
                   .HasMajorGridlines = False
                   .HasMinorGridlines = False

            End With
            With .PlotArea
                 .Interior.ColorIndex = xlNone
                 .ClearFormats
            End With
            .HasLegend = False
    End With
End Sub
```

Bibliography

Benston GJ (1966) Multiple regression analysis of cost behavior. Account Rev 41(4):657–672

Kvanli A, Guynes C, Pavur R (1992) Introduction to business statistics: a computer integrated approach. West, New York

Lee CF, Lee JC, Lee AC (2013) Statistics for business and financial economics. Springer, New York

Microsoft Inc., Excel 2013. Microsoft Inc., Redmond

Minitab Inc. Minitab 17. Minitab Inc., State College

SAS Institute Inc. SAS 2014. SAS Institute Inc., Cary

Stevens J (1986) Applied multivariate statistics for the social sciences. Lawrence Erlbaum, Hillsdale

Chapter 17
Nonparametric Statistics

17.1 Introduction

There are three types of statistical data. The three types of statistical data are numerical, categorical, and ordinal. Numerical data is used for measurement, for example, the height of a person and miles to New York City. You can make mathematical operations on numerical data and the resulting number has meaning. Categorical data represents characteristics, for example, male or female and true or false. Ordinal data has ranking. For example, an experience could be poor, fair, and excellent. In this chapter we will do statistical tests on ordinal data.

In most of the previous chapters, we have been interested in estimating distribution parameters like the mean and variance and have assumed that we knew which distribution we were dealing with. In nonparametric statistics, these objectives and assumptions have been changed or cannot be achieved. Now we will be doing the test without making restrictive distribution assumptions about the parameters of the populations.

Below is a model showing how we will proceed in studying nonparametric statistics.

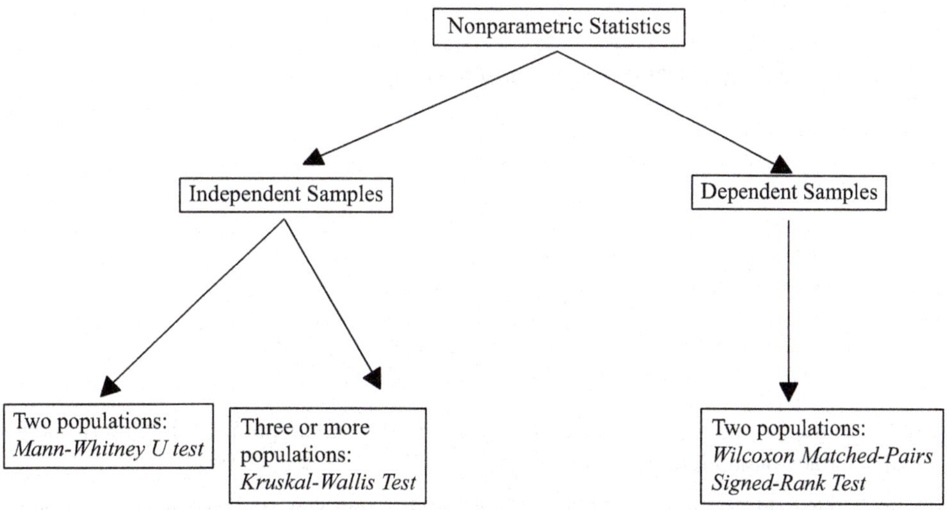

In general, relative rankings are used to test whether the probability distributions are equal or not. We will first look at a test called the Wilcoxon rank-sum test, also known as the Mann–Whitney U test.

C.-F. Lee et al., *Essentials of Excel, Excel VBA, SAS and Minitab for Statistical and Financial Analyses*, DOI 10.1007/978-3-319-38867-0_17

17.2 Mann–Whitney *U* Test

A Mann–Whitney *U* test is a nonparametric test of independent samples from two populations when we cannot assume that the sample is from a population that has a normal distribution.

Example 17.1 A company wishes to compare typing accuracy on two kinds of computer keyboards. Fifteen experienced typists type the same 600 words. Keyboard A is used seven times and keyboard B eight times, with the following results:

	Number of Errors							
Board A	13	9	16	15	10	11	12	
Board B	15	9	18	12	14	17	20	19

Use the Mann–Whitney *U* test to see if the two kinds of computer keyboards have the same type of accuracy at a 0.05 alpha level.

In previous chapters we used the *t* test for problems like this, but we had to assume that we were sampling from a normal distribution. If we assume that this keyboard problem does not follow a normal distribution, we cannot use a *t* test. We will use the Mann–Whitney *U* nonparametric test instead because it does not assume that the sample will come from a normal distribution.

Our null and alternative hypotheses are:

H_0: Accuracy on the two types of keyboards is equal.
H_1: Accuracy on the two types of keyboards is not equal.

We will use the *mann–whitney* Minitab command to solve this problem. We will put the board A data in column c1 and board B data in column c2 of the Minitab worksheet.

```
MTB > name c1 'Board A'
MTB > name c2 'Board B'
MTB > set 'Board A'
DATA> 13 9 16 15 10 11 12
DATA> end
MTB > set 'Board B'
DATA> 15 9 18 12 14 17 20 19
DATA> end
MTB > mann-whitney 'Board A' 'Board B'
```

Mann–Whitney *U* Test and CI: Board A and Board B

```
         N  Median
Board A  7  12.000
Board B  8  16.000


Point estimate for η1 - η2 is -3.500
95.7 Percent CI for η1 - η2 is (-6.998,1.000)
W = 41.5
Test of η1 = η2 vs η1 ≠ η2 is significant at 0.1052
The test is significant at 0.1043 (adjusted for ties)
```

From the Minitab output, we can see that we cannot reject the hypothesis that the two keyboards have equal accuracy at an alpha value of 0.05 because the *p*-value is 0.1052.

Example 17.2 The table below shows the research and development expenditures of 15 companies in each of two major industries, A and B. Use the Mann–Whitney *U* test to see if the two industries' expenditures are equal.

R&D Expenditures of Two Major Industries (in millions of dollars)	
Industry A	Industry B
40	8
41	6
43	15
46	16
47	17
53	18
55	19
56	20
61	21
63	22
64	25
68	26
79	27
80	29
85	30

Our null and alternative hypotheses would be:

H_0: The R&D expenditures of the two industries are equal.
H_1: The R&D expenditures of the two industries are not equal.

We put the data on industry A in column c1 and the data on industry B in column c2 of the Minitab worksheet. The *mann–whitney* command is illustrated below.

```
MTB > name c1 'IndustA'
MTB > name c2 'IndustB'
MTB > read 'IndustA' - 'IndustB'
DATA> 40 8
DATA> 41 6
DATA> 43 15
DATA> 46 16
DATA> 47 17
DATA> 53 18
DATA> 55 19
DATA> 56 20
DATA> 61 21
DATA> 63 22
DATA> 64 25
DATA> 68 26
DATA> 79 27
DATA> 80 29
DATA> 85 30
DATA> end
15 rows read.
MTB > mann-whitney 'IndustA' 'IndustB'
```

Mann–Whitney Test and CI: IndustA and IndustB

```
          N  Median
IndustA  15   56.00
IndustB  15   20.00


Point estimate for η1 - η2 is 37.00
95.4 Percent CI for η1 - η2 is (28.00,48.00)
W = 345.0
Test of η1 = η2 vs η1 ≠ η2 is significant at 0.0000
```

From this Minitab output, we can see that the calculated *p*-value is 0.0000. Since this is less than the alpha value of 0.05, we will reject the null hypothesis that the two industries spend the same amount on R&D.

Example 17.3 The producer commodity price indexes for January 1985 and January 1986 for six product categories are shown below. The data are from *Standard & Poor's Statistical Service, Currency Statistics,* Jan. 1987, pp 12–13. Use the Mann–Whitney U test to see if the probability distribution of these economic indexes was the same in January 1985 and January 1986.

Product Category	Jan-85	Jan-86
Processed poultry	198.80	192.40
Concrete ingredients	331.00	339.00
Lumber	343.00	329.60
Gas fuels	1,073.00	1,034.30
Drugs and pharmaceuticals	247.40	265.90
Synthetic fibers	157.60	151.10

Our null and alternative hypotheses are:

H_0: The economic indexes are the same in January 1985 and January 1986.
H_1: The economic indexes are not the same in January 1985 and January 1986.

We will put the January 1985 data in column c1 and the January 1986 data in column c2.

```
MTB > name c1 'Jan-85'
MTB > name c2 'Jan-86'
MTB > read 'Jan-85' 'Jan-86'
DATA> 198.8 192.40
DATA> 331.00 339.00
DATA> 343.00 329.60
DATA> 1073.00 1034.3
DATA> 247.40 265.90
DATA> 157.60 151.10
DATA> end
6 rows read.
MTB > mann-whitney c1 c2
```

Mann–Whitney Test and CI: Jan-85 and Jan-86

```
         N  Median
Jan-85   6   289.2
Jan-86   6   297.8


Point estimate for η1 - η2 is 5.2
95.5 Percent CI for η1 - η2 is (-181.4,192.0)
W = 41.0
Test of η1 = η2 vs η1 ≠ η2 is significant at 0.8102
```

Minitab calculates the p-value as 0.8102. Since this is greater than an alpha value of 0.05, we cannot reject the null hypothesis that the economic indexes are the same in January 1985 and January 1986.

17.3 Kruskal–Wallis Test

The Kruskal–Wallis test is a one-factor analysis of variance by ranks. It is a nonparametric test that represents a generalization of the two-sample Mann–Whitney U test to situations where more than two populations are involved. Unlike one-factor analyses of variance, the Kruskal–Wallis test makes no assumptions about the population distributions.

Example 17.4

Shaving Cream Sales (cases per thousand of population)		
A	B	C
38	26	40
42	30	36
27	18	32
60	42	37
36	24	42
54	30	46
40	26	38

The manufacturer of a new shaving cream tests three new advertising campaigns in a total of 21 markets. Sales in the third week after introduction are given in the above table.

Our task is to determine whether the median sales levels for the three campaigns are different at the 10 % alpha level. For this we will use the *kruskal–wallis* command.

To use the *kruskal–wallis* Minitab command, we have to input the data in a special way. The data will have to go into column c1 and the group for each data in column c2. To do this we will represent group A with a 1, group B with a 2, and group C with a 3.

Our hypotheses are:

H_0: The different advertising campaigns produce no difference in sales.
H_1: The different advertising campaigns do produce different sales.

```
MTB >  read c1 c2
DATA> 38 1
DATA> 42 1
DATA> 27 1
DATA> 60 1
DATA> 36 1
DATA> 54 1
DATA> 40 1
DATA> 26 2
DATA> 30 2
DATA> 18 2
DATA> 42 2
DATA> 24 2
DATA> 30 2
DATA> 26 2
DATA> 40 3
DATA> 36 3
DATA> 32 3
DATA> 37 3
DATA> 42 3
DATA> 46 3
DATA> 38 3
DATA> end
21 rows read.
MTB > kruskal-wallis c1 c2
```

Kruskal–Wallis Test: C1 Versus C2

```
Kruskal-Wallis Test on C1

C2         N  Median  Ave Rank      Z
1          7   40.00      14.2   1.68
2          7   26.00       5.7  -2.76
3          7   38.00      13.1   1.08
Overall   21              11.0

H = 7.74  DF = 2  P = 0.021
H = 7.78  DF = 2  P = 0.020  (adjusted for ties)
```

From the Minitab output, we can see that the calculated p-value is 0.021. Since this is less than the alpha value of 0.10, we can reject the null hypothesis that there is no difference among the different advertising campaigns and accept the alternative hypothesis that the different advertising campaigns do produce different sales levels.

Example 17.5 Assume that samples of executive vice presidents in a certain industry were drawn from firms classified into three size categories. After being assured of the confidentiality of their replies, the 20 executives were asked to rate the overall performance quality of their board of directors in setting general corporate policy during the past 3-year period on a scale from 0 to 100. The scores are shown below.

Firms Classified by Size		
Large	Medium	Small
79	69	83
96	78	66
86	85	51
88	62	94
76	63	71
91	73	61
81		74

Our hypotheses would be:

H_0: There is no difference in the rankings among the three groups of companies.
H_1: There is a difference in the rankings among the three groups of companies.

We have to enter the data in the same fashion as in the previous example. We will call the large firms group 1, the medium firms group 2, and the small firms group 3. Then, when we execute the *kruskal–wallis* command, we get the following output.

```
MTB > read c1 c2
DATA> 79 1
DATA> 96 1
DATA> 86 1
DATA> 88 1
DATA> 76 1
DATA> 91 1
DATA> 81 1
DATA> 69 2
DATA> 78 2
DATA> 85 2
DATA> 62 2
DATA> 63 2
DATA> 73 2
DATA> 83 3
DATA> 66 3
DATA> 51 3
DATA> 94 3
DATA> 71 3
DATA> 61 3
DATA> 74 3
DATA> end
20 rows read.
MTB > kruskal-wallis c1 c2
```

Kruskal–Wallis Test: C1 Versus C2

```
Kruskal-Wallis Test on C1

C2         N  Median  Ave Rank      Z
1          7   86.00      15.1   2.58
2          6   71.00       7.8  -1.32
3          7   71.00       8.1  -1.31
Overall   20              10.5

H = 6.64  DF = 2  P = 0.036
```

Minitab calculates the p-value as 0.036. This means that we will reject the null hypothesis that the rankings are the same among the different firm sizes at the 0.05 alpha level and accept the alternative that the rankings are not the same among the different size firms.

17.4 Wilcoxon Matched-Pairs Signed-Rank Test

The previous two tests have been nonparametric tests for independent samples. We will now look at a nonparametric test for two dependent samples, the Wilcoxon matched-pairs signed-rank test.

Example 17.6 Suppose the president of a company is interested in learning if there is any difference in customer satisfaction in two different stores, A and B. The ranking is from 1 to 10. A higher ranking is better. Below is a sample of ten customers from each store.

Customer Satisfaction

	A	B
1	6	7
2	5	8
3	7	8
4	8	9
5	9	4
6	5	5
7	4	7
8	9	6
9	8	6
10	7	6

Our hypotheses would be:

H_0: There is no difference in customer satisfaction between stores A and B.
H_1: There is a difference in customer satisfaction between stores A and B.

We will use the *wtest* command to carry out the Wilcoxon matched-pairs signed-rank test.

```
MTB > read c1 c2
DATA> 6 7
DATA> 5 8
DATA> 7 8
DATA> 8 9
DATA> 9 4
DATA> 5 5
DATA> 4 7
DATA> 9 6
DATA> 8 6
DATA> 7 6
DATA> end
10 rows read.
MTB > let c3 = c2-c1
MTB > wtest c3
```

Wilcoxon Signed-Rank Test: C3

```
Test of median = 0.000000 versus median ≠ 0.000000

          N for   Wilcoxon                Estimated
      N   Test   Statistic         P         Median
C3   10      9        21.5     0.953    0.000000000
```

Notice that we take the difference of the two samples before we use the *wtest* command. The output shows that the *p*-value 0.953 is greater than an alpha value of either 0.10 or 0.05. Therefore, we cannot conclude that there is any difference in satisfaction between stores A and B.

Example 17.7 We have data on net income from Lawrence, Inc., before and after a corporate reorganization. We are interested in testing to see if net income differs significantly before and after the reorganization. The data are shown below.

Market Region	Net Income Before Reorganization	Net Income After Reorganization
1	41	62
2	34	49
3	43	39
4	29	28
5	55	55
6	63	66
7	35	47
8	42	72
9	57	84
10	45	42

We will put the "before" data in column c1 and the "after" data in column c2. Our hypotheses would be:

H_0: There is no difference in income after reorganization.
H_1: There is a difference in income after reorganization.

The Minitab commands and output are shown below.

```
MTB > read c1 c2
DATA> 41 62
DATA> 34 49
DATA> 43 39
DATA> 29 28
DATA> 55 55
DATA> 63 66
DATA> 35 47
DATA> 42 72
DATA> 57 84
DATA> 45 42
DATA> end
10 rows read.
MTB > let c3 = c2-c1
MTB > wtest c3
```

Wilcoxon Signed-Rank Test: C3

```
Test of median = 0.000000 versus median ≠ 0.000000

          N for   Wilcoxon            Estimated
     N    Test   Statistic       P       Median
C3  10       9        37.5   0.086        10.00
```

From the Minitab output, the calculated *p*-value for the data is 0.086. This means that at a 0.05 alpha level, we cannot reject the null hypothesis that there is no difference between income before and after the reorganization.

17.5 Ranking

If we did the tests examined in this chapter manually, we would have to rank the data. For this or other reasons, we are often interested in ranking data. Minitab has a command that can rank data in a column. But suppose we want to rank data from more than one column. In this section we will see how to do this.

Product Category	Jan-92	Jan-93
Processed poultry	198.80	192.40
Concrete ingredients	331.00	339.00
Lumber	343.00	329.60
Gas fuels	1,073.00	1,034.30
Drugs and pharmaceuticals	247.40	265.90
Synthetic fibers	157.60	151.10

Suppose we are interested in ranking the above index figures from January 1992 and January 1993. Let us assume that we had previously stored the 1992 data in column c1 and the 1993 data in column c2 on a Minitab worksheet. The *rank* command will rank data in one column, but we want to make a single ranking of the data in both columns. We have the choice of either typing the above data again in a single column or telling Minitab to put it in one column. Letting Minitab put all the data in one column is called stacking the data.

The worksheet would look like the following before anything is done.

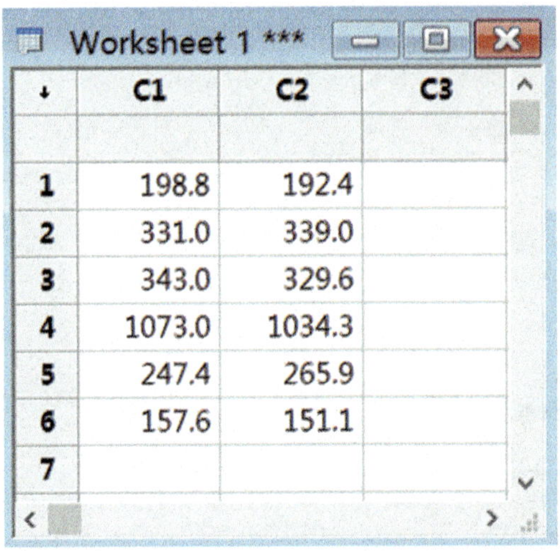
Worksheet 1 ***

↓	C1	C2	C3
1	198.8	192.4	
2	331.0	339.0	
3	343.0	329.6	
4	1073.0	1034.3	
5	247.4	265.9	
6	157.6	151.1	
7			

Our first step, before we can rank the above data, is to stack the data in column c1 onto c2 and put the results into c3. We will use the *stack* command to stack data.

```
MTB > stack c1 c2 c3
```

The first argument, "c1," is the location of the first data set. The second argument, "c2," is the location of the second data set. The third argument, "c3," is the location where we are going to stack the two data sets. The worksheet will look like the following after stacking the two data sets.

↓	C1	C2	C3	C4
1	198.8	192.4	198.8	
2	331.0	339.0	331.0	
3	343.0	329.6	343.0	
4	1073.0	1034.3	1073.0	
5	247.4	265.9	247.4	
6	157.6	151.1	157.6	
7			192.4	
8			339.0	
9			329.6	
10			1034.3	
11			265.9	
12			151.1	
13				

We can now tell Minitab to rank the stacked data in column c3 and put the rankings in column c4.

```
MTB > rank c3 c4
```

The first argument, "c3," is the location of the data to be ranked. The second argument, "c4," is the location of the ranking of the data. The worksheet should look like the following after the *rank* command is issued.

↓	C1	C2	C3	C4	C5
1	198.8	192.4	198.8	4	
2	331.0	339.0	331.0	8	
3	343.0	329.6	343.0	10	
4	1073.0	1034.3	1073.0	12	
5	247.4	265.9	247.4	5	
6	157.6	151.1	157.6	2	
7			192.4	3	
8			339.0	9	
9			329.6	7	
10			1034.3	11	
11			265.9	6	
12			151.1	1	
13					

17.6 SAS Programming Code Instructions and Examples

17.6.1 Mann–Whitney U Test

Example 17.2A We will use the **PROC NPARLWAY** command to solve this problem.

```
Data Typing;
input Error Board $ @@;
cards;
13 A 9 A 16 A 15 A 10 A 11 A 12 A
15 B 9 B 18 B 12 B 14 B 17 B 20 B 19 B
;
Run;
```

The following statements request a Wilcoxon test of the null hypothesis that there is no difference between the errors of the two keyboards. Board is the class variable, and Error is the analysis variable. The WILCOXON option[1] requests an analysis of Wilcoxon scores.

```
proc npar1way data = Typing wilcoxon;
  class Board;
  var Error;
run;
```

Wilcoxon scores (rank sums) for variable Error classified by variable Board					
Board	N	Sum of scores	Expected under H_0	Std. dev. under H_0	Mean score
A	7	41.50	56.0	8.617811	5.928571
B	8	78.50	64.0	8.617811	9.812500
Average scores were used for ties					

Wilcoxon two-sample test	
Statistic	41.5000
Normal approximation	
Z	−1.6245
One-sided Pr < Z	0.0521
Two-sided Pr > |Z|	0.1043
t Approximation	
One-sided Pr < Z	0.0633
Two-sided Pr > |Z|	0.1266
Z includes a continuity correction of 0.5	

To interpret the first table of the output, we give the rank table of boards A and B below:

Board A : 13 9 16 15 10 11 12
Board B : 15 9 18 12 14 17 20 19

[1] The detailed information can be found in http://support.sas.com/documentation/cdl/en/statug/63033/HTML/default/viewer.htm#statug_npar1way_sect022.htm.

$$\begin{array}{ll} \text{Board}: & \underline{13}\ \underline{9}\ \underline{16}\ \underline{15}\ \underline{10}\ \underline{11}\ \underline{12}\ 15\ 9\ 18\ 12\ 14\ 17\ 20\ 19 \\ \text{Rank}: & \underline{7}\ \underline{1.5}\ \underline{11}\ \underline{9.5}\ \underline{3}\ \underline{4}\ \underline{5.5}\ 9.5\ 1.5\ 13\ 5.5\ 8\ 12\ 15\ 14 \end{array}$$

The number with underscore belongs to board A. The sum of sores is the sum of ranks from boards A and B. Take A, for example. The sum of scores equals 7 + 1.5 + 11 + 9.5 + 3 + 4 + 5.5 = 41.5. Suppose group A has N_a records and group B has N_b records. The expected under H_0 for group A is $N_a{}^*(N_a + N_b + 1)/2$. For board A, the expected under H_0 is 7 * (7 + 8 + 1)/2 = 56. The std. dev. under H_0 is $\sqrt{N_a{}^*N_b{}^*(N_a + N_{b+1})/24}$. The Z value can be calculated as (sum of sores − expected under H_0)/std. dev. under H_0.

The results suggest that there is no statistically significant difference between the accuracy of two keyboards (z = −1.6245, p = 0.1043).

Example 17.8

```
Data RandD;
input Expenditure Industry $ @@;
cards;
40 A 8 B
41 A 6 B
43 A 15 B
46 A 16 B
47 A 17 B
53 A 18 B
55 A 19 B
56 A 20 B
61 A 21 B
63 A 22 B
64 A 25 B
68 A 26 B
79 A 27 B
80 A 29 B
85 A 30 B
;
Run;

proc npar1way data = RandD wilcoxon;
  class Industry;
  var Expenditure;
run;
```

Wilcoxon scores (rank sums) for variable Expenditure classified by variable Industry					
Industry	N	Sum of scores	Expected under H_0	Std. dev. under H_0	Mean score
A	15	345.0	232.50	24.109127	23.0
B	15	120.0	232.50	24.109127	8.0

Wilcoxon two-sample test	
Statistic	345.0000
Normal approximation	
Z	4.6455
One-sided Pr > Z	<0.0001
Two-sided Pr > |Z|	<0.0001
t Approximation	
One-sided Pr > Z	<0.0001
Two-sided Pr > |Z|	<0.0001
Z includes a continuity correction of 0.5	

Example 17.9 We put the data under Jan-85 as A in column Category and Jan-86 as B. The **PROC NPARLWAY** command is illustrated below.

```
Data Product;
input Jan Category $ @@;
cards;
198.80 A 192.40 B
331.00 A 339.00 B
343.00 A 329.60 B
1073.00 A 1034.30 B
247.40 A 265.90 B
157.60 A 151.10 B
;
Run;

proc npar1way data = Product wilcoxon;
  class Category;
  var Jan;
run;
```

Wilcoxon scores (rank sums) for variable Jan classified by variable Category					
Category	N	Sum of scores	Expected under H_0	Std. dev. under H_0	Mean score
A	6	41.0	39.0	6.244998	6.833333
B	6	37.0	39.0	6.244998	6.166667

Wilcoxon two-sample test	
Statistic	41.0000
Normal approximation	
Z	0.2402
One-sided Pr > Z	0.4051
Two-sided Pr > |Z|	0.8102
t Approximation	
One-sided Pr > Z	0.4073
Two-sided Pr > |Z|	0.8146
Z includes a continuity correction of 0.5	

17.6.2 Kruskal–Wallis Test

Example 17.10 **PROC NPARLWAY** command is used to do the Kruskal–Wallis test in SAS.

```
Data Advertising;
input Sale Group $ @@;
cards;
38 A 26 B 40 C
42 A 30 B 36 C
27 A 18 B 32 C
60 A 42 B 37 C
36 A 24 B 42 C
54 A 30 B 46 C
40 A 26 B 38 C
;
Run;

proc npar1way data = Advertising wilcoxon;
  class Group;
  var Sale;
run;
```

Wilcoxon scores (rank sums) for variable Sale classified by variable Group					
Group	N	Sum of scores	Expected under H_0	Std. dev. under H_0	Mean score
A	7	99.50	77.0	13.364755	14.214286
B	7	40.00	77.0	13.364755	5.714286
C	7	91.50	77.0	13.364755	13.071429
Average scores were used for ties					

Kruskal–Wallis test	
Chi-square	7.7839
DF	2
Pr > chi-square	0.0204

Example 17.11 We have to enter the data in the same fashion as in the previous example. We will call the large firms group Large, the medium firms group Medium, and the small firms group Small. Then, when we execute the **PROC NPARLWAY** command, we get the following output.

```
Data Firm;
input Rate Size $ @@;
cards;
79 Large 69 Medium 83 Small
96 Large 78 Medium 66 Small
86 Large 85 Medium 51 Small
88 Large 62 Medium 94 Small
76 Large 63 Medium 71 Small
91 Large 73 Medium 61 Small
81 Large 74 Small
;
Run;

proc npar1way data = Firm wilcoxon;
  class Size;
  var Rate;
run;
```

Wilcoxon scores (rank sums) for variable Rate classified by variable Size					
Size	N	Sum of scores	Expected under H_0	Std. dev. under H_0	Mean score
Large	7	106.0	73.50	12.619429	15.142857
Medium	6	47.0	63.00	12.124356	7.833333
Small	7	57.0	73.50	12.619429	8.142857

Kruskal–Wallis test	
Chi-square	6.6415
DF	2
Pr > chi-square	0.0361

17.6.3 Wilcoxon Matched-Pairs Signed-Rank Test

Example 17.12 The Wilcoxon matched-pairs signed-rank test is calculated by using **PROC UNIVARIATE**.

```
Data Company;
input A B @@;
Diff=A-B;
cards;
6 7
5 8
7 8
8 9
9 4
5 5
4 7
9 6
8 6
7 6
;
Run;

PROC UNIVARIATE DATA = Company;
    VAR Diff;
RUN;
```

Tests for location: Mu0 = 0				
Test	Statistic		p-Value	
Student's t	t	0.24577	Pr > |t|	0.8114
Sign	M	−0.5	Pr ≥ |M|	1.0000
Signed rank	S	1	Pr ≥ |S|	0.9883

The Wilcoxon signed-rank statistic gives $p = 0.9883$, indicating that there is no significant difference between treatments.

Example 17.13 **PROC UNIVARIATE** is used to calculate the Wilcoxon matched-pairs signed-rank test.

```
Data Income;
input Before After @@;
Diff=Before-After;
cards;
41 62
34 49
43 39
29 28
55 55
63 66
35 47
42 72
57 84
45 42
;
Run;

PROC UNIVARIATE DATA = Income;
    VAR Diff;
RUN;
```

Tests for location: Mu0 = 0				
Test	Statistic		p-Value	
Student's t	t	−2.471	Pr > |t|	0.0355
Sign	M	−1.5	Pr ≥ |M|	0.5078
Signed rank	S	−15	Pr ≥ |S|	0.0820

The Wilcoxon signed-rank statistic gives p = 0.0820, indicating that there is no significant difference between treatments.

17.6.4 Ranking

The RANK procedure computes ranks for one or more numeric variables across the observations of a SAS data set and outputs the ranks to a new SAS data set. PROC RANK by itself produces no printed output.

```
Data Product;
input Jan @@;
cards;
198.80 192.40
331.00 339.00
343.00 329.60
1073.00 1034.30
247.40 265.90
157.60 151.10
;
Run;
```

The following output shows the results of ranking the values of one variable with a simple **PROC RANK** step.[2]

```
proc rank data=Product out=rankings;
   var Jan;
```

Obs	Jan	Jan_Rank
1	198.8	4
2	192.4	3
3	331.0	8
4	339.0	9
5	343.0	10
6	329.6	7
7	1073.0	12
8	1034.3	11
9	247.4	5
10	265.9	6
11	157.6	2
12	151.1	1

[2] The detailed information can be found in http://support.sas.com/documentation/cdl/en/proc/61895/HTML/default/viewer.htm#rank-overview.htm.

17.7 Statistical Summary

In this chapter, we discussed three statistical tests that do not require the assumption of normality in the distribution of the population. These tests are called nonparametric statistical tests. The three tests were the Mann–Whitney U test, the Kruskal–Wallis test, and the Wilcoxon matched-pairs signed-rank test.

Bibliography

Lee CF, Lee JC, Lee AC (2013) Statistics for business and financial economics. Springer, New York
Microsoft Inc., Excel 2013. Microsoft Inc., Redmond
Minitab Inc. Minitab 17. Minitab Inc., State College, Pennsylvania
SAS Institute Inc. SAS 2014. SAS Institute Inc., Cary

Chapter 18
Time Series: Analysis, Model, and Forecasting

18.1 Introduction

In statistics there are two kinds of data, cross-section data and time-series data. Time-series data are those recorded over time. Cross-section data pertain to a particular time. In this chapter we will look at specific issues of time-series data.

Time-series data can be broken down into four components:

1. Trend component, T
2. Cyclical component, C
3. Seasonal component, S
4. Irregular component, I

These four components can be specified as an additive model

$$x = T + C + S + I$$

or as a multiplicative model

$$x = T^{*}C^{*}S^{*}I$$

Time-series data can fluctuate wildly, mostly because of these four components. In the next section, we discuss how we can use the moving average method to reduce the effects of fluctuations associated with seasonality and irregularity.

18.2 Moving Averages in Minitab

We will use the quarterly earnings per share data of Johnson & Johnson (J&J) to show how moving averages can be obtained.

C.-F. Lee et al., *Essentials of Excel, Excel VBA, SAS and Minitab for Statistical and Financial Analyses*, DOI 10.1007/978-3-319-38867-0_18

Period, t	Earnings per Share	Period, t	Earnings per Share	Period, t	Earnings per Share	Period, t	Earnings per Share	Period, t	Earnings per Share
1	0.3	11	0.255	21	0.47	31	0.61	41	0.73
2	0.2717	12	0.21	22	0.425	32	0.475	42	1.06
3	0.2967	13	0.35	23	0.435	33	0.81		
4	0.215	14	0.37	24	0.35	34	0.785		
5	0.325	15	0.395	25	-0.375	35	0.71		
6	0.345	16	0.17	26	0.52	36	0.555		
7	0.345	17	0.39	27	0.505	37	0.95		
8	0.24	18	0.315	28	0.275	38	0.89		
9	0.415	19	0.375	29	0.68	39	0.8		
10	0.38	20	0.295	30	0.65	40	0.61		

The negative earnings in period 25 were the result of the Tylenol poisoning tragedy.

We will put the earnings per share data in column c1 of the Minitab worksheet and name column c1 *eps*.

```
MTB > name c1 'eps'
MTB > SET 'EPS'
DATA> .3000 .2717 .2967 .2150 .3250 .3450 .3450 .2400 .4150 .3800
DATA> .2550 .2100 .3500 .3700 .3950 .1700 .3900 .3150 .3750 .2950
DATA> .4700 .4250 .4350 .3500 -0.3750 .5200 .5050 .2750 .6800 .6500
DATA> .6100 .4750 .8100 .7850 .7100 .5550 .9500 .8900 .8000 .6100
DATA> .7300 1.0600
DATA> end
```

The formula to calculate a four-point moving average is

$$M_t = \frac{Y_{t-2} + Y_{t-1} + Y_t + Y_{t+1}}{4}$$

where Y_t is the earnings for period t.

To follow this formula, we have to start our calculation at period 3 and end it one period before the end of the data set. The calculation for period 3 is

$$M_3 = \frac{Y_1 + Y_2 + Y_3 + Y_4}{4} = \frac{0.3 + 0.2717 + 0.2967 + 0.215}{4} = 0.27085$$

We will run the *Ma.mac* macro that comes with Minitab. This macro is stored in the *macro* folder under the *mtbwin* folder. The first parameter of the *Ma* macro indicates the column that contains the data to be average. The second parameter of *Ma* macro indicates the moving average length.

```
MTB > MA 'eps' 4;
SUBC> brief 2.
```

Moving Average for eps

```
Data      eps
Length    42
NMissing  0

Moving Average

Length  4

Accuracy Measures

MAPE  29.1715
MAD    0.1336
MSD    0.0384
```

Time	eps	MA	Predict	Error
1	0.3000	*	*	*
2	0.2717	*	*	*
3	0.2967	*	*	*
4	0.2150	0.270850	*	*
5	0.3250	0.277100	0.270850	0.054150
6	0.3450	0.295425	0.277100	0.067900
7	0.3450	0.307500	0.295425	0.049575
8	0.2400	0.313750	0.307500	-0.067500
9	0.4150	0.336250	0.313750	0.101250
10	0.3800	0.345000	0.336250	0.043750
11	0.2550	0.322500	0.345000	-0.090000
12	0.2100	0.315000	0.322500	-0.112500
13	0.3500	0.298750	0.315000	0.035000
14	0.3700	0.296250	0.298750	0.071250
15	0.3950	0.331250	0.296250	0.098750
16	0.1700	0.321250	0.331250	-0.161250
17	0.3900	0.331250	0.321250	0.068750
18	0.3150	0.317500	0.331250	-0.016250
19	0.3750	0.312500	0.317500	0.057500
20	0.2950	0.343750	0.312500	-0.017500
21	0.4700	0.363750	0.343750	0.126250
22	0.4250	0.391250	0.363750	0.061250
23	0.4350	0.406250	0.391250	0.043750
24	0.3500	0.420000	0.406250	-0.056250
25	-0.3750	0.208750	0.420000	-0.795000
26	0.5200	0.232500	0.208750	0.311250
27	0.5050	0.250000	0.232500	0.272500
28	0.2750	0.231250	0.250000	0.025000
29	0.6800	0.495000	0.231250	0.448750
30	0.6500	0.527500	0.495000	0.155000
31	0.6100	0.553750	0.527500	0.082500
32	0.4750	0.603750	0.553750	-0.078750
33	0.8100	0.636250	0.603750	0.206250
34	0.7850	0.670000	0.636250	0.148750
35	0.7100	0.695000	0.670000	0.040000
36	0.5550	0.715000	0.695000	-0.140000
37	0.9500	0.750000	0.715000	0.235000
38	0.8900	0.776250	0.750000	0.140000
39	0.8000	0.798750	0.776250	0.023750
40	0.6100	0.812500	0.798750	-0.188750
41	0.7300	0.757500	0.812500	-0.082500
42	1.0600	0.800000	0.757500	0.302500

Moving Average Plot for eps

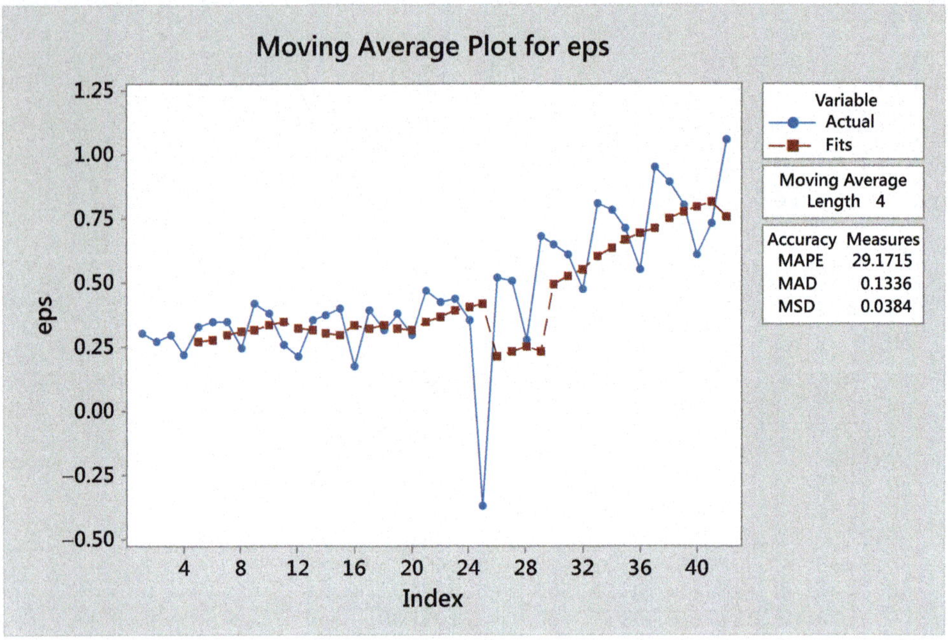

18.3 Moving Averages in Excel: Trend Line Method

In Excel there are two ways to produce a moving average chart. The first way is to produce a line chart of J&J earning data and then use Excel's Chart *Add Trendline* feature. The second way is to use the moving average tool in Excel's Data Analysis menu item under the *Data* menu. We will first show how to create a moving average chart using Excel's Chart *Add Trendline* feature.

The first thing to do is to add J&J's EPS data into column A as shown below.

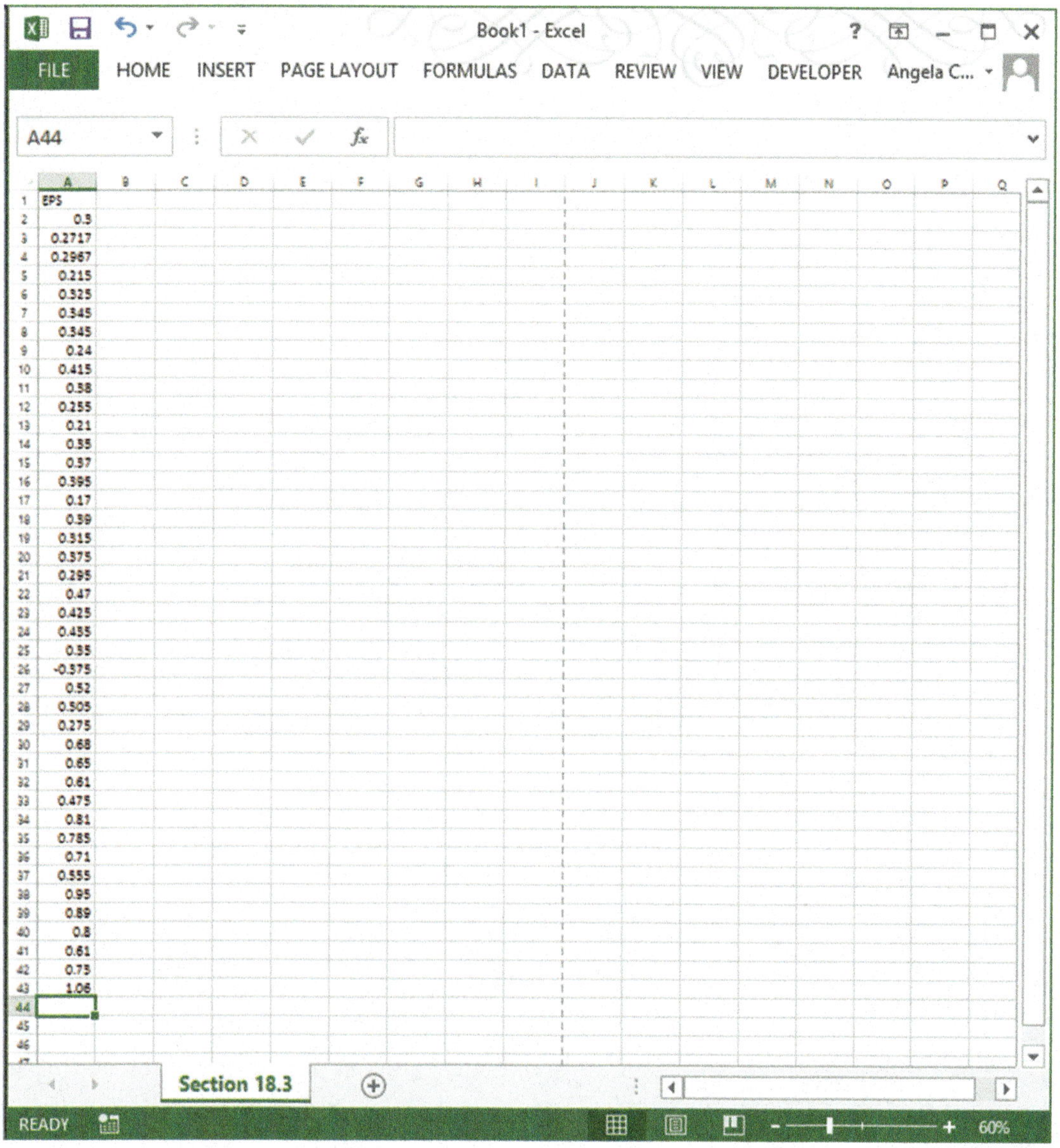

The next thing to do is create a line chart of J&J's EPS data. This line chart is shown below.

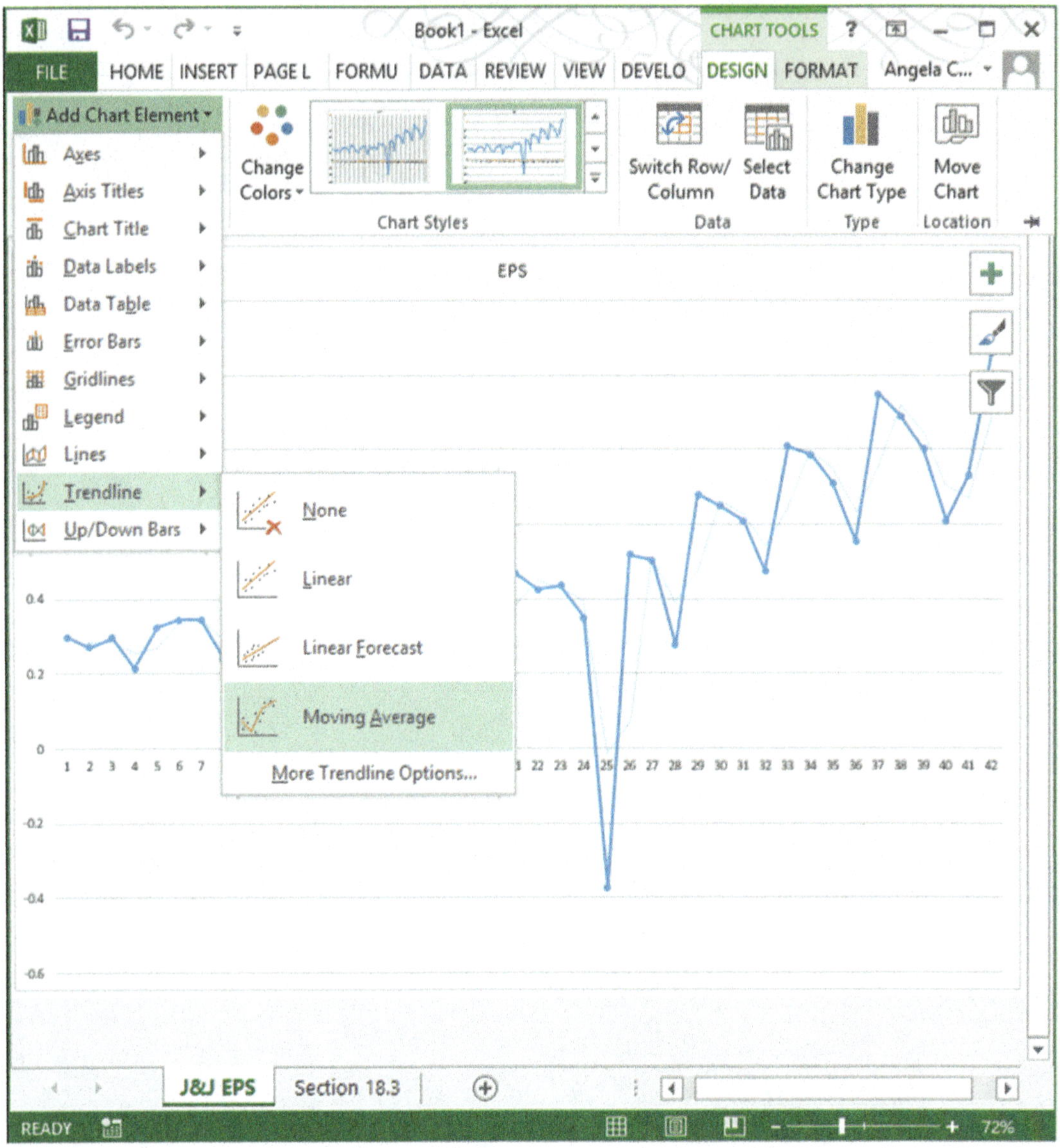

From the *Chart* menu, choose *Add Chart Element* and then *Trendline* as shown above. Choosing the *Moving Average* option, the Excel automatically assumes that the moving average period is 2. However, we need to change it to 4. To change it, double-click on the EPS line and a format chart will appear on the right-hand side. Click on the triangle on the side and select *Trendline*. Options will be shown as below.

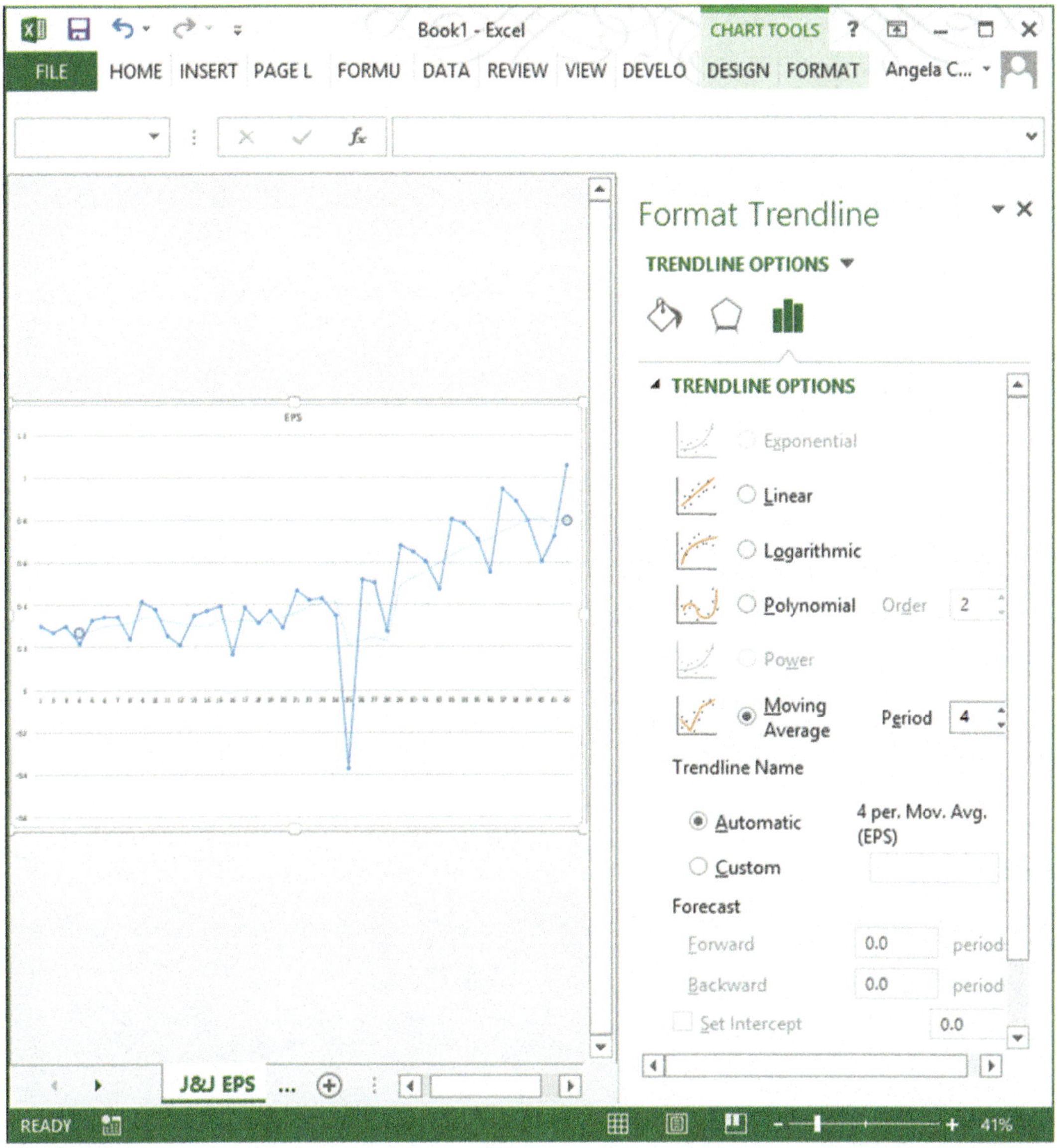
Book1 - Excel
CHART TOOLS
FILE
HOME
INSERT
PAGE L
FORMU
DATA
REVIEW
VIEW
DEVELO
DESIGN
FORMAT
Angela C...
Format Trendline
TRENDLINE OPTIONS
TRENDLINE OPTIONS
Exponential
Linear
Logarithmic
Polynomial
Order
2
Power
Moving Average
Period
4
Trendline Name
Automatic
4 per. Mov. Avg. (EPS)
Custom
Forecast
Forward
0.0
period
Backward
0.0
period
Set Intercept
0.0
J&J EPS
READY
41%

Change the moving average period to 4. Below is the final chart.

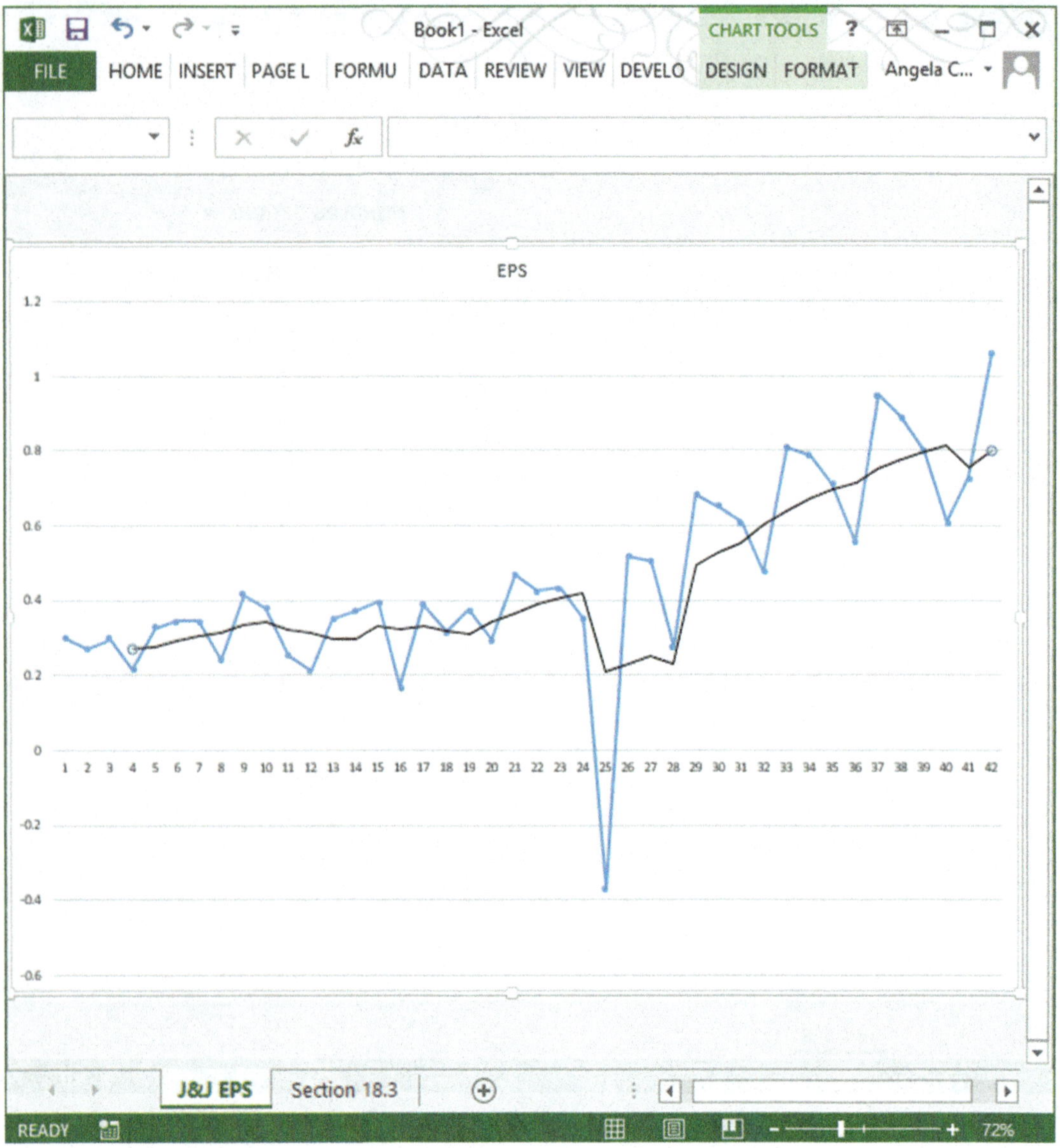

18.4 Moving Averages in Excel: Data Analysis Method

The other way to produce a moving average chart is to choose the *Moving Average* option from the *Data Analysis* as shown below.

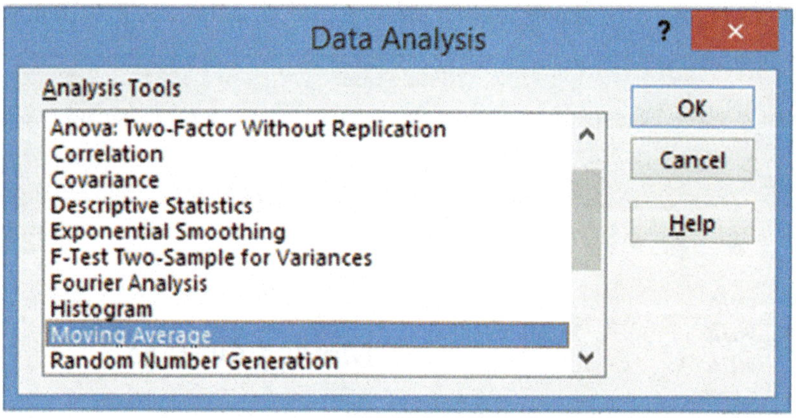

After pressing the *OK* button, the *Moving Average* dialog box is displayed as shown below.

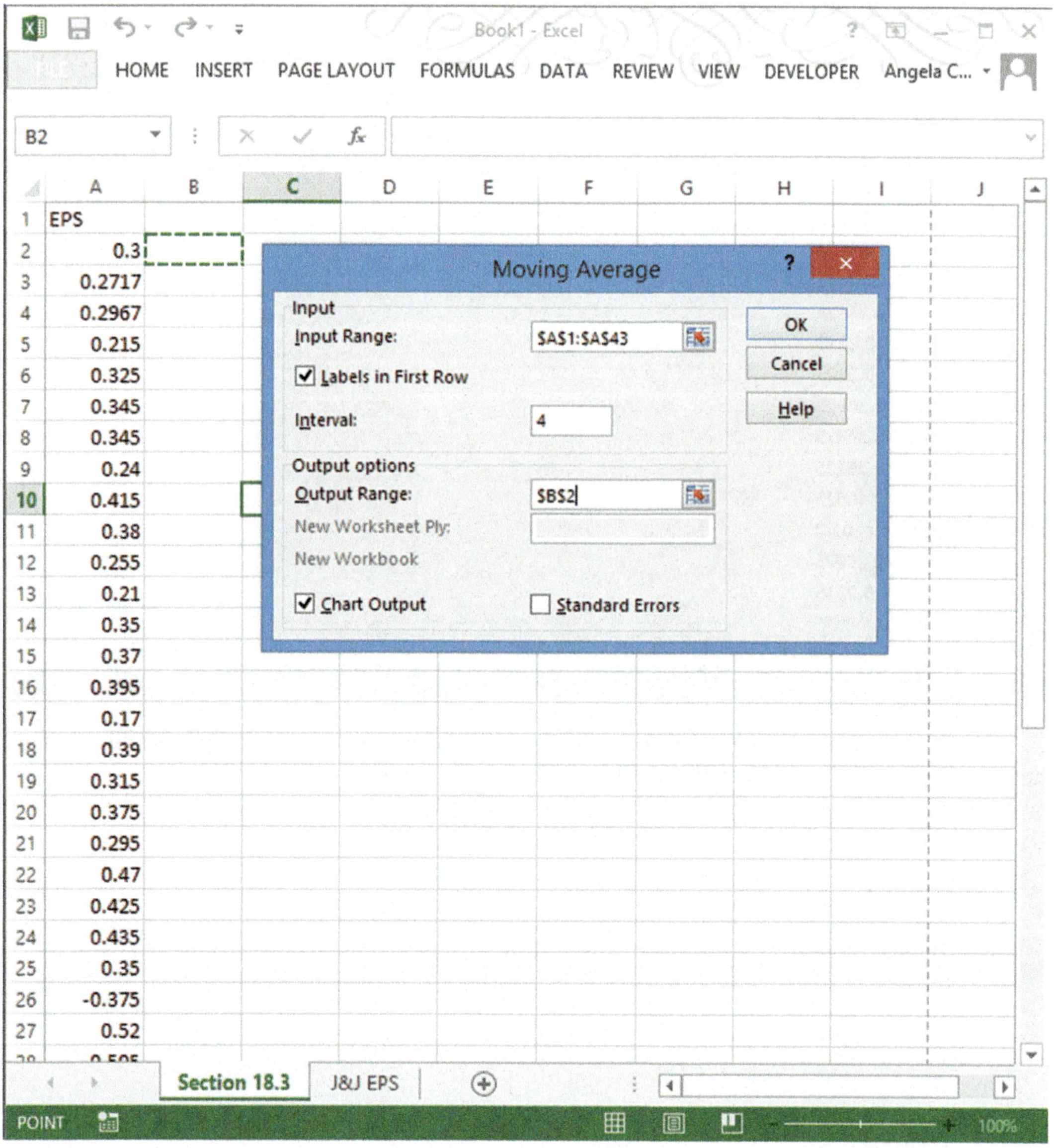

Filling the *Moving Average* dialog box and pressing the *OK* will result in the following.

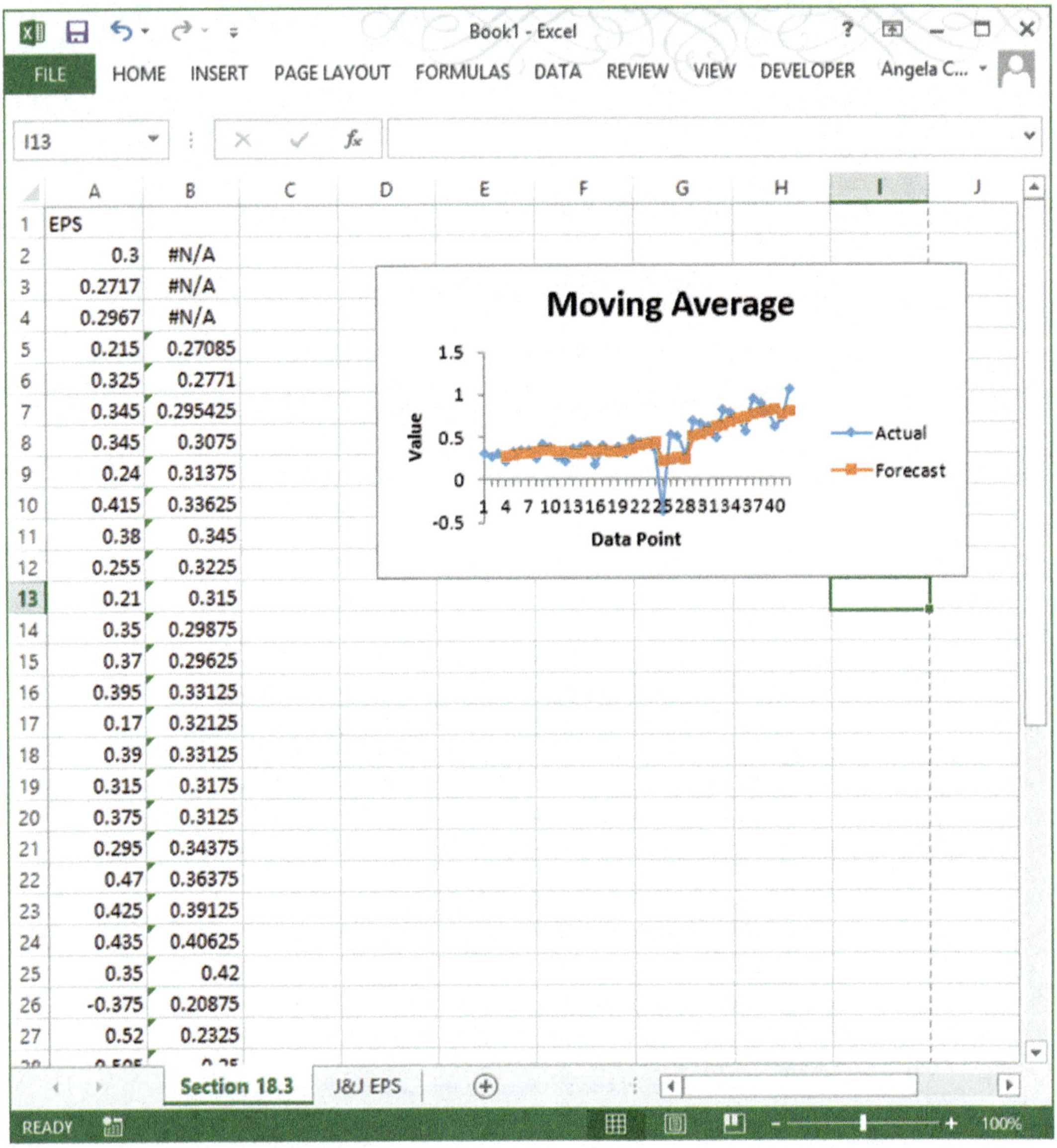

As shown above, the *Moving Average* dialog box puts the moving average calculation in column B and puts the moving average chart at cell D2.

To make the chart bigger, click on the chart once, and click on the right mouse button to get the shortcut menu as shown below.

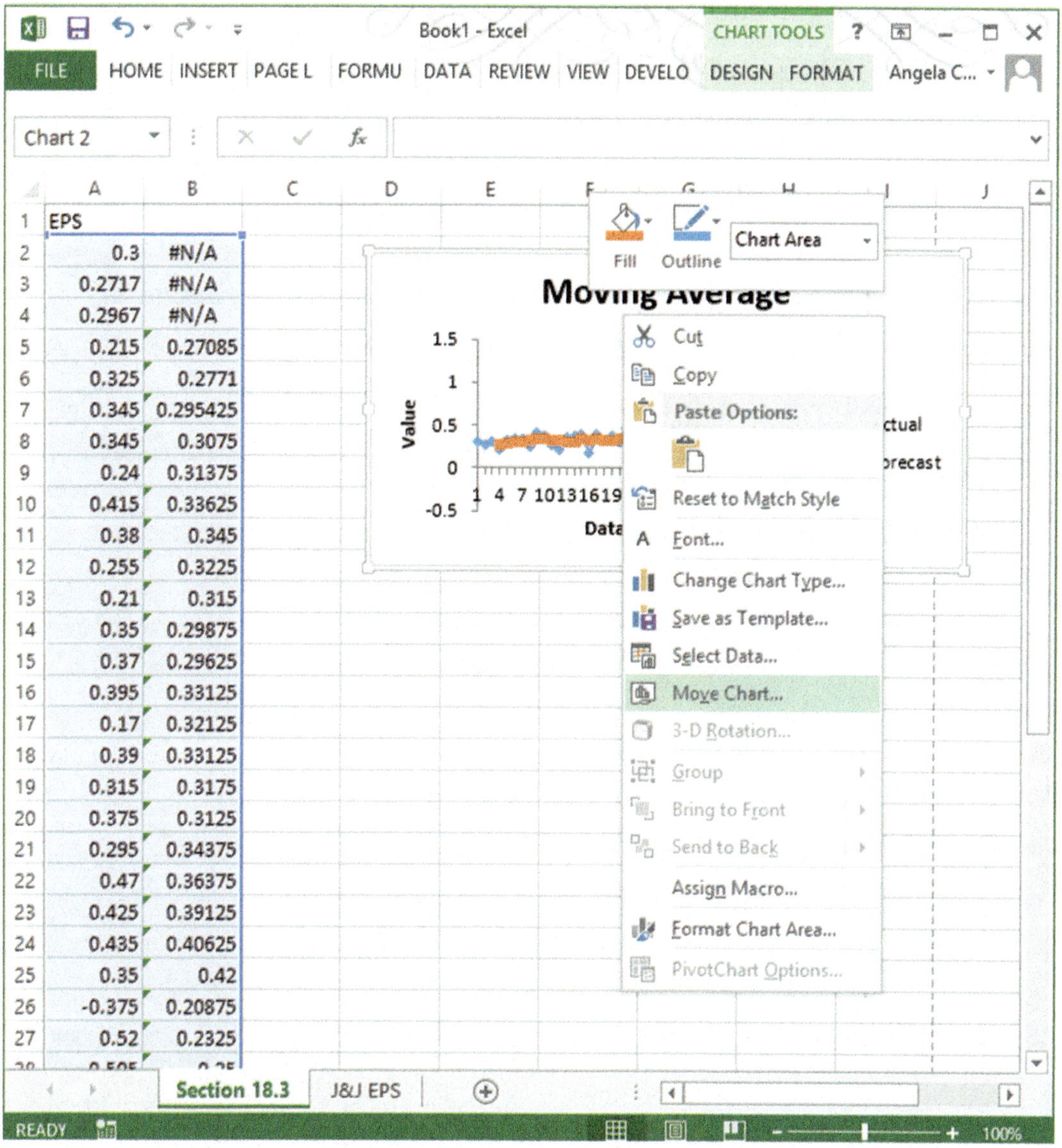

Choosing the *Move Chart* menu item as shown above will get the *Move Chart* dialog box as shown below.

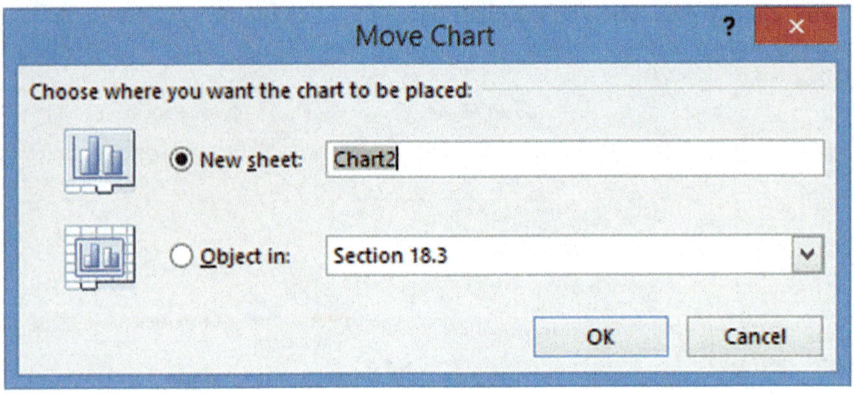

Pressing the *OK* on the *Move Chart* dialog box will produce the following chart.

18.5 Linear Time Trend Regression

If a time series is expected to change linearly over time, regression analysis can be used. We will use the annual sales of Ford to study the regression of a time series.

Year	Sales (millions)	*t*
1968	$14,075.10	1
1969	$14,755.60	2
1970	$14,979.90	3
1971	$16,433.00	4
1972	$20,194.40	5
1973	$23,015.10	6
1974	$23,620.60	7
1975	$24,009.11	8
1976	$28,839.61	9
1977	$37,841.51	10
1978	$42,784.11	11
1979	$43,513.71	12
1980	$37,085.51	13
1981	$38,247.11	14
1982	$37,067.21	15
1983	$44,454.61	16
1984	$52,366.41	17
1985	$52,774.41	18
1986	$62,868.30	19
1987	$72,797.20	20
1988	$82,193.00	21
1989	$82,879.40	22
1990	$81,844.00	23

We will put the sales data in column c1 and name the column *sales*.

```
MTB > name c1 'sales'
MTB > set 'sales'
DATA> 14075.10 14755.60 14979.90 16433.00 20194.40
DATA> 23015.10 23620.60 24009.11 28839.61 37841.51
DATA> 42784.11 43513.71 37085.51 38247.11 37067.21
DATA> 44454.61 52366.41 52774.41 62868.30 72797.20
DATA> 82193.00 82879.40 81844.00
DATA> end
MTB > print 'sales'
```

Data Display

```
sales
   14075.1   14755.6   14979.9   16433.0   20194.4   23015.1   23620.6
   24009.1   28839.6   37841.5   42784.1   43513.7   37085.5   38247.1
   37067.2   44454.6   52366.4   52774.4   62868.3   72797.2   82193.0
   82879.4   81844.0
```

We will run the *Trend.mac* macro that comes with Minitab. This macro is stored in the *macro* folder under the *mtbwin* folder. The first parameter of the *Trend* macro indicates the column that contains the data to be analyzed.

```
MTB > Trend 'sales';
SUBC>   Brief 2.
```

Trend Analysis for Sales

```
Data      sales
Length    23
NMissing  0


Fitted Trend Equation

Yt = 3240 + 3167×t


Accuracy Measures

MAPE         15
MAD        5373
MSD    42731943


Time     sales     Trend   Detrend
1      14075.1    6407.1    7668.0
2      14755.6    9574.2    5181.4
3      14979.9   12741.3    2238.6
4      16433.0   15908.4     524.6
5      20194.4   19075.5    1118.9
6      23015.1   22242.6     772.5
7      23620.6   25409.7   -1789.1
8      24009.1   28576.8   -4567.7
9      28839.6   31743.9   -2904.3
10     37841.5   34911.0    2930.5
11     42784.1   38078.1    4706.0
12     43513.7   41245.2    2268.5
13     37085.5   44412.3   -7326.8
14     38247.1   47579.4   -9332.3
15     37067.2   50746.5  -13679.3
16     44454.6   53913.6   -9459.0
17     52366.4   57080.7   -4714.3
18     52774.4   60247.8   -7473.4
19     62868.3   63414.9    -546.6
20     72797.2   66582.0    6215.2
21     82193.0   69749.1   12443.9
22     82879.4   72916.2    9963.2
23     81844.0   76083.3    5760.7
```

Trend Analysis Plot for Sales

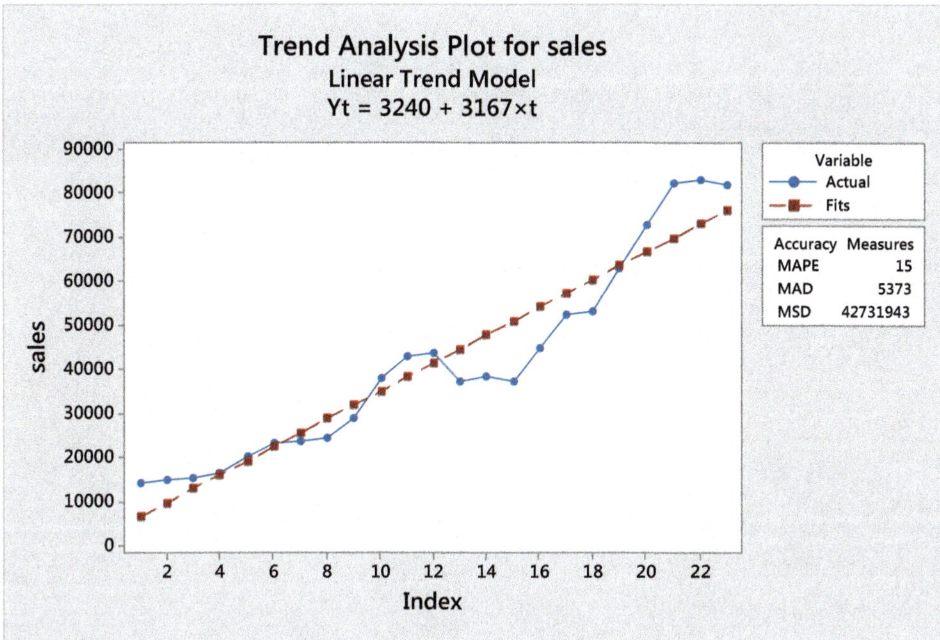

We will now regress sales over time. To do this, we will put the period in column c2 and name it "period."

```
MTB > Name c2 'period'
MTB > Set 'period'
DATA>   1( 1 : 23 / 1 )1
DATA>   End.
MTB >  Regress;
SUBC>   Response 'sales';
SUBC>   Nodefault;
SUBC>   Continuous 'period';
SUBC>   Terms period;
SUBC>   Constant;
SUBC>   Unstandardized;
SUBC>   Tmethod;
SUBC>   Tanova;
SUBC>   Tsummary;
SUBC>   Tcoefficients;
SUBC>   Tequation;
SUBC>   TDiagnostics 0.
```

Regression Analysis: Sales Versus Period

```
Analysis of Variance

Source        DF       Adj SS       Adj MS  F-Value  P-Value
Regression     1  10150900144  10150900144   216.89    0.000
  period       1  10150900144  10150900144   216.89    0.000
Error         21    982834678     46801651
Total         22  11133734822


Model Summary

      S    R-sq  R-sq(adj)  R-sq(pred)
6841.17  91.17%     90.75%      89.28%


Coefficients

Term       Coef  SE Coef  T-Value  P-Value   VIF
Constant   3240     2949     1.10    0.284
period     3167      215    14.73    0.000  1.00


Regression Equation

sales = 3240 + 3167 period


Fits and Diagnostics for Unusual Observations

Obs  sales    Fit   Resid  Std Resid
 15  37067  50746  -13679      -2.05  R

R  Large residual
```

We can see from the adjusted R-square of 90.75 % that time explains sales pretty well. This indicates that Ford's sales have a linear trend. This linear trend is indicated by the coefficient of the independent variable, which is 3167. This means that the trend of Ford's sales is an increase of $3.167 billion every year.

18.6 Exponential Smoothing in Minitab

Smoothing techniques are often used to forecast future values of a time series. One problem that arises in using a moving average to forecast time series is that values in the series are lost. To solve this problem, we can use exponential smoothing.

Exponential smoothing is calculated as follows:

$$\begin{aligned}
s_1 &= x_1 \\
s_2 &= \alpha x_2 + (1-\alpha)s_1 \\
s_3 &= \alpha x_3 + (1-\alpha)s_2 \\
&\cdots \\
s_t &= \alpha x_t + (1-\alpha)s_{t-1}
\end{aligned}$$

where α is the smoothing constant.

We will do exponential smoothing on the annual earnings per share for J&J and IBM.

J&J

YEAR	EPS
1980	1.0834
1981	1.255
1982	1.26
1983	1.285
1984	1.375
1985	1.68
1986	0.925
1987	2.415
1988	2.86
1989	3.25

IBM

YEAR	EPS
1980	6.1
1981	5.63
1982	7.39
1983	9.04
1984	10.77
1985	10.67
1986	7.81
1987	8.72
1988	9.8
1989	6.47

We will put J&J's EPS before the exponential smoothing calculations in column c2 and name it "j&j*eps*." We will put the years in column c1 and name it *year*.

```
MTB > name c1 'year'
MTB > name c2 'j&jeps'
MTB > read c1-c2
DATA> 1980 1.0834
DATA> 1981 1.2550
DATA> 1982 1.2600
DATA> 1983 1.2850
DATA> 1984 1.3750
DATA> 1985 1.6800
DATA> 1986 0.9250
DATA> 1987 2.4150
DATA> 1988 2.8600
DATA> 1989 3.2500
DATA> end
10 rows read.
MTB > Name c3 "SMOO1"
MTB > SES 'j&jeps';
SUBC>   Weight 0.3;
SUBC>   Initial 1;
SUBC>   Smoothed 'SMOO1';
SUBC>   Smplot;
SUBC>   Brief 2.
```

Single Exponential Smoothing for j&jeps

```
Data     j&jeps
Length   10


Smoothing Constant

α  0.3


Accuracy Measures

MAPE  25.1203
MAD    0.5166
MSD    0.5061


Time  j&jeps   Smooth   Predict     Error
1     1.0834  1.08340  1.08340   0.00000
2     1.2550  1.13488  1.08340   0.17160
3     1.2600  1.17242  1.13488   0.12512
4     1.2850  1.20619  1.17242   0.11258
5     1.3750  1.25683  1.20619   0.16881
6     1.6800  1.38378  1.25683   0.42317
7     0.9250  1.24615  1.38378  -0.45878
8     2.4150  1.59680  1.24615   1.16885
9     2.8600  1.97576  1.59680   1.26320
10    3.2500  2.35803  1.97576   1.27424
```

Single Exponential Smoothing Plot for j&jeps

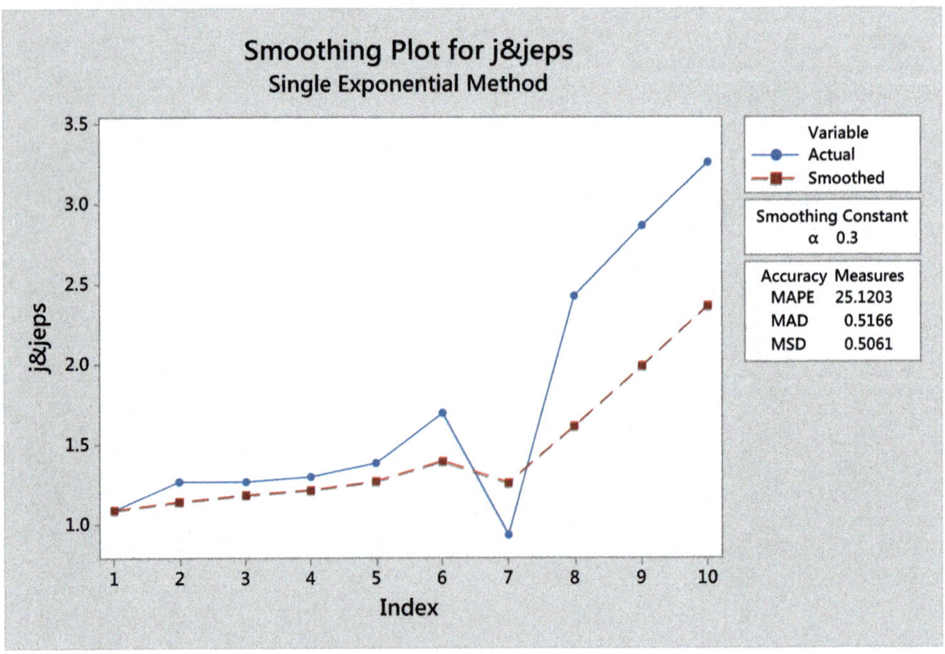

We will now illustrate the macro for the IBM data. We will put the EPS before the exponential smoothing calculations in column c4 and name it *ibmeps*. Column c5, *smooth2*, will contain the EPS after the exponential smoothing calculations.

```
MTB > name c4 'ibmeps'
MTB > set 'ibmeps'
DATA> 6.1 5.63 7.39 9.04 10.77 10.67 7.81 8.72 9.8 6.47
DATA> end
MTB > Name c5 "SMOO2"
MTB > SES 'ibmeps';
SUBC>   Weight 0.3;
SUBC>   Initial 1;
SUBC>   Smoothed 'SMOO2';
SUBC>   Smplot;
SUBC>   Brief 2.
```

Single Exponential Smoothing for ibmeps

```
Data    ibmeps
Length  10


Smoothing Constant

α  0.3


Accuracy Measures

MAPE  17.9736
MAD    1.5484
MSD    3.7062


Time  ibmeps    Smooth   Predict      Error
1       6.10  6.10000  6.10000    0.00000
2       5.63  5.95900  6.10000   -0.47000
3       7.39  6.38830  5.95900    1.43100
4       9.04  7.18381  6.38830    2.65170
5      10.77  8.25967  7.18381    3.58619
6      10.67  8.98277  8.25967    2.41033
7       7.81  8.63094  8.98277   -1.17277
8       8.72  8.65766  8.63094    0.08906
9       9.80  9.00036  8.65766    1.14234
10      6.47  8.24125  9.00036   -2.53036
```

Single Exponential Smoothing Plot for ibmeps

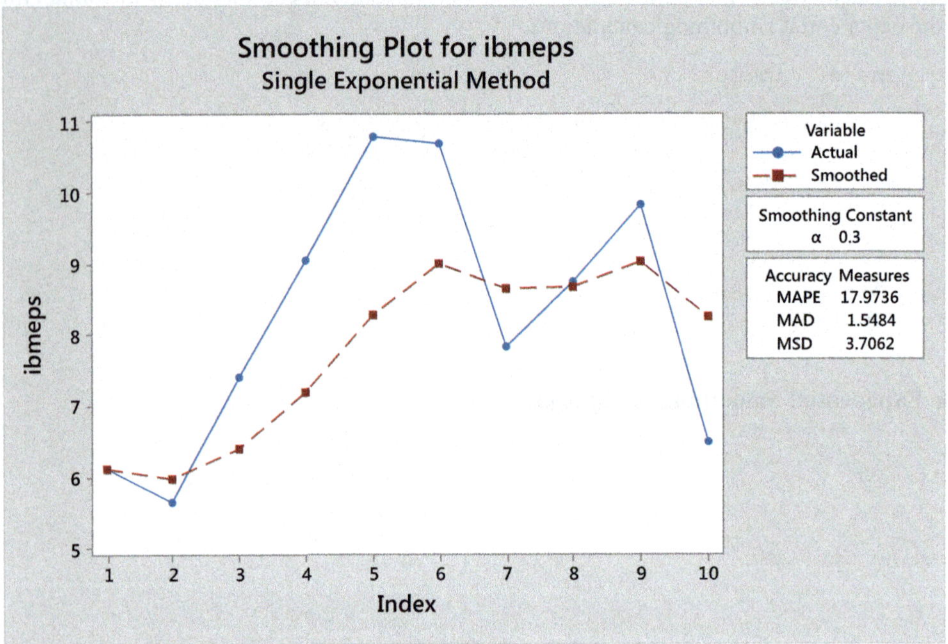

18.7 Exponential Smoothing in Excel

It is important to note that Excel uses a different equation for exponential smoothing. The formula that Excel uses for exponential smoothing is

$$F_{t+1} = F_t + a(A_t - F_t) = F_t + (1 - \text{damp Fact})(A_t - F_t)$$

We use Excel's *Analysis Tool Pack* to do our exponential smoothing of J&J's and IBM's earning per share.

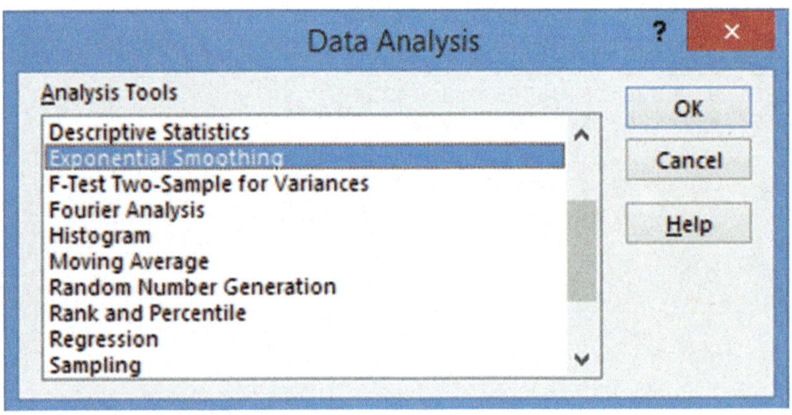

Below shows the *Exponential Smoothing* dialog box for the J&J earnings per share.

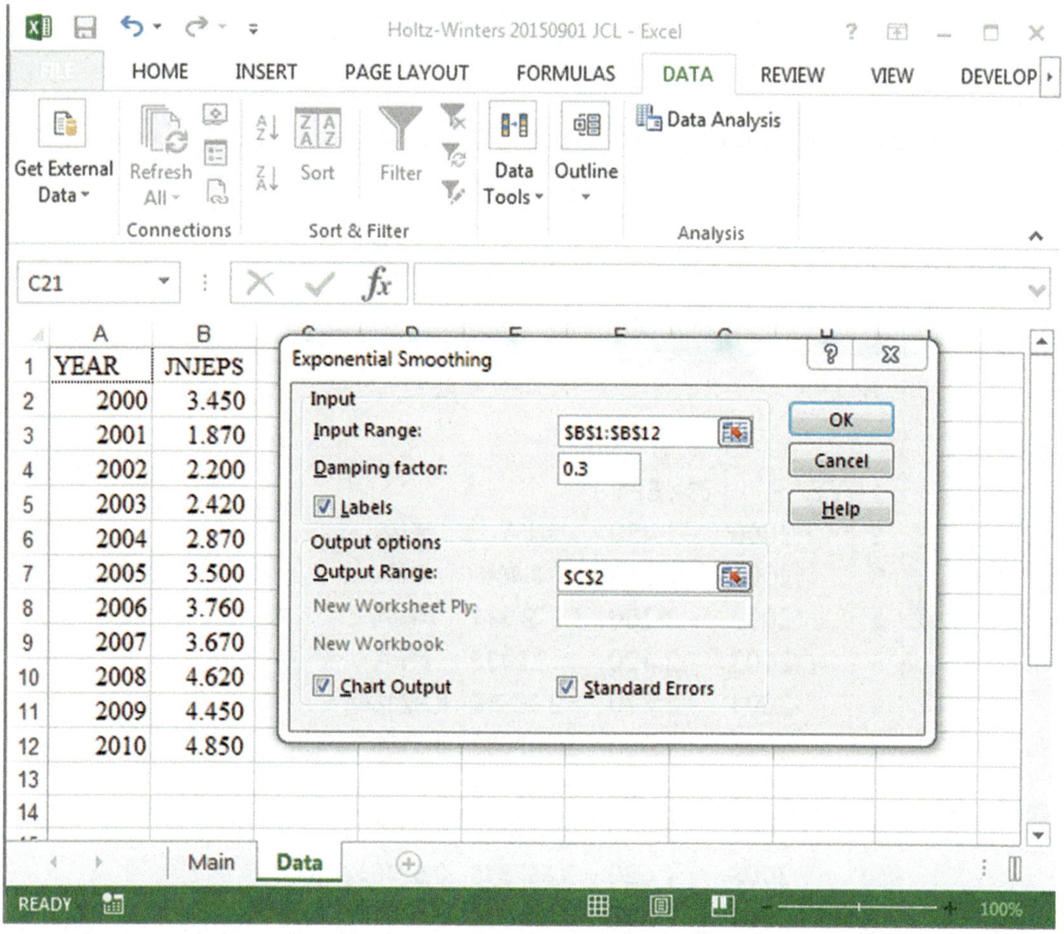

Below are the calculated exponential smoothing numbers and exponential smoothing chart for J&J's EPS numbers.

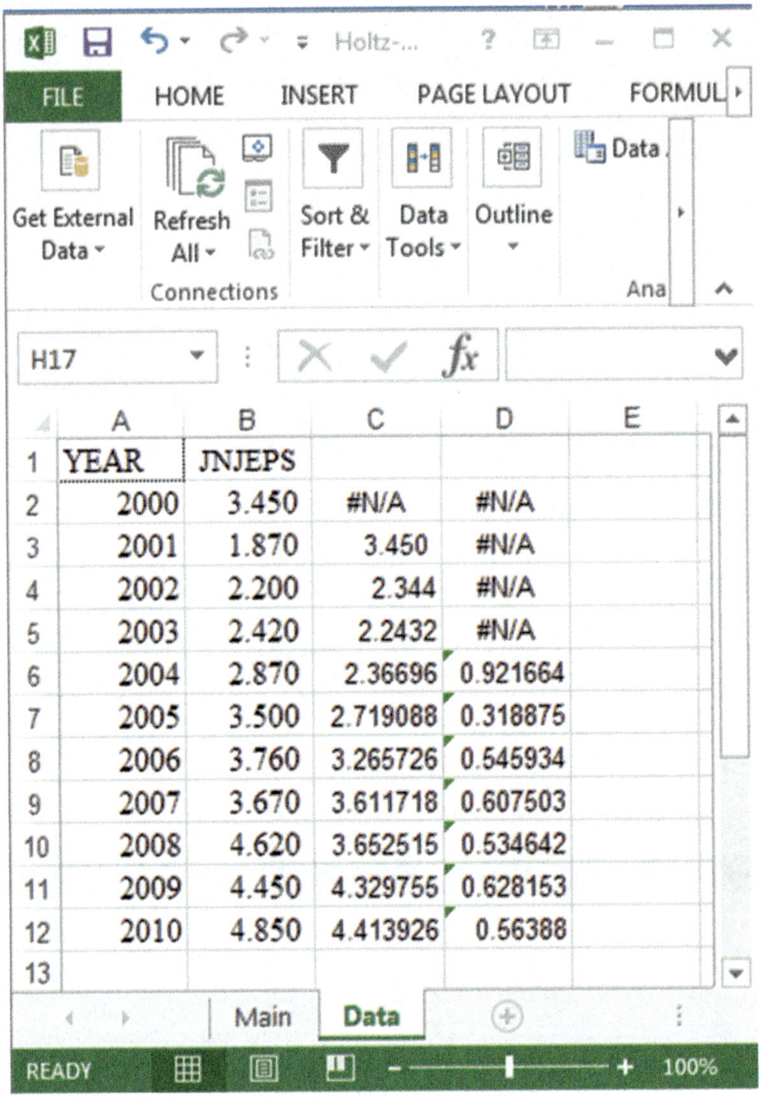

	A	B	C	D
1	YEAR	JNJEPS		
2	2000	3.450	#N/A	#N/A
3	2001	1.870	3.450	#N/A
4	2002	2.200	2.344	#N/A
5	2003	2.420	2.2432	#N/A
6	2004	2.870	2.36696	0.921664
7	2005	3.500	2.719088	0.318875
8	2006	3.760	3.265726	0.545934
9	2007	3.670	3.611718	0.607503
10	2008	4.620	3.652515	0.534642
11	2009	4.450	4.329755	0.628153
12	2010	4.850	4.413926	0.56388

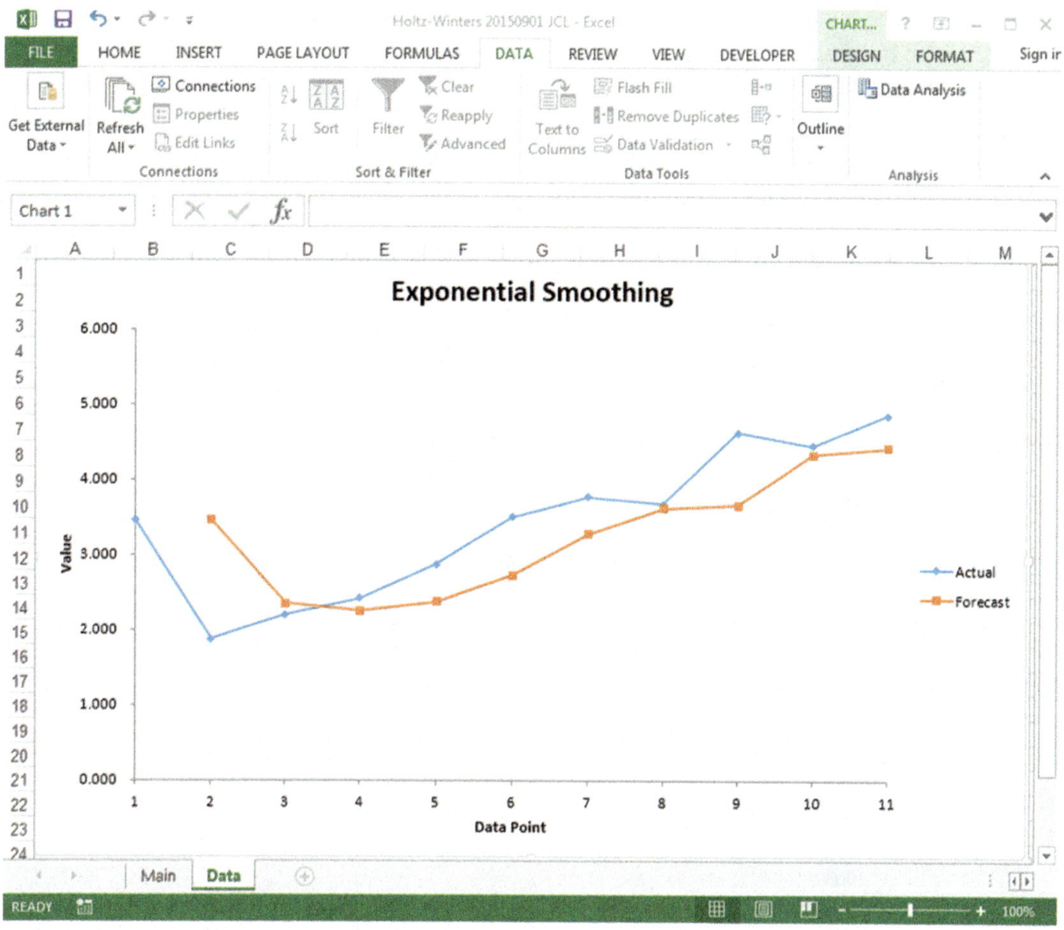

18.8 Holt–Winters Forecasting Model for Nonseasonal Series

Exponential smoothing does not recognize the trend in time-series data, but the Holt–Winters model does. The Holt–Winters model contains a smoothing component, s, and a trend component, T. The smoothing and trend components can be described as follows:

$$s_t = \alpha x_t + (1-\alpha)(s_{t-1} + T_{t-1})$$
$$T_t = \beta(s_t - s_{t-1}) + (1-\beta)T_{t-1}$$

where α and β are the exponential and trend smoothing constants, respectively.

The procedure for calculating the Holt–Winters components is as follows:

1. Choose an exponential smoothing constant α between 0 and 1. A small value for α gives less weight to the current values of the time series and more weight to the past. A large value for α gives more weight to the most recent trend of the series.
2. Choose a trend smoothing constant β between 0 and 1. Small values of β give less weight to the current changes in the level of the series and more weight to the past trend. Larger choices assign more weight to the most recent trend series.
3. Estimate the first observation of the trend T_1 by one of the following alternative methods.

Method 1:
Let $T_1 = 0$. If there are a large number of observations in the time series, this method provides an adequate initial estimate of the trend.

Method 2:
Use the first five (or so) observations to estimate the initial trend by following the linear time trend regression line. Then use the estimated slope b as the first trend observation: that is, $T_1 = b$.

We will use the J&J and IBM EPS data again to calculate the Holt–Winters numbers. The data are reproduced below for convenience.

Year	JNJ EPS
2000	3.450
2001	1.870
2002	2.200
2003	2.420
2004	2.870
2005	3.500
2006	3.760
2007	3.670
2008	4.620
2009	4.450
2010	4.850

Year	IBM EPS
2000	4.580
2001	4.450
2002	3.130
2003	4.420
2004	5.040
2005	4.990
2006	6.150
2007	7.320
2008	9.070
2009	10.120
2010	11.690

18.9 Excel Macro: Holt–Winters Forecasting Model for Nonseasonal Series

To perform the Holt–Winters calculations in Excel, we will use the *Holt–Winters.xls* workbook that is included in the CD that accompanies this book. The Holt–Winters.xls workbook will calculate the Holt–Winters calculations twice. It will perform the Holt–Winters calculations where $T_1 = 0$, and it will perform the Holt–Winters calculations where T_1 is the b of the regression of the first five time-series observations. The Holt–Winters.xls will produce five sheets. The five sheets are:

(1) Holt–Winters calculations where $T_1 = 0$
(2) Holt–Winters calculations where $T_1 = b$
(3) The regression calculations of the first five time-series data
(4) A chart of the Holt–Winters calculations where $T_1 = 0$
(5) A chart of the Holt–Winters calculations where $T_1 = b$.

We will first perform the Holt–Winters calculation on J&J's EPS. Below shows J&J's EPS data in the data sheet of the *Holt–Winters.xls*.

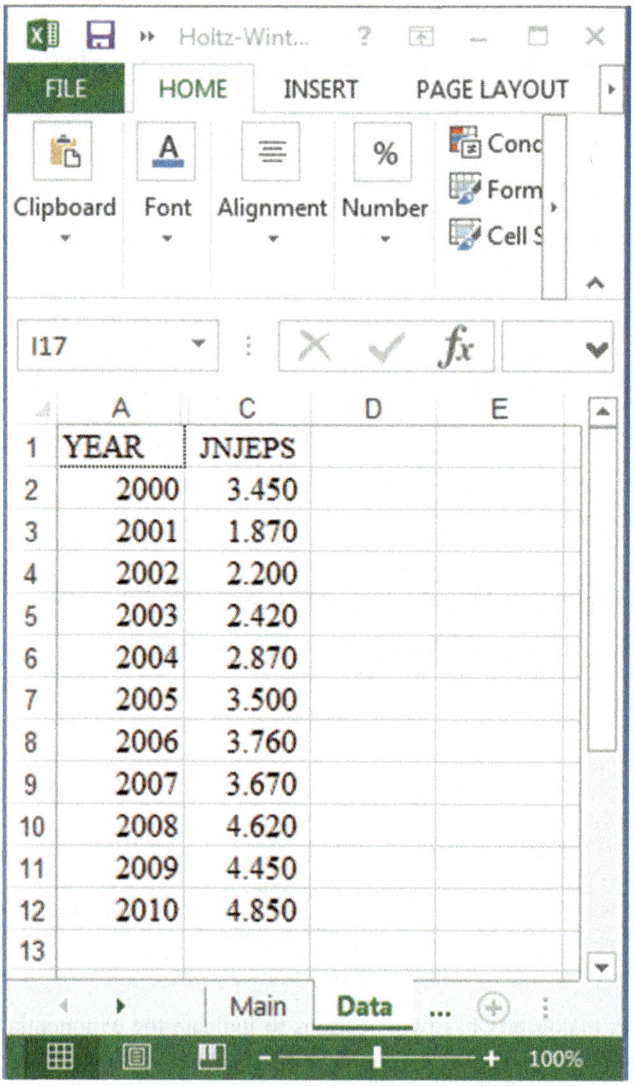

	A	C
1	YEAR	JNJEPS
2	2000	3.450
3	2001	1.870
4	2002	2.200
5	2003	2.420
6	2004	2.870
7	2005	3.500
8	2006	3.760
9	2007	3.670
10	2008	4.620
11	2009	4.450
12	2010	4.850
13		

We then go to the *Main* sheet and click on the *Holtz–Winters* button as shown below.

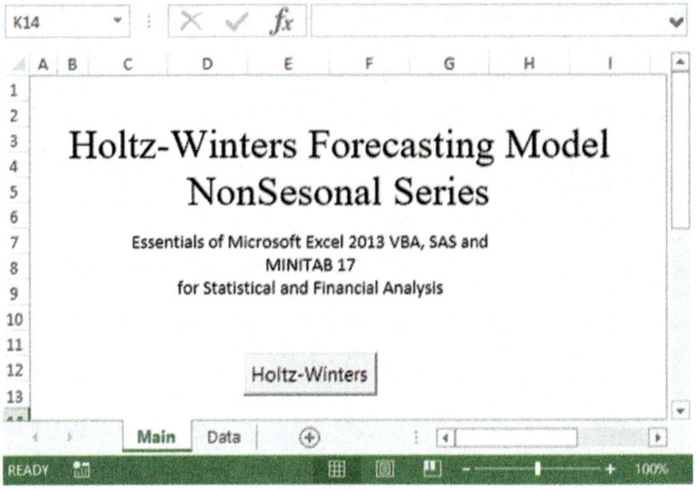

Pushing the *Holtz–Winters* button shows the *Holtz–Winters* dialog box as shown below.

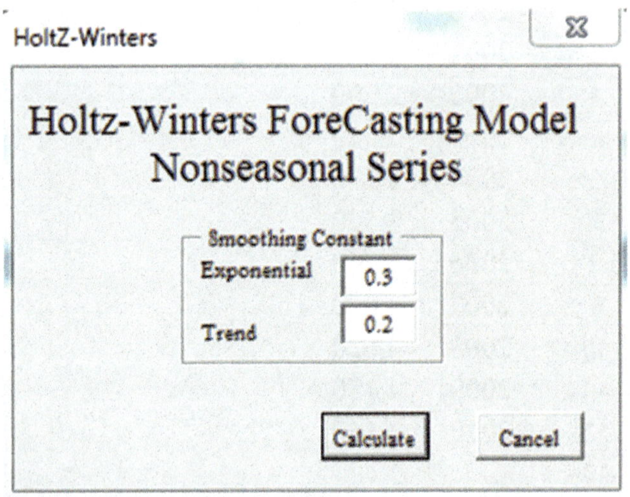

The *Holtz–Winters* dialog box has two locations to indicate the exponential smoothing constant and the trend constant. The *Holtz–Winters* dialog box defaults the exponential smoothing constant as 0.3 and the trend constant as 0.2. The following are the five sheets that are produce after pressing the *Calculate* button on the *Holtz–Winters* dialog box.

The first sheet that is produced is a chart with J&J's EPS actual data and J&J's smoothed EPS data and the J&J's trend data. The first observation for the trend data is the regression of the first five observed EPS.

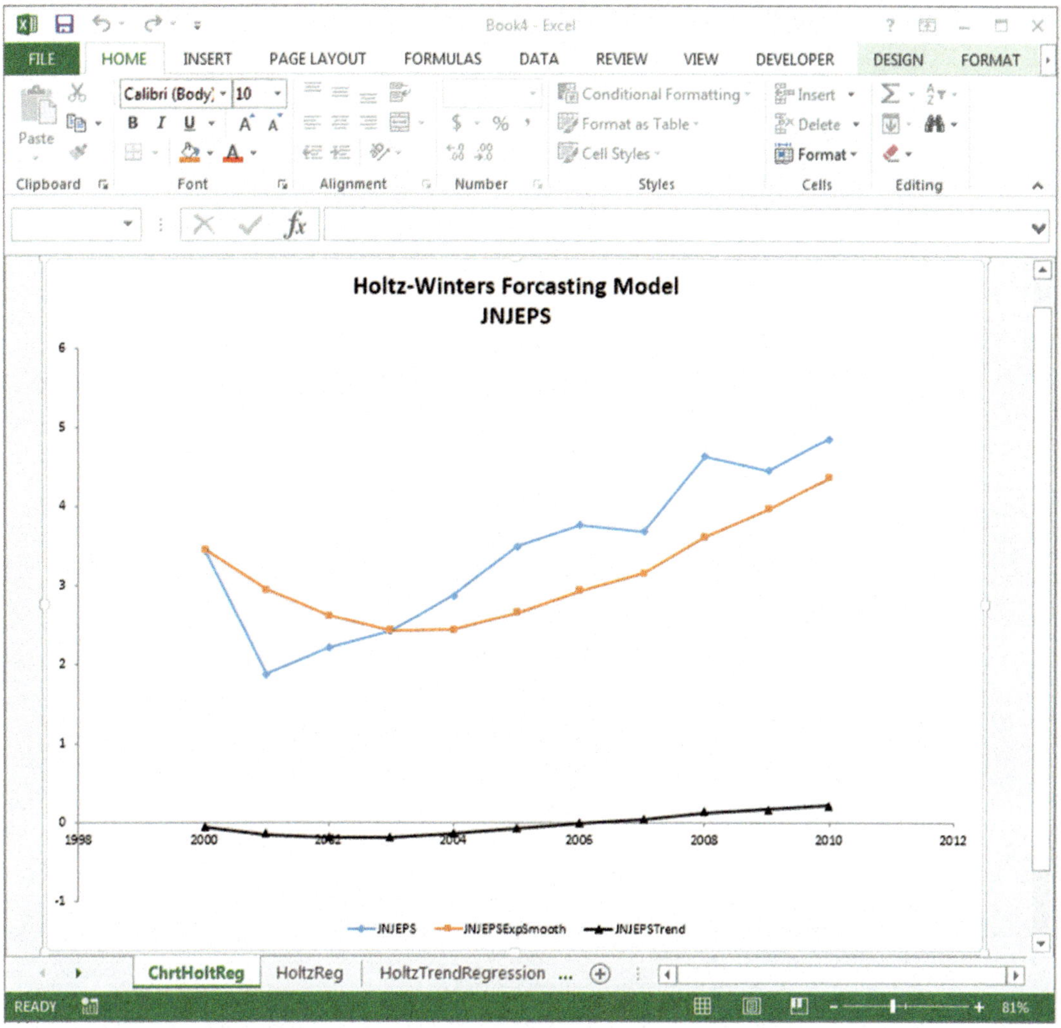
Book4 - Excel
FILE
HOME
INSERT
PAGE LAYOUT
FORMULAS
DATA
REVIEW
VIEW
DEVELOPER
DESIGN
FORMAT
Holtz-Winters Forcasting Model
JNJEPS
JNJEPS
JNJEPSExpSmooth
JNJEPSTrend
ChrtHoltReg
HoltzReg
HoltzTrendRegression
READY

The second sheet *HoltzReg* contains the smooth and trend calculation.

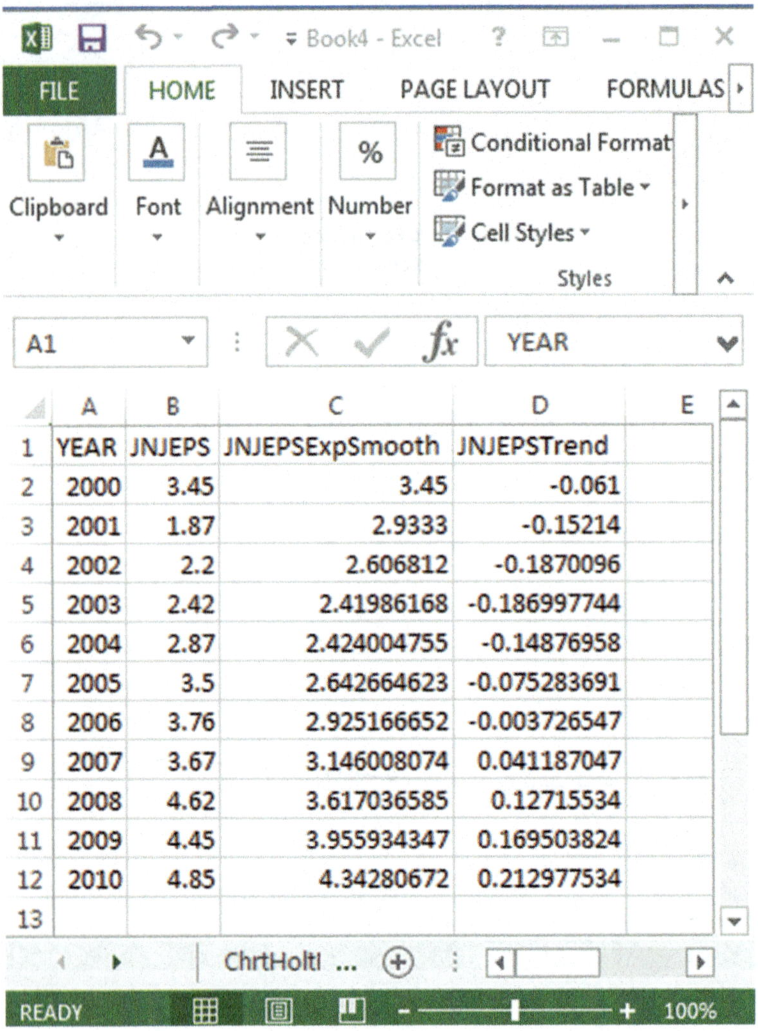

	A	B	C	D	E
1	YEAR	JNJEPS	JNJEPSExpSmooth	JNJEPSTrend	
2	2000	3.45	3.45	-0.061	
3	2001	1.87	2.9333	-0.15214	
4	2002	2.2	2.606812	-0.1870096	
5	2003	2.42	2.41986168	-0.186997744	
6	2004	2.87	2.424004755	-0.14876958	
7	2005	3.5	2.642664623	-0.075283691	
8	2006	3.76	2.925166652	-0.003726547	
9	2007	3.67	3.146008074	0.041187047	
10	2008	4.62	3.617036585	0.12715534	
11	2009	4.45	3.955934347	0.169503824	
12	2010	4.85	4.34280672	0.212977534	
13					

The second sheet *HoltzTrendRegression* contains the regress for the first five J&J EPS. The coefficient for "Year" will be used as the first observation value for the trend.

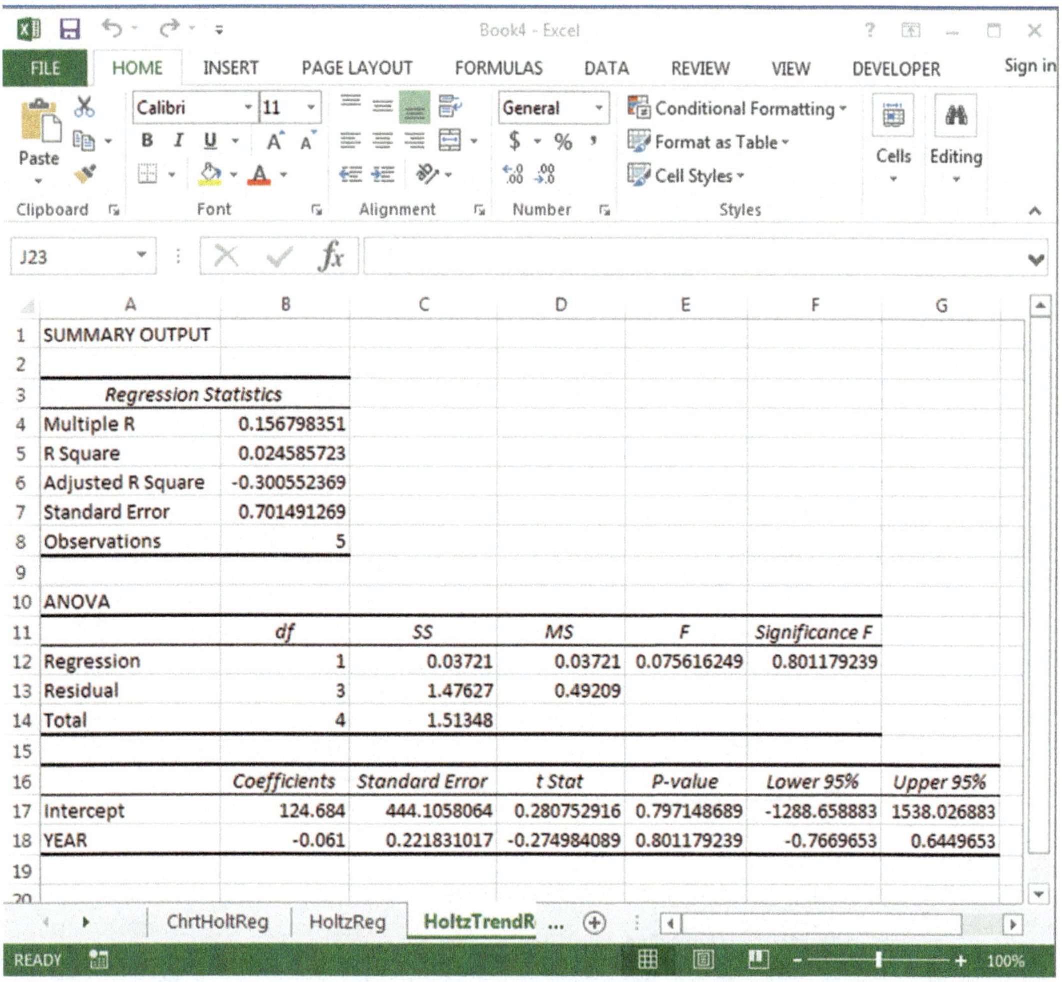

SUMMARY OUTPUT						
Regression Statistics						
Multiple R	0.156798351					
R Square	0.024585723					
Adjusted R Square	-0.300552369					
Standard Error	0.701491269					
Observations	5					
ANOVA						
	df	*SS*	*MS*	*F*	*Significance F*	
Regression	1	0.03721	0.03721	0.075616249	0.801179239	
Residual	3	1.47627	0.49209			
Total	4	1.51348				
	Coefficients	*Standard Error*	*t Stat*	*P-value*	*Lower 95%*	*Upper 95%*
Intercept	124.684	444.1058064	0.280752916	0.797148689	-1288.658883	1538.026883
YEAR	-0.061	0.221831017	-0.274984089	0.801179239	-0.7669653	0.6449653

The third sheet *chrtHoltZero* that is produced is a chart with J&J's EPS actual data and J&J's smoothed EPS data and the J&J's trend data. The first observation for the trend data is zero.

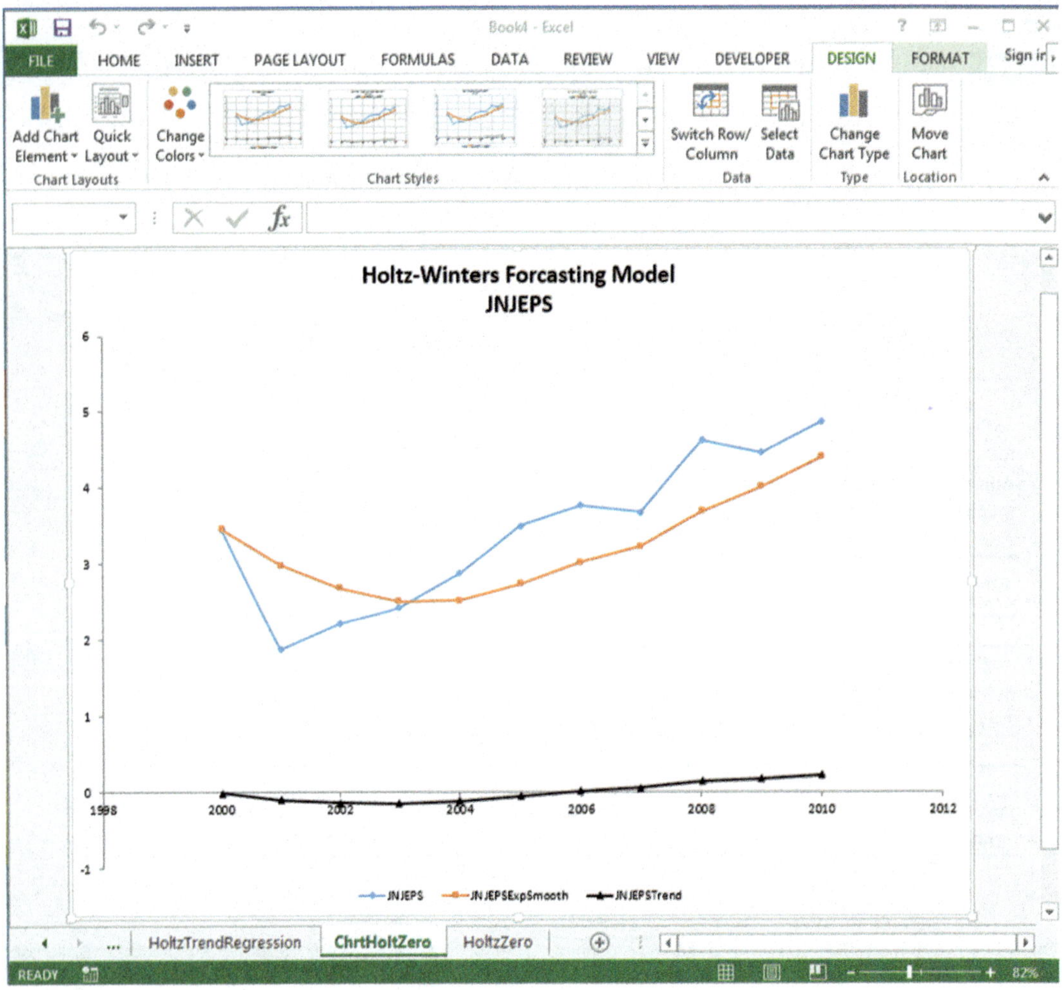

We will now perform the Holt–Winters calculation on IBM's EPS. Below show IBM's EPS data in the data sheet of the *Holt–Winters.xlsm*.

YEAR	IBMEPS
2000	4.580
2001	4.450
2002	3.130
2003	4.420
2004	5.040
2005	4.990
2006	6.150
2007	7.320
2008	9.070
2009	10.120
2010	11.690

Below are the five sheets that are produced by the *Holt–Winters.xlsm* workbook on IBM's EPS data.

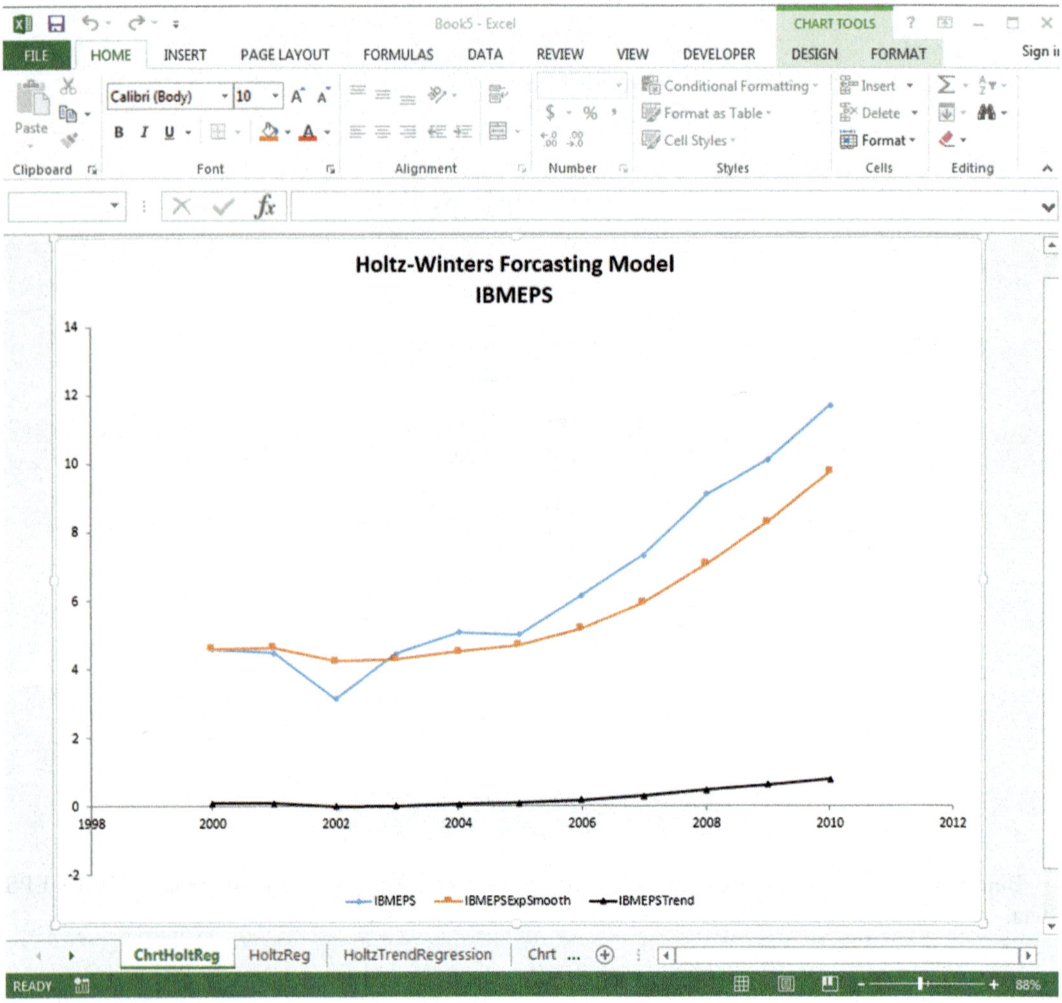
Book5 - Excel
Holtz-Winters Forcasting Model
IBMEPS
14
12
10
8
6
4
2
0
-2
1998
2000
2002
2004
2006
2008
2010
2012
IBMEPS
IBMEPSExpSmooth
IBMEPSTrend
ChrtHoltReg
HoltzReg
HoltzTrendRegression
Chrt ...
READY

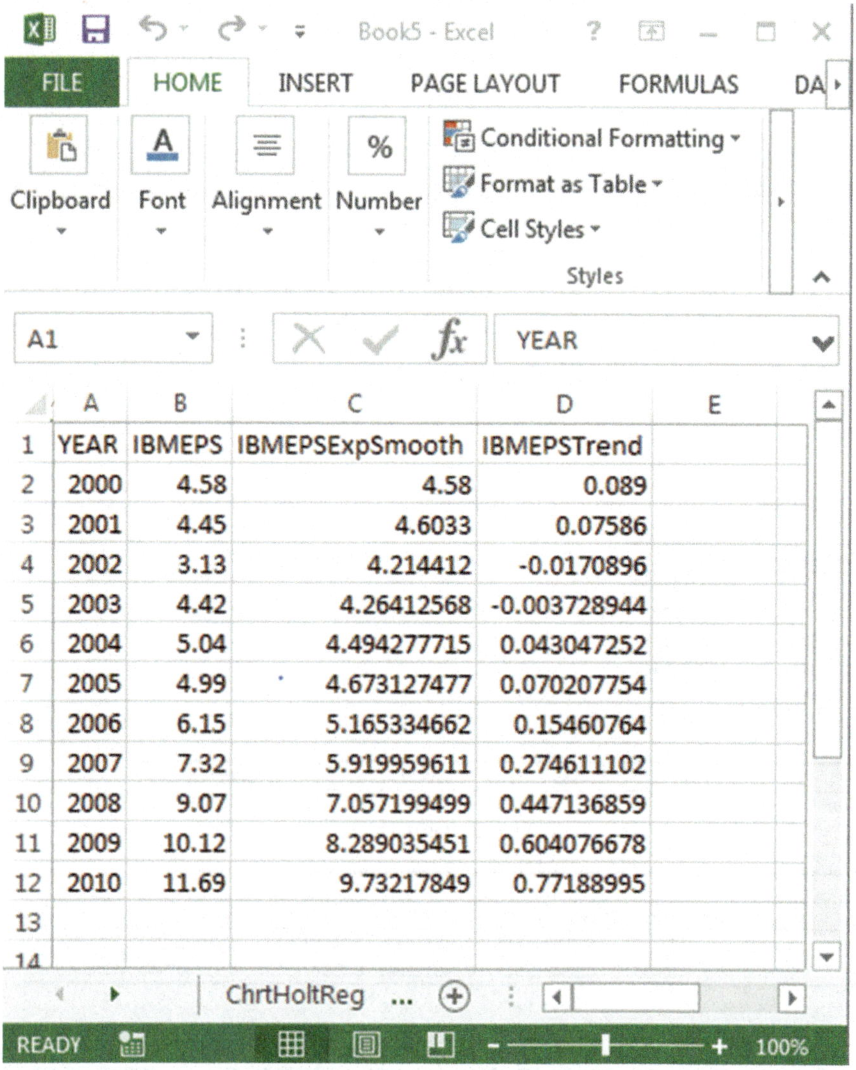

YEAR	IBMEPS	IBMEPSExpSmooth	IBMEPSTrend
2000	4.58	4.58	0.089
2001	4.45	4.6033	0.07586
2002	3.13	4.214412	-0.0170896
2003	4.42	4.26412568	-0.003728944
2004	5.04	4.494277715	0.043047252
2005	4.99	4.673127477	0.070207754
2006	6.15	5.165334662	0.15460764
2007	7.32	5.919959611	0.274611102
2008	9.07	7.057199499	0.447136859
2009	10.12	8.289035451	0.604076678
2010	11.69	9.73217849	0.77188995

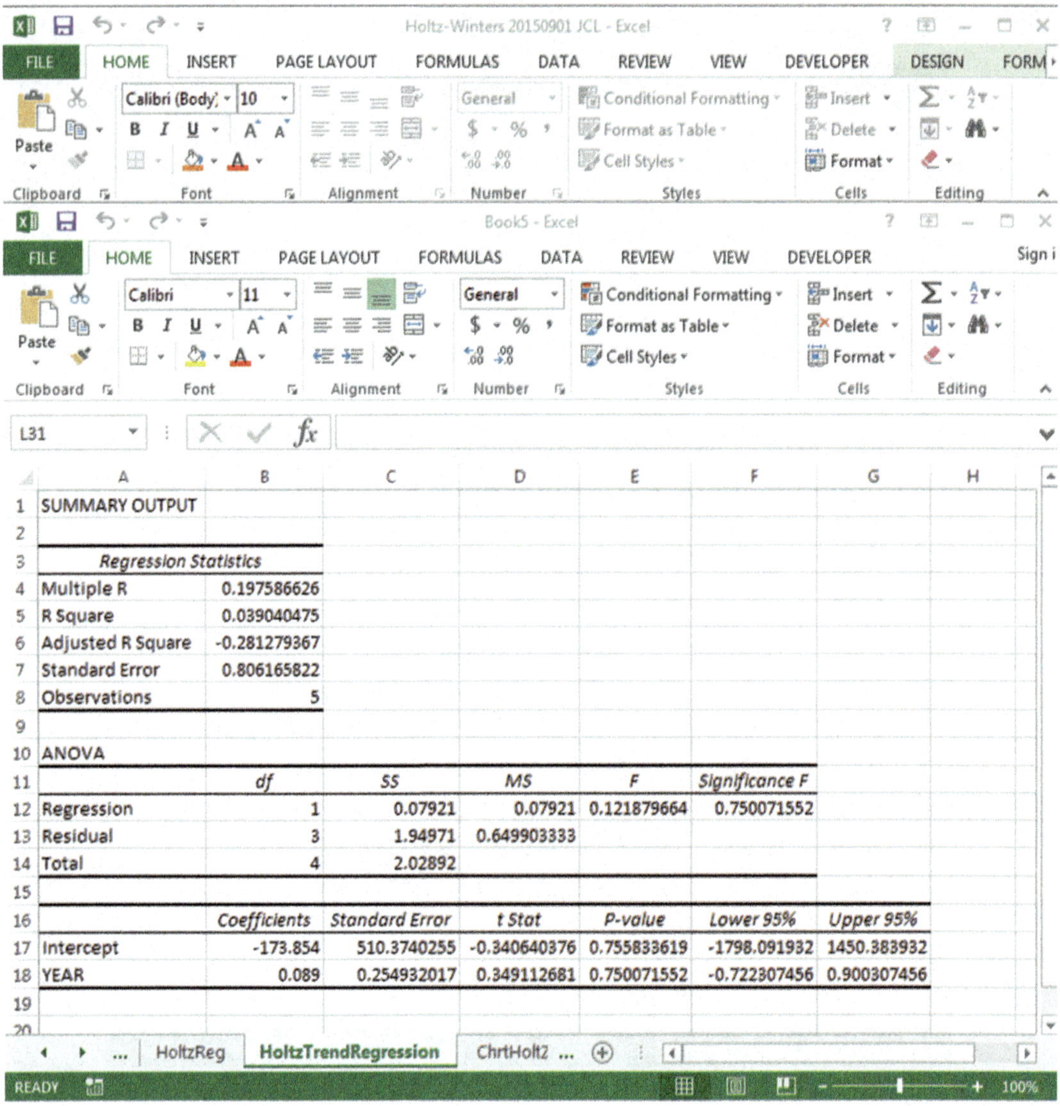

SUMMARY OUTPUT

Regression Statistics	
Multiple R	0.197586626
R Square	0.039040475
Adjusted R Square	-0.281279367
Standard Error	0.806165822
Observations	5

ANOVA

	df	*SS*	*MS*	*F*	*Significance F*
Regression	1	0.07921	0.07921	0.121879664	0.750071552
Residual	3	1.94971	0.649903333		
Total	4	2.02892			

	Coefficients	*Standard Error*	*t Stat*	*P-value*	*Lower 95%*	*Upper 95%*
Intercept	-173.854	510.3740255	-0.340640376	0.755833619	-1798.091932	1450.383932
YEAR	0.089	0.254932017	0.349112681	0.750071552	-0.722307456	0.900307456

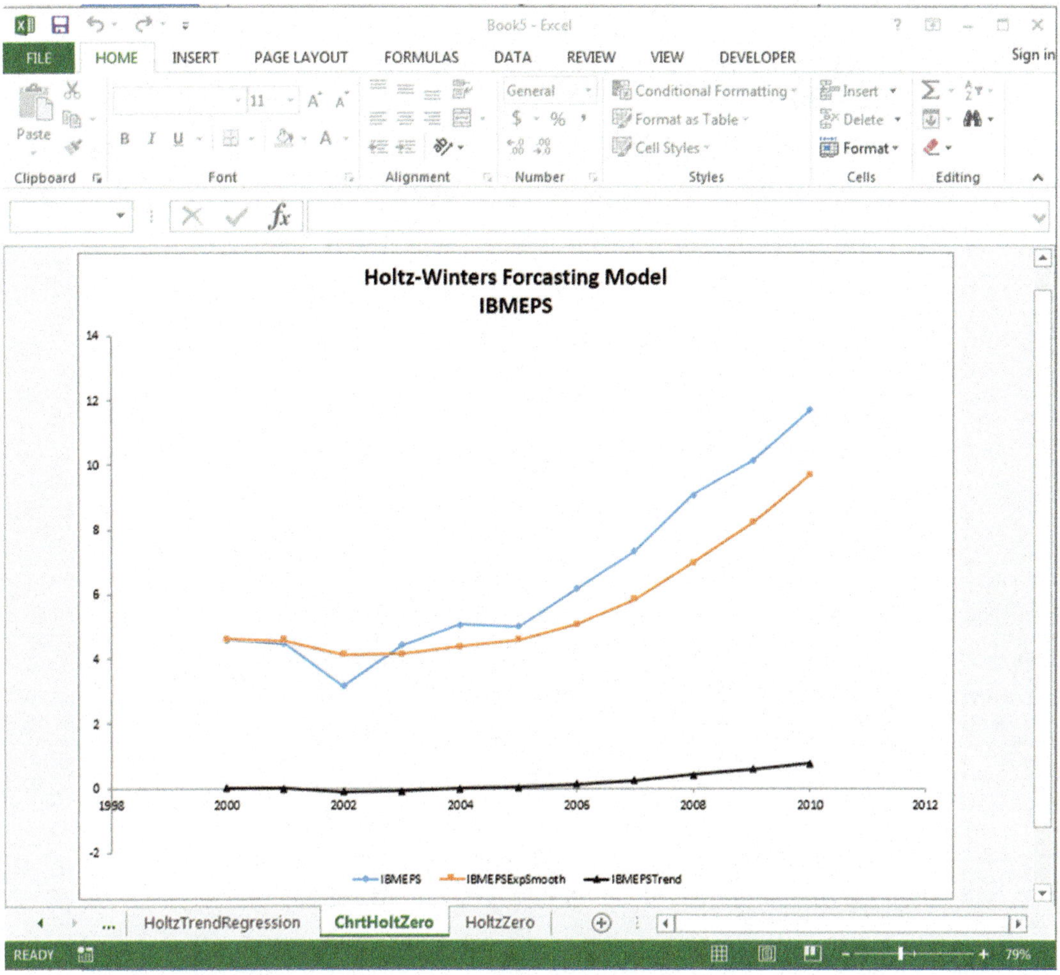
Holtz-Winters Forcasting Model
IBMEPS
14
12
10
8
6
4
2
0
-2
1998
2000
2002
2004
2006
2008
2010
2012
IBMEPS
IBMEPSExpSmooth
IBMEPSTrend
HoltzTrendRegression
ChrtHoltZero
HoltzZero

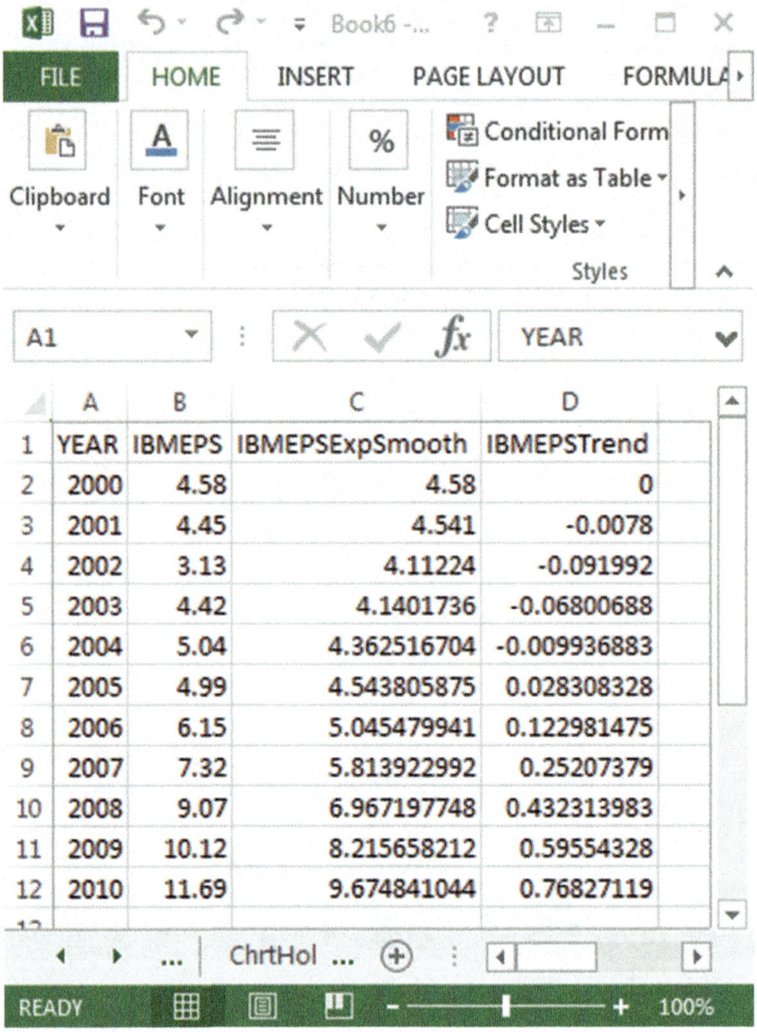

	A	B	C	D
1	YEAR	IBMEPS	IBMEPSExpSmooth	IBMEPSTrend
2	2000	4.58	4.58	0
3	2001	4.45	4.541	-0.0078
4	2002	3.13	4.11224	-0.091992
5	2003	4.42	4.1401736	-0.06800688
6	2004	5.04	4.362516704	-0.009936883
7	2005	4.99	4.543805875	0.028308328
8	2006	6.15	5.045479941	0.122981475
9	2007	7.32	5.813922992	0.25207379
10	2008	9.07	6.967197748	0.432313983
11	2009	10.12	8.215658212	0.59554328
12	2010	11.69	9.674841044	0.76827119

18.10 SAS Programming Code Instructions and Examples

18.10.1 Moving Averages in SAS

We use the same data in Sect. 18.2 to show how moving averages can be obtained. In the following commands, we read the data into SAS. The counter _N_ counts the iterations through the implicit loop of the DATA step.

```
Data JNJ_EPS;
Row=_n_;
period=_n_;
input eps @@;
cards;
.3000 .2717 .2967 .2150 .3250 .3450 .3450 .2400 .4150 .3800
.2550 .2100 .3500 .3700 .3950 .1700 .3900 .3150 .3750 .2950
.4700 .4250 .4350 .3500 -0.3750 .5200 .5050 .2750 .6800 .6500
.6100 .4750 .8100 .7850 .7100 .5550 .9500 .8900 .8000 .6100
.7300 1.0600
;
Run;
```

The formula to calculate a four-point moving average is

$$M_i = \frac{Y_{i-3} + Y_{i-2} + Y_{i-1} + Y_i}{4}$$

where Yt is the earnings for period t.

To calculate the four-point moving average, we mainly use function SUM and LAGn. As we mentioned before, a LAGn (n = 1–100) function returns the value of the nth previous execution of the function. It is easy to assume that the LAGn functions return values of the nth previous observation. For the ith observation, $\text{sum} = Y_1 + Y_2 + \ldots + Y_{i-1} + Y_i$ and $\text{lag4(sum)} = Y_1 + Y_2 + \ldots + Y_{i-4}$. Therefore, $\text{lag4(sum)} - \text{sum} = Y_{i-3} + Y_{i-2} + Y_{i-1} + Y_i$.

```
data JNJ_AVG;
set JNJ_EPS;
sum+eps;
sum1=lag1(sum);
if sum1='.' then sum1=0;
sum1=lag3(sum1);
MovAvg4=(sum-sum1)/4;
Predict=lag1(MovAvg4);
Error= eps - Predict;
drop sum sum1;
run;
```

Obs	Row	Period	eps	MovAvg4	Predict	Error
1	1	1	0.3000			
2	2	2	0.2717			
3	3	3	0.2967			
4	4	4	0.2150	0.27085		
5	5	5	0.3250	0.27710	0.27085	0.05415
6	6	6	0.3450	0.29543	0.27710	0.06790
7	7	7	0.3450	0.30750	0.29543	0.04957
8	8	8	0.2400	0.31375	0.30750	−0.06750
9	9	9	0.4150	0.33625	0.31375	0.10125
10	10	10	0.3800	0.34500	0.33625	0.04375
11	11	11	0.2550	0.32250	0.34500	−0.09000
12	12	12	0.2100	0.31500	0.32250	−0.11250
13	13	13	0.3500	0.29875	0.31500	0.03500
14	14	14	0.3700	0.29625	0.29875	0.07125
15	15	15	0.3950	0.33125	0.29625	0.09875
16	16	16	0.1700	0.32125	0.33125	−0.16125

(continued)

Obs	Row	Period	eps	MovAvg4	Predict	Error
17	17	17	0.3900	0.33125	0.32125	0.06875
18	18	18	0.3150	0.31750	0.33125	−0.01625
19	19	19	0.3750	0.31250	0.31750	0.05750
20	20	20	0.2950	0.34375	0.31250	−0.01750
21	21	21	0.4700	0.36375	0.34375	0.12625
22	22	22	0.4250	0.39125	0.36375	0.06125
23	23	23	0.4350	0.40625	0.39125	0.04375
24	24	24	0.3500	0.42000	0.40625	−0.05625
25	25	25	−0.3750	0.20875	0.42000	−0.79500
26	26	26	0.5200	0.23250	0.20875	0.31125
27	27	27	0.5050	0.25000	0.23250	0.27250
28	28	28	0.2750	0.23125	0.25000	0.02500
29	29	29	0.6800	0.49500	0.23125	0.44875
30	30	30	0.6500	0.52750	0.49500	0.15500
31	31	31	0.6100	0.55375	0.52750	0.08250
32	32	32	0.4750	0.60375	0.55375	−0.07875
33	33	33	0.8100	0.63625	0.60375	0.20625
34	34	34	0.7850	0.67000	0.63625	0.14875
35	35	35	0.7100	0.69500	0.67000	0.04000
36	36	36	0.5550	0.71500	0.69500	−0.14000
37	37	37	0.9500	0.75000	0.71500	0.23500
38	38	38	0.8900	0.77625	0.75000	0.14000
39	39	39	0.8000	0.79875	0.77625	0.02375
40	40	40	0.6100	0.81250	0.79875	−0.18875
41	41	41	0.7300	0.75750	0.81250	−0.08250
42	42	42	1.0600	0.80000	0.75750	0.30250

18.10.2 Create a Line Chart of J&J's EPS Data

We will use **PROC GPLOT** to create a line chart of J&J's EPS data, though the same approach can be taken with any of the other SAS/GRAPH procedures. In this example, one PLOT statement plots both the EPS and Predict variables against the variable PERIOD using two plot requests. The OVERLAY option on the PLOT statement determines that both plot lines appear on the same graph. There are a few options that allow us to control and adjust the graphic output:

Plot Options

- Legend = or no legend: specifies figure legend options
- Overlay: allows overlay of more than one Y variable
- Skipmiss: breaks the plotting line where Y values are missing

Appearance Option

- Axis: specifies axis label and value options
- Symbol: specifies symbol options
- href and vref: draw vertical or horizontal reference lines on the plot
- frame/fr or noframe/nofr: specifies whether or not to frame the plot
- caxis/ca, cframe/cfr, chref/ch, cvref/cv, and ctext/c: specify colors used for axis, frame, text, or reference lines

```
Goptions ftext='Times New Roman' htext=4 gunit=pct;
title  height=6 "Johnson&Johnson's Moving Average Calculation";
legend1 label=none value=(j=left 'Actual' j=left 'Predicted')
mode=protect position=(bottom inside right) cborder=black across=1
shape=line(10);
symbol1 i=join  v=dot  c=black L=1 h=2;
symbol2 i=join  v=square c=gray L=1 line=2 w=4;
axis1  label=('eps')
       order=(-0.5 to 1 by 0.5);
axis2  label=('Time')
       order=(0 to 40 by 10);
Proc gplot data=JNJ_AVG;
plot  eps*period predict*period /vaxis=axis1 haxis=axis2 overlay legend=legend1;
Run;
```

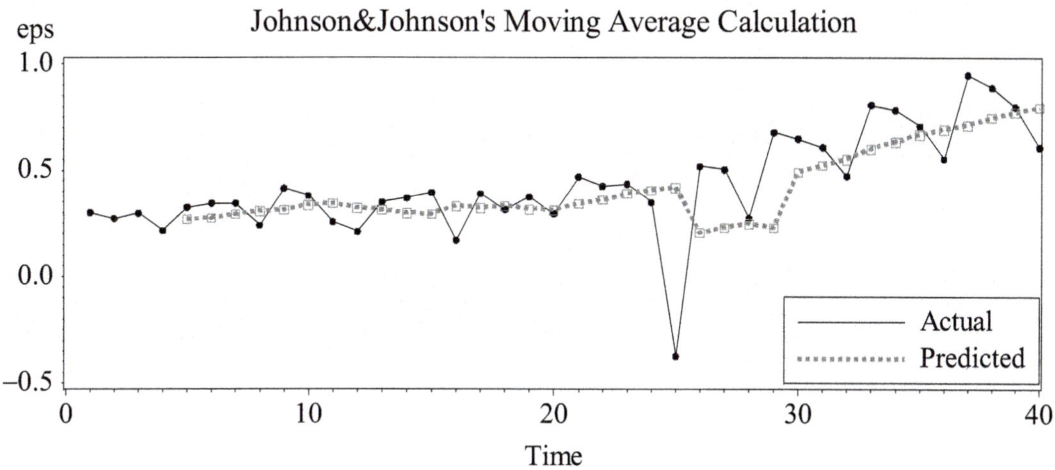

18.10.3 Linear Time Trend Regression

We will put the Ford sales data in SAS and name the column "Sales."

```
Data Ford_Sales;
t=_n_;
input Sales @@;
cards;
14075.10 14755.60 14979.90 16433.00 20194.40
23015.10 23620.60 24009.11 28839.61 37841.51
42784.11 43513.71 37085.51 38247.11 37067.21
44454.61 52366.41 52774.41 62868.30 72797.20
82193.00 82879.40 81844.00
;
Run;

Goptions ftext='Times New Roman' htext=5 gunit=pct;
title  height=7 'Ford Linear Time Trend';
symbol1 v=dot  c=black L=1 h=2 i=rl;
axis1  label=('Sales')
       order=(10000 to 90000 by 10000);
axis2  label=('T')
       order=(1 to 23 by 1);
Proc gplot data=ford_sales;
plot  Sales*t /vaxis=axis1 haxis=axis2 regeqn;
Run;
```

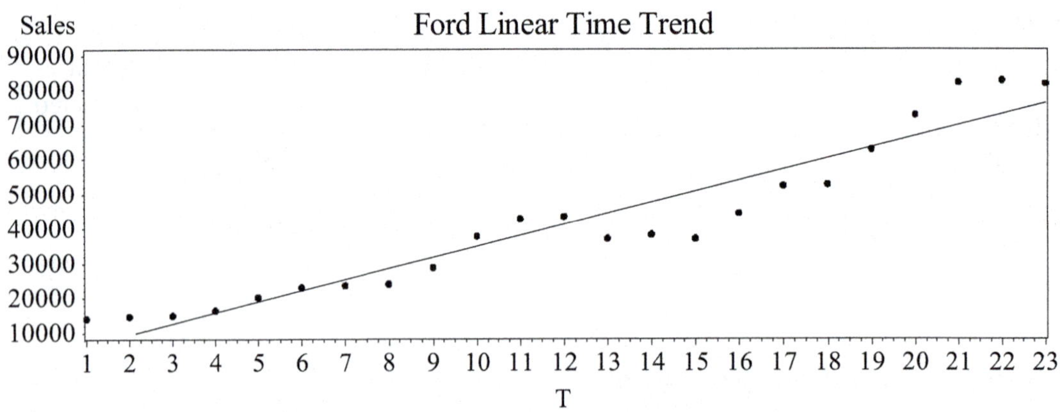

Regression Equation:
Sales = 3239.949 + 3167.102*t

We will now regress sales over time in SAS.

```
PROC REG DATA = ford_sales;
    MODEL Sales = t;
RUN;
```

Analysis of variance					
Source	DF	Sum of squares	Mean square	F value	Pr > F
Model	1	10150900144	10150900144	216.89	<.0001
Error	21	982834678	46801651		
Corrected total	22	11133734822			

Parameter estimates					
Variable	DF	Parameter estimate	Standard error	t Value	Pr > |t|
Intercept	1	3239.94854	2948.62305	1.10	0.2843
t	1	3167.10179	215.05044	14.73	<.0001

Root MSE	6841.17324	R-Square	0.9117
Dependent mean	41245	Adj. R-sq.	0.9075
Coeff. var.	16.58660		

18.10.4 Exponential Smoothing in SAS

We will put JNJ's EPS before the exponential smoothing calculations and name it "jnj_eps." jnj_smootheps will contain the EPS after the exponential smoothing calculations. We take $\alpha = 0.3$.

```
Data JNJ_smooth;
input Time jnj_eps;
cards;
1       1.0834
2       1.2550
3       1.2600
4       1.2850
5       1.3750
6       1.6800
7       0.9250
8       2.4150
9       2.8600
10      3.2500
;
Run;
```

According to the exponential smoothing calculation formula, we get jnj_smootheps.

```
Data JNJ_smooth;
set JNJ_smooth;
jnj_smootheps=0.3*jnj_eps+(1-0.3)*(lag(jnj_eps));
if jnj_smootheps=. then jnj_smootheps=1.08340;
Predict=lag(jnj_smootheps);
if Predict=. then Predict=1.08340;
Error=jnj_eps-Predict;
run;
```

Obs	Time	jnj_eps	jnj_smootheps	Predict	Error
1	1	1.0834	1.08340	1.08340	0.00000
2	2	1.2550	1.13488	1.08340	0.17160
3	3	1.2600	1.25650	1.13488	0.12512
4	4	1.2850	1.26750	1.25650	0.02850
5	5	1.3750	1.31200	1.26750	0.10750
6	6	1.6800	1.46650	1.31200	0.36800
7	7	0.9250	1.45350	1.46650	−0.54150
8	8	2.4150	1.37200	1.45350	0.96150
9	9	2.8600	2.54850	1.37200	1.48800
10	10	3.2500	2.97700	2.54850	0.70150

```
Goptions ftext='Times New Roman' htext=4 gunit=pct;
title  height=6 'J&J EPS Eponential Smoothing(alpha=0.3)';
legend1 label=none value=(j=left 'Actual' j=left 'Predicted')
mode=protect position=(bottom inside right) cborder=black across=1
shape=line(10);
symbol1 i=join  v=dot  c=black L=1 h=2;
symbol2 i=join  v=square c=gray L=1 line=2 w=4;
axis1  label=(a=90 'JNJ_EPS')
       order=(0.5 to 3.5 by 0.5);
axis2  label=('Time')
       order=(0 to 10 by 1);
Proc gplot data=JNJ_smooth;
plot  jnj_eps*Time Predict*Time /vaxis=axis1 haxis=axis2 overlay legend=legend1;
Run;
```

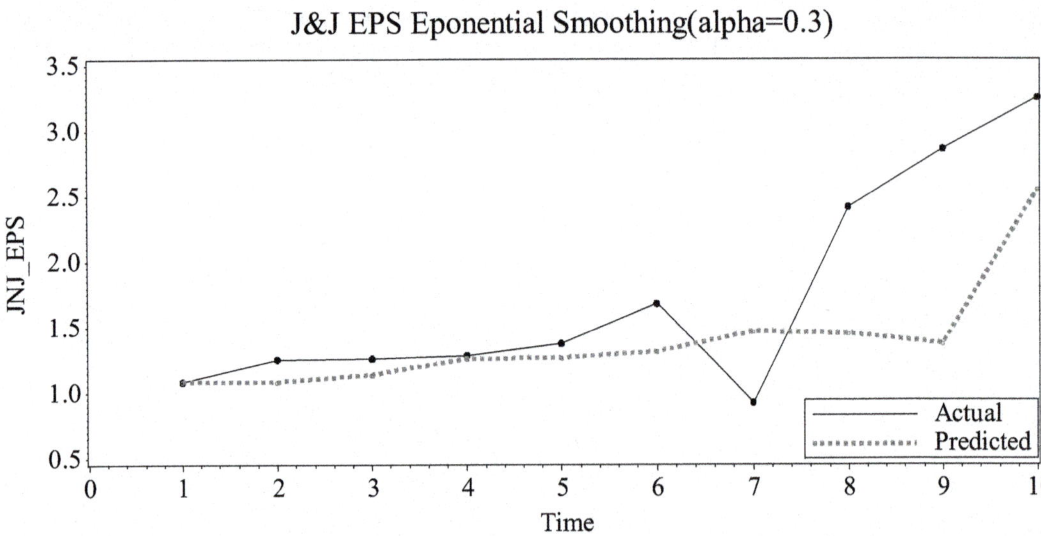

We will now illustrate the IBM data. The same procedure is used as we introduced before.

```
Goptions ftext='Times New Roman' htext=6 gunit=pct;
title  height=7 "IBM EPS Exponential Smoothing";
legend1 label=none value=(j=left 'Actual' j=left 'Smoothed')
mode=protect position=(bottom inside right) cborder=black across=1 shape=line(10);
symbol1 i=join  v=dot  c=black L=1 h=2;
symbol2 i=join  v=square c=gray L=1 line=2 w=4;
axis1  label=('ibmeps')
       order=(5 to 11 by 1);
axis2  label=('Time')
       order=(0 to 10 by 5);
Proc gplot data=IBM_smooth;
plot  ibmeps*Time ibmexpsmth*Time /vaxis=axis1 haxis=axis2 overlay legend=legend1;
Run;
```

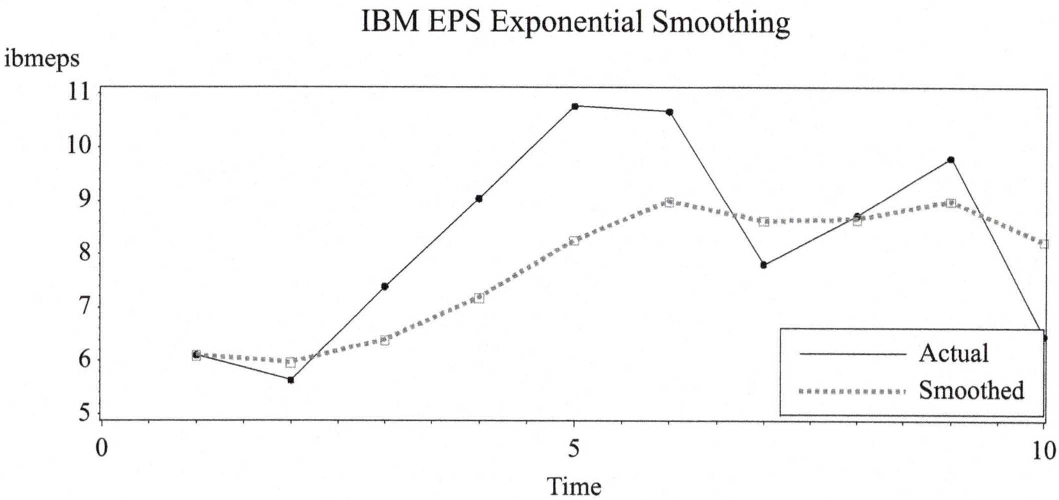

18.10.5 Exponential Smoothing in SAS

The formula that Excel uses for exponential smoothing is

$$F_{t+1} = F_t + a(A_t - F_t) = F_t + (1 - \text{damp Fact})(A_t - F_t)$$

We take damping factor equal to 0.3. Then, the equation above can be expressed as

$$F_{t+1} = 0.3{*}F_t + 0.7{*}A_t$$

Below are the calculated exponential smoothing numbers.

```
data Exponential_smooth;
set JNJ_smooth;
retain Forecast 1.0834;
Forecast=0.3*Forecast + 0.7*jnj_eps;
run;
data Exponential_smooth;
set Exponential_smooth;
Forecast=lag(Forecast);
run;
```

Obs	Time	jnj_eps	Forecast
1	1	1.0834	
2	2	1.2550	1.08340
3	3	1.2600	1.20352
4	4	1.2850	1.24306
5	5	1.3750	1.27242
6	6	1.6800	1.34423
7	7	0.9250	1.57927
8	8	2.4150	1.12128
9	9	2.8600	2.02688
10	10	3.2500	2.61007

```
Goptions ftext='Times New Roman' htext=6 gunit=pct;
title  height=7 "Exponential Smoothing";
legend1 label=none value=(j=left 'Actual' j=left 'Forecast')
mode=protect position=(bottom inside right) cborder=black across=1
shape=line(10);
symbol1 i=join  v=dot  c=black L=1 h=2;
symbol2 i=join  v=square c=gray L=1 line=2 w=4;
axis1  label=('Value')
       order=(0 to 3.5 by 0.5);
axis2  label=('Data Point')
       order=(1 to 10 by 1);
Proc gplot data=Exponential_smooth;
plot  jnj_eps*Time Forecast*Time /vaxis=axis1 haxis=axis2 overlay
legend=legend1;
Run;
```

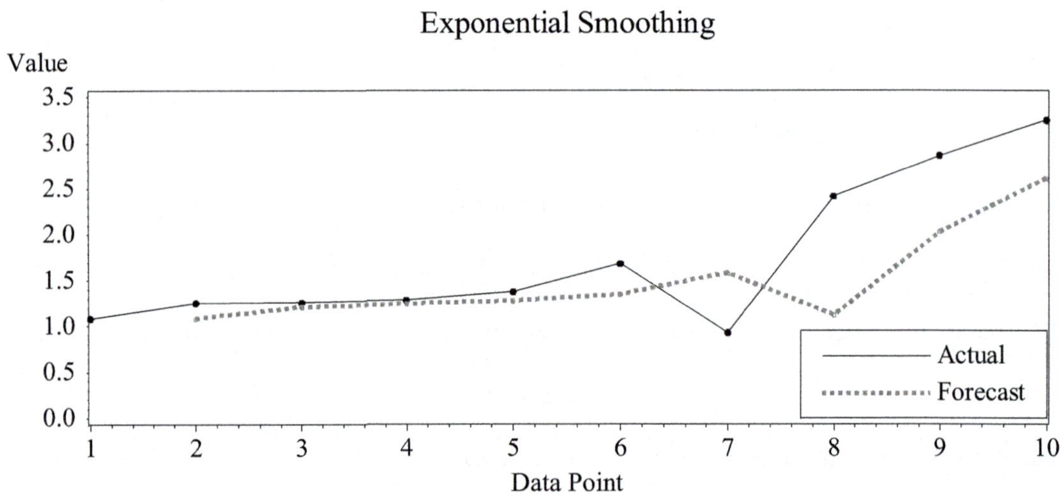

18.10.6 Holt–Winters Forecasting Model for Nonseasonal Series

As it is introduced in the previous section, we first do the regression for the first five J&J EPS.

```
Data JNJ;
input Time jnj_eps;
cards;
1980      1.0834
1981      1.2550
1982      1.2600
1983      1.2850
1984      1.3750
1985      1.6800
1986      0.9250
1987      2.4150
1988      2.8600
1989    3.2500
;
Run;
```

A macro is designed to calculate the values of smooth and trend according to the formula introduced in the previous section. There are four parameters in this macro. a is an exponential smoothing constant between 0 and 1. b is a trend smoothing constant between 0 and 1. Filename is the data to deal with. T1 is the initial value of the trend component. There are two options for T1: "zero" and "reg." The initial value of trend component will be zero if choosing option "zero." We use the estimated slope b as the first trend observation with choosing option "reg."

```
%macro Forecast(a,b,filename,T1);

%if "&T1."="zero" %then %do;

Data &filename._Reg;
set &filename. (obs=5);
run;

proc reg data=&filename._Reg;
model &filename._eps=Time;
ods output ParameterEstimates=PE;
run;

data PE1; set PE;
where Variable='Time';
run;

proc sql;
create table &filename.1 as
select  &filename..*, PE1.Estimate as Initial
from PE1, &filename.;
quit;

data &filename.2;
set &filename.1;
num=_N_;
if num=1 then do;
        ss=&filename._eps;
        tt=Initial;
end;
%do i=2 %to 10;
if num<=&i then do;
              oldss=lag(ss);
              oldtt=lag(tt);
              if num=&i then do;
                     ss=&a*&filename._eps+(1-&a)*(oldss+oldtt);
                     tt=&b*(ss-oldss)+(1-&b)*oldtt;
              end;
end;
%end;
drop oldss oldtt;
run;

%end;
```

```
%if "&T1."="reg" %then %do;

data &filename.2;
set &filename.;
num=_N_;
if num=1 then do;
        ss=&filename._eps;
        tt=0;
end;
%do i=2 %to 10;
if num<=&i then do;
                oldss=lag(ss);
                oldtt=lag(tt);
                if num=&i then do;
                        ss=&a*&filename._eps+(1-&a)*(oldss+oldtt);
                        tt=&b*(ss-oldss)+(1-&b)*oldtt;
                end;
end;

%end;
drop oldss oldtt;
run;

%end;

proc print data=&filename.2;
run;

Goptions ftext='Times New Roman' htext=6 gunit=pct;
title  height=7 "Holt-Winters Forceasting Model &Filename.EPS";
legend1 label=none value=(h=4 "&filename.EPS" "&filename.EPSExpSmooth"
"&filename.EPSTrend")
mode=protect position=(top outside right) cborder=black across=1 shape=line(4));
symbol1 i=join  v=dot  c=black L=1 w=1;
symbol2 i=join  v=dot c=red L=1 w=1;
symbol3 i=join  v=dot c=blue L=1 w=1;
axis1  label=('Value');
axis2  label=('Data Point');
Proc gplot data=&filename.2;
plot  (&filename._eps ss tt)*Time /vaxis=axis1 haxis=axis2 overlay
legend=legend1;
Run;
%mend test;
```

Let α and β equal to 0.3 and 0.2, respectively. The first observation for the trend data is the regression of the first five observed EPS. We call the macro and get the table of smooth and trend.

```
%Forecast (0.3, 0.2, jnj, reg);
```

The results are shown in the following:

Analysis of variance					
Source	DF	Sum of squares	Mean square	F value	Pr > F
Model	1	0.03760	0.03760	15.85	0.0283
Error	3	0.00712	0.00237		
Corrected total	4	0.04472			

Root MSE	0.04870	R-Square	0.8409
Dependent mean	1.25168	Adj. R-sq.	0.7878
Coeff. var.	3.89076		

Parameter estimates					
Variable	DF	Parameter estimate	Standard error	t Value	Pr > |t|
Intercept	1	−120.28456	30.52329	−3.94	0.0291
Time	1	0.06132	0.01540	3.98	0.0283

Obs	Time	jnj_eps	num	ss	tt
1	1980	1.0834	1	1.08340	0.06132
2	1981	1.2550	2	1.17780	0.06794
3	1982	1.2600	3	1.25002	0.06879
4	1983	1.2850	4	1.30867	0.06676
5	1984	1.3750	5	1.37530	0.06674
6	1985	1.6800	6	1.51343	0.08102
7	1986	0.9250	7	1.39361	0.04085
8	1987	2.4150	8	1.72862	0.09968
9	1988	2.8600	9	2.13781	0.16158
10	1989	3.2500	10	2.58458	0.21862

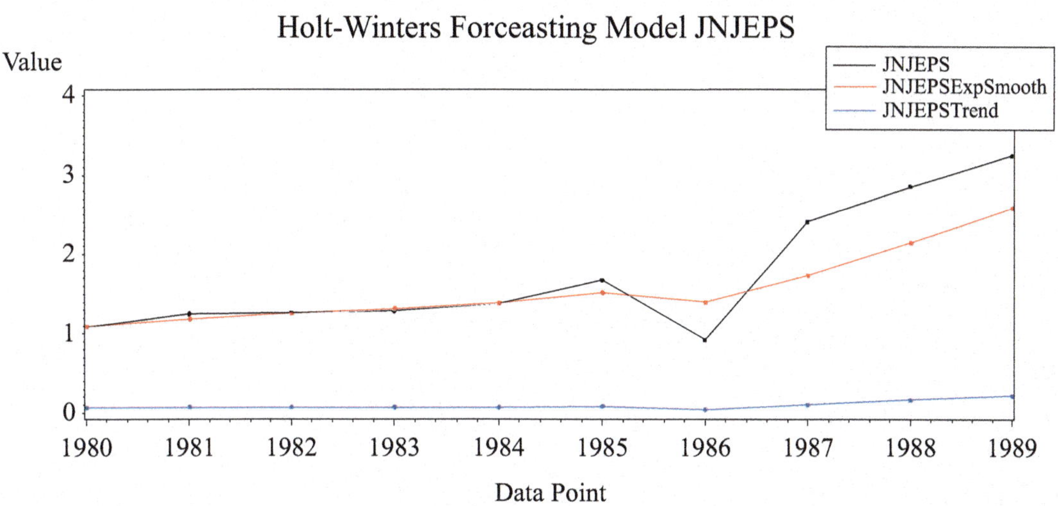

If the first observation for the trend data is zero, we just give T1 a value of b and call the macro again.

```
%Forecast(0.3, 0.2, JNJ, zero);
```

The result is shown below.

Obs	Time	jnj_eps	num	ss	tt
1	1980	1.0834	1	1.08340	0.00000
2	1981	1.2550	2	1.13488	0.01030
3	1982	1.2600	3	1.17962	0.01719
4	1983	1.2850	4	1.22327	0.02248
5	1984	1.3750	5	1.28452	0.03023
6	1985	1.6800	6	1.42433	0.05215
7	1986	0.9250	7	1.31103	0.01906
8	1987	2.4150	8	1.65556	0.08415
9	1988	2.8600	9	2.07580	0.15137
10	1989	3.2500	10	2.53402	0.21274

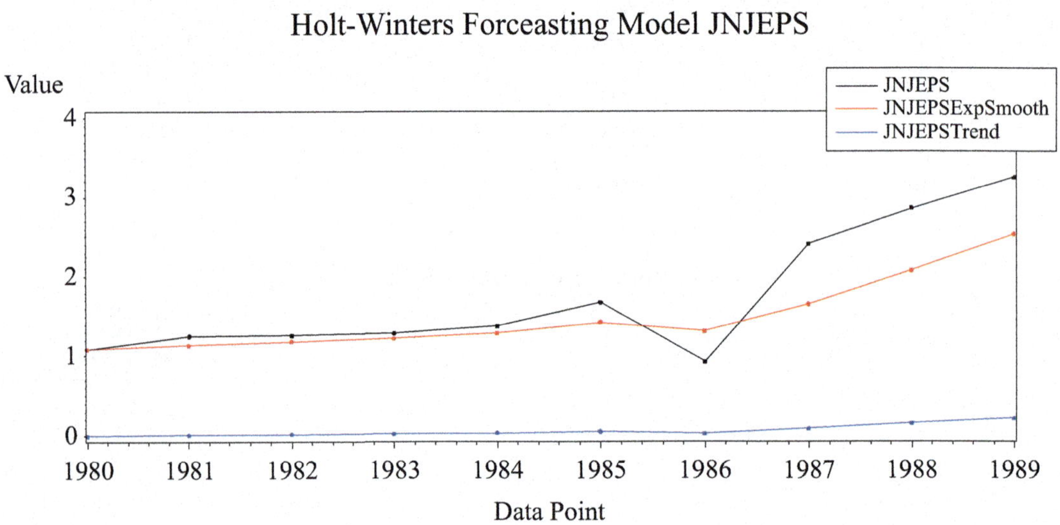

We will now perform the Holt–Winters calculation on IBM's EPS. The same macro will be used.

```
Data IBM;
input Time IBM_eps;
cards;
1980 6.1
1981 5.63
1982 7.39
1983 9.04
1984 10.77
1985 10.67
1986 7.81
1987 8.72
1988 9.8
1989 6.47
;
Run;
```

We call macro Forecast first.

```
%Forecast(0.3, 0.2, IBM, reg);
```

The result is shown below.

Analysis of variance					
Source	DF	Sum of squares	Mean square	F value	Pr > F
Model	1	16.25625	16.25625	26.10	0.0145
Error	3	1.86827	0.62276		
Corrected Total	4	18.12452			

Root MSE	0.78915	R-Square	0.8969
Dependent mean	7.78600	Adj. R-Sq.	0.8626
Coeff. var.	10.13549		

Parameter estimates					
Variable	DF	Parameter estimate	Standard error	t Value	Pr > |t|
Intercept	1	−2519.26400	494.61007	−5.09	0.0146
Time	1	1.27500	0.24955	5.11	0.0145

Obs	Time	IBM_eps	num	ss	tt
1	1980	6.10	1	6.1000	1.27500
2	1981	5.63	2	6.8515	1.17030
3	1982	7.39	3	7.8323	1.13239
4	1983	9.04	4	8.9873	1.13691
5	1984	10.77	5	10.3179	1.17566
6	1985	10.67	6	11.2465	1.12625
7	1986	7.81	7	11.0039	0.85248
8	1987	8.72	8	10.9155	0.66430
9	1988	9.80	9	11.0458	0.55751
10	1989	6.47	10	10.0634	0.24951

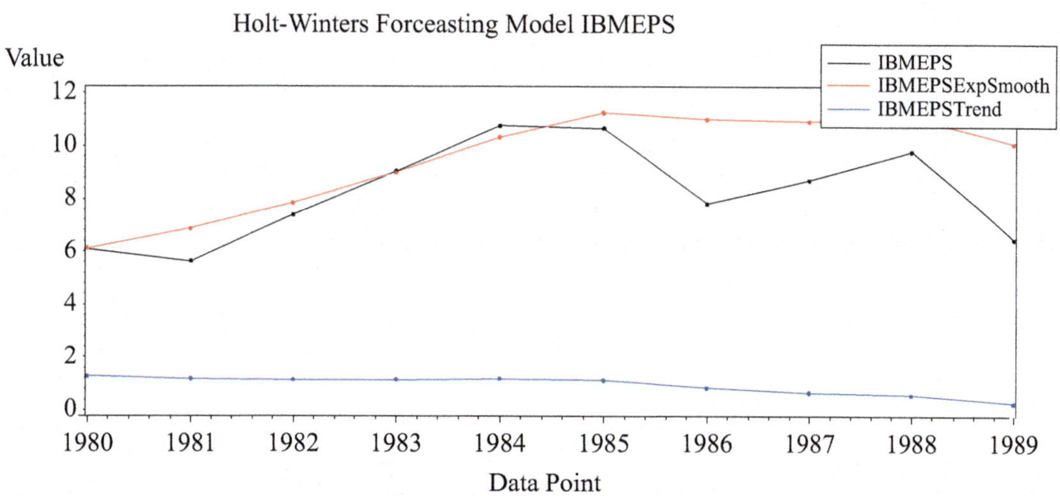

If the first observation for the trend data is zero, we get the smooth and trend of IBM below.

```
%Forecast(0.3, 0.2, IBM, zero);
```

Obs	Time	IBM_eps	num	ss	tt
1	1980	6.10	1	6.10000	0.00000
2	1981	5.63	2	5.95900	−0.02820
3	1982	7.39	3	6.36856	0.05935
4	1983	9.04	4	7.21154	0.21608
5	1984	10.77	5	8.43033	0.41662
6	1985	10.67	6	9.39387	0.52600
7	1986	7.81	7	9.28691	0.39941
8	1987	8.72	8	9.39642	0.34143
9	1988	9.80	9	9.75650	0.34516
10	1989	6.47	10	9.01216	0.12726

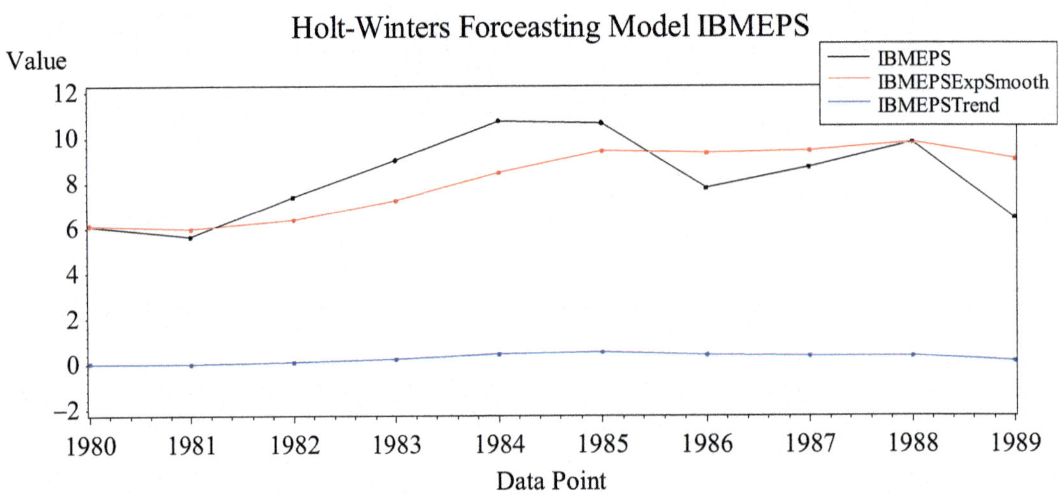

18.11 Statistical Summary

In this chapter, we first discuss how the moving average method can be used to remove both seasonal and irregular components of time-series data. Then, we discussed how linear time trend regression and exponential smoothing can be used to analyze the time-series data.

Appendix 18.1: Excel Code (Holtz–Winters)

The Holtz–Winters Excel macros below are in the form *frmHolzWinters* in the workbook *Holt–Winters.xlsm* workbook.

```
'/**************************************************************************
'/Essentials of Microsoft Excel 2013 VBA, SAS, and
'/MINITAB 17
'/for Statistical and Financial Analysis
'/**************************************************************************

Option Explicit
Dim mwsData As Worksheet
Dim mrData As Range
Dim mrDataLabel As Range
Dim mwbReport As Workbook
Dim mwsReportSmoothData As Worksheet
Dim mrExponentialSmooth As Range
Dim mrTrendSmooth As Range
Dim mChrtWintersZero As Chart
Property Set ChartWintersZero(c As Chart)
    Set mChrtWintersZero = c
End Property

Property Get ChartWintersZero() As Chart
    Set ChartWintersZero = mChrtWintersZero
End Property


Property Set ReportSmoothData(ws As Worksheet)
    Set mwsReportSmoothData = ws
End Property

Property Get ReportSmoothData() As Worksheet
    Set ReportSmoothData = mwsReportSmoothData
End Property
Property Set ReportWorkbook(wb As Workbook)
    Set mwbReport = wb
End Property

Property Get ReportWorkbook() As Workbook
    Set ReportWorkbook = mwbReport
End Property

Property Set DataWorksheet(ws As Worksheet)
    Set mwsData = ws
End Property

Property Get DataWorksheet() As Worksheet
    Set DataWorksheet = mwsData
End Property
Property Set DataRange(r As Range)
    'Set it so the data range does not include the labels
    Set r = r.Resize(r.Rows.Count - 1, r.Columns.Count).Offset(1, 0)
    Set mrData = r
End Property
Property Get DataRange() As Range
    Set DataRange = mrData
End Property
'/****************************************************************************
'/Purpose:  Somtimes you want data range with Label and sometimes you do not
'/          Use this propoerty of the form if you want the data range with label
'/****************************************************************************
Property Set DataRangeLabel(r As Range)
    Set mrDataLabel = r
End Property
```

```
Property Get DataRangeLabel() As Range
    Set DataRangeLabel = mrDataLabel
End Property
Property Set ExponentialSmooth(r As Range)
    Set mrExponentialSmooth = r
End Property

Property Get ExponentialSmooth() As Range
    Set ExponentialSmooth = mrExponentialSmooth
End Property
Property Set TrendSmooth(r As Range)
    Set mrTrendSmooth = r
End Property

Property Get TrendSmooth() As Range
    Set TrendSmooth = mrTrendSmooth
End Property

'/*********************************************************************
'/call by:  the command button cmdCalculate
'/*********************************************************************
Private Sub cmdCalculate_Click()
    Dim wb As Workbook

    Set wb = Workbooks.Add

    With Me
        Set .ReportWorkbook = wb
        Set .ReportSmoothData = .ReportWorkbook.Worksheets.Add
        .ReportSmoothData.Cells(1, 1). _
            Resize(.DataRangeLabel.Rows.Count, .DataRangeLabel.Columns.Count).Value = _
            .DataRangeLabel.Value
        Set .ExponentialSmooth = .ReportSmoothData.Cells(1,
1).CurrentRegion.Columns(1).Offset(0, 2)
        Set .TrendSmooth = .ExponentialSmooth.Offset(0, 1)
        .ExponentialSmooth.Cells(1).Value = .ExponentialSmooth.Cells(1).Offset(0, -
1).Value & "ExpSmooth"
        .TrendSmooth.Cells(1).Value = .TrendSmooth.Cells(1).Offset(0, -2).Value & "Trend"
    End With

    HoltzWintersZero
    HoltzWingersReg

    CleanUp
    Unload UserForm1
End Sub
'/****************************************************************
'/Purpose:  Caculates trend for T1 as the regression of the first 5
'/          observations
'/****************************************************************
Sub HoltzWingersReg()
    Dim rX As Range
    Dim rY As Range
    Dim ws As Worksheet
    Dim rRegress As Range
    Dim rSmooth As Range
    Dim rTrend As Range
    Dim dblBeta As Double     'Regression Beta number
    Dim rBeta As Range        'Location of Beta variable on regression report

    Const OBSERVATION = 5     ' This is the number of observation to regress
                              ' Change here for a different num of ob to regress
```

```
    Set ws = Me.ReportWorkbook.Worksheets.Add
    Set rX = Me.ReportSmoothData.Cells(1, 1).CurrentRegion
    Set rX = rX.Columns(1).Resize(6, 1)
    Set rY = rX.Offset(0, 1)

    Application.Run "ATPVBAEN.XLAM!Regress", rY, rX, False, True, , ws.Cells(1, 1), _
        False, False, False, False, , False

    Set rRegress = Selection
    With rRegress
         .EntireColumn.AutoFit
        Set rBeta = .Columns(1).Find(Me.ReportSmoothData.Cells(1))
    End With

    Dim wsHoltzReg As Worksheet

    With Me

        Set wsHoltzReg = .ReportWorkbook.Worksheets.Add
        .ReportSmoothData.Cells(1).CurrentRegion.Columns(1).Resize(, 2).Copy
wsHoltzReg.Cells(1)
    End With

    Set rSmooth = wsHoltzReg.Cells(1).CurrentRegion.Columns(2).Offset(0, 1)
    Set rSmooth = rSmooth.Resize(rSmooth.Rows.Count - 1, 1).Offset(1, 0)
    Set rTrend = rSmooth.Offset(0, 1)


    rSmooth.Cells(1).Offset(-1, 0) = Me.ReportSmoothData.Cells(1, 3)
    rTrend.Cells(1).Offset(-1, 0) = Me.ReportSmoothData.Cells(1, 4)

    rSmooth.Formula = "=" & Me.txtConstantExp & "*RC[-1]+(1-" & Me.txtConstantExp &
")*(+R[-1]C+R[-1]C[1])"
    rTrend.Formula = "=" & Me.txtConstantTrend & "*(RC[-1]-R[-1]C[-1])+(1-" &
Me.txtConstantTrend & ")*R[-1]C"

    rSmooth.Cells(1) = rSmooth.Cells(1).Offset(0, -1)
    rTrend.Cells(1) = rBeta.Offset(0, 1)

    rTrend.CurrentRegion.EntireColumn.AutoFit
    ws.Name = "HoltzTrendRegression"
    wsHoltzReg.Name = "HoltzReg"

    Me.ChartHoltzWinters rData:=wsHoltzReg.Cells(1).CurrentRegion, sName:="ChrtHoltReg"

End Sub

'/*******************************************************************************
'/Purpose: To chart the Holtz-Winters Data
'/Call by: cmdCalculate_Click
'/********************************************************************************
Sub ChartHoltzWinters(rData As Range, sName As String)
    Dim seriesCurrent As Series
    Dim chrt As Chart

    CreatChart chrtBox:=chrt, _
               seriesCurrent:=seriesCurrent, _
               wbWorkbook:=Me.ReportWorkbook, _
               sChartSubType:=xlXYScatterLines

    Set Me.ChartWintersZero = chrt
```

```
    With Me.ChartWintersZero
        .SetSourceData Source:=rData, PlotBy:=xlColumns
        .HasLegend = True
        .Legend.Position = xlLegendPositionBottom
        With .ChartTitle.Characters
            .Text = "Holtz-Winters Forcasting Model" & _
                     Chr(10) & Me.DataRangeLabel.Cells(1, 2)
            .Font.Size = 18
        End With
        'Third series defualt to yellow, change to b lack
        With .SeriesCollection(3)
                .Border.ColorIndex = 1
                .MarkerBackgroundColorIndex = 1
                .MarkerForegroundColorIndex = 1
        End With
        .Name = sName
    End With
End Sub

'/*****************************************************************************
'/Purpose: To clean up the report workbook
'/Call by: cmdCalculate_Click
'/*****************************************************************************
Sub CleanUp()
    Dim ws As Worksheet

    Application.DisplayAlerts = False
    For Each ws In Me.ReportWorkbook.Worksheets
        If Left(ws.Name, 5) = "Sheet" Then
            ws.Delete
        End If
    Next

    Me.ReportSmoothData.Cells(1).CurrentRegion.EntireColumn.AutoFit

    Application.DisplayAlerts = True
End Sub

'/*****************************************************************************
'/Call by: cmdCalculate_Click
'/*****************************************************************************
Sub HoltzWintersZero()
    Dim rSmooth As Range
    Dim rTrend As Range

    With Me
        .ExponentialSmooth.Cells(2) = .ExponentialSmooth.Cells(2).Offset(0, -1)
        .TrendSmooth.Cells(2) = 0
        .ReportSmoothData.Name = "HoltzZero"
        Set rSmooth = .ExponentialSmooth
        Set rTrend = .TrendSmooth
    End With


    Set rSmooth = rSmooth.Resize(rSmooth.Rows.Count - 2, 1).Offset(2, 0)

    Set rTrend = rTrend.Resize(rTrend.Rows.Count - 2, 1).Offset(2, 0)

    rSmooth.Formula = "=" & Me.txtConstantExp & "*RC[-1]+(1-" & Me.txtConstantExp &
")*(+R[-1]C+R[-1]C[1])"
    rTrend.Formula = "=" & Me.txtConstantTrend & "*(RC[-1]-R[-1]C[-1])+(1-" &
Me.txtConstantTrend & ")*R[-1]C"
```

```
    ChartHoltzWinters rData:=Me.ReportSmoothData.Cells(1).CurrentRegion,
sName:="ChrtHoltZero"

End Sub


'/**************************************************************************
'/Purpose: To close the form without performing any calculations
'/**************************************************************************
Private Sub cmdCancel_Click()
    Unload Me
End Sub

'/**************************************************************************
'/Purpose:  This procedure is autimatically runned when the form is loaded.
'/          Mainly use this procedure to set form display
'/**************************************************************************
Private Sub UserForm_Initialize()
    With Me
        .txtConstantExp = 0.3
        .txtConstantTrend = 0.2
        Set .DataWorksheet = wsData
        Set .DataRange = wsData.Cells(1, 1).CurrentRegion
        Set .DataRangeLabel = wsData.Cells(1, 1).CurrentRegion
    End With
End Sub
'/******************************************************************************
'/Purpose:  To create and format the chart before indicating the data series.
'/******************************************************************************
Sub CreatChart(chrtBox As Chart, _
               seriesCurrent As Series, _
               wbWorkbook As Workbook, _
               sChartSubType As String)
    Set chrtBox = wbWorkbook.Charts.Add

    If chrtBox.SeriesCollection.Count = 0 Then
        Set seriesCurrent = chrtBox.SeriesCollection.NewSeries
    Else
        Set seriesCurrent = chrtBox.SeriesCollection(1)
    End If

    With chrtBox
            'xlXYScatterSmooth
            .ChartType = sChartSubType
            .HasTitle = True

            With .Axes(xlCategory)
                  .HasMajorGridlines = False
                  .HasMinorGridlines = False
            End With
            With .Axes(xlValue)
                  .HasMajorGridlines = False
                  .HasMinorGridlines = False

            End With

            With .PlotArea
                .Interior.ColorIndex = xlNone
                .ClearFormats
            End With
            .HasLegend = False
    End With

End Sub
```

Bibliography

Lee CF, Lee JC, Lee AC (2013) Statistics for business and financial economics. Springer, New York
Microsoft Inc., Excel 2013. Microsoft Inc., Redmond
Minitab Inc. Minitab 17. Minitab Inc., State College
SAS Institute Inc. SAS 2014. SAS Institute Inc., Cary

Chapter 19
Index Numbers and Stock Market Indexes

19.1 Introduction

In this chapter we will look at indexes. Index numbers are numbers that compare an activity in one time or place to a similar activity in a specific base period or place. The first index that we will look at is the simple price index of a single item. When used this way, we can measure the relative price change over time.

19.2 Simple Price Index

The formula for the simple price index is

$$I_t = \frac{P_t}{P_0} * 100$$

where P_t is the price in period t, P_0 is the base period price, and I_t is the simple price index at time t.

Example 19.1 Suppose we would like to create a simple price index for eggs. We know the price of eggs for the following years.

PRICE OF EGGS

YEAR	PRICE
1991	\$1.00
1992	\$1.20
1993	\$1.50

In creating an index, we have to decide on a base year. Let us choose 1991 as the base year in this case. To use Minitab to create the index, we should put the years in column c1 and the prices in column c2. When finished the worksheet should look like the following.

C.-F. Lee et al., *Essentials of Excel, Excel VBA, SAS and Minitab for Statistical and Financial Analyses*, DOI 10.1007/978-3-319-38867-0_19

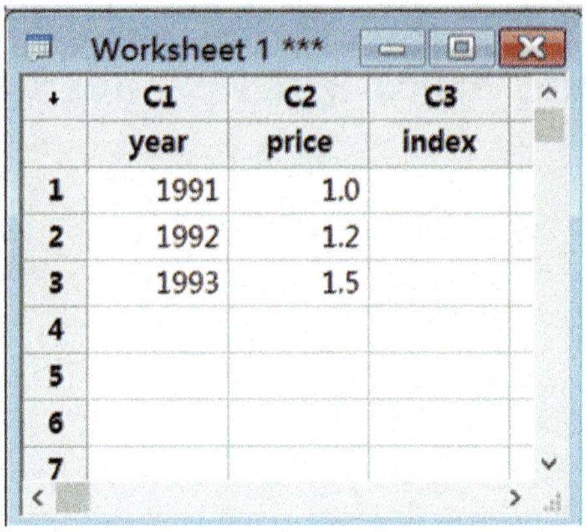

The index numbers will go in column c3. The index numbers can be easily created by using the *let* command as illustrated below.

```
MTB > let c3=c2/c2(1)  * 100
MTB > print c1-c3
```

Data Display

```
Row    year  price  index

  1    1991    1.0    100
  2    1992    1.2    120
  3    1993    1.5    150
```

In the above expression, c2 is the numerator of the simple price index and c2(1) is the denominator or the simple price index.

The index numbers tell us that the price of eggs in 1992 was 20 % greater than it was in 1991 or 120 % of the price of eggs in 1991. The index numbers also tell us that the price of eggs in 1993 was 50 % greater than it was in 1991 or 150 % of the price of eggs in 1991.

We will create custom functions in Excel to calculate the indexes in this chapter. The function to calculate the simple price index is shown below.

```
'/*****************************************************************************
'/Purpose: Calculates the price index given the current year and base year
'/Parameter: CurrentYear: Range that contains current year.
'/           BaseYear: Range that contains the base year.
'/*****************************************************************************

Function PriceIndex(CurrentYear As Range, BaseYear As Range) As Double
            PriceIndex = (WorksheetFunction.Sum(CurrentYear) / _
                          WorksheetFunction.Sum(BaseYear)) * 100
End Function
```

In Excel input the price of eggs data and enter the *PriceIndex* function as shown below.

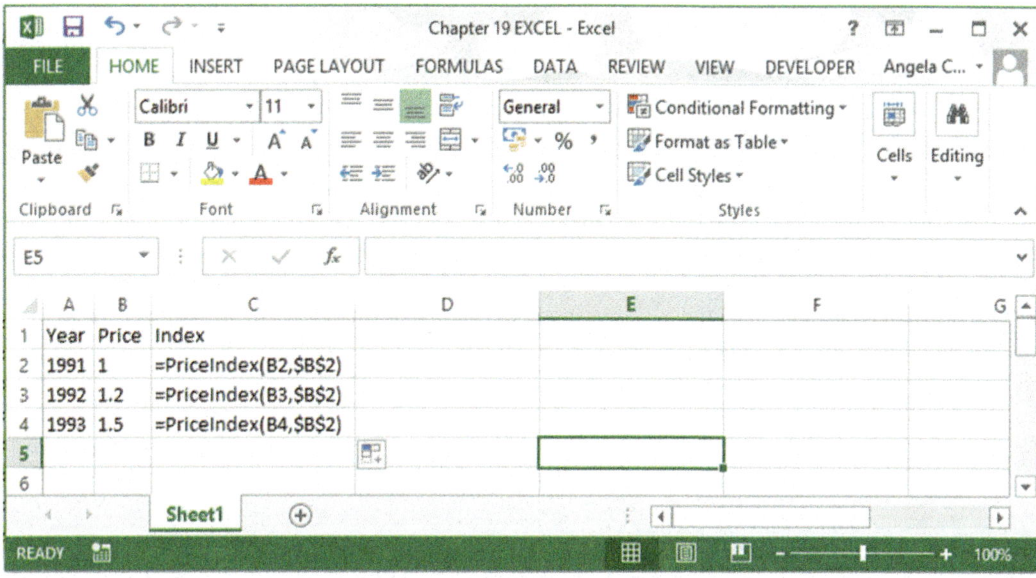

After entering the *PriceIndex* function, your spreadsheet should look like the following.

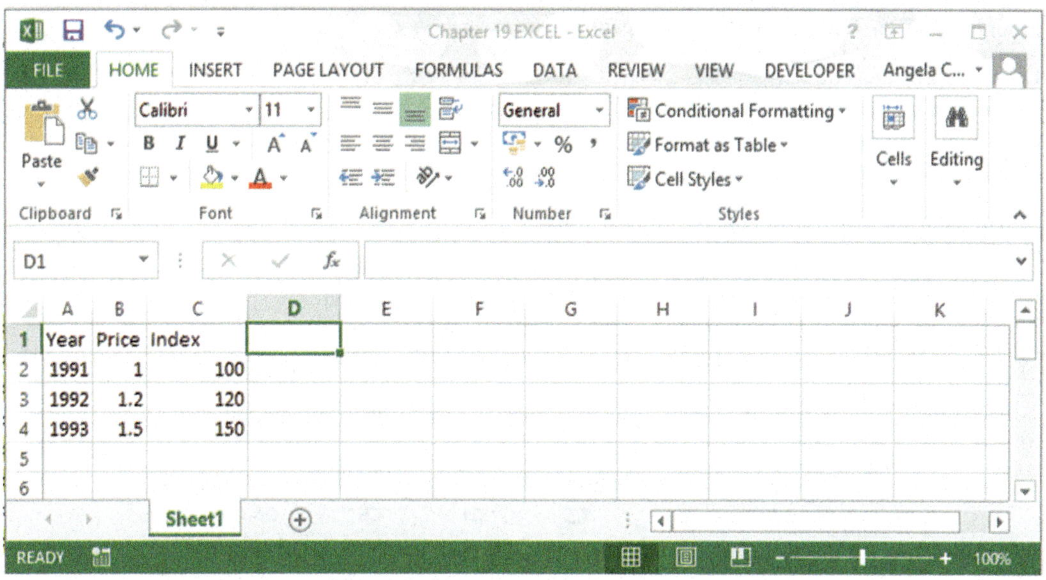

In studying index numbers, it is important to remember what the base period is. Otherwise, we may end up with misinformation.

Example 19.2 Instead of choosing 1991 as the base year for indexing the eggs, let us choose 1992 as the base year.

```
MTB > let c3=c2/c2(2) * 100
MTB > print c1-c3
```

Data Display

```
Row     year  price      index

  1     1991    1.0     83.333
  2     1992    1.2    100.000
  3     1993    1.5    125.000
```

We can see that after choosing a different base year, the index number for every year is different. The index for the year 1991 was 100 when the base year was 1991, but now it is 83.333. This example also shows that an index number can be less than 100, indicating that the price (or other variables) is less than in the base year.

Instead of having an index for a single item, we can have an index for a group of items. The mathematical formula for a group of items is shown below:

$$I_t = \frac{\sum_{i=1}^{n} P_{ti}}{\sum_{i=1}^{n} P_{0i}} * 100$$

where P_{ti} is the price at period t for item i, P_{0i} is the base period price for item i, and I_t is the simple price index at time t.

The consumer price index (CPI) is an example of a simple index for a group of items. The CPI has around 2000 items but we will use only four:

One dozen of eggs
One gallon of milk
One pound of butter
One loaf of bread

The data for the four items are the following.

Year	Eggs	Milk	Butter	Bread
1991	1.00	1.50	1.10	0.40
1992	1.20	1.75	1.35	0.70
1993	1.50	2.00	1.60	0.90

The Minitab worksheet should look like the following after the data are entered.

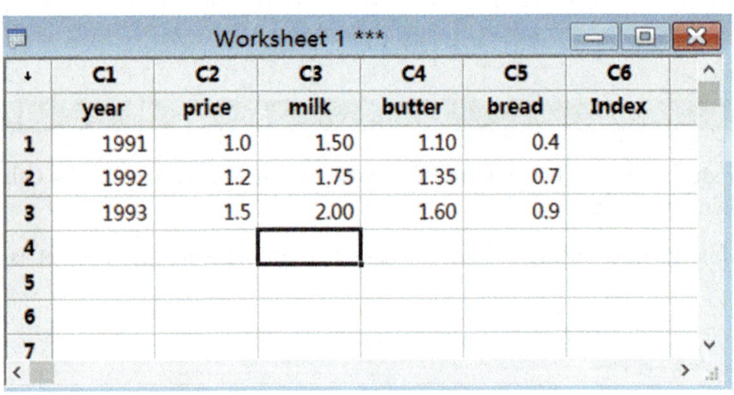
Worksheet 1 ***

↓	C1	C2	C3	C4	C5	C6
	year	price	milk	butter	bread	Index
1	1991	1.0	1.50	1.10	0.4	
2	1992	1.2	1.75	1.35	0.7	
3	1993	1.5	2.00	1.60	0.9	
4						
5						
6						
7						

We will again use 1991 as the base year and create the index numbers with only the *let* command. This is illustrated below.

```
MTB > let c6=((c2+c3+c4+c5)/(c2(1)+c3(1)+C4(1)+c5(1)))*100
MTB > print c1-c6
```

Data Display

```
Row     year    eggs    milk  butter  bread  Index

  1     1991     1.0    1.50    1.10    0.4    100
  2     1992     1.2    1.75    1.35    0.7    125
  3     1993     1.5    2.00    1.60    0.9    150
```

In the *let* expression, `(c2+c3+c4+c5)` represents the numerator of the simple index of a group of items. The `(c2(1)+c3(1)+C4(1)+c5(1))` represents the denominator of the simple index of a group of items.

The Minitab output tells us that the 1992 group of goods is 25 % more expensive than the same group of goods in 1991 or 125 % of the price of the same group in 1991. The Minitab output also tells us that the group of goods in 1993 is 50 % more expensive than the same group in 1991 or 150 % of the price of the goods in 1991.

19.3 Laspeyres Price Index

Previously we have only looked at price in our index. The Laspeyres price index takes into consideration both price and quantity. The base period quantity is in both the numerator and the denominator of the Laspeyres index. The formula for the Laspeyres index is

$$I_t = \frac{\sum_{i=1}^{n} P_{ti}Q_{0i}}{\sum_{i=1}^{n} P_{0i}Q_{0i}} * 100$$

where P_{0i} is the base price of item i, Q_{0i} is the base quantity of item i, P_{ti} is the price at time t of item i, and I_t is the Laspeyres price index.

We will use the following items to calculate the Laspeyres price index.

	Eggs		Milk		Butter		Bread		Shirts	
Year	Price	Quantity	Price	Quantity	Price	Quantity	Price	Quantity	Price	Quantity
1991	1.00	150	1.50	300	1.10	200.00	0.40	1100	16.00	10
1992	1.20	160	1.75	250	1.35	180.00	0.70	1000	20.00	15
1993	1.50	180	2.00	300	1.60	250.00	0.90	1050	10.00	20

The Minitab worksheet should look like the following after inputting the data.

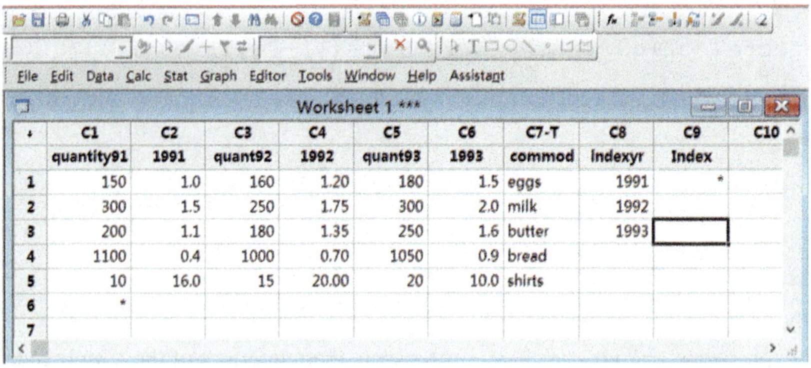

↓	C1	C2	C3	C4	C5	C6	C7-T	C8	C9	C10
	quantity91	1991	quant92	1992	quant93	1993	commod	indexyr	Index	
1	150	1.0	160	1.20	180	1.5	eggs	1991	*	
2	300	1.5	250	1.75	300	2.0	milk	1992		
3	200	1.1	180	1.35	250	1.6	butter	1993		
4	1100	0.4	1000	0.70	1050	0.9	bread			
5	10	16.0	15	20.00	20	10.0	shirts			
6	*									
7										

Note that the quantity of the base year, 1991, is entered in column c1. We then put the prices for each year in a separate column. Column c5 contains descriptions of the content, of each row. The first row is the data for eggs, the second for milk, the third for butter, the fourth for bread, and the fifth for shirts. The *indexyr* column is where we put the index for each year.

Note that in the worksheet, only the quantity for the base year is entered because the Laspeyres price index only has the base-year quantity, Q_{0i}, in its formula.

The Minitab commands for the Laspeyres price index are illustrated below.

```
MTB > let c7(1)=(sum(c1*c2)/sum(c1*c2)) * 100
MTB > let c7(2)=(sum(c1*c3)/sum(c1*c2)) * 100
MTB > let c7(3)=(sum(c1*c4)/sum(c1*c2)) * 100
MTB > print c1-c7
```

Data Display

```
Row  quantity   1991    1992   1993  commod   indexyr     index

  1       150    1.0    1.20    1.5  eggs        1991   100.000
  2       300    1.5    1.75    2.0  milk        1992   136.972
  3       200    1.1    1.35    1.6  butter      1993   157.394
  4      1100    0.4    0.70    0.9  bread
  5        10   16.0   20.00   10.0  shirts
```

From the Minitab output, we can see that the 1992 basket of goods is 37 % more expensive than the 1991 basket of goods and that the 1993 basket of goods is 57 % more expensive than the 1991 basket of goods.

The following exhibit shows the various parts of the *let* command expression that correspond to the Laspeyres price index.

```
1991 index ——→ MTB > let c7(1)=(sum(c1*c2)/sum(c1*c2)) * 100
1992 index ——→ MTB > let c7(2)=(sum(c1*c3)/sum(c1*c2)) * 100
1993 index ——→ MTB > let c7(3)=(sum(c1*c4)/sum(c1*c2)) * 100
```

Numerator of Laspeyres Price Index   Denominator of Laspeyres Price Index

We will now create a custom Laspeyres price index function in Excel. The Excel function code to calculate the Laspeyres price index is shown below.

```
'/**************************************************************************
'/Purpose: Calculates the Laspeyres price index  given the current year and base year
'/         and the quantity of the base year.
'/Parameter: CurrentYear: Range that contains current year.
'/           BaseYear: Range that contains the base year.
'/           BaseQuantity: Range that contains the base year quantity of the base year.
'/**************************************************************************
Function LaspeyresPriceIndex(BaseQuantity As Range, BaseYear As Range, _
                             CurrentYear As Range) As Double
         'the sumproduct is a worksheet function
         LaspeyresPriceIndex = (WorksheetFunction.SumProduct(BaseQuantity, CurrentYear)/
_
                                WorksheetFunction.SumProduct(BaseQuantity, BaseYear)) _
                                * 100

End Function
```

Enter the data as shown below in Excel.

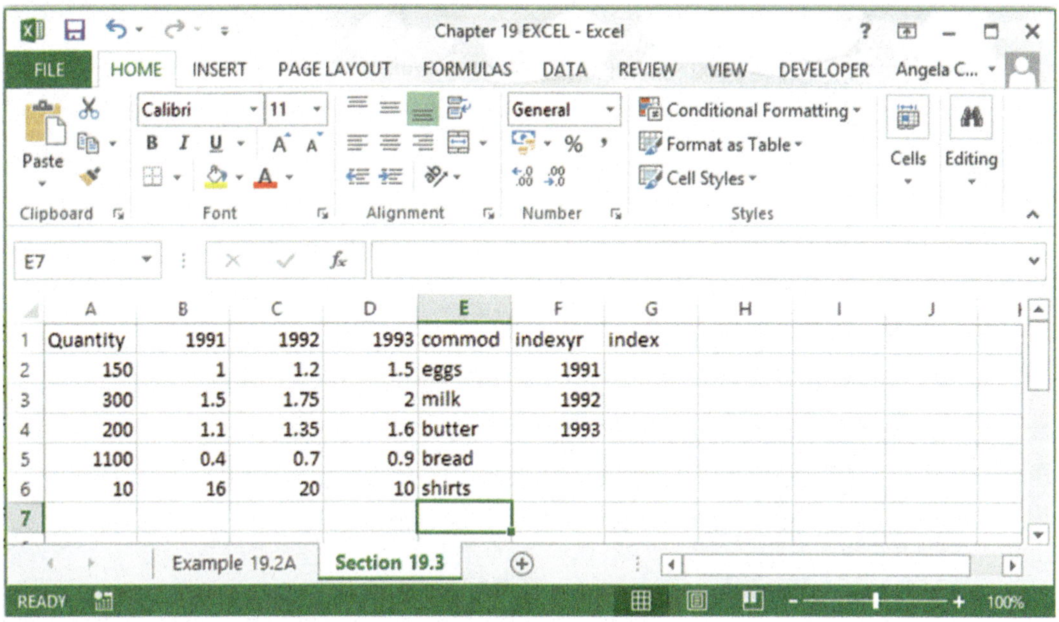

	A	B	C	D	E	F	G
1	Quantity	1991	1992	1993	commod	indexyr	index
2	150	1	1.2	1.5	eggs	1991	
3	300	1.5	1.75	2	milk	1992	
4	200	1.1	1.35	1.6	butter	1993	
5	1100	0.4	0.7	0.9	bread		
6	10	16	20	10	shirts		
7							

In column G, enter the LaspeyresPriceIndex function as shown below.

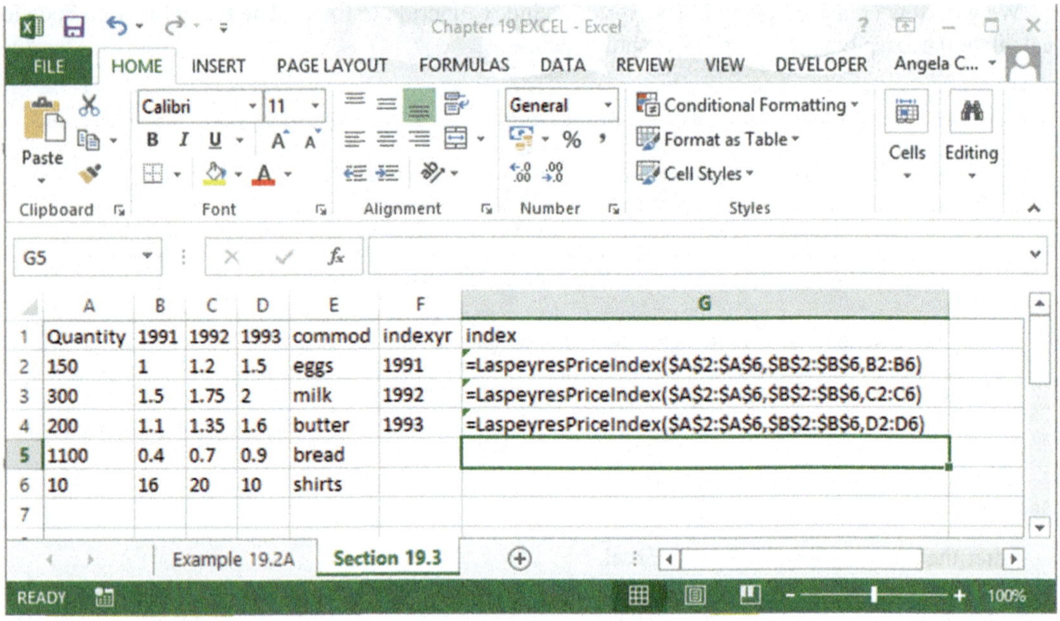

After entering the data and function properly, your Excel worksheet should look like the following.

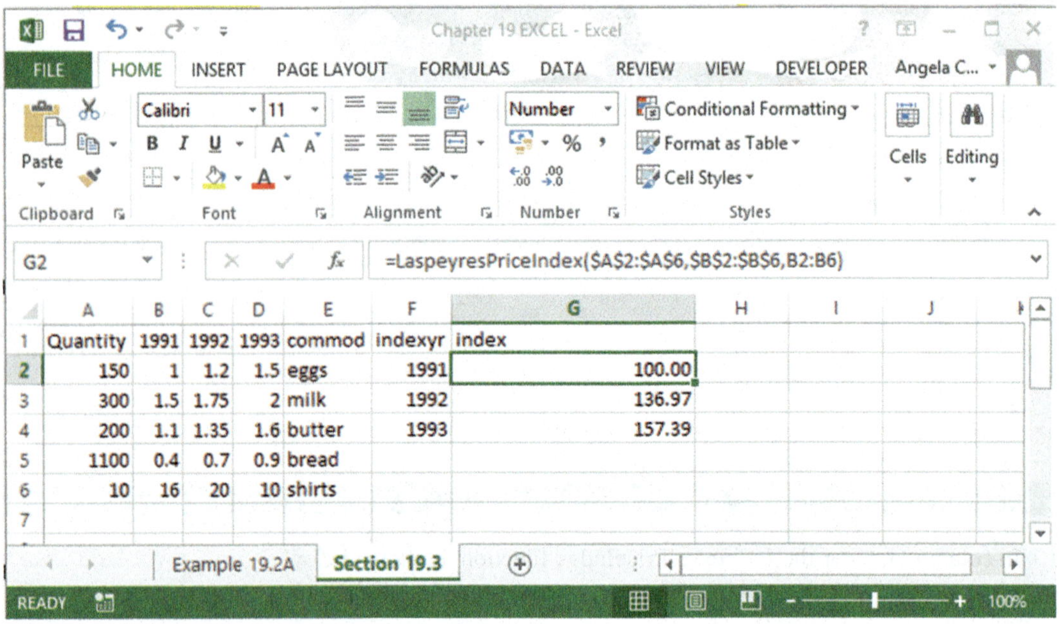

19.4 Paasche Price Index

The Paasche price index is like the Laspeyres index except that it uses the current-year quantities instead of the base-year quantity. The formula for the Paasche price index is

$$I_t = \frac{\sum_{i=1}^{n} P_{ti}Q_{ti}}{\sum_{i=1}^{n} P_{0i}Q_{ti}} * 100$$

We will use the same data that we did for the Laspeyres price index. It is reproduced below for convenience.

	Eggs		Milk		Butter		Bread		Shirts	
Year	Price	Quantity	Price	Quantity	Price	Quantity	Price	Quantity	Price	Quantity
1991	1.00	150	1.50	300	1.10	200.00	0.40	1100	16.00	10
1992	1.20	160	1.75	250	1.35	180.00	0.70	1000	20.00	15
1993	1.50	180	2.00	300	1.60	250.00	0.90	1050	10.00	20

The Minitab worksheet should look like the following after inputting the data.

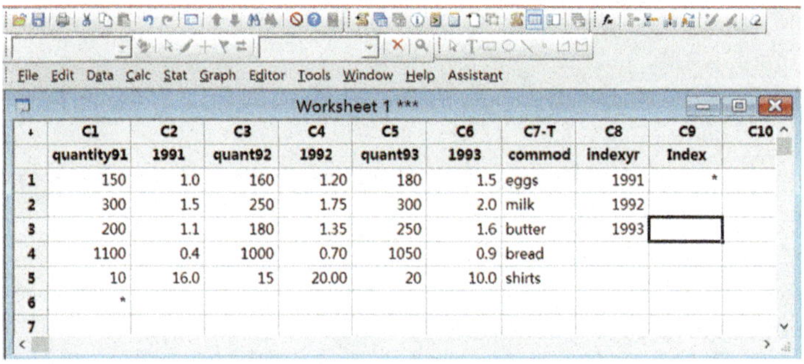

↓	C1	C2	C3	C4	C5	C6	C7-T	C8	C9	C10
	quantity91	1991	quant92	1992	quant93	1993	commod	indexyr	Index	
1	150	1.0	160	1.20	180	1.5	eggs	1991	*	
2	300	1.5	250	1.75	300	2.0	milk	1992		
3	200	1.1	180	1.35	250	1.6	butter	1993		
4	1100	0.4	1000	0.70	1050	0.9	bread			
5	10	16.0	15	20.00	20	10.0	shirts			
6	*									
7										

```
MTB > let c9(1)=(sum(c1*c2)/sum(c1*c2)) * 100
MTB > let c9(2)=(sum(c3*c4)/sum(c3*c2)) * 100
MTB > let c9(3)=(sum(c5*c6)/sum(c5*c2)) * 100
MTB > print c1-c9
```

Data Display

```
 Row  quant91   1991  quant92    1992  quant93   1993  commod  indexyr

   1      150    1.0      160    1.20      180    1.5  eggs       1991
   2      300    1.5      250    1.75      300    2.0  milk       1992
   3      200    1.1      180    1.35      250    1.6  butter     1993
   4     1100    0.4     1000    0.70     1050    0.9  bread
   5       10   16.0       15   20.00       20   10.0  shirts

index
   100.000   136.380   146.809
```

From the computer output, we can see that the 1992 price of goods is 37 % greater than the price of goods in 1991. We can see that the 1993 price of goods is 47 % greater than the price of goods in 1991.

The following exhibit shows the various parts of the *let* command expression that correspond to the Paasche price index.

```
MTB > let c9(1)=(sum(c1*c2)/sum(c1*c2)) * 100
MTB > let c9(2)=(sum(c3*c4)/sum(c3*c2)) * 100
MTB > let c9(3)=(sum(c5*c6)/sum(c5*c2)) * 100
```

Numerator of Paasche Price Index   Denominator of Paasche Price Index

We will now create a custom Paasche price index in Excel. The Excel function code to calculate the Paasche price index is shown below.

```
'/*****************************************************************************
'/Purpose: Calculates the Paasche price index  given the current year and base year
'/         and the current quantity
'/Parameter: CurrentYear: Range that contains current year.
'/           BaseYear: Range that contains the base year.
'/           CurrentQuantity: Range that contains the current year quantity.
'/*****************************************************************************
Function PaaschePriceIndex(CurrentQuantity As Range, CurrentYear As Range, _
                              BaseYear As Range) As Double


         PaaschePriceIndex =(WorksheetFunction.SumProduct(CurrentQuantity, CurrentYear)/
_
                               WorksheetFunction.SumProduct(CurrentQuantity, BaseYear)) _
                               * 100

End Function
```

Enter the data in Excel as shown below.

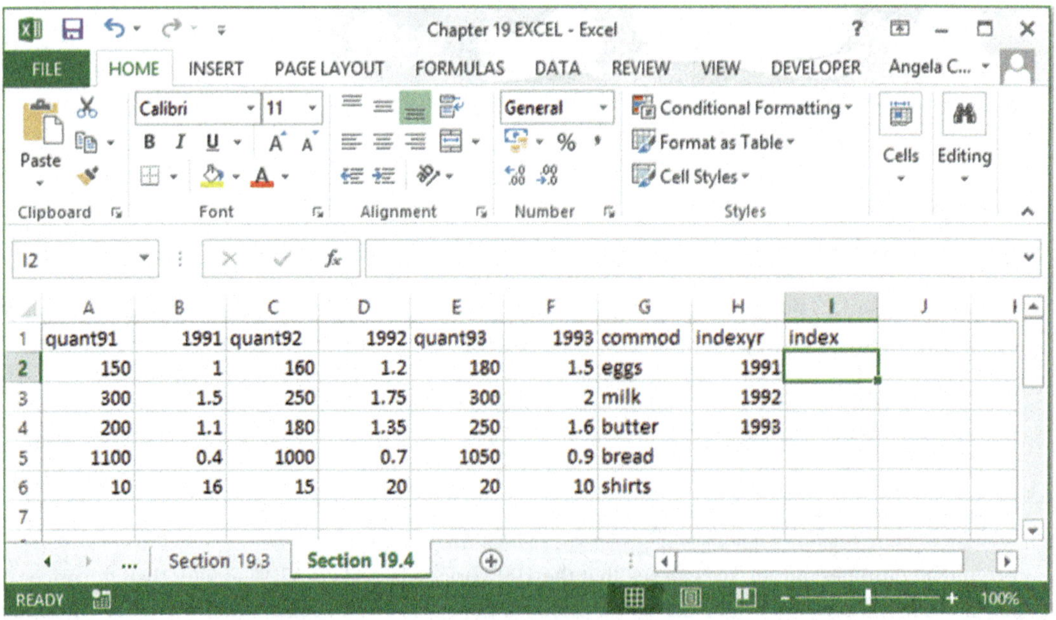

Enter the Paasche price index function as shown below.

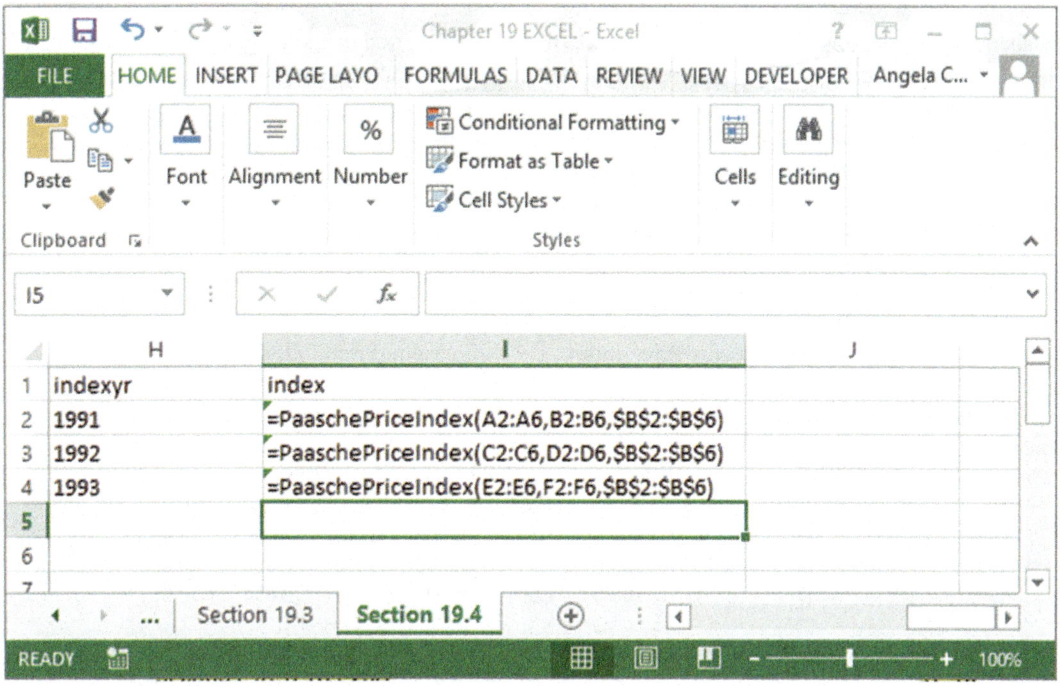

After entering the Paasche price index function, the Excel worksheet should look like the following.

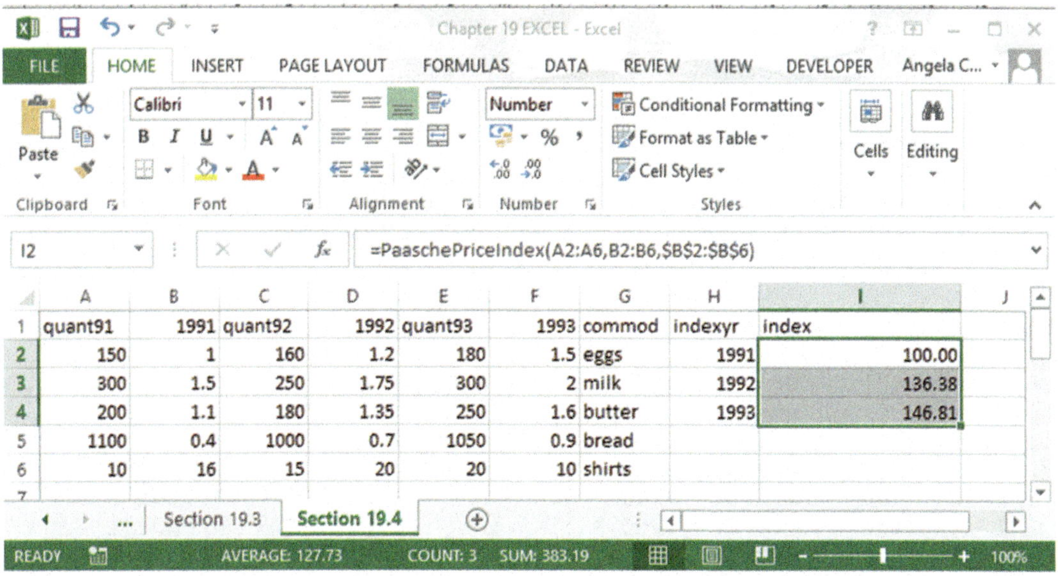

19.5 Fisher's Ideal Price Index

Fisher's ideal price index, which contains both the Laspeyres price index and the Paasche price index, has a value between those two indexes. The mathematical formula for the Fisher's ideal price index is

$$FI = \sqrt{\frac{\sum_{i=1}^{n} P_{ti}Q_{0i}}{\sum_{i=1}^{n} P_{0i}Q_{0i}} * \frac{\sum_{i=1}^{n} P_{ti}Q_{ti}}{\sum_{i=1}^{n} P_{0i}Q_{ti}}}$$

Laspeyres Price Index   Paasche Price Index

We will use the same data as the Laspeyres price index. It is reproduced below for convenience.

	Eggs		Milk		Butter		Bread		Shirts	
Year	Price	Quantity	Price	Quantity	Price	Quantity	Price	Quantity	Price	Quantity
1991	1.00	150	1.50	300	1.10	200.00	0.40	1100	16.00	10
1992	1.20	160	1.75	250	1.35	180.00	0.70	1000	20.00	15
1993	1.50	180	2.00	300	1.60	250.00	0.90	1050	10.00	20

The Minitab worksheet should look like the following after inputting the data.

Worksheet 1 ***

↓	C1	C2	C3	C4	C5	C6	C7-T	C8	C9	C10
	quantity91	1991	quant92	1992	quant93	1993	commod	indexyr	Index	
1	150	1.0	160	1.20	180	1.5	eggs	1991	*	
2	300	1.5	250	1.75	300	2.0	milk	1992		
3	200	1.1	180	1.35	250	1.6	butter	1993		
4	1100	0.4	1000	0.70	1050	0.9	bread			
5	10	16.0	15	20.00	20	10.0	shirts			

```
MTB > let c9(1)=sqrt((sum(c1*c2)/sum(c1*c2))*(sum(c1*c2)/sum(c1*c2))) * 100
MTB > let c9(2)=sqrt((sum(c1*c4)/sum(c1*c2))*(sum(c3*c4)/sum(c3*c2))) * 100
MTB > let c9(3)=sqrt((sum(c1*c6)/sum(c1*c2))*(sum(c5*c6)/sum(c5*c2))) * 100
MTB > print c1-c9
```

Data Display

```
Row  quant91   1991  quant92    1992  quant93   1993  commod  indexyr

  1      150    1.0      160    1.20      180    1.5  eggs       1991
  2      300    1.5      250    1.75      300    2.0  milk       1992
  3      200    1.1      180    1.35      250    1.6  butter     1993
  4     1100    0.4     1000    0.70     1050    0.9  bread
  5       10   16.0       15   20.00       20   10.0  shirts


index
  100.000   136.676   152.009
```

From the Minitab output, we can see that the price of goods in 1992 is 36 % greater than the price of goods in 1991. We can also see that the price of goods in 1993 is 52 % greater than the price of goods in 1991.

The following exhibit shows the various parts of the *let* command expression that correspond to the Fisher's ideal price index.

```
let c9(1)=sqrt((sum(c1*c2)/sum(c1*c2))*(sum(c1*c2)/sum(c1*c2)))
let c9(2)=sqrt((sum(c1*c4)/sum(c1*c2))*(sum(c3*c4)/sum(c3*c2)))
let c9(3)=sqrt((sum(c1*c6)/sum(c1*c2))*(sum(c5*c6)/sum(c5*c2)))
```

Laspeyres Price Index   Paasche Price Index

We will now create a custom Fisher's ideal price index in Excel. The Excel function code to calculate the Fisher's ideal price index is shown below.

```
'/****************************************************************************
'/Purpose: Calculates the Fisher Ideal Price Index. Calculates the index by
'          calling the Laseyres Price Index and Paasche Price Index.
'/Parameter: CurrentYear: Range that contains current year.
'/           BaseYear: Range that contains the base year.
'/           BaseQuantity: Range that contains the base year quantity of the base year.
'/           CurrentQuantity: Range that contains the current year quantity
'/****************************************************************************
Function FisherIdealPriceIndex(CurrentQuantity As Range, CurrentYear As Range, _
                               BaseQuantity As Range, BaseYear As Range) As Double

 FisherIdealPriceIndex = Sqr(LaspeyresPriceIndex(BaseQuantity, BaseYear, CurrentYear) _
                             * PaaschePriceIndex(CurrentQuantity, CurrentYear, BaseYear))

End Function
```

To illustrate the Fisher ideal price index, we will use the Paasche price index data that we entered in Excel. We will put the Fisher index in column J as shown below.

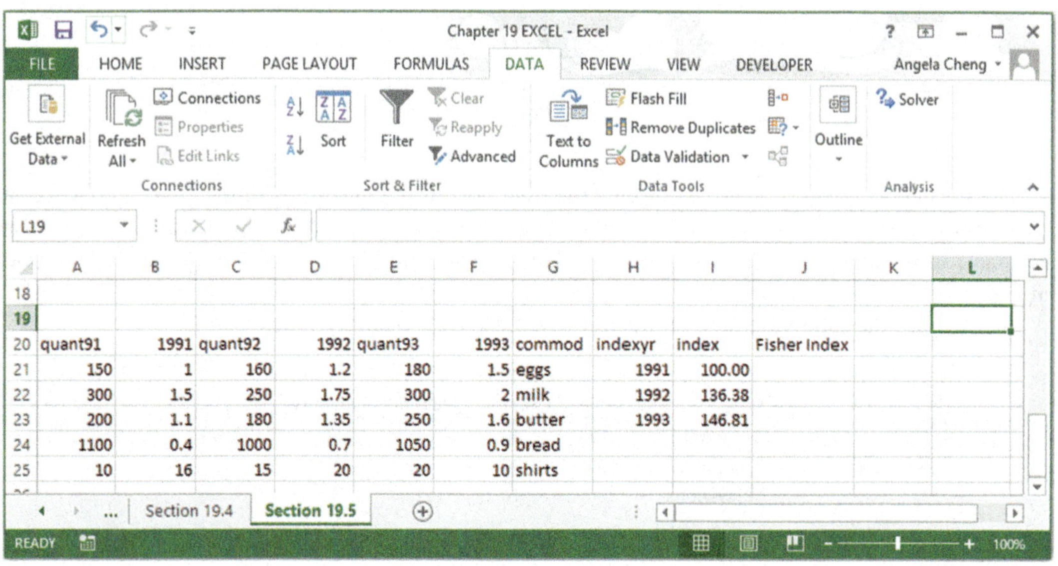

Enter the Fisher function in column J as shown below.

J
Fisher Index
=FisherIdealPriceIndex(A21:A25,B21:B25,A21:A25,B21:B25)
=FisherIdealPriceIndex(C21:C25,D21:D25,A21:A25,B21:B25)
=FisherIdealPriceIndex(E21:E25,F21:F25,A21:A25,B21:B25)

The Excel worksheet should look like the following after inputting the Fisher index.

	A	B	C	D	E	F	G	H	I	J
18										
19										
20	quant91	1991	quant92	1992	quant93	1993	commod	indexyr	index	Fisher Index
21	150	1	160	1.2	180	1.5	eggs	1991	100.00	100.00
22	300	1.5	250	1.75	300	2	milk	1992	136.38	136.68
23	200	1.1	180	1.35	250	1.6	butter	1993	146.81	152.01
24	1100	0.4	1000	0.7	1050	0.9	bread			
25	10	16	15	20	20	10	shirts			

19.6 Laspeyres Quantity Index

The Laspeyres quantity index is very much like the Laspeyres price index. The difference between the two is that the Laspeyres quantity index keeps the price in the numerator constant, while the Laspeyres price index keeps the quantity in the numerator constant. The formula for the Laspeyres quantity index is shown below:

$$I_t = \frac{\sum_{i=1}^{n} P_{0i} Q_{ti}}{\sum_{i=1}^{n} P_{0i} Q_{0i}} * 100$$

The Laspeyres quantity index represents the total cost of the quantities in the year in question at base-year prices as a percentage of the total cost of the base-year quantities. Because prices are kept constant, any change in the index is due to the change in quantities between the base year and the year in question.

We will use the following data to illustrate the Laspeyres quantity index.

Year	1991 Price	1991 Quantity	1992 Quantity	1993 Quantity
Automobiles	1000	40	50	60
Computers	500	30	40	50
Televisions	200	10	20	30

The Minitab worksheet should look like the following.

Worksheet 1 ***

↓	C1	C2	C3	C4	C5-T	C6	C7
	quantity	1991	1992	1993	command	indexyr	index
1	1000	40	50	60	auto	1991	
2	500	30	40	50	computer	1992	
3	200	10	20	30	tele	1993	
4							
5							

```
MTB > let c7(1)=(sum(c1*c2)/sum(c1*c2)) * 100
MTB > let c7(2)=(sum(c1*c3)/sum(c1*c2)) * 100
MTB > let c7(3)=(sum(c1*c4)/sum(c1*c2)) * 100
MTB > print c1-c7
```

Data Display

```
Row  quantity   1991   1992   1993   command   indexyr     index

  1      1000     40     50     60  auto          1991   100.000
  2       500     30     40     50  computer      1992   129.825
  3       200     10     20     30  tele          1993   159.649
```

Notice that the commands for the Laspeyres quantity index are the same as for the Laspeyres price index. This is possible because the worksheet format is the same for both.

In the Minitab output, the indexes for 1992 and 1993 indicate that the cost of the three commodities increased 30 % and 60 %, respectively. The price has been held constant, so changes in the indexes are due to changes in the quantities of the commodities for each period. In other words, the 1992 and 1993 indexes show that the quantities of the goods increased 30 % and 60 %, respectively, from the 1991 base year.

We will now create a custom Laspeyres quantity index function in Excel. The Excel function code to calculate the Laspeyres price index is shown below.

```
'/*********************************************************************************
'/Purpose: Calculates the Laspeyres price index  given the current year and base year
'/         and the quantity of the base year.
'/Parameter: CurrentQuantity: Range that contains current year price.
'/           BasePrice: Range that contains the base year price.
'/           BaseQuantity: Range that contains the base year quantity of the base year.
'/*********************************************************************************
Function LaspeyresQuantityIndex(BaseQuantity As Range, BasePrice As Range, _
                             CurrentQuantity As Range) As Double

    LaspeyresQuantityIndex = (WorksheetFunction.SumProduct(BasePrice, CurrentQuantity) /
_
                               WorksheetFunction.SumProduct(BaseQuantity, BasePrice)) _
                               * 100

End Function
```

To illustrate the Laspeyres quantity index, enter data in Excel as shown below.

	A	B	C	D	E	F	G
1	Price	1991	1992	1993	commod	indexyr	index
2	1000	40	50	60	auto	1991	
3	500	30	40	50	computer	1992	
4	200	10	20	30	tele	1993	
5							

	A	B	C	D	E	F	G
30	Price	1991	1992	1993	commod	indexyr	index
31	1000	40	50	60	auto	1991	
32	500	30	40	50	computer	1992	
33	200	10	20	30	tele	1993	

In column G, enter the Laspeyres quantity index function as shown below.

G
index
=LaspeyresQuantityIndex(B2:B4,B2:B4,A2:A4)
=LaspeyresQuantityIndex(B2:B4,C2:C4,A2:A4)
=LaspeyresQuantityIndex(B2:B4,D2:D4,A2:A4)

	G
30	index
31	=LaspeyresQuantityIndex(B31:B33,B31:B33,A31:A33)
32	=LaspeyresQuantityIndex(B31:B33,C31:C33,A31:A33)
33	=LaspeyresQuantityIndex(B31:B33,D31:D33,A31:A33)

After entering the Laspeyres quantity index function, the Excel data sheet should look like the following.

	A	B	C	D	E	F	G
1	Price	1991	1992	1993	commod	indexyr	index
2	1000	40	50	60	auto	1991	100
3	500	30	40	50	computer	1992	129.8246
4	200	10	20	30	tele	1993	159.6491

19.7 Paasche Quantity Index

The Paasche quantity index is like the Paasche price index. The only difference is that the Paasche price index denominator holds the price constant, while the Paasche quantity index holds the quantity constant in the denominator. The formula for the Paasche quantity index is

$$I_t = \frac{\sum_{i=1}^{n} P_{ti}Q_{ti}}{\sum_{i=1}^{n} P_{ti}Q_{0i}} * 100$$

The Minitab worksheet should look like the following after inputting the data.

Worksheet 1 ***

↓	C1	C2	C3	C4	C5	C6	C7-T	C8	C9
	1991	quant91	1992	quant92	1993	quant93	commod	indexyr	index
1	2000	40	2100	50	2500	60	auto	1991	
2	600	30	600	40	700	50	comput	1992	
3	250	10	300	20	300	20	tele	1993	
4									
5									

```
MTB > let c9(1)=sum(c1*c2)/sum(c1*c2) * 100
MTB > let c9(2)=sum(c3*c4)/sum(c3*c2) * 100
MTB > let c9(3)=sum(c5*c6)/sum(c5*c2) * 100
MTB > print c1-c9
```

Data Display

```
 Row    1991  quant91    1992  quant92    1993  quant93  commod
indexyr

   1    2000       40    2100       50    2500       60  auto
1991
   2     600       30     600       40     700       50  comput
1992
   3     250       10     300       20     300       20  tele
1993


index
   100.000   128.571   154.032
```

Prices are held fixed in the equation, so a change between numerator and denominator reflects a change in the quantities over 3 years. In other words, 1992 and 1993 saw a 29 % and 54 % increase in quantity, respectively, over the 1991 quantity.

We will now create a custom Paasche quantity index function in Excel. The Excel function code to calculate the Paasche quantity index is shown below.

```
'/*******************************************************************************
'/Purpose: Calculates the Paasche quantity index  given the current price and base
'/         quantity and the current quantity
'/Parameter: CurrentPrice: Range that contains current year price.
'/           BaseQuantity: Range that contains the base quantity.
'/           CurrentQuantity: Range that contains the current year quantity.
'/*******************************************************************************
Function PaascheQuantityIndex(CurrentQuantity As Range, CurrentPrice As Range, _
                              BaseQuantity As Range) As Double


  PaascheQuantityIndex = (WorksheetFunction.SumProduct(CurrentQuantity, CurrentPrice) / _
                         WorksheetFunction.SumProduct(BaseQuantity, CurrentPrice)) _
                         * 100

End Function
```

To illustrate the Paasche quantity index, enter data in Excel as shown below.

	A	B	C	D	E	F	G	H	I
1	1991	quant91	1992	quant92	1993	quant93	commod	indexyr	index
2	2000	40	2100	50	2500	60	auto	1991	
3	600	30	600	40	700	50	computer	1992	
4	250	10	300	20	300	20	tele	1993	

In column I, enter the Paasche quantity index function as shown below.

	I
1	index
2	=PaascheQuantityIndex(F2:F4,E2:E4,B2:B4)
3	=PaascheQuantityIndex(D2:D4,C2:C4,B2:B4)
4	=PaascheQuantityIndex(F2:F4,E2:E4,B2:B4)

	I
40	index
41	=FaascheQuantityIndex(F41:F43,E41:E43,B41:B43)
42	=FaascheQuantityIndex(D41:D43,C41:C43,B41:B43)
43	=FaascheQuantityIndex(F41:F43,E41:E43,B41:B43)
44	

After entering the Paasche quantity index function, the Excel data sheet should look like the following.

	A	B	C	D	E	F	G	H	I
1	1991	quant91	1992	quant92	1993	quant93	commod	indexyr	index
2	2000	40	2100	50	2500	60	auto	1991	154.0323
3	600	30	600	40	700	50	computer	1992	128.5714
4	250	10	300	20	300	20	tele	1993	154.0323

19.8 Fisher's Ideal Quantity Index

Fisher's ideal quantity index is defined as

$$FIQ = \sqrt{(\text{Laspeyres quantity index})(\text{Paasche quantity index})}$$

We can use Minitab to calculate the FIQ for 1992 and 1993.

```
MTB > let k1=sqrt(129.825*128.571)
MTB > print k1
```

Data Display

```
K1      129.196
MTB > let k2=sqrt(159.649*154.032)
MTB > print k2
```

Data Display

K2 156.815

From the Minitab output, the 1992 FIQ which is contained in $k1$ is 129.196, and the 1993 FIQ, contained in $k2$, is 156.815.

19.9 Stock Indexes: S&P 500 Index

We will now look at two popular stock indexes. A stock market index is a statistical measure that shows how the price of a group of stocks changes over time. A stock market index encompasses either all or only a portion of stocks in its market. Stock market indexes employ different weighting schemes, so we can use this basis to categorize the indexes by type. The three most common types of stock market indexes are market-value-weighted indexes, price-weighted indexes, and equally weighted indexes. Price per share in the current period (P_0), price per share in the next period (P_1), number of shares outstanding in the current period (Q_0), and number of shares outstanding in the next period (Q_1) are listed below. These data are used to illustrate the various weighting schemes and to provide information about the weights applied to the market stock indexes.

Price Per Share and Outstanding Shares for Stocks, A,B,C

Stock	P_0	P_1	Q_0	Q_1
A	100	100	60	55
B	50	60	90	100
C	20	40	250	300

The first stock market index we will look at is the *market-value-weighted index,* which is shown below:

$$I = \frac{\sum_{i=1}^{n} Q_{ti}P_{ti}}{\sum_{i=1}^{n} Q_{0i}P_{0i}}$$

The following is the market-value-weighted index of the table above:

$$\begin{aligned} I &= \frac{55(100)+100(60)+300(40)}{60(100)+90(50)+250(20)}(100) \\ &= \frac{23,500}{15,500}(100) = 152 \end{aligned}$$

The market-value-weighted index implies that there was a 52 % increase from the base period to the current period.

Standard and Poor's 500 composite index is an example of a market-value-weighted index. The S&P 500 index comprises industrial firms, utilities, transportation firms, and financial firms. Changes in the index are based on changes in the firm's total market value with respect to a base year. Currently, the base period (1941–1943 = 10) for the S&P 500 index is stated formally as follows:

$$S\&P\,500\,\text{index} = \frac{\sum_{i=1}^{n} Q_{ti}P_{ti}}{\sum_{i=}^{n} Q_{0i}P_{0i}}(10)$$

It is interesting to note that the S&P 500 contains more than 500 stocks. It actually contains 505 stocks because it includes two share classes of stock from five of its component companies. On October 14, 2015, the following Chicago Board Options Exchange Internet URL lists all the 505 stocks of the S&P 500: http://www.cboe.com/products/snp500.aspx.

19.10 Stock Indexes: Dow Jones Industrial Average (DJIA)

The next stock market index that we will look at is the *price-weighted index*. The price-weighted index shows the change in the average price of the stocks that are included in the index. Using the table in Sect. 19.9, we can compute the price-weighted index as follows:

$$\begin{aligned} I &= \frac{(100+60+40)/3}{(100+50+20)/3}(100) \\ &= \frac{200/3}{170/3}(100) = 117.65 \end{aligned}$$

The closest thing to a true price-weighted stock market index is the *Dow Jones industrial average (DJIA)*. Simply stated, the DJIA is an arithmetic average of the stock prices that make up the index. The DJIA originally assumed a single share of each stock in the index, and the total of stock prices was divided by the number of stocks that made up of the index:

$$DJIA_t = \frac{\frac{\sum_{i=1}^{30} P_{ti}}{30}}{\frac{\sum_{i=1}^{30} P_{0i}}{30}}$$

Today the index is adjusted for stock splits and the issuance of stock dividends:

$$DJIA_t = \frac{\frac{\sum_{i=1}^{30} \frac{P_{ti}}{AD_t}}{30}}{\frac{\sum_{i=1}^{30} P_{0i}}{30}}$$

where P_{ti} = the closing price of stock i on day t and AD_t = the adjusted divisor on the day t. The adjustment process is shown below.

Adjustment of DJIA Divisor to Allow for a Stock Split

Before Split		After 2-for-1 Stock Split by Stock A
Stock	*Price*	*Price*
A	60	30
B	30	30
C	20	20
D	10	10
	120	90

Average before split = 120/4 = 30

Adjust divisor = (sum of prices after the split)/(Average before split)=90/30=3

Average after split = 90/30 = 30

Before split divisor = 4 After split divisor = 30

As shown above, the adjustment process is designed to keep the index value the same as it would have been if the split had not occurred. Similar adjustments have been made when it has been found necessary to replace one of the component stocks with the stock of another company, thus preserving the consistency and the comparability of index values at different points in time.

Nevertheless, the adjustment process used for the DJIA has its share of critics. Because price weighting itself causes high-priced stocks to dominate the series, the same effect can cause a shift in this balance when rapidly growing firms split their stock. For example, a 20 % increase in the price of stock A in the table above would in itself have caused a 10 % increase in the value of the sample index before the split, whereas a 20 % increase in the price of stock B would have caused only a 5 % increase in the index value. After the 2-for-1 split of stock A, a 20 % increase in either stock A or stock B would have the same effect on the index value (a 6.7 % increase); a downward shift in the importance of stock A relative to that of the other stocks in the sample has occurred. This effect would relegate the stock of the fastest-growing companies to a position of *least* importance in the determining index values.

19.11 Components of the Dow Jones Industrial Average (DJIA)

One frequent question about the Dow Jones industrial average is "What are the stocks that compose the DJIA?" On October 14, 2015, the following CNBC Internet URL lists all the 30 stock components of the DJIA: http://www.cnbc.com/dow-components/.

	Symbol	Name	Price	Change	%Change	Previous
1	AXP	American Express Co	76.17	-0.43	-0.56%	76.6
2	AAPL	Apple Inc	110.21	-1.58	-1.41%	111.79
3	BA	Boeing Co	134.22	-6.07	-4.33%	140.25
4	CAT	Caterpillar Inc	70.71	0.54	0.77%	70.17
5	CSCO	Cisco Systems Inc	27.82	-0.03	-0.11%	27.85
6	CVX	Chevron Corp	89.76	1.38	1.56%	88.38
7	DD	E I du Pont de Nemours and Co	56.4	0.64	1.15%	55.76
8	XOM	Exxon Mobil Corp	80.16	1	1.26%	79.16
9	GE	General Electric Co	27.6	-0.27	-0.97%	27.92
10	GS	Goldman Sachs Group Inc	179.51	-1.46	-0.81%	180.97
11	HD	Home Depot Inc	120.27	-1.34	-1.10%	121.61
12	IBM	International Business Machines Corp	150.01	0.39	0.26%	149.62
13	INTC	Intel Corp	32.8	0.76	2.37%	32.04
14	JNJ	Johnson & Johnson	94.53	-0.92	-0.96%	95.48
15	KO	Coca-Cola Co	41.68	0.03	0.07%	41.65
16	JPM	JPMorgan Chase and Co	59.99	-1.56	-2.53%	61.55
17	MCD	McDonald's Corp	102.82	-0.56	-0.54%	103.38
18	MMM	3M Co	148.32	-0.79	-0.53%	149.11
19	MRK	Merck & Co Inc	49.54	0.07	0.14%	49.45
20	MSFT	Microsoft Corp	46.68	-0.21	-0.45%	46.88
21	NKE	Nike Inc	125.84	0.03	0.02%	125.81
22	PFE	Pfizer Inc	33.04	0.06	0.18%	32.98
23	PG	Procter & Gamble Co	74.21	0.1	0.13%	74.11
24	TRV	Travelers Companies Inc	102.77	-0.54	-0.52%	103.31
25	UNH	UnitedHealth Group Inc	122.07	-1.92	-1.55%	124
26	UTX	United Technologies Corp	92.17	-2.11	-2.24%	94.28
27	VZ	Verizon Communications Inc	43.99	-0.37	-0.83%	44.37
28	V	Visa Inc	74.2	-0.8	-1.07%	74.98
29	WMT	Wal Mart Stores Inc	60.03	-6.7	-10.04%	66.73
30	DIS	Walt Disney Co	105.73	-0.86	-0.81%	106.59

19.12 SAS Programming Code Instructions and Examples

19.12.1 Simple Price Index

Example 19.3 The index numbers can be easily created by using the **Data** command as illustrated below.

```
Data Egg;
input Year Price @@;
cards;
1991 1.00
1992 1.20
1993 1.50
;
Run;

Data Egg_index;
Set Egg;
P0=1.00;
Index=(Price/P0)*100;
run;
```

Obs	Year	Price	P_0	Index
1	1991	1.0	1	100
2	1992	1.2	1	120
3	1993	1.5	1	150

Example 19.4 Instead of choosing 1991 as the base year for indexing the eggs, let us choose 1992 as the base year.

```
Data Egg_index;
Set Egg;
P0=1.20;
Index=(Price/P0)*100;
run;
```

Obs	Year	Price	P_0	Index
1	1991	1.0	1.2	83.333
2	1992	1.2	1.2	100.000
3	1993	1.5	1.2	125.000

We will again use 1991 as the base year and create the index numbers with the same command in SAS.

```
Data Price;
input Year Eggs Milk Butter Bread @@;
cards;
1991 1.00 1.50 1.10 0.40
1992 1.20 1.75 1.35 0.70
1993 1.50 2.00 1.60 0.90
;
Run;

Data Price_index;
Set Price;
P0=1.00+1.50+1.10+0.40;
Sum=Eggs+Milk+Butter+Bread;
Index=(Sum/P0)*100;
run;
```

Obs	Year	Eggs	Milk	Butter	Bread	P_0	Sum	Index
1	1991	1.0	1.50	1.10	0.4	4	4	100
2	1992	1.2	1.75	1.35	0.7	4	5	125
3	1993	1.5	2.00	1.60	0.9	4	6	150

19.12.2 Laspeyres Price Index

To get the Laspeyres price index, we first read the data into SAS as shown below.

```
Data Laspeyres;
input quantity Year_1991 Year_1992 Year_1993 commod $;
sum_1991=quantity*Year_1991;
sum_1992=quantity*Year_1992;
sum_1993=quantity*Year_1993;
cards;
150     1.0     1.20     1.5  eggs
300     1.5     1.75     2.0  milk
200     1.1     1.35     1.6  butter
1100     0.4     0.70     0.9  bread
10  16.0    20.00    10.0  shirts
;
run;
```

We use **PROC SQL** to calculate the Laspeyres price index and the other entire indexes in the following sections. **PROC SQL** is a perfect function to summarize data. **PROC SQL** is more intuitive than **PROC MEANS** or **PROC SUMMARY**, where SAS will create an output table that always contains more rows and columns than you need and you have to choose the right _TYPE_ value. PROC SQL below gives you the Laspeyres price index directly.

```
proc sql;
     create table Laspeyres_Index as
     select sum(quantity*Year_1991)/sum(quantity*Year_1991)*100 as Index_1991,
          sum(quantity*Year_1992)/sum(quantity*Year_1991)*100 as   Index_1992,
          sum(quantity*Year_1993)/sum(quantity*Year_1991)*100 as Index_1993
            from Laspeyres;
  quit;
```

Obs	Index_1991	Index_1992	Index_1993
1	100	136.972	157.394

19.12.3 Paasche Price Index

We use the same procedure as we introduced in the previous section.

```
Data PAASCHE;
input quantity91 Year_1991 quantity92 Year_1992 quantity93 Year_1993 commod $;
cards;
150 1.0 160 1.20 180 1.5  eggs
300 1.5 250 1.75 300 2.0  milk
200 1.1 180 1.35 250 1.6  butter
1100 0.4 1000 0.70 1050 0.9  bread
10 16.0 15 20.00 20 10.0  shirts
;
run;

proc sql;
  create table PAASCHE_Index as
  select sum(quantity91*Year_1991)/sum(quantity91*Year_1991)*100 as Index_1991,
         sum(quantity92*Year_1992)/sum(quantity92*Year_1991)*100 as Index_1992,
         sum(quantity93*Year_1993)/sum(quantity93*Year_1991)*100 as Index_1993
  from PAASCHE;
 quit;
```

Obs	Index_1991	Index_1992	Index_1993
1	100	136.380	146.809

19.12.4 Fisher's Ideal Price Index

Read the table into SAS and calculate Fisher's ideal price index using **PROC SQL**.

```
Data Fisher;
input quantity91 Year_1991 quantity92 Year_1992 quantity93 Year_1993 commod $;
cards;
150 1.0 160 1.20 180 1.5  eggs
300 1.5 250 1.75 300 2.0  milk
200 1.1 180 1.35 250 1.6  butter
1100 0.4 1000 0.70 1050 0.9  bread
10 16.0 15 20.00 20 10.0  shirts
;
run;

proc sql;
     create table Fisher_Index as
     select
sqrt(sum(quantity91*Year_1991)/sum(quantity91*Year_1991)*sum(quantity91*Year_1991)/sum
(quantity91*Year_1991))*100 as Index_1991,

sqrt(sum(quantity91*Year_1992)/sum(quantity91*Year_1991)*sum(quantity92*Year_1992)/sum
(quantity92*Year_1991))*100 as Index_1992,

sqrt(sum(quantity91*Year_1993)/sum(quantity91*Year_1991)*sum(quantity93*Year_1993)/sum
(quantity93*Year_1991))*100 as Index_1993
            from Fisher;
 quit;
```

Obs	Index_1991	Index_1992	Index_1993
1	100	136.676	152.009

19.12.5 Laspeyres Quantity Index

Based on the equation of Laspeyres quantity index, we calculate it using **PROC SQL**.

```
Data Laspeyres_Quantity;
input quantity Year_1991 Year_1992 Year_1993 command $;
cards;
1000 40 50 60 auto
500 30 40 50 computer
200 10 20 30 tele
;
run;

proc sql;
   create table Laspeyres_Quantity_Index as
   select sum(quantity*Year_1991)/sum(quantity*Year_1991)*100 as Index_1991,
          sum(quantity*Year_1992)/sum(quantity*Year_1991)*100 as Index_1992,
          sum(quantity*Year_1993)/sum(quantity*Year_1991)*100 as Index_1993
   from Laspeyres_Quantity;
 quit;
```

The result is shown below:

Obs	Index_1991	Index_1992	Index_1993
1	100	129.825	159.649

19.12.6 Paasche Quantity Index

```
Data PAASCHE_Quantity;
input quantity91 Year_1991 quantity92 Year_1992 quantity93 Year_1993 commod $;
cards;
2000 40 2100 50 2500 60  auto
600 30 600 40 700 50  comput
250 10 300 20 300 20  tele

;
run;


proc sql;
  create table PAASCHE_Quantity_Index as
  select sum(quantity91*Year_1991)/sum(quantity91*Year_1991)*100 as Index_1991,
         sum(quantity92*Year_1992)/sum(quantity92*Year_1991)*100 as Index_1992,
         sum(quantity93*Year_1993)/sum(quantity93*Year_1991)*100 as Index_1993
             from PAASCHE_Quantity;
 quit;
```

Obs	Index_1991	Index_1992	Index_1993
1	100	128.571	154.032

19.12.7 Fisher's Ideal Quantity Index

```
proc sql;
     create table FISHER_IDEAL_Quantity as
     select
sqrt(Laspeyres_Quantity_Index.Index_1992*PAASCHE_Quantity_Index.Index_1992) as
K1,

sqrt(Laspeyres_Quantity_Index.Index_1993*PAASCHE_Quantity_Index.Index_1993) as
K2
        from Laspeyres_Quantity_Index,PAASCHE_Quantity_Index;
 quit;
```

Obs	K1	K2
1	129.196	156.816

19.13 Statistical Summary

In this chapter, we discussed several price and quantity indexes that are used to analyze economic activities. We also looked at two popular stock market indexes. We saw how to find the components for the S&P 500 and the DJIA.

Bibliography

Lee CF, Lee JC, Lee AC (2013) Statistics for business and financial economics. Springer, New York

Microsoft Inc., Excel 2013. Microsoft Inc., Redmond

Minitab Inc. Minitab 17. Minitab Inc., State College

SAS Institute Inc. SAS 2014. SAS Institute Inc., Cary

Chapter 20
Sampling Surveys: Methods and Applications

20.1 Introduction

In previous chapters we investigated sampling but only in terms of simple random sampling, in which each potential sample of n members has an equal chance of being chosen. Most of the time, this requirement is satisfied due to the fact that the sample size is small compared to the population. But when the sample size becomes a large part of the population, some adjustments must be made.

20.2 Random Number Tables

Random tables are very often used to obtain random samples. Minitab can easily create random tables in response to the *random* command. As shown in Chaps. 8 and 9, the *random* command can randomly generate numbers from many different kinds of distributions. The distribution that we will use here to generate random tables is the uniform distribution, because in many applications we are interested in every number having an equal chance of occurring. The uniform distribution has this property.

Suppose we are interested in creating a 10-by-10 table that has numbers randomly generated from 0 to 100. The creation of this table is illustrated below.

```
MTB > random 10 c1-c10;
SUBC> integer 0 100.
MTB > print c1-c10
```

Data Display

Row	C1	C2	C3	C4	C5	C6	C7	C8	C9	C10
1	44	59	3	51	72	78	85	95	72	95
2	8	20	89	40	73	26	49	28	30	15
3	18	90	30	14	62	48	45	0	39	19
4	46	0	28	2	45	73	73	50	48	3
5	44	73	36	93	30	46	31	6	89	46
6	6	81	73	6	64	53	12	13	94	24
7	32	41	96	66	13	46	77	63	13	91
8	32	5	79	45	89	8	90	96	73	15
9	48	11	26	66	75	5	90	73	69	63
10	41	8	48	45	37	55	45	83	84	83

C.-F. Lee et al., *Essentials of Excel, Excel VBA, SAS and Minitab for Statistical and Financial Analyses*, DOI 10.1007/978-3-319-38867-0_20

We used the *integer* subcommand instead of the *uniform* subcommand because the *integer* subcommand is a uniform distribution that has only integer numbers. Now let us create another random table, using exactly the same commands, to illustrate that random tables are in fact random.

```
MTB > random 10 c1-c10;
SUBC> integer 0 100.
MTB > print c1-c10
```

Data Display

Row	C1	C2	C3	C4	C5	C6	C7	C8	C9	C10
1	31	39	30	28	7	10	43	0	57	97
2	81	6	68	97	67	18	77	26	60	56
3	91	69	97	46	69	50	20	66	88	53
4	92	53	18	84	93	98	12	90	79	50
5	93	55	89	1	48	84	2	71	45	88
6	1	76	22	53	48	78	81	2	29	74
7	85	49	41	46	3	60	81	22	59	43
8	48	80	95	71	66	5	41	24	15	67
9	32	18	24	79	25	15	18	0	84	76
10	52	11	14	67	4	93	79	47	50	99

Below is the Excel code to create a 10-by-10 table that has numbers randomly generated from 0 to 100.

```
'/********************************************************
'/Purpose:  To generate 100 numbers from 0 to 100 on the
'/          active sheet
'/********************************************************
Sub Random100()
    Dim rRandom As Range
    Dim rCell As Range

    Set rRandom = ActiveSheet.Range("a1:j10")

    For Each rCell In rRandom.Cells
        Randomize   ' Initialize random-number generator.
        rCell.Value = Rnd * 100
    Next

    rRandom.NumberFormat = "0"
End Sub
```

Below shows a 10-by-10 random table generated by the above Excel code.

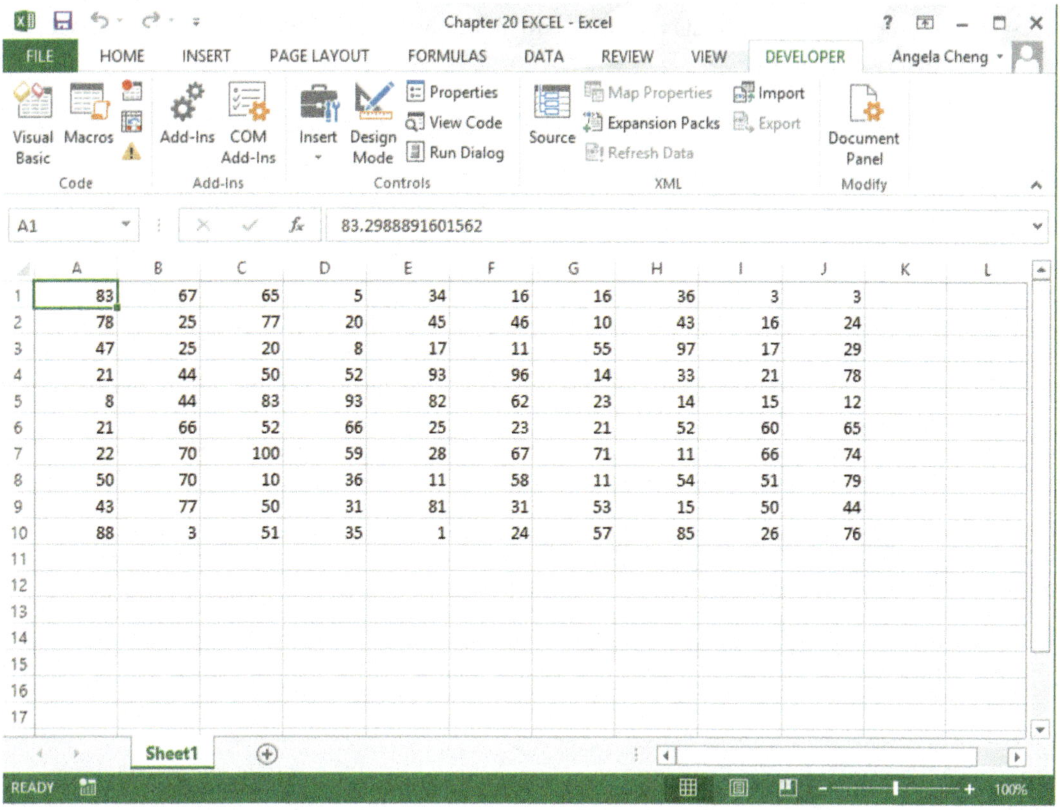

	A	B	C	D	E	F	G	H	I	J
1	83	67	65	5	34	16	16	36	3	3
2	78	25	77	20	45	46	10	43	16	24
3	47	25	20	8	17	11	55	97	17	29
4	21	44	50	52	93	96	14	33	21	78
5	8	44	83	93	82	62	23	14	15	12
6	21	66	52	66	25	23	21	52	60	65
7	22	70	100	59	28	67	71	11	66	74
8	50	70	10	36	11	58	11	54	51	79
9	43	77	50	31	81	31	53	15	50	44
10	88	3	51	35	1	24	57	85	26	76

20.3 Confidence Interval for the Population Mean

In previous chapters we assumed either that the population size was sufficiently large or the sample was small enough compared to the population for the sample to be random. When this is not the case, we have to make adjustments. One such adjustment is the variance of the sample mean. The adjusted variance of the sample mean is defined as

$$\hat{\sigma}_{\bar{x}}^2 = \frac{S^2}{n} * \frac{N-n}{N-1}$$

As a rule of thumb, if $n/N > 0.05$, we should use the above adjusted sample variance.
If $n/N < 0.05$, we should use the calculated sample variance, s^2.

Example 20.1 Suppose an investment adviser is trying to decide whether a small retirement community consisting of 1000 residents represents a promising source of potential clients. To determine the potential business, the investment adviser decides to analyze the size of the residents' investment portfolios. A random sample of 75 residents, who were able to respond anonymously, produces a sample mean of \$375,000 with a sample standard deviation of \$120,000. Calculate the 95% confidence interval for the mean value of the investment portfolio.

The ratio of the sample size to the population size is $n/N = 75/1000 = 0.08$. Since the ratio of the sample size to the population size is greater than 0.05, we should use the adjusted sample variance.

When we do not have to make an adjustment to the variance of the sample mean, the formula for the confidence interval would be

$$\bar{x} - z_{\alpha/2}(\sigma/\sqrt{n}) < \mu < \bar{x} + z_{\alpha/2}(\sigma/\sqrt{n})$$

With the adjustment to the variance of the sample mean, the confidence interval formula is

$$\bar{x} - z_{\alpha/2}\hat{\sigma}_{\bar{x}} < \mu < \bar{x} + z_{\alpha/2}\hat{\sigma}_{\bar{x}}$$

We will create a macro to calculate the adjusted variance of the sample mean and the confidence interval for a population mean that came from a large sample. To do this, we will borrow a lot from the *zint* macro in Chap. 10 to program the macro *adjzint*. The analysis of the *adjzint* macro is very similar to that of the *zint* macro.

The use of the macro *adjzint* to solve Example 20.3A is shown below. The Minitab code of the macro *adjzint* is shown in Appendix 20.1.

```
MTB > %adjzint
Executing from file: C:\Program Files\Mtbwin\MACROS\adjzint.MAC
this macro calculates the adjusted sample variance
and then finds a confidence interval for a sample mean
based on the adjusted sample variance


What is the population size?
DATA> 1000
What is the sample size?
DATA> 75
What is the sample standard deviation?
DATA> 120000
The adjusted sample variance is
```

Data Display

```
KadjVar     177777778
The adjusted standard deviation is
```

Data Display

```
Kadjstdv    13333.3

What is the mean?
DATA> 375000
What is the desired percentage confidence interval?
DATA> .95
The confidence interval is:
```

Data Display

```
Row   Clowint  Chighint

  1    348867    401133
```

From the *adjzint* macro, we can see that the 95% confidence interval is from \$348,867 to \$401,133. The set of commands in the *adjzint* macro is shown below.

20.4 Confidence Interval for the Population Proportion

We will also need to adjust the variance of the sample proportion when the sample relative to the population is not so small. The adjusted variance for a sample proportion is

$$\hat{\sigma}\overline{x}^2 = \frac{\hat{p}\,(1-\hat{p}\,)}{n} * \frac{N-n}{N-1}$$

Example 20.2 Suppose we want to determine the proportion of college-bound high school seniors in a class of 500. A survey of 30 randomly selected students reveals that 19 will be attending college. Create a 90% confidence interval for the population proportion p.

The ratio of the sample size to the population size is 30/500 = 0.06. Since this is greater than 0.05, we will have to make an adjustment to the sample variance. To do this, we will create a macro called *adjpint* to calculate both the adjusted variance and the confidence interval for a population proportion. Because we will borrow a lot from the *pint* macro in Chap. 10 to program the macro *adjpint*, the analysis of the *adjpint* macro is very similar to that of the *pint* macro.

```
MTB > %adjpint
Executing from file: C:\Program Files\Mtbwin\MACROS\adjpint.MAC
this macro calculates the adjusted sample variance
and then finds a confidence interval for a sample proportion
based on the adjusted sample variance


What is the population size?
DATA> 500
What is the sample size?
DATA> 30
What is proportion?
DATA> .6333
The adjusted sample variance is
```

Data Display

```
KadjVar    0.00729116

The adjusted standard deviation is
```

Data Display

```
Kadjstdv    0.0853883

What is the desired percentage confidence interval?
DATA> .90

The confidence interval is:
```

Data Display

```
Row    Clowint    Chighint

  1   0.492849    0.773751
```

From the *adjpint* macro, we see that the 90% confidence interval is from 49.2849% to 77.3751%. The Minitab *adjpint* macro is shown in Appendix 20.2.

20.5 Determining Sample Size

Up to now we have obtained the sample size first and then found the standard deviation. What if we did the reverse and knew a specific standard deviation but did not know the sample size? The formula to find the required sample size for a required standard deviation would be

$$n = \frac{N\sigma^2}{(N-1)\hat{\sigma}_{\bar{x}}^2 + \sigma^2}$$

Example 20.3 Crow Company's accountant decides that the best way to determine the company's mean accounts receivable is to take a simple random sample of the 1025 accounts. Assume that the population variance, σ^2, is \$2425. What sample size should the accountant take if she would like to have a level of precision of \$75?

Now let's see how we would use *sampsize* in the Crow Company case. The Minitab code for the macro *sampsize* is shown in Appendix 20.3.

```
MTB > %sampsize
Executing from file: C:\Program Files\Mtbwin\MACROS\sampsize.MAC
This macro calculates the required sample size
for a specific standard deviation.
What is the population size?
DATA> 1025
What is the sample variance?
DATA> 75
What is the population variance?
DATA> 2425

The required sample size is
```

Data Display

```
Ksize    31.3743
```

From the *sampsize* macro, we can see that a sample of 32 accounts will produce the desired result.

Example 20.4 Use the information in Example 20.5A, but assume the accountant would like a sample variance of \$50.

```
MTB > %sampsize
Executing from file: C:\Program Files\Mtbwin\MACROS\sampsize.MAC
This macro calculates the required sample size
for a specific standard deviation.
What is the population size?
DATA> 1025
What is the sample variance?
DATA> 50
What is the population variance?
DATA> 2425

The required sample size is
```

Data Display

```
Ksize     46.3520
```

From the *sampsize* macro output, we can see that 47 is the required sample size for a sample variance of $50. Notice that as the sample variance decreases, the sample size increases.

20.6 SAS Programming Code Instructions and Examples

20.6.1 Random Number Tables

In this section, we will use two loops to create a p by q table that has numbers randomly generated from 0 to p * q. The first loop will go through all the p columns, and the second loop will define q random values for each column. Function **ranuni(seed)** gives out the random number between 0 and 1 with uniform distribution. Function **ROUND()** makes the value integer.

In macro Random_Table, the parameter "row" and "col" will decide the table size and then generate a random table with size row*col.

```
%macro Random_Table(row, col);
data Random_Table (drop=row col);
/*Generate ten columns named C1 to C10*/
 ARRAY C(&col);
 do row=1 to &row;
    do col=1 to &col;
    C(col)=round(ranuni(123)*100);
       end;
       output;
  end;
 run;
 proc print data=Random_Table;
 run;
 %mend;

 %Random_Table(10, 10);
```

Obs	C1	C2	C3	C4	C5	C6	C7	C8	C9	C10
1	75	32	18	91	36	22	79	40	12	19
2	78	44	97	26	71	55	53	86	14	86
3	65	77	70	33	52	93	5	61	66	37
4	59	31	54	28	53	56	10	11	58	67
5	48	15	75	96	76	96	9	68	89	28
6	66	80	79	32	78	55	70	83	10	7
7	25	23	79	69	26	86	47	69	70	55
8	8	63	75	82	9	51	35	20	11	22
9	100	1	24	2	84	68	1	34	85	12
10	75	96	95	22	29	35	83	67	31	61

20.6.2 Confidence Interval for the Population Mean

Example 20.5 The SAS code to calculate the adjusted variance of the sample mean and the confidence interval for a population mean is shown below.

```
%macro CI(N_p, n_s, s, x_mean);
data CI;
     sigma=sqrt(&s.**2/&n_s.*(&N_p.-&n_s.)/(&N_p.-1));
     Clowint=&x_mean.-probit(0.975)*sigma;
     Chighint=&x_mean.+probit(0.975)*sigma;
run;
proc print data=CI;
run;
%mend;

%CI(1000,75,120000,375000);
```

Obs	Sigma	Clowint	Chighint
1	13333.33	348867.15	401132.85

From the *adjzint* macro, we can see that the 95% confidence interval is from \$348,867 to \$401,133.

20.6.3 Confidence Interval for the Population Proportion

Example 20.6 This macro is very similar to that of the macro CIProp().

```
%macro CIProp(N_p, n_s, p);
data StatTable;
       sigma=sqrt(&p.*(1-&p.)/&n_s.*(&N_p.-&n_s.)/(&N_p.-1));
       Clowint=&p.-probit(0.95)*sigma;
          Chighint=&p.+probit(0.95)*sigma;
run;
proc print data=StatTable;
run;
%mend;

%CIProp(500, 30, 0.6333);
```

Obs	Sigma	Clowint	Chighint
1	0.085388	0.49285	0.74273

20.6.4 Determining Sample Size

Example 20.7 Macro **SamSize()** is used to calculate the sample size.

```
%macro SamSize(N_p, var, sam_var);
data StatTable;
       n=&N_p.*&var./((&N_p.-1)*&sam_var.+&var.);
run;
proc print data=StatTable;
run;
%mend;

%SamSize(1025, 2425, 75);
```

Obs	n
1	31.3743

Example 20.8 The same macro is used in this example.

```
%SamSize(1025, 2425, 50);
```

Obs	n
1	46.3520

20.7 Statistical Summary

In this chapter we first discussed how random number tables can be generated. Then we discussed samples that have a ratio of sample size to population size greater than 0.05. This is important because in such a case, we have to make some adjustments to make the estimate unbiased.

We also learned how to determine the required sample size for a desired sample variance.

Appendix 20.1: Minitab Code—*Adjzint*

```
macro
adjZint
mconstant Kmean Kstdv Ksize Kpercent Kinvcdf KinvcdfP KinvcdfS Kupper Klower
mconstant Kpop KadjVar Kadjstdv
mcolumn   Cmean Cstdv Csize CPercent Chighint Clowint Cpop
noecho
#
note this macro calculates the adjusted sample variance
note and then finds a confidence interval for a sample mean
note based on the adjusted sample variance
note
note What is the population size?
set 'terminal' Cpop;
nobs 1.
end

note What is the sample size?
set 'terminal' Csize;
nobs 1.
End

note What is the sample standard deviation?
set 'terminal' Cstdv;
nobs 1.
End

let Kpop =  Cpop
let Kstdv = Cstdv
let Ksize = Csize

#calculate variance
let Kadjvar = ((Kstdv**2)/Ksize) * ((Kpop-Ksize)/(Kpop - 1))
let Kadjstdv= sqrt(Kadjvar)

note The adjusted sample variance is
print Kadjvar
```

```
note
note The adjusted standard deviation is
print Kadjstdv

note
note What is the mean?
set 'terminal' Cmean;
nobs 1.
end


note What is the desired percentage confidence interval?
set 'terminal' Cpercent;
nobs 1.
End

let Kmean = Cmean
let Kpercent = Cpercent
let KinvcdfP = (1 + Kpercent)/2
invcdf kinvcdfP KinvcdfS;
normal 0 1.


#calculating the upper and lower end of interval
let Chighint= kmean + Kinvcdfs *Kadjstdv
let Clowint= kmean - Kinvcdfs *Kadjstdv


note The confidence interval is:
print clowint chighint
```

Appendix 20.2: Minitab Code—*Adjpint*

```
macro
adjPint
mconstant Ksize Kpercent Kinvcdf KinvcdfP KinvcdfS
mconstant Kupper Klower Kpop KadjVar Kadjstdv Kpro
mcolumn   Csize CPercent Chighint Clowint Cpop Cpro
noecho
#
note this macro calculates the adjusted sample variance
note and then finds a confidence interval for a sample proportion
note based on the adjusted sample variance
note
note
note What is the population size?
set 'terminal' Cpop;
nobs 1.
end
note What is the sample size?
set 'terminal' Csize;
nobs 1.
End
```

```
note What is proportion?
set 'terminal' Cpro;
nobs 1.
End

let Kpop =  Cpop
let Kpro = Cpro
let Ksize = Csize

#calculate variance
let Kadjvar = ((Kpro * (1-Kpro))/Ksize) * ((Kpop-Ksize)/(Kpop-1))
let Kadjstdv= sqrt(Kadjvar)

note The adjusted sample variance is
print Kadjvar
note
note The adjusted standard deviation is
print Kadjstdv


note
note What is the desired percentage confidence interval?
set 'terminal' Cpercent;
nobs 1.
End


let Kpercent = Cpercent
let KinvcdfP = (1 + Kpercent)/2
invcdf kinvcdfP KinvcdfS;
normal 0 1.


#calculating the upper and lower end of interval
let Chighint= Kpro+ Kinvcdfs *Kadjstdv
let Clowint= Kpro- Kinvcdfs *Kadjstdv

note
note The confidence interval is:
print clowint chighint

endmacro
```

Appendix 20.3: Minitab Macro—*SampSize*

```
macro
sampsize
mconstant Ksize
mcolumn Cpop Csampvar CpopVar
noecho
note This macro calculates the required sample size
note for a specific standard deviation.
note What is the population size?
set 'terminal' Cpop;
nobs 1.
end
note What is the sample variance?
set 'terminal' Csampvar;
nobs 1.
end
note What is the population variance?
set 'terminal' CpopVar;
nobs 1.
end
let Ksize=(Cpop * CpopVar)/((Cpop-1)* Csampvar + CpopVar)
note
note The required sample size is
print Ksize

endmacro
```

Bibliography

Lee CF, Lee JC, Lee AC (2013) Statistics for business and financial economics. Springer, New York
Microsoft Inc., Excel 2013. Microsoft Inc., Redmond
Minitab Inc. Minitab 17. Minitab Inc., State College
SAS Institute Inc. SAS 2014. SAS Institute Inc., Cary

Chapter 21
Statistical Decision Theory

21.1 Introduction

This chapter was originally in Lee et al. (2013). It discusses the statistical decision theory. The main topics covered in this chapter are (1) Four Key Elements of a Decision, (2) Decisions Based on Extreme Values, (3) Expected Monetary Value and Utility Analysis, (4) Bayes Strategies, (5) Decision Trees and Expected Monetary Values, (6) Mean and Variance Trade-Off Analysis (Optional), and (7) The Mean and Variance Method for Capital Budgeting Decisions.

The Excel applications will be emphasized in this chapter. Section 21.2 discusses the Excel applications in the decision tree approach to calculate the expected monetary values. Section 21.3 presents Excel applications for net present value (NPV) and internal rate of return (IRR) methods for capital budgeting decision under certainty. Section 21.4 discusses Excel application of z-statistic method for capital budgeting decision under uncertainty. Section 21.5 summarizes the chapter.

21.2 Decision Trees and Expected Monetary Values

Following the example from Sect. 21.6 of Lee et al. (2013), we assume that an oil company is trying to decide whether to test for the presence of oil or to drill for oil. If oil is struck, the revenues are $1 million, with a cost of $100,000 to drill. The firm has to decide whether to test first for the presence of oil. Without testing, the probability of striking oil is .1. Thus the firm's expected value without testing is 0 ($1 million times .1, less drilling fees of $100,000). The cost of the oil test is $50,000. If the test is positive, there is a .6 probability that oil will be struck; if the test is negative, the probability of striking oil is .05.

The expected gain for a negative test result is .05 (1,000,000) − 100,000 − 50,000 = −$100,000. If the test is positive, the expected profit is (1,000,000) (.6) − 100,000 − 50,000 = $450,000. If the firm's test is negative, the firm should not drill; however, if the test is positive, it should drill. This analysis makes it clear that the test should be done. The abovementioned information can be summarized as in Table 21.1. The Excel functions of decision trees in terms of Table 21.1 are presented in Fig. 21.1. The results of decision trees in terms of Fig. 21.1 are presented in Fig. 21.2.

C.-F. Lee et al., *Essentials of Excel, Excel VBA, SAS and Minitab for Statistical and Financial Analyses*, DOI 10.1007/978-3-319-38867-0_21

Table 21.1 The inputs of the example above

Numbers for the oil drilling problems	
Test cost	50,000
Drilling cost	100,000
Payoff for successful drilling	1,000,000
Probability for test results	
Positive	0.5
Negative	0.5
Probability for drilling results	
Test positive	
Successful	0.6
Fail	0.4
Test negative	
Successful	0.05
Fail	0.95
No test	
Successful	0.1
Fail	0.9

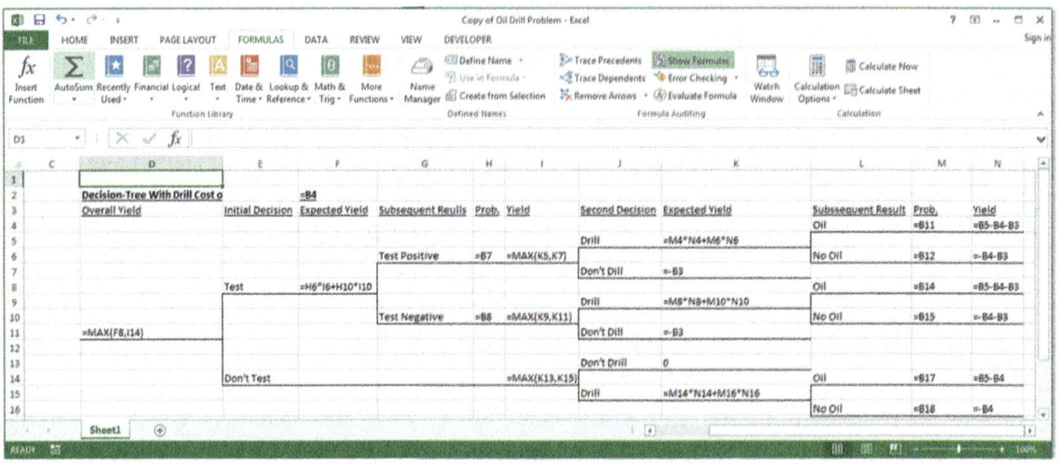

Fig. 21.1 The Excel functions of decision trees in terms of Table 21.1

21.3 NPV and IRR Method for Capital Budgeting Decision Under Certainty

Both net present value (NPV) method and internal rate of return (IRR) method can be used to do the capital budgeting decision. For example, project A and project B, the initial outlays, and net cash inflow for year 0 to year 4 are presented in Table 21.2. In Table 21.2, we know that the initial outlays at year 0 for projects A and B are $80,000 and $50,000, respectively. In year 1, additional investments for projects A and B are $20,000 and $50,000, respectively. The net cash inflows of project A for the next 4 years are $20,000, $30,000, $50,000, and $50,000, respectively. The net cash inflows of project B for the next 4 years are $40,000, $60,000, $30,000, and $10,000, respectively.

The net present value of a project is computed by discounting the project's cash flows to the present by the appropriate cost of capital. The formula used to calculate NPV can be defined as follows:

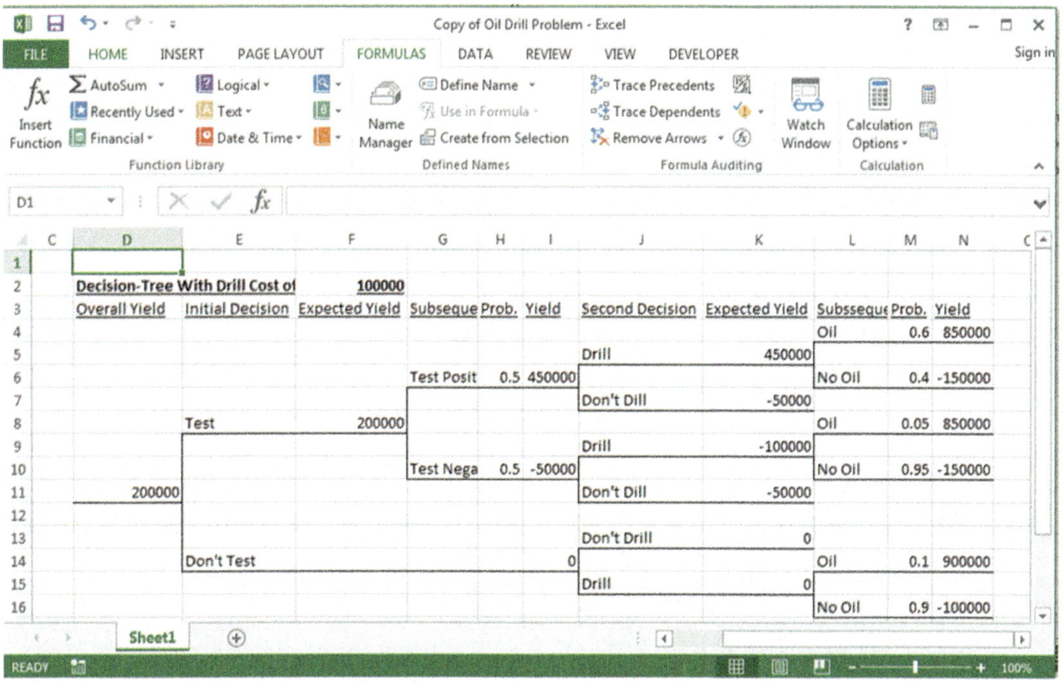

Fig. 21.2 The results of decision trees in terms of Fig. 21.1

Table 21.2 Partial input information for NPV method and IRR method

		Year 0	Year 1	Year 2	Year 3	Year 4
Project A	Cost	−$80,000	−$20,000	0	0	0
	Return	0	$20,000	$30,000	$50,000	$50,000
Project B	Cost	−$50,000	−$50,000	0	0	0
	Return	0	$40,000	$60,000	$30,000	$10,000

$$NPV = \sum_{t=1}^{N} \frac{CF_t}{(1 + k)^t} - I \tag{21.1}$$

where:

k = the appropriate discount rate
CFt = net cash flow (positive or negative) in period t
I = initial outlay
N = life of the project

Using the Excel function, we can calculate NPV for both projects A and B. We also can calculate the example above using Excel NPV function. NPV is a function to calculate the net present value of an investment by using a discount rate and a series of future payments (negative values) and income (positive values). The NPV function in Cell H10 is equal to

$$= \mathrm{NPV}(\mathrm{C2}, \mathrm{D10} : \mathrm{G10}) + \mathrm{C10}$$

Based upon the NPV function on Fig. 21.3, the NPV results are shown in Fig. 21.4.

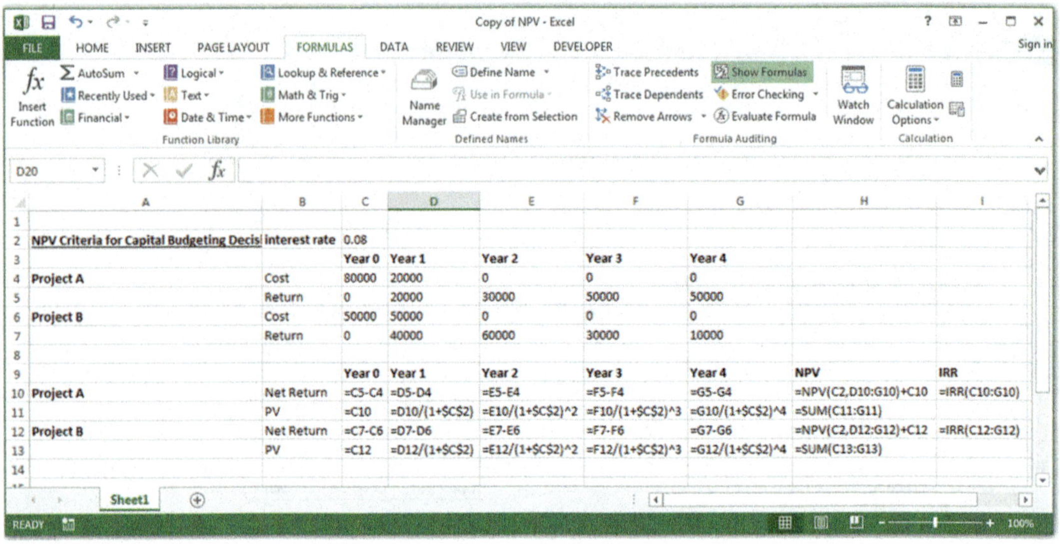

	A	B	C	D	E	F	G	H	I
1									
2	NPV Criteria for Capital Budgeting Decis	interest rate	0.08						
3			Year 0	Year 1	Year 2	Year 3	Year 4		
4	Project A	Cost	80000	20000	0	0	0		
5		Return	0	20000	30000	50000	50000		
6	Project B	Cost	50000	50000	0	0	0		
7		Return	0	40000	60000	30000	10000		
8									
9			Year 0	Year 1	Year 2	Year 3	Year 4	NPV	IRR
10	Project A	Net Return	=C5-C4	=D5-D4	=E5-E4	=F5-F4	=G5-G4	=NPV(C2,D10:G10)+C10	=IRR(C10:G10)
11		PV	=C10	=D10/(1+$C$2)	=E10/(1+$C$2)^2	=F10/(1+$C$2)^3	=G10/(1+$C$2)^4	=SUM(C11:G11)	
12	Project B	Net Return	=C7-C6	=D7-D6	=E7-E6	=F7-F6	=G7-G6	=NPV(C2,D12:G12)+C12	=IRR(C12:G12)
13		PV	=C12	=D12/(1+$C$2)	=E12/(1+$C$2)^2	=F12/(1+$C$2)^3	=G12/(1+$C$2)^4	=SUM(C13:G13)	

Fig. 21.3 Excel calculation functions for NPV method

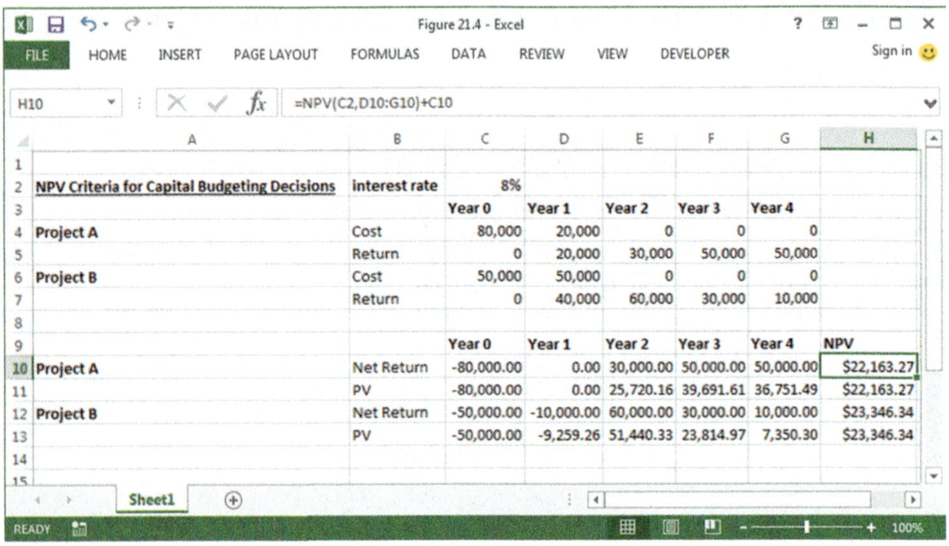

	A	B	C	D	E	F	G	H
1								
2	NPV Criteria for Capital Budgeting Decisions	interest rate	8%					
3			Year 0	Year 1	Year 2	Year 3	Year 4	
4	Project A	Cost	80,000	20,000	0	0	0	
5		Return	0	20,000	30,000	50,000	50,000	
6	Project B	Cost	50,000	50,000	0	0	0	
7		Return	0	40,000	60,000	30,000	10,000	
8								
9			Year 0	Year 1	Year 2	Year 3	Year 4	NPV
10	Project A	Net Return	-80,000.00	0.00	30,000.00	50,000.00	50,000.00	$22,163.27
11		PV	-80,000.00	0.00	25,720.16	39,691.61	36,751.49	$22,163.27
12	Project B	Net Return	-50,000.00	-10,000.00	60,000.00	30,000.00	10,000.00	$23,346.34
13		PV	-50,000.00	-9,259.26	51,440.33	23,814.97	7,350.30	$23,346.34

Fig. 21.4 Excel calculation results for NPV method

The internal rate of return (IRR, r) is the discount rate which equates the discounted cash flows from a project to its investment. Thus, one must solve iteratively for the r in Eq. (21.2):

$$\sum_{t=1}^{N} \frac{CF_t}{(1+r)^t} = I \tag{21.2}$$

where:

CFt = net cash flow (positive or negative) in period t
I = initial investment

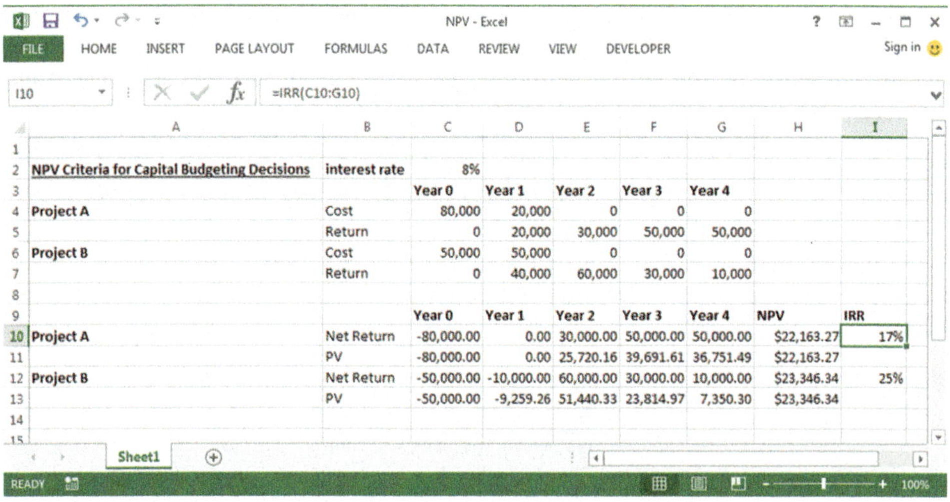

	A	B	C	D	E	F	G	H	I
1									
2	NPV Criteria for Capital Budgeting Decisions	interest rate	8%						
3			Year 0	Year 1	Year 2	Year 3	Year 4		
4	Project A	Cost	80,000	20,000	0	0	0		
5		Return	0	20,000	30,000	50,000	50,000		
6	Project B	Cost	50,000	50,000	0	0	0		
7		Return	0	40,000	60,000	30,000	10,000		
8									
9			Year 0	Year 1	Year 2	Year 3	Year 4	NPV	IRR
10	Project A	Net Return	-80,000.00	0.00	30,000.00	50,000.00	50,000.00	$22,163.27	17%
11		PV	-80,000.00	0.00	25,720.16	39,691.61	36,751.49	$22,163.27	
12	Project B	Net Return	-50,000.00	-10,000.00	60,000.00	30,000.00	10,000.00	$23,346.34	25%
13		PV	-50,000.00	-9,259.26	51,440.33	23,814.97	7,350.30	$23,346.34	
14									
15									

Fig. 21.5 The Excel calculation results for IRR

N = life of the project
r = the internal rate of return

In addition, we can use Excel function IRR to calculate the internal rate of return. IRR is a function to calculate the internal rate of return which is the rate of return received for an investment consisting of payments (negative values) and income (positive values) that occur at regular periods. The IRR function in Cell I10 is

$$= \mathrm{IRR}(\mathrm{C10} : \mathrm{G10})$$

Based upon the IRR function in Fig. 21.3, the IRR results in terms of Excel calculations are shown in Fig. 21.5.

21.4 The Statistical Distribution Method for Capital Budgeting Decision Under Uncertainty

21.4.1 Methodology

The riskiness of an investment proposal is generally defined as the variability of its possible returns. Decision situations can generally be classified into three types: certainty, risk, and uncertainty. The distinction between risk and uncertainty is that risk involves situations in which the probabilities of a particular event occurring are known, whereas with uncertainty these probabilities themselves are not known. Throughout this chapter, the terms "risk" and "uncertainty" will be used interchangeably, although the strict difference between them is not a matter to be taken lightly.

Economic factors peculiar to investment, competition, technological development, consumer preferences, and labor market conditions are a few of the factors that make it virtually impossible to foretell the future. Consequently, the revenues, costs, and economic life of investment projects are less than certain.

With the introduction of risk, a firm is no longer indifferent between two investment proposals having equal net present values or internal rates of return. Both net present value and its standard deviation (σ_{NPV}) should be estimated in performing capital budgeting analyses under uncertainty. An NPV approach under uncertainty can be defined as

$$\widetilde{NPV} = \sum_{t=1}^{N} \frac{\widetilde{C}_t}{(1+k)^t} + \frac{S_n}{(1+k)^N} - I_0 \tag{21.3}$$

where C_t is the uncertain net cash flow in period t, k is the risk-adjusted discount rate, S_n is the salvage value, and I_0 is the initial outlay. The mean of the NPV distribution and its standard deviation can be defined as

$$\overline{NPV} = \sum_{t=1}^{N} \frac{\overline{C}_t}{(1+k)^t} + \frac{S_n}{(1+k)^N} - I_0 \tag{21.4}$$

$$\sigma_{NPV} = \left[\sum_{t=1}^{N} \frac{\sigma_t^2}{(1+k)^{2t}} + \sum_{\tau=1}^{N} \sum_{t=1}^{N} w_t w_\tau Cov(C_\tau, C_t) \right]^{1/2} \quad (\tau \neq t) \tag{21.5}$$

where:

σ_t^2 = variance of cash flows in the tth period
W_t, W_τ = discount factor for the tth and the τth period, respectively, that is

$$W_t = \frac{1}{(1+K_f)^t} \quad \text{and} \quad W_\tau = \frac{1}{(1+K_f)^\tau}$$

$\text{Cov}(C_t, C_\tau)$ = covariability between cash flows C_τ and C_t

Cash flows between periods t and τ are generally related. Therefore, $\text{Cov}(C_t, C_\tau)$ is an important factor in the estimation of σ_{NPV}. The magnitude, sign, and degree of the relationships of these cash flows depend on the economic operating conditions and on the nature of the product or service being produced. If there are only three periods, then all terms within parenthesis in Eq. (21.5) can be presented as in Table 21.3. The summation of the diagonal elements $[W_1^2\sigma_1^2, W_2^2\sigma_2^2, W_3^2\sigma_3^2]$ of Table 21.3 results in the first part of Eq. (21.5), or

$$\sum_{t=1}^{N} \frac{\sigma_t^2}{(1+K_f)^{2t}}$$

The summation of all other elements in Table 21.3 gives the second portion of Eq. (21.5), or

$$\sum_{t=1}^{N} \sum_{\tau=1}^{N} W_t W_\tau \text{Cov}(C_\tau, C_t), \quad t \neq \tau$$

Table 21.3 Variance-covariance matrix

$W_1^2\sigma_1^2$	$W_1W_2\text{Cov}(C_1, C_2)$	$W_1W_3\text{Cov}(C_1, C_3)$
$W_2W_1\text{Cov}(C_2, C_1)$	$W_2^2\sigma_2^2$	$W_1W_3\text{Cov}(C_2, C_3)$
$W_3W_1\text{Cov}(C_3, C_1)$	$W_2W_3\text{Cov}(C_2, C_3)$	$W_3^2\sigma_3^2$

Equation (21.5) is the general equation for σ_{NPV}. Both Eq. (21.7) for σ_{NPV} under perfectly positively correlated cash flow and Eq. (21.8) for independent cash flows are special cases derived from the general Eq. (21.5). If $\rho_{12} = \rho_{13} = \rho_{23} = 1$, then $\mathrm{Cov}(C_1, C_2) = \sigma_1\sigma_2$, $\mathrm{Cov}(C_1, C_2) = \sigma_1\sigma_3$, and $\mathrm{Cov}(C_2, C_3) = \sigma_1\sigma_3$. Therefore, Eq. (21.5) reduces to

$$\sigma_{NPV} = \left(\frac{\sigma_1^2}{(1+K_f)^2} + \frac{\sigma_2^2}{(1+K_f)^4} + \frac{\sigma_3^2}{(1+K_f)^6} + \frac{2\sigma_1\sigma_2}{(1+K_f)^3} + \frac{2\sigma_1\sigma_3}{(1+K_f)^4} + \frac{2\sigma_2\sigma_3}{(1+K_f)^5}\right)^{\frac{1}{2}} = \sum_{t=1}^{3} \frac{\sigma_t}{(1+K_f)^t}$$

which is Eq. (21.7).

$\mathrm{Cov}(C_\tau, C_t)$ is used to measure the covariability between the cash flow in the tth and τth periods. It is well known that

$$\mathrm{Cov}(C_t, C_\tau) = \rho_{\tau t}\sigma_\tau\sigma_t \tag{21.6}$$

where $\rho_{\tau t}$ is the correlation coefficient between the cash flow in the ith period and that in the τth period. In one special case, the cash flows are perfectly correlated over time. If such is the case, then the deviation of the actual cash flow from the mean of its corresponding profitability distribution of expected cash flows for the period gives us the information that cash flows in future periods will deviate in a similar fashion. In other words, Eq. (21.5) can be rewritten as[3]

$$\sigma_{NPV} = \sum_{t-1}^{N} \frac{\sigma_t}{(1+k)^t} \tag{21.7}$$

If the cash flows are mutually independent, then Eq. (21.5) reduces to

$$\sigma_{NPV} = \sqrt{\sum_{t=1}^{N} \frac{\sigma_t^2}{(1+k)^{2t}}} \tag{21.8}$$

Hillier (1963) combines the assumption of mutual independence and perfect correlation in developing a model to deal with mixed situations. The model as defined in Eq. (21.9) can be used to analyze investment proposals in which some of the expected cash flows are closely related, and others are fairly independent:

$$\sigma = \sqrt{\sum_{t=1}^{N} \frac{\sigma_{yt}^2}{(1+k)^{2t}} + \sum_{h}^{m} \left(\sum_{t=1}^{N} \frac{\sigma_{zt}^h}{(1+k)^t}\right)^2} \tag{21.9}$$

where $\sigma_{yt}{}^2$ is the variance for an independent net cash flow in period t and $\sigma_{zt}{}^h$ is the standard deviation for stream h of a perfectly correlated stream of cash flows t. It is clear that Eq. (21.9) is a combination of Eqs. (21.7) and (21.8). Chen and Moore (1982) have generalized this model by introducing the estimation risk.

They argue that Hillier's (1963) approach is unrealistic due to the assumption of parameter certainty. In reality, the parameters of the probability distribution are never known with certainty, and thus the capital budgeting decisions based on Hillier's approach may be potentially misleading because this approach ignores estimation risk. Chen and Moore (1982) hence extend Hillier's approach with a Bayesian framework which leads to the derivation of a predicative distribution that is not conditioned on unknown parameters. Therefore, the effect of estimation risk in capital budgeting decisions can be eliminated in Chen and Moore (1982) approach.

Table 21.4 Expected cash flow for new product

Year	Source	Expected value of net cash flow (in thousands)	Standard deviation (in thousands)
0	Initial investment	−$600	$50
1	Production cash outflow	−250	20
2	Production cash outflow	−200	10
3	Production cash outflow	−200	10
4	Production cash outflow	−200	10
5	Production outflow—salvage value	−100	15
1	Marketing	300	50
2	Marketing	600	100
3	Marketing	500	100
4	Marketing	400	100
5	Marketing	300	100

Reprinted with permission from Hillier, F., "The derivation of probabilistic information for the evaluation of risky investments," *Management Science* 9 (April 1963). Copyright 1963 by the Institute of Management Sciences

In addition, Wang and Lee (2010) provide the fuzzy real option valuation approach to solve the capital budgeting decisions under uncertainty environment. In Wang and Lee's model framework, the concept of probability is employed in describing fuzzy events, and the estimated cash flows are based on the fuzzy numbers which can better reflect the uncertainty in the project. By using a fuzzy real option valuation, the managers can select fuzzy projects and determine the optimal time to abandon the project under the assumption of limited capital budget.

To illustrate the model of Eq. (21.9), suppose that a firm is considering the introduction of a new product with returns expected over the next 5 years. Because the product's market reception is uncertain, management feels that, if initial reception exceeds expectations, actual acceptance in later years will also exceed expectations, in approximately the same proportions. For simplicity, it is believed that the net marketing cash flow (sales less marketing and advertising expenses) can be treated as perfectly correlated over time.

On the other hand, estimates of the initial investment in the project and the production costs are reasonably reliable, so that any deviation from expectations is assumed to be attributable to random fluctuations. Consequently, initial investment and net production cash flows are regarded as being mutually independent over time. The probability information for the introduction of the new product is shown in Table 21.4. We assume that each of the probability distributions involved can be regarded as normal. If 10 % is used as the risk-free rate, the expected value of the net present value for the proposal is

$$\begin{aligned} NPV &= \sum_{t=0}^{5} \frac{C_t}{(1.04)^t} = -600 + \frac{300-250}{(1.04)} + \frac{600-200}{(1.04)^2} + \frac{500-200}{(1.04)^3} \\ &\quad + \frac{400-200}{(1.04)^4} + \frac{300-100}{(1.04)^5} \\ &= \$419.95 = \$420.00 \end{aligned}$$

Following Eq. (21.5), we can calculate the standard deviation as

$$\sigma_{NPV} = \left[50^2 + \frac{20^2}{(1.04)^2} + \cdots + \frac{15^2}{(1.04)^{10}} + \left(\frac{50}{1.04} + \cdots + \frac{100}{(1.04)^5} \right)^2 \right]^{1/2} = \$398$$

Thus, the expected net present value of the proposal is \$420, and the standard deviation of the distribution relating to NPV is \$398. Using this information, we can express an interval inference for NPV as:

Pr (\$420 ± \$398) = 68.27 %
Pr (\$420 ± (2)(\$398)) = 95.45 %
Pr (\$420 ± (3)(\$398)) = 99.73 %

Hillier's approach takes us a long way toward coping with the correlation of cash flows over time, but his method is not always practically applicable. For many investment proposals, cash flows fall into neither of these categories, since in practice they show less than perfect correlation over time. If the correlation is moderate, then the classification suggested by Hillier will not be appropriate. The method for handling the problem of moderate correlation is with a series of conditional probability distributions. Wang and Lee (2010) have expended Bonini's result in terms of fuzzy set and real option approach.

To illustrate the use of this model, we refer to Table 21.5, where we consider a project requiring an initial outlay of \$10,000 (column 1). In each of the following time periods (columns 3, 5, and 7), the cash flow to be received is not known with perfect certainty, but the probabilities associated with each cash flow in each period are assumed to be known (columns 2, 4, and 6), so that we are dealing with a case of risk and not strict uncertainty. We note that columns (4) and (6) are conditional probability

Table 21.5 Illustration of conditional probability distribution approach

Initial outlay period 0 (1)	Initial probability P(1) (2)	Net cash flow (3)	Conditional probability P(3|2,1) (4)	Net cash flow (5)	Conditional probability P(3|2,1) (6)	Cash flow (7)	Joint probability (8)
					0.2	1000	0.015
			0.25	2000	0.5	2000	0.0375
					0.3	3000	0.0225
					0.2	2000	0.03
10,000	0.3	2000	0.5	3000	0.5	3000	0.075
					0.3	4000	0.045
					0.2	3000	0.015
			0.25	4000	0.5	4000	0.0375
					0.3	5000	0.0225
					0.3	2000	0.045
			0.3	3000	0.4	4000	0.06
					0.3	6000	0.045
					0.3	4000	0.06
	0.5	4000	0.4	5000	0.4	5000	0.08
					0.3	6000	0.06
					0.3	5000	0.045
			0.3	7000	0.4	7000	0.06
					0.3	9000	0.045
					0.25	4000	0.0125
			0.25	5000	0.5	6000	0.025
					0.25	8000	0.0125
					0.25	5000	0.025
	0.2	6000	0.5	7000	0.5	7000	0.05
					0.25	9000	0.025
					0.25	7000	0.0125
			0.25	8000	0.5	9000	0.025
					0.25	11,000	0.0125

Table 21.6 NPV and joint probability

PV	NPV	Probability
4661.2	−5338.8	0.015
5550.2	−4449.8	0.0375
6439.2	−3560.8	0.0225
6474.76	−3525.24	0.03
7363.76	−2636.24	0.075
8252.76	−1747.24	0.045
8288.32	−1711.68	0.015
9177.32	−822.68	0.0375
10,066.32	66.32	0.0225
8297.84	−1602.16	0.045
10,175.84	175.84	0.06
11,953.84	1953.84	0.045
12,024.96	2024.96	0.06
12,913.96	2913.96	0.08
13,802.96	3802.96	0.06
14,763.08	4763.08	0.045
16,541.08	6541.08	0.06
18,319.08	8319.08	0.045
13,948.04	3948.04	0.0125
15,726.04	5726.04	0.025
17,504.04	7504.04	0.0125
16,686.16	6686.16	0.025
18,464.16	8464.16	0.05
20,242.16	10,242.16	0.025
19,388.72	9388.72	0.0125
21,166.72	11,166.72	0.025
22,944.72	12,944.72	0.0125

Discount rate = 4 %
NPV = $2517.182
Variance = $20,359,090.9093
Standard deviation = $4512.105

figures, where the later periods' expected cash flows depend highly on what occurs in earlier time periods. Given our cash flow and simple probability estimates for the periods 1, 2, and 3, we find 27 possible joint probabilities (column 8), each of which corresponds to a cash flow series. Thus, in the uppermost path, we find the joint probability of 0.015 being associated with a cash flow of $2000 in period 1, followed by cash flows of $2000 and $1000 in periods 2 and 3, respectively. This particular path is the worst possible result; in nondiscounted dollar terms, there is a 50 % loss on the investment. At the other end of the spectrum, we find the best possible outcome offers a return of $12,945 (the last line of Table 21.6) with a joint probability of 0.0125. The returns and joint probabilities can be calculated similarly for the other 25 possible patterns.

While the primary hurdle in this process is in estimating the cash flows and the probabilities, we recognize that it is the NPV figures we are ultimately interested in. Table 21.6 gives the present value of the cash flows and the net present value of the project for each cash flow series, where a constant 4 % discount rate was employed. When we multiply each joint probability by the expected net present value associated with that probability, we obtain the expected net present value for the project as a whole, there shown to be $2517.18. Also, from all 27 expected net present value figures, we can calculate the variance and the standard deviation for the project as a whole, shown in the lower portion of Table 21.8b as $20,359,090.91 and $4512.11, respectively, where these figures can be used to assess the riskiness of the project.

The use of this form of conditional probability distribution enables us to take account of the dependence or correlation in the cash flows over time. In the above example, the cash flow in period 3 depends upon the size of the cash flow in periods 2 and 1. Given the cash flow in period 1, the cash flow in period 2 can vary within a range, but more of a known range than if we did not know the cash flow in period 1. Similarly, the cash flows in period 3 are allowed to vary within a range of values, the range being determined in part by the realized cash flow in the prior period. Most importantly, it should be noted that we assume we know what the distribution of each cash flow figure is, and the probabilities associated with each flow, and, further, that once we make the investment decision we more or less have locked ourselves into the project. In the following section, we relax this latter assumption and allow sequential decisions to be introduced into the conditional probability decision framework.

21.4.2 Excel and VBA Application

In this section, we will show how Excel and VBA can be used to calculate Eqs. (21.4), (21.5), and (21.7). First, we will show how Excel program can be used to calculate the mean and standard deviation in a continuous data present in Table 21.6. In Table 21.6, the first column PV represents the summation of present value of future net cash inflow. The second column represents net present value which is PV minus initial outlay. Finally, the third column represents the probability of occurrence.

Now, we discuss how we can use either Excel program or VBA to calculate the mean and variance in terms of the data present in Fig. 21.6. In this table, column A presents state of economy, column B presents probability of occurrence, column C presents the rate of return, and column D presents

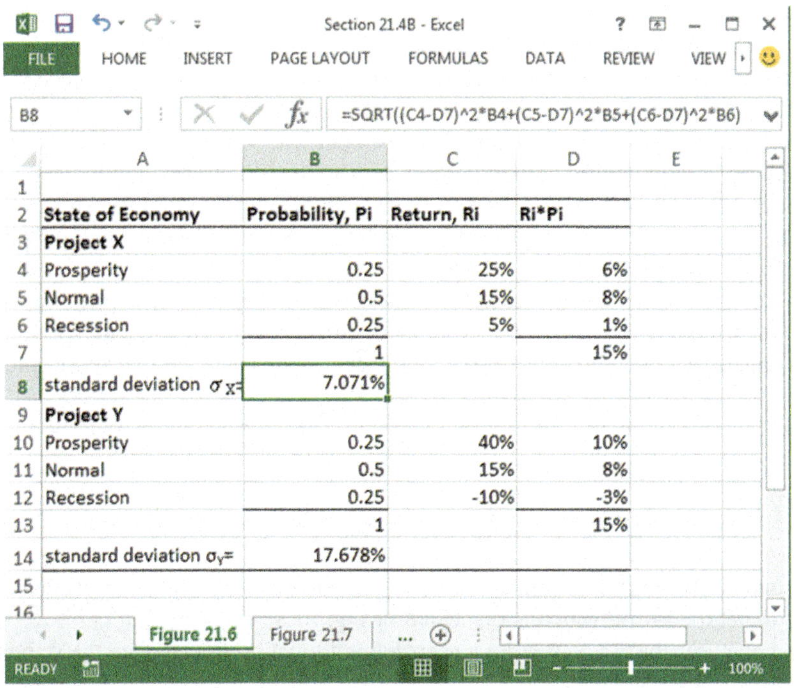

	A	B	C	D
1				
2	State of Economy	Probability, Pi	Return, Ri	Ri*Pi
3	Project X			
4	Prosperity	0.25	25%	6%
5	Normal	0.5	15%	8%
6	Recession	0.25	5%	1%
7		1		15%
8	standard deviation σ_X=	7.071%		
9	Project Y			
10	Prosperity	0.25	40%	10%
11	Normal	0.5	15%	8%
12	Recession	0.25	-10%	-3%
13		1		15%
14	standard deviation σ_Y=	17.678%		
15				

Fig. 21.6 Input data

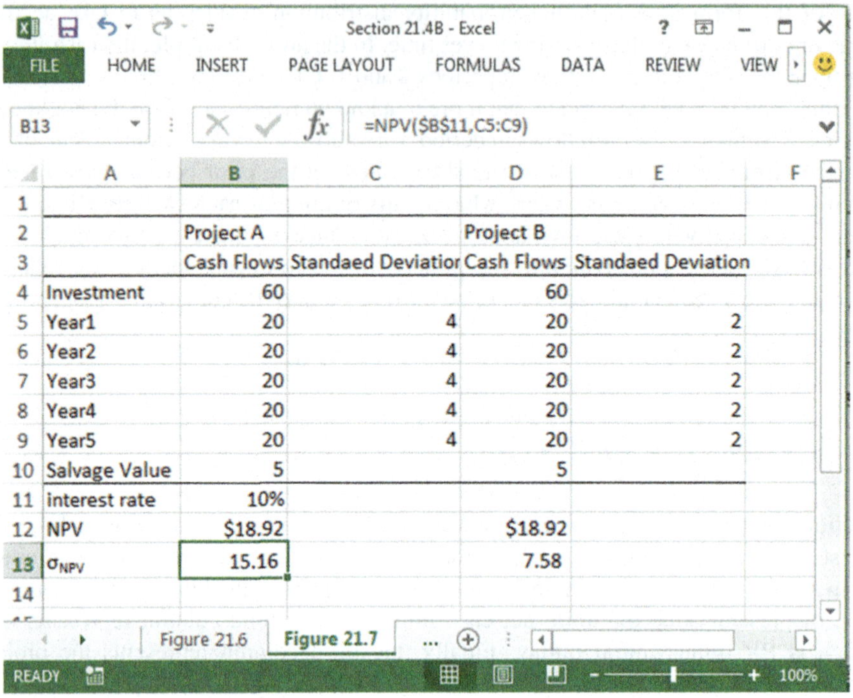

	A	B	C	D	E
1					
2		Project A		Project B	
3		Cash Flows	Standaed Deviatior	Cash Flows	Standaed Deviation
4	Investment	60		60	
5	Year1	20	4	20	2
6	Year2	20	4	20	2
7	Year3	20	4	20	2
8	Year4	20	4	20	2
9	Year5	20	4	20	2
10	Salvage Value	5		5	
11	interest rate	10%			
12	NPV	$18.92		$18.92	
13	σ_{NPV}	15.16		7.58	
14					

Fig. 21.7 Information for capital budgeting decision

column B times column C. By using this information, we can use Excel program to calculate both mean and standard deviation for projects X and Y. From Fig. 21.6, we found both rates of return for projects X and Y are 15 %; however, standard deviation for project X is 7.071 % and for project Y is 17.678 %. Therefore, we can conclude that we should prefer project X than project Y.

Figure 21.7 presents how Excel program can be used to calculate Eqs. (21.4) and (21.7). In Fig. 21.7, we have the information of initial investments, net cash inflow, and standard deviation associated with net cash inflow for both projects A and B. From the Excel output, we found that NPVs for projects A and B are both $18.92. However, the standard deviation for project A is 15.16 which is higher than the standard deviation of project B which is 7.58. Therefore, we will conclude project B will perform better than project A.

Now we will generalize the previous example by allowing the cash flow between different periods are not positively and perfectly correlated. In other words, we will use Eqs. (21.4) and (21.5) to perform capital budgeting decision under uncertainty. There are two approaches to calculate σ_{NPV}.

The first one uses Excel matrix to multiply and transpose to calculate σ_{NPV}. The function in Cell B10 is

$$= \mathrm{SQRT}(\mathrm{MMULT}(\mathrm{MMULT}(\mathrm{TRANSPOSE}(G4:G8), B4:F8), G4:G8))$$

Because it is one kind of matrix arithmetic, after typing the function in the cell, we have to press Ctrl +Shift+Enter to get the result, which is presented in Fig. 21.8.

The second one uses the VBA to write a user-defined function to calculate the standard deviation of NPV. Here is the VBA code.

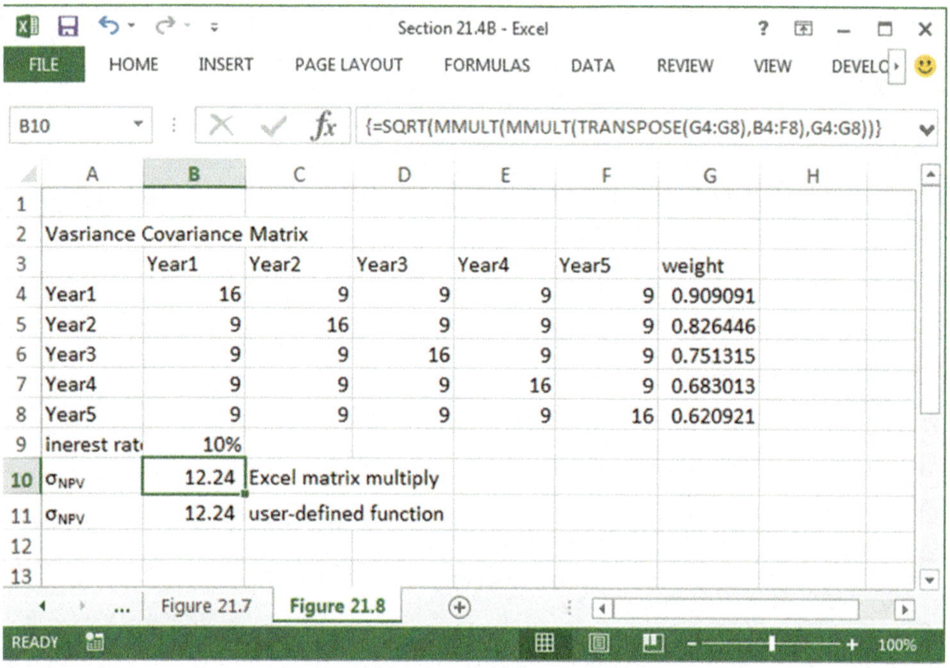

	A	B	C	D	E	F	G
1							
2	Vasriance Covariance Matrix						
3		Year1	Year2	Year3	Year4	Year5	weight
4	Year1	16	9	9	9	9	0.909091
5	Year2	9	16	9	9	9	0.826446
6	Year3	9	9	16	9	9	0.751315
7	Year4	9	9	9	16	9	0.683013
8	Year5	9	9	9	9	16	0.620921
9	inerest rat	10%					
10	σ_{NPV}	12.24	Excel matrix multiply				
11	σ_{NPV}	12.24	user-defined function				
12							
13							

Fig. 21.8 A general formula for the standard deviation of NPV

```
Function SigmaNPV(vcvm, rate)
  Dim n, i, j
  Dim sigmasum() As Variant
  n = vcvm.Columns.Count
  ReDim sigmasum(n, n)
  For i = 1 To n
      For j = 1 To n
        sigmasum(i, j) = vcvm(i, j) / (1 + rate) ^ (i + j)
      Next j
  Next i
  SigmaNPV = Sqr(Application.Sum(sigmasum))
End Function
```

Figure 21.9 shows the result of using user-defined VBA function to calculate the standard deviation (sigma) of NPV. Below in the Cell B11

$$= \text{sigmaNPV}(\text{B4} : \text{F8}, \text{B9})$$

In Fig. 21.9, we input the information of variance-covariance matrix among five periods. This information is generally calculated by input data of cash flow information related to the five periods of a project.

By either method, we find the standard deviation of NPV is 12.24. By following the same procedure discussed in the previous example, we can perform capital budgeting under uncertainty.

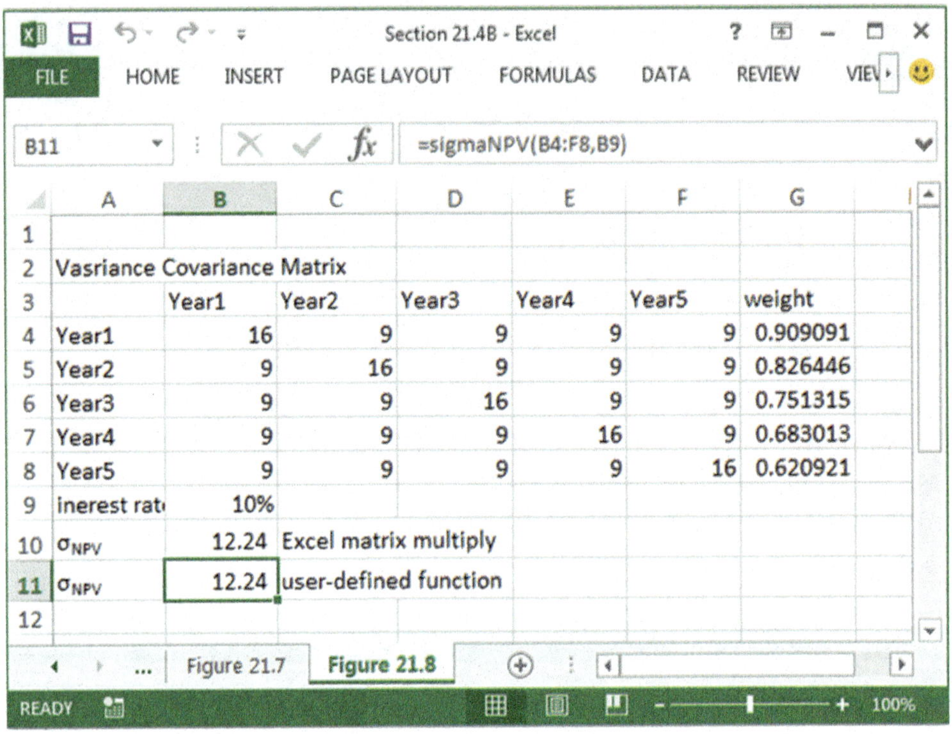

Section 21.4B - Excel

B11 =sigmaNPV(B4:F8,B9)

	A	B	C	D	E	F	G
1							
2	Vasriance Covariance Matrix						
3		Year1	Year2	Year3	Year4	Year5	weight
4	Year1	16	9	9	9	9	0.909091
5	Year2	9	16	9	9	9	0.826446
6	Year3	9	9	16	9	9	0.751315
7	Year4	9	9	9	16	9	0.683013
8	Year5	9	9	9	9	16	0.620921
9	inerest rat	10%					
10	σ_{NPV}	12.24	Excel matrix multiply				
11	σ_{NPV}	12.24	user-defined function				
12							

Fig. 21.9 A general formula for the standard deviation of NPV by using VBA

21.5 Summary

In this chapter, we have shown how Excel program can be used to do (1) decision trees and expected monetary values and (2) NPV and IRR methods for capital budgeting decision under certainty. (3) We use either Excel program or VBA program to perform statistical distribution method for capital budgeting under uncertainty. In Part B of this book, we will discuss Excel and VBA programs in further detail.

Bibliography

Chen SN, Moore WT (1982) Investment decision under uncertainty: application of estimation risk in the Hillier approach. J Financ Quant Anal 17:425–440

Microsoft Inc. Excel 2013. Microsoft Inc., Redmond

Hillier F (1963) The derivation of probabilistic information for the evaluation of risky investments. Manag Sci 9:443–457

Lee CF, Lee JC, Lee AC (2013) Statistics for business and financial economics. Springer, New York

Minitab Inc. Minitab 17. Minitab Inc., State College

SAS Institute Inc. SAS 2014. SAS Institute Inc., Cary

Wang SW, Lee CF (2010) Fuzzy set and real option approach for capital budget decision. Int J Inf Technol Decis Mak 9(5):695–714

Part B
Advanced Applications of Microsoft Excel Programs in Financial Analysis

Chapter 22
Introduction to Excel Programming

22.1 Introduction

A lot of the work done by an Excel user is repetitive and time consuming. Fortunately for an Excel user, Excel offers a powerful and professional programming language and powerful and professional programming environment to automate a big portion of their work. This book has illustrated some of the things that can be accomplished by Excel's programming language called Visual Basic for Applications or more commonly known as VBA.

This chapter will introduce programming in Excel.

22.2 Excel's Macro Recorder

There is one common question that both a novice and an experienced Excel VBA programmer will ask about Excel VBA programming: "How do I program this in Excel VBA?" The reason that the novice VBA programmer will ask this question is because of his or her lack of experience. To understand why the experienced VBA programmer will ask this question, we need to realize that Excel has an enormous amount of features. It is virtually impossible for anybody to remember how to program every feature of Excel. Interestingly, the answer to the question is the same for both the novice and the experienced programmer. The answer is Excel's macro recorder. Excel's macro recorder will record any action done by the user. The recorded result is the Excel VBA code. This is important because both the novice and experienced VBA programmer can study the resulting Excel VBA code.

Suppose that we have a great need to do the following to any cell(s) that we selected:

1. Bold the words in a/the cell(s) that we selected.
2. Italicize the words in a/the cell(s) that we selected.
3. Underline the words in a/the cell(s) that we selected.
4. Center the words in a/the cells(s) that we selected.

What is the Excel VBA code to accomplish the above list? The thing for both the novice and the experienced VBA programmer to do is to use Excel's macro recorder to record manually the actions required to get the results desired. This process is shown below.

C.-F. Lee et al., *Essentials of Excel, Excel VBA, SAS and Minitab for Statistical and Financial Analyses*, DOI 10.1007/978-3-319-38867-0_22

Before we do anything, let's type in the words as shown in worksheet "Sheet1" as shown below.

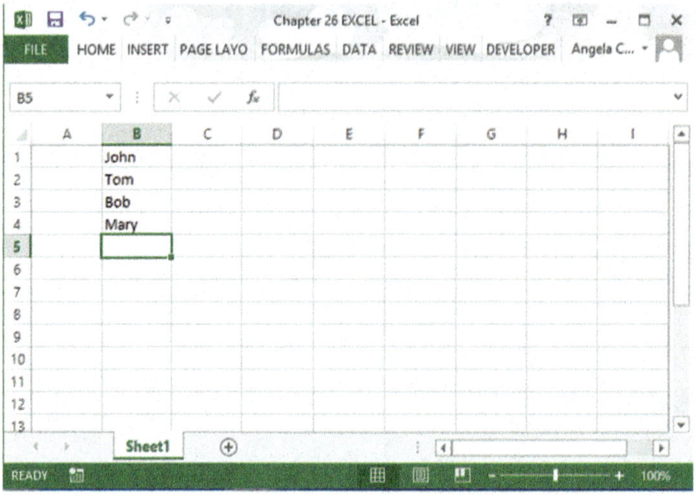

Next, highlight the words above before we start using Excel's macro recorder to generate the VBA code. To highlight the list, first select the word "John" then press the *Shift* key on the keyboard and while pressing the *Shift* key press the ↓ key on the keyboard three times. The end result is shown below.

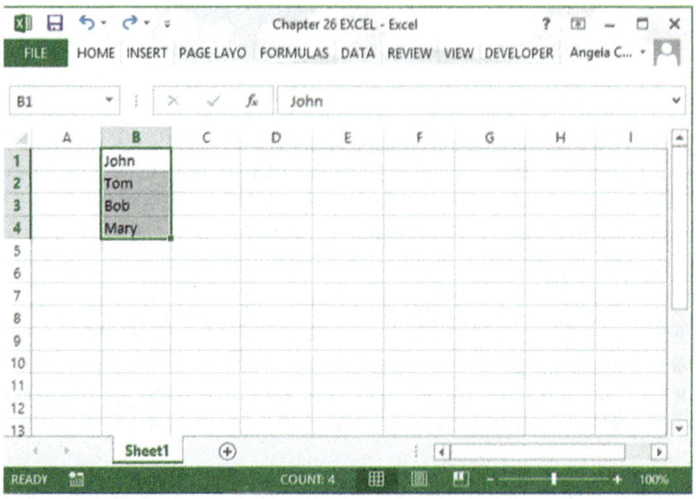

Now let's turn on Excel's macro recorder. To do this, we would choose DEVELOPER → Record Macro. This is shown below.

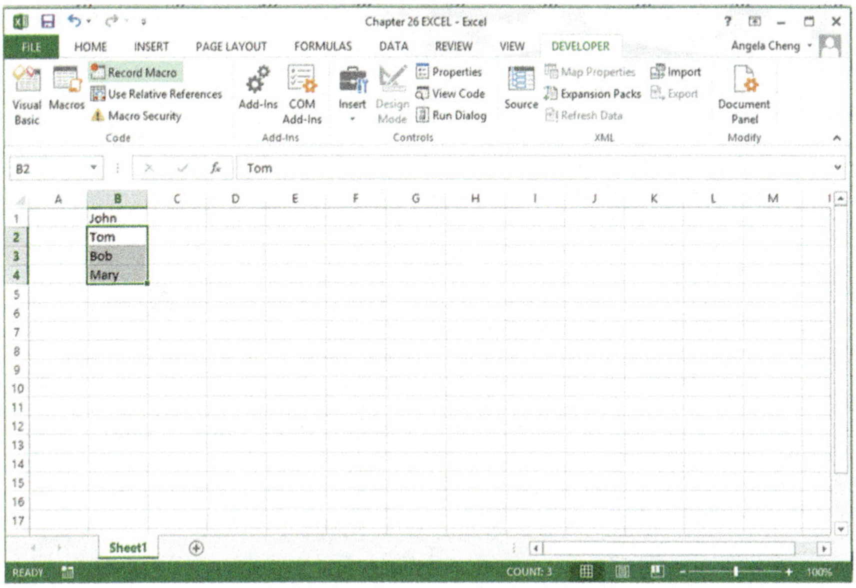

Choosing the *Record Macro* menu item would result in the *Record Macro* dialog box shown below.

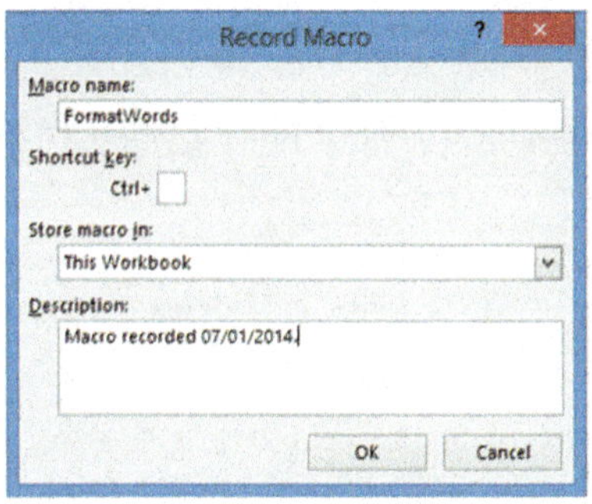

Next type "FormatWords" in the *Macro name*: option to indicate the name of our macro. After doing this, press the *OK* button.

Let's first bold the words by pressing *Ctrl+B* or press B under *HOME*. This is shown below.

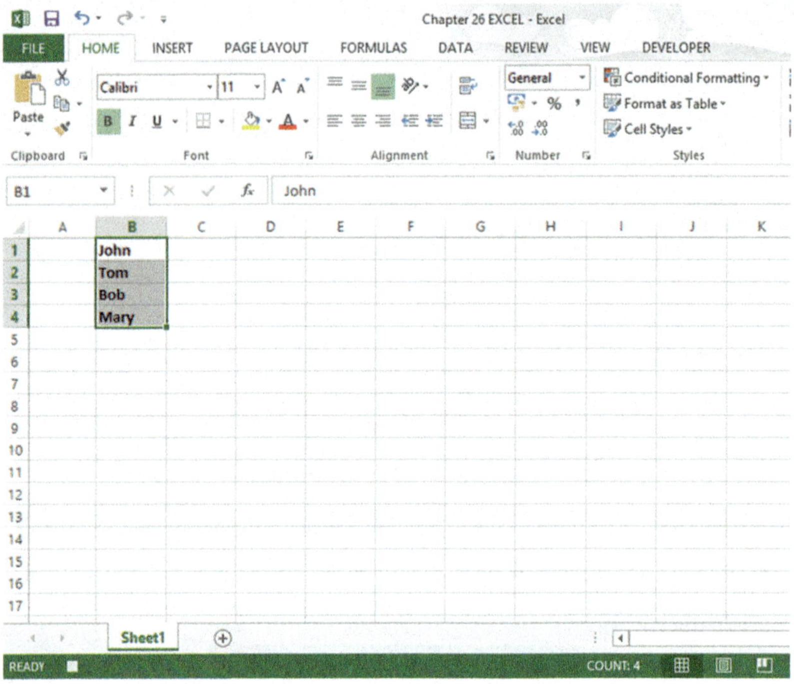

Next italicize the words by pressing *Ctrl+I* or press I under *HOME*. This is shown below.

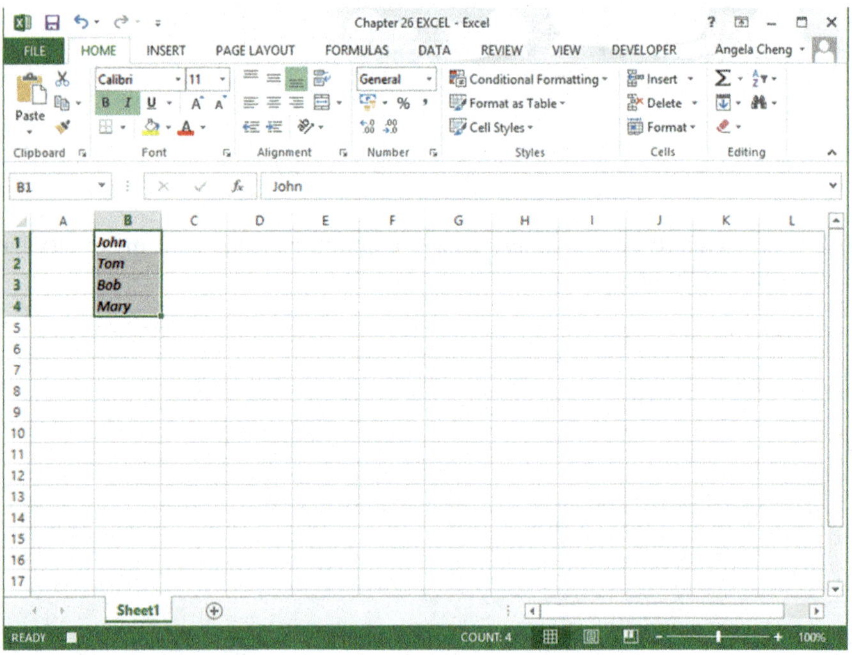

Next underline the words by pressing *Ctrl+U* or press U under *HOME*. This is shown below.

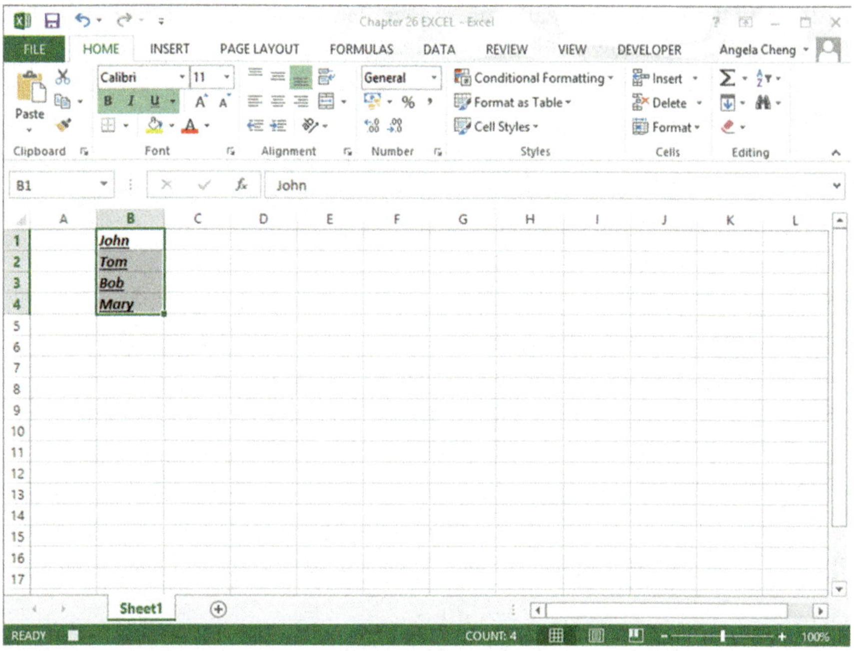

Next center the words by pressing the "Center" under *HOME*. This is shown below.

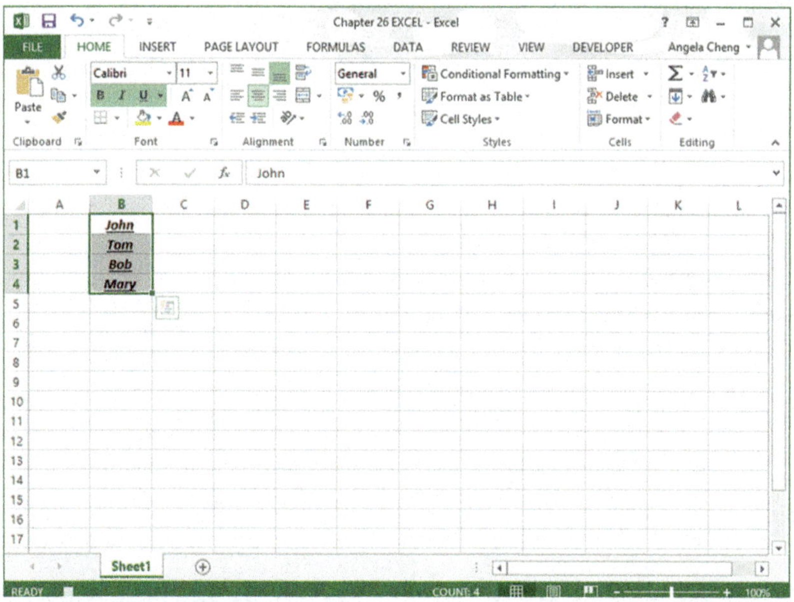

The next thing to do is to stop Excel's macro recorder. Go to *DEVELOPER* and click on "Stop Recording." As highlighted below.

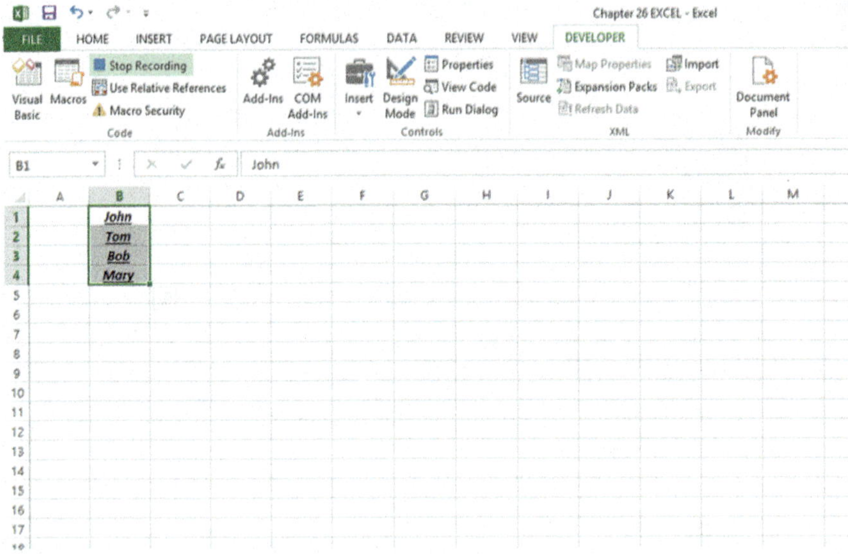

Below shows the end result of our exercise.

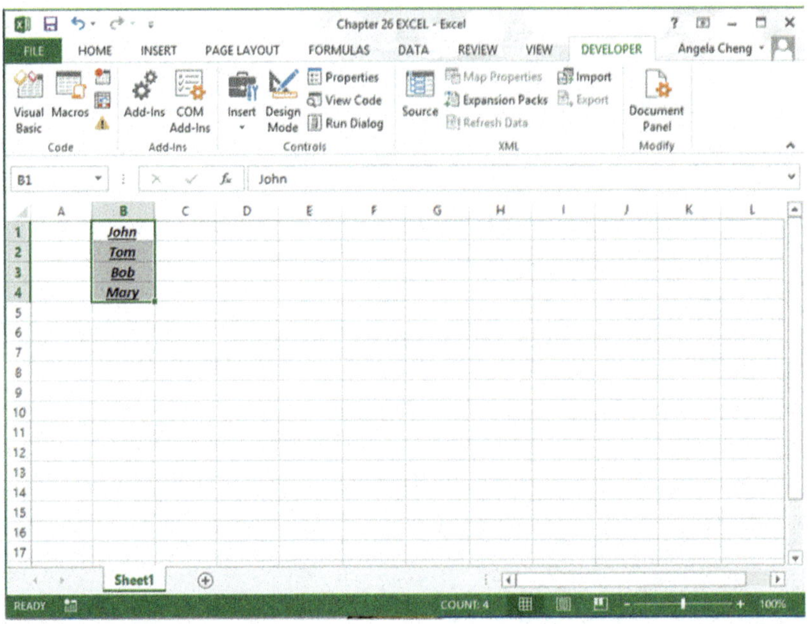

Let's look at the resulting VBA code by choosing DEVELOPER → Macros or press *Alt+F8*. Doing this would result in the *Macro* dialog box shown below.

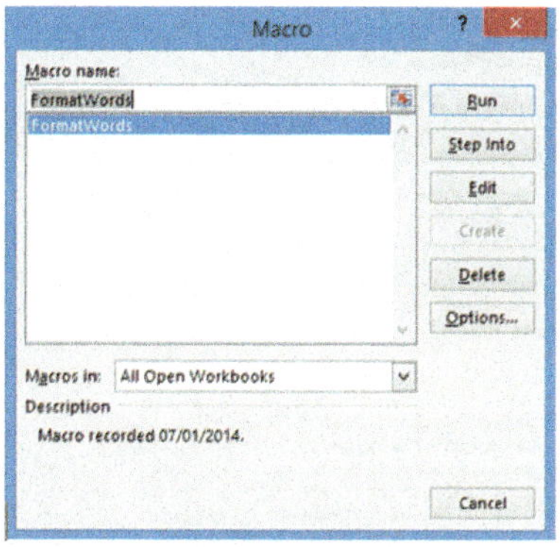

The *Macro* dialog box shows all the available macros in a workbook. The *Macro* dialog box shows one macro, the macro that we created. Let's now look at the "FormatWords" macro that we created. To look at this macro, highlight the macro name and then press the *Edit* button on the *Macro* dialog box. Pushing the *Edit* button would result in the *Microsoft Visual Basic Editor* (VBE). The next section discusses the Visual Basic Editor.

22.3 Excel's Visual Basic Editor

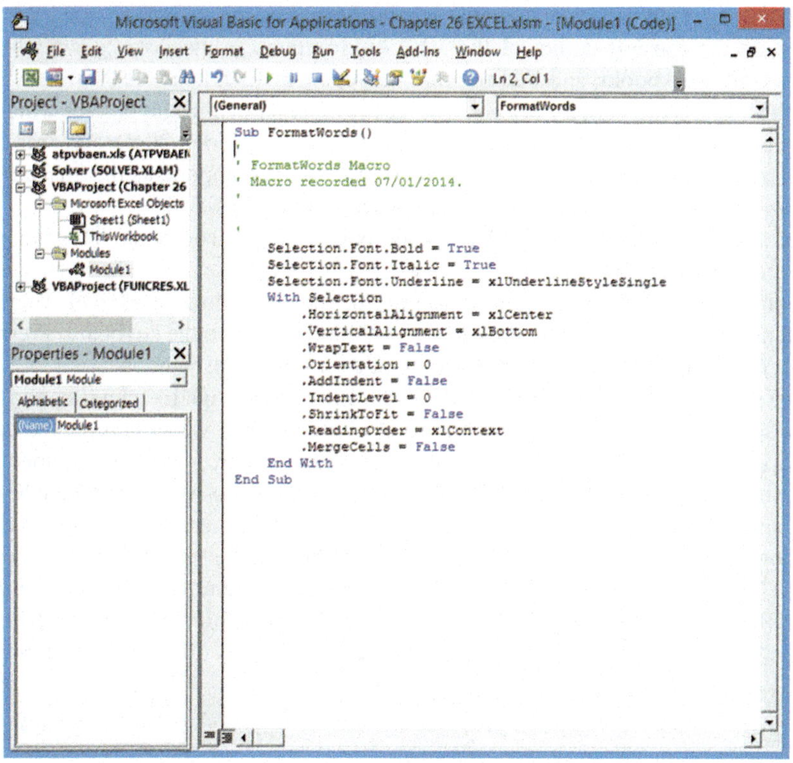

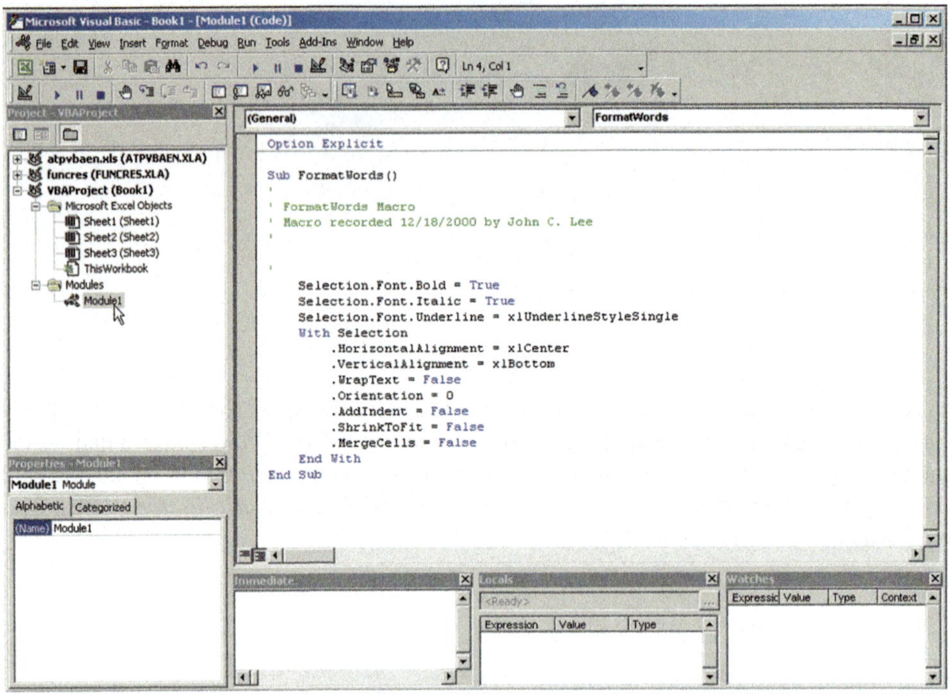

The VBE is Excel's programming environment. This programming environment is very similar to Visual Basic's programming environment. Visual Basic is the language used by professional programmers. At the top left corner of the VBE environment is the *project window*. The project window shows all workbooks and add-ins that are open in Excel. In the VBE environment, the workbooks and add-ins are called *projects*. The last *project* shown in the project window is for our workbook "Chapter 22 Excel.xlsm." This is where we created our "FormatWords" macro. The "Chapter 22 Excel.xlsm" project is expanded and shows all the components that make up the project.

The *module* component is where our "FormatWords" macro resides.

It is important to remember that the VBA code is part of the workbook. This is unlike Minitab where the macro code is a separate file.

The VBE environment is presented to the user in a different window than Excel. To go to the Excel window from the VBE window, press the *Alt* key and the *F11* key on the keyboard. Pressing *Alt+F11* keys will also navigate the user from the Excel window to the VBE window.

It should be noted that Excel writes bad VBA code. But even though Excel writes bad VBA code, it is valuable to the experienced VBA programmer. As noted above, we should realize that Excel is a feature-rich application. It is almost impossible for even an expert VBA programmer to remember how to program every feature in VBA. The above-recorded macro would be valuable to an experienced programmer that has never or has forgotten how to program the "Bold" or "Italic" or "Underline" or "Center" features of Excel. This is where Excel's macro recorder comes to play. The end result helps guide the experienced and expert VBA programmer in how to program an Excel feature in VBA. The way that an experienced VBA programmer would write the macro "Formatwords" is shown below. We name it "Formatwords2" to distinguish it from the recorded macro.

```
Sub Formatwords2()
    With Selection
        With .Font
            .Bold = True
            .Italic = True
            .Underline = xlUnderlineStyleSingle
        End With
        .HorizontalAlignment = xlCenter
    End With
End Sub
```

Note how much more efficient "Formatwords2" is compared to "Formatwords."

22.4 Running an Excel Macro

The previous section recorded the macro "Formatwords." This section will show how to run that macro. Before we do this, we will need to set up worksheet "Sheet2." The "Sheet2" format is shown below.

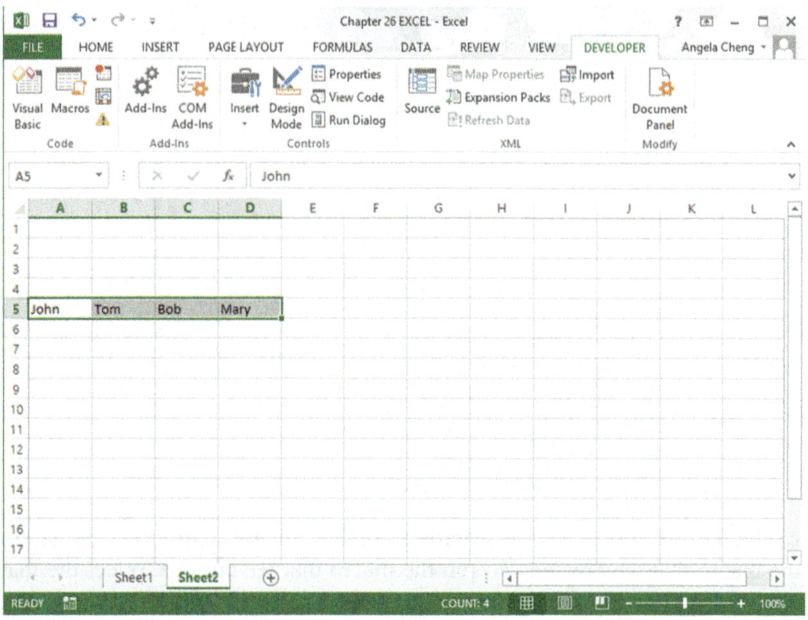

We will use the "FormatWords" macro to format the names in worksheet "Sheet2." To do this, we will need to select the names as shown above and then choose DEVELOPER → Macros or press *Alt* +*F8*. This is shown below.

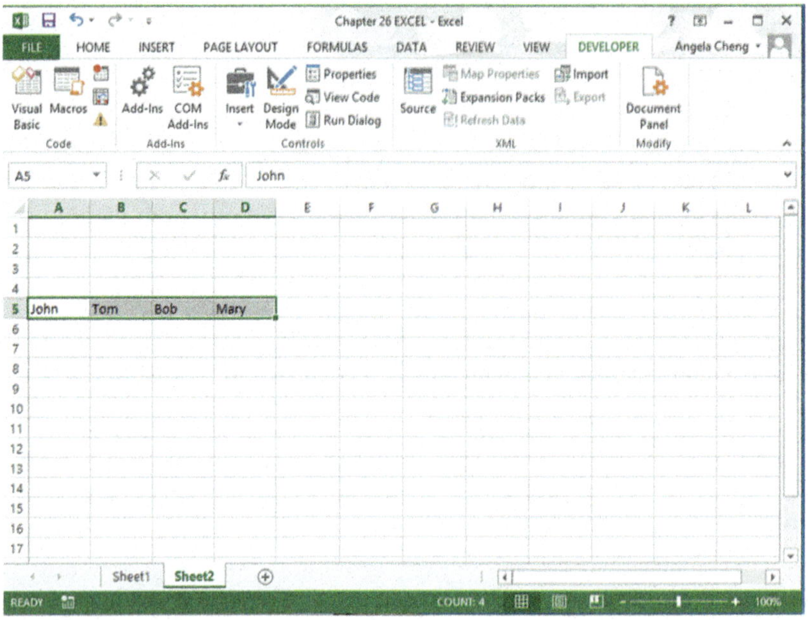

Choosing the *Macros* menu item will display the *Macros* dialog box shown below.

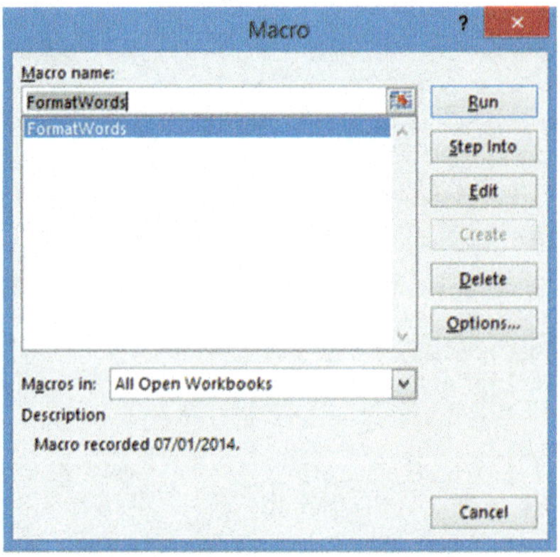

The *Macro* dialog box shows all the macros available for use. Currently the *macro* dialog box shows only the macro that we created. To run the macro that we created, select the macro and then press the *Run* button as shown as shown above. Below shows the end result after pressing the *Run* button.

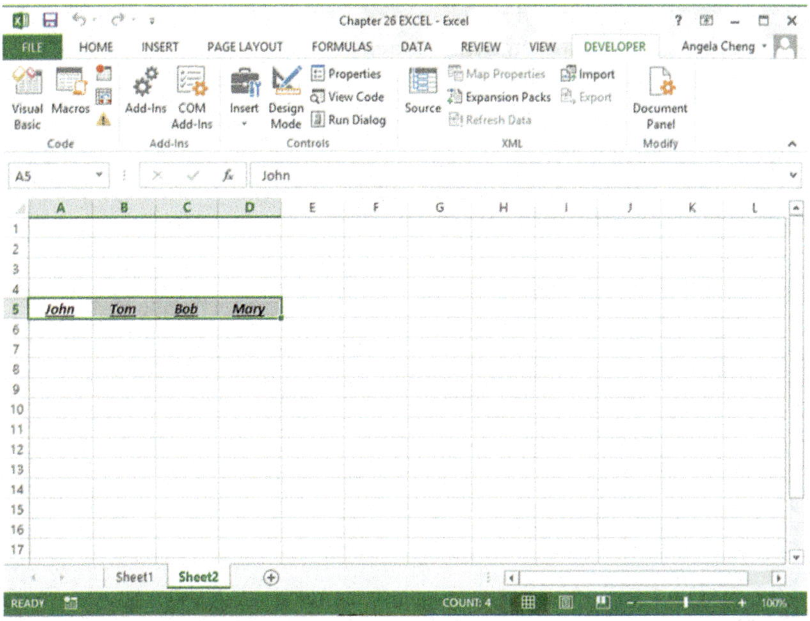

22.5 Adding Macro Code to a Workbook

Let's now add the "Formatwords2" created in Sect. 22.3. The first thing that we need to do is to go to the VBE editor by pressing the *Alt+F11*. Let's put this macro in another module. To do this we will need to add another module. Below shows how to add another module.

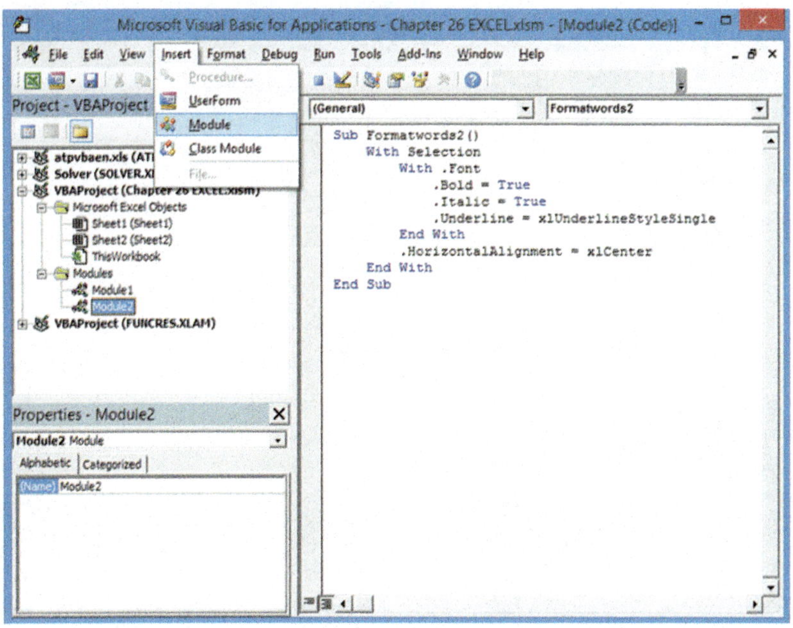

Choosing Insert → Module will add a new module called "Module2." In "Module2," type in the macro "FormatWords2." Below shows the two modules and the macro "FormatWords2" in the VBE. Below also indicates that "Module2" is the active component in the project.

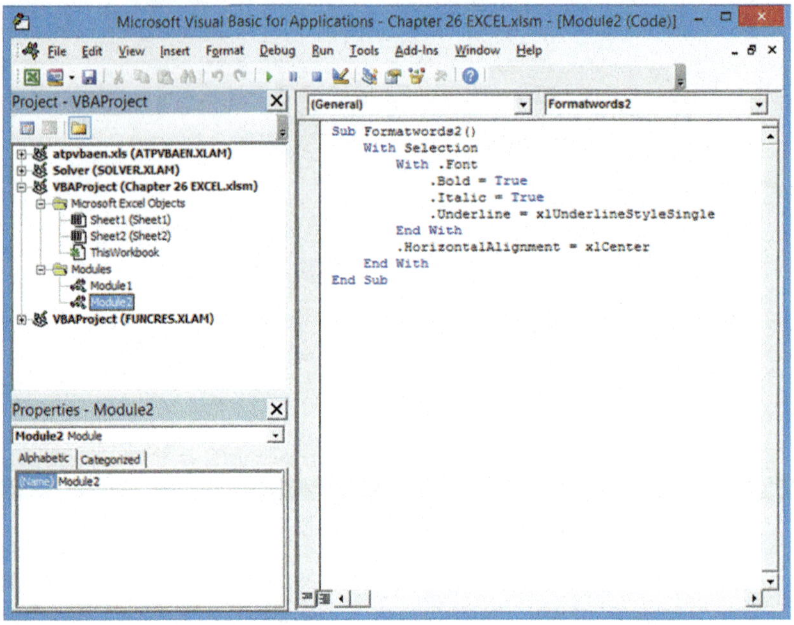

When the VBA program gets larger, it might make sense to name the modules to a more meaningful name. In the bottom left of the VBE window, there is a properties window for "Module2." Shown in the properties window (left bottom corner) is the name property for "Module2." Let's change the name to "Format." Below shows the end result. Notice in the project window that it now shows a "Format" module.

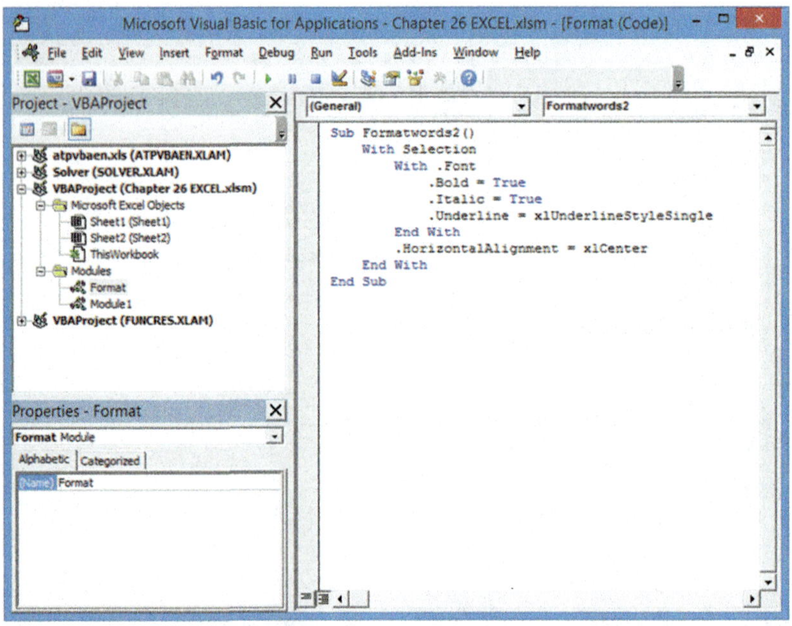

Now let's go back and look at the *Macro* dialog box. To go from the VBE window to the Excel window, press *Alt+F11*. Below shows the *Macro* dialog box after typing in the macro "FormatWords2" into the VBE editor.

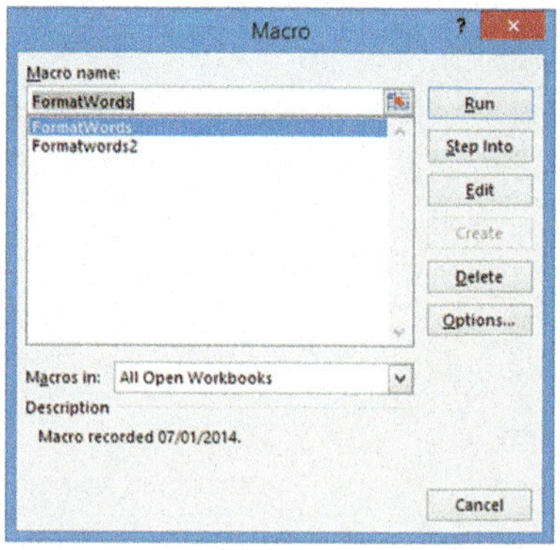

The *Macro* dialog box now shows the two macros that were created.

22.6 Push Button

In the sections above, we used menu items to run macros. In this section, we will use *push* buttons to execute a specific macro. Push buttons are used when a specific macro is used frequently. Before we illustrate *push buttons*, let's set up worksheet "Sheet3" as shown below.

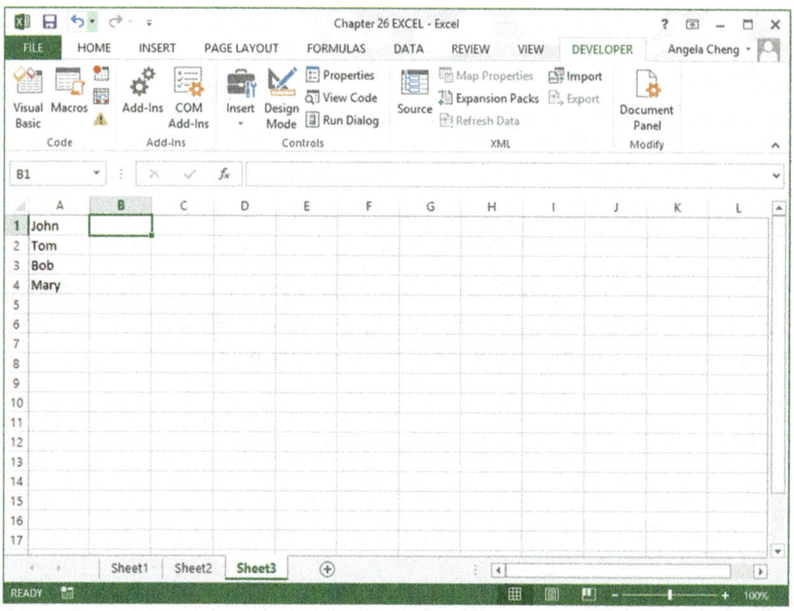

We will need the *Forms* toolbar to create push button. Go to DEVELOPER → Insert → Button.

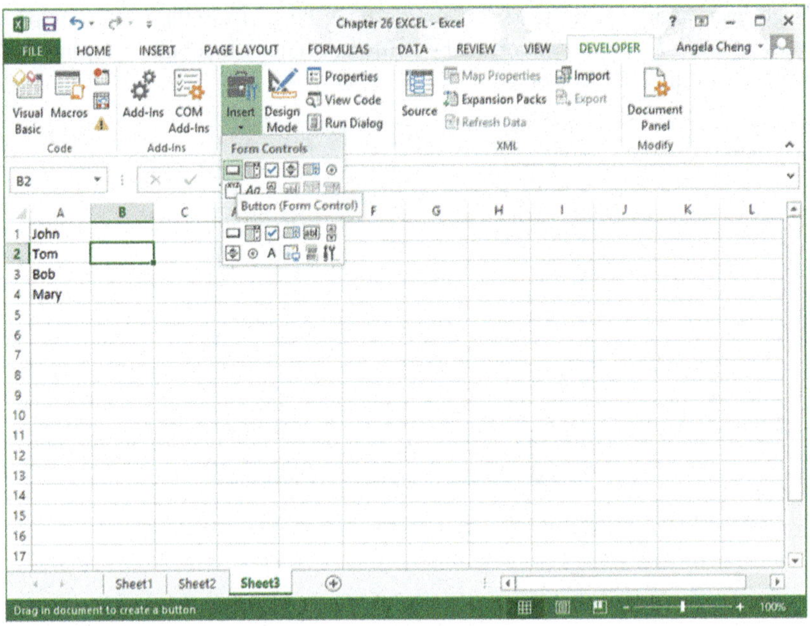

After that, click on the cell where we want the button to be located. This will bring up the *Assign Macro* dialog box.

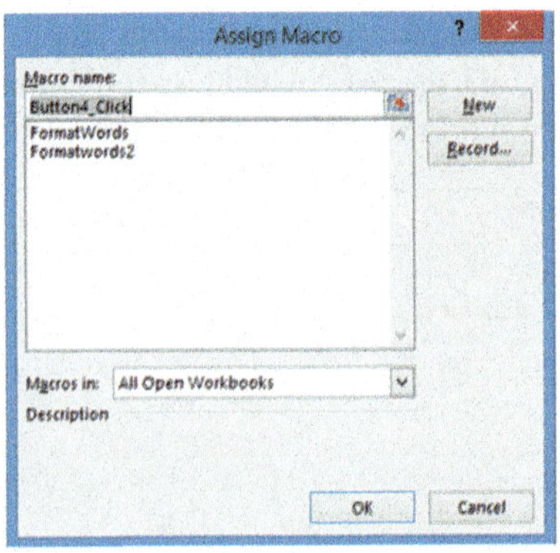

The *Assign Macro* dialog box shows all the available to be assigned to the button. Choose the macro "FormatWord2" as shown above and press the *OK* button. Pressing the *OK* button will assign the macro "FormatWord2" to button. The end result is shown below.

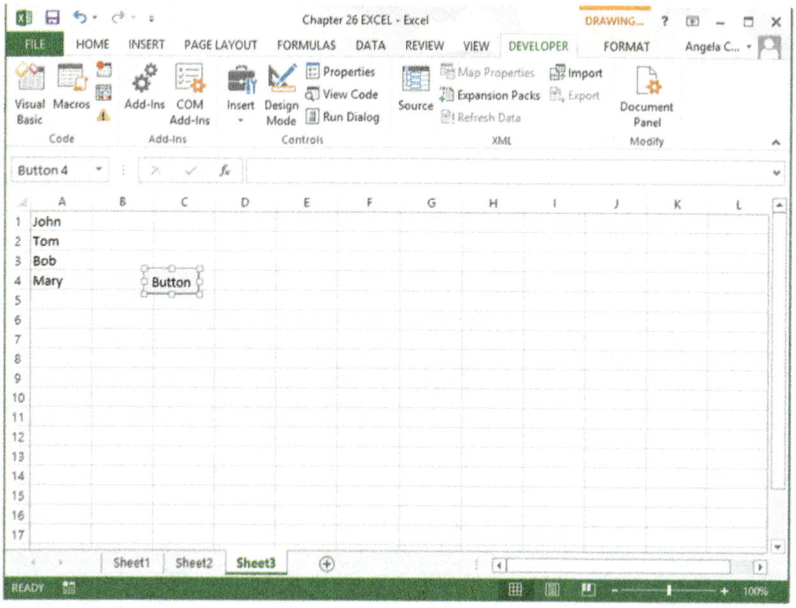

Next, move the mouse cursor over the button "Button 1." Next click on the left mouse button. This will result in cell A1 to be formatted. The end result is shown below.

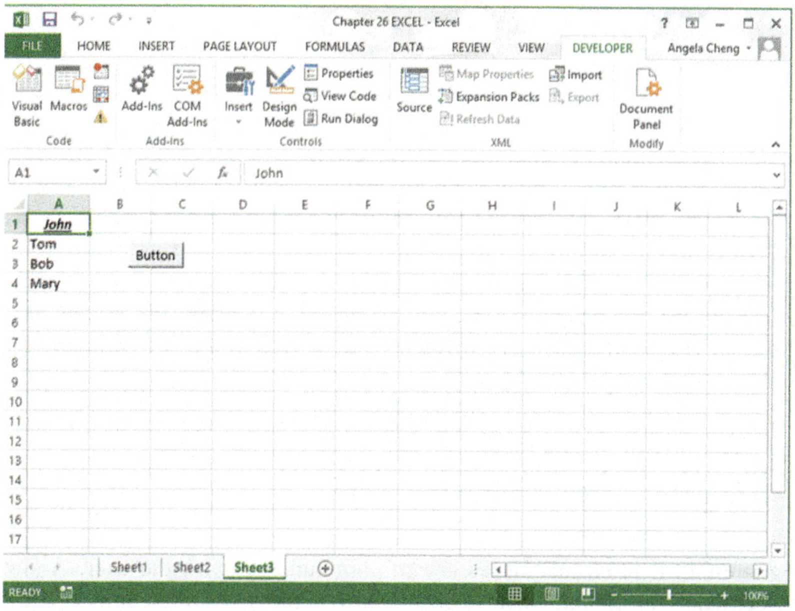

The name "Button" for the button is probably not a good name. To change the name "Button," move the mouse pointer over the button. After doing this, click on the right mouse button. This will display a shortcut menu for the button and select *Edit Text*. Change the name to "Format" The end result is shown below.

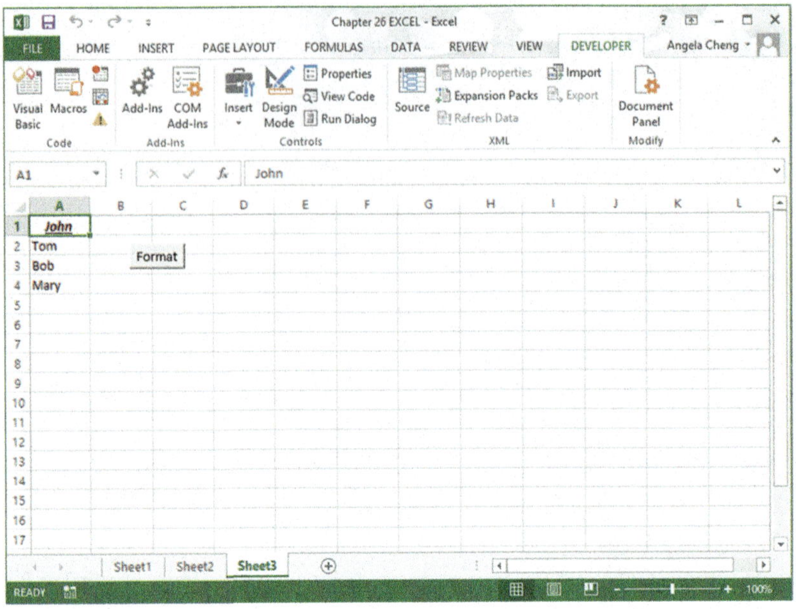

22.7 Subprocedures

In the previous sections we dealt with two logic groups of Excel VBA code. One group was called "FormatWords" and the other group of VBA code was called "FormatWords2" In both groups there was a word *sub* to distinguish the beginning of the group of VBA code and the words *end sub* to indicate the end of a group VBA code. Both *sub* and *end sub* are called *keywords*. Keywords are words that are part of the VBA programming language. In a basic sense, a program is an accumulation of groups of VBA code.

We saw in the previous section that subprocedures in modules are all listed in the *Macro* dialog box. Modules are not the only place where subprocedures are. Subprocedures can all be put in class modules and forms. These subprocedures will not be displayed in the *Macro* dialog box.

22.8 Message Box and Programming Help

In Excel programming it is usually necessary to communicate with the user. A simple but very popular VBA command to communicate with the user is the *msgbox* command. This command is used to display a message to the user. Below shows the very popular "Hello World" subprocedures in VBA.

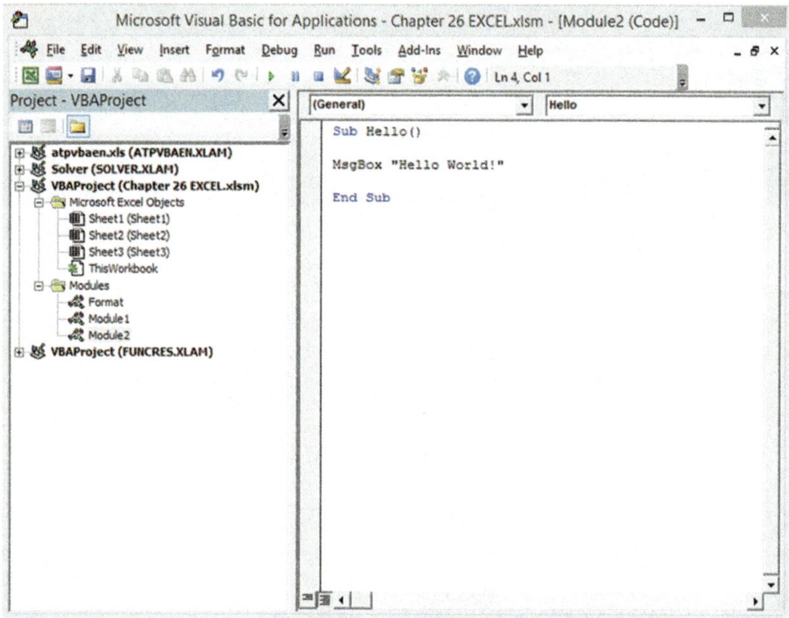

It is not necessary as indicated in the previous section to go to the *Macros* dialog box to run the "Hello" subprocedure shown above. To run this macro, place the cursor inside the procedure and press the *F5* key on the keyboard. Pressing the *F5* key will result in the following:

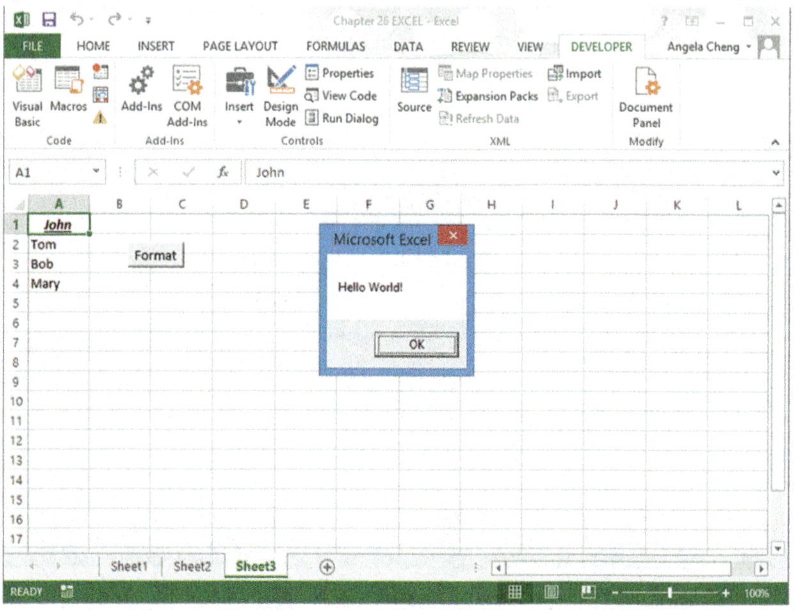

Notice that in the message box above, the title of the message box is "Microsoft Excel." Suppose we want the title of the message box to be "Hello." Below shows the VBA code to accomplish this.

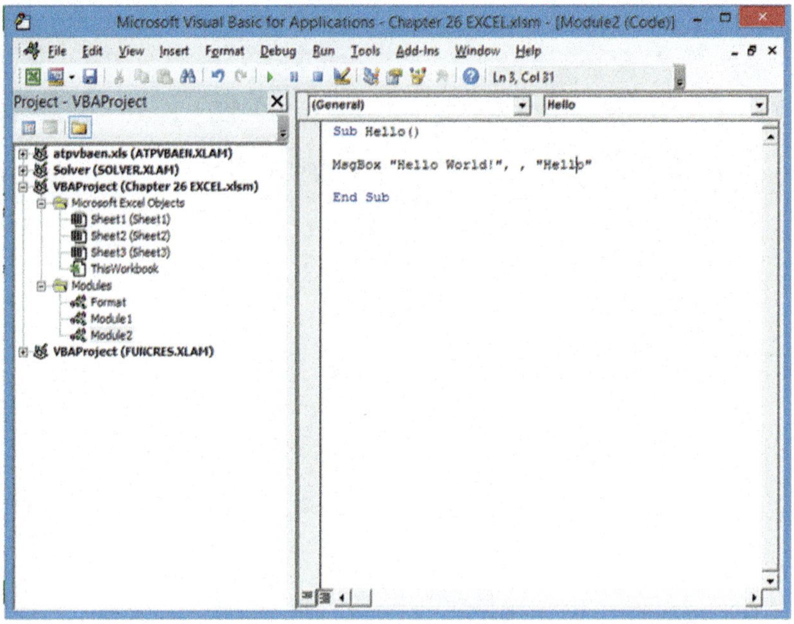

Below shows the result of running the above code. Notice that the title of the message box is "Hello."

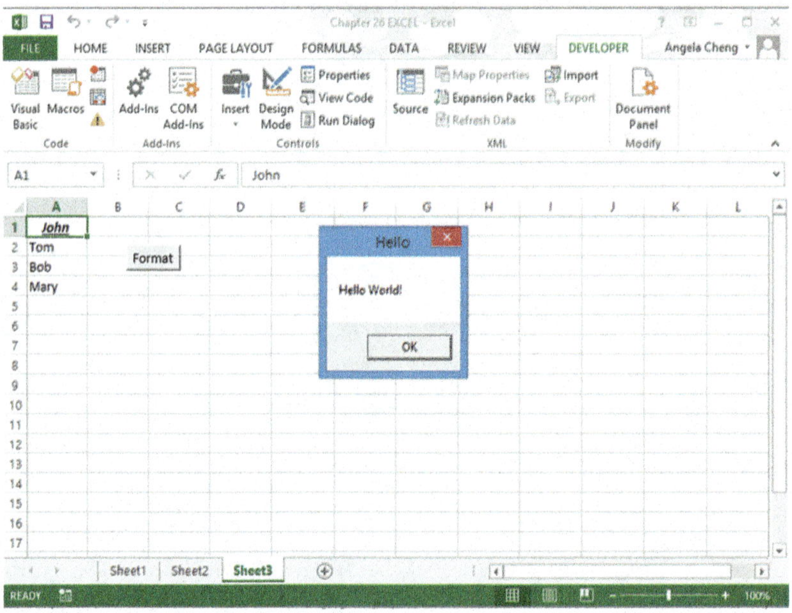

The *msgbox* command can do a lot of things. But one problem is remembering how to program all the features. This is one issue that the VBE editor is very good at dealing with. Notice in the above code that commas separate the arguments to set the *msgbox* command. This then brings up the question: How many arguments does the VBA *msgbox* have? Below shows how the VBE editor assists the programmer in programming the *msgbox* command.

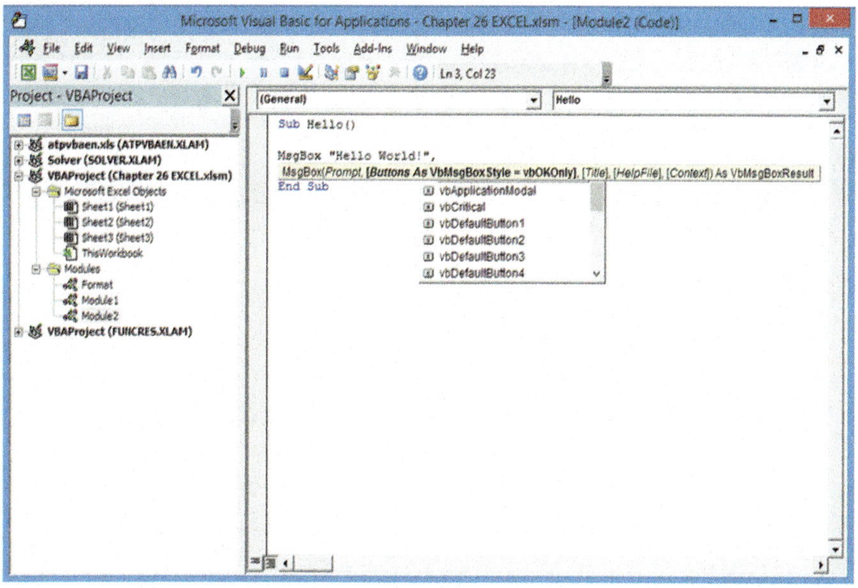

We see above that after typing the first comma, the VBE editor shows two things. The first thing is a horizontal list that shows and names all the arguments of the *msgbox* command and in that list it bolds the argument that we are working on. The second thing that the VBE editor shows is a vertical list that lists all the possible values of the arguments that we are currently working on. A list is only shown when an argument has a set of predefined values.

If the above two features are not sufficient in aiding in how to program the *msgbox* command, we can move the cursor on the *msgbox* command as shown below and press the *F1* key on the keyboard.

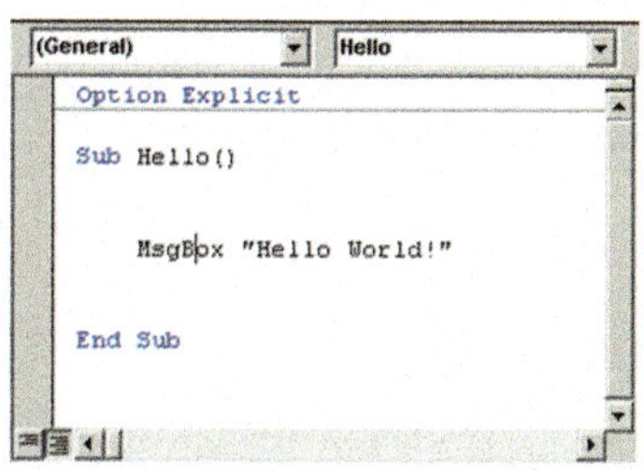

Pressing the *F1* key would open up the web browser to a detailed explanation of the *msgbox* command. This is shown below.

Bibliography

Walkenbach J (2013) Excel 2013 power programming with VBA. Wiley, Hoboken

Bullen S (2007) Excel 2007 VBA programmer's reference. Wiley, Indianapolis

Bovey R (2009) Professional excel development. Pearson Education, Boston

Hillier F (1963) The derivation of probabilistic information for the evaluation of risky investments. Management Science, 9, 443–457

Wang SW and CF Lee (2010) Fuzzy set and real option approach for capital budget decision. International Journal of Information Technology and Decision Making, 09, 695–714

Chapter 23
Introduction to VBA Programming

23.1 Introduction

In the previous chapter, we mentioned that VBA was Excel's programming language. It turns out that VBA is the programming language for all the Microsoft Office applications. Also VBA is becoming the programming language of other popular applications. Some of the other applications that are using VBA as its programming language are Visio and Business Objects. This is very exciting news for people who automate applications. The main advantage is that a person does not have to constantly learn a new language when he or she deals with a new application. Another important advantage of an application adopting VBA as a programming language is it is a language that uses most of the programming concepts that professional programmers use; therefore, VBA is a powerful language. Now, it is easier for an Excel programmer to do Access VBA programming, Word VBA, or even PowerPoint VBA programming. The finance people would say that there is a return in the investment of time in learning VBA.

In this chapter we will study VBA and specific Excel VBA issues.

23.2 Excel's Object Model

There is one thing that is frequently done with an Excel VBA program; it sets a value to a cell or a range of cells. For example, suppose we are interested in setting the cell A5 in worksheet "Sheet1" to the value of 100. Below is a common way that a novice would program a VBA program to set the cell A5 to 100.

```
Sub Chp23Example1()

    Range("A5").Value = 100

End Sub
```

The *range* command above is used to reference specific cells of a worksheet. So, if the worksheet "Sheet1" is the active worksheet, cell A5 of worksheet "Sheet1" will be populated with the value of 100. This is shown below.

C.-F. Lee et al., *Essentials of Excel, Excel VBA, SAS and Minitab for Statistical and Financial Analyses*, DOI 10.1007/978-3-319-38867-0_23

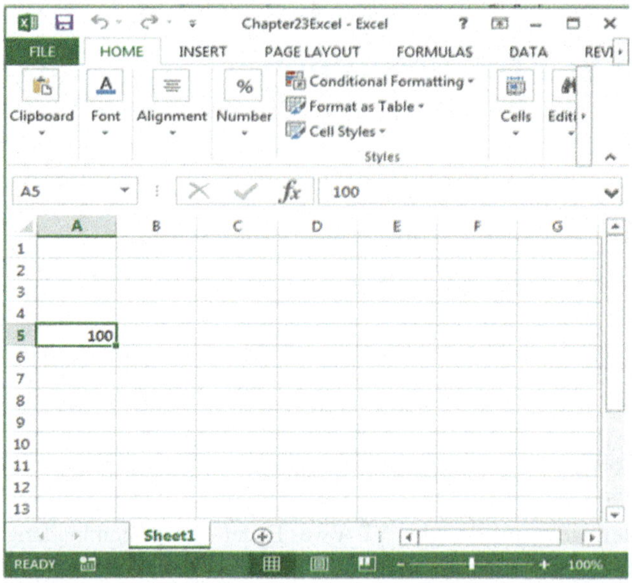

Many times, it is important that cell A5 in worksheet "Sheet1" has the value of 100 and not cell A5 in the other worksheets of the workbook. But if we run the above macro when worksheet "Sheet2" is active, cell A5 in worksheet "Sheet2" will be populated with the value of 100. To solve this issue, experienced programmers will rewrite the above VBA procedure as shown below.

```
Sub Chp23Example2()

   ThisWorkbook.Worksheets("Sheet1").Range("A5").Value = 100

End Sub
```

Notice that the VBA code line is longer in the procedure "Chp23Example2" than in the procedure "Chp23Example1." To understand why, we will need to look at Excel's object model. We can think of Excel's object as an upside-down tree. A lot of Excel VBA programming is basically traversing the tree. In VBA programming, moving from one level of a tree to another level is indicated by a period. The VBA code in the procedure "Chp23Example2" traverses the Excel's object model through three levels.

Among all the Microsoft Office products, Excel has the most detailed object model. When we are talking about object models, we are talking about concepts that a professional programmer would talk about. When we are talking about object models, there are three words that even a novice must know. Those three words are *objects*, *properties*, and *methods*. These words can take up chapters or even books to explain. I have found that a very crude but somewhat effective way to think about what these words mean is to think about the English grammar. We can crudely equate *objects* as a noun. We can crudely equate *properties* as an adjective. We can crudely equate *methods* as an adverb. In Excel some examples of objects are worksheets, workbooks, and charts. These objects have properties that describe them or have methods that act on them. The main point is that every novice VBA programmer should right away think about object models, objects of the object models, properties, and methods of an object when programming in VBA. Not doing so will cause the novice much unnecessary headaches and pain.

In the Excel object model, there is a parent and child relationship between objects. The topmost object is the Excel object. A frequently used object and a child of the Excel object is the *workbook* object. Another frequently used object and a child of the *workbook* object is the *worksheet*. Another frequently used object and a child of the *worksheet* object is the *range* object. If we look at the Excel object model, we will be able to see the relationship between the Excel object, the workbook object, the worksheet object, and the range object.

We can use the help in the VB Editor (VBE) to look at the Excel object model. To do this we would need to choose Help → Microsoft Visual Basic Help.

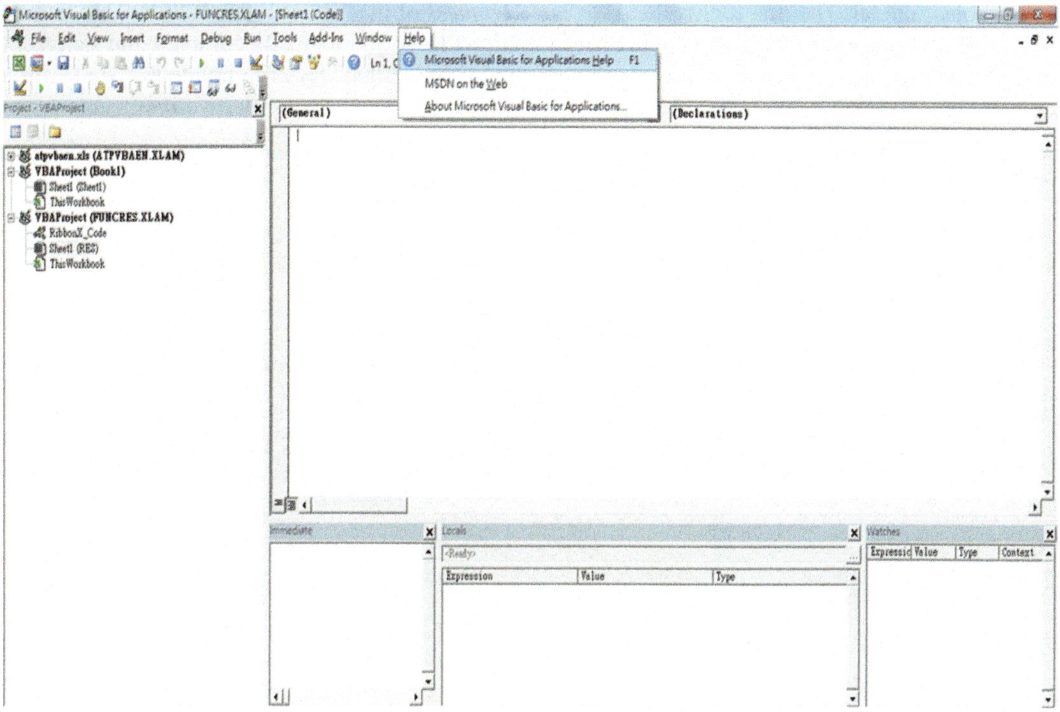

In the Excel 2013, there is no off-line help, you need to go online to explore the Microsoft website. This will result in the following help screen:

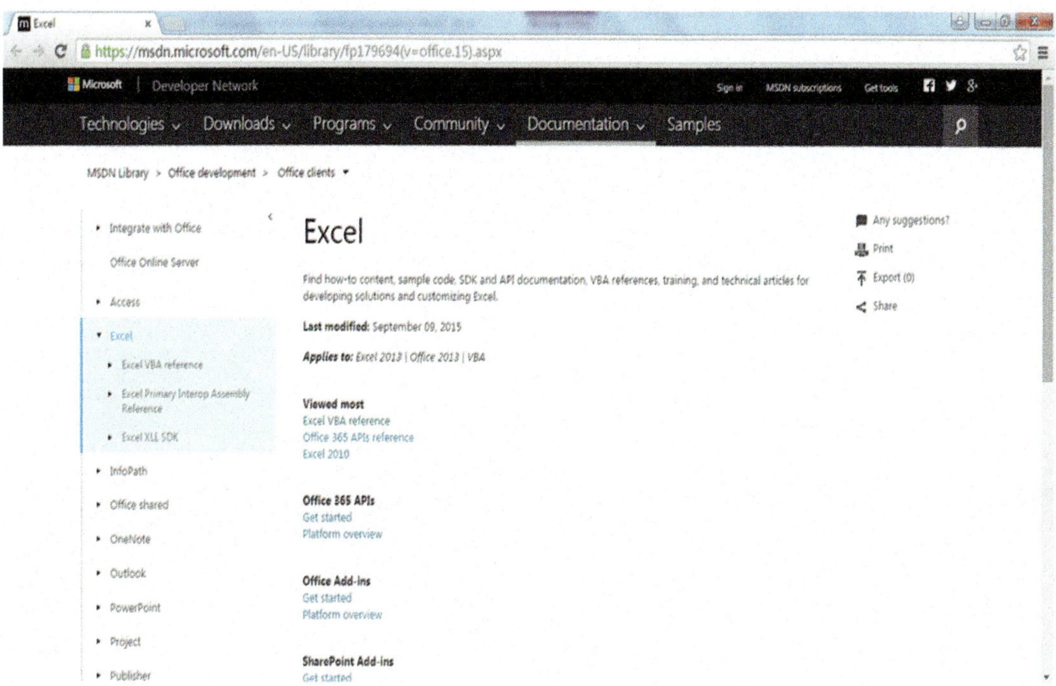

23.3 Auto List Members

The Excel VBA programmer should always be thinking about the Excel object model. Because of the importance of the Excel object model, the VBE has tools to aid the VBA programmer in dealing with Excel object model. The first tool is the *Auto List Members* of the Visual Basic Editor. This feature will display for an object a list that contains information that would logically complete the statement at the current insertion point. For example, below shows the list that would complete the *Application* object. This list contains the properties, methods, and child objects of the *Application* object.

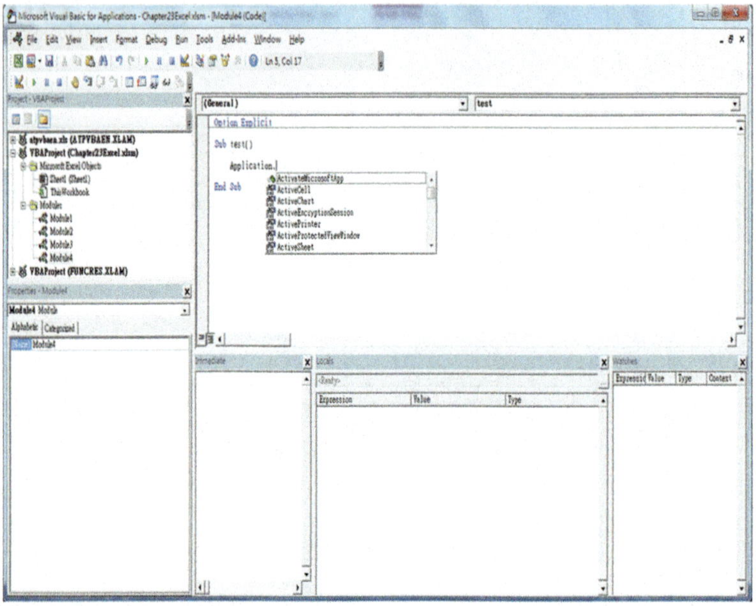

The *Auto List* is a great aid in helping the VBA programmer in dealing with the methods, properties, and child objects of each object.

23.4 Object Browser

Another tool to aid the VBA programmer in dealing with the Excel object model is the *Object Browser*. To view the *Object Browser*, choose View → Object Browser. This is shown below.

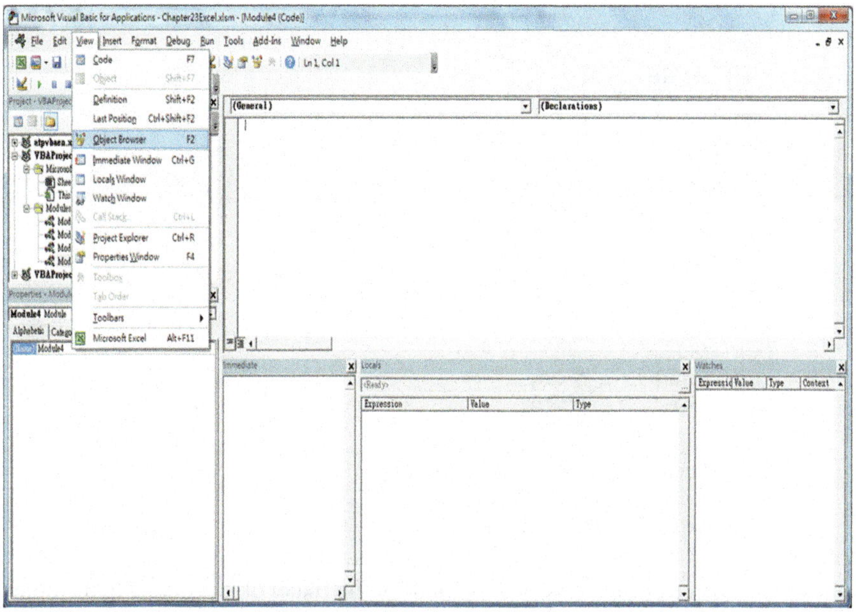

The default display for the *Object Browser* is shown below.

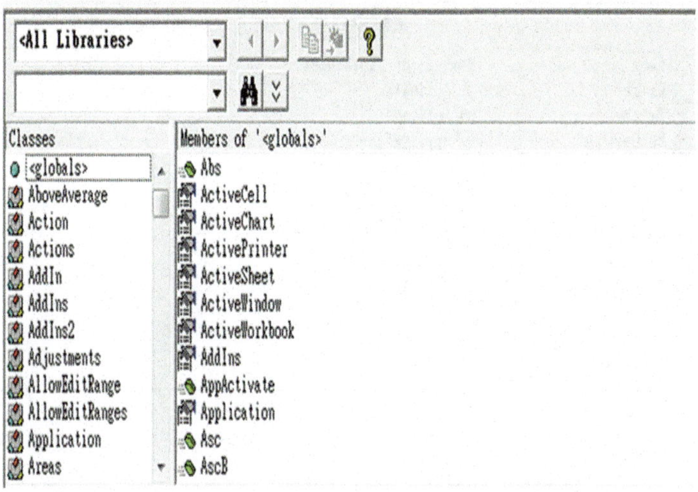

Below shows how to view the Excel object model from the *Object Browser*.

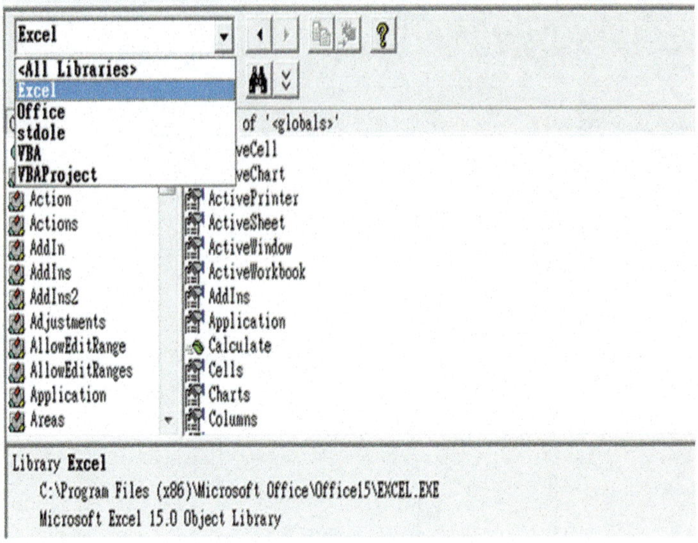

Below shows the objects, properties, and methods of the Excel object model in the object browser.

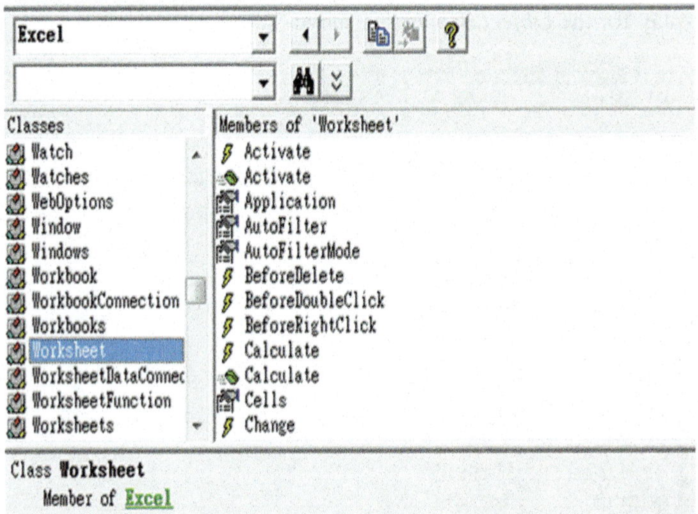

In the *Object Browser* above, the object *worksheets* is chosen on the left side of the object browser and on the right side all the properties, methods, and child objects of the *worksheets* object.

It is important to note that the Excel object model is not the only object model that the VBE handles. This issue was alluded to above. The default display for the *Object Browser* shows "<All Libraries>" This suggests that other object models were available. Above we also saw the following list in the object browser:

The list above indicates the object models used by the Visual Basic Editor. Of all the object models shown above, the VBA object model is used most after the Excel object model. Below shows the VBA object model in the *object browser*.

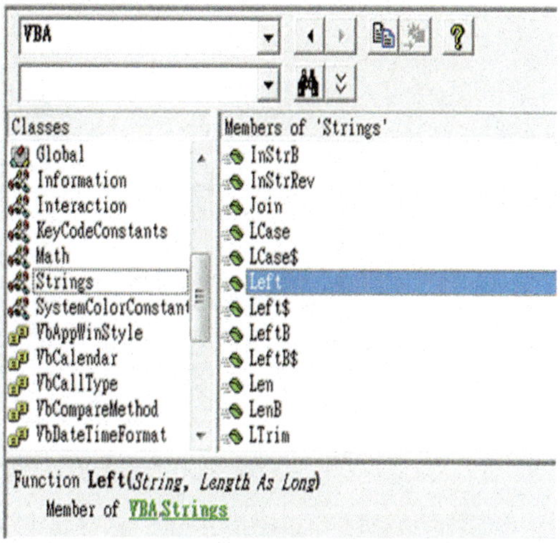

The main reason that an Excel VBA programmer uses the VBA object model is that the VBA object model provides a lot of useful function. Professional programmers will say that functions of an object model are properties of an object model. For example, for the *Left* function shown above, we can say that the *Left* function is a property of the VBA object model. Below shows an example on using the property *Left* of the VBA object model.

```
    MsgBox "The first word is: " & VBA.Strings.Left("New York", 3)

End Sub
```

Below shows the result of executing the “Chp23Example3” macro.

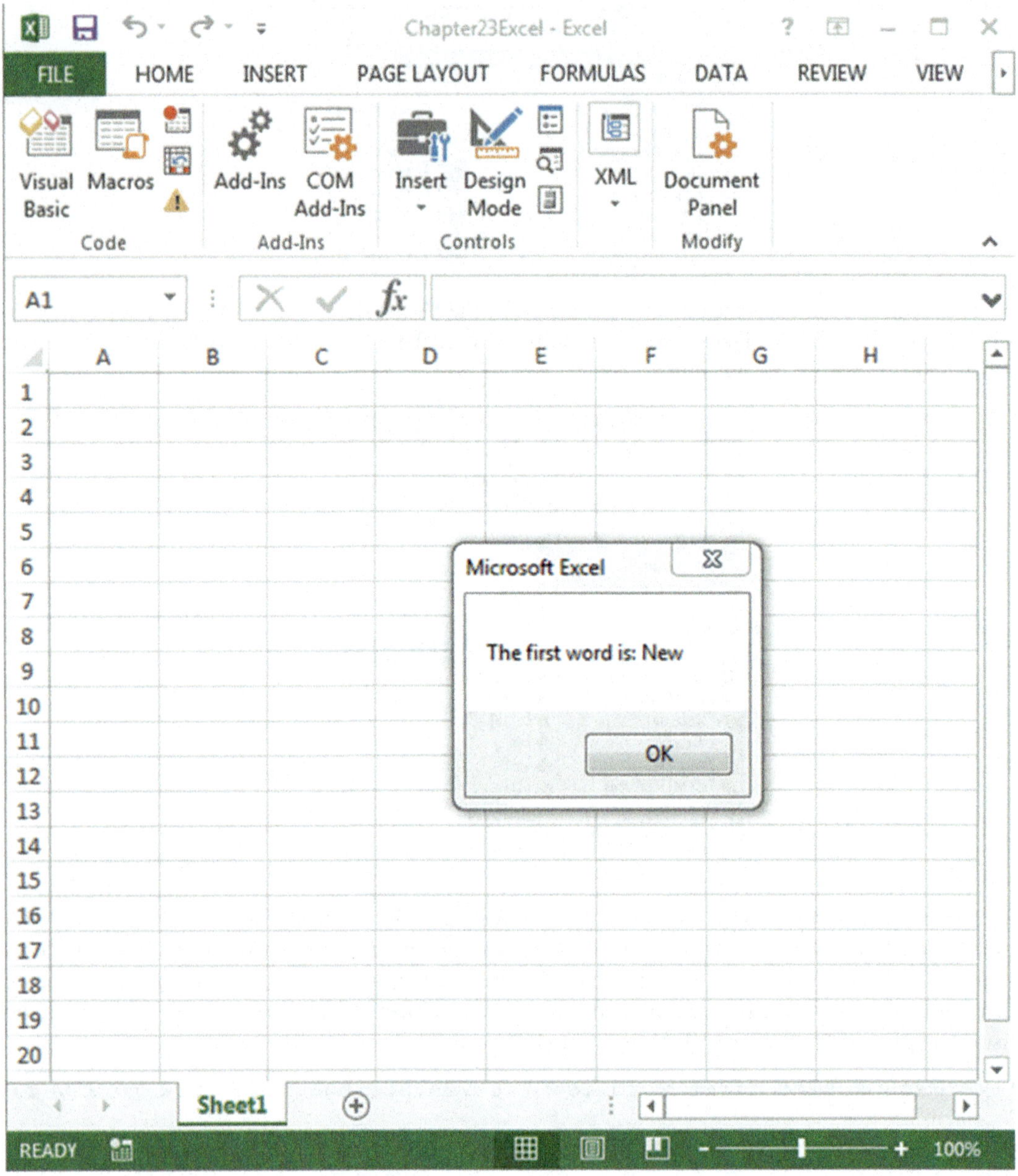

Many times an Excel VBA programmer will write macros that use both Microsoft Excel and Microsoft Access. To do this we would need to setup the VBE so it can also use Access' object model. To do this we would first have to choose Tools → Reference in the VBE. This is shown below.

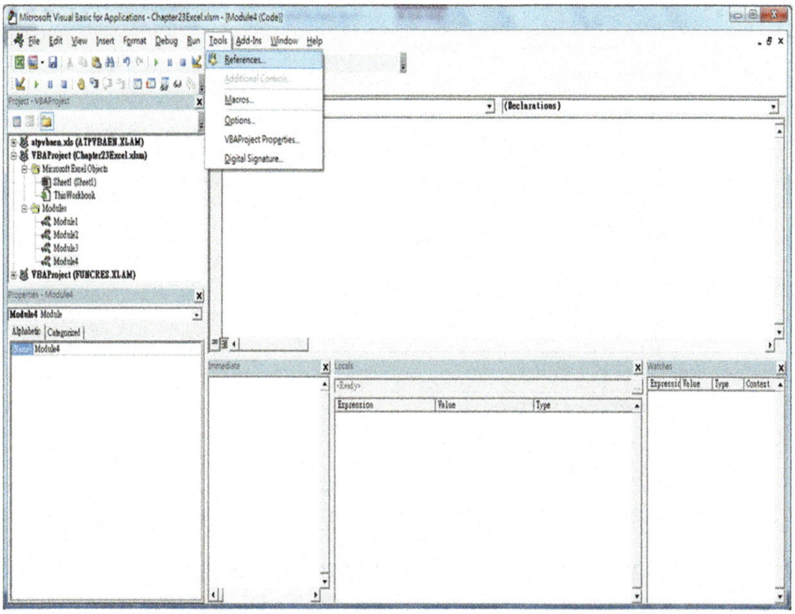

The resulting *References* dialog box is shown below.

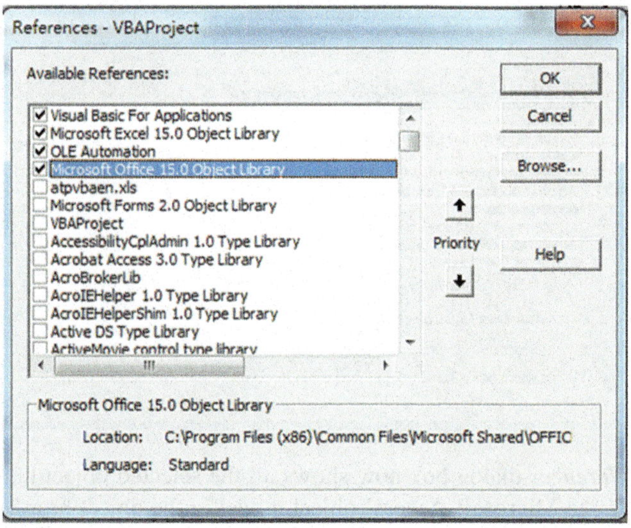

In the above *References* dialog box, the Excel object model is selected. The bottom of the dialog box shows the location of the file that contains Excel's object model. The file that contains an object model is called a *type library*.

To program Microsoft Access while programming Excel, we will need to find the type library for Microsoft Access. Below shows the Microsoft Access object model being selected.

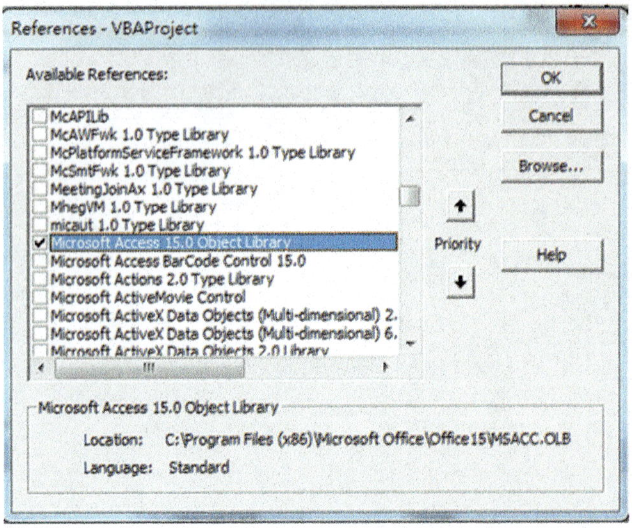

If we press the *OK* button and go back to the *References* dialog box, we will see the following:

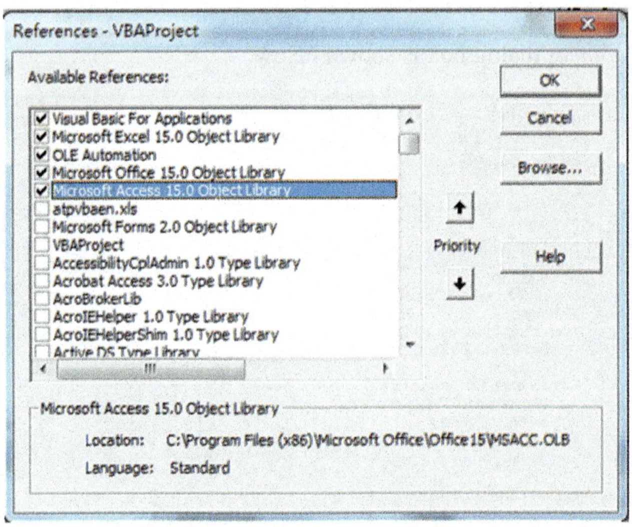

Notice that the *References* dialog box now shows all the selected object libraries on the top. We now should be able to see Microsoft Access' object model in the object browser. Below shows that Microsoft Access' object model is included in the object browser's list.

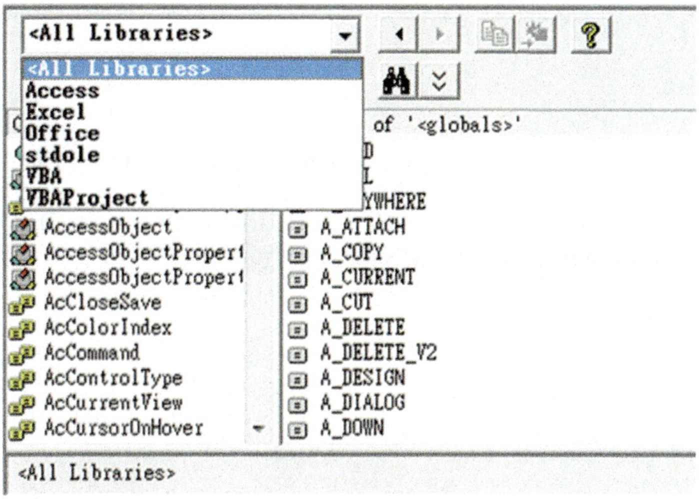

Below shows Microsoft Access' object model in the object browser.

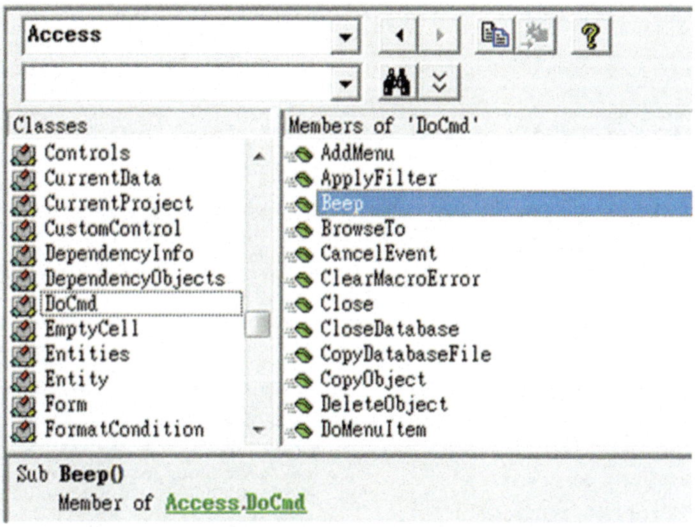

The Excel object model does not have a method to make the PC make a beep sound. Fortunate it turns out that the Access object does have a method to make the PC make a beep sound. Below is a macro that will make the PC make a beep sound. The *Access* keyword indicates that we are using the Access object model. The keyword *DoCmd* is a child object of the *Access* object. The keyword *Beep* is a method of the *DoCmd* object.

```
Sub Chp24Example3a()

    Access.DoCmd.Beep

End Sub
```

It turns out that in the VBA object model, there is also a *beep* method. Below shows a macro using the VBA object model to make the PC make a beep sound.

```
Sub Chp24Example3b()

    VBA.Interaction.Beep

End Sub
```

23.5 Variables

In VBA programming *variables* are used to store and manipulate data during macro execution. When dealing with data, it is often useful when processing data to only deal with a specific type of data. In VBA it is possible to define a specific type for specific variables. Below is a data type summary of what variables can be defined as. This list was obtained from the VBE help.

Data Type Summary

Data Type Summary

The following table shows the supported data types, including storage sizes and ranges.

Data type	Storage size	Range
Byte	1 byte	0 to 255
Boolean	2 bytes	**True** or **False**
Integer	2 bytes	-32,768 to 32,767
Long (long integer)	4 bytes	-2,147,483,648 to 2,147,483,647
Single (single-precision floating-point)	4 bytes	-3.402823E38 to -1.401298E-45 for negative values; 1.401298E-45 to 3.402823E38 for positive values
Double (double-precision floating-point)	8 bytes	-1.79769313486231E308 to -4.94065645841247E-324 for negative values; 4.94065645841247E-324 to 1.79769313486232E308 for positive values
Currency (scaled integer)	8 bytes	-922,337,203,685,477.5808 to 922,337,203,685,477.5807
Decimal	14 bytes	+/-79,228,162,514,264,337,593,543,950,335 with no decimal point; +/-7.9228162514264337593543950335 with 28 places to the right of the decimal; smallest non-zero number is +/-0.0000000000000000000000000001
Date	8 bytes	January 1, 100 to December 31, 9999
Object	4 bytes	Any **Object** reference
String (variable-length)	10 bytes + string length	0 to approximately 2 billion
String (fixed-length)	Length of string	1 to approximately 65,400
Variant (with numbers)	16 bytes	Any numeric value up to the range of a **Double**
Variant (with characters)	22 bytes + string length	Same range as for variable-length **String**
User-defined (using **Type**)	Number required by elements	The range of each element is the same as the range of its data type.

Note Arrays of any data type require 20 bytes of memory plus 4 bytes for each array dimension plus the number of bytes occupied by the data itself. The memory occupied by the data can be calculated by multiplying the number of data elements by the size of each element. For example, the data in a single-dimension array consisting of 4 **Integer** data elements of 2 bytes each occupies 8 bytes. The 8 bytes required for the data plus the 24 bytes of overhead brings the total memory requirement for the array to 32 bytes.
A **Variant** containing an array requires 12 bytes more than the array alone.
Note Use the **StrConv** function to convert one type of string data to another.

Below shows how to define and use variables in VBA.

```
Sub Chp23Example4()
    Dim iNum As Integer
    Dim sString As String
    Dim iLength As Integer

    'Inputbox is used to prompt the user for data
    iNum = 4 + InputBox("Please enter an integer to be added by four.")
    sString = InputBox("Please enter a word. " & _
              "The length of the word will be calculated.")
    iLength = VBA.Strings.Len(sString)

    ThisWorkbook.ActiveSheet.Range("a1") = iNum
    ThisWorkbook.ActiveSheet.Range("a2") = sString
    ThisWorkbook.ActiveSheet.Range("a3") = "The length of the word " & _
                                           sString & " is " & iLength

End Sub
```

Running the above will result in the following:

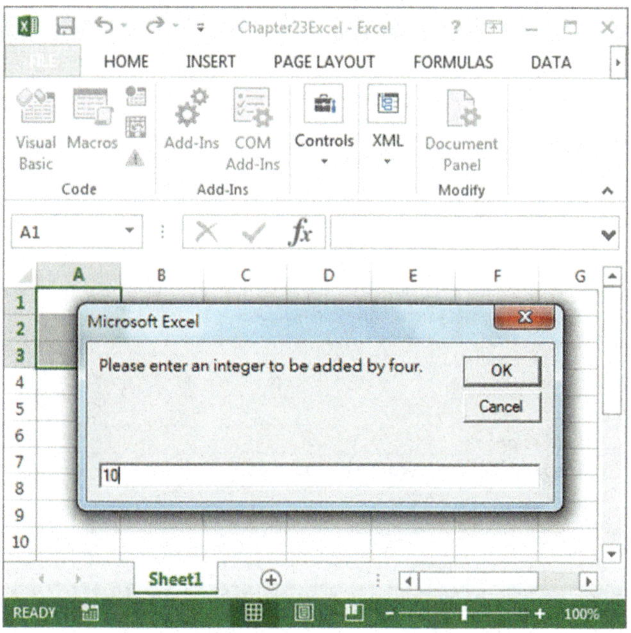

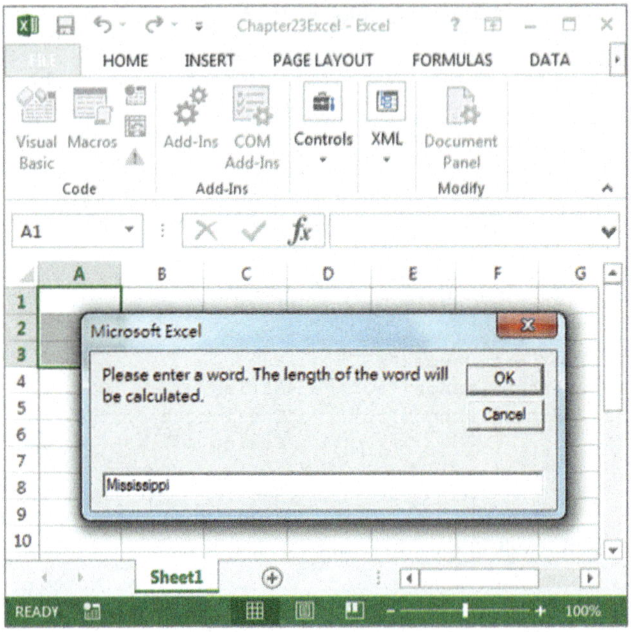

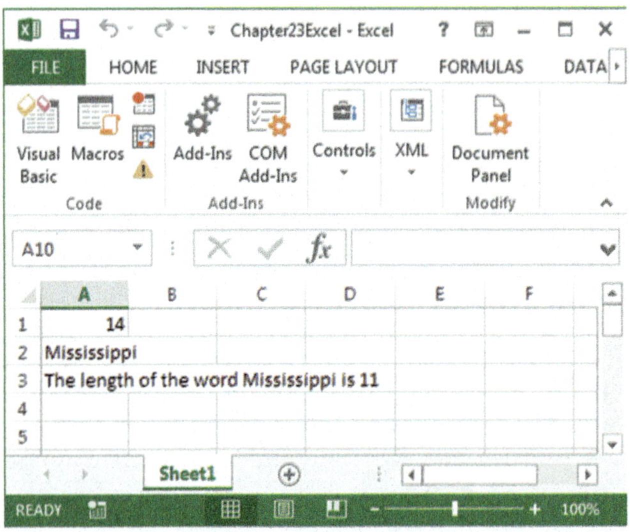

There are a lot of things happening in the macro "Chp23Example4":

1. In this macro, we used the keyword *Dim* to define two variables to hold *integer* data and one variable to hold *string* data.
2. In this macro, we used the keyword *inputbox* to prompt the user for data.
3. We used the single apostrophe to tell the VBE to ignore everything to the right. Programmers use the single apostrophe to comment about the VBA code.
4. Double quotes are used to hold string values.
5. "&" is used to put together two strings.
6. The character "_" is used to indicate that the VBA command line is continued in the next line.
7. We calculated the data we received and put the calculated result in ranges A1–A3.

We will now show why data typing a variable is important. The first input box requested an integer so can four to that number. Suppose that by accident, we enter a word instead. Below shows what happens when we do this.

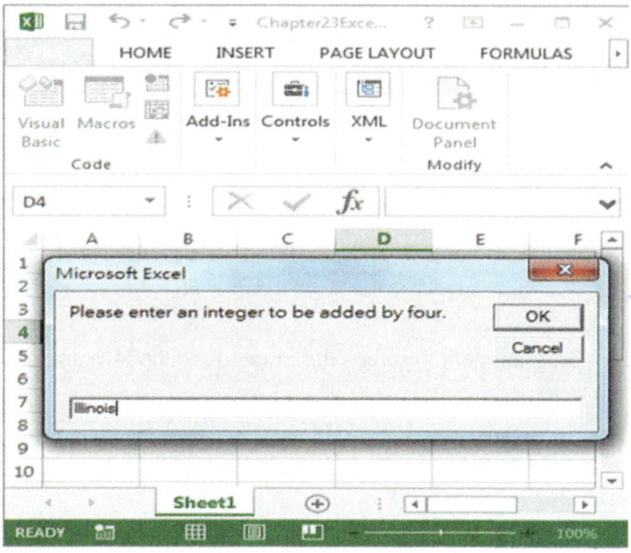

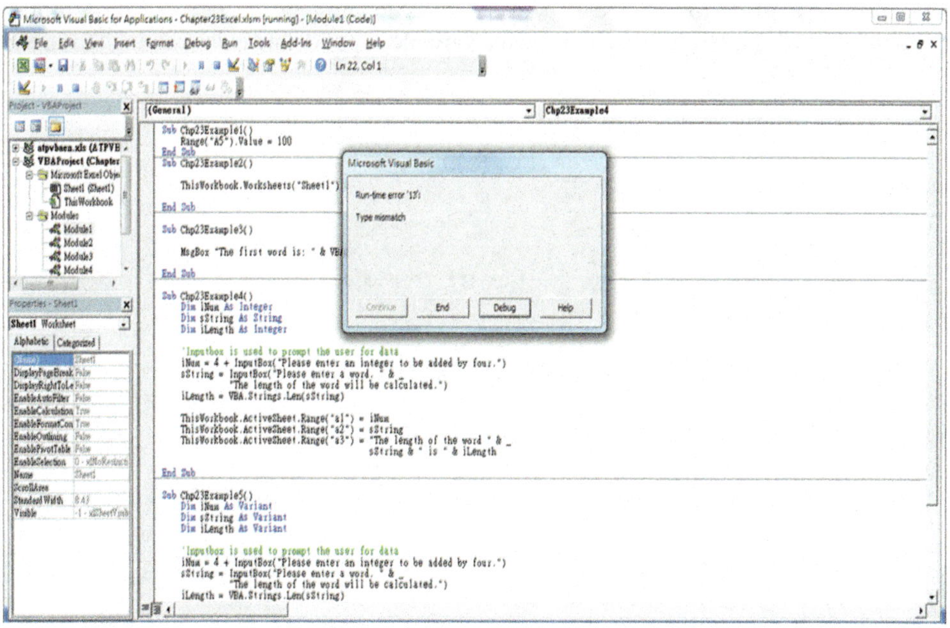

Above shows that the VBE will complain about having a wrong data type for the variable "iNum." There are VBA techniques to handle for this type of situation so the user will not have to see the above VBA error message.

From the data type list, it important to note that the data type *variant* is a data type that can be any type. The type of a variable is determined during run time (when the macro is running). The macro "Chp23Example4" can be rewritten as follows:

```
Sub Chp24Example5()
    Dim iNum As Variant
    Dim sString As Variant
    Dim iLength As Variant

    'Inputbox is used to prompt the user for data
    iNum = 4 + InputBox("Please enter an integer to be added by four.")
    sString = InputBox("Please enter a word. " & _
              "The length of the word will be calculated.")
    iLength = VBA.Strings.Len(sString)

    ThisWorkbook.ActiveSheet.Range("a1") = iNum
    ThisWorkbook.ActiveSheet.Range("a2") = sString
    ThisWorkbook.ActiveSheet.Range("a3") = "The length of the word " & _
                                           sString & " is " & iLength
End Su
```

Experienced VBA programmers prefer the macro "Chp24Example4" over the macro "Chp24Example5."

23.6 Option Explicit

In VBA programming, it is actually possible to use variables without first being defined, but good programming practice dictates that every variable should be defined. Excel VBA has the two keywords *Option Explicit* to indicate that every variable must be declared. Below shows what happens when *Option Explicit* is used and when a variable is not defined when trying to run a macro.

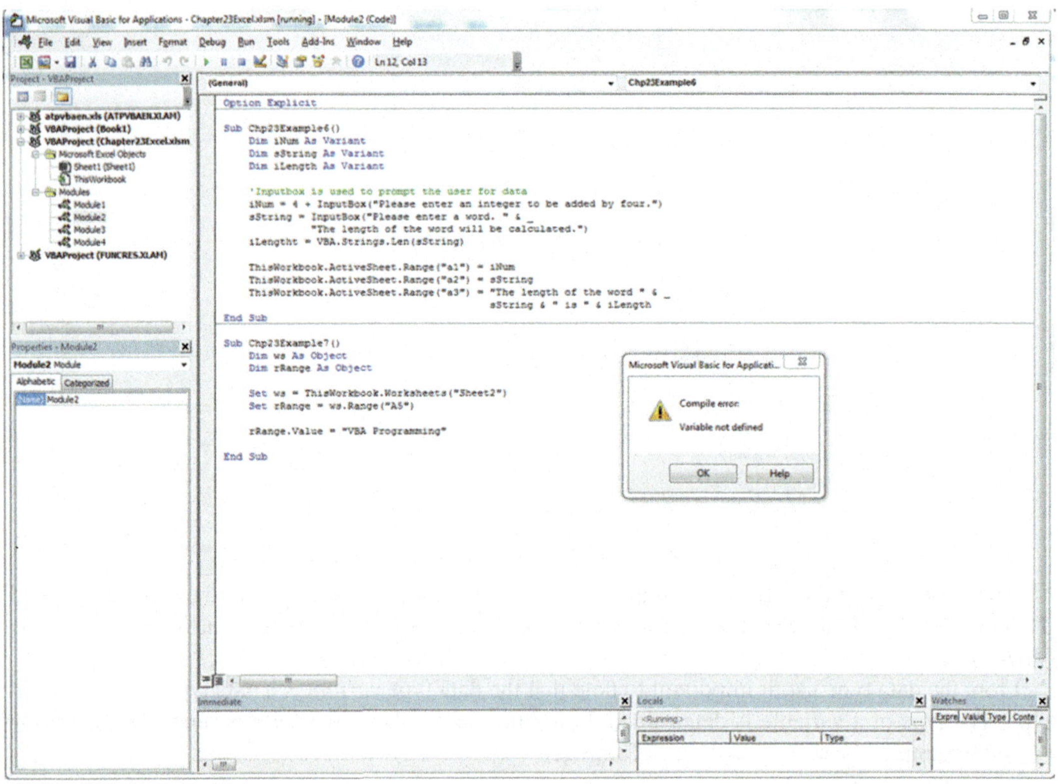

Notice that using the *Option Explicit* keywords result in the following:

1. The variable that is not defined is highlighted.
2. A message indicating that a variable is not defined is displayed.

When a new module is inserted into a project, the keywords *Option Explicit* by default are not inserted into the new module. This can cause problems especially in bigger macros. The VBE has a feature where the keywords *Option Explicit* are automatically included in a new module. To do this Choose Tools → Options. This is shown below.

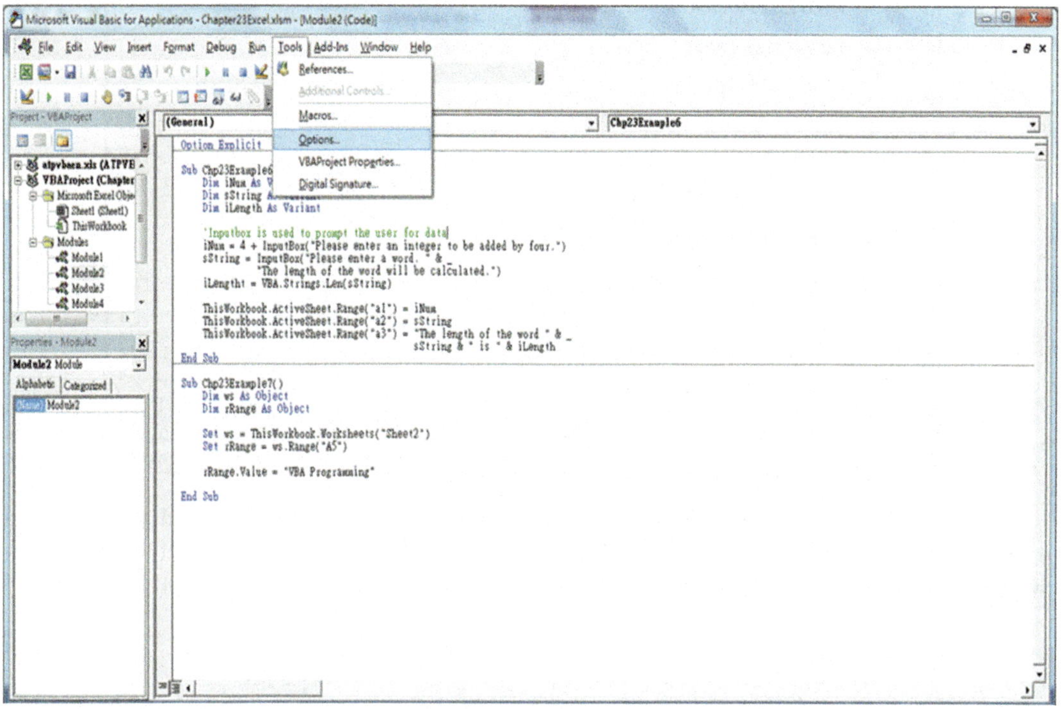

This will result in the following *Options* dialog box:

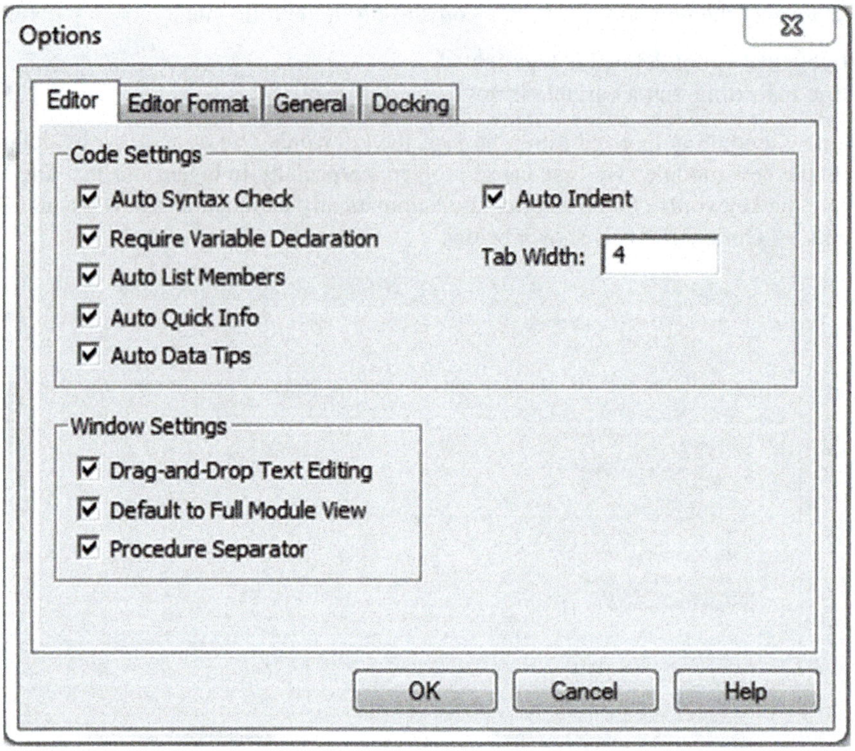

Choose the *Required Variable Declaration* option in the *Editor* tab of the *Options* dialog box to set it so the keywords *Options Explicit* are included with every new module. It important to note that by default the *Require Variable Declaration* option is not selected.

23.7 Object Variables

The data type *Object* is used to define variable to "point" to objects in the Excel object model. Like the data type *Variant*, the specific object data type for the data type *Object* is determined at run time. The macro below will set the cell A5 in the worksheet "Sheet2" to the value "VBA Programming." This macro is not sensitive to which worksheet is active.

```
Sub Chp24Example7()
    Dim ws As Object
    Dim rRange As Object

    Set ws = ThisWorkbook.Worksheets("Sheet2")
    Set rRange = ws.Range("A5")

    rRange.Value = "VBA Programming"

End Sub
```

Below rewrites the macro "Chp24Example7" by defining the variable "ws" as a *worksheet* data type and the variable "rRange" as a *range* data types.

```
Sub Chp24Example8()
    Dim ws As Worksheet
    Dim rRange As Range

    Set ws = ThisWorkbook.Worksheets("Sheet2")
    Set rRange = ws.Range("A5")

    rRange.Value = "VBA Programming"

End Sub
```

Experienced VBA programmers prefer the macro “Chp24Examples8” over the macro “Chp24Examples7.”

One reason to use specific data object types over the generic *object* data type is that the auto list member feature will not work with variables that are defined as an *object* data type. The auto list member feature will work with variables that are defined as a specific defined data types. This is shown below.

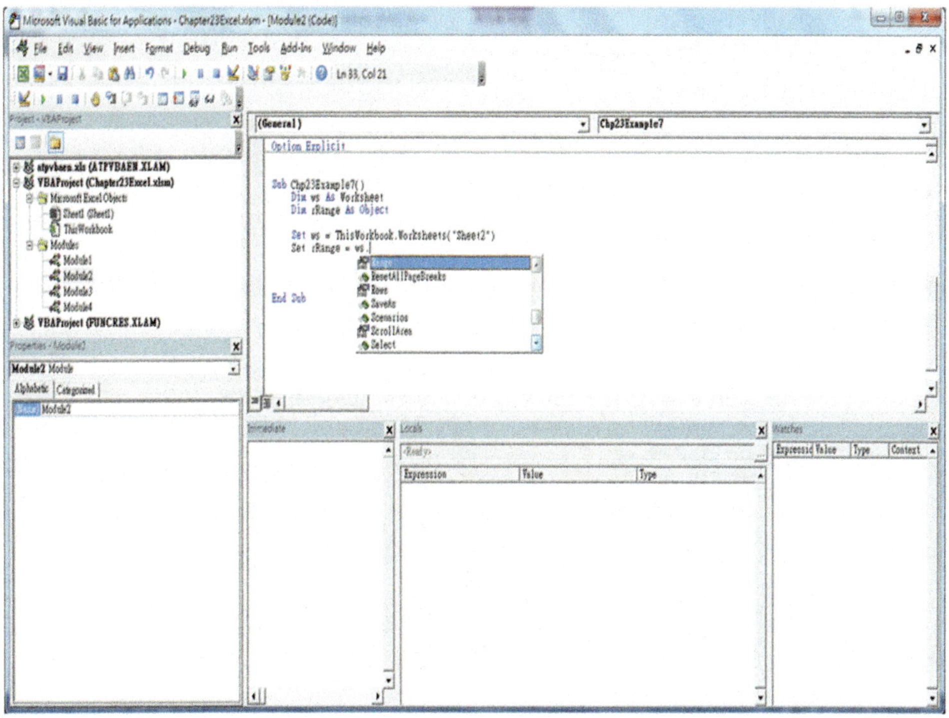

23.8 Functions

Functions in VBA act very much like functions in math. For example, below is a function that multiplies every number by .10.

$$f(x) = x^*.1$$

So in the above function, if x is 1000, then $f(x)$ is 100. The above function can be used in a bank that has a CD that pays 10 %. So if a customer opens a $1000 CD, a banker can use the above function to calculate the interest which according to the function is $100. Below is a VBA function that creates the above mathematical function.

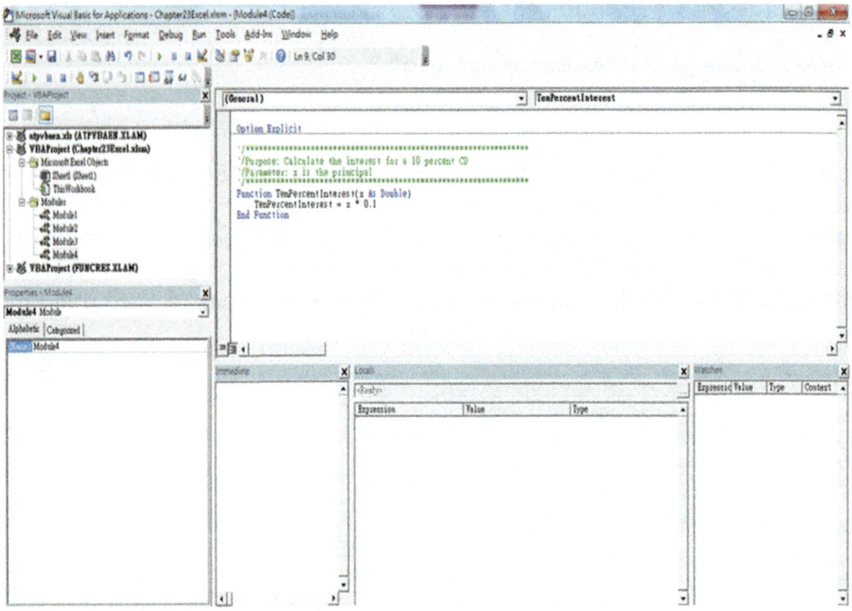

Functions created in Excel VBA can be used in the workbook that contains the function. To demonstrate this, go to the function wizard as shown below.

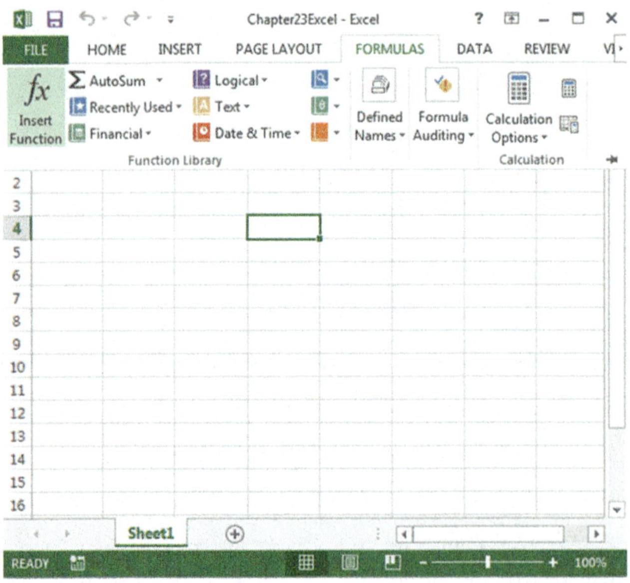

Next in the *Paste Function* dialog box, scroll down to the user section as shown below.

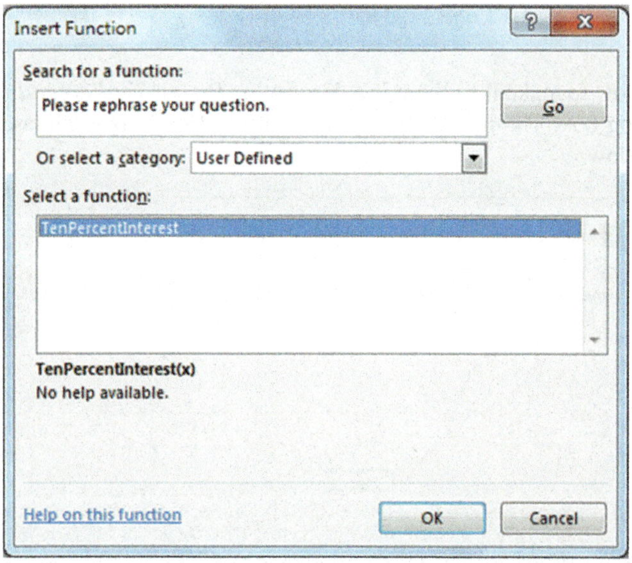

Notice that the function *TenPercentInterest* is listed in the *Paste Function* dialog box. To use the function that we created, highlight the function that we created as shown above and then press the *OK* button. Pressing the *OK* button will result in the following:

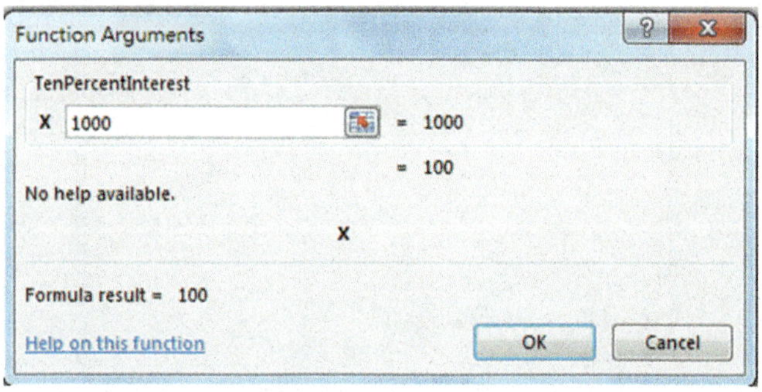

Notice that the above dialog box displays the parameter of the function. The above dialog box shows that entering the value 1000 for the parameter x will result in a value of 100. In functions that come with Excel, this dialog box will describe the function of interest. We can also do this for our *TenPercentInterest* function.

The following is the result after pressing the *OK* button in the above dialog box:

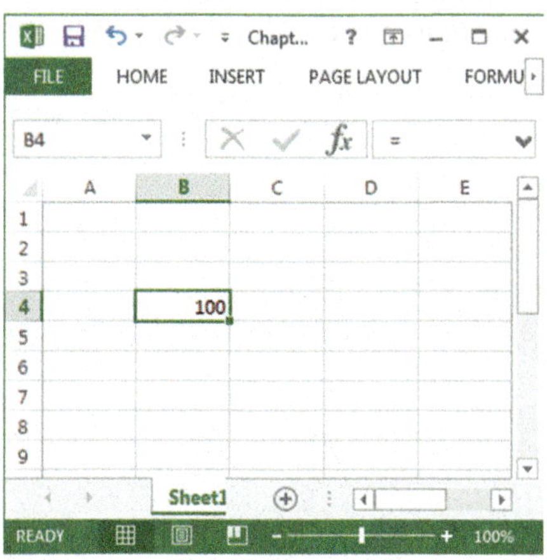

23.9 Adding a Function Description

We will now show how to make it so there is a description for our *TenPercentInterest* function in the *Paste Function* dialog box. The first thing that we will need to do is to choose Tools → Macro → Macros as shown below.

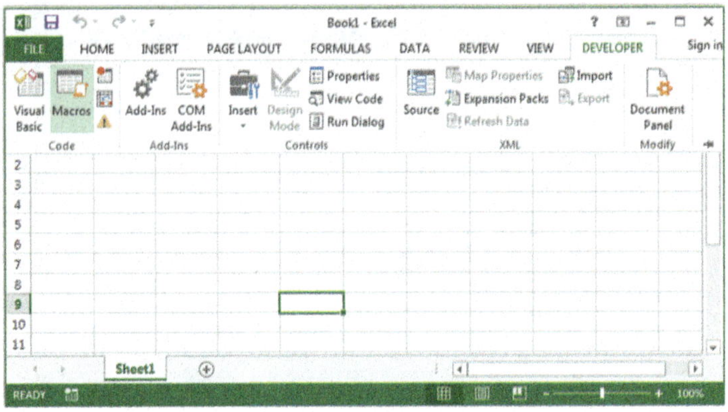

The resulting dialog box *Macro* dialog box is shown below.

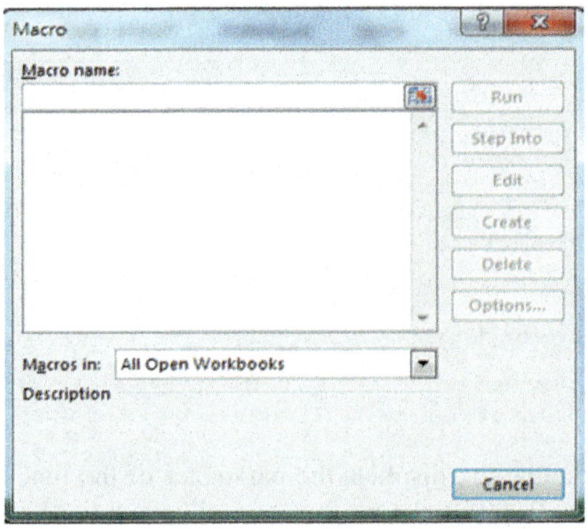

Notice that in the above *Macro* dialog box that no macro name is displayed and the only button active is the *Cancel* button. The reason for this is that the *Macro* dialog box only shows sub procedures. We did not include any sub procedures in our workbook. To write a description for a function, we would type in our functions name in the *Macro name:* option of the *Macro* dialog box as shown below.

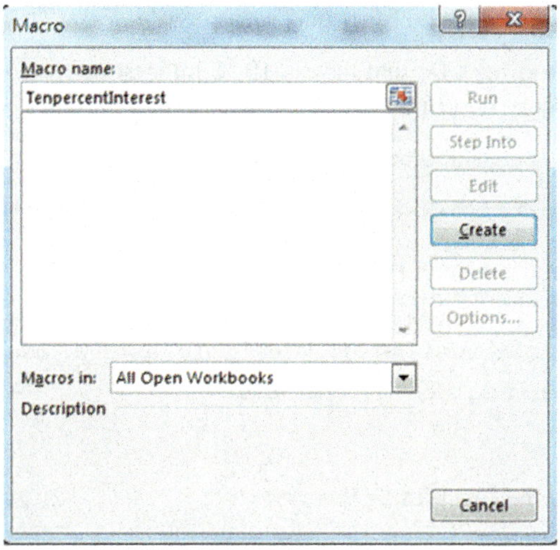

The next thing to do would be to press the *Options* button of the *Macro* dialog box to get the *Macro Options* dialog box shown below.

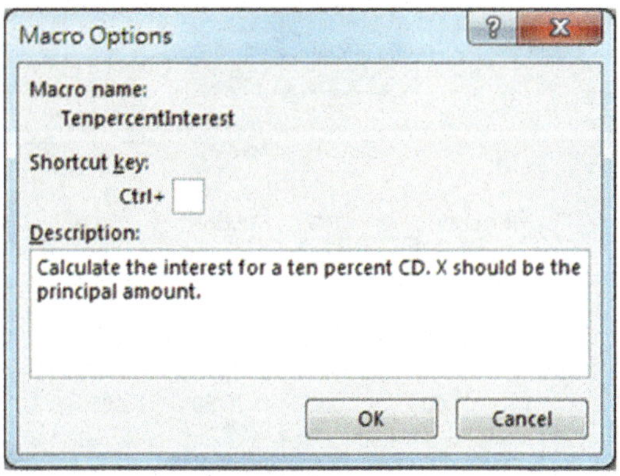

The next thing to do is to type the description for the function in the *Description:* option of the *Macro Options* dialog box. After you finish typing in the description, press the *OK* button.

If we now go back to the function wizard, we should now see the description that we typed in for our function. This is shown below.

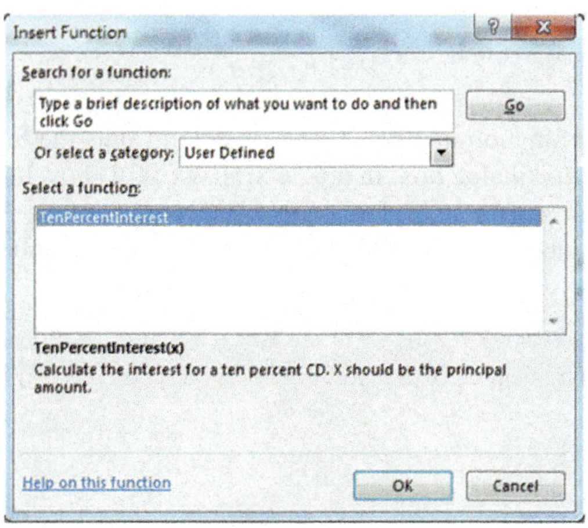

There are a few limitations with the function *TenPercentInterest*. The limitations are:

1. This function is only good for CDs that have a 10 % interest rate.
2. The parameter *x* is not very descriptive.

The function *CDInterest* address this issue.

```
'/***********************************************************
'/Purpose: Calculates the interest for a given CD
'/Parameter: Principal is the principal amount of CD
'/           Interest is the inert percent of CD
'/***********************************************************
Function CDInterest(Principal As Double, Interest As Double)
    CDInterest = Principal* Interest
End Function
```

Below shows the use of the function *CDInterest*.

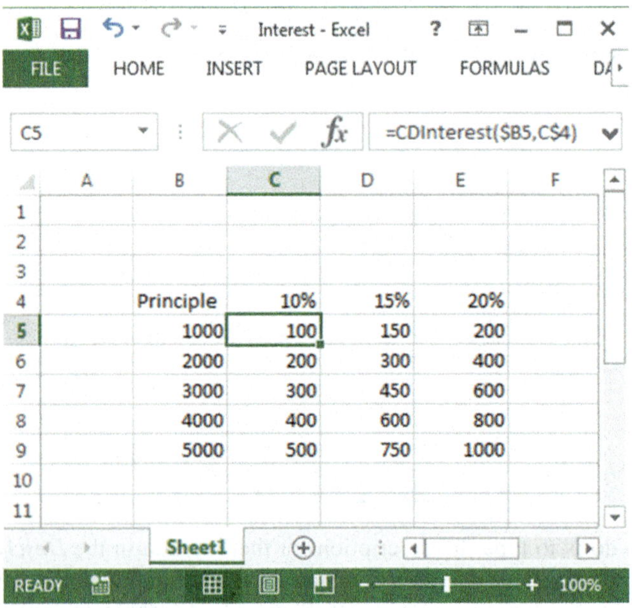

23.10 Specifying a Function Category

When you create a custom function in VBA, Excel by default puts the function in the *User Defined* category of the *Paste Function* dialog box. In this section, we will show how through VBA to set it so that the function *CDInterest* shows up in the "financial" category of the *Paste Function* dialog box.

Below is the VBA procedure to set it so the *CDInterest* function will be categorized as "financial" category.

```
Sub Auto_Open()
    Application.MacroOptions Macro:="CDInterest", Category:=1
End Sub
```

The *MacroOptions* method of the *Application* object sets function *CDInterest* to the "finance" category of the *Paste Function* dialog box. The *MacroOptions* method must be executed every time when we open the workbook that contains the function *CDInterest*.

This task is done by the procedure *Auto_Open* because VBA will execute any procedure called "Auto_Open" when a workbook is opened. Below shows the procedure *Auto_Open* and the function *CDInterest* in the workbook "CDInterest.xls."

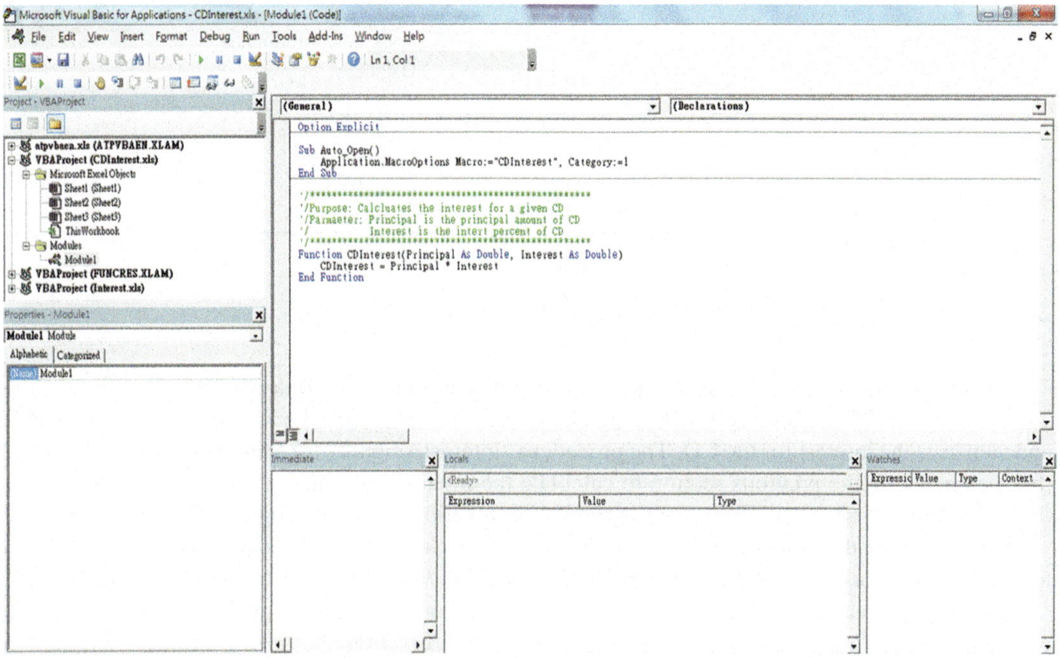

Below shows the function *CDInterest* in the "Financial" category in the *Paste Function* dialog box.

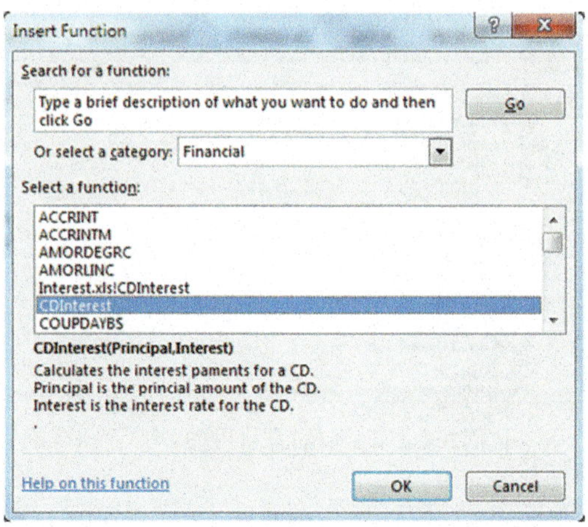

Below is a table showing the category number for the other categories of the *Paste Function* dialog box.

Function Categories

Category number	Category name
0	All
1	Financial
2	Date and time
3	Math and trig
4	Statistical
5	Lookup and reference
6	Database
7	Text
8	Logical
9	Information
14	User defined
15	Engineering

23.11 Conditional Programming with the IF Statement

The VBA *IF* statement is used to do conditional programming. Below shows the procedure "InterestAmount" This procedure will assign an interest rate based on the amount of the CD balance and then give the interest for the CD. The procedure "InterestAmount" uses the function "CDInterest" that we created in the previous section to calculate the interest amount.

It is possible to use most of the built-in worksheet functions in VBA programming. The procedure "CDInterest" uses the worksheet function "Isnumber" to check if the principal amount entered is a number or not. Worksheet functions belong to the *worksheetfunction* object of the Excel object model.

We can say that the module "module1" in the workbook "CDInterst.xls" is a program. It is a program because "module1" has two procedures and one function. A VBA program is basically a grouping of procedures and functions.

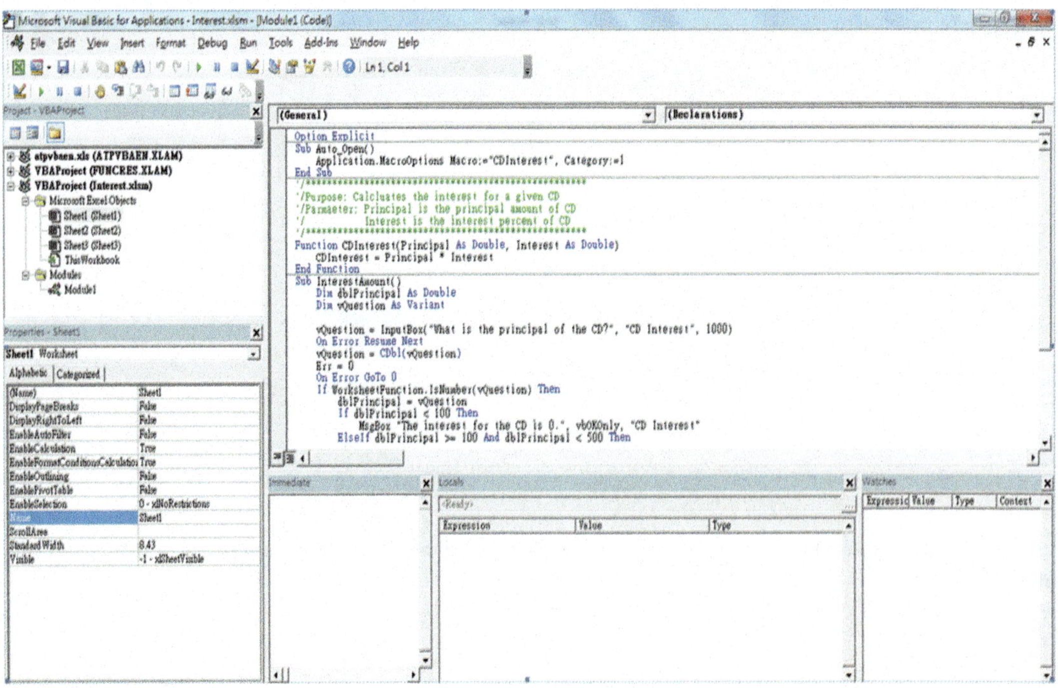

Below demonstrates the procedure "InterestAmount."

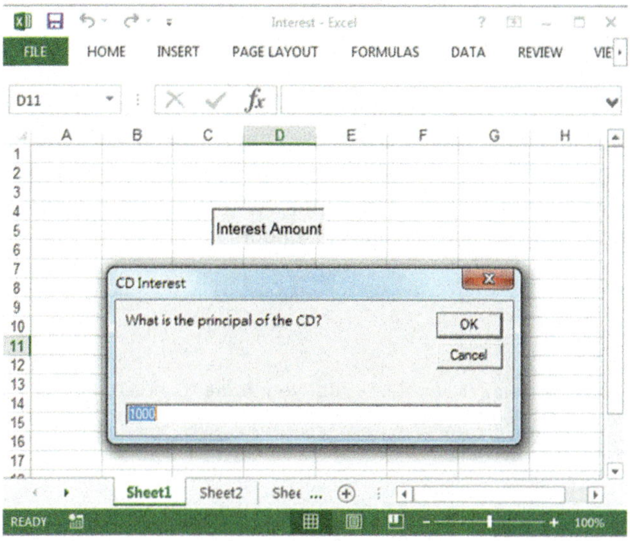

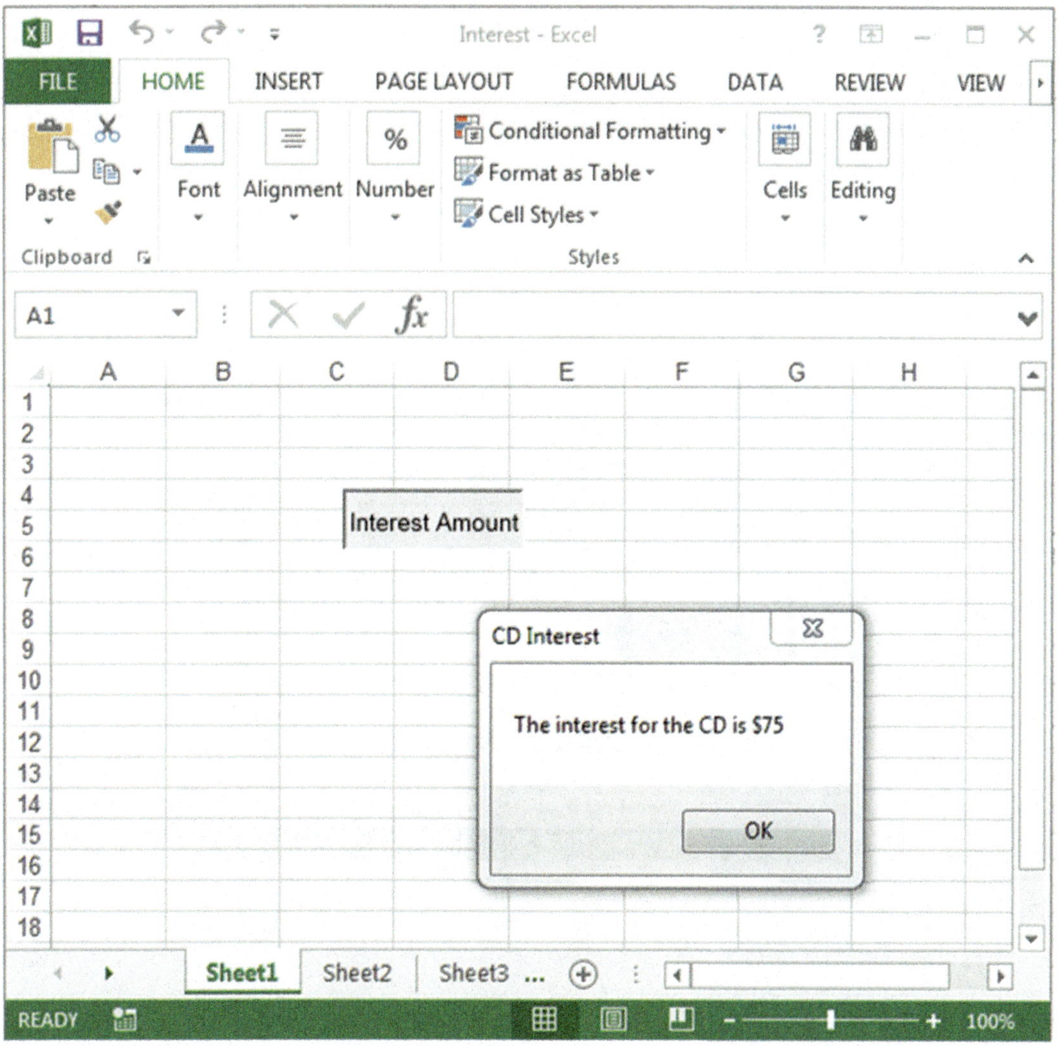

23.12 For Loop

Up to this point, the VBA code that we have been writing is executed sequentially from top to bottom. When the VBA code reaches the bottom it stops. We will now look at looping, the concept of where VBA code is executed more than once. The first looping code that we will look at is the *For* loop. The *For* loop is used when it can be determined how many times the loop should be. To demonstrate the *For* loop, we will extend our CD program in our previous section. We will add the procedure below to ask how many CD's we want to calculate. The improved CD program is in the workbook *InterestFor.xls* that is included in the CD that accompanies this book.

```
Sub MultiplyLoopFor()
    Dim vQuestion As Variant
    Dim i As Long

    vQuestion = InputBox("How many CD's do you want to calculate?", "CD Interest", 1)
    On Error Resume Next
    vQuestion = CDbl(vQuestion)
    Err = 0
    On Error GoTo 0
    If WorksheetFunction.IsNumber(vQuestion) Then
        For i = 1 To vQuestion
            InterestAmount
        Next
    End If
End Sub
```

Below demonstrates the *InterestFor.xls* workbook.

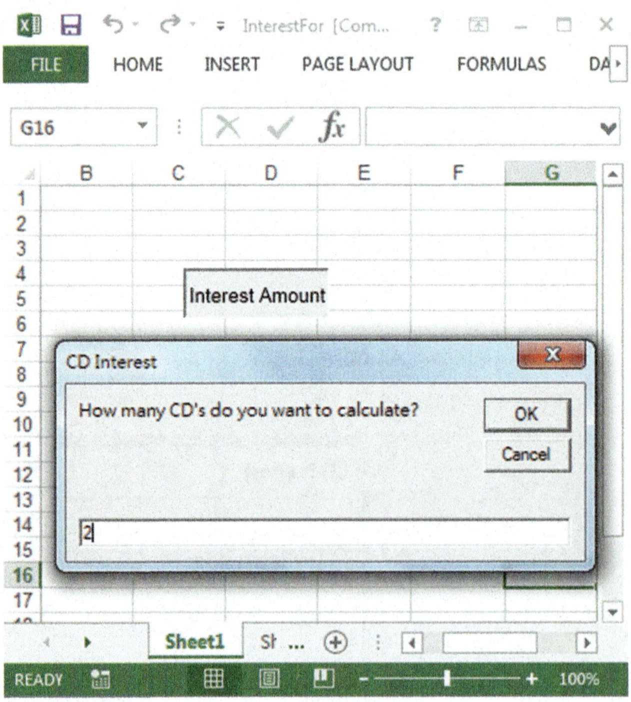

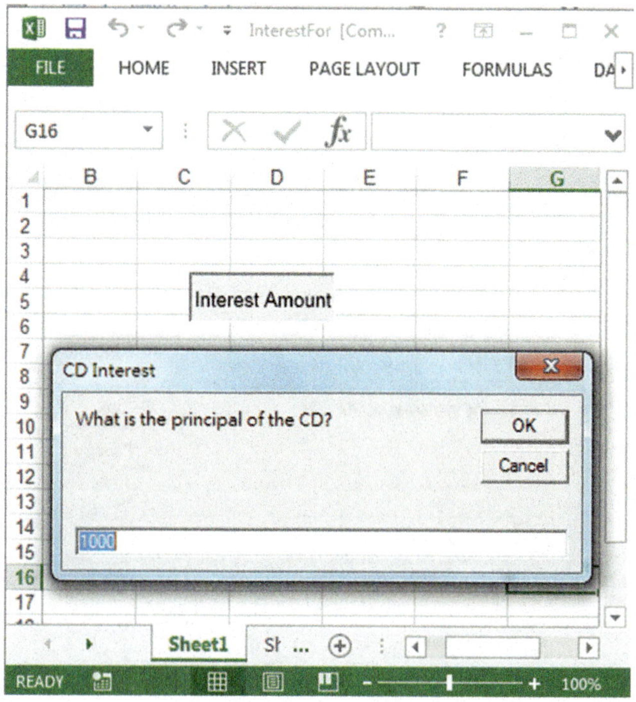
InterestFor [Com...
FILE
HOME
INSERT
PAGE LAYOUT
FORMULAS
G16
Interest Amount
CD Interest
What is the principal of the CD?
OK
Cancel
1000
Sheet1
READY
100%

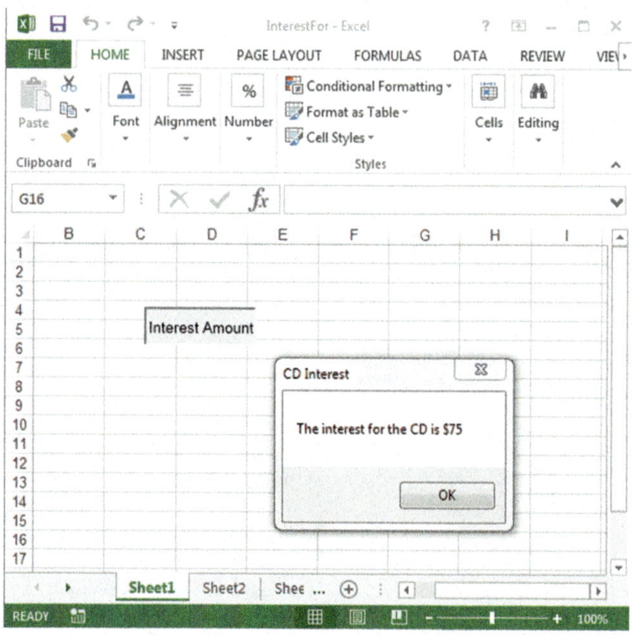
InterestFor - Excel
FILE
HOME
INSERT
PAGE LAYOUT
FORMULAS
DATA
REVIEW
Paste
Font
Alignment
Number
Conditional Formatting
Format as Table
Cell Styles
Cells
Editing
Clipboard
Styles
G16
Interest Amount
CD Interest
The interest for the CD is $75
OK
Sheet1
Sheet2
READY
100%

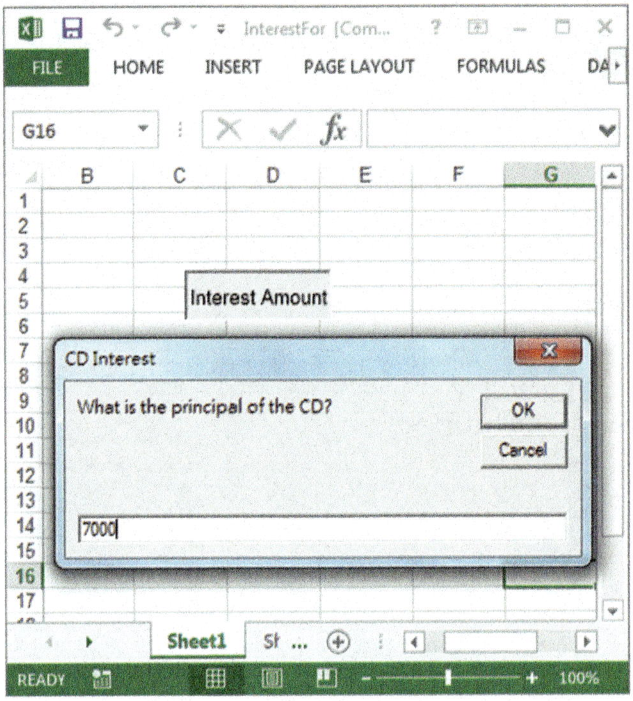
InterestFor [Com...
FILE
HOME
INSERT
PAGE LAYOUT
FORMULAS
G16
Interest Amount
CD Interest
What is the principal of the CD?
OK
Cancel
7000
Sheet1
READY
100%

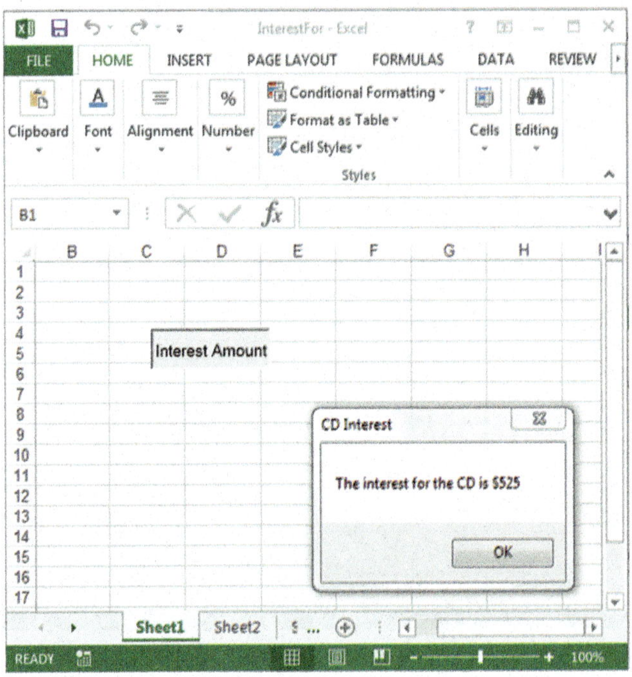
InterestFor - Excel
FILE
HOME
INSERT
PAGE LAYOUT
FORMULAS
DATA
REVIEW
Clipboard
Font
Alignment
Number
Conditional Formatting
Format as Table
Cell Styles
Styles
Cells
Editing
B1
Interest Amount
CD Interest
The interest for the CD is $525
OK
Sheet1
Sheet2
READY
100%

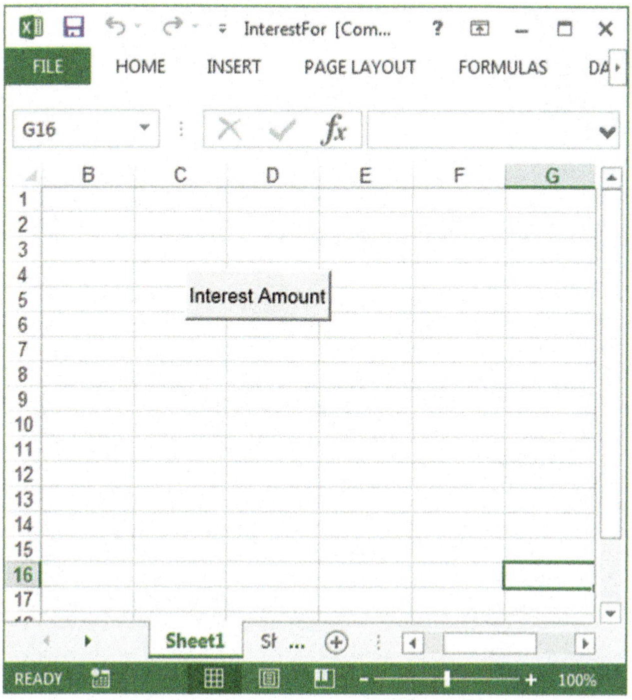

23.13 While Loop

Many times we do not know how many loops beforehand. In this case, the *While* loop is used instead. The *While* loop does a conditional test during each loop determine if a loop should be continued or not. To demonstrate the *While* loop, we will rewrite the program in the workbook *InterestFor.xls* to use the *While* loop instead of the *For* loop. The resulting workbook will be called *InterestWhile.xls*, which is included in the CD that accompanies this book. The following procedure that contains the *While* loop is in the *InterestWhile.xls*:

```
Sub MultiplyLoopWhile()

    InterestAmount
    While MsgBox("Do you want to calculate another CD?", _
        vbYesNo, "CDInterest") = vbYes
        InterestAmount
    Wend
End Sub
```

Below illustrates the program in the workbook *InterestWhile.xls*.

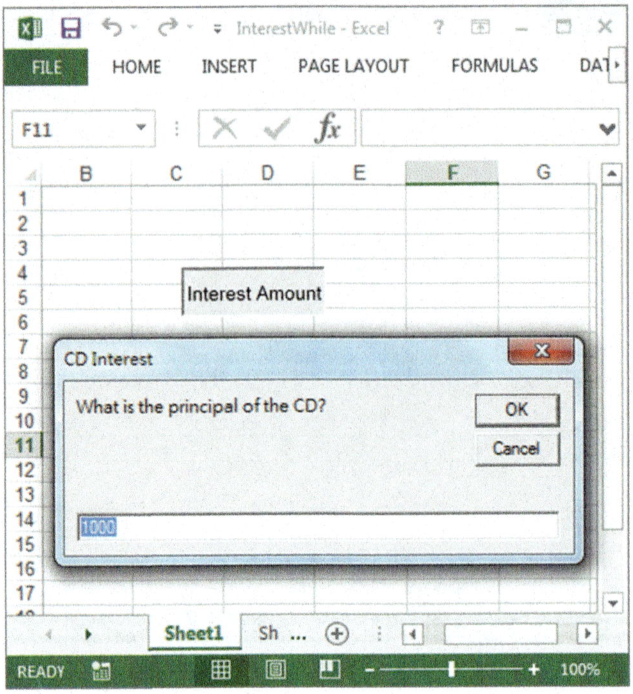

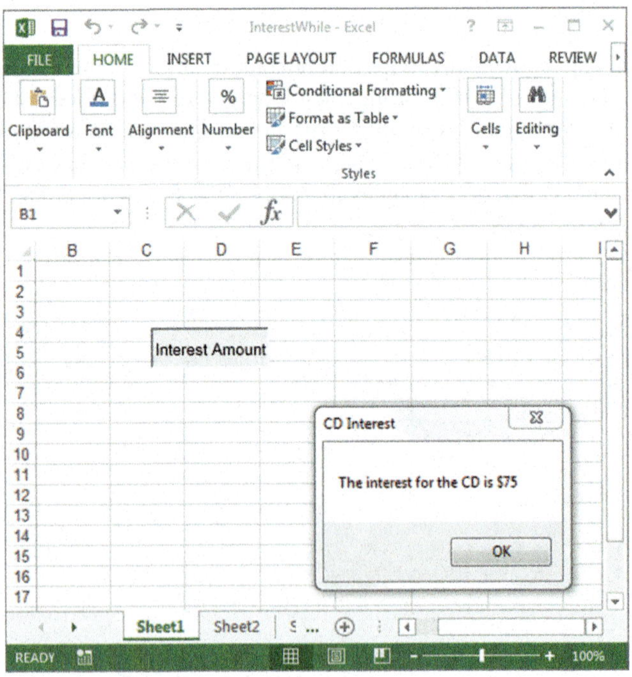

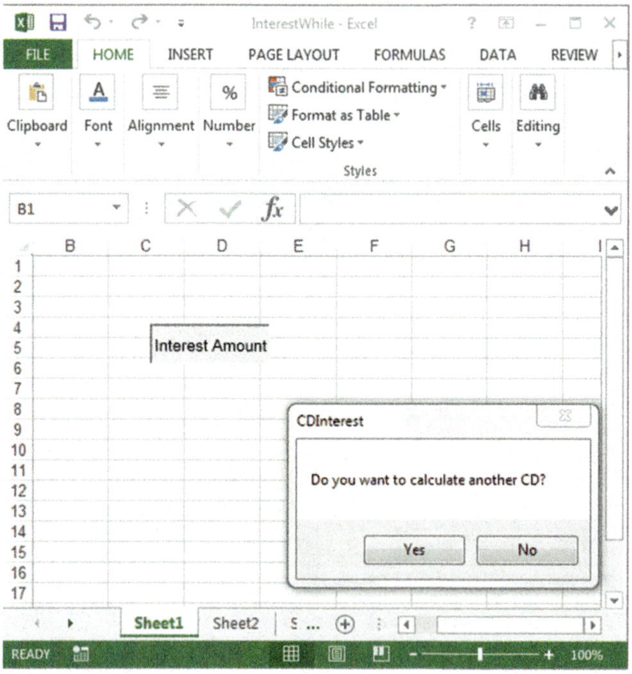
InterestWhile - Excel
FILE
HOME
INSERT
PAGE LAYOUT
FORMULAS
DATA
REVIEW
Clipboard
Font
Alignment
Number
Conditional Formatting
Format as Table
Cell Styles
Styles
Cells
Editing
B1
Interest Amount
CDInterest
Do you want to calculate another CD?
Yes
No
Sheet1
Sheet2
READY
100%

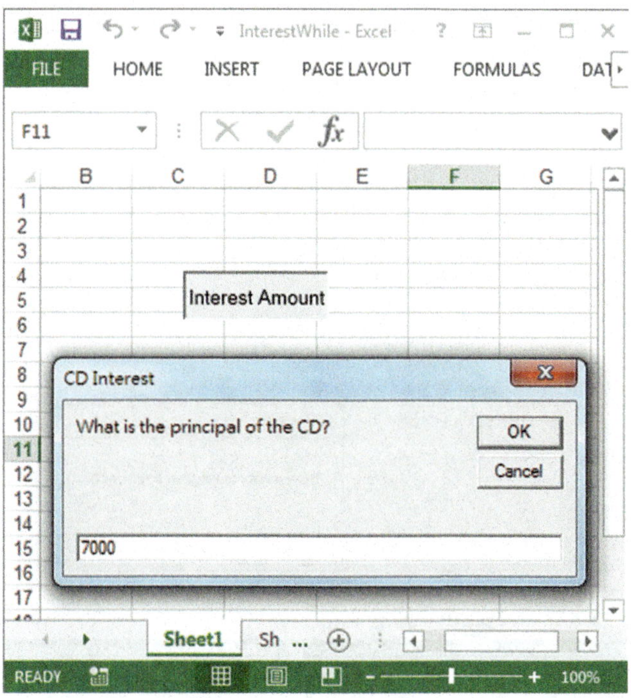
InterestWhile - Excel
FILE
HOME
INSERT
PAGE LAYOUT
FORMULAS
F11
Interest Amount
CD Interest
What is the principal of the CD?
OK
Cancel
7000
Sheet1
READY
100%

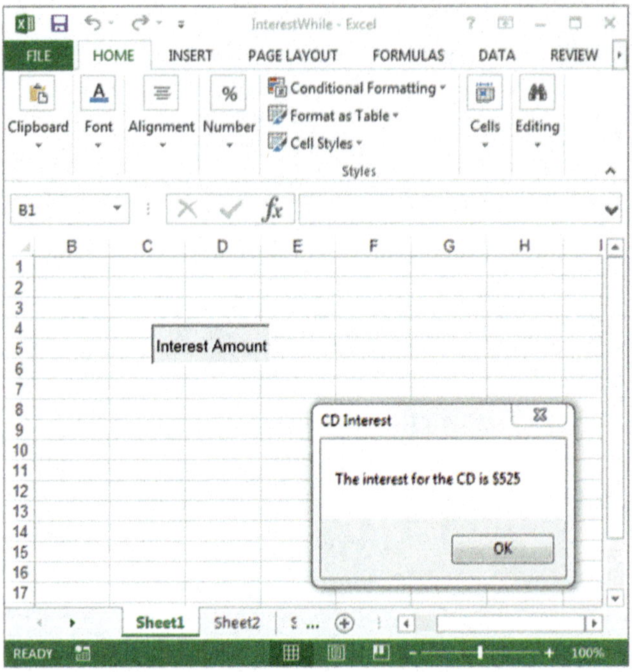
InterestWhile - Excel
Interest Amount
CD Interest
The interest for the CD is $525
OK

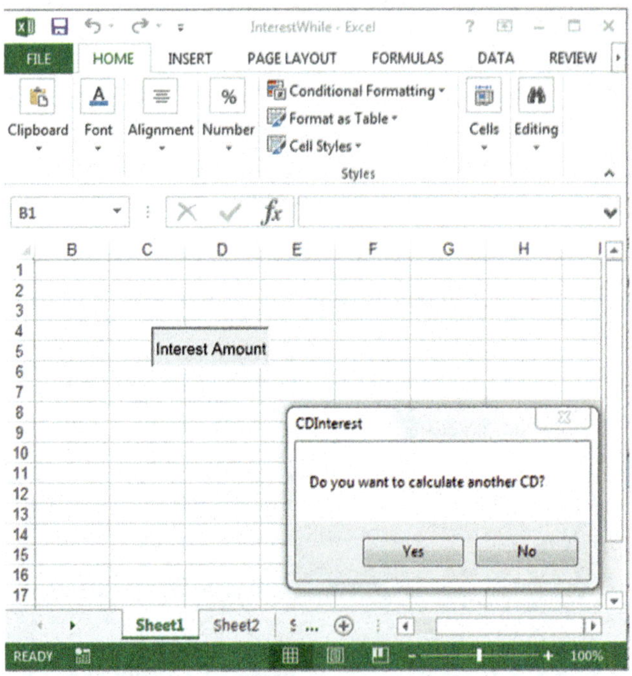
InterestWhile - Excel
Interest Amount
CDInterest
Do you want to calculate another CD?
Yes
No

23.14 Arrays

Most of the time when we are analyzing a data set, the data set contains data of the same data type. For example, we may have a data set of accounting salaries, a data set of GM stock prices, a data set of accounts receivables, or a data set of certificates of deposits. We might define 50 variables if we are processing a data set of salaries that have 50 data items. We might define the variables as "Salary1," "Salary2," "Salary3," . . . "Salary50." Another alternative is to define an *Array* of salaries. An *Array* is a group or collection of like data items. We would reference a particular salary through an index. The following is how to define our salary array variable of 50 elements:

```
Dim Salary(1 To 50) As Double
```

The following shows how to assign 15,000 to the 20th salary item:

```
Salary(20) = 15000
```

Suppose we need to calculate every 2 weeks the income tax to be withheld from 30 employees. This situation is very similar to our example in calculating the interest of certificate of deposits. When we calculate the certificate of deposits, we prompted the user for the principal amount. This process is very time consuming and very tedious. In the business world, it is common that the information of interest is already in an application. The procedure would then be to extract the information to a file to be processed. For our salary example, we will extract the salary data to a *csv* file format. A CSV file format is basically a text file that is separated by commas. A common tool to read CSV files is Notepad. Below shows the "salary.csv" that we are interested in processing.

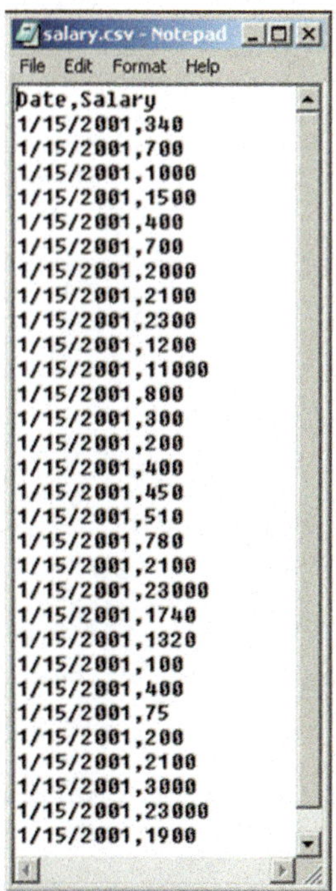
salary.csv - Notepad
File Edit Format Help

```
Date,Salary
1/15/2001,340
1/15/2001,700
1/15/2001,1000
1/15/2001,1500
1/15/2001,400
1/15/2001,700
1/15/2001,2000
1/15/2001,2100
1/15/2001,2300
1/15/2001,1200
1/15/2001,11000
1/15/2001,800
1/15/2001,300
1/15/2001,200
1/15/2001,400
1/15/2001,450
1/15/2001,510
1/15/2001,780
1/15/2001,2100
1/15/2001,23000
1/15/2001,1740
1/15/2001,1320
1/15/2001,100
1/15/2001,400
1/15/2001,75
1/15/2001,200
1/15/2001,2100
1/15/2001,3000
1/15/2001,23000
1/15/2001,1900
```

The thing to note about the *csv* file is that the first row is usually the header. This is the row that describes the columns of a data set. In the salary file above, we can say that the header contains two fields. One field is the date field and the other field is the salary field.

Below shows code on how to process the salary file. The procedure below is in the workbook "SalaryInterest.xls" which is in the CD that accompanies this book. Make sure the files "salary.csv" and "SalaryInterest.xls" are in the same directory or folder.

```
'Option Explicit indicates that all variable must be declared.
Option Explicit
'Option base 1 indicates that the lower array index is 1
Option Base 1
'/*********************************************************************
'/Purpose: To cacuclate the weekly interst to be withheld
'/         The tax table will be as follows.
'/              1% < 500
'/              2% < 1000
'/              3% < 1750
'/              4% > 1749
'/*********************************************************************
Sub SalaryTax()
    'To dynamically size the arrays we will need to use variant variables
    Dim vSalary As Variant
    Dim vTax As Variant
    Dim rSalary As Range
    Dim wbSalary As Workbook
    Dim wbTax As Workbook
    Dim lCount As Long
    Dim lTax As Long

    'Open salary file.  It is assumed in same location as this workbook
    Set wbSalary = Workbooks.Open(ThisWorkbook.Path & "\" & "salary.csv")

    'Figure out salary range
    'CurrentRegion Method will find row and columns that are completely
    'surronded by blank cells
    Set rSalary = ActiveSheet.Cells(1).CurrentRegion
    Set rSalary = rSalary.Resize(rSalary.Rows.Count - 1, rSalary.Columns.Count)
    Set rSalary = rSalary.Offset(1, 0)

    'Import information into array
    vSalary = WorksheetFunction.Transpose(rSalary.Columns(2))
    'The UBound function will find the upper index of an array
    lTax = UBound(vSalary)
    'Size vTax to the same size as vSalary
    'Redim statement will reset an array to a specific size
    ReDim vTax(lTax)

    wbSalary.Close False

    'Loop through Salary array to calculate the taxes to be with held
    For lCount = 1 To lTax
        If vSalary(lCount) < 500 Then
            vTax(lCount) = vSalary(lCount) * 1.01
        ElseIf vSalary(lCount) < 1000 Then
            vTax(lCount) = vSalary(lCount) * 1.02
        ElseIf vSalary(lCount) < 1750 Then
            vTax(lCount) = vSalary(lCount) * 1.03
        ElseIf vSalary(lCount) > 1749 Then
            vTax(lCount) = vSalary(lCount) * 1.04
        End If
    Next

    'Create a resulting workbook and have the variable wbTax point to it
    Set wbTax = Workbooks.Add
    With wbTax.Worksheets(1)
        .Cells(1, 1) = "Salary"
        .Cells(1, 2) = "Tax"
        'Put resulting array to range
        .Cells(2, 1).Resize(lTax, 1) = WorksheetFunction.Transpose(vSalary)
        .Cells(2, 2).Resize(lTax, 1) = WorksheetFunction.Transpose(vTax)
        .Name = "Tax Withholding"
    End With
End Sub
```

Below illustrates the above procedure.

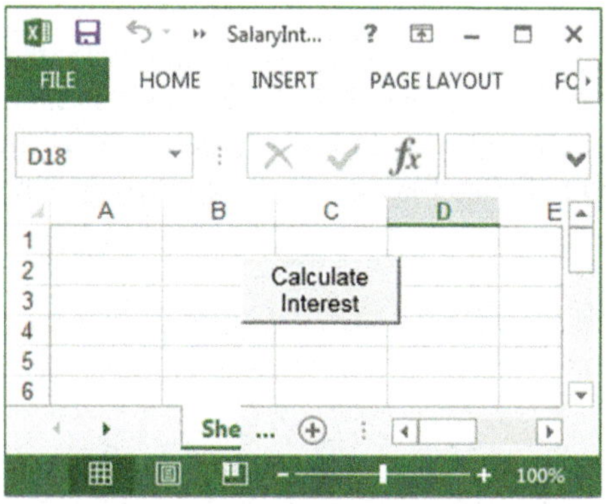

Pushing the *Calculate Interest* button will result in the following workbook:

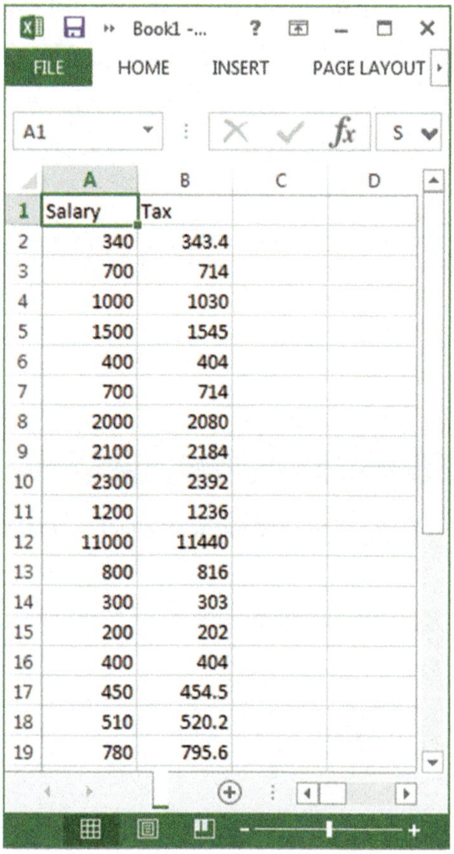

23.15 Option Base 1

When normal people think about lists, they usually start with the number 1. A lot of times programmers begin a list with the number 0. In VBA programming the beginning of any array index is 0. To set it so the beginning of array index is 1, we would use the statement "Option Base 1." This was done in the procedure "SalaryTax" in the previous procedure.

23.16 Collections

In VBA programming there is a lot of programming with a group of like items. Groups of like items are called *Collections*. Examples are collections of workbooks, worksheets, cells, charts, and names. There are two ways to reference a collection. The first way is through an index. The second way is by name. For example, suppose we have the following workbook that contains three worksheets:

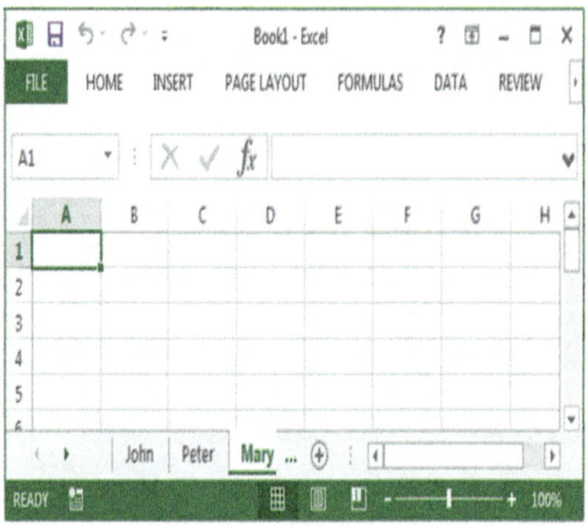

Below is a procedure that references the second worksheet by index.

```
Sub PeterIndex()
    MsgBox ThisWorkbook.Worksheets(2).Name
End Sub
```

Below demonstrates the procedure "PeterIndex."

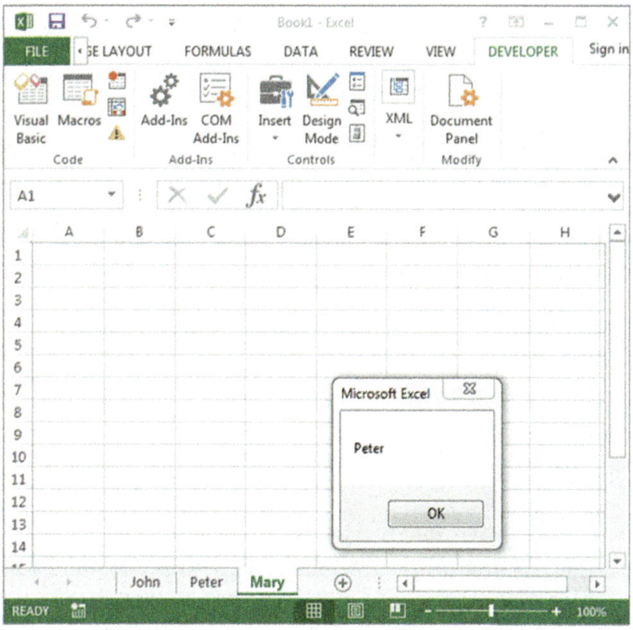

Below is a procedure that references the second worksheet by name.

```
Sub PeterName()
    MsgBox ThisWorkbook.Worksheets("Peter").Name
End Sub
```

It is important to note what the effect of removing an item from a collection to VBA code is. Below shows the workbook without the worksheet "John."

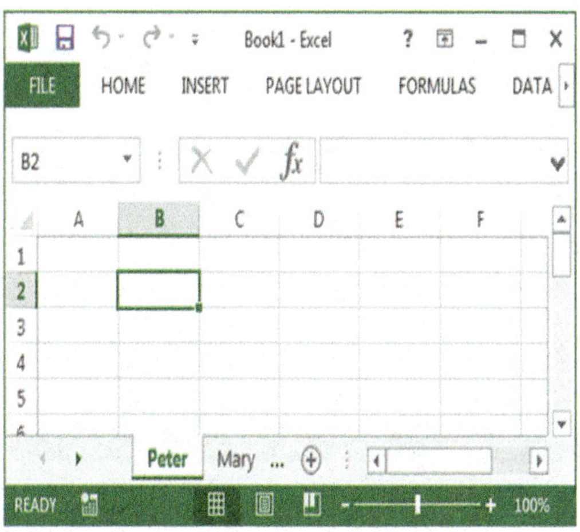

Below is the result when executing the procedure "PeterIndex."

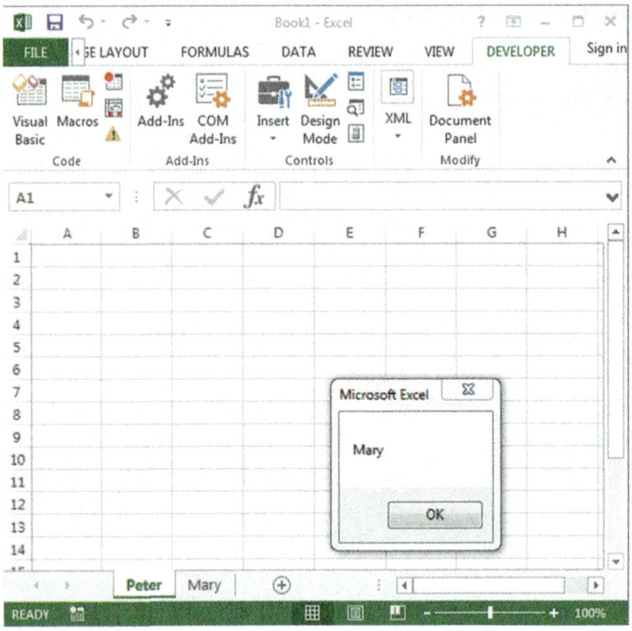

Below is the result when executing the procedure "PeterName."

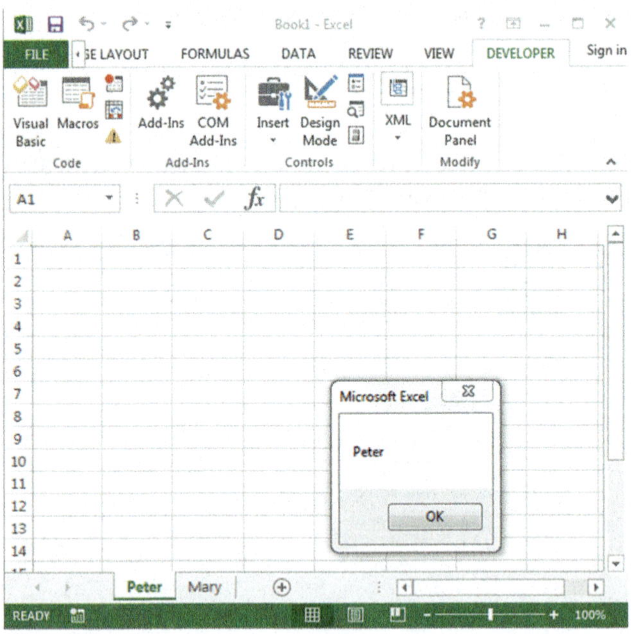

The above demonstrates that referencing an item in a collection by name is preferable when there are additions or deletions to a collection.

23.17 Looping a Collection: For Each

If we are interested in looping through all the elements of a collection we can use the *For Each* loop.

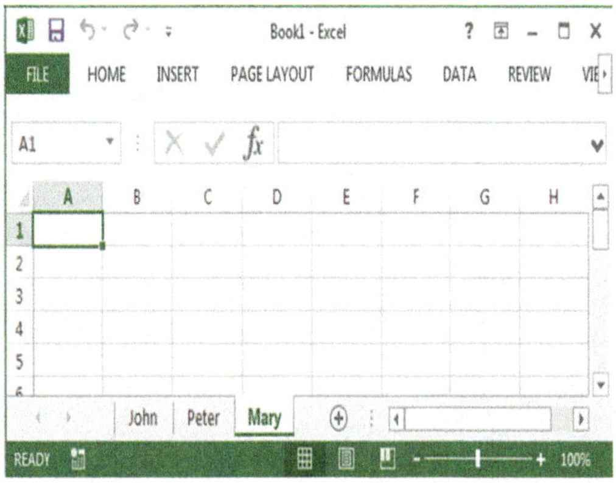

Below is a procedure that will display the name of every worksheet in the above workbook.

```
Sub WorksheetName()
  Dim ws As Worksheet
     For Each ws In ThisWorkbook.Worksheets
         MsgBox ws.Name
     Next
End Sub
```

Below demonstrates the procedure "WorksheetName."

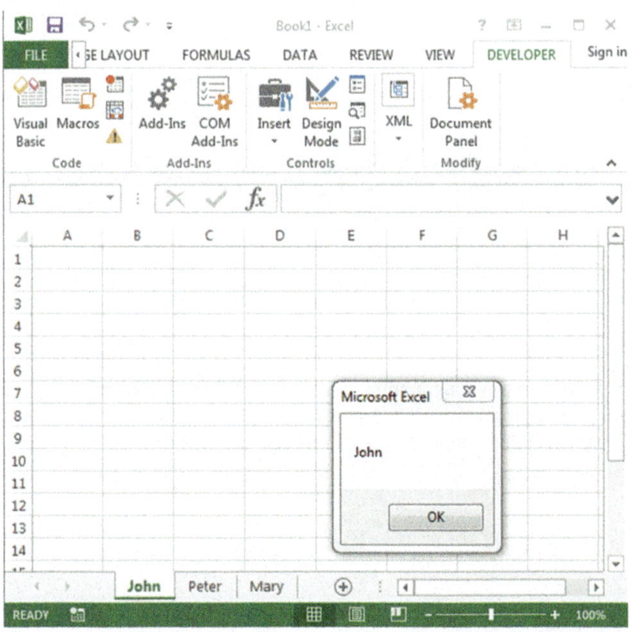

We can also use the *For Each* loop to loop through arrays. The procedure "ArraySalaryDemo" shown below demonstrates this.

```
Sub ArraySalaryDemo()
    Dim dblSalary(1 To 3) As Double
    Dim vnt As Variant

    dblSalary(1) = 1000
    dblSalary(2) = 2000
    dblSalary(3) = 3000

    For Each vnt In dblSalary
        MsgBox vnt
    Next

End Sub
```

Below demonstrates the procedure "ArraySalaryDemo."

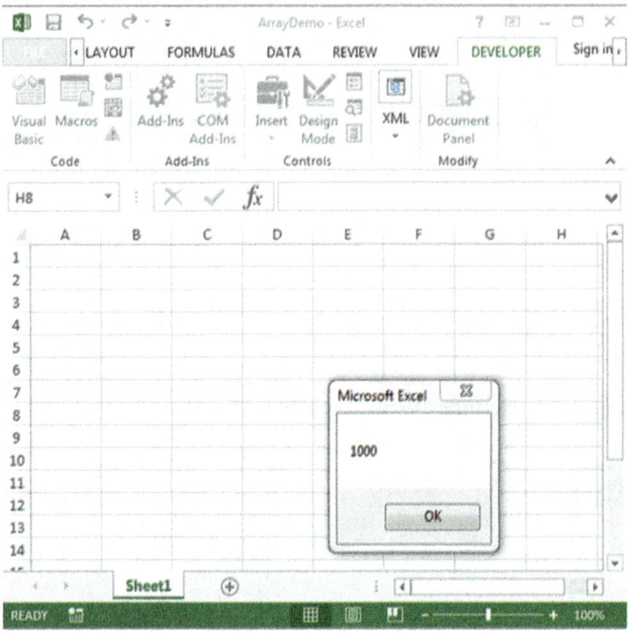

Bibliography

Walkenbach J (2013) Excel 2013 power programming with VBA. Wiley, Hoboken
Bullen S (2007) Excel 2007 VBA programmer's reference. Wiley, Indianapolis
Bovey R (2009) Professional excel development. Pearson Education, Boston

Chapter 24
Professional Techniques Used in Excel and Excel VBA Techniques

24.1 Introduction

In this chapter we will discuss Excel and Excel VBA techniques that I find useful and are not usually discussed or pointed out in Excel and Excel VBA books. These techniques come from my experience as an Excel professional.

24.2 Finding the Range of a Table: CurrentRegion Property

Many times we are interested in finding the range or an address of a table. My favorite way to do this is to use the *CurrentRegion* property of the range object. One common situation where there is a need to do this is when we import data files. Usually Excel places the data in the upper left-hand corner of the first worksheet. As an example, let's open the "CD.csv" file that is included in the CD that accompanies this book.

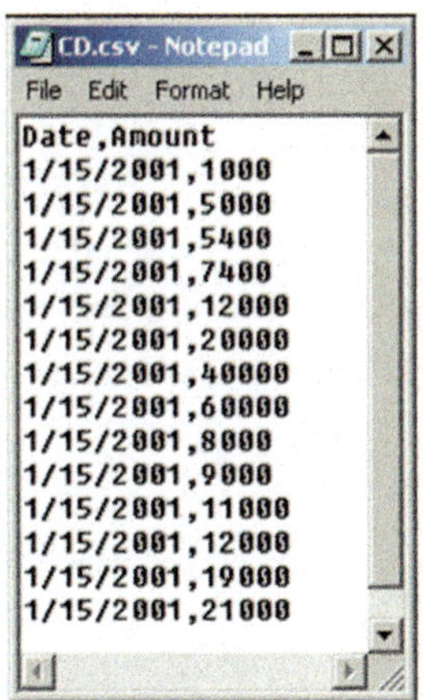

C.-F. Lee et al., *Essentials of Excel, Excel VBA, SAS and Minitab for Statistical and Financial Analyses*, DOI 10.1007/978-3-319-38867-0_24

```
'/*****************************************************************************
'/Purpose:  To find the data range of an imported file
'/****************************************************************************
Sub FindCurrentRegion()
    Dim rCD As Range
    Dim wbCD As Workbook

    On Error Resume Next
    'Open CD file.  It is assumed in same location as this workbook
    Set wbCD = Workbooks.Open(ThisWorkbook.Path & "\" & "CD.csv")

    If wbCD Is Nothing Then
        MsgBox "Could not find the file CD.csv in the path " _
                    & ThisWorkbook.Path, vbCritical

        End
    End If
    'Figure out salary range
    'CurrentRegion Method will find row and columns that are completely
    'surronded by blank cells
    Set rCD = ActiveSheet.Cells(1).CurrentRegion

    rCD.Select
    MsgBox "The address of the data is " & rCD.Address

    wbCD.Close False
End Sub
```

The above procedure will open the "CD.csv" file and then select the data range by using the *CurentRegion* property of the range object and also display the address of the data range. Below demonstrates the "FindCurrentRegion" procedure.

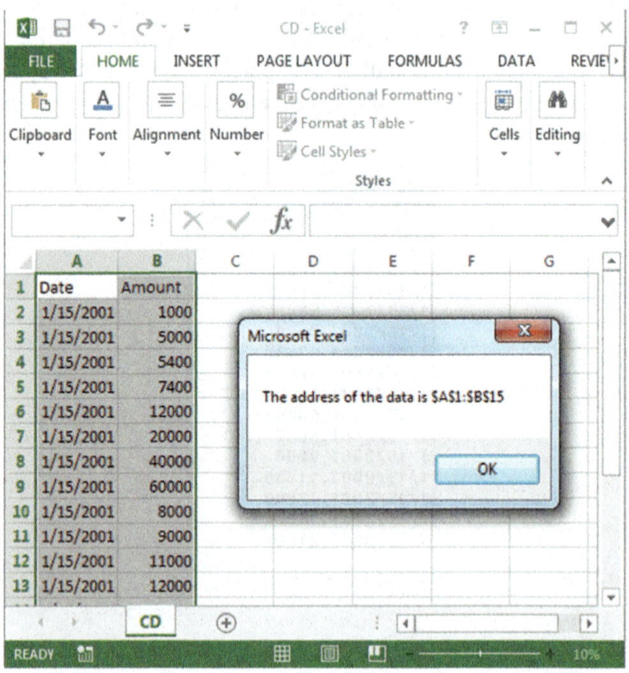

Notice that the currentregion area contains the header or row 1. Many times when data is imported, we will want to exclude the header row. To solve this problem we will look at the *offset* property of the range object in the next section.

24.3 Offset Property of the Range Object

The *offset property* is one of those properties and methods that are usually mentioned in passing in most books. From my experience it is one of the properties that I use a lot. The *offset* property has two arguments. The first argument is for the row offset. The second argument is for the column offset. Below is a procedure that illustrates the *offset* property. This procedure is in the workbook *CurrentRegionOffset.xls* that is included in the CD that accompanies this book.

```
'/*************************************************************************************
'/Purpose:  To find the data range of an imported file
'/***********************************************************************************
Sub CurrentRegionOffset()
    Dim rCD As Range
    Dim wbCD As Workbook

    On Error Resume Next
    'Open CD file.  It is assumed in same location as this workbook
    Set wbCD = Workbooks.Open(ThisWorkbook.Path & "\" & "CD.csv")

    If wbCD Is Nothing Then
        MsgBox "Could not find the file CD.csv in the path " _
                    & ThisWorkbook.Path, vbCritical
        End
    End If
    'Figure out salary range
    'CurrentRegion Method will find row and columns that are completely
    'surronded by blank cells
    Set rCD = ActiveSheet.Cells(1).CurrentRegion

    'Offset the current region by one row.
    'The offset property has row offset argument and column offset argument
    Set rCD = rCD.Offset(rowoffset:=1, columnoffset:=0)
    rCD.Select

    MsgBox "The address of the data is " & rCD.Address

    wbCD.Close False
End Sub
```

Below demonstrates the "CurrentRegionOffset" procedure.

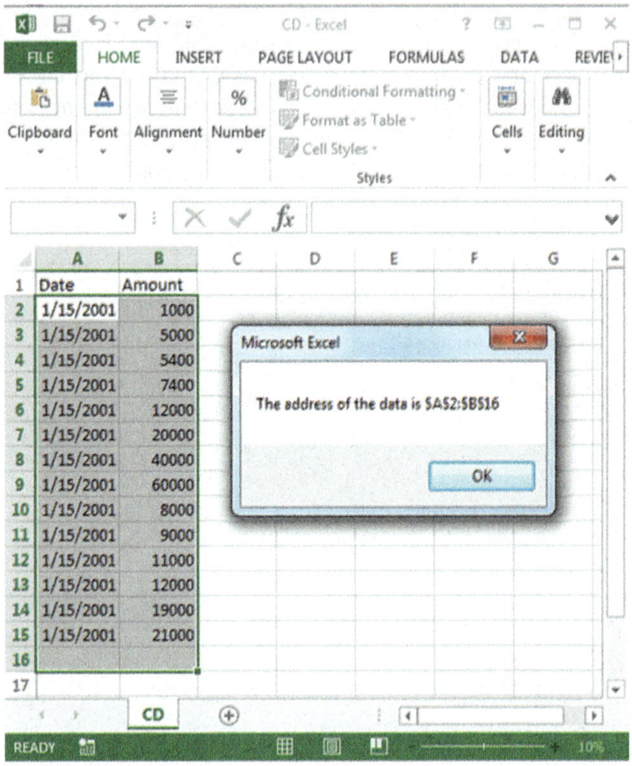

Notice that when we used the *offset* property, we shifted the whole currentregion by one row. As shown above, offsetting the currentregion by one row causes the blank row 16 to be included. To solve this problem, we will use the *resize* property of the range object. The *resize* property is discussed in the next section.

24.4 Resize Property of the Range Object

Like the *offset* property, the *resize* property is one of those properties and methods that are usually mentioned in passing in most books but is one of those properties that I use a lot.

The *resize* property has two arguments. The first argument is to resize the row to a certain size. The second argument is to resize the column to a certain size. Below is a procedure that illustrates the *resize* property. This procedure is in the workbook *CurrentRegionOffsetResize.xls* that is included in the CD that accompanies this book.

```
'/*************************************************************************************
'/Purpose:  To find the data range of an imported file
'/***********************************************************************************
Sub CurrentRegionOffsetResize()
    Dim rCD As Range
    Dim wbCD As Workbook

    On Error Resume Next
    'Open CD file.  It is assumed in same location as this workbook
    Set wbCD = Workbooks.Open(ThisWorkbook.Path & "\" & "CD.csv")

    If wbCD Is Nothing Then
        MsgBox "Could not find the file CD.csv in the path " _
                    & ThisWorkbook.Path, vbCritical
        End
    End If
    'Figure out salary range
    'CurrentRegion Method will find row and columns that are completely
    'surronded by blank cells
    Set rCD = ActiveSheet.Cells(1).CurrentRegion

    'Offset the current region by one row.
    'The offset property has row offset argument and column offset argument
    Set rCD = rCD.Offset(rowoffset:=1, columnoffset:=0)

    'resize the range by the amount previous rows -1
    'resize the columns to same number columns as previously
    Set rCD = rCD.Resize(rowsize:=rCD.Rows.Count - 1, columnsize:=rCD.Columns.Count)
    rCD.Select
    MsgBox "The address of the data is " & rCD.Address

    wbCD.Close False
End Sub
```

Below demonstrates the *resize* property.

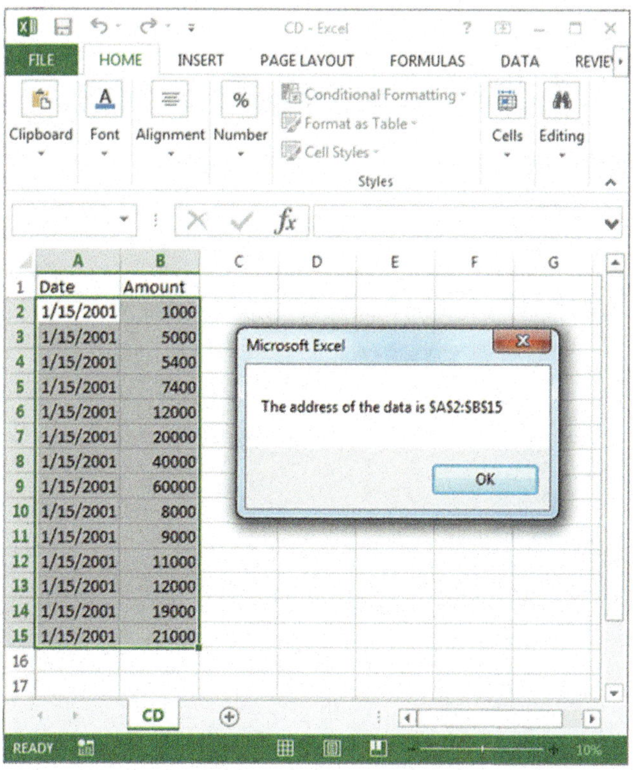

24.5 UsedRange Property of the Range Object

Another useful property to know is the *UsedRange* property of the range object. The VBA Help file defines the usedrange as the used range of a specific worksheet. Below demonstrates the difference between the *usedrange* and the *currentregion*. To demonstrate both concepts, let's first select cell F11 as shown below.

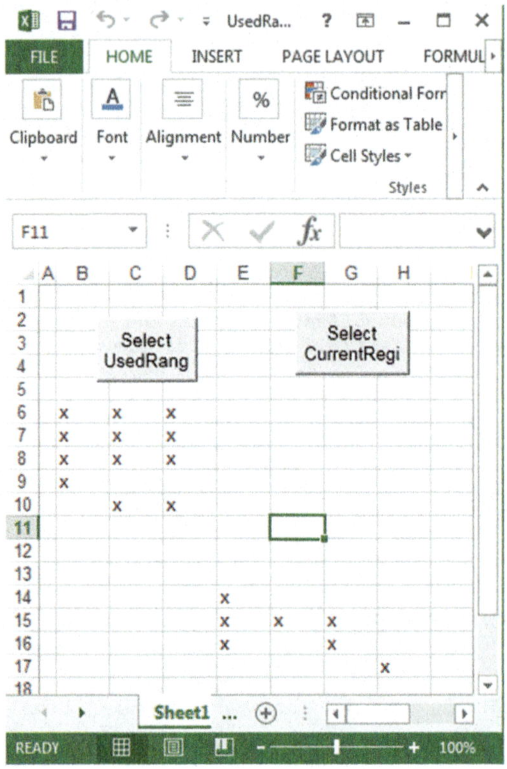

Below shows what happens after pushing the *Select UsedRange* button.

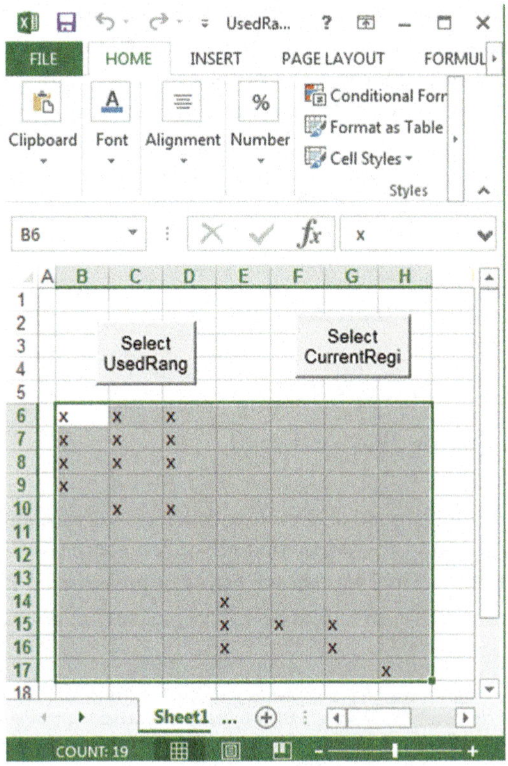

Below shows what happens after pushing the *Select CurrentRegion* button.

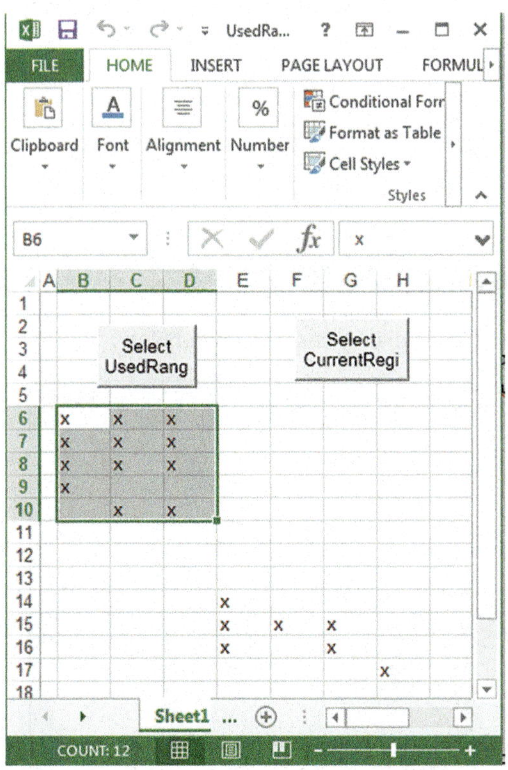

To understand the difference between the *usedrange* and the *currentregion*, it is important to know how the help file defines the *currentregion*. The VBA Help file defines the currentregion as "a range bounded by any combination of blank rows and blank columns."

Below are the procedures to find the usedrange and the currentregion.

```
Sub FindUsedRange()
    ActiveCell.Parent.UsedRange.Select
End Sub

Sub FindCurrentRegion()
    ActiveCell.CurrentRegion.Select
End Sub
```

24.6 Go To Special Dialog Box of Excel

As we have seen so far, navigating around an Excel worksheet is very important. A tool to help navigate around an Excel worksheet is the *Go To Special* dialog box. To get to the *Go To Special* dialog box, we would need to first choose Edit → Go To as shown below or press the *F5* key on the keyboard.

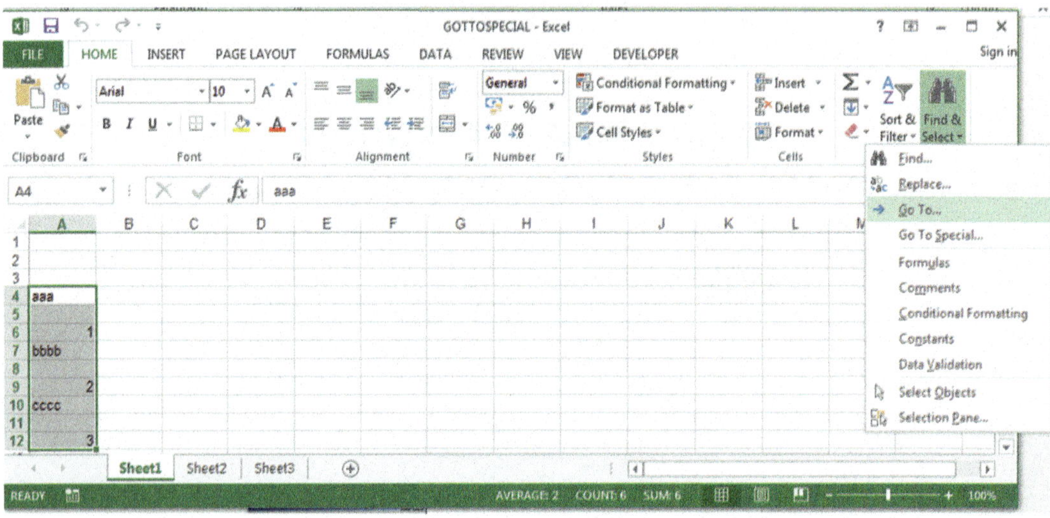

Doing this will show the *Go To* dialog box as shown below.

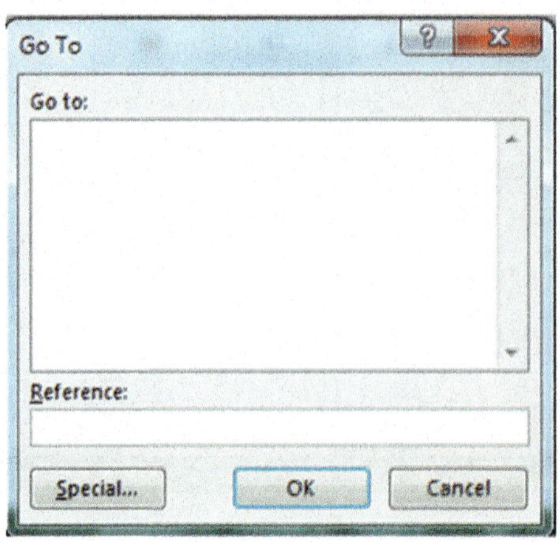

Next press the *Special* button as shown above to get the *Go To Special* button shown below.

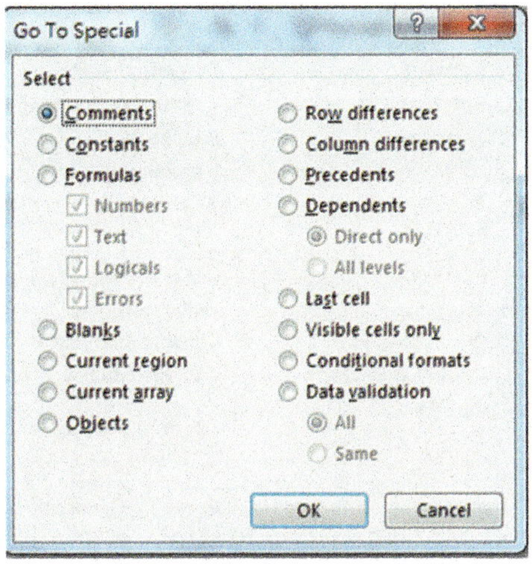

Below illustrates the *Go To Special* dialog box. Suppose we are interested in finding the blank cells inside the selected range shown below.

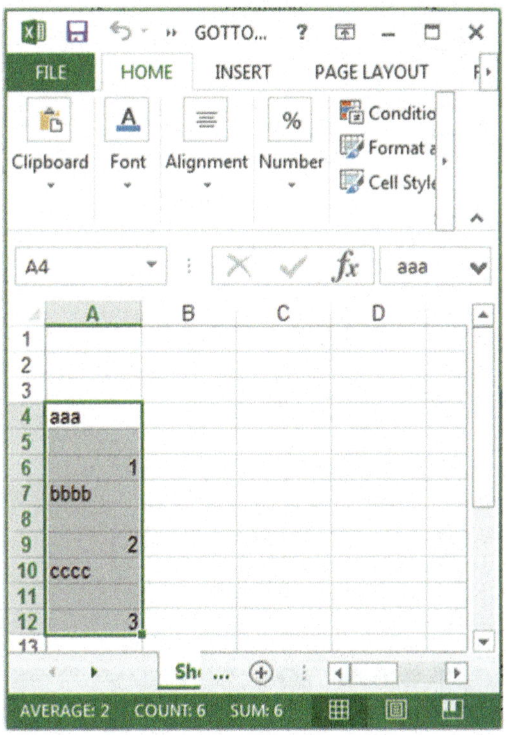

To find the blank cells, we would go to the *Go To Special* dialog box and then choose the *Blanks* options as shown below.

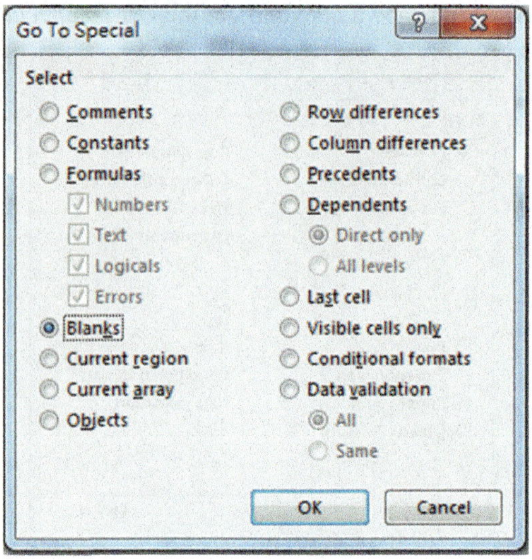

The following is the result after pressing the *OK* button on the *Go To Special* dialog box.

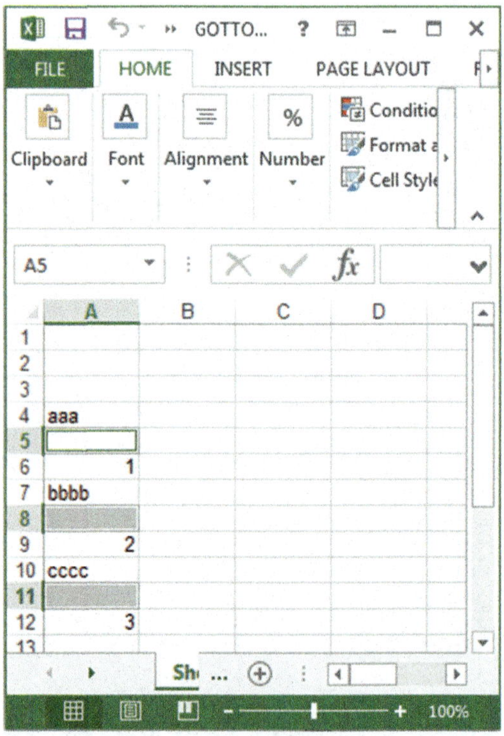

24.7 Importing Column Data into Arrays

Many times we are interested in importing data into arrays. The main and only reason to do this is speed. When the data set is large, there is a noticeable difference between manipulating data in arrays versus manipulating arrays in a worksheet. One way to get data into an array is to loop through every cell and put each data element individually into an array. The other way to get data into an array is shown below.

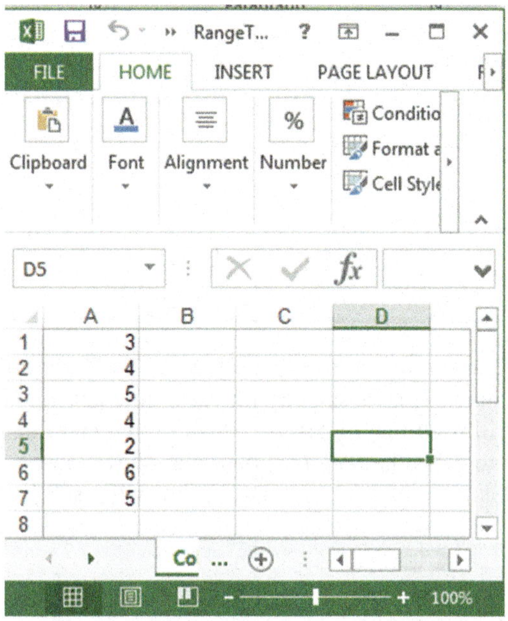

Suppose we are interested in importing the data in column A in the above worksheet to an array. This can be accomplished by the procedure shown below.

```
Sub IntoArrayColumnNoTransPose()
    Dim vNum As Variant
    vNum = Worksheets("Column").Range("a1").CurrentRegion
End Sub
```

Notice in the above procedure, it requires only one line of VBA code to bring data into an array from a worksheet. It is important to note that for the above technique to work, the array variable "vNum" must be defined as a variant.

To illustrate that the above technique works, we will have to use the professional programming tools provided by Excel. The tools are in the Visual Basic Editor. The Visual Basic Editor is shown below.

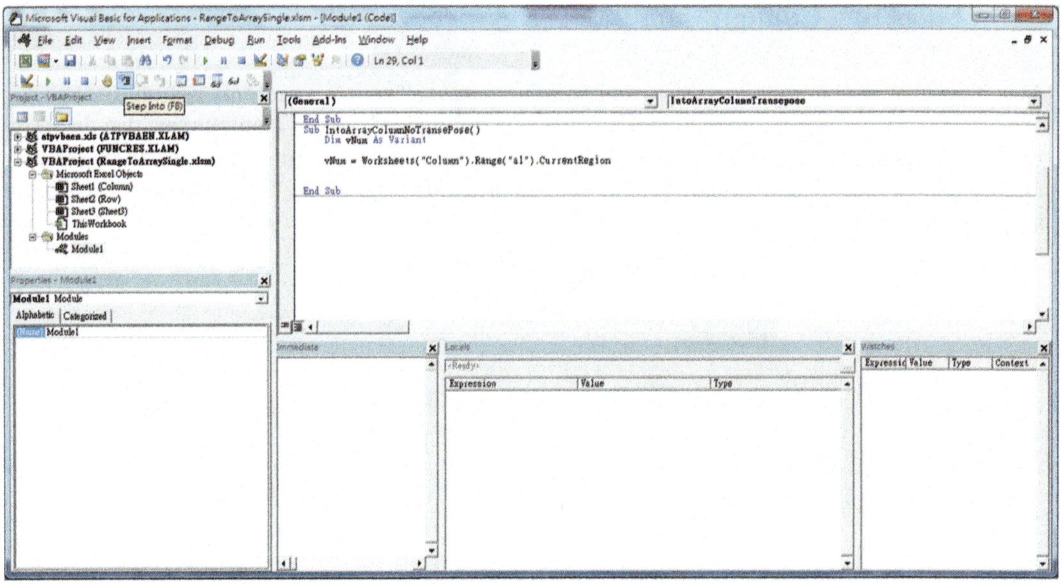

To illustrate the technique discussed in this section, we will need to run the VBA code in the procedure "IntoArrayColumnNoTranspose" one line at a time and look at value of variables after each line. To do this we will need first to put the cursor on the first line of the procedure as shown above. Then we will need to press the *Step Into* button as shown above or press the *F8* key on the keyboard. Pressing the *F8* is my preferred way. Doing this will result in the following:

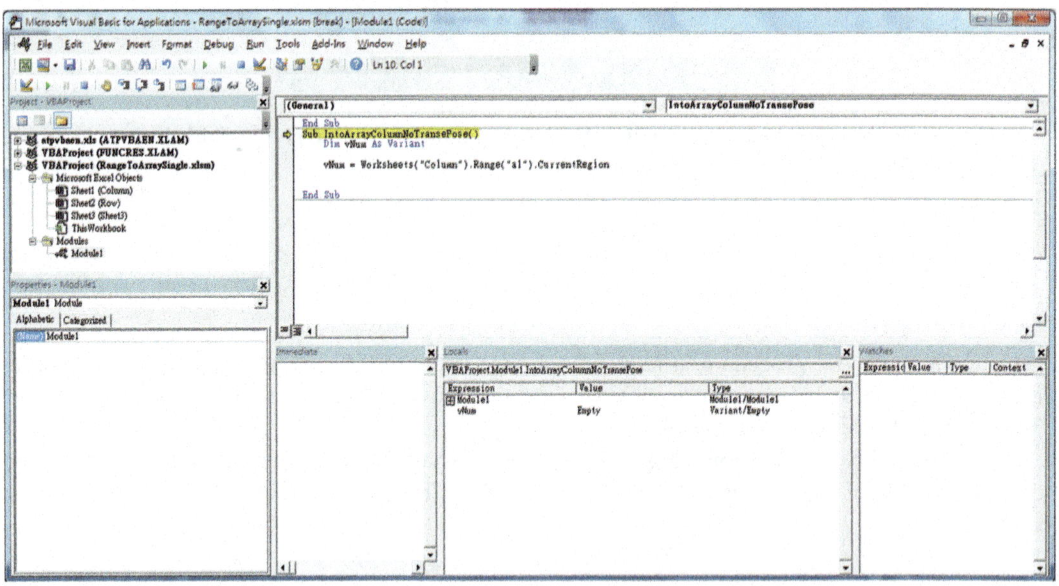

The first thing to notice is that in the code window, the first line of the procedure is highlighted in yellow. This is shown above. The other thing to note is in the *Locals* window. It shows the value of all the variables. At this point it indicates that the variable "vNum" has an empty value, which means no value.

The next thing that we need to do now is to press the *F8* key on the keyboard to move to the next VBA line. Below shows what happens after pressing the *F8* key.

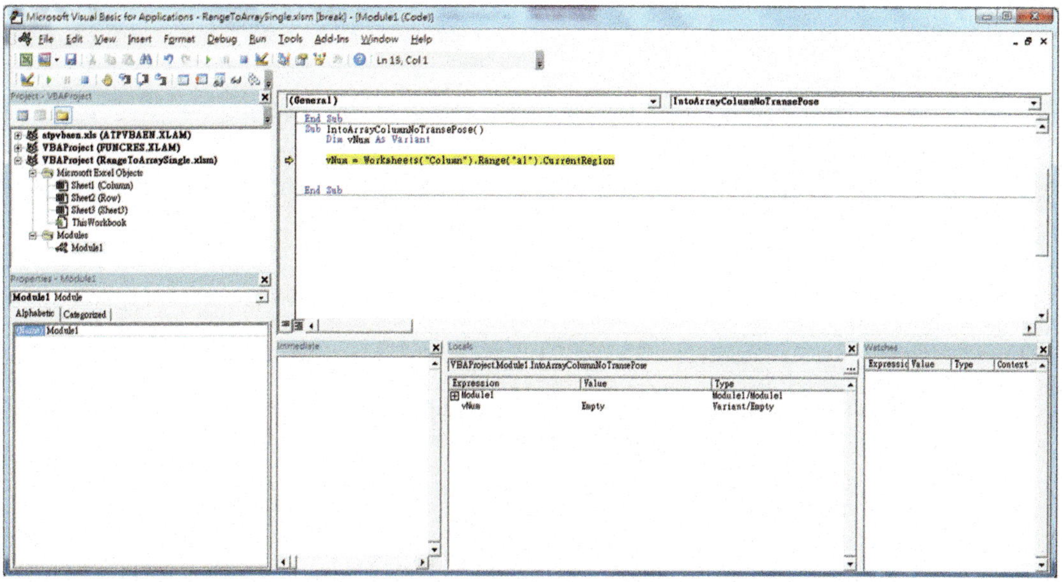

Notice at this point the variable "vNum" still has no value. Let's press the *F8* key one more time.

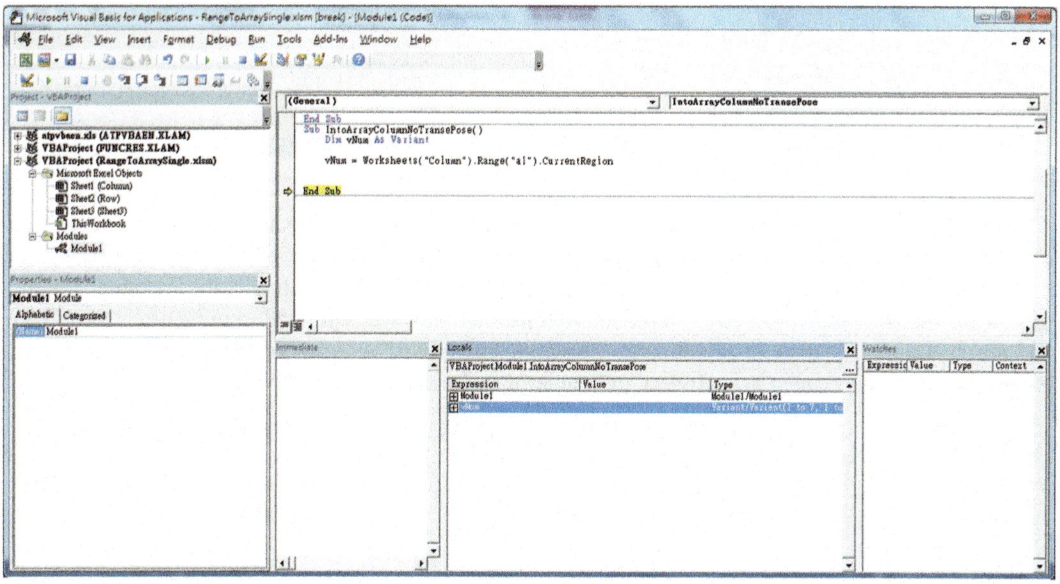

Notice at this point the variable "vNum" no longer indicates empty. There is also a symbol next to the variable. This symbol indicates that there are values for the array. We will need to click on the symbol to look at the values of the array. The following shows the result of clicking on the symbol.

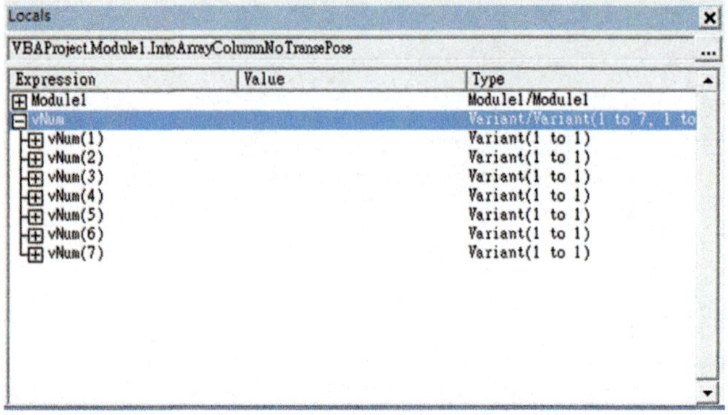

The *Locals* window now shows that there are seven elements in the array "vNum." Let's now click on each element of the array. The end result is shown below.

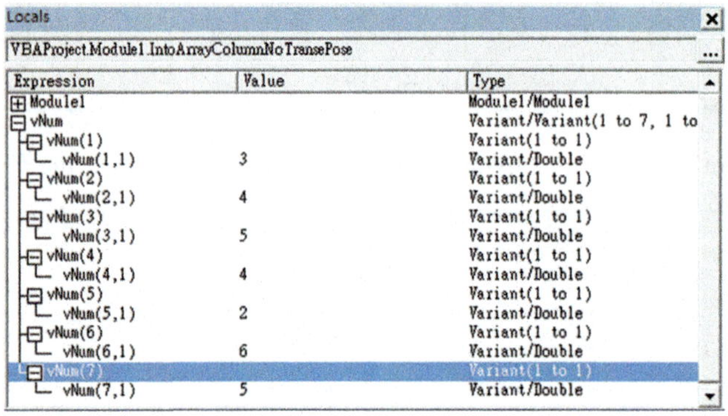

The *Locals* window indicates that the first element of the array is 3. The values of the rest of the elements agree with the values in the worksheet. Note that in the *Locals* window, the third element has a reference of "vNum(3,1)." This indicates that VBA has automatically set the variable "vNum" to a two-dimensional array. So to reference the third element, we will need to indicate "vNum(3,1)" and not "vNum(3)." This can be illustrated with the *Immediate* window of the Visual Basic Editor. Below shows in the *Immediate* window the value of the array element "vNum(3,1)."

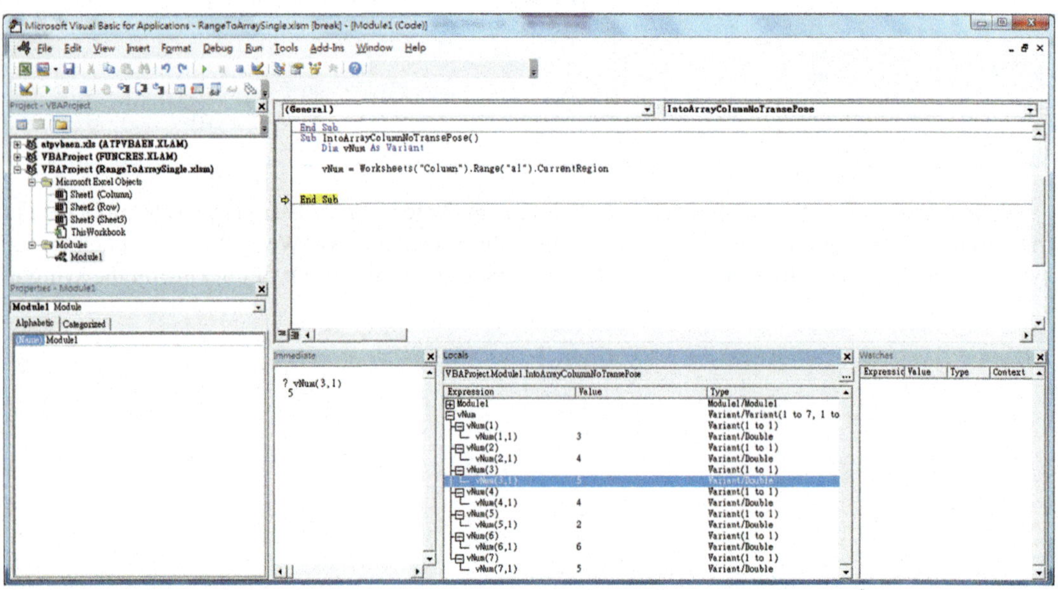

Below shows what happens when we try to reference the third element as "vNum(3)."

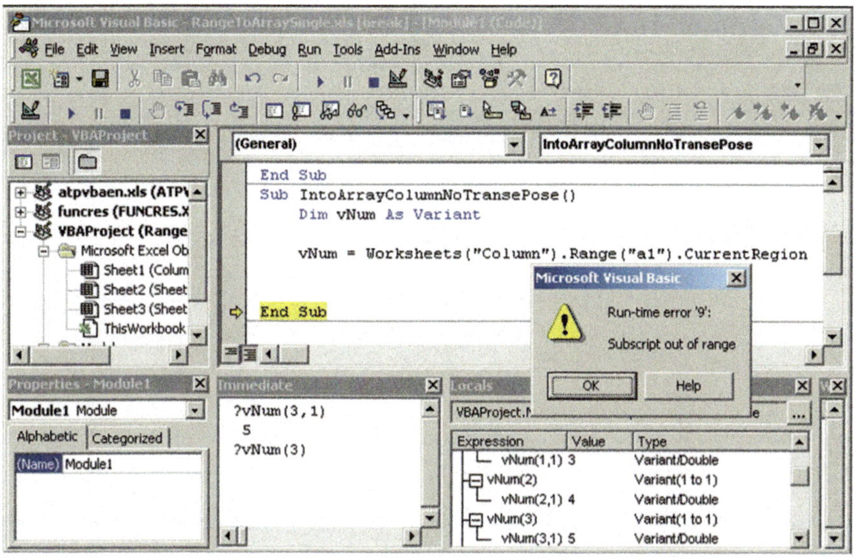

The Visual Basic Editor complains when we try to reference the third element as "vNum(3)."

Many times we are interested in the variable being a one-dimensional array. To do this we, will use the *Transpose* method of the *worksheetfunction* object. The procedure "IntoArrayColumnTranspose" accomplishes this.

```
Sub IntoArrayColumnTransepose()
    Dim vNum As Variant

    vNum = WorksheetFunction.Transpose(Worksheets("Columnm") _
           .Range("a1").CurrentRegion)

End Sub
```

Instead of stepping through the code line by line, we can tell the VBE to run the VBA code and stop at a certain point. To indicate where to stop, put the cursor at the "end Sub" VBA line as shown below. Then, press the *ToggleBreakPoint* button as shown below or press the *F9* key on the keyboard. My preferred method is to press the *F9* key.

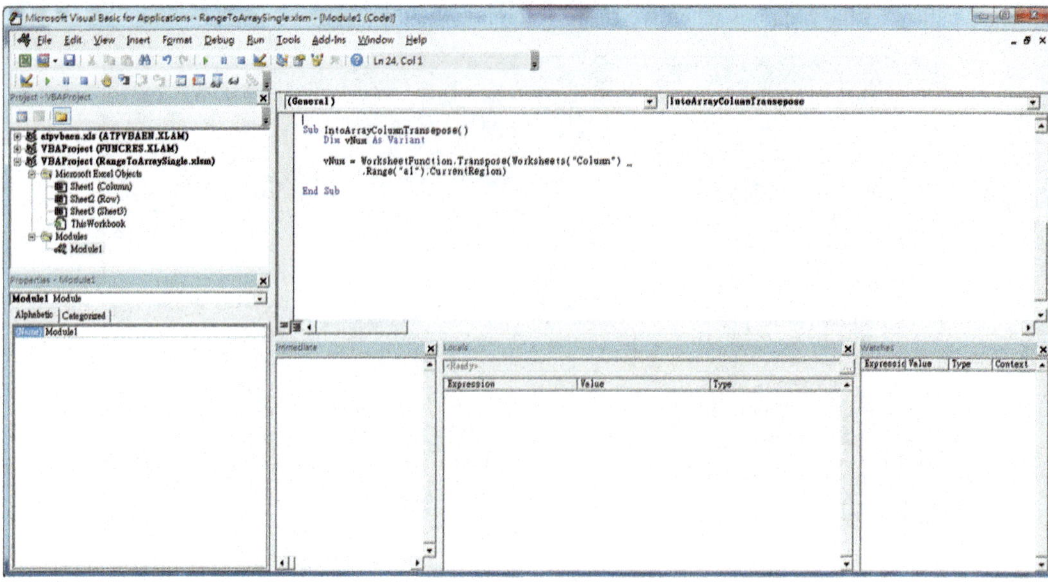

Below shows what happens after pressing the *F9* key.

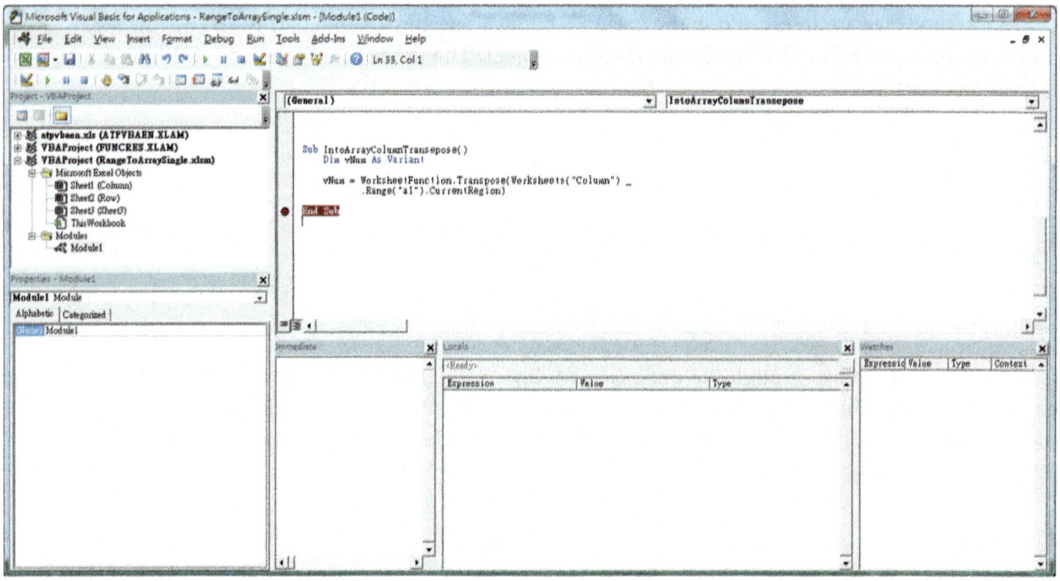

Pressing the *F5* key will run the VBA code until the break point. Below shows the state of the VBE after pressing the *F5* key.

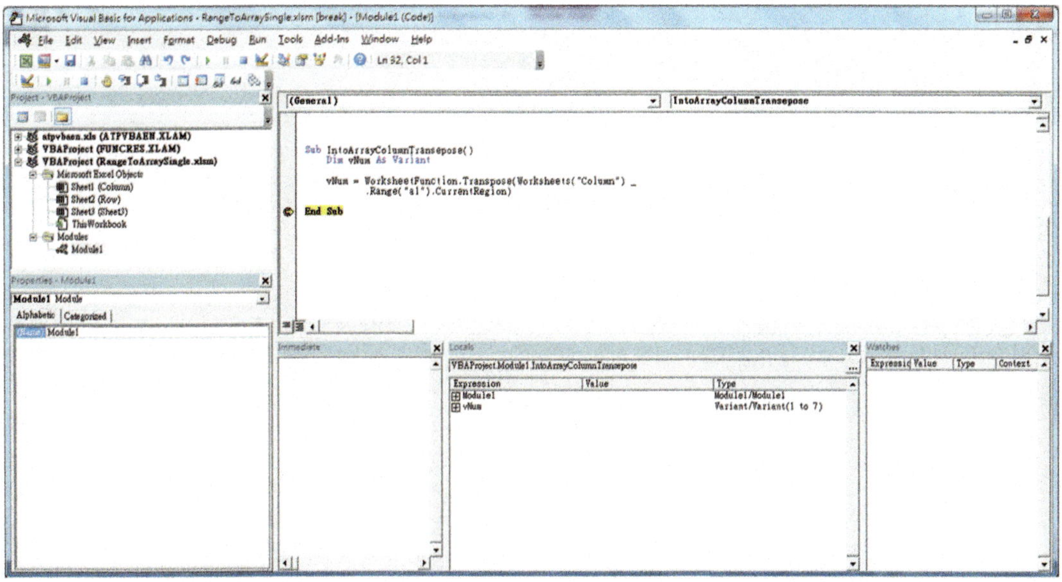

Let's now expand the variable "vNum" in the *Locals* window. Below shows the state of the *Locals* window after expanding the "vNum" variable.

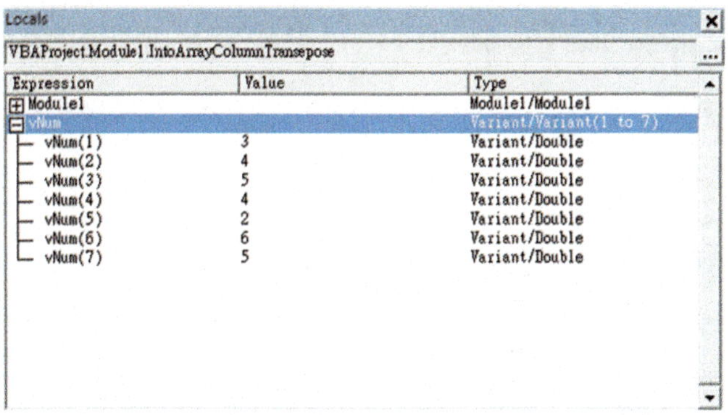

The above *Locals* window shows that the variable "vNum" is one-dimensional. Below shows the *Immediate* pane referencing the third element of the variable "vNum" as a one-dimensional variable.

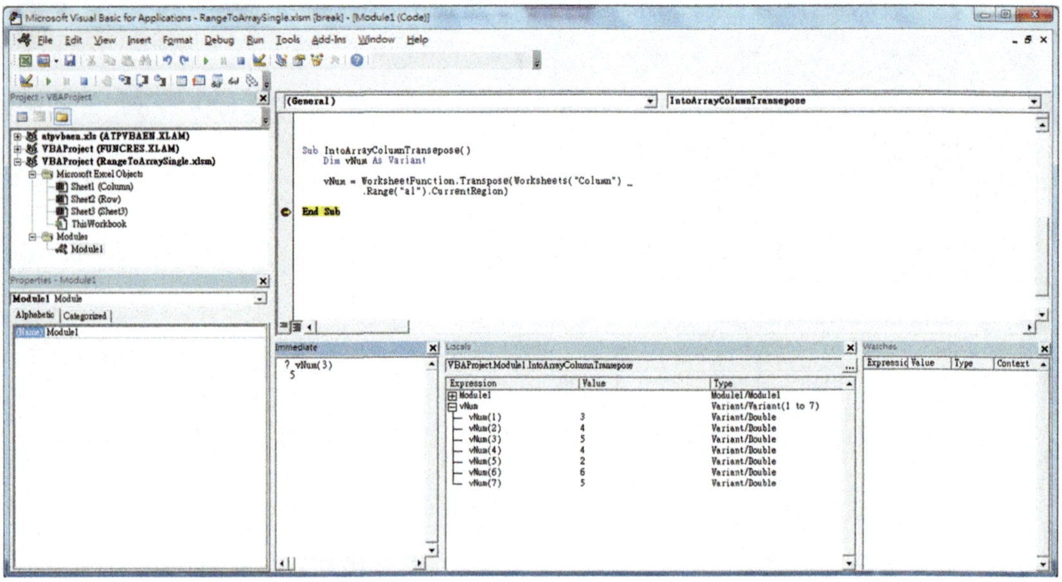

When you are finished analyzing the procedure above, choose Debug → Clear All Breakpoints as shown below. This will clear out all the break points.

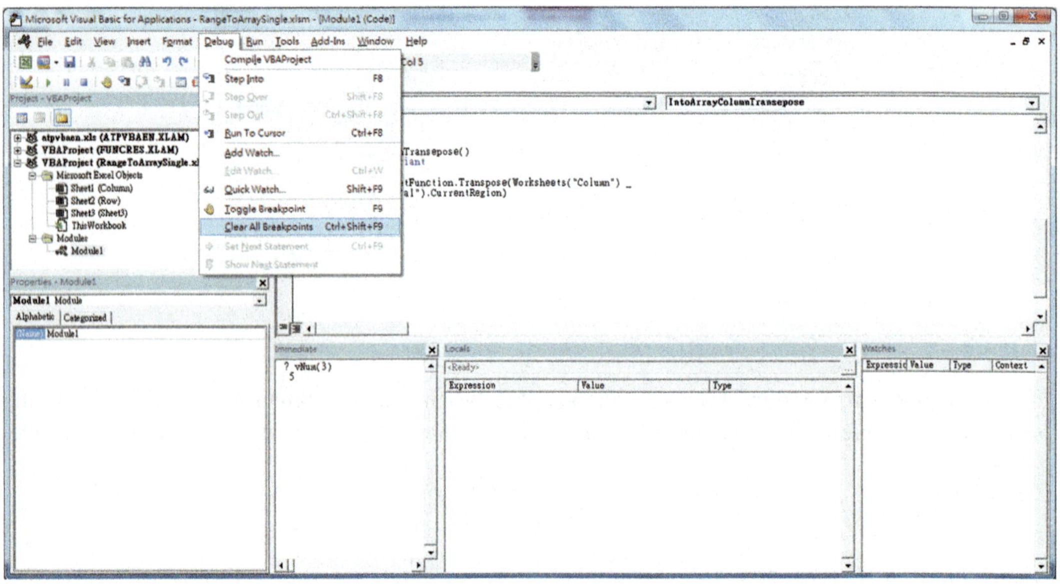

Not clearing out the break points will cause the macro to stop at this point after you reopen the workbook and then rerun the macro!

24.8 Importing Row Data into an Array

In the previous section we used the *Transpose* property (function) to transpose the column data. For some reasons, which I don't understand, we need to use the *Transpose* property twice for row data. Let's import the row data shown below to an array.

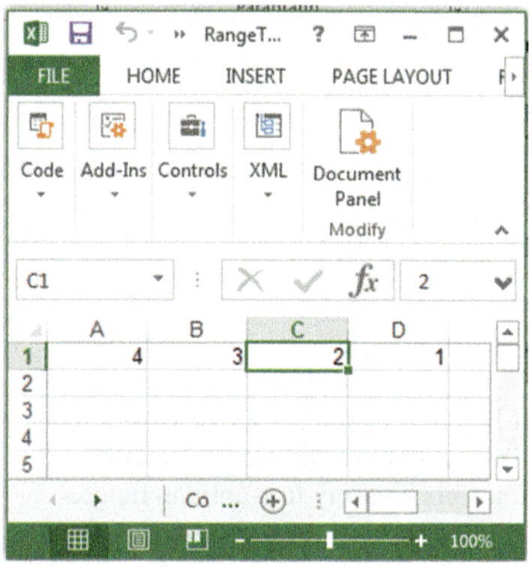

The procedure below will import the row data above to an array. Notice the two *Transpose methods.*

```
Sub IntoArrayRow()
    Dim vNum As Variant

    vNum = WorksheetFunction.Transpose(WorksheetFunction. _
            Transpose(Worksheets("Row"). _
            Range("a1").CurrentRegion.Value))
End Sub
```

Below demonstrates the above procedure.

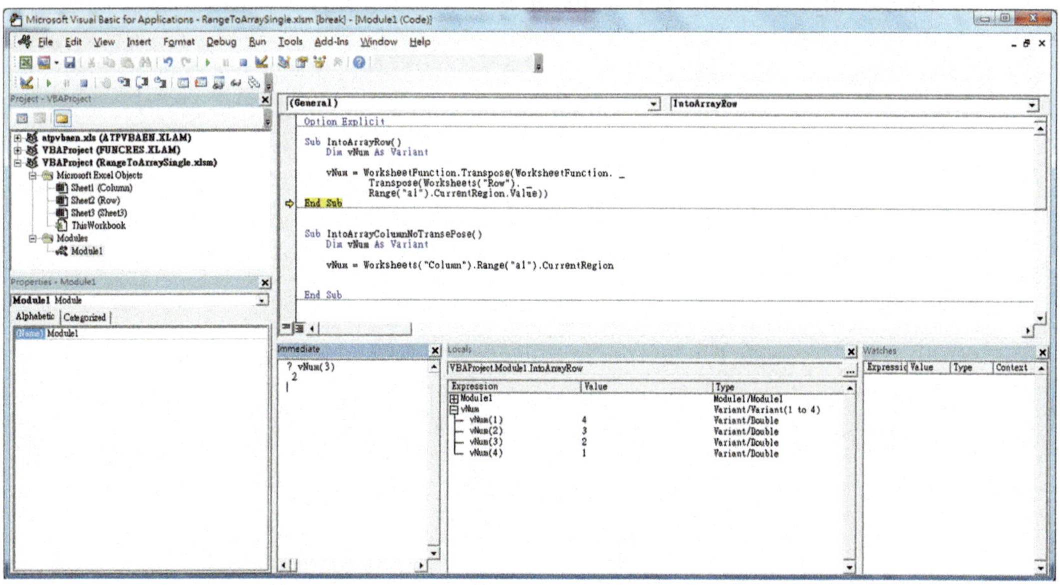

24.9 Transferring Data from an Array to a Range

In this section we will illustrate how to transfer an array to a range. We will first illustrate how to transfer an array to a row range, and then we will illustrate how to transfer an array to a column range. The following procedure transfers an array to a row range:

```
Sub TransferToRow()

    Dim v As Variant

    v = Array(1, 2, 3, 4)

    With ActiveSheet.Range("a1")
        .CurrentRegion.ClearContents
        'Need to resize range to row first
        'Need to transpose array
        .Resize(4, 1) = WorksheetFunction.Transpose(v)
    End With
End Sub
```

It is important to note that before transferring an array to a row range, it is necessary to set the range to a properly sized row range and then transpose the array.

The following procedure transfers an array to a column range.

```
Sub TransfertoColumn()

    Dim v As Variant

    v = Array(1, 2, 3, 4)

    With ActiveSheet.Range("a1")
        .CurrentRegion.ClearContents
        'Need to resize range to column first
        .Resize(1, 4) = (v)
    End With

End Sub
```

It is important to note that before transferring an array to a column range, it is necessary to set the range to a column range and then transpose the array.

Below demonstrates the "TransfertoRow" procedure.

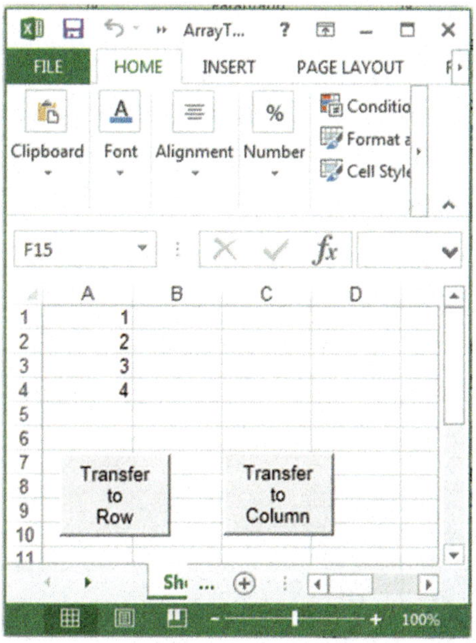

Below demonstrates the "TransfertoColumn" procedure.

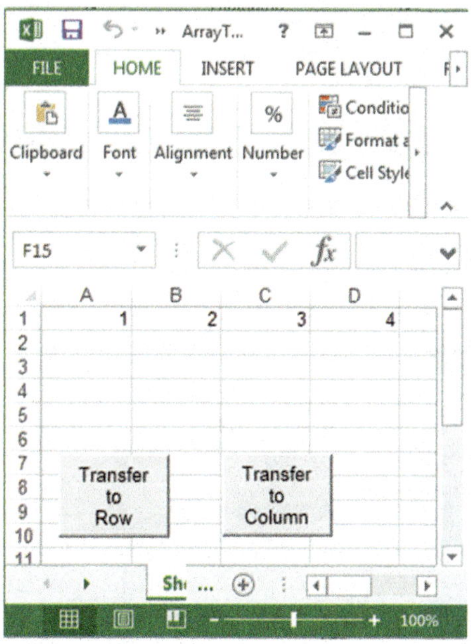

24.10 Workbook Names

We can do a lot of things with workbook names. The first thing that we will do is to assign names to worksheet ranges. It is common to set a range name by first selecting the range and then assigning a name to the selected range in the *Name Box*. This is shown below.

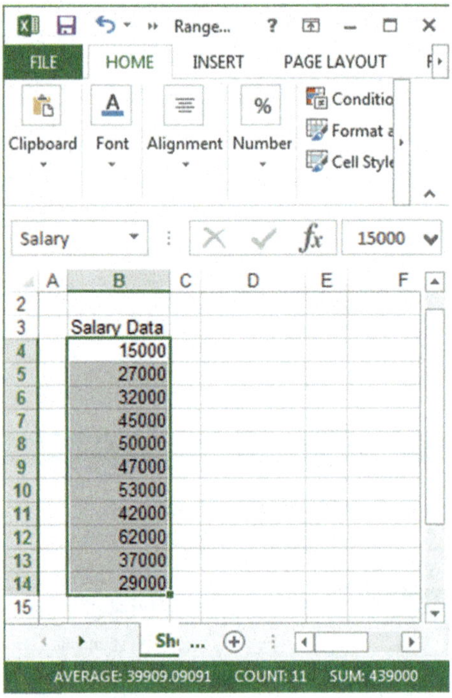

Notice as shown above that Excel will automatically sum any range selected.

One thing that can be done with workbook names is range navigation. As an illustration let's choose cell E5 as shown below and then press the *F5* key.

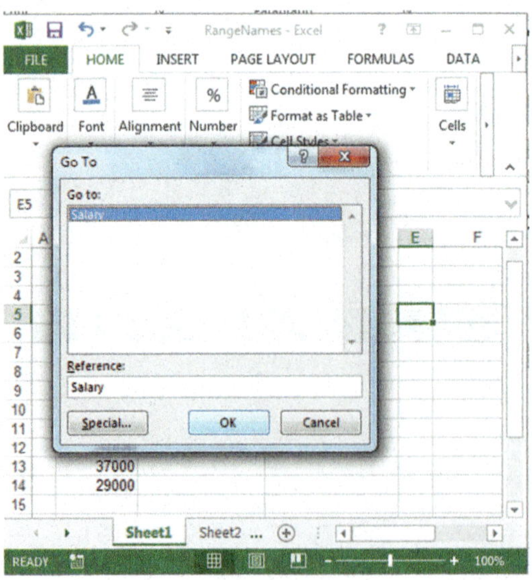

Notice that the *Go To* dialog box shows all workbook names. The next thing that we should do is highlight the *Salary* range and press the *OK* button as shown above. Pressing the *OK* button caused Excel to select the *Salary* range as shown below.

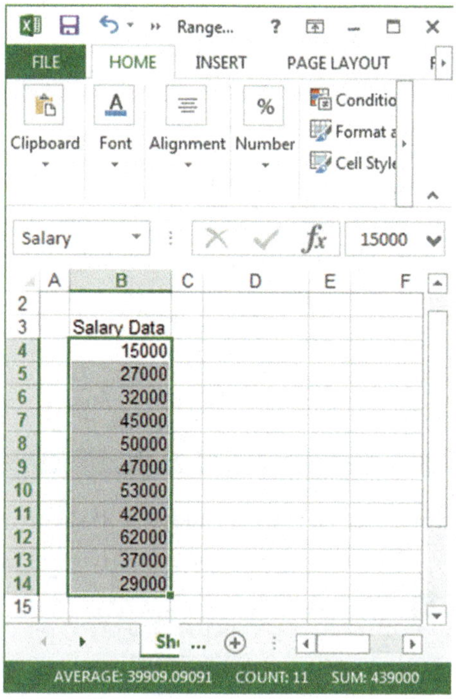

Another thing that we can do with *Names* is to use *Name* in worksheet equations. Suppose we want to increase the salary data by 10 %. The most common technique is to reference each individual cell and multiply it by .10. This is shown below.

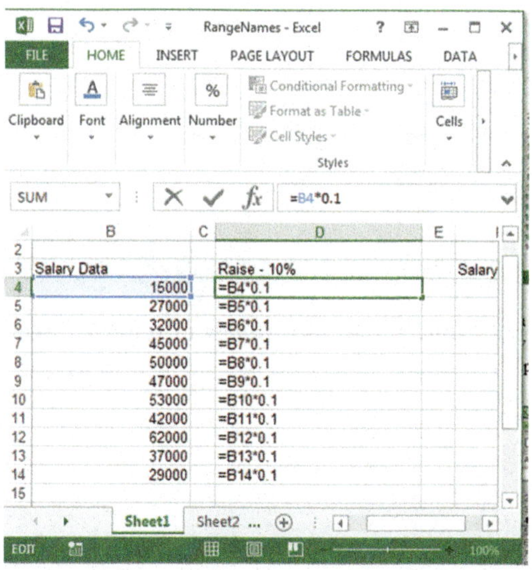

Above shows the formula view of a worksheet. To switch between the formula view and the regular worksheet view, press *ctrl*.

Alternatively we can use workbook *Names*. This is shown below.

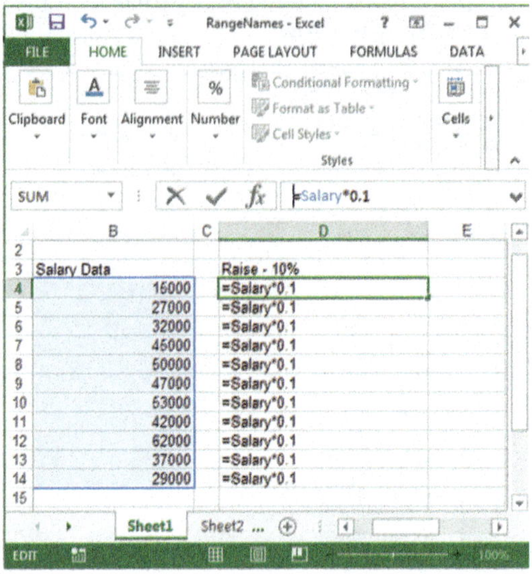

Below shows the regular worksheet view of the above worksheet.

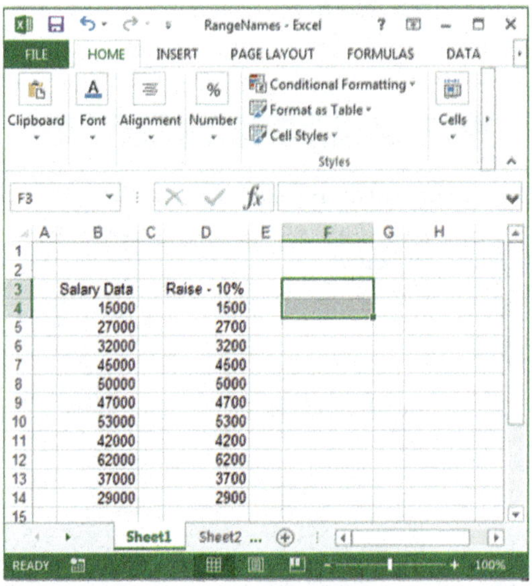

Another use for workbook *Names* is to use *Names* to reference worksheet expressions. Below demonstrates this concept.

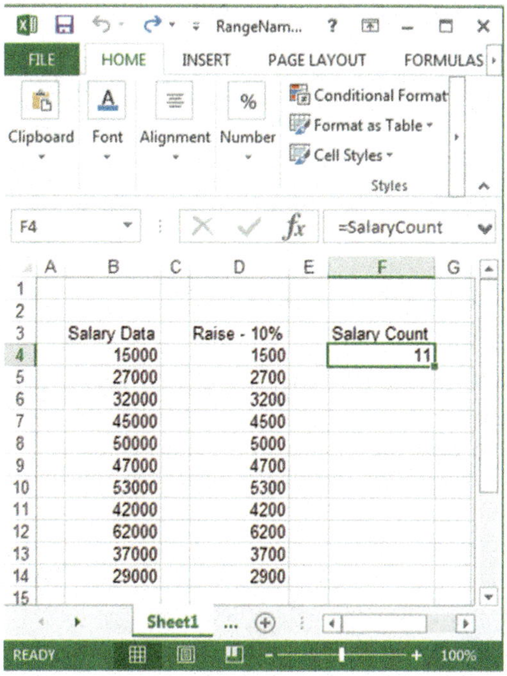

Above shows that cell F4 is referencing a workbook name called “SalaryCount.” The workbook name “SalaryCount” returns a value of 11. This number is the number of salaries in column B. We can confirm that this number is correct by manually counting the number of salaries in column B. To see how the name *SalaryCount* is defined, let’s choose Insert → Name → Define as shown below.

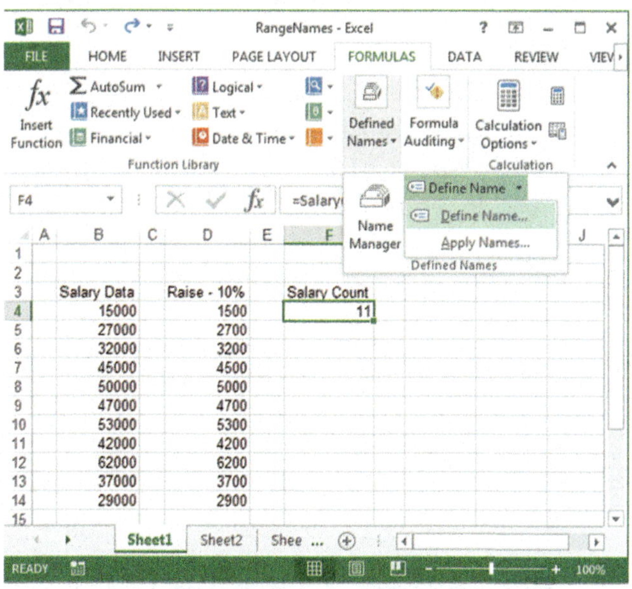

This shows the *Defined Name* dialog box shown below.

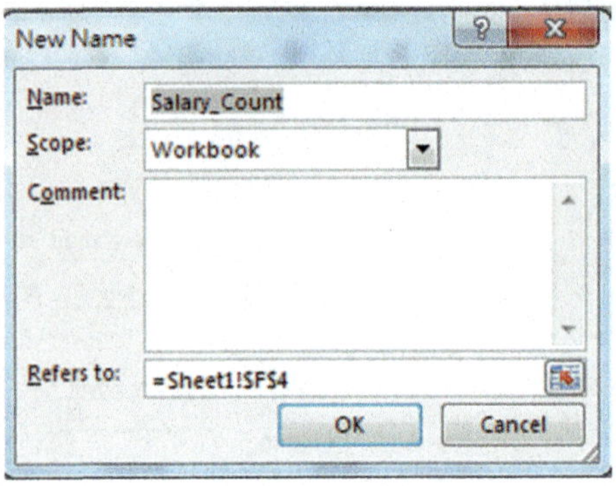

The *Define Name* dialog box shows that the workbook name "SalaryCount" is defined as "=Counta(Sheet1!$B:B)-1." Notice that workbook names can contain worksheet functions.

24.11 Dynamic Ranges

In this section we will illustrate how to create dynamic ranges. Dynamic ranges use the worksheet function *counta* and the worksheet function *offset.*

The function *counta* counts the number of cells that are not empty. This concept is illustrated below.

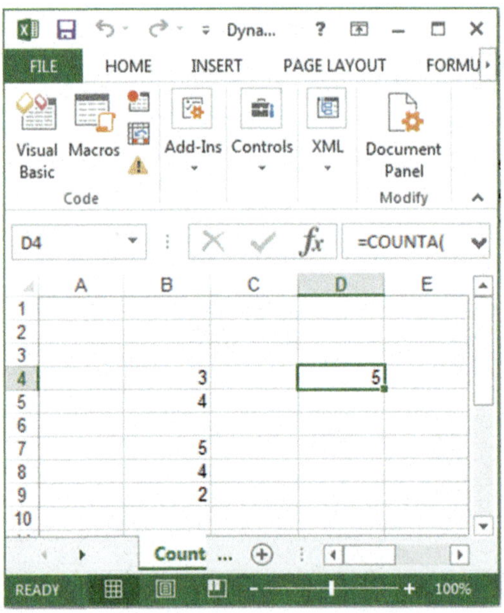

Now we will look at the worksheet function *offset*. This function works very much like the VBA *offset* and the VBA resize property. The worksheet function *offset* takes five parameters. The first parameter is where to anchor off. The second and third parameter works like the VBA *offset* property. The fourth and fifth parameter works very much like the VBA *resize* property.

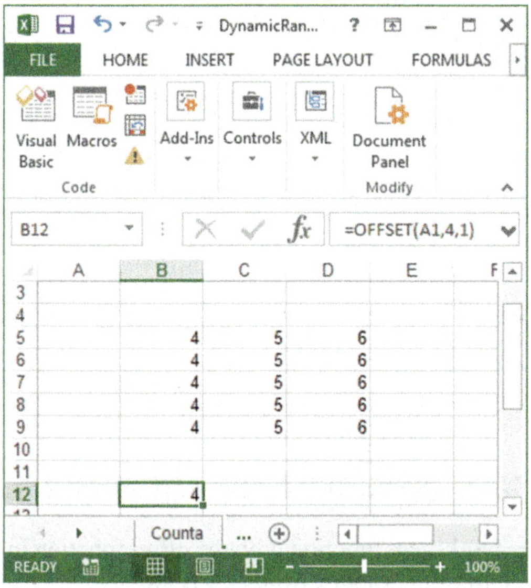

The *offset* function requires that at least first three parameters be used. The *offset* function shown above indicates to start at cell A1 then offset four rows and one column. This would bring us to cell B5. The *offset* function above returns a value of 4, which agrees with the cell value of B5.

Below shows the *offset* function with all five parameters being used.

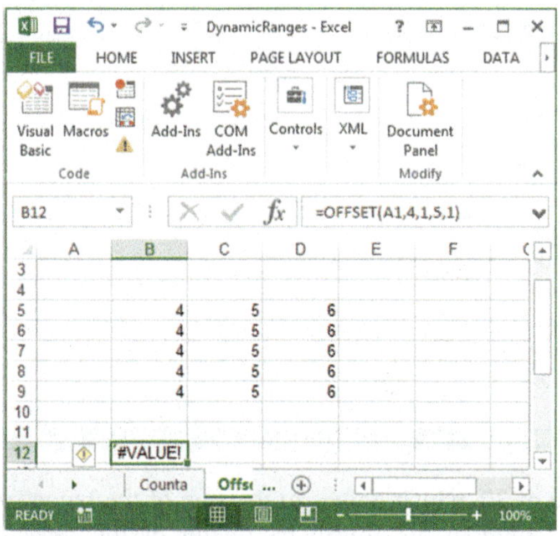

The fourth parameter indicates how many rows to resize, which in this case is 5. The fifth parameter indicates how many columns to resize to, which in this case is 1. The *offset* function returns a range. Based on the arguments in the *offset* function shown above, the *offset* function is supposed to return the range from B5 to B10. This is why we see a #VALUE! error in cell B12.

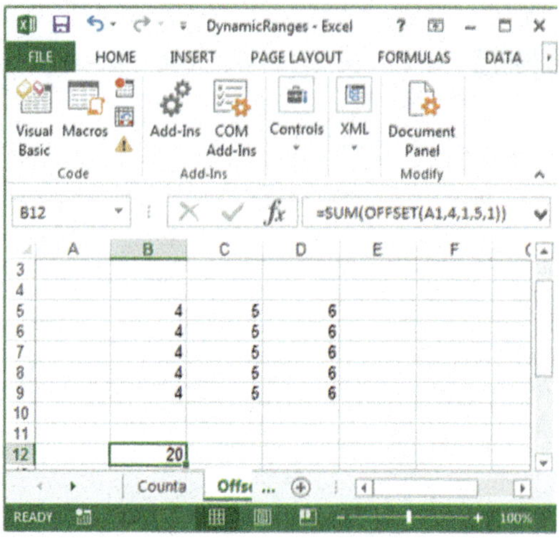

The above shows the *sum* worksheet function with the *offset* function in cell B12. The above shows a value of 20, which is the sum of the ranges from cell B5 to B9.

Next we will illustrate how to dynamically sum column D in the above workbook. We do this by inserting a *counta* function to the fourth parameter of the *offset* function. Since we are adding column D, we change the fourth parameter of the *offset* to 3, which means to offset three columns to the right. This is shown below.

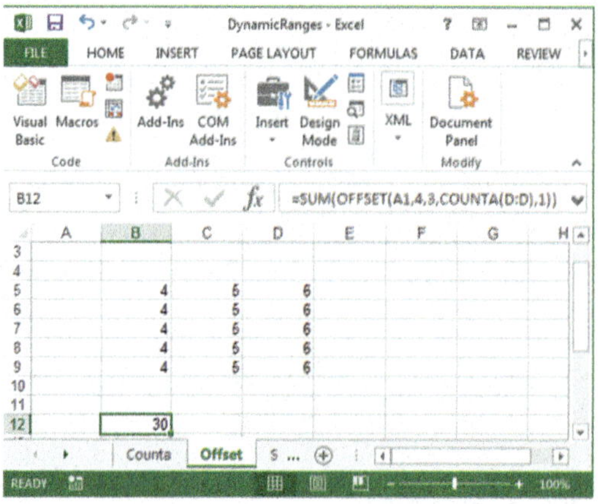

Cell B12 shows a value of 30, which agrees to the sum of the range from cell D5 to D9.

We put the function *counta* in the fourth parameter of the *offset* function. This causes the expressing in cell B12 to be dynamic. We can demonstrate this dynamic concept by entering a value of 6 in cell D10. Entering the value 6 in cell D10 will cause B12 to have a value of 36. This is shown below.

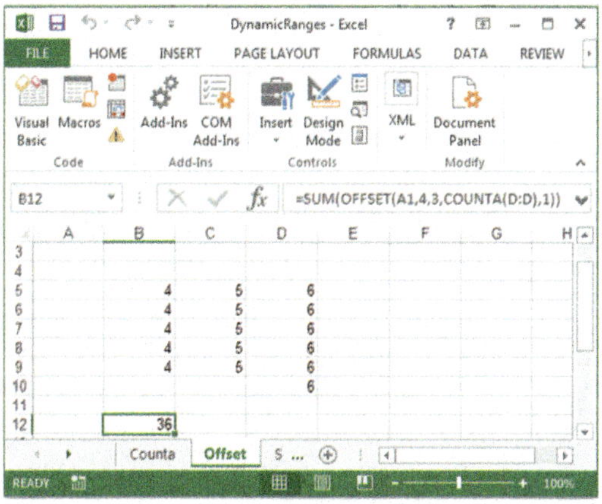

24.12 Global Versus Local Workbook Names

With workbook names there is a distinction between "global" names versus "local" names. Not knowing the distinctions can cause a lot of problems and confusion. We will in this section look at many scenarios for "global" names and "local" names. By default all workbook names are created as "global" names. Below demonstrates the consequences of names being "global" names. The first thing that we will do is define the cell A1 in worksheet "Sheet1" as "Salary" through the *Name Box*. This is illustrated below.

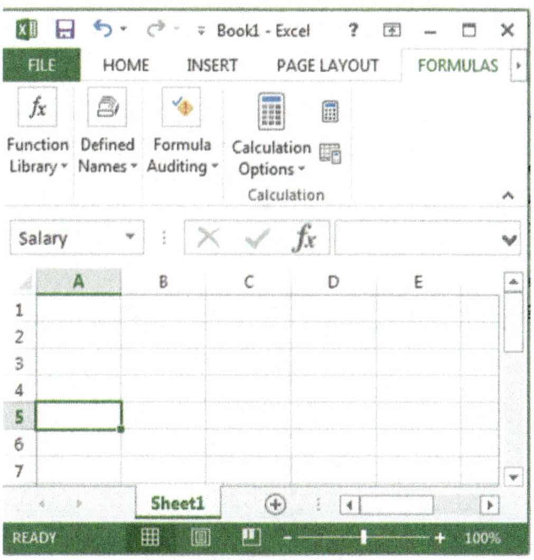

Now suppose we are also interested in defining cell A5 in worksheet "Sheet2" also as "Salary." What we will find out is when we try to define cell A5 in worksheet "Sheet2" through the *Name Box* is Excel will jump to cell A5 in worksheet "Sheet1," our first definition of "Salary." This illustrates the concept that there can only be one "global" workbook name.

It is also possible to define names by selecting Insert → Name → Define as shown below.

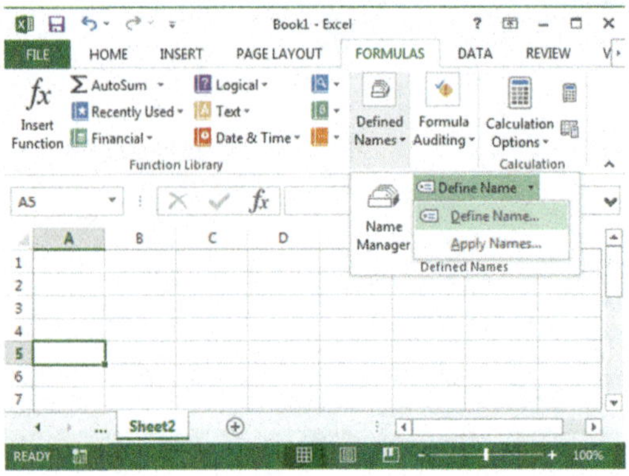

If we first choose the cell A5 in worksheet "Sheet2" and then select Insert → Name → Define, we will get the following *Names* dialog box.

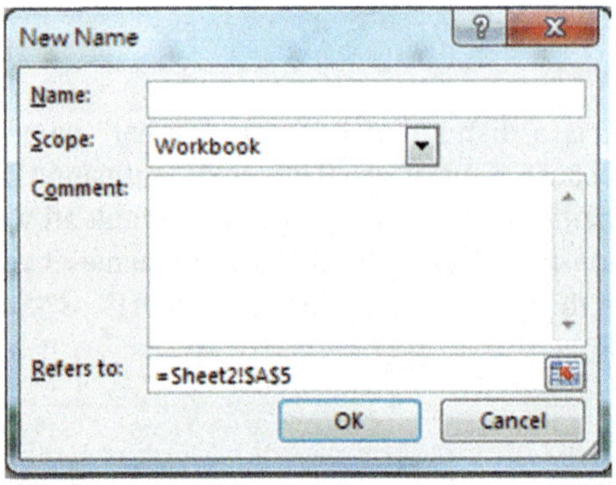

The *Define Name* dialog box shows in the *Refers to:* textbox the address of the active cell. Let's now type in "Salary" in the *Names in workbook:* textbox and then press the *Add* button. This is shown below.

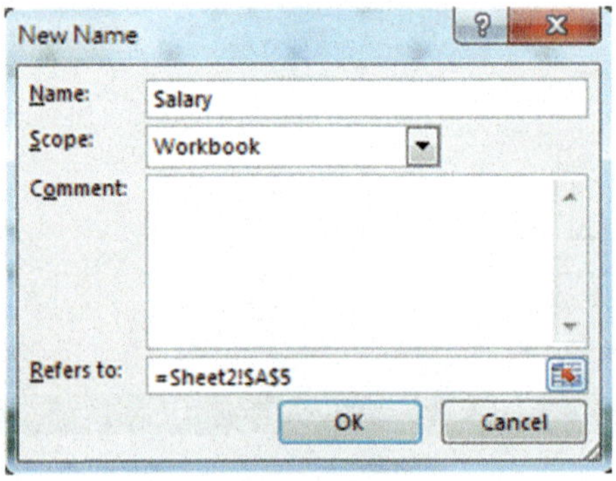

Below shows what the *Define Name* dialog box looks like after clicking on the *Add* button.

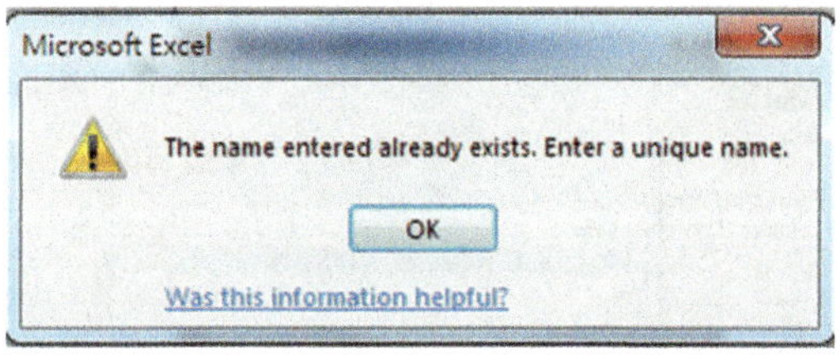

Let's now press the *Close* button to get back to the workbook and select cell A11 in worksheet "Sheet1." What we will find out from the result of the above action is that when we try to navigate to the range "Salary," we will go to cell A5 in worksheet "Sheet 2." This also illustrates that we can have only one global name.

Let's now illustrate how we can cell A5 in worksheet "Sheet1" and cell A5 in worksheet "Sheet2" both be defined as "Salary." To do this let's first go to cell A5 in worksheet "Sheet1" and then go to the *Define Name* dialog box.

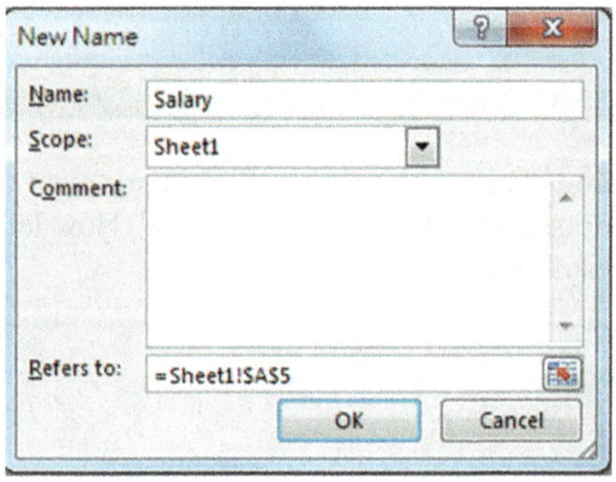

In the *Names in workbook:* textbox "Sheet!Salary" and then press the *Add* button. This is shown above. Below shows what the *Define Name* dialog box will show after clicking on the *Add* button.

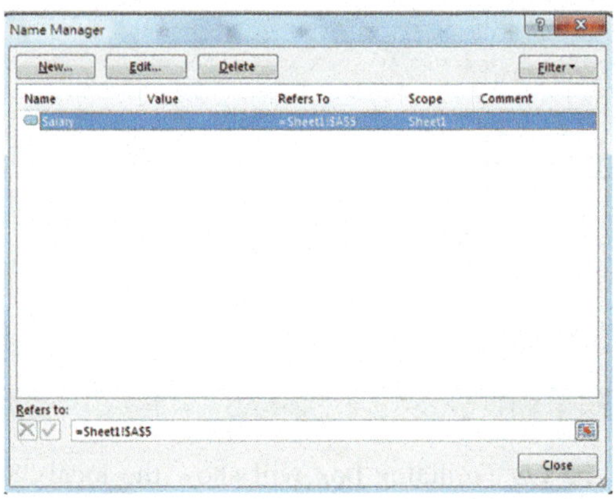

Notice that the name *Salary* is highlighted and that in the same row, the worksheet name "Sheet1" is indicated. This indicates that there is a local "Salary" range name defined for worksheet "Sheet1." Let's now go back to worksheet "Sheet1" and select cell D11 and then press the *F5*. The end result is shown below.

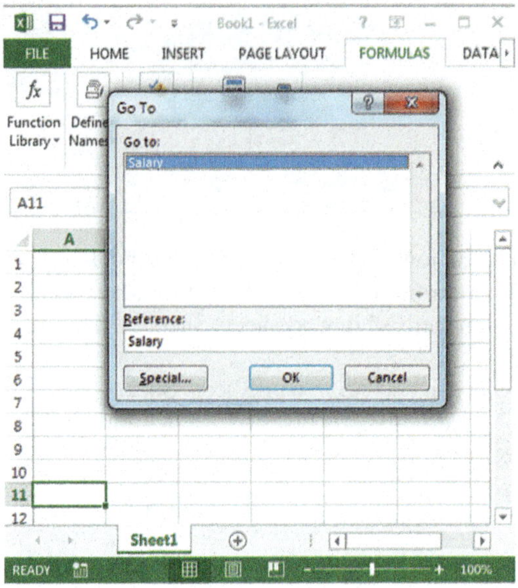

Notice that the *Go To* dialog shows one range called salary even though there is a range name called "salary" in both the worksheets "Sheet1" and "Sheet2." Now let's press the *OK* button on the *Go To* dialog box. The end result is shown below.

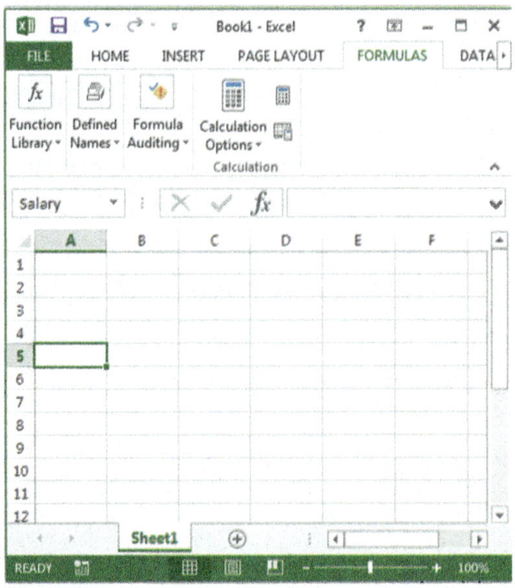

This demonstrates that the *Go To* dialog box will show the local "Salary" name of worksheet "Sheet1" and not the global "Salary" name that is defined as cell A5 of worksheet "Sheet2." Let's now select cell A11 of worksheet "Sheet3" and press the *F5* key. The end result is shown below.

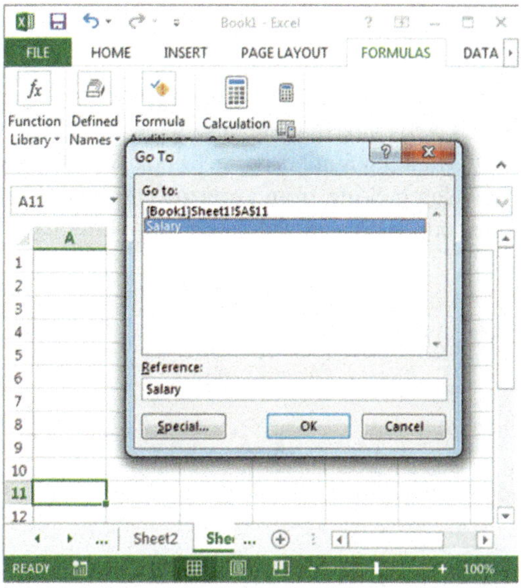

Let's now select the "salary" as shown above and then press the *OK* button. The end result is shown below.

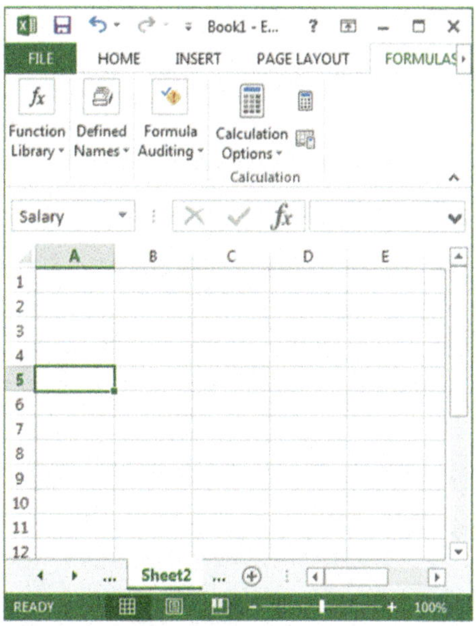

The above exercise illustrates that when displaying the *Go To* dialog box on a worksheet that does not have a "local" name, it will show the "global" name.

24.13 Dynamic Charting

I have been daily keeping track of how well the sales has been to the previous edition of this book by looking up the sales ranking on www.amazon.com and www.barnesandnoble.com. Below shows a workbook that contains the sales ranking number for my book from October 2000 to December 2000 and one chart that charts the sales ranking from Amazon.com and one chart that charts the sales ranking from BarnesandNoble.com. Below shows the sales ranking data.

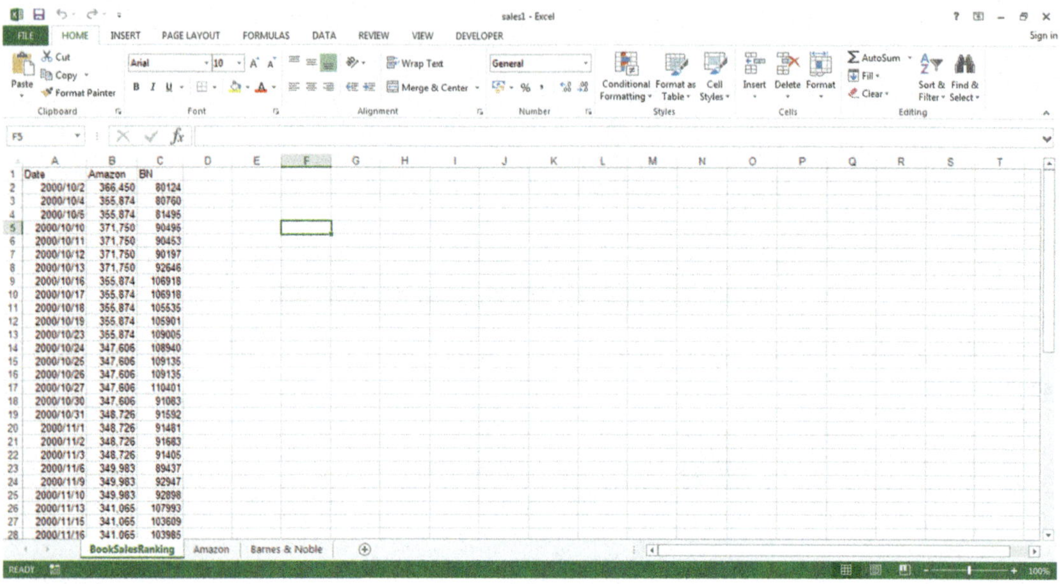

Using the Excel chart wizard on the sales data would result in the two sales charts shown below.

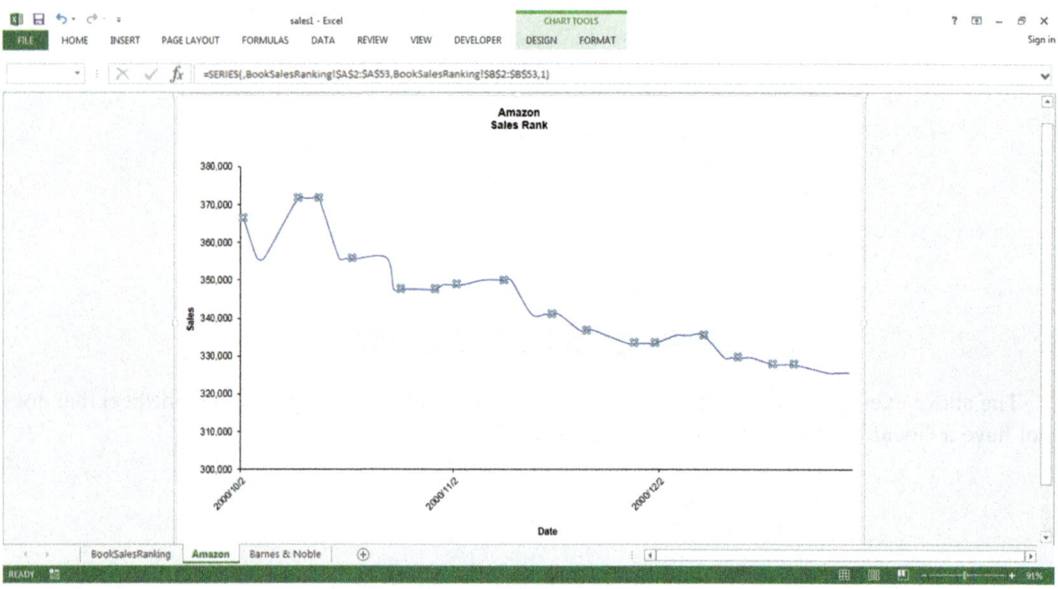

Both charts have a series chart function that indicates that sales data is from row 2 to row 53. The first argument of the function indicates the name for the series. A chart can have more than one series. The second argument indicates the data source for the *x*-axis of the chart. The third argument indicates the *y*-axis of the chart. The fourth indicates the order of the series. In the above charts, BookSalesRanking!$A$2:$A$53 is the data source for the *x*-axis of the chart. BookSalesRanking!$B $2:$B$53 is the data source for the *x*-axis for the Amazon chart, and BookSalesRanking!$C$2:$C$53 is the data source for the *x*-axis for the Barnes and Noble chart. Row 53 indicates the sales data for 12/29/2000. On January 2, 2001, I would then look up the sales ranking and put the information on row 54. Then I would have to adjust the data source for the charts. I did this routine for a while, but it got pretty annoying to have to constantly adjust the charts' data source.

Luckily I found a technique that would not require me to adjust the data source for the charts every time I entered new information. I applied workbook names in charts. The workbook *sales2.xls* that is included in the CD that accompanies this book demonstrates workbook names in charts.

Below shows the charts using workbook names. It is important to note that it is necessary to use local names as data sources. Local names were discussed in the previous section.

Below shows the definition of the workbook names used in the charts.

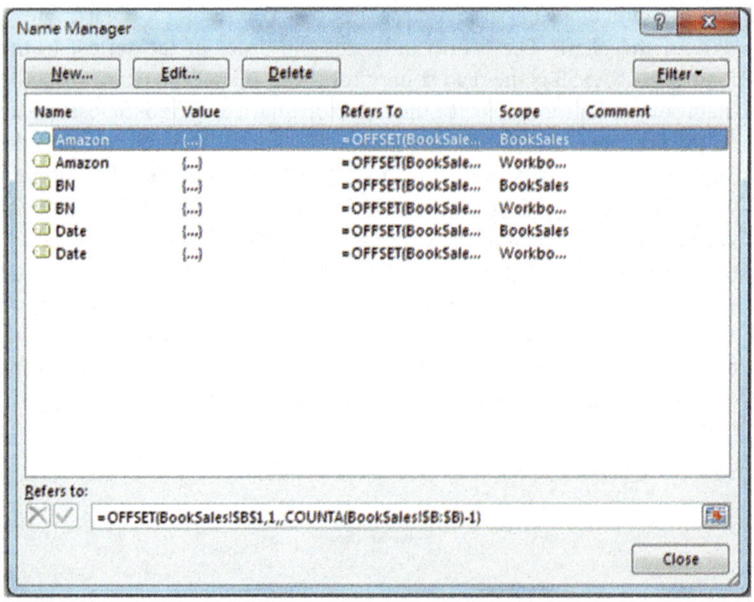

Using workbook *sales2.xls*, I now no longer have to adjust the charts after I enter new sales information.

24.14 Searching All File in a Directory

Every now and then there is interest to search all files in a directory. Before learning the procedure shown below, I wrote a complicated macro that did the same thing. The secret to the macros' simplicity is using the *FileSystemObject* object.

```
Option Explicit

'Creating a FileSystemObject
Public FSO As New FileSystemObject

Sub ListFiles()
'Declaring variables
Dim objFolder As Folder
    Dim objFile As File
    Dim strPath As String
    Dim NextRow As Long
    Dim wb As Workbook
    Dim ws As Worksheet
    Dim wsMain As Worksheet

    Set wsMain = ThisWorkbook.Worksheets("Main")

    'Specify the path of the folder
    strPath = wsMain.Range("Directory")

    If Not FSO.FolderExists(strPath) Then
        MsgBox "The folder " & strPath & " does not exits."
        Exit Sub
    End If

    'Create the object of this folder
    Set objFolder = FSO.GetFolder(strPath)

    'Check if the folder is empty or not
    If objFolder.Files.Count = 0 Then
      MsgBox "No files were found...", vbExclamation
      Exit Sub
    End If

    Set wb = Workbooks.Add
    Set ws = wb.Worksheets(1)

    ws.Cells(2, 1).Select
    ActiveWindow.FreezePanes = True

    'Adding Column names for A
    ws.Cells(1, "A").Value = "File Name"
    ws.Cells(1, "B").Value = "Size"
    ws.Cells(1, "C").Value = "Modified Date/Time"
    ws.Cells(1, "D").Value = "User Name"
    ws.Cells(1, 1).Resize(1, 4).Font.Bold = True

    'Find the next available row
    NextRow = ws.Cells(2, 1).Row
    'Loop through each file in the folder
    For Each objFile In objFolder.Files

        'List the name of the current file
        ws.Cells(NextRow, 1).Value = objFile.Name
        ws.Cells(NextRow, 2).Value = Format(objFile.Size, "#,##0")
        ws.Cells(NextRow, 3).Value = Format(objFile.DateLastModified, "mmm-dd-yyyy
hh:mm:ss am/pm")
        ws.Cells(NextRow, 4).Value = Application.UserName

        'Find the next row
        NextRow = NextRow + 1
    Next objFile

    With ws
        .Cells.EntireColumn.AutoFit
    End With

End Sub
```

Below demonstrates the above procedure using the file *FileSearch.xlsm*. This file is in the CD that accompanies this book.

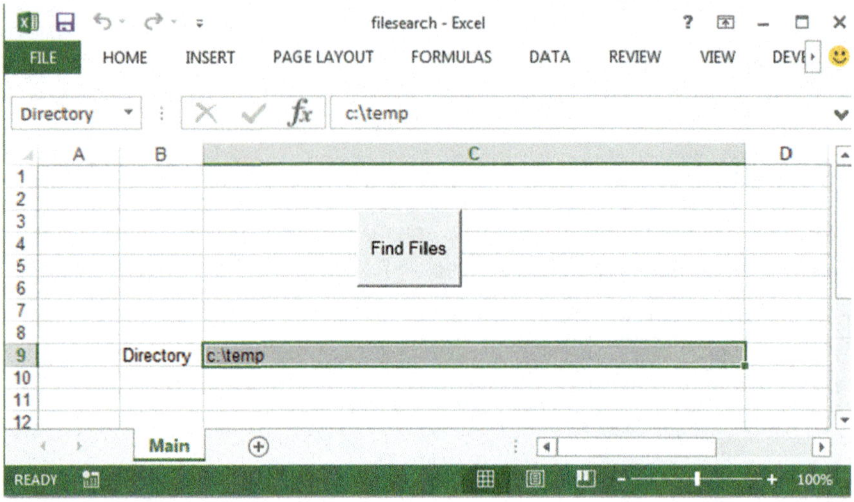

Below shows all the files that are in my "temp" folder.

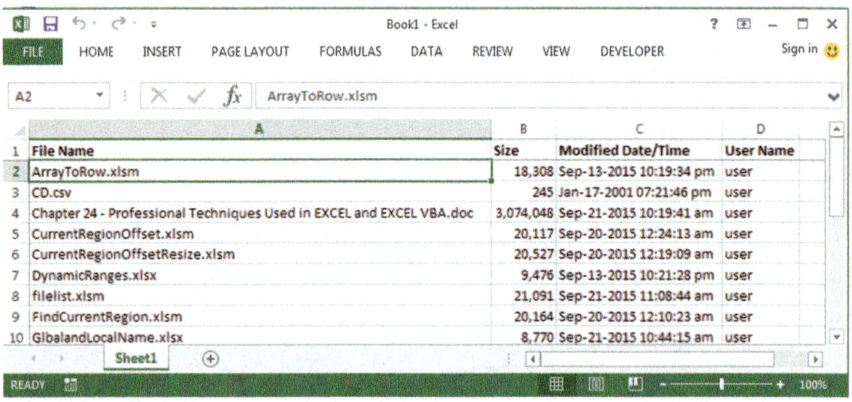

Bibliography

Walkenbach J (2013) Excel 2013 power programming with VBA. Wiley, Hoboken
Bullen S (2007) Excel 2007 VBA programmer's reference. Wiley, Indianapolis
Bovey R (2009) Professional excel development. Pearson Education, Boston

Chapter 25
Binomial Option Pricing Model Decision Tree Approach

25.1 Introduction

Microsoft Excel is one of the most powerful and valuable tools available to the business users. The financial industry in New York City has recognized this value. We can see this by going to one of the many job sites on the Internet. Two Internet sites that demonstrate the value of someone who knows Microsoft Excel very well are www.dice.com and www.indeed.com. For both of these Internet sites, search for New York City and VBA, which is Microsoft Excel's programming language, and you will see many job postings requiring VBA.

The academic world has begun to realize the value of Microsoft Excel. There are now many books that use Microsoft Excel to do statistical analysis and financial modeling. This can be shown by going to the Internet site www.amazon.com, and search for books like "Data Analysis Microsoft Excel" and "Financial Modeling Microsoft Excel."

The binomial option pricing model is one the most famous models used to price options. Only the Black-Scholes model is more famous. One problem with learning the binomial option pricing model is that it is computationally intensive. This results in a very complicated formula to price an option.

The complexity of the binomial option pricing model makes it a challenge to learn the model. Most books teach the Binomial Option model by describing the formula. This is a not very effective because it usually requires the learner to mentally keep track of many details, many times to the point of information overload. There is a well-known principle in psychology that the average number of things that a person can remember at one time is seven.

Many books are used as teaching aids which include Decision Trees. Because of the computational intensity of the model, most books do not present Decision Trees with more than three periods. One problem with this is that the Binomial Option model is best when the periods are large.

This chapter will do two things. It will first demonstrate the power of Microsoft Excel. It will do this by demonstrating that it is possible to create large Decision Trees for the binomial pricing model using Microsoft Excel. A 10-period Decision Tree would require 2047 call calculations and 2047 put calculations. This paper will also show the Decision Tree for the price of a stock and the price of a bond, each requiring 2047 calculation. Therefore, there would be 2047*4 = 8188 calculations for a complete set of 10-period Decision Trees.

The second thing that this paper will do is present the Binomial Option model in a less mathematical matter. It will try to make it so that the reader will not have to keep track of many things at one time. It will do this by using Decision Trees to price call and put options

In this chapter we show how the binomial distribution is combined with some basic finance concepts to generate a model for determining the price stock of options.

C.-F. Lee et al., *Essentials of Excel, Excel VBA, SAS and Minitab for Statistical and Financial Analyses*, DOI 10.1007/978-3-319-38867-0_25

25.2 Call and Put Options

A *call option* gives the owner the right but not the obligation to buy the underlying security at a specified price. The price in which the owner can buy the underlying price is called the *exercise price*. A call option becomes valuable when the exercise price is less than the current price of the underlying stock price.

For example, a call option on an IBM stock with an exercise price of $100 when the stock price of an IBM stock is $110 is worth $10. The reason it is worth $10 is because a holder of the call option can buy the IBM stock at $100 and then sell the IBM stock at the prevailing price of $110 for a profit of $10. Also, a call option on an IBM stock with an exercise price of $100 when the stock price of an IBM stock is $90 is worth $0.

A *put option* gives the owner the right but not the obligation to sell the underlying security at a specified price. A put option becomes valuable when the exercise price is more than the current price of the underlying stock price.

For example, a put option on an IBM stock with an exercise price of $100 when the stock price of an IBM stock is $90 is worth $10. The reason it is worth $10 is because a holder of the put option can buy the IBM stock at the prevailing price of $90 and then sell the IBM stock at the put price of $100 for a profit of $10. Also, a put option on an IBM stock with an exercise price of $100 when the stock price of the IBM stock is $110 is worth $0.

Below are charts showing the value of call and put options of the above IBM stock at varying prices.

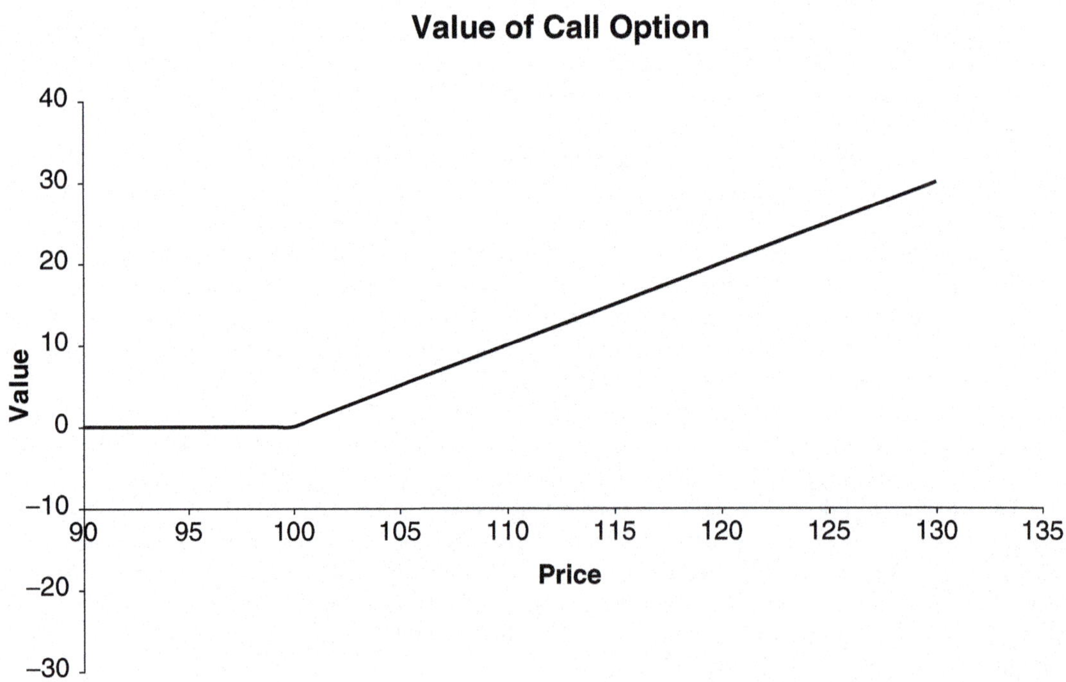

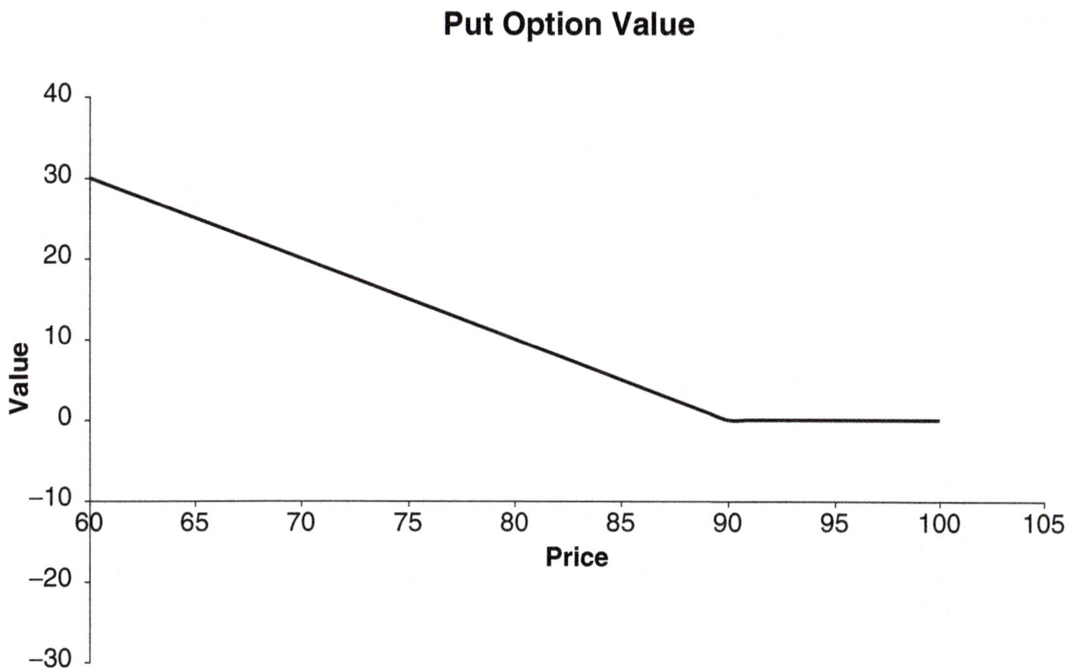

25.3 Option Pricing: One Period

What should be the value of these options? Let's look at a case where we are only concern with the value of options for one period. In the next period, a stock price can either go up or go down. Let's look at a case where we know for certain that a stock with a price of \$100 will either go up 10 % or go down 10 % in the next period and the exercise after one period is \$100. Below shows the Decision Tree for the stock price, the call option price, and the put option price.

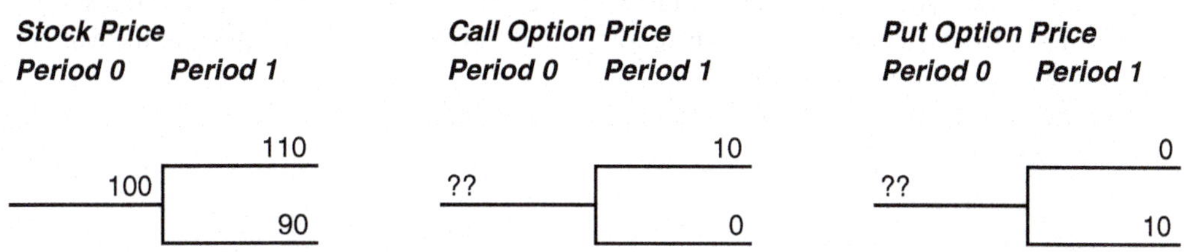

Let's first consider the issue of pricing a call option. Using a one-period Decision Tree, we can illustrate the price of a stock if it goes up and the price of a stock if it goes down. Since we know the possible ending values of a stock, we can derive the possible ending values of a call option. If the stock price increases to \$110, the price of the call option will then be \$10 (\$110 − \$100). If the stock price decreases to \$90, the value of the call option will worth \$0 because it would be below the exercise price of \$100. We have just discussed the possible ending value of a call option in period 1. But, what we are really interested is what is the value now of the call option knowing the two resulting value of a call option.

To help determine the value of a one-period call option, it's useful to know that it is possible to replicate the resulting two state of the value of the call option by buying a combination of stocks and bonds. Below is the formula to replicate the situation where the price increases to \$110. We will assume that the interest rate for the bond is 7 %.

$$110\text{S} + 1.07\text{B} = 10$$
$$90\text{S} + 1.07\text{B} = 0$$

We can use simple algebra to solve for both S and B. The first thing that we need to do is to rearrange the second equation as follows:

$$1.07\text{B} = -90\text{S}$$

With the above equation, we can rewrite the first equation as

$$110\text{S} + (-90\text{S}) = 10$$
$$20\text{S} = 10$$
$$\text{S} = .5$$

We can solve for B by substituting the value .5 for S in the first equation.

$$110(.5) + 1.07\text{B} = 10$$
$$55 + 1.07\text{B} = 10$$
$$1.07\text{B} = -45$$
$$\text{B} = -42.05607$$

Therefore, from the above simple algebraic exercise, we should at period 0 buy .5 shares of IBM stock and borrow 42.05607 at 7 % to replicate the payoff of the call option. This means the value of a call option should be 5*100 – 42.05607 = 7.94393.

If this were not the case, there would then be arbitrage profits. For example, if the call option were sold for \$8, there would be a profit of .056607. This would result in the increase in the selling of the call option. The increase in the supply of call options would push the price down for the call options. If the call option were sold for \$7, there would be a saving of .94393. This saving would result in the increase demand for the call option. This increase demand would result in the price of the call option to increase. The equilibrium point would be 7.94393.

Using the abovementioned concept and procedure, Benninga (2000) has derived a one-period call option model as

$$C = q_u \text{Max}[\text{S}(1+\text{u}) - \text{X}, 0] + q_d \text{Max}[\text{S}(1+\text{d}) - \text{X}, 0] \tag{25.1}$$

where

$$q_u = \frac{i-d}{(1+i)(u-d)}$$
$$q_d = \frac{u-i}{(1+i)(u-d)}$$

u = increase factor
d = down factor
i = interest rate

If we let $i = r, p = (r - d)/(u - d)$, $1 - p = (u - r)/(u - d)$, $\text{R} = 1/(1 + r)$, $C_u = \text{Max}[\text{S}(1 + \text{u}) - \text{X}, 0]$, and $C_d = \text{Max}[\text{S}(1 + \text{d}) - \text{X}, 0]$, then we have

$$C = [pC_u + (1 - p)C_d]/\text{R} \tag{25.2}$$

where

C_u = call option price after increase
C_d = call option price after decrease

Equation (25.2) represents one-period call option value.

Below calculates the value of the above one-period call option where the strike price, X, is \$100 and the risk-free interest rate is 7 %. We will assume that the price of a stock for any given period will either increase or decrease by 10 %.

$$X = \$100$$
$$\text{S} = \$100$$
$$u = 1.10$$
$$\text{d} = .9$$
$$\text{R} = 1 + r + .07$$
$$\text{p} = (1.07 - .90)/(1.10 - .90)$$
$$C = [.85(10) + .15(0)]/1.07 = \$7.94$$

Therefore, from the above calculations, the value of the call option is \$7.94.

From the above calculations, the call option pricing Decision Tree should look like the following.

Call Option Price
Period 0 Period 1

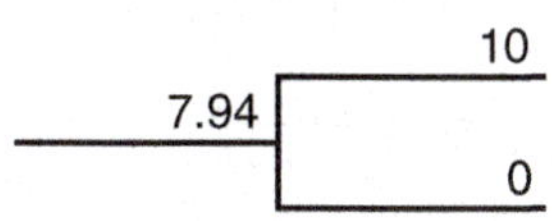

25.4 Put Option Pricing: One Period

Like the call option, it is possible to replicate the resulting two state of the value of the put option by buying a combination of stocks and bonds. Below is the formula to replicate the situation where the price decreases to \$90:

$$110\text{S} + 1.07\text{B} = 0$$
$$90\text{S} + 1.07\text{B} = 10$$

We will use simple algebra to solve for both S and B. The first thing we will do is to rewrite the second equation as follows:

$$1.07\text{B} = 10 - 90\text{S}$$

The next thing to do is to substitute the above equation to the first put option equation. Doing this would result in the following:

$$110\mathrm{S} + 10 - 90\mathrm{S} = 0$$

The following solves for S:

$$20\mathrm{S} = -10$$
$$\mathrm{S} = -.5$$

Now let's solve for B by putting the value of S into the first equation. This is shown below.

$$110(.5) + 1.07\mathrm{B} = 0$$
$$1.07\mathrm{B} = 55$$
$$\mathrm{B} = 51.04$$

From the above simple algebra exercise, we have S = −.5 and B = 51.04. This tells us that we should in period 0 lend $51.04 at 7 % and sell .5 shares of stock to replicate the put option payoff for period 1. And the value of the put option should be 100*(−.5) + 51.40 = −50 + 51.40 = 1.40.

Using the same arbitrage argument that we used in the discussing of the call option, 1.40 has to be the equilibrium price of the put option.

As with the call option, Benninga (2000) has derived a one-period put option model as

$$P = q_u \mathrm{Max}[\mathrm{X} - \mathrm{S}(1 + \mathrm{u}), 0] + q_d \mathrm{Max}[\mathrm{X} - \mathrm{S}(1 + \mathrm{d}), 0] \tag{25.3}$$

where

$$q_u = \frac{i - d}{(1 + i)(u - d)}$$
$$q_d = \frac{u - i}{(1 + i)(u - d)}$$

u = increase factor
d = down factor
i = interest rate

If we let $i = r, p = (r - d)/(u - d)$, $1 - p = (u - r)/(u - d)$, $\mathrm{R} = 1/(1 + r)$, $P_u = \mathrm{Max}[\mathrm{X} - \mathrm{S}(1 + \mathrm{u}), 0]$, and $P_d = \mathrm{Max}[\mathrm{X} - \mathrm{S}(1 + \mathrm{d}), 0]$, then we have

$$P = [pP_u + (1 - p)P_d]/\mathrm{R} \tag{25.4}$$

where

P_u = put option price after increase
P_d = put option price after decrease

Below calculates the value of the above one-period call option where the strike price, *X*, is $100 and the risk-free interest rate is 7 %.

$$P = [.85(0) + .15(10)]/1.07 = \$1.40$$

From the above calculation, the put option pricing Decision Tree would look like the following.

Put Option Price
Period 0 Period 1

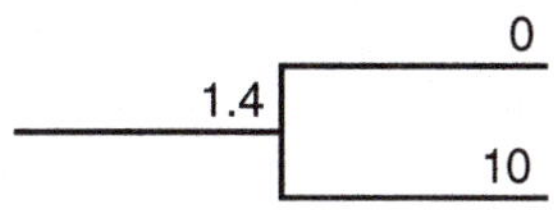

25.5 Option Pricing: Two Periods

We now will look at pricing options for two periods. Below shows the stock price Decision Tree based on the parameters indicated in the last section.

Stock Price
Period 0 Period 1 Period 2

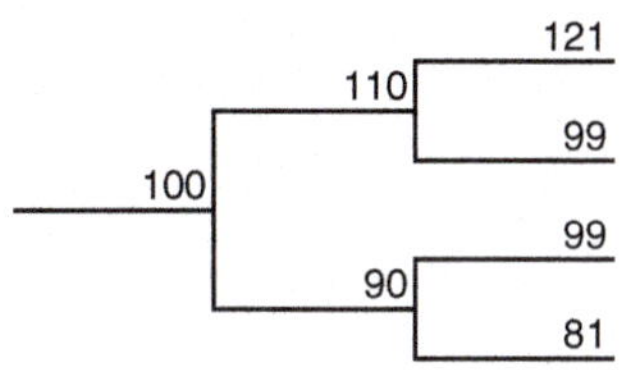

This Decision Tree was created based on the assumption that a stock price will either increase by 10 % or decrease by 10 %.

How do we price the value of a call and put option for two periods?

The highest possible value for our stock based on our assumption is \$121. We get this value first by multiplying the stock price at period 0 by 110 % to get the resulting value of \$110 of period 1. We then again multiply the stock price in period 1 by 110 % to get the resulting value of \$121. In period 2, the value of a call option when a stock price is \$121 is the stock price minus the exercise price, \$121 − 100, or \$21 dollars. In period 2, the value of a put option when a stock price \$121 is the exercise price minus the stock price, \$100 − \$121, or −\$21. A negative value has no value to an investor so the value of the put option would be \$0.

The lowest possible value for our stock based on our assumptions is \$81. We get this value first by multiplying the stock price at period 0 by 90 % (decreasing the value of the stock by 10 %) to get the resulting value of \$90 of period 1. We then again multiply the stock price in period 1 by 90 % to get the resulting value of \$81. In period 2, the value of a call option, when a stock price is \$81, is the stock price minus the exercise price, \$81 − \$100 or −\$19. A negative value has no value to an investor so the value of a call option would be \$0. In period 2, the value of a put option when a stock price is \$81 is the exercise price minus the stock price, \$100 − \$ 81 or \$19. We can derive the call and put option value for the other possible value of the stock in period 2 in the same fashion.

The following shows the possible call and put option values for period 2.

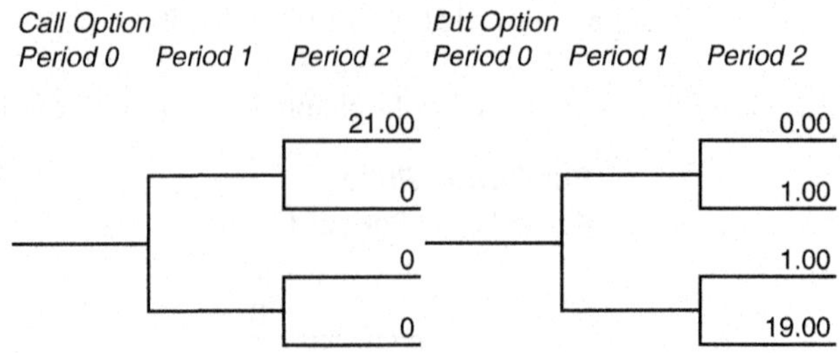

We cannot calculate the value of the call and put option in period 1 the same way we did in period 2 because it's not the ending value of the stock. In period 1 there are two possible call values. One value is when the stock price increased and one value is when the stock price decreased. The call option Decision Tree shown above shows two possible values for a call option in period 1. If we just focus on the value of a call option when the stock price increases from period one, we will notice that it is like the Decision Tree for a call option for one period. This is shown below.

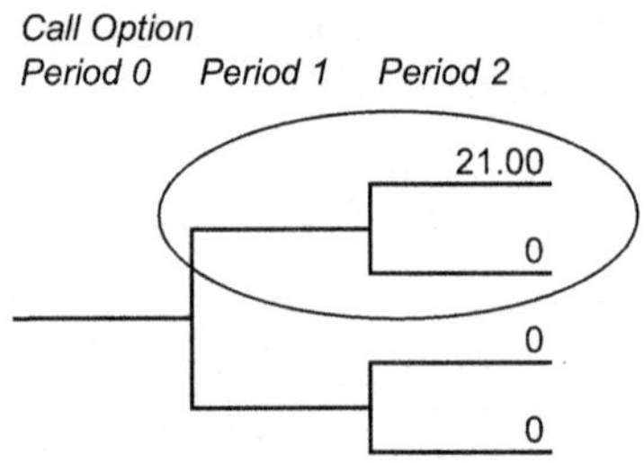

Using the same method for pricing a call option for one period, the price of a call option when stock price increase from period 0 will be $16.68. The resulting Decision Tree is shown below.

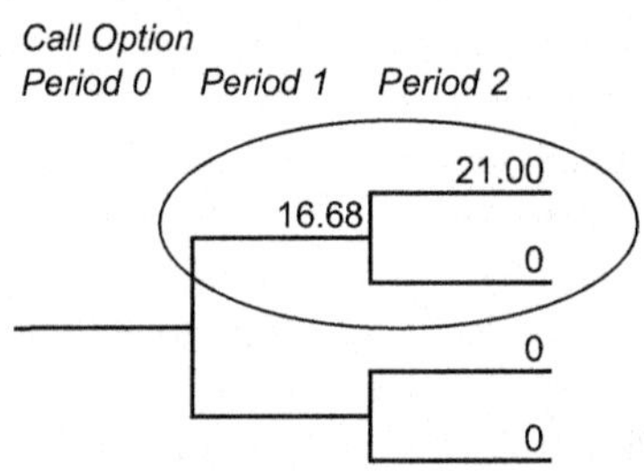

In the same fashion, we can price the value of a call option when a stock price decreases. The price of a call option when a stock price decreases from period 0 is $0. The resulting Decision Tree is shown below.

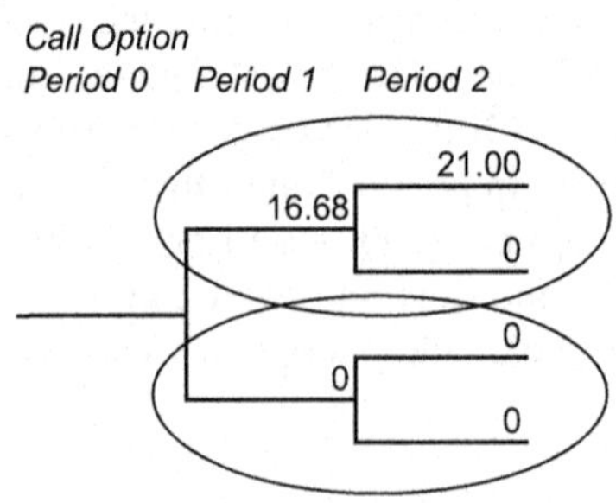

In the same fashion, we can price the value of a call option in period 0. The resulting Decision Tree is shown below.

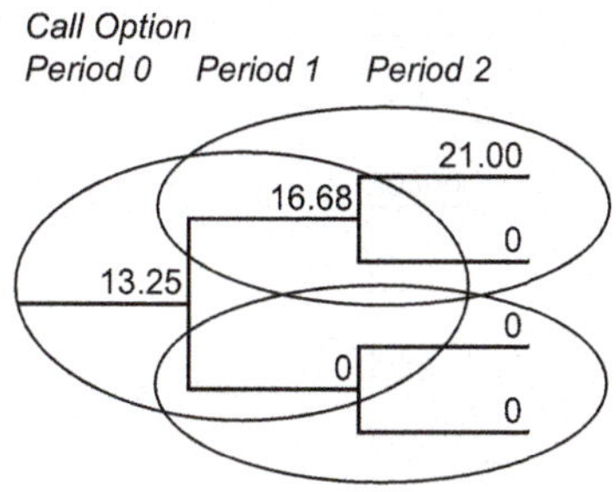

We can calculate the value of a put option in the same manner as we did in calculating the value of a call option. The Decision Tree for a put option is shown below.

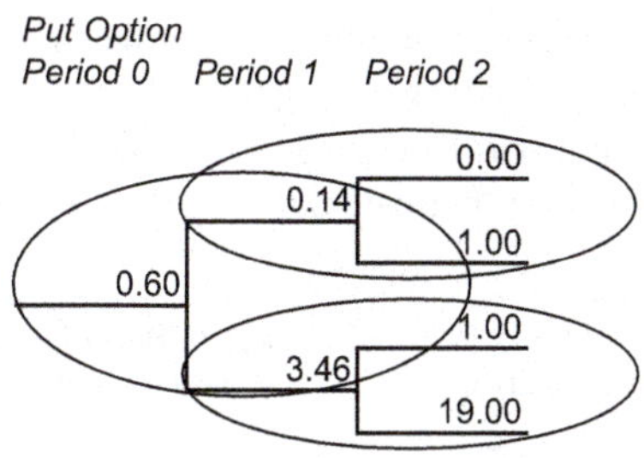

25.6 Option Pricing: Four Periods

We now will look at pricing options for three periods. Below shows the stock price Decision Tree based on the parameters indicated in the last section.

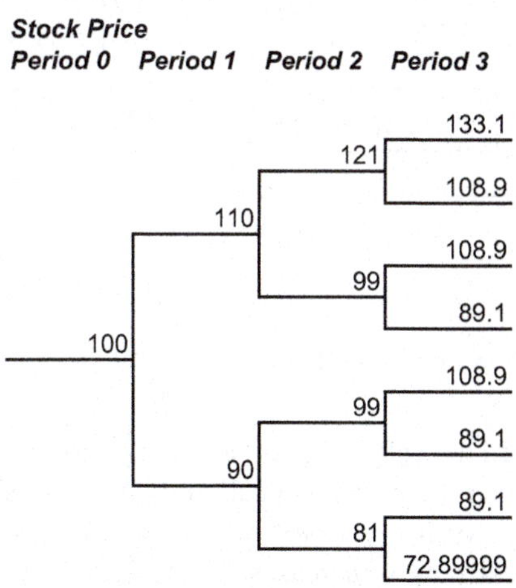

From the above stock price decision tree, we can figure out the values for the call and put options for period 3. The values for the call and put options are shown below.

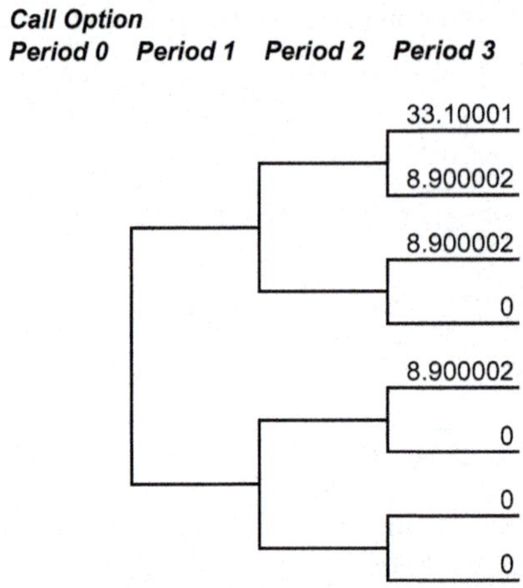

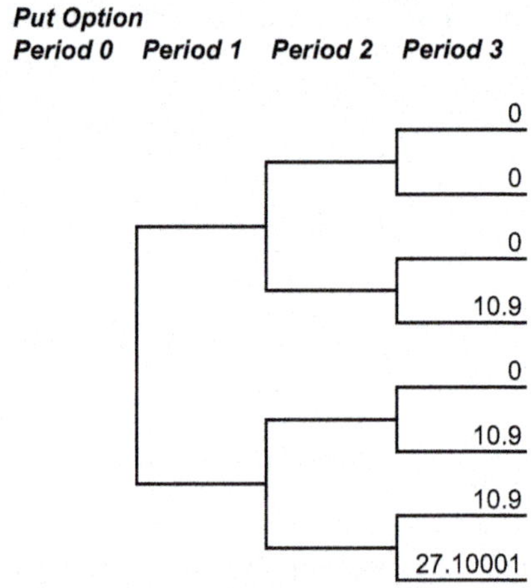

The value is $33.10 for the top most call option because the stock price is $133.1 and the exercise price is $100. In other words $133.1 − $100 = $33.10.

To get the price of the call and put option at period 0, we will need to price backward from period 3 to period 0 as shown below. Each circled calculation below is basically a one-period calculation shown in the previous section.

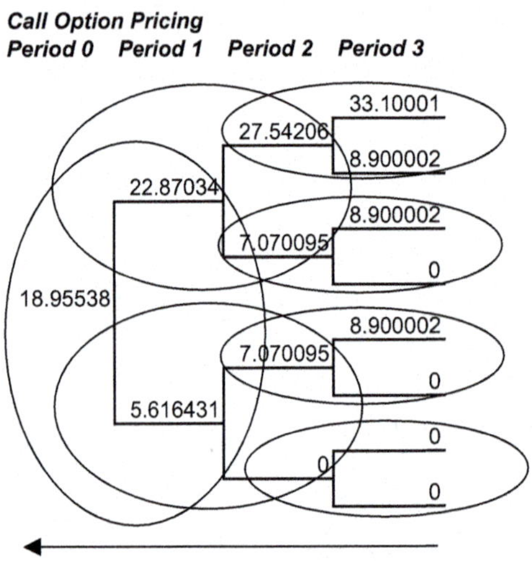

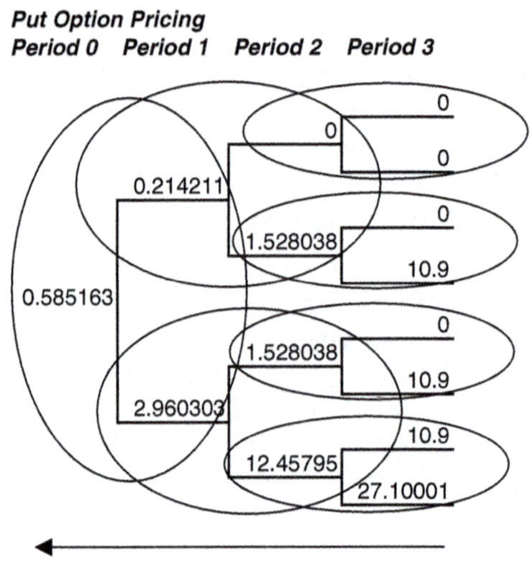

25.7 Using Microsoft Excel to Create the Binomial Option Call Trees

In the previous section, we priced the value of a call and put option by pricing backward, from the last period to the first period. This method of pricing call and put options will work for any *n* period. To price the value of a call options for two periods required seven sets of calculations. The number of calculations increases dramatically as *n* increases. Table 25.1 lists the number of calculations for specific number of periods.

After two periods, it becomes very cumbersome to calculate and create the Decision Trees for a call and put option. In the previous section, we saw that calculations were very repetitive and mechanical. To solve this problem, this paper will use Microsoft Excel to do the calculations and

Table 25.1 Number of calculations for specific number of periods

Periods	Calculations
1	3
2	7
3	15
4	31
5	63
6	127
7	255
8	511
9	1023
10	2047
11	4065
12	8191

create the Decision Trees for the call and put options. We will also use Microsoft Excel to calculate and draw the related Decision Trees for the underlying stock and bond.

To solve this repetitive and mechanical calculation of the binomial option pricing model, we will look at a Microsoft Excel file called *binomialoptionpricingmodel.xlsm.*

We will use this Excel file to produce four Decision Trees for the IBM stock that was discussed in the previous sections. The four Decision Trees are:

1. Stock price
2. Call option price
3. Put option price
4. Bond price

This section will demonstrate how to use the *binomialoptionpricingmodel.xlsm* Excel file to create the four Decision Trees.

The following shows the Excel file *binomialoptionpricingmodel.xlsm* after the file is opened.

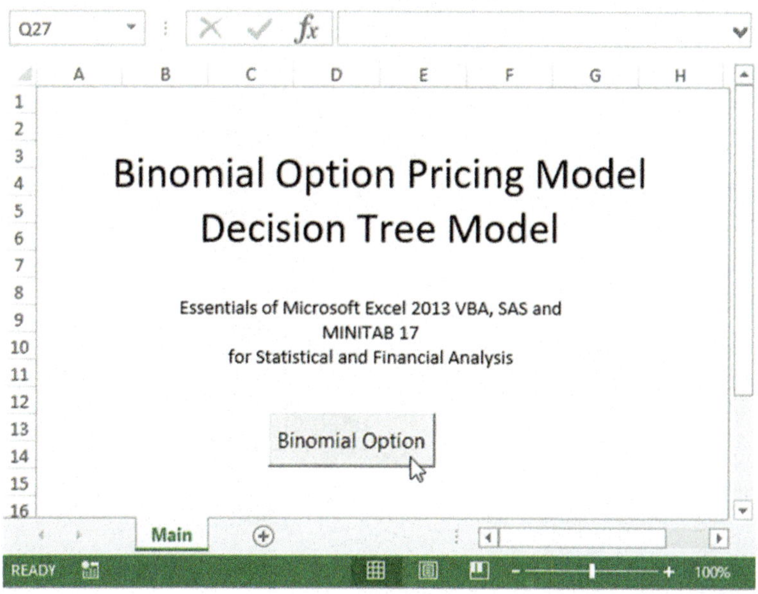

Pushing the Binomial Option button shown above will get the dialog box shown below.

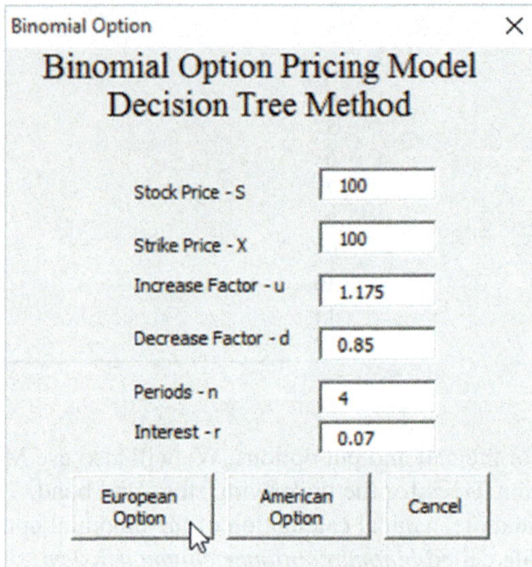

The dialog box shown above shows the parameters for the *binomial option pricing model*. These parameters are changeable. The dialog box shows the default values.

Pushing the *European Option* button produces four binomial option decision trees:

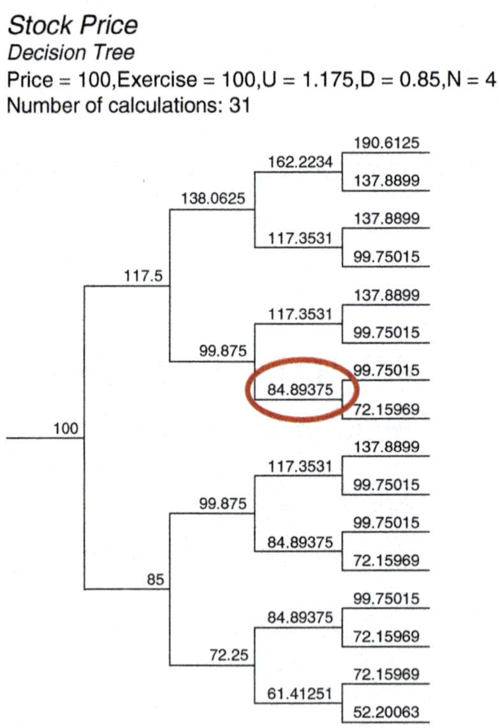

European Call Option Pricing
Decision Tree
Price = 100,Exercise = 100,U = 1.175,D = 0.85,N = 4
Number of calculations: 31

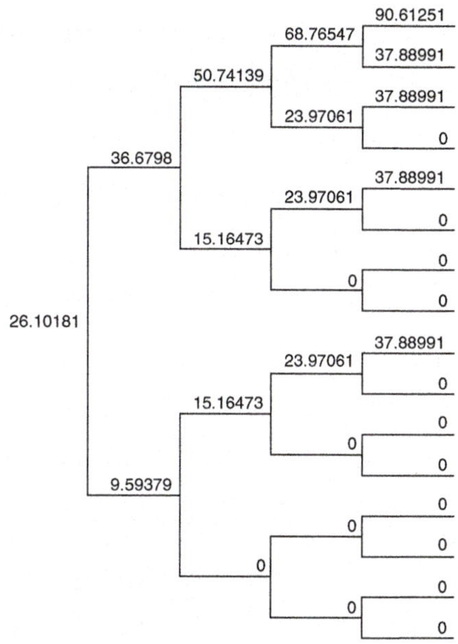

European Put Option Pricing
Decision Tree
Price = 100,Exercise = 100,U = 1.175,D = 0.85,N = 4
Number of calculations: 31

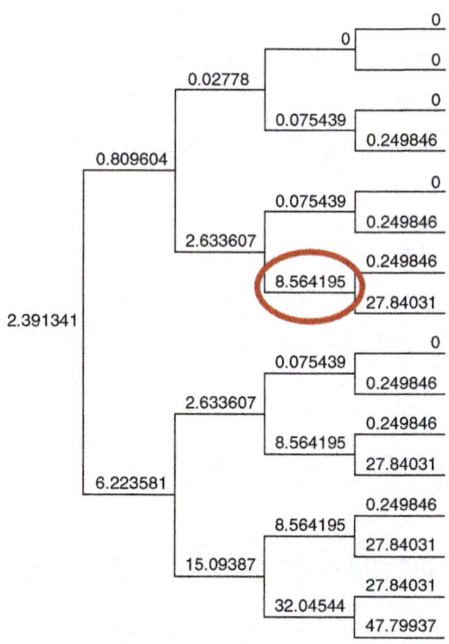

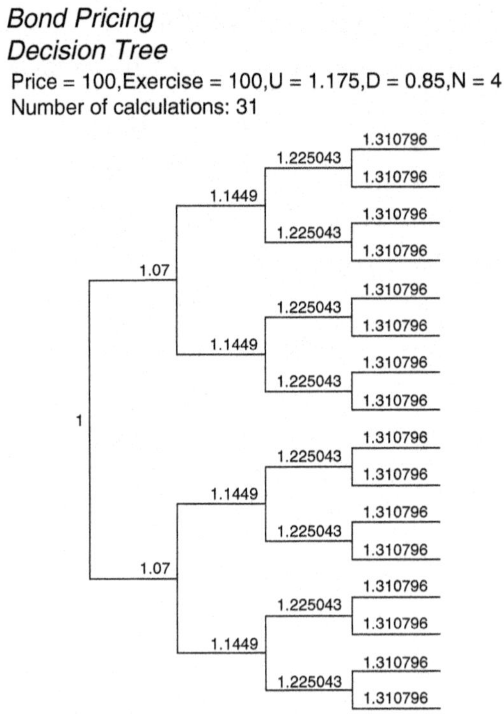

The table at the beginning of this section indicated 31 calculations were required to create a Decision Tree that has four periods. This section showed four Decision Trees. Therefore, the Excel file did 31*4 = 121 calculations to create the four Decision Trees.

Benninga (2000, p. 260) has defined the price of a call option in a binomial option pricing model with n periods as

$$C = \sum_{i=0}^{n} \binom{n}{i} q_u^i q_d^{n-i} \ \max\left[S(1+u)^i(1+d)^{n-i} - X, 0\right] \tag{25.5}$$

and the price of a put option in a binomial option pricing model with n periods as

$$P = \sum_{i=0}^{n} \binom{n}{i} q_u^i q_d^{n-i} \ \max\left[X - S(1+u)^i(1+d)^{n-i}, 0\right] \tag{25.6}$$

Lee et al. (2000, p. 237) has defined the pricing of a call option in a binomial option pricing model with n period as

$$C = \frac{1}{R^n} \sum_{k=0}^{n} \frac{n!}{k!(n-k!)} p^k (1-p)^{n-k} \ \max\left[0, (1+u)^k(1+d)^{n-k} S - X\right] \tag{25.7}$$

The definition of the pricing of a put option in a binomial option pricing model with n period would then be defined as

$$P = \frac{1}{R^n} \sum_{k=0}^{n} \frac{n!}{k!(n-k!)} p^k (1-p)^{n-k} \ \max\left[0, X - (1+u)^k(1+d)^{n-k} S\right] \tag{25.8}$$

25.8 SAS Programming Code Instructions to Implement the Binomial Option Trees

The following SAS macro is used to implement binomial trees and calculate the price of a stock, a call option, a put option, and a risk-free bond price. The parameters of this macro are:

S: stock price
X: strike price
U: increase factor
D: decrease factor
N: periods
r: interest

```
%macro test(S, X, U, D, N, r);
data a;
%do i=1 %to &N.;
    output;
%end;
run;

/*Stock Price*/
data b;
set a;
array S(&N.);
num=_N_;
if num=1 then do;
    S1=&S;
end;

%do j=1 %to (&N.-1);
    if num=1 then do;
        S(&j.+1)=&U*S(&j.);
    end;
    newS=lag(S(&j.));
    if num>1 then do;
        S(&j.+1)=&D*newS;
    end;

%end;
drop newS num;
run;

proc print data=b;
run;

/*Call Option Pricing*/
data cpre;
set b;
v=&N.+1-_N_;
run;

proc sort data=cpre;
by v;
run;
```

```
data c;
set cpre;
array C(&N.);
C&N.=max(S&N.-&X.,0);
prob=(1+&r.-&D.)/(&U.-&D.);

%do k=&N. %to 2 %by -1;
     newC=lag(C&k.);
     C(&k.-1)=((1-prob)*newC+prob*C(&k.))/(1+&r.);
%end;
drop newC prob S1-S&N.;
run;

proc sort data=c;
by descending v;
run;

data c;
set c;
drop v;
run;

proc print data=c;
run;

/*Put Option Pricing*/
data dpre;
set b;
u=&N.+1-_N_;
run;

proc sort data=dpre;
by u;
run;

data d;
set dpre;
array P(&N.);
P&N.=max(&X.-S&N.,0);
prob=(1+&r.-&D.)/(&U.-&D.);

%do k=&N. %to 2 %by -1;
     newP=lag(P&k.);
     P(&k.-1)=((1-prob)*newP+prob*P(&k.))/(1+&r.);
%end;
drop newP prob S1-S&N.;
run;

proc sort data=d;
by descending u;
run;

data d;
set d;
drop u;
run;

proc print data=d;
run;
```

```
/*Bond Pricing*/
data e;
set a;
array B(&N.);
num=_N_;
if num=1 then do;

B1=1;
end;

%do q=1 %to (&N.-1);
     if num=1 then do;
          B(&q.+1)=(1+&r.)*B(&q.);
     end;
     newB=lag(B(&q.));
     if num>1 then do;
          B(&q.+1)=(1+&r.)*newB;
end;
%end;
drop newB num;
run;

proc print data=e;
run;

%mend;
```

The current stock price is $100 and the strike price of call and put option is $100. The increase and decrease factors are 1.175 and 0.85, respectively. The constant interest rate is 7 %. The five-period binomial trees can be created by following SAS macro code:

```
%test(100,100,1.175,0.85,5,0.07);
```

The output is shown below:

Stock Price

Obs	S1	S2	S3	S4	S5
1	100	117.5	138.063	162.223	190.613
2	.	85.0	99.875	117.353	137.890
3	.	.	72.250	84.894	99.750
4	.	.	.	61.413	72.160
5	.	.	.	.	52.201

Call Option Pricing

Obs	C1	C2	C3	C4	C5
1	26.1018	36.6798	50.7414	68.7655	90.6125
2	.	9.5938	15.1647	23.9706	37.8899
3	.	.	0.0000	0.0000	0.0000
4	.	.	.	0.0000	0.0000
5	.	.	.	.	0.0000

Put Option Pricing

Obs	P1	P2	P3	P4	P5
1	2.39134	0.80960	0.0228	0.0000	0.0000
2	.	6.22358	2.6336	0.0754	0.0000
3	.	.	15.0939	8.5642	0.2498
4	.	.	.	32.0454	27.8403
5	.	.	.	.	47.7994

Bond Pricing'

Obs	B1	B2	B3	B4	B5
1	1	1.07	1.1449	1.22504	1.31080
2	.	1.07	1.1449	1.22504	1.31080
3	.	.	1.1449	1.22504	1.31080
4	.	.	.	1.22504	1.31080
5	.	.	.	.	1.31080

25.9 American Options

An American option is an option that the holder may exercise at any time between the start date and the maturity date. Therefore, the holder of an American option faces with the dilemma of deciding when to exercise. Binomial tree valuation can be adapted to include the possibility of exercise at intermediate dates and not just the maturity date. This feature needs to be incorporated into the pricing of American options.

The first step of pricing an American option is the same as a European option. For an American put option, the second step is to take the max of the difference between the strike price of the stock and the price of the stock at that node N and the value of the European put option at node N. The value of a European put option is shown in Eq. (25.4).

Below shows the American put option binomial tree. This American put option has the same parameters as the European put option.

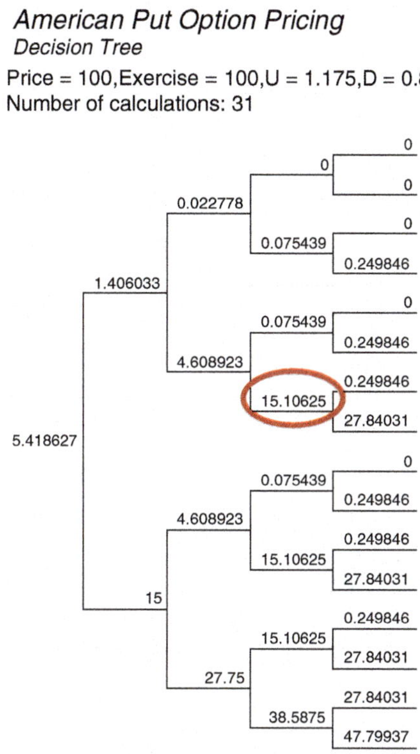

With the same input parameters, we can see that the value of the European put option and the value of the American put option are different. The value of the European put option is 2.391341, while the value of the American put option is 5.418627.

The red circle in the American put option binomial tree is one reason why. At this node the American put option has a value of 15.10625, while at the same node the European put option has a value of 8.564195. At this node the value of the put option is the max of the difference between the strike stock's strike price and stock price at this node and the value of the European put option at this node. At this node the stock price is 84.09375 and the stock strike price is 100.

Mathematically the price of the American put option at this node is

$$\text{Max}(X - St,\ 8.564195) = \text{Max}(100 - 84.09375,\ 8.56195) = 15.10625$$

25.10 Alternative Tree Methods

In this section we will introduce three binomial tree methods and one trinomial tree method to price option values. Three binomial tree methods include Cox et al. (1979), Jarrow and Rudd (1983), and Leisen and Reimer (1996). These methods will generate different kinds of underlying asset trees to represent different trends of asset movement. Kamrad and Ritchken (1991) extend binomial tree method to multinomial approximation models. Trinomial tree method is one of multinomial models.

25.10.1 Cox, Ross, and Rubinstein

Cox, Ross, and Rubinstein (1979) (hereafter CRR) propose an alternative choice of parameters that also create a risk-neutral valuation environment. The price multipliers, u and d, depend only on volatility σ and on δt, not on drift

$$u = e^{\sigma\sqrt{\delta t}}$$

$$d = \frac{1}{u}$$

To offset the absence of a drift component in u and d, the probability of an up move in the CRR tree is usually greater than 0.5 to ensure that the expected value of the price increases by a factor of exp [(r − q)δt] on each step. The formula for p is:

$$p = \frac{e^{(r-q)\delta t} - d}{u - d}$$

Below is the asset price tree base on CRR binomial tree model.

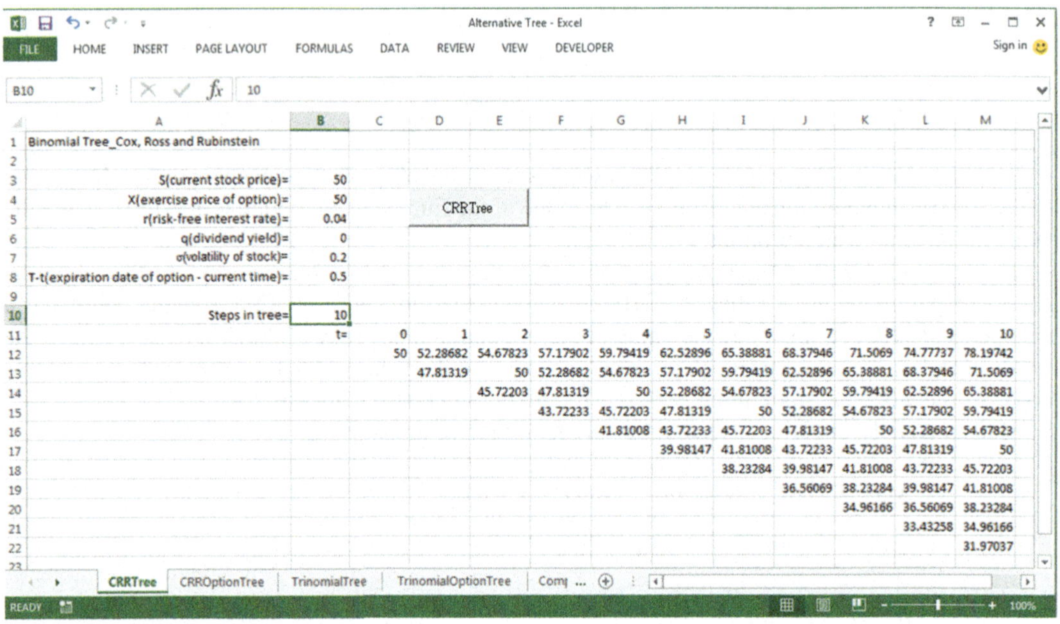

	A	B	C	D	E	F	G	H	I	J	K	L	M
1	Binomial Tree_Cox, Ross and Rubinstein												
2													
3	S(current stock price)=	50											
4	X(exercise price of option)=	50		CRRTree									
5	r(risk-free interest rate)=	0.04											
6	q(dividend yield)=	0											
7	σ(volatility of stock)=	0.2											
8	T-t(expiration date of option - current time)=	0.5											
9													
10	Steps in tree=	10											
11		t=	0	1	2	3	4	5	6	7	8	9	10
12			50	52.28682	54.67823	57.17902	59.79419	62.52896	65.38881	68.37946	71.5069	74.77737	78.19742
13				47.81319	50	52.28682	54.67823	57.17902	59.79419	62.52896	65.38881	68.37946	71.5069
14					45.72203	47.81319	50	52.28682	54.67823	57.17902	59.79419	62.52896	65.38881
15						43.72233	45.72203	47.81319	50	52.28682	54.67823	57.17902	59.79419
16							41.81008	43.72233	45.72203	47.81319	50	52.28682	54.67823
17								39.98147	41.81008	43.72233	45.72203	47.81319	50
18									38.23284	39.98147	41.81008	43.72233	45.72203
19										36.56069	38.23284	39.98147	41.81008
20											34.96166	36.56069	38.23284
21												33.43258	34.96166
22													31.97037

We can see that CRR tree is symmetric to its initial asset price, in this case is 50. Next, we want to create option tree in the worksheet. For example, a call option value is on this asset price. Let $f_{i,j}$ denotes the option value in node (i, j), where j refers to period j (j = 0, 1, 2,. . ., N) and i denotes the *i*th node in period j (in the binomial tree model, node numbers increase going up in the lattice, so i = 0,. . ., j). With these assumptions, the underlying asset price in node (i, j) is $Su^j d^{i-j}$. At the expiration we have

$$f_{i,N} = \max\left[Su^i d^{N-i} - X, 0\right] \quad i = 0, 1, \ldots, N$$

Going backward in time (decreasing j), we get

$$f_{i,j} = e^{-r\delta t}\left[pf_{i+1,j+1} + (1-p)f_{i,j+1}\right]$$

The CRR option value tree is shown below.

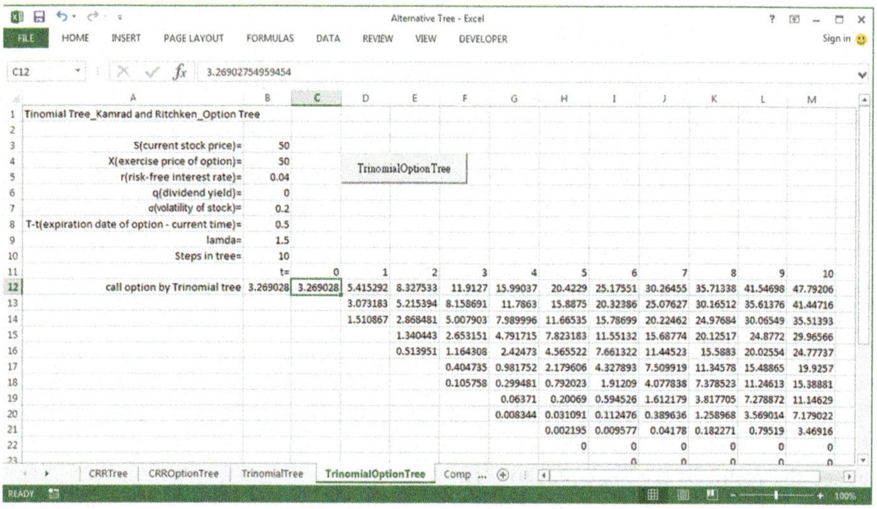

	A	B	C	D	E	F	G	H	I	J	K	L	M
1	Tinomial Tree_Kamrad and Ritchken_Option Tree												
2													
3	S(current stock price)=	50											
4	X(exercise price of option)=	50		TrinomialOptionTree									
5	r(risk-free interest rate)=	0.04											
6	q(dividend yield)=	0											
7	σ(volatility of stock)=	0.2											
8	T-t(expiration date of option - current time)=	0.5											
9	lamda=	1.5											
10	Steps in tree=	10											
11		t=	0	1	2	3	4	5	6	7	8	9	10
12	call option by Trinomial tree	3.269028	3.269028	5.415292	8.327533	11.9127	15.99037	20.4229	25.17551	30.26455	35.71338	41.54698	47.79206
13				3.073183	5.215394	8.158691	11.7863	15.8875	20.32386	25.07627	30.16512	35.61376	41.44716
14				1.510867	2.868481	5.007903	7.989996	11.66535	15.78699	20.22462	24.97684	30.06549	35.51393
15					1.340443	2.653151	4.791715	7.823183	11.55132	15.68774	20.12517	24.8772	29.96566
16					0.513951	1.164308	2.42473	4.565522	7.661322	11.44523	15.5883	20.02554	24.77737
17						0.404735	0.981752	2.179606	4.327893	7.509919	11.34578	15.48865	19.9257
18						0.105758	0.299481	0.792023	1.91209	4.077838	7.378523	11.24613	15.38881
19							0.06371	0.20069	0.594526	1.612179	3.817705	7.278872	11.14629
20							0.008344	0.031091	0.112476	0.389636	1.258968	3.569014	7.179022
21								0.002195	0.009577	0.04178	0.182271	0.79519	3.46916
22								0	0	0	0	0	0

We can see the call option value at time zero is equal to 3.244077 in the cell C12. We also can write a VBA function to price call option. Below is the function:

```
'   Returns CRR Binomial Option Value
Function CRRBinCall(S, X, r, q, T, sigma, Nstep)
    Dim dt, erdt, ermqdt, u, d, p
    Dim i As Integer, j As Integer
    Dim vvec() As Variant
    ReDim vvec(Nstep)
    dt = T / Nstep
    erdt = Exp(r * dt)
    ermqdt = Exp((r - q) * dt)
    u = Exp(sigma * Sqr(dt))
    d = 1 / u
    p = (ermqdt - d) / (u - d)
    For i = 0 To Nstep
      vvec(i) = Application.Max(S * (u ^ i) * (d ^ (Nstep - i)) - X, 0)
    Next i
        For j = Nstep - 1 To 0 Step -1
            For i = 0 To j
                vvec(i) = (p * vvec(i + 1) + (1 - p) * vvec(i)) / erdt
            Next i
        Next j
    CRRBinCall = vvec(0)
End Function
```

Using this function and put parameters in the function, we can get call option value under different steps. This result is shown below.

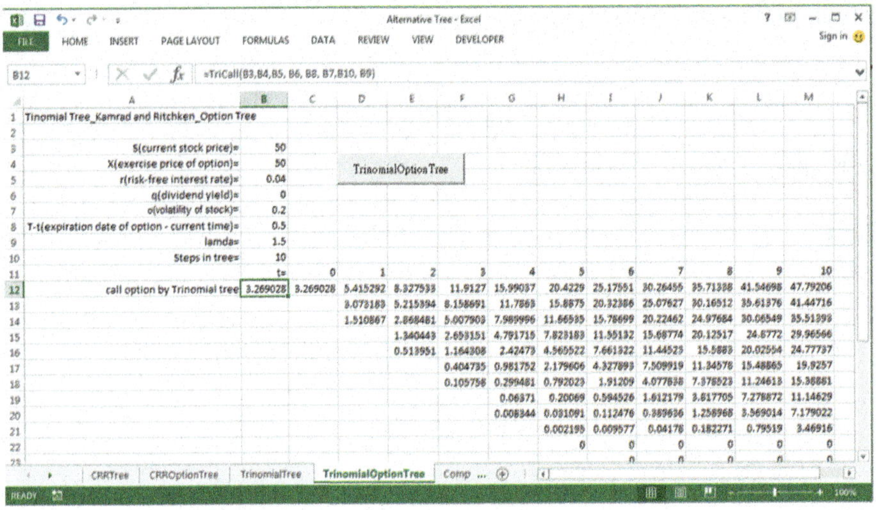

The function in cell B12 is

$$= \text{CRRBinCall}\,(\text{B3}, \text{B4}, \text{B5}, \text{B6}, \text{B8}, \text{B7}, \text{B10})$$

We can see the result in B12 is equal to C12.

25.10.2 Trinomial Tree

Because binomial tree methods are computationally expensive, Kamrad and Ritchken (1991) propose multinomial models. New multinomial models include as special cases existing models. The more general models are shown to be computationally more efficient.

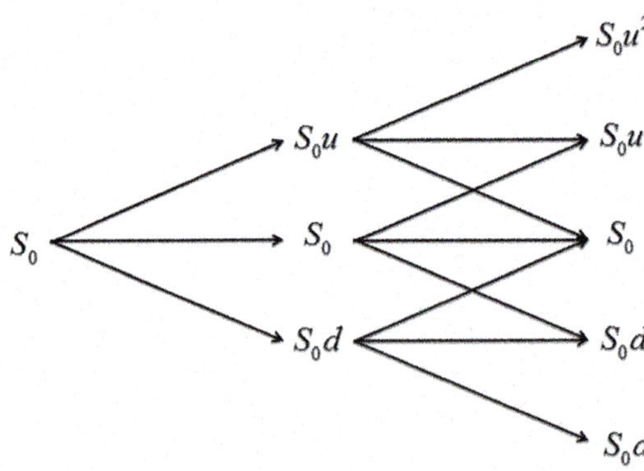

Expressed algebraically, the trinomial tree parameters are:

$$u = e^{\lambda\sigma\sqrt{\delta t}}$$

$$d = \frac{1}{u}$$

The formula for probability p:

$$p_u = \frac{1}{2\lambda^2} + \frac{(r - \sigma^2/2)\sqrt{\delta t}}{2\lambda\sigma}$$

$$p_m = 1 - \frac{1}{\lambda^2}$$

$$p_d = 1 - p_u - p_m$$

If parameter λ is equal 1, then trinomial tree model reduce to a binomial tree model. Below is the underlying asset price pattern base on trinomial tree model.

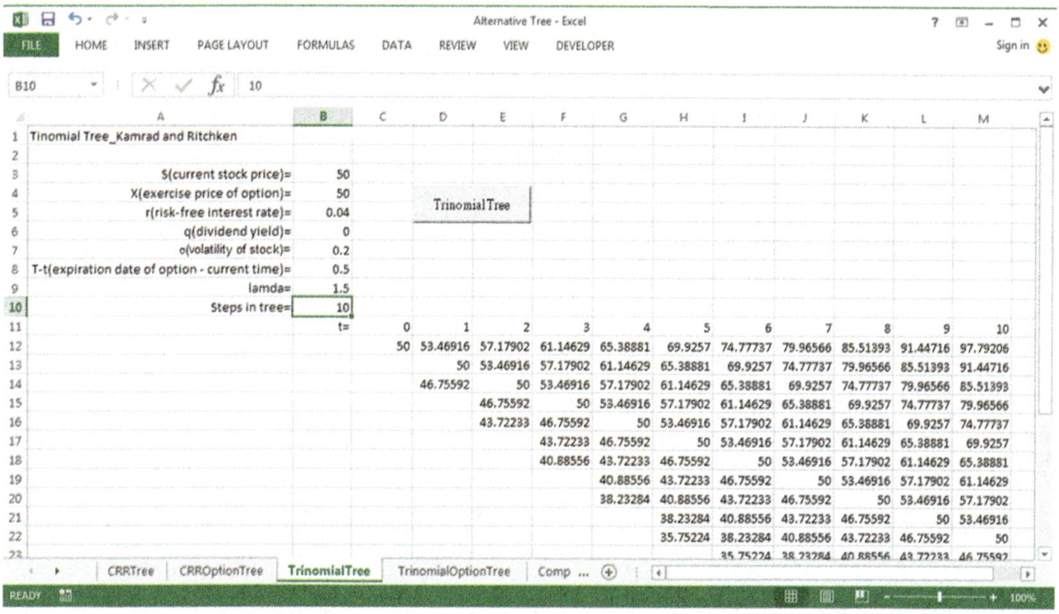

	A	B	C	D	E	F	G	H	I	J	K	L	M
1	Tinomial Tree_Kamrad and Ritchken												
2													
3	S(current stock price)=	50											
4	X(exercise price of option)=	50											
5	r(risk-free interest rate)=	0.04											
6	q(dividend yield)=	0											
7	σ(volatility of stock)=	0.2											
8	T-t(expiration date of option - current time)=	0.5											
9	lamda=	1.5											
10	Steps in tree=	10											
11		t=	0	1	2	3	4	5	6	7	8	9	10
12			50	53.46916	57.17902	61.14629	65.38881	69.9257	74.77737	79.96566	85.51393	91.44716	97.79206
13				50	53.46916	57.17902	61.14629	65.38881	69.9257	74.77737	79.96566	85.51393	91.44716
14				46.75592	50	53.46916	57.17902	61.14629	65.38881	69.9257	74.77737	79.96566	85.51393
15					46.75592	50	53.46916	57.17902	61.14629	65.38881	69.9257	74.77737	79.96566
16					43.72233	46.75592	50	53.46916	57.17902	61.14629	65.38881	69.9257	74.77737
17						43.72233	46.75592	50	53.46916	57.17902	61.14629	65.38881	69.9257
18						40.88556	43.72233	46.75592	50	53.46916	57.17902	61.14629	65.38881
19							40.88556	43.72233	46.75592	50	53.46916	57.17902	61.14629
20							38.23284	40.88556	43.72233	46.75592	50	53.46916	57.17902
21								38.23284	40.88556	43.72233	46.75592	50	53.46916
22								35.75224	38.23284	40.88556	43.72233	46.75592	50
23									35.75224	38.23284	40.88556	43.72233	46.75592

We can see this trinomial tree model is also a symmetric tree. The middle price in each period is the same as initial asset price, 50.

Through the similar rule, we can use this tree to price a call option. At first, we can draw the option tree based on trinomial underlying asset price tree. The result is shown below.

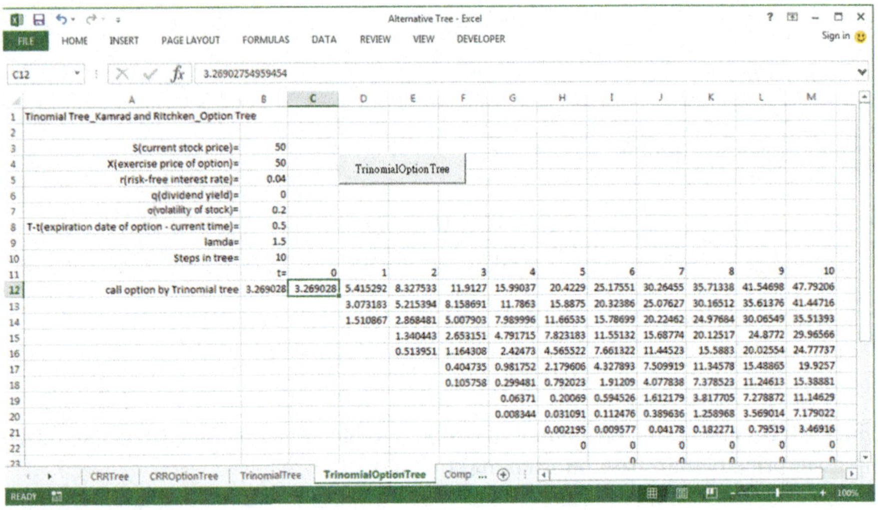

The call option value at time zero is 3.269028 in cell C12. In addition, we also can write a function to price a call option based on trinomial tree model. The function is shown below.

```
'   Returns Trinomial Option Value
Function TriCall(S, X, r, q, T, sigma, Nstep, lamda)
    Dim dt, erdt, ermqdt, u, d, pu, pm, pd
    Dim i As Integer, j As Integer
    Dim vvec() As Variant
    ReDim vvec(2 * Nstep)
    dt = T / Nstep
    erdt = Exp(r * dt)
    ermqdt = Exp((r - q) * dt)
    u = Exp(lamda * sigma * Sqr(dt))
    d = 1 / u
    pu = 1 / (2 * lamda ^ 2) + (r - sigma ^ 2 / 2) * Sqr(dt) / (2 * lamda * sigma)
    pm = 1 - 1 / (lamda ^ 2)
    pd = 1 - pu - pm
    For i = 0 To 2 * Nstep
      vvec(i) = Application.Max(S * (d ^ Nstep) * (u ^ i) - X, 0)
    Next i
    For j = Nstep - 1 To 0 Step -1
        For i = 0 To 2 * j
            vvec(i) = (pu * vvec(i + 2) + pm * vvec(i + 1) + pd * vvec(i)) / erdt
        Next i
    Next j
    TriCall = vvec(0)
End Function
```

Similar data can be used in this function and get the same call option of today's price.

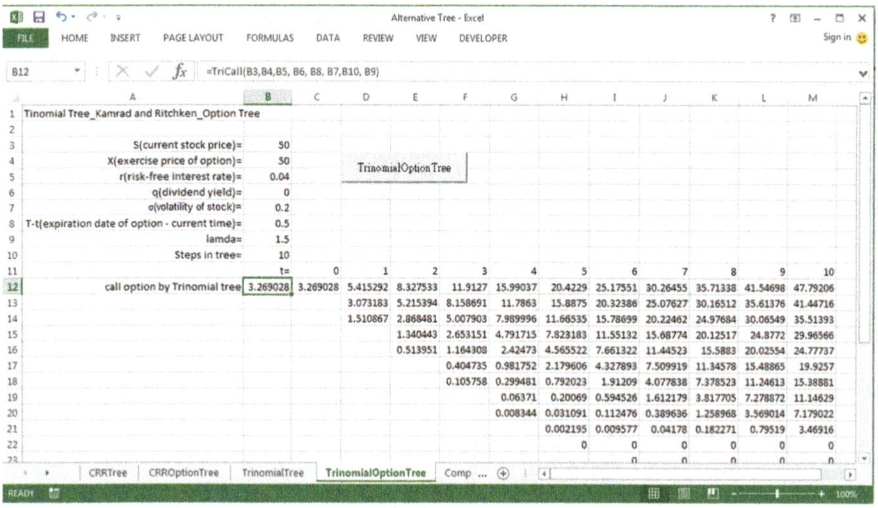

The function in cell B12 is equal to

$$= \text{TriCall}\,(\text{B3}, \text{B4}, \text{B5}, \text{B6}, \text{B8}, \text{B7}, \text{B10}, \text{B9})$$

25.10.3 Compare the Option Price Efficiency

In this section, we would like to compare the efficient between these two methods. In the table below, we represent different numbers of step, 1, 2,. . ., 50. And we represent Black and Scholes, CRR binominal tree, and trinomial tree method results. The following figure is the result.

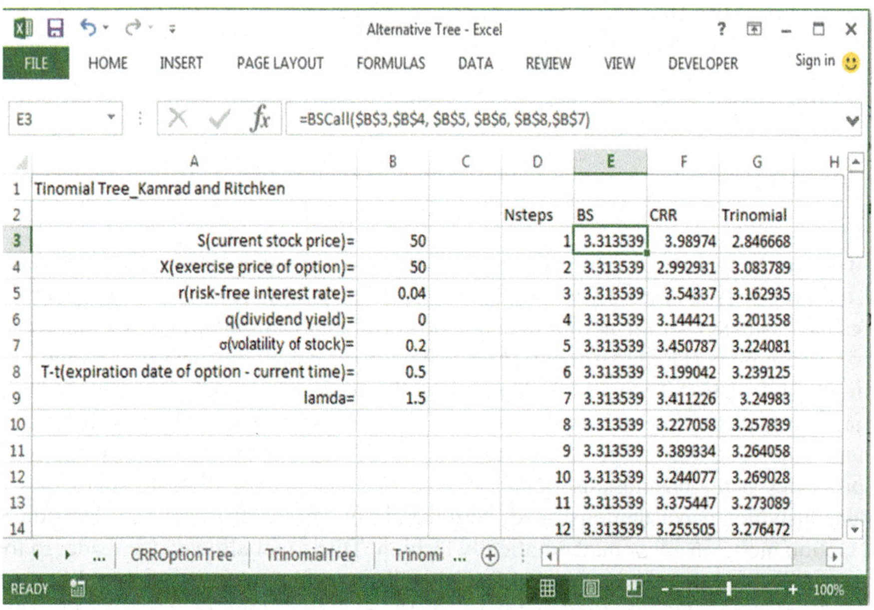

In order to see the result more deeply, we draw the result in the picture. The picture is shown below.

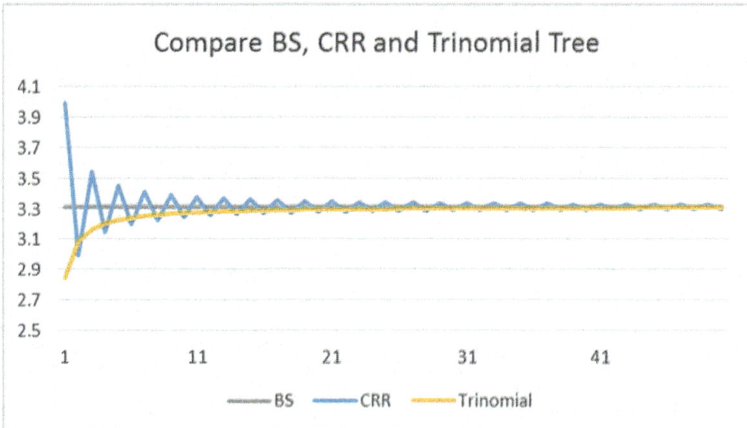

After we increase the number of steps, we can see that trinomial tree method is more quickly convergence to Black and Scholes than CRR binomial tree method.

25.11 Retrieving Option Prices from Yahoo Finance

The following is the URL to retrieve Coca-Cola's option prices:

http://finance.yahoo.com/q/op?s=KO+Options

The following is the URL to retrieve Home Depot's option prices:

http://finance.yahoo.com/q/op?s=HD+Options

The following is the URL to retrieve Microsoft's option prices:

http://finance.yahoo.com/q/op?s=MSFT+Options

25.12 Summary

In this paper we demonstrated why Microsoft Excel is a very powerful application and why the Financial Industry in New York City value people that know Microsoft Excel very well. Microsoft Excel gives the business user the ability to create powerful application quickly without relying on the Information Technology (IT) department. Prior to Microsoft Excel, business user would have to rely heavily on the Information Technology department. There are two problems with relying on the IT department. The first problem is that the tools that the IT department was using resulted in longer development time. The second problem was that the IT department was not as familiar with the business processes as the business users.

Simultaneously this paper demonstrated, with the aid of Microsoft Excel and Decision Trees, the Binomial Option model in a less mathematically fashion. This paper allowed the reader to focus more on the concepts by studying the associated Decision Trees, which were created by Microsoft Excel. This paper also demonstrates that using Microsoft Excel releases the reader from the computation burden of the Binomial Option model.

This paper also published the Microsoft Excel VBA code that created the Binomial Option Decision Trees. This allows for those who are interested to study the many advance Microsoft Excel VBA programming concepts that were used to create the Decision Trees. One major computer science programming concept used by Microsoft Excel VBA is recursive programming. Recursive programming is the ideal of a procedure calling itself many times. Inside the procedure, there are statements to decide when not to call itself.

Appendix 25.1: Excel Code—Binomial Option Pricing Model

```
'/*************************************************************************
'/Essentials of Microsoft Excel 2013 VBA, SAS
'/               and MINITAB 17
'/   for Statistical and Financial Analysis
'/
'/*************************************************************************
Option Explicit
Dim mwbTreeWorkbook As Workbook
Dim mwsTreeWorksheet As Worksheet
Dim mwsCallTree As Worksheet
Dim mwsPutTree As Worksheet
Dim mwsBondTree As Worksheet
Dim mdblPFactor As Double
Dim mBinomialCalc As Long
Dim mOptionType As String
'/*************************************************
'/Purpose:  Keep track the numbers of binomial calc
'/*************************************************
Property Let OptionType(t As String)
    mOptionType = t
End Property
Property Get OptionType() As String
    OptionType = mOptionType
End Property
Property Let BinomialCalc(l As Long)
    mBinomialCalc = l
End Property
Property Get BinomialCalc() As Long
    BinomialCalc = mBinomialCalc
End Property
Property Set TreeWorkbook(wb As Workbook)
    Set mwbTreeWorkbook = wb
End Property
Property Get TreeWorkbook() As Workbook
    Set TreeWorkbook = mwbTreeWorkbook
End Property
Property Set TreeWorksheet(ws As Worksheet)
    Set mwsTreeWorksheet = ws
End Property
Property Get TreeWorksheet() As Worksheet
    Set TreeWorksheet = mwsTreeWorksheet
End Property
Property Set CallTree(ws As Worksheet)
    Set mwsCallTree = ws
End Property
Property Get CallTree() As Worksheet
    Set CallTree = mwsCallTree
End Property
Property Set PutTree(ws As Worksheet)
    Set mwsPutTree = ws
End Property
Property Get PutTree() As Worksheet
```

```
    Set PutTree = mwsPutTree
End Property
Property Set BondTree(ws As Worksheet)
    Set mwsBondTree = ws
End Property
Property Get BondTree() As Worksheet
    Set BondTree = mwsBondTree
End Property

Property Let PFactor(r As Double)
    Dim dRate As Double

    dRate = ((1 + r) - Me.txtBinomialD) / (Me.txtBinomialU - Me.txtBinomialD)
    Let mdblPFactor = dRate
End Property
Property Get PFactor() As Double
    Let PFactor = mdblPFactor
End Property
Private Sub cmdCalculate_Click()
    Me.Hide
    BinomialOption
    Unload Me
End Sub

Private Sub cmdCalculateAmerican_Click()

    Me.Hide
    Me.OptionType = "American"
    BinomialOption
    Unload Me
End Sub

Private Sub cmdCalculateEuropean_Click()
    Me.Hide
    Me.OptionType = "European"
    BinomialOption
    Unload Me
End Sub

Private Sub cmdCancel_Click()
    Unload Me
End Sub


Private Sub UserForm_Initialize()
   With Me
        .txtBinomialS = 100
        .txtBinomialX = 100
        .txtBinomialD = 0.85
        .txtBinomialU = 1.175
        .txtBinomialN = 4
        .txtBinomialr = 0.07
    End With
    Me.Hide
End Sub

Sub BinomialOption()
    Dim wbTree As Workbook
    Dim wsTree As Worksheet
    Dim rColumn As Range
    Dim ws As Worksheet

    Set Me.TreeWorkbook = Workbooks.Add
```

```
    Set Me.BondTree = Me.TreeWorkbook.Worksheets.Add
    Set Me.PutTree = Me.TreeWorkbook.Worksheets.Add
    Set Me.CallTree = Me.TreeWorkbook.Worksheets.Add
    Set Me.TreeWorksheet = Me.TreeWorkbook.Worksheets.Add

    Set rColumn = Me.TreeWorksheet.Range("a1")

    With Me
        .BinomialCalc = 0
        .PFactor = Me.txtBinomialr
        .CallTree.Name = "American Call Option Price"
        .PutTree.Name = "American Put Option Price"
        .TreeWorksheet.Name = "Stock Price"
        .BondTree.Name = "Bond"
    End With

    DecitionTree rCell:=rColumn, nPeriod:=Me.txtBinomialN + 1, _
             dblPrice:=Me.txtBinomialS, sngU:=Me.txtBinomialU, _
             sngD:=Me.txtBinomialD

    DecitionTreeFormat
    TreeTitle wsTree:=Me.TreeWorksheet, sTitle:="Stock Price "
    TreeTitle wsTree:=Me.CallTree, sTitle:=Me.OptionType & " Call Option Pricing"
    TreeTitle wsTree:=Me.PutTree, sTitle:=Me.OptionType & " Put Option Pricing"
    TreeTitle wsTree:=Me.BondTree, sTitle:="Bond Pricing"

    Application.DisplayAlerts = False
    For Each ws In Me.TreeWorkbook.Worksheets
        If Left(ws.Name, 5) = "Sheet" Then
            ws.Delete
        Else
            ws.Activate
            ActiveWindow.DisplayGridlines = False
        End If
    Next
    Application.DisplayAlerts = True
    Me.TreeWorksheet.Activate
End Sub
Sub TreeTitle(wsTree As Worksheet, sTitle As String)
    wsTree.Range("A1:a5").EntireRow.Insert (xlShiftDown)
    With wsTree
        With .Cells(1)
               .Value = sTitle
               .Font.Size = 20
               .Font.Italic = True
        End With
        With .Cells(2, 1)
             .Value = "Decision Tree"
             .Font.Size = 16
             .Font.Italic = True
        End With
        With .Cells(3, 1)
             .Value = "Price = " & Me.txtBinomialS & _
                    ",Exercise = " & Me.txtBinomialX & _
                    ",U = " & Me.txtBinomialU & _
                    ",D = " & Me.txtBinomialD & _
                    ",N = " & Me.txtBinomialN
             .Font.Size = 14
        End With
        With .Cells(4, 1)
            .Value = "Number of calculations: " & Me.BinomialCalc
            .Font.Size = 14
        End With
```

```
    End With
End Sub
Sub BondDecisionTree(rPrice As Range, arCell As Variant, iCount As Long)
    Dim rBond As Range
    Dim rPup As Range
    Dim rPDown As Range

    Set rBond = Me.BondTree.Cells(rPrice.Row, rPrice.Column)
    Set rPup = Me.BondTree.Cells(arCell(iCount - 1).Row, arCell(iCount - 1).Column)
    Set rPDown = Me.BondTree.Cells(arCell(iCount).Row, arCell(iCount).Column)

    If rPup.Column = Me.TreeWorksheet.UsedRange.Columns.Count Then
        rPup.Value = (1 + Me.txtBinomialr) ^ (rPup.Column - 1)
        rPDown.Value = rPup.Value
    End If

    rBond.Value = (1 + Me.txtBinomialr) ^ (rBond.Column - 1)

            rPDown.Borders(xlBottom).LineStyle = xlContinuous
            With rPup
                .Borders(xlBottom).LineStyle = xlContinuous
                .Offset(1, 0).Resize((rPDown.Row - rPup.Row), 1). _
                    Borders(xlEdgeLeft).LineStyle = xlContinuous
            End With

End Sub


Sub PutDecisionTree(rPrice As Range, arCell As Variant, iCount As Long)
    Dim rCall As Range
    Dim rPup As Range
    Dim rPDown As Range

    Set rCall = Me.PutTree.Cells(rPrice.Row, rPrice.Column)
    Set rPup = Me.PutTree.Cells(arCell(iCount - 1).Row, arCell(iCount - 1).Column)
    Set rPDown = Me.PutTree.Cells(arCell(iCount).Row, arCell(iCount).Column)

    If rPup.Column = Me.TreeWorksheet.UsedRange.Columns.Count Then
        rPup.Value = WorksheetFunction.Max(Me.txtBinomialX - arCell(iCount - 1), 0)
        rPDown.Value = WorksheetFunction.Max(Me.txtBinomialX - arCell(iCount), 0)
    End If


    If Me.OptionType = "American" Then
        'American Option
        'Striket price - put option price for perion N
        rCall.Value = WorksheetFunction.Max(Me.txtBinomialX - arCell(iCount - 1) /
Me.txtBinomialU, _
                        (Me.PFactor * rPup + (1 - Me.PFactor) * rPDown) / (1 +
Me.txtBinomialr))
    Else
        'European Option
        rCall.Value = (Me.PFactor * rPup + (1 - Me.PFactor) * rPDown) / (1 +
Me.txtBinomialr)

    End If

            rPDown.Borders(xlBottom).LineStyle = xlContinuous
            With rPup
                .Borders(xlBottom).LineStyle = xlContinuous
                .Offset(1, 0).Resize((rPDown.Row - rPup.Row), 1). _
                    Borders(xlEdgeLeft).LineStyle = xlContinuous
            End With
```

```
End Sub
Sub CallDecisionTree(rPrice As Range, arCell As Variant, iCount As Long)
    Dim rCall As Range
    Dim rCup As Range
    Dim rCDown As Range

    Set rCall = Me.CallTree.Cells(rPrice.Row, rPrice.Column)
    Set rCup = Me.CallTree.Cells(arCell(iCount - 1).Row, arCell(iCount - 1).Column)
    Set rCDown = Me.CallTree.Cells(arCell(iCount).Row, arCell(iCount).Column)

    If rCup.Column = Me.TreeWorksheet.UsedRange.Columns.Count Then
        rCup.Value = WorksheetFunction.Max(arCell(iCount - 1) - Me.txtBinomialX, 0)
        rCDown.Value = WorksheetFunction.Max(arCell(iCount) - Me.txtBinomialX, 0)
    End If


    If Me.OptionType = "American" Then
            'Call option price for Period N - strike price
            rCall.Value = WorksheetFunction.Max(arCell(iCount - 1) / Me.txtBinomialU -
Me.txtBinomialX, _
                            (Me.PFactor * rCup + (1 - Me.PFactor) * rCDown) / (1 +
Me.txtBinomialr))

    Else
            'European
            rCall.Value = (Me.PFactor * rCup + (1 - Me.PFactor) * rCDown) / (1 +
Me.txtBinomialr)
    End If

            rCDown.Borders(xlBottom).LineStyle = xlContinuous
            With rCup
                .Borders(xlBottom).LineStyle = xlContinuous
                .Offset(1, 0).Resize((rCDown.Row - rCup.Row), 1). _
                    Borders(xlEdgeLeft).LineStyle = xlContinuous
            End With

End Sub

Sub DecitionTreeFormat()
    Dim rTree As Range
    Dim nColumns As Integer
    Dim rLast As Range
    Dim rCell As Range
    Dim lCount As Long
    Dim lCellSize As Long
    Dim vntColumn As Variant
    Dim iCount As Long
    Dim lTimes As Long
    Dim arCell() As Range
    Dim sFormatColumn As String
    Dim rPrice As Range


    Application.StatusBar = "Formatting Tree.. "
    Set rTree = Me.TreeWorksheet.UsedRange
    nColumns = rTree.Columns.Count


    Set rLast = rTree.Columns(nColumns).EntireColumn.SpecialCells(xlCellTypeConstants,
23)
    lCellSize = rLast.Cells.Count
    For lCount = nColumns To 2 Step -1
```

```
        sFormatColumn = rLast.Parent.Columns(lCount).EntireColumn.Address
        Application.StatusBar = "Formatting column " & sFormatColumn
        ReDim vntColumn(1 To (rLast.Cells.Count / 2), 1)

        Application.StatusBar = "Assigning values to array for column " & _
            rLast.Parent.Columns(lCount).EntireColumn.Address
        vntColumn = rLast.Offset(0, -1).EntireColumn.Cells(1).Resize(rLast.Cells.Count /
2, 1)
        rLast.Offset(0, -1).EntireColumn.ClearContents

        ReDim arCell(1 To rLast.Cells.Count)
        lTimes = 1
        Application.StatusBar = "Assigning cells to arrays. Total number of cells: " &
lCellSize
        For Each rCell In rLast.Cells
            Application.StatusBar = "Array to column " & sFormatColumn & " Cells " &
rCell.Row
            Set arCell(lTimes) = rCell
            lTimes = lTimes + 1
        Next

        lTimes = 1

        Application.StatusBar = "Formatting leaves for column " & sFormatColumn
        For iCount = 2 To lCellSize Step 2

            Application.StatusBar = "Formatting leaves for cell " & arCell(iCount).Row
            If rLast.Cells.Count <> 2 Then
                Set rPrice = arCell(iCount).Offset(-1 * ((arCell(iCount).Row -
arCell(iCount - 1).Row) / 2), -1)
                rPrice.Value = vntColumn(lTimes, 1)
            Else
                Set rPrice = arCell(iCount).Offset(-1 * ((arCell(iCount).Row -
arCell(iCount - 1).Row) / 2), -1)
                rPrice.Value = vntColumn
            End If


            arCell(iCount).Borders(xlBottom).LineStyle = xlContinuous
            With arCell(iCount - 1)
                .Borders(xlBottom).LineStyle = xlContinuous
                .Offset(1, 0).Resize((arCell(iCount).Row - arCell(iCount - 1).Row), 1). _
                    Borders(xlEdgeLeft).LineStyle = xlContinuous
            End With
            lTimes = 1 + lTimes

            CallDecisionTree rPrice:=rPrice, arCell:=arCell, iCount:=iCount
            PutDecisionTree rPrice:=rPrice, arCell:=arCell, iCount:=iCount
            BondDecisionTree rPrice:=rPrice, arCell:=arCell, iCount:=iCount
        Next

        Set rLast = rTree.Columns(lCount -
1).EntireColumn.SpecialCells(xlCellTypeConstants, 23)
        lCellSize = rLast.Cells.Count
    Next ' / outer next

    rLast.Borders(xlBottom).LineStyle = xlContinuous
    Application.StatusBar = False
End Sub

'/*********************************************************************
'/Purpse: To calculate the price value of every state of the binomial
'/       decision tree
```

```
'/***********************************************************************
Sub DecitionTree(rCell As Range, nPeriod As Integer, _
                dblPrice As Double, sngU As Single, sngD As Single)
    Dim lIteminColumn As Long


    If Not nPeriod = 1 Then
        'Do Up
        DecitionTree rCell:=rCell.Offset(0, 1), nPeriod:=nPeriod - 1, _
                dblPrice:=dblPrice * sngU, sngU:=sngU, _
                sngD:=sngD
        'Do Down
        DecitionTree rCell:=rCell.Offset(0, 1), nPeriod:=nPeriod - 1, _
                dblPrice:=dblPrice * sngD, sngU:=sngU, _
                sngD:=sngD
    End If


    lIteminColumn = WorksheetFunction.CountA(rCell.EntireColumn)


    If lIteminColumn = 0 Then
        rCell = dblPrice
    Else
        If nPeriod <> 1 Then
            rCell.EntireColumn.Cells(lIteminColumn + 1) = dblPrice
        Else
            rCell.EntireColumn.Cells(((lIteminColumn + 1) * 2) - 1) = dblPrice
        End If

    End If
    Me.BinomialCalc = Me.BinomialCalc + 1
    Application.StatusBar = "The number of binomial calcs are : " & Me.BinomialCalc
End Sub
```

Bibliography

Benninga S (2000) Financial modeling. MIT Press, Cambridge

Benninga S (2008) Financial modeling. MIT Press, Cambridge

Black F, Scholes M (1973) The pricing of options and corporate liabilities. J Polit Econ 31:637–659

Cox J, Ross SA, Rubinstein M (1979) Option pricing: a simplified approach. J Financ Econ 7:229–263

Daigler RT (1994) Financial futures and options markets concepts and strategies. Harper Collins, New York

Jarrow R, Rudd A (1983) Option pricing. Irwin, Chicago

Jarrow R, TurnBull S (1996) Derivative securities. South-Western College, Cincinnati

Kamrad B, Ritchken P (1991) Multinomial approximating models for options with k state variables. Manage Sci 37:1640–1652

Lee CF, Lee AC, Lee J (2010) Handbook of quantitative finance and risk management. Springer, New York

Lee CF, Lee AC (2013) Encyclopedia of finance, 2nd edn. Springer, New York

Lee CF, Lee JC, Lee AC (2000) Statistics for business and financial economics, 3rd edn. Springer, New York

Lee JC, Lee CF, Wang RS, Lin TI (2004) On the limit properties of binomial and multinomial option pricing models: review and integration. In: Advances in quantitative analysis of finance and accounting, vol 1. World Scientific, Singapore, pp 271–295

Lee CF, Tsai CM, Lee AC (2013) Asset pricing with disequilibrium price adjustment: theory and empirical evidence. Quant Finance 13(2):227–240

Lee JC (2001) Using Microsoft excel and decision trees to demonstrate the binomial option pricing model. Adv Invest Anal Portfolio Manage 8:303–329

Leisen DPJ, Reimer M (1996) Binomial models for option valuation—examining and improving convergence. Appl Math Finance 3(4):319–346

Lo AW, Wang J (2000) Trading volume: definition, data analysis, and implications of portfolio theory. Rev Financ Stud 13:257–300

Rendleman RJ Jr, Barter BJ (1979) Two-state option pricing. J Finance 34(5):1093–1110

Wells E, Harshbarger S (1997) Microsoft Excel 97 developer's handbook. Microsoft Press, Redmond

Walkenbach J (2003) Excel 2003 power programming with VBA. Wiley, Indianapolis

Chapter 26
Microsoft Excel Approach to Estimating Alternative Option Pricing Models

26.1 Introduction

This chapter shows how Microsoft Excel can be used to estimate call and put options for (a) Black–Scholes model for individual stock, (b) Black–Scholes model for stock indices, and (c) Black–Scholes model for currencies. In addition, we are going to present how an Excel program can be used to estimate American Options. Section 26.2 presents an option pricing model for Individual Stocks, Sect. 26.3 presents an option pricing model for Stock Indices, Sect. 26.4 presents option pricing model for Currencies, Sect. 26.5 presents Bivariate Normal Distribution Approach to calculate American Call Options, Sect. 26.6 presents the Black's approximation method to calculate American Call Options, Sect. 26.7 presents how to evaluate American Call option when dividend yield is known, and Sect. 26.8 summarizes this chapter. Appendix 26.1 defines the Bivariate Normal probability density function, and Appendix 26.2 presents the Excel program to calculate the American call option when dividend payments are known.

26.2 Option Pricing Model for Individual Stock

The call option formula for an individual stock can be defined as

$$C = SN(d_1) - Xe^{-r(T)}N(d_2) \tag{26.1}$$

where

$$d_1 = \frac{\ln\left(\frac{S}{X}\right) + \left(r + \frac{1}{2}\sigma^2\right)T}{\sigma\sqrt{T}}$$

$$d_2 = \frac{\ln\left(\frac{S}{X}\right) + \left(r - \frac{1}{2}\sigma^2\right)T}{\sigma\sqrt{T}} = d_1 - \sigma\sqrt{T}$$

This chapter was written by Professor Cheng F. Lee and Dr. Ta-Peng Wu of Rutgers University.

C.-F. Lee et al., *Essentials of Excel, Excel VBA, SAS and Minitab for Statistical and Financial Analyses*, DOI 10.1007/978-3-319-38867-0_26

C = price of the call option
S = current price of the stock
X = exercise price of the option
$e = 2.71828\ldots$
r = short-term interest rate (T-Bill rate) = R_f
T = time to expiration of the option, in years
$N(d_i)$ = value of the cumulative standard normal distribution ($i = 1,2$)
σ^2 = variance of the stock rate of return

The put option formula can be defined as

$$P = Xe^{-r(T)}N(-d_2) - SN(-d_1) \tag{26.2}$$

where

P = price of the put option.

The other notations have been defined in Eq. (26.1)

Assume S = 42, X = 40, r = 0.1, σ = 0.2, and T = 0.5. The following shows how to set up Microsoft Excel to solve the problem (Fig. 26.1).

The following shows the answer to the problem in Microsoft Excel (Fig. 26.2).

From the Excel output, we find that the price of a call option and a put option is \$4.76 and \$0.81 respectively

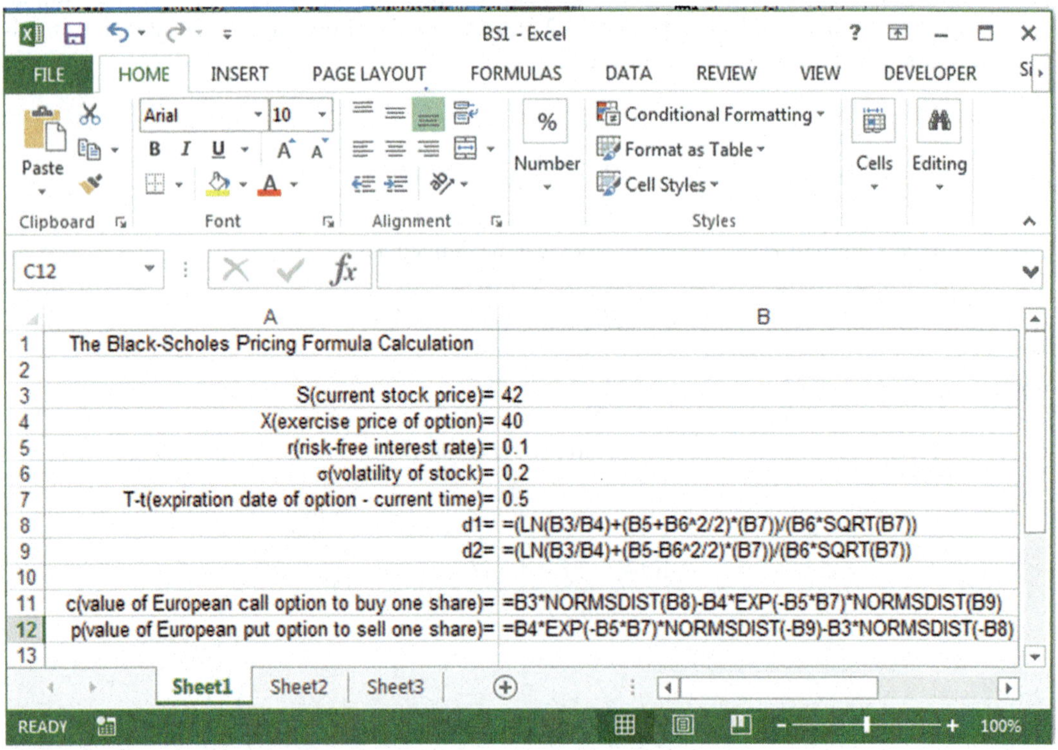

Fig. 26.1 The inputs and Excel functions of European call and put options

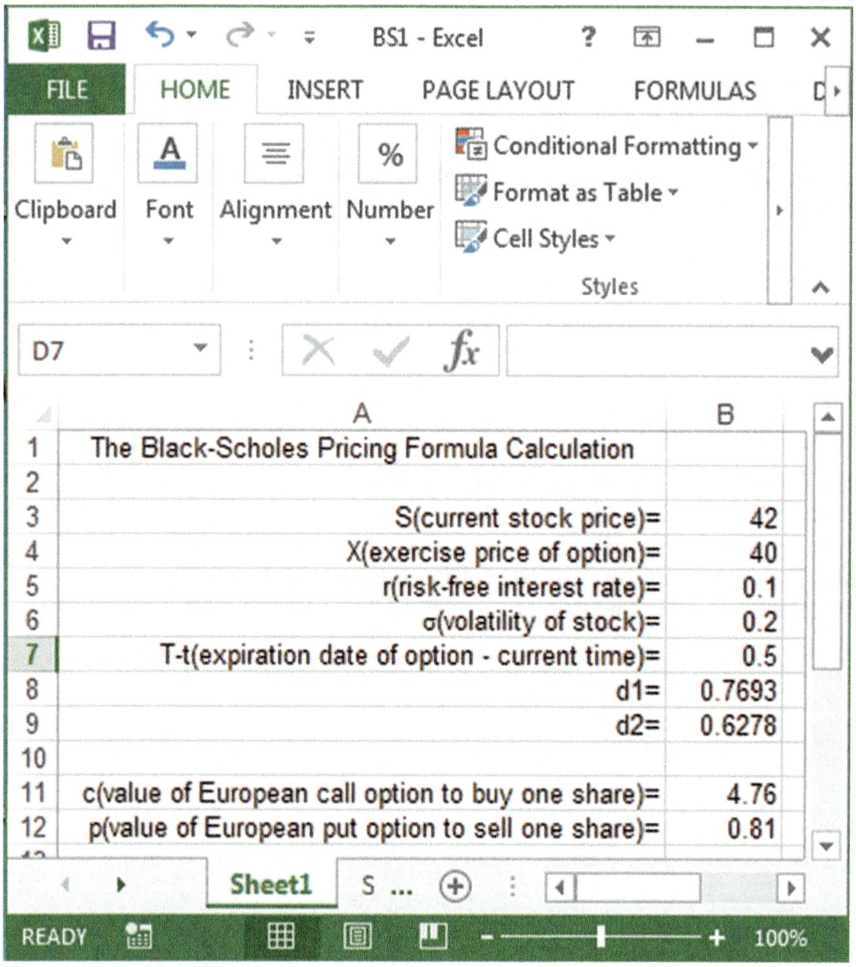

Fig. 26.2 Results for functions contained in Fig. 26.1

26.3 Option Pricing Model for Stock Indices

The call option formula for a stock index can be defined as

$$C = Se^{-q(T)}N(d_1) - Xe^{-r(T)}N(d_2) \tag{26.3}$$

where

$$d_1 = \frac{\ln(S/X) + \left(r - q + \frac{\sigma^2}{2}\right)(T)}{\sigma\sqrt{T}}$$

$$d_2 = \frac{\ln(S/X) + \left(r - q - \frac{\sigma^2}{2}\right)(T)}{\sigma\sqrt{T}} = d_1 - \sigma\sqrt{T}$$

q = dividend yield
S = value of index
X = exercise price
r = short-term interest rate (T-Bill rate) = R_f
T = time to expiration of the option, in years
$N(d_i)$ = value of the cumulative standard normal distribution (i = 1,2)
σ^2 = variance of the stock rate of return

The put option formula for a stock index can be defined as

$$P = Xe^{-r(T)}N(-d_2) - Se^{-q(T)}N(-d_1) \tag{26.4}$$

where

P = the price of the put option.

The other notations have been defined in Eq. (26.3).

Assume that S = 950, X = 900, r = 0.06, σ = 0.15, q = 0.03, and T = 2/12. The following shows how to set up Microsoft Excel to solve the problem (Fig. 26.3).

The following shows the answer to the problem in Microsoft Excel (Fig. 26.4).

From the Excel output, we find that the price of a call option and a put option is $59.26 and $5.01 respectively.

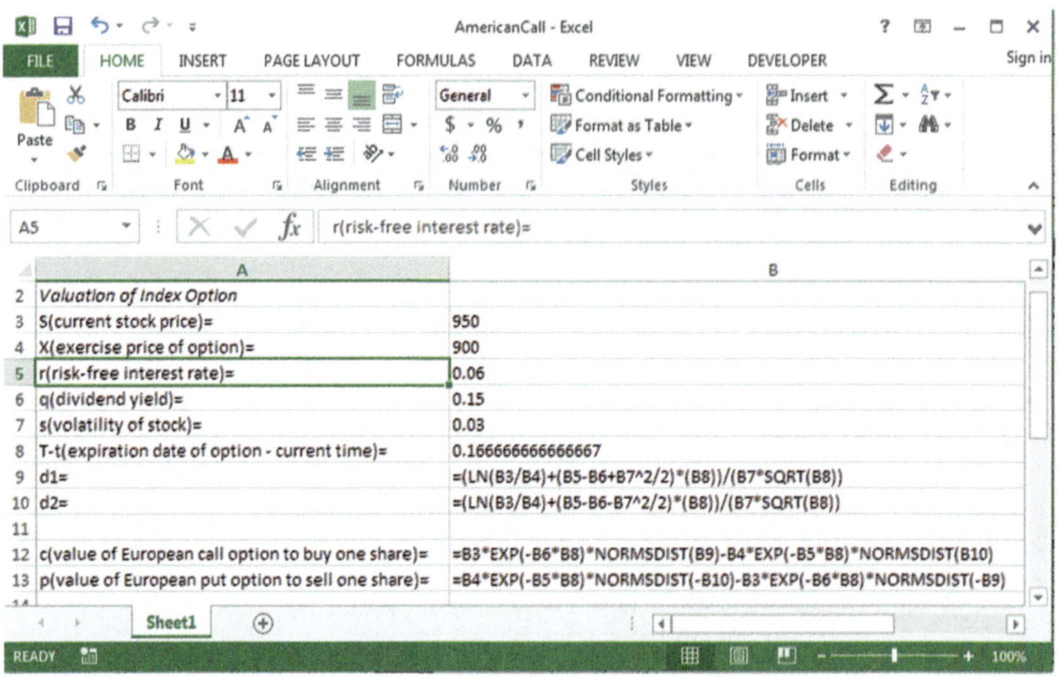

Fig. 26.3 The inputs and Excel functions of European call and put options

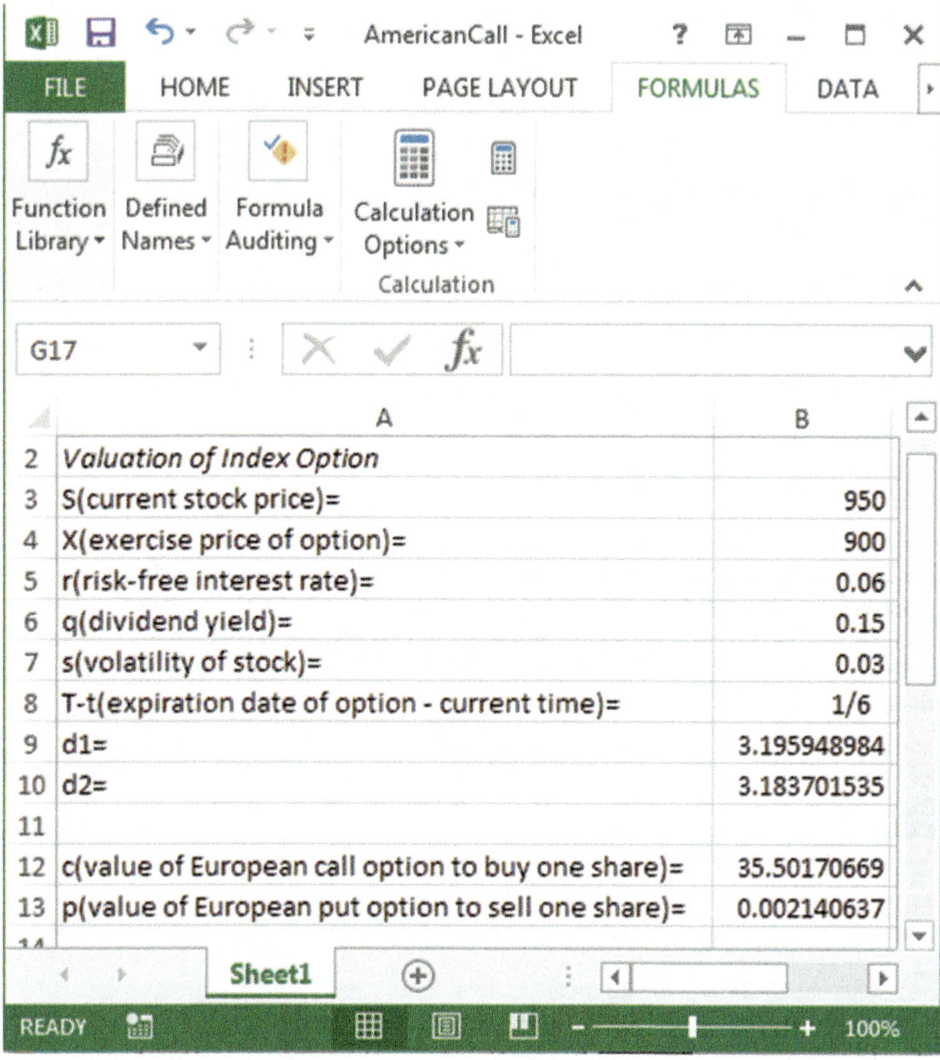

	A	B
2	*Valuation of Index Option*	
3	S(current stock price)=	950
4	X(exercise price of option)=	900
5	r(risk-free interest rate)=	0.06
6	q(dividend yield)=	0.15
7	s(volatility of stock)=	0.03
8	T-t(expiration date of option - current time)=	1/6
9	d1=	3.195948984
10	d2=	3.183701535
11		
12	c(value of European call option to buy one share)=	35.50170669
13	p(value of European put option to sell one share)=	0.002140637

Fig. 26.4 Results for functions contained in Fig. 26.3

26.4 Option Pricing Model for Currencies

The call option formula for a currency can be defined as

$$C = Se^{-r_f(T)}N(d_1) - Xe^{-r(T)}N(d_2) \tag{26.5}$$

where

$$d_1 = \frac{\ln(S/X) + \left(r - r_f + \frac{\sigma^2}{2}\right)(T)}{\sigma\sqrt{T}}$$

$$d_2 = \frac{\ln(S/X) + \left(r - r_f - \frac{\sigma^2}{2}\right)(T)}{\sigma\sqrt{T}} = d_1 - \sigma\sqrt{T}$$

S = spot exchange rate
r = risk free rate for domestic country
X = exercise price
T = time to expiration of the option, in years
$N(d_i)$ = value of the cumulative standard normal distribution (i = 1,2)
σ = standard deviation of spot rate

The put option formula for a currency can be defined as

$$P = Xe^{-r(T)}N(-d_2) - Se^{-r_f(T)}N(-d_1) \tag{26.6}$$

where

P = the price of the put option.

The other notations have been defined in Eq. (26.5).

Assume that S = 130, X = 125, r = 0.06, r_f = 0.02, σ = 0.15, and T = 4/12. The following shows how to set up Microsoft Excel to solve the problem (Fig. 26.5).

The following shows the answer to the problem in Microsoft Excel (Fig. 26.6).

From the Excel output, we find that the price of a call option and a put option is \$8.43 and \$1.82 respectively.

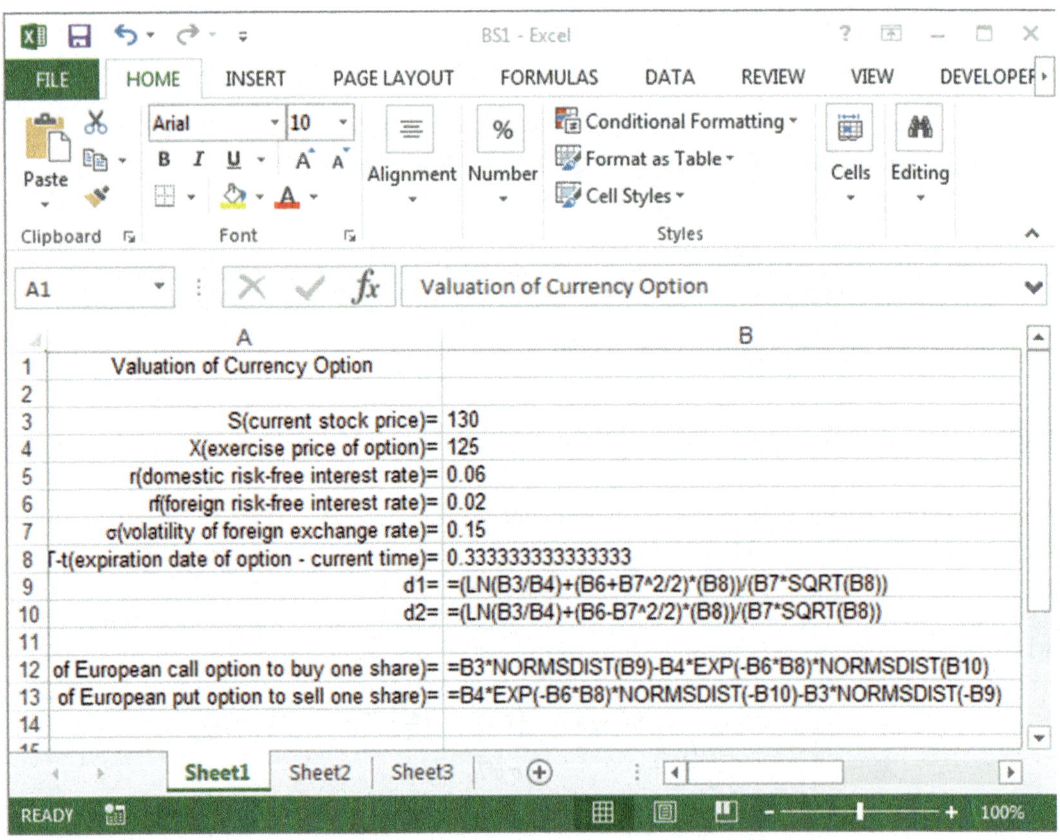

Fig. 26.5 The inputs and Excel functions of European call and put options

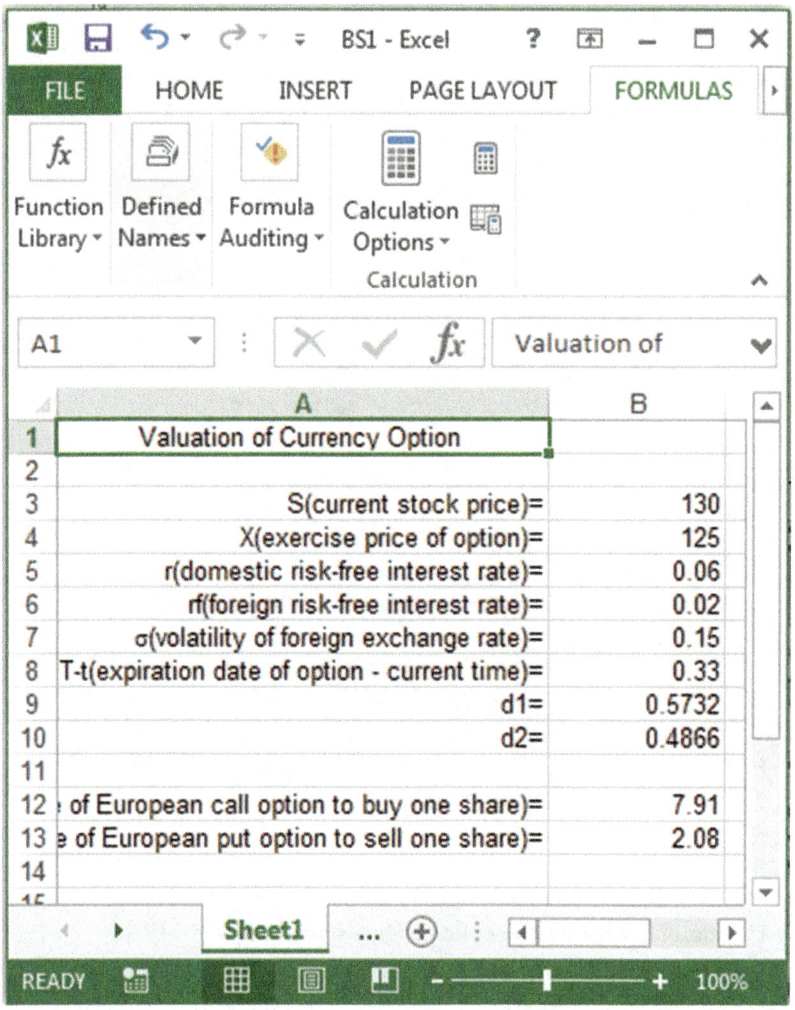

Fig. 26.6 Results for functions contained in Fig. 26.5

26.5 Future Option

Black (1976) showed that the original call-option formula for stocks can be easily modified to be used in pricing call options on futures. The formula is

$$C(T, F, \sigma^2, X, r) = e^{-rT}[FN(d_1) - XN(d_2)], \tag{26.7}$$

$$d_1 = \frac{\ln(F/X) + \frac{1}{2}\sigma^2 T}{\sigma\sqrt{T},} \tag{26.8}$$

$$d_2 = \frac{\ln(F/X) - \frac{1}{2}\sigma^2 T}{\sigma\sqrt{T}.} \tag{26.9}$$

In Eq. (26.7), F now denotes the current futures price. The other four variables are as before—time to maturity, volatility of the underlying futures price, exercise price, and risk-free rate. Note that

Eq. (26.7) differs from Eq. (26.1) only in one respect: by substituting $e^{-rT}F$ for S in the original Eq. (26.1), Eq. (26.7) is obtained. This holds because the investment in a futures contract is zero, which causes the interest rate in Eqs. (26.8) and (26.9) to drop out. The following excel results are obtained by substituting F = 42, X = 40, r = 0.1, σ = 0.2, T-t = 0.5, d1 = 0.4157, and d2 = 0.2743 into Eqs. (26.7)–(26.9).

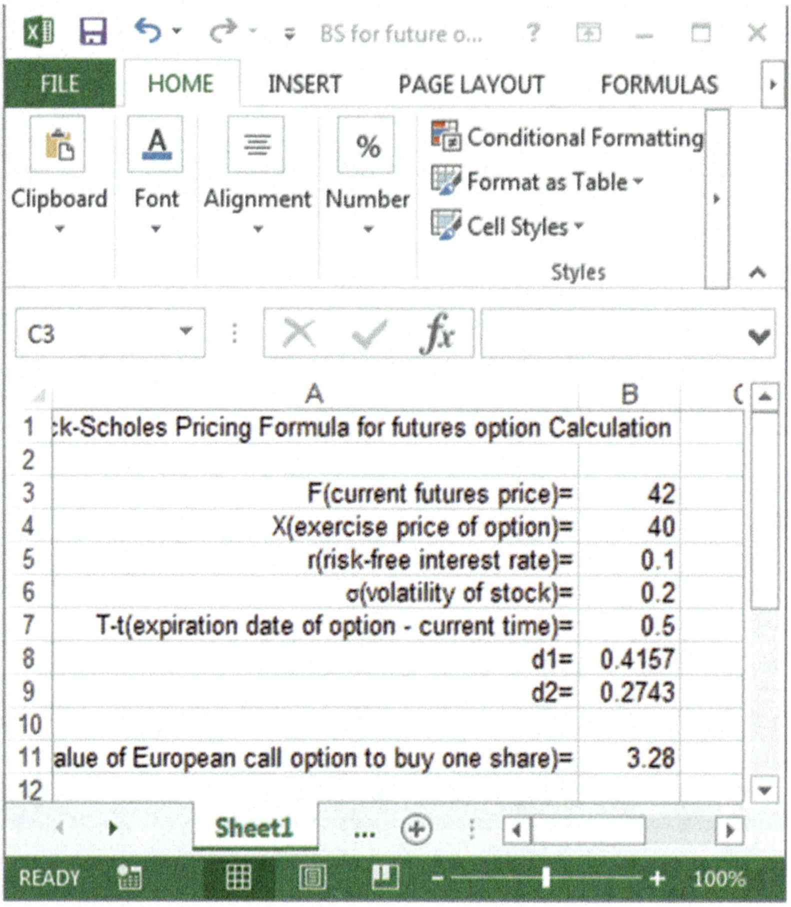

26.6 SAS Programming Code Instructions and Examples

In this section, we write SAS macro function code to price options for individual stock, stock indices, and currencies. We use same examples in previous sections to show the results in SAS.

26.6.1 Option Model for Individual Stock

```
%macro Individual_Stock(S, X, r, sigma, T);


  data Individual;
        d1=(log(&S./&X.)+(&r.+1/2*&sigma.**2)*&T.)/(&sigma.*sqrt(&T.));
           d2=(log(&S./&X.)+(&r.-1/2*&sigma.**2)*&T.)/(&sigma.*sqrt(&T.));
           C=&S.*CDF('NORMAL',d1)-&X.*exp(-&r.*&T.)*CDF('NORMAL',d2);
           P=&X.*exp(-&r.*&T.)*CDF('NORMAL',-d2)-&S.*CDF('NORMAL',-d1);
  run;

  proc print data=Individual label;
      var d1 d2 C P;
      label C="C(the European call option price)"
             P="P(the European put option price)";
  run;
  %mend;

  %Individual_Stock(42,40,0.1,0.2,0.5);
```

Obs	d1	d2	C(the European call option price)	P(the European put option price)
1	0.76926	0.62784	4.75942	0.80860

26.6.2 Option Model for Stock Indices

```
%macro Stock_Indices(S, X, r, sigma, q, T);
data Indices;
     d1=(log(&S./&X.)+(&r.-&q.+1/2*&sigma.**2)*&T.)/(&sigma.*sqrt(&T.));
        d2=d1-&sigma.*sqrt(&T.);
        C=&S.*exp(-&q.*&T.)*CDF('NORMAL',d1)-&X.*exp(-&r.*&T.)*CDF('NORMAL',d2);
        P=&X.*exp(-&r.*&T.)*CDF('NORMAL',-d2)-&S.*exp(-&q.*&T.)*CDF('NORMAL',-d1);
run;
proc print data=Indices label;
    var d1 d2 C P;
    label C="C(the European call option price)"
           P="P(the European put option price)";
run;
%mend;

%Stock_Indices(950,900,0.06,0.15,0.03,1/6);
```

Obs	d1	d2	C(the European call option price)	P(the European put option price)
1	0.99518	0.93395	59.2225	5.00552

26.6.3 Option for Currencies

```
%macro Currencies(S, X, r, rf, sigma, T);
data Individual;
     d1=(log(&S./&X.)+(&r.-&rf.+1/2*&sigma.**2)*&T.)/(&sigma.*sqrt(&T.));
        d2=d1-&sigma.*sqrt(&T.);
        C=&S.*exp(-&rf.*&T.)*CDF('NORMAL',d1)-&X.*exp(-&r.*&T.)*CDF('NORMAL',d2);
        P=&X.*exp(-&r.*&T.)*CDF('NORMAL',-d2)-&S.*exp(-&rf.*&T.)*CDF('NORMAL',-d1);
run;
proc print data=Individual label;
    var d1 d2 C P;
    label C="C(the European call option price)"
           P="P(the European put option price)";
run;

%mend;

%Currencies(130,125,0.06,0.02,0.15,4/12);
```

Obs	d1	d2	C(the European call option price)	P(the European put option price)
1	0.65014	0.56354	8.42754	1.81616

26.7 Using Bivariate Normal Distribution Approach to Calculate American Call Options

Following Chap. 19 of Lee et al. (2013), the call option formula for American options for a stock that pays a dividend, and there is at least one known dividend, can be defined as

$$\begin{aligned} C(S,T,X) = {} & S^x\left[N_1(b_1) + N_2\left(a_1,\, -b_1; -\sqrt{t/T}\right)\right] \\ & - Xe^{-rt}\left[N_1(b_2)e^{r(T-t)} + N_2\left(a_2,\, -b_2; -\sqrt{t/T}\right)\right] \\ & + De^{-rt}N_1(b_2), \end{aligned} \tag{26.10}$$

where

$$a1 = \frac{\ln\left(\frac{S^x}{X}\right) + \left(r + \frac{1}{2}\sigma^2\right)T}{\sigma\sqrt{T}}, a2 = a1 - \sigma\sqrt{T}, \tag{26.11}$$

$$b1 = \frac{\ln\left(\frac{S^x}{S_t^*}\right) + \left(r + \frac{1}{2}\sigma^2\right)t}{\sigma\sqrt{t}}, b2 = b1 - \sigma\sqrt{t} \tag{26.12}$$

$$S^x = S - De^{-rT} \tag{26.13}$$

S^x represents the corrected stock net price of the present value of the promised dividend per share (D); t represents the time dividend to be paid.

S_t^* is the ex-dividend stock price for which

$$C\left(S_t^*, T-t\right) = S_t^* + D - X$$

Both $N_1(b_1)$ and $N_2(b_2)$ represent the cumulative univariate normal density function. $N_2(a,b;\rho)$ is the cumulative bivariate normal density function with upper integral limits a and b and correlation coefficient $\rho = -\sqrt{t/T}$

If we want to calculate the call option value of the American Option, we need first to calculate a_1 and b_1. For calculating a_1 and b_1, we need to first calculate S^x and S_t^*. The calculation of S^x can be found in Eq. (26.9). The calculation will be explained in the following example from Chap. 19 of Lee et al. (2013).

An American call option whose exercise price is $48 has an expiration time of 90 days. Assume the risk-free rate of interest is 8 % annually, the underlying price is $50, the standard deviation of the rate of return of the stock is 20 %, and the stock pays a dividend of $2 exactly for 50 days. (a) What is the European call value? (b) Can the early exercise price predicted? (c) What is the value of the American call?

(a) The current stock net price of the present value of the promised dividend is

$$S^x = 50 - 2e^{-0.08\left(50/365\right)} = 48.0218.$$

The European call value can be calculated as

$$C = (48.0218)N(d_1) - 48e^{-0.08\left(90/365\right)}N(d_2),$$

where

$$d_1 = \frac{\left[\ln(48.208/48) + \left(0.08 + 0.5(0.20)^2\right)(90/365)\right]}{.20\sqrt{90/365}} = 0.25285$$

$$d_2 = 0.292 - 0.0993 = 0.15354.$$

From the standard normal table, we obtain

$$N(0.25285) = 0.5 + .3438 = 0.599809$$

$$N(0.15354) = 0.5 + .3186 = 0.561014.$$

So the European call value is
$C = (48.516)(0.599809) - 48(0.980)(0.561014) = 2.40123.$

(b) The present value of the interest income that would be earned by deferring exercise until expiration is

$$X\left(1 - e^{-r(T-t)}\right) = 48\left(1 - e^{-0.08(90-50)/365}\right) = 48(1 - 0.991) = 0.432.$$

Since $d = 2 > 0.432$, therefore, the early exercise is not precluded.

(c) The value of the American call is now calculated as

$$\begin{aligned} C = 48.208&\left[N_1(b_1) + N_2\left(a_1, \ -b_1; \sqrt{50/90}\right)\right] \\ &- 48e^{-0.08(90/365)}\left[N_1(b_2)e^{-0.08(40/365)} + N_2\left(a_2, \ -b_2; -\sqrt{50/90}\right)\right] \\ &+ 2e^{-0.08(50/365)}N_1(b_2) \end{aligned} \tag{26.14}$$

since both b_1 and b_2 depend on the critical ex-dividend stock price S_t^*, which can be determined by

$$C(S_t^*, 40/365; 48) = S_t^* + 2 - 48.$$

By using trial and error, we find that $S_t^* = 46.9641$. An Excel program used to calculate this value is presented in Fig. 26.7.

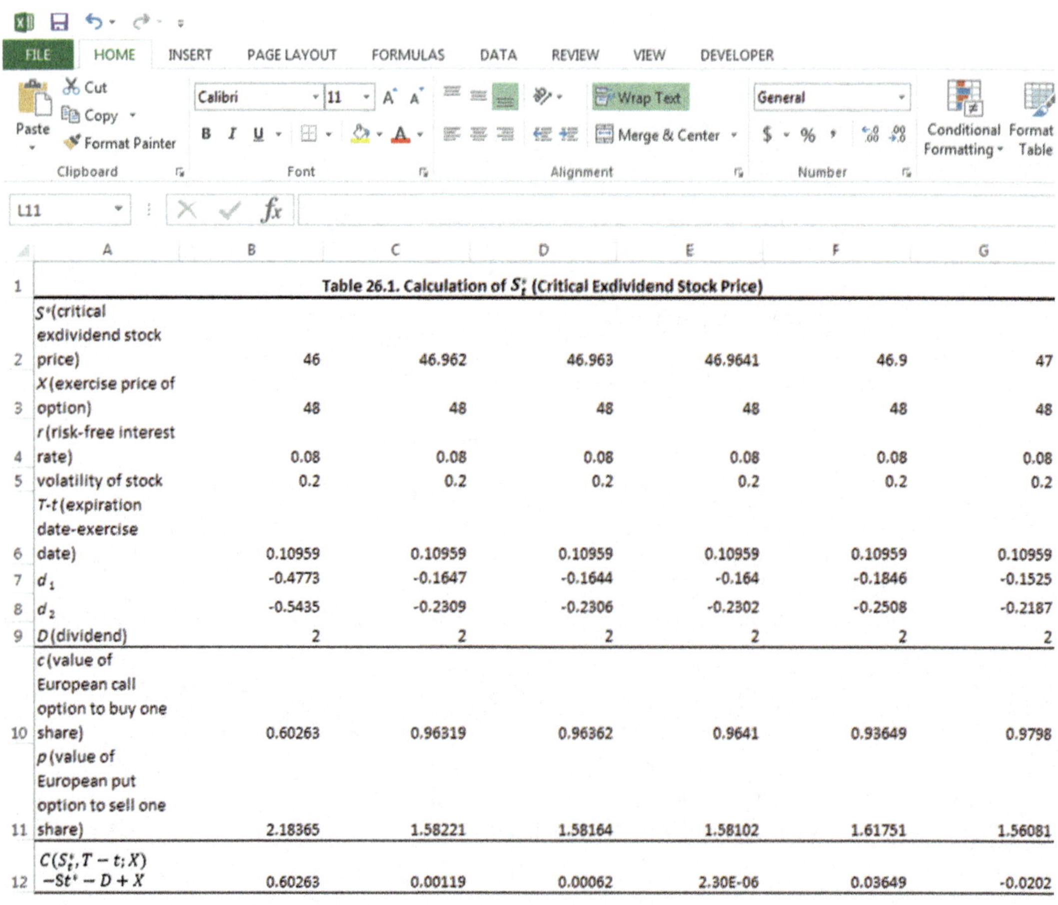

	A	B	C	D	E	F	G
1	Table 26.1. Calculation of S_t^* (Critical Exdividend Stock Price)						
2	S*(critical exdividend stock price)	46	46.962	46.963	46.9641	46.9	47
3	X(exercise price of option)	48	48	48	48	48	48
4	r(risk-free interest rate)	0.08	0.08	0.08	0.08	0.08	0.08
5	volatility of stock	0.2	0.2	0.2	0.2	0.2	0.2
6	T-t(expiration date-exercise date)	0.10959	0.10959	0.10959	0.10959	0.10959	0.10959
7	d_1	-0.4773	-0.1647	-0.1644	-0.164	-0.1846	-0.1525
8	d_2	-0.5435	-0.2309	-0.2306	-0.2302	-0.2508	-0.2187
9	D(dividend)	2	2	2	2	2	2
10	c(value of European call option to buy one share)	0.60263	0.96319	0.96362	0.9641	0.93649	0.9798
11	p(value of European put option to sell one share)	2.18365	1.58221	1.58164	1.58102	1.61751	1.56081
12	$C(S_t^*, T-t; X) - St^* - D + X$	0.60263	0.00119	0.00062	2.30E-06	0.03649	-0.0202

Fig. 26.7 Calculation of S_t^* (critical ex-dividend stock price)

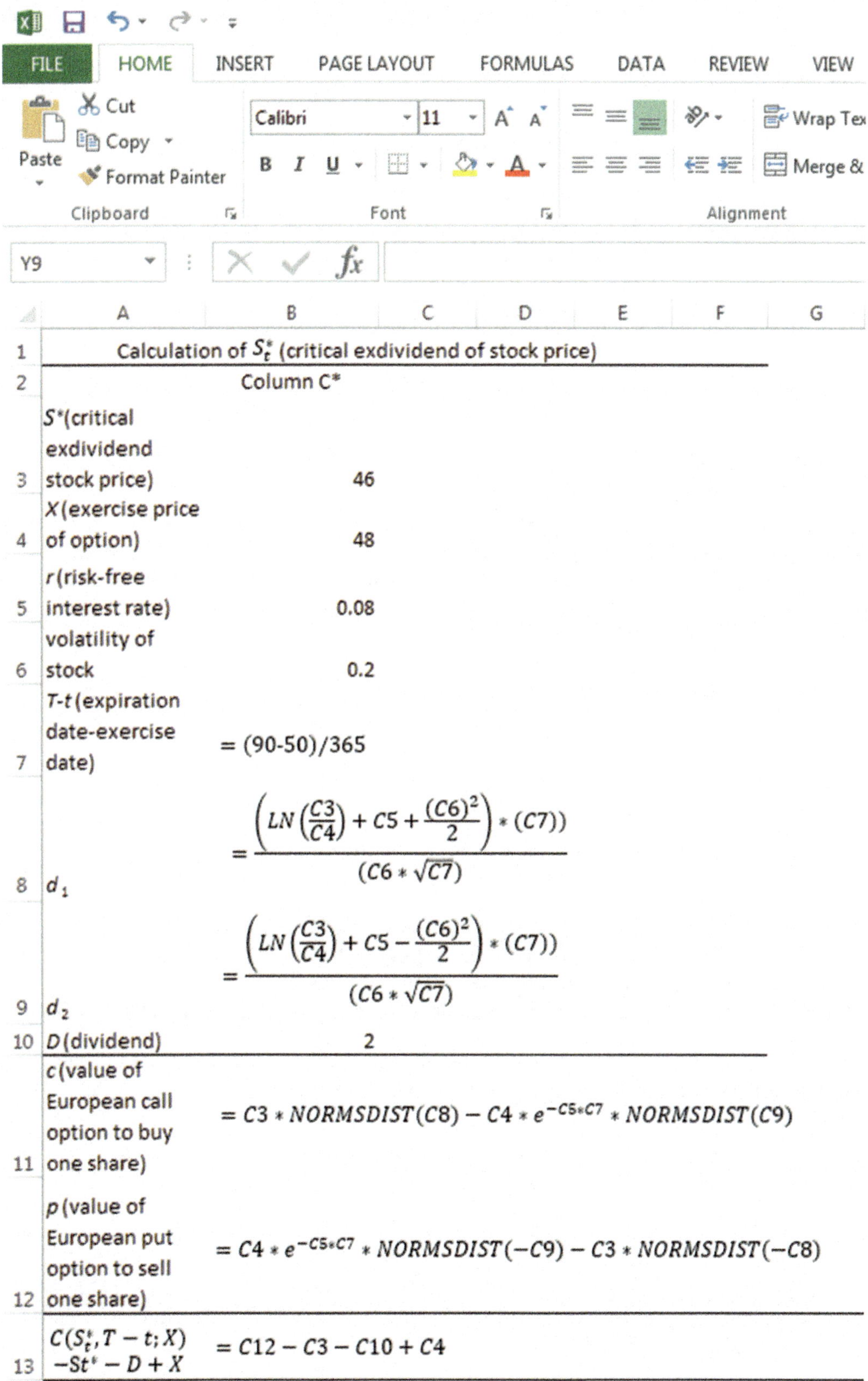

	A	B	C
1	Calculation of S_t^* (critical exdividend of stock price)		
2		Column C*	
3	S^*(critical exdividend stock price)		46
4	X(exercise price of option)		48
5	r(risk-free interest rate)		0.08
6	volatility of stock		0.2
7	T-t(expiration date-exercise date)	= (90-50)/365	
8	d_1	$=\dfrac{\left(LN\left(\frac{C3}{C4}\right)+C5+\frac{(C6)^2}{2}\right)*(C7))}{(C6*\sqrt{C7})}$	
9	d_2	$=\dfrac{\left(LN\left(\frac{C3}{C4}\right)+C5-\frac{(C6)^2}{2}\right)*(C7))}{(C6*\sqrt{C7})}$	
10	D(dividend)		2
11	c(value of European call option to buy one share)	$= C3*NORMSDIST(C8)-C4*e^{-C5*C7}*NORMSDIST(C9)$	
12	p(value of European put option to sell one share)	$= C4*e^{-C5*C7}*NORMSDIST(-C9)-C3*NORMSDIST(-C8)$	
13	$C(S_t^*, T-t; X)$ $-St^*-D+X$	$= C12-C3-C10+C4$	

Fig. 26.7 continued

Substituting $S^x = 48.208$, $X = \$48$ and S_t^* into Eqs. (26.8) and (26.9), we can calculate a_1, a_2, b_1, and b_2:

$$a_1 = d_1 = 0.25285.$$

$$a_2 = d_2 = 0.15354.$$

$$b_1 = \frac{\ln\left(\frac{48.208}{46.9641}\right) + \left(0.08 + \frac{0.2^2}{2}\right)\left(\frac{50}{365}\right)}{(.20)\sqrt{50/365}} = 0.4859.$$

$$b_2 = 0.485931 - 0.074023 = 0.4119.$$

In addition, we also know $\rho = -\sqrt{50/90} = -\ 0.7454$.

From the above information, we now calculate related normal probability as follows:

$N_1(b_1) = N_1(0.4859) = 0.6865$
$N_1(b_2) = N_1(0.7454) = 0.6598$

Following Eq. (26.13), we now calculate the value of $N_2(0.25285, -0.4859; -0.7454)$ and $N_2\ (0.15354, -0.4119; -0.7454)$ as follows:

Since $ab\rho > 0$ for both cumulative bivariate normal density function, therefore, we can use Equation $N_2(a,\ b;\rho) = N_2(\mathrm{a},0;\rho_{ab}) + N_2(b,\ 0;\rho_{\mathrm{ba}}) - \delta$ to calculate the value of both $N_2(a,b;\rho)$ as follows:

$$\rho_{ab} = \frac{\left[(-0.7454)(0.25285) + 0.4859\right](1)}{\sqrt{(0.25285)^2 - 2(0.7454)(0.25285)(-0.4859) + (0.4859)^2}} = 0.87002$$

$$\rho_{ba} = \frac{\left[(-0.7454)(-0.4859) - 0.25285\right](1)}{\sqrt{(0.25285)^2 - 2(0.7454)(0.25285)(-0.4859) + (0.4859)^2}} = -0.31979$$

$$\delta = (1 - (1)(-1))/4 = 1/2$$

$N_2(0.292,\ -0.4859;\ -0.7454) = N_2(0.292,\ 0.0844) + N_2(-0.5377,\ 0.0656) - 0.5 = N_1(0) + N_1(-0.5377) - \Phi(-0.292, 0; -0.0844) - \Phi(-0.5377, 0; -0.0656) - 0.5 = 0.07525$

Using a Microsoft Excel programs presented in Appendix 19A of Lee et al. (2013), we obtain $N_2(0.1927, -0.4119; -0.7454) = 0.06862$.

Then substituting the related information into the Eq. (26.11), we obtain $C = \$3.08238$ and all related results are presented in Appendix 26.2.

The following is the VBA code necessary for Microsoft excel to run the Bivariate Normal Distribution approach to calculate an American Call Option:

```
Option Explicit

Public Function Bivarncdf(a As Double, b As Double, rho As Double) As Double

    Dim rho_ab As Double, rho_ba As Double
    Dim delta As Double

    If (a * b * rho) <= 0 Then

        If (a <= 0 And b <= 0 And rho <= 0) Then
            Bivarncdf = Phi(a, b, rho)
        End If

        If (a <= 0 And b >= 0 And rho > 0) Then
            Bivarncdf = Application.WorksheetFunction.NormSDist(a) - Phi(a, -b, -rho)
        End If

        If (a >= 0 And b <= 0 And rho > 0) Then
            Bivarncdf = Application.WorksheetFunction.NormSDist(b) - Phi(-a, b, -rho)
        End If

        If (a >= 0 And b >= 0 And rho <= 0) Then
            Bivarncdf = Application.WorksheetFunction.NormSDist(a) + Application.WorksheetFunction.NormSDist(b) - 1 + Phi(-a, -b, rho)
        End If

    Else

        rho_ab = ((rho * a - b) * IIf(a >= 0, 1, -1)) / Sqr(a ^ 2 - 2 * rho * a * b + b ^ 2)
        rho_ba = ((rho * b - a) * IIf(b >= 0, 1, -1)) / Sqr(a ^ 2 - 2 * rho * a * b + b ^ 2)
        delta = (1 - IIf(a >= 0, 1, -1) * IIf(b >= 0, 1, -1)) / 4
        Bivarncdf = Bivarncdf(a, 0, rho_ab) + Bivarncdf(b, 0, rho_ba) - delta

    End If
End Function
```

```
Public Function Phi(a As Double, b As Double, rho As Double) As Double

    Dim a1 As Double, b1 As Double
    Dim w(5) As Double, x(5) As Double
    Dim i As Integer, j As Integer
    Dim doublesum As Double

    a1 = a / Sqr(2 * (1 - rho ^ 2))
    b1 = b / Sqr(2 * (1 - rho ^ 2))

    w(1) = 0.24840615
    w(2) = 0.39233107
    w(3) = 0.21141819
    w(4) = 0.03324666
    w(5) = 0.00082485334

    x(1) = 0.10024215
    x(2) = 0.48281397
    x(3) = 1.0609498
    x(4) = 1.7797294
    x(5) = 2.6697604

    doublesum = 0

    For i = 1 To 5
        For j = 1 To 5
            doublesum = doublesum + w(i) * w(j) * Exp(a1 * (2 * x(i) - a1) + b1 * (2 * x(j) - b1) + 2 * rho * (x(i) - a1) * (x(j) - b1))
            Next j
        Next i

    Phi = 0.31830989 * Sqr(1 - rho ^ 2) * doublesum

End Function
```

26.8 Black's Approximation Method for American Option With One Dividend Payment

By using the same data as the Bivariate Normal Distribution (from Sect. 26.4), we will show how Black's Approximation method can be used to calculate the value of an American option. The first step is to calculate the stock price minus the current value of the dividend and then calculate d_1 and d_2 to calculate the call price at time T (the time of maturity).

$$2e^{-0.13699(0.08)} + 2e^{-0.24658(0.08)} = 0 = 1.9782$$

- The option price can therefore be calculated from the Black–Scholes formula with $S_0 = 48.0218$, $K = 48$, $r = 0.08$, $\sigma = 0.2$, and $T = 0.24658$. We have

$$\mathrm{d}_1 = \frac{\ln\left(\frac{48.0218}{48}\right) + \left(0.08 + \frac{0.2^2}{2}\right)(0.24658)}{0.8\sqrt{0.24658}} = 0.2529$$

$$d_1 = \frac{\ln\left(\frac{48.0218}{48}\right) + \left(0.08 - \frac{0.2^2}{2}\right)(0.24658)}{0.8\sqrt{0.24658}} = 0.1535$$

- We can get from the normal table

$$\mathrm{N}(d_1) = 0.5998, \ N(d_2) = 0.5610$$

- And the call price is

$$48.0218(0.5998) - 48e^{-0.08(0.24658)}(0.5610) = \$2.40$$

You then calculate the call price at time t (the time of the dividend payment) using the current stock price.

$$\mathrm{d}_1 = \frac{\ln\left(\frac{50}{48}\right) + \left(0.08 + \frac{0.2^2}{2}\right)(0.13699)}{0.8\sqrt{0.13699}} = 0.7365$$

$$d_1 = \frac{\ln\left(\frac{50}{48}\right) + \left(0.08 - \frac{0.2^2}{2}\right)(0.13699)}{0.8\sqrt{0.13699}} = 0.6625$$

- We can get from the normal table

$$\mathrm{N}(d_1) = 0.7693, \ N(d_2) = 0.7462$$

- And the call price is

$$50(0.7693) - 48e^{-0.08(0.24658)}(0.7462) = \$3.04$$

Comparing the greater of the two call option values will show if it is worth waiting until the time to maturity or exercising at the dividend payment.

$$\$3.04 > \$2.40$$

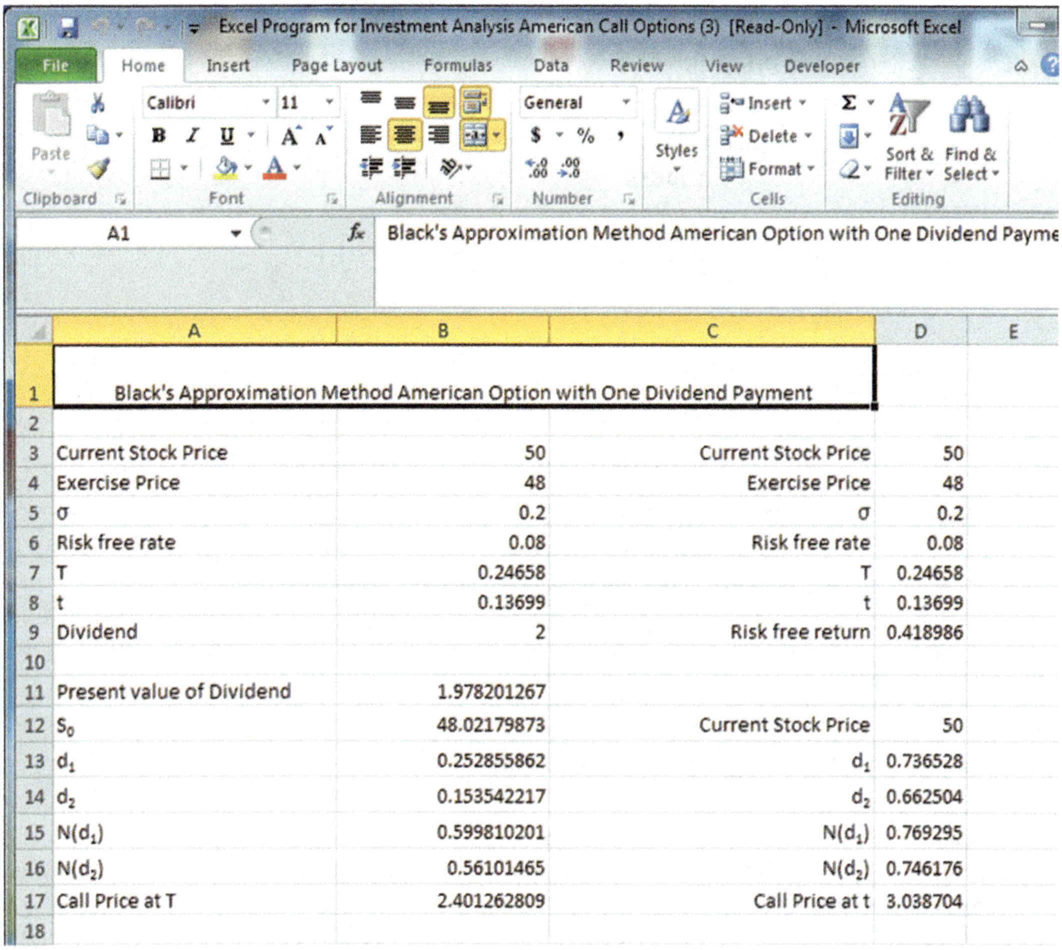

	A	B	C	D
1	Black's Approximation Method American Option with One Dividend Payment			
2				
3	Current Stock Price	50	Current Stock Price	50
4	Exercise Price	48	Exercise Price	48
5	σ	0.2	σ	0.2
6	Risk free rate	0.08	Risk free rate	0.08
7	T	0.24658	T	0.24658
8	t	0.13699	t	0.13699
9	Dividend	2	Risk free return	0.418986
10				
11	Present value of Dividend	1.978201267		
12	S_0	48.02179873	Current Stock Price	50
13	d_1	0.252855862	d_1	0.736528
14	d_2	0.153542217	d_2	0.662504
15	$N(d_1)$	0.599810201	$N(d_1)$	0.769295
16	$N(d_2)$	0.56101465	$N(d_2)$	0.746176
17	Call Price at T	2.401262809	Call Price at t	3.038704
18				

Current Stock Price	50
Exercise Price	48
σ	0.2
Risk free rate	0.08
T	0.24658
t	0.13699
Dividend	2
Present value of Dividend	=B9*EXP(-B8*B6)
S_0	=B3-B11
d_1	=(LN(B12/B4)+(B6+((B5^2)/2))*(B7))/(B5*SQRT(B7))
d_2	=(LN(B12/B4)+(B6-((B5^2)/2))*(B7))/(B5*SQRT(B7))
$N(d_1)$	=NORMSDIST(B13)
$N(d_2)$	=NORMSDIST(B14)
Call Price at T	=B12*(B15)-(B4*(EXP(-B6*B7))*B16)

Current Stock Price	50
Exercise Price	48
σ	0.2
Risk free rate	0.08
T	0.24658
t	0.13699
Risk free return	=$B$4*(1-EXP(-$B$6*($B$7-$B$8)))
Current Stock Price	=B3
d_1	=(LN(D12/D4)+(D6+((D5^2)/2))*(D8))/(D5*SQRT(D8))
d_2	=(LN(D12/D4)+(D6-((D5^2)/2))*(D8))/(D5*SQRT(D8))
$N(d_1)$	=NORMSDIST(D13)
$N(d_2)$	=NORMSDIST(D14)
Call Price at t	=D12*(D15)-(D4*(EXP(-D6*D8))*D16)

26.9 American Call Option When Dividend Yield Is Known

In Sections 26.5 and 26.6, we discuss American option valuation procedure when the dividend payment amounts are known. In this section, we discuss the American option valuation when dividend yield instead of dividend payment is known.

Following Technical Note No. 8* named "Options, Futures, and Other Derivatives, Ninth Edition" by John Hull, we use the following procedures to calculate the American call options value. Hull' method is derived from Barone-Adesi and Whaley (1987). In our words, Hull replaces Barone-Adesi and Whaley's commodity option model in terms of stock option model. They use a quadratic approximation to get an analytic approximation for American option.

26.9.1 Theory and Method

Consider an option written on a stock providing a dividend yield equal to q. The European call prices at time t will be denoted by c(S, t), where S is the stock price, and the corresponding American call will be denoted by C(S, t). The relationship between American option and European option can be represented as

$$C(S,t) = \begin{cases} c(S,t) + A_2\left(\frac{S}{S^*}\right)^{\gamma_2} & \text{when } S < S^* \\ S - K & \text{when } S \geq S^* \end{cases}$$

where

$$A_2 = \frac{S^*}{\gamma_2}\left\{1 - e^{-q(T-t)}N\left[d_1\left(S^*\right)\right]\right\}$$

$$\gamma_2 = \left[-(\beta - 1) + \sqrt{(\beta - 1)^2 + \frac{4\alpha}{h}}\right]/2$$

$$d_1 = \frac{\ln\left(\frac{S}{K}\right) + \left(r - q + \frac{\sigma^2}{2}\right)(T - t)}{\sigma\sqrt{T - t}}$$

$$\alpha = \frac{2r}{\sigma^2}$$

$$\beta = \frac{2(r - q)}{\sigma^2}$$

$$h = 1 - e^{-r(T-t)}$$

To find the critical stock price, S*, it is necessary to solve

$$S^* - K = c\left(S^*, t\right) + \frac{S^*}{\gamma_2}\left\{1 - e^{-q(T-t)}N\left[d_1\left(S^*\right)\right]\right\}$$

Since this cannot be done directly, an iterative procedure must be developed.

26.9.2 VBA Program for Calculating American Option When Dividend Yield Is Known

We can use Excel GoalSeek tool to develop the iterative process. We set Cell F7 equal to zero by changing Cell B3 to find *S**. The function in Cell F7 is

$$= B12 + (1 - EXP(-B6{*}B8){*}NORMSDIST(B9)){*}\frac{B3}{F6} - B3 + B4$$

After doing the iterative procedure, the result shows that *S** is equal to 44.82072.

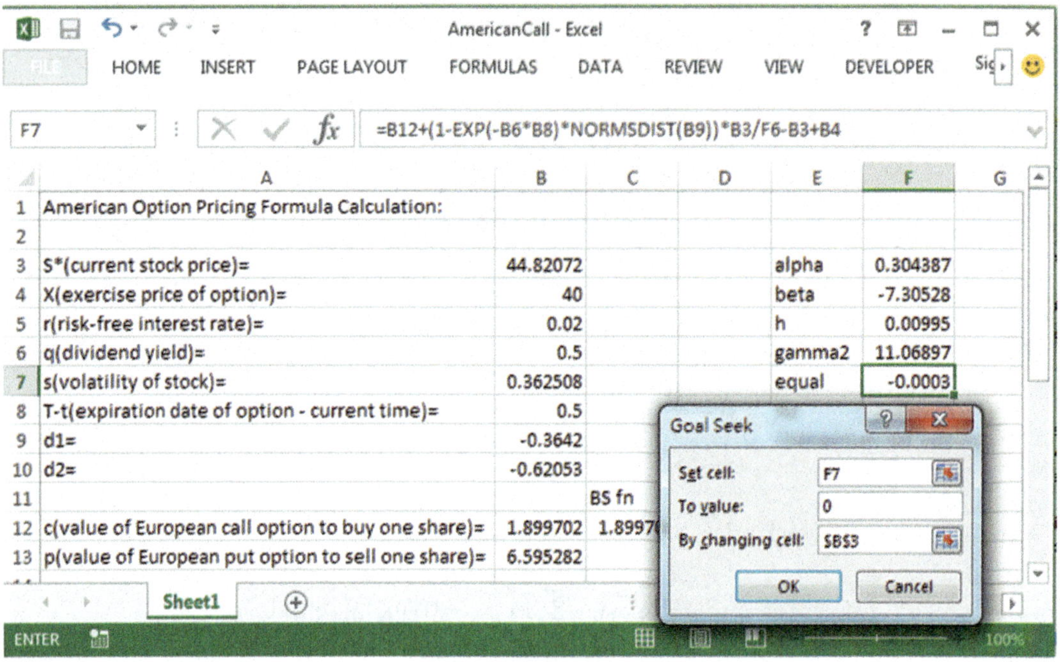

After we get S^*, we can calculate the value of American call option when S is equal to 42 in Cell B15. The function to calculate American call option in Cell H9 is

$$= \mathrm{IF}(\mathrm{B}15 < B3,\ B24 + F8^*(B15/B3)^\wedge F6, B15 - B4)$$

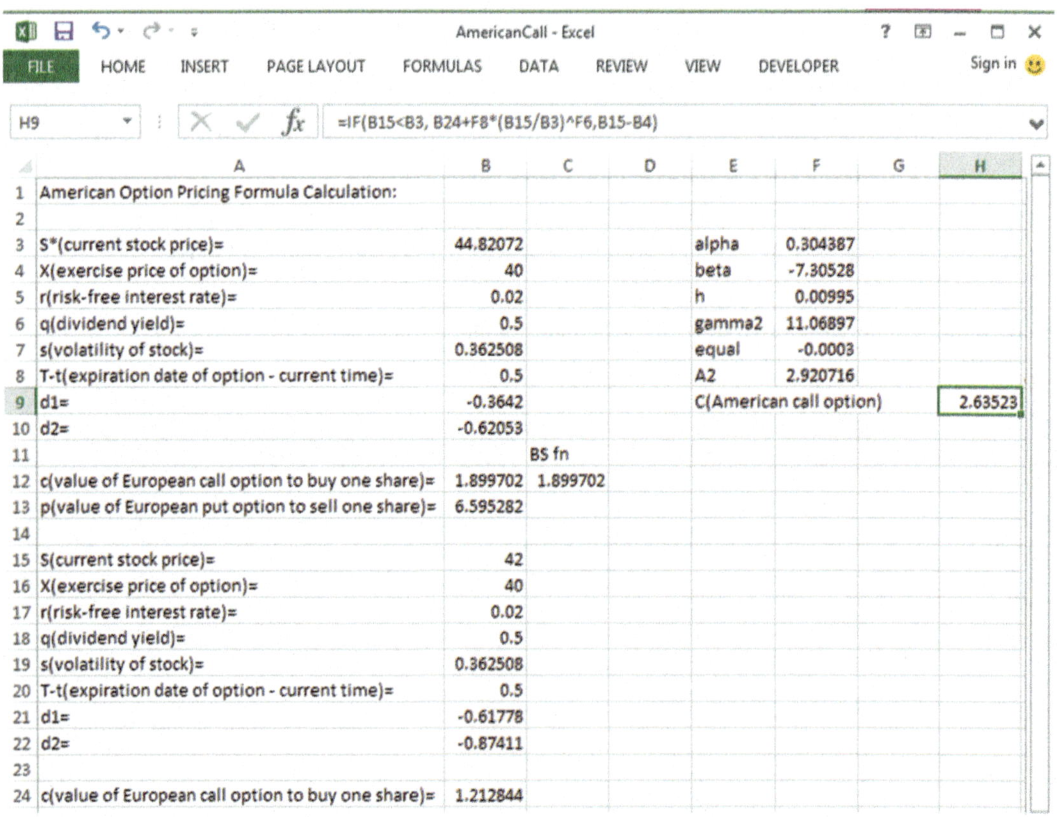

In addition to GoSeek tool, we also can write a user-defined function to calculate this value of American call option. Below is the VBA function

```
Function AmericanCall(S, X, r, q, T, sigma, a, b)
' Estimate implied volatility by Bisection
' Uses BSCall fn
  Dim yb, ya, c, yc, alpha, beta, h, gamma2, d1, A2, Sa
  alpha = 2 * r / sigma ^ 2
  beta = 2 * (r - q) / sigma ^ 2
  h = 1 - Exp(-r * T)
  gamma2 = (-(beta - 1) + Sqr((beta - 1) ^ 2 + 4 * alpha / h)) / 2
  d1 = (Log(b / X) + (r - q + 0.5 * sigma ^ 2) * T) / (sigma * Sqr(T))
  yb = BSCall(b, X, r, q, T, sigma) + (1 - Exp(-q * T) * Application.NormSDist(d1)) * b / gamma2 - b + X
  d1 = (Log(a / X) + (r - q + 0.5 * sigma ^ 2) * T) / (sigma * Sqr(T))
  ya = BSCall(a, X, r, q, T, sigma) + (1 - Exp(-q * T) * Application.NormSDist(d1)) * a / gamma2 - a + X
  If yb * ya > 0 Then
   BSIVBisection = CVErr(xlErrValue)
  Else
    Do While Abs(a - b) > 0.000000001
     c = (a + b) / 2
     d1 = (Log(c / X) + (r - q + 0.5 * sigma ^ 2) * T) / (sigma * Sqr(T))
     yc = BSCall(c, X, r, q, T, sigma) + (1 - Exp(-q * T) * Application.NormSDist(d1)) * c / gamma2 - c + X
     d1 = (Log(a / X) + (r - q + 0.5 * sigma ^ 2) * T) / (sigma * Sqr(T))
     ya = BSCall(a, X, r, q, T, sigma) + (1 - Exp(-q * T) * Application.NormSDist(d1)) * a / gamma2 - a + X
     If ya * yc < 0 Then
       b = c
     Else
       a = c
     End If
    Loop
    Sa = (a + b) / 2
  End If
    d1 = (Log(Sa / X) + (r - q + 0.5 * sigma ^ 2) * T) / (sigma * Sqr(T))
    A2 = (Sa / gamma2) * (1 - Exp(-q * T) * Application.NormSDist(d1))
  If S < Sa Then
    AmericanCall = BSCall(S, X, r, q, T, sigma) + A2 * (S / Sa) ^ gamma2
  Else
    AmericanCall = S - X
  End If
End Function
```

The function in Cell I9 is

$$= \text{AmericanCall}(\text{B15},\ \text{B4}, \text{B5},\ \text{B6}, \text{B8}, \text{B7},\ 0.0001,\ 1000)$$

After putting the parameters in function of the Cell I9, the result is similar to the value of American call option calculated by GoalSeek in Cell H9.

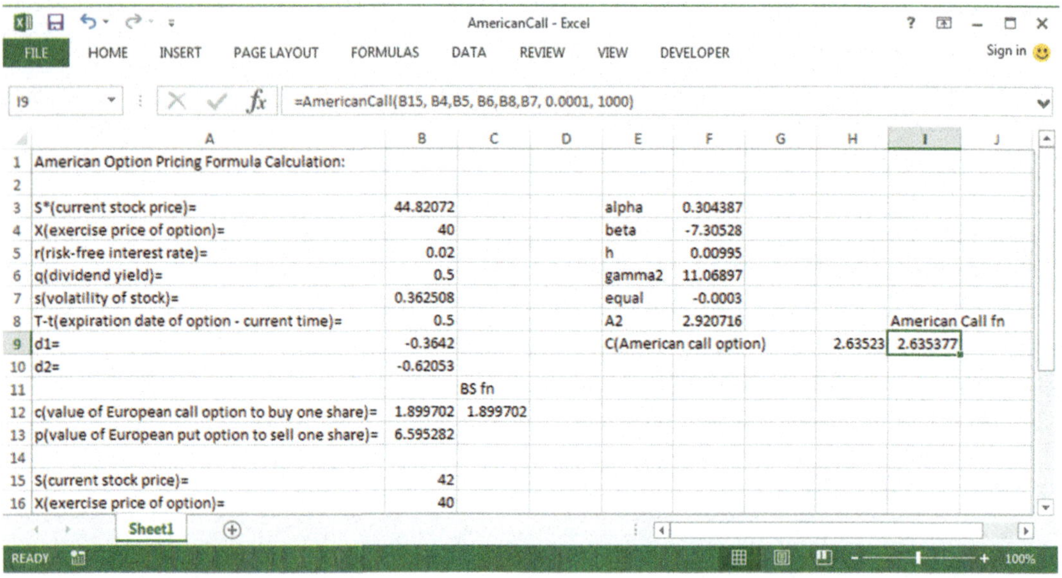

	A	B	C	D	E	F	G	H	I	J
1	American Option Pricing Formula Calculation:									
2										
3	S*(current stock price)=	44.82072			alpha	0.304387				
4	X(exercise price of option)=	40			beta	-7.30528				
5	r(risk-free interest rate)=	0.02			h	0.00995				
6	q(dividend yield)=	0.5			gamma2	11.06897				
7	s(volatility of stock)=	0.362508			equal	-0.0003				
8	T-t(expiration date of option - current time)=	0.5			A2	2.920716			American Call fn	
9	d1=	-0.3642			C(American call option)			2.63523	2.635377	
10	d2=	-0.62053								
11			BS fn							
12	c(value of European call option to buy one share)=	1.899702	1.899702							
13	p(value of European put option to sell one share)=	6.595282								
14										
15	S(current stock price)=	42								
16	X(exercise price of option)=	40								

26.10 Summary

This chapter showed how Microsoft Excel can be used to estimate European call and put options for (a) Black–Scholes model for Individual Stock, (b) Black–Scholes model for Stock Indices, and (c) Black–Scholes model for Currencies. In addition, we also discuss alternative methods to evaluate American call option when either dividend payment or dividend yield is known.

Appendix 26.1: Bivariate Normal Distribution

We have shown how the cumulative univariate normal density function can be used to evaluate a European call option in the previous sections of this chapter. If a common stock pays a discrete dividend during the option's life, the American call option valuation equation requires the evaluation of a cumulative bivariate normal density function. While there are many available approximations for the cumulative bivariate normal distribution, the approximation provided here relies on Gaussian quadratures. The approach is straightforward and efficient, and its maximum absolute error is 0.00000055.

The probability that $x^{'}$ is less than a and that $y^{'}$ is less than b for the standardized cumulative bivariate normal distribution can be defined as

$$P\left(X^{'} < a, Y^{'} < b\right) = \frac{1}{2\pi\sqrt{1-\rho^2}} \int_{-\infty}^{a} \int_{-\infty}^{b} \exp\left[\frac{2x^{'2} - 2\rho x^{'} y^{'} + y^{'2}}{2(1-\rho^2)}\right] dx^{'} dy^{'},$$

where $x^{'} = \frac{x-\mu_x}{\sigma_x}$, $y^{'} = \frac{y-\mu_y}{\sigma_y}$, and p is the correlation between the random variables $x^{'}$ and $y^{'}$.

The first step in the approximation of the bivariate normal probability $N_2(a, b; \rho)$ is as follows:

$$\phi(a,b;\rho) \approx .31830989\sqrt{1-\rho^2}\sum_{i=1}^{5}\sum_{j=1}^{5} w_i w_j f\left(x_i', x_j'\right), \tag{26.15}$$

where $f\left(x_i', x_j'\right) = \exp\left[a_1\left(2x_i' - a_1\right) + b_1\left(2x_j' - b_1\right) + 2\rho\left(x_i' - a_1\right)\left(x_j' - b_1\right)\right]$.

The pairs of weights (w) and corresponding abscissa values (x') are

i,j	w	x'
1	0.24840615	0.10024215
2	0.39233107	0.48281397
3	0.21141819	1.0609498
4	0.033246660	1.7797294
5	0.00082485334	2.6697604

(This portion is based upon Appendix 13.1 of Stoll H. R. and R. E Whaley. *Futures and Options.* Cincinnati, OH: South Western Publishing, 1993.)

and the coefficients a_1 and b_1 are computed using $a_1 = \frac{a}{\sqrt{2(1-\rho^2)}}$ and $b_1 = \frac{b}{\sqrt{2(1-\rho^2)}}$.

The second step in the approximation involves computing the product $ab\rho$; if $ab\rho \leq 0$, compute the bivariate normal probability, $N_2(a, b; \rho)$, using the following rules:

$$\begin{aligned}
&(1)\ \text{If } a \leq 0,\ b \leq 0,\ \text{and } \rho \leq 0,\ \text{then } N_2(a,b;\rho) = \phi(a,b;\rho);\\
&(2)\ \text{If } a \leq 0,\ b \geq 0,\ \text{and } \rho > 0,\ \text{then } N_2(a,b;\rho) = N_1(a) - \phi(a,-b;-\rho);\\
&(3)\ \text{If } a \geq 0,\ b \leq 0,\ \text{and } \rho > 0,\ \text{then } N_2(a,b;\rho) = N_1(b) - \phi(-a,b;-\rho);\\
&(4)\ \text{If } a \geq 0,\ b \geq 0,\ \text{and } \rho \leq 0,\ \text{then } N_2(a,b;\rho) = N_1(a) + N_1(b) - 1 + \phi(-a,-b;\rho).
\end{aligned} \tag{26.16}$$

If $ab\rho > 0$, compute the bivariate normal probability, $N_2(a, b; \rho)$, as

$$N_2(a,b;\rho) = N_2(a,0;\rho_{ab}) + N_2(b,0;\rho_{ab}) - \delta, \tag{26.17}$$

where the values of $N_2(\bullet)$ on the right-hand side are computed from the rules, for $ab\rho \leq 0$

$$\rho_{ab} = \frac{(\rho a - b)Sgn(a)}{\sqrt{a^2 - 2\rho ab + b^2}},\ \rho_{ba} = \frac{(\rho b - a)Sgn(b)}{\sqrt{a^2 - 2\rho ab + b^2}},\ \delta = \frac{1 - Sgn(a) \times Sgn(b)}{4},$$

and

$$Sgn(x) = \begin{cases} 1 & x \geq 0 \\ -1 & x < 0 \end{cases},$$

$N_1(d)$ is the cumulative univariate normal probability.

Appendix 26.2: Excel Program to Calculate the American Call Option When Dividend Payments Are Known

The following is a Microsoft Excel Program which can be used to calculate the price of an American Call Option using the Bivariate Normal Distribution method (Table 26.1):

Table 26.1 Microsoft Excel program for calculating the American call options

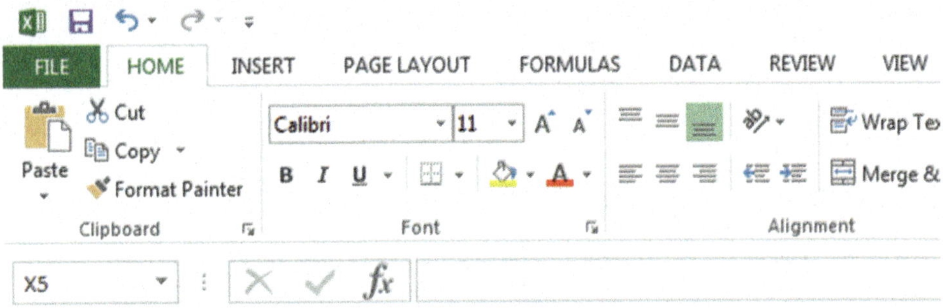

	A	B	C
1	A	B	C*
2			
3	Option Pricing Calculation		
4			
5	*S* (current stock price) =	50	50
6	S_t^* (critical exdividend stock price) =	46.9641	46.9641
7	*S* (current stock price NPV of promised dividend) =	x=B5-B13*EXP(-B9*B12)	48.0218
8	X (exercise price of option) =	48	48
9	*r* (risk-free interest rate) =	0.08	0.08
10	σ (volatility of stock) =	0.2	0.2
11	*T (expiration date)* =	0.24658	0.24658
12	*t* (Exercise date) =	0.13699	0.13699
13	*D* (Dividend) =	2	2
14	d_1 (nondividend-paying)=	x=(LN(B5/B8)+(B9+0.5*B10^2)*B11)/(B10*SQRT(B11))	0.659325
15	d_2 (nondividend-paying)=	x=B14-B10*SQRT(B11)	0.560012
16	d_1* (critical exdividend stock price) =	x=(LN(B6/B8)+(B9+0.5*B10^2)*(B11-B12))/(B10*SQRT(B11-B12))	-0.164
17	d_2 *(critical exdividend stock price) =	x=B16 - B10*SQRT(B11-B12)	-0.23021
18	d_1 (dividend- paying) =	x=(LN(B7/B8)+(B9+0.5*B10^2)*(B11))/(B10*SQRT(B11))	0.252856
19	d_2 (dividend- paying)=	x=B18-B10*SQRT(B11)	0.153542
20	a_1 =	x=(LN((B5-B13*EXP(-B9*B12))/B8)+(B9+0.5*B10^2)*(B11))/(B10*SQRT((B11)))	0.252856
21	a_2 =	x=B20-B10*SQRT(B11)	0.153542
22	b_1 =	x=(LN((B5-B13*EXP(-B9*B12))/B6)+(B9+0.5*B10^2)*(B12))/(B10*SQRT(B12))	0.485929
23	b_2 =	x=B22-B10*SQRT(B12)	0.411905
24			
25	$C(S_t^*, T - t; X)$	x=B6*NORMSDIST(B16)-B8*EXP(-B9*(B11-B12))*NORMSDIST(B17)	0.964112

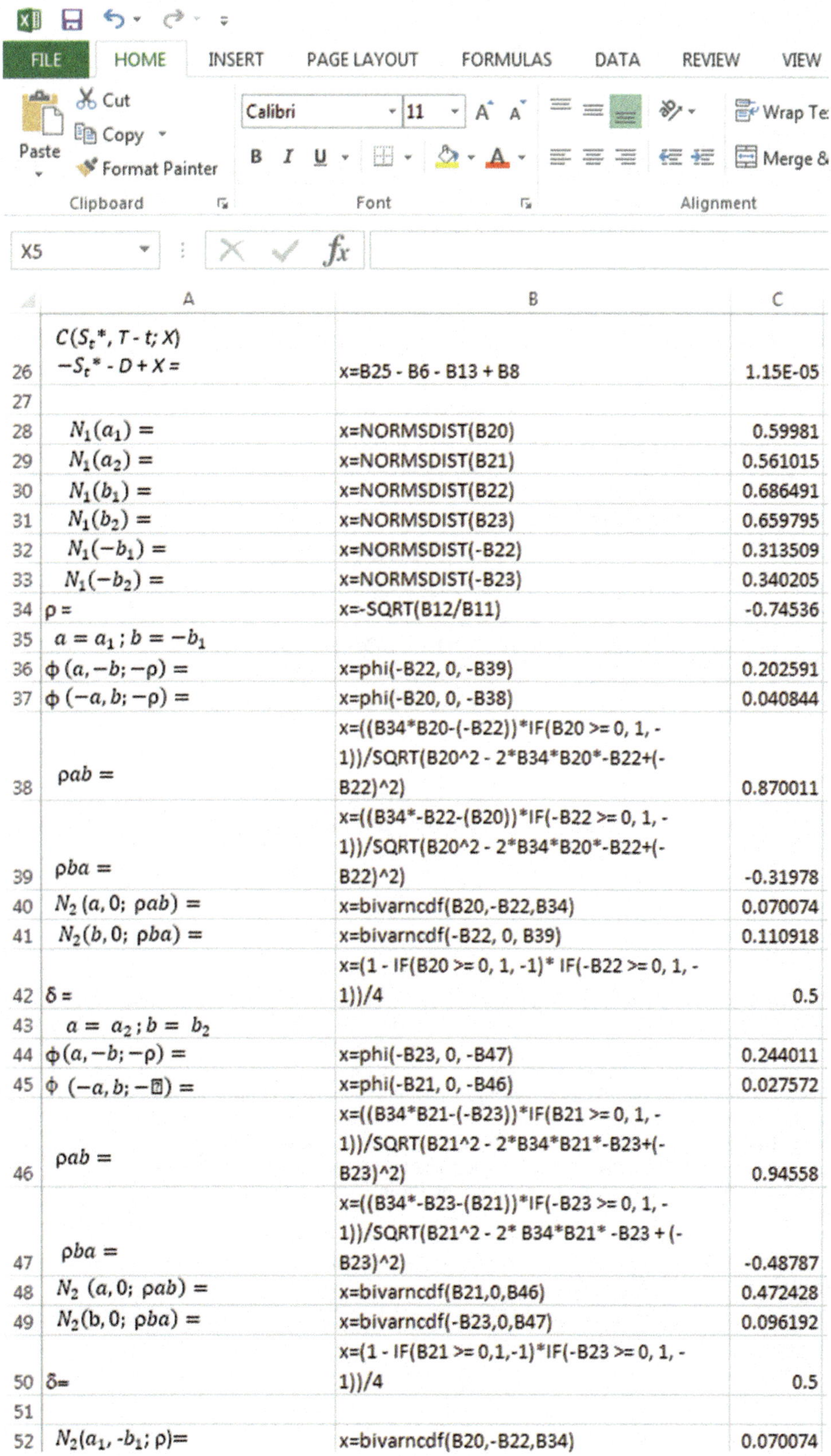

	A	B	C
26	$C(S_t^*, T-t; X)$ $-S_t^* - D + X =$	x=B25 - B6 - B13 + B8	1.15E-05
27			
28	$N_1(a_1) =$	x=NORMSDIST(B20)	0.59981
29	$N_1(a_2) =$	x=NORMSDIST(B21)	0.561015
30	$N_1(b_1) =$	x=NORMSDIST(B22)	0.686491
31	$N_1(b_2) =$	x=NORMSDIST(B23)	0.659795
32	$N_1(-b_1) =$	x=NORMSDIST(-B22)	0.313509
33	$N_1(-b_2) =$	x=NORMSDIST(-B23)	0.340205
34	ρ =	x=-SQRT(B12/B11)	-0.74536
35	$a = a_1 ; b = -b_1$		
36	$\phi(a, -b; -\rho) =$	x=phi(-B22, 0, -B39)	0.202591
37	$\phi(-a, b; -\rho) =$	x=phi(-B20, 0, -B38)	0.040844
38	$\rho ab =$	x=((B34*B20-(-B22))*IF(B20 >= 0, 1, -1))/SQRT(B20^2 - 2*B34*B20*-B22+(-B22)^2)	0.870011
39	$\rho ba =$	x=((B34*-B22-(B20))*IF(-B22 >= 0, 1, -1))/SQRT(B20^2 - 2*B34*B20*-B22+(-B22)^2)	-0.31978
40	$N_2(a, 0; \rho ab) =$	x=bivarncdf(B20,-B22,B34)	0.070074
41	$N_2(b, 0; \rho ba) =$	x=bivarncdf(-B22, 0, B39)	0.110918
42	δ =	x=(1 - IF(B20 >= 0, 1, -1)* IF(-B22 >= 0, 1, -1))/4	0.5
43	$a = a_2 ; b = b_2$		
44	$\phi(a, -b; -\rho) =$	x=phi(-B23, 0, -B47)	0.244011
45	$\phi(-a, b; -$□$) =$	x=phi(-B21, 0, -B46)	0.027572
46	$\rho ab =$	x=((B34*B21-(-B23))*IF(B21 >= 0, 1, -1))/SQRT(B21^2 - 2*B34*B21*-B23+(-B23)^2)	0.94558
47	$\rho ba =$	x=((B34*-B23-(B21))*IF(-B23 >= 0, 1, -1))/SQRT(B21^2 - 2* B34*B21* -B23 + (-B23)^2)	-0.48787
48	$N_2(a, 0; \rho ab) =$	x=bivarncdf(B21,0,B46)	0.472428
49	$N_2(b, 0; \rho ba) =$	x=bivarncdf(-B23,0,B47)	0.096192
50	δ=	x=(1 - IF(B21 >= 0,1,-1)*IF(-B23 >= 0, 1, -1))/4	0.5
51			
52	$N_2(a_1, -b_1; \rho)=$	x=bivarncdf(B20,-B22,B34)	0.070074

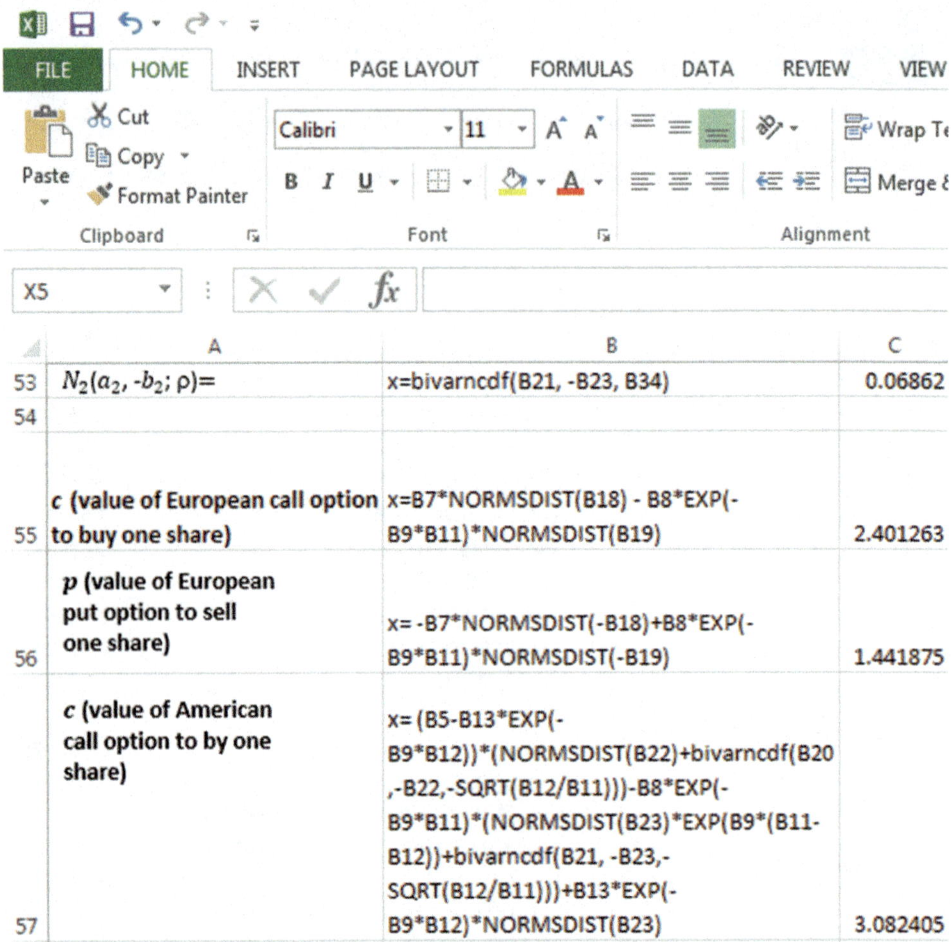

	A	B	C
53	$N_2(a_2, -b_2; \rho)$=	x=bivarncdf(B21, -B23, B34)	0.06862
54			
55	**c (value of European call option to buy one share)**	x=B7*NORMSDIST(B18) - B8*EXP(-B9*B11)*NORMSDIST(B19)	2.401263
56	**p (value of European put option to sell one share)**	x= -B7*NORMSDIST(-B18)+B8*EXP(-B9*B11)*NORMSDIST(-B19)	1.441875
57	**c (value of American call option to by one share)**	x= (B5-B13*EXP(-B9*B12))*(NORMSDIST(B22)+bivarncdf(B20,-B22,-SQRT(B12/B11)))-B8*EXP(-B9*B11)*(NORMSDIST(B23)*EXP(B9*(B11-B12))+bivarncdf(B21, -B23,-SQRT(B12/B11)))+B13*EXP(-B9*B12)*NORMSDIST(B23)	3.082405

Bibliography

Anderson TW (2003) An introduction to multivariate statistical analysis, 3rd edn. Wiley-Interscience, New York

Barone-Adesi G, Whaley RE (1987) Efficient analytic approximation of American option values. J Financ 42:301–320

Black F (1976) The pricing of commodity contracts. J Financ Econ 3:167–178

Cox JC, Ross SA (1976) The valuation of options for alternative stochastic processes. J Financ Econ 3:145–166

Cox J, Ross S, Rubinstein M (1979) Option pricing: a simplified approach. J Financ Econ 7:229–263

Johnson NL, Kotz S (1970) Distributions in statistics: continuous univariate distributions 2. Wiley, New York

Johnson NL, Kotz S (1972) Distributions in statistics: continuous multivariate distributions. Wiley, New York

Lee CF, Lee JC, Lee AC (2013) Statistics for business and financial economics. Springer, New York

Rubinstein M (1976) The valuation of uncertain income streams and the pricing of options. Bell J Econ Manage Sci 7:407–425

Stoll HR (1969) The relationship between put and call option prices. J Financ 24:801–824

Whaley RE (1981) On the valuation of American call options on stocks with known dividends. J Financ Econ 9:207–211

Chapter 27
Alternative Methods to Estimate Implied Variance

27.1 Introduction

In this chapter we will introduce how to use Excel to estimate implied volatility. First, we use approximate linear function to derive the volatility implied by Black–Merton–Scholes model. Second, we use nonlinear method, which includes goal seek and bisection method, to calculate implied volatility. Third, we demonstrate how to get the volatility smile using IBM data. Fourth, we introduce constant elasticity volatility (CEV) model and use bisection method to calculate the implied volatility of CEV model. Finally, we calculate the 52 weeks historical volatility of a stock. We used the Excel function *webservice* to retrieve the 52 historical stock prices.

27.2 Excel Program to Estimate Implied Variance with Black–Scholes Option Pricing Model

27.2.1 Black, Scholes, and Merton Model

In a classic option pricing developed by Black and Scholes (1973) and Merton (1973), the value of a European call option on a stock is stated:

$$c = Se^{-qT}N(d) + Xe^{-rT}N\left(d - \sigma\sqrt{T}\right)$$

$$d = \frac{\ln\left(\frac{S}{X}\right) + \left(r - q + \frac{\sigma^2}{2}\right)T}{\sigma\sqrt{T}}$$

where the stock price, exercise price, interest rate, dividend yield, and time until option expiration are denoted by S, K, r, q, and T, respectively. The instantaneous standard deviation of the log stock price is represented by σ, and $N(.)$ is the standard normal distribution function. If we can get the parameter in the model, we can calculate the option price. Below shows the Black–Scholes formula in the spreadsheet.

C.-F. Lee et al., *Essentials of Excel, Excel VBA, SAS and Minitab for Statistical and Financial Analyses*, DOI 10.1007/978-3-319-38867-0_27

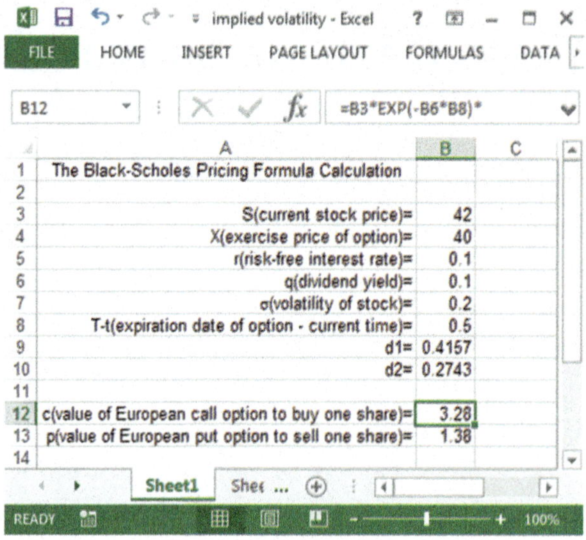

For a call option on a stock, the Black–Scholes formula in cell B12 is:

$$= \mathrm{B3} * \mathrm{EXP}(-\mathrm{B6} * \mathrm{B8}) * \mathrm{NORMSDIST}(\mathrm{B9}) - \mathrm{B4} * \mathrm{EXP}(-\mathrm{B5} * \mathrm{B8}) * \mathrm{NORMSDIST}(\mathrm{B10})$$

where NORMSDIST takes care of the cumulative distribution function of standard normal distribution.

It is easy to write a function to price a call function using Black and Scholes formula. The VBA function program is below:

```
'   BS Call Option Value

Function BSCall(S, X, r, q, T, sigma)

    Dim d1, d2, Nd1, Nd2

    d1 = (Log(S / X) + (r - q + 0.5 * sigma ^ 2) * T) / (sigma * Sqr(T))

    d2 = d1 - sigma * Sqr(T)

    Nd1 = Application.NormSDist(d1)

    Nd2 = Application.NormSDist(d2)

    BSCall = Exp(-q * T) * S * Nd1 - Exp(-r * T) * X * Nd2

End Function
```

If we use this function to calculate, we just put the parameters into the function. And we can get the result. We don't need to write the Black and Scholes formula again. This is shown below.

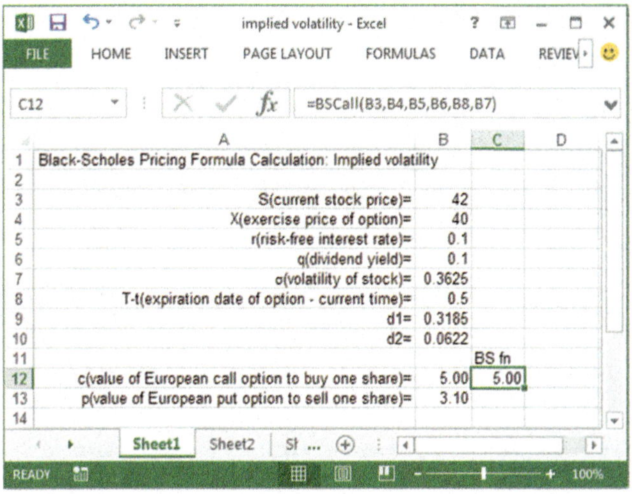

The user-defined VBA function in cell C12 is:

$$= \text{BSCall}(\text{B3}, \text{B4}, \text{B5}, \text{B6}, \text{B8}, \text{B7})$$

The call value in cell C12 is 5.00 which is equal to B12 calculated by spreadsheet.

27.2.2 Approximating Linear Function for Implied Volatility

All model parameters except the log stock price standard deviation are directly observable from market data. This allows a market-based estimate of a stock's future price volatility to be obtained by inverting (27.1), thereby yielding an implied volatility.

Unfortunately, there is no closed-form solution for an implied standard deviation from (27.1). We have to solve a nonlinear equation. Corrado and Miller (1996) have suggested an analytic formula that produces an approximation for the implied volatility. They start by approximating N(z) as a linear function

$$\mathrm{N(z)} = \frac{1}{2} + \frac{1}{\sqrt{2\pi}}\left(z - \frac{z^3}{6} + \frac{z^5}{40} + \ldots\right)$$

Substituting expansions of the normal cumulative probabilities N(d) and $\mathrm{N}\left(\mathrm{d} - \sigma\sqrt{T}\right)$ into Black–Scholes call option price,

$$\mathrm{c} = \mathrm{S}e^{-qT}\left(\frac{1}{2} + \frac{d}{\sqrt{2\pi}}\right) + Xe^{-rT}\left(\frac{1}{2} + \frac{d - \sigma\sqrt{T}}{\sqrt{2\pi}}\right)$$

After solving the quadratic equation and some approximations, we can get

$$\sigma = \frac{\sqrt{2\pi/T}}{M + K}\left(c - \frac{M - K}{2} + \sqrt{\left(c - \frac{M - K}{2}\right)^2 - \frac{(M - K)^2}{\pi}}\right)$$

where $\mathrm{M} = \mathrm{S}e^{-qT}$ and $\mathrm{K} = \mathrm{X}e^{-\mathrm{r}\mathrm{T}}$.

After typing Corrado and Miller's formula into excel worksheet, we can get the approximation of implied volatility easily. This is shown below.

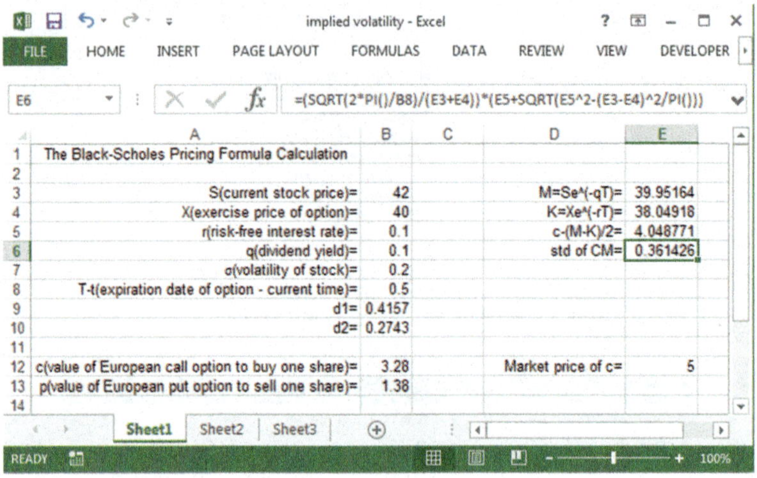

If the market price of call option is E12, the approximation value of implied volatility using the Corrado and Miller's formula shown in E6 is:

$$= (\mathrm{SQRT}(2 * \mathrm{PI}()/\mathrm{B8})/(\mathrm{F3} + \mathrm{F4})) * \left(\mathrm{F5} + \mathrm{SQRT}\left(\mathrm{F5}^2 - (\mathrm{F3} - \mathrm{F4})^2/\mathrm{PI}()\right)\right)$$

If we want to write a function to calculate implied volatility of Corrado and Miller, here is the VBA function:

```
'   Estimate implied volatility by Corrando and Miller

Function BSIVCM(S, X, r, q, T, callprice)

    Dim M, K, p, diff, sqrtest

    M = S * Exp(-q * T)

    K = X * Exp(-r * T)

    p = Application.Pi()

    diff = callprice - 0.5 * (M - K)

    sqrtest = (diff ^ 2) - ((M - K) ^ 2) / p

    If sqrtest < 0 Then

        BSIVCM = -1

    Else

        BSIVCM = (Sqr(2 * p / T) / (M + K)) * (diff + Sqr(sqrtest))

    End If

End Function
```

Using this function, it's easy to calculate an approximation of implied volatility. The output is shown below.

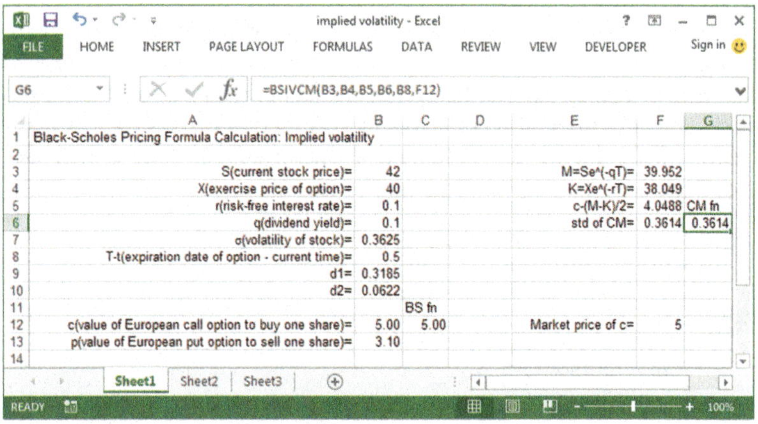

	A	B	C	D	E	F	G
1	Black-Scholes Pricing Formula Calculation: Implied volatility						
2							
3	S(current stock price)=	42			M=Se^(-qT)=	39.952	
4	X(exercise price of option)=	40			K=Xe^(-rT)=	38.049	
5	r(risk-free interest rate)=	0.1			c-(M-K)/2=	4.0488	CM fn
6	q(dividend yield)=	0.1			std of CM=	0.3614	0.3614
7	σ(volatility of stock)=	0.3625					
8	T-t(expiration date of option - current time)=	0.5					
9	d1=	0.3185					
10	d2=	0.0622					
11			BS fn				
12	c(value of European call option to buy one share)=	5.00	5.00		Market price of c=	5	
13	p(value of European put option to sell one share)=	3.10					
14							

The Corrado and Miller implied volatility formula in G6 is:

$$= \text{BSIVCM}(\text{B3}, \text{B4}, \text{B5}, \text{B6}, \text{B8}, \text{F12})$$

The approximation value in G6 is 0.3614 which is equal to F6.

27.2.3 Nonlinear Method for Implied Volatility

There are two nonlinear methods for implied volatility. The first one is Newton–Raphson method. The second one is bisection. Using the slope to improve, the accuracy of subsequent guesses is known as the Newton–Raphson method.

27.2.3.1 Newton–Raphson Method

Newton–Raphson method is a method for finding successively better approximations to the roots of a nonlinear function.

$$\mathrm{x} : \mathrm{f}(\mathrm{x}) = 0$$

The Newton–Raphson method in one variable is accomplished as follows:

Given a function *f*(*x*) and its derivative $f'(x)$, we begin with a first guess x_0 for a root of the function *f*. The process is iterated as

$$x_{n+1} = x_n - \frac{f(x_n)}{f'(x_n)}$$

until a sufficiently accurate value is approached.

In order to use Newton–Raphson to estimate implied volatility, we need f′(.), if option pricing model is Vega:

$$\mathrm{v} = \frac{\partial \mathrm{C}}{\partial \sigma} = \mathrm{S}e^{-\mathrm{qT}}\sqrt{T}N'(d_1)$$

Goal Seek is a procedure in Excel. It uses the Newton–Raphson method to solve the root of nonlinear equation. In the figure below, we would like to show how to use Goal Seek procedure to find the implied volatility. The details of our vanilla option are set out (cell B3 to B8). Suppose the observed call option market value is 5. 00. Our work is to choose a succession of volatility estimates in cell B6 until the BSM call option value in cell B11 equals to the observed price, 5.00. This can be done by applying the Goal Seek command in the Data part of Excel's menu.

[Data] → [What If Analysis] → [Goal Seek]

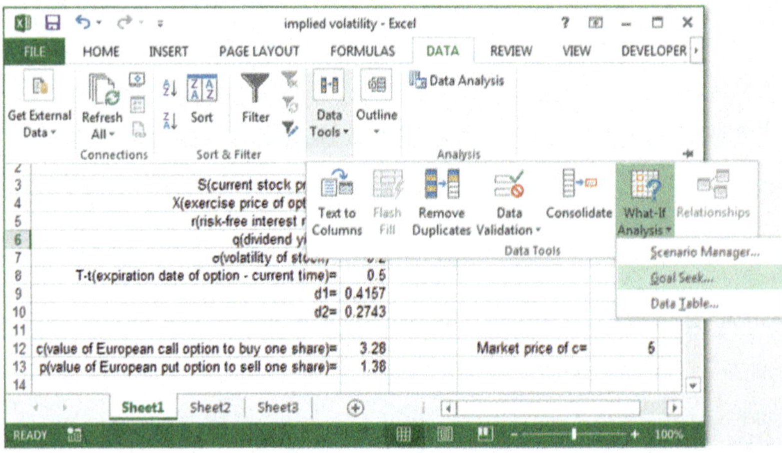

Insert the following data into [Goal Seek] dialogue box:

Set cell: B12
To value: 5.00
By changing cell: $B$7

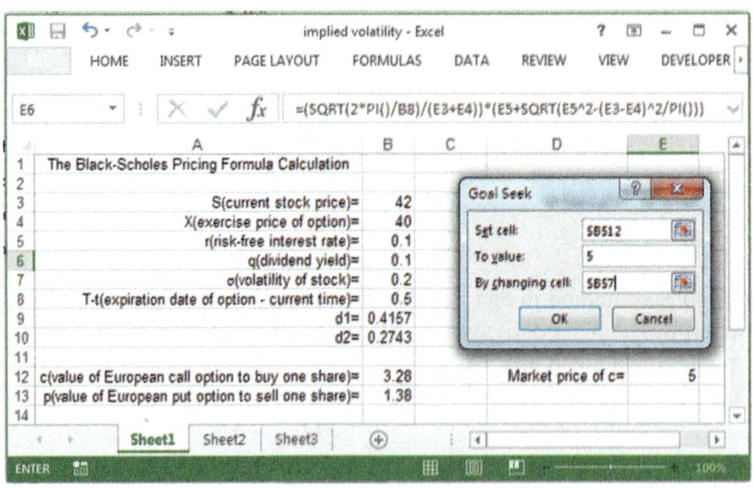

After we press OK button, we should find that the true implied volatility is 36.3 %.

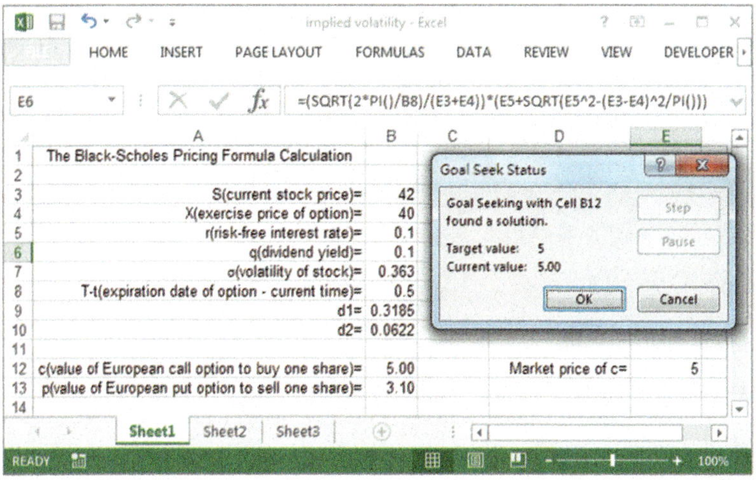

We can find that Corrado and Miller (1996) analytical, 0.361, which is near the Goal Seek solution 0.363.

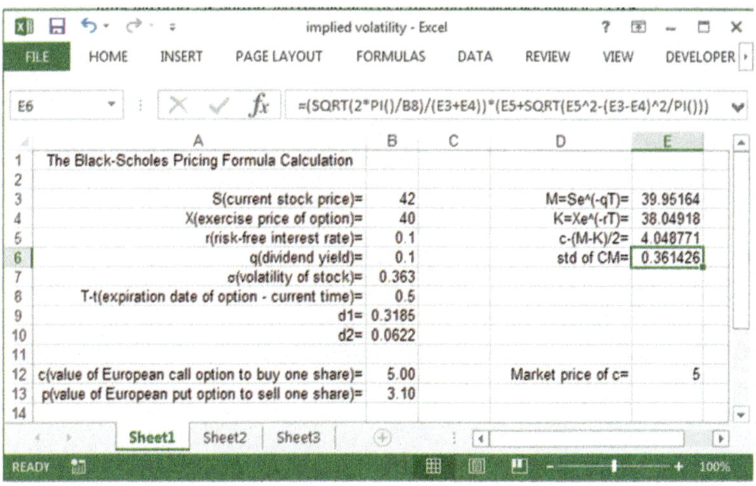

27.2.3.2 Bisection Method

In addition to Newton–Raphson method, we have another method to solve the root of nonlinear equation. This is bisection method. Start with two numbers, a and b, where a < b and f(a)*f(b) < 0. If we evaluate f and midpoint c = (a + b)/2, then

1. f(c) = 0.
2. f(a)*f(c) < 0.
3. f(c)*f(b) < 0.

In call option example, f(.) = BSCall(.)—market price of call option and a, b, and c are the candidates of implied volatility.

Although this method is a little slower than Newton–Raphson method, it will not run down when we give a bad initial value like Newton–Raphson method. We also can create a function to estimate implied volatility by using bisection method. The VBA function is shown below.

```
'   Estimate implied volatility by Bisection

'   Uses BSCall fn

Function BSIVBisection(S, X, r, q, T, callprice, a, b)

    Dim yb, ya, c, yc

    yb = BSCall(S, X, r, q, T, b) - callprice

    ya = BSCall(S, X, r, q, T, a) - callprice

    If yb * ya > 0 Then

      BSIVBisection = CVErr(xlErrValue)

    Else

       Do While Abs(a - b) > 0.000000001

         c = (a + b) / 2

         yc = BSCall(S, X, r, q, T, c) - callprice

         ya = BSCall(S, X, r, q, T, a) - callprice

         If ya * yc < 0 Then

            b = c

         Else

            a = c

         End If

        Loop

       BSIVBisection = (a + b) / 2

    End If

End Function
```

When we use this function to estimate implied volatility, the result is shown below.

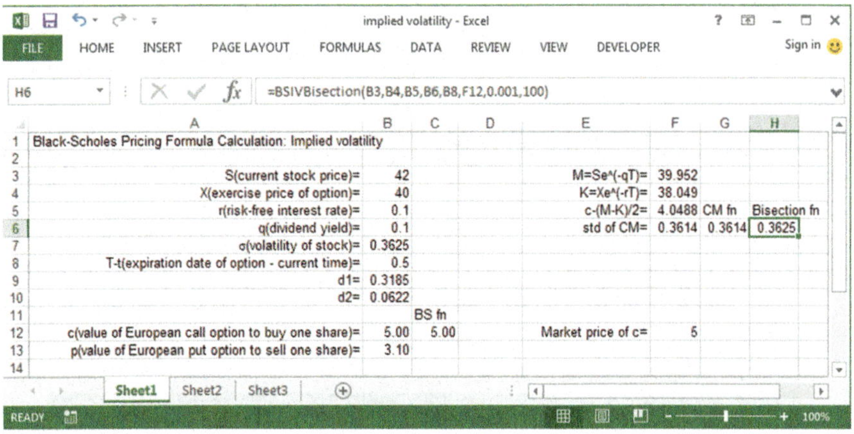

The bisection formula of implied volatility in H6 is:

$$= \text{BSIVBisection}(\text{B3}, \text{B4}, \text{B5}, \text{B6}, \text{B8}, \text{F12}, 0.001, 100)$$

Implied volatility, 0.3625, estimated from bisection method is much closer to Newton–Raphson method of Goal Seek, 0.3625, than Corrado and Miller's approximation, 0.3614.

27.2.3.3 Compare Newton–Raphson Method and Bisection Method

Before we write a user-defined function for Newton–Raphson method, we need a Vega function for vanilla call option. Below is the function for Vega.

```
'    BS Call Option Vega

Function BSCallVega(S, X, r, q, T, sigma)

    Dim d1, Ndash1

    d1 = (Log(S / X) + (r - q + 0.5 * sigma ^ 2) * T) / (sigma * Sqr(T))

    Ndash1 = (1 / Sqr(2 * Application.Pi)) * Exp(-d1 ^ 2 / 2)

    BSCallVega = Exp(-q * T) * S * Sqr(T) * Ndash1

End Function
```

In the figure below, we can see in cell B15 the function to calculate Vega:

$$= \text{BSCallVega}(\text{B3}, \text{B4}, \text{B5}, \text{B6}, \text{B8}, \text{B7})$$

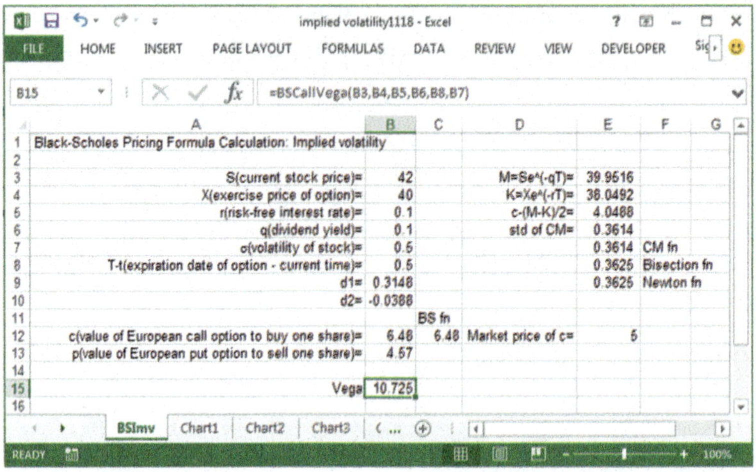

In order to compare Newton–Raphson method and Bisection method, we have to write a user-defined function of Newton–Raphson. According to the methodology in Sect. 27.2.3.1, below is the VBA function.

```
'    Estimate implied volatility by Newton
'    Uses BSCall fn & BSCallVega

Function BSIVNewton(S, X, r, q, T, callprice, initial)

    Dim bias, iv, ya, ydasha

    bias = 0.0001

    iv = initial

    Do

      ya = BSCall(S, X, r, q, T, iv) - callprice

      ydasha = BSCallVega(S, X, r, q, T, iv)

      iv = iv - ya / ydasha

    Loop While Abs(ya / ydasha) > bias

    BSIVNewton = iv

End Function
```

Using this function we can calculate the implied volatility by Newton–Raphson method.

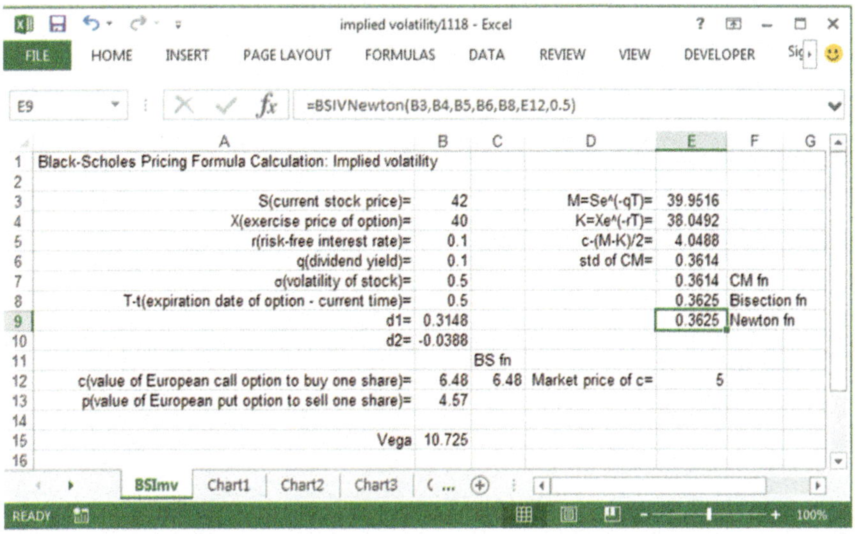

In the cell E9, we can see the function is

$$= \text{BSIVNewton}(\text{B3}, \text{B4}, \text{B5}, \text{B6}, \text{B8}, \text{E12}, 0.5)$$

And the output is 0.3625 which is equal to output of bisection method. The last input, 0.5, is the initial value. The most important input in the Newton–Raphson method is initial value. If we change the initial value to 0.01 or 5, we can find that the output is #VALUE! This is the biggest problem of Newton–Raphson method. If the initial is not suitable, we will not find the correct result. However, if there exists a solution and initial value covers this solution, we can get a correct solution no matter how big or small the initial value. The figure below shows the F(σ) = Cbs − Cmarket. We can find that there exists a unit solution at F(σ) = 0.

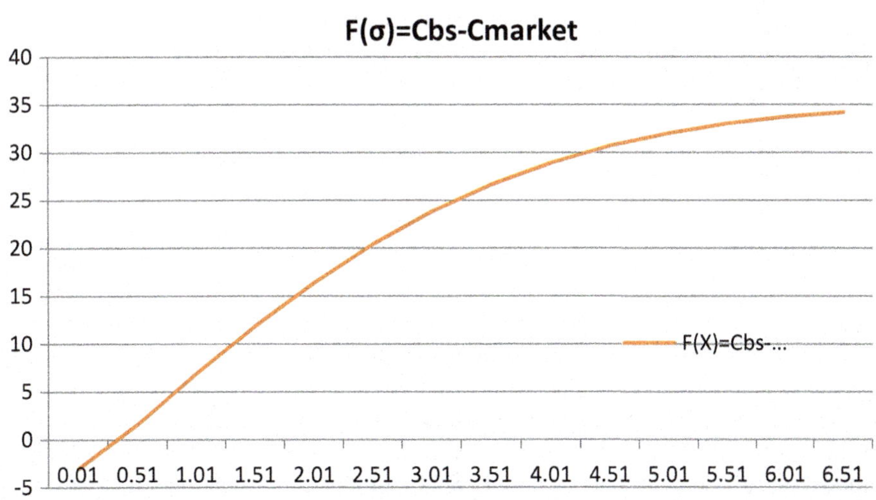

Although bisection method has less initial value problem, it still has a problem of more iterations. We calculate iterations and errors for these two methods and plot the figures below.

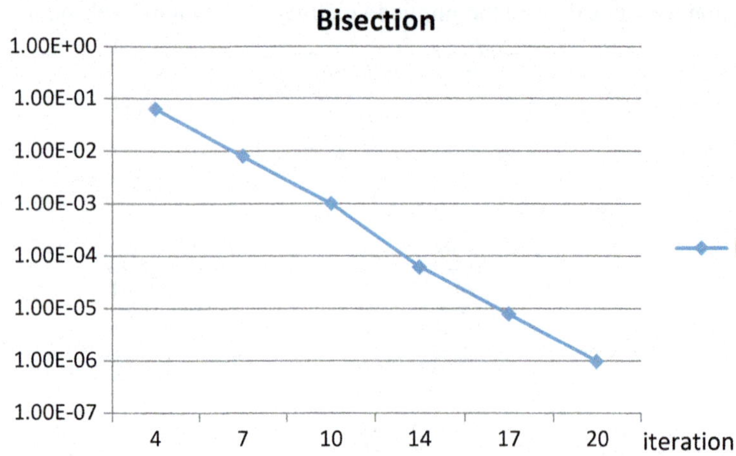

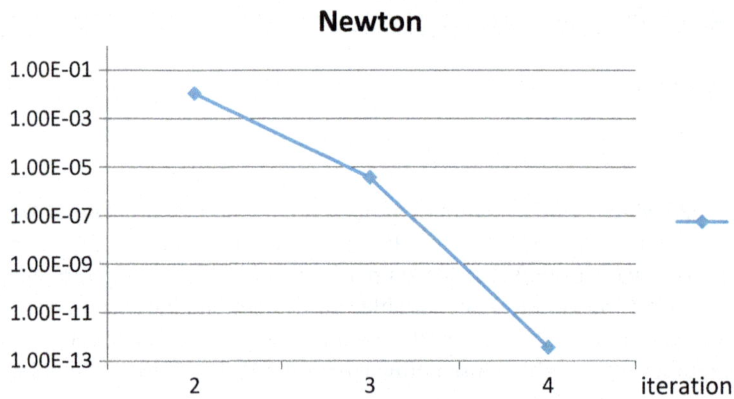

We can find that bisection method needs 20 iterations to reduce an error of around 10^{-6}. However, Newton–Raphson method only needs four iterations to produce an error of around 10^{-13}. This problem may occur in the past but today's computer is more efficient. So, we don't need to care about this problem too much now.

27.3 Volatility Smile

The existence of volatility smile is because Black–Scholes formula cannot precisely evaluate the either call or put option value. The main reason is that the Black–Scholes formula assumes the stock price per share is log-normally distributed. If we introduce extra distribution parameters into the option pricing determination formula, we can obtain the constant elasticity volatility (CEV) option pricing formula. This formula can be found in Sect. 27.4 of this Chapter. Lee et al. (2004) show that the CEV model performs better than the Black–Scholes model in evaluating either call or put option value.

A plot of the implied volatility of an option as a function of its strike price is known as a *volatility smile*. Now we use IBM's data to show the volatility smile. The call option data listed in the table below can be found in Yahoo Finance: http://finance.yahoo.com/q/op?s=IBM&date=1450396800.

We use the IBM option contract with the expiration date of November 18.

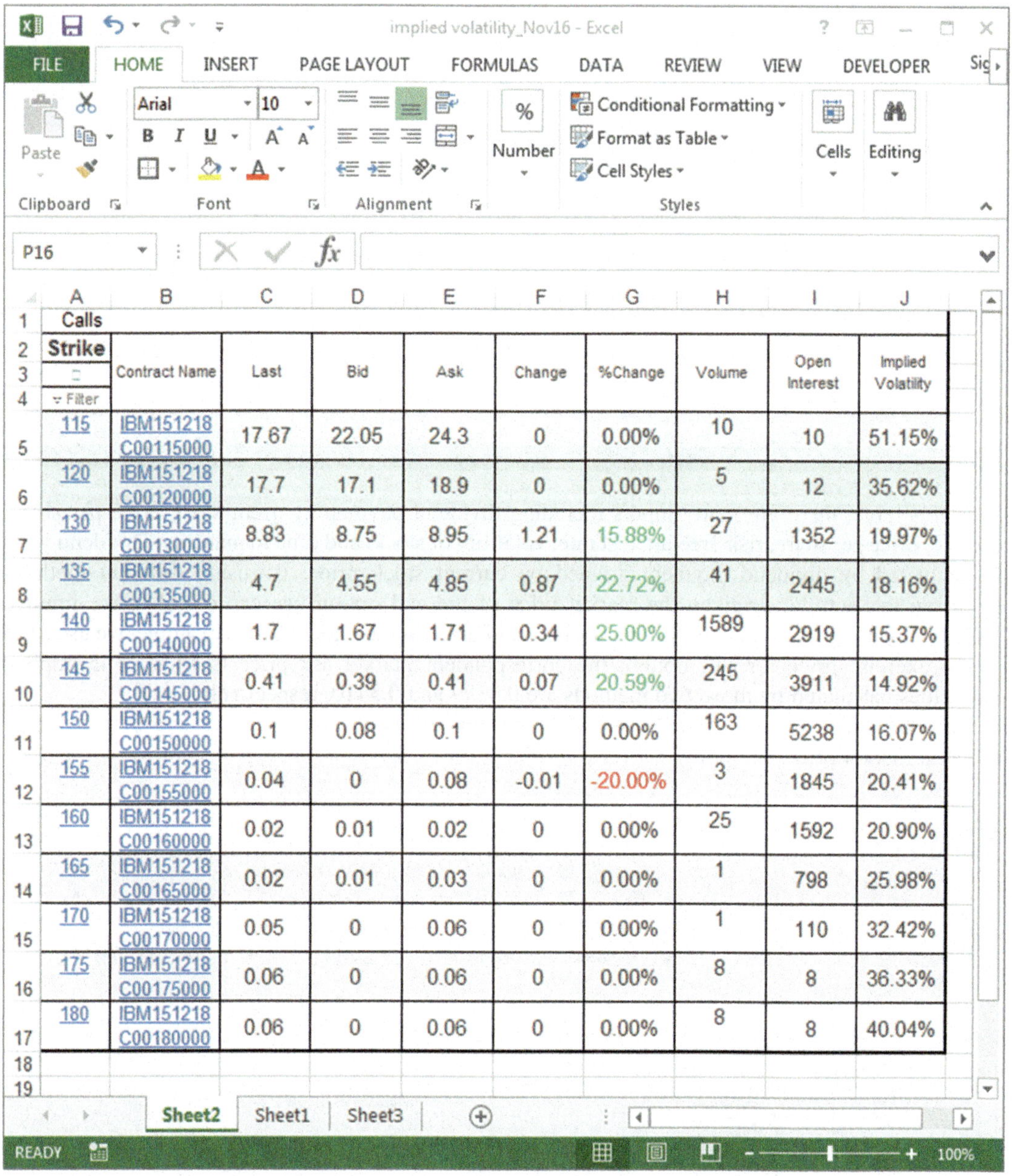

Calls									
Strike	Contract Name	Last	Bid	Ask	Change	%Change	Volume	Open Interest	Implied Volatility
115	IBM151218C00115000	17.67	22.05	24.3	0	0.00%	10	10	51.15%
120	IBM151218C00120000	17.7	17.1	18.9	0	0.00%	5	12	35.62%
130	IBM151218C00130000	8.83	8.75	8.95	1.21	15.88%	27	1352	19.97%
135	IBM151218C00135000	4.7	4.55	4.85	0.87	22.72%	41	2445	18.16%
140	IBM151218C00140000	1.7	1.67	1.71	0.34	25.00%	1589	2919	15.37%
145	IBM151218C00145000	0.41	0.39	0.41	0.07	20.59%	245	3911	14.92%
150	IBM151218C00150000	0.1	0.08	0.1	0	0.00%	163	5238	16.07%
155	IBM151218C00155000	0.04	0	0.08	-0.01	-20.00%	3	1845	20.41%
160	IBM151218C00160000	0.02	0.01	0.02	0	0.00%	25	1592	20.90%
165	IBM151218C00165000	0.02	0.01	0.03	0	0.00%	1	798	25.98%
170	IBM151218C00170000	0.05	0	0.06	0	0.00%	1	110	32.42%
175	IBM151218C00175000	0.06	0	0.06	0	0.00%	8	8	36.33%
180	IBM151218C00180000	0.06	0	0.06	0	0.00%	8	8	40.04%

Then we use the implied volatility excel program in the last section to calculate the implied volatility with a specific exercise price list in the table above.

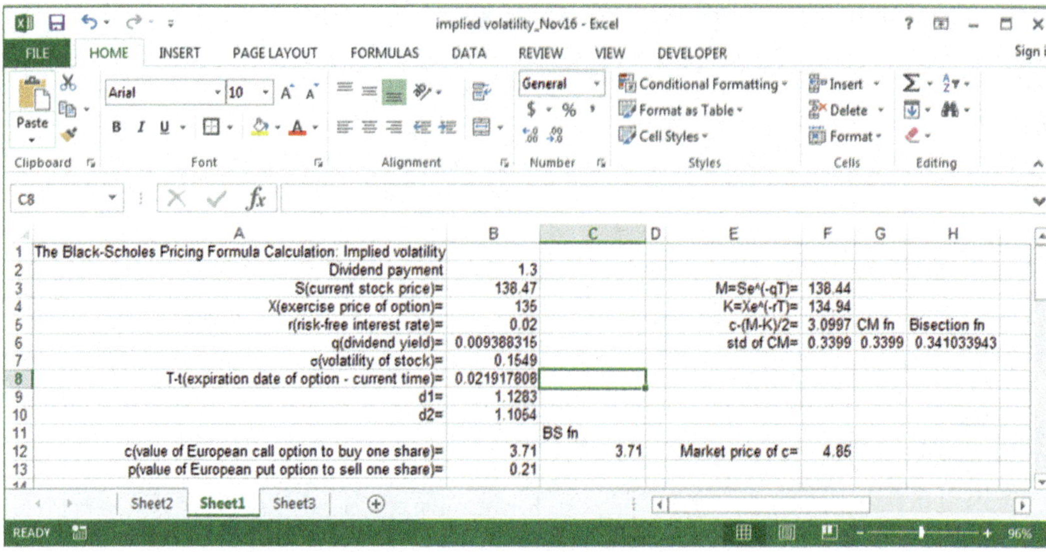

In this table, there are many inputs including dividend payment, current stock price per share, exercise price per share, risk-free interest rate, volatility of stock, and time to maturity. Dividend yield is calculated by dividend payment divided by current stock price. By using different methods discussed in Sect. 27. 2, given the market price of the call option, we can calculate the implied volatility by using Corrado and Miller's formula and bisection methods. In this example, we use $135 as our exercise price for call option; the correspondent market ask price is $4.85. The implied volatilities calculated by those two methods are 0.3399 and 0.3410, respectively.

Now we calculate the implied volatility by using different exercise price and correspondent different market price.

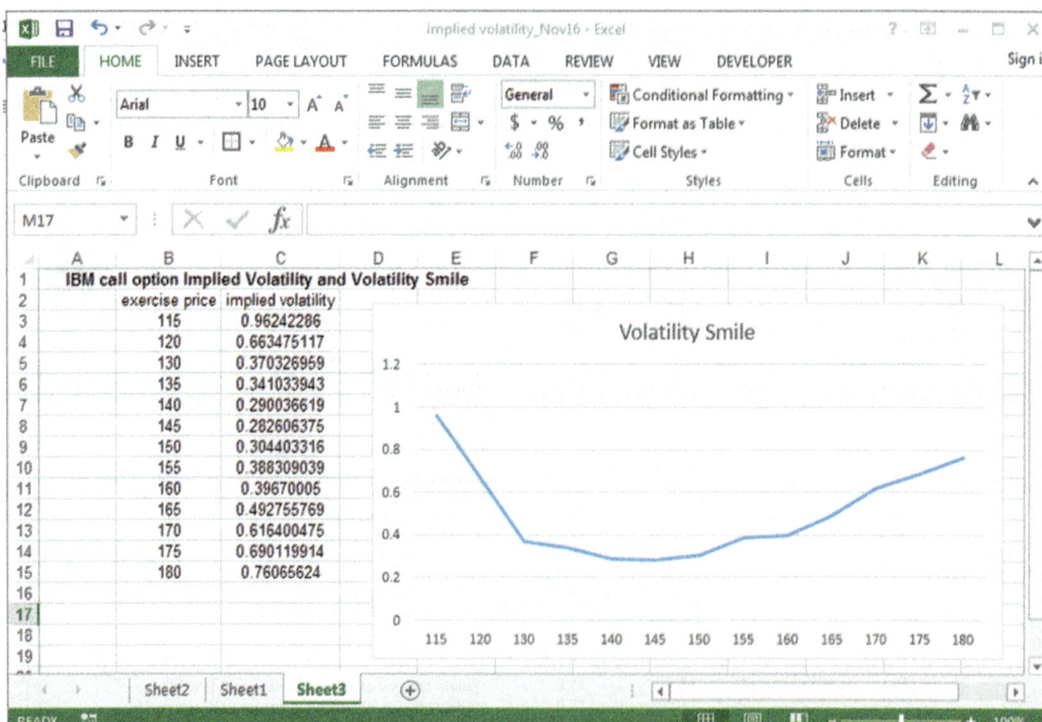

In the Excel table above, we calculate the implied volatility for correspondent of different exercise prices by using bisection method. Then plot the implied volatility; we can get the volatility smile as the above.

27.4 Excel Program to Estimate Implied Variance with CEV Model

In order to price a European option under a CEV model, we need a noncentral chi-square distribution. The following figure shows the charts of the noncentral chi-square distribution with 5 degrees of freedom for noncentral parameter δ = 0, 2, 4, 6.

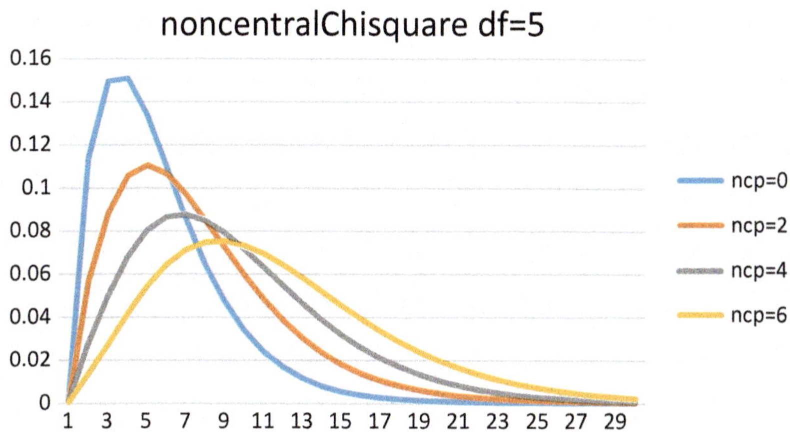

Under the theory in this chapter, we can write a call option price under CEV model. Below is the figure to do this:

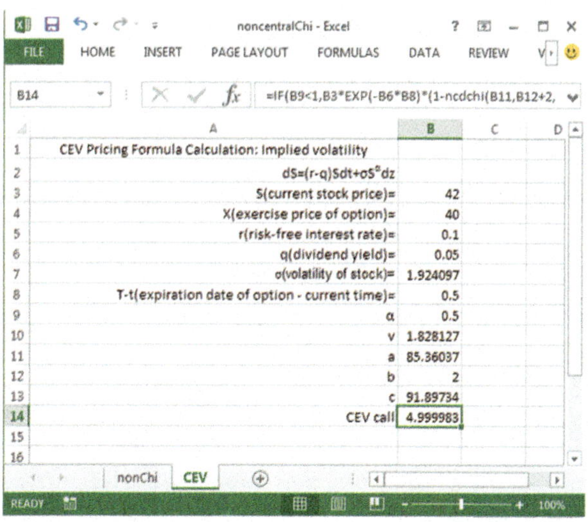

Hence, the formula for CEV call option in B14 is:

$$
\begin{aligned}
= \mathrm{IF}\big(&B9 < 1, B3 * EXP(-B6 * B8) * (1 - ncdchi(B11, B12 + 2, B13)) \\
&- \mathrm{B4} * \mathrm{EXP}(-\mathrm{B5} * \mathrm{B8}) * \mathrm{ncdchi}(\mathrm{B13}, \mathrm{B12}, \mathrm{B11}), \\
&\mathrm{B3} * \mathrm{EXP}(-\mathrm{B6} * \mathrm{B8}) * (1 - \mathrm{ncdchi}(\mathrm{B13}, -\mathrm{B12}, \mathrm{B11})) \\
&- \mathrm{B4} * \mathrm{EXP}(-\mathrm{B5} * \mathrm{B8}) * \mathrm{ncdchi}(\mathrm{B11}, 2 - \mathrm{B12}, \mathrm{B13})\big)
\end{aligned}
$$

The ncdchi is the noncentral chi-square cumulative distribution function. The function, IF, is used to separate the two conditions for this formula, $0 < \alpha < 1$ and $\alpha > 1$.

We can write a function to price the call option under CEV model. Below is the code to accomplish this:

```
'    CEV Call Option Value

Function CEVCall(S, X, r, q, T, sigma, alpha)

    Dim v As Double

    Dim aa As Double

    Dim bb As Double

    Dim cc As Double

    v = (Exp(2 * (r - q) * (alpha - 1) * T) - 1) * (sigma ^ 2) / (2 * (r - q) * (alpha
- 1))

    aa = ((X * Exp(-(r - q) * T)) ^ (2 * (1 - alpha))) / (((1 - alpha) ^ 2) * v)

    bb = 1 / (1 - alpha)

    cc = (S ^ (2 * (1 - alpha))) / (((1 - alpha) ^ 2) * v)
    If alpha < 1 Then

    CEVCall = Exp(-q * T) * S * (1 - ncdchi(aa, bb + 2, cc)) - Exp(-r * T) * X *
ncdchi(cc, bb, aa)

    Else

    CEVCall = Exp(-q * T) * S * (1 - ncdchi(cc, -bb, aa)) - Exp(-r * T) * X *
ncdchi(aa, 2 - bb, cc)

    End If

End Function
```

Use this function to value the call option which is shown below.

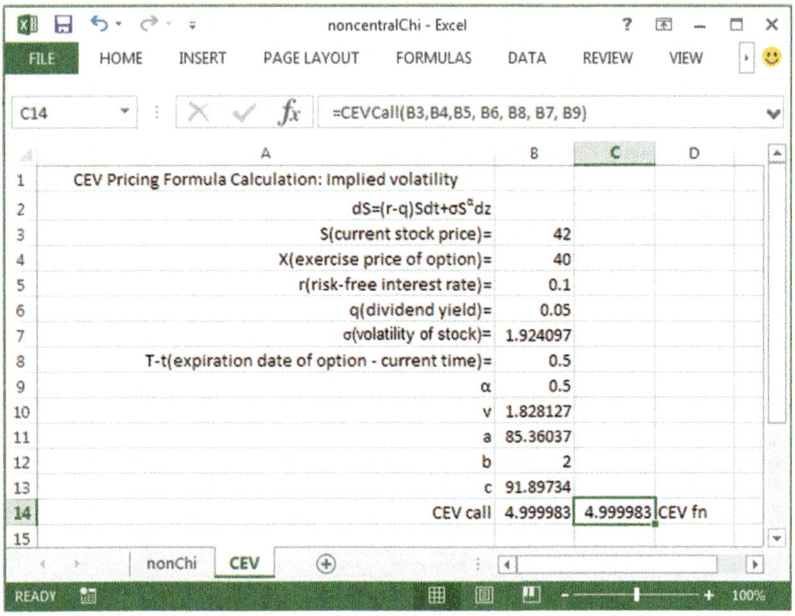

The CEV call option formula in C14 is:

$$= \text{CEVCall}(\text{B3}, \text{B4}, \text{B5},\ \text{B6},\ \text{B8},\ \text{B7},\ \text{B9})$$

The value of CEV call option in C14 is equal to B14.

Next, we want to use Goal Seek procedure to calculate the implied volatility. To do this we can see the figure below.

Set cell: B14
To value: 4
By changing cell: $B$7

After pressing the OK button, we can get the sigma value in B7.

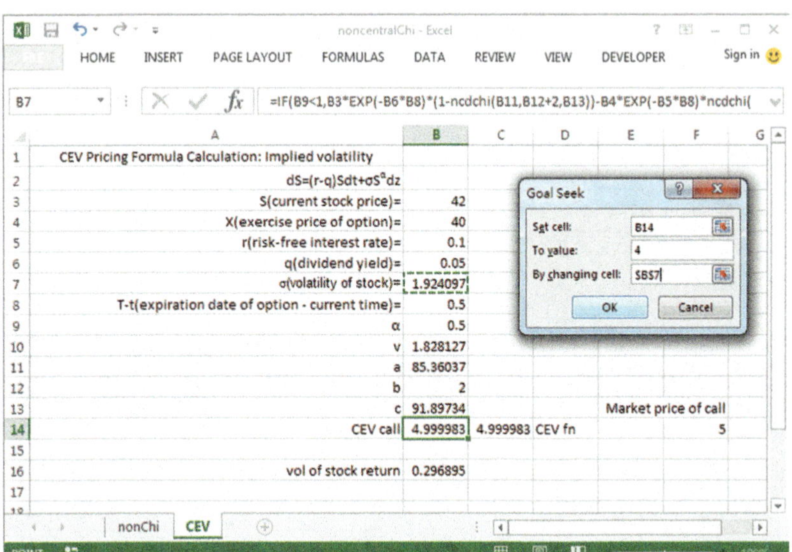

If we want to calculate implied volatility of stock return, we show this result in B16 of the figure below. The formula of implied volatility of stock return in B16 is:

$$= \mathrm{B7}^{*}\mathrm{B3}^{(\mathrm{B9}-1)}$$

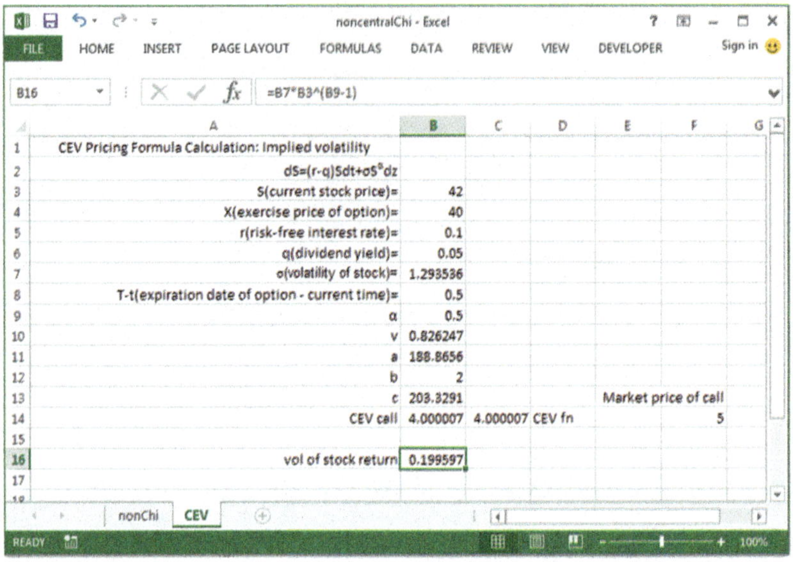

We use bisection method to write a function to calculate the implied volatility of CEV model. The following code can accomplish this task:

```
'   Estimate implied volatility by Bisection
'   Uses BSCall fn
Function CEVIVBisection(S, X, r, q, T, alpha, callprice, a, b)
    Dim yb, ya, c, yc
    yb = CEVCall(S, X, r, q, T, b, alpha) - callprice
    ya = CEVCall(S, X, r, q, T, a, alpha) - callprice
    If yb * ya > 0 Then
      CEVIVBisection = CVErr(xlErrValue)
    Else
       Do While Abs(a - b) > 0.000000001
         c = (a + b) / 2
         yc = CEVCall(S, X, r, q, T, c, alpha) - callprice
         ya = CEVCall(S, X, r, q, T, a, alpha) - callprice
         If ya * yc < 0 Then
            b = c
         Else
            a = c
         End If
        Loop
       CEVIVBisection = (a + b) / 2
    End If
End Function
```

After typing the parameters in the above function, we can get the sigma and implied volatility of stock return. The result is shown below.

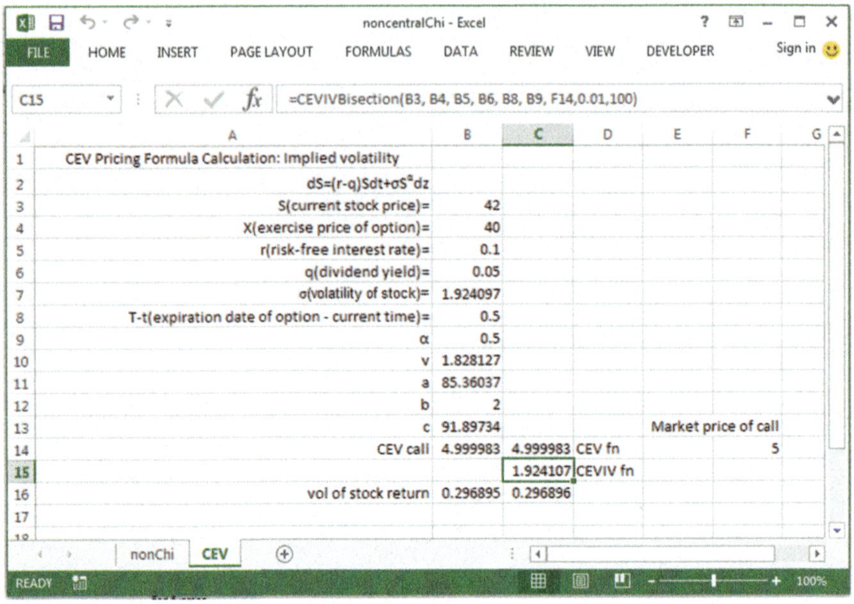

	A	B	C	D	E	F
1	CEV Pricing Formula Calculation: Implied volatility					
2	dS=(r-q)Sdt+σS^αdz					
3	S(current stock price)=	42				
4	X(exercise price of option)=	40				
5	r(risk-free interest rate)=	0.1				
6	q(dividend yield)=	0.05				
7	σ(volatility of stock)=	1.924097				
8	T-t(expiration date of option - current time)=	0.5				
9	α	0.5				
10	v	1.828127				
11	a	85.36037				
12	b	2				
13	c	91.89734			Market price of call	
14	CEV call	4.999983	4.999983	CEV fn		5
15			1.924107	CEVIV fn		
16	vol of stock return	0.296895	0.296896			

The formula of sigma in CEV model in C15 is:

$$= \text{CEVIVBisection}(\text{B3, B4, B5, B6, B8, B9, F14}, 0.01, 100)$$

The value of sigma in C15 is similar to B7 calculated by Goal Seek procedure. In the same way, we can calculate volatility of stock return in C16. The value of volatility of stock return in C16 is also near B16.

27.5 Webservice Excel Function

An URL is a request and response Internet convention. A user would request a URL by typing the URL in the Internet browser and the browser will respond to the request. For example, the user would request the USA Today website by typing in http://www.usatoday.com/ in the browser and the browser would display the USA Today website. A lot of information is returned to the user. The browser would return a lot of text and graphical information and the browser will format text and graphical information.

There are URLs that are constructed to return only data. One popular thing to do is retrieve stock prices from Yahoo.

The following URL will return the last stock price of Microsoft:

http://finance.yahoo.com/d/quotes.csv?s=MSFT&f=l1

The following URL will return the last stock price of IBM:

http://finance.yahoo.com/d/quotes.csv?s=IBM&f=l1

The following URL will return the last stock price of GM:

http://finance.yahoo.com/d/quotes.csv?s=GM&f=l1

The following URL will return the last stock price of Ford:

http://finance.yahoo.com/d/quotes.csv?s=F&f=l1

It's also possible to retrieve the ticker name from Yahoo.

The following URL will return the ticker name of Microsoft:

http://finance.yahoo.com/d/quotes.csv?s=MSFT&f=N

The following URL will return the ticker name of IBM:

http://finance.yahoo.com/d/quotes.csv?s=IBM&f=N

The following URL will return the ticker of GM:

http://finance.yahoo.com/d/quotes.csv?s=GM&f=N

The following URL will return the name of Ford:

http://finance.yahoo.com/d/quotes.csv?s=F&f=N

A person would use the Excel *webservice* function to use an Internet URL.

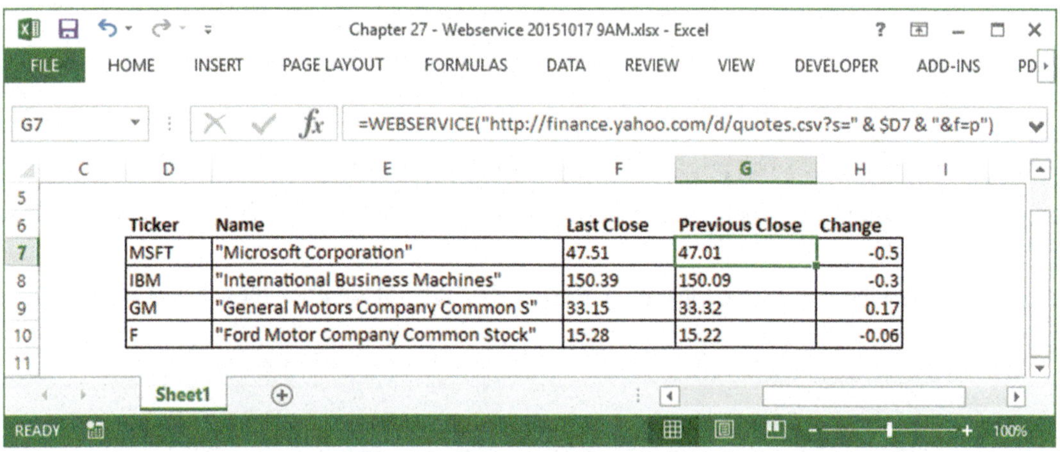

Press Ctrl-~ to show the formulas of a worksheet.

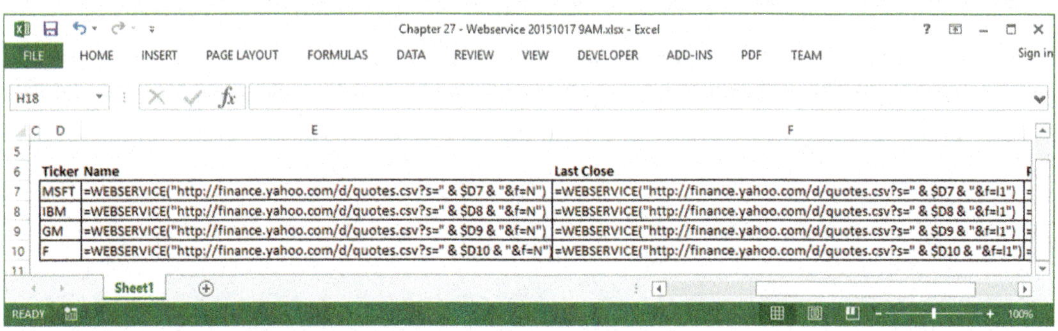

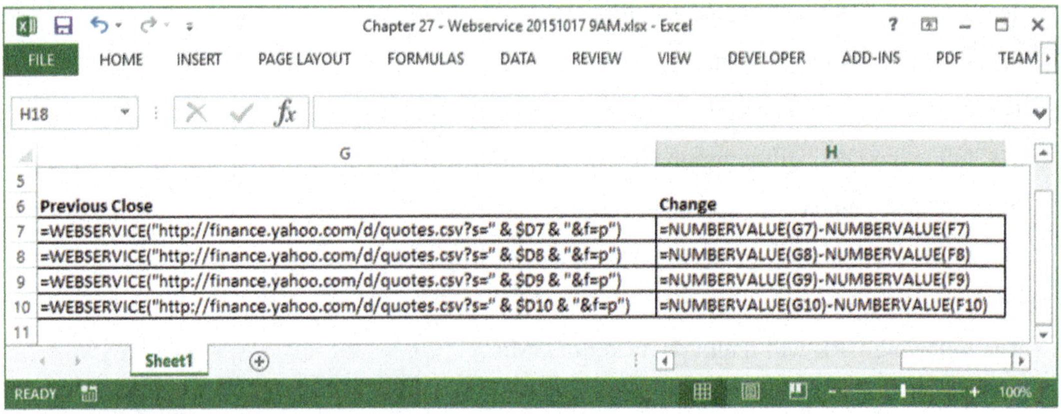

There are two things to notice in the above formulas. To retrieve the previous closing price, the last character in the URL is a "p." The *webservice* function returns everything as a text. The *numbervalue* function in column H is used to convert the *webservice* function from a tel0xt to a number.

27.6 Retrieving a Stock Price for a Specific Date

The URL to retrieve the stock price for MSFT for October 15, 2015, is

http://real-chart.finance.yahoo.com/table.csv?s=MSFT&a=09&b=15&c=2015&d=09&e=15&f=2015&g=d&ignore=.csv

Below shows the *webservice* function retrieving the MSFT stock price for October 15, 2015.

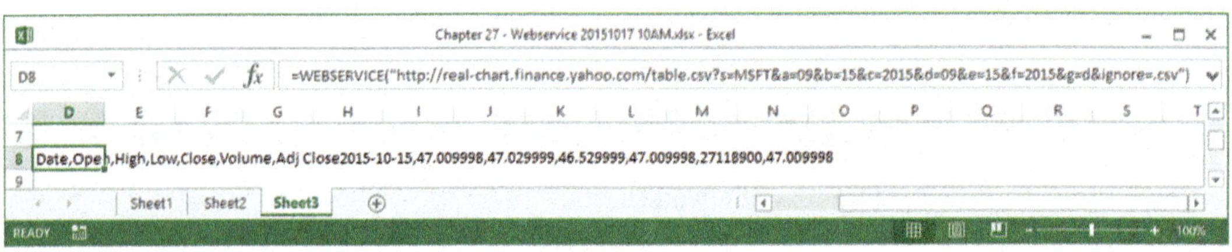

The number after the last comma is the stock price. To get this number, it is necessary to create a custom function. Below is the custom function *TickerInfo* to retrieve a stock price for a specific day.

```
'Example: =TickerInfo("MSFT","Adj Close","10/15/2015")
'sTicker: Ticker symbol
'sAttribute: Date,Open,High,Low,Close,Volume,Adj Close
Function TickerInfo(sTicker As String, sAttribute, sDate As Date) As Variant

    Dim vArrayTicker As Variant

    Dim sURL As String
    Dim sInfo As String
    Dim dDate As Date

    dDate = DateSerial(Year(sDate), Month(sDate), Day(sDate))

    sURL = "http://real-chart.finance.yahoo.com/table.csv?s="
    sURL = sURL & "<Ticker>&a=<Month>&b=<Day>&c=<Year>"
    sURL = sURL & "&d=<Month>&e=<Day>&f=<Year>&g=d&ignore=.csv"

    sURL = VBA.Replace(sURL, "<Ticker>", sTicker)
    sURL = VBA.Replace(sURL, "<Month>", Format(Month(dDate) - 1, "00"))
    sURL = VBA.Replace(sURL, "<Day>", Format(Day(dDate), "00"))
    sURL = VBA.Replace(sURL, "<Year>", Format(Year(dDate), "0000"))

    sInfo = Application.WorksheetFunction.WebService(sURL)

    vArrayTicker = Split(sInfo, ",")

    If Trim(UCase(sAttribute)) = "ADJ CLOSE" Then
        TickerInfo = CDbl(vArrayTicker(12))
    ElseIf Trim(UCase(sAttribute)) = "VOLUME" Then
        TickerInfo = CDbl(vArrayTicker(11))
    ElseIf Trim(UCase(sAttribute)) = "CLOSE" Then
        TickerInfo = CDbl(vArrayTicker(10))
    ElseIf Trim(UCase(sAttribute)) = "LOW" Then
        TickerInfo = CDbl(vArrayTicker(9))
    ElseIf Trim(UCase(sAttribute)) = "HIGH" Then
        TickerInfo = CDbl(vArrayTicker(8))
    ElseIf Trim(UCase(sAttribute)) = "OPEN" Then
        TickerInfo = CDbl(vArrayTicker(7))
    ElseIf Trim(UCase(sAttribute)) = "DATE" Then
        TickerInfo = Format(CDate(Mid(vArrayTicker(6), 11, 10)), "MMM-DD-YYYY")
    End If
End Function
```

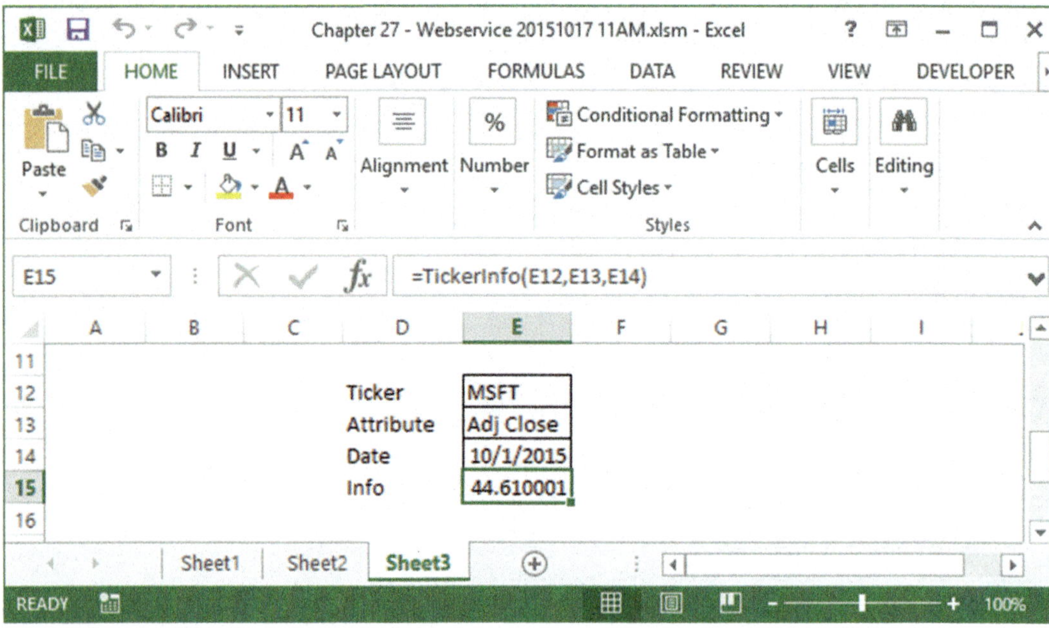

Using the custom function *TickerInfo*, we were able to get the price change from August 2015 to September 30, 2015.

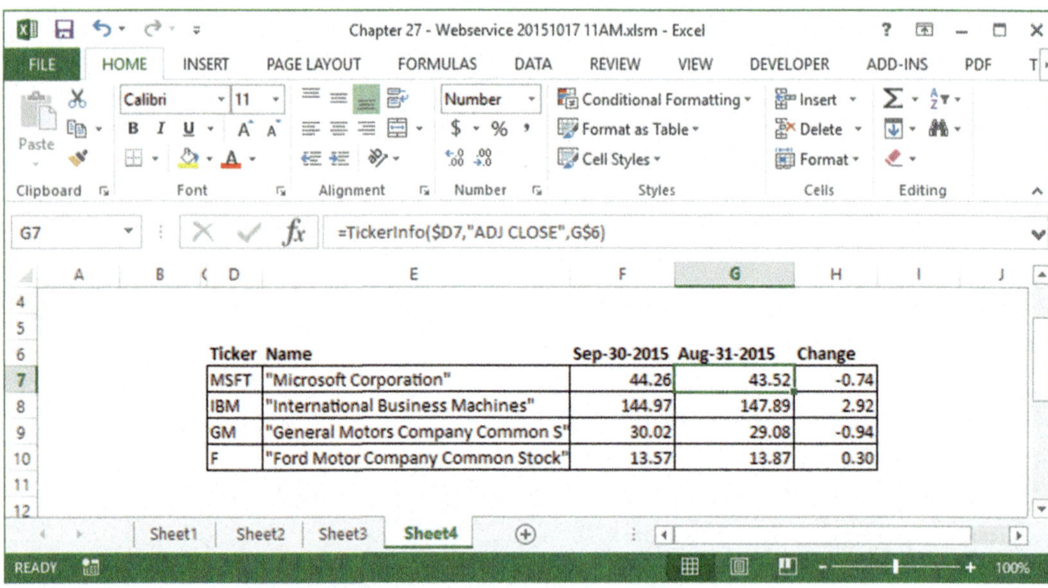

Press Ctrl-~ to show the formulas of the above worksheet.

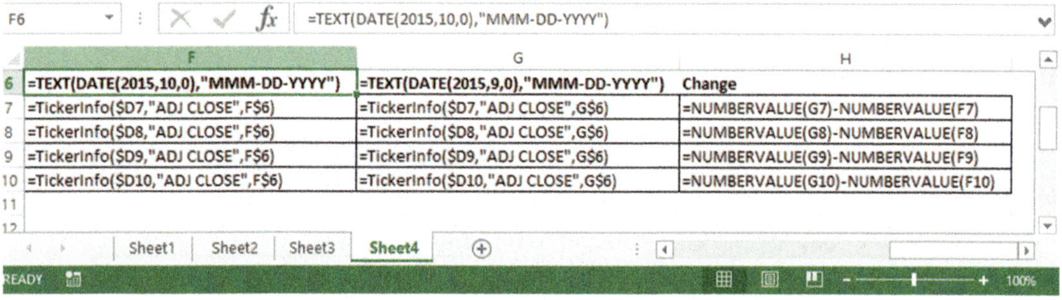

F6 =TEXT(DATE(2015,10,0),"MMM-DD-YYYY")

	F	G	H
6	=TEXT(DATE(2015,10,0),"MMM-DD-YYYY")	=TEXT(DATE(2015,9,0),"MMM-DD-YYYY")	Change
7	=TickerInfo($D7,"ADJ CLOSE",F$6)	=TickerInfo($D7,"ADJ CLOSE",G$6)	=NUMBERVALUE(G7)-NUMBERVALUE(F7)
8	=TickerInfo($D8,"ADJ CLOSE",F$6)	=TickerInfo($D8,"ADJ CLOSE",G$6)	=NUMBERVALUE(G8)-NUMBERVALUE(F8)
9	=TickerInfo($D9,"ADJ CLOSE",F$6)	=TickerInfo($D9,"ADJ CLOSE",G$6)	=NUMBERVALUE(G9)-NUMBERVALUE(F9)
10	=TickerInfo($D10,"ADJ CLOSE",F$6)	=TickerInfo($D10,"ADJ CLOSE",G$6)	=NUMBERVALUE(G10)-NUMBERVALUE(F10)

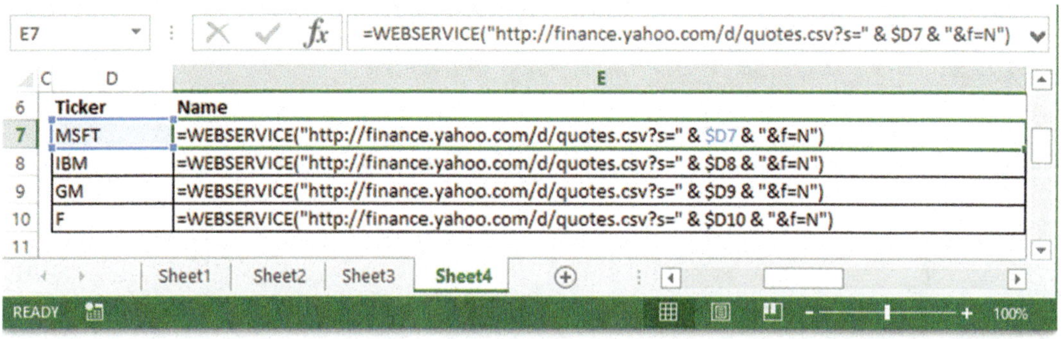

E7 =WEBSERVICE("http://finance.yahoo.com/d/quotes.csv?s=" & $D7 & "&f=N")

	D	E
6	Ticker	Name
7	MSFT	=WEBSERVICE("http://finance.yahoo.com/d/quotes.csv?s=" & $D7 & "&f=N")
8	IBM	=WEBSERVICE("http://finance.yahoo.com/d/quotes.csv?s=" & $D8 & "&f=N")
9	GM	=WEBSERVICE("http://finance.yahoo.com/d/quotes.csv?s=" & $D9 & "&f=N")
10	F	=WEBSERVICE("http://finance.yahoo.com/d/quotes.csv?s=" & $D10 & "&f=N")

There are two things to note. The *text* function in cell F6 is used to format a date. The 0 in the *date* function date(2015,10,0) causes the function to return to the last day of September.

27.7 Calculated Holiday List

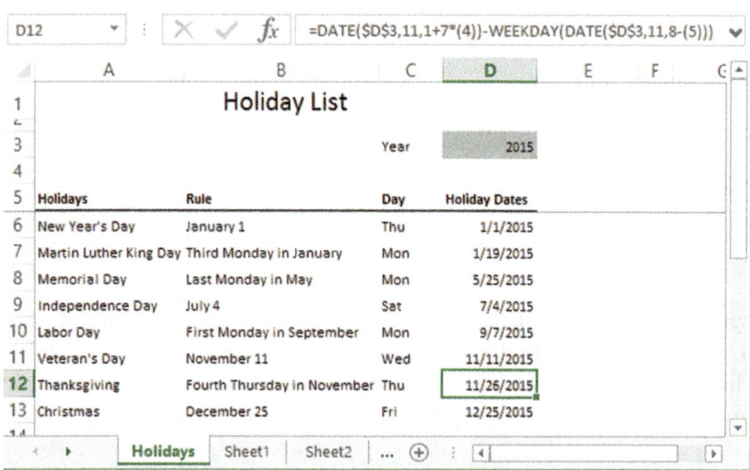

D12 =DATE($D$3,11,1+7*(4))-WEEKDAY(DATE($D$3,11,8-(5)))

Holiday List

Year 2015

Holidays	Rule	Day	Holiday Dates
New Year's Day	January 1	Thu	1/1/2015
Martin Luther King Day	Third Monday in January	Mon	1/19/2015
Memorial Day	Last Monday in May	Mon	5/25/2015
Independence Day	July 4	Sat	7/4/2015
Labor Day	First Monday in September	Mon	9/7/2015
Veteran's Day	November 11	Wed	11/11/2015
Thanksgiving	Fourth Thursday in November	Thu	11/26/2015
Christmas	December 25	Fri	12/25/2015

Financial calculation often needs to take into consideration of holidays. Above is a list of holidays for the year of 2015 that is dynamically calculated using Excel functions. Below shows how each holiday is calculated.

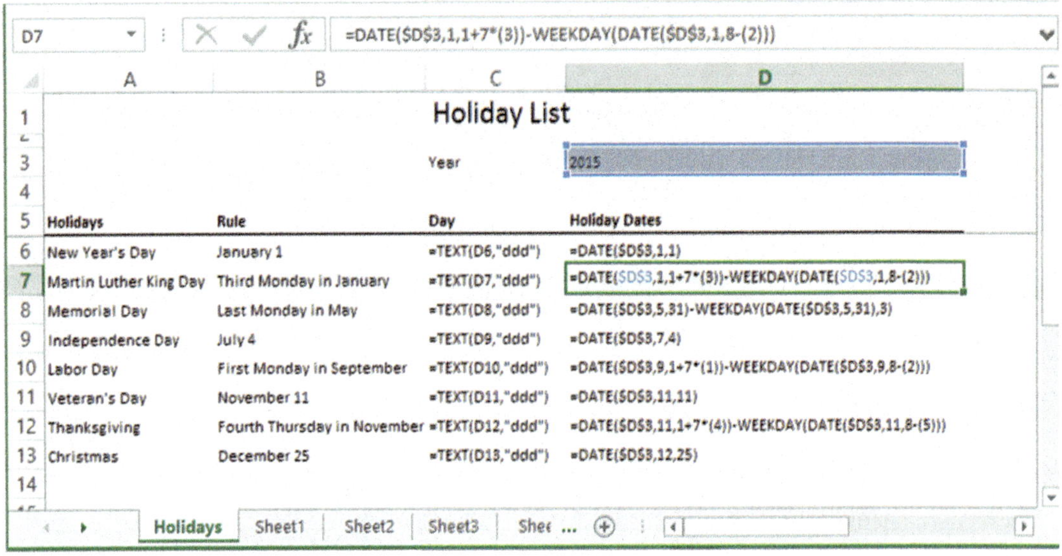

27.8 Calculating Historical Volatility

Another way to get the volatility value is to calculate historical volatility. It's a lot of effort to do this because it takes a lot of effort to get the historical price of a stock for each specific day. We will use our custom Excel function *TickerInfo* to solve this problem.

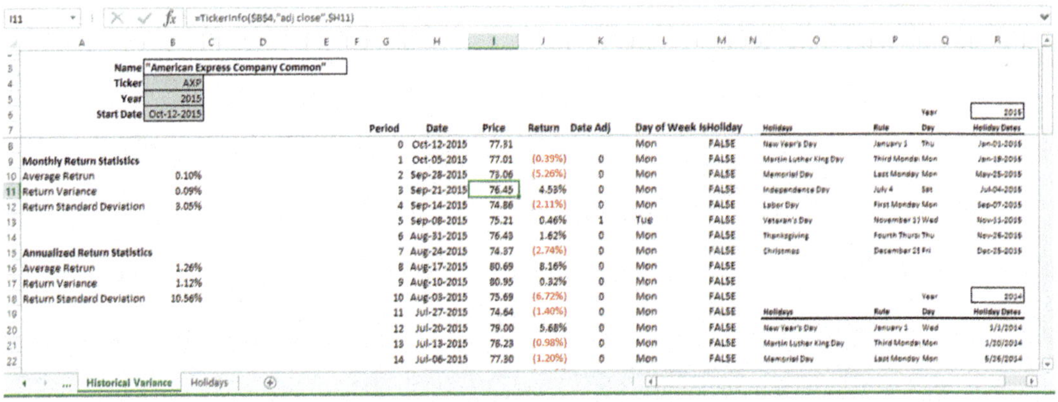

Above is a spreadsheet that calculates a 52-week historical variance for any stock. There are three input values to the spreadsheet. The three input values are "Ticker," "Year," and "Start Date." In calculating the historical variance, we have to be concerned about holidays because there are no stock prices on holidays. The "Year" input is used by the calculated calendars in columns O to R.

Below shows the formulas for the spreadsheet.

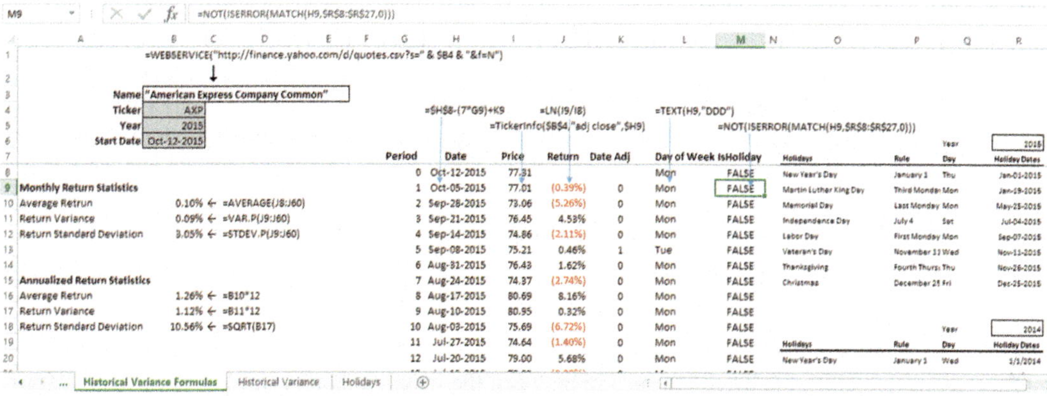

Every row in the date column is 7 days prior to the previous row. In cell H13 the date should have been Sep-07-2015. In the 2015 holiday calendar, column O shows that Sep-07-2015 is a holiday; therefore, there is no stock price. Because of this we have to push the date forward by one day. Pushing the day forward is done in column K.

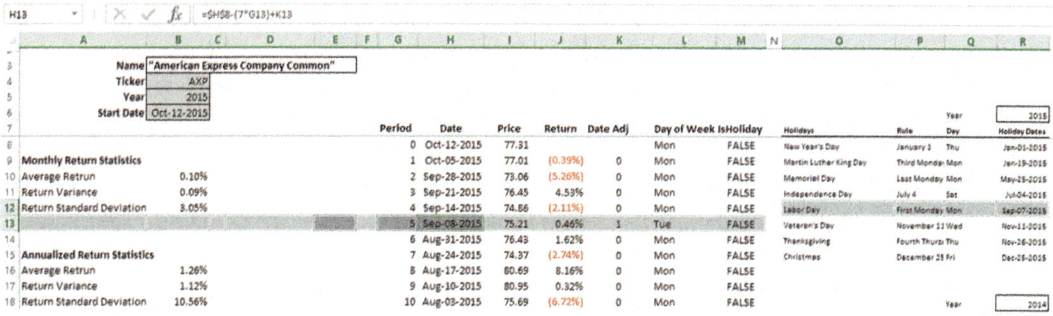

27.9 Summary

In the inputs of Black and Scholes formula, only the volatility can't be measured directly. If we use the market price of an option, we can estimate the volatility implied by option market price. In this chapter, we introduce Corrado and Miller's approximation to estimate implied volatility. Next, we use the Goal Seek facility of Excel to solve the root of nonlinear equation which is based on Newton–Raphson method. We apply a VBA function to calculate implied volatility by using bisection method.

We also calculated a 52-week volatility of stock. This is a very difficult task because it is very labor intensive to get the stock price for all 52 weeks. To make it more difficult, we have to take into consideration the holidays. To reduce the burden of getting the 52 historical stock prices, we created a custom function called *TickerInfo*. This custom function uses the Excel function *webservice*, which goes to Yahoo to retrieve each of the 52 historical stock prices.

We also showed the Excel equations to dynamically calculate holidays for any particular year.

Appendix 27.1: Application of CEV Model to Forecasting Implied Volatilities for Options on Index Futures

In this appendix, we use CEV model to forecast implied volatility (called IV hereafter) of options on index futures. Cox (1975) and Cox and Ross (1976) developed the "constant elasticity of variance (CEV) model" which incorporates an observed market phenomenon that the underlying asset variance tends to fall as the asset price increases (and vice versa). The advantage of CEV model is that it can describe the interrelationship between stock prices and its volatility. The constant elasticity of variance (CEV) model for a stock price, S, can be represented as follows:

$$dS = (r - q)Sdt + \delta S^{\alpha}dZ \tag{27.1}$$

where r is the risk-free rate, q is the dividend yield, dZ is a Wiener process, δ is a volatility parameter, and α is a positive constant. The relationship between the instantaneous volatility of the asset return, $\sigma(S, t)$, and parameters in CEV model can be represented as:

$$\sigma(S, t) = \delta S^{\alpha-1} \tag{27.2}$$

When $\alpha = 1$, the CEV model is the geometric Brownian motion model we have been using up to now. When $\alpha < 1$, the volatility increases as the stock price decreases. This creates a probability distribution similar to that observed for equities with a heavy left tail and a less heavy right tail. When $\alpha > 1$, the volatility increases as the stock price increases, giving a probability distribution with a heavy right tail and a less left tail. This corresponds to a volatility smile where the implied volatility is an increasing function of the strike price. This type of volatility smile is sometimes observed for options on futures.

The formula for pricing a European call option in CEV model is:

$$C_t = \begin{cases} S_t e^{-q\tau}[1 - \chi^2(a,\ b+2,\ c)] - Ke^{-r\tau}\chi^2(c,\ b,\ a) \ \ when\ \ \alpha < 1 \\ S_t e^{-q\tau}[1 - \chi^2(c,\ -b,\ a)] - Ke^{-r\tau}\chi^2(a,\ 2-b,\ c) \ \ when\ \ \alpha > 1 \end{cases} \tag{27.3}$$

where $a = \frac{\left[Ke^{-(r-q)\tau}\right]^{2(1-\alpha)}}{(1-\alpha)^2 \upsilon}$, $b = \frac{1}{1-\alpha}$, $c = \frac{S_t^{2(1-\alpha)}}{(1-\alpha)^2 \upsilon}$, $\upsilon = \frac{\delta^2}{2(r-q)(\alpha-1)}\left[e^{2(r-q)(\alpha-1)\tau} - 1\right]$ and $\chi^2(z, k, v)$ is the cumulative probability that a variable with a noncentral χ^2 distribution[1] with noncentrality parameter v and k degrees of freedom is less than z. Hsu, Lin, and Lee (2008) provided the detailed derivation of approximative formula for CEV model. Based on the approximated formula, CEV model can reduce computational and implementation costs rather than the complex models such as jump-diffusion stochastic volatility model. Therefore, CVE model with one more parameter than Black–Scholes–Merton option pricing model (BSM) can be a better choice to improve the performance of predicting implied volatilities of index options (Singh and Ahmad 2011).

Beckers (1980) investigate the relationship between the stock price and its variance of returns by using an approximative closed-form formulas for CEV model based on two special cases of the constant elasticity class ($\alpha = 1 or 0$). Based on the significant relationship between the stock price and its volatility in the empirical results, Beckers (1980) claimed that CEV model in terms of noncentral chi-square distribution performs better than BC model in terms of log-normal distribution in description of stock price behavior. MacBeth and Merville (1980) is the first paper to empirically test the

[1] The calculation process of $\chi^2(z, k, v)$ value can be referred to Ding (1992). The complementary noncentral chi-square distribution function can be expressed as an infinite double sum of gamma function, which can be referred to Benton and Krishnamoorthy (2003).

performance of CEV model. Their empirical results show the negative relationship between stock prices and its volatility of returns; that is, the elasticity class is less than 2 (i.e., $\alpha < 2$). Jackwerth and Rubinstein (2001) and Lee et al. (2004) used S&P 500 index options to do empirical work and found that CEV model performed well because it took account the negative correlation between the index level and volatility into model assumption. Pun and Wong (2013) combine asymptotic approach with CEV model to price American options. Larguinho et al. (2013) compute Greek letters under CEV model to measure different dimension to the risk in option positions and investigate leverage effects in option markets.

Since the future price equals the expected future spot price in a risk-neutral measurement, the S&P 500 index futures prices have same distribution property of S&P 500 index prices. Therefore, for a call option on index futures can be given by (27.3) with S_t replaced by F_t and $q = r$ as (27.4)[2]:

$$C_t^F = \begin{cases} e^{-r\tau}(F_t[1-\chi^2(a,\ b+2,\ c)] - K\chi^2(c,\ b,\ a)) & when \quad \alpha < 1 \\ \\ e^{-r\tau}(F_t[1-\chi^2(c,\ -b,\ a)] - K\chi^2(a,\ 2-b,\ c)) & when \quad \alpha > 1 \end{cases} \tag{27.4}$$

where $a = \frac{K^{2(1-\alpha)}}{(1-\alpha)^2\upsilon}$, $b = \frac{1}{1-\alpha}$, $c = \frac{F_t^{2(1-\alpha)}}{(1-\alpha)^2\upsilon}$, $\upsilon = \delta^2\tau$

The MATLAB code to price European call option on the future price using CEV model is as follows:

[2] When substituting $q = r$ into $\upsilon = \frac{\delta^2}{2(r-q)(\alpha-1)}\left[e^{2(r-q)(\alpha-1)\tau} - 1\right]$, we can use L'Hospital's rule to obtain υ. Let $x = r - q$, and then

$$\lim_{x\to 0}\frac{\delta^2\left[e^{2x(\alpha-1)\tau}-1\right]}{2x(\alpha-1)} = \lim_{x\to 0}\frac{\dfrac{\partial\delta^2\left[e^{2x(\alpha-1)\tau}-1\right]}{\partial x}}{\dfrac{\partial 2x(\alpha-1)}{\partial x}} = \lim_{x\to 0}\frac{(2(\alpha-1)\tau)\delta^2\left[e^{2x(\alpha-1)\tau}\right]}{2(\alpha-1)} = \lim_{x\to 0}\frac{\tau\delta^2\left[e^{2x(\alpha-1)\tau}\right]}{1}$$

$$= \tau\delta^2$$

```
function [ call ] = CevFCall(F,K,T,r,sigma,alpha)
% Compute European Call option on future price using CEV Model
% F is future price
% K is vector for options with different strike price on the same day
% Scaling S & K in the next tree line to enable
% APE=@(x) sum(abs(data(:,4)-CevFCall(data(:,7), data(:,1), data(:, 10)/360,
data(:,11), x(1),x(2))))
% PPE=@(x) sum(abs(data(:,4)-CevFCall(data(:,7), data(:,1), data(:, 10)/360,
data(:,11), x(1),x(2)))./data(:,4))
% SSE=@(x) sum(abs(data(:,4)-CevFCall(data(:,7), data(:,1), data(:, 10)/360,
data(:,11), x(1),x(2))).^2)
% [x,fval,exitflag, output] = fminsearch(SSE,[0.27,-1])
% Volatility = blsimpv(Price, Strike, Rate, Time, Value, Limit, Tolerance,
% Class)
% ctrl+c to stop Matlab when it is busy
KK = K;
F = F./K;
K = ones(size(K));

if (alpha ~= 1)

    v = (sigma^2)*T;

    a = K.^(2*(1-alpha))./(v*(1-alpha)^2);

    b = ones(size(K)).*(1/(1-alpha));

    c = (F.^ (2 *(1-alpha)))./(v*(1-alpha)^2);
    % Multiplying the call price by KK enable us to scale back
    % if (0 < alpha && alpha < 1)
    if (alpha < 1)
        call = KK.*( F.*( ones(size(K)) - ncx2cdf( a,b + 2,c)) -
K.*(ncx2cdf(c,b,a))).*exp(-r.*T);
    elseif (alpha > 1)
        call = KK.*( F.*ncx2cdf(c,-b,a) - K.*ncx2cdf(a,2-b,c)).*exp(-r.*T);
    end
else
    call = 0; % function not defined for alpa < 0 or = 1
end

end
```

The procedures to obtain estimated parameters of CEV model are as follows:

1. Let $C^F_{i,n,t}$ be the market price of the nth option contract in category i, $C^{\hat{F}}_{i,n,t}(\delta_0, \alpha_0)$ be the model option price determined by CEV model in (27.4) with the initial value of parameters, $\delta = \delta_0$ and $\alpha = \alpha_0$. For nth option contract in category i at date t, the difference between market price and model option price can be described as:

$$\varepsilon^F_{i,n,t} = C^F_{i,n,t} - C^{\hat{F}}_{i,n,t}(\delta_0, \alpha_0) \tag{27.5}$$

The MATLAB code to find initial value of parameters in CEV model is as follows:

```
function STradingTM=cevslpine(TradingTM,TM)
sigma=[0.1:0.05:0.7];
alpha=[-0.5; -0.3; -0.1; 0.1; 0.3; 0.5; 0.7; 0.9];
LA=length(alpha);
LB=length(sigma);

L=length(TradingTM);
Tn=ones(L,1);
Tr=ones(L,1);
y=ones(L,length(alpha),length(sigma));


a=ones(L,1);
b=ones(L,1);
iniError=ones(L,1);
inisigmaplace=ones(L,1);
iniaplhaplace=ones(L,1);
inisigma=ones(L,1);
inialpha=ones(L,1);

for i=1:L

    Tn(i)=Tr(i)+TradingTM(i,1)-1;
    if(i<L) Tr(i+1)=Tn(i)+1; end
end


for k=1:L
for i=1:LA
  for j=1:LB
    y(k,i,j)= sum(abs(TM(Tr(k):Tn(k),2)-CevFCall(TM(Tr(k):Tn(k),3), TM(Tr(k):Tn(k),1),
TM(Tr(k):Tn(k),4)/360.0, TM(Tr(k):Tn(k),5), sigma(j), alpha(i))));
  end
end
[~,b]=min(y(k,:,:));
  [iniError(k),inisigmaplace(k)]=min(min(y(k,:,:)));
  inialphaplace(k)=b(inisigmaplace(k));
  inisigma(k)=sigma(inisigmaplace(k));
  inialpha(k)=alpha(inialphaplace(k));
  disp(sprintf('iteration %d contract %d alpha and %d sigma', k, i,j));
end

STradingTM=[TradingTM Tr Tn inisigma inialpha];
end
```

2. For each date t, we can obtain the optimal parameters in each group by solving the minimum value of absolute pricing errors (minAPE) as:

$$\mathrm{minAPE}_{i,t} = \min_{\delta_0,\,\alpha_0} \sum_{n=1}^{N} \left| \varepsilon_{i,n,t}^{F} \right| \tag{27.6}$$

where N is the total number of option contracts in group i at time t.

3. Using optimization function in MATLAB to find a minimum value of the unconstrained multivariable function, the function code is as follows:

$$[\mathrm{x},\ \mathrm{fval}] = \mathrm{fminunc}(\mathrm{fun},\ \mathrm{x}_0) \tag{27.7}$$

where x is the optimal parameters of CEV model, fval is the local minimum value of minAPE, fun is the specified MATLAB function of (27.4), and x_0 is the initial points of parameters obtained in step (1). The algorithm of fminunc function is based on quasi-Newton method. The MATLAB code is as follows:

```
function [ call ] = CevFCalltr(F,K,T,r,sigma,alpha)
% Compute European Call option on future price using CEV Model
% F is future price
% K is vector for options with different strike price on the same day
% Scaling S & K in the next tree line to enable
% APE=@(x) sum(abs(data(:,4)-CevFCall(data(:,7), data(:,1), data(:, 10)/360,
data(:,11), x(1),x(2))))
% PPE=@(x) sum(abs(data(:,4)-CevFCall(data(:,7), data(:,1), data(:, 10)/360,
data(:,11), x(1),x(2)))./data(:,4))
% SSE=@(x) sum(abs(data(:,4)-CevFCall(data(:,7), data(:,1), data(:, 10)/360,
data(:,11), x(1),x(2))).^2)
% [x,fval,exitflag, output] = fminsearch(SSE,[0.27,-1])
% Volatility = blsimpv(Price, Strike, Rate, Time, Value, Limit, Tolerance,
% Class)
% ctrl+c to stop Matlab when it is busy

if (alpha ~= 1)

    v = (sigma^2)*T;

    a = K.^(2*(1-alpha))./(v*(1-alpha)^2);

    b = ones(size(K)).*(1/(1-alpha));

    c = (F.^ (2 *(1-alpha)))./(v*(1-alpha)^2);
    % Multiplying the call price by KK enable us to scale back
    % if (0 < alpha && alpha < 1)
    if (alpha < 1)
        call = ( F.*( ones(size(K)) - ncx2cdf( a,b + 2,c)) -
K.*(ncx2cdf(c,b,a))).*exp(-r.*T);
    elseif (alpha > 1)
        call =( F.*ncx2cdf(c,-b,a) - K.*ncx2cdf(a,2-b,c)).*exp(-r.*T);
    end
else
    call = 0; % function not defined for alpa < 0 or = 1
end

end

function EstCev=CevIVIA(Ini_id, Ini_ed, STradingTM,TM)

L=Ini_ed-Ini_id+1;
Tr=STradingTM(:,3);
Tn=STradingTM(:,4);
x_1=STradingTM(:,5);
x_2=STradingTM(:,6);
```

```
 EstCev=ones(L,9);
 CIVAPE=ones(L,1);
 CIAAPE=ones(L,1);
 CErrorAPE=ones(L,1);
 CIVPPE=ones(L,1);
 CIAPPE=ones(L,1);
 CErrorPPE=ones(L,1);
 CIVSSE=ones(L,1);
 CIASSE=ones(L,1);
 CErrorSSE=ones(L,1);

 %countforloop=0;

fileID=fopen('EstCev.txt', 'w');
%parfor i=1:L
parfor i=1:L

  Id_global=Ini_id+i-1;
      APE=@(x) sum(abs(TM(Tr(i):Tn(i),2)-CevFCall(TM(Tr(i):Tn(i),3), TM(Tr(i):Tn(i),1),
TM(Tr(i):Tn(i),4)/360.0, TM(Tr(i):Tn(i),5), x(1), x(2))));
   % [x,fval] =fminsearch(APE, [x0(i), 0.5], options);
     % using fmincon will cause error because the warning :Warning: Large-scale (trust
region) method does not currently solve this type of problem,
  %                                                                  switching to medium-
scale (line search).
%  disp(sprintf('fminunc doing %d contract with initial sigma %d  and alpha %d sigma',
i, x_1(i), x_2(i)));
   options = psoptimset('UseParallel', 'always', 'CompletePoll', 'on', 'Vectorized',
'off', 'TimeLimit', 30, 'TolFun', 1e-2, 'TolX', 1e-4);
  [x,fval] =fminunc(APE, [x_1(i), x_2(i)],options);
    disp(sprintf('%d Id_global contract, %d contract local minimum IV is %d and aplha
is %d, minAPE is %d with initial sigma %d  and alpha %d',Id_global, i, x(1), x(2),
fval, x_1(i), x_2(i)));

    fprintf(fileID, '%d Id_global contract, %d contract local minimum IV is %d and
aplha is %d, minAPE is %d where initial sigma %d  and alpha %d',Id_global, i, x(1),
x(2), fval, x_1(i), x_2(i));

    CIVAPE(i)=x(1);
    CIAAPE(i)=x(2);
     CErrorAPE(i)=fval;
      CErrorPPE(i)=abs(sum((TM(Tr(i):Tn(i),2)-CevFCall(TM(Tr(i):Tn(i),3),
TM(Tr(i):Tn(i),1), TM(Tr(i):Tn(i),4)/360.0,TM(Tr(i):Tn(i),5), x(1),
x(2)))./TM(Tr(i):Tn(i),2)));
   CErrorSSE(i)=sum(abs(TM(Tr(i):Tn(i),2)-CevFCall(TM(Tr(i):Tn(i),3),
TM(Tr(i):Tn(i),1), TM(Tr(i):Tn(i),4)/360.0,TM(Tr(i):Tn(i),5), x(1), x(2))).^2);

end
disp(sprintf('farloop is over'));
fclose(fileID);
EstCev=[CIVAPE CIAAPE CErrorAPE CIVPPE CIAPPE CErrorPPE CIVSSE CIASSE CErrorSSE];

%matlabpool close
end
```

The Data is the options on S&P 500 index futures expired within January 1, 2010, to December 31, 2013, which are traded at the Chicago Mercantile Exchange (CME).[3] The reason for using options on S&P 500 index futures instead of S&P 500 index is to eliminate from nonsimultaneous price

[3] Nowadays Chicago Mercantile Exchange (CME), Chicago Board of Trade (CBOT), New York Mercantile Exchange (NYMEX), and Commodity Exchange (COMEX) are merged and operate as designated contract markets (DCM) of the CME Group which is the world's leading and most diverse derivatives marketplace. Website of CME group: http://www.cmegroup.com/

effects between options and its underlying assets (Harvey and Whaley 1991). The option and future markets are closed at 3:15 pm central time (CT), while stock market is closed at 3 pm CT. Therefore, using closing option prices to estimate the volatility of underlying stock return is problematic even though the correct option pricing model is used. In addition to no nonsynchronous price issue, the underlying assets, S&P 500 index futures, do not need to be adjusted for discrete dividends. Therefore, we can reduce the pricing error in accordance with the needless dividend adjustment. According to the suggestions in Harvey and Whaley (1991, 1992a, b), we select simultaneous index option prices and index future prices to do empirical analysis.

The risk-free rate is based on 1-year Treasury Bill from Federal Reserve Bank of St. Louis.[4] Daily closing price and trading volumes of options on S&P 500 index futures and its underlying asset can be obtained from Datastream.

The future options expired on March, June, and September in both 2010 and 2011 are selected because they have over 1 year trading date (above 252 observations), while other options only have more or less 100 observations. Studying futures option contracts with same expired months in 2010 and 2011 will allow the examination of IV characteristics and movements over time as well as the effects of different market climates.

In order to ensure reliable estimation of IV, we estimate market volatility by using multiple option transactions instead of a single contract. For comparing prediction power of Black model and CEV model, we use all future options expired in 2010 and 2013 to generate implied volatility surface. Here we exclude the data based on the following criteria:

1. IV cannot be computed by Black model.
2. Trading volume is lower than ten for excluding minuscule transactions.
3. Time to maturity is less than 10 days for avoiding liquidity-related biases.
4. Quotes not satisfying the arbitrage restriction: excluding option contact if its price larger than the difference between S&P 500 index future and exercise price.
5. Deep-in/out-of-money contacts where the ratio of S&P 500 index future price to exercise price is either above 1.2 or below 0.8.

After arranging data based on these criteria, we still have 30,364 observations of future options which are expired within the period of 2010–2013. The period of option prices is from March 19, 2009, to November 5, 2013.

To deal with moneyness- and maturity-related biases, we use the "implied-volatility matrix" to find proper parameters in CEV model. The option contracts are divided into nine categories by moneyness and time to maturity. Option contracts are classified by moneyness level as at-the-money (ATM), out-of-the-money (OTM), or in-the-money (ITM) based on the ratio of underlying asset price, S, to exercise price, K. If an option contract with S/K ratio is between 0.95 and 1.01, it belongs to ATM category. If its S/K ratio is higher (lower) than 1.01 (0.95), the option contract belongs to ITM (OTM) category.

According to the large observations in ATM and OTM, we divide moneyness-level group into five levels: ratio above 1.01, ratio between 0.98 and 1.01, ratio between 0.95 and 0.98, ratio between 0.90 and 0.95, and ratio below 0.90. By expiration day, we classified option contracts into short term (less than 30 trading days), medium term (between 30 and 60 trading days), and long term (more than 60 trading days).

In Fig. 27.1, we find that each option on index future contract's IV estimated by Black model varies across moneyness and time to maturity. This graph shows volatility skew (or smile) in options

[4] Website of Federal Reserve Bank of St. Louis: http://research.stlouisfed.org/

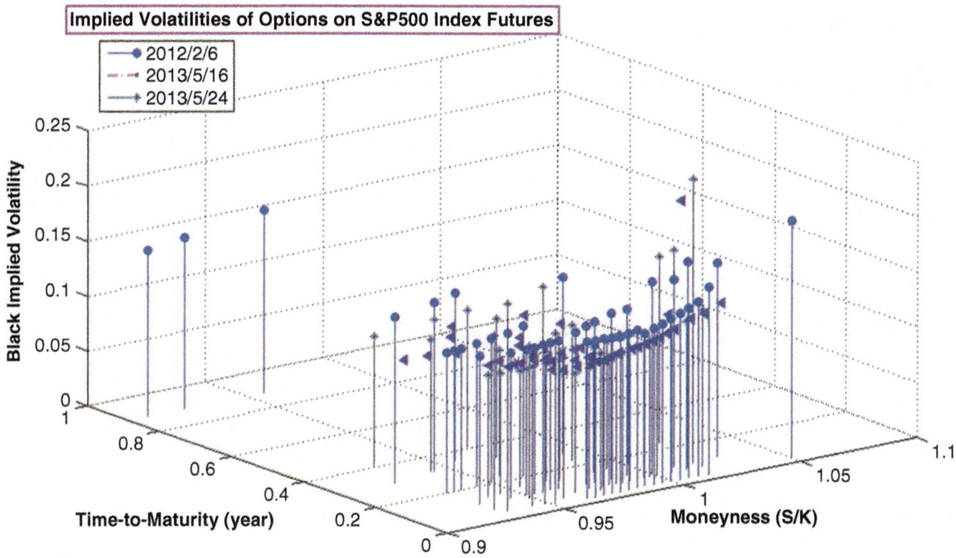

Fig. 27.1 Implied volatilities in black model

Table 27.1 Average daily and total number of observations in each group

Time to maturity (TM)	TM < 30		30 ≦ TM ≦ 60		TM > 60		All TM	
Moneyness (S/K ratio)	Daily obs.	Total obs.	Daily obs.	Total obs.	Daily obs.	Total obs.	Daily obs.	Total obs.
S/K ratio >1.01	1.91	844	1.64	499	1.53	462	2.61	1805
0.98≦S/K ratio ≦1.01	4.26	3217	2.58	1963	2.04	1282	6.53	6462
0.95≦S/K ratio <0.98	5.37	4031	3.97	3440	2.58	1957	9.32	9428
0.9≦S/K ratio <0.95	4.26	3194	4.37	3825	3.27	2843	9.71	9862
S/K ratio <0.9	2.84	764	2.68	798	2.37	1244	4.42	2806
All ratio	12.59	12,050	10.78	10,526	7.45	7788	27.62	30,364

The whole period of option prices is from March 19, 2009, to November 5, 2013. Total observations is 30, 364. The lengths of period in groups are various. The range of lengths is from 260 (group with ratio below 0.90 and time to maturity within 30 days) to 1100 (whole samples)

on S&P 500 index futures, i.e., the implied volatilities decrease as the strike price increases (the moneyness level decreases).

Even though everyday implied volatility surface changes, this characteristic still exists. Therefore, we divided future option contracts into a six by four matrix based on moneyness and time-to-maturity levels when we estimate implied volatilities of future options in CEV model framework in accordance with this character. The whole option samples expired within the period of 2010–2013 contains 30,364 observations. The whole period of option prices is from March 19, 2009, to November 5, 2013. The observations for each group are presented in Table 27.1.

Since most trades are in the future options with short time to maturity, the estimated implied volatility of the option samples in 2009 may be significantly biased because we didn't collect the future options expired in 2009. Therefore, we only use option prices in the period between January 1, 2010, and November 5, 2013, to estimate parameters of CEV model. In order to find global optimization instead of local minimum of absolute pricing errors, the ranges for searching suitable δ_0 and α_0 are set as $\delta_0 \in [0.01, 0.81]$ with interval 0.05 and $\alpha_0 \in [-0.81, 1.39]$ with interval 0.1, respectively. First, we find the value of parameters, (δ_0, α_0), within the ranges such that minimize value of absolute pricing errors in (27.5). Then we use this pair of parameters, (δ_0, α_0), as optimal

Table 27.2 Initial parameters of CEV model for estimation procedure

Time to maturity (TM)	TM < 30		30 ≦ TM ≦ 60		TM > 60		All TM	
Moneyness (S/K ratio)	α_0	δ_0	α_0	δ_0	α_0	δ_0	α_0	δ_0
S/K ratio >1.01	0.677	0.400	0.690	0.433	0.814	0.448	0.692	0.429
0.98≦S/K ratio≦1.01	0.602	0.333	0.659	0.373	0.567	0.361	0.647	0.345
0.95≦S/K ratio <0.98	0.513	0.331	0.555	0.321	0.545	0.349	0.586	0.343
0.9≦S/K ratio <0.95	0.502	0.344	0.538	0.332	0.547	0.318	0.578	0.321
S/K ratio <0.9	0.777	0.457	0.526	0.468	0.726	0.423	0.709	0.423
All ratio	0.854	0.517	0.846	0.512	0.847	0.534	0.835	0.504

The sample period of option prices is from January 1, 2010, to November 5, 2013. During the estimating procedure for initial parameters of CEV model, the volatility for S&P 500 index futures equals to $\delta_0 S^{\alpha_0 - 1}$

Table 27.3 Total number of observations and trading days in each group

Time to maturity (TM)	TM < 30		30 ≦ TM ≦ 60		TM > 60		All TM	
Moneyness (S/K ratio)	Days	Total obs.	Days	Total obs.	Days	Total obs.	Days	Total obs.
S/K ratio >1.01	172	272	104	163	81	122	249	557
0.98≦S/K ratio≦1.01	377	1695	354	984	268	592	448	3271
0.95≦S/K ratio <0.98	362	1958	405	1828	349	1074	457	4860
0.9≦S/K ratio <0.95	315	919	380	1399	375	1318	440	3636
S/K ratio <0.9	32	35	40	73	105	173	134	281
All ratio	441	4879	440	4447	418	3279	461	12,605

The subsample period of option prices is from January 1, 2012, to November 5, 2013. Total observations is 13, 434. The lengths of period in groups are various. The range of lengths is from 47 (group with ratio below 0.90 and time to maturity within 30 days) to 1100 (whole samples). The range of daily observations is from 1 to 30

initial estimates in the procedure of estimating local minimum minAPE based on steps (1)–(3). The initial parameter setting of CEV model is presented in Table 27.2.

In Table 27.2, the average sigma is almost the same, while the average alpha value in either each group or whole sample is less than one. This evidence implies that the alpha of CEV model can capture the negative relationship between S&P 500 index future prices and its volatilities shown in Fig. 27.1. The instant volatility of S&P 500 index future prices equals to $\delta_0 S^{\alpha_0 - 1}$ where S is S&P 500 index future prices and δ_0 and α_0 are the parameters in CEV model. The estimated parameters in Table 27.2 are similar across time-to-maturity level but volatile across moneyness.

Because of the implementation and computational costs, we select the subperiod from January 2012 to November 2013 to analyze the performance of CEV model. The total number of observations and the length of trading days in each group are presented in Table 27.3. The estimated parameters in Table 27.2 are similar across time-to-maturity level but volatile across moneyness. Therefore, we investigate the performance of all groups except the groups on the bottom row of Table 27.3. The performance of models can be measured by either the implied volatility graph or the average absolute pricing errors (AveAPE). The implied volatility graph should be flat across different moneyness levels and time to maturity. We use subsample like what Bakshi et al. (1997) and Chen et al. (2009) did to test implied volatility consistency among moneyness-maturity categories. Using the subsample data from January 2012 to May 2013 to test in-the-sample fitness, the average daily implied volatility of both CEV and Black models and average alpha of CEV model are computed in Table 27.4. The fitness performance is shown in Table 27.5. The implied volatility graphs for both models are shown in Fig. 27.2. In Table 27.4, we estimate the optimal parameters of CEV model by using a more efficient program. In this efficient program, we scale the strike price and future price to speed up the program where the implied volatility of CEV model equals to $\delta(\text{ratio}^{\alpha-1})$, ratio is the moneyness level, and δ and α are the optimal parameters of program

Table 27.4 Average daily parameters of in sample

Time to maturity (TM)	TM < 30				30 ≦ TM ≦ 60				TM > 60				All TM			
Moneyness (S/K ratio)	CEV			Black	CEV			Black	CEV			Black	CEV			Black
Parameters	α	δ	IV	IV	α	δ	IV	IV	α	δ	IV	IV	α	δ	IV	IV
S/K ratio >1.01	0.29	0.19	0.188	0.200	0.14	0.18	0.183	0.181	0.29	0.21	0.204	0.196	0.25	0.19	0.1890	0.1882
0.98 ≦ S/K ratio ≦1.01	0.34	0.16	0.162	0.1556	0.30	0.16	0.154	0.147	0.14	0.16	0.155	0.155	0.39	0.17	0.151	0.150
0.95≦S/K ratio <0.98	0.22	0.13	0.137	0.135	0.30	0.13	0.134	0.131	0.24	0.14	0.141	0.139	0.37	0.14	0.136	0.132
0.9≦S/K ratio <0.95	0.05	0.15	0.159	0.152	0.25	0.13	0.133	0.128	0.26	0.14	0.136	0.131	0.38	0.14	0.135	0.129
S/K ratio <0.9	−0.23	0.22	0.252	0.243	−1.67	0.14	0.193	0.159	0.25	0.15	0.145	0.142	0.23	0.15	0.157	0.152

The in-sample period of option prices is from January 1, 2012, to May 30, 2013. In the in-sample estimating procedure, CEV implied volatility for S&P 500 index futures (CEV IV) equals to $\delta(S/K\ \text{ratio})^{\alpha-1}$ according to reduce computational costs. The optimization setting of finding CEV IV and Black IV is under the same criteria

Table 27.5 AveAPE performance for in-sample fitness

Time to maturity (TM)	TM < 30			30 ≦ TM ≦ 60			TM > 60			All TM		
Moneyness (S/K ratio)	CEV	Black	Obs.	CEV	Black	Obs.	CEV	Black	Obs.	CEV	Black	Obs.
S/K ratio >1.01	**1.65**	1.88	202	1.81	**1.77**	142	5.10	**5.08**	115	**5.80**	6.51	459
0.98 ≦ S/K ratio ≦1.01	**6.63**	7.02	1290	**4.00**	4.28	801	4.59	**4.53**	529	**18.54**	18.90	2620
0.95≦S/K ratio <0.98	2.38	**2.34**	1560	4.25	**4.14**	1469	3.96	**3.89**	913	14.25	**14.15**	3942
0.9≦S/K ratio <0.95	0.69	**0.68**	710	1.44	**1.43**	1094	3.68	**3.62**	1131	**7.08**	7.10	2935
S/K ratio <0.9	0.01	0.01	33	**0.13**	0.18	72	0.61	**0.60**	171	0.69	**0.68**	276

The in-sample period of option prices is from January 1, 2012, to May 30, 2013

which are not the parameters of CEV model in (27.4). In Table 27.5, we found that CEV model perform well at in-the-money group.

Figure 27.2 shows the IV computed by CEV and Black models. Although their implied volatility graphs are similar in each group, the reasons to cause volatility smile are totally different. In Black model, the constant volatility setting is misspecified. The volatility parameter of Black model in Fig. 27.2b varies across moneyless and time-to-maturity levels, while the IV in CEV model is a function of the underlying price and the elasticity of variance (alpha parameter). Therefore, we can image that the prediction power of CEV model will be better than Black model because of the explicit function of IV in CEV model. We can use alpha to measure the sensitivity of relationship between option price and its underlying asset. For example, in Fig. 27.2c, the in-the-money future options near expired date have significantly negative relationship between future price and its volatility.

The better performance of CEV model may result from the over-fitting issue that will hurt the forecastability of CEV model. Therefore, we use out-of-sample data from June 2013 to November 2013 to compare the prediction power of Black and CEV models. We use the estimated parameters in previous day as the current day's input variables of model. Then, the theoretical option price computed by either Black or CEV model can calculate bias between theoretical price and market price. Thus, we can calculate the average absolute pricing errors (AveAPE) for both models. The lower the value of a model's AveAPE, the higher the pricing prediction power of the model. The pricing errors of out-of-sample data are presented in Table 27.6. Here we find that CEV model can predict options on S&P 500 index futures more precisely than Black model. Based on the better performance in both in sample and out of sample, we claim that CEV model can describe the options of S&P 500 index futures more precisely than Black model.

With regard to generate implied volatility surface to capture whole prediction of the future option market, the CEV model is the better choice than Black model because it not only captures the skewness and kurtosis effects of options on index futures but also has less computational costs than other jump-diffusion stochastic volatility models.

In sum, we show that CEV model performs better than Black model in aspects of either in-sample fitness or out-of-sample prediction. The setting of CEV model is more reasonable to depict the negative relationship between S&P 500 index future price and its volatilities. The elasticity of variance parameter in CEV model captures the level of this characteristic. The stable volatility parameter in CEV model in our empirical results implies that the instantaneous volatility of index future is mainly determined by current future price and the level of elasticity of variance parameter.

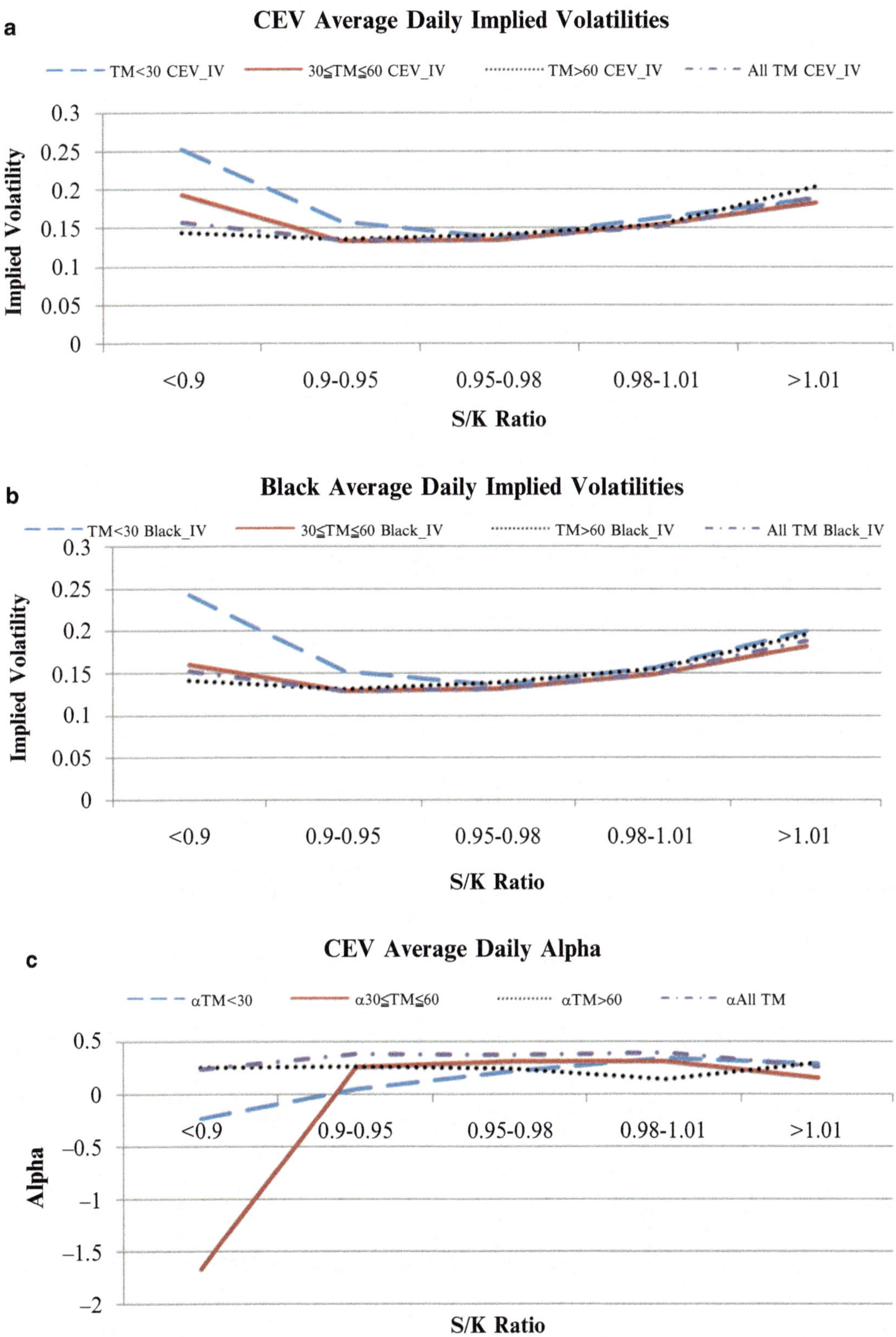

Fig. 27.2 Implied volatilities and CEV alpha graph

Table 27.6 AveAPE performance for out of sample

Time to maturity (TM)	TM < 30		30 ≦ TM ≦ 60		TM > 60		All TM	
Moneyness (S/K ratio)	CEV	Black	CEV	Black	CEV	Black	CEV	Black
S/K ratio >1.01	**3.22**	3.62	**3.38**	4.94	**8.96**	13.86	**4.25**	5.47
0.98≦S/K ratio ≦1.01	**2.21**	2.35	2.63	**2.53**	**3.47**	3.56	**2.72**	2.75
0.95≦S/K ratio <0.98	**0.88**	1.04	**1.42**	1.46	1.97	**1.95**	**1.44**	1.45
0.9≦S/K ratio <0.95	**0.34**	0.53	**0.61**	0.62	1.40	1.40	**0.88**	0.90
S/K ratio <0.9	**0.23**	0.79	**0.25**	0.30	1.28	**1.27**	**1.03**	1.66

References

Bakshi G, Cao C, Chen Z (1997) Empirical performance of alternative option pricing models. J Finance 52:2003–2049

Beckers S (1980) The constant elasticity of variance model and its implications for option pricing. J Finance 35:661–673

Black F, Scholes M (1973) The pricing of options and corporate liabilities. J Polit Econ 81(3):637–654

Chen R, Lee CF, Lee H (2009) Empirical performance of the constant elasticity variance option pricing model. Rev Pacific Basin Financ Markets Policies 12(2):177–217

Corrado CJ, Miller TW (1996) A note on a simple, accurate formula to compute implied standard deviations. J Bank Financ 20:595–603

Cox JC (1975) Notes on option pricing I: constant elasticity of variance diffusions. Working paper, Stanford University

Cox JC, Ross SA (1976) The valuation of options for alternative stochastic processes. J Financ Econ 3:145–166

Harvey CR, Whaley RE (1991) S&P 100 index option volatility. J Finance 46:1551–1561

Harvey CR, Whaley RE (1992a) Market volatility prediction and the efficiency of the S&P 100 index option market. J Financ Econ 31:43–73

Harvey CR, Whaley RE (1992b) Dividends and S&P 100 index option valuation. J Futures Mark 12:123–137

Hsu YL, Lin TI, Lee CF (2008) Constant elasticity of variance (CEV) option pricing model: integration and detailed derivation. Math Comput Simul 79(1):60–71

Jackwerth JC, Rubinstein M (2001) Recovering stochastic processes from option prices. Working paper, London Business School

Larguinho M, Dias JC, Braumann CA (2013) On the computation of option prices and Greeks under the CEV model. Quant Finance 13(6):907–917

Lee CF, Wu T, Chen R (2004) The constant elasticity of variance models: new evidence from S&P 500 index options. Rev Pacific Basin Financ Markets Policies 7(2):173–190

MacBeth JD, Merville LJ (1980) Tests of the Black-Scholes and Cox call option valuation models. J Finance 35:285–301

Merton RC (1973) A rational theory of option pricing. Bell J Econ Manag Sci 4:141–183

Pun CS, Wong HY (2013) CEV asymptotics of American options. J Math Anal Appl 403(2):451–463

Singh VK, Ahmad N (2011) Forecasting performance of constant elasticity of variance model: empirical evidence from India. Int J Appl Econ Financ 5:87–96

Chapter 28
Greek Letters and Portfolio Insurance

28.1 Introduction

In Chap. 26, we have discussed how the call option value can be affected by stock price per share, exercise price per share, the contract period of the option, the risk-free rate, and the volatility of the stock return. In this chapter, we will mathematically analyze these kinds of relationships. Parts of these mathematical relationships are called "Greek letters" by finance professionals. Here we specifically derive Greek letters for call (put) options on non-dividend stock and dividend-paying stock. Some examples will be provided to explain applications of these Greek letters. Sections 28.1–28.5 discuss the formula, Excel function, and applications of delta, theta, gamma, vega, and rho, respectively. Section 28.6 derives the partial derivative of stock options with respect to their exercise prices. Section 28.7 describes the relationship between delta, theta, and gamma and their implication in delta-neutral portfolio. Section 28.8 presents a portfolio insurance example. Finally in Sect. 28.9, we summarize and conclude this chapter.

28.2 Delta

The delta of an option, Δ, is defined as the rate of change of the option price with respect to the rate of change of underlying asset price:

$$\Delta = \frac{\partial \Pi}{\partial S},$$

where Π is the option price and S is underlying asset price. We next show the derivation of delta for various kinds of stock option.

28.2.1 Formula of Delta for Different Kinds of Stock Options

From Black–Scholes option pricing model, we know the price of call option on a non-dividend stock can be written as

$$C_t = S_t N(d_1) - X e^{-r\tau} N(d_2),$$

C.-F. Lee et al., *Essentials of Excel, Excel VBA, SAS and Minitab for Statistical and Financial Analyses*, DOI 10.1007/978-3-319-38867-0_28

and the price of put option on a non-dividend stock can be written as

$$P_t = Xe^{-r\tau}N(-d_2) - S_tN(-d_1),$$

where

$$d_1 = \frac{\ln\left(\frac{S_t}{X}\right) + \left(r + \frac{\sigma_s^2}{2}\right)\tau}{\sigma_s\sqrt{\tau}},$$

$$d_2 = \frac{\ln\left(\frac{S_t}{X}\right) + \left(r - \frac{\sigma_s^2}{2}\right)\tau}{\sigma_s\sqrt{\tau}} = d_1 - \sigma_s\sqrt{\tau},$$

$$\tau = T - t,$$

$N(\cdot)$ is the cumulative density function of normal distribution.

$$N(d_1) = \int_{-\infty}^{d_1} f(u)du = \int_{-\infty}^{d_1} \frac{1}{\sqrt{2\pi}} e^{-\frac{u^2}{2}} du$$

For a European call option on a non-dividend stock, delta can be shown as

$$\Delta = N(d_1)$$

For a European put option on a non-dividend stock, delta can be shown as

$$\Delta = N(d_1) - 1$$

If the underlying asset is a dividend-paying stock providing a dividend yield at rate q, Black–Scholes formulas for the prices of a European call option on a dividend-paying stock and a European put option on a dividend-paying stock are

$$C_t = S_te^{-q\tau}N(d_1) - Xe^{-r\tau}N(d_2),$$

and

$$P_t = Xe^{-r\tau}N(-d_2) - S_te^{-q\tau}N(-d_1),$$

where

$$d_1 = \frac{\ln\left(\frac{S_t}{X}\right) + \left(r - q + \frac{\sigma_s^2}{2}\right)\tau}{\sigma_s\sqrt{\tau}},$$

$$d_2 = \frac{\ln\left(\frac{S_t}{X}\right) + \left(r - q - \frac{\sigma_s^2}{2}\right)\tau}{\sigma_s\sqrt{\tau}} = d_1 - \sigma_s\sqrt{\tau},$$

For a European call option on a dividend-paying stock, delta can be shown as

$$\Delta = e^{-q\tau} N(d_1).$$

For a European put option on a dividend-paying stock, delta can be shown as

$$\Delta = e^{-q\tau} [N(d_1) - 1].$$

28.2.2 Excel Function of Delta for European Call Options

We can write a function to calculate the delta of call options. Below is the Visual Basic Application (VBA) function

```
' BS Call Option Delta
Function BSCallDelta(S, X, r, q, T, sigma)
    Dim d1, Nd1
    d1 = (Log(S / X) + (r - q + 0.5 * sigma ^ 2) * T) / (sigma *
Sqr(T))
    Nd1 = Application.NormSDist(d1)
    BSCallDelta = Exp(-q * T) * Nd1
End Function
```

With this function we can use it in the Excel to calculate delta.

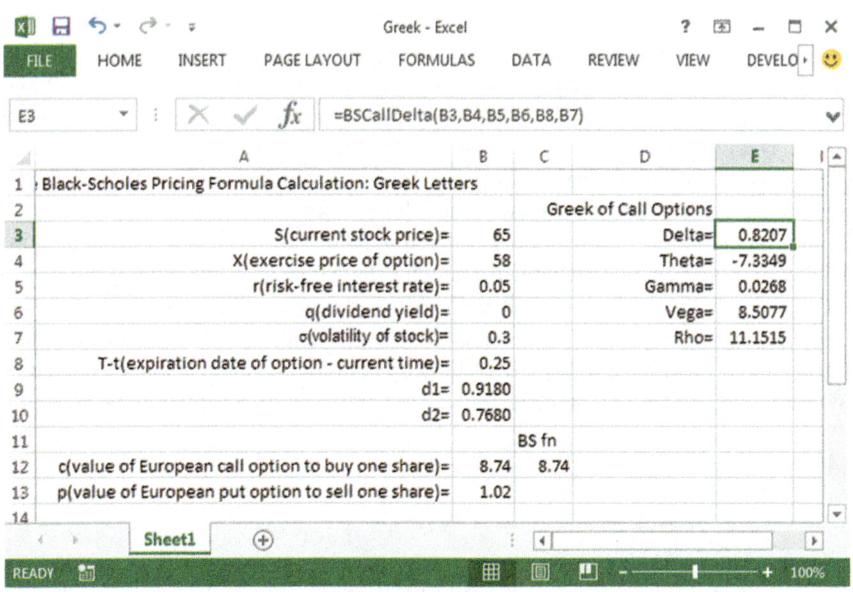

The formula for delta of a call option in Cell E3 is

$$= \text{BSCallDelta}(B3, B4, B5, B6, B8, B7)$$

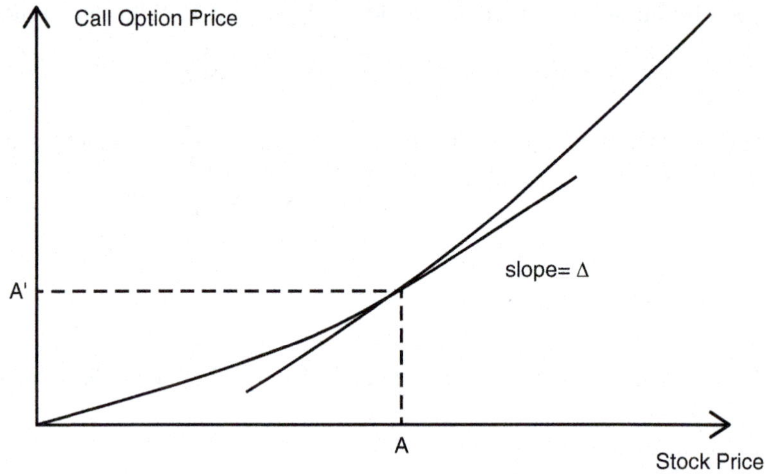

Fig. 28.1 The relationship between the price of a call option and the price of its underlying asset

28.2.3 Application of Delta

Figure 28.1 shows the relationship between the price of a call option and the price of its underlying asset. The delta of this call option is the slope of the line at the point of A corresponding to current price of the underlying asset.

By calculating delta ratio, a financial institution that sells option to a client can make a delta-neutral position to hedge the risk of changes of the underlying asset price. Suppose that the current stock price is \$100, the call option price on stock is \$10, and the current delta of the call option is 0.4. A financial institution sold 10 call option to its client, so that the client has right to buy 1000 shares at time to maturity. To construct a delta hedge position, the financial institution should buy $0.4 \times 1000 = 400$ shares of stock. If the stock price goes up to \$1, the option price will go up by \$0.40. In this situation, the financial institution has a \$400 (\$1 $\times$ 400 shares) gain in its stock position and a \$400 (\$0.40 $\times$ 1000 shares) loss in its option position. The total payoff of the financial institution is zero. On the other hand, if the stock price goes down by \$1, the option price will go down by \$0.40. The total payoff of the financial institution is also zero.

However, the relationship between option price and stock price is not linear, so delta changes over different stock price. If an investor wants to remain his portfolio in delta neutral, he should adjust his hedged ratio periodically. The more frequent adjustments he does, the better delta hedging he gets.

Figure 28.2 exhibits the change in delta affecting the delta hedges. If the underlying stock has a price equal to \$20, then the investor who uses only delta as risk measure will consider that his or her portfolio has no risk. However, as the underlying stock prices changes, either up or down, the delta changes as well, and thus he or she will have to use different delta hedging. Delta measure can be combined with other risk measures to yield better risk measurement. We will discuss it further in the following sections.

28.3 Theta

The theta of an option, Θ, is defined as the rate of change of the option price with respect to the passage of time:

$$\Theta = \frac{\partial \Pi}{\partial t},$$

where Π is the option price and t is the passage of time.

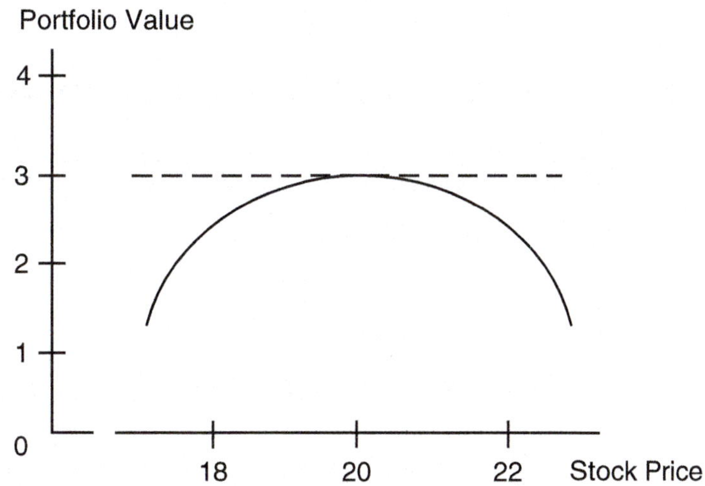

Fig. 28.2 Changes of delta hedge

If $\tau = T - t$, theta (Θ) can also be defined as minus one timing the rate of change of the option price with respect to time to maturity. The derivation of such transformation is easy and straightforward:

$$\Theta = \frac{\partial \Pi}{\partial t} = \frac{\partial \Pi}{\partial \tau}\frac{\partial \tau}{\partial t} = (-1)\frac{\partial \Pi}{\partial \tau},$$

where $\tau = T - t$ is time to maturity. For the derivation of theta for various kinds of stock option, we use the definition of negative differential on time to maturity.

28.3.1 Formula of Theta for Different Kinds of Stock Options

For a European call option on a non-dividend stock, theta can be written as

$$\Theta = -\frac{S_t \sigma_s}{2\sqrt{\tau}} \cdot N'(d_1) - rX \cdot e^{-r\tau} N(d_2).$$

For a European put option on a non-dividend stock, theta can be shown as

$$\Theta = -\frac{S_t \sigma_s}{2\sqrt{\tau}} \cdot N'(d_1) + rX \cdot e^{-r\tau} N(-d_2).$$

For a European call option on a dividend-paying stock, theta can be shown as

$$\Theta = q \cdot S_t e^{-q\tau} N(d_1) - \frac{S_t e^{-q\tau} \sigma_s}{2\sqrt{\tau}} \cdot N'(d_1) - rX \cdot e^{-r\tau} N(d_2).$$

For a European put option on a dividend-paying stock, theta can be shown as

$$\Theta = rX \cdot e^{-r\tau} N(-d_2) - qS_t e^{-q\tau} N(-d_1) - \frac{S_t e^{-q\tau} \sigma_s}{2\sqrt{\tau}} \cdot N'(d_1).$$

28.3.2 Excel Function of Theta of European Call Option

We also can write a function to calculate theta. The VBA function can be written as

```
' BS Call Option Theta
Function BSCallTheta(S, X, r, q, T, sigma)
    Dim d1, d2, Nd1, Nd2, Ndash1
    d1 = (Log(S / X) + (r - q + 0.5 * sigma ^ 2) * T) / (sigma * Sqr(T))
    d2 = d1 - sigma * Sqr(T)
    Nd1 = Application.NormSDist(d1)
    Nd2 = Application.NormSDist(d2)
    Ndash1 = (1 / Sqr(2 * Application.Pi)) * Exp(-d1 ^ 2 / 2)
    BSCallTheta = q * Exp(-q * T) * S * Nd1 - S * Ndash1 * sigma * Exp(-q * T) / (2 * Sqr(T)) - r * Exp(-r * T) * X * Nd2
End Function
```

Using this function we can value the theta of a call option.

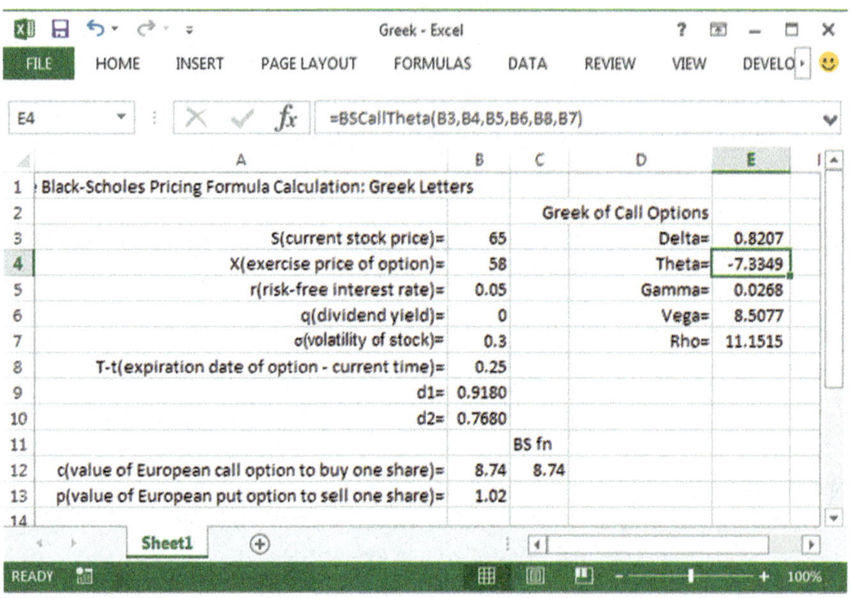

The function of theta for a European call option in Cell E4 is

$$= \text{BSCallTheta}(\text{B3}, \text{B4}, \text{B5}, \text{B6}, \text{B8}, \text{B7})$$

28.3.3 Application of Theta

The value of option is the combination of time value and stock value. When time passes, the time value of the option decreases. Thus, the rate of change of the option price with respect to the passage of time, theta, is usually negative.

Because the passage of time on an option is not uncertain, we do not need to make a theta-hedge portfolio against the effect of the passage of time. However, we still regard theta as a useful parameter, because it is a proxy of gamma in the delta-neutral portfolio. For the specific detail, we will discuss in the following sections.

28.4 Gamma

The gamma of an option, Γ, is defined as the rate of change of delta with respect to the rate of change of underlying asset price:

$$\Gamma = \frac{\partial \Delta}{\partial S} = \frac{\partial^2 \Pi}{\partial S^2},$$

where Π is the option price and S is the underlying asset price.

Because the option is not linearly dependent on its underlying asset, delta-neutral hedge strategy is useful only when the movement of underlying asset price is small. Once the underlying asset price moves wider, gamma-neutral hedge is necessary. We next show the derivation of gamma for various kinds of stock option.

28.4.1 Formula of Gamma for Different Kinds of Stock Options

For a European call option on a non-dividend stock, gamma can be shown as

$$\Gamma = \frac{1}{S_t \sigma_s \sqrt{\tau}} N'(d_1).$$

For a European put option on a non-dividend stock, gamma can be shown as

$$\Gamma = \frac{1}{S_t \sigma_s \sqrt{\tau}} N'(d_1).$$

For a European call option on a dividend-paying stock, gamma can be shown as

$$\Gamma = \frac{e^{-q\tau}}{S_t \sigma_s \sqrt{\tau}} N'(d_1).$$

For a European put option on a dividend-paying stock, gamma can be shown as

$$\Gamma = \frac{e^{-q\tau}}{S_t \sigma_s \sqrt{\tau}} N'(d_1).$$

28.4.2 Excel Function of Gamma for European Call Options

In addition, we can write a code to price gamma of a call option. Here is the VBA function to calculate gamma.

```
'   BS Call Option Gamma
Function BSCallGamma(S, X, r, q, T, sigma)
    Dim d1, Ndash1
    d1 = (Log(S / X) + (r - q + 0.5 * sigma ^ 2) * T) / (sigma *
Sqr(T))
    Ndash1 = (1 / Sqr(2 * Application.Pi)) * Exp(-d1 ^ 2 / 2)
    BSCallGamma = Exp(-q * T) * Ndash1 / (S * sigma * Sqr(T))
End Function
```

We can use the function in Excel spreadsheet to calculate gamma.

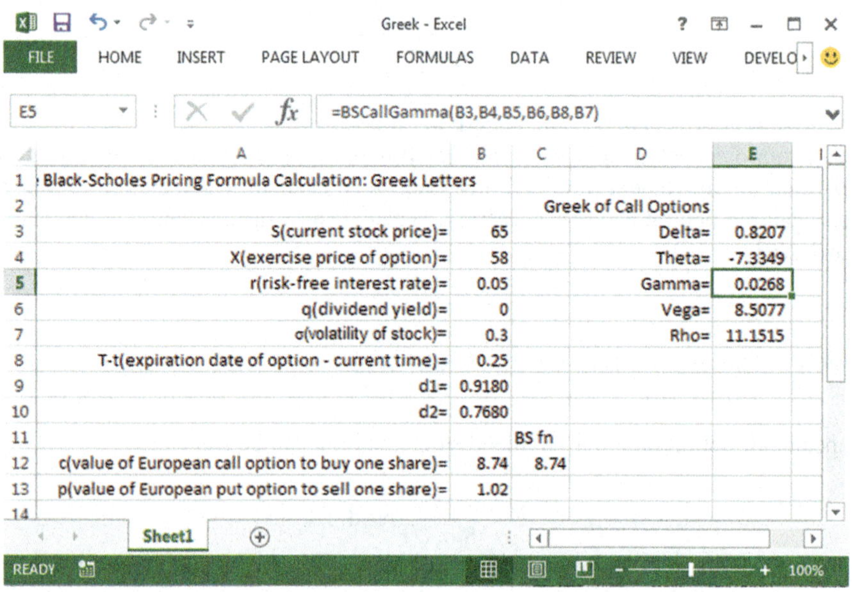

The function of gamma for a European call option in Cell E5 is

$$= \text{BSCallGamma}(\text{B3}, \text{B4}, \text{B5}, \text{B6}, \text{B8}, \text{B7})$$

28.4.3 Application of Gamma

One can use delta and gamma together to calculate the changes of the option due to changes in the underlying stock price. This change can be approximated by the following relation:

$$\text{change in option value} \approx \Delta \times \text{change in stock price} + \frac{1}{2} \times \Gamma \times (\text{change in stock price})^2.$$

From the above relation, one can observe that the gamma makes the correction for the fact that the option value is not a linear function of underlying stock price. This approximation comes from the Taylor series expansion near the initial stock price. If we let V be option value, S be stock price, and S_0 be initial stock price, then the Taylor series expansion around S_0 yields the following:

$$\begin{aligned} V(S) &\approx V(S_0) + \frac{\partial V(S_0)}{\partial S}(S - S_0) + \frac{1}{2!}\frac{\partial^2 V(S_0)}{\partial S^2}(S - S_0)^2 + \cdots + \frac{1}{2!}\frac{\partial^n V(S_0)}{\partial S^n}(S - S_0)^n \\ &\approx V(S_0) + \frac{\partial V(S_0)}{\partial S}(S - S_0) + \frac{1}{2!}\frac{\partial^2 V(S_0)}{\partial S^2}(S - S_0)^2 + o(S) \end{aligned}$$

If we only consider the first three terms, the approximation is then

$$\begin{aligned} V(S) - V(S_0) &\approx \frac{\partial V(S_0)}{\partial S}(S - S_0) + \frac{1}{2!}\frac{\partial^2 V(S_0)}{\partial S^2}(S - S_0)^2 \\ &\approx \Delta(S - S_0) + \frac{1}{2}\Gamma(S - S_0)^2. \end{aligned}$$

For example, if a portfolio of options has a delta equal to \$10,000 and a gamma equal to \$5,000, the change in the portfolio value if the stock price drop to \$34 from \$35 is approximately

$$\begin{aligned} \text{change in portfolio value} &\approx (\$10,000) \times (\$34 - \$35) + \frac{1}{2} \times (\$5000) \times (\$34 - \$35)^2 \\ &\approx -\$7500 \end{aligned}$$

The above analysis can also be applied to measure the price sensitivity of interest rate-related assets or portfolio to interest rate changes. Here we introduce *modified duration* and *convexity* as risk measure corresponding to the above delta and gamma. Modified duration measures the percentage change in asset or portfolio value resulting from a percentage change in interest rate.

$$\begin{aligned} \text{Modified Duration} &= \left(\frac{\text{Change in price}}{\text{Change in interest rate}}\right)/\text{Price} \\ &= -\Delta/P \end{aligned}$$

Using the modified duration,

$$\begin{aligned} \text{Change in Portfolio Value} &= \Delta \times \text{Change in interest rate} \\ &= (-\mathit{Duration} \times \text{P}) \times \text{Change in interest rate}, \end{aligned}$$

we can calculate the value changes of the portfolio. The above relation corresponds to the previous discussion of delta measure. We want to know how the price of the portfolio changes given a change in interest rate. Similar to delta, modified duration only shows the first-order approximation of the changes in value. In order to account for the nonlinear relation between the interest rate and portfolio value, we need a second-order approximation similar to the gamma measure before; this is then the convexity measure. Convexity is the interest rate gamma divided by price,

$$Convexity = \Gamma/P,$$

and this measure captures the nonlinear part of the price changes due to interest rate changes. Using the modified duration and convexity together allow us to develop first- as well as second-order approximation of the price changes similar to previous discussion.

$$\text{Change in Portfolio Value} \approx -Duration \times P \times (\text{change in rate}) + \frac{1}{2} \times Convexity \times P \times (\text{change in rate})^2$$

As a result, (−Duration × P) and (Convexity × P) act like the delta and gamma measure, respectively, in the previous discussion. This shows that these Greek letters can also be applied in measuring risk in interest-rate-related assets or portfolio.

Next we discuss how to make a gamma-neutral portfolio. Suppose the gamma of a delta-neutral portfolio is Γ, the gamma of the option in this portfolio is Γ_o, and ω_o is the number of options added to the delta-neutral portfolio. Then, the gamma of this new portfolio is

$$\omega_o\Gamma_o + \Gamma.$$

To make a gamma-neutral portfolio, we should trade $\omega_o{}^* = -\Gamma/\Gamma_o$ options. Because of the position of option changes, the new portfolio is not in the delta neutral. We should change the position of the underlying asset to maintain delta neutral.

For example, the delta and gamma of a particular call option are 0.7 and 1.2. A delta-neutral portfolio has a gamma of −2400. To make a delta-neutral and gamma-neutral portfolio, we should add a long position of 2400/1.2 = 2000 shares and a short position of 2000 × 0.7 = 1400 shares in the original portfolio.

28.5 Vega

The vega of an option, ν, is defined as the rate of change of the option price respective to the volatility of the underlying asset:

$$\nu = \frac{\partial \Pi}{\partial \sigma},$$

where Π is the option price and σ is volatility of the stock price. We next show the derivation of vega for various kinds of stock option.

28.5.1 *Formula of Vega for Different Kinds of Stock Options*

For a European call option on a non-dividend stock, vega can be shown as

$$\nu = S_t\sqrt{\tau} \cdot N'(d_1).$$

For a European put option on a non-dividend stock, vega can be shown as

$$\nu = S_t \sqrt{\tau} \cdot N'(d_1).$$

For a European call option on a dividend-paying stock, vega can be shown as

$$\nu = S_t e^{-q\tau} \sqrt{\tau} \cdot N'(d_1).$$

For a European put option on a dividend-paying stock, vega can be shown as

$$\nu = S_t e^{-q\tau} \sqrt{\tau} \cdot N'(d_1).$$

28.5.2 Excel Function of Vega for European Call Options

We can write a function to calculate vega. Below is the VBA function of vega for European call options.

```
'    BS Call Option Vega
Function BSCallVega(S, X, r, q, T, sigma)
    Dim d1, Ndash1
    d1 = (Log(S / X) + (r - q + 0.5 * sigma ^ 2) * T) / (sigma *
Sqr(T))
    Ndash1 = (1 / Sqr(2 * Application.Pi)) * Exp(-d1 ^ 2 / 2)
    BSCallVega = Exp(-q * T) * S * Sqr(T) * Ndash1
End Function
```

Using this function we can calculate vega for a European call option in the Excel spreadsheet.

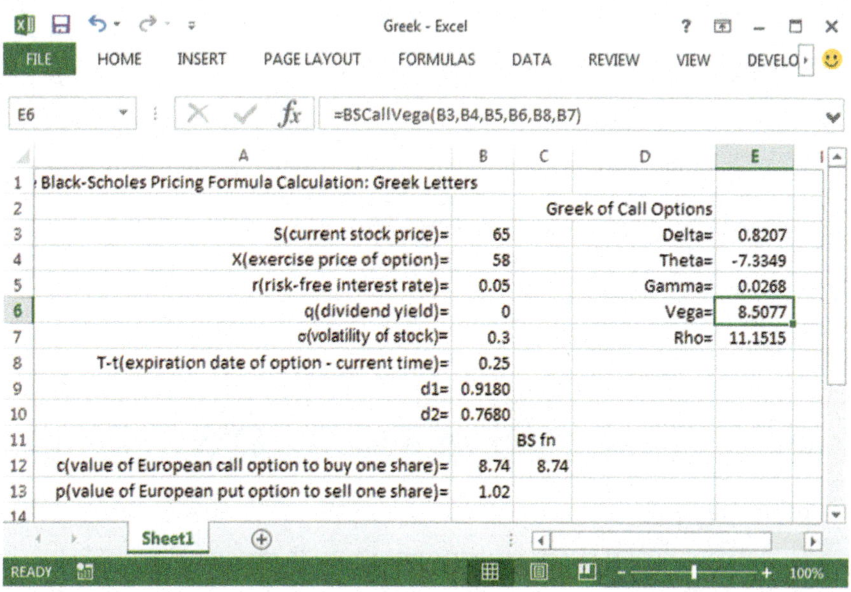

	A	B	C	D	E
1	Black-Scholes Pricing Formula Calculation: Greek Letters				
2			Greek of Call Options		
3	S(current stock price)=	65		Delta=	0.8207
4	X(exercise price of option)=	58		Theta=	-7.3349
5	r(risk-free interest rate)=	0.05		Gamma=	0.0268
6	q(dividend yield)=	0		Vega=	8.5077
7	σ(volatility of stock)=	0.3		Rho=	11.1515
8	T-t(expiration date of option - current time)=	0.25			
9	d1=	0.9180			
10	d2=	0.7680			
11			BS fn		
12	c(value of European call option to buy one share)=	8.74	8.74		
13	p(value of European put option to sell one share)=	1.02			
14					

The function of vega for a European call option in Cell E5 is

$$= \text{BSCallVega}(\text{B3}, \text{B4}, \text{B5}, \text{B6}, \text{B8}, \text{B7})$$

28.5.3 Application of Vega

Suppose a delta-neutral and gamma-neutral portfolio has a vega equal to ν and the vega of a particular option is ν_o, similar to gamma, we can add a position of $-\nu/\nu_o$ in option to make a vega-neutral portfolio. To maintain delta neutral, we should change the underlying asset position. However, when we change the option position, the new portfolio is not gamma neutral. Generally, a portfolio with one option cannot maintain its gamma neutral and vega neutral at the same time. If we want a portfolio to be both gamma neutral and vega neutral, we should include at least two kinds of option on the same underlying asset in our portfolio.

For example, a delta-neutral and gamma-neutral portfolio contains option A, option B, and underlying asset. The gamma and vega of this portfolio are −3200 and −2500, respectively. Option A has a delta of 0.3, gamma of 1.2, and vega of 1.5. Option B has a delta of 0.4, gamma of 1.6, and vega of 0.8. The new portfolio will be both gamma neutral and vega neutral when adding ω_A of option A and ω_B of option B into the original portfolio.

$$\text{Gamma Neutral}: \quad -3200 + 1.2\omega_A + 1.6\omega_B = 0.$$

$$\text{Vega Neutral}: \quad -2500 + 1.5\omega_A + 0.8\omega_B = 0.$$

From two equations shown above, we can get the solution that $\omega_A = 1000$ and $\omega_B = 1250$. The delta of new portfolio is $1000 \times .3 + 1250 \times 0.4 = 800$. To maintain delta neutral, we need to short 800 shares of the underlying asset.

We can use Excel matrix function to solve these linear equations.

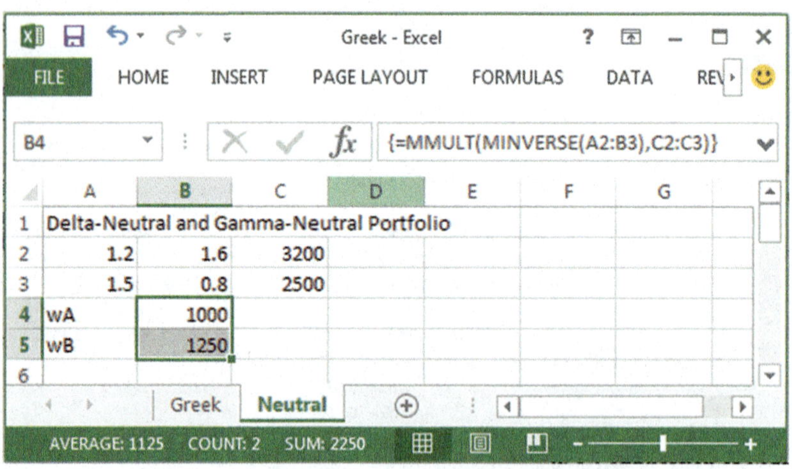

The function in Cell B4:B5 is

$$= \text{MMULT}(\text{MINVERSE}(\text{A2}:\text{B3}), \text{C2}:\text{C3})$$

Because this is matrix function, we need to use [ctrl] + [shift] + [enter] to get our result.

28.6 Rho

The rho of an option is defined as the rate of change of the option price with respect to the interest rate:

$$\text{rho} = \frac{\partial \Pi}{\partial r},$$

where Π is the option price and r is interest rate. The rho for an ordinary stock call option should be positive because higher interest rate reduces the present value of the strike price which in turn increases the value of the call option. Similarly, the rho of an ordinary put option should be negative by the same reasoning. We next show the derivation of rho for various kinds of stock option.

28.6.1 Formula of Rho for Different Kinds of Stock Options

For a European call option on a non-dividend stock, rho can be shown as

$$\text{rho} = X\tau \cdot e^{-r\tau} N(d_2).$$

For a European put option on a non-dividend stock, rho can be shown as

$$\text{rho} = -X\tau \cdot e^{-r\tau} N(-d_2).$$

For a European call option on a dividend-paying stock, rho can be shown as

$$\text{rho} = X\tau \cdot e^{-r\tau} N(d_2).$$

For a European put option on a dividend-paying stock, rho can be shown as

$$\text{rho} = -X\tau \cdot e^{-r\tau} N(-d_2).$$

28.6.2 Excel Function of Rho for European Call Options

We can write a function to calculate rho. Here is the VBA function to calculate rho for European call options.

```
'   BS Call Option Rho
Function BSCallRho(S, X, r, q, T, sigma)
    Dim d1, d2, Nd2
    d1 = (Log(S / X) + (r - q + 0.5 * sigma ^ 2) * T) / (sigma *
Sqr(T))
    d2 = d1 - sigma * Sqr(T)
    Nd2 = Application.NormSDist(d2)
    BSCallRho = T * Exp(-r * T) * X * Nd2
End Function
```

Then we can use this function to calculate rho in the Excel worksheet.

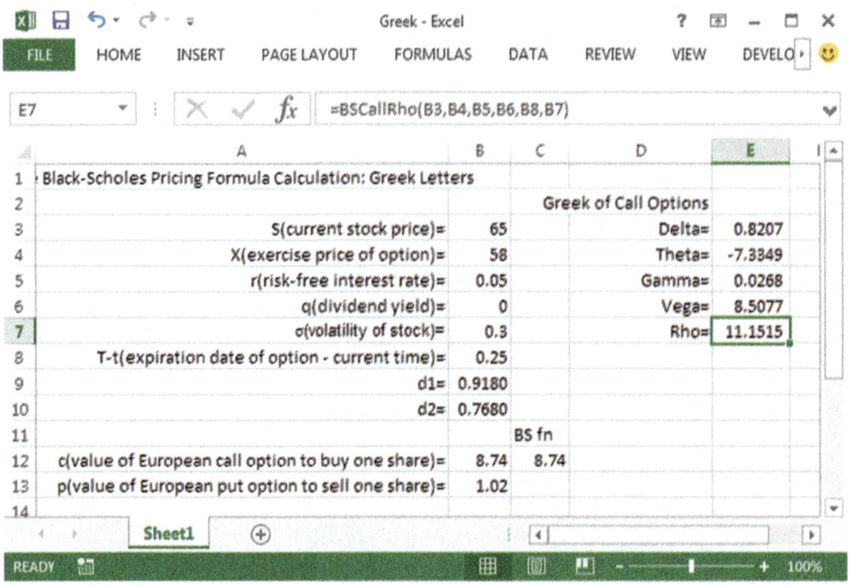

	A	B	C	D	E
1	Black-Scholes Pricing Formula Calculation: Greek Letters				
2			Greek of Call Options		
3	S(current stock price)=	65		Delta=	0.8207
4	X(exercise price of option)=	58		Theta=	-7.3349
5	r(risk-free interest rate)=	0.05		Gamma=	0.0268
6	q(dividend yield)=	0		Vega=	8.5077
7	σ(volatility of stock)=	0.3		Rho=	11.1515
8	T-t(expiration date of option - current time)=	0.25			
9	d1=	0.9180			
10	d2=	0.7680			
11			BS fn		
12	c(value of European call option to buy one share)=	8.74	8.74		
13	p(value of European put option to sell one share)=	1.02			

The function of rho in the Cell E7 is

$$= \text{BSCallRho}(\text{B3}, \text{B4}, \text{B5}, \text{B6}, \text{B8}, \text{B7})$$

28.6.3 Application of Rho

Assume that an investor would like to see how interest rate changes affect the value of a 3-month European call option she holds with the following information. The current stock price is \$65 and the strike price is \$58. The interest rate and the volatility of the stock are 5 % and 30 % per annum, respectively. The rho of this European call can be calculated as follows:

$$Rho_{put} = X\tau e^{-r\tau}N(d_2) = 11.1515$$

This calculation indicates that given 1 % change increase in interest rate, say from 5 to 6 %, the value of this European call option will decrease 0.111515 (0.01 × 11.1515). This simple example can be further applied to stocks that pay dividends using the derivation results shown previously.

28.7 Formula of Sensitivity for Stock Options with Respect to Exercise Price

For a European call option on a non-dividend stock, the sensitivity can be shown as

$$\frac{\partial C_t}{\partial X} = -e^{-r\tau}N(d_2)$$

For a European put option on a non-dividend stock, the sensitivity can be shown as

$$\frac{\partial P_t}{\partial X} = e^{-r\tau}N(-d_2)$$

For a European call option on a dividend-paying stock, the sensitivity can be shown as

$$\frac{\partial C_t}{\partial X} = -e^{-r\tau}N(d_2).$$

For a European put option on a dividend-paying stock, the sensitivity can be shown as

$$\frac{\partial P_t}{\partial X} = e^{-r\tau}N(-d_2).$$

28.8 Relationship Between Delta, Theta, and Gamma

So far, the discussion has introduced the derivation and application of each individual Greek letters and how they can be applied in portfolio management. In practice, the interaction or trade-off between these parameters is of concern as well. For example, recall the Black–Scholes–Merton differential equation with non-dividend paying stock can be written as

$$\frac{\partial \Pi}{\partial t} + rS\frac{\partial \Pi}{\partial S} + \frac{1}{2}\sigma^2 S^2 \frac{\partial^2 \Pi}{\partial S^2} = r\Pi,$$

where Π is the value of the derivative security contingent on stock price, S is the price of stock, r is the risk-free rate, σ is the volatility of the stock price, and t is the time to expiration of the derivative. Given the earlier derivation, we can rewrite the Black–Scholes partial differential equation (PDE) as

$$\Theta + rS\Delta + \frac{1}{2}\sigma^2 S^2 \Gamma = r\Pi.$$

This relation gives us the trade-off between delta, gamma, and theta. For example, suppose there are two delta-neutral ($\Delta = 0$) portfolios, one with positive gamma ($\Gamma > 0$) and the other one with negative gamma ($\Gamma < 0$) and they both have value of \$1 ($\Pi = 1$). The trade-off can be written as

$$\Theta + \frac{1}{2}\sigma^2 S^2 \Gamma = r.$$

For the first portfolio, if gamma is positive and large, then theta is negative and large. When gamma is positive, changes in stock prices result in higher value of the option. This means that when there is no change in stock prices, the value of the option declines as we approach the expiration date. As a result, the theta is negative. On the other hand, when gamma is negative and large, changes in stock prices result in lower option value. This means that when there is no stock price change, the value of the option increases as we approach the expiration and theta is positive. This gives us a trade-off between gamma and theta, and they can be used as proxy for each other in a delta-neutral portfolio.

28.9 Portfolio Insurance

Portfolio insurance is a strategy of hedging a portfolio of stocks against the market risk by using a synthetic put option. What is a synthetic put option? A synthetic put option is like to buy a put option to hedge a portfolio. That is a protective put strategy. Although this strategy uses short stocks or futures to construct a delta which is like buying a put option, the risk of this strategy is not the same as buying a put option.

Consider two strategies. The first one we long 1 index portfolio and long 1 put, then the delta in this strategy is $1 + \Delta p$, where Δp is the delta of put and the value is negative. The second one we long 1 index portfolio, short $-\Delta p$ amount of index, and invest the money that short index to riskless asset, then the delta of this strategy is $1-(-\Delta p*1) = 1 + \Delta p$, which is equal to the first strategy.

The second strategy is the so-called portfolio insurance. The dynamic adjustment in this strategy is like below. As the value of the index portfolio increases, Δp becomes less negative, and some of the index portfolio is repurchased. As the value of the index portfolio decreases, Δp becomes more negative, and more of the index portfolio have to be sold.

However, the portfolio insurance strategy did not work well on October 19, 1987. That day stock market declined very quickly. The managers using portfolio insurance strategy should short index portfolio. This action increased the pace of decline in the stock market. Therefore, synthetic put cannot create the same payoff like buying a put option. There is no any effect of insurance in the crash market.

28.10 Summary

In this chapter, we have shown the partial derivatives of stock option with respect to five variables. Delta (Δ), the rate of change of option price to change in price of underlying asset, is first derived. After delta is obtained, gamma (Γ) can be derived as the rate of change of delta with respect to underlying asset price. Another two risk measures are theta (Θ) and rho (ρ); they measure the change in option value with respect to passing time and interest rate, respectively. Finally, one can also measure the change in option value with respect to the volatility of the underlying asset, and this gives us the vega (v). The applications of these Greek letters in the portfolio management have also been discussed. In addition, we use the Black–Scholes PDE to show the relationship between these risk measures. In sum, risk management is one of the important topics in finance for both academics and practitioners. Given the recent credit crisis, one can observe that it is crucial to properly measure the risk related to the even more complicated financial assets. The comparative static analysis of option pricing models gives an introduction to the portfolio risk management.

Bibliography

Bjork T (1998) Arbitrage theory in continuous time. Oxford University Press, New York
Boyle PP, Emanuel D (1980) Discretely adjusted option hedges. J Financ Econ 8(3):259–282
Duffie D (2001) Dynamic asset pricing theory. Princeton University Press, Princeton, NJ
Fabozzi FJ (2007) Fixed income, analysis, 2nd edn. Wiley, New York
Figlewski S (1989) Options arbitrage in imperfect markets. J Finance 44(5):1289–1311
Galai D (1983) The components of the return from hedging options against stocks. J Bus 56(1):45–54
Hull J (2011) Options, futures, and other derivatives, 8th edn. Pearson, Upper Saddle River, NJ
Hull J, White A (1987) Hedging the risks from writing foreign currency options. J Int Money Finance 6(2):131–152
Karatzas I, Shreve SE (2000) Brownian motion and stochastic calculus. Springer, Berlin
Klebaner FC (2005) Introduction to stochastic calculus with applications. Imperial College Press, London
McDonald RL (2005) Derivatives markets, 2nd edn. Addison-Wesley, Boston, MA
Shreve SE (2004) Stochastic, calculus for finance II: continuous time model. Springer, New York
Tuckman B (2002) Fixed income securities: tools for today's markets, 2nd edn. Wiley, New York

Chapter 29
Portfolio Analysis and Option Strategies

29.1 Introduction

The main purposes of this chapter are to show how Excel programs can be used to perform portfolio selection decisions and to construct option strategies. In Sect. 29.2, we demonstrate how Microsoft Excel can be used to inverse the matrix. In Sect. 29.3, we discuss how Excel programs can be used to estimate the Markowitz portfolio models. In Sect. 29.4, we discuss alternative option strategies. In Sect. 29.5, we summarize the results of this chapter.

29.2 Three Alternative Methods to Solve Simultaneous Equation

In this section, we discuss four alternative methods to solve the system of linear equations including Sect. 29.2.1 Substitution Method, Sect. 29.2.2 Cramer's Rule, Sect. 29.2.3 Matrix Method, and Sect. 29.2.4 Excel Matrix Inversion and Multiplication.

29.2.1 Substitution Method (Reference: Wikipedia)

The simplest method for solving a system of linear equations is to repeatedly eliminate variables. This method can be described as follows:

1. In the first equation, solve for one of the variables in terms of the others.
2. Substitute this expression into the remaining equations. This yields a system of equations with one fewer equation and one fewer unknown.
3. Continue until you have reduced the system to a single linear equation.
4. Solve this equation and then back-substitute until the entire solution is found.

For example, consider the following system:

$$\begin{aligned} x + 3y - 2z &= 5 \\ 3x + 5y + 6z &= 7 \\ 2x + 4y + 3z &= 8. \end{aligned}$$

C.-F. Lee et al., *Essentials of Excel, Excel VBA, SAS and Minitab for Statistical and Financial Analyses*, DOI 10.1007/978-3-319-38867-0_29

Solving the first equation for x gives x = 5 + 2z − 3y, and plugging this into the second and third equation yields

$$-4y + 12z = -8$$
$$-2y + 7z = -2.$$

Solving the first of these equations for y yields y = 2 + 3z, and plugging this into the second equation yields z = 2. We now have

$$x = 5 + 2z - 3y$$
$$y = 2 + 3z$$
$$z = 2.$$

Substituting z = 2 into the second equation gives y = 8, and substituting z = 2 and y = 8 into the first equation yields x = −15. Therefore, the solution set is the single point (x, y, z) = (−15, 8, 2).

29.2.2 Cramer's Rule

Cramer's rule is an explicit formulas for small systems (Reference: Wikipedia)

Consider the linear system $\begin{cases} a_1x + b_1y = c_1 \\ a_2x + b_2y = c_2 \end{cases}$ which in matrix format is $\begin{bmatrix} a_1 & b_1 \\ a_2 & b_2 \end{bmatrix}\begin{bmatrix} x \\ y \end{bmatrix} = \begin{bmatrix} c_1 \\ c_2 \end{bmatrix}$.

Assume $a_1b_2 - b_1a_2$ is nonzero. Then, x and y can be found with Cramer's rule as

$$x = \begin{vmatrix} c_1 & b_1 \\ c_2 & b_2 \end{vmatrix} \Big/ \begin{vmatrix} a_1 & b_1 \\ a_2 & b_2 \end{vmatrix} = \frac{c_1b_2 - b_1c_2}{a_1b_2 - b_1a_2}$$

and

$$y = \begin{vmatrix} a_1 & c_1 \\ a_2 & c_2 \end{vmatrix} \Big/ \begin{vmatrix} a_1 & b_1 \\ a_2 & b_2 \end{vmatrix} = \frac{a_1c_2 - c_1a_2}{a_1b_2 - b_1a_2}.$$

The rules for 3 × 3 matrices are similar. Given $\begin{cases} a_1x + b_1y + c_1z = d_1 \\ a_2x + b_2y + c_2z = d_2 \\ a_3x + b_3y + c_3z = d_3 \end{cases}$ which in matrix format is $\begin{bmatrix} a_1 & b_1 & c_1 \\ a_2 & b_2 & c_2 \\ a_3 & b_3 & c_3 \end{bmatrix}\begin{bmatrix} x \\ y \\ z \end{bmatrix} = \begin{bmatrix} d_1 \\ d_2 \\ d_3 \end{bmatrix}$.

Then the values of x, y, and z can be found as follows:

$$x = \frac{\begin{vmatrix} d_1 & b_1 & c_1 \\ d_2 & b_2 & c_2 \\ d_3 & b_3 & c_3 \end{vmatrix}}{\begin{vmatrix} a_1 & b_1 & c_1 \\ a_2 & b_2 & c_2 \\ a_3 & b_3 & c_3 \end{vmatrix}}, \quad y = \frac{\begin{vmatrix} a_1 & d_1 & c_1 \\ a_2 & d_2 & c_2 \\ a_3 & d_3 & c_3 \end{vmatrix}}{\begin{vmatrix} a_1 & b_1 & c_1 \\ a_2 & b_2 & c_2 \\ a_3 & b_3 & c_3 \end{vmatrix}}, \quad \text{and} \quad z = \frac{\begin{vmatrix} a_1 & b_1 & d_1 \\ a_2 & b_2 & d_2 \\ a_3 & b_3 & d_3 \end{vmatrix}}{\begin{vmatrix} a_1 & b_1 & c_1 \\ a_2 & b_2 & c_2 \\ a_3 & b_3 & c_3 \end{vmatrix}}.$$

And then you need to use determinant calculation; the calculation for determinant is as follows:

For example, for 3 × 3 matrices, the determinant of a 3 × 3 matrix is defined by

$$\begin{vmatrix} a & b & c \\ d & e & f \\ g & h & i \end{vmatrix} = a\begin{vmatrix} e & f \\ h & i \end{vmatrix} - b\begin{vmatrix} d & f \\ g & i \end{vmatrix} + c\begin{vmatrix} d & e \\ g & h \end{vmatrix}$$
$$= a(ei - fh) - b(di - fg) + c(dh - eg)$$
$$= aei + bfg + cdh - ceg - bdi - afh.$$

We use the same example as we did in the first method:

$$\text{x} = \frac{\begin{bmatrix} 5 & 3 & -2 \\ 7 & 5 & 6 \\ 8 & 4 & 3 \end{bmatrix}}{\begin{bmatrix} 1 & 3 & -2 \\ 3 & 5 & 6 \\ 2 & 4 & 3 \end{bmatrix}},\ y = \frac{\begin{bmatrix} 1 & 5 & -2 \\ 3 & 7 & 6 \\ 2 & 8 & 3 \end{bmatrix}}{\begin{bmatrix} 1 & 3 & -2 \\ 3 & 5 & 6 \\ 2 & 4 & 3 \end{bmatrix}},\ z = \frac{\begin{bmatrix} 1 & 3 & 5 \\ 3 & 5 & 7 \\ 2 & 4 & 8 \end{bmatrix}}{\begin{bmatrix} 1 & 3 & -2 \\ 3 & 5 & 6 \\ 2 & 4 & 3 \end{bmatrix}}$$

$$\text{x} = \frac{5*5*3+3*6*8+(-2)*7*4-(-2)*5*8-3*7*3-5*6*4}{1*5*3+3*6*2+(-2)*3*4-(-2)*5*2-3*3*3-1*6*4}$$
$$= \frac{75+144-28+80-63-120}{15+36-24+20-27-24} = \frac{60}{-4} = -15$$

$$\text{y} = \frac{1*7*3+5*6*2+(-2)*3*8-(-2)*7*2-5*3*3-1*6*8}{1*5*3+3*6*2+(-2)*3*4-(-2)*5*2-3*3*3-1*6*4}$$
$$= \frac{21+60-48+28-45-48}{15+36-24+20-27-24} = \frac{-32}{-4} = 8$$

$$\text{z} = \frac{1*5*8+3*7*2+5*3*4-5*5*2-3*3*8-1*7*4}{1*5*3+3*6*2+(-2)*3*4-(-2)*5*2-3*3*3-1*6*4}$$
$$= \frac{40+42+60-50-72-28}{15+36-24+20-27-24} = \frac{-8}{-4} = 2.$$

29.2.3 Matrix Method

Using the example in the last two sections above, we can derive the following matrix equation:

$$\begin{bmatrix} x \\ y \\ z \end{bmatrix} = \begin{bmatrix} 1 & 3 & -2 \\ 3 & 5 & 6 \\ 2 & 4 & 3 \end{bmatrix}^{-1} * \begin{bmatrix} 5 \\ 7 \\ 8 \end{bmatrix}.$$

The inversion of matrix A is, by the definition,

$$A^{-1} = \frac{1}{\det A} * (AdjA).$$

The Adjoint A is defined by the transpose of the cofactor matrix. First we need to calculate the cofactor matrix of A. Suppose the cofactor matrix is

$$\text{cofactor matrix} = \begin{bmatrix} A_{11} & A_{12} & A_{13} \\ A_{21} & A_{22} & A_{23} \\ A_{31} & A_{32} & A_{33} \end{bmatrix},$$

$$A_{11} = \begin{bmatrix} 5 & 6 \\ 4 & 3 \end{bmatrix} = -9,\ A_{12} = -\begin{bmatrix} 3 & 6 \\ 2 & 3 \end{bmatrix} = 3,\ A_{13} = \begin{bmatrix} 3 & 5 \\ 2 & 4 \end{bmatrix} = 2,$$

$$A_{21} = -\begin{bmatrix} 3 & -2 \\ 4 & 3 \end{bmatrix} = -17,\ A_{22} = \begin{bmatrix} 1 & -2 \\ 2 & 3 \end{bmatrix} = 7,\ A_{23} = -\begin{bmatrix} 1 & 3 \\ 2 & 4 \end{bmatrix} = 2,$$

$$A_{31} = \begin{bmatrix} 3 & -2 \\ 5 & 6 \end{bmatrix} = 28,\ A_{32} = -\begin{bmatrix} 1 & -2 \\ 3 & 6 \end{bmatrix} = -12,\ A_{33} = \begin{bmatrix} 1 & 3 \\ 3 & 5 \end{bmatrix} = -4.$$

Therefore,

$$\text{Cofactor matrix} = \begin{bmatrix} -9 & 3 & 2 \\ -17 & 7 & 2 \\ 28 & -12 & -4 \end{bmatrix}.$$

Then, we can get the Adjoint A:

$$\text{Adj A} = \begin{bmatrix} -9 & -17 & 28 \\ 3 & 7 & -12 \\ 2 & 2 & -4 \end{bmatrix}.$$

The determinant of A we have calculated in Cramer's rule:

$$\text{Det A} = \begin{bmatrix} 1 & 3 & -2 \\ 3 & 5 & 6 \\ 2 & 4 & 3 \end{bmatrix} = -4,$$

$$A^{-1} = \frac{1}{(-4)} * \begin{bmatrix} -9 & -17 & 28 \\ 3 & 7 & -12 \\ 2 & 2 & -4 \end{bmatrix} = \begin{bmatrix} \frac{9}{4} & \frac{17}{4} & -\frac{28}{4} \\ -\frac{3}{4} & -\frac{7}{4} & 3 \\ -\frac{1}{2} & -\frac{1}{2} & 1 \end{bmatrix}.$$

Therefore,

$$\begin{bmatrix} x \\ y \\ z \end{bmatrix} = \begin{bmatrix} \frac{9}{4} & \frac{17}{4} & -\frac{28}{4} \\ -\frac{3}{4} & -\frac{7}{4} & 3 \\ -\frac{1}{2} & -\frac{1}{2} & 1 \end{bmatrix} * \begin{bmatrix} 5 \\ 7 \\ 8 \end{bmatrix} = \begin{bmatrix} \frac{9}{4}*5 + \frac{17}{4}*7 + \left(-\frac{28}{4}\right)*8 \\ \left(-\frac{3}{4}\right)*5 + \left(-\frac{7}{4}\right)*7 + 3*8 \\ \left(-\frac{1}{2}\right)*5 + \left(-\frac{1}{2}\right)*7 + 1*8 \end{bmatrix} = \begin{bmatrix} -15 \\ 8 \\ 2 \end{bmatrix}.$$

29.2.4 Excel Matrix Inversion and Multiplication

1. Using minverse () function to get the A inverse, type "Ctrl + Shift + Enter" together and you will get the inverse of A.

Matrix A
1
3
2

Matrix A Inver
2.25
-0.75
-0.5

Matrix A		
1	3	-2
3	5	6
2	4	3

Matrix A Inverse		
=minverse(B20:D22)		

2. Using mmult () function to do the matrix multiplication, type "ctrl + shift + enter" together and you will get the answers for x, y, and z.

Matrix A Inverse				Vector D		Vector_(x,y,z)
2.25	4.25	-7		5		=MMULT(B25:D27,F25:F27)
-0.75	-1.75	3	*	7	=	
-0.5	-0.5	1		8		

Matrix A Inverse				Vector D		Vector_(x,y,z)
2.25	4.25	-7		5		-15
-0.75	-1.75	3	*	7	=	8
-0.5	-0.5	1		8		2

Excel matrix inversion and multiplication method discussed in this section is identical to the method discussed in the previous section.

29.3 Markowitz Model for Portfolio Selection

The Markowitz model of portfolio selection is a mathematical approach for deriving optimal portfolios. There are two methods to obtain optimal weights for portfolio selection; these two methods are (a) the least risk for a given level of expected return and (b) the greatest expected return for a given level of risk.

How does a portfolio manager apply these techniques in the real world?

The process would normally begin with a universe of securities available to the fund manager. These securities would be determined by the goals and objectives of the mutual fund. For example, a portfolio manager who runs a mutual fund specializing in health-care stocks would be required to select securities from the universe of health-care stocks. This would greatly reduce the analysis of the fund manager by limiting the number of securities available.

The next step in the process would be to determine the proportions of each security to be included in the portfolio. To do this, the fund manager would begin by setting a target rate of return for the portfolio. After determining the target rate of return, the fund manager can determine the different proportions of each security that will allow the portfolio to reach this target rate of return.

The final step in the process would be for the fund manager to find the portfolio with the lowest variance given the target rate of return.

The optimal portfolio can be obtained mathematically through the use of the Lagrangian multipliers. The Lagrangian method allows the minimization or maximization of an objective function when the objective function is subject to some constraints. One of the goals of portfolio analysis is minimizing the risk or variance of the portfolio, subject to the portfolio's attaining some target expected rate of return and also subject to the portfolio weights' summing to one. The problem can be stated mathematically:

$$\text{Min } \sigma_p^2 = \sum_{i=1}^{n}\sum_{j=1}^{n} W_i W_j \sigma_{ij} \tag{29.1}$$

subject to

$$\text{(i)} \qquad \sum_{i=1}^{n} W_i E(R_i) = E^*,$$

where E^* is the target expected return and

$$\text{(ii)} \qquad \sum_{i=1}^{n} W_i = 1.0.$$

The first constraint simply says that the expected return on the portfolio should equal the target return determined by the portfolio manager. The second constraint says that the weights of the securities invested in the portfolio must sum to one.

The Lagrangian objective function can be written:

$$C = \sum_{i=1}^{n} \sum_{j=1}^{n} W_i W_j \operatorname{Cov}(R_i R_j) + \lambda_1 \left(1 - \sum_{i=1}^{n} W_i\right) + \lambda_2 \left[E^* - \sum_{i=1}^{n} W_i E(R_i)\right]. \tag{29.2}$$

For three-security case, the Lagrangian objective function is as follows:

$$\begin{aligned} C = {} & W_1^2\sigma_1^2 + W_2^2\sigma_2^2 + W_3^2\sigma_3^2 + 2W_1W_2\sigma_{12} + 2W_1W_3\sigma_{13} + 2W_2W_3\sigma_{23} \\ & + \lambda_1(1 - W_1 - W_2 - W_3) + \lambda_2 E^* - W_1E(R_1) - W_2E(R_2) - W_3E(R_3). \end{aligned}$$

Taking the partial derivatives of (29.3) with respect to each of the variables, W_1, W_2, W_3, λ_1, λ_2, and setting the resulting five equations equal to zero yield the minimization of risk subject to the Lagrangian constraints. We can obtain the following equations:

$$\begin{aligned} \frac{\partial C}{\partial W_1} &= 2W_1\sigma_1{}^2 + 2W_2\sigma_{12} + 2W_3\sigma_{13} - \lambda_1 - \lambda_2 E(R_1) = 0 \\ \frac{\partial C}{\partial W_2} &= 2W_2\sigma_2{}^2 + 2W_1\sigma_{12} + 2W_3\sigma_{23} - \lambda_1 - \lambda_2 E(R_2) = 0 \\ \frac{\partial C}{\partial W_3} &= 2W_3\sigma_3{}^2 + 2W_1\sigma_{13} + 2W_2\sigma_{23} - \lambda_1 - \lambda_2 E(R_3) = 0 \end{aligned} \tag{29.3}$$

$$\frac{\partial C}{\partial \lambda_1} = 1 - W_1 - W_2 - W_3 = 0$$

$$\frac{\partial C}{\partial \lambda_2} = E^* - W_1E(R_1) - W_2E(R_2) - W_3E(R_3) = 0.$$

This system of five equations and five unknowns can be solved by the use of matrix algebra. Briefly, the Jacobian matrix of these equations is

Table 29.1 Data for three securities

Company	$E(r_i)$	σ_i^2	$\mathrm{Cov}(R_i, R_j)$
JNJ	0.0080	0.0025	$\sigma_{12} = 0.0007$
IBM	0.0050	0.0071	$\sigma_{23} = 0.0006$
BA	0.0113	0.0083	$\sigma_{13} = 0.0007$

$$\begin{bmatrix} 2\sigma_{11} & 2\sigma_{12} & 2\sigma_{13} & -1 & -E(R_1) \\ 2\sigma_{21} & 2\sigma_{22} & 2\sigma_{23} & -1 & -E(R_2) \\ 2\sigma_{31} & 2\sigma_{32} & 2\sigma_{33} & -1 & -E(R_3) \\ 1 & 1 & 1 & 0 & 0 \\ E(R_1) & E(R_2) & E(R_3) & 0 & 0 \end{bmatrix} \times \begin{bmatrix} W_1 \\ W_2 \\ W_3 \\ \lambda_1 \\ \lambda_2 \end{bmatrix} = \begin{bmatrix} 0 \\ 0 \\ 0 \\ 1 \\ E^* \end{bmatrix}. \tag{29.4a}$$

Equation (29.4a) can be redefined as

$$\mathrm{AW} = \mathrm{K} \tag{29.4b}$$

To solve for the unknown W of (29.4b), we can premultiply both sides of the equation (29.4b) by the inverse of A (denoted A^{-1}) and solve for the W column. This procedure can be found in Sect. 29.2.3.

Following the example from Lee et al. (2013), this example uses the information of returns and risk of Johnson & Johnson (JNJ), International Business Machines Corp. (IBM), and Boeing Co. (BA), for the period April 2001 to April 2010. The data used are tabulated in Table 29.1.

Plugging the data listed in Table 29.1 and $E^* = 0.00106$ into the matrix defined in equation (29.4a) yields

$$\begin{bmatrix} 0.0910 & 0.0018 & 0.0008 & -1 & -0.0053 \\ 0.0036 & 0.1228 & 0.0020 & -1 & -0.0055 \\ 0.0008 & 0.0020 & 0.1050 & -1 & -0.0126 \\ 1 & 1 & 1 & 0 & 0 \\ 0.0053 & 0.0055 & 0.0126 & 0 & 0 \end{bmatrix} \times \begin{bmatrix} W_1 \\ W_2 \\ W_3 \\ \lambda_1 \\ \lambda_2 \end{bmatrix} = \begin{bmatrix} 0 \\ 0 \\ 0 \\ 1 \\ 0.00106 \end{bmatrix}. \tag{29.5}$$

When matrix A is properly inverted and post multiplied by K, the solution vector $\mathrm{A}^{-1}\mathrm{K}$ is derived:

$$\begin{matrix} W & & \mathrm{A}^{-1}\mathrm{K} \\ \begin{bmatrix} W_1 \\ W_2 \\ W_3 \\ \lambda_1 \\ \lambda_2 \end{bmatrix} & = & \begin{bmatrix} 0.9442 \\ 0.6546 \\ -0.5988 \\ 0.1937 \\ -20.1953 \end{bmatrix} \end{matrix} \tag{29.6}$$

with the knowledge of the efficient portfolio weights given that $E(R_p)$ is equal to 0.00106, 0.00212, and 0.00318.

Now we use data of IBM, Microsoft (MSFT), and S&P500 as an example to calculate optimal weights of Markowitz model. The monthly rates of return for these three companies from 2009 to 2013 for all three stocks can be found in Appendix 1. The means, variances, and variance-covariance matrices for these three companies are presented in Fig. 29.1. By using Excel program, we can calculate optimal Markowitz portfolio model and its results are presented in Fig. 29.2.

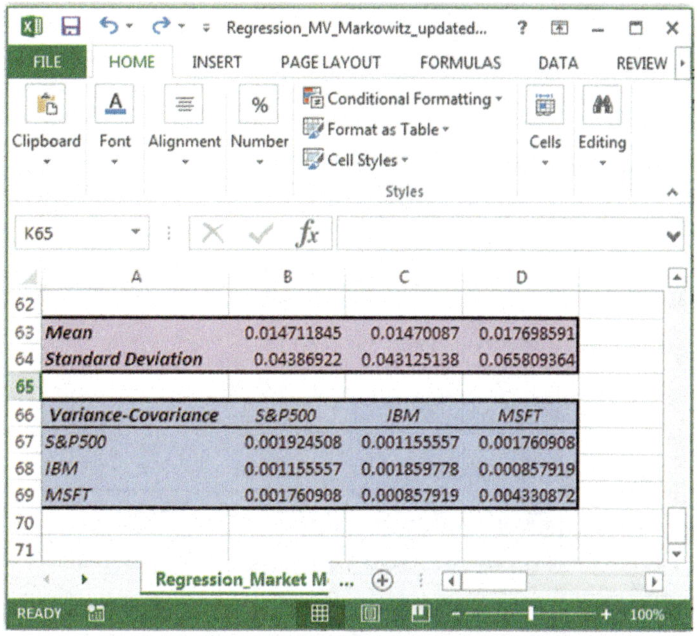

Fig. 29.1 The mean, standard deviation, and variance-covariance matrix for company S&P 500, IBM, and MSFT

In Fig. 29.2, the top portion is the equation system used to calculate optimal weights, which was discussed previously. Then we use the input data and calculate related information for equation system as presented in Step 1. Step 2 presents the procedure of calculating optimal weights. Finally, the lower portion of this figure, we present the expected rate of return and the variance for this optimal portfolio.

There is a special case in terms of Markowitz model. This case is minimum variance model. The only difference between these two models is that we exclude the expected return constraint that is

$$\sum_{i=1}^{n} W_i E(R_i) = E^*.$$

For calculating the optimal expected return of the specific portfolio, we need first to calculate the mean, standard deviation, and variance-covariance matrix for companies. In this chapter, we use that shown in Fig. 29.3.

29.4 Option Strategies

In this section, we will discuss how Excel can be used to calculate seven different option strategies. The seven strategies will include a long straddle, a short straddle, a long vertical spread, a short vertical spread, a protective put, a covered call, and a collar. The IBM option data on October 12, 2015, as presented in Appendix 2 is used to do the following seven option strategies.

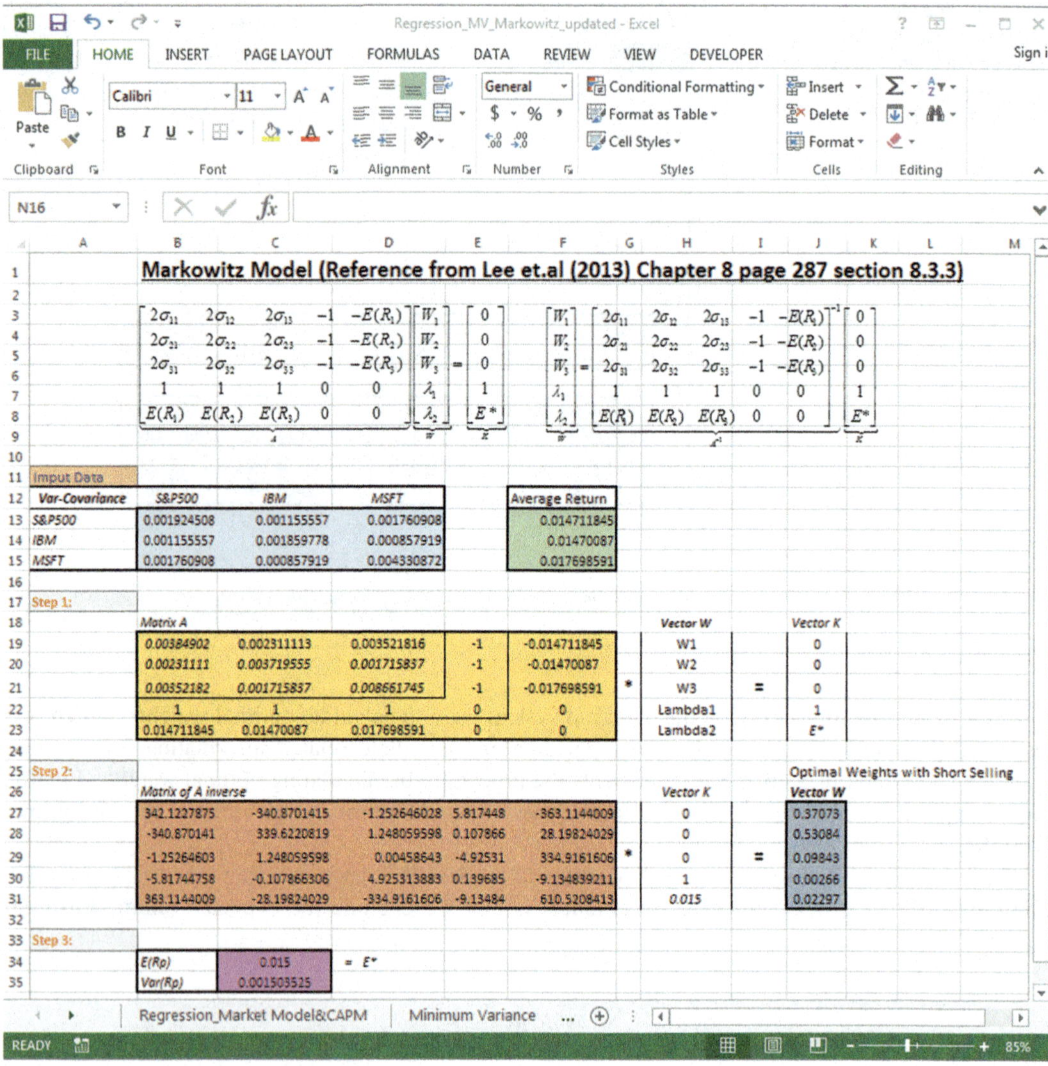

Fig. 29.2 Excel application of Markowitz model

29.4.1 Long Straddle

Assume that an investor expects the volatility of IBM stock to increase in the future and then uses a long straddle to profit. The investor can purchase a call option and a put option with the same exercise price of \$150. The investor will profit on this type of position as long as the price of the underlying asset moves sufficiently up or down to more than cover the original cost of the option premiums. Let S_T and X denote the stock purchase price, future stock price at the expiration time T, and the strike price, respectively. Given $X(\mathrm{E}) = \$150$, S_T (you can find the value for S_T in the first column of the table in Fig. 29.4) and the premium for the call option is \$4.60 and put option is \$4.40, Fig. 29.4 shows the values for long straddle at different stock prices at time T. For information in details, you can find the Excel function in Fig. 29.5 for calculations of the numbers in Fig. 29.4. The profit profile of the long straddle position is constructed in Fig. 29.6. The break-even point means when the profit equals to zero. The formula for calculating the upper break-even point is strike price of long call + net premium paid

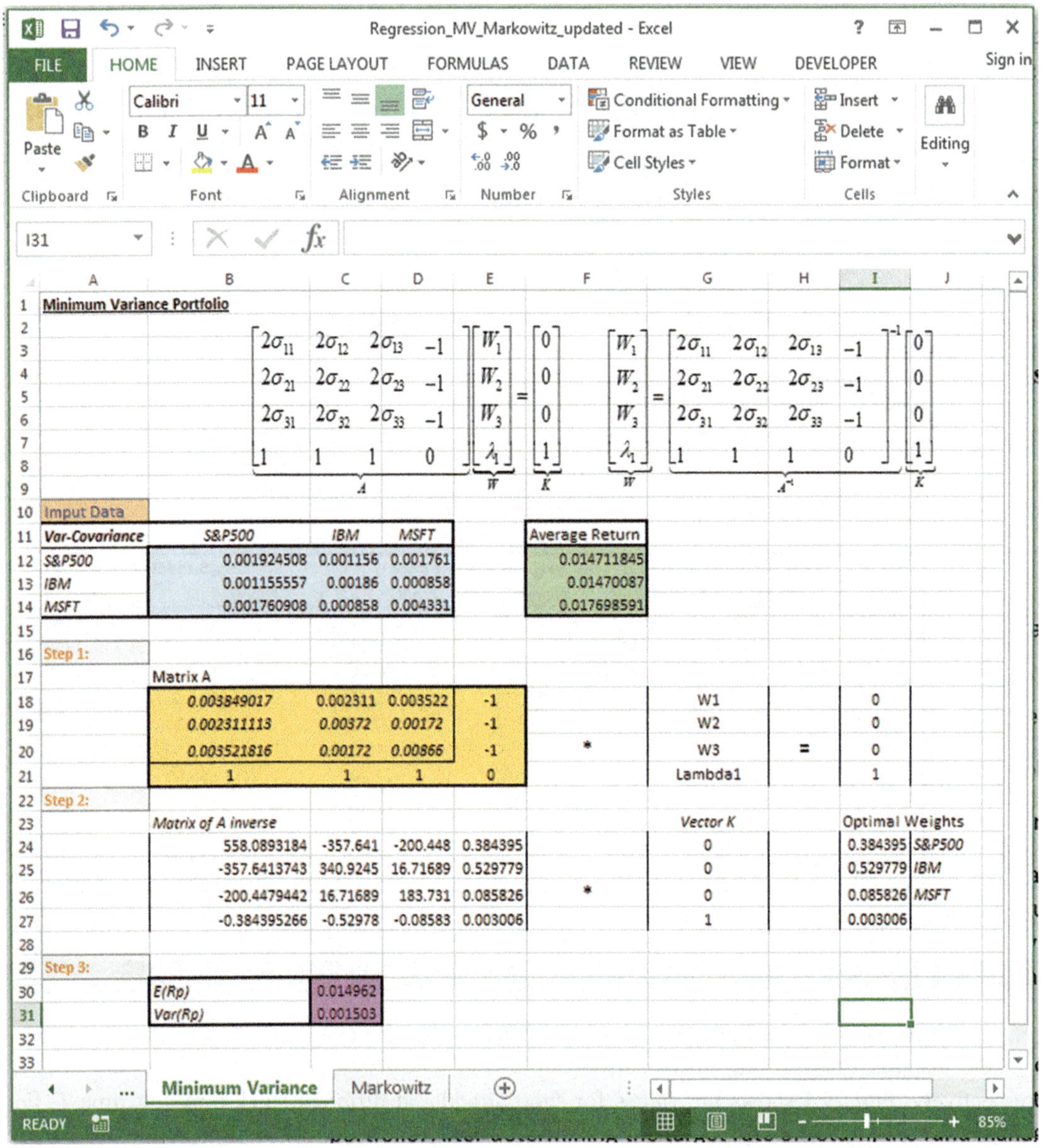

Fig. 29.3 Excel application for minimum variance model

and the lower break-even point can be calculated as strike price of long put − net premium paid. For this example, the upper break-even point is $159 and the lower break-even point is $141.

29.4.2 Short Straddle

Contrary to the long straddle strategy, an investor will use a short straddle via a short call and a short put on IBM stock with the same exercise price of $150, when he or she expects little or no movement in the price of IBM stock. Given $X(E) = \$150$, S_T (you can find the value for S_T in the first column of the table in Fig. 29.7) and the premiums for the call option and put option are $4.35 and $4.15,

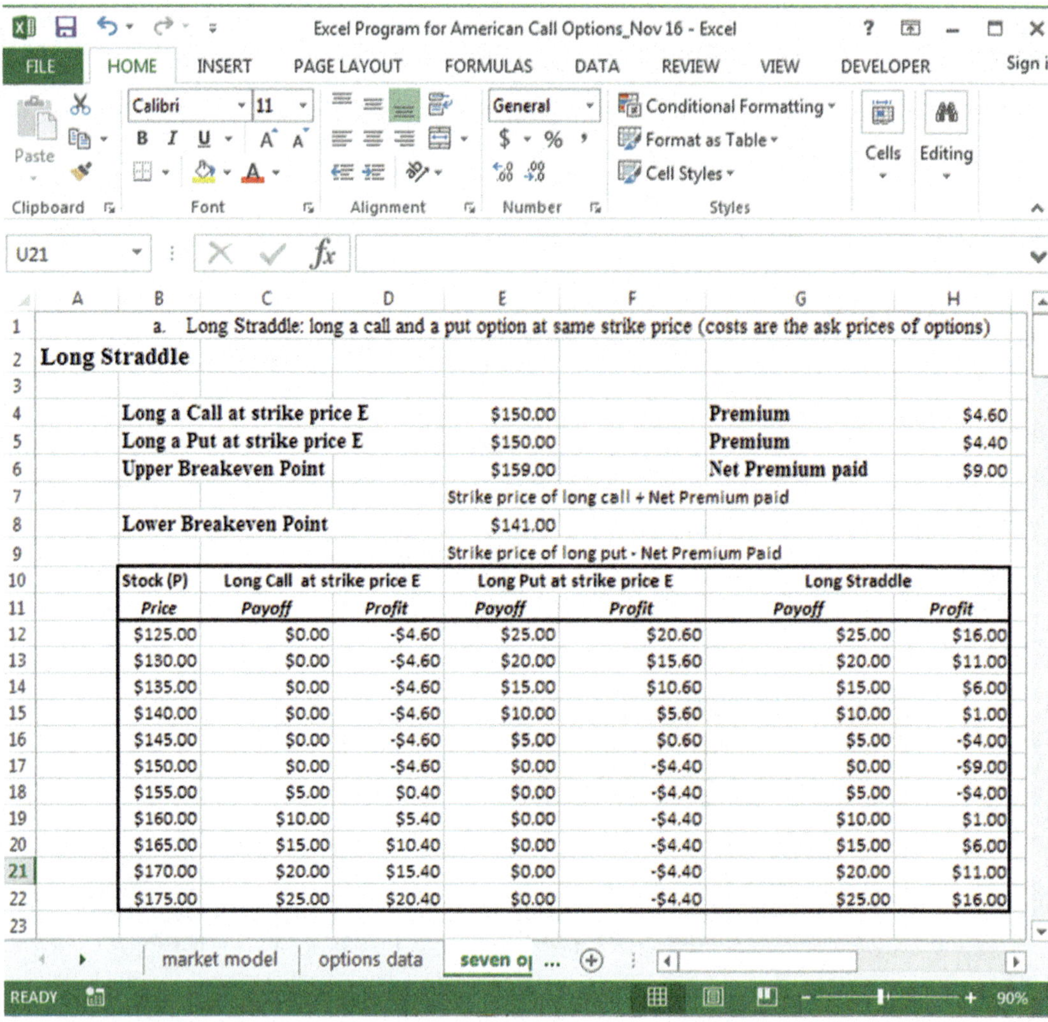

a. Long Straddle: long a call and a put option at same strike price (costs are the ask prices of options)

Long Straddle

Long a Call at strike price E	$150.00	Premium	$4.60
Long a Put at strike price E	$150.00	Premium	$4.40
Upper Breakeven Point	$159.00	Net Premium paid	$9.00
	Strike price of long call + Net Premium paid		
Lower Breakeven Point	$141.00		
	Strike price of long put - Net Premium Paid		

Stock (P)	Long Call at strike price E		Long Put at strike price E		Long Straddle	
Price	Payoff	Profit	Payoff	Profit	Payoff	Profit
$125.00	$0.00	-$4.60	$25.00	$20.60	$25.00	$16.00
$130.00	$0.00	-$4.60	$20.00	$15.60	$20.00	$11.00
$135.00	$0.00	-$4.60	$15.00	$10.60	$15.00	$6.00
$140.00	$0.00	-$4.60	$10.00	$5.60	$10.00	$1.00
$145.00	$0.00	-$4.60	$5.00	$0.60	$5.00	-$4.00
$150.00	$0.00	-$4.60	$0.00	-$4.40	$0.00	-$9.00
$155.00	$5.00	$0.40	$0.00	-$4.40	$5.00	-$4.00
$160.00	$10.00	$5.40	$0.00	-$4.40	$10.00	$1.00
$165.00	$15.00	$10.40	$0.00	-$4.40	$15.00	$6.00
$170.00	$20.00	$15.40	$0.00	-$4.40	$20.00	$11.00
$175.00	$25.00	$20.40	$0.00	-$4.40	$25.00	$16.00

Fig. 29.4 Value of a long straddle position at option expiration

respectively, Fig. 29.7 shows the values for short straddle at different stock prices at time T. For information in details, you can find the Excel function in Fig. 29.8 for calculations of the numbers in Fig. 29.7. The profit profile of the short straddle position is constructed in Fig. 29.9. The break-even point means when the profit equals to zero. The upper break-even point for short straddle can be calculated as strike price of short call + net premium received and the lower break-even point can be calculated as strike price of short put − net premium received. For this example, the upper break-even point is $158.50 and the lower break-even point is $141.50.

29.4.3 Long Vertical Spread

This strategy combines a long call (or put) with a low strike price and a short call (or put) with a high strike price. For example, an investor purchases a call with the exercise price of $155 and sells a call with the exercise price of $150. Given X_1(E1) = $155, X_2(E2) = $150, S_T (you can find the value for

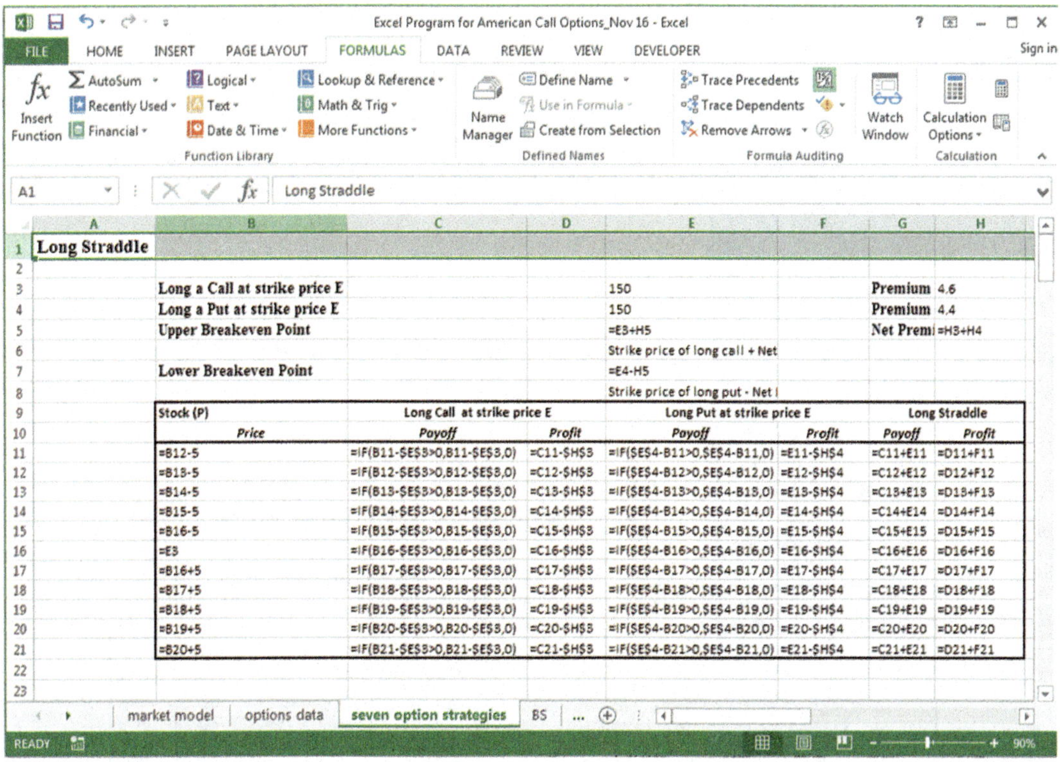

	A	B	C	D	E	F	G	H
1	**Long Straddle**							
2								
3		**Long a Call at strike price E**			150		**Premium**	4.6
4		**Long a Put at strike price E**			150		**Premium**	4.4
5		**Upper Breakeven Point**			=E3+H5		**Net Premi**	=H3+H4
6					Strike price of long call + Net			
7		**Lower Breakeven Point**			=E4-H5			
8					Strike price of long put - Net I			
9		Stock (P)	Long Call at strike price E		Long Put at strike price E		Long Straddle	
10		*Price*	*Payoff*	*Profit*	*Payoff*	*Profit*	*Payoff*	*Profit*
11		=B12-5	=IF(B11-$E$3>0,B11-$E$3,0)	=C11-$H$3	=IF($E$4-B11>0,$E$4-B11,0)	=E11-$H$4	=C11+E11	=D11+F11
12		=B13-5	=IF(B12-$E$3>0,B12-$E$3,0)	=C12-$H$3	=IF($E$4-B12>0,$E$4-B12,0)	=E12-$H$4	=C12+E12	=D12+F12
13		=B14-5	=IF(B13-$E$3>0,B13-$E$3,0)	=C13-$H$3	=IF($E$4-B13>0,$E$4-B13,0)	=E13-$H$4	=C13+E13	=D13+F13
14		=B15-5	=IF(B14-$E$3>0,B14-$E$3,0)	=C14-$H$3	=IF($E$4-B14>0,$E$4-B14,0)	=E14-$H$4	=C14+E14	=D14+F14
15		=B16-5	=IF(B15-$E$3>0,B15-$E$3,0)	=C15-$H$3	=IF($E$4-B15>0,$E$4-B15,0)	=E15-$H$4	=C15+E15	=D15+F15
16		=E3	=IF(B16-$E$3>0,B16-$E$3,0)	=C16-$H$3	=IF($E$4-B16>0,$E$4-B16,0)	=E16-$H$4	=C16+E16	=D16+F16
17		=B16+5	=IF(B17-$E$3>0,B17-$E$3,0)	=C17-$H$3	=IF($E$4-B17>0,$E$4-B17,0)	=E17-$H$4	=C17+E17	=D17+F17
18		=B17+5	=IF(B18-$E$3>0,B18-$E$3,0)	=C18-$H$3	=IF($E$4-B18>0,$E$4-B18,0)	=E18-$H$4	=C18+E18	=D18+F18
19		=B18+5	=IF(B19-$E$3>0,B19-$E$3,0)	=C19-$H$3	=IF($E$4-B19>0,$E$4-B19,0)	=E19-$H$4	=C19+E19	=D19+F19
20		=B19+5	=IF(B20-$E$3>0,B20-$E$3,0)	=C20-$H$3	=IF($E$4-B20>0,$E$4-B20,0)	=E20-$H$4	=C20+E20	=D20+F20
21		=B20+5	=IF(B21-$E$3>0,B21-$E$3,0)	=C21-$H$3	=IF($E$4-B21>0,$E$4-B21,0)	=E21-$H$4	=C21+E21	=D21+F21

Fig. 29.5 Excel formula for calculating the value of a long straddle position at option expiration

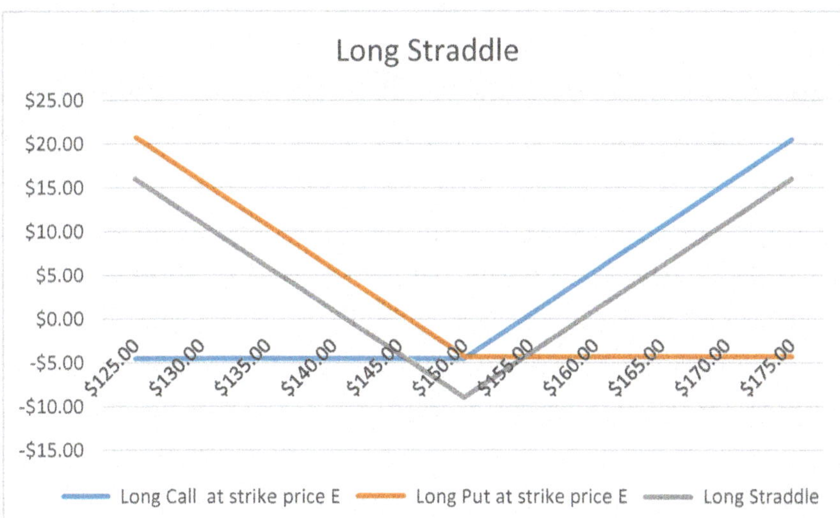

Fig. 29.6 Profit profile for long straddle

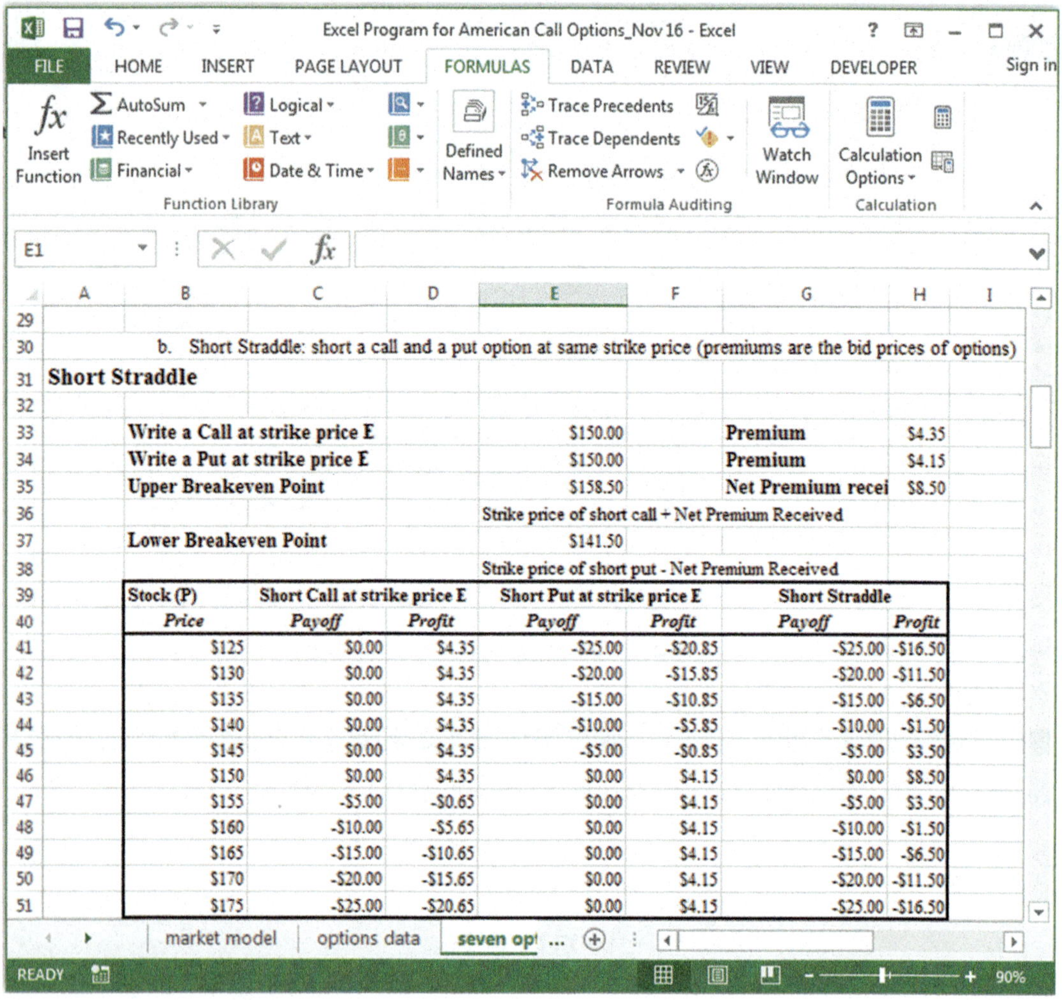

b. Short Straddle: short a call and a put option at same strike price (premiums are the bid prices of options)

Short Straddle

Write a Call at strike price E	$150.00	**Premium**	$4.35
Write a Put at strike price E	$150.00	**Premium**	$4.15
Upper Breakeven Point	$158.50	**Net Premium recei**	$8.50
	Strike price of short call + Net Premium Received		
Lower Breakeven Point	$141.50		
	Strike price of short put - Net Premium Received		

Stock (P)	**Short Call at strike price E**		**Short Put at strike price E**		**Short Straddle**	
Price	***Payoff***	***Profit***	***Payoff***	***Profit***	***Payoff***	***Profit***
$125	$0.00	$4.35	-$25.00	-$20.85	-$25.00	-$16.50
$130	$0.00	$4.35	-$20.00	-$15.85	-$20.00	-$11.50
$135	$0.00	$4.35	-$15.00	-$10.85	-$15.00	-$6.50
$140	$0.00	$4.35	-$10.00	-$5.85	-$10.00	-$1.50
$145	$0.00	$4.35	-$5.00	-$0.85	-$5.00	$3.50
$150	$0.00	$4.35	$0.00	$4.15	$0.00	$8.50
$155	-$5.00	-$0.65	$0.00	$4.15	-$5.00	$3.50
$160	-$10.00	-$5.65	$0.00	$4.15	-$10.00	-$1.50
$165	-$15.00	-$10.65	$0.00	$4.15	-$15.00	-$6.50
$170	-$20.00	-$15.65	$0.00	$4.15	-$20.00	-$11.50
$175	-$25.00	-$20.65	$0.00	$4.15	-$25.00	-$16.50

Fig. 29.7 Value of a short straddle position at option expiration

S_T in the first column of the table in Fig. 29.10) and the premium for the long call option is $1.97 and the short call option is $4.60, Fig. 29.10 shows the values for long vertical spread at different stock prices at time T. For information in details, you can find the Excel function in Fig. 29.11 for calculations of the numbers in Fig. 29.10. The profit profile of the long vertical spread is constructed in Fig. 29.12. The break-even point means when the profit equals to zero. The break-even point for long vertical spread can be calculated as strike price of long call + net premium paid.

For this example, the break-even point is $152.63.

29.4.4 Short Vertical Spread

Contrary to a long vertical spread, this strategy combines a long call (or put) with a high strike price and a short call (or put) with a low strike price. For example, an investor purchases a call with the exercise price of $150 and sells a call with the exercise price of $155. Given X_1(E1) = $150,

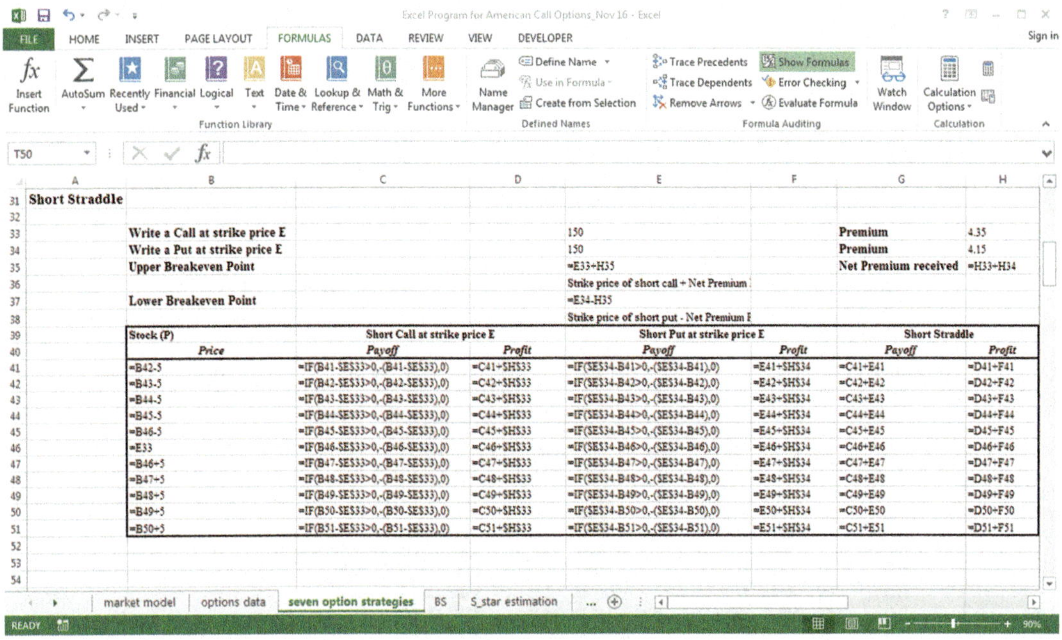

	A	B	C	D	E	F	G	H
31	**Short Straddle**							
32								
33		**Write a Call at strike price E**			150		**Premium**	4.35
34		**Write a Put at strike price E**			150		**Premium**	4.15
35		**Upper Breakeven Point**			=E33+H35		**Net Premium received**	=H33+H34
36					Strike price of short call + Net Premium			
37		**Lower Breakeven Point**			=E34-H35			
38					Strike price of short put - Net Premium F			
39		**Stock (P)**	**Short Call at strike price E**		**Short Put at strike price E**		**Short Straddle**	
40		***Price***	***Payoff***	***Profit***	***Payoff***	***Profit***	***Payoff***	***Profit***
41		=B42-5	=IF(B41-$E$33>0,-(B41-$E$33),0)	=C41+$H$33	=IF($E$34-B41>0,-($E$34-B41),0)	=E41+$H$34	=C41+E41	=D41+F41
42		=B43-5	=IF(B42-$E$33>0,-(B42-$E$33),0)	=C42+$H$33	=IF($E$34-B42>0,-($E$34-B42),0)	=E42+$H$34	=C42+E42	=D42+F42
43		=B44-5	=IF(B43-$E$33>0,-(B43-$E$33),0)	=C43+$H$33	=IF($E$34-B43>0,-($E$34-B43),0)	=E43+$H$34	=C43+E43	=D43+F43
44		=B45-5	=IF(B44-$E$33>0,-(B44-$E$33),0)	=C44+$H$33	=IF($E$34-B44>0,-($E$34-B44),0)	=E44+$H$34	=C44+E44	=D44+F44
45		=B46-5	=IF(B45-$E$33>0,-(B45-$E$33),0)	=C45+$H$33	=IF($E$34-B45>0,-($E$34-B45),0)	=E45+$H$34	=C45+E45	=D45+F45
46		=E33	=IF(B46-$E$33>0,-(B46-$E$33),0)	=C46+$H$33	=IF($E$34-B46>0,-($E$34-B46),0)	=E46+$H$34	=C46+E46	=D46+F46
47		=B46+5	=IF(B47-$E$33>0,-(B47-$E$33),0)	=C47+$H$33	=IF($E$34-B47>0,-($E$34-B47),0)	=E47+$H$34	=C47+E47	=D47+F47
48		=B47+5	=IF(B48-$E$33>0,-(B48-$E$33),0)	=C48+$H$33	=IF($E$34-B48>0,-($E$34-B48),0)	=E48+$H$34	=C48+E48	=D48+F48
49		=B48+5	=IF(B49-$E$33>0,-(B49-$E$33),0)	=C49+$H$33	=IF($E$34-B49>0,-($E$34-B49),0)	=E49+$H$34	=C49+E49	=D49+F49
50		=B49+5	=IF(B50-$E$33>0,-(B50-$E$33),0)	=C50+$H$33	=IF($E$34-B50>0,-($E$34-B50),0)	=E50+$H$34	=C50+E50	=D50+F50
51		=B50+5	=IF(B51-$E$33>0,-(B51-$E$33),0)	=C51+$H$33	=IF($E$34-B51>0,-($E$34-B51),0)	=E51+$H$34	=C51+E51	=D51+F51

Fig. 29.8 Excel formula for calculating the value of a short straddle position at option expiration

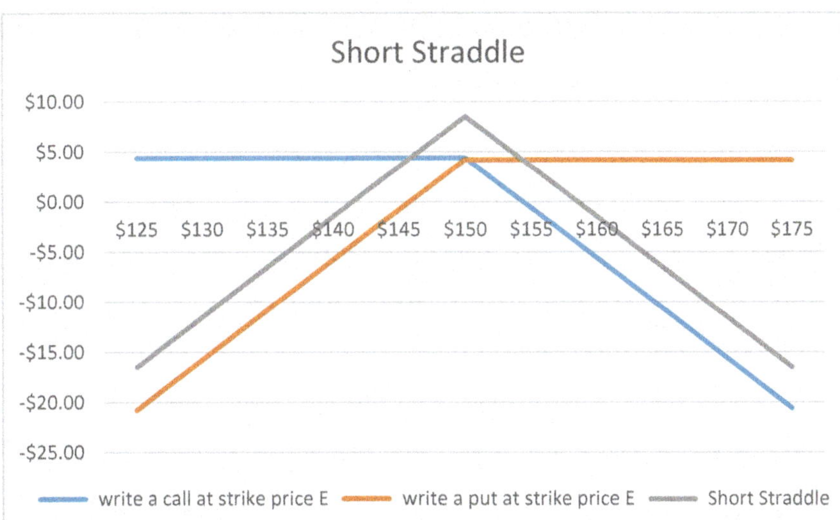

Fig. 29.9 Profit profile for short straddle

X_2(E2) = \$155, S_T (you can find the value for S_T in the first column of the table in Fig. 29.13) and the premium for the long call option is \$4.35 and the short call option is \$2.13, Fig. 29.13 shows the values for short vertical spread at different stock prices at time T. For information in details, you can find the Excel function in Fig. 29.14 for calculations of the numbers in Fig. 29.13. The profit profile of the short vertical spread is constructed in Fig. 29.15. The break-even point means when the profit equals to zero. The break-even point for short vertical spread can be calculated as strike price of short call + net premium received. For this example, the break-even point is \$152.22.

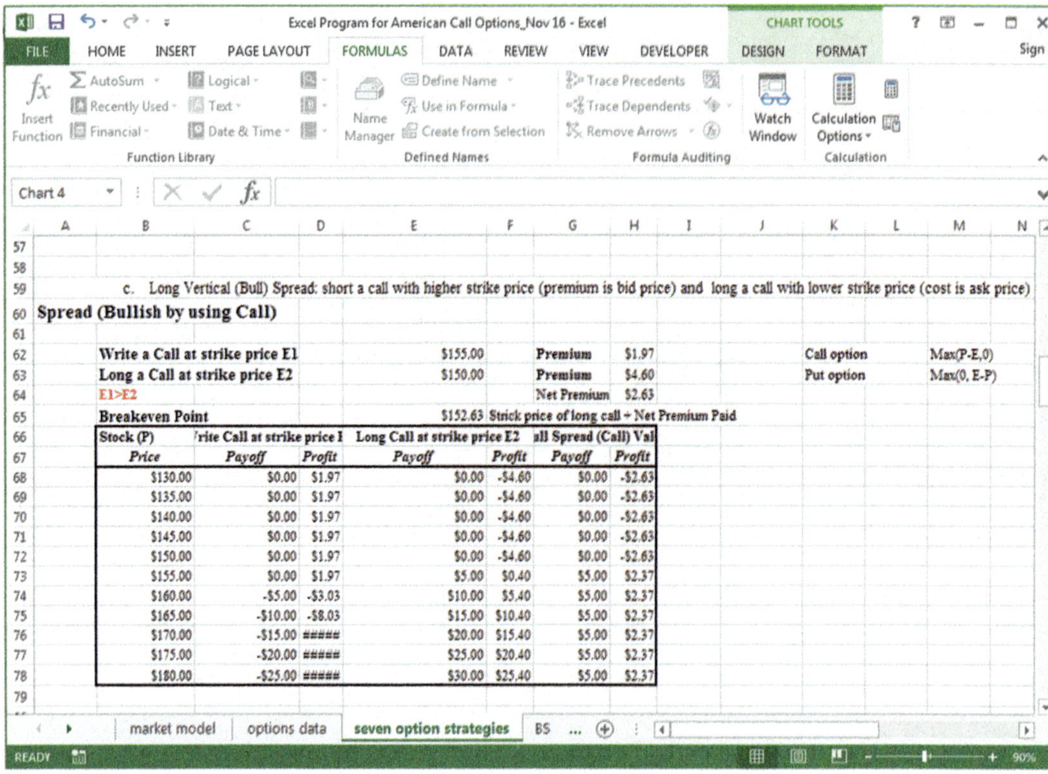

Fig. 29.10 Value of a long vertical spread position at option expiration

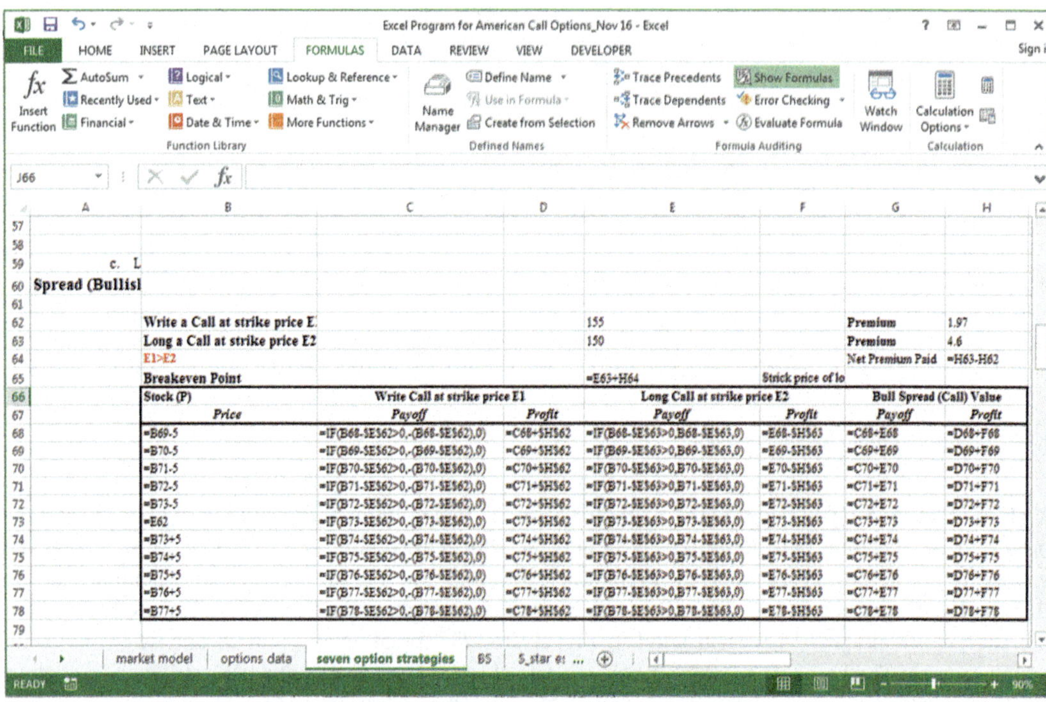

Fig. 29.11 Excel formula for calculating the value of a long vertical spread position at option expiration

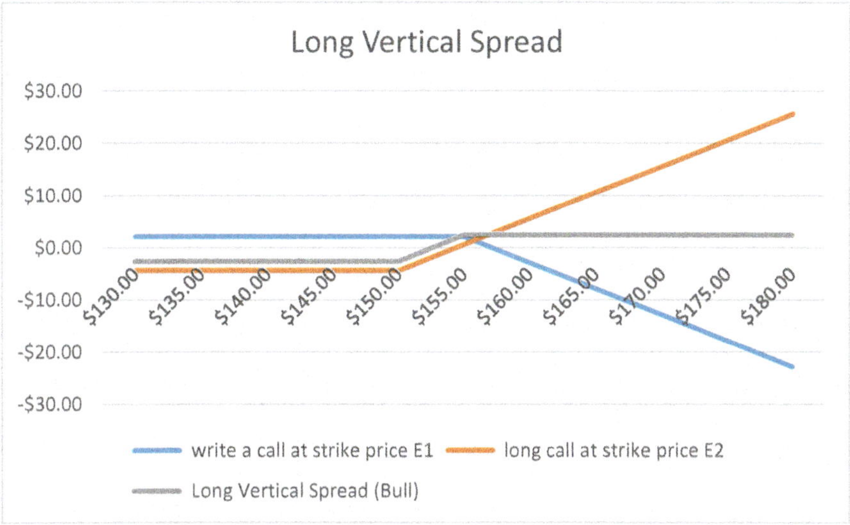

Fig. 29.12 Profit profile for long vertical spread

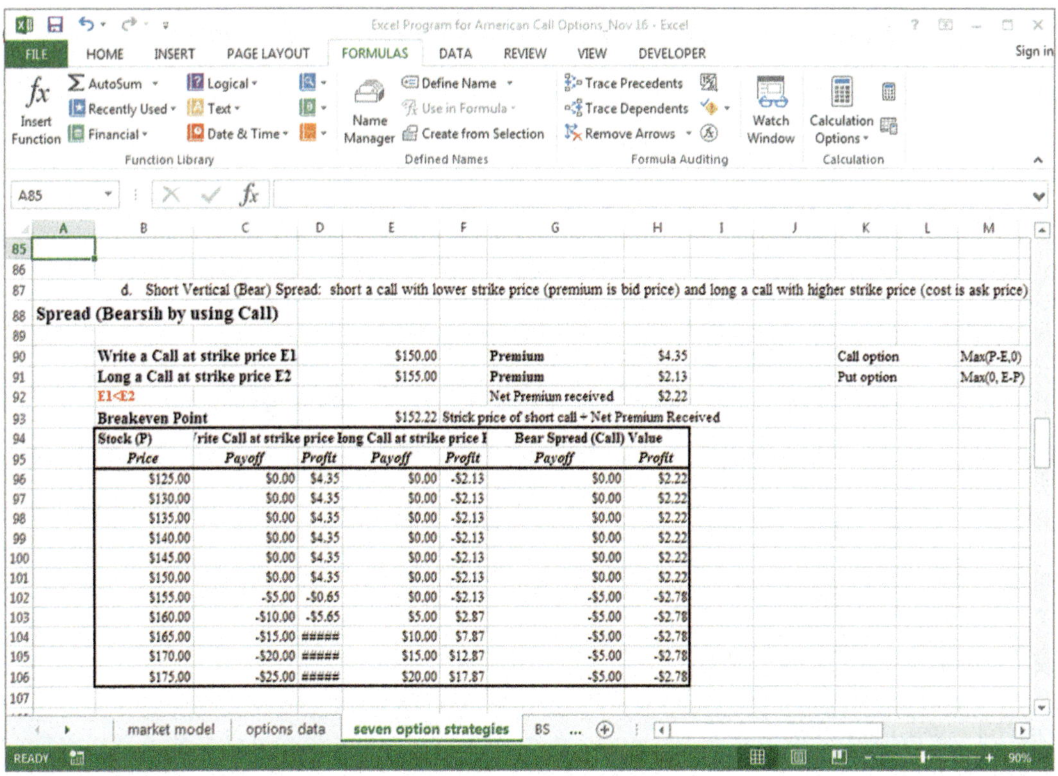

d. Short Vertical (Bear) Spread: short a call with lower strike price (premium is bid price) and long a call with higher strike price (cost is ask price)

Spread (Bearsih by using Call)

Write a Call at strike price E1	$150.00	Premium	$4.35	Call option	Max(P-E,0)
Long a Call at strike price E2	$155.00	Premium	$2.13	Put option	Max(0, E-P)
E1<E2		Net Premium received	$2.22		
Breakeven Point	$152.22	Strick price of short call + Net Premium Received			

Stock (P)	'rite Call at strike price		long Call at strike price I		Bear Spread (Call) Value	
Price	*Payoff*	*Profit*	*Payoff*	*Profit*	*Payoff*	*Profit*
$125.00	$0.00	$4.35	$0.00	-$2.13	$0.00	$2.22
$130.00	$0.00	$4.35	$0.00	-$2.13	$0.00	$2.22
$135.00	$0.00	$4.35	$0.00	-$2.13	$0.00	$2.22
$140.00	$0.00	$4.35	$0.00	-$2.13	$0.00	$2.22
$145.00	$0.00	$4.35	$0.00	-$2.13	$0.00	$2.22
$150.00	$0.00	$4.35	$0.00	-$2.13	$0.00	$2.22
$155.00	-$5.00	-$0.65	$0.00	-$2.13	-$5.00	-$2.78
$160.00	-$10.00	-$5.65	$5.00	$2.87	-$5.00	-$2.78
$165.00	-$15.00	#####	$10.00	$7.87	-$5.00	-$2.78
$170.00	-$20.00	#####	$15.00	$12.87	-$5.00	-$2.78
$175.00	-$25.00	#####	$20.00	$17.87	-$5.00	-$2.78

Fig. 29.13 Value of a short vertical spread position at option expiration

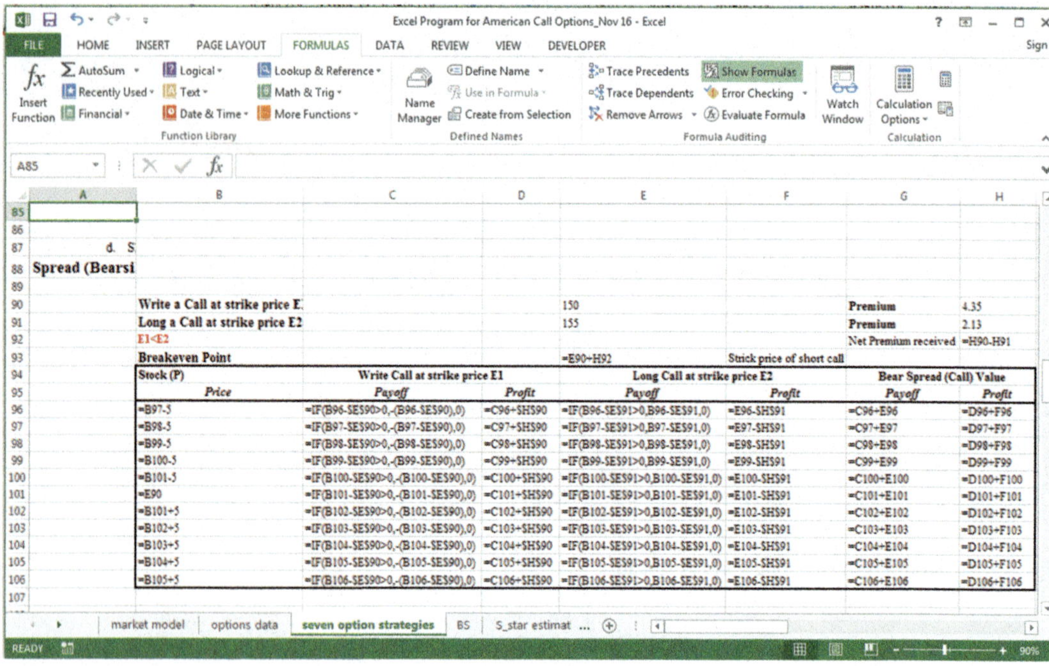

	B	C	D	E	F	G	H
88	Spread (Bearsi						
90	Write a Call at strike price E1			150		Premium	4.35
91	Long a Call at strike price E2			155		Premium	2.13
92	E1<E2					Net Premium received	=H90-H91
93	Breakeven Point			=E90+H92	Strick price of short call		
94	Stock (P)	Write Call at strike price E1		Long Call at strike price E2		Bear Spread (Call) Value	
95	Price	Payoff	Profit	Payoff	Profit	Payoff	Profit
96	=B97-5	=IF(B96-$E$90>0,-(B96-$E$90),0)	=C96+$H$90	=IF(B96-$E$91>0,B96-$E$91,0)	=E96-$H$91	=C96+E96	=D96+F96
97	=B98-5	=IF(B97-$E$90>0,-(B97-$E$90),0)	=C97+$H$90	=IF(B97-$E$91>0,B97-$E$91,0)	=E97-$H$91	=C97+E97	=D97+F97
98	=B99-5	=IF(B98-$E$90>0,-(B98-$E$90),0)	=C98+$H$90	=IF(B98-$E$91>0,B98-$E$91,0)	=E98-$H$91	=C98+E98	=D98+F98
99	=B100-5	=IF(B99-$E$90>0,-(B99-$E$90),0)	=C99+$H$90	=IF(B99-$E$91>0,B99-$E$91,0)	=E99-$H$91	=C99+E99	=D99+F99
100	=B101-5	=IF(B100-$E$90>0,-(B100-$E$90),0)	=C100+$H$90	=IF(B100-$E$91>0,B100-$E$91,0)	=E100-$H$91	=C100+E100	=D100+F100
101	=E90	=IF(B101-$E$90>0,-(B101-$E$90),0)	=C101+$H$90	=IF(B101-$E$91>0,B101-$E$91,0)	=E101-$H$91	=C101+E101	=D101+F101
102	=B101+5	=IF(B102-$E$90>0,-(B102-$E$90),0)	=C102+$H$90	=IF(B102-$E$91>0,B102-$E$91,0)	=E102-$H$91	=C102+E102	=D102+F102
103	=B102+5	=IF(B103-$E$90>0,-(B103-$E$90),0)	=C103+$H$90	=IF(B103-$E$91>0,B103-$E$91,0)	=E103-$H$91	=C103+E103	=D103+F103
104	=B103+5	=IF(B104-$E$90>0,-(B104-$E$90),0)	=C104+$H$90	=IF(B104-$E$91>0,B104-$E$91,0)	=E104-$H$91	=C104+E104	=D104+F104
105	=B104+5	=IF(B105-$E$90>0,-(B105-$E$90),0)	=C105+$H$90	=IF(B105-$E$91>0,B105-$E$91,0)	=E105-$H$91	=C105+E105	=D105+F105
106	=B105+5	=IF(B106-$E$90>0,-(B106-$E$90),0)	=C106+$H$90	=IF(B106-$E$91>0,B106-$E$91,0)	=E106-$H$91	=C106+E106	=D106+F106

Fig. 29.14 Excel formula for calculating the value of a short vertical spread position at option expiration

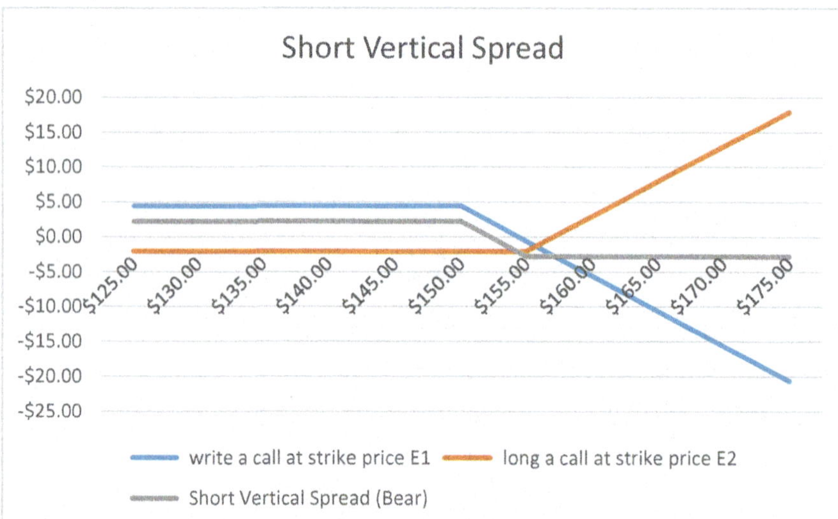

Fig. 29.15 Profit profile for short vertical spread

29.4.5 *Protective Put*

Assume that an investor wants to invest in the IBM stock on March 29, 2011, but does not desire to bear any potential loss for prices below \$150. The investor can purchase IBM stock and at the same time buy the put option with a strike price of \$150. Given current stock $S_0 = \$155.54$, exercise price $X(E) = \$150$, S_T (you can find the value for S_T in the first column of the table in Fig. 29.16) and the

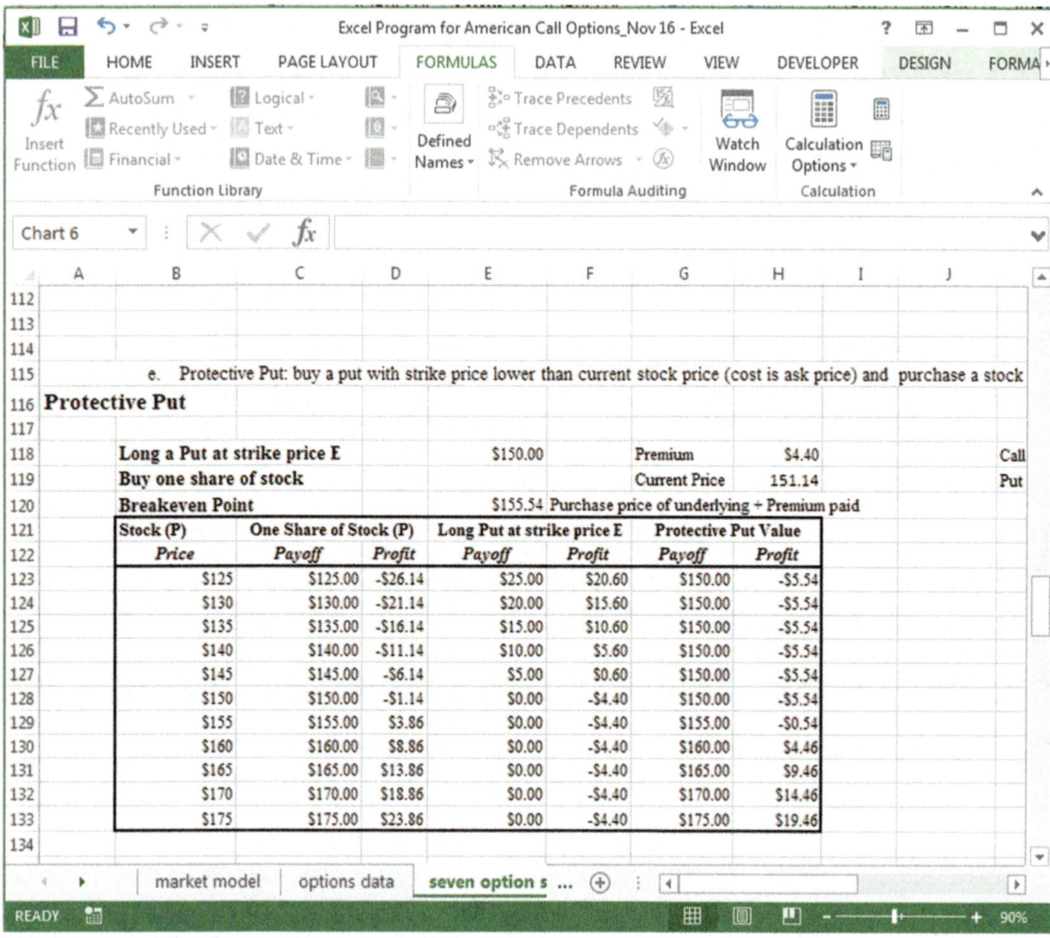

e. Protective Put: buy a put with strike price lower than current stock price (cost is ask price) and purchase a stock

Protective Put

Long a Put at strike price E			$150.00	Premium	$4.40	
Buy one share of stock				Current Price	151.14	
Breakeven Point			$155.54	Purchase price of underlying + Premium paid		

Stock (P)	One Share of Stock (P)		Long Put at strike price E		Protective Put Value	
Price	*Payoff*	*Profit*	*Payoff*	*Profit*	*Payoff*	*Profit*
$125	$125.00	-$26.14	$25.00	$20.60	$150.00	-$5.54
$130	$130.00	-$21.14	$20.00	$15.60	$150.00	-$5.54
$135	$135.00	-$16.14	$15.00	$10.60	$150.00	-$5.54
$140	$140.00	-$11.14	$10.00	$5.60	$150.00	-$5.54
$145	$145.00	-$6.14	$5.00	$0.60	$150.00	-$5.54
$150	$150.00	-$1.14	$0.00	-$4.40	$150.00	-$5.54
$155	$155.00	$3.86	$0.00	-$4.40	$155.00	-$0.54
$160	$160.00	$8.86	$0.00	-$4.40	$160.00	$4.46
$165	$165.00	$13.86	$0.00	-$4.40	$165.00	$9.46
$170	$170.00	$18.86	$0.00	-$4.40	$170.00	$14.46
$175	$175.00	$23.86	$0.00	-$4.40	$175.00	$19.46

Fig. 29.16 Value of a protective put position at option expiration

premium for the put option is $4.40 (the ask price), Fig. 29.16 shows the values for protective put at different stock prices at time T. For information in details, you can find the Excel function in Fig. 29.17 for calculations of the numbers in Fig. 29.16. The profit profile of the protective put position is constructed in Fig. 29.18. The break-even point means when the profit equals to zero. The break-even point for protective put can be calculated as purchase price of underlying + premium paid. For this example, the break-even point is $155.54.

29.4.6 Covered Call

This strategy involves investing in a stock and selling a call option on the stock at the same time. The value at the expiration of the call will be the stock value minus the value of the call. The call is "covered" because the potential obligation of delivering the stock is covered by the stock held in the portfolio. In essence, the sale of the call sold the claim to any stock value above the strike price in return for the initial premium. Suppose a manager of a stock fund holds a share of IBM stock on October 12, 2015 and she plans to sell the IBM stock if its price hits $155. Then she can write a share

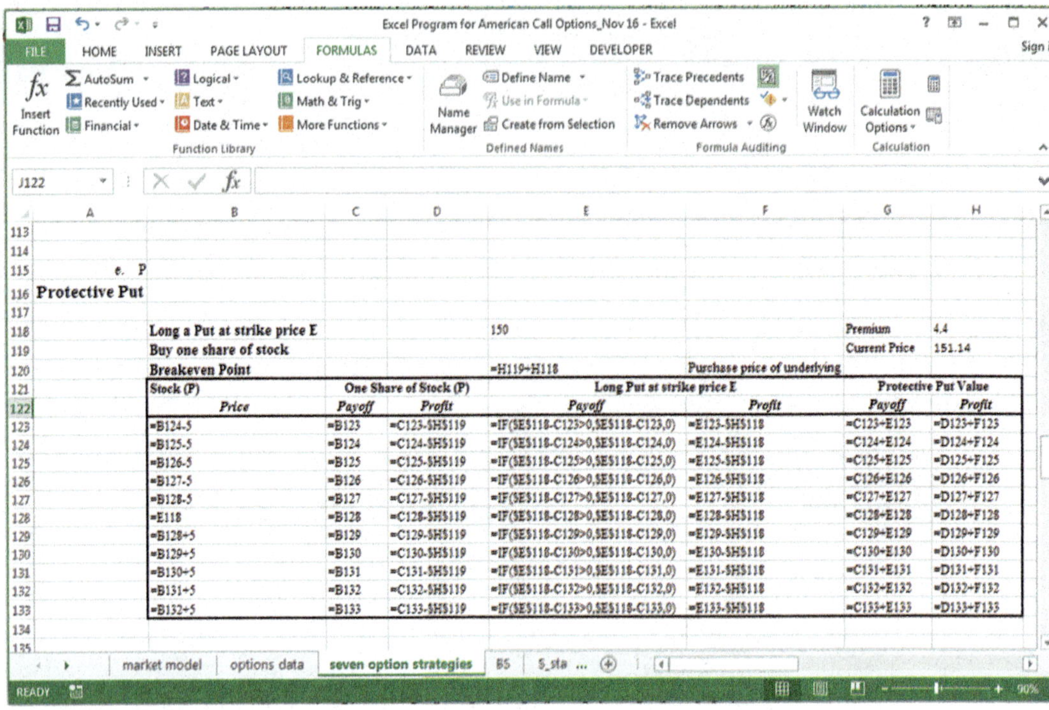

	B	C	D	E	F	G	H
e. P							
Protective Put							
	Long a Put at strike price E			150		Premium	4.4
	Buy one share of stock					Current Price	151.14
	Breakeven Point			=H119+H118	Purchase price of underlying		
	Stock (P)	One Share of Stock (P)		Long Put at strike price E		Protective Put Value	
	Price	Payoff	Profit	Payoff	Profit	Payoff	Profit
	=B124-5	=B123	=C123-$H$119	=IF($E$118-C123>0,$E$118-C123,0)	=E123-$H$118	=C123+E123	=D123+F123
	=B125-5	=B124	=C124-$H$119	=IF($E$118-C124>0,$E$118-C124,0)	=E124-$H$118	=C124+E124	=D124+F124
	=B126-5	=B125	=C125-$H$119	=IF($E$118-C125>0,$E$118-C125,0)	=E125-$H$118	=C125+E125	=D125+F125
	=B127-5	=B126	=C126-$H$119	=IF($E$118-C126>0,$E$118-C126,0)	=E126-$H$118	=C126+E126	=D126+F126
	=B128-5	=B127	=C127-$H$119	=IF($E$118-C127>0,$E$118-C127,0)	=E127-$H$118	=C127+E127	=D127+F127
	=E118	=B128	=C128-$H$119	=IF($E$118-C128>0,$E$118-C128,0)	=E128-$H$118	=C128+E128	=D128+F128
	=B128+5	=B129	=C129-$H$119	=IF($E$118-C129>0,$E$118-C129,0)	=E129-$H$118	=C129+E129	=D129+F129
	=B129+5	=B130	=C130-$H$119	=IF($E$118-C130>0,$E$118-C130,0)	=E130-$H$118	=C130+E130	=D130+F130
	=B130+5	=B131	=C131-$H$119	=IF($E$118-C131>0,$E$118-C131,0)	=E131-$H$118	=C131+E131	=D131+F131
	=B131+5	=B132	=C132-$H$119	=IF($E$118-C132>0,$E$118-C132,0)	=E132-$H$118	=C132+E132	=D132+F132
	=B132+5	=B133	=C133-$H$119	=IF($E$118-C133>0,$E$118-C133,0)	=E133-$H$118	=C133+E133	=D133+F133

Fig. 29.17 Excel formula for calculating the value of a protective put position at option expiration

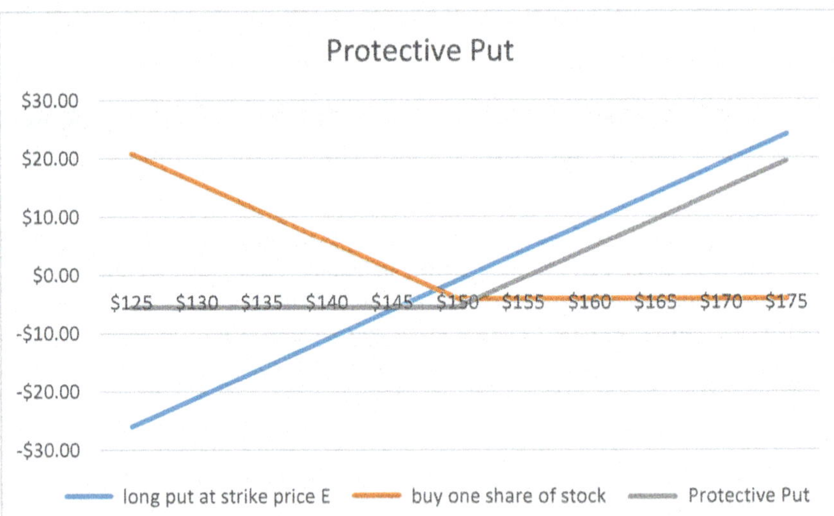

Fig. 29.18 Profit profile for protective put

of a call option with a strike price of $155 to establish the position. She shorts the call and collects premiums. Given that current stock price $S_0 = \$151.14$, $X(E) = \$155$, S_T (you can find the value for S_T in the first column of the table in Fig. 29.19) and the premium for the call option is $1.97(the bid price), Fig. 29.19 shows the values for covered call at different stock prices at time T. For information in details, you can find the Excel function in Fig. 29.20 for calculations of the numbers in Fig. 29.19.

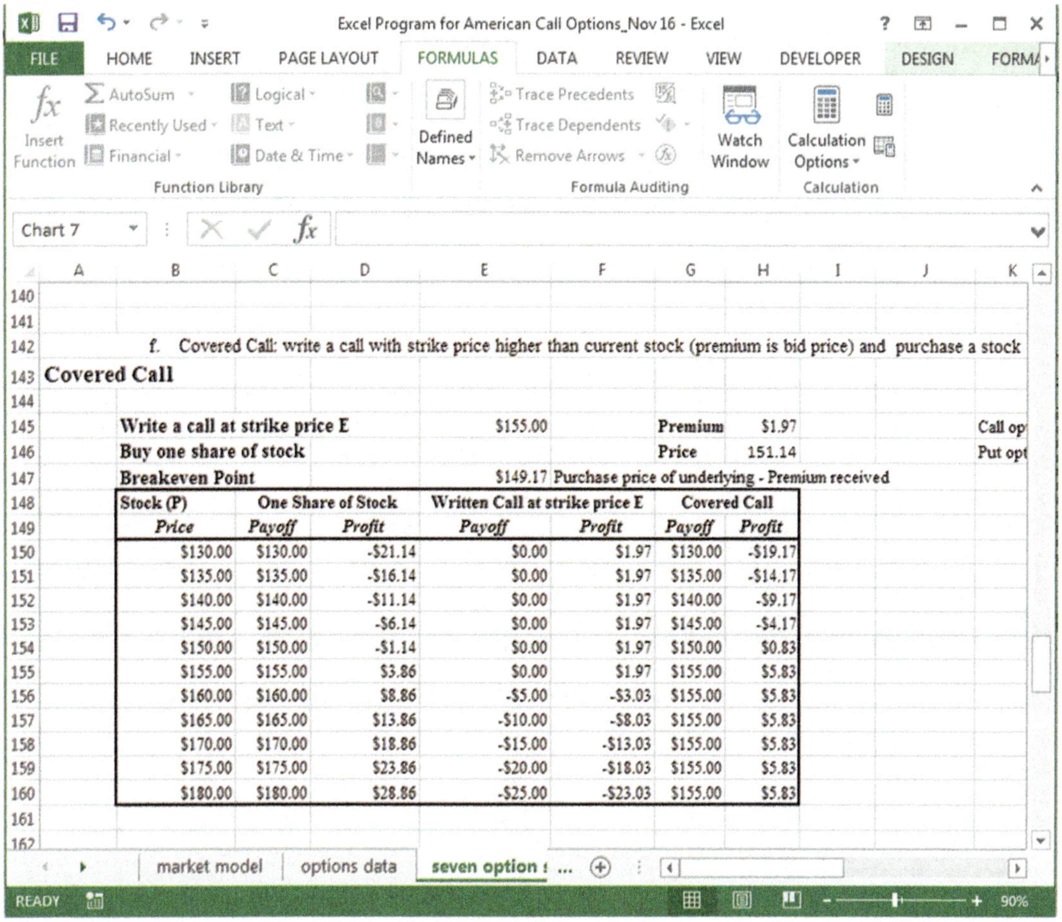

f. Covered Call: write a call with strike price higher than current stock (premium is bid price) and purchase a stock

Covered Call

Write a call at strike price E			$155.00		Premium	$1.97	
Buy one share of stock					Price	151.14	
Breakeven Point			$149.17	Purchase price of underlying - Premium received			
Stock (P)	One Share of Stock		Written Call at strike price E		Covered Call		
Price	Payoff	Profit	Payoff	Profit	Payoff	Profit	
$130.00	$130.00	-$21.14	$0.00	$1.97	$130.00	-$19.17	
$135.00	$135.00	-$16.14	$0.00	$1.97	$135.00	-$14.17	
$140.00	$140.00	-$11.14	$0.00	$1.97	$140.00	-$9.17	
$145.00	$145.00	-$6.14	$0.00	$1.97	$145.00	-$4.17	
$150.00	$150.00	-$1.14	$0.00	$1.97	$150.00	$0.83	
$155.00	$155.00	$3.86	$0.00	$1.97	$155.00	$5.83	
$160.00	$160.00	$8.86	-$5.00	-$3.03	$155.00	$5.83	
$165.00	$165.00	$13.86	-$10.00	-$8.03	$155.00	$5.83	
$170.00	$170.00	$18.86	-$15.00	-$13.03	$155.00	$5.83	
$175.00	$175.00	$23.86	-$20.00	-$18.03	$155.00	$5.83	
$180.00	$180.00	$28.86	-$25.00	-$23.03	$155.00	$5.83	

Fig. 29.19 Value of a covered call position at option expiration

The profit profile of the covered call position is constructed in Fig. 29.21. It can be shown that the payoff pattern of covered call is exactly equal to shorting a put. Therefore, the covered call has frequently been used to replace shorting a put in dynamic hedging practice. The break-even point means when the profit equals to zero. The break-even point for covered call can be calculated as purchase price of underlying + premium received. For this example, the break-even point is $149.17.

29.4.7 Collar

A collar combines a protective put and a short call option to bracket the value of a portfolio between two bounds. For example, an investor holds the IBM stock selling at $151.10. Buying a protective put using the put option with an exercise price of $150 places a lower bound of $150 on the value of the portfolio. At the same time, the investor can write a call option with an exercise price of $155. You can find the S_T, which is the value for S_T in the first column of the table in Fig. 29.22. The call and the put sell at $1.97 (the bid price) and $4.40 (the ask price), respectively, make the net outlay for the two options to be only $2.43. Figure 29.22 shows the values of the collar position at different stock prices

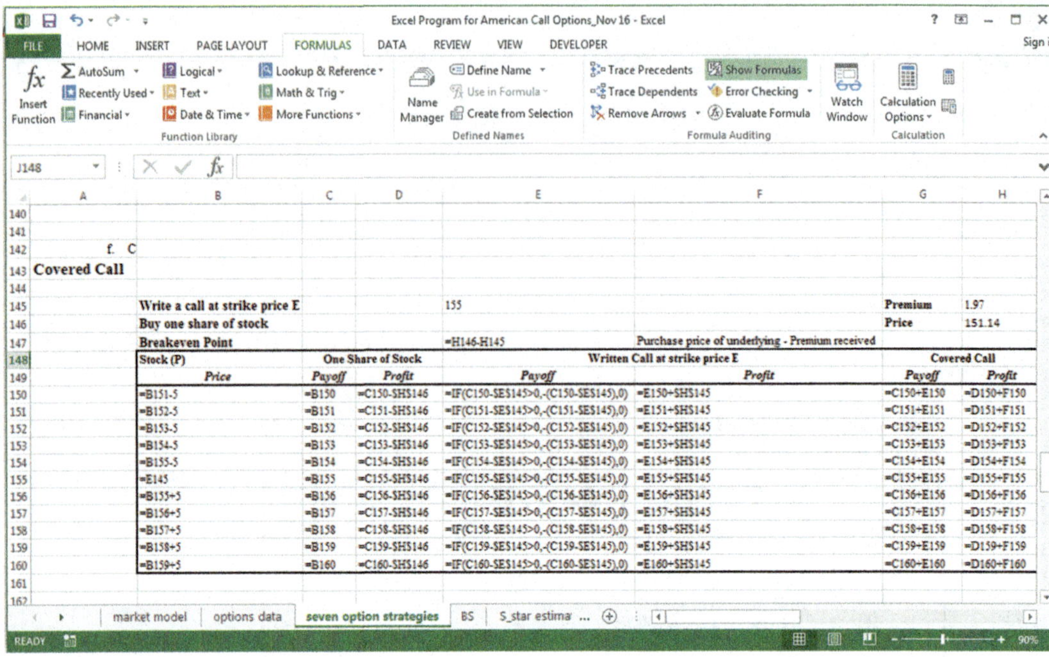

	A	B	C	D	E	F	G	H
142	f. C							
143	**Covered Call**							
145		**Write a call at strike price E**			155		**Premium**	1.97
146		**Buy one share of stock**					**Price**	151.14
147		**Breakeven Point**			=H146-H145	Purchase price of underlying - Premium received		
148		**Stock (P)**	**One Share of Stock**		**Written Call at strike price E**		**Covered Call**	
149		*Price*	*Payoff*	*Profit*	*Payoff*	*Profit*	*Payoff*	*Profit*
150		=B151-5	=B150	=C150-$H$146	=IF(C150-$E$145>0,-(C150-$E$145),0)	=E150+$H$145	=C150+E150	=D150+F150
151		=B152-5	=B151	=C151-$H$146	=IF(C151-$E$145>0,-(C151-$E$145),0)	=E151+$H$145	=C151+E151	=D151+F151
152		=B153-5	=B152	=C152-$H$146	=IF(C152-$E$145>0,-(C152-$E$145),0)	=E152+$H$145	=C152+E152	=D152+F152
153		=B154-5	=B153	=C153-$H$146	=IF(C153-$E$145>0,-(C153-$E$145),0)	=E153+$H$145	=C153+E153	=D153+F153
154		=B155-5	=B154	=C154-$H$146	=IF(C154-$E$145>0,-(C154-$E$145),0)	=E154+$H$145	=C154+E154	=D154+F154
155		=E145	=B155	=C155-$H$146	=IF(C155-$E$145>0,-(C155-$E$145),0)	=E155+$H$145	=C155+E155	=D155+F155
156		=B155+5	=B156	=C156-$H$146	=IF(C156-$E$145>0,-(C156-$E$145),0)	=E156+$H$145	=C156+E156	=D156+F156
157		=B156+5	=B157	=C157-$H$146	=IF(C157-$E$145>0,-(C157-$E$145),0)	=E157+$H$145	=C157+E157	=D157+F157
158		=B157+5	=B158	=C158-$H$146	=IF(C158-$E$145>0,-(C158-$E$145),0)	=E158+$H$145	=C158+E158	=D158+F158
159		=B158+5	=B159	=C159-$H$146	=IF(C159-$E$145>0,-(C159-$E$145),0)	=E159+$H$145	=C159+E159	=D159+F159
160		=B159+5	=B160	=C160-$H$146	=IF(C160-$E$145>0,-(C160-$E$145),0)	=E160+$H$145	=C160+E160	=D160+F160

Fig. 29.20 Excel formula for calculating the value of a covered call position at option expiration

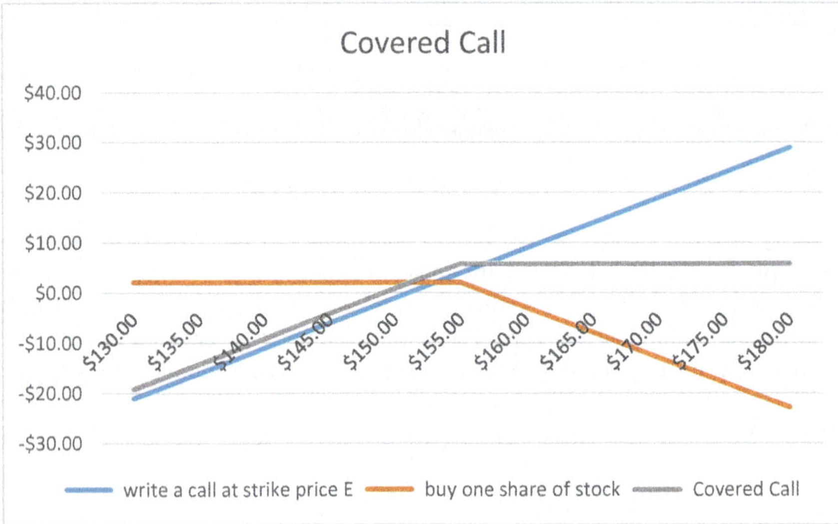

Fig. 29.21 Profit profile for covered call

at time T. For information in details, you can find the Excel function in Fig. 29.23 for calculations of the numbers in Fig. 29.22. The profit profile of the collar position is shown in Fig. 29.24. The break-even point means when the profit equals to zero. The break-even point for collar can be calculated as purchase price of underlying + net premium paid. For this example, the break-even point is \$153.57.

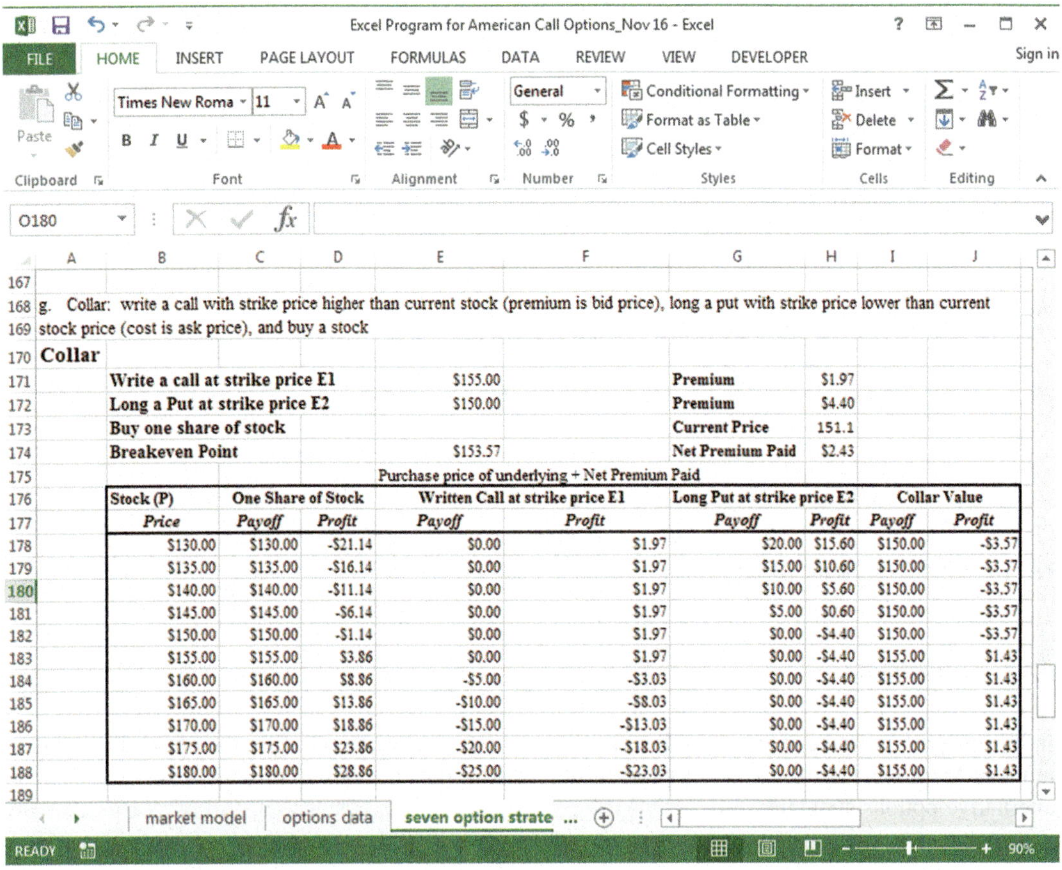

g. Collar: write a call with strike price higher than current stock (premium is bid price), long a put with strike price lower than current stock price (cost is ask price), and buy a stock

Collar

Write a call at strike price E1	$155.00	**Premium**	$1.97
Long a Put at strike price E2	$150.00	**Premium**	$4.40
Buy one share of stock		**Current Price**	151.1
Breakeven Point	$153.57	**Net Premium Paid**	$2.43
	Purchase price of underlying + Net Premium Paid		

Stock (P)	One Share of Stock		Written Call at strike price E1		Long Put at strike price E2		Collar Value	
Price	*Payoff*	*Profit*	*Payoff*	*Profit*	*Payoff*	*Profit*	*Payoff*	*Profit*
$130.00	$130.00	-$21.14	$0.00	$1.97	$20.00	$15.60	$150.00	-$3.57
$135.00	$135.00	-$16.14	$0.00	$1.97	$15.00	$10.60	$150.00	-$3.57
$140.00	$140.00	-$11.14	$0.00	$1.97	$10.00	$5.60	$150.00	-$3.57
$145.00	$145.00	-$6.14	$0.00	$1.97	$5.00	$0.60	$150.00	-$3.57
$150.00	$150.00	-$1.14	$0.00	$1.97	$0.00	-$4.40	$150.00	-$3.57
$155.00	$155.00	$3.86	$0.00	$1.97	$0.00	-$4.40	$155.00	$1.43
$160.00	$160.00	$8.86	-$5.00	-$3.03	$0.00	-$4.40	$155.00	$1.43
$165.00	$165.00	$13.86	-$10.00	-$8.03	$0.00	-$4.40	$155.00	$1.43
$170.00	$170.00	$18.86	-$15.00	-$13.03	$0.00	-$4.40	$155.00	$1.43
$175.00	$175.00	$23.86	-$20.00	-$18.03	$0.00	-$4.40	$155.00	$1.43
$180.00	$180.00	$28.86	-$25.00	-$23.03	$0.00	-$4.40	$155.00	$1.43

Fig. 29.22 Value of a collar position at option expiration

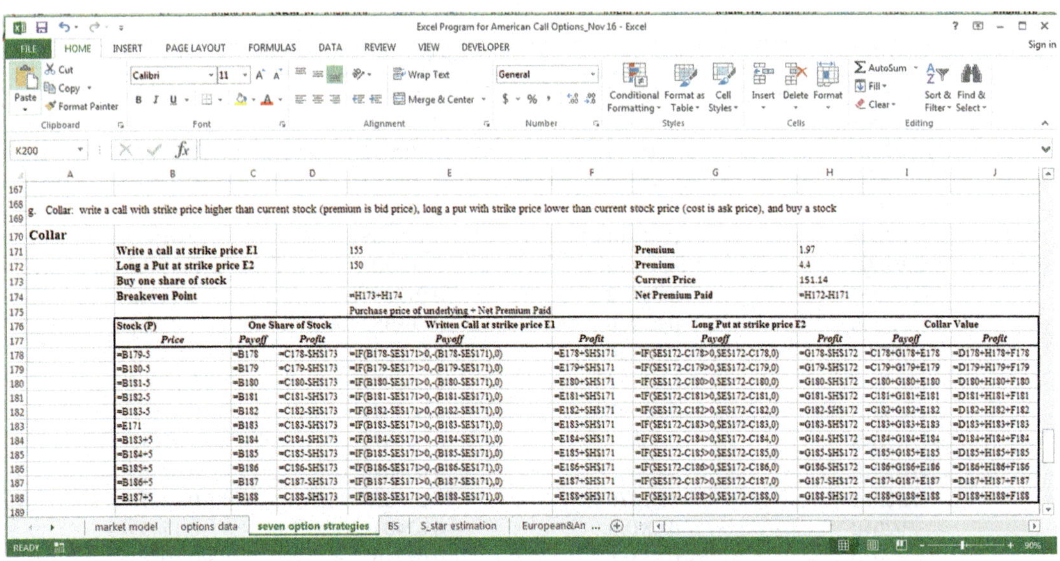

g. Collar: write a call with strike price higher than current stock (premium is bid price), long a put with strike price lower than current stock price (cost is ask price), and buy a stock

Collar

Write a call at strike price E1	155	**Premium**	1.97
Long a Put at strike price E2	150	**Premium**	4.4
Buy one share of stock		**Current Price**	151.14
Breakeven Point	=H173+H174	**Net Premium Paid**	=H172-H171
	Purchase price of underlying + Net Premium Paid		

Stock (P)	One Share of Stock		Written Call at strike price E1		Long Put at strike price E2		Collar Value	
Price	*Payoff*	*Profit*	*Payoff*	*Profit*	*Payoff*	*Profit*	*Payoff*	*Profit*
=B179-5	=B178	=C178-$H$173	=IF(B178-$E$171>0,-(B178-$E$171),0)	=E178+$H$171	=IF($E$172-C178>0,$E$172-C178,0)	=G178-$H$172	=C178+G178+E178	=D178+H178+F178
=B180-5	=B179	=C179-$H$173	=IF(B179-$E$171>0,-(B179-$E$171),0)	=E179+$H$171	=IF($E$172-C179>0,$E$172-C179,0)	=G179-$H$172	=C179+G179+E179	=D179+H179+F179
=B181-5	=B180	=C180-$H$173	=IF(B180-$E$171>0,-(B180-$E$171),0)	=E180+$H$171	=IF($E$172-C180>0,$E$172-C180,0)	=G180-$H$172	=C180+G180+E180	=D180+H180+F180
=B182-5	=B181	=C181-$H$173	=IF(B181-$E$171>0,-(B181-$E$171),0)	=E181+$H$171	=IF($E$172-C181>0,$E$172-C181,0)	=G181-$H$172	=C181+G181+E181	=D181+H181+F181
=B183-5	=B182	=C182-$H$173	=IF(B182-$E$171>0,-(B182-$E$171),0)	=E182+$H$171	=IF($E$172-C182>0,$E$172-C182,0)	=G182-$H$172	=C182+G182+E182	=D182+H182+F182
=E171	=B183	=C183-$H$173	=IF(B183-$E$171>0,-(B183-$E$171),0)	=E183+$H$171	=IF($E$172-C183>0,$E$172-C183,0)	=G183-$H$172	=C183+G183+E183	=D183+H183+F183
=B183+5	=B184	=C184-$H$173	=IF(B184-$E$171>0,-(B184-$E$171),0)	=E184+$H$171	=IF($E$172-C184>0,$E$172-C184,0)	=G184-$H$172	=C184+G184+E184	=D184+H184+F184
=B184+5	=B185	=C185-$H$173	=IF(B185-$E$171>0,-(B185-$E$171),0)	=E185+$H$171	=IF($E$172-C185>0,$E$172-C185,0)	=G185-$H$172	=C185+G185+E185	=D185+H185+F185
=B185+5	=B186	=C186-$H$173	=IF(B186-$E$171>0,-(B186-$E$171),0)	=E186+$H$171	=IF($E$172-C186>0,$E$172-C186,0)	=G186-$H$172	=C186+G186+E186	=D186+H186+F186
=B186+5	=B187	=C187-$H$173	=IF(B187-$E$171>0,-(B187-$E$171),0)	=E187+$H$171	=IF($E$172-C187>0,$E$172-C187,0)	=G187-$H$172	=C187+G187+E187	=D187+H187+F187
=B187+5	=B188	=C188-$H$173	=IF(B188-$E$171>0,-(B188-$E$171),0)	=E188+$H$171	=IF($E$172-C188>0,$E$172-C188,0)	=G188-$H$172	=C188+G188+E188	=D188+H188+F188

Fig. 29.23 Excel formula for calculating the value of a collar position at option expiration

Fig. 29.24 Profit profile for collar

29.5 Summary

In this chapter, we have shown how Excel programs can be used to calculate the optimal weights in terms of Markowitz portfolio model. In addition, we also show how Excel programs can be used to do alternative options strategies.

Appendix 29.1: Monthly Rates of Returns for S&P 500, IBM, and MSFT

Date	S&P 500	IBM	MSFT
2/2/2009	−0.10993	0.009608	−0.0488
3/2/2009	0.0854	0.052765	0.13695
4/1/2009	0.09393	0.065225	0.10289
5/1/2009	0.05308	0.035127	0.03811
6/1/2009	0.0002	−0.01743	0.13753
7/1/2009	0.07414	0.129383	−0.0106
8/3/2009	0.03356	0.005705	0.05404
9/1/2009	0.03572	0.013204	0.04342
10/1/2009	−0.01976	0.008352	0.07835
11/2/2009	0.05736	0.052244	0.06527
12/1/2009	0.01777	0.03607	0.03622
1/4/2010	−0.03697	−0.06504	−0.0755
2/1/2010	0.02851	0.043575	0.02212
3/1/2010	0.0588	0.008642	0.02165
4/1/2010	0.01476	0.005853	0.04276
5/3/2010	−0.08198	−0.02404	−0.1515
6/1/2010	−0.05388	−0.01426	−0.108

(continued)

Date	S&P 500	IBM	MSFT
7/1/2010	0.06878	0.039888	0.12152
8/2/2010	−0.04745	−0.03633	−0.0857
9/1/2010	0.08755	0.089493	0.04331
10/1/2010	0.03686	0.070499	0.08896
11/1/2010	−0.00229	−0.0105	−0.0469
12/1/2010	0.0653	0.037447	0.10505
1/3/2011	0.02265	0.103829	−0.0068
2/1/2011	0.03196	0.003244	−0.0356
3/1/2011	−0.00105	0.007324	−0.0449
4/1/2011	0.0285	0.046047	0.02087
5/2/2011	−0.0135	−0.00526	−0.0285
6/1/2011	−0.01826	0.015548	0.03946
7/1/2011	−0.02147	0.060001	0.05399
8/1/2011	−0.05679	−0.05052	−0.0232
9/1/2011	−0.07176	0.017245	−0.0643
10/3/2011	0.10772	0.055761	0.07005
11/1/2011	−0.00506	0.022365	−0.0323
12/1/2011	0.00853	−0.02193	0.0148
1/3/2012	0.04358	0.047428	0.1375
2/1/2012	0.04059	0.02546	0.08205
3/1/2012	0.03133	0.060548	0.01625
4/2/2012	−0.0075	−0.0075	−0.0073
5/1/2012	−0.06265	−0.06457	−0.0826
6/1/2012	0.03955	0.013882	0.04828
7/2/2012	0.0126	0.002032	−0.0366
8/1/2012	0.01976	−0.00149	0.05252
9/4/2012	0.02424	0.064682	−0.0344
10/1/2012	−0.01979	−0.06231	−0.041
11/1/2012	0.00285	−0.01863	−0.0595
12/3/2012	0.00707	0.007802	0.00316
1/2/2013	0.05043	0.060094	0.02796
2/1/2013	0.01106	−0.00684	0.02107
3/1/2013	0.03599	0.062066	0.02927
4/1/2013	0.01809	−0.05045	0.15676
5/1/2013	0.02076	0.031919	0.06177
6/3/2013	−0.015	−0.08133	−0.0104
7/1/2013	0.04946	0.0206	−0.078
8/1/2013	−0.0313	−0.06082	0.05628
9/3/2013	0.02975	0.015992	−0.0037
10/1/2013	0.0446	−0.03225	0.06399
11/1/2013	0.02805	0.00799	0.08512
12/2/2013	0.02356	0.043936	−0.019

Appendix 29.2: Options Data for IBM on October 12, 2015

Calls

Strike ∵ Filter	Contract Name	Last	Bid	Ask	Change	%Change	Volume	Open Interest	Implied Volatility
139	IBM151113C00139000	8.05	12.45	14.8	0	0	103	203	0.4113
140	IBM151113C00140000	6.88	11.65	13.9	0	0	32	32	0.3988
141	IBM151113C00141000	6.65	10.75	13	0	0	50	50	0.3857
142	IBM151113C00142000	5.5	10	12.15	0	0	5	10	0.3755
143	IBM151113C00143000	5	9.1	11	0	0	20	20	0.345
144	IBM151113C00144000	4.45	8.4	9.45	0	0	25	29	0.2885
145	IBM151113C00145000	7.26	7.6	8.6	0	0	24	32	0.2778
146	IBM151113C00146000	6.35	6.95	7.6	0	0	5	11	0.257
147	IBM151113C00147000	6.66	6.3	6.85	0	0	12	18	0.2505
148	IBM151113C00148000	7.16	5.65	6	0	0	37	45	0.2367
149	IBM151113C00149000	6.54	5	5.35	0	0	12	51	0.2329
150	IBM151113C00150000	4.64	4.35	4.6	-0.36	-0.072	42	58	0.2217
152.5	IBM151113C00152500	3.8	3	3.25	-0.05	-0.013	7	82	0.2138
155	IBM151113C00155000	2.02	1.97	2.13	-0.64	-0.2406	5	219	0.2041
157.5	IBM151113C00157500	1.33	1.23	1.33	-0.62	-0.3179	123	16	0.198
160	IBM151113C00160000	0.84	0.69	0.8	-0.26	-0.2364	28	275	0.1948
162.5	IBM151113C00162500	0.69	0.37	0.58	0	0	1	1	0.2061
165	IBM151113C00165000	0.27	0.19	0.34	0	0	1	1	0.2051

Puts

Strike ∵ Filter	Contract Name	Last	Bid	Ask	Change	%Change	Volume	Open Interest	Implied Volatility
115	IBM151113P00115000	0.41	0.02	0.23	0	0	6	6	0.4863
125	IBM151113P00125000	0.86	0.08	0.56	0	0	17	17	0.4302
130	IBM151113P00130000	0.75	0.27	0.73	0	0	100	111	0.3843
133	IBM151113P00133000	0.92	0.44	0.87	0	0	1	15	0.3572
134	IBM151113P00134000	1.1	0.51	0.93	0	0	50	62	0.3489
135	IBM151113P00135000	1.28	0.59	1.05	0	0	5	5	0.3462
136	IBM151113P00136000	1.24	0.7	1.13	0	0	2	5	0.3384
137	IBM151113P00137000	1.54	0.79	1.16	0	0	20	30	0.3251
138	IBM151113P00138000	1.51	0.9	1.25	0	0	20	83	0.3171
139	IBM151113P00139000	1.25	1.02	1.42	0	0	10	22	0.3152
140	IBM151113P00140000	1.37	1.15	1.55	0	0	11	60	0.3086
141	IBM151113P00141000	2.02	1.36	1.67	0	0	4	16	0.3002
142	IBM151113P00142000	2.22	1.69	1.84	0	0	6	32	0.2947
143	IBM151113P00143000	2.55	1.88	2.06	0	0	4	20	0.2913
144	IBM151113P00144000	2.36	2.12	2.29	0	0	6	16	0.2872
145	IBM151113P00145000	3.15	2.37	2.56	0	0	8	22	0.2843
146	IBM151113P00146000	2.65	2.63	2.87	0	0	1	7	0.2822
147	IBM151113P00147000	2.89	2.99	3.2	0	0	2	16	0.2797
148	IBM151113P00148000	3.19	3.3	3.55	0	0	1	25	0.2767
149	IBM151113P00149000	3.55	3.7	3.95	0.1	0.029	5	18	0.2749
150	IBM151113P00150000	4	4.15	4.4	0	0	6	20	0.2742
152.5	IBM151113P00152500	5.25	5.3	5.7	0.39	0.0802	5	100	0.2742
157.5	IBM151113P00157500	8.3	7.7	9.15	0	0	14	23	0.291
160	IBM151113P00160000	10.29	9.5	11.2	0	0	1	1	0.3062
162.5	IBM151113P00162500	21.51	11.4	14.05	0	0	5	5	0.3694

Bibliography

Alexander GJ, Francis JC (1986) Portfolio analysis. Prentice-Hall, New York

Amram M, Kulatilaka N (2001) Real options. Oxford University Press, New York

Ball C, Torous W (1983) Bond prices dynamics and options. J Financ Quant Anal 18:517–532

Baumol WJ (1963) An expected gain-confidence limit criterion for portfolio selection. Manage Sci 10:171–182

Bertsekas D (1974) Necessary and sufficient conditions for existence of an optimal portfolio. J Econ Theory 8:235–247

Bhattacharya M (1980) Empirical properties of the Black–Scholes formula under ideal conditions. J Financ Quant Anal 15:1081–1106

Black F (1972) Capital market equilibrium with restricted borrowing. J Bus 45:444–455

Black F (1985) Fact and fantasy in the use of options. Financ Anal J 31:36–72

Black F, Scholes M (1973) The pricing of options and corporate liabilities. J Polit Econ 31:637–654

Blume M (1970) Portfolio theory: a step toward its practical application. J Bus 43:152–173

Bodhurta J, Courtadon G (1986) Efficiency tests of the foreign currency options market. J Finance 41:151–162

Bodie Z, Kane A, Marcus A (2010) Investments, 9th edn. McGraw-Hill Book, New York

Bookstaber RM (1981) Option, pricing and strategies in investing. Addison-Wesley, Reading, MA

Bookstaber RM, Clarke R (1983) Option strategies for institutional investment management. Addison-Wesley, Reading, MA

Brealey RA, Hodges SD (1975) Playing with portfolios. J Finance 30:125–134

Breen W, Jackson R (1971) An efficient algorithm for solving large-scale portfolio problems. J Financ Quant Anal 6:627–637

Brennan MJ (1975) The optimal number of securities in a risky asset portfolio where there are fixed costs of transaction: theory and some empirical results. J Financ Quant Anal 10:483–496

Brennan M, Schwartz E (1977) The valuation of American put options. J Finance 32:449–462

Cohen K, Pogue J (1967) An empirical evaluation of alternative portfolio-selection models. J Bus 46:166–193

Cox JC (1979) Option pricing: a simplified approach. J Financ Econ 8:229–263

Cox JC, Rubinstein M (1985) Option markets. Prentice-Hall, Englewood Cliffs, NJ

Dyl EA (1975) Negative betas: the attractions of selling short. J Portf Manage 1:74–76

Eckardt W, Williams S (1984) The complete options indexes. Financ Anal J 40:48–57

Elton EJ, Gruber M (1974) Portfolio theory when investment relatives are log normally distributed. J Finance 29:1265–1273

Elton EJ, Padberg ME (1976) Simple criteria for optimal portfolio selection. J Finance 11:1341–1357

Elton EJ, Padberg ME (1978) Simple criteria for optimal portfolio selection: tracing out the efficient frontier. J Finance 13:296–302

Elton EJ, Gruber MJ, Brown SJ, Goetzmann WN (2006) Modern portfolio theory and investment analysis, 7th edn. Wiley, New York

Ervine J, Rudd A (1985) Index options: the early evidence. J Finance 40:743–756

Evans J, Archer S (1968) Diversification and the reduction of dispersion: an empirical analysis. J Finance 3:761–767

Fama EF (1970) Efficient capital markets: a review of theory and empirical work. J Finance 25:383–417

Feller W (1968) An introduction to probability theory and its application, vol 1. Wiley, New York

Finnerty J (1978) The Chicago board options exchange and market efficiency. J Financ Quant Anal 13:28–38

Francis JC, Archer SH (1979) Portfolio analysis. Prentice-Hall, New York

Galai D, Masulis RW (1976) The option pricing model and the risk factor of stock. J Financ Econ 3:53–81

Galai D, Geske R, Givots S (1988) Option markets. Addison-Wesley, Reading, MA

Gastineau G (1979) The stock options manual. McGraw-Hill, New York

Geske R, Shastri K (1985) Valuation by approximation: a comparison of alternative option valuation techniques. J Financ Quant Anal 20:45–72

Gressis N, Philiippatos G, Hayya J (1976) Multiperiod portfolio analysis and the inefficiencies of the market portfolio. J Finance 31:1115–1126

Guerard JB (2010) Handbook of portfolio and construction: contemporary applications of Markowitz techniques. Springer, New York

Henderson J, Quandt R (1980) Microeconomic theory: a mathematical approach, 3rd edn. McGraw-Hill, New York

Hull J (2005) Options, futures, and other derivatives, 6th edn. Prentice Hall, Upper Saddle River, NJ

Jarrow RA, Rudd A (1983) Option pricing. Richard D. Irwin, Homewood, IL

Jarrow R, Turnbull S (1999) Derivatives securities, 2nd edn. South-Western College, Cincinnati, OH

Lee CF (2009) Handbook of quantitative finance and risk management. Springer, New York

Lee CF, Lee AC (2006) Encyclopedia of finance. Springer, New York

Lee CF, Lee JC, Lee AC (2000) Statistics for business and financial economics. World Scientific, Singapore

Lee CF, Lee AC, Lee JC (2010) Handbook of quantitative finance and risk management. Springer, New York
Lee CF, Lee JC, Lee AC (2013) Statistics for business and financial economics. Springer, New York
Levy H, Sarnat M (1971) A note on portfolio selection and investors' wealth. J Financ Quant Anal 6:639–642
Lewis AL (1988) A simple algorithm for the portfolio selection problem. J Finance 43:71–82
Liaw KT, Moy RL (2000) The Irwin guide to stocks, bonds, futures, and options. McGraw-Hill, New York
Lintner J (1965) The valuation of risk assets and the selection of risky investments in stock portfolio and capital budgets. Rev Econ Stat 47:13–27
Macbeth J, Merville L (1979) An empirical examination of the Black–Scholes call option pricing model. J Finance 34:J173–J186
Maginn JL, Tuttle DL, Pinto JE, McLeavey DW (2007) Managing investment portfolios: a dynamic process, 3rd edn, CFA Institute investment series. Wiley, New York
Mao JCF (1969) Quantitative, analysis of financial decisions. Macmillan, New York
Markowitz HM (1952) Portfolio selection. J Finance 1:77–91
Markowitz HM (1959) Portfolio selection; Cowles Foundation monograph 16. Wiley, New York
Markowitz HM (1976) Markowitz revisited. Financ Anal J 32:47–52
Markowitz HM (1987) Mean-variance, analysis in portfolio choice and capital markets. Blackwell, New York
Martin AD Jr (1955) Mathematical programming of portfolio selections. Manage Sci 1:152–166
McDonald RL (ed) (2005) Derivatives markets, 2nd edn. Addison Wesley, Boston, MA
Merton R (1972) An analytical derivation of efficient portfolio frontier. J Financ Quant Anal 7:1851–1872
Merton R (1973) Theory of rational option pricing. Bell J Econ Manage Sci 4:141–183
Mossin J (1968) Optimal multiperiod portfolio policies. J Bus 41:215–229
Rendleman RJ Jr, Barter BJ (1979) Two-state option pricing. J Finance 34:1093–1110
Ritchken P (1987) Options: theory, strategy and applications. Scott, Foresman, Glenview, IL
Ross SA (1982) On the general validity of the mean-variance approach in large markets. In: Sharpe WF, Cootner CM (eds) Financial economics: essays in honor of Paul Cootner. Prentice Hall, New York, pp 52–84
Rubinstein M, Leland H (1981) Replicating options with positions in stock and cash. Financ Anal J 37:63–72
Sears S, Trennepohl G (1982) Measuring portfolio risk in options. J Financ Quant Anal 17:391–410
Sharpe WF (1970) Portfolio, theory and capital markets. McGraw-Hill, New York
Simkowitz MA, Beedles WL (1978) Diversification in a three-moment world. J Financ Quant Anal 13:927–941
Smith C (1976) Option pricing: a review. J Financ Econ 3:3–51
Stoll H (1969) The relationships between put and call option prices. J Finance 24:801–824
Summa JF, Lubow JW (2001) Options on futures. Wiley, New York
Trennepohl G (1981) A comparison of listed option premium and Black–Scholes model prices: 1973–1979. J Financ Res 4:11–20
Von Neumann J, Morgenstern O (1947) Theory of games and economic behavior, 2nd edn. Princeton University Press, Princeton, NJ
Wackerly D, Mendenhall W, Scheaffer RL (2007) Mathematical statistics with applications, 7th edn. Duxbury Press, California
Weinstein M (1983) Bond systematic risk and the options pricing model. J Finance 38:1415–1430
Welch W (1982) Strategies for put and call option trading. Winthrop, Cambridge, MA
Whaley R (1982) Valuation of American call options on dividend paying stocks: empirical tests. J Financ Econ 10:29–58
Zhang PG (1998) Exotic options: a guide to second generation options, 2nd edn. World Scientific, Singapore

Chapter 30
Simulation and Its Application

30.1 Introduction

In this chapter, we will introduce Monte Carlo simulation which is a problem-solving technique. This technique can approximate the probability of certain outcomes by using random variables, called simulations. Monte Carlo simulation is named after the city in Monaco. The primary attractions in this place are casinos having gambling games, like dice, roulette, and slot machines. In these games of chance, there exist random behavior.

In option pricing methods, we can use Monte Carlo simulation to generate underlying asset price process, then to value today's option price. At first, we will introduce how to use Excel to simulate stock price and get the option price. Next, we also introduce different methods to improve the efficiency of simulation. These include antithetic variates and Quasi-Monte Carlo simulation. Finally, we apply Monte Carlo simulation to path-depend option.

30.2 Monte Carlo Simulation

The advantages of Monte Carlo simulation are its generality and relative easy to use. For instance, it may take many complicating features of exotic options into account and it lends itself to treating high-dimensional problems. However, it is difficult to apply simulation to American options. Simulation goes forward in time, but establishing an optimal exercise policy requires going backward in time.

At first, we generate asset price paths by Monte Carlo simulation. For convenience, we recall the geometric Brownian motion for asset price. Geometric Brownian motion is the standard assumption for stock price process. This stock process is plausibly explained in John Hull text book. Mathematically speaking, the asset price S(t), with drift μ and volatility σ:

$$dS = \mu S dt + \sigma S dz,$$

where $dz\left(=\varepsilon\sqrt{dt}\right)$ is Brownian motion, ε is standard normal random variable, dt is in a very short time.

Using Ito's lemma, we can get the stochastic process under a logarithm stock price:

$$d\ln S = \left(\mu - 0.5\sigma^2\right)dt + \sigma dz.$$

C.-F. Lee et al., *Essentials of Excel, Excel VBA, SAS and Minitab for Statistical and Financial Analyses*, DOI 10.1007/978-3-319-38867-0_30

Because there is no stock price in the drift and diffusion term, we can discretize the time period and get the stock price process like this:

$$\ln S(t + \delta t) - \ln S(t) = (\mu - 0.5\sigma^2)\delta t + \sigma\sqrt{\delta t}\varepsilon.$$

We also can use another form to represent the stock price process:

$$S(t + \delta t) = S(t)\exp\left[(\mu - 0.5\sigma^2)\delta t + \sigma\sqrt{\delta t}\varepsilon\right].$$

In order to generate a stock price process, we can use the below subroutine:

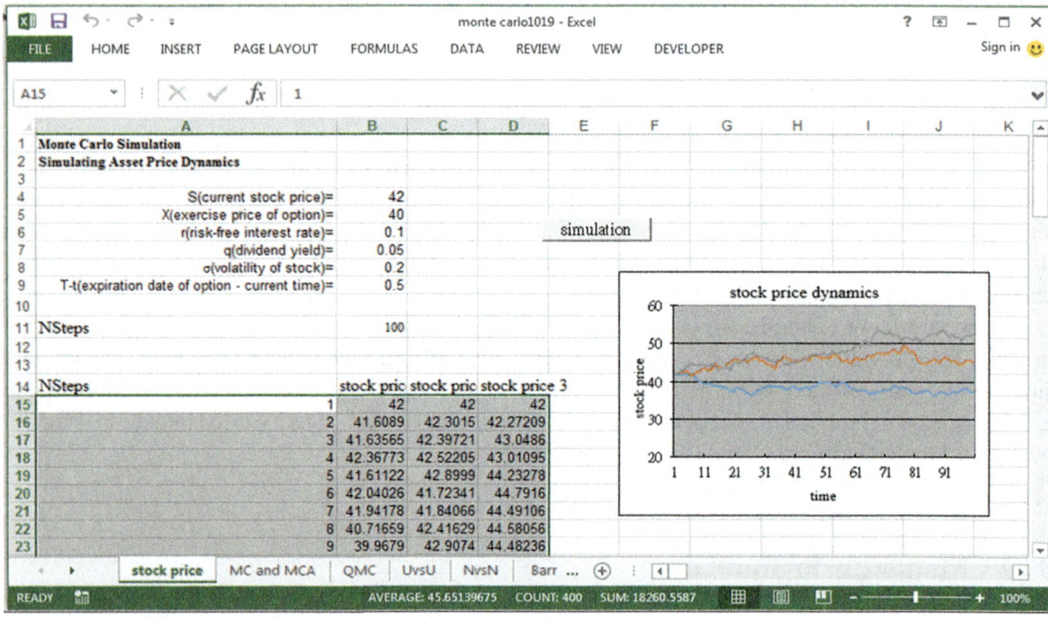

At the heart of Monte Carlo simulation for option valuation is the stochastic process that generates the share price. The stochastic equation for the underlying share price at time T when the option on the share expires was given as:

$$S_T = S_0\, exp\left[(\mu - 0.5\sigma^2)T + \sigma\sqrt{T}\epsilon\right].$$

The associated European option payoff depends on the expectation of S_T in the risk-neutral world. Thus the stochastic equation for S_T for risk neutral valuation takes the form:

$$S_T = S_0\, exp\left[(r - q - 0.5\sigma^2)T + \sigma\sqrt{T}\epsilon\right].$$

The share price process outlined above is the same as that assumed for binomial tree valuation. RAND gives random numbers uniformly distributed in the range [0, 1]. Regarding its outputs as cumulative probabilities, the NORMSINV function converts them into standard normal variate values, mostly between −3 and 3. The random normal samples (value of ε) are then used to generate share prices and the corresponding option payoff.

In European option pricing, we need to estimate the expected value of the discounted payoff of the option:

$$
\begin{aligned}
f &= e^{-rT}E(f_T) \\
&= e^{-rT}E[\max(S_T - X, 0)] \\
&= e^{-rT}E\left[\max\left(S_0 exp\left[(r - q - 0.5\sigma^2)T + \sigma\sqrt{T}\epsilon\right] - X, 0\right)\right].
\end{aligned}
$$

The standard deviation of the simulated payoffs divided by the square root of the number of trials is relatively large. To improve the precision of the Monte Carlo value estimate, the number of simulation trial must be increased.

We can replicate many stock prices at option maturity date in the Excel worksheet.

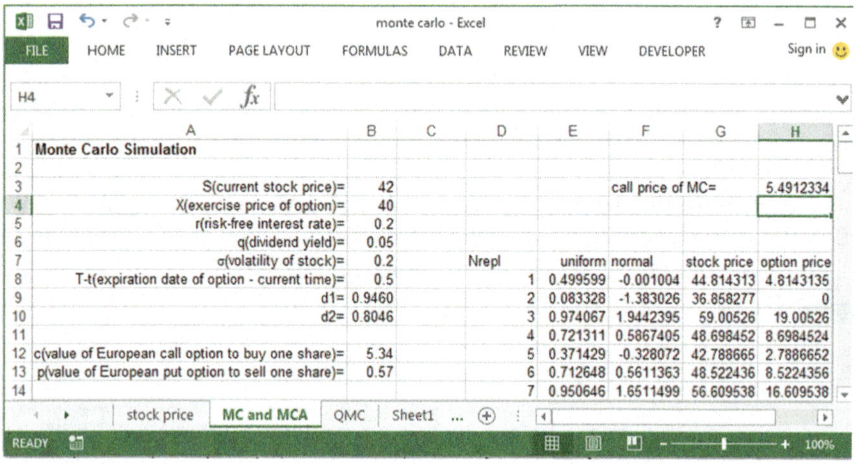

	A	B	C	D	E	F	G	H
1	**Monte Carlo Simulation**							
2								
3	S(current stock price)=	42				call price of MC=		5.4912334
4	X(exercise price of option)=	40						
5	r(risk-free interest rate)=	0.2						
6	q(dividend yield)=	0.05						
7	σ(volatility of stock)=	0.2		Nrepl	uniform	normal	stock price	option price
8	T-t(expiration date of option - current time)=	0.5		1	0.499599	-0.001004	44.814313	4.8143135
9	d1=	0.9460		2	0.083328	-1.383026	36.858277	0
10	d2=	0.8046		3	0.974067	1.9442395	59.00526	19.00526
11				4	0.721311	0.5867405	48.698452	8.6984524
12	c(value of European call option to buy one share)=	5.34		5	0.371429	-0.328072	42.788665	2.7886652
13	p(value of European put option to sell one share)=	0.57		6	0.712648	0.5611363	48.522436	8.5224356
14				7	0.950646	1.6511499	56.609538	16.609538

Using Excel RAND() function, we can generate uniform random number, like cell E8. We simulate 100 random numbers in the sheet. Next, we use the NORMSINV function to transfer uniform random number to standard normal random number in cell F8. Then the random normal samples are used to generate stock prices from the formula, like G8. The stock price formula in G8 is:

$$= \$B\$3 * EXP\left((\$B\$5 - \$B\$6 - 0.5 * \$B\$7^2) * \$B\$8 + \$B\$7 * SQRT(\$B\$8) * F8\right)$$

Finally, the corresponding call option payoff in cell H8 is:

$$= MAX(G8 - \$B\$4, 0)$$

The discount value of the average of the 100 simulated option payoffs is the estimate call value from Monte Carlo simulation. Pressing the F9 in Excel, we can generate a further 100 trials and another Monte Carlo simulation. The formula for call option estimated by Monte Carlo simulation in H3 is:

$$= EXP(-\$B\$5 * \$B\$8) * AVERAGE(H8 : H107)$$

The value in H3 is 5.49. Compared with the true Black and Scholes call value, there exist some differences. To improve the precision of the Monte Carlo estimate, the number of simulation trials has to be increased.

We can write a function for crude Monte Carlo simulation.

```
'   Monte-Carlo simulation for Call option
Function MCCall(S, X, r, q, T, sigma, NRepl)
    Dim nuT, siT, Sum, randns, ST, payoff
    Dim i As Integer
    Randomize
    nuT = (r - q - 0.5 * sigma ^ 2) * T
    siT = sigma * Sqr(T)
    Sum = 0
    For i = 1 To NRepl
        randns = Application.NormSInv(Rnd)
        ST = S * Exp(nuT + randns * siT)
        payoff = Application.Max((ST - X), 0)
        Sum = Sum + payoff
    Next i
    MCCall = Exp(-r * T) * Sum / NRepl
End Function
```

Using this function we can get option price calculated by Monte Carlo simulation. We can also change different numbers of replication to get a more efficient option price. This is shown below.

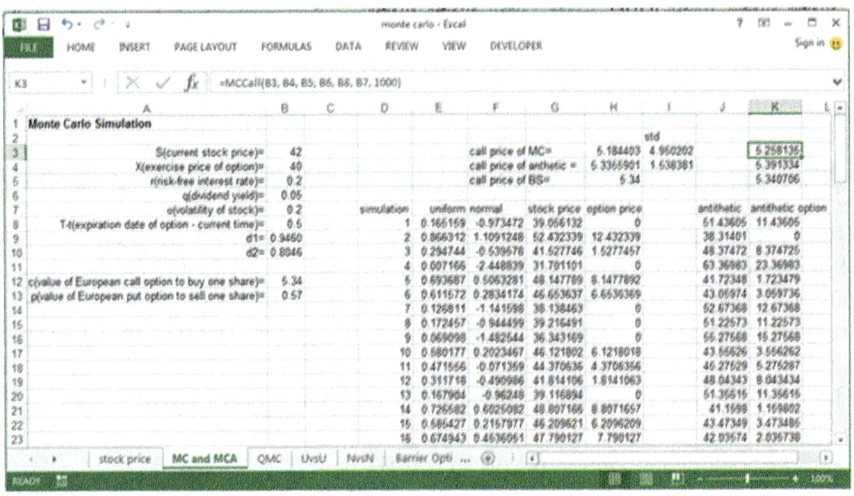

The Monte Carlo simulation for European call option in K3 is:

$$= \text{MCCall(B3, B4, B5, B6, B8, B7, 1000)}$$

In this case, we replicate 1000 times to get the call option value. The value in K3 is equal to 5.2581 which is a little more near the value of Black–Scholes, 5.34.

30.3 Antithetic Variables

In addition to increasing the number of trials, we have another way of improving the precision of the Monte Carlo estimate is antithetic variables. The antithetic variates method is a variance reduction technique used in Monte Carlo methods. The standard error of the Monte Carlo estimate (with antithetic variables) is substantially lower than that for the uncontrolled sampling approach. Therefore, the antithetic variates method reduces the variance of the simulation results and improve the efficiency of simulation.

The antithetic variates technique consists, for every sample path obtained, in taking its antithetic path. Suppose that we have two random samples X_1 and X_2:

$$X_1(1),\ X_1(2),\ \ldots,\ X_1(n)$$

$$X_2(1),\ X_2(2),\ \ldots,\ X_2(n)$$

We would like to estimate

$$\theta = \mathrm{E}[\mathrm{h}(\mathrm{X})]$$

An unbiased estimator is given by

$$\mathrm{X(i)} = \frac{X_1(i) + X_2(i)}{2}$$

Therefore,

$$\begin{aligned} \mathrm{var}\left[\frac{\sum X(i)}{n}\right] &= \frac{var[X(i)]}{n} \\ &= \frac{var[X_1(i)] + var[X_2(i) + 2cov[X_1(i), X_2(i)]}{4n} < \frac{var[X_1(i)]}{n} \end{aligned}$$

In order to reduce the sample mean variance, we should take $\mathrm{cov}[X_1(i),\ X_2(i)] < var[X_i(i)]$. In the antithetic method, we will choose the second sample in such a way that X_1 and X_2 are not i.i.d., but cov(X_1, X_2) is negative. As a result, variance is reduced.

There are two advantages in the antithetic method. First, it reduces the number of normal samples to be taken to generate N paths. Second, it reduces the variance of the sample paths, improving the accuracy. An important point to bear in mind is that antithetic sampling may not yield a variance reduction when some monotonicity condition is not satisfied.

We use a spreadsheet to implement the antithetic method which is shown below. The stock price in G8 is:

$$= \$B\$3 * \mathrm{EXP}\left((\$B\$5 - \$B\$6 - 0.5 * \$B\$7^2) * \$B\$8 + \$B\$7 * \mathrm{SQRT}(\$B\$8) * \mathrm{F8}\right)$$

The antithetic variable method generates the other stock price in J8 which is equal to:

$$= \$B\$3 * \mathrm{EXP}\left((\$B\$5 - \$B\$6 - 0.5 * \$B\$7^2) * \$B\$8 + \$B\$7 * \mathrm{SQRT}(\$B\$8) * (-\mathrm{F8})\right)$$

The most important in these two formulas are random variables. The first one uses F8 and the other one use −F8.

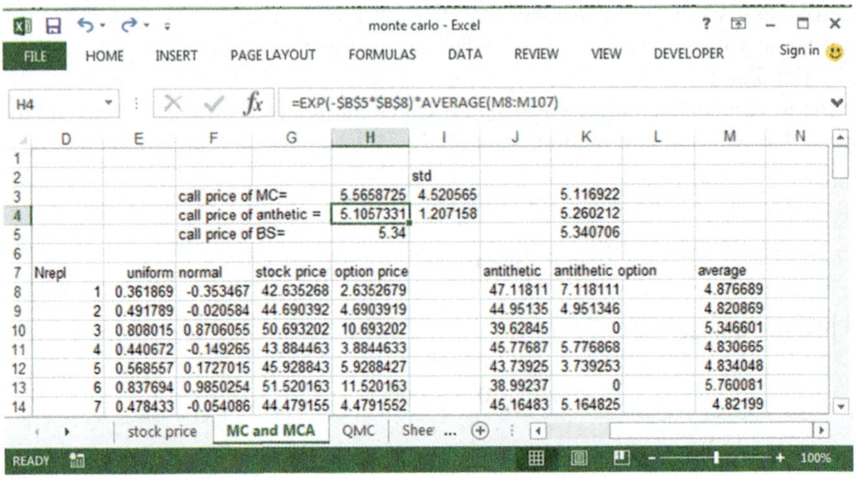

	D	E	F	G	H	I	J	K	L	M
1										
2						std				
3			call price of MC=		5.5658725	4.520565		5.116922		
4			call price of anthetic =		5.1057331	1.207158		5.260212		
5			call price of BS=		5.34			5.340706		
6										
7	Nrepl	uniform	normal	stock price	option price		antithetic	antithetic option		average
8	1	0.361869	-0.353467	42.635268	2.6352679		47.11811	7.118111		4.876689
9	2	0.491789	-0.020584	44.690392	4.6903919		44.95135	4.951346		4.820869
10	3	0.808015	0.8706055	50.693202	10.693202		39.62845	0		5.346601
11	4	0.440672	-0.149265	43.884463	3.8844633		45.77687	5.776868		4.830665
12	5	0.568557	0.1727015	45.928843	5.9288427		43.73925	3.739253		4.834048
13	6	0.837694	0.9850254	51.520163	11.520163		38.99237	0		5.760081
14	7	0.478433	-0.054086	44.479155	4.4791552		45.16483	5.164825		4.82199

The call option value estimated by antithetic method in H4 is:

$$= \mathrm{EXP}(-\$B\$5 * \$B\$8) * \mathrm{AVERAGE}(M8 : M107)$$

We also calculate the standard deviations of Monte Carlo simulation and antithetic variates method in I3 and I4. The standard deviation of antithetic variates method is smaller than that of Monte Carlo simulation.

In addition, we can write a function for antithetic method to improve the precision of Monte Carlo estimate. Below is the code:

```
'   Monte-Carlo simulation and antithetic variates for Call option
Function MCCallAnti(S, X, r, q, T, sigma, NRepl)
    Dim nuT, siT, Sum, randns, ST1, ST2, payoff1, payoff2
    Dim i As Integer
    Randomize
    nuT = (r - q - 0.5 * sigma ^ 2) * T
    siT = sigma * Sqr(T)
    Sum = 0
    For i = 1 To NRepl
        randns = Application.NormSInv(Rnd)
        ST1 = S * Exp(nuT + randns * siT)
        ST2 = S * Exp(nuT - randns * siT)
        payoff1 = Application.Max((ST1 - X), 0)
        payoff2 = Application.Max((ST2 - X), 0)
        Sum = Sum + 0.5 * (payoff1 + payoff2)
    Next i
    MCCallAnti = Exp(-r * T) * Sum / NRepl
End Function
```

We can directly use this function in the worksheet to get the estimate of antithetic method. After changing the numbers of replication, we can get the option prices in different numbers of replication.

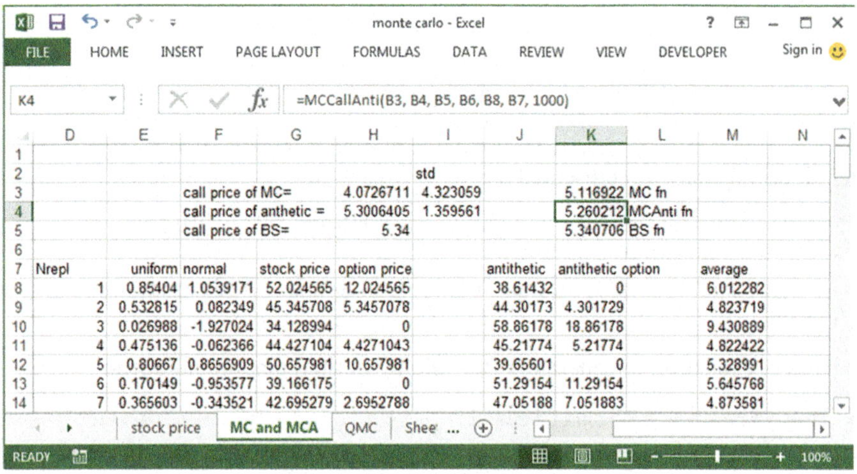

K4 =MCCallAnti(B3, B4, B5, B6, B8, B7, 1000)

	D	E	F	G	H	I	J	K	L	M	N
1											
2						std					
3			call price of MC=		4.0726711	4.323059		5.116922	MC fn		
4			call price of anthetic =		5.3006405	1.359561		5.260212	MCAnti fn		
5			call price of BS=		5.34			5.340706	BS fn		
6											
7	Nrepl		uniform	normal	stock price	option price	antithetic	antithetic option		average	
8		1	0.85404	1.0539171	52.024565	12.024565	38.61432	0		6.012282	
9		2	0.532815	0.082349	45.345708	5.3457078	44.30173	4.301729		4.823719	
10		3	0.026988	-1.927024	34.128994	0	58.86178	18.86178		9.430889	
11		4	0.475136	-0.062366	44.427104	4.4271043	45.21774	5.21774		4.822422	
12		5	0.80667	0.8656909	50.657981	10.657981	39.65601	0		5.328991	
13		6	0.170149	-0.953577	39.166175	0	51.29154	11.29154		5.645768	
14		7	0.365603	-0.343521	42.695279	2.6952788	47.05188	7.051883		4.873581	

The formula for call value of antithetic variates method in K4 is:

$$= \text{MCCallAnti(B3, B4, B5, B6, B8, B7, 1000)}$$

The value in K4 is closer to Black–Scholes, K5, than K3 estimated by Monte Carlo in 100 times replication.

30.4 Quasi-Monte Carlo Simulation

Quasi-Monte Carlo simulation is another way to improve the efficiency of Monte Carlo. This simulation method is a method for solving some other problems using low-discrepancy sequences (also called quasi-random sequences or sub-random sequences). This is in contrast to the regular Monte Carlo simulation, which are based on sequences of pseudorandom numbers. To generate U(0, 1) variables, the standard method is based on linear congruential generators (LCGs). LCG is a process that given an initial z_0 and through a formula to generate the next number. The formula is

$$z_i = (a * z_{i-1} + c)\ (mod\, m)$$

For example, 15 mod 6 = 3 (remainder of integer division). Then the uniform random number is

$$U_i = \frac{z_i}{m}$$

There is nothing random in this sequence. First, it must start from an initial number z_0, seed. Secondly, the generator is periodic.

The inverse transform is a general approach to transform uniform variates to normal variates. Since no analytical form for it is known, we cannot invert the normal distribution function efficiently. One old-fashioned possibility, which is still suggested in some textbooks is to exploit the central limit theorem to generate normal random number by summing a suitable number of uniform variates. Computational efficiency would restrict the number of uniform variates. An alternative method is the Box-Muller approach. Consider two independent variables X, Y ~ N(0, 1), and let (R, θ) be the polar coordinates of the point of Cartesian coordinates (X, Y) in the planes, so that

$$d = R^2 = X^2 + Y^2$$

$$\theta = \tan^{-1}\frac{Y}{X}$$

Box-Muller algorithm can be represented as:

1. Generate two independent uniform random variates U1 and U2 ~ U(0, 1).
2. Set $R^2 = -2$*log(U1) and $\theta = 2\pi$*U2.
3. Set X = R*cosθ and Y = R*sinθ, then X ~ N(0, 1) and Y ~ N(0, 1) are independent standard normal variates.

Here is the VBA function to generate a Box-Muller normal random numbers:

```
'   Box Muller transformation 1
Function BMNormSInv1(x1 As Double, x2 As Double) As Double
    Dim vlog, norm1
    vlog = Sqr(-2 * Log(x1))
    norm1 = vlog * Cos(2 * Application.Pi() * x2)
    BMNormSInv1 = norm1
End Function


'   Box Muller transformation 2
 Function BMNormSInv2(x1 As Double, x2 As Double) As Double
    Dim vlog, norm2
    vlog = Sqr(-2 * Log(h1))
    norm2 = vlog * Sin(2 * Application.Pi() * h2)
    BMNormSInv2 = norm2
End Function
```

The random numbers produced by a LCG or by more sophisticated algorithms are not random at all. So one could try to devise alternative deterministic sequences of numbers which are in some sense evenly distributed. This idea may be made more precise by defining the discrepancy of a sequence of number. The only trick in the selection process is to remember the values of all the previous numbers chosen as each new number is selected. Using quasi-random sampling means that the error in any estimate based on the samples is proportional to 1/n rather than 1/sqrt(n).

There are many quasi-random sequence (low-discrepancy sequence), like Halton's sequence, Sobol's sequence, Faure's sequence, and Niederreiter's sequence. For instance, the Halton's sequence is constructed according to a deterministic method that uses a prime number as its base. Here is a simple example to create Halton's sequence which base is 2:

1. Representing an integer number n in a base b, where b is a prime number:

$$\begin{aligned} n &= (\ldots d_4\, d_3\, d_2\, d_1 d_0)_b \\ &= \sum\nolimits_{k=0}^{m} d_k b^k \end{aligned}$$

   For example, $4 = (100)_2 = 1 * 2^2 + 0 * 2^1 + 0 * 2^0$.
2. Reflecting the digits and adding a radix point to obtain a number with the unit interval:

$$\begin{aligned} h(n, b) &= (0.d_0\, d_1\, d_2\, d_3 d_4 \ldots)_b \\ &= \sum\nolimits_{k=0}^{m} d_k b^{-k+1} \end{aligned}$$

For example, $(0.001)_2 = 1 * \frac{1}{2^3} + 0 * \frac{1}{2^2} + 0 * \frac{1}{2} = \frac{1}{8}$

Therefore, we get a Halton's sequence:

$$\begin{array}{cccccccc} n: & 1 & 2 & 3 & 4 & 5 & 6 & 7\ldots \\ h(n,2): & 1/2 & 1/4 & 3/4 & 1/8 & 5/8 & 3/8 & 7/8\ldots \end{array}$$

Below is a function to generate a Halton's sequence:

```
' Helton's sequence
Function Halton(n, b) As Double
  Dim h As Double, f As Double
  Dim n1 As Integer, n0 As Integer, r As Integer
  n0 = n
  h = 0
  f = 1 / b
  Do While n0 > 0
     n1 = Int(n0 / b)
     r = n0 - n1 * b
     h = h + f * r
     f = f / b
     n0 = n1
  Loop
  Halton = h
End Function
```

Using this function in the worksheet, we can get a sequence number generated by Halton's function. In addition, we can change the prime number to get a Halton's number form different base.

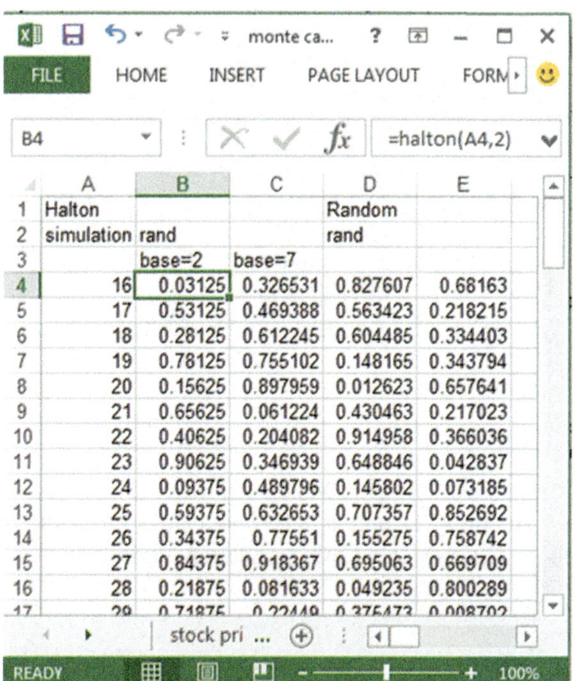

	A	B	C	D	E
1	Halton			Random	
2	simulation	rand		rand	
3		base=2	base=7		
4	16	0.03125	0.326531	0.827607	0.68163
5	17	0.53125	0.469388	0.563423	0.218215
6	18	0.28125	0.612245	0.604485	0.334403
7	19	0.78125	0.755102	0.148165	0.343794
8	20	0.15625	0.897959	0.012623	0.657641
9	21	0.65625	0.061224	0.430463	0.217023
10	22	0.40625	0.204082	0.914958	0.366036
11	23	0.90625	0.346939	0.648846	0.042837
12	24	0.09375	0.489796	0.145802	0.073185
13	25	0.59375	0.632653	0.707357	0.852692
14	26	0.34375	0.77551	0.155275	0.758742
15	27	0.84375	0.918367	0.695063	0.669709
16	28	0.21875	0.081633	0.049235	0.800289
17	29	0.71875	0.22449	0.375473	0.008702

The formula for Halton's number in B4 is:

$$= \text{halton}(A4, 2)$$

which is the 16th number under the base is equal to 2. We can change the base to 7 as shown in C4.

Two independent numbers generated by Halton or random generator can construct a join distribution. The results are shown in the below figures. We can see that the numbers generated from Halton's sequence is more discrepant than the numbers generated from random generator in Excel.

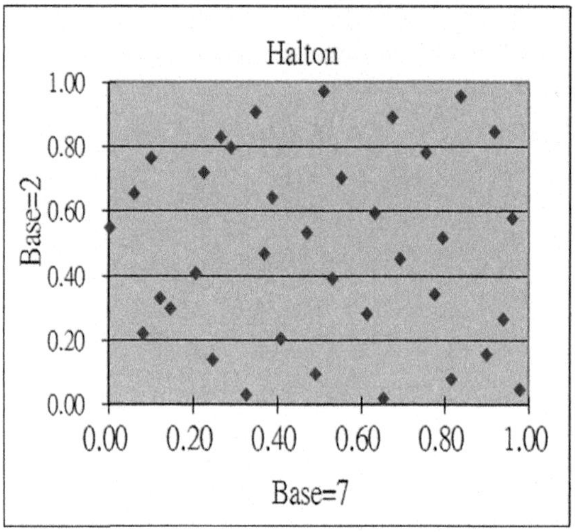

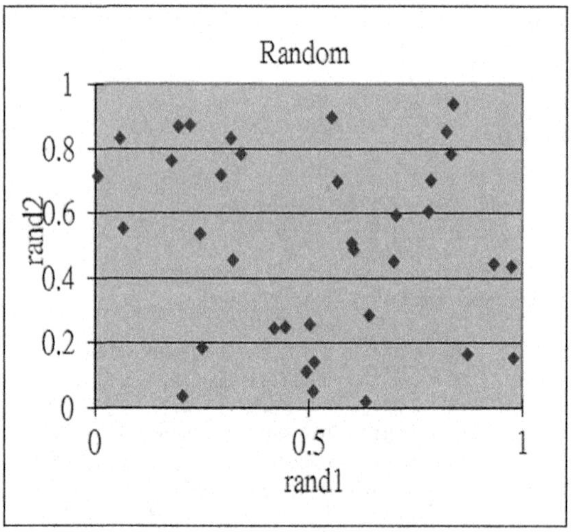

We can use Halton's sequences and Box-Muller approach to generate normal random number. And create stock prices at maturity of option. Then we can estimate the option price today. This estimate process is called the Quasi-Monte Carlo simulation. Following function can accomplish this task:

```
'   Quasi Monte-Carlo simulation for Call option
Function QMCCallBM(S, X, r, q, T, sigma, NRepl)
    Dim nuT, siT, sum, ST1, qrandns1, ST2, qrandns2, NRepl1
    Dim i As Integer, iskip As Integer
    nuT = (r - q - 0.5 * sigma ^ 2) * T
    siT = sigma * Sqr(T)
    iskip = (2 ^ 4) - 1
    sum = 0
    NRepl1 = Application.Ceiling(NRepl / 2, 1)
   For i = 1 To NRepl1
        qrandns1 = BMNormSInv1(Halton(i + iskip, 2), Halton(i + iskip,
3))
        ST1 = S * Exp(nuT + qrandns1 * siT)
        qrandns2 = BMNormSInv2(Halton(i + iskip, 2), Halton(i + iskip,
3))
        ST2 = S * Exp(nuT + qrandns2 * siT)
        sum = sum + 0.5 * (Application.Max((ST1 - X), 0) +
Application.Max((ST2 - X), 0))
    Next i
    QMCCallBM = Exp(-r * T) * sum / NRepl1
End Function
```

Halton's sequence can have the desirable property. The error in any estimate based on the samples is proportional to 1/M rather than $1/\sqrt{M}$, where M is the number of samples. We compare Monte Carlo estimates, Antithetic variates, and Quasi-Monte Carlo estimates in different simulation numbers. In the table below, we represent different replication numbers, 100, 200, …, 2000, to price option. Following figure is the result.

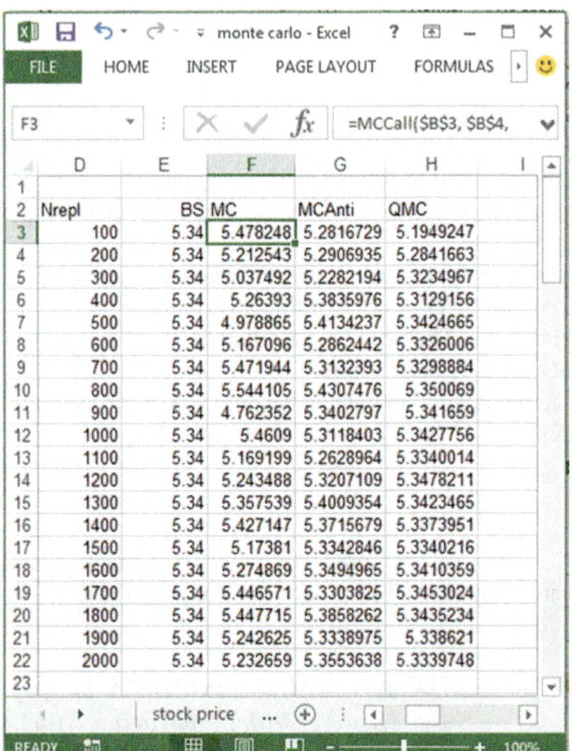

	D	E	F	G	H
1					
2	Nrepl	BS	MC	MCAnti	QMC
3	100	5.34	5.478248	5.2816729	5.1949247
4	200	5.34	5.212543	5.2906935	5.2841663
5	300	5.34	5.037492	5.2282194	5.3234967
6	400	5.34	5.26393	5.3835976	5.3129156
7	500	5.34	4.978865	5.4134237	5.3424665
8	600	5.34	5.167096	5.2862442	5.3326006
9	700	5.34	5.471944	5.3132393	5.3298884
10	800	5.34	5.544105	5.4307476	5.350069
11	900	5.34	4.762352	5.3402797	5.341659
12	1000	5.34	5.4609	5.3118403	5.3427756
13	1100	5.34	5.169199	5.2628964	5.3340014
14	1200	5.34	5.243488	5.3207109	5.3478211
15	1300	5.34	5.357539	5.4009354	5.3423465
16	1400	5.34	5.427147	5.3715679	5.3373951
17	1500	5.34	5.17381	5.3342846	5.3340216
18	1600	5.34	5.274869	5.3494965	5.3410359
19	1700	5.34	5.446571	5.3303825	5.3453024
20	1800	5.34	5.447715	5.3858262	5.3435234
21	1900	5.34	5.242625	5.3338975	5.338621
22	2000	5.34	5.232659	5.3553638	5.3339748
23					

In the column E, we use the Black–Scholes function BSCall(S, X, r, q, T, sigma) which is used as a benchmark. Monte Carlo simulation function, MCCall(S, X, r, q, T, sigma, NRepl), is used in the column F. In the column G, the call value is evaluated by antithetic variates function, MCCallAnti (S, X, r, q, T, sigma, NRepl). Quasi-Monte Carlo simulation function, QMCCallBM(S, X, r, q, T, sigma, NRepl) is used in column H.

The relative convergence of different Monte Carlo simulation can be compared. The data in range E3:H22 can be used to chart. The result is shown below.

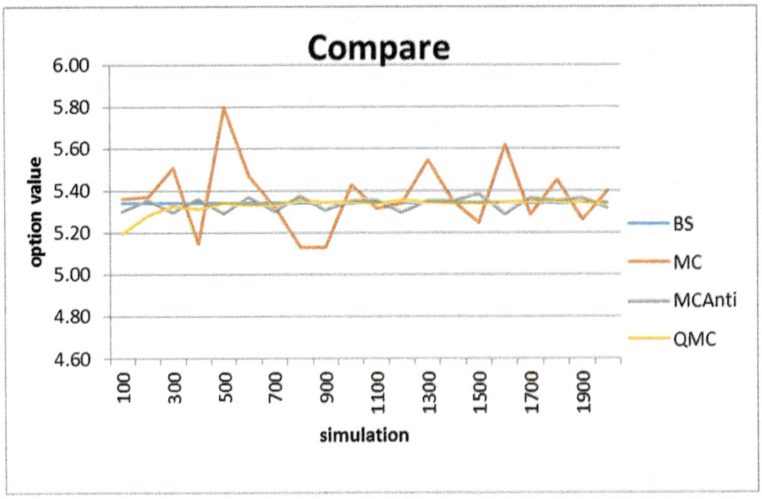

In the figure above, we can see that Monte Carlo estimate is more volatile than Antithetic variates and Quasi-Monte Carlo estimates.

30.5 Application

Binomial tree method is well suited to price American option. However, Monte Carlo simulation is suitable to value path-dependent options. In this section, we introduce the application of Monte Carlo simulation in path-depend option.

Barrier options are one kind of path-depend options where the payoff depends on whether the price of underlying reaches a certain level of price during a certain period of time. There are a number of different types of barrier options. They can be classified as knock-out or knock-in options. Here, we give a down-and-out put option as an example.

A down-and-out put option is a put option that become void if the asset price falls below the barrier S_b ($S_b < S_0$ and $S_b < X$)

$$\mathrm{P} = P_{di} + P_{do}$$

In principle, the barrier might be monitored continuously; in practice periodic monitoring may be applied. If the barrier can be monitored continuously, analytical pricing formulas are available for certain barrier option

$$\begin{aligned}\mathrm{P}_{do} = {} & Xe^{-rT}\{N(d_4) - N(d_2) - a[N(d_7) - N(d_5)]\} \\ & - \mathrm{S}_0 e^{-qT}\{N(d_3) - N(d_1) - b[N(d_8) - N(d_6)]\}\end{aligned}$$

$$a = (S_b/S_0)^{-1+2r/\sigma^2}$$

$$b = (S_b/S_0)^{1+2r/\sigma^2}$$

$$\begin{aligned}
d_1 &= [\ln(S_0/X) + (r - q + \sigma^2/2)T]/(\sigma\sqrt{T}) \\
d_2 &= d_1 - \sigma\sqrt{T} \\
d_3 &= [\ln(S_0/S_b) + (r - q + \sigma^2/2)T]/(\sigma\sqrt{T}) \\
d_4 &= d_3 - \sigma\sqrt{T} \\
d_5 &= [\ln(S_0/S_b) - (r - q - \sigma^2/2)T]/(\sigma\sqrt{T}) \\
d_6 &= d_5 - \sigma\sqrt{T} \\
d_7 &= [\ln(S_0X/S_b^2) - (r - q - \sigma^2/2)T]/(\sigma\sqrt{T}) \\
d_8 &= d_7 - \sigma\sqrt{T}
\end{aligned}$$

As an example, down-and-out put option with strike price X, expiring in T time units, with a barrier set to S_b. S_0, r, q, σ have the usual meaning.

To accomplish this we can use the below code generate a function:

```
'Down-and-out put option
Function DOPut(S, X, r, q, T, sigma, Sb)
    Dim NDOne, NDTwo, NDThree, NDFour, NDFive, NDSix, NDSeven, NDEight,
a, b, DOne, DTwo, DThree, DFour, DFive, DSix, DSeven, DEight
    a = (Sb / S) ^ (-1 + (2 * r / sigma ^ 2))
    b = (Sb / S) ^ (1 + (2 * r / sigma ^ 2))
    DOne = (Log(S / X) + (r - q + 0.5 * sigma ^ 2) * T) / (sigma *
Sqr(T))
    DTwo = (Log(S / X) + (r - q - 0.5 * sigma ^ 2) * T) / (sigma *
Sqr(T))
    DThree = (Log(S / Sb) + (r - q + 0.5 * sigma ^ 2) * T) / (sigma *
Sqr(T))
    DFour = (Log(S / Sb) + (r - q - 0.5 * sigma ^ 2) * T) / (sigma *
Sqr(T))
    DFive = (Log(S / Sb) - (r - q - 0.5 * sigma ^ 2) * T) / (sigma *
Sqr(T))
    DSix = (Log(S / Sb) - (r - q + 0.5 * sigma ^ 2) * T) / (sigma *
Sqr(T))
    DSeven = (Log(S * X / Sb ^ 2) - (r - q - 0.5 * sigma ^ 2) *T) /
(sigma * Sqr(T))
    DEight = (Log(S * X / Sb ^ 2) - (r - q + 0.5 * sigma ^ 2) *T) /
(sigma * Sqr(T))
    NDOne = Application.NormSDist(DOne)
    NDTwo = Application.NormSDist(DTwo)
    NDThree = Application.NormSDist(DThree)
    NDFour = Application.NormSDist(DFour)
    NDFive = Application.NormSDist(DFive)
    NDSix = Application.NormSDist(DSix)
    NDSeven = Application.NormSDist(DSeven)
    NDEight = Application.NormSDist(DEight)
    DOPut = X * Exp(-r * T) * (NDFour - NDTwo - a * (NDSeven - NDFive))
- S * Exp(-q * T) * (NDThree - NDOne - b * (NDEight - NDSix))
End Function
```

Barrier options often have a very different properties from plain vanilla option. For instance, sometimes Greek letter, vega, is negative. Below is the spreadsheet to show this phenomenon.

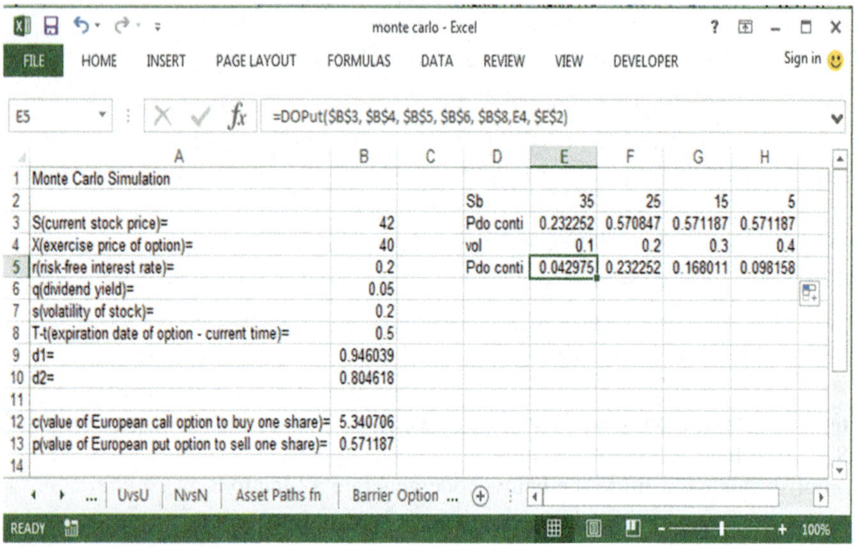

	A	B	C	D	E	F	G	H
1	Monte Carlo Simulation							
2				Sb	35	25	15	5
3	S(current stock price)=	42		Pdo conti	0.232252	0.570847	0.571187	0.571187
4	X(exercise price of option)=	40		vol	0.1	0.2	0.3	0.4
5	r(risk-free interest rate)=	0.2		Pdo conti	0.042975	0.232252	0.168011	0.098158
6	q(dividend yield)=	0.05						
7	s(volatility of stock)=	0.2						
8	T-t(expiration date of option - current time)=	0.5						
9	d1=	0.946039						
10	d2=	0.804618						
11								
12	c(value of European call option to buy one share)=	5.340706						
13	p(value of European put option to sell one share)=	0.571187						
14								

The formula for down-and-out put option in cell E5 is:

$$= \text{DOPut}(\$B\$3,\ \$B\$4,\ \$B\$5,\ \$B\$6,\ \$B\$8, E4,\ \$E\$2)$$

As volatility increases, the price of down-and-out put option may decrease because stock is easy to drop down across the barrier. We can see this effect in the below figure. As the volatility increases from 0.1 to 0.2, barrier option price increases. However, as the volatility increases from 0.2 to 0.3, barrier option price decreases.

However, monitored continuously barrier option is theoretical. In practice, we can only consider a down-and-out put option periodically, under the assumption that the barrier is checked at the end for each trading day. In order to price the barrier option, we have to generate a stock price process not only the maturity price. Below are the functions to generate two asset price processes under random number and Halton's sequence:

```
'Random Asset Paths
Function AssetPaths(S, r, q, T, sigma, NSteps, NRepl)
     Dim dt, nut, sit
     Dim i, j As Integer
     Dim spath()
     Randomize
     dt = T / NSteps
     nut = (r - q - 0.5 * sigma ^ 2) * dt
     sit = sigma * Sqr(dt)
     ReDim spath(NSteps, 1 To NRepl)
     For j = 1 To NRepl
        spath(0, j) = S
        For i = 1 To NSteps
            randns = Application.NormSInv(Rnd)

            spath(i, j) = spath(i - 1, j) * Exp(nut + randns * sit)
        Next i
     Next j
     AssetPaths = spath
End Function

' Halton Asset Paths
Function AssetPathsHalton(S, r, q, T, sigma, NSteps, NRepl)
     Dim dt, nut, sit
     Dim i, j As Integer
     Dim spath()
     Randomize
     dt = T / NSteps
     nut = (r - q - 0.5 * sigma ^ 2) * dt
     sit = sigma * Sqr(dt)
     ReDim spath(NSteps, 1 To NRepl)
     For j = 1 To NRepl
        spath(0, j) = S
        For i = 1 To NSteps
            randns = Application.NormSInv(Halton((j - 1) * NStpes + i +
16, 13))
            spath(i, j) = spath(i - 1, j) * Exp(nut + randns * sit)
        Next i
     Next j
     AssetPathsHalton = spath
 End Function
```

where NSteps is the number of time interval from now to option maturity, NRepl is how many replications to simulate. After we input the parameters, we can get the stock price process. Below we replicate three stock price process for each method. Each process with 20 time intervals.

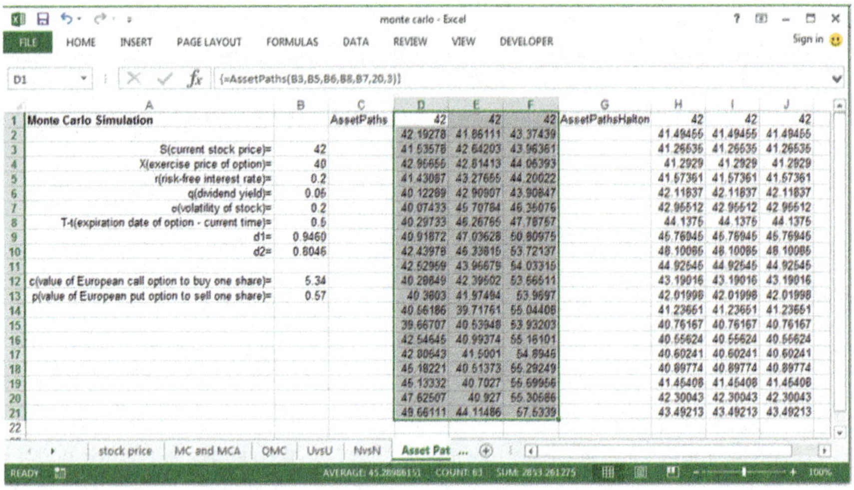

Because the output of this function is a matrix, we should follow the step below to generate the outcome. First, select the range of cells in which you want to enter the array formula, in this example D1:F21. Second, enter the formula that you want to use, in this example, AssetPaths(B3,B5,B6,B8, B7,20,3). Finally, press Ctrl + Shift + Enter.

Now, we can use a Monte Carlo simulation to compute price of the down-and-out put option. Following the function can help us accomplish this task:

```
'Down-and-out put Monte Carlo Simulation
Function DOPutMC(S, X, r, q, T, sigma, Sb, NSteps, NRepl)
    Dim payoff, sum
    Dim spath()
    ReDim spath(NSteps, 1 To NRepl)
    sum = 0
    spath = AssetPaths(S, r, q, T, sigma, NSteps, NRepl)
    For j = 1 To NRepl
        payoff = Application.Max(X - spath(NSteps, j), 0)
        For i = 1 To NSteps
            If spath(i, j) <= Sb Then
            payoff = 0
            i = NSteps
            End If
        Next i
        sum = sum + payoff
    Next j
    DOPutMC = Exp(-r * tyr) * sum / NRepl
End Function
```

Using this function, we can enter parameters into the function at the worksheet. Or we can generate stock price process in the worksheet directly. Below is the figure to show this two results.

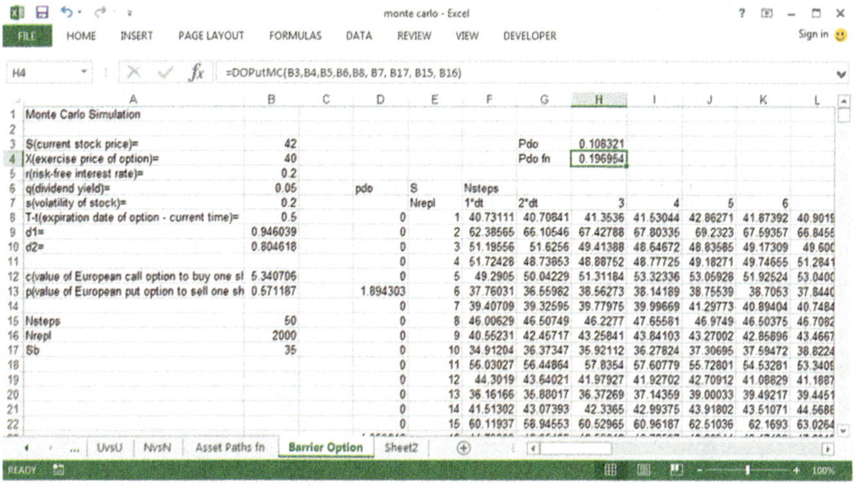

The formula in cell H3 estimated by worksheet is:

$$= \text{AVERAGE(D13 : D1012)} * \text{EXP}(-\$B\$5 * \$B\$8)$$

The formula in cell H4 estimated by user-defined VBA function is:

$$= \text{DOPutMC(B3, B4, B5, B6, B8, B7, B17, B15, B16)}$$

If you want to know how many replications the stock price cross the barrier, below is the function to complete this job:

```
'Down-and-out put Monte Carlo Simulation and cross times
Function DOPutMC_2(S, X, r, q, T, sigma, Sb, NSteps, NRepl)
    Dim payoff, Sum, cross
    Dim temp(1)
    Dim spath()
    ReDim spath(NSteps, 1 To NRepl)
    Sum = 0
    cross = 0
    spath = AssetPaths(S, r, q, T, sigma, NSteps, NRepl)
    For j = 1 To NRepl
        payoff = Application.Max(X - spath(NSteps, j), 0)
        For i = 1 To NSteps
            If spath(i, j) <= Sb Then
            payoff = 0
            i = NSteps
            cross = cross + 1
            End If
        Next i
        Sum = Sum + payoff
    Next j
    temp(0) = Exp(-r * T) * Sum / NRepl
    temp(1) = cross
    DOPutMC_2 = temp
End Function
```

Using the above function, we can get two outcomes in the cells, H5:I5. H5 is down-and-out put option value and I5 is the times that price crosses the barrier. The formula for option price and crossed number in cells, H5:I5 is

$$= \text{DOPutMC\_2(B3, B4, B5, B6, B8, B7, B17, B15, B16)}$$

We should mark the range H5:I5, then type the formula. Finally, press the [ctrl] + [shift] + [enter]. Then we can get the result.

In order to see the different crossed number, we set two barrier, Sb. In the first case Sb is equal to 5. Because barrier Sb in this case is 5, much below exercise and stock price, there is no price crosses the barrier.

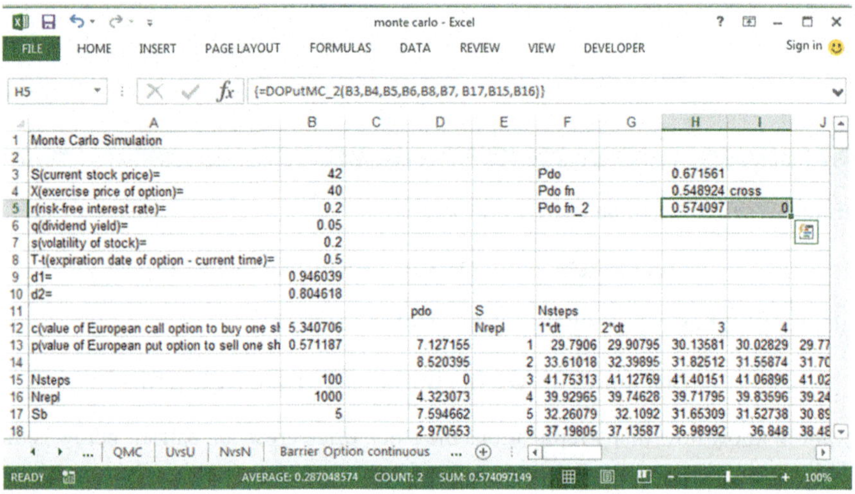

In the second case Sb is equal to 35. We can see this case, Sb = 35 is near strike price 40. Hence, there are 95 times that stock price crosses the barrier.

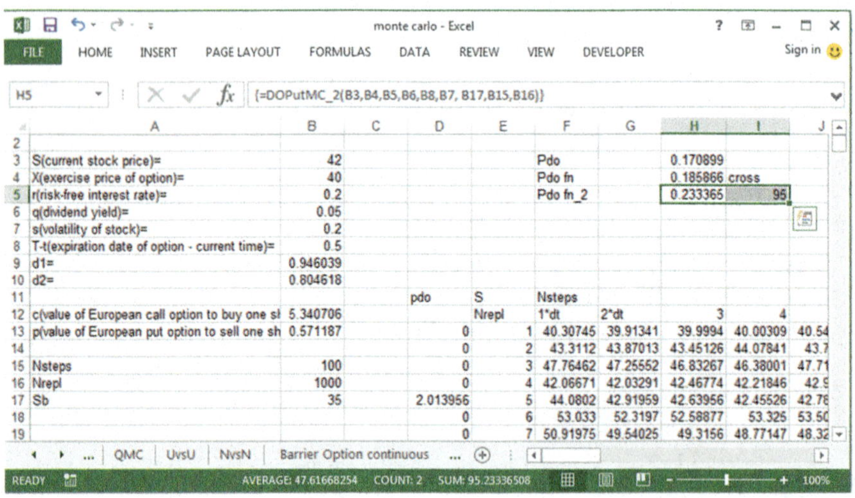

30.6 Summary

Monte Carlo simulation consists of using random numbers to generate stochastic stock price. Traditionally, we use the random generator in the Excel, rand(). However, it takes a lot of time to run a Monte Carlo simulation. In this chapter, we introduce antithetic variates to improve the efficiency in the simulation. In addition, random number generated from random generator is not discrepancy. We generate Halton's sequence, a non-random number, and use Box-Muller to generate normal samples. Then we can run a Quasi-Monte Carlo simulation, which produce a smaller error of estimation. In the application, we apply Monte Carlo simulation to the path-depend option. We simulate all the underlying asset price process to price barrier option which is one kind of path-depend option.

Appendix 30.1: Excel Code—Share Price Paths

```
'Native code to generate share price paths by Monte Carlo simulation
Sub shareprice()
    Dim nudt, sidt, Sum, randns
    Dim i As Integer
    Randomize
    Range("A15:d200").Select
    Selection.ClearContents
    S = Cells(4, 2)
    X = Cells(5, 2)
    r = Cells(6, 2)
    q = Cells(7, 2)
    T = Cells(9, 2)
    sigma = Cells(8, 2)
    NSteps = Cells(11, 2)
    nudt = (r - q - 0.5 * sigma ^ 2) * (T / NSteps)
    sidt = sigma * Sqr(T / NSteps)
    Sum = 0
    Cells(14, 1) = Cells(11, 1)
    Cells(14, 2) = "stock price 1"
    Cells(14, 3) = "stock price 2"
    Cells(14, 4) = "stock price 3"
    Cells(15, 1) = 1
    Cells(15, 2) = Cells(4, 2)
    Cells(15, 3) = Cells(4, 2)
    Cells(15, 4) = Cells(4, 2)
    For i = 2 To NSteps
        randns = Application.NormSInv(Rnd)
        Cells(14 + i, 2) = Cells(14 + i - 1, 2) * Exp(nudt + randns *
sidt)
        randns = Application.NormSInv(Rnd)
        Cells(14 + i, 3) = Cells(14 + i - 1, 3) * Exp(nudt + randns *
sidt)
        randns = Application.NormSInv(Rnd)
        Cells(14 + i, 4) = Cells(14 + i - 1, 4) * Exp(nudt + randns *
sidt)
        Cells(14 + i, 1) = i
    Next i
End Sub
```

References

On the Web

http://roth.cs.kuleuven.be/wiki/Main_Page

Further Reading

Boyle PP (1977) Options: a Monte Carlo approach. J Financ Econ 4(3):323–338

Boyle P, Broadie M, Glasserman P (1997) Monte Carlo methods for security pricing. J Econ Dyn Control 21(8):1267–1321

Hull JC (2015) Options, futures, and other derivatives. Prentice Hall, Upper Saddle River, NJ

Joy C, Boyle PP, Tan KS (1996) Quasi-Monte Carlo methods in numerical finance. Manage Sci 42(6):926–938

Wilmott P (2013) Paul Wilmott on quantitative finance. Wiley, Chichester

Part C
Applications of SAS Programs to Financial Analysis

Chapter 31
Application of Simultaneous Equation in Finance Research: Methods and Empirical Results

31.1 Introduction

Based upon the paper by Chen and Lee (2010) and Lee et al. (2015), we will first discuss the development of 2SLS, 3SLS, and generalized method of moments (GMM). Then, we will use GE as an example to show how SAS program can be used to estimate simultaneous equations system in terms of 2SLS, 3SLS, and GMM. In the second section, we will develop the model specification for 2SLS, 3SLS, and GMM. In the third section, we will discuss the simultaneous equation system for investment policy, financing policy, and dividend policy. In the fourth section, we will use GE data to show how simultaneous equation system can be estimated by 2SLS, 3SLS, and GMM in terms of SAS program. The GE data is presented in Appendix 31.1. The SAS program used to estimate these different methods will be presented in Appendix 31.2. Finally, in the fifth section, we summarize the results.

31.2 Model Development for 2SLS, 3SLS, and GMM

31.2.1 Introduction

Empirical finance research often employs a single equation for estimation and testing. However, single equation rarely happens in the economic or financial theory. Using OLS method to estimate equation(s), which should otherwise be treated as a simultaneous equation system, is likely to produce biased and inconsistent parameter estimators. To illustrate, let's start with a simple Keynesian consumption function specified as follows:

$$C_t = \alpha + \beta Y_t + \mu_t \tag{31.1}$$

$$Y_t = C_t + I_t \tag{31.2}$$

$$I_t = I_0 \tag{31.3}$$

Where C_t is the consumption expenditure at time t, Y_t is the national income at time t, I_t is the investment expenditure at time t, which is assumed fixed at I_0, and μ_t is the stochastic disturbance

C.-F. Lee et al., *Essentials of Excel, Excel VBA, SAS and Minitab for Statistical and Financial Analyses*, DOI 10.1007/978-3-319-38867-0_31

term at time t. Equation (31.1) is the consumption function; Eq. (31.2) is the equilibrium condition (national income accounting identity); and Eq. (31.3) is the investment function. Some of the variables in the model are endogenous, others are exogenous. For example, C_t and Y_t are endogenous, meaning they are determined within the model. On the other hand, I_t is the exogenous variable, which is not determined in the model, hence not correlated with μ_t.

A simultaneous equation bias arises when OLS is applied to estimate the consumption function because Y_t is correlated with the disturbance term μ_t, which violates the OLS assumption that independent variables are orthogonal to the disturbance term. To see this, through a series of substitutions, we can obtain the reduced form equations from the structural Eqs. (31.1) through (31.3) as:

$$C_t = \frac{\alpha}{1-\beta} + \frac{\beta}{1-\beta} I_0 + \frac{\mu_t}{1-\beta} \tag{31.4}$$

$$Y_t = \frac{\alpha}{1-\beta} + \frac{1}{1-\beta} I_0 + \frac{\mu_t}{1-\beta} \tag{31.5}$$

Based upon Eq. (31.5), clearly Y_t is correlated with the disturbance term, μ_t, which violates the OLS assumption that independent variables and the disturbance term are uncorrelated. This bias is commonly referred to in the literature as the simultaneous equation bias. Furthermore, the OLS estimate of β in Eq. (31.1) is not only bias, but also inconsistent, meaning β estimate does not converge to the true β when the sample size increases to very large.

31.2.2 Two-Stage and Three-Stage Least Squares Method

To resolve the simultaneous equation bias problem as illustrated in Sect. 31.1, in this section we discuss two popular simultaneous equation estimation methods. Different methods are available to handle the estimation of a simultaneous equation model: Indirect least squares, instrumental variable procedure, two-stage least squares, three-stage least squares, limited information likelihood method, and full information maximum likelihood method, just to name a few. In this section, we will focus on two methods that popular statistical and/or econometric software are readily available.

31.2.2.1 Identification Problem

Before getting into the estimation methods, it is necessary to discuss the "identification problem". Identification problem arises when we cannot identify the difference between, say, two functions. Consider the demand and supply model of gold. The structure equations can be written as:

$$Q_d = \alpha_0 + \alpha_1 P + \varepsilon \tag{31.6}$$

$$Q_s = \beta_0 + \beta_1 P + e \tag{31.7}$$

$$Q_d = Q_s = Q \tag{31.8}$$

Equation (31.6) is the demand for gold function, where the demand Q_d is determined by the price of gold, P; Eq. (31.7) is the supply of gold, and it is a function of gold price; Eq. (31.8) is an identity stating the market equilibrium. Can we apply the OLS method to Eqs. (31.6) and (31.7) to obtain parameter estimates? To answer this question, we first obtain the "reduced form" equations for P and Q through substitutions.

$$P = \frac{\beta_0 - \alpha_0}{\alpha_1 - \beta_1} + \frac{\varepsilon - e}{\alpha_1 - \beta_1} \tag{31.9}$$

$$Q = \frac{\alpha_1\beta_0 - \alpha_0\beta_1}{\alpha_1 - \beta_1} + \frac{\alpha_1\varepsilon - \beta_1 e}{\alpha_1 - \beta_1} \tag{31.10}$$

Obviously, it is impossible to estimate Eqs. (31.9) and (31.10) using OLS method because there are four parameters (α_0, α_1, β_0, *and* β_1) to be estimated, but there are only two equations. Therefore, we cannot estimate the parameters in the structure equations. This is the situation called "under-identification".

To differentiate demand equation from supply equation, now suppose we assume that demand curve for gold may shift due to the changes in economic uncertainty, which can be proxied by, say, stock market volatility, V, which is assumed to be exogenous. Hence, Eqs. (31.6) and (31.7) can be modified as:

$$Q_d = \alpha_0 + \alpha_1 P + \alpha_2 V + \varepsilon \tag{31.11}$$

$$Q_s = \beta_0 + \beta_1 P + e \tag{31.12}$$

The reduced form becomes

$$P = \frac{\beta_0 - \alpha_0}{\alpha_1 - \beta_1} - \frac{\alpha_2}{\alpha_1 - \beta_1} V + \frac{\varepsilon - e}{\alpha_1 - \beta_1}$$

$$= \gamma_0 + \gamma_1 V + \pi_1 \tag{31.13}$$

$$Q = (\beta_0 + \beta_1\gamma_0) + \beta_1\gamma_1 V + \frac{\alpha_1\varepsilon - \beta_1 e}{\alpha_1 - \beta_1}$$

$$= \lambda_0 + \lambda_1 V + \pi_2 \tag{31.14}$$

Because V is assumed exogenous and uncorrelated with residuals π_1 and π_2, OLS can be applied to the reduced form Eqs. (31.13) and (31.14) and obtain estimators of γ_0, γ_1, λ_0, and λ_1. Examining Eq. (31.14), we find that $\lambda_0 = \beta_0 + \beta_1\gamma_0$, and $\lambda_1 = \beta_1\gamma_1$. Since γ_0, γ_1, λ_0, and λ_1 are all obtained from the OLS estimates, β_0 and β_1 can be solved. Therefore, the supply function (Eq. (31.14)) is said to be identified. However, from the demand function, we find that

$$\gamma_0 = \frac{\beta_0 - \alpha_0}{\alpha_1 - \beta_1}, \text{ and } \gamma_1 = \frac{\alpha_2}{\alpha_1 - \beta_1}.$$

Since there are only two equations, we cannot possibly estimate three unknowns, α_0, α_1, *and* α_2, hence the demand function is not identified. Based upon the discussions of Eqs. (31.6) and (31.7), we thus know that in a two-equation model, if one variable is omitted from one equation, then this equation is identified. On the other hand, there is no omitted variable in Eq. (31.6), hence the demand function is not identified.

Now let's further modify Eqs. (31.6) and (31.7) as follow:

$$Q_d = \alpha_0 + \alpha_1 P + \alpha_2 V + \varepsilon \tag{31.15}$$

$$Q_s = \beta_0 + \beta_1 P + \beta_2 D + e \tag{31.16}$$

All variables are defined as before except now we have added a new variable D in the supply equation. Let D be the government deficit of Russia, which is assumed exogenous to the system. When Russia's budget deficit deteriorates, the government increases the gold production for cash. The reduced form of P and Q based upon Eqs. (31.11) and (31.12) is

$$P = \frac{\beta_0 - \alpha_0}{\alpha_1 - \beta_1} - \frac{\alpha_2}{\alpha_1 - \beta_1} V + \frac{\beta_2}{\alpha_1 - \beta_1} D + \frac{\varepsilon - e}{\alpha_1 - \beta_1}$$

$$= \gamma_0 + \gamma_1 V + \gamma_2 D + \pi_1 \tag{31.17}$$

$$Q = (\alpha_0 + \alpha_1 \gamma_0) + (\alpha_1 \gamma_1 + \gamma_2) V + \alpha_1 \gamma_2 D + \frac{\alpha_1 \varepsilon - \beta_1 e}{\alpha_1 - \beta_1}$$

$$= \lambda_0 + \lambda_1 V + \lambda_2 D + \pi_2 \tag{31.18}$$

Based upon the OLS estimates of Eqs. (31.17) and (31.18), we can obtain unique estimates for the structure parameters α_0, α_1, α_2, β_0, β_1, and β_2, hence both the demand and supply functions are identified. In this case, we call the situation as "exactly identified".

In a scenario when there are multiple solutions to the structure parameters, the equation is said to be "over-identified". For example, in Eqs. (31.19) and (31.20), we modify the supply equation by adding another exogenous variable, q, representing lagged quantity of gold produced (i.e., supply of gold in the last period), which is predetermined.

$$Q_d = \alpha_0 + \alpha_1 P + \alpha_2 V + \varepsilon \tag{31.19}$$

$$Q_s = \beta_0 + \beta_1 P + \beta_2 D + \beta_3 q + e \tag{31.20}$$

The reduced form becomes

$$P = \frac{\beta_0 - \alpha_0}{\alpha_1 - \beta_1} - \frac{\alpha_2}{\alpha_1 - \beta_1} V + \frac{\beta_2}{\alpha_1 - \beta_1} D + \frac{\beta_3 q}{\alpha_1 - \beta_1} + \frac{\varepsilon - e}{\alpha_1 - \beta_1}$$

$$= \gamma_0 + \gamma_1 V + \gamma_2 D + \gamma_3 q + \pi_1 \tag{31.21}$$

$$Q = (\alpha_0 + \alpha_1 \gamma_0) + (\alpha_1 \gamma_1 + \gamma_2) V + \alpha_1 \gamma_2 D + \alpha_1 \gamma_3 + \frac{\alpha_1 \varepsilon - \beta_1 e}{\alpha_1 - \beta_1}$$

$$= \lambda_0 + \lambda_1 V + \lambda_2 D + \lambda_3 q + \pi_2 \tag{31.22}$$

Based upon Eqs. (31.21) and (31.22), we find $\alpha_1 \gamma_2 = \lambda_2$, hence structure equation parameter α_1 can be estimated as γ_2/λ_2. However, we also find $\alpha_1 \gamma_3 = \lambda_3$, hence α_1 can also take another value, γ_3/λ_3. Therefore, α_1 does not have a unique solution, and we say the model is "over-identified".

The condition we employ in the above discussions for model identification is the so-called "order condition of identification". To summarize the order condition of model identification, a general rule is that the number of variables excluded from an equation must be $\geq$ the number of structural equations.[1]

Although "order condition" is a popular way of model identification, it provides only a necessary condition for model identification, not a sufficient condition. Alternatively, "rank condition" provides both necessary and sufficient conditions for model identification. An equation satisfies the rank condition if and only if at least one determinant of rank (M-1) can be constructed from the column coefficients corresponding to the variables that have been excluded from the equation, where M is the number of equations in the system. However, "rank condition" is more complicated than "order condition" and it is difficult to determine in a large simultaneous equation model. The following example based upon Eqs. (31.23) through (31.25) provides some basic ideas about "rank condition". Note, Eqs. (31.23) through (31.25) are similar to Eqs. (31.11), (31.12), and (31.8) with terms rearranged.

$$Q_d - \alpha_0 - \alpha_1 P - \alpha_2 V = \varepsilon \tag{31.23}$$

$$Q_s - \beta_0 - \beta_1 P = e \tag{31.24}$$

$$Q_d - Q_s = 0 \tag{31.25}$$

In the following table, all structural parameters are stripped from the equations and placed in a matrix.

	Coefficients of the variables			
Equations	Intercept	Q	P	V
Eq. (31.23)	$-\alpha_0$	1	$-\alpha_1$	$-\alpha_2$
Eq. (31.24)	$-\beta_0$	1	β_1	0

Since only variable V is excluded from Eq. (31.24), the rank of remaining parameter in column V is

$$\text{rank}(-\alpha_2) = 1.$$

Because this has a rank of one, which is equal to the number of endogenous variables subtracts one, Eq. (31.24) is identified. On the other hand, rank condition of remaining parameter matrix cannot be constructed for Eq. (31.23) because none of the variables are excluded. Hence, Eq. (31.23) is "under-identified".[2]

31.2.2.2 Two-Stage Least Squares

Two-stage least squares (2SLS) method is easy to apply and can be applied to a model that is exactly or over-identified. To illustrate, let's use Eqs. (31.15) and (31.16) for demonstration and rewrite them as the follows:

$$Q_d = \alpha_0 + \alpha_1 P + \alpha_2 V + \varepsilon \tag{31.26}$$

[1] The discussions of the order condition draw heavily from Ramanathan (1995).

[2] For more detailed discussions of the rank condition, see econometric books such as Greene (2003), Judge et al. (1985), Fisher (1966), Blalock (1969), and Fogler and Ganapathy (1982).

$$Q_s = \beta_0 + \beta_1 P + \beta_2 D + e \tag{31.27}$$

Based upon the "order condition", Eqs. (31.26) and (31.27) each has one variable excluded from the other equation, which is equal to the number of equations minus one. Hence, the model is identified.

Since endogenous variable P is correlated with the disturbance term, the first stage for the 2SLS calls for the estimation of "predicted P" ($\hat{P}$) using a reduced form containing all exogenous variables. To do this, we can apply the OLS to the following equation:

$$P = \eta_0 + \eta_1 V + \eta_2 D + \tau \tag{31.28}$$

OLS will yield unbiased and consistent estimation because both V and D are exogenous, hence not correlated with the disturbance term τ. With the parameters in Eq. (31.28) estimated, the "predicted P can be calculated as:

$$\hat{P} = \hat{\eta}_0 + \hat{\eta}_1 V + \hat{\eta}_2 D.$$

This $\hat{P}$ is the instrumental variable to be used in the second stage estimation and is not corrected with the structure equation disturbance term. Substituting $\hat{P}$ into Eqs. (31.26) and (31.27), we have

$$Q_d = \alpha_0 + \alpha_1 \hat{P} + \alpha_2 V + \varepsilon \tag{31.29}$$

$$Q_s = \beta_0 + \beta_1 \hat{P} + \beta_2 D + e \tag{31.30}$$

Since $\hat{P}$ is not correlated with the disturbance terms, OLS method can be applied to Eqs. (31.29) and (31.30).

31.2.2.3 Three-Stage Least Squares

The 2SLS method is a limited information method. On the other hand, the three-stage least squares (3SLS) method is a full information method. A full information method takes into account the information from the complete system, hence it is more efficient than the limited information method. Simply put, 3SLS method incorporates information obtained from the variance-covariance matrix of the system disturbance terms to estimate structural equation parameters. On the other hand, 2SLS method assumes that ε and e in Eqs. (31.29) and (31.30) are independent and estimates structural equation parameters separately, thus it might lose some information when in fact the disturbance terms are not independent. This section briefly explains a 3SLS estimation method.

For example, we can rewrite the models Eqs. (31.26) and (31.27) in matrix form as an illustration of 3SLS estimation. Let

$$\mathbf{Y} = \begin{bmatrix} \boldsymbol{Q} \\ \boldsymbol{P} \end{bmatrix}, \ \boldsymbol{Z} = \begin{bmatrix} \boldsymbol{Z}_1 & 0 \\ 0 & \boldsymbol{Z}_2 \end{bmatrix} = \begin{bmatrix} (\boldsymbol{P}\ 1\ \boldsymbol{V}) & 0 \\ 0 & (\boldsymbol{Q}\ 1\ \boldsymbol{D}) \end{bmatrix}, \boldsymbol{\psi} = \begin{bmatrix} \psi_1 \\ \psi_2 \end{bmatrix}$$

$$= \begin{bmatrix} (\alpha_1 \quad \alpha_0 \quad \alpha_2)' \\ \left(\frac{1}{\beta_1} \quad \frac{-\beta_0}{\beta_1} \quad \frac{-\beta_2}{\beta_1}\right)' \end{bmatrix}, \boldsymbol{\varepsilon} = \begin{bmatrix} \varepsilon \\ \frac{e}{-\beta_1} \end{bmatrix} = \begin{bmatrix} \varepsilon_1 \\ \varepsilon_2 \end{bmatrix}$$

Then, we have

$$\mathbf{Y} = \boldsymbol{Z\psi} + \boldsymbol{\varepsilon} \tag{31.31}$$

here $\mathbf{Y}$ is a vector of $2n$ observations on the left-hand side endogenous variables; $\mathbf{Z}_\mathrm{j}$ is a matrix consisting of the right-hand side endogenous and exogenous variables in jth equation; and $\boldsymbol{\psi}$ is a vector of structural equation parameters of interest. Similar to 2SLS estimation, we use all exogenous variables as instrumental variables in each equation of the system. Thus, we can multiply both sides of Eq. (31.31) by a matrix $\mathbf{X}'$, defined by

$$\mathbf{X}' = \begin{bmatrix} (1\ \boldsymbol{V}\,\boldsymbol{D})' & 0 \\ 0 & (\ 1\ \boldsymbol{V}\,\boldsymbol{D})' \end{bmatrix} \tag{31.32}$$

i.e.,

$$\mathbf{X}'\mathbf{Y} = \mathbf{X}'\boldsymbol{Z\psi} + \mathbf{X}'\boldsymbol{\varepsilon} \tag{31.33}$$

Then the variance-covariance matrix of the disturbance term in Eq. (31.33), $\mathbf{X}'\ \boldsymbol{\varepsilon}$ will be

$$\mathbf{E}\left(\mathbf{X}'\boldsymbol{\varepsilon\varepsilon}'\mathbf{X}\right) = \boldsymbol{\Omega} \otimes \left(\mathbf{X}'\mathbf{X}\right), \boldsymbol{\Omega} = \begin{bmatrix} \sigma_{11} & \sigma_{12} \\ \sigma_{21} & \sigma_{22} \end{bmatrix} \tag{31.34}$$

The 3SLS estimator of structural equation parameters can thus be obtained as

$$\widehat{\boldsymbol{\psi}} = \left\{Z'X\left[\boldsymbol{\Omega}^{-1} \otimes \left(\mathbf{X}'\mathbf{X}\right)^{-1}\right]X'Z\right\}^{-1} Z'X\left[\boldsymbol{\Omega}^{-1} \otimes \left(\mathbf{X}'\mathbf{X}\right)^{-1}\right]X'Y \tag{31.35}$$

A question arises in the estimation process because the σ ' s are unknown, hence the matrix Ω^{-1} is also unknown. This problem can be resolved by using the residuals from the structural equations estimated by the 2SLS to form the mean sum of residual squares and use them to estimate Ω^{-1}. Standard econometric software such as SAS can be easily used to estimate Eq. (31.35). In sum, 3SLS takes three stages to estimate the structural parameters. The first stage is to estimate the reduce form system; the second stage uses 2SLS to estimate the $\hat{\Omega}$ matrix; and the third stage completes the estimation using Eq. (31.35). Since Ω contains information pertinent to the correlations between disturbance terms in the structural equations, 3SLS is called a full information method.[3]

Since 3SLS is a full information estimation method, the parameters estimated are asymptotically more efficient than the 2SLS estimates. However, this statement is correct only if the model is correctly specified. In effect, 3SLS is quite vulnerable to model misspecifications. This is because model misspecification in a single equation could easily propagate itself into the entire system.

31.2.3 GMM Methodology

Suppose that a set of observations on a variable y is drawn independently from probability distribution and depends on an unknown vector of parameters $\boldsymbol{\beta}$ of interest. One general approach for estimating parameters $\boldsymbol{\beta}$ is based on maximum likelihood (ML) estimation. The intuition behind ML estimation is to specify a probability distribution for it and then find an estimate $\hat{\boldsymbol{\beta}}$ in which the data would be most likely to have been observed. The drawback with maximum likelihood methods is that we have

[3] For more detailed discussions, see Ghosh (1991), Judge et al. (1985), and Greene (2003).

to specify a full probability distribution for the data. Here, we introduce an alternative approach for parameter estimation known as generalized method of moments (GMM). The GMM estimation was formalized by Hansen (1982) and since has become one of the most widely used methods of estimation in economics and finance. Hansen won his Nobel Prize in 2013 in economics for deriving the GMM estimation. In contrast to ML estimation, the GMM estimation only requires the specification of certain moment conditions rather than the form of likelihood function.

The idea behind GMM estimation is to choose parameter estimate so as to make the sample moment conditions as close as possible to the population moment of zero according to the measure of Euclidean distance. The GMM estimation proposed a weighting matrix reflecting the importance given to matching each of the moments. Alternative weighting matrix is associated with alternative estimator. Many standard estimators, including ordinary least squares (OLS), method of moments (MM), ML, instrumental variable (IV), two-stage least squares (2SLS), and three-stage least squares (3SLS), can be seen as special cases of GMM estimators. For example, when the number of moment conditions and unknown parameters is the same, solving the quadratic criterion yields the GMM estimator, which is the same as the method of moment (MM) estimator sets the sample moment condition exactly equal to zero. The weighting matrix does not matter in this case. In particular, in models for which there are more moment conditions than model parameters, GMM estimation provides a straightforward way to test the specification of the proposed model. This is an important feature that is unique to GMM estimation.

Recently, the endogeneity concern has received much attention in empirical corporate finance research. There are at least three generally recognized sources of endogeneity: omitted explanatory variables, simultaneity bias, and errors in variables. Whenever there is endogeneity, the application of ordinary least squares (OLS) estimation yields biased and inconsistent estimates. In literature, the instrumental variable (IV) methods are commonly used to deal with this endogeneity problem. The basic motivation for the instrumental variable method was to deal with equations that exhibited both simultaneity and measurement errors in exogenous variables. The idea behind IV estimation is to select suitable instruments that are orthogonal to the disturbance while sufficiently correlated with the regressors. The IV estimator makes the linear combinations of sample orthogonality conditions as possible to zeros. Sargan (1958, 1959) established a fully developed theory of IV estimation. The GMM estimator proposed by Hansen (1982) is also based on orthogonality conditions and provides an alternative solution. Hansen (1982)'s GMM estimator generalized Sargan (1958, 1959)'s linear and nonlinear IV estimators based on optimal weighing matrix for the moment conditions. In contrast to traditional IV class estimators such as 2SLS and 3SLS estimators, the GMM estimator uses a weighting matrix and takes into account temporal dependence, heteroscedasticity, or autocorrelation. The detailed specification and estimation methods for GMM will be discussed in Appendix 31.1.

31.3 Model Specification for Investment, Financing, and Dividend Policies

The investment, dividend, and debt financing are major decisions of a firm. Past studies argue about some relations among investment, dividend, and debt financing.[4] To control for the possible endogenous problems among investment, dividend, and debt financing decisions, we apply two-stage least

[4] Higgins (1972), Fama (1974), Morgan and Saint-Pierre (1978), Smirlock and Marshall (1929), Lee et al. (2011) and Chen et al. (2013) investigate the relationship between investment decision and dividend decision. Fama and French (2002) consider the interaction between dividend and financing decisions. Dhrymes and Kurz (1967), McDonald et al. (1975), McCabe (1979), Peterson and Benesh (1929), Switzer (1984), and Pruitt and Gitman (1991) argue that the investment decision is related to financing decision and dividend decision. Chava and Roberts (2008) show how financing impacts corporate investment via debt covenants. Lambrecht and Myers (2012) develop a combined theory of payout, debt, and investment.

squares (2SLS), three-stage least squares (3SLS), and generalized method of moments (GMM) method to estimate the simultaneous-equations model that considers the interaction of the three policies. There are three equations in our simultaneous-equations system; each equation contains the remaining two endogenous variables as explanatory variables along with other exogenous variables.

The structural equations are estimated as follows:

$$Inv_{it} = \alpha_{1i} + \alpha_{2i}Div_{it} + \alpha_{3i}Leverage_{it} + \alpha_{4i}Inv_{i,t-1} + \alpha_{5i}Q_{it} + \varepsilon_{it}, \tag{31.36}$$

$$Div_{it} = \beta_{1i} + \beta_{2i}Inv_{it} + \beta_{3i}Leverage_{it} + \beta_{4i}Div_{i,t-1} + \beta_{5i}P_{it} + \eta_{it}, \tag{31.37}$$

$$\begin{aligned} Leverage_{it} = {} & \gamma_{1i} + \gamma_{2i}Inv_{it} + \gamma_{3i}Div_{it} + \gamma_{4i}Leverage_{i,t-1} + \gamma_{5i}ln\,A_{i,t-1} \\ & + \gamma_{6i}(E_{i,t-1}/A_{i,t-1}) + \xi_{it}. \end{aligned} \tag{31.38}$$

Where Inv_{it}—the investment of firm *i* at year *t*

Div_{it}—the dividend of firm *i* at year *t*
$Debt_{it}$—change of the debt of firm *i* at year *t*
Q_{it}—sales plus the change in inventories of firm *i* at year *t*
P_{it}—net income before extraordinary items plus depreciation minus preferred dividends of firm *i* at year *t*
$\ln A_{i,t-1}$—natural logarithm of lagged total assets of firm *i* at year *t-1*
$E_{i,t-1}/A_{i,t-1}$—the lag of earnings before interest and taxes divided by total assets of firm *i* at year *t-1*

Here, Inv_{it}, Div_{it} and $Debt_{it}$ are the three endogenous variables. Q_{it}, P_{it}, $\ln A_{i,t-1}$, and $E_{i,t-1}/A_{i,t-1}$ are the four exogenous variables.

We use GE as an example to show how SAS program can be used to estimate 2SLS, 3SLS, and GMM. The sample period for all the variables as specified in Eqs. (31.36)–(31.38) is from 1966 to 2012. The data for all these variables is presented in Appendix 31.2. The SAS programs used to estimate this equation system for 2SLS, 3SLS, and GMM are presented in Appendix 31.3. Then, the empirical results are presented in Tables 31.1, 31.2, and 31.3.

Empirical results for GE company.

Table 31.1 2SLS results for GE

Dependent variables			
Independent variables	Inv_{it}	Div_{it}	$Leverage_{it}$
Constant	−10.382***	0.552*	−0.131
	(−4.57)	(1.88)	(−0.77)
$Inv_{i,t-1}$	0.230***		
	(3.95)		
$Div_{i,t-1}$		0.123	
		(1.68)	
$Leverage_{i,t-1}$			0.788***
			(7.06)
Inv_{it}		0.033	−0.005
		(0.84)	(−0.86)
Div_{it}	5.805***		0.050
	(4.00)		(0.91)
$Leverage_{it}$	11.047***	−0.540*	
	(4.15)	(−1.88)	
Q_{it}	0.069**		
	(2.05)		
P_{it}		0.146	
		(1.45)	
$ln\ A_{i,t-1}$			0.027*
			(1.78)
$E_{i,t-1}/A_{i,t-1}$			−0.240
			(−0.36)
$Industry_{i,t-1}$			0.007
			(0.80)
Adjusted R-squares	0.95	0.94	0.92

This table presents the 2SLS regression results of a simultaneous equation system model for investment, dividend, and debt financing:

$$Inv_{it} = \alpha_{1i} + \alpha_{2i}Div_{it} + \alpha_{3i}Leverage_{it} + \alpha_{4i}Inv_{i,t-1} + \alpha_{5i}Q_{it} + \varepsilon_{it},$$

$$Div_{it} = \beta_{1i} + \beta_{2i}Inv_{it} + \beta_{3i}Leverage_{it} + \beta_{4i}Div_{i,t-1} + \beta_{5i}P_{it} + \eta_{it},$$

$$Leverage_{it} = \gamma_{1i} + \gamma_{2i}Inv_{it} + \gamma_{3i}Div_{it} + \gamma_{4i}Leverage_{i,t-1} + \gamma_{5i}ln\ A_{i,t-1} + \gamma_{6i}(E_{i,t-1}/A_{i,t-1}) + \gamma_{7i}Industry_{i,t-1} + \xi_{it}.$$

Regressions are based on annual data of General Electric (GE) from 1966 to 2012. The three endogenous variables are Inv_{it}, Div_{it}, and $Leverage_{it}$, which are net plant and equipment, dividends, and book leverage ratio, respectively. The independent variables in the investment regression are lagged investment ($Inv_{i,t-1}$), and sales plus change in inventories (Q_{it}). The independent variables in the dividend regression are lagged dividends ($Div_{i,t-1}$), and net income minus preferred dividends (P_{it}). All the variables in both of investment and dividend equations are measured on a per share basis. The independent variables in the debt financing regression are lagged book leverage ($Leverage_{i,t-1}$), natural logarithm of lagged total assets ($ln\ A_{i,t-1}$), the lag of earnings before interest and taxes divided by total assets ($E_{i,t-1}/A_{i,t-1}$), and lagged industry averages of book leverage ($Industry_{i,t-1}$). Numbers in the parentheses are t-statistics. ***, **, and * denote statistical significance at the 1 %, 5 %, and 10 % levels, respectively.

Table 31.2 3SLS results for GE

Dependent variables			
Independent variables	Inv_{it}	Div_{it}	$Leverage_{it}$
Constant	−9.908***	0.799***	−0.153
	(−4.55)	(2.91)	(−0.91)
$Inv_{i,t-1}$	0.124**		
	(2.62)		
$Div_{i,t-1}$		−0.011	
		(−0.18)	
$Leverage_{i,t-1}$			0.818***
			(7.38)
Inv_{it}		0.065**	−0.008
		(2.07)	(−1.51)
Div_{it}	7.871***		0.084
	(6.77)		(1.55)
$Leverage_{it}$	10.016***	−0.761***	
	(4.07)	(−2.80)	
Q_{it}	0.038		
	(1.38)		
P_{it}		0.091	
		(1.14)	
$ln\ A_{i,t-1}$			0.026*
			(1.74)
$E_{i,t-1}/A_{i,t-1}$			−0.256
			(−0.39)
$Industry_{i,t-1}$			0.007
			(0.85)
Adjusted R-squares	0.98		

This table presents the 3SLS regression results of a simultaneous equation system model for investment, dividend, and debt financing:

$$Inv_{it} = \alpha_{1i} + \alpha_{2i}Div_{it} + \alpha_{3i}Leverage_{it} + \alpha_{4i}Inv_{i,t-1} + \alpha_{5i}Q_{it} + \varepsilon_{it},$$

$$Div_{it} = \beta_{1i} + \beta_{2i}Inv_{it} + \beta_{3i}Leverage_{it} + \beta_{4i}Div_{i,t-1} + \beta_{5i}P_{it} + \eta_{it},$$

$$Leverage_{it} = \gamma_{1i} + \gamma_{2i}Inv_{it} + \gamma_{3i}Div_{it} + \gamma_{4i}Leverage_{i,t-1} + \gamma_{5i}\ln A_{i,t-1} + \gamma_{6i}(E_{i,t-1}/A_{i,t-1}) + \gamma_{7i}Industry_{i,t-1} + \xi_{it}.$$

Regressions are based on annual data of General Electric (GE) from 1966 to 2012. The three endogenous variables are Inv_{it}, Div_{it}, and $Leverage_{it}$, which are net plant and equipment, dividends, and book leverage ratio, respectively. The other variables are the same as in Table 31.1. Numbers in the parentheses are t-statistics. ***, **, and * denote statistical significance at the 1 %, 5 % and 10 % levels, respectively.

Table 31.3 GMM results for GE

Dependent variables Independent variables	Inv_{it}	Div_{it}	$Leverage_{it}$
Constant	−9.855***	0.533**	−0.112*
	(−6.44)	(2.47)	(−1.81)
$Inv_{i,t-1}$	0.193***		
	(4.75)		
$Div_{i,t-1}$		0.079	
		(1.28)	
$Leverage_{i,t-1}$			0.867***
			(13.08)
Inv_{it}		0.041	−0.003
		(1.68)	(−1.15)
Div_{it}	7.031***		0.036
	(5.87)		(1.68)
$Leverage_{it}$	10.019***	−0.459**	
	(5.62)	(−2.21)	
Q_{it}	0.047***		
	(1.94)		
P_{it}		0.133**	
		(2.21)	
$ln\ A_{i,t-1}$			0.019**
			(2.35)
$E_{i,t-1}/A_{i,t-1}$			−0.182
			(−0.67)
$Industry_{i,t-1}$			0.006*
			(1.72)
Adjusted R-squares	0.94	0.94	0.92

This table presents the GMM regression results of a simultaneous equation system model for investment, dividend, and debt financing:

$$Inv_{it} = \alpha_{1i} + \alpha_{2i}Div_{it} + \alpha_{3i}Leverage_{it} + \alpha_{4i}Inv_{i,t-1} + \alpha_{5i}Q_{it} + \varepsilon_{it},$$

$$Div_{it} = \beta_{1i} + \beta_{2i}Inv_{it} + \beta_{3i}Leverage_{it} + \beta_{4i}Div_{i,t-1} + \beta_{5i}P_{it} + \eta_{it},$$

$$Leverage_{it} = \gamma_{1i} + \gamma_{2i}Inv_{it} + \gamma_{3i}Div_{it} + \gamma_{4i}Leverage_{i,t-1} + \gamma_{5i}\ln A_{i,t-1} + \gamma_{6i}(E_{i,t-1}/A_{i,t-1}) + \gamma_{7i}Industry_{i,t-1} + \xi_{it}.$$

Regressions are based on annual data of General Electric (GE) from 1966 to 2012. The three endogenous variables are Inv_{it}, Div_{it}, and $Leverage_{it}$, which are net plant and equipment, dividends, and book leverage ratio, respectively. The other variables are the same as in Table 31.1. Numbers in the parentheses are t-statistics. The sign in bracket is the expected sign of each variable of regressions. ***, **, and * denote statistical significance at the 1 %, 5 %, and 10 % levels, respectively.

Appendix 31.1: Application of GMM Estimation in the Linear Regression Model

31.1.1 Consider the Following Linear Regression Model

$$y_t = \mathbf{x}_t\boldsymbol{\beta} + \varepsilon_t, \quad t = 1, \ldots, T \tag{31.39}$$

where y is the endogenous variable, $\mathbf{x}_t$ is a $1 \times K$ regressor vector and includes constant term, and ε_t is the error term. Here, $\boldsymbol{\beta}$ denotes a $K \times 1$ parameter vector of interest. The critical assumption made for the OLS estimation is that the disturbance ε_t is uncorrelated with the regressors $\mathbf{x}_t$, $E(\mathbf{x}_t{}'\varepsilon_t) = \mathbf{0}$. The T observations in model Eq. (31.39) can be written in matrix form as

$$\mathbf{Y} = \mathbf{X}\boldsymbol{\beta} + \boldsymbol{\varepsilon} \tag{31.40}$$

here $\mathbf{Y}$ denotes the $T \times 1$ data vector for the endogenous variable and $\mathbf{X}$ is a $T \times K$ data matrix for all regressors. In this matrix notation, we have the OLS estimator for $\boldsymbol{\beta}$ as follows:

$$\hat{\boldsymbol{\beta}}_{\mathrm{OLS}} = (\mathbf{X}'\mathbf{X})^{-1}\mathbf{X}'\mathbf{Y} \tag{31.41}$$

If the disturbance term is correlated with at least some components of regressors, we say that the regressors are endogenous. Whenever there is endogeneity, the application of ordinary least squares (OLS) estimation to Eq. (31.40) yields biased and inconsistent estimates. The instrumental variable (IV) methods are commonly used to deal with this endogeneity problem. In a typical IV application, the researcher first chooses a set of variables as instruments that are exogenous and applies two stage least squares (2SLS) method to estimate the parameter $\boldsymbol{\beta}$. A good instrument should be highly correlated with the endogenous regressors while uncorrelated with the disturbance in the structural equation. The IV estimator for $\boldsymbol{\beta}$ can be regarded as the solution to following moment conditions of the form

$$E[\mathbf{z}_t{}'\varepsilon_t] = E[\mathbf{z}_t{}'(y_t - \mathbf{x}_t{}'\boldsymbol{\beta})] = \mathbf{0} \tag{31.42}$$

where $\mathbf{z}_t$ is a $1 \times L$ vector of instrumental variables which are uncorrelated with disturbance but correlated with $\mathbf{x}_t$, and the sample moment conditions are

$$\frac{1}{T}\sum_{t=1}^{T}\mathbf{z}_t{}'\left(y_t - \mathbf{x}_t\hat{\boldsymbol{\beta}}\right) = \mathbf{0} \tag{31.43}$$

Assume $\mathbf{Z}$ denotes a $T \times L$ instruments matrix. If the system is just identified ($L = K$) and $\mathbf{Z}'\ \mathbf{X}$ is invertible, the system of sample moment conditions in Eq. (31.43) has a unique solution. We have an IV estimator $\hat{\boldsymbol{\beta}}_{IV}$ as follows:

$$\hat{\boldsymbol{\beta}}_{\mathrm{IV}} = (\mathbf{Z}'\mathbf{X})^{-1}\mathbf{Z}'\mathbf{Y} \tag{31.44}$$

Suppose that the number of instruments exceeds the number of explanatory variables ($L > K$), the system in Eq. (31.43) is over-identified. Then there the question arises that how to select or combine more than enough moment conditions to get K equations. Here, the two-stage least squares (2SLS) estimator, which is the most efficient IV estimator out of all possible linear combinations of the valid instruments under homoscedasticity, is employed in this case. The first stage of 2SLS estimator is regressing each endogenous regressor on all instruments to get its OLS prediction, expressed in matrix notation as $\hat{\mathbf{X}} = \mathbf{Z}(\mathbf{Z}'\mathbf{Z})^{-1}\mathbf{Z}'\mathbf{X}$. The second stage is regressing dependent variable on $\hat{\mathbf{X}}$ to obtain the 2SLS estimator for $\boldsymbol{\beta}$, $\hat{\boldsymbol{\beta}}_{2SLS} = \left(\hat{\mathbf{X}}'\hat{\mathbf{X}}\right)^{-1}\hat{\mathbf{X}}'\mathbf{Y}$. Substitute $\mathbf{Z}(\mathbf{Z}'\mathbf{Z})^{-1}\mathbf{Z}'\mathbf{X}$ for $\hat{\mathbf{X}}$, the 2SLS estimator $\hat{\boldsymbol{\beta}}_{2SLS}$ can be written as

$$\hat{\boldsymbol{\beta}}_{2SLS} = \left[(\mathbf{X}'\mathbf{Z})(\mathbf{Z}'\mathbf{Z})^{-1}\mathbf{Z}'\mathbf{X}\right]^{-1}(\mathbf{X}'\mathbf{Z})(\mathbf{Z}'\mathbf{Z})^{-1}\mathbf{Z}'\mathbf{Y} \tag{31.45}$$

Hansen (1982)'s (GMM) estimation provides an alternative approach for parameter estimation in this over-identified model. The idea behind GMM estimation is to choose parameter estimate so as to make the sample moment conditions in Eq. (31.43) as close as possible to the population moment of zero. The GMM estimator is constructed based on the moment conditions Eq. (31.43) minimizes in the following quadratic function:

$$\left[\sum_{t=1}^{T}\mathbf{z}_t'(y_t - \mathbf{x}_t\boldsymbol{\beta})\right]'\mathbf{W}_T^{-1}\left[\sum_{t=1}^{T}\mathbf{z}_t'(y_t - \mathbf{x}_t\boldsymbol{\beta})\right] \tag{31.46}$$

for some $L \times L$ positive definite weighting matrix $\mathbf{W}_T^{-1}$. If the system is just identified and $\mathbf{Z}'\ \mathbf{X}$ is invertible, we can solve for the parameter vector which makes the sample moment conditions of zero in Eq. (31.43). In this case, the weighting matrix is irrelevant. The corresponding GMM estimator is just as the IV estimator $\hat{\boldsymbol{\beta}}_{IV}$ in Eq. (31.44). If the model is over-identified, we cannot set the sample moment conditions in Eq. (31.43) exactly equal to zero. The GMM estimator for $\boldsymbol{\beta}$ can be obtained by minimizing the quadratic function in Eq. (31.46) as follows:

$$\hat{\boldsymbol{\beta}}_{GMM} = \left[(\mathbf{X}'\mathbf{Z})\mathbf{W}_T^{-1}\mathbf{Z}'\mathbf{X}\right]^{-1}(\mathbf{X}'\mathbf{Z})\mathbf{W}_T^{-1}\mathbf{Z}'\mathbf{Y} \tag{31.47}$$

Alternative weighting matrices $\mathbf{W}_T$ are associated with alternative estimators. The question in GMM estimation is which $\mathbf{W}_T$ to use in Eq. (31.46). Hansen (1982) shows that the optimal weighting matrix $\mathbf{W}_T$ for the resulting estimator is

$$\mathbf{W}_T = \mathrm{Var}[\mathbf{z}'\varepsilon] \equiv E\left[\mathbf{z}\mathbf{z}'\varepsilon^2\right] = E_{\mathbf{z}}\left\{\mathbf{z}\mathbf{z}'\left[E\left(\varepsilon^2|\mathbf{z}\right)\right]\right\} \tag{31.48}$$

Under conditional homoscedasticity $E\left(\varepsilon^2|\mathbf{z}\right) = \sigma^2$, the optimal weighting matrix in which case is

$$\mathbf{W}_T = \left(\frac{\mathbf{Z}'\mathbf{Z}}{T}\right)\sigma^2 \tag{31.49}$$

Hence, any scalar in $\mathbf{W}_T$ will be canceled in this case yields

$$\hat{\boldsymbol{\beta}}_{GMM} = \left[(\mathbf{X}'\mathbf{Z})(\mathbf{Z}'\mathbf{Z})^{-1}\mathbf{Z}'\mathbf{X}\right]^{-1}(\mathbf{X}'\mathbf{Z})(\mathbf{Z}'\mathbf{Z})^{-1}\mathbf{Z}'\mathbf{Y} \tag{31.50}$$

Thus, the GMM estimator is simply the 2SLS estimator under conditional homoscedasticity. However, if the conditional variance of ε_t given $\mathbf{z}_t$ depends on $\mathbf{z}_t$, the optimal weighting matrix $\mathbf{W}_T$ should be estimated by

$$\mathbf{W}_T = \frac{1}{T}\sum_{t=1}^{T} \mathbf{z}_t{}'\mathbf{z}_t\hat{\varepsilon}_t^2 = \frac{1}{T}\mathbf{Z}'\mathbf{D}\mathbf{Z} \tag{31.51}$$

where $\hat{\varepsilon}_t$ is sample residuals and $\mathbf{D} = \text{diag}\left(\hat{\varepsilon}_1^2, \ \ldots, \ \hat{\varepsilon}_T^2\right)$. Here, we can apply the two-stage least-squares (2SLS) estimator in Eq. (31.45) to obtain the sample residuals by $\hat{\varepsilon}_t = y_t - \mathbf{x}_t\hat{\boldsymbol{\beta}}_{\text{2SLS}}$, then the GMM estimator $\hat{\boldsymbol{\beta}}_{\text{GMM}}$ is

$$\hat{\boldsymbol{\beta}}_{\mathbf{GMM}} = \left[(\mathbf{X}'\mathbf{Z})(\mathbf{Z}'\mathbf{D}\mathbf{Z})^{-1}\mathbf{Z}'\mathbf{X}\right]^{-1}(\mathbf{X}'\mathbf{Z})(\mathbf{Z}'\mathbf{D}\mathbf{Z})^{-1}\mathbf{Z}'\mathbf{Y} \tag{31.52}$$

Note that the GMM estimator is obtained by two-step procedure under heteroscedasticity. First use the 2SLS estimator as an initial estimator since it is consistent to get the residuals by $\hat{\varepsilon}_t = y_t - \mathbf{x}_t\hat{\boldsymbol{\beta}}_{\text{2SLS}}$. Then substitute $\sum_{t=1}^{T} \mathbf{z}_t{}'\mathbf{z}_t\hat{\varepsilon}_t^2$ into $\mathbf{W}_T$ as the weighting matrix to obtain the GMM estimator. For this reason, the GMM estimator sometimes called a two-stage instrumental variables estimator.

31.1.2 Application of GMM Estimation in the Simultaneous Equations Model

Consider the following linear simultaneous equations model:

$$\begin{aligned} y_{1t} &= \delta_{12}y_{2t} + \delta_{13}y_{3t} + \cdots + \delta_{1J}y_{Jt} + \mathbf{x}_{1t}\boldsymbol{\gamma}_1 + \varepsilon_{1t} \\ y_{2t} &= \delta_{21}y_{1t} + \delta_{23}y_{3t} + \cdots + \delta_{2J}y_{Jt} + \mathbf{x}_{2t}\boldsymbol{\gamma}_2 + \varepsilon_{2t} \\ &\vdots \\ y_{Jt} &= \delta_{J1}y_{1t} + \delta_{J2}y_{2t} + \cdots + \delta_{J(J-1)}y_{(J-1)t} + \mathbf{x}_{Jt}\boldsymbol{\gamma}_J + \varepsilon_{Jt} \end{aligned} \tag{31.53}$$

Here $t = 1, \ 2, \ \ldots, \ T$. Define that $\mathbf{y}_t = \begin{bmatrix} y_{1t} & y_{2t} & \cdots & y_{Jt} \end{bmatrix}'$ is an $J \times 1$ vector for endogenous variables, $\mathbf{x}_t = \begin{bmatrix} \mathbf{x}_{1t} & \mathbf{x}_{2t} & \cdots & \mathbf{x}_{Jt} \end{bmatrix}'$ is a vector for all exogenous variables in this system and includes constant term. $\boldsymbol{\varepsilon}_t = \begin{bmatrix} \varepsilon_{1t} & \varepsilon_{2t} & \cdots & \varepsilon_{Jt} \end{bmatrix}$ is an $J \times 1$ vector for the disturbances. Here, $\boldsymbol{\delta}$ and $\boldsymbol{\gamma}$ are the parameters matrices of interest defined as

$$\boldsymbol{\delta} = \begin{bmatrix} \delta_{12} & \delta_{13} & \cdots & \delta_{1J} \\ \delta_{21} & \delta_{23} & \cdots & \delta_{2J} \\ \vdots & \vdots & \vdots & \vdots \\ \delta_{J1} & \delta_{J2} & \cdots & \delta_{J(J-1)} \end{bmatrix} = \begin{bmatrix} \boldsymbol{\delta}_1 \\ \boldsymbol{\delta}_2 \\ \vdots \\ \boldsymbol{\delta}_J \end{bmatrix} \text{ and } \boldsymbol{\gamma} = \begin{bmatrix} \boldsymbol{\gamma}_1 \\ \boldsymbol{\gamma}_2 \\ \vdots \\ \boldsymbol{\gamma}_J \end{bmatrix} \tag{31.54}$$

There are two approaches to estimate the structural parameters $\boldsymbol{\delta}$ and $\boldsymbol{\gamma}$ of the system, one is the single equation estimation and the other is the system estimation. Firstly, we introduce the single equation estimation as below. We can rewrite the j-th equation in our simultaneous equation model in terms of the full set of T observations:

$$\mathbf{y}_j = \mathbf{Y}_j\boldsymbol{\delta}_j + \mathbf{X}_j\boldsymbol{\gamma}_j + \ \boldsymbol{\varepsilon}_j = \mathbf{Z}_j\boldsymbol{\beta}_j + \ \boldsymbol{\varepsilon}_j, \quad j = 1, \ldots, J \tag{31.55}$$

where $\mathbf{y}_j$ denotes the $T \times 1$ vector of observations for the endogenous variables on left-hand side of j-th equation. $\mathbf{Y}_j$ denotes the $T \times (J-1)$ data matrix for the endogenous variables on right-hand side of this equation. $\mathbf{X}_j$ is a data matrix for all exogenous variables in this equation. Since these jointly determined variables $\mathbf{y}_j$ and $\mathbf{Y}_j$ are determined within the system, they are correlated with the disturbance terms. This correlation usually creates estimation difficulties because the OLS estimator would be biased and inconsistent (e.g., Johnston and DiNardo 1997; Greene 2011).

As discussed above, the application of OLS estimation to Eq. (31.55) yields biased and inconsistent estimates because of the correlation of $\mathbf{Z}_j$ and $\boldsymbol{\varepsilon}_j$. The two-stage least squares (2SLS) approach is the most common method used to deal with this endogeneity problem resulting from the correlation of $\mathbf{Z}_j$ and $\boldsymbol{\varepsilon}_j$. The 2SLS estimation uses all the exogenous variables in this system as instruments to obtain the predictions of $\mathbf{Y}_j$. In the first stage, we regress $\mathbf{Y}_j$ on all exogenous variables in the system to receive the predictions of the endogenous variables on right-hand side of this equation, $\hat{\mathbf{Y}}_j$. In the second stage, we regress $\mathbf{y}_j$ on $\hat{\mathbf{Y}}_j$ and $\mathbf{X}_j$ to obtain the estimator of $\boldsymbol{\beta}_j$ in Eq. (31.55). Thus, the 2SLS estimator for $\boldsymbol{\beta}_j$ in Eq. (31.55) is,

$$\hat{\boldsymbol{\beta}}_{j,2SLS} = \left[\left(\mathbf{Z}_j'\mathbf{X}\right)\left(\mathbf{X}'\mathbf{X}\right)^{-1}\mathbf{X}'\mathbf{Z}_j\right]^{-1}\left(\mathbf{Z}_j'\mathbf{X}\right)\left(\mathbf{X}'\mathbf{X}\right)^{-1}\mathbf{X}'\mathbf{y}_j \tag{31.56}$$

where $\mathbf{X} = [\mathbf{X}_1 \quad \mathbf{X}_2 \quad \cdots \quad \mathbf{X}_J]$ is a matrix for all exogenous variables in this system.

The GMM estimation provides an alternative approach to deal with this simultaneity bias problem. As for the GMM estimator with instruments $\mathbf{X}$, the moment conditions in the Eq. (31.55) is,

$$E_t\left(\mathbf{x}_t'\varepsilon_{jt}\right) = E_t\left[\mathbf{x}_t'\left(y_{jt} - \mathbf{Z}_{jt}\boldsymbol{\beta}_j\right)\right] = \mathbf{0} \tag{31.57}$$

We can apply the two-stage least-squares (2SLS) estimator in Eq. (31.56) with instruments $\mathbf{X}$ to estimate $\boldsymbol{\beta}_j$ and obtain the sample residuals $\hat{\boldsymbol{\varepsilon}}_j = \mathbf{y}_j - \mathbf{Z}_j\hat{\boldsymbol{\beta}}_{j,2SLS}$. Then, compute the weight matrix $\mathbf{W}_j$ for GMM estimator based on those residuals as follows:

$$\mathbf{W}_j = \left[\frac{1}{T^2}\left(\sum_{t=1}^{T}\mathbf{x}_t'\hat{\varepsilon}_{jt}\hat{\varepsilon}_{jt}'\mathbf{x}_t\right)\right] \tag{31.58}$$

The GMM estimator based on the moment conditions Eq. (31.57) minimizes the following quadratic function:

$$\left[\sum_{t=1}^{T}\mathbf{x}_t'\left(y_{jt} - \mathbf{Z}_{jt}\boldsymbol{\beta}_j\right)\right]\mathbf{W}_j^{-1}\left[\sum_{t=1}^{T}\mathbf{x}_t'\left(y_{jt} - \mathbf{Z}_{jt}\boldsymbol{\beta}_j\right)\right] \tag{31.59}$$

The GMM estimator that minimizes this quadratic function Eq. (31.59) is obtained as

$$\hat{\boldsymbol{\beta}}_{GMM} = \left[\left(\mathbf{Z}_j'\mathbf{X}\right)\hat{\mathbf{W}}_j^{-1}\left(\mathbf{X}'\mathbf{Z}_j\right)\right]^{-1}\left[\left(\mathbf{Z}_i'\mathbf{X}\right)\hat{\mathbf{W}}_j^{-1}\left(\mathbf{X}'\mathbf{y}_j\right)\right] \tag{31.60}$$

In the homoscedastic and serially independent case, a good estimate of the weight matrix $\hat{\mathbf{W}}_j$ would merely be

$$\hat{\mathbf{W}} = \left[\frac{\hat{\sigma}^2}{T}\left(\mathbf{X}'\mathbf{X}\right)\right] \tag{31.61}$$

Given the estimate of $\hat{\sigma}^2$ is obtained, then rearrange terms in Eq. (31.60), which yields

$$\hat{\boldsymbol{\beta}}_{GMM} = \left[\left(\mathbf{Z}_j'\mathbf{X}\right)\left(\mathbf{X}'\mathbf{X}\right)^{-1}\mathbf{X}'\mathbf{Z}_j\right]^{-1}\left(\mathbf{Z}_j'\mathbf{X}\right)\left(\mathbf{X}'\mathbf{X}\right)^{-1}\left(\mathbf{X}'\mathbf{y}_j\right) \tag{31.62}$$

Thus, the two-stage least-squares (2SLS) estimator is a special case of GMM estimator.

As Chen and Lee (2010) pointed out, the 2SLS estimation is a limited information method. The three-stage least squares (3SLS) estimation is a full information method. The 3SLS estimation takes into account the information from full system of equations. Thus, it is more efficient than the 2SLS estimation. The 3SLS method estimates all structural parameters of this system jointly. This allows the possibility of contemporaneous correlation between the disturbances in different structural equation. We introduce the 3SLS estimation as below. Rewrite our full system of equations in Eq. (31.56) as

$$\mathbf{Y} = \mathbf{Z}\boldsymbol{\beta} + \boldsymbol{\varepsilon} \tag{31.63}$$

where $\mathbf{Y}$ is a vector defined as $[\mathbf{y}_1 \quad \mathbf{y}_2 \quad \cdots \quad \mathbf{y}_J]'$. $\mathbf{Z} = \text{diag}[\mathbf{Z}_1 \quad \mathbf{Z}_2 \quad \cdots \quad \mathbf{Z}_J]$ is a block diagonal data matrix for all variables on right-hand side of this system with the form $\mathbf{Z}_j = [\mathbf{Y}_j \quad \mathbf{X}_j]$ as defined in Eq. (31.56). $\boldsymbol{\beta}$ is a vector of interest parameters defined as $[\boldsymbol{\beta}_1 \quad \boldsymbol{\beta}_2 \quad \cdots \quad \boldsymbol{\beta}_J]'$. $\boldsymbol{\varepsilon}$ is a vector of disturbances defined as $[\boldsymbol{\varepsilon}_1 \quad \boldsymbol{\varepsilon}_2 \quad \cdots \quad \boldsymbol{\varepsilon}_J]'$ with $\text{E}(\boldsymbol{\varepsilon}) = \mathbf{0}$ and $\text{E}(\boldsymbol{\varepsilon}\boldsymbol{\varepsilon}') = \boldsymbol{\Sigma} \otimes \mathbf{I}_T$ where $\otimes$ signifies the Kroneker product. Here, $\boldsymbol{\Sigma}$ is defined as

$$\boldsymbol{\Sigma} = \begin{bmatrix} \sigma_{11} & \sigma_{12} & \cdots & \sigma_{1J} \\ \sigma_{21} & \sigma_{22} & \cdots & \sigma_{2J} \\ \vdots & \vdots & \ddots & \vdots \\ \sigma_{J1} & \sigma_{J2} & \cdots & \sigma_{JJ} \end{bmatrix} \tag{31.64}$$

The three-stage least squares (3SLS) approach is the most common method used to estimate the structural parameters of this system simultaneously. Basically, the 3SLS estimator is a generalized least square (GLS) estimator in the entire system taking account of the covariance matrix in Eq. (31.64). The 3SLS estimator is equivalent to using the all exogenous variables as instruments and estimating the entire system using GLS estimation (Intriligator et al. 1996). The 3SLS estimation uses all exogenous variables $\mathbf{X} = [\mathbf{X}_1 \quad \mathbf{X}_2 \quad \cdots \quad \mathbf{X}_J]$ as instruments in each equation of this system, pre-multiplying the model Eq. (31.63) by $\mathbf{X}_I' = \text{diag}[\mathbf{X}' \quad \cdots \quad \mathbf{X}'] = \mathbf{X} \otimes \mathbf{I}_J$ yields the model

$$\mathbf{X}_I'\mathbf{Y} = \mathbf{X}_I'\mathbf{Z}\boldsymbol{\beta} + \mathbf{X}_I'\mathbf{u} \tag{31.65}$$

The covariance matrix from Eq. (31.64) is

$$Cov\left(\mathbf{X}_I'\boldsymbol{\varepsilon}\right) = \mathbf{X}_I'Cov(\boldsymbol{\varepsilon})\mathbf{X}_I = \mathbf{X}_I'(\boldsymbol{\Sigma} \otimes \mathbf{I}_T)\mathbf{X}_I \tag{31.66}$$

The GLS estimator of the Eq. (31.65) is the 3SLS estimator. Thus, the 3SLS estimator is given as follows:

$$\hat{\boldsymbol{\beta}}_{3SLS} = \left\{\mathbf{Z}'\mathbf{X}_I\left[\mathbf{X}_I'(\boldsymbol{\Sigma} \otimes \mathbf{I}_T)\mathbf{X}_I\right]^{-1}\mathbf{X}_I'\mathbf{Z}\right\}^{-1}\mathbf{Z}'\mathbf{X}_I\left[\mathbf{X}_I'(\boldsymbol{\Sigma} \otimes \mathbf{I}_T)\mathbf{X}_I\right]^{-1}\mathbf{X}_I'\mathbf{Y} \tag{31.67}$$

In the case, $\boldsymbol{\Sigma}$ is a diagonal matrix, the 3SLS estimator is equivalent to the 2SLS estimator. As discussed above, the GMM estimator with all exogenous variables $\mathbf{X} = [\mathbf{X}_1 \quad \mathbf{X}_2 \quad \cdots \quad \mathbf{X}_J]$ as instruments, the moment conditions of this system Eq. (31.63) is,

$$\begin{aligned} E(\mathbf{X}_I'\boldsymbol{\varepsilon}) &= E[\mathbf{X}_I'(\mathbf{Y} - \mathbf{Z}\boldsymbol{\beta})] \\ &= \left(E[\mathbf{X}_I'(\mathbf{y}_1 - \mathbf{Z}_1\boldsymbol{\beta}_1)] \quad E[\mathbf{X}_I'(\mathbf{y}_2 - \mathbf{Z}_2\boldsymbol{\beta}_2)] \quad \cdots \quad E[\mathbf{X}_I'(\mathbf{y}_J - \mathbf{Z}_J\boldsymbol{\beta}_J)]\right)' = \mathbf{0} \end{aligned} \tag{31.68}$$

We can apply the 2SLS estimator with instruments $\mathbf{X}$ to estimate $\boldsymbol{\beta}_j$ and obtain the sample residuals $\hat{\varepsilon}_j = y_j\text{-}Z_j\hat{\beta}_{j,2SLS}$. Then, compute the weighting matrix $\widehat{W}_{jl}$ for GMM estimator based on those residuals as follows:

$$\widehat{W}_{jl} = \left[\frac{1}{T}\left(\sum_{t=1}^{T} x_{jt}'\hat{\varepsilon}_{jt}\hat{\varepsilon}_{lt}'x_{jt}\right)\right] \tag{31.69}$$

The system GMM estimator based on the moment conditions Eq. (31.68) minimizes the quadratic function:

$$\begin{bmatrix} X'(y_1\text{-}Z_1\beta_1) \\ X'(y_2\text{-}Z_2\beta_2) \\ \vdots \\ X'(y_J - Z_J\beta_J) \end{bmatrix}' \begin{bmatrix} \widehat{W}_{11} & \widehat{W}_{12} & \cdots & \widehat{W}_{1J} \\ \widehat{W}_{21} & \widehat{W}_{22} & \cdots & \widehat{W}_{2J} \\ \vdots & \vdots & \ddots & \vdots \\ \widehat{W}_{J1} & \widehat{W}_{J2} & \cdots & \widehat{W}_{JJ} \end{bmatrix}^{-1} \begin{bmatrix} X'(y_1\text{-}Z_1\beta_1) \\ X'(y_2\text{-}Z_2\beta_2) \\ \vdots \\ X'(y_J - Z_J\beta_J) \end{bmatrix} \tag{31.70}$$

The GMM estimator that minimizes this quadratic function Eq. (31.70) is obtained as

$$\begin{bmatrix} \hat{\beta}_{1,GMM} \\ \hat{\beta}_{2,GMM} \\ \vdots \\ \hat{\beta}_{J,GMM} \end{bmatrix} = \begin{bmatrix} Z_1'X\widehat{\mathbf{W}}_{11}{-}1X'Z_1 & \cdots & Z_1'X\widehat{\mathbf{W}}_{1J}{-}1X'Z_J \\ Z_2'X\widehat{\mathbf{W}}_{21}{-}1X'Z_1 & \cdots & Z_2'X\widehat{\mathbf{W}}_{2J}{-}1X'Z_J \\ \vdots & \ddots & \vdots \\ Z_J'X\widehat{\mathbf{W}}_{J1}{-}1X'Z_1 & \cdots & Z_J'X\widehat{\mathbf{W}}_{JJ}{-}1X'Z_J \end{bmatrix}^{-1} \begin{bmatrix} \sum_{l=1}^{J} Z_1'X\widehat{\mathbf{W}}_{1J}{-}1y_1 \\ \sum_{l=1}^{J} Z_2'X\widehat{\mathbf{W}}_{2J}{-}1y_1 \\ \vdots \\ \sum_{l=1}^{J} Z_J'X\widehat{\mathbf{W}}_{JJ}{-}1y_1 \end{bmatrix} \cdots \tag{31.71}$$

Here, $\widehat{W}_{ij}{-}1$ is the *ij*-th block in the inverse of center matrix in Eq. (31.70). The 2SLS and 3SLS estimators are the special cases of system GMM estimator. If $\widehat{W}_{jl} = \left[\frac{\hat{\sigma}_{jj}}{T}\left(\sum_{t=1}^{T} x_t'x_t\right)\right]$ and $\widehat{W}_{jl} = 0$ for $j \neq l$, then the system GMM estimator is equivalent to the 2SLS estimator. In the case that $\widehat{W}_{jl} = \left[\frac{\hat{\sigma}_{jl}}{T}\left(\sum_{t=1}^{T} x_t'x_t\right)\right]$, the system GMM estimator is equivalent to the 3SLS estimator.

Appendix 31.2: Data for GE

ff48	industry	gvkey	datadate	pstar	div	inv	q	debtrati	deltainv	deltadiv	deltadeb	divlag1	invlag	debtrati	etlag	lnalag	fyear	invlag_1	divlag_1	debtrati	deltainv_	deltadiv_	deltadeb	period
	0.52097	5047	2E+07	6.31905	2.58913	13.8999	82.9562	0.53495	2.47652	0.20234	0.03455	2.38679	11.4234	0.50041	0.15008	8.36646	1966							1960
	0.52401	5047	2E+07	7.08571	2.58551	16.5051	85.2219	0.55487	2.60525	-0.00362	0.01991	2.58913	13.8999	0.53495	0.13027	8.48708	1967	13.8999	2.58913	0.53495	2.60525	-0.00362	0.01991	1960
	0.47032	5047	2E+07	7.25462	2.59148	18.5177	92.8585	0.5589	2.01262	0.00597	0.00403	2.58551	16.5051	0.55487	0.11986	8.58433	1968	16.5051	2.58551	0.55487	2.01262	0.00597	0.00403	1960
	0.44329	5047	2E+07	6.96077	2.60193	20.0757	94.6804	0.58113	1.55794	0.01045	0.02224	2.59148	18.5177	0.5589	0.11279	8.65587	1969	18.5177	2.59148	0.5589	1.55794	0.01045	0.02224	1960
	0.46414	5047	2E+07	7.30243	2.59186	19.2632	95.6263	0.58136	-0.81243	-0.01007	0.00022	2.60193	20.0757	0.58113	0.08256	8.6817	1970	20.0757	2.60193	0.58113	-0.81243	-0.01007	0.00022	1970
	0.44949	5047	2E+07	4.09284	1.37105	11.1227	43.4756	0.58707	-8.14054	-1.22081	0.00571	2.59186	19.2632	0.58136	0.08856	8.73206	1971	19.2632	2.59186	0.58136	-8.14054	-1.22081	0.00571	1970
	0.4432	5047	2E+07	4.63016	1.39733	11.7172	56.9505	0.5774	0.59445	0.02628	-0.00967	1.37105	11.1227	0.58707	0.107	8.83751	1972	11.1227	1.37105	0.58707	0.59445	0.02628	-0.00967	1970
	0.41211	5047	2E+07	5.04593	1.49824	12.9593	64.8072	0.58885	1.24216	0.10091	0.01145	1.39733	11.7172	0.5774	0.11007	8.90948	1973	11.7172	1.39733	0.5774	1.24216	0.10091	0.01145	1970
	0.50253	5047	2E+07	5.38897	1.5943	14.3202	74.8882	0.59703	1.36089	0.09606	0.00818	1.49824	12.9593	0.58885	0.1147	9.02692	1974	12.9593	1.49824	0.58885	1.36089	0.09606	0.00818	1970
	0.57188	5047	2E+07	5.42095	1.58983	13.899	71.7941	0.57466	-0.42124	-0.00446	-0.02237	1.5943	14.3202	0.59703	0.1062	9.14517	1975	14.3202	1.5943	0.59703	-0.42124	-0.00446	-0.02237	1970
	0.54375	5047	2E+07	6.25185	1.59209	14.8106	68.1843	0.55419	0.91166	0.00225	-0.02047	1.58983	13.899	0.57466	0.09433	9.18641	1976	13.899	1.58983	0.57466	0.91166	0.00225	-0.02047	1970
	0.5591	5047	2E+07	7.05774	2.09019	15.7086	77.8069	0.55652	0.898	0.49811	0.00233	1.59209	14.8106	0.55419	0.1268	9.3968	1977	14.8106	1.59209	0.55419	0.898	0.49811	0.00233	1970
	0.54726	5047	2E+07	7.92024	2.49873	17.6402	87.9433	0.55191	1.93157	0.40854	-0.00461	2.09019	15.7086	0.55652	0.12397	9.52492	1978	15.7086	2.09019	0.55652	1.93157	0.40854	-0.00461	1970
	0.53897	5047	2E+07	8.92253	2.73702	20.2458	99.285	0.54856	2.60563	0.23829	-0.00335	2.49873	17.6402	0.55191	0.13021	9.6182	1979	17.6402	2.49873	0.55191	2.60563	0.23829	-0.00335	1970
	0.64517	5047	2E+07	9.75128	2.94163	25.377	110.385	0.5487	5.13121	0.20461	0.00014	2.73702	20.2458	0.54856	0.12796	9.71984	1980	20.2458	2.73702	0.54856	5.13121	0.20461	0.00014	1980
	0.59083	5047	2E+07	11.1257	3.13926	30.049	120.117	0.5562	4.67201	0.19763	0.0075	2.94163	25.377	0.5487	0.12117	9.82612	1981	25.377	2.94163	0.5487	4.67201	0.19763	0.0075	1980
	0.5352	5047	2E+07	12.2942	3.33579	32.0763	114.413	0.52056	2.02722	0.19654	-0.03564	3.13926	30.049	0.5562	0.11685	9.94951	1982	30.049	3.13926	0.5562	2.02723	0.19654	-0.03564	1980
	0.60659	5047	2E+07	6.83631	1.87405	16.9302	52.5937	0.50885	-15.1461	-1.25496	-0.01172	3.33579	32.0763	0.52056	0.11127	9.98114	1983	32.0763	3.33579	0.52056	-15.1461	-1.46174	-0.01172	1980
	0.50335	5047	2E+07	7.4306	2.04451	16.9057	62.5606	0.48637	-0.02451	0.17047	-0.02247	1.87405	16.9302	0.50885	0.10946	10.0557	1984	16.9302	1.87405	0.50885	-0.02451	0.17047	-0.02247	1980
	0.46848	5047	2E+07	7.81182	2.23696	18.2641	62.6242	0.46913	1.35841	0.19245	-0.01724	2.04451	16.9057	0.48637	0.11504	10.1158	1985	16.9057	2.04451	0.48637	1.35841	0.19245	-0.01724	1980
	0.92819	5047	2E+07	8.66866	2.37116	21.5861	79.895	0.55775	3.32199	0.1342	0.08862	2.23696	18.2641	0.46913	0.11104	10.1823	1986	18.2641	2.23696	0.46913	3.32199	0.1342	0.08862	1980
	1.53518	5047	2E+07	4.05669	1.33894	10.2497	39.1582	0.57169	-11.3364	-1.03222	0.01394	2.37116	21.5861	0.55775	0.08849	10.4514	1987	21.5861	2.37116	0.55775	-11.3364	-1.03222	0.01394	1980
	1.59229	5047	2E+07	6.26527	1.45658	15.0879	55.343	0.82459	4.83816	0.11764	0.2169	1.33894	10.2497	0.57169	0.09453	10.5693	1988	10.2497	1.33894	0.57169	4.83816	0.11764	0.2529	1980
	1.27219	5047	2E+07	6.84696	1.69875	17.2926	59.8972	0.82943	2.20472	0.24218	0.00485	1.45658	15.0879	0.82459	0.08049	11.2891	1989	15.0879	1.45658	0.82459	2.20472	0.24218	0.00485	1980
	1.52682	5047	2E+07	7.80076	1.94246	19.0478	67.3698	0.85125	1.7552	0.24371	0.02181	1.69875	17.2926	0.82943	0.09107	11.2891	1990	17.2926	1.69875	0.82943	1.7552	0.24371	0.02181	1990
	1.09599	5047	2E+07	8.40969	2.0923	21.9599	68.8561	0.86389	2.91208	0.14984	0.01264	1.94246	19.0478	0.85125	0.08363	11.2891	1991	19.0478	1.94246	0.85125	2.91208	0.14984	0.01264	1990
	0.76459	5047	2E+07	8.32682	2.32047	23.8325	66.506	0.8714	1.87262	0.22818	0.00752	2.0923	21.9599	0.86389	0.07759	11.2891	1992	21.9599	2.0923	0.86389	1.87263	0.22818	0.00752	1990
	0.57601	5047	2E+07	9.00251	2.61114	24.8673	74.8415	0.89074	1.03482	0.29066	0.01933	2.32047	23.8325	0.8714	0.06422	11.2891	1993	23.8325	2.32047	0.8714	1.03482	0.29066	0.01933	1990
	0.5615	5047	2E+07	5.34711	1.49241	13.7547	25.321	0.85483	-11.1126	-1.11873	-0.03591	2.61114	24.8673	0.89074	0.05161	11.2891	1994	24.8673	2.61114	0.89074	-11.1126	-1.11873	-0.03591	1990
	0.75684	5047	2E+07	6.10077	1.70296	15.4088	42.0646	0.85719	1.65417	0.21055	0.00237	1.49241	13.7547	0.85483	0.06678	11.2891	1995	13.7547	1.49241	0.85483	1.65417	0.21055	0.00237	1990
	1.28677	5047	2E+07	6.72832	1.90813	17.5094	47.2622	0.8747	2.1006	0.20517	0.01751	1.70296	15.4088	0.85719	0.07225	11.2891	1996	15.4088	1.70296	0.85719	2.1006	0.20517	0.01751	1990
	0.48724	5047	2E+07	3.7631	1.08283	9.89894	25.7359	0.87461	-7.61049	-0.8253	-9E-05	1.90813	17.5094	0.8747	0.06733	11.2891	1997	17.5094	1.90813	0.8747	-7.61049	-0.8253	-9E-05	1990
	0.49522	5047	2E+07	4.63303	1.24752	10.9223	31.0938	0.87876	1.02334	0.16469	0.00415	1.08283	9.89894	0.87461	0.06521	11.2891	1998	9.89894	1.08283	0.87461	1.02334	0.16469	0.00415	1990
	3.75715	5047	2E+07	5.29949	1.457	12.4883	33.1761	0.88211	1.56599	0.20948	0.00335	1.24752	10.9223	0.87876	0.06419	11.2891	1999	10.9223	1.24752	0.87876	1.56599	0.20948	0.00335	1990
	3.75715	5047	2E+07	2.06111	0.56857	4.02889	10.6497	0.87316	-8.45937	-0.88843	-0.00894	1.457	12.4883	0.88211	0.06209	11.2891	2000	12.4883	1.457	0.88211	-8.45937	-0.88843	-0.00894	2000
	3.75715	5047	2E+07	2.13758	0.66039	4.24544	12.7715	0.87871	0.21655	0.09183	0.00555	0.56857	4.02889	0.87316	0.06775	11.2891	2001	4.02889	0.56857	0.87316	0.21655	0.09183	0.00555	2000
	3.75715	5047	2E+07	2.11948	0.72879	4.73465	13.2995	0.87974	0.48921	0.0684	0.00103	0.66039	4.24544	0.87871	0.06237	11.2891	2002	4.24544	0.66039	0.87871	0.48921	0.0684	0.00103	2000
	2.72696	5047	2E+07	2.24036	0.77103	5.30472	13.0984	0.86878	0.57006	0.04224	-0.01096	0.72879	4.73465	0.87974	0.04941	11.2891	2003	4.73465	0.72879	0.87974	0.57006	0.04224	-0.01096	2000
	1.98364	5047	2E+07	2.3808	0.8118	5.98261	14.8253	0.83051	0.67789	0.04077	-0.03828	0.77103	5.30472	0.86878	0.04637	11.2891	2004	5.30472	0.77103	0.86878	0.67789	0.04077	-0.03827	2000
	0.5845	5047	2E+07	2.72637	0.92014	6.44089	14.4488	0.82563	0.45828	0.10834	-0.00488	0.8118	5.98261	0.83051	0.04314	11.2891	2005	5.98261	0.8118	0.83051	0.45828	0.10834	-0.00488	2000
	0.60619	5047	2E+07	3.07598	1.03869	7.29428	16.1624	0.82805	0.85339	0.11855	0.00241	0.92014	6.44089	0.82563	0.05519	11.2891	2006	6.44089	0.92014	0.82563	0.85339	0.11855	0.00241	2000
	0.65427	5047	2E+07	3.48692	1.17275	7.79917	17.0267	0.84464	0.5049	0.13406	0.01659	1.03869	7.29428	0.82805	0.06063	11.2891	2007	7.29428	1.03869	0.82805	0.5049	0.13406	0.01659	2000
	0.7272	5047	2E+07	2.9987	1.20045	7.45286	16.5852	0.85759	-0.34631	0.02769	0.01295	1.17275	7.79917	0.84464	0.06335	11.2891	2008	7.79917	1.17275	0.84464	-0.34631	0.02769	0.01295	2000
	0.66106	5047	2E+07	2.21831	0.63631	6.49081	14.3051	0.83994	-0.96205	-0.56414	-0.01765	1.20045	7.45286	0.85759	0.05817	11.2891	2009	7.45286	1.20045	0.85759	-0.96205	-0.56414	-0.01765	2000
	0.55963	5047	2E+07	2.26888	0.49099	6.23756	13.9893	0.83467	-0.25325	-0.14532	-0.00527	0.63631	6.49081	0.83994	0.03595	11.2891	2010	6.49081	0.63631	0.83994	-0.25325	-0.14532	-0.00527	2000
	0.47726	5047	2E+07	2.26615	0.70916	6.21762	13.7007	0.83529	-0.01994	0.21818	0.00062	0.49099	6.23756	0.83467	0.03866	11.2891	2011	6.23756	0.49099	0.83467	-0.01994	0.21818	0.00062	2000
	0.57959	5047	2E+07	2.46405	0.70846	6.70243	14.0643	0.81254	0.48481	-0.0007	-0.02275	0.70916	6.21762	0.83529	0.04028	11.2891	2012	6.21762	0.70916	0.83529	0.48481	-0.0007	-0.02275	2000

Appendix 31.3: SAS Program

```
/*=========================================
 * 2SLS
 *=========================================*/
PROC SYSLIN DATA=GE 2SLS;      /*Analyze data "GE" and employ 2SLS*/
ENDOGENOUS inv div leverage;                /*Endogenous variables*/
INSTRUMENTS invlag_1 q divlag_1 p leveragelag_1 lnalag etlag industry;                    /*Instrumental
variables*/
MODEL inv = div leverage invlag_1 q;/*Simultaneous equations models*/
MODEL div = inv leverage divlag_1 p;
MODEL leverage = inv div leveragelag_1 lnalag etlag industry;
RUN;

/*=========================================
 * 3SLS
 *=========================================*/
PROC SYSLIN DATA=GE 3SLS;      /*Analyze data "GE" and employ 3SLS*/
ENDOGENOUS inv div leverage;                /*Endogenous variables*/
INSTRUMENTS invlag_1 q divlag_1 p leveragelag_1 lnalag etlag industry;                    /*Instrumental
variables*/
MODEL inv = div leverage invlag_1 q;/*Simultaneous equations models*/
MODEL div = inv leverage divlag_1 p;
MODEL leverage = inv div leveragelag_1 lnalag etlag industry;
RUN;

/*=========================================
 * GMM
 *=========================================*/
PROC MODEL DATA=GE;                         /*Analyze data "GE"*/
ENDOGENOUS inv div leverage;                /*Endogenous variables*/
INSTRUMENTS invlag_1 q divlag_1 p leveragelag_1 lnalag etlag industry;                    /*Instrumental
variables*/
inv = alpha_1+ alpha_2*div + alpha_3*leverage + alpha_4*invlag_1 + alpha_5*q;
/*Simultaneous equations models*/
div = beta_1+beta_2*inv+beta_3*leverage+beta_4*divlag_1+beta_5*p;
leverage = gamma_1+gamma_2*inv+gamma_3*div
+gamma_4*leveragelag_1+gamma_5*lnalag+gamma_6*etlag+gamma_7*industry;
FIT inv div leverage /GMM;                  /*Employ GMM*/
RUN;
```

Bibliography

Aggarwal R, Jacques KT (2001) The impact of FDICIA and prompt corrective action on bank capital and risk: estimates using a simultaneous equations model. J Bank Financ 25:1139–1160

Aggarwal R, Kyaw NA (2010) Capital structure, dividend policy, and multinationality: theory versus empirical evidence. Int Rev Financ Anal 19:140–150

Agrawal A, Knoeber CR (1996) Firm performance and mechanisms to control agency problems between managers and shareholders. J Financ Quant Anal 31:377–397

Aivazian VA, Booth L, Cleary S (2006) Dividend smoothing and debt ratings. J Financ Quant Anal 41:439–453

Antle R, Gordon E, Narayanamoorthy G, Zhou L (2006) The joint determination of audit fees, non-audit fees, and abnormal accruals. Rev Quant Financ Acc 27:235–266

Aunon-Nerin D, Ehling P (2008) Why firms purchase property insurance. J Financ Econ 90:298–312

Babenko I (2009) Share repurchases and pay-performance sensitivity of employee compensation contracts. J Financ 64:117–150
Berger AN, Bonaccorsi di Patti E (2006) Capital structure and firm performance: a new approach to testing agency theory and an application to the banking industry. J Bank Financ 30:1065–1102
Betton S, Eckbo BE (2000) Toeholds, bid jumps, and expected payoffs in takeovers. Rev Financ Stud 13:841–882
Bhagat S, Black BS (2002) The non-correlation between board independence and long-term firm performance. J Corp Law 27:231–273
Billett MT, Xue H (2007) The takeover deterrent effect of open market share repurchases. J Financ 62:1827–1850
Billett MT, King THD, Mauer DC (2007) Growth opportunities and the choice of leverage, debt maturity, and covenants. J Financ 62:697–730
Blalock HM (1969) Theory construction: from verbal to mathematical formations. Prentice-Hall, Englewood Cliffs
Boone AL, Field LC, Karpoff JM, Raheja CG (2007) The determinants of corporate board size and composition: an empirical analysis. J Financ Econ 85:66–101
Bound J, Jaeger D, Baker R (1995) Problems with instrumental variable estimation when the correlation between the instruments and the endogenous explanatory variables is weak. J Am Stat Assoc 90:443–450
Brav A, Graham JR, Harvey CR, Michaely R (2005) Payout policy in the 21st century. J Financ Econ 77:483–527
Chava S, Roberts MR (2008) How does financing impact investment? The role of debt covenants. J Financ 63:2085–2121
Chen CR, Lee CF (2010) Application of simultaneous equation in finance research. In: Lee CF et al (eds) Handbook of quantitative finance and risk management. Springer, New York, pp 1301–1306
Chen CR, Steiner TL, Whyte AM (2006) Does stock option-based executive compensation induce risk-taking? An analysis of the banking industry. J Bank Financ 30:915–945
Chen WP, Chung H, Lee CF, Liao WL (2007) Corporate governance and equity liquidity: analysis of S&P transparency and disclosure rankings. Corp Govern Int Rev 15:644–660
Chen HY, Gupta MC, Lee AC, Lee CF (2013) Sustainable growth rate, optimal growth rate, and optimal payout ratio: a joint optimization approach. J Bank Financ 37:1205–1222
Coles JL, Daniel ND, Naveen L (2006) Managerial incentives and risk-taking. J Financ Econ 79:431–468
Cook DO, Tang T (2010) Macroeconomic conditions and capital structure adjustment speed. J Corp Finance 16:73–87
Core JE, Guay WR (2001) Stock option plans for non-executive employees. J Financ Econ 61:253–287
De La Bruslerie H (2013) Crossing takeover premiums and mix of payment: an empirical test of contractual setting in M&A Transactions. J Bank Financ 37:2106–2123
DeAngelo H, Roll R (2015) How stable are corporate capital structure? J Financ 70:373–418
Demsetz H, Villalonga B (2001) Ownership structure and corporate performance. J Corp Finance 7:209–233
Deng X, Kang JK, Low BS (2013) Corporate social responsibility and stakeholder value maximization: Evidence from mergers. J Financ Econ 110:87–109
Dhrymes PJ, Kurz M (1967) Investment, dividend, and external finance behavior of firms. In: Ferber R (ed) Determinants of investment behavior. NBER, Cambridge, pp 427–486
Edmans A, Goldstein I, Jiang W (2012) The real effects of financial markets: the impact of prices on takeovers. J Financ 67:933–971
Fama EF (1974) The empirical relationships between the dividend and investment decisions of firms. Am Econ Rev 64:304–318
Fama EF, French KR (2001) Disappearing dividends: changing firm characteristics or lower propensity to pay? J Financ Econ 60:3–43
Fama EF, French KR (2002) Testing trade-off and pecking order predictions about dividends and debt. Rev Financ Stud 15:1–33
Ferreira MA, Matos P (2008) The colors of investors' money: the role of institutional investors around the world. J Financ Econ 88:499–533
Fich EM, Shivdasani A (2007) Financial fraud, director reputation, and shareholder wealth. J Financ Econ 86:306–336
Fidrmuc JP, Roosenboom P, Paap R, Teunissen T (2012) One size does not fit all: selling firms to private equity versus strategic acquirers. J Corp Finance 18:828–848
Fisher FM (1966) The identification problem in econometrics. McGraw-Hill, New York
Flannery MJ, Rangan KP (2006) Partial adjustment toward target capital structures. J Financ Econ 79:469–506
Fogler HR, Ganapathy S (1982) Financial econometrics. Prentice-Hall, Englewood Cliffs
Frank MZ, Goyal VK (2009) Capital structure decisions: which factors are reliably important? Financ Manag 38:1–37
Froot KA, Scharfstein DS, Stein JC (1993) Risk management: coordinating corporate investment and financing policies. J Financ 48:1629–1658
Ghosh SK (1991) Econometrics: theory and application. Prentice Hall, Englewood Cliffs
Gong G, Louis H, Sun AX (2008) Earnings management and firm performance following open-market repurchases. J Financ 63:947–986

Grabowski HG, Mueller DC (1972) Managerial and stockholder welfare models of firm expenditures. Rev Econ Stat 54:9–24
Graham JR, Rogers DA (2002) Do firms hedge in response to tax incentives? J Financ 57:815–839
Greene WH (2003) Econometric analysis, Prentice Hall. Upper Saddle River, NJ
Greene WH (2011) Econometric analysis, 7th edn. Prentice Hall, New Jersey
Grullon G, Michaely R (2002) Dividends, share repurchases, and the substitution hypothesis. J Financ 57:1649–1684
Grundy BD, Lim B, Verwijmeren P (2012) Do option markets undo restrictions on short sales? Evidence from the 2008 short-sale ban. J Financ Econ 106:331–348
Gugler K (2003) Corporate governance, dividend payout policy, and the interrelation between dividends, R&D, and capital investment. J Bank Financ 27:1297–1321
Hahn J, Hausman J (2005) Instrumental variable estimation with valid and invalid instruments. Working Paper.
Hansen LP (1982) Large sample properties of generalized method of moments estimators. Econometrica 50:1029–1054
Harford J, Klasa S, Maxwell WF (2014) Refinancing risk and cash holdings. J Financ 69:975–1012
Harris M, Raviv A (1990) Capital structure and the informational role of debt. J Financ 45:321–349
Harvey CR, Lins KV, Roper AH (2004) The effect of capital structure when expected agency costs are extreme. J Financ Econ 74:3–30
Hayashi F (2000) Econometrics. Princeton University, New Jersey
Higgins RC (1972) The corporate dividend-saving decision. J Financ Quant Anal 7:1527–1541
Huang R, Ritter JR (2009) Testing theories of capital structure and estimating the speed of adjustment. J Financ Quant Anal 44:237–271
Intriligator MD, Bodkin RG, Hsiao C (1996) Econometric models, techniques, and applications, 2nd edn. Prentice Hall, New Jersey
Jacques K, Nigro P (1997) Risk-based capital, portfolio risk, and bank capital: A simultaneous equations approach. J Econ Bus 49:533–547
Jensen GR, Solberg DP, Zorn TS (1992) Simultaneous determination of insider ownership, debt, and dividend policies. J Financ Quant Anal 27:247–263
John K, Nachman DC (1985) Risky debt, investment incentives, and reputation in a sequential equilibrium. J Financ 40:863–878
Johnston J, DiNardo J (1997) Econometric methods. McGraw-Hill, New York
Judge GG, Griffiths WE, Hill RC, Lutkepohl H, Lee T (1985) The theory and practice of econometrics. Wiley, New York
Lambrecht BM, Myers SC (2012) A Lintner model of payout and managerial rents. J Financ 67:1761–1810
Leary MT, Roberts MR (2014) Do peer firms affect corporate financial policy? J Financ 69:139–178
Lee CF (1976) A note on the interdependent structure of security returns. J Financ Quant Anal 11:73–86
Lee CF, Lin FL, Chen ML (2010) International hedge ratios for index futures market: a simultaneous equations approach. Rev Pac Basin Financ Mark Policies 13:203–213
Lee CF, Gupta MC, Chen HY, Lee AC (2011) Optimal payout ratio under uncertainty and the flexibility hypothesis: theory and empirical evidence. J Corp Finance 17:483–501
Lee C-F, Lee JC, Lee AC (eds) (2013) Statistics for business and financial economics, 3rd edn. New York, Springer Academic. ISBN 978-1-4614-5896-8
Lee JC (ed) (2014) Handbook of financial econometrics and statistics. Springer Academic, New York
Loderer C, Martin K (1997) Executive stock ownership and performance tracking faint traces. J Financ Econ 45:223–255
Long MS, Malitz IB (1985) Investment patterns and financial leverage. In: Friedman BM (ed) Corporate capital structures in the United States. University of Chicago, Chicago, pp 353–377
MacKay P, Moeller SB (2007) The value of corporate risk management. J Financ 62:1379–1419
MacKay P, Phillips GM (2005) How does industry affect firm financial structure? Rev Financ Stud 18:1433–1466
McCabe GM (1979) The empirical relationship between investment and financing: a new look. J Financ Quant Anal 14:119–135
McDonald JG, Jacquillat B, Nussenbaum M (1975) Dividend, investment and financing decisions: empirical evidence on French firms. J Financ Quant Anal 10:741–755
Morgan IG, Saint-Pierre J (1978) Dividend and investment decisions of Canadian firms. Can J Econ 11:20–37
Myers SC, Majluf NS (1984) Corporate financing and investment decisions when firms have information that investors do not have. J Financ Econ 13:187–221
Officer MS (2003) Termination fees in mergers and acquisitions. J Financ Econ 69:431–467
Pagan AR, Hall D (1983) Diagnostic tests as residual analysis. Econom Rev 2:159–218
Pesaran MH, Taylor LW (1999) Diagnostics for IV regressions. Oxf Bull Econ Stat 61:255–281
Peterson PP, Benesh GA (1983) A reexamination of the empirical relationship between investment and financing decisions. J Financ Quant Anal 18(4):439–453

Prevost AK, Rao RP, Hossain M (2002) Determinants of board composition in New Zealand: a simultaneous equations approach. J Empir Finance 9:373–397

Pruitt SW, Gitman LJ (1991) The interactions between the investment, financing, and dividend decisions of major US firms. Financ Rev 26:409–430

Rajan RG, Zingales L (1995) What do we know about capital structure? Some evidence from international data. J Financ 50:1421–1460

Ramanathan R (1995) Introductory econometrics with application. Dryden, Fort Worth

Ross SA (1977) The determination of financial structure: the incentive-signaling approach. Bell J Econ 8:23–40

Ruland W, Zhou P (2005) Debt, diversification, and valuation. Rev Quant Financ Acc 25:277–291

Sargan JD (1958) The estimation of economic relationships using instrumental variables. Econometrica 26:393–415

Sargan JD (1959) The estimation of relationships with autocorrelated residuals by the use of instrumental variables. J Roy Stat Soc B Stat Meth 21:91–105

Setia-Atmaja L, Tanewski GA, Skully M (2009) The role of dividends, debt and board structure in the governance of family controlled firms. J Bus Financ Acc 36:863–898

Shrieves RE, Dahl D (1992) The relationship between risk and capital in commercial banks. J Bank Financ 16:439–457

Simkowitz MA, Logue DE (1973) The interdependent structure of security returns. J Financ Quant Anal 8:259–272

Smirlock M, Marshall W (1983) An examination of the empirical relationship between the dividend and investment decisions: a note. J Financ 38:1659–1667

Staiger D, Stock JH (1997) Instrumental variables regression with weak instruments. Econometrica 65:557–586

Stock JH, Yogo M (2005) Testing for weak instruments in linear IV regression. In: Andrews DWK (ed) Identification and inference for econometric models. Cambridge University, New York, pp 80–108

Stock JH, Wright JH, Yogo M (2002) A survey of weak instruments and weak identification in generalized method of moments. J Bus Econ Stat 20:518–529

Switzer L (1984) The determinants of industrial R&D: a funds flow simultaneous equation approach. Rev Econ Stat 66:163–168

Velury U, Reisch JT, O'Reilly DM (2003) Institutional ownership and the selection of industry specialist auditors. Rev Quant Financ Acc 21:35–48

Wang CJ (2015) Instrumental variables approach to correct for endogeneity in finance. In: Lee CF, Lee J (eds) Handbook of financial econometrics and statistics. Springer, New York, pp 2577–2600

Woidtke T (2002) Agents watching agents?: evidence from pension fund ownership and firm value. J Financ Econ 63:99–131

Ye P (2012) The value of active investing: Can active institutional investors remove excess comovement of stock returns? J Financ Quant Anal 47:667–688

Yoon PS, Starks LT (1995) Signaling, investment opportunities, and dividend announcements. Rev Financ Stud 8:995–1018

Chapter 32
Hedge Ratios: Theory and Applications

32.1 Introduction

One of the best uses of derivative securities such as futures contracts is in hedging. In the past, both academicians and practitioners have shown great interest in the issue of hedging with futures. This is quite evident from the large number of articles written in this area.

One of the main theoretical issues in hedging involves the determination of the optimal hedge ratio. However, the optimal hedge ratio depends on the particular objective function to be optimized. Many different objective functions are currently being used. For example, one of the most widely used hedging strategies is based on the minimization of the variance of the hedged portfolio (e.g., see Johnson 1960; Ederington 1979; Myers and Thompson 1989). This so-called minimum-variance (MV) hedge ratio is simple to understand and estimate. However, the MV hedge ratio completely ignores the expected return of the hedged portfolio. Therefore, this strategy is in general inconsistent with the mean-variance framework unless the individuals are infinitely risk-averse or the futures price follows a pure martingale process (i.e., expected futures price change is zero).

Other strategies that incorporate both the expected return and risk (variance) of the hedged portfolio have been recently proposed (e.g., see Howard and D'Antonio 1984; Cecchetti et al. 1988; Hsin et al. 1994). These strategies are consistent with the mean-variance framework. However, it can be shown that if the futures price follows a pure martingale process, then the optimal mean-variance hedge ratio will be the same as the MV hedge ratio.

Another aspect of the mean-variance based strategies is that even though they are an improvement over the MV strategy, for them to be consistent with the expected utility maximization principle, either the utility function needs to be quadratic or the returns should be jointly normal. If neither of these assumptions is valid, then the hedge ratio may not be optimal with respect to the expected utility maximization principle. Some researchers have solved this problem by deriving the optimal hedge ratio based on the maximization of the expected utility (e.g., see Cecchetti et al. 1988; Lence 1995, 1996). However, this approach requires the use of specific utility function and specific return distribution.

Attempts have been made to eliminate these specific assumptions regarding the utility function and return distributions. Some of them involve the minimization of the mean extended-Gini (MEG) coefficient, which is consistent with the concept of stochastic dominance (e.g., see Cheung et al. 1990; Kolb and Okunev 1992, 1993; Lien and Luo 1993a; Shalit 1995; Lien and Shaffer 1999). Shalit (1995) shows that if the prices are normally distributed, then the MEG-based hedge ratio will be the same as the MV hedge ratio.

C.-F. Lee et al., *Essentials of Excel, Excel VBA, SAS and Minitab for Statistical and Financial Analyses*, DOI 10.1007/978-3-319-38867-0_32

Recently, hedge ratios based on the generalized semivariance (GSV) or lower partial moments have been proposed (e.g., see De Jong et al. 1997; Lien and Tse 1998, 2000; Chen et al. 2001). These hedge ratios are also consistent with the concept of stochastic dominance. Furthermore, these GSV-based hedge ratios have another attractive feature whereby they measure portfolio risk by the GSV, which is consistent with the risk perceived by managers, because of its emphasis on the returns below the target return (see Crum et al. 1981; Lien and Tse 2000). Lien and Tse (1998) show that if the futures and spot returns are jointly normally distributed and if the futures price follows a pure martingale process, then the minimum-GSV hedge ratio will be equal to the MV hedge ratio. Finally, Hung et al. (2006) have proposed a related hedge ratio that minimizes the value-at-risk associated with the hedged portfolio when choosing hedge ratio. This hedge ratio will also be equal to MV hedge ratio if the futures price follows a pure martingale process.

Most of the studies mentioned above (except Lence (1995, 1996)) ignore transaction costs as well as investments in other securities. Lence (1995, 1996) derives the optimal hedge ratio where transaction costs and investments in other securities are incorporated in the model. Using a CARA utility function, Lence finds that under certain circumstances the optimal hedge ratio is zero; i.e., the optimal hedging strategy is not to hedge at all.

In addition to the use of different objective functions in the derivation of the optimal hedge ratio, previous studies also differ in terms of the dynamic nature of the hedge ratio. For example, some studies assume that the hedge ratio is constant over time. Consequently, these static hedge ratios are estimated using unconditional probability distributions (e.g., see Ederington 1979; Howard and D'Antonio 1984; Benet 1992; Kolb and Okunev 1992, 1993; Ghosh 1993). On the other hand, several studies allow the hedge ratio to change over time. In some cases, these dynamic hedge ratios are estimated using conditional distributions associated with models such as ARCH (Autoregressive conditional heteroscedasticity) and GARCH (Generalized Autoregressive conditional heteroscedasticity) (e.g., see Cecchetti et al. 1988; Baillie and Myers 1991; Kroner and Sultan 1993; Sephton 1993a). The GARCH-based method has recently been extended by Lee and Yoder (2007) where regime-switching model is used. Alternatively, the hedge ratios can be made dynamic by considering a multi-period model where the hedge ratios are allowed to vary for different periods. This is the method used by Lien and Luo (1993b).

When it comes to estimating the hedge ratios, many different techniques are currently being employed, ranging from simple to complex ones. For example, some of them use such a simple method as the ordinary least squares (OLS) technique (e.g., see Ederington 1979; Malliaris and Urrutia 1991; Benet 1992). However, others use more complex methods such as the conditional heteroscedastic (ARCH or GARCH) method (e.g., see Cecchetti et al. 1988; Baillie and Myers 1991; Sephton 1993a), the random coefficient method (e.g., see Grammatikos and Saunders 1983), the cointegration method (e.g., see Ghosh 1993; Lien and Luo 1993b; Chou et al. 1996), or the cointegration-heteroscedastic method (e.g., see Kroner and Sultan 1993). Recently, Lien and Shrestha (2007) have suggested the use of wavelet analysis to match the data frequency with the hedging horizon. Finally, Lien and Shrestha (2010) also suggest the use of multivariate skew-normal distribution in estimating the minimum-variance hedge ratio.

It is quite clear that there are several different ways of deriving and estimating hedge ratios. In the chapter we review these different techniques and approaches and examine their relations.

The chapter is divided into five sections. In Sect. 32.2, alternative theories for deriving the optimal hedge ratios are reviewed. Various estimation methods are discussed in Sect. 32.3. This section is divided into four subsections. Section 32.3.1 discusses estimation of minimum-variance hedge ratio, Sect. 32.3.2 discusses estimation of the optimum mean-variance and Sharpe Hedge ratios, and Sect. 32.3.3 discusses the estimation of the maximum expected unity hedge ratio. Section 32.3.4 discusses the estimation of mean extended-Gini coefficient-based hedge ratio and Sect. 32.3.5 discusses the estimation of generalized semivariance (GSV) based hedge ratio. Section 32.4 presents a discussion on the relationship among lengths of hedging horizon, maturity of futures contract, data frequency, and hedging effectiveness. Finally, in Sect. 32.5, we provide a summary and conclusion.

32.2 Alternative Theories for Deriving the Optimal Hedge Ratio

The basic concept of hedging is to combine investments in the spot market and futures market to form a portfolio that will eliminate (or reduce) fluctuations in its value. Specifically, consider a portfolio consisting of C_s units of a long spot position and C_f units of a short futures position.[1] Let S_t and F_t denote the spot and futures prices at time t, respectively. Since the futures contracts are used to reduce the fluctuations in spot positions, the resulting portfolio is known as the hedged portfolio. The return on the hedged portfolio, R_h, is given by:

$$R_h = \frac{C_s S_t R_s - C_f F_t R_f}{C_s S_t} = R_s - hR_f, \tag{32.1a}$$

where $h = \frac{C_f F_t}{C_s S_t}$ is the so-called hedge ratio, and $R_s = \frac{S_{t+1} - S_t}{S_t}$ and $R_f = \frac{F_{t+1} - F_t}{F_t}$ are so-called one-period returns on the spot and futures positions, respectively. Sometimes, the hedge ratio is discussed in terms of price changes (profits) instead of returns. In this case the profit on the hedged portfolio, ΔV_H, and the hedge ratio, H, are respectively given by:

$$\Delta V_H = C_s \Delta S_t - C_f \Delta F_t \quad \text{and} \quad H = \frac{C_f}{C_s}, \tag{32.1b}$$

where $\Delta S_t = S_{t+1} - S_t$ and $\Delta F_t = F_{t+1} - F_t$.

The main objective of hedging is to choose the optimal hedge ratio (either h or H). As mentioned above, the optimal hedge ratio will depend on a particular objective function to be optimized. Furthermore, the hedge ratio can be static or dynamic. In Sects. 32.2.1 and 32.2.2, we will discuss the static hedge ratio and then the dynamic hedge ratio.

It is important to note that in the above setup, the cash position is assumed to be fixed and we only look for the optimum futures position. Most of the hedging literature assumes that the cash position is fixed, a setup that is suitable for financial futures. However, when we are dealing with commodity futures, the initial cash position becomes an important decision variable that is tied to the production decision. One such setup considered by Lence (1995, 1996) will be discussed in Sect. 32.2.3.

32.2.1 Static Case

We consider here that the hedge ratio is static if it remains the same over time. The static hedge ratios reviewed in this chapter can be divided into eight categories, as shown in Table 32.1. We will discuss each of them in the chapter.

[1] Without loss of generality, we assume that the size of the futures contract is one.

Table 32.1 A list of different static hedge ratios

Hedge ratio	Objective function
• Minimum-variance (MV) hedge ratio	Minimize variance of R_h
• Optimum mean-variance hedge ratio	Maximize $E(R_h) - \frac{A}{2}Var(R_h)$
• Sharpe hedge ratio	Maximize $\frac{E(R_h)-R_F}{\sqrt{Var(R_h)}}$
• Maximum expected utility hedge ratio	Maximize $E[U(W_1)]$
• Minimum mean extended-Gini (MEG) coefficient hedge ratio	Minimize $\Gamma_v(R_h v)$
• Optimum mean-MEG hedge ratio	Maximize $E[R_h] - \Gamma_v(R_h v)$
• Minimum generalized semivariance (GSV) hedge ratio	Minimize $V_{\delta,\alpha}(R_h)$
• Maximum mean-GSV hedge ratio	Maximize $E[R_h] - V_{\delta,\alpha}(R_h)$
• Minimum VaR hedge ratio over a given time period τ	Minimize $Z_\alpha \sigma_h \sqrt{\tau} - E[R_h]\tau$

Notes

1. R_h = return on the hedged portfolio
   $E(R_h)$ = expected return on the hedged portfolio
   $Var(R_h)$ = variance of return on the hedged portfolio
   σ_h = standard deviation of return on the hedged portfolio
   Z_α = negative of left percentile at α for the standard normal distribution
   A = risk aversion parameter,
   R_F = return on the risk-free security
   $E(U(W_1))$ = expected utility of end-of-period wealth
   $\Gamma_v(R_h v)$ = mean extended-Gini coefficient of R_h
   $V_{\delta,\alpha}(R_h)$ = generalized semivariance of R_h
2. With W_1 given by (32.17), the maximum expected utility hedge ratio includes the hedge ratio considered by Lence (1995, 1996)

32.2.1.1 Minimum-Variance Hedge Ratio

The most widely used static hedge ratio is the minimum-variance (MV) hedge ratio. Johnson (1960) derives this hedge ratio by minimizing the portfolio risk, where the risk is given by the variance of changes in the value of the hedged portfolio as follows:

$$Var(\Delta V_H) = C_s^2 Var(\Delta S) + C_f^2 Var(\Delta F) - 2C_s C_f Cov(\Delta S, \Delta F).$$

The MV hedge ratio, in this case, is given by:

$$H_J^* = \frac{C_f}{C_s} = \frac{Cov(\Delta S, \Delta F)}{Var(\Delta F)}. \tag{32.2a}$$

Alternatively, if we use definition (32.1a) and use $Var(R_h)$ to represent the portfolio risk, then the MV hedge ratio is obtained by minimizing $Var(R_h)$ which is given by:

$$Var(R_h) = Var(R_s) + h^2 Var(R_f) - 2hCov(R_s, R_f).$$

In this case, the MV hedge ratio is given by:

$$h_J^* = \frac{Cov(R_s, R_f)}{Var(R_f)} = \rho \frac{\sigma_s}{\sigma_f}, \tag{32.2b}$$

where ρ is the correlation coefficient between R_s and R_f, and σ_s and σ_f are standard deviations of R_s and R_f, respectively.

The attractive features of the MV hedge ratio are that it is easy to understand and simple to compute. However, in general, the MV hedge ratio is not consistent with the mean-variance framework since it ignores the expected return on the hedged portfolio. For the MV hedge ratio to be consistent with the mean-variance framework, either the investors need to be infinitely risk-averse or the expected return on the futures contract needs to be zero.

32.2.1.2 Optimum Mean-Variance Hedge Ratio

Various studies have incorporated both risk and return in the derivation of the hedge ratio. For example, Hsin et al. (1994) derive the optimal hedge ratio that maximizes the following utility function:

$$\underset{C_f}{Max}\, V(E(R_h), \sigma; A) = E(R_h) - 0.5A\sigma_h^2, \tag{32.3}$$

where A represents the risk aversion parameter. It is clear that this utility function incorporates both risk and return. Therefore, the hedge ratio based on this utility function would be consistent with the mean-variance framework. The optimal number of futures contract and the optimal hedge ratio are respectively given by:

$$h_2 = -\frac{C_f^* F}{C_s S} = -\left[\frac{E(R_f)}{A\sigma_f^2} - \rho\frac{\sigma_s}{\sigma_f}\right]. \tag{32.4}$$

One problem associated with this type of hedge ratio is that in order to derive the optimum hedge ratio, we need to know the individual's risk aversion parameter. Furthermore, different individuals will choose different optimal hedge ratios, depending on the values of their risk aversion parameter.

Since the MV hedge ratio is easy to understand and simple to compute, it will be interesting and useful to know under what condition the above hedge ratio would be the same as the MV hedge ratio. It can be seen from (32.2b) and (32.4) that if $A \to \infty$ or $E(R_f) = 0$, then h_2 would be equal to the MV hedge ratio h_J^*. The first condition is simply a restatement of the infinitely risk-averse individuals. However, the second condition does not impose any condition on the risk-averseness, and this is important. It implies that even if the individuals are not infinitely risk-averse, then the MV hedge ratio would be the same as the optimal mean-variance hedge ratio if the expected return on the futures contract is zero (i.e., futures prices follow a simple martingale process). Therefore, if futures prices follow a simple martingale process, then we do not need to know the risk aversion parameter of the investor to find the optimal hedge ratio.

32.2.1.3 Sharpe Hedge Ratio

Another way of incorporating the portfolio return in the hedging strategy is to use the risk-return trade-off (Sharpe measure) criteria. Howard and D'Antonio (1984) consider the optimal level of futures contracts by maximizing the ratio of the portfolio's excess return to its volatility:

$$\underset{C_f}{Max}\ \theta = \frac{E(R_h) - R_F}{\sigma_h}, \tag{32.5}$$

where $\sigma_h^2 = Var(R_h)$ and R_F represents the risk-free interest rate. In this case the optimal number of futures positions, C_f^*, is given by:

$$C_f^* = -C_s \frac{\left(\frac{S}{F}\right)\left(\frac{\sigma_s}{\sigma_f}\right)\left[\frac{\sigma_s}{\sigma_f}\left(\frac{E(R_f)}{E(R_s)-R_F}\right) - \rho\right]}{\left[1 - \frac{\sigma_s}{\sigma_f}\left(\frac{E(R_f)\rho}{E(R_s)-R_F}\right)\right]}. \tag{32.6}$$

From the optimal futures position, we can obtain the following optimal hedge ratio:

$$h_3 = -\frac{\left(\frac{\sigma_s}{\sigma_f}\right)\left[\frac{\sigma_s}{\sigma_f}\left(\frac{E(R_f)}{E(R_s)-R_F}\right) - \rho\right]}{\left[1 - \frac{\sigma_s}{\sigma_f}\left(\frac{E(R_f)\rho}{E(R_s)-R_F}\right)\right]}. \tag{32.7}$$

Again, if $E(R_f) = 0$, then h_3 reduces to:

$$h_3 = \left(\frac{\sigma_s}{\sigma_f}\right)\rho, \tag{32.8}$$

which is the same as the MV hedge ratio h_J^*.

As pointed out by Chen et al. (2001), the Sharpe ratio is a highly nonlinear function of the hedge ratio. Therefore, it is possible that (32.7), which is derived by equating the first derivative to zero, may lead to the hedge ratio that would minimize, instead of maximizing, the Sharpe ratio. This would be true if the second derivative of the Sharpe ratio with respect to the hedge ratio is positive instead of negative. Furthermore, it is possible that the optimal hedge ratio may be undefined as in the case encountered by Chen et al. (2001), where the Sharpe ratio monotonically increases with the hedge ratio.

32.2.1.4 Maximum Expected Utility Hedge Ratio

So far we have discussed the hedge ratios that incorporate only risk as well as the ones that incorporate both risk and return. The methods, which incorporate both the expected return and risk in the derivation of the optimal hedge ratio, are consistent with the mean-variance framework. However, these methods may not be consistent with the expected utility maximization principle unless either the utility function is quadratic or the returns are jointly normally distributed. Therefore, in order to make the hedge ratio consistent with the expected utility maximization principle, we need to derive the hedge ratio that maximizes the expected utility. However, in order to maximize the expected utility, we need to assume a specific utility function. For example, Cecchetti et al. (1988) derive the hedge ratio that maximizes the expected utility where the utility function is assumed to be the logarithm of terminal wealth. Specifically, they derive the optimal hedge ratio that maximizes the following expected utility function:

$$\int_{R_s}\int_{R_f} \log\left[1 + R_s - hR_f\right] f(R_s, R_f)\, dR_s dR_f,$$

where the density function $f(R_s, R_f)$ is assumed to be bivariate normal. A third-order linear bivariate ARCH model is used to get the conditional variance and covariance matrix, and a numerical procedure is used to maximize the objective function with respect to the hedge ratio.[2]

32.2.1.5 Minimum Mean Extended-Gini Coefficient Hedge Ratio

This approach of deriving the optimal hedge ratio is consistent with the concept of stochastic dominance and involves the use of the mean extended-Gini (MEG) coefficient. Cheung et al. (1990), Kolb and Okunev (1992), Lien and Luo (1993a), Shalit (1995), and Lien and Shaffer (1999) all consider this approach. It minimizes the MEG coefficient $\Gamma_\nu(R_h)$ defined as follows:

$$\Gamma_\nu(R_h) = -\nu Cov\left(R_h, (1 - G(R_h))^{\nu-1}\right), \tag{32.9}$$

where G is the cumulative probability distribution and ν is the risk aversion parameter. Note that $0 \leq \nu < 1$ implies risk seekers, $\nu = 1$ implies risk-neutral investors, and $\nu > 1$ implies risk-averse investors. Shalit (1995) has shown that if the futures and spot returns are jointly normally distributed, then the minimum MEG hedge ratio would be the same as the MV hedge ratio.

32.2.1.6 Optimum Mean-MEG Hedge Ratio

Instead of minimizing the MEG coefficient, Kolb and Okunev (1993) alternatively consider maximizing the utility function defined as follows:

$$U(R_h) = E(R_h) - \Gamma_\nu(R_h). \tag{32.10}$$

The hedge ratio based on the utility function defined by (32.10) is denoted as the M-MEG hedge ratio. The difference between the MEG and M-MEG hedge ratios is that the MEG hedge ratio ignores the expected return on the hedged portfolio. Again, if the futures price follows a martingale process (i.e., $E(R_f) = 0$), then the MEG hedge ratio would be the same as the M-MEG hedge ratio.

32.2.1.7 Minimum Generalized Semivariance Hedge Ratio

In recent years, a new approach for determining the hedge ratio has been suggested (see De Jong et al. 1997; Lien and Tse 1998, 2000; Chen et al. 2001). This new approach is based on the relationship between the generalized semivariance (GSV) and expected utility as discussed by Fishburn (1977) and Bawa (1978). In this case, the optimal hedge ratio is obtained by minimizing the GSV given below:

$$V_{\delta,\alpha}(R_h) = \int_{-\infty}^{\delta} (\delta - R_h)^\alpha dG(R_h), \quad \alpha > 0, \tag{32.11}$$

[2] Lence (1995) also derives the hedge ratio based on the expected utility. We will discuss it later in Sect. 32.2.3.

where $G(R_h)$ is the probability distribution function of the return on the hedged portfolio R_h. The parameters δ and α (which are both real numbers) represent the target return and risk aversion, respectively. The risk is defined in such a way that the investors consider only the returns below the target return (δ) to be risky. It can be shown (see Fishburn 1977) that $\alpha < 1$ represents a risk-seeking investor and $\alpha > 1$ represents a risk-averse investor.

The GSV, due to its emphasis on the returns below the target return, is consistent with the risk perceived by managers (see Crum et al. 1981; Lien and Tse 2000). Furthermore, as shown by Fishburn (1977) and Bawa (1978), the GSV is consistent with the concept of stochastic dominance. Lien and Tse (1998) show that the GSV hedge ratio, which is obtained by minimizing the GSV, would be the same as the MV hedge ratio if the futures and spot returns are jointly normally distributed and if the futures price follows a pure martingale process.

32.2.1.8 Optimum Mean-Generalized Semivariance Hedge Ratio

Chen et al. (2001) extend the GSV hedge ratio to a Mean-GSV (M-GSV) hedge ratio by incorporating the mean return in the derivation of the optimal hedge ratio. The M-GSV hedge ratio is obtained by maximizing the following mean-risk utility function, which is similar to the conventional mean-variance based utility function (see (32.3)):

$$U(R_h) = E[R_h] - V_{\delta,\alpha}(R_h). \tag{32.12}$$

This approach to the hedge ratio does not use the risk aversion parameter to multiply the GSV as done in conventional mean-risk models (see Hsin et al. 1994, and (32.3)). This is because the risk aversion parameter is already included in the definition of the GSV, $V_{\delta,\alpha}(R_h)$. As before, the M-GSV hedge ratio would be the same as the GSV hedge ratio if the futures price follows a pure martingale process.

32.2.1.9 Minimum Value-at-Risk Hedge Ratio

Hung et al. (2006) suggest a new hedge ratio that minimizes the value-at-risk of the hedged portfolio. Specifically, the hedge ratio h is derived by minimizing the following value-at-risk of the hedged portfolio over a given time period τ:

$$VaR(R_h) = Z_\alpha \sigma_h \sqrt{\tau} - E[R_h]\tau \tag{32.13}$$

The resulting optimal hedge ratio, which Hung et al. (2006) refer to as zero-VaR hedge ratio, is given by

$$h^{VaR} = \rho\frac{\sigma_s}{\sigma_f} - E\left[R_f\right]\frac{\sigma_s}{\sigma_f}\sqrt{\frac{1-\rho^2}{Z_\alpha^2\sigma_f^2 - E\left[R_f\right]^2}} \tag{32.14}$$

It is clear that, if the futures price follows martingale process, the zero-VaR hedge ratio would be the same as the MV hedge ratio.

32.2.2 Dynamic Case

We have up to now examined the situations in which the hedge ratio is fixed at the optimum level and is not revised during the hedging period. However, it could be beneficial to change the hedge ratio over time. One way to allow the hedge ratio to change is by recalculating the hedge ratio based on the current (or conditional) information on the covariance (σ_{sf}) and variance (σ_f^2). This involves calculating the hedge ratio based on conditional information (i.e., $\sigma_{sf}|\Omega_{t-1}$ and $\sigma_f^2|\Omega_{t-1}$) instead of unconditional information. In this case, the MV hedge ratio is given by:

$$h_1|\Omega_{t-1} = -\frac{\sigma_{sf}|\Omega_{t-1}}{\sigma_f^2|\Omega_{t-1}}.$$

The adjustment to the hedge ratio based on new information can be implemented using such conditional models as ARCH and GARCH (to be discussed later) or using the moving window estimation method.

Another way of making the hedge ratio dynamic is by using the regime-switching GARCH model (to be discussed later) as suggested by Lee and Yoder (2007). This model assumes two different regimes where each regime is associated with different set of parameters and the probabilities of regime switching must also be estimated when implementing such method. Alternatively, we can allow the hedge ratio to change during the hedging period by considering multi-period models, which is the approach used by Lien and Luo (1993b).

Lien and Luo (1993b) consider hedging with T periods' planning horizon and minimize the variance of the wealth at the end of the planning horizon, W_T. Consider the situation where $C_{s,t}$ is the spot position at the beginning of period t and the corresponding futures position is given by $C_{f,t} = -b_t C_{s,t}$. The wealth at the end of the planning horizon, W_T, is then given by:

$$\begin{aligned} W_T &= W_0 + \sum_{t=0}^{T-1} C_{s,t}[S_{t+1} - S_t - b_t(F_{t+1} - F_t)] \\ &= W_0 + \sum_{t=0}^{T-1} C_{s,t}[\Delta S_{t+1} - b_t \Delta F_{t+1}]. \end{aligned} \tag{32.15}$$

The optimal b_ts are given by the following recursive formula:

$$b_t = \frac{Cov(\Delta S_{t+1}, \Delta F_{t+1})}{Var(\Delta F_{t+1})} + \sum_{i=t+1}^{T-1} \left(\frac{C_{s,i}}{C_{s,t}}\right) \frac{Cov(\Delta F_{t+1}, \Delta S_{i+1} + b_i \Delta F_{t+i})}{Var(\Delta F_{t+1})}. \tag{32.16}$$

It is clear from (32.16) that the optimal hedge ratio b_t will change over time. The multi-period hedge ratio will differ from the single-period hedge ratio due to the second term on the right-hand side of equation (32.16). However, it is interesting to note that the multi-period hedge ratio would be different from the single-period one if the changes in current futures prices are correlated with the changes in future futures prices or with the changes in future spot prices.

32.2.3 Case with Production and Alternative Investment Opportunities

All the models considered in Sects. 32.2.1 and 32.2.2 assume that the spot position is fixed or predetermined, and thus production is ignored. As mentioned earlier, such an assumption may be appropriate for financial futures. However, when we consider commodity futures, production should be considered in which case the spot position becomes one of the decision variables. In an important chapter, Lence (1995) extends the model with a fixed or predetermined spot position to a model where production is included. In his model, Lence (1995) also incorporates the possibility of investing in a risk-free asset and other risky assets, borrowing, as well as transaction costs. We will briefly discuss the model considered by Lence (1995) below.

Lence (1995) considers a decision maker whose utility is a function of terminal wealth $U(W_1)$, such that $U' > 0$ and $U'' < 0$. At the decision date $(t = 0)$, the decision maker will engage in the production of Q commodity units for sale at terminal date $(t = 1)$ at the random cash price P_1. At the decision date, the decision maker can lend L dollars at the risk-free lending rate $(R_L - 1)$ and borrow B dollars at the borrowing rate $(R_B - 1)$, invest I dollars in a different activity that yields a random rate of return $(R_I - 1)$ and sell X futures at futures price F_0. The transaction cost for the futures trade is f dollars per unit of the commodity traded to be paid at the terminal date. The terminal wealth (W_1) is therefore given by:

$$W_1 = W_0 R = P_1 Q + (F_0 - F_1)X - f|X| - R_B B + R_L L + R_I I, \tag{32.17}$$

where R is the return on the diversified portfolio. The decision maker will maximize the expected utility subject to the following restrictions:

$$W_0 + B \geq v(Q)Q + L + I, \quad 0 \leq B \leq k_B v(Q)Q, \; k_B \geq 0,$$

$$L \geq k_L F_0 |X|, \; k_L \geq 0, \; I \geq 0,$$

where $v(Q)$ is the average cost function, k_B is the maximum amount (expressed as a proportion of his initial wealth) that the agent can borrow, and k_L is the safety margin for the futures contract.

Using this framework, Lence (1995) introduces two opportunity costs: opportunity cost of alternative (suboptimal) investment (c_{alt}) and opportunity cost of estimation risk (e^{Bayes}).[3] Let R_{opt} be the return of the expected utility maximizing strategy and let R_{alt} be the return on a particular alternative (suboptimal) investment strategy. The opportunity cost of alternative investment strategy c_{alt} is then given by:

$$E\left[U\left(W_0 R_{opt}\right)\right] = E[U(W_0 R_{alt} + c_{alt})]. \tag{32.18}$$

In other words, c_{alt} is the minimum certain net return required by the agent to invest in the alternative (suboptimal hedging) strategy rather than in the optimum strategy. Using the CARA utility function and some simulation results, Lence (1995) finds that the expected utility maximizing hedge ratios are substantially different from the minimum-variance hedge ratios. He also shows that under certain conditions, the optimal hedge ratio is zero; i.e., the optimal strategy is not to hedge at all.

Similarly, the opportunity cost of the estimation risk (e^{Bayes}) is defined as follows:

[3] Our discussion of the opportunity costs is very brief. We would like to refer interested readers to Lence (1995) for a detailed discussion. We would also like to point to the fact that production can be allowed to be random as is done in Lence (1996).

$$E_\rho\left[E\left(U\left\{W_0\left[R_{opt}(\rho) - e_\rho^{Bayes}\right]\right\}\right)\right] = E_\rho\left[E\left(U\left(W_0 R_{opt}^{Bayes}\right)\right)\right], \tag{32.19}$$

where $R_{opt}(\rho)$ is the expected utility maximizing return where the agent knows with certainty the value of the correlation between the futures and spot prices (ρ), ${R_{opt}}^{Bayes}$ is the expected utility maximizing return where the agent only knows the distribution of the correlation ρ, and $E_\rho[.]$ is the expectation with respect to ρ. Using simulation results, Lence (1995) finds that the opportunity cost of the estimation risk is negligible and thus the value of the use of sophisticated estimation methods is negligible.

32.3 Alternative Methods for Estimating the Optimal Hedge Ratio

In Sect. 32.2, we discussed different approaches to deriving the optimum hedge ratios. However, in order to apply these optimum hedge ratios in practice, we need to estimate these hedge ratios. There are various ways of estimating them. In this section, we briefly discuss these estimation methods.

32.3.1 Estimation of the Minimum-Variance (MV) Hedge Ratio

32.3.1.1 OLS Method

The conventional approach to estimating the MV hedge ratio involves the regression of the changes in spot prices on the changes in futures price using the OLS technique (e.g., see Junkus and Lee 1985). Specifically, the regression equation can be written as:

$$\Delta S_t = a_0 + a_1 \Delta F_t + e_t, \tag{32.20}$$

where the estimate of the MV hedge ratio, H_j, is given by a_1. The OLS technique is quite robust and simple to use. However, for the OLS technique to be valid and efficient, assumptions associated with the OLS regression must be satisfied. One case where the assumptions are not completely satisfied is that the error term in the regression is heteroscedastic. This situation will be discussed later.

Another problem with the OLS method, as pointed out by Myers and Thompson (1989), is the fact that it uses unconditional sample moments instead of conditional sample moments, which use currently available information. They suggest the use of the conditional covariance and conditional variance in (32.2a). In this case, the conditional version of the optimal hedge ratio (32.2a) will take the following form:

$$H_J^* = \frac{C_f}{C_s} = \frac{Cov(\Delta S, \Delta F)\big|\Omega_{t-1}}{Var(\Delta F)\big|\Omega_{t-1}}. \tag{32.2a*}$$

Suppose that the current information (Ω_{t-1}) includes a vector of variables (X_{t-1}) and the spot and futures price changes are generated by the following equilibrium model:

$$\Delta S_t = X_{t-1}\alpha + u_t,$$
$$\Delta F_t = X_{t-1}\beta + v_t.$$

In this case, the maximum likelihood estimator of the MV hedge ratio is given by (see Myers and Thompson (1989)):

$$\hat{h}\,|X_{t-1} = \frac{\hat{\sigma}_{uv}}{\hat{\sigma}_v^2}, \tag{32.21}$$

where $\hat{\sigma}_{uv}$ is the sample covariance between the residuals u_t and v_t, and $\hat{\sigma}_v^2$ is the sample variance of the residual v_t. In general, the OLS estimator obtained from (32.20) would be different from the one given by (32.21). For the two estimators to be the same, the spot and futures prices must be generated by the following model:

$$\Delta S_t = \alpha_0 + u_t, \quad \Delta F_t = \beta_0 + v_t.$$

In other words, if the spot and futures prices follow a random walk, then with or without drift, the two estimators will be the same. Otherwise, the hedge ratio estimated from the OLS regression (32.18) will not be optimal. Now we show how SAS can be used to estimate the hedge ratio in terms of OLS method.

SAS Applications:

Based upon monthly S&P 500 index and its futures as presented in Appendix 32.3, we use OLS method in term of equation* (32.20) *to estimated hedge ratio in this section. By using SAS regression procedure which has been discussed in Sect. 13.10 of Chap. 13 and Sect. 14.13 of Chap. 14, we obtain the following program code and empirical results.

```
proc reg data = sp500monthly;
model C_spot = C_future;
ods output ParameterEstimates = estimates_ols;
run;
```

Variable	DF	Estimate	StdErr	tValue	Probt
Intercept	1	0.10961	0.28169	0.39	0.6976
C_future	1	0.98795	0.00535	184.51	<.0001

C_spot—the change of the monthly spot returns from Jan 1st, 2000 to July 31st, 2015.
C_future—the change of the monthly futures returns from Jan 1st, 2000 to July 31st, 2015.

32.3.1.2 Multivariate Skew-Normal Distribution Method

An alternative way of estimating the MV hedge ratio involves the assumption that the spot price and futures price follow a multivariate skew-normal distribution as suggested by Lien and Shrestha (2010). The estimate of covariance matrix under skew-normal distribution can be different from the estimate of covariance matrix under the usual normal distribution resulting in different estimates of MV hedge ratio. Let Y be a k-dimensional random vector. Then Y is said to have skew-normal distribution if its probability density function is given as follows:

$$f_Y(y) = 2\phi_k(y, \Omega_Y)\Phi(\alpha^t y)$$

where α is a k-dimensional column vector, $\phi_k(y, \Omega_Y)$ is the probability density function of a k-dimensional standard normal random variable with zero mean and correlation matrix Ω_Y, and $\Phi(\alpha^t y)$ is the probability distribution function of a one-dimensional standard normal random variable evaluated at $\alpha^t y$.

32.3.1.3 ARCH and GARCH Methods

Ever since the development of ARCH and GARCH models, the OLS method of estimating the hedge ratio has been generalized to take into account the heteroscedastic nature of the error term in (32.20). In this case, rather than using the unconditional sample variance and covariance, the conditional variance and covariance from the GARCH model are used in the estimation of the hedge ratio. As mentioned above, such a technique allows an update of the hedge ratio over the hedging period.

Consider the following bivariate GARCH model (see Cecchetti et al. 1988; Baillie and Myers 1991):

$$\begin{bmatrix} \Delta S_t \\ \Delta F_t \end{bmatrix} = \begin{bmatrix} \mu_1 \\ \mu_2 \end{bmatrix} + \begin{bmatrix} e_{1t} \\ e_{2t} \end{bmatrix} \Leftrightarrow \Delta Y_t = \mu + e_t,$$

$$e_t | \Omega_{t-1} \tilde{} N(0, H_t), \quad H_t = \begin{bmatrix} H_{11,t} & H_{12,t} \\ H_{12,t} & H_{22,t} \end{bmatrix},$$

$$vec(H_t) = C + A\,vec\left(e_{t-1}e_{t-1}^{'}\right) + B\,vec(H_{t-1}). \tag{32.22}$$

The conditional MV hedge ratio at time t is given by $h_{t-1} = H_{12,t}/H_{22,t}$. This model allows the hedge ratio to change over time, resulting in a series of hedge ratios instead of a single hedge ratio for the entire hedging horizon. Equation (32.22) represents a GARCH model. This GARCH model will reduce to ARCH if B is equal to zero.

SAS Applications:

Based upon monthly S&P 500 index and its futures as presented in Appendix 32.3, we use ARCH and GARCH method in term of equation (32.22) ***to estimated hedge ratio in this section.***

```
proc autoreg data = sp500monthly;
model C_spot = C_future/nlag=2 garch=(q=1,p=1) maxit=50;
ods output ParameterEstimates=estimate_autoreg;
run;
```

Variable	DF	Estimate	StdErr	tValue	Probt
Intercept	1	0.0563	0.009424	5.97	<.0001
C_future	1	0.9997	0.000461	2167.28	<.0001
AR1	1	0.9944	0.007702	129.12	<.0001
AR2	1	1.0042	0.015	66.86	<.0001
ARCH0	1	0.005013	0.005881	0.85	0.394
ARCH1	1	0.8427	0.1154	7.3	<.0001
GARCH1	1	0.589	0.0371	15.88	<.0001

C_spot—the change of the monthly spot returns from Jan 1st, 2000 to July 31st, 2015.

C_future—the change of the monthly futures returns from Jan 1st, 2000 to July 31st, 2015.

The model can be extended to include more than one type of cash and futures contracts (see Sephton 1993a). For example, consider a portfolio that consists of spot wheat ($S_{1,t}$), spot canola (S_{2t}), wheat futures (F_{1t}), and canola futures (F_{2t}). We then have the following multivariate GARCH model:

$$\begin{bmatrix} \Delta S_{1t} \\ \Delta S_{2t} \\ \Delta F_{1t} \\ \Delta F_{2t} \end{bmatrix} = \begin{bmatrix} \mu_1 \\ \mu_2 \\ \mu_3 \\ \mu_4 \end{bmatrix} + \begin{bmatrix} e_{1t} \\ e_{2t} \\ e_{3t} \\ e_{4t} \end{bmatrix} \Leftrightarrow \Delta Y_t = \mu + e_t, \quad e_t | \Omega_{t-1} \sim N(0, H_t).$$

The MV hedge ratio can be estimated using a similar technique as described above. For example, the conditional MV hedge ratio is given by the conditional covariance between the spot and futures price changes divided by the conditional variance of the futures price change. Now we show how SAS can be used to estimate ratio in terms of ARCH and GARCH models.

32.3.1.4 Regime-Switching GARCH Model

The GARCH model discussed above can be further extended by allowing regime switching as suggested by Lee and Yoder (2007). Under this model, the data generating process can be in one of two states or regime denoted by state variable $s_t = \{1, 2\}$, which is assumed to follow a first-order Markov process. The state transition probabilities are assumed to follow a logistic distribution where the transition probabilities are given by

$$\Pr(s_t = 1 | s_{t-1} = 1) = \frac{e^{p_0}}{1 + e^{p_0}} \quad \& \quad \Pr(s_t = 2 | s_{t-1} = 2) = \frac{e^{q_0}}{1 + e^{q_0}}.$$

The conditional covariance matrix is given by

$$H_{t,s_t} = \begin{bmatrix} h_{1,t,s_t} & 0 \\ 0 & h_{2,t,s_t} \end{bmatrix} \begin{bmatrix} 1 & \rho_{t,s_t} \\ \rho_{t,s_t} & 1 \end{bmatrix} \begin{bmatrix} h_{1,t,s_t} & 0 \\ 0 & h_{2,t,s_t} \end{bmatrix}$$

where

$$h^2_{1,t,s_t} = \gamma_{1,s_t} + \alpha_{1,s_t} e^2_{1.t-1} + \beta_{1,s_t} h^2_{1,t-1}$$

$$h^2_{2,t,s_t} = \gamma_{2,s_t} + \alpha_{2,s_t} e^2_{2.t-1} + \beta_{2,s_t} h^2_{2,t-1}$$

$$\rho_{t,s_t} = (1 - \theta_{1,s_t} - \theta_{2,s_t})\rho + \theta_{1,s_t}\rho_{t-1} + \theta_{2,s_t}\phi_{t-1}$$

$$\phi_{t-1} = \frac{\sum_{j=1}^{2} \varepsilon_{1,t-j}\varepsilon_{2,t-j}}{\sqrt{\left(\sum_{j=1}^{2} \varepsilon^2_{1,t-j}\right)\left(\sum_{j=1}^{2} \varepsilon^2_{2,t-j}\right)}}, \quad \varepsilon_{i,t} = \frac{e_{i,t}}{h_{it}}, \quad \theta_1, \theta_2 \geq 0 \ \& \ \theta_1 + \theta_2 \leq 1$$

Once the conditional covariance matrix is estimated, the time varying conditional MV hedge ratio is given by the ratio of the covariance between the spot and futures returns to the variance of the futures return.

32.3.1.5 Random Coefficient Method

There is another way to deal with heteroscedasticity. This involves use of the random coefficient model as suggested by Grammatikos and Saunders (1983). This model employs the following variation of (32.20):

$$\Delta S_t = \beta_0 + \beta_t \Delta F_t + e_t, \tag{32.23}$$

where the hedge ratio $\beta_t = \overline{\beta} + v_t$ is assumed to be random. This random coefficient model can, in some cases, improve the effectiveness of hedging strategy. However, this technique does not allow for the update of the hedge ratio over time even though the correction for the randomness can be made in the estimation of the hedge ratio.

32.3.1.6 Cointegration and Error-Correction Method

The techniques described so far do not take into consideration the possibility that spot price and futures price series could be nonstationary. If these series have unit roots, then this will raise a different issue. If the two series are cointegrated as defined by Engle and Granger (1987), then the regression (32.20) will be mis-specified and an error-correction term must be included in the equation. Since the arbitrage condition ties the spot and futures prices, they cannot drift far apart in the long run. Therefore, if both series follow a random walk, then we expect the two series to be cointegrated in which case we need to estimate the error-correction model. This calls for the use of the cointegration analysis.

The cointegration analysis involves two steps. First, each series must be tested for a unit root (e.g., see Dickey and Fuller 1981; Phillips and Perron 1988). Second, if both series are found to have a single unit root, then the cointegration test must be performed (e.g., see Engle and Granger 1987; Johansen and Juselius 1990; Osterwald-Lenum 1992).

If the spot price and futures price series are found to be cointegrated, then the hedge ratio can be estimated in two steps (see Ghosh 1993; Chou et al. 1996). The first step involves the estimation of the following cointegrating regression:

$$S_t = a + bF_t + u_t. \tag{32.24}$$

The second step involves the estimation of the following error-correction model:

$$\Delta S_t = \rho u_{t-1} + \beta \Delta F_t + \sum_{i=1}^{m} \delta_i \Delta F_{t-i} + \sum_{j=1}^{n} \theta_i \Delta S_{t-j} + e_j, \tag{32.25}$$

where u_t is the residual series from the cointegrating regression. The estimate of the hedge ratio is given by the estimate of β. Some researchers (e.g., see Lien and Luo 1993b) assume that the long-run cointegrating relationship is $(S_t - F_t)$, and estimate the following error-correction model:

$$\Delta S_t = \rho(S_{t-1} - F_{t-1}) + \beta \Delta F_t + \sum_{i=1}^{m} \delta_i \Delta F_{t-i} + \sum_{j=1}^{n} \theta_i \Delta S_{t-j} + e_j. \tag{32.26}$$

Alternatively, Chou et al. (1996) suggest the estimation of the error-correction model as follows:

$$\Delta S_t = \alpha \hat{u}_{t-1} + \beta \Delta F_t + \sum_{i=1}^{m} \delta_i \Delta F_{t-i} + \sum_{j=1}^{n} \theta_i \Delta S_{t-j} + e_j, \tag{32.27}$$

where $\hat{u}_{t-1} = S_{t-1} - (a + bF_{t-1})$; i.e., the series $\hat{u}_t$ is the estimated residual series from (32.24). The hedge ratio is given by β in (32.26).

Kroner and Sultan (1993) combine the error-correction model with the GARCH model considered by Cecchetti et al. (1988) and Baillie and Myers (1991) in order to estimate the optimum hedge ratio. Specifically, they use the following model:

$$\begin{bmatrix} \Delta \log_e(S_t) \\ \Delta \log_e(F_t) \end{bmatrix} = \begin{bmatrix} \mu_1 \\ \mu_2 \end{bmatrix} + \begin{bmatrix} \alpha_s(\log_e(S_{t-1}) - \log_e(F_{t-1})) \\ \alpha_f(\log_e(S_{t-1}) - \log_e(F_{t-1})) \end{bmatrix} + \begin{bmatrix} e_{1t} \\ e_{2t} \end{bmatrix}, \tag{32.28}$$

where the error processes follow a GARCH process. As before, the hedge ratio at time $(t-1)$ is given by $h_{t-1} = H_{12,t}/H_{22,t}$.

32.3.2 Estimation of the Optimum Mean-Variance and Sharpe Hedge Ratios

The optimum mean-variance and Sharpe hedge ratios are given by (32.4) and (32.7), respectively. These hedge ratios can be estimated simply by replacing the theoretical moments by their sample moments. For example, the expected returns can be replaced by sample average returns, the standard deviations can be replaced by the sample standard deviations, and the correlation can be replaced by sample correlation.

32.3.3 Estimation of the Maximum Expected Utility Hedge Ratio

The maximum expected utility hedge ratio involves the maximization of the expected utility. This requires the estimation of distributions of the changes in spot and futures prices. Once the distributions are estimated, one needs to use a numerical technique to get the optimum hedge ratio. One such method is described in Cecchetti et al. (1988) where an ARCH model is used to estimate the required distributions.

32.3.4 Estimation of Mean Extended-Gini (MEG) Coefficient-Based Hedge Ratios

The MEG hedge ratio involves the minimization of the following MEG coefficient:

$$\Gamma_v(R_h) = -vCov\left(R_h, (1 - G(R_h))^{v-1}\right).$$

In order to estimate the MEG coefficient, we need to estimate the cumulative probability density function $G(R_h)$. The cumulative probability density function is usually estimated by ranking the observed return on the hedged portfolio. A detailed description of the process can be found in Kolb and Okunev (1992), and we briefly describe the process here.

The cumulative probability distribution is estimated by using the rank as follows:

$$G(R_{h,i}) = \frac{Rank(R_{h,i})}{N},$$

where N is the sample size. Once we have the series for the probability distribution function, the MEG is estimated by replacing the theoretical covariance by the sample covariance as follows:

$$\Gamma_v^{sample}(R_h) = -\frac{v}{N}\sum_{i=1}^{N}\left(R_{h,i} - \overline{R}_h\right)\left((1 - G(R_{h,i}))^{v-1} - \Theta\right), \tag{32.29}$$

where

$$\overline{R}_h = \frac{1}{N}\sum_{i=1}^{N} R_{h,i} \quad \text{and} \quad \Theta = \frac{1}{N}\sum_{i=1}^{N}(1 - G(R_{h,i}))^{v-1}.$$

The optimal hedge ratio is now given by the hedge ratio that minimizes the estimated MEG. Since there is no analytical solution, the numerical method needs to be applied in order to get the optimal hedge ratio. This method is sometimes referred to as the empirical distribution method.

Alternatively, the instrumental variable (IV) method suggested by Shalit (1995) can be used to find the MEG hedge ratio. Shalit's method provides the following analytical solution for the MEG hedge ratio:

$$h^{IV} = \frac{Cov\left(S_{t+1}, [1 - G(F_{t+1})]^{v-1}\right)}{Cov\left(F_{t+1}, [1 - G(F_{t+1})]^{v-1}\right)}.$$

It is important to note that for the IV method to be valid, the cumulative distribution function of the terminal wealth (W_{t+1}) should be similar to the cumulative distribution of the futures price (F_{t+1}); i.e., $G(W_{t+1}) = G(F_{t+1})$. Lien and Shaffer (1999) find that the IV-based hedge ratio (h^{IV}) is significantly different from the minimum MEG hedge ratio.

Lien and Luo (1993a) suggest an alternative method of estimating the MEG hedge ratio. This method involves the estimation of the cumulative distribution function using a nonparametric kernel function instead of using a rank function as suggested above.

Regarding the estimation of the M-MEG hedge ratio, one can follow either the empirical distribution method or the nonparametric kernel method to estimate the MEG coefficient. A numerical method can then be used to estimate the hedge ratio that maximizes the objective function given by (32.10).

32.3.5 Estimation of Generalized Semivariance (GSV) Based Hedge Ratios

The GSV can be estimated from the sample by using the following sample counterpart:

$$V_{\delta,\alpha}^{sample}(R_h) = \frac{1}{N}\sum_{i=1}^{N}(\delta - R_{h,i})^{\alpha}U(\delta - R_{h,i}), \tag{32.30}$$

where

$$U(\delta - R_{h,i}) = \begin{cases} 1 & for \quad \delta \geq R_{h,i} \\ 0 & for \quad \delta < R_{h,i} \end{cases}.$$

Similar to the MEG technique, the optimal GSV hedge ratio can be estimated by choosing the hedge ratio that minimizes the sample GSV, $V_{\delta,\alpha}{}^{sample}(R_h)$. Numerical methods can be used to search for the optimum hedge ratio. Similarly, the M-GSV hedge ratio can be obtained by minimizing the mean-risk function given by (32.12), where the expected return on the hedged portfolio is replaced by the sample average return and the GSV is replaced by the sample GSV.

One can instead use the kernel density estimation method suggested by Lien and Tse (2000) to estimate the GSV, and numerical techniques can be used to find the optimum GSV hedge ratio. Instead of using the kernel method, one can also employ the conditional heteroscedastic model to estimate the density function. This is the method used by Lien and Tse (1998).

32.4 Hedging Horizon, Maturity of Futures Contract, Data Frequency, and Hedging Effectiveness

In this section, we discuss the relationship among the length of hedging horizon (hedging period), maturity of futures contracts, data frequency (e.g., daily, weekly, monthly, or quarterly), and hedging effectiveness.

Since there are many futures contracts (with different maturities) that can be used in hedging, the question is whether the minimum-variance (MV) hedge ratio depends on the time to maturity of the futures contract being used for hedging. Lee et al. (1987) find that the MV hedge ratio increases as the maturity is approached. This means that if we use the nearest to maturity futures contracts to hedge, then the MV hedge ratio will be larger compared to the one obtained using futures contracts with a longer maturity.

Aside from using futures contracts with different maturities, we can estimate the MV hedge ratio using data with different frequencies. For example, the data used in the estimation of the optimum hedge ratio can be daily, weekly, monthly, or quarterly. At the same time, the hedging horizon could be from a few hours to more than a month. The question is whether a relationship exists between the data frequency used and the length of the hedging horizon.

Malliaris and Urrutia (1991) and Benet (1992) utilize (32.20) and weekly data to estimate the optimal hedge ratio. According to Malliaris and Urrutia (1991), the *ex ante* hedging is more effective when the hedging horizon is 1 week compared to a hedging horizon of 4 weeks. Benet (1992) finds that a shorter hedging horizon (4-weeks) is more effective (in *ex ante* test) compared to a longer hedging horizon (8- and 12-weeks). These empirical results seem to be consistent with the argument that when estimating the MV hedge ratio, the hedging horizon's length must match the data frequency being used.

There is a potential problem associated with matching the length of the hedging horizon and the data frequency. For example, consider the case where the hedging horizon is 3 months (one quarter). In this case, we need to use quarterly data to match the length of the hedging horizon. In other words, when estimating (32.20) we must employ quarterly changes in spot and futures prices. Therefore, if we have 5 years' worth of data, then we will have 19 nonoverlapping price changes, resulting in a sample size of 19. However, if the hedging horizon is 1 week, instead of 3 months, then we will end up with approximately 260 nonoverlapping price changes (sample size of 260) for the same 5 years' worth of data. Therefore, the matching method is associated with a reduction in sample size for a longer hedging horizon.

One way to get around this problem is to use overlapping price changes. For example, Geppert (1995) utilizes k-period differencing for a k-period hedging horizon in estimating the regression-based MV hedge ratio. Since Geppert (1995) uses approximately 13 months of data for estimating the hedge ratio, he employs overlapping differencing in order to eliminate the reduction in sample size caused by differencing. However, this will lead to correlated observations instead of independent observations and will require the use of a regression with autocorrelated errors in the estimation of the hedge ratio.

In order to eliminate the autocorrelated errors problem, Geppert (1995) suggests a method based on cointegration and unit-root processes. We will briefly describe his method. Suppose that the spot and futures prices, which are both unit-root processes, are cointegrated. In this case, the futures and spot prices can be described by the following processes (see Stock and Watson 1988; Hylleberg and Mizon 1989):

$$S_t = A_1 P_t + A_2 \tau_t, \tag{32.31a}$$

$$F_t = B_1 P_t + B_2 \tau_t, \tag{32.31b}$$

$$P_t = P_{t-1} + w_t, \tag{32.31c}$$

$$\tau_t = \alpha_1 \tau_{t-1} + v_t, \quad 0 \le |\alpha_1| < 1, \tag{32.31d}$$

where P_t and τ_t are permanent and transitory factors that drive the spot and futures prices and w_t and v_t are white noise processes. Note that P_t follows a pure random walk process and τ_t follows a stationary process. The MV hedge ratio for a k-period hedging horizon is then given by (see Geppert 1995):

$$H_J^* = \frac{A_1 B_1 k \sigma_w^2 + 2A_2 B_2 \left(\frac{(1-\alpha^k)}{1-\alpha^2}\right) \sigma_v^2}{B_1^2 k \sigma_w^2 + 2B_2^2 \left(\frac{(1-\alpha^k)}{1-\alpha^2}\right) \sigma_v^2}. \tag{32.32}$$

One advantage of using (32.32) instead of a regression with nonoverlapping price changes is that it avoids the problem of a reduction in sample size associated with nonoverlapping differencing.

An alternative way of matching the data frequency with the hedging horizon is by using the wavelet to decompose the time series into different frequencies as suggested by Lien and Shrestha (2007). The decomposition can be done without the loss of sample size (see Lien and Shrestha (2007) for detail). For example, the daily spot and future returns series can be decomposed using the maximal overlap discrete wavelet transform (MODWT) as follows:

$$R_{s,t} = B^s_{J,t} + D^s_{J,t} + D^s_{J-1,t} + \cdots + D^s_{1,t}$$

$$R_{f,t} = B^f_{J,t} + D^f_{J,t} + D^f_{J-1,t} + \cdots + D^f_{1,t}$$

where $D_{j,t}{}^s$ and $D_{j,t}{}^f$ are the spot and futures returns series with changes on the time scale of length 2^{j-1} days, respectively.[4] Similarly, $B_{J,t}{}^s$ and $B_{J,t}{}^2$ represent spot and futures returns series corresponding to time scale of 2^J days and longer. Now, we can run the following regression to find the hedge ratio corresponding to hedging horizon equal to 2^{j-1} days:

$$D^s_{j,t} = \theta_{j,0} + \theta_{j,1} D^f_{j,t} + e_j \tag{32.33}$$

where the estimate of the hedge ratio is given by the estimate of $\theta_{j,1}$.

32.5 Summary and Conclusions

In this chapter, we have reviewed various approaches to deriving the optimal hedge ratio, as summarized in Appendix 32.1. These approaches can be divided into the mean-variance based approach, the expected utility maximizing approach, the mean extended-Gini coefficient-based approach, and the generalized semivariance-based approach. All these approaches will lead to the same hedge ratio as the conventional minimum-variance (MV) hedge ratio if the futures price follows a pure martingale process and if the futures and spot prices are jointly normal. However, if these conditions do not hold, then the hedge ratios based on the various approaches will be different.

The MV hedge ratio is the most understood and most widely used hedge ratio. Since the statistical properties of the MV hedge ratio are well known, statistical hypothesis testing can be performed with the MV hedge ratio. For example, we can test whether the optimal MV hedge ratio is the same as the naïve hedge ratio. Since the MV hedge ratio ignores the expected return, it will not be consistent with the mean-variance analysis unless the futures price follows a pure martingale process. Furthermore, if the martingale and normality condition do not hold, then the MV hedge ratio will not be consistent with the expected utility maximization principle. Following the MV hedge ratio is the mean-variance hedge ratio. Even if this hedge ratio incorporates the expected return in the derivation of the optimal hedge ratio, it will not be consistent with the expected maximization principle unless either the normality condition holds or the utility function is quadratic.

In order to make the hedge ratio consistent with the expected utility maximization principle, we can derive the optimal hedge ratio by maximizing the expected utility. However, to implement such approach, we need to assume a specific utility function and we need to make an assumption regarding the return distribution. Therefore, different utility functions will lead to different optimal hedge ratios. Furthermore, analytic solutions for such hedge ratios are not known and numerical methods need to be applied.

New approaches have recently been suggested in deriving optimal hedge ratios. These include the mean-Gini coefficient-based hedge ratio, semivariance-based hedge ratios, and value-at-risk based hedge ratios. These hedge ratios are consistent with the second-order stochastic dominance principle. Therefore, such hedge ratios are very general in the sense that they are consistent with the expected utility maximization principle and make very few assumptions on the utility function. The only requirement is that the marginal utility be positive and the second derivative of the utility function be

[4] For example, $D_{1,t}{}^s$ represents daily time scale and $D_{4,t}{}^s$ represents 8-day time scale.

negative. However, both of these hedge ratios do not lead to a unique hedge ratio. For example, the mean-Gini coefficient-based hedge ratio depends on the risk aversion parameter (ν) and the semivariance-based hedge ratio depends on the risk aversion parameter (α) and target return (δ). It is important to note, however, that the semivariance-based hedge ratio has some appeal in the sense that the semivariance as a measure of risk is consistent with the risk perceived by individuals. The same argument can be applied to value-at-risk based hedge ratio.

So far as the derivation of the optimal hedge ratio is concerned, almost all of the derivations do not incorporate transaction costs. Furthermore, these derivations do not allow investments in securities other than the spot and corresponding futures contracts. As shown by Lence (1995), once we relax these conventional assumptions, the resulting optimal hedge ratio can be quite different from the ones obtained under the conventional assumptions. Lence's (1995) results are based on a specific utility function and some other assumption regarding the return distributions. It remains to be seen if such results hold for the mean extended-Gini coefficient-based as well as semivariance-based hedge ratios.

In this chapter, we have also reviewed various ways of estimating the optimum hedge ratio, as summarized in Appendix 32.2. As far as the estimation of the conventional MV hedge ratio is concerned, there are a large number of methods that have been proposed in the literature. These methods range from a simple regression method to complex cointegrated heteroscedastic methods with regime switching, and some of the estimation methods include a kernel density function method as well as an empirical distribution method. Except for many of mean-variance based hedge ratios, the estimation involves the use of a numerical technique. This has to do with the fact that most of the optimal hedge ratio formulae do not have a closed-form analytic expression. Again, it is important to mention that based on his specific model, Lence (1995) finds that the value of complicated and sophisticated estimation methods is negligible. It remains to be seen if such a result holds for the mean extended-Gini coefficient-based as well as semivariance-based hedge ratios.

In this chapter, we have also discussed about the relationship between the optimal MV hedge ratio and the hedging horizon. We feel that this relationship has not been fully explored and can be further developed in the future. For example, we would like to know if the optimal hedge ratio approaches the naïve hedge ratio when the hedging horizon becomes longer.

The main thing we learn from this review is that if the futures price follows a pure martingale process and if the returns are jointly normally distributed, then all different hedge ratios are the same as the conventional MV hedge ratio, which is simple to compute and easy to understand. However, if these two conditions do not hold, then there are many optimal hedge ratios (depending on which objective function one is trying to optimize) and there is no single optimal hedge ratio that is distinctly superior to the remaining ones. Therefore, further research needs to be done to unify these different approaches to the hedge ratio.

For those who are interested in research in this area, we would like to finally point out that one requires a good understanding of financial economic theories and econometric methodologies. In addition, a good background in data analysis and computer programming would also be helpful.

Appendix 32.1: Theoretical Models

References	Return definition and objective function	Summary
Johnson (1960)	Ret_1 O_1	The chapter derives the minimum-variance hedge ratio. The hedging effectiveness is defined as E_1, but no empirical analysis is done

(continued)

References	Return definition and objective function	Summary
Hsin et al. (1994)	Ret_2 O_2	The chapter derives the utility function-based hedge ratio. A new measure of hedging effectiveness E_2 based on a certainty equivalent is proposed. The new measure of hedging effectiveness is used to compare the effectiveness of futures and options as hedging instruments
Howard and D'Antonio (1984)	Ret_2 O_3	The chapter derives the optimal hedge ratio based on maximizing the Sharpe ratio. The proposed hedging effectiveness E_3 is based on the Sharpe ratio
Cecchetti et al. (1988)	Ret_2 O_4	The chapter derives the optimal hedge ratio that maximizes the expected utility function: $\int_{R_s}\int_{R_f} \log\left[1 + R_s(t) - h(t)R_f(t)\right] f_t(R_s, R_f)\, dR_s dR_f$, where the density function is assumed to be bivariate normal. A third-order linear bivariate ARCH model is used to get the conditional variance and covariance matrix. A numerical procedure is used to maximize the objective function with respect to hedge ratio. Due to ARCH, the hedge ratio changes over time. The chapter uses certainty equivalent (E_2) to measure the hedging effectiveness
Cheung et al. (1990)	Ret_2 O_5	The chapter uses mean-Gini ($v = 2$, not mean extended-Gini coefficient) and mean-variance approaches to analyze the effectiveness of options and futures as hedging instruments
Kolb and Okunev (1992)	Ret_2 O_5	The chapter uses mean extended-Gini coefficient in the derivation of the optimal hedge ratio. Therefore, it can be considered as a generalization of the mean-Gini coefficient method used by Cheung et al. (1990)
Kolb and Okunev (1993)	Ret_2 O_6	The chapter defines the objective function as O_6, but in terms of wealth (W) $U(W) = E[W] - \Gamma_v(W)$ and compares with the quadratic utility function $U(W) = E[W] - m\sigma^2$. The chapter plots the EMG efficient frontier in W and $\Gamma_v(W)$ space for various values of risk aversion parameters (v)
Lien and Luo (1993b)	Ret_1 O_9	The chapter derives the multi-period hedge ratios where the hedge ratios are allowed to change over the hedging period. The method suggested in the chapter still falls under the minimum-variance hedge ratio
Lence (1995)	O_4	This chapter derives the expected utility maximizing hedge ratio where the terminal wealth depends on the return on a diversified portfolio that consists of the production of a spot commodity, investment in a risk-free asset, investment in a risky asset, as well as borrowing. It also incorporates the transaction costs
De Jong et al. (1997)	Ret_2 O_7 (also uses O_1 and O_3)	The chapter derives the optimal hedge ratio that minimizes the generalized semivariance (GSV). The chapter compares the GSV hedge ratio with the minimum-variance (MV) hedge ratio as well as the Sharpe hedge ratio. The chapter uses E_1 (for the MV hedge ratio), E_3 (for the Sharpe hedge ratio), and E_4 (for the GSV hedge ratio) as the measures of hedging effectiveness
Chen et al. (2001)	Ret_1 O_8	The chapter derives the optimal hedge ratio that maximizes the risk-return function given by $U(R_h) = E[R_h] - V_{\delta,\alpha}(R_h)$. The method can be considered as an extension of the GSV method used by De Jong et al. (1997)
Hung et al. (2006)	Ret_2 O_{10}	The chapter derives the optimal hedge ratio that minimizes the value-at-risk for a hedging horizon of length τ given by $Z_\alpha \sigma_h \sqrt{\tau} - E[R_h]\tau$

Notes:

A. Return Model:

(Ret$_1$) $\Delta V_H = C_s \Delta P_s + C_f \Delta P_f \Rightarrow$ hedge ratio $= H = \frac{C_f}{C_s}$, C_s = units of spot commodity and C_f = units of futures contract

(Ret$_2$) $R_h = R_s + hR_f,\quad R_s = \dfrac{S_t - S_{t-1}}{S_{t-1}}$ (a) $R_f = \dfrac{F_t - F_{t-1}}{F_{t-1}} \Rightarrow$ hedge ratio: $h = \dfrac{C_f F_{t-1}}{C_s S_{t-1}}$.

(b) $R_f = \dfrac{F_t - F_{t-1}}{S_{t-1}} \Rightarrow$ hedge ratio: $h = \dfrac{C_f}{C_s}$

B. Objective Function:

(O$_1$) Minimize $Var(R_h) = C_s^2\sigma_s^2 + C_f^2\sigma_f^2 + 2C_sC_f\sigma_{sf}$ or $Var(R_h) = \sigma_s^2 + h^2\sigma_f^2 + 2h\sigma_{sf}$

(O$_2$) Maximize $E(R_h) - \dfrac{A}{2}Var(R_h)$

(O$_3$) Maximize $\dfrac{E(R_h) - R_F}{Var(R_h)}$ (Sharpe ratio), R_F = risk-free interest rate

(O$_4$) Maximize $E[U(W)]$, $U(.)$ = utility function, W = terminal wealth

(O$_5$) Minimize $\Gamma_v(R_h),\quad \Gamma_v(R_h) = -vCov\left(R_h, (1 - F(R_h))^{v-1}\right)$

(O$_6$) Maximize $E[R_h] - \Gamma_v(R_h v)$

(O$_7$) Minimize $V_{\delta,\alpha}(R_h) = \int_{-\infty}^{\delta} (\delta - R_h)^\alpha dG(R_h),\quad \alpha > 0$

(O$_8$) Maximize $U(R_h) = E[R_h] - V_{\delta,\alpha}(R_h)$

(O$_9$) Minimize $Var(W_t) = Var\left(\sum_{t=1}^{T} C_{st}\Delta S_t + C_{ft}\Delta F_t\right)$.

(O$_{10}$) Minimize $Z_\alpha \sigma_h \sqrt{\tau} - E[R_h]\tau$

C. Hedging Effectiveness:

(E$_1$) $e = \left(1 - \frac{Var(R_h)}{Var(R_s)}\right)$

(E$_2$) $e = R_h^{ce} - R_{ss}^{ce}$, R_h^{ce} $\left(R_s^{ce}\right)$ = certainty equivalent return of hedged (unhedged) portfolio

(E$_3$) $e = \dfrac{\frac{(E[R_h] - R_F)}{Var(R_h)}}{\frac{(E[R_s] - R_F)}{Var(R_s)}}$ or $e = \frac{(E[R_h] - R_F)}{Var(R_h)} - \frac{(E[R_s] - R_F)}{Var(R_s)}$

(E$_4$) $e = 1 - \frac{V_{\delta,\alpha}(R_h)}{V_{\delta,\alpha}(R_s)}$.

Appendix 32.2: Empirical Models

References	Commodity	Summary
Ederington (1979)	GNMA futures (1/1976-12/1977), Wheat (1/1976-12/1977), Corn (1/1976-12/1977), T-bill futures (3/1976-12/1977) [weekly data]	The chapter uses the Ret$_1$ definition of return and estimates the minimum-variance hedge ratio (O$_1$). E$_1$ is used as a hedging effectiveness measure. The chapter uses nearby contracts (3–6 months, 6–9 months, and 9–12 months) and a hedging period of 2 and 4 weeks. OLS (M$_1$) is used to estimate the parameters. Some of the hedge ratios are found not to be different from zero and the hedging effectiveness increases with

(continued)

References	Commodity	Summary
		the length of hedging period. The hedge ratio also increases (closer to unity) with the length of hedging period
Grammatikos and Saunders (1983)	Swiss franc, Canadian dollar, British pound, DM, Yen (1/1974-6/1980) [weekly data]	The chapter estimates the hedge ratio for the whole period and moving window (2-year data). It is found that the hedge ratio changes over time. Dummy variables for various subperiods are used, and shifts are found. The chapter uses a random coefficient (M_3) model to estimate the hedge ratio. The hedge ratio for Swiss franc is found to follow a random coefficient model. However, there is no improvement in effectiveness when the hedge ratio is calculated by correcting for the randomness
Junkus and Lee (1985)	Three stock index futures for Kansas City Board of Trade, New York Futures Exchange, and Chicago Mercantile Exchange (5/82-3/83) [daily data]	The chapter tests the applicability of four futures hedging models: a variance-minimizing model introduced by Johnson (1960), the traditional one to one hedge, a utility maximization model developed by Rutledge (1972), and a basis arbitrage model suggested by Working (1953). An optimal ratio or decision rule is estimated for each model, and measures for the effectiveness of each hedge are devised. Each hedge strategy performed best according to its own criterion. The Working decision rule appeared to be easy to use and satisfactory in most cases. Although the maturity of the futures contract used affected the size of the optimal hedge ratio, there was no consistent maturity effect on performance. Use of a particular ratio depends on how closely the assumptions underlying the model approach a hedger's real situation
Lee et al. (1987)	S&P 500, NYSE, Value Line (1983) [daily data]	The chapter tests for the temporal stability of the minimum-variance hedge ratio. It is found that the hedge ratio increases as maturity of the futures contract nears. The chapter also performs a functional form test and finds support for the regression of rate of change for discrete as well as continuous rates of change in prices
Cecchetti et al. (1988)	Treasury bond, Treasury bond futures (1/1978-5/1986) [monthly data]	The chapter derives the hedge ratio by maximizing the expected utility. A third-order linear bivariate ARCH model is used to get the conditional variance and covariance matrix. A numerical procedure is used to maximize the objective function with respect to the hedge ratio. Due to ARCH, the hedge ratio changes over time. It is found that the hedge ratio changes over time and is significantly less (in absolute value) than the minimum-variance (MV) hedge ratio (which also changes over time). E_2 (certainty equivalent) is used to measure the performance effectiveness. The proposed utility-maximizing hedge ratio performs better than the MV hedge ratio
Cheung et al. (1990)	Swiss franc, Canadian dollar, British pound, German mark, Japanese yen (9/1983-12/1984) [daily data]	The chapter uses mean-Gini coefficient ($v = 2$) and mean-variance approaches to analyze the effectiveness of options and futures as hedging instruments. It considers both mean-variance and expected return mean-Gini coefficient frontiers.

(continued)

References	Commodity	Summary
		It also considers the minimum-variance (MV) and minimum mean-Gini coefficient hedge ratios. The MV and minimum mean-Gini approaches indicate that futures is a better hedging instrument. However, the mean-variance frontier indicates futures to be a better hedging instrument whereas the mean-Gini frontier indicates options to be a better hedging instrument
Baillie and Myers (1991)	Beef, Coffee, Corn, Cotton, Gold, Soybean (contracts maturing in 1982 and 1986) [daily data]	The chapter uses a bivariate GARCH model (M_2) in estimating the minimum-variance (MV) hedge ratios. Since the models used are conditional models, the time series of hedge ratios are estimated. The MV hedge ratios are found to follow a unit-root process. The hedge ratio for beef is found to be centered around zero. E_1 is used as a hedging effectiveness measure. Both in-sample and out-of-sample effectiveness of the GARCH-based hedge ratios is compared with a constant hedge ratio. The GARCH-based hedge ratios are found to be significantly better compared to the constant hedge ratio
Malliaris and Urrutia (1991)	British pound, German mark, Japanese yen, Swill franc, Canadian dollar (3/1980–12/1988) [weekly data]	The chapter uses regression autocorrelated errors model to estimate the minimum-variance (MV) hedge ratio for the five currencies. Using overlapping moving windows, the time series of the MV hedge ratio and hedging effectiveness are estimated for both ex post (in-sample) and ex ante (out-of-sample) cases. E_1 is used to measure the hedging effectiveness for the ex post case whereas average return is used to measure the hedging effectiveness. Specifically, the average return close to zero is used to indicate a better performing hedging strategy. In the ex post case, the 4-week hedging horizon is more effective compared to the 1-week hedging horizon. However, for the ex ante case the opposite is found to be true
Benet (1992)	Australian dollar, Brazilian cruzeiro, Mexican peso, South African rand, Chinese yuan, Finnish markka, Irish pound, Japanese yen (8/1973–12/1985) [weekly data]	This chapter considers direct and cross hedging, using multiple futures contracts. For minor currencies, the cross hedging exhibits a significant decrease in performance from ex post to ex ante. The minimum-variance hedge ratios are found to change from one period to the other except for the direct hedging of Japanese yen. On the ex ante case, the hedging effectiveness does not appear to be related to the estimation period length. However, the effectiveness decreases as the hedging period length increases
Kolb and Okunev (1992)	Corn, Copper, Gold, German mark, S&P 500 (1989) [daily data]	The chapter estimates the mean extended-Gini (MEG) hedge ratio (M_9) with v ranging from 2 to 200. The MEG hedge ratios are found to be close to the minimum-variance hedge ratios for a lower level of risk parameter v (for v from 2 to 5). For higher values of v, the two hedge ratios are found to be quite different. The hedge ratios are found to increase with the risk aversion parameter for

(continued)

References	Commodity	Summary
		S&P 500, Corn, and Gold. However, for Copper and German mark, the hedge ratios are found to decrease with the risk aversion parameter. The hedge ratio tends to be more stable for higher levels of risk
Kolb and Okunev (1993)	Cocoa (3/1952–1976) for four cocoa-producing countries (Ghana, Nigeria, Ivory Coast, and Brazil) [March and September data]	The chapter estimates the Mean-MEG (M-MEG) hedge ratio (M_{12}). The chapter compares the M-MEG hedge ratio, minimum-variance hedge ratio, and optimum mean-variance hedge ratio for various values of risk aversion parameters. The chapter finds that the M-MEG hedge ratio leads to reverse hedging (buy futures instead of selling) for v less than 1.24 (Ghana case). For high-risk aversion parameter values (high v), all hedge ratios are found to converge to the same value
Lien and Luo (1993a)	S&P 500 (1/1984–12/1988) [weekly data]	The chapter points out that the mean extended-Gini (MEG) hedge ratio can be calculated either by numerically optimizing the MEG coefficient or by numerically solving the first-order condition. For $v = 9$, the hedge ratio of -0.8182 is close to the minimum-variance (MV) hedge ratio of -0.8171. Using the first-order condition, the chapter shows that for a large v the MEG hedge ratio converges to a constant. The empirical result shows that the hedge ratio decreases with the risk aversion parameter v. The chapter finds that the MV and MEG hedge ratio (for low v) series (obtained by using a moving window) are more stable compared to the MEG hedge ratio for a large v. The chapter also uses a nonparametric Kernel estimator to estimate the cumulative density function. However, the kernel estimator does not change the result significantly
Lien and Luo (1993b)	British pound, Canadian dollar, German mark, Japanese yen, Swiss franc (3/1980–12/1988), MMI, NYSE, S&P (1/1984–12/1988) [weekly data]	This chapter proposes a multi-period model to estimate the optimal hedge ratio. The hedge ratios are estimated using an error-correction model. The spot and futures prices are found to be cointegrated. The optimal multi-period hedge ratios are found to exhibit a cyclical pattern with a tendency for the amplitude of the cycles to decrease. Finally, the possibility of spreading among different market contracts is analyzed. It is shown that hedging in a single market may be much less effective than the optimal spreading strategy
Ghosh (1993)	S&P futures, S&P index, Dow Jones Industrial average, NYSE composite index (1/1990-12/1991) [daily data]	All the variables are found to have a unit root. For all three indices the same S&P 500 futures contracts are used (cross hedging). Using the Engle-Granger two-step test, the S&P 500 futures price is found to be cointegrated with each of the three spot prices: S&P 500, DJIA, and NYSE. The hedge ratio is estimated using the error-correction model (ECM) (M_4). Out-of-sample performance is better for the hedge ratio from the ECM compared to the Ederington model

(continued)

References	Commodity	Summary
Sephton (1993a)	Feed wheat, Canola futures (1981–82 crop year) [daily data]	The chapter finds unit roots on each of the cash and futures (log) prices, but no cointegration between futures and spot (log) prices. The hedge ratios are computed using a four-variable GARCH(1,1) model. The time series of hedge ratios are found to be stationary. Reduction in portfolio variance is used as a measure of hedging effectiveness. It is found that the GARCH-based hedge ratio performs better compared to the conventional minimum-variance hedge ratio
Sephton (1993b)	Feed wheat, Feed barley, Canola futures (1988/89) [daily data]	The chapter finds unit roots on each of the cash and futures (log) prices, but no cointegration between futures and spot (log) prices. A univariate GARCH model shows that the mean returns on the futures are not significantly different from zero. However, from the bivariate GARCH canola is found to have a significant mean return. For canola the mean-variance utility function is used to find the optimal hedge ratio for various values of the risk aversion parameter. The time series of the hedge ratio (based on bivariate GARCH model) is found to be stationary. The benefit in terms of utility gained from using a multivariate GARCH decreases as the degree of risk aversion increases
Kroner and Sultan (1993)	British pound, Canadian dollar, German mark, Japanese yen, Swiss franc (2/1985–2/1990) [weekly data]	The chapter uses the error-correction model with a GARCH error (M_5) to estimate the minimum-variance (MV) hedge ratio for the five currencies. Due to the use of conditional models, the time series of the MV hedge ratios are estimated. Both within-sample and out-of-sample evidence shows that the hedging strategy proposed in the chapter is potentially superior to the conventional strategies
Hsin et al. (1994)	British pound, German mark, Yen, Swiss franc (1/1986–12/1989) [daily data]	The chapter derives the optimum mean-variance hedge ratio by maximizing the objective function O_2. The hedging horizons of 14, 30, 60, 90, and 120 calendar days are considered to compare the hedging effectiveness of options and futures contracts. It is found that the futures contracts perform better than the options contracts
Shalit (1995)	Gold, Silver, Copper, Aluminum (1/1977–12/1990) [daily data]	The chapter shows that if the prices are jointly normally distributed, the mean extended-Gini (MEG) hedge ratio will be same as the minimum-variance (MV) hedge ratio. The MEG hedge ratio is estimated using the instrumental variable method. The chapter performs normality tests as well as the tests to see if the MEG hedge ratios are different from the MV hedge ratios. The chapter finds that for a significant number of futures contracts the normality does not hold and the MEG hedge ratios are different from the MV hedge ratios
Geppert (1995)	German mark, Swiss franc, Japanese yen, S&P 500, Municipal Bond Index (1/1990–1/1993) [weekly data]	The chapter estimates the minimum-variance hedge ratio using the OLS as well as the cointegration methods for various lengths of hedging horizon. The in-sample results indicate

(continued)

References	Commodity	Summary
		that for both methods the hedging effectiveness increases with the length of the hedging horizon. The out-of-sample results indicate that in general the effectiveness (based on the method suggested by Malliaris and Urrutia (1991)) decreases as the length of the hedging horizon decreases. This is true for both the regression method and the decomposition method proposed in the chapter. However, the decomposition method seems to perform better than the regression method in terms of both mean and variance
De Jong et al. (1997)	British pound (12/1976–10/1993), German mark (12/1976–10/1993), Japanese yen (4/1977–10/1993) [daily data]	The chapter compares the minimum-variance, generalized semivariance, and Sharpe hedge ratios for the three currencies. The chapter computes the out-of-sample hedging effectiveness using nonoverlapping 90-day periods where the first 60 days are used to estimate the hedge ratio and the remaining 30 days are used to compute the out-of-sample hedging effectiveness. The chapter finds that the naïve hedge ratio performs better than the model-based hedge ratios
Lien and Tse (1998)	Nikkei Stock Average (1/1989–8/1996) [daily data]	The chapter shows that if the rates of change in spot and futures prices are bivariate normal and if the futures price follows a martingale process, then the generalized semivariance (GSV) (referred to as lower partial moment) hedge ratio will be same as the minimum-variance (MV) hedge ratio. A version of the bivariate asymmetric power ARCH model is used to estimate the conditional joint distribution, which is then used to estimate the time varying GSV hedge ratios. The chapter finds that the GSV hedge ratio significantly varies over time and is different from the MV hedge ratio
Lien and Shaffer (1999)	Nikkei (9/86–9/89), S&P (4/82–4/85), TOPIX (4/90–12/93), KOSPI (5/96–12/96), Hang Seng (1/87–12189), IBEX (4/93–3/95) [daily data]	This chapter empirically tests the ranking assumption used by Shalit (1995). The ranking assumption assumes that the ranking of futures prices is the same as the ranking of the wealth. The chapter estimates the mean extended-Gini (MEG) hedge ratio based on the instrumental variable (IV) method used by Shalit (1995) and the true MEG hedge ratio. The true MEG hedge ratio is computed using the cumulative probability distribution estimated employing the kernel method instead of the rank method. The chapter finds that the MEG hedge ratio obtained from the IV method to be different from the true MEG hedge ratio. Furthermore, the true MEG hedge ratio leads to a significantly smaller MEG coefficient compared to the IV-based MEG hedge ratio
Lien and Tse (2000)	Nikkei Stock Average (1/1988–8/996) [daily data]	The chapter estimates the generalized semivariance (GSV) hedge ratios for different values of parameters using a nonparametric kernel estimation method. The kernel method is compared with the empirical distribution method. It is found that the hedge ratio from one method is not different from the hedge ratio from another.

(continued)

References	Commodity	Summary
		The Jarque–Bera (1987) test indicates that the changes in spot and futures prices do not follow normal distribution
Chen et al. (2001)	S&P 500 (4/1982–12/1991) [weekly data]	The chapter proposes the use of the M-GSV hedge ratio. The chapter estimates the minimum-variance (MV), optimum mean-variance, Sharpe, mean extended-Gini (MEG), generalized semivariance (GSV), mean-MEG (M-MEG), and mean-GSV (M-GSV) hedge ratios. The Jarque–Bera (1987) Test and D'Agostino (1971) D Statistic indicate that the price changes are not normally distributed. Furthermore, the expected value of the futures price change is found to be significantly different from zero. It is also found that for a high level of risk aversion, the M-MEG hedge ratio converges to the MV hedge ratio whereas the M-GSV hedge ratio converges to a lower value
Hung et al. (2006)	S&P 500 (01/1997–12/1999) [daily data]	The chapter proposes minimization of value-at-risk in deriving the optimum hedge ratio. The chapter finds cointegrating relationship between the spot and futures returns and uses bivariate constant correlation GARCH(1,1) model with error-correction term. The chapter compares the proposed hedge ratio with MV hedge ratio and hedge ratio (HKL hedge ratio) proposed by Hsin et al. (1994). The chapter finds the performance of the proposed hedge ratio to be similar to the HKL hedge ratio. Finally, the proposed hedge ratio converges to the MV hedge ratio for high risk-averse levels
Lee and Yoder (2007)	Nikkei 225 and Hang Send index futures (01/1989-12/2003) [weekly data]	The chapter proposes regime-switching time varying correlation GARCH model and compares the resulting hedge ratio with constant correlation GARCH and time varying correlation GARCH. The proposed model is found to outperform the other two hedge ratios in both in-sample and out-of-sample for both contracts
Lien, D and Shrestha, K (2007)	23 different futures contracts (sample period depends on contracts) [daily data]	This chapter proposes wavelet-based hedge ratio to compute the hedge ratios for different hedging horizons (1-day, 2-day, 4-day, 8-day, 16-day, 32-day, 64-day, 128-day; and 256-day and longer). It is found that the wavelet-based hedge ratio and the error-correction based hedge ratio are larger than MV hedge ratio. The performance of wavelet-based hedge ratio improves with the length of hedging horizon
Lien, D and Lien and Shrestha (2010)	22 different futures contracts (sample period depends on contracts) [daily data]	The chapter proposes the hedge ratio based on skew-normal distribution (SKN hedge ratio). The chapter also estimates the semivariance (lower partial moment (LPM)) hedge ratio and MV hedge ratio among other hedge ratios. SKN hedge ratios are found to be different from the MV hedge ratio based on normal distribution. SKN hedge ratio performs better than LPM hedge ratio for long hedger especially for the out-of-sample cases

Notes:
A. Minimum-Variance Hedge Ratio:
A.1. OLS:

$$(\mathrm{M}_1): \quad \Delta S_t = a_0 + a_1 \Delta F_t + e_t: \quad \textit{Hedge ratio} = a_1$$
$$R_s = a_0 + a_1 R_f + e_t: \quad \textit{Hedge ratio} = a_1$$

A.2.Multivariate Skew-Normal:

(M_2): The return vector $Y = \begin{bmatrix} R_s \\ R_f \end{bmatrix}$ is assumed to have skew-normal distribution with covariance matrix V:
Hedge ration $= H_{skn} = \frac{V(1,2)}{V(2,2)}$

A.3. ARCH/GARCH:

$$(\mathrm{M}_3): \begin{bmatrix} \Delta S_t \\ \Delta F_t \end{bmatrix} = \begin{bmatrix} \mu_1 \\ \mu_2 \end{bmatrix} + \begin{bmatrix} e_{1t} \\ e_{2t} \end{bmatrix}, \; e_t|\Omega_{t-1} \sim N(0, H_t), \; H_t = \begin{bmatrix} H_{11,t} & H_{12,t} \\ H_{12,t} & H_{22,t} \end{bmatrix} \; \textit{Hedge ratio} = H_{12,t}/H_{22,t}$$

A.4. Regime-Switching GARCH :

(M_4): The transition probabilities are given by:

$$\Pr(s_t = 1 | s_{t-1} = 1) = \frac{e^{p_0}}{1 + e^{p_0}} \quad \& \quad \Pr(s_t = 2 | s_{t-1} = 2) = \frac{e^{q_0}}{1 + e^{q_0}}$$

The GARCH model: Two-series GARCH model with first series as return on futures.

$$H_{t,s_t} = \begin{bmatrix} h_{1,t,s_t} & 0 \\ 0 & h_{2,t,s_t} \end{bmatrix} \begin{bmatrix} 1 & \rho_{t,s_t} \\ \rho_{t,s_t} & 1 \end{bmatrix} \begin{bmatrix} h_{1,t,s_t} & 0 \\ 0 & h_{2,t,s_t} \end{bmatrix}$$

$$h^2_{1,t,s_t} = \gamma_{1,s_t} + \alpha_{1,s_t} e^2_{1.t-1} + \beta_{1,s_t} h^2_{1,t-1}, \quad h^2_{2,t,s_t} = \gamma_{2,s_t} + \alpha_{2,s_t} e^2_{2.t-1} + \beta_{2,s_t} h^2_{2,t-1}$$

$$\rho_{t,s_t} = (1 - \theta_{1,s_t} - \theta_{2,s_t})\rho + \theta_{1,s_t}\rho_{t-1} + \theta_{2,s_t}\phi_{t-1}$$

$$\phi_{t-1} = \frac{\sum_{j=1}^{2} \varepsilon_{1,t-j}\varepsilon_{2,t-j}}{\sqrt{\left(\sum_{j=1}^{2} \varepsilon^2_{1,t-j}\right)\left(\sum_{j=1}^{2} \varepsilon^2_{2,t-j}\right)}}, \quad \varepsilon_{i,t} = \frac{e_{i,t}}{h_{it}}, \; \theta_1, \theta_2 \geq 0 \; \& \; \theta_1 + \theta_2 \leq 1, \; \textit{Hedge ratio} = \frac{H_{t,s_t}(1,2)}{H_{t,s_t}(2,2)}$$

A.5. Random Coefficient:

$$(\mathrm{M}_5): \quad \Delta S_t = \beta_0 + \beta_t \Delta F_t + e_t$$
$$\beta_t = \overline{\beta} + v_t, \qquad \textit{Hedge ratio} = \overline{\beta}$$

A.6. Cointegration and Error Correction :

$$(\mathrm{M}_6): \quad S_t = a + bF_t + u_t$$

$$\Delta S_t = \rho u_{t-1} + \beta \Delta F_t + \sum_{i=1}^{m} \delta_i \Delta F_{t-i} + \sum_{j=1}^{n} \theta_i \Delta S_{t-j} + e_j, \quad EC\,Hedge\ ratio\ = \beta$$

A.7. Error Correction with GARCH:

$$(\mathrm{M}_7): \begin{bmatrix} \Delta \log_e(S_t) \\ \Delta \log_e(F_t) \end{bmatrix} = \begin{bmatrix} \mu_1 \\ \mu_2 \end{bmatrix} + \begin{bmatrix} \alpha_s(\log_e(S_{t-1}) - \log_e(F_{t-1})) \\ \alpha_f(\log_e(S_{t-1}) - \log_e(F_{t-1})) \end{bmatrix} + \begin{bmatrix} e_{1t} \\ e_{2t} \end{bmatrix}, \; e_t|\Omega_{t-1} \sim N(0, H_t), \quad H_t = \begin{bmatrix} H_{11,t} & H_{12,t} \\ H_{12,t} & H_{22,t} \end{bmatrix} Hedge\ ratio\ = h_{t-1} = H_{12,t}/H_{22,t}$$

A.8. Common Stochastic Trend:

$$(\mathrm{M}_8): \quad S_t = A_1 P_t + A_2 \tau_t, \; F_t = B_1 P_t + B_2 \tau_t, \; P_t = P_{t-1} + w_t \; \tau_t = \alpha_1 \tau_{t-1} + v_t, \; 0 \le |\alpha_1| < 1,$$

$$Hedge\ ratio\,for\ k - period\ investment\ horizon\ = H_J^* = \frac{A_1 B_1 k \sigma_w^2 + 2A_2 B_2 \left(\frac{(1-\alpha^k)}{1-\alpha^2} \right) \sigma_v^2}{B_1^2 k \sigma_w^2 + 2B_2^2 \left(\frac{(1-\alpha^k)}{1-\alpha^2} \right) \sigma_v^2},$$

B. Optimum Mean-Variance Hedge Ratio:

(M_9): $Hedge\ ratio = h_2 = -\frac{C_f^* F}{C_s S} = -\left[\frac{E(R_f)}{A\sigma_f^2} - \rho \frac{\sigma_s}{\sigma_f} \right]$, where the moments $E[R_f]$, σ_s and σ_f are estimated by sample moments.

C. Sharpe Hedge Ratio:

(M_{10}): $Hedge\ ratio\ = h_3 = -\frac{\left(\frac{\sigma_s}{\sigma_f}\right)\left[\frac{\sigma_s}{\sigma_f}\left(\frac{E(R_f)}{E(R_s)-i}\right)-\rho\right]}{\left[1-\frac{\sigma_s}{\sigma_f}\left(\frac{E(R_f)\rho}{E(R_s)-i}\right)\right]}$, where the moments and correlation are estimated by their sample counterparts.

D. Mean-Gini Coefficient-Based Hedge Ratios:

(M_{11}): The hedge ratio is estimated by numerically minimizing the following mean extended-Gini coefficient, where the cumulative probability distribution function is estimated using the rank function:

$$\hat{\Gamma}_v(R_h) = -\frac{v}{N} \sum_{i=1}^{N} \left(R_{h,i} - \overline{R}_h \right) \left((1 - G(R_{h,i}))^{v-1} - \Theta \right).$$

(M_{12}): The hedge ratio is estimated by numerically solving the first-order condition, where the cumulative probability distribution function is estimated using the rank function.

(M_{13}): The hedge ratio is estimated by numerically solving the first-order condition, where the cumulative probability distribution function is estimated using the kernel-based estimates.

(M_{14}): The hedge ratio is estimated by numerically maximizing the following function:

$$U(R_h) = E(R_h) - \Gamma_v(R_h),$$

where the expected values and the mean extended-Gini coefficient are replaced by their sample counterparts and the cumulative probability distribution function is estimated using the rank function.

E. *Generalized Semivariance-Based Hedge Ratios:*

(M_{15}): The hedge ratio is estimated by numerically minimizing the following sample generalized hedge ratio:

$$V_{\delta,\alpha}^{sample}(R_h) = \frac{1}{N}\sum_{i=1}^{N}(\delta - R_{h,i})^{\alpha}U(\delta - R_{h,i}), \text{ where } U(\delta - R_{h,i}) = \begin{cases} 1\, for\ \delta \geq R_{h,i} \\ 0\, for\ \delta < R_{h,i} \end{cases}.$$

(M_{16}): The hedge ratio is estimated by numerically maximizing the following function:

$$U(R_h) = R_h - V_{\delta,\alpha}^{sample}(R_h).$$

E. *Minimum Value-at-Risk Hedge Ratio:*

(M_{17}): The hedge ratio is estimated by minimizing the following value-at-risk:

$$VaR(R_h) = Z_\alpha \sigma_h \sqrt{\tau} - E[R_h]\tau$$

The resulting hedge ratio is given by

$$h^{VaR} = \rho\frac{\sigma_s}{\sigma_f} - E[R_f]\frac{\sigma_s}{\sigma_f}\sqrt{\frac{1-\rho^2}{Z_\alpha^2\sigma_f^2 - E[R_f]^2}}$$

Appendix 32.3: Monthly Data of S&P 500 Index and Its Futures (January 2000–June 2015)

Date	SPOT	FUTURES	C_spot	C_future
2000M01	1394.46	1401	−28.04	−29
2000M02	1366.42	1372	132.16	143.3
2000M03	1498.58	1515.3	−46.15	−55.3
2000M04	1452.43	1460	−31.83	−37.8
2000M05	1420.6	1422.2	34	45.9
2000M06	1454.6	1468.1	−23.77	−29.2
2000M07	1430.83	1438.9	86.85	82.3
2000M08	1517.68	1521.2	−81.17	−67.5
2000M09	1436.51	1453.7	−7.11	−13.5
2000M10	1429.4	1440.2	−114.45	−118.7
2000M11	1314.95	1321.5	5.33	13.5
2000M12	1320.28	1335	45.73	37.9
2001M01	1366.01	1372.9	−126.07	−130.9
2001M02	1239.94	1242	−79.61	−72.8
2001M03	1160.33	1169.2	89.13	85.1
2001M04	1249.46	1254.3	6.36	3.1
2001M05	1255.82	1257.4	−31.4	−25.7
2001M06	1224.42	1231.7	−13.19	−16.4
2001M07	1211.23	1215.3	−77.65	−80.2
2001M08	1133.58	1135.1	−92.64	−91.4
2001M09	1040.94	1043.7	18.84	17

(continued)

Date	SPOT	FUTURES	C_spot	C_future
2001M10	1059.78	1060.7	79.67	79.3
2001M11	1139.45	1140	8.63	9.2
2001M12	1148.08	1149.2	−17.88	−18.8
2002M01	1130.2	1130.4	−23.47	−23.5
2002M02	1106.73	1106.9	40.66	42.3
2002M03	1147.39	1149.2	−70.47	−72
2002M04	1076.92	1077.2	−9.78	−9.7
2002M05	1067.14	1067.5	−77.33	−77.4
2002M06	989.81	990.1	−78.19	−78.6
2002M07	911.62	911.5	4.45	4.6
2002M08	916.07	916.1	−100.79	−101.1
2002M09	815.28	815	70.48	70.4
2002M10	885.76	885.4	50.55	50.6
2002M11	936.31	936	−56.49	−57.1
2002M12	879.82	878.9	−24.12	−24.2
2003M01	855.7	854.7	−14.55	−13.8
2003M02	841.15	840.9	7.03	6.1
2003M03	848.18	847	68.74	69.1
2003M04	916.92	916.1	46.67	47.2
2003M05	963.59	963.3	10.91	10
2003M06	974.5	973.3	15.81	16
2003M07	990.31	989.3	17.7	18.4
2003M08	1008.01	1007.7	−12.04	−13.6
2003M09	995.97	994.1	54.74	55.4
2003M10	1050.71	1049.5	7.49	8.3
2003M11	1058.2	1057.8	53.72	52.8
2003M12	1111.92	1110.6	19.21	19.3
2004M01	1131.13	1129.9	13.81	14.7
2004M02	1144.94	1144.6	−18.73	−19.7
2004M03	1126.21	1124.9	−18.91	−18.8
2004M04	1107.3	1106.1	13.38	14.2
2004M05	1120.68	1120.3	20.16	20.1
2004M06	1140.84	1140.4	−39.12	−39.3
2004M07	1101.72	1101.1	2.52	3
2004M08	1104.24	1104.1	10.34	10.8
2004M09	1114.58	1114.9	15.62	15.4
2004M10	1130.2	1130.3	43.62	43.8
2004M11	1173.82	1174.1	38.1	39.6
2004M12	1211.92	1213.7	−30.65	−32
2005M01	1181.27	1181.7	22.33	22.4
2005M02	1203.6	1204.1	−23.01	−20.2
2005M03	1180.59	1183.9	−23.74	−25.4
2005M04	1156.85	1158.5	34.65	33.8
2005M05	1191.5	1192.3	−0.17	3.2
2005M06	1191.33	1195.5	42.85	41.3
2005M07	1234.18	1236.8	−13.85	−15.4
2005M08	1220.33	1221.4	8.48	12.9
2005M09	1228.81	1234.3	−21.8	−24.5
2005M10	1207.01	1209.8	42.47	41.3
2005M11	1249.48	1251.1	−1.19	3.7
2005M12	1248.29	1254.8	31.79	28.8
2006M01	1280.08	1283.6	0.58	−1.2

(continued)

Date	SPOT	FUTURES	C_spot	C_future
2006M02	1280.66	1282.4	14.17	20.9
2006M03	1294.83	1303.3	15.78	12.6
2006M04	1310.61	1315.9	−40.52	−43.8
2006M05	1270.09	1272.1	0.11	7.3
2006M06	1270.2	1279.4	6.46	2.4
2006M07	1276.66	1281.8	27.16	23.8
2006M08	1303.82	1305.6	32.03	39.8
2006M09	1335.85	1345.4	42.09	37.8
2006M10	1377.94	1383.2	22.69	19.7
2006M11	1400.63	1402.9	17.67	25.5
2006M12	1418.3	1428.4	19.94	14.6
2007M01	1438.24	1443	−31.42	−34.1
2007M02	1406.82	1408.9	14.04	22.3
2007M03	1420.86	1431.2	61.51	57.2
2007M04	1482.37	1488.4	48.25	44.5
2007M05	1530.62	1532.9	−27.27	−17.5
2007M06	1503.35	1515.4	−48.08	−53.5
2007M07	1455.27	1461.9	18.72	14.8
2007M08	1473.99	1476.7	52.76	61.4
2007M09	1526.75	1538.1	22.63	16.8
2007M10	1549.38	1554.9	−68.24	−71.2
2007M11	1481.14	1483.7	−12.79	−6.5
2007M12	1468.35	1477.2	−89.8	−97.6
2008M01	1378.55	1379.6	−47.92	−48.3
2008M02	1330.63	1331.3	−7.93	−7.3
2008M03	1322.7	1324	62.89	62
2008M04	1385.59	1386	14.79	14.6
2008M05	1400.38	1400.6	−120.38	−119.5
2008M06	1280	1281.1	−12.62	−14
2008M07	1267.38	1267.1	15.45	15.5
2008M08	1282.83	1282.6	−116.47	−113.6
2008M09	1166.36	1169	−197.61	−201.7
2008M10	968.75	967.3	−72.51	−72
2008M11	896.24	895.3	7.01	4.8
2008M12	903.25	900.1	−77.37	−77.6
2009M01	825.88	822.5	−90.79	−88.3
2009M02	735.09	734.2	62.78	60.6
2009M03	797.87	794.8	74.94	75.2
2009M04	872.81	870	46.33	48.1
2009M05	919.14	918.1	0.18	−2.6
2009M06	919.32	915.5	68.16	68.9
2009M07	987.48	984.4	33.14	35.3
2009M08	1020.62	1019.7	36.46	33.2
2009M09	1057.08	1052.9	−20.89	−19.9
2009M10	1036.19	1033	59.44	61.8
2009M11	1095.63	1094.8	19.47	15.9
2009M12	1115.1	1110.7	−41.23	−40.3
2010M01	1073.87	1070.4	30.62	33
2010M02	1104.49	1103.4	64.94	61.8
2010M03	1169.43	1165.2	17.26	18.2
2010M04	1186.69	1183.4	−97.28	−94.9
2010M05	1089.41	1088.5	−58.7	−61.9

(continued)

Date	SPOT	FUTURES	C_spot	C_future
2010M06	1030.71	1026.6	70.89	71.7
2010M07	1101.6	1098.3	−52.27	−50
2010M08	1049.33	1048.3	91.87	88.4
2010M09	1141.2	1136.7	42.06	43
2010M10	1183.26	1179.7	−2.71	−0.1
2010M11	1180.55	1179.6	77.09	73.4
2010M12	1257.64	1253	28.48	29.4
2011M01	1286.12	1282.4	41.1	43.7
2011M02	1327.22	1326.1	−1.39	−5.1
2011M03	1325.83	1321	37.78	38.7
2011M04	1363.61	1359.7	−18.41	−15.8
2011M05	1345.2	1343.9	−24.56	−28.4
2011M06	1320.64	1315.5	−28.36	−27.1
2011M07	1292.28	1288.4	−73.39	−70.7
2011M08	1218.89	1217.7	−87.47	−91.7
2011M09	1131.42	1126	121.88	123.3
2011M10	1253.3	1249.3	−6.34	−3.3
2011M11	1246.96	1246	10.64	6.6
2011M12	1257.6	1252.6	54.81	55.6
2012M01	1312.41	1308.2	53.27	56.2
2012M02	1365.68	1364.4	42.79	38.8
2012M03	1408.47	1403.2	−10.56	−9.6
2012M04	1397.91	1393.6	−87.58	−84.4
2012M05	1310.33	1309.2	51.83	47.2
2012M06	1362.16	1356.4	17.16	18.2
2012M07	1379.32	1374.6	27.26	30.5
2012M08	1406.58	1405.1	34.09	29.1
2012M09	1440.67	1434.2	−28.51	−27.4
2012M10	1412.16	1406.8	4.02	7.6
2012M11	1416.18	1414.4	10.01	5.7
2012M12	1426.19	1420.1	71.92	73.2
2013M01	1498.11	1493.3	16.57	20
2013M02	1514.68	1513.3	54.51	49.4
2013M03	1569.19	1562.7	28.38	29.5
2013M04	1597.57	1592.2	33.17	36.8
2013M05	1630.74	1629	−24.46	−29.7
2013M06	1606.28	1599.3	79.44	81.2
2013M07	1685.72	1680.5	−52.75	−49.2
2013M08	1632.97	1631.3	48.58	43
2013M09	1681.55	1674.3	74.99	76.7
2013M10	1756.54	1751	49.27	53.1
2013M11	1805.81	1804.1	42.55	37
2013M12	1848.36	1841.1	−65.77	−64.5
2014M01	1782.59	1776.6	76.86	81
2014M02	1859.45	1857.6	12.88	7
2014M03	1872.33	1864.6	11.62	13.3
2014M04	1883.95	1877.9	39.62	43.6
2014M05	1923.57	1921.5	36.66	30.9
2014M06	1960.23	1952.4	−29.56	−27.6
2014M07	1930.67	1924.8	72.7	76.6
2014M08	2003.37	2001.4	−31.09	−35.9
2014M09	1972.28	1965.5	45.77	45.9
2014M10	2018.05	2011.4	49.51	54.9

(continued)

Date	SPOT	FUTURES	C_spot	C_future
2014M11	2067.56	2066.3	−8.66	−13.9
2014M12	2058.9	2052.4	−63.91	−64
2015M01	1994.99	1988.4	109.51	114.4
2015M02	2104.5	2102.8	−36.61	−42
2015M03	2067.89	2060.8	17.62	18.1
2015M04	2085.51	2078.9	21.88	27.1
2015M05	2107.39	2106	−44.28	−51.6
2015M06	2063.11	2054.4	40.73	44

References

Baillie RT, Myers RJ (1991) Bivariate Garch estimation of the optimal commodity futures hedge. J Appl Econom 6:109–124

Bawa VS (1978) Safety-first, stochastic dominance, and optimal portfolio choice. J Financ Quant Anal 13:255–271

Benet BA (1992) Hedge period length and ex-ante futures hedging effectiveness: the case of foreign-exchange risk cross hedges. J Futures Mark 12:163–175

Cecchetti SG, Cumby RE, Figlewski S (1988) Estimation of the optimal futures hedge. Rev Econ Stat 70:623–630

Chen SS, Lee CF, Shrestha K (2001) On a mean-generalized semivariance approach to determining the hedge ratio. J Futures Mark 21:581–598

Cheung CS, Kwan CCY, Yip PCY (1990) The hedging effectiveness of options and futures: a mean-Gini approach. J Futures Mark 10:61–74

Chou WL, Fan KK, Lee CF (1996) Hedging with the Nikkei index futures: the conventional model versus the error correction model. Q Rev Econ Finance 36:495–505

Crum RL, Laughhunn DL, Payne JW (1981) Risk-seeking behavior and its implications for financial models. Financ Manage 10:20–27

D'Agostino RB (1971) An omnibus test of normality for moderate and large size samples. Biometrika 58:341–348

De Jong A, De Roon F, Veld C (1997) Out-of-sample hedging effectiveness of currency futures for alternative models and hedging strategies. J Futures Mark 17:817–837

Dickey DA, Fuller WA (1981) Likelihood ratio statistics for autoregressive time series with a unit root. Econometrica 49:1057–1072

Ederington LH (1979) The hedging performance of the new futures markets. J Finance 34:157–170

Engle RF, Granger CW (1987) Co-integration and error correction: representation, estimation and testing. Econometrica 55:251–276

Fishburn PC (1977) Mean-risk analysis with risk associated with below-target returns. Am Econ Rev 67:116–126

Geppert JM (1995) A statistical model for the relationship between futures contract hedging effectiveness and investment horizon length. J Futures Mark 15:507–536

Ghosh A (1993) Hedging with stock index futures: estimation and forecasting with error correction model. J Futures Mark 13:743–752

Grammatikos T, Saunders A (1983) Stability and the hedging performance of foreign currency futures. J Futures Mark 3:295–305

Howard CT, D'Antonio LJ (1984) A risk-return measure of hedging effectiveness. J Financ Quant Anal 19:101–112

Hsin CW, Kuo J, Lee CF (1994) A new measure to compare the hedging effectiveness of foreign currency futures versus options. J Futures Mark 14:685–707

Hung JC, Chiu CL, Lee MC (2006) Hedging with zero-value at risk hedge ratio. Appl Financ Econ 16:259–269

Hylleberg S, Mizon GE (1989) Cointegration and error correction mechanisms. Econ J 99:113–125

Jarque CM, Bera AK (1987) A test for normality of observations and regression residuals. Int Stat Rev 55:163–172

Johansen S, Juselius K (1990) Maximum likelihood estimation and inference on cointegration—with applications to the demand for money. Oxford Bull Econ Stat 52:169–210

Johnson LL (1960) The theory of hedging and speculation in commodity futures. Rev Econ Stud 27:139–151

Junkus JC, Lee CF (1985) Use of three index futures in hedging decisions. J Futures Mark 5:201–222

Kolb RW, Okunev J (1992) An empirical evaluation of the extended mean-Gini coefficient for futures hedging. J Futures Mark 12:177–186

Kolb RW, Okunev J (1993) Utility maximizing hedge ratios in the extended mean Gini framework. J Futures Mark 13:597–609

Kroner KF, Sultan J (1993) Time-varying distributions and dynamic hedging with foreign currency futures. J Financ Quant Anal 28:535–551

Lee HT, Yoder J (2007) Optimal hedging with a regime-switching time-varying correlation GARCH model. J Futures Mark 27:495–516

Lee CF, Bubnys EL, Lin Y (1987) Stock index futures hedge ratios: test on horizon effects and functional form. Adv Futures Opt Res 2:291–311

Lence SH (1995) The economic value of minimum-variance hedges. Am J Agric Econ 77:353–364

Lence SH (1996) Relaxing the assumptions of minimum variance hedging. J Agric Resourc Econ 21:39–55

Lien D, Luo X (1993a) Estimating the extended mean-Gini coefficient for futures hedging. J Futures Mark 13:665–676

Lien D, Luo X (1993b) Estimating multiperiod hedge ratios in cointegrated markets. J Futures Mark 13:909–920

Lien D, Shaffer DR (1999) Note on estimating the minimum extended Gini hedge ratio. J Futures Mark 19:101–113

Lien D, Shrestha K (2007) An empirical analysis of the relationship between hedge ratio and hedging horizon using wavelet analysis. J Futures Mark 27:127–150

Lien D, Shrestha K (2010) Estimating optimal hedge ratio: a multivariate skew-normal distribution. Appl Financ Econ 20:627–636

Lien D, Tse YK (1998) Hedging time-varying downside risk. J Futures Mark 18:705–722

Lien D, Tse YK (2000) Hedging downside risk with futures contracts. Appl Financ Econ 10:163–170

Malliaris AG, Urrutia JL (1991) The impact of the lengths of estimation periods and hedging horizons on the effectiveness of a hedge: evidence from foreign currency futures. J Futures Mark 3:271–289

Myers RJ, Thompson SR (1989) Generalized optimal hedge ratio estimation. Am J Agric Econ 71:858–868

Osterwald-Lenum M (1992) A note with quantiles of the asymptotic distribution of the maximum likelihood cointegration rank test statistics. Oxford Bull Econ Stat 54:461–471

Phillips PCB, Perron P (1988) Testing unit roots in time series regression. Biometrika 75:335–46

Rutledge DJS (1972) Hedgers' demand for futures contracts: a theoretical framework with applications to the United States soybean complex. Food Res Inst Stud 11:237–256

Sephton PS (1993a) Hedging wheat and canola at the Winnipeg commodity exchange. Appl Financ Econ 3:67–72

Sephton PS (1993b) Optimal hedge ratios at the Winnipeg commodity exchange. Can J Econ 26:175–193

Shalit H (1995) Mean-Gini hedging in futures markets. J Futures Mark 15:617–635

Stock JH, Watson MW (1988) Testing for common trends. J Am Stat Assoc 83:1097–1107

Working H (1953) Hedging reconsidered. J Farm Econ 35:544–561

Author Index

C.-F. Lee et al., *Essentials of Excel, Excel VBA, SAS and Minitab for Statistical and Financial Analyses*, DOI 10.1007/978-3-319-38867-0

Subject Index

C.-F. Lee et al., *Essentials of Excel, Excel VBA, SAS and Minitab for Statistical and Financial Analyses*, DOI 10.1007/978-3-319-38867-0

W

Z

CPSIA information can be obtained
at www.ICGtesting.com
Printed in the USA
BVOW07s0831211216
471498BV00004B/47/P

9 783319 388656